INTEGRATED BIOREFINERIES

DESIGN, ANALYSIS, AND OPTIMIZATION

GREEN CHEMISTRY AND CHEMICAL ENGINEERING

Series Editor: Sunggyu Lee
Ohio University, Athens, Ohio, USA

Proton Exchange Membrane Fuel Cells: Contamination and Mitigation Strategies
Hui Li, Shanna Knights, Zheng Shi, John W. Van Zee, and Jiujun Zhang

Proton Exchange Membrane Fuel Cells: Materials Properties and Performance
David P. Wilkinson, Jiujun Zhang, Rob Hui, Jeffrey Fergus, and Xianguo Li

Solid Oxide Fuel Cells: Materials Properties and Performance
Jeffrey Fergus, Rob Hui, Xianguo Li, David P. Wilkinson, and Jiujun Zhang

Efficiency and Sustainability in the Energy and Chemical Industries: Scientific Principles and Case Studies, Second Edition
Krishnan Sankaranarayanan, Jakob de Swaan Arons, and Hedzer van der Kooi

Nuclear Hydrogen Production Handbook
Xing L. Yan and Ryutaro Hino

Magneto Luminous Chemical Vapor Deposition
Hirotsugu Yasuda

Carbon-Neutral Fuels and Energy Carriers
Nazim Z. Muradov and T. Nejat Veziroğlu

Oxide Semiconductors for Solar Energy Conversion: Titanium Dioxide
Janusz Nowotny

Lithium-Ion Batteries: Advanced Materials and Technologies
Xianxia Yuan, Hansan Liu, and Jiujun Zhang

Process Integration for Resource Conservation
Dominic C. Y. Foo

Chemicals from Biomass: Integrating Bioprocesses into Chemical Production Complexes for Sustainable Development
Debalina Sengupta and Ralph W. Pike

Hydrogen Safety
Fotis Rigas and Paul Amyotte

Biofuels and Bioenergy: Processes and Technologies
Sunggyu Lee and Y. T. Shah

Integrated Biorefineries: Design, Analysis, and Optimization
Paul R. Stuart and Mahmoud M. El-Halwagi

GREEN CHEMISTRY AND CHEMICAL ENGINEERING

INTEGRATED BIOREFINERIES

DESIGN, ANALYSIS, AND OPTIMIZATION

Edited by

Paul R. Stuart
Mahmoud M. El-Halwagi

CRC Press
Taylor & Francis Group
Boca Raton London New York

CRC Press is an imprint of the
Taylor & Francis Group, an **informa** business

CRC Press
Taylor & Francis Group
6000 Broken Sound Parkway NW, Suite 300
Boca Raton, FL 33487-2742

Version Date: 20121030

International Standard Book Number-13: 978-1-4398-0346-2 (Hardback)

Visit the Taylor & Francis Web site at
http://www.taylorandfrancis.com

and the CRC Press Web site at
http://www.crcpress.com

Contents

Section III Biorefinery Process Analysis

Section IV Biorefinery Feedstocks

Section V Biorefinery Technologies

Series Preface

Green Chemistry and Chemical Engineering

The subjects and disciplines of chemistry and chemical engineering have reached a new landmark in the way of thinking about, developing, and designing chemical products and processes. This revolutionary philosophy, termed "green chemistry and chemical engineering," focuses on the designs of products and processes that are conducive to reducing or eliminating the use and generation of hazardous substances. In dealing with hazardous or potentially hazardous substances, there may be some overlaps and interrelationships between environmental chemistry and green chemistry. While environmental chemistry is the chemistry of the natural environment and the pollutant chemicals in nature, green chemistry proactively aims to reduce and prevent pollution at its very source. In essence, the philosophies of green chemistry and chemical engineering tend to focus more on industrial application and practice rather than on academic principles and phenomenological science. However, as both chemistry and chemical engineering philosophy, green chemistry and chemical engineering derive from and build upon organic chemistry, inorganic chemistry, polymer chemistry, fuel chemistry, biochemistry, analytical chemistry, physical chemistry, environmental chemistry, thermodynamics, chemical reaction engineering, transport phenomena, chemical process design, separation technology, automatic process control, and more. In short, green chemistry and chemical engineering is the rigorous use of chemistry and chemical engineering for pollution prevention and environmental protection.

The Pollution Prevention Act of 1990 in the United States established a national policy to prevent or reduce pollution at its source whenever feasible. And adhering to the spirit of this policy, the Environmental Protection Agency (EPA) launched its Green Chemistry Program to promote innovative chemical technologies that reduce or eliminate the use or generation of hazardous substances in the design, manufacture, and use of chemical products. Global efforts in green chemistry and chemical engineering have recently gained a substantial amount of support from the international community of science, engineering, academia, industry, and governments in all phases and aspects.

Some of the successful examples and key technological developments include the use of supercritical carbon dioxide as a green solvent in separation technologies, application of supercritical water oxidation for the destruction of harmful substances, process integration with carbon dioxide sequestration steps, solvent-free synthesis of chemicals and polymeric materials, exploitation of biologically degradable materials, use of aqueous hydrogen peroxide for efficient oxidation, development of hydrogen proton exchange membrane (PEM) fuel cells for a variety of power generation needs, advanced biofuel productions, devulcanization of spent tire rubber, avoidance of the use of chemicals and processes that generate volatile organic compounds (VOCs), replacement of traditional petrochemical processes by microorganism-based bioengineering processes, replacement of chlorofluorocarbons (CFCs) with nonhazardous alternatives, advances in the design of energy-efficient processes, use of clean, alternative, and renewable energy sources in manufacturing, and much more. This list, even though it is only a partial compilation, is undoubtedly growing exponentially.

This book series on Green Chemistry and Chemical Engineering by CRC Press/Taylor & Francis is designed to meet the new challenges of the twenty-first century in the chemistry and chemical engineering disciplines by publishing books and monographs based on cutting-edge research and development that aim at reducing adverse impacts on the environment by chemical enterprises. And in achieving this, the series will detail the development of alternative sustainable technologies that will minimize the hazard and maximize the efficiency of any chemical choice. The series aims at delivering readers in academia and industry with an authoritative information source in the field of green chemistry and chemical engineering. The publisher and the series editor are fully aware of the rapidly evolving nature of the subject and its long-lasting impact on the quality of human life in both the present and the future. As such, the team is committed to making this series the most comprehensive and accurate literary source in the field of green chemistry and chemical engineering.

Sunggyu Lee

Preface

We read each day about rapid advances in processes that can potentially transform biomass into bioproducts—biorefinery innovation is all around us. However, understanding how to create competitive advantage by designing, synthesizing, and optimizing the processes into existing processes and infrastructure is complex. The book assembles a well-coordinated set of contributions from leading experts in the field of integrated biorefineries from around the world, working in the areas of design, synthesis, and optimization of integrated biorefineries. It captures today's state-of-the-art in the rapidly advancing bioeconomy.

The *integrated biorefinery* we have considered is any processing facility that transforms biomass into value-added products. Focus is given to second-generation feedstocks, which include forestry resources, energy crops, agricultural residues, oils, and various waste materials. The products from biorefining span a broad spectrum from commodity through to specialty. Examples of these products include biofuels, forestry products, biochemicals, food, and pharmaceuticals. With the increasing interest in sustainability and with the need to enhance the utilization of renewable resources in industrial facilities, biorefineries are destined to play an instrumental role in the process industries and have significant economic, environmental, and societal implications.

An integrated biorefinery may involve one or more of the following types of integration:

- **Process Integration** is based on a holistic approach to design and operation, thus focusing on the unity of the process, and may be in the form of mass, energy, and property integration.
- **Infrastructure Integration** links the process units to the existing infrastructure. Examples include the integration of a biorefinery with a petroleum refinery or pulp and paper mill, the sequestration of carbon dioxide from the gaseous emissions of an industrial process to grow algae, which is subsequently used to produce biodiesel, or the injection of syngas produced from biomass gasification into commercial gas pipelines or into gas-to-liquid plants.
- **Feedstock and Product Integration** exploits complementary characteristics of raw materials or products for a biorefinery with other materials. Examples include the addition of bioethanol to gasoline, the blending of biodiesel with petrodiesel, and the mixing of bioethylene produced from ethanol with ethylene produced from natural gas to form a feedstock for a polyethylene plant.
- **Supply-Chain Integration** coordinates the resources and sequence of activities associated with the life cycle of a product, from feedstock operations through to delivery of the product portfolio (e.g., biomass growth, harvesting, pre-processing, transportation, processing, usage, recycle, storage, etc.).
- **Policy and Environmental Integration** integrates the environmental performance of a biorefinery with the surroundings or with other processing facilities in the same region, and meets environmental regulations as well as other environmental targets, for example, greenhouse gas emission reduction for the new biorefinery product portfolio.

The potential for synthesizing and integrating novel process–product pathways for biorefineries is tremendous. The candidate feedstocks provide various precursors that may be processed via an ever increasing number of technology platforms to yield diverse products. While much work has been undertaken in the area of experimental development of biorefinery technology, there is relatively little information in the literature that can effectively guide engineers and decision makers in the synthesis, selection, design, analysis, and optimization of biorefineries. The objective of this book is to fill this gap by providing state-of-the-art advances in the design, analysis, and optimization of integrated biorefineries. For example, the book addresses the following questions:

- What is the current state of affairs for biorefineries?
- How can strategic opportunities for biorefineries be identified from the many available?
- What are the various processing platforms (e.g., biological, chemical, thermal, mechanical, etc.)? What are their key characteristics?
- How can the design and analysis of a biorefinery be made to maximize different opportunities for integration and create competitive advantage?
- How can the techno-economics of many biorefinery processes be done, and those with the most interesting returns be considered further?
- How should future market changes and uncertainties in the design and operation of biorefineries be considered in biorefinery selection?
- How can the life cycle impact of biorefinery products and processes be evaluated?
- What is the role of discovery and innovation in generating integrated biorefineries?
- What is the impact of energy and other policies on biorefinery design and selection?

The chapters have been clustered into six categories:

- Biorefinery product and process design
- Biorefinery supply chains
- Biorefinery process analysis
- Biorefinery feedstocks
- Biorefinery technologies
- Biorefinery policy and environmental analysis

We are grateful to our contributing authors who, with their enormous expertise, have provided state-of-the-art coverage of diverse and critical topics in the field of integrated biorefineries. Much appreciation is due to the numerous reviewers who have provided valuable feedback and recommendations. We thank Allison Shatkin, senior editor, and her team at CRC Press/Taylor & Francis for being a wonderful partner throughout the phases of producing this book. We also appreciate and extend our thanks to Syed Mohamad Shajahan and his colleagues at Techset Composition Ltd for typesetting the book. We gratefully acknowledge the fine editorial work of Diana Bouchard. The coeditors of this book have a passion for what they do! Paul Stuart would like to thank his talented graduate students who enthusiastically supported many book chapters, and his family for their patience and support during the production of this book. Mahmoud M. El-Halwagi is thankful to his collaborators for their excellent contributions and is grateful to his parents, wife Amal, and sons Omar and Ali for their continued support and love.

There are tremendous opportunities in the field of biorefining. We hope that this book will provide readers with the essential concepts, tools, and applications to foster innovation, and to facilitate and enable the sustainable development of integrated biorefineries around the world.

Paul R. Stuart
École Polytechnique de Montréal, Quebec, Canada

Mahmoud M. El-Halwagi
Texas A&M University, College Station, Texas

Editors

Paul R. Stuart, PhD, is a professor in the Department of Chemical Engineering at École Polytechnique of the University of Montréal, where he holds the National Sciences and Engineering Research Council of Canada (NSERC) Environmental Design Engineering Chair. As an academic and consultant, Dr. Stuart employs product and process design methodologies to develop business and technology plans for companies seeking to implement them in the biorefinery. Prior to joining academia, he held positions in leading consulting firms, working as a partner, company associate, as well as the director of process and environmental engineering. Dr. Stuart is a fellow of The Canadian Academy of Engineering and a past president of the Canadian Society for Chemical Engineering.

Mahmoud M. El-Halwagi, PhD, is a professor and holds the McFerrin Professorship at the Artie McFerrin Department of Chemical Engineering, Texas A&M University, College Station. His main research interests include process integration, biofuels and integrated biorefineries, and sustainable process design, operation, and optimization. Dr. El-Halwagi is the author/coeditor of seven books and monographs and more than 200 refereed papers and book chapters. He has served as a consultant in the biofuels, chemical, energy, petrochemical, petroleum, pharmaceutical, and metal finishing industries.

Contributors

Thomas E. Amidon
Department of Paper and Bioprocess Engineering
SUNY College of Environmental Science and Forestry
Syracuse, New York

Youngsun Baek
Oak Ridge National Laboratory
Oak Ridge, Tennessee

Dieudonné R. Batsy
Ecole Polytechnique Montreal
Montreal, Québec, Canada

Thore Berntsson
Department of Energy and Environment
Chalmers University of Technology
Gothenburg, Sweden

David L. Bilhartz, II
The Artie McFerrin Department of Chemical Engineering
Texas A&M University
College Station, Texas

Marilyn A. Brown
School of Public Policy
Georgia Institute of Technology
Atlanta, Georgia

Tom Browne
FPInnovations–Pulp & Paper Division
Pointe-Claire, Québec, Canada

John P. Buyondo
Department of Paper and Bioprocess Engineering
SUNY College of Environmental Science and Forestry
Syracuse, New York

Agustin Jaime Castro-Montoya
Facultad de Ingeniería Química
Universidad Michoacana de San Nicolás de Hidalgo
Morelia, Michoacán, México

Virginie Chambost
Ecole Polytechnique Montreal
Montreal, Québec, Canada

Hou-min Chang
North Carolina State University
Raleigh, North Carolina

Sophie D'Amours
Département de génie mécanique
Université Laval
Laval, Québec, Canada

Louis Patrick Dansereau
Department of Chemical Engineering
Ecole Polytechnique Montreal
Montreal, Québec, Canada

Mario R. Eden
Department of Chemical Engineering
Auburn University
Auburn, Alabama

Mahmoud M. El-Halwagi
Department of Chemical Engineering
Texas A&M University
College Station, Texas

Yan Feng
Département de génie mécanique
Université Laval
Laval, Québec, Canada

W. J. Frederick, Jr.
Table Mountain Consulting, LLC
Boulder, Colorado

Gil Garnier
Department of Chemical Engineering
Monash University
Victoria, Australia

Michael Garver
Department of Paper and Bioprocess Engineering
SUNY College of Environmental Science and Forestry
Syracuse, New York

Ignacio E. Grossmann
Department of Chemical Engineering
Carnegie-Mellon University
Pittsburgh, Pennsylvania

Mark Holtzapple
Department of Chemical Engineering
Texas A&M University
College Station, Texas

Damon Honnery
Department of Mechanical and Aerospace Engineering
Monash University
Victoria, Australia

Eemeli Hytönen
VTT Technical Research Centre of Finland
Espoo, Finland

Hasan Jameel
Department of Forest Biomaterials
North Carolina State University
Raleigh, North Carolina

Matty Janssen
Department of Energy and Environment
Chalmers University of Technology
Göteborg, Sweden

Naceur Jemaa
FPInnovations–Pulp & Paper Division
Pointe-Claire, Québec, Canada

Arturo Jiménez-Gutiérrez
Departamento de Ingeniería Química
Instituto Tecnológico de Celaya
Celaya, Guanajuato, México

Yongcan Jin
North Carolina State University
Raleigh, North Carolina

Vyacheslav Kafarov
Department of Chemical Engineering
Industrial University of Santander
Bucaramanga, Santander, Colombia

Ibrahim Karidio
Advanced Energy Research Facility
Edmonton Waste Management Centre
Edmonton, Alberta, Canada

Antonis C. Kokossis
School of Chemical Engineering
National Technical University of Athens
Athens, Greece

Bon Wook Koo
Department of Forest Biomaterials
North Carolina State University
Raleigh, North Carolina

Milan Korbel
Department of Chemical Engineering
Ecole Polytechnique Montreal
Montreal, Québec, Canada

Apostolis A. Koutinas
Department of Food Science and Technology
Agricultural University of Athens
Athens, Greece

Andrzej Kraslawski
Lappeenranta University of Technology
Lappeenranta, Finland

Luc LeBel
Départment des sciences du bois et de la forêt
Université Laval
Laval, Québec, Canada

Yichao Lei
State Key Laboratory of Pulp and Paper Engineering
South China University of Technology
Guangzhou, China

Bin Liang
College of Chemical Engineering
Sichuan University
Chengdu, China

Ta-Chen Lin
Faculty of Engineering and Physical Sciences
University of Surrey
Surrey, United Kingdom

Shijie Liu
Department of Paper and Bioprocess Engineering
SUNY College of Environmental Science and Forestry
Syracuse, New York

and

State Key Laboratory of Pulp and Paper Engineering
South China University of Technology
Guangzhou, China

Houfang Lu
College of Chemical Engineering
Sichuan University
Chengdu, China

Warren E. Mabee
Energy and Environmental Policy
Kingston, Queen's University
Kingston, Ontario, Canada

Talat Mahmood
FPInnovations–Pulp & Paper Division
Pointe-Claire, Québec, Canada

Maryam Mahmoudkhani
Department of Energy and Environment
Chalmers University of Technology
Gothenburg, Sweden

Behrang Mansoornejad
Department of Chemical Engineering
Ecole Polytechnique Montreal
Montreal, Québec, Canada

Mariano Martín
Department of Chemical Engineering
Carnegie-Mellon University
Pittsburgh, Pennsylvania

José Meléndez
Department of Chemical Engineering
Ecole Polytechnique Montreal
Montreal, Québec, Canada

Patrick Moriarty
Department of Design
Monash University
Victoria, Australia

Denny K. S. Ng
Department of Chemical and Environmental Engineering
University of Nottingham Malaysia
Selangor, Malaysia

Daryl Nichol
Alberta Pacific Forest Industries Inc.
Boyle, Alberta, Canada

Mustapha Nourelfath
Département de génie mécanique
Université Laval
Laval, Québec, Canada

Karina Ojeda
Department of Chemical Engineering
University of San Buenaventura
Cartagena, Bolivar, Colombia

Mike Paice
Michael Paice and Associates
Richmond, British Columbia, Canada

Mike Paleologou
FPInnovations–Pulp & Paper Division
Pointe-Claire, Québec, Canada

Viet Pham
Department of Chemical Engineering
Texas A&M University
College Station, Texas

Richard B. Phillips
Department of Forest Biomaterials
North Carolina State University
Raleigh, North Carolina

Grace Pokoo-Aikins
The Artie McFerrin Department of Chemical Engineering
Texas A&M University
College Station, Texas

Norm E. Sammons
Department of Chemical Engineering
Auburn University
Auburn, Alabama

Shabnam Sanaei
Department of Chemical Engineering
Ecole Polytechnique Montreal
Montreal, Québec, Canada

Eduardo Sánchez-Tuirán
Department of Chemical Engineering
Industrial University of Santander
Bucaramanga, Santander, Colombia

and

Department of Chemical Engineering
University of San Buenaventura
Cartagena, Bolivar, Colombia

Gary M. Scott
Department of Paper and Bioprocess Engineering
SUNY College of Environmental Science and Forestry
Syracuse, New York

Doug Singbeil
FPInnovations–Pulp & Paper Division
Vancouver, British Columbia, Canada

Charles C. Solvason
Department of Chemical Engineering
Auburn University
Auburn, Alabama

Paul R. Stuart
Department of Chemical Engineering
Ecole Polytechnique Montreal
Montreal, Québec, Canada

Runcang Sun
State Key Laboratory of Pulp and Paper Engineering
South China University of Technology
Guangzhou, China

and

Institute of Biomass Chemistry and Technology
Beijing Forestry University
Beijing, China

Douglas H. S. Tay
Department of Chemical and Environmental Engineering
University of Nottingham Malaysia
Selangor, Malaysia

Trevor H. Treasure
Department of Forest Biomaterials
North Carolina State University
Raleigh, North Carolina

Marinella Tsakalova
School of Chemical Engineering
National Technical University of Athens
Athens, Greece

Alexey Vishtal
TK 5103 Biofibre Processing
VIT—Technical Research Centre of Finland
Jyväskylä, Finland

Yang Wang
Department of Paper and Bioprocess Engineering
SUNY College of Environmental Science and Forestry
Syracuse, New York

Yuanzhen Wang
Department of Paper and Bioprocess Engineering
SUNY College of Environmental Science and Forestry
Syracuse, New York

Al Ward
Alberta Pacific Forest Industries Inc.
Boyle, Alberta, Canada

Christopher D. Wood
Department of Paper and Bioprocess Engineering
SUNY College of Environmental Science and Forestry
Syracuse, New York

Aidong Yang
Faculty of Engineering and Physical Sciences
University of Surrey
Surrey, United Kingdom

Xiao Zhang
School of Chemical Engineering and Bioengineering
Washington State University
Richland, Washington

Section I

Biorefinery Product and Process Design

1

Product Portfolio Selection and Process Design for the Forest Biorefinery

Dieudonné R. Batsy, Charles C. Solvason, Norm E. Sammons, Virginie Chambost, David L. Bilhartz, II, Mario R. Eden, Mahmoud M. El-Halwagi, and Paul R. Stuart

CONTENTS

This chapter provides the reader with an overview of the design process. It accounts for the special features of the biorefinery and illustrates how these may affect the design process. Here, the biorefinery is presented, not as a project, but as a strategy. The reader will obtain answers to the following questions: (a) how product and process design tools can support the definition of successful business strategies, (b) how the technology strategy can best serve the business model, and (c) how the model should be incorporated into the decision-making process.

The chapter is organized into seven sections. Key concepts of the biorefinery and its characteristics are introduced in the first section. Objectives are presented in the second. A literature review for the concepts of Product Design and Process Design is presented in the third section, highlighting the general aspects applicable in the biorefinery context. A review of classical methodologies is carried out, and the limitations of conventional approaches are presented. Design methods are reviewed in the fourth section, and a systematic approach to evaluating the product portfolio is proposed. A case study which illustrates the integration tools and concepts is presented in the fifth section. The sixth and seventh sections summarize and discuss the concepts presented and their influence on the future of sustainability and scenario planning.

1.1 Introduction

Economic development in today's world demands energy. However, a favorable response to the demand for energy must respect sustainability. Sustainable development can be defined as, "development that meets present needs without compromising the ability of future generations to meet their own needs" (p. 16) [1]. To align the world's manufacturing activities with these new environmental objectives, governments, nongovernmental organizations (NGOs), scientists, researchers, and authorities will need to be involved in a convergence of resources. As part of *Agenda 21*, an action program for the twenty-first century adopted by the summit participants in Rio de Janeiro in 1992, the intelligent use of constrained resources will be of paramount importance, and a number of principles, objectives, and policy instruments that emphasize renewable resources will need to be implemented to achieve this goal [2]. Fortunately, many types of primary renewable resources are available and can be used to produce energy. The rise to prominence of renewable energy sources can be attributed in part to the increased visibility of climate change. Climate change has been correlated with massive emissions of greenhouse gases (GHGs) from fuels derived from fossil resources, which are used in energy production. Today, several options are being considered to replace existing fuels and fossil-based products with biofuels and bioproducts. Bioproducts can have molecular structures similar or dissimilar to those of conventional fossil-based products. They are, respectively, called *replacement* (similar structure) and *substitution* (dissimilar structure) products and are produced from renewable biomass. As a result, biomass-derived products and fuels have the capability to respond positively to increasing energy demand while potentially considerably reducing environmental impact (e.g., GHG emissions). The exploitation of this primary raw material resource involves the concept of *biorefining*.

As the demand for sustainable products has risen, an unrelated rapid decline in the North American forestry industry has occurred, which has freed up biomass-handling equipment and manufacturing capability. The decline in the industry is due to the rising cost of energy and raw materials and has been compounded by strong competition on the international market from emerging countries in Asia and Latin America. To overcome this crisis while remaining competitive in the global market, pulp and paper companies have opted to combine or optimize their technologies for the production of their current product portfolios [3]. Unfortunately, these strategies cannot be effective over the long term. For these companies to thrive, new strategies based on sustainable products must be developed. Among these strategies, the forest biorefinery (FBR) platform represents an excellent option.

A biorefinery is defined by the National Renewable Energy Laboratory (NREL) as a "facility that integrates biomass conversion processes and equipment to produce fuels, power, and chemicals from biomass" [4]. There are currently four categories of biorefinery; they are classified as first, second, third, and fourth generation and are shown in Table 1.1 [5].

The first-generation biorefinery produces biofuels from agricultural biomass (e.g., corn, starch, vegetable oils, sugar cane). This kind of biomass is rich in sugar, which results in high production yields. Although it presents substantial advantages, this technology raises

TABLE 1.1

Classification of Biorefineries Based on Their Technology Generations

Generation	Feedstocks	Examples
First-generation biorefineries	Sugars, starch, vegetable oils, or animal fats	Bioalcohols, vegetable oil, biodiesel, biosyngas, biogas
Second-generation biorefineries	Nonfood crops, wheat straw, corn wood, solid waste, energy crop	Bioalcohols, bio-oil, bio-DMF, biohydrogen, bio-Fischer–Tropsch diesel
Third-generation biorefineries	Algae	Vegetable oil, biodiesel
Fourth-generation biorefineries	Vegetable oil, biodiesel	Biogasoline

environmental and social concerns such as: (1) the risk of creating a competition between food consumption and biorefinery processes for the same biomass, and (2) the risk of deforestation by overuse of land as well as environmental risk by the widespread use of fertilizers and pesticides. Furthermore, the life-cycle assessment (LCA) cost of a first-generation biorefinery exceeds, in many cases, that of biofuels from fossil petroleum [6]. The second-generation biorefinery uses mainly lignocellulosic biomass and is commonly called the *forest biorefinery*. Unlike the first generation, the second-generation biorefinery improves the environmental balance, and the biomass used does not compete with food production. Moreover, this biomass is so abundant that the input purchase cost is low, and therefore the production costs are reduced. The caveat for this generation is that the technologies needed to execute it are still under development. The third-generation biorefinery uses aquatic biomass like algae. This category has advantages in terms of yield and land use compared to a first-generation biorefinery. The Aquatic Species Program (ASP) reported that, "two hundred thousand hectares (less than 0.1% of climatically suitable land areas in the United States) could produce one quad of fuel" (p. 13) [7]. Like the second-generation technology, this technology is also still under development. The fourth-generation biorefinery uses biomass from municipal waste vegetable oils. It solves the recurring problem of treatment and management of this waste. The technology for this category is thermochemical, and most of the thermal and physical processes involved are already in use on an industrial scale.

A company seeking to switch to a biorefinery platform will also need to decide whether the biorefinery will be *brownfield, greenfield*, or *retrofit*. A project is described as *brownfield* when it uses infrastructure, equipment, or land abandoned by another manufacturer. The risk of a brownfield is the presence of potentially hazardous substances, pollutants, and contaminants from its previous use [8]. However, because this type of project addresses environmental issues, revitalizes life in its neighborhood, and brings new jobs and higher tax revenues to the local community, it is generally supported in the United States by the Environmental Protection Agency (EPA) and other federal partners. With the right incentives, these former economic engines can once again generate value for both private and public sectors [8]. A *greenfield* project is a project developed and implemented from scratch, and a *retrofit* project is an integration of new technology within an existing operating plant structure. For example, retrofitting biorefinery technology within an existing manufacturing facility can improve energy efficiency while at the same time diversifying the product portfolio and reducing emissions. Such integration presents many advantages, including the use of the existing supply chain (leading to synergy in feedstock supply and sales) and of existing facilities, which will undoubtedly lead to a reduction in initial capital costs [9].

This chapter will focus on the second-generation biorefinery, FBR, in a retrofit context. This approach enables pulp and paper manufacturers to continue to produce traditional forestry products while also diversifying their portfolio to make other value-added products such as biofuels, specialty chemicals, and pharmaceuticals [10]. There is a real opportunity to consider the FBR as an alternative to improving the existing pulp and paper business model, but some risks and difficulties exist. Biorefinery technologies are still under development and have not yet reached maturity, which represents a significant challenge for developers. These technological challenges inevitably lead developers into competition, which results in the protection of information and nondisclosure of relevant advances in the field for intellectual property reasons. As for new product development, the options for diversification are also considerable. Diversification of the existing product portfolio to range from commodities to specialty products can lead to potential revenue creation and margin improvement. However, the identification of the most promising products and related market strategies is not obvious given the current economic stalemate in the industry. Researchers developing design tools to assist decision-making are often forced to make assumptions based on heuristics to fill up the information gap when data are limited [11], which often creates uncertainty in the solution. There are many sources of uncertainty (e.g., future prices of energy, products, and feedstocks, climate policies, and process scaleup). Some of the uncertainties are outside the manufacturer's control and may include one or more of the following:

- What product/process combination offers the best value proposition?
- Does it lead to value creation and potential value retention over the long term?
- What are the associated competitive advantages over the long term?
- How might market dynamics, that is, the business cycle, impact on the profitability potential?
- What are the business strategies associated with the targeted value chain with respect to the market?
- Who would be the best collaboration partner to penetrate existing and mature value chains?
- Will the technology be efficient enough to compete?
- Is there enough capital available for investment?

Other uncertainties are within the manufacturer's control. These may include one or more of the following:

- Does the manufacturer have an appropriate business model?
- Does the manufacturer have the appropriate management leadership and expertise to execute the business model?
- Which product or family of products should the manufacturer produce?
- Which technology platform and process should the manufacturer use to produce the product?

Simultaneously managing these uncertainties is best performed using a method to screen out unsustainable technologies early in the development process. A review of one such method is presented in this chapter.

1.2 Objective

The objective of this chapter is to present a systematic approach to evaluating the appropriate product portfolio for biomass-derived products for an industry seeking to diversify from an ethanol-centric market. To meet this objective, this chapter first presents an overview of product and process design. This is followed by a discussion on the limitations of the current state of the art in managing portfolio design. A method is then proposed for integrating a portfolio design approach into an integrated product/process design framework. Special attention is paid to defining a business model for the forest biorefinery (FBR).

1.3 Literature Review

To consider a new business model aiming at new product/process introduction into the existing forestry industry portfolio, a review of both chemical product and process designs is required. Generic elements are first presented before extrapolating key issues to the context of the FBR.

1.3.1 Product Design

Across all fields of engineering, product design is defined according to the *Encyclopedia of Science and Technology* as, "the determination and specification of the parts of a product and their inter-relationship so that they become a unified whole. The design must satisfy a broad array of requirements in a condition of balanced effectiveness. A product is designed:

- To perform a particular function or set of functions effectively and reliably;
- To be economically manufacturable;
- To be profitably saleable;
- To suit the purposes and the attitudes of the consumer; and
- To be durable, safe, and economical to operate.

For instance, the design must take into consideration the particular manufacturing facilities, available materials, know-how, and economic resources of the manufacturer. The product should also appear significant, effective, compatible with the culture, and appear to be worth more than the price" [12].

1.3.1.1 Classical Approach

In the realm of chemical engineering, product design is thought to consist of three pillars (property function, process function, and usage function) in a chemical product pyramid, as shown in Figures 1.1 and 1.2 [13].

Optimizing the three pillars to develop new products and processes that use known technologies and meet customer needs is then the focus. The complexity of this problem has prevented a specific classical and comprehensive methodology for chemical product design from being developed [11]. Rather, general, holistic systems approaches for specific categories of problems have been developed.

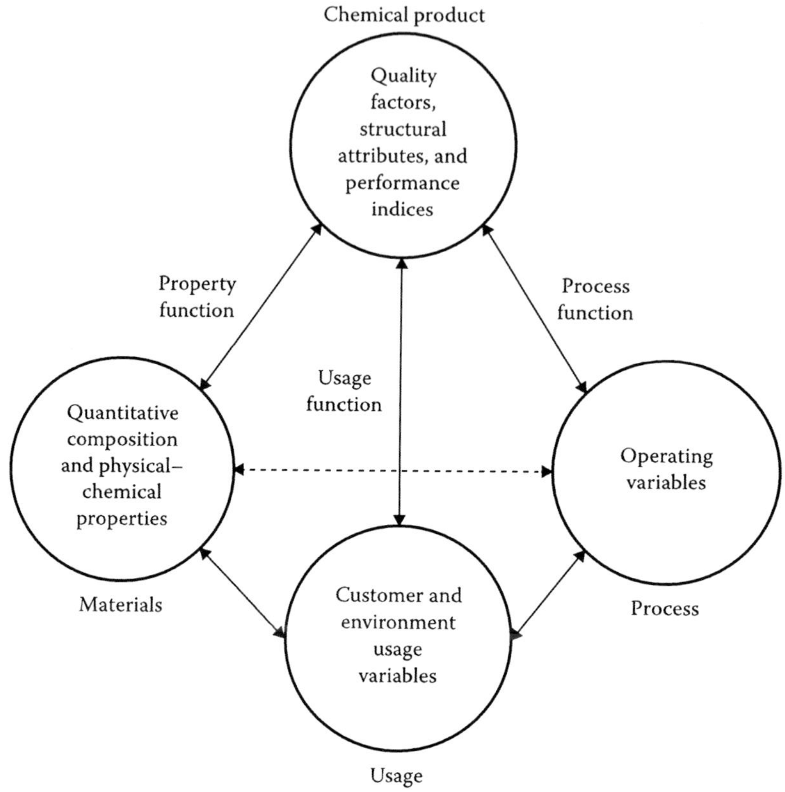

FIGURE 1.1
Structure for chemical product engineering. (Adapted from R. Costa et al., *AIChE Journal*, 52, 1981, 2006.)

One basic holistic approach consisting of seven steps has been proposed by Ulrich and Eppinger [14] and adapted to chemical product design by Costa et al. [13]. Beginning with step 1 of the seven steps shown in Figure 1.3, the needs that should be met by the product are identified. Next, target specifications for the properties that influence identified consumer needs are defined. In the third step, product ideas which potentially satisfy the identified needs are generated, followed by the fourth step, the selection of the most

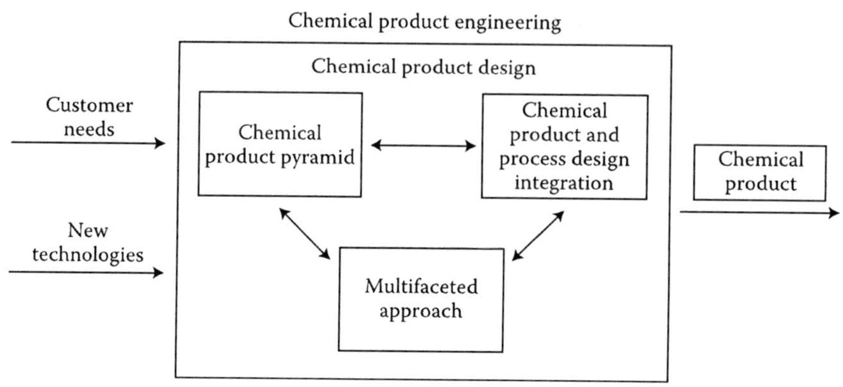

FIGURE 1.2
Structure for chemical product engineering. (Adapted from R. Costa et al., *AIChE Journal*, 52, 1978, 2006.)

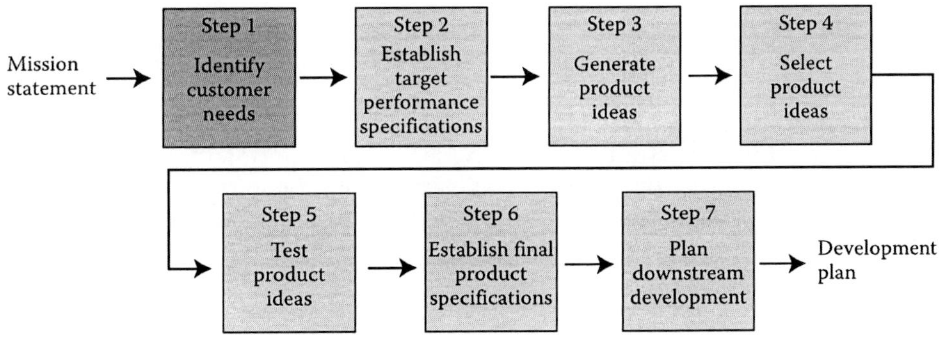

FIGURE 1.3
Product design steps. (Adapted from K.T. Ulrich and S.D. Eppinger, *Product Design and Development*, 2008.)

promising product idea. Before the final step of manufacturing the desired product, the product ideas must undergo rigorous testing and centerlining to determine the final product and process specifications, as shown in steps 5–7.

Other approaches have been proposed in the literature and are generally tailored to specific product architectures. For example, Hill [11] suggests an eight-step methodology for designing homogeneous products: (1) product definition; (2) technical product requirements; (3) product performance relationships; (4) product candidate generation; (5) product candidate selection; (6) process design; (7) risk analysis; and (8) financial (business case) analysis [11]. The methodology for the design of structured products is more difficult because of its complexity and the lack of comprehensive data describing the interdependent nature of its molecular architecture. Other authors like Westerberg and Subrahmanian [15] have suggested an alternative approach incorporating experience gained during the teaching of product design courses [15].

Most chemical product design problems can be adapted from the problem formulation proposed by Gani [16]. This formulation provides a useful tool for integrating the first four steps shown in Figure 1.3 into a mathematical program which can be solved efficiently. In this formulation, an overall product and process design problem would be described as follows:

$$F_{\mathrm{OBJ}} = \max\{C^{\mathrm{T}}\mathbf{y} + f(\mathbf{x})\} \tag{1.1}$$

$$h_1(\mathbf{x}) = 0 \tag{1.2}$$

$$h_2(\mathbf{x}) = 0 \tag{1.3}$$

$$h_3(\mathbf{x},\mathbf{y}) = 0 \tag{1.4}$$

$$l_1 \le g_1(\mathbf{x}) \le u_1 \tag{1.5}$$

$$l_2 \le g_2(\mathbf{x},\mathbf{y}) \le u_2 \tag{1.6}$$

$$l_3 \le \mathbf{B}y + \mathbf{C}x \le u_3 \tag{1.7}$$

where **x** is a vector of continuous variables representing mixture compositions and **y** is a vector of binary integer variables of molecular descriptors identifying the presence of atoms, molecular groups, and other types of product architectures. The product attributes (or consumer needs) and goals are defined by a list of objectives and constraints. $h_1(\mathbf{x})$ is the set of attribute equality constraints related to the process design parameters (e.g., pressure, reflux ratio) and explicitly described in terms of their chemical composition [16]. $h_2(\mathbf{x})$ is the set of attribute equality constraints explicitly described in terms of process models like mass and energy balances. $h_3(\mathbf{x},\mathbf{y})$ are a set of attribute equality constraints related to molecular structure, composition, and other information regarding product architecture. $g_1(\mathbf{x})$ is a set of compositionally dependent inequality bounds on the process design specifications. $g_2(\mathbf{x},\mathbf{y})$ is a set of composition and molecular architecture inequality bounds on the product. It should be noted that any linear inequality constraints are described explicitly by Equation 1.7. Finally, the term $f(\mathbf{x})$ represents a vector of linear or nonlinear objective functions. Using only the structure-attribute models (Equation 1.4), the constraints on the attributes (Equation 1.6), and the molecular feasibility logical constraints (Equation 1.7) results in a problem formulation for chemical product design that is useful as a screening tool before rigorous design [17]. For example, Solvason et al. [17] used this approach to screen for environmentally benign candidate additives to combine with R-125 to replace R-134a in a refrigerant design problem. If desired, the best candidate solution can also be found by including the objective function in the formulation and solving the system as a mixed integer nonlinear program (MINLP) using a branch-and-bound algorithm [18]. For example, in a similar refrigerant design problem, Sahindis et al. [19] used an MINLP to select nitrosyl fluoride as the best refrigerant candidate to replace R-12, while also identifying, through various relaxations, eight other previously unknown candidates.

With the decline of natural resources, volatility of oil prices, and growth of environmental awareness, there is an opportunity to utilize product design algorithms such as these to aide in the search for new, bioderived products manufactured by environmentally friendly processes. Bioderived products can be generally categorized as *substitutes*, *replacements*, or *novel*.

The term *replacement* refers to a product that satisfies an existing consumer need with the same molecular architecture as an existing product, but using new, green raw materials in an environmentally friendly manufacturing process. The term *substitute* refers to a product that satisfies an existing consumer need or improves an existing product functionality using a different molecular architecture manufactured from green raw materials by an environmentally friendly process. Note that both replacement and substitute chemical products are intended to respond to market demands in terms of quality and performance. Occasionally, a new manufacturing process makes it possible to produce a product with attributes significantly beyond what consumers believe is possible. Under these conditions, when a product has no preexisting consumer need or market, it may have any type of molecular functionality and may be manufactured from any type of biomass raw material [10], making it *novel*. An example of a novel chemical product includes tooth-whitening strips which use a specially configured chemical structure [20]. At this time, novel configured chemical products are beyond the processing technology currently available for bioderived products, and therefore the designs discussed in this chapter will be limited to replacement and substitute chemical products.

1.3.1.2 Adaptation to the Biorefinery

A forest biorefinery (FBR) creates a special set of conditions for which the existing approach to chemical product design must be adjusted. In particular, the capacity of the FBR is such that merely switching to a new fine or specialty chemical product may saturate the market in that area, driving down prices and ultimately making the switch unsustainable.

A methodology was developed by the NREL [21] for bio-based product analysis and was adapted by the Pacific Northwest National Laboratory (PNNL) [22] to identify the potential of lignin by-products. Based on the definition of a carbohydrate platform consisting of more than 300 chemicals, the NREL analysis targeted a group of promising value-added chemicals, taking into account (1) preliminary economic and technical criteria, (2) chemical functionality and technical screening, (3) technical barriers based on the best available technical pathways, and (4) the potential for each building-block chemical to produce a range of derivatives.

A more market-driven approach has been proposed in various reports such as Penner [23] to select promising building blocks for the biorefinery considering product technical feasibility and a thorough market potential analysis. Figure 1.4 shows the necessity of combining both market and technology-driven approaches to determine the right slate of products for the biorefinery.

The schematic in Figure 1.4 also illustrates the importance of a market-centric approach for defining new products and the technological impacts needed for successful design. The choice of product architecture is influenced not only by product functionalities, but also by technological constraints and product strategies, as shown in Figure 1.5.

One of the most important challenges facing forestry companies considering the biorefinery concerns the successful diversification of their existing product portfolio toward the production of products showing promising market potential and leading to competitive advantage over the long term. In general, a *product portfolio* is defined as a set of multiple chemical products related through a common molecular architecture or manufacturing process. A product portfolio is usually based on a central platform commodity chemical that can be easily converted to the other commodity or specialty chemicals in its family. This definition leads to a codependency of products on the processes used to manufacture them and complicates the chemical product design process. For this reason, it is common to differentiate product portfolio design from product design because when designing

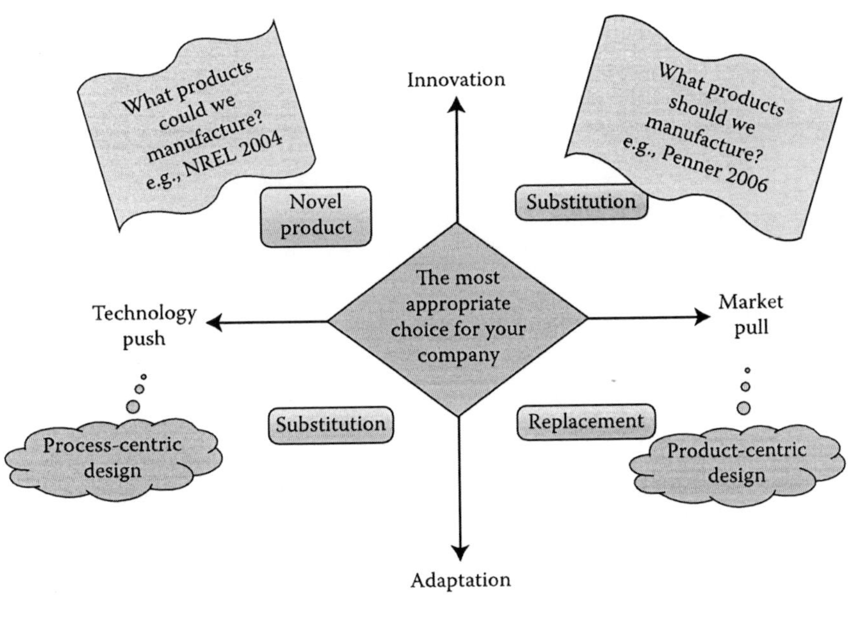

FIGURE 1.4
Analytical approaches for examining FBR product portfolios. (Adapted from V. Chambost et al., *Pulp and Paper Canada*, 109, 1–9, 2008.)

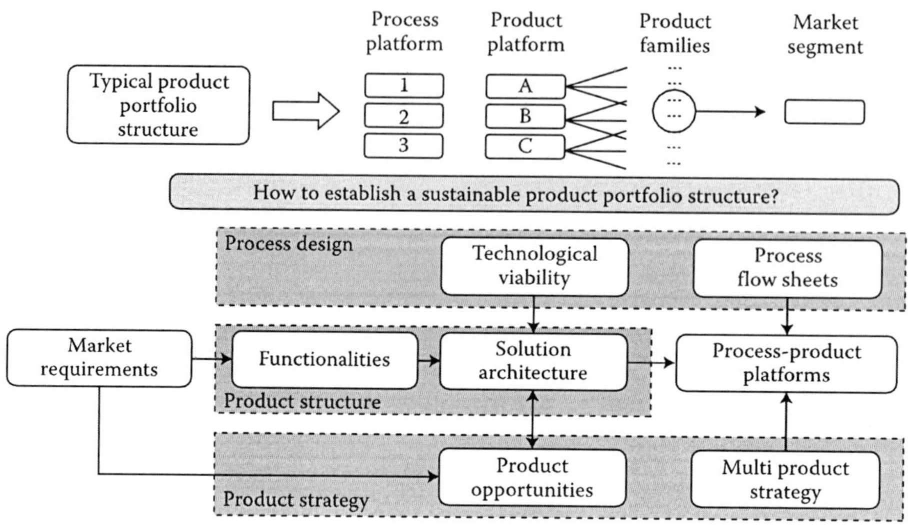

FIGURE 1.5
Product portfolio structure.

chemical product portfolios, any new product portfolio designed must take into account issues related both to its manufacture and to its product family's manufacture.

In addition, the choice of product portfolios should be driven by the optimization of profit margins through advanced supply chain strategies beyond the manufacturing facility. Critical questions will need to be addressed by companies establishing biorefinery strategies, such as the following: (1) who are the best business partners and business models for successful implementation, (2) what emerging production processes (biochemical, thermochemical, or chemical) enable product diversification while providing the targeted return on investment, and (3) what would be the best business model for outsourcing some of the key market delivery functions.

Incorporation of market constraints within the design of the FBR product portfolio can occur in three phases. The objective of the first phase is to lower the cost of the operation within the existing plant by changing the consumption pattern. Attempts should be made to replace any source of nonrenewable energy such as fuel oil with a green, renewable alternative so that the plant can become more "green." Fortunately, many options for the use of renewable energy, including bark, sludge, and other manufacturing by-products, exist within conventional pulp and paper mills. After streamlining the synthesis routes within the FBR, the focus switches to the identification and production of commodities, the building-block chemicals from which other specialty, high-value products can be made. Ethanol has been the conventional choice as a commodity for most FBRs. Other commodities such as levulinic acid may be an option, but the intrinsic risks associated with switching to a relatively unknown commodity must be mitigated.

The second phase is characterized by the development of derivatives and their production processes. Diversification into the commodities identified in phase I can help to increase revenues. Partnerships can also be considered to minimize technical and commercial risks. In general, forest product derivatives involve considerable change in both the company's business plan and its new market development strategy.

The last phase consists of improving profit margins by adopting certain strategies: promoting supply flexibility based on knowledge of market demand, using new optimization methods to identify novel products, changing the business models, developing new partnerships, and restructuring the supply chain. Although this approach provides the guidelines for implementing an FBR, a systematic methodology still needs to be developed and applied.

In conclusion, the creation of a systematic approach to the design of chemical products for the FBR will require a highly integrated approach that relies on manufacturing, market, and environmental constraints. Managing the complexity of this design will require the integration of the process and product design problems with screening heuristics from, market and the environmental constraints.

1.3.2 Process Design

In its general form, process design is a broad concept which can be defined as the approach that engineering disciplines use to specify how to create or do something [24]. Successful designs must generally satisfy a functional specification and meet implicit or explicit requirements on performance and resource usage [24]. A number of important elements must be considered when designing a process: a clear definition of problems and objectives, a process design framework, and a hierarchical stepwise approach to meet the objectives. This section will focus on a review of chemical process design. Fortunately, like product design, process design can be mathematically formulated using an objective function (Equation 1.1) followed by a series of constraints (Equations 1.2 through 1.7) on the process configurations and property domains. This formulation is ideally suited for computer-based solution. The next section discusses the classical approach to process design, followed by its extension to the design of an FBR portfolio.

1.3.2.1 Classical Approach

Process design is considered to be the cornerstone of the chemical engineering curriculum. The design process is the set of activities involved in developing and producing a chemical product. These activities are grouped into four major stages: preliminary design, basic process design, detailed engineering, and startup and operation [25]. As shown in Table 1.2 (Kaibel and Schoenmakers [26]), the lifecycle steps of an industrial process design begin with product design.

In fact, it should be noted that these steps are generally intertwined with the greater product design process, representing steps 4 through 7 of Figure 1.3 and the subsequent development steps. The most common use of process design is to provide a quick estimate of the cost of a particular processing route when investigating a particular chemical product. One design framework recently developed by Sammons et al. [27] is especially capable of handling process designs with biomass feedstocks. In the context of the FBR, such a framework for optimizing process routes and evaluating the profitability of different possible production pathways while maximizing stakeholder value could be used to build or consolidate the existing business model.

As shown in Figure 1.6, this methodology begins with a preliminary superstructure derived from a rigorous scientific review of the state of the art. Gross profit analysis is conducted on all feasible steps, and unprofitable processing routes are discarded. Next, black-box models are configured to evaluate yield, conversion, and energy efficiency. Any unprofitable routes are again discarded. Subsequent process modeling, energy integration, and mass integration are performed, resulting in a final superstructure from which economic and environmental optimization (Figure 1.7) can proceed. Through global supply chain optimization, the framework enables decision-makers to decide which products

TABLE 1.2
Overview of All Steps to Commercial Operation and End of Life

Life cycle step	Involves
Chemical route synthesis	Development of chemical synthesis steps Selection of best chemical synthesis steps
Conceptual process design	Function integration Heuristic selecting unit operations and recycle structure Superstructure optimization
Process development	Experiments for kinetic, physical data Reaction and separation tests Pilot plant Cold flow scale-up tests
Process engineering	Definition of all equipment and control for accurate economic evaluation
Site integration	Connect energy and mass flows with other processes and utilities
Detailed engineering	Definition of all process details to allow purchasing and construction
Plant operation	
End of life	Find second use Deconstruct and reuse parts

Source: Reprinted from G. Kaibel and H. Schoenmakers, Process synthesis and design in industrial practice, pp. 9–22, Copyright (2002), In: J. Grievink, J. Van Schijndel (Eds.), *Proceedings of the European Symposium on Computer-Aided Process Engineering (CAPE-12)*, with permission from Elsevier.

and pathways to pursue to maximize net present value while measuring and minimizing environmental impact [27]. Note that the method proposed by Sammons et al. [27] is a strong technique for evaluating chemical processes, but requires *a priori* knowledge of what products and processing routes to include in the superstructure. As such, its solutions are only as good as the raw materials, processing routes, and product options identified. A more detailed description of process design can be found in the works of Biegler et al. [25] and Douglas [28].

1.3.2.2 Adaptation to the Biorefinery

The process design needs for the biorefinery concept have been covered in the introduction of this chapter. This section focuses on how to meet those needs using conventional process design tools adapted to the FBR concept. In general, several manufacturing routes can be envisaged for the FBR, depending on the type of process involved: physical, chemical, thermochemical, biochemical, or biological (Figure 1.8) [29].

The process superstructure, on the basis of which rigorous optimization can be performed, will ultimately vary from one plant to another depending on the existing equipment, available capital, desired markets, and species of forest raw material. It is then essential to identify the advantages and disadvantages of different scenarios rather than the selection of particular products, which adds some uncertainty to the design. Because the design will vary depending on the variability of the local raw materials (agricultural, forest, marine, or symbiotic biomass, etc.), the number of potential product scenarios will become quite large. In this situation, it is useful to apply a set of heuristics to narrow down the products and product pathways associated with each scenario. Useful production heuristics may include the number of conversion steps or reactions required, the types of conversions, public domain knowledge of the reaction

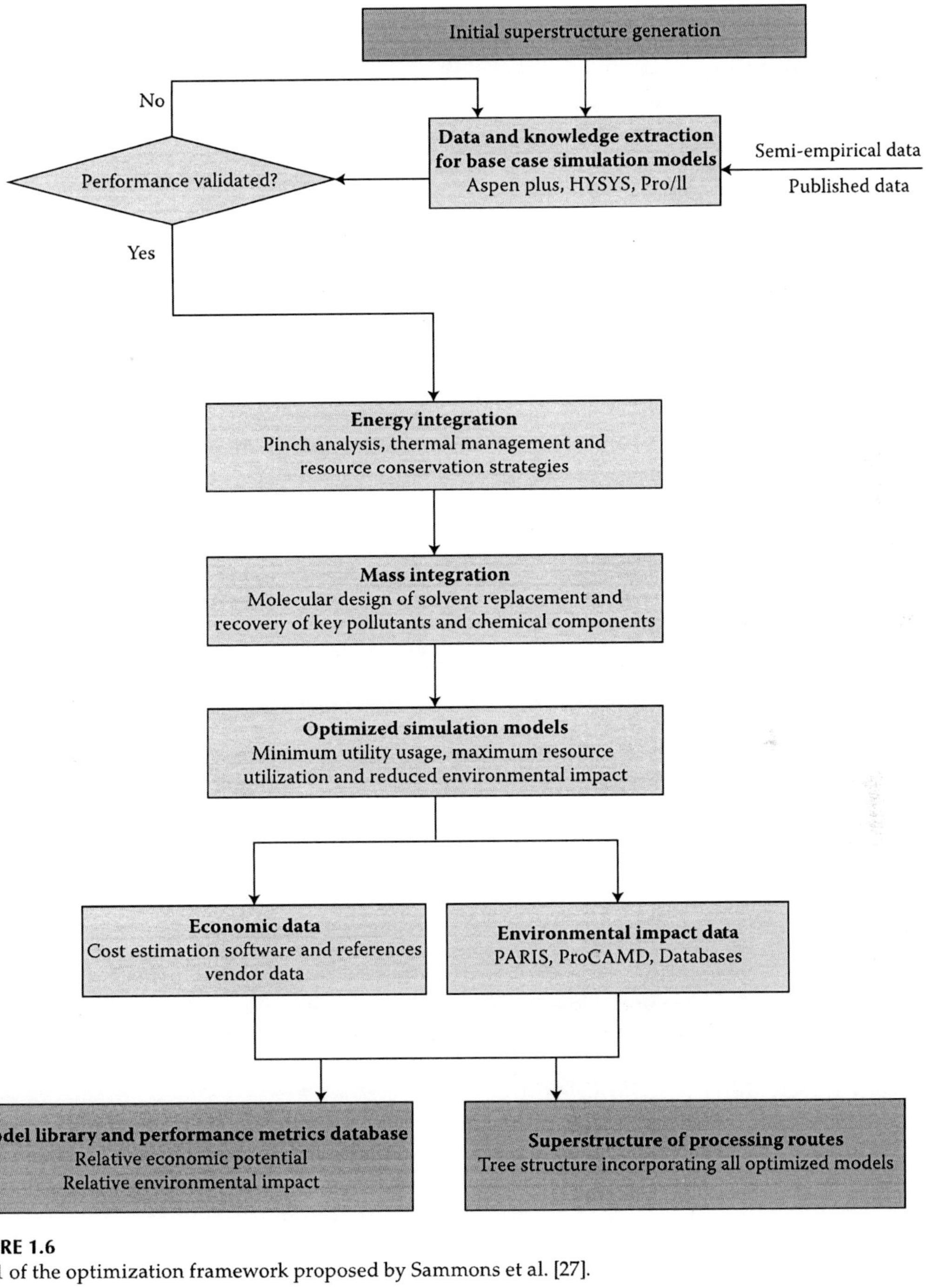

FIGURE 1.6
Part 1 of the optimization framework proposed by Sammons et al. [27].

chemistries, a minimum selling price, and others. Market heuristics, such as specialty market size or consumer appeal, and environmental heuristics like toxicity or sustainability can also be used to limit the superstructure. Once the superstructure has been developed, the optimization method developed by Sammons et al. [27] can be used to select the best product and processing route. The cumulative environmental impact of

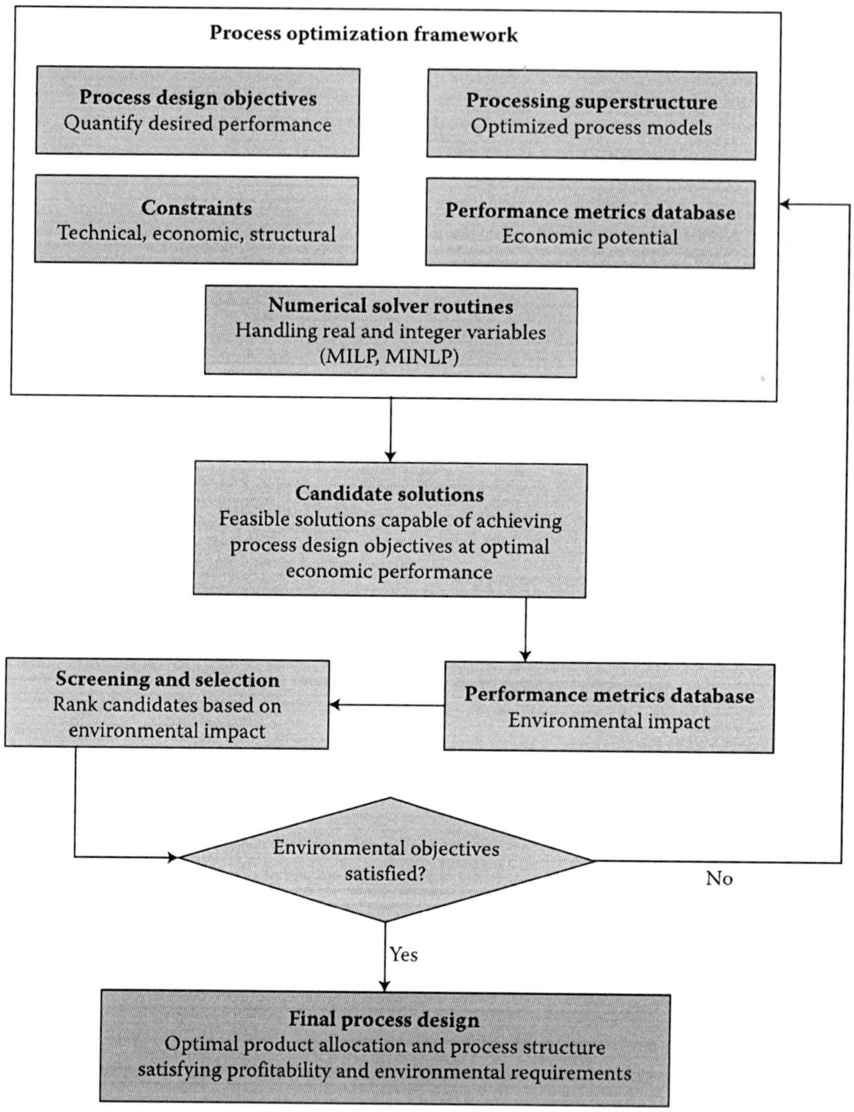

FIGURE 1.7
Part 2 of the optimization framework proposed by Sammons et al. [27].

the economically desirable products and their pathways can then be measured using the Waste Reduction (WAR) algorithm proposed by Young and Cabezas [30].

Critical to the economic and commercial success of the FBR is the identification and management of a "biorefinery product platform" (Figure 1.9). The FBR platform definition involves the determination of building-block commodities (i.e., platform chemicals) and value-added derivatives (i.e., a platform chemical's product family). This platform-based approach is typical of the petrochemical industry, where building blocks like naphtha are produced from crude oil and natural gas and then transformed into primary and secondary chemicals. However, unlike the history of the petrochemical industry, the time available to incrementally investigate biorefinery process pathways is short. Computational

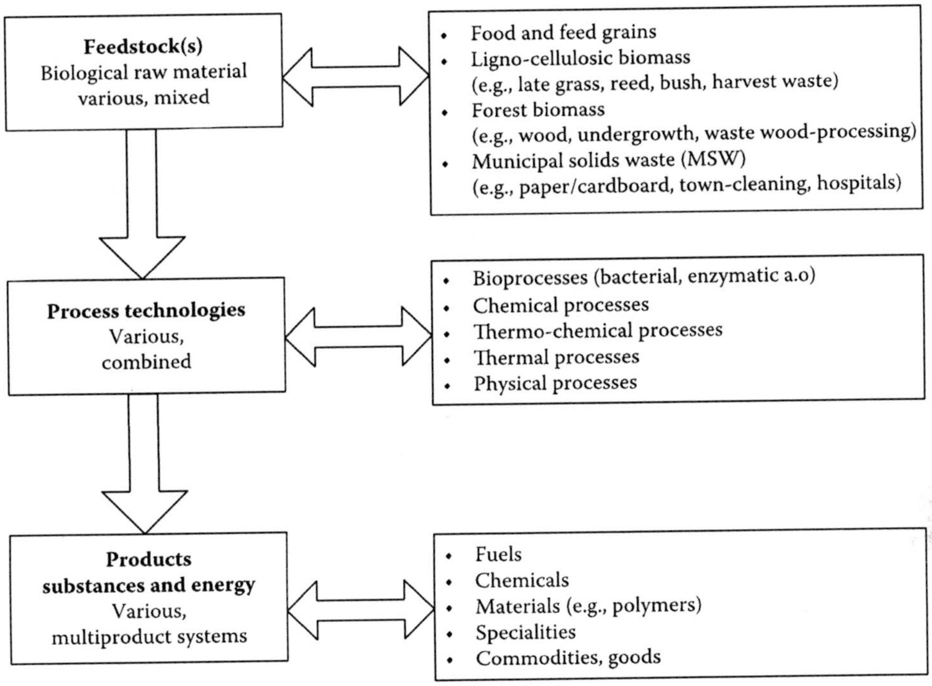

FIGURE 1.8
A generic picture of the biorefinery. (B. Kamm and P. Gruber, R., *Lignocellulosic Feedstock, Biorefineries—Industrial Processes and Products*. Vol. 1, pp. 139–164. 2006. Copyright Wiley-VCH Verlag GmbH & Co. KgaA. Reproduced with permission.)

techniques like process design optimization will be needed to identify process pathways quickly. To do so, several changes to the conventional process design framework will need to be made. First, heuristics will need to be applied individually to each of the product platform stages shown in Figure 1.10 to limit the size of the superstructure. Significantly limiting the number of platform chemicals or building blocks and their subsequent families of derivatives will be vital to ensure that timely optimization can proceed. Second, optimization may need to be conducted in stages, first on the platform chemicals, and then on their derivatives, to limit the complexity of the design. Finally, because of the uncertainty introduced by the use of heuristics, it will be beneficial to rewrite the optimization framework as a Monte Carlo optimization problem, introducing risk and variability in the parameters and measuring their impact on the product families. Other techniques intended to restrict the size of the optimization problem to ensure that it can be solved efficiently are discussed in Solvason et al. [17].

1.3.3 Limitations of Conventional Approaches

The previous sections presented two design stages, product design and process design, which must be integrated for successful application of the FBR concept. The three-phase approach (Figure 1.10) gives an insight into how the FBR can strategically be implemented by lowering operating cost, increasing revenue, and improving margins. To choose a successful set of products and processing routes, it is important to recognize that state-of-the-art product and process design tools can be used in a contextual

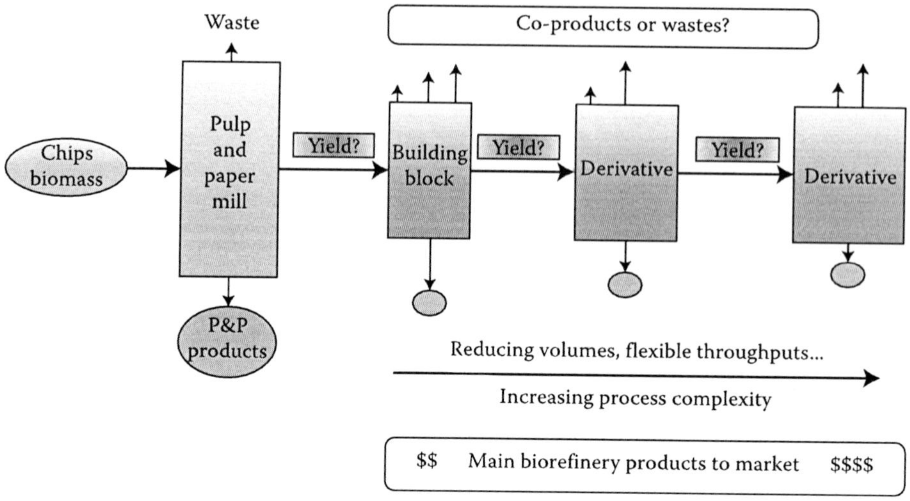

FIGURE 1.9
Biorefinery platform definition. (Adapted from V. Chambost and P. Stuart, *Industrial Biotechnology*, 3, 112–119, 2007.)

approach. In particular, chemical product and process design methodologies and formulations seldom extend beyond developing single chemical products, whereas the decision to produce bioderived chemical products will require the simultaneous design of multiple products, which is a by-product of the significant capital investment required.

For the pulp and paper industries, product portfolios will also need to be grafted onto business strategies (such as market demand, product performance, consumer needs,

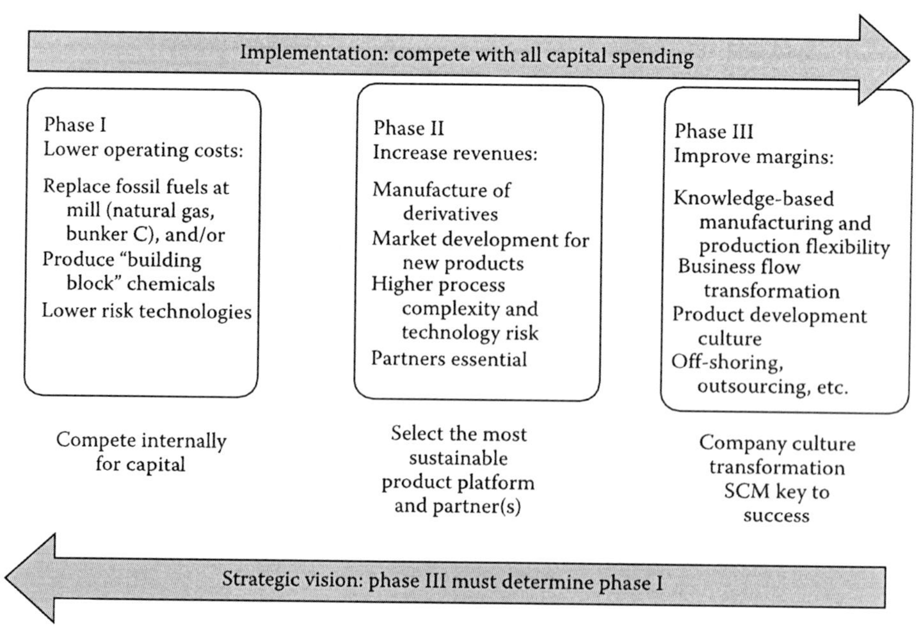

FIGURE 1.10
A three-phase approach. (Adapted from V. Chambost et al., *Pulp and Paper Canada*, 109, 1–9, 2008.)

environmental constraints, economic viability, climate policies, and government incentives). For this reason, product portfolio design has traditionally used an empirical "trial and error" approach, which costs time and money to perfect and bring to maturity [13]. The National Renewable Energy Laboratory (NREL) has recognized this limitation and has performed a preliminary search that identifies the top platform chemicals around which to build product families [21]. While promising, this report is only a starting point for the design of a biorefinery and its product portfolio. In today's world of increasing competition, there is a need to reach the market quickly. A new, rapid, and efficient approach for developing new product portfolios is required.

Recent advances in the systems community by Gani [16] and many others [11,13,17–19,25,32,40] have shown that the same programming techniques used to solve process design problems can be used to solve product design problems. Eden et al. [32] showed that these two problems can be integrated into a reverse problem formulation that enables simultaneous design. Approaches that make use of these innovations will be necessary for the pulp and paper industries to move from a commodity focus (which stresses process optimization for low operating costs and high production volumes) to an FBR with specialty markets (which stresses first-to-market principles and high-value, low-volume products).

1.4 Design Methodology

Product portfolio design in chemical engineering requires the integration of consumer attributes, market constraints, environmental constraints, economic models, and product and process design tools. In particular, the design of the FBR is sensitive to the choice of raw material. Complexity and lack of comprehensive data is a recurring problem, and therefore it is not possible to solve product portfolio design problems completely at the early stage of design without high risk. Instead, it is proposed to develop a universal method that incorporates market (consumer needs, competition), environmental, and economic (process and product design) constraints in a multicriterion decision matrix from which conclusions can be drawn [33,34]. This has the added benefit of giving a voice to various experts whose intrinsic knowledge may not be well represented in the parameters used in the various models. Note also that the market, environmental, and economic factors calculated for the analysis do not need to be completely independent of one another. For instance, Smith and Ierapepritou [35] used consumer preference, market needs, competition, and labor contracts to constrain product options in a multiobjective optimization problem. As proposed in this section, the results of such a technique would serve as inputs to a multicriterion decision-making (MCDM) tool (see Section 4.2.1) because the models used in an FBR design are poorly defined. Therefore, the methodologies developed and used within this approach will be applicable only on a case-by-case basis.

1.4.1 Product Portfolio Design

Chambost et al. [10] raised key issues related to the identification of a product portfolio for the biorefinery, as shown in Figure 1.11.

To consider those issues in a systematic manner, Chambost and Stuart [36], developed a classical "design funnel" approach (Figure 1.12) to define a set of promising product options and then eliminate the less promising ones through market analyses, called "evaluation of entry point" in existing or new value chains. Bio-based replacement or substitution of existing products in the market requires a fundamental understanding of market dynamics, the potential for penetrating existing and mature value chains, and the related potential value propositions. Each product within a portfolio must be screened using a systematic assessment of its market potential which takes into account a set of market, technology, and technoeconomic criteria such as market growth potential, product revenue potential, product yield potential to match market volume, and margin creation. The definition of the value-chain point of entry is closely linked with the potential for partnering with a "quality" third party [37]. A major effort should be devoted to competitive analysis of the overall portfolio to identify a unique value proposition for product delivery to a value chain, involving trade-offs that are distinct from those of the competition. For highly competitive markets such as the commodity market, product manufacturing and delivery cost-competitiveness are critical. On the other hand, for specialty products, differentiation and first-to-market strategies will drive competitive advantage.

This methodology can be adapted to the solution of a product portfolio design for the FBR by creating an MCDM that includes three inputs: market, economic, and environmental. Market inputs include various heuristics from the NREL as well as inputs from pulp and paper manufacturers that will serve to determine consumer needs, the presence of specialty or commodity markets, and the capacity of the FBR to meet these needs.

Economic inputs form the backbone of the MCDM and largely rely on existing product and process design methodologies, but as adapted to product portfolios. This adaptation is achieved by formulating a two-stage process, first discovering the most

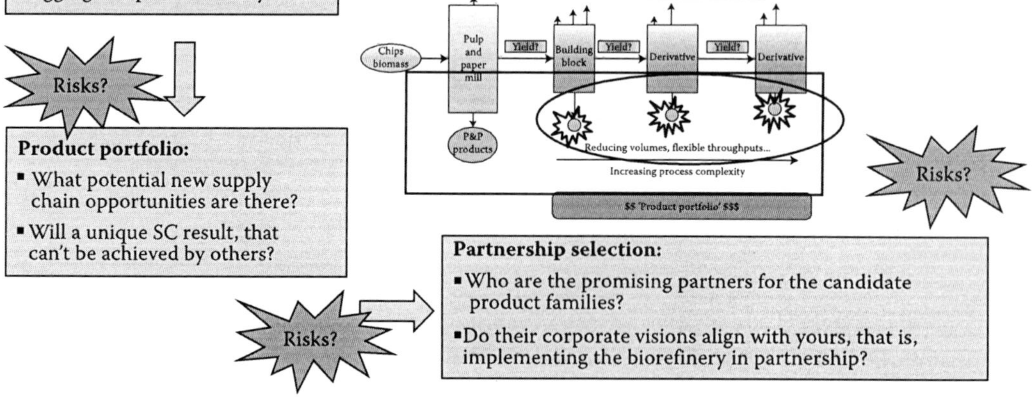

FIGURE 1.11
Key questions and the three-stage methodology proposed by V. Chambost and P. Stuart. (Adapted from V. Chambost and P. Stuart in *Design for Energy and the Environment*: CRC Press, 2009, pp. 907–917.)

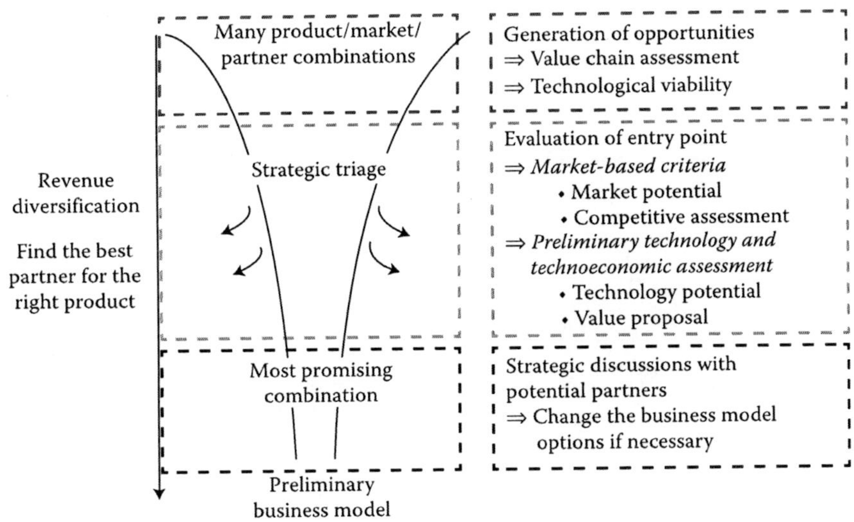

FIGURE 1.12
Value approach for preliminary business model definition. (Adapted from V. Chambost et al., *21st European Symposium on Computer Aided Process Engineering–ESCAPE 21*, 2010.)

appropriate platform chemical that can be sold as a commodity, followed by a second stage to determine the family of specialty and commodity products associated with that platform chemical. It has been proposed by Solvason et al. [17] that the superstructure is mathematically generated using reaction networks and enzymatic networks as proposed by Ugi et al. [39] and Broadbelt et al. [40]. These networks categorize reactions based on the reactive electron donor and acceptor sites, which are often common across different molecular architectures. Using generation algorithms then enables synthesis routes (that have not yet been published) to be included in the initial superstructure generation. Next, the pathways can be constrained using various heuristics such as the NREL report on the top platform chemicals [21]. Once the first-stage superstructure has been developed, the conventional integrated product and process design techniques proposed by Gani [16] or Sammons et al. [27] can be used to select the best options. Note that the platform chemical MUST be a commodity to support the production capacity of an FBR.

In the second stage, the process is repeated, using the platform products as the raw materials and mathematically generating the product family. Again, heuristics are used, but this time the commodity stipulation is removed. The result is a two-tiered superstructure which can then be evaluated using the technoeconomic assessment proposed by Sammons et al. [27]. Due to the uncertainty of market prices and process parameters, the method of Sammons et al. is modified to perform a Monte Carlo simulation, which gives the production ratio of potential products within the family. The resulting combinations are then entered into the MCDM as potential options.

Once the product pathways have been designed and developed, the WAR algorithm is used to estimate environmental impact [30]. The WAR algorithm consists of a set of health and environmental measures combined into a single potential environmental impact (PEI) value for a particular scenario. This forms a third group of parameters used in the MCDM to evaluate options. A summary of the proposed method is presented in Figure 1.13.

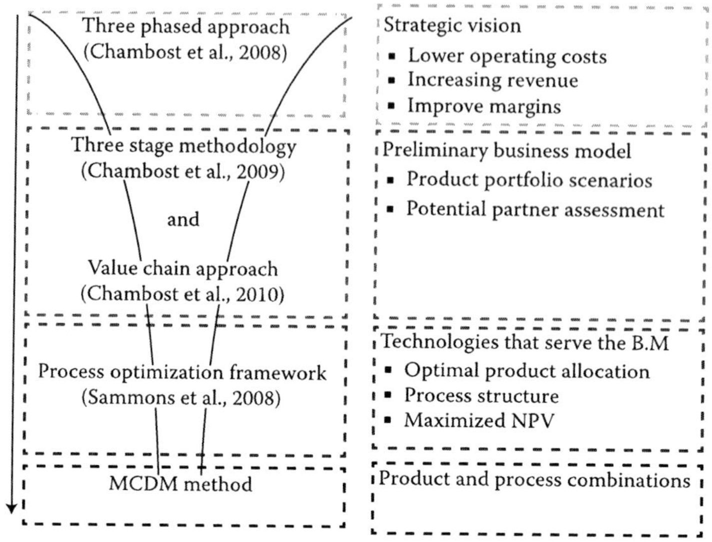

FIGURE 1.13
A systematic approach for evaluating a product portfolio. (Adapted from R. L. Keeney, *Operations Research*, 30, 803–838, 1982.)

1.4.2 Multicriterion Decision Making

The importance of the multicriterion decision-making (MCDM) tool can be characterized by observing the need for the three groups of parameters used to make a strategic decision: market, economic, and environmental. Although the FBR is an opportunity for the North American forestry sector, its implementation still presents risks and uncertainties related to early-stage design issues, the embryonic state of emerging technologies on the market, scaleup issues, biomass procurement strategies, and new market penetration. Because the industries must meet various objectives including profitability, environmental, and socioeconomic objectives, these risks and uncertainties should be assessed, and uncertain sustainable criteria which reflect all these aspects should be developed. Obviously, the interpretation of these criteria can be challenged because some of them may be in conflict when analyzed simultaneously. Therefore, the best choice of strategies cannot be made using simple common sense.

The decision-making process is complex and structured. Keeney [41] illustrated this complexity as shown in Figure 1.14 and proposed a four-step approach [41].

The biggest challenge is the implementation of the MCDM method within an organization's decision-making process. In this context, Janssen et al. [34] proposed a series of steps to conduct dealings with an MCDM panel.

The panel consists of a group of experts involved in the decision-making process. The two-phased approach shown in Figure 1.15 demonstrates how an organization can handle the decision process once the alternatives are known and completely defined. In the first phase, the decision structure, the utility function, attributes, and decision criteria are established, while in the second phase, decision problems and a weighting procedure are introduced to the panel. The members are made aware of the dependencies that may occur among criteria, and consequently, members are better equipped to address decisions under uncertainties. Once all criterion interpretations have been validated by panel

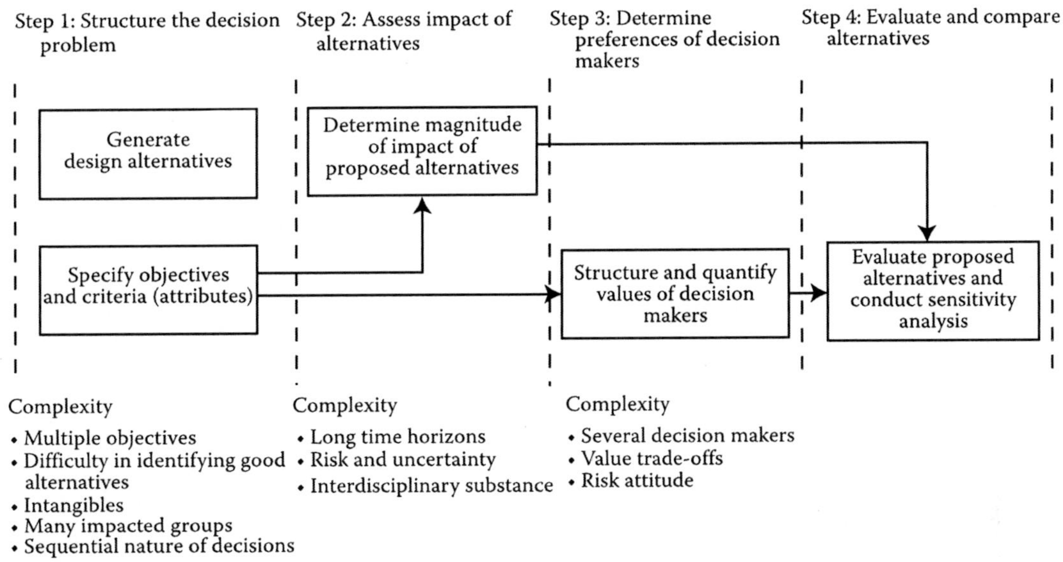

FIGURE 1.14
Schematic representation of decision analysis steps. (Adapted from R. L. Keeney, *Operations Research*, 30, 803–838, 1982.)

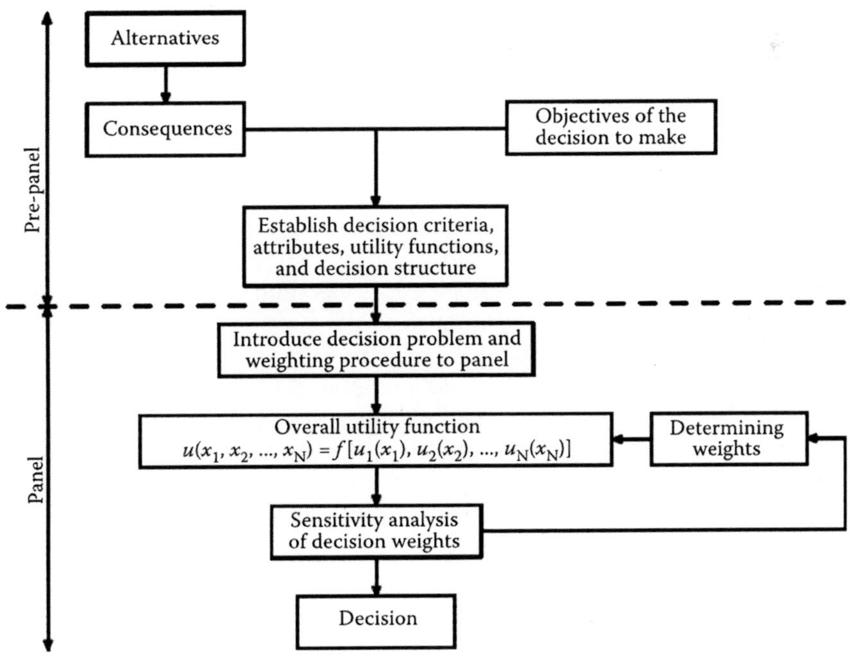

FIGURE 1.15
General procedure for working with a multicriterion decision-making (MCDM) panel. (Adapted from Janssen Matty, Retrofit design methodology based on process and product modeling, Unpublished doctoral dissertation, Chemical engineering, Université de montréal, 2007; M. Janssen et al., Eds., *Design for Energy and the Environment* 2009, FOCAPD 2009.)

members, the weighting method is used to assign a relative importance to each criterion. This generic procedure has been applied in several case studies involving key decision-makers in the forest industry [9,33].

1.5 Case Study

To illustrate the use of the proposed methodology, potential process platforms and chemicals based on cellulosic feedstock were enumerated, designed, and evaluated in a two-stage process. The first stage enumerated possibilities from cellulose to various chemical platforms, and the second stage enumerated possibilities from the optimal chemical platform to various members of the platform family. In this case study, a proposed commercial cellulose-to-ethanol plant was used as the base case for all new process designs [42]. The report from which the base case was obtained contains detailed process information, including process equipment, mass and energy balances, and a complete, scalable economic analysis. In the case study, the simplifying assumption was made that the feedstock was composed purely of cellulose to obtain a much simpler superstructure than would have been obtained by including both lignin and hemicellulose.

Twelve possible building-block value-added chemicals were presented in work published by the PNNL and the NREL [21]. The report identified 1,4 diacids (further classified as succinic, fumaric, and malic), 2,5-furan dicarboxylic acid (2,5-FDA), 3-hydroxypropionic acid (3-HPA), aspartic acid, glucaric acid, glutamic acid, itaconic acid, levulinic acid, 3-hydroxybutyrolactone, glycerol, sorbitol, and xylitol/arabinitol as potential candidates for value-added production in biorefineries. Each of the products in this list was examined using a heuristic approach to determine its profitability upper bound (PUB). For the base-case study, it was assumed that the feedstock of the biorefinery consisted of wood chips. The wood chips were processed to produce a mixture of cellulose, hemicellulose, lignin, and other by-products. For the case study, it was assumed that pure cellulose was separated from the mixture and fed into the bioreactor. For production of the value-added products, the cellulose was converted into glucose at 80% conversion using enzymatic hydrolysis via cellulase [43]. *SciFinder Scholar 2007* was then used to validate the reactions of glucose to value-added products by obtaining peer-reviewed publications of reaction mechanisms and percentage yields. Biotransformation mechanisms for 3-hydroxybutyrolactone, xylitol/arabinitol, fumaric acid, and malic acid from glucose could not be validated using *SciFinder Scholar* because of patent protection and lack of available data.

Figure 1.16 shows the resulting superstructure after validation of the proposed reaction pathways. During research on the chemical reaction pathways of glucose to the NREL building-block materials, a common chemical intermediate product, 5-(hydroxymethyl) furfural (5-HMF), was determined to link several building-block chemicals to glucose. Where possible, direct reaction pathways from glucose to product were preferred over the 5-HMF intermediate route because of the additional loss of product due to conversion in the latter pathway. Rudimentary mass balances were determined using stoichiometrically balanced chemical reactions and yield data. A basis of 100 kg of dry cellulose was defined for Stage 1 and 100 kg of levulinic acid for Stage 2 reactions, and it was assumed that no side reactions occurred. Results for product mass outputs and effective percentage yields, where effective percentage yields was defined by multiplying the percentage yields for each reaction step, can be found in Solvason et al. [17]. Market research was

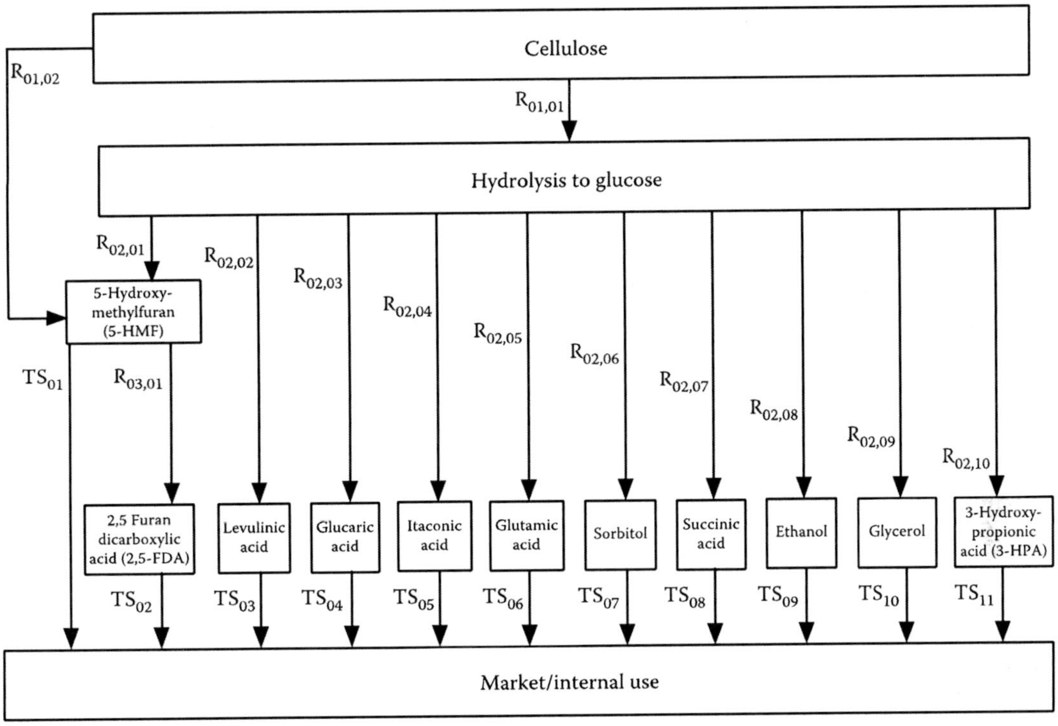

FIGURE 1.16
Validated building-block superstructure.

performed to determine the prices of the chemicals in interest in the bulk market; however, prices for some products were not available in bulk quantities. The bulk prices for these products were determined using price information for laboratory quantities of the substance. The price was converted into an estimated bulk quantity price using a price correlation factor and a scaling factor of −0.75. The purchase price of pure cellulose was then estimated using the following assumptions: 50% biomass moisture content, 1.2 variable cost penalty factor, 50% composition of cellulose in biomass, \$28.51 per m^3 [44], and an average spruce–pine–fir (SPF) density of 500 kg/m^3. The calculated prices and market prices from research are tabulated and listed in Solvason et al. [17].

PUB analysis was then performed by multiplying the mass of the cellulose entering the system by its price per unit mass. The selling price of each product per unit mass was then multiplied by the mass of product leaving the system. The "in" amount was then subtracted from the "out" amount to obtain the PUB. The products and their PUB values were then ranked in descending order. Extreme PUB values were obtained for the products 2,5-FDA, glucaric acid, and 5-HMF, primarily due to the price estimate per unit mass for these chemicals. Because of the nonexistence of a bulk market for these products and the use of the price-scaling correlation factor to determine the bulk price, these PUBs were deemed infeasible and were disregarded in the recommendation.

Succinic acid, levulinic acid, and ethanol had the highest realistic PUB values and were recommended for investigation. As a result, the original superstructure in Figure 1.16 was refined to that shown in Figure 1.17. These three potential processes were then relayed to the process synthesis and design group for further investigation of potential production.

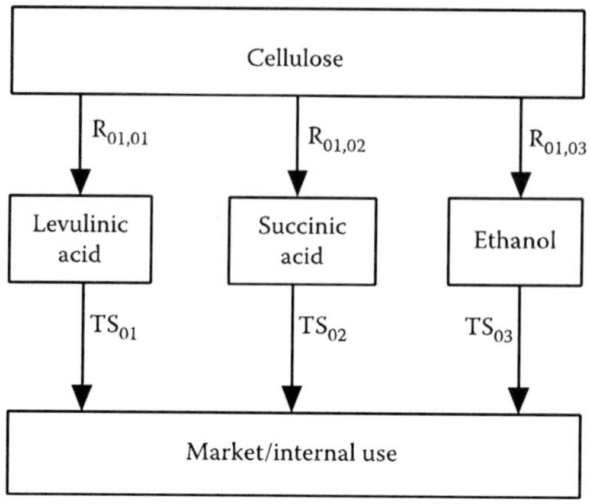

FIGURE 1.17
Modified superstructure for Stage 1 process selection.

These teams synthesized, designed, and estimated the plant cost; details of their work are described in Section 1.3. This information was then entered into the optimization-based framework presented by Sammons et al. [27] and used in Section 1.4.

A similar analysis was performed for each potential platform chemical in Stage 2. Beginning with levulinic acid, a literature from the NREL and the PNNL was again consulted for possible derivatives. Nine derivatives were suggested: diphenolic acid, g-valerolactone, methyltetrahydrofurfural (MTHF), acetylacrylic acid, 1,4-pentanediol, d-aminolevulinate (DALA), a-angelica lactone, methyl levulinate, and ethyl levulinate [21]. For each product, *SciFinder Scholar* was used to determine a reaction pathway from levulinic acid. The reactions for these products were verified, and the percentage yields were obtained or calculated from the literature. The verified superstructure was then constructed and is shown in Figure 1.18. The constraint of including only bio-enzymatic reactions was lifted for the reactions of levulinic acid, and chemical reactions were allowed in the superstructure. As a result, most of the reactions from levulinic acid to products contain additional reactants in the reaction mechanisms. For the mass balance calculations, it was assumed that 100 kg of pure levulinic acid along with stoichiometric amounts of the additional reactants were fed into the reactor.

Price values for potential products as well as additional reactants were obtained from market research or laboratory-to-bulk scaleup. In most cases, the price-scaling correlation factor was used to estimate a bulk quantity price for each of the products. Extremely high-price values were obtained for acetylacrylic acid, 1,4-pentanediol, and d-aminolevulinate; therefore, the PUB values for these products were also the highest among all product candidates. These products, however, were not disregarded in the recommendation because there is evidence that a bulk market for these products exists. PUB analysis results were tabulated for the levulinic acid product platform as shown in Solvason et al. [17], and again the top three most promising product candidates were further evaluated for process synthesis and design in the following section [45]. The refined superstructure illustrating only the top three candidates is presented in Figure 1.19.

The initial prescreening of Stage 1 chemical platform possibilities showed that ethanol, levulinic acid, and succinic acid had high levels of promise in terms of profitability. Process designs were then synthesized for the conversion of cellulose into levulinic acid and

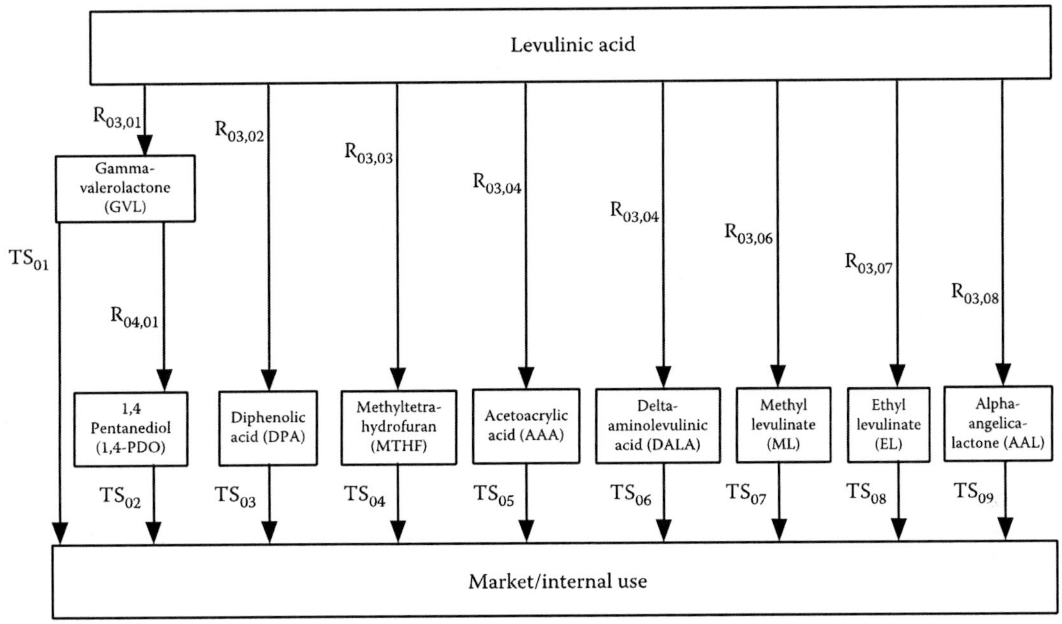

FIGURE 1.18
Process superstructure for Stage 2 from levulinic acid.

succinic acid; the process for converting cellulose into ethanol has already been well defined by Wooley et al. [42]. Figure 1.20 shows a large block diagram overview of the base-case cellulose-to-ethanol plant, and the footnotes for Figure 1.20 explain the changes made to the base case for the other processes in the first stage.

For each synthesized process, necessary steps for synthesizing each chemical were identified, and block diagrams were constructed which represented these subprocesses. The process for converting cellulose into levulinic acid is defined in the literature at the laboratory scale [46], and the process conditions and parameters were scaled up to be incorporated into the changes necessary to the ethanol base case. Because the conversion of cellulose into levulinic acid was modeled using the Biofine process, which involves dilute acid catalysis, a set of parallel reactors were added to the base-case process. Moreover, formic acid is a by-product of the production of levulinic acid, which required the use of an amine separation unit to separate a formic-acid and water mixture [47].

The succinic acid process is also defined in the literature at the laboratory scale, and a similar scaleup procedure was performed. Fermentation is also used to convert cellulose into succinic acid, and it was assumed that existing equipment for fermentation to ethanol could also be used in this synthesized process.

The large block diagram for cellulose-to-succinic acid conversion is not shown, but can be assumed to be nearly identical because a similar fermentation process is used for conversion. The large block diagram for cellulose-to-levulinic acid is also not shown, but additional process blocks include a large glass-lined reactor for acid-catalyzed conversion of cellulose into levulinic acid using the Biofine process and an amine separation unit for the separation of water and formic-acid by-product. During scaleup, the incoming cellulose feed rate from the base case was kept constant, and the large blocks were modified to reflect the conversion rates found in the literature. Overall and large block mass balances were performed to determine

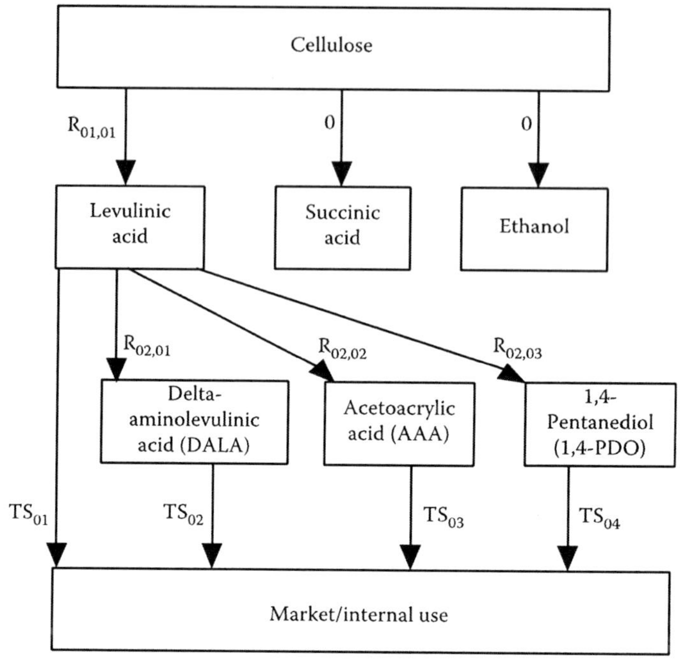

FIGURE 1.19
Modified superstructure for Stage 2 process selection.

overall conversion rates and separation requirements for subsequent detailed design. In the design stage, equipment materials, equipment size, and total installed capital cost were determined for the synthesized process. For large block components present in the ethanol base case, a scaleup depending on mass flow was performed. For block components not found in the base case, process blocks were synthesized and designed manually.

For (Figure 1.7) levulinic acid, and initial prescreening reduced the derivative search space to delta-aminolevulinic acid (DALA), 1,4-pentanediol (PDO), and acetoacrylic acid (AAA). Figure 1.21 illustrates the large block diagram for conversion of glucose into DALA via levulinic acid, and the footnotes of Figure 1.21 explain the differences between the DALA process and the PDO and AAA processes. The literature was once again reviewed for these three processes to synthesize and design the processes in a similar fashion to Stage 1. The entering flowrates for these Stage 2 processes were set to the existing flow rate of the Stage 1 process for conversion of cellulose into levulinic acid. Because the DALA process also produces formic acid as a valuable by-product, the amine extraction unit in the levulinic acid process was replicated as a second identical unit. Distillation columns were used to purify the product streams to approximately 98% purity. All reactors were overdesigned by 30% to ensure safe yet efficient operation.

In the case of DALA production, a second amine separation unit is assumed to be present because DALA production results in additional formic-acid by-product. However, the second amine separation unit is not present in other processes. The large block diagram for conversion of cellulose into PDO via levulinic acid is not shown, but can be assumed to be similar with the addition of a flash tank and a distillation column. The large block diagram for conversion of cellulose into AAA via levulinic acid is also not shown, but can be assumed to be similar to the PDO block diagram in its overall unit operations.

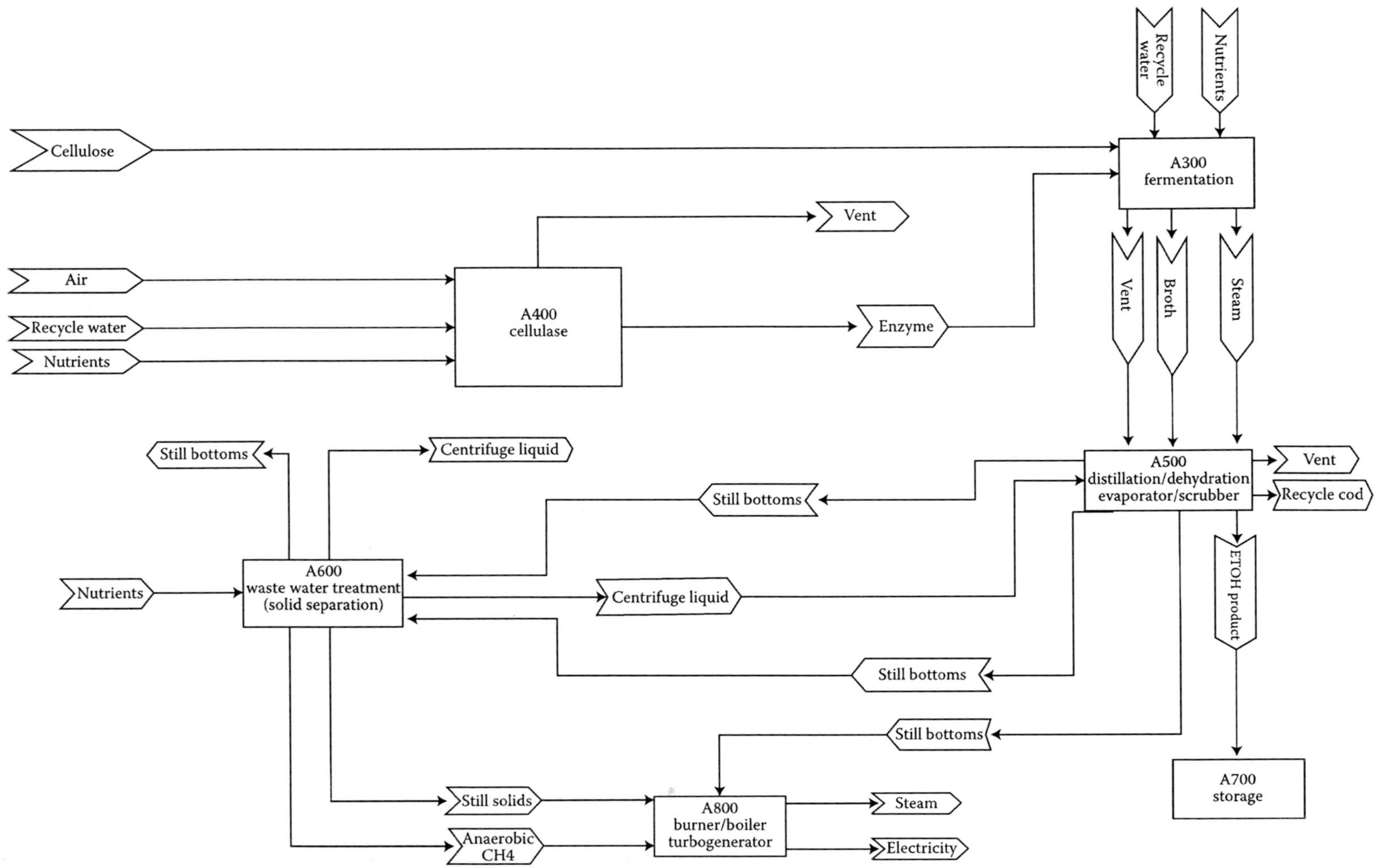

FIGURE 1.20
Large block diagram for cellulose-to-ethanol conversion. (Adapted from Robert Wooley et al., Lignocellulosic biomass to ethanol process design and economics utilizing co-current dilute acid prehydrolysis and enzymatic hydrolysis current and futuristic scenarios. National Renewable Energy Laboratory, 1999.)

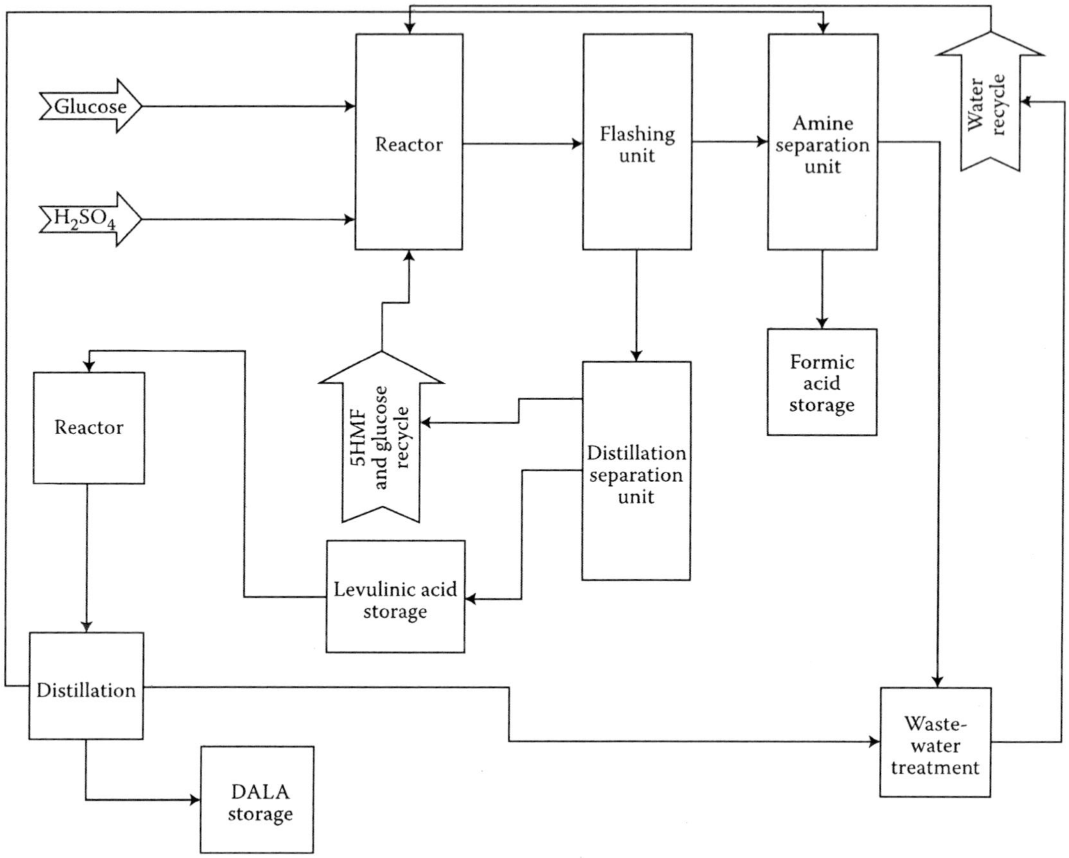

FIGURE 1.21
Large block diagram for cellulose to DALA via levulinic acid.

In the product allocation framework presented by Sammons et al. [27], mathematical optimization was used to determine which processing routes should be pursued to optimize the objective function at hand. In the interests of simplicity, in Stage 1, the framework sought to maximize annual gross profit as its measure to determine what chemical product platform should be pursued. The optimization program maximized the following objective function:

$$\max Profit = \sum_{m}\left(\sum_{k} TS_{mk}C_k^s - \sum_{i}\sum_{j} R_{mij}C_{mij}^P - C_m^{BM}\sum_{j} R_{m1j}\right) \tag{1.8}$$

The first set of terms in the objective function represents the revenue realized by selling a given product on the market. The second set of terms represent the accumulated costs, both fixed and variable, that are incurred when a certain production pathway is pursued, and the last set of terms denote feedstock cost. This optimization is subject to constraints on mass balances around production points of the superstructure and on maximum processing capacity for both feedstock and products.

In qualitative order, the optimization framework selected levulinic acid as its most profitable platform in Stage 1, followed by succinic acid and ethanol. Although PUB calculations suggested that succinic acid should be produced due to its higher market price, the coproduction of formic acid enhanced the profitability of levulinic-acid production. The execution of the optimization framework for Stage 2 processes suggested that DALA should be produced, followed by AAA and PDO. As can be seen in the figure, the market prices for all three of these products are much higher than those of commodity-grade chemicals due to the lack of available market data for these products and the need to scaleup prices from laboratory quantities.

Instead of focusing on the absolute profitability of pursuing these process pathways, a more interesting result might be obtained using Monte Carlo simulation of price movement. A normal distribution for price movement was assumed, where the mean was the calculated bulk price and the standard deviation was set to half of this mean. Under these conditions, approximately 98% of product prices will fall into the range between zero and double the calculated price. The levulinic-acid price was kept constant as a failsafe in the rare event that the prices returned from the normal distribution were all zero or less, which would mean that levulinic acid should be produced and sold directly to the market. Figure 1.22 illustrates the distribution of the optimal product selected by the mathematical optimization procedure with this price movement. Although DALA is the predominantly chosen solution for optimal profitability in production, it should be noted that there is market risk involved that could possibly shift the optimal answer to another product in the family.

These results can then form the product options for an MCDM, preferably weighted by their likelihood of production. Environmental impact for each option can be calculated using the potential environmental impact (PEI) of the WAR algorithm [30]. Although environmental impact was not measured as a part of this case study, it should be noted that growing environmental concerns are key to decision-making with regard to biorefining processes, and therefore an environmental impact assessment will be necessary once the full MCDM matrix has been developed. Other items such as risk can also be entered using various augmentations of price volatility as identified by Chambost et al. [10] and Solvason et al. [17]. Such an approach is crucial when accounting for other metrics in addition to profitability and environmental impact, such as process risk, market

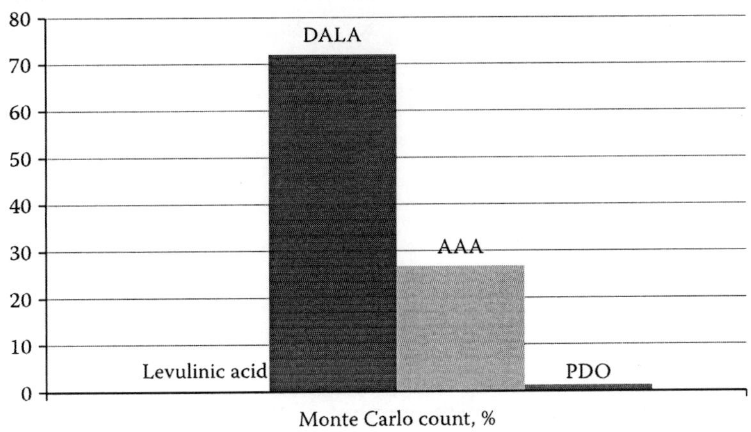

FIGURE 1.22
Monte Carlo price movement simulation and histogram of selected products from the optimization program.

risk, process flexibility, and other unforeseen metrics. One of the weaknesses encountered in the case study was lack of accurate price information and its subsequent impact within the optimization framework. More research needs to be performed to determine better methods of approximating bulk chemical prices. In addition, market risk in this case study was tied to price movement alone, but the price of a chemical is the result of many factors, and these factors may not be the same ones that drive the risk of a chemical within the market. This type of risk should also be accounted for, and using MCDM as the tool to do so would enable buy-in among all stakeholders in the decision-making process.

1.6 Conclusions

The objective of this chapter was to present a systematic approach for evaluating the appropriate product portfolio for biomass-derived products for an industry seeking to diversify from an ethanol-centric market. Special attention was paid to defining the business model for the FBR, focusing on the second generation of biorefinery, called the *forest biorefinery* (FBR), in the retrofit context. This approach enabled the presentation of the FBR as a real opportunity for the North American forest industry. A literature search was carried out to provide an overview of product and process design. The classical approaches of product and process design were reviewed, and their adaptation to the biorefinery was described. This step enabled an analysis and a discussion on the limitations of the current state of the art for performing portfolio design. Based on the limitations discussed, a method for integrating portfolio design and multicriterion decision-making (MCDM) into the integrated product and process design framework was proposed. The proposed methodology was illustrated using a case study in which biorefining possibilities were generated using systematic methods, prescreened by basic economic calculations, synthesized and designed, and then screened further by means of an established optimization-based framework to determine which product and process combination resulted in maximum profitability. One of the challenges encountered in the case study was lack of accurate price information and its subsequent impact within the optimization framework. More research needs to be performed to determine better methods of approximating bulk chemical prices. In addition, market risk in the case studies was tied to price movement that may not be risks of the chemical within the market.

1.7 Future Directions: Sustainability and Scenario Planning

On one hand, given the current crisis in the North American forestry industry, it is imperative for companies to consider the improvement of their business model by integrating new and value-added products into their existing product portfolios [10]. The new business model should accommodate process constraints as well as environmental standards and government policies. Today, for example, the product environmental declaration (PED), carbon footprint, and government policies are key elements to take into account in decision-making about the final business model. Strategic decisions should consider sustainability in the early stages of product and process design by defining economic, social, and environmental objectives instead of only an economic objective as is usual in conventional process design [33]. On the other, given the unpredictable nature of future policy, the North American forest industry need to adopt scenario planning, an approach called a tool for strategic thinking by Schoemaker [48] and proposed as a tool for survival in an uncertain world by Peterson et al. [49], to help the industry scan the future and take advantage of the unexpected opportunities that will come along. The success and advantages of scenario planning have been proven in various cases, including the classic stories of Shell Oil in the 1970s (during the oil crisis) and in the 1980s (with the declining price of oil), which enabled Shell to outperform its competitors and changed it from one of the smallest multinational oil companies to the second largest [50,51]. These dynamic changes offer substantial opportunities for the biorefining sector and highlight the importance of the tools presented in this chapter.

References

1. G. Brundtland, *Our Common Future: The World Commission on Environment and Development*. Oxford: Oxford University Press, p. 16, 1987.
2. United Nations General Assembly, Rapport de la conférence des nations unies sur l'environnement et le développement à Rio de Janeiro. United Nations, 1992.
3. U. Wising and P. Stuart, Identifying the Canadian forest biorefinery. *Pulp and Paper Canada*, 107, 25–30, 2006.
4. National Renewable Energy Laboratory. What is a biorefinery? Available online from: http://www.nrel.gov/biomass/biorefinery.html, Accessed on July 9, 2012.
5. A. Demirbas, *Biorefineries: For Biomass Upgrading Facilities*. Springer, Trabzon, Turkey, 2010.
6. Agence de l'environnement et de la maîtrise de l'energie (ADEME), Life cycle assessments applied to first generation biofuels used in France, Agence de l'environnement et de la Maîtrise de l'Energie, February, 2010.
7. J. Sheehan, T. Dunahay, J. Benemann, and P. Roessler, *A Look Back at the U.S. Department of Energy's Aquatic Species Program: Biodiesel from Algae*, National Renewable Energy Laboratory, US DEO, p. 13, 1998.
8. EPA, *Brownfields Tax Incentive Guidelines*. US EPA, 2008.
9. J. Cohen, M. Janssen, V. Chambost, and P. Stuart, Critical analysis of emerging forest biorefinery (FBR) technologies for ethanol production. *Pulp & Paper Canada*, 111, 24–30, 2010.
10. V. Chambost, J. McNutt, and P.R. Stuart, Guided tour: Implementing the forest biorefinery (FBR) at existing pulp and paper mills. *Pulp and Paper Canada*, 109, 1–9, 2008.
11. M. Hill, Chemical product engineering—The third paradigm. *Computers & Chemical Engineering*, 33, 947–953, 2009.
12. Richard I. Felver, "Product design," in AccessScience, ©McGraw-Hill Companies, 2008, Available online at http://www.accessscience.com.lib-ezproxy.tamu.edu:2048, July 9, 2012.
13. R. Costa, G. D. Moggridge, and P. M. Saraiva, Chemical product engineering: An emerging paradigm within chemical engineering. *AIChE Journal*, 52, 1976–1986, 2006.
14. K. T. Ulrich and S. D. Eppinger, *Product Design and Development*. New York: McGraw-Hill, 2000.
15. A. Westerberg and E. Subrahmanian, Product design. *Computer and Chemical Engineering*, 24, 959–966, 2000.
16. R. Gani, Chemical product design: Challenges and opportunities. *Computers & Chemical Engineering*, 28, 2441–2457, 2004.

17. C. C. Solvason, N. E. Sammons Jr., D. Bilhartz, J. A. Pack, J. Harsimran, V. Chambost, P. R. Stuart, M. R. Eden, and M. M. El-Halwagi, Chemical product portfolio design: A case study for the forest biorefinery. *Chemical Engineering Research & Design* (submitted), 2012.
18. A. P. Duvedi and L. E. K. Achenie, Designing environmentally safe refrigerants using mathematical programming. *Chemical Engineering Science*, 51, 3727–3739, 1996.
19. N. V. Sahnidis, M. Tawarmalani, and M. Yu, Design of alternative refrigerants via global optimization. *AIChE Journal*, 49, 1761–1775, 2003.
20. W. D. Seider, J. D. Seader, D. R. Lewin, and S. Widagdo, *Product and Process Design Principles: Synthesis, Analysis, and Evaluation*. New York: Wiley, 2009.
21. T. Werpy and G. Petersen, *Top Value Added Chemicals from Biomass*. National Renewable Energy Laboratory and Pacific Northwest National Laboratory, US DOE, 2004.
22. J. E. Holladay, J. J. Bozell, J. F. White, and D. Johnson, *Top Value-Added Chemicals from Biomass-Volume II—Results of Screening for Potential Candidates from Biorefinery Lignin*. PNNL-16983; Other: BM0102070, 2007.
23. G. Penner, The future for bioproducts. *BlueWater Sustainability Initiative Conference*, Sarnia, 2007.
24. The Free On-line Dictionary of Computing, © Denis Howe 2010 available at http://foldoc.org.
25. L. Biegler, I. Grossman, and A. Westerberg, *Systematic Methods of Chemical Process Design*. New Jersey: Prentice-Hall PTR, 1997.
26. G. Kaibel and H. Schoenmakers, Process synthesis and design in industrial practice. *Computer Aided Chemical Engineering*, 10, 9–22, 2002.
27. N. E. Sammons Jr., W. Yuan, M. R. Eden, B. Askoy, and H.T. Cullinan, Optimal biorefinery product allocation by combining process and economic modeling. *Chemical Engineering Research and Design*, 86, 800–808, 2008.
28. J. Douglas, *Conceptual Design of Chemical Processs*. New York: McGraw-Hill, 1988.
29. B. Kamm, P. R. Gruber, and M. Kamm, Biorefineries – Industrial Processes and Products. Ullmann's Encyclopedia of Industrial Chemistry. 2007.
30. D. M. Young and H. Cabezas, The waste reduction (WAR) algorithm: Environmental impacts, energy consumption, and engineering economics. *Waste Management*, 20, 605–615, 2000.
31. V. Chambost and P. Stuart, Selecting the most appropriate products for the forest biorefinery. *Industrial Biotechnology*, 3, 112–119, 2007.
32. M. R. Eden, S. B. Jorgensen, R. Gani, and M. M. El-Halwagi, A novel framework for simultaneous separation process and product design. *Chemical Engineering and Processing*, 43, 595–608, 2004.
33. M. Janssen, Retrofit design methodology based on process and Product modeling, Unpublished doctoral dissertation, Chemical engineering, Université de montréal, 2007.
34. M. Janssen, V. Chambost, and P. Stuart, Choice of a sustainable forest biorefinery product platform using an mcdm method. *Design for Energy and the Environment*, 389–404, 2009.
35. B. V. Smith and M. Ierapepritou, Framework for consumer-integrated optimal product design. *Industrial & Engineering Chemistry Research*, 48, 8566–8574, 2009.
36. V. Chambost and P. Stuart, Product portfolio design for forest biorefinery implementation at an existing pulp and paper mill. *Design for Energy and the Environment*, 907–917, 2009.
37. M. Janssen, V. Chambost, and P. R. Stuart, Successful partnerships for the forest biorefinery. *Industrial Biotechnology*, 4, 352–362, 2008.
38. V. Chambost, B. Mansoornejad, and P. Stuart, The role of supply chain analysis in market-driven product portfolio selection for the forest biorefinery. *Computer Aided Chemical Engineering*, 29, 1030–1034, 2011.
39. I. Ugi, J. Bauer, C. Blomberger, J. Barndt, A. Dietz, E. Fontain, B. Gruber, A. von Scholley-Pfab, A. Senff, and N. Stein, Models, concepts, theories, and formal languages in chemistry and their use as a basis for computer assistance in chemistry. *Journal of Chemical Information and Computer Sciences*, 34, 3–16, 1994.
40. L. J. Broadbelt, C. S. Henry, and V. Hatzimanikatis, Discovery of novel routes for the production of fuels and chemicals. *Design for Energy and the Environment*, 141–148, 2009.

41. R. L. Keeney, Feature article—Decision analysis: An overview. *Operations Research*, 30, 803–838, 1982.
42. R. Wooley, M. Ruth, J. Sheehan, and K. Ibsen, Lignocellulosic biomass to ethanol process design and economics utilizing co-current dilute acid prehydrolysis and enzymatic hydrolysis current and futuristic scenarios. National Renewable Energy Laboratory, US DOE, 1999.
43. C. N. Hamelinck, G. van Hooijdonk, and A. P. C. Faajj, Ethanol from lignocellulosic biomass: Techno-economic performance in short-, middle- and long-term. *Biomass and Bioenergy*, 28, 384–410, 2005.
44. Ministry of Forests, Lands, and Natural Resource Operations, British Columbia Interior Log Market Report for the 3 month period July 1, 2009 to September 30, 2009 8, Revenue Branch, Ministry of Forests and Range, 2009.
45. M. Paster, J. L. Pellegrino, T. M. Carole, *Industrial Bioproducts: Today and Tomorrow*," U.S. Department of Energy, US DOE, 2003.
46. P. Lee, S. Lee, S. Hong, and H. Chang, Isolation and characterization of a new succinic acid-producing bacterium. *Applied Microbiology and Biotechnology*, 58, 663–668, 2002.
47. H. Uslu, Reactive extraction of formic acid by using tri octyl amine (TOA). *Separation Science and Technology*, 44, 1784–1798, 2009.
48. P. J. H. Schoemaker, Scenario planning: A tool for strategic thinking, *Sloan Management Review*, 36, 25–40, 1995.
49. G. D. Peterson, G. S. Cumming, and S. R. Carpenter, Scenario planning: A tool for conservation in an uncertain world. *Conservation Biology*, 17, 358–366, 2003.
50. P. Wack, Scenarios: Uncharted waters ahead. *Harvard Business Review*, 63, 72–89, 1985a.
51. P. Wack, Scenarios: Shooting the rapids. *Harvard Business Review*, 63, 139–150, 1985b.

2

Systematic Screening of Multiple Processing Paths in Biorefineries: The ABC (Assessing Biomass to Chemicals) Project and Its Potential to Build Process Synthesis Capabilities

Antonis C. Kokossis, Aidong Yang, Marinella Tsakalova, and Ta-Chen Lin

CONTENTS

2.1 Introduction

The use of renewable biomass feedstocks has recently received increased attention in the process industry, induced by the gradual depletion of oil, uncertainties in energy supplies, and a commanding requirement to reduce greenhouse gas (GHG) emissions. In the renewable energy sector, biofuels and bioenergy already play an important role. In the chemical and biochemical sector, McKinsey (Riese, 2006) predicted an annual growth rate of 5% for (biomass-derived) fermentation products, with all bio-based products to account for 10% of the chemical industry (96€ billion) by 2010. On the other hand, according to the statistics published by Eurostat (http://ec.europe.eu/eurostat) in 2005, the chemical industry currently still remains highly dependent on oil and gas which account for 80% of its feedstock and energy consumption. In contrast to the alarming limitations of the availability of fossil feedstock, most of the annual production of biomass in the world is currently left underexploited (220 billion tons per year), offering an abundant renewable source of chemicals and carbon-based fuels. This contrast indicates a great potential of biomass utilization which is still to be explored.

On the way of utilizing biomass as renewable feedstock, process industry has been experiencing a steady growth in the production of biofuels (Demirbas, 2007; Huber et al., 2006) that is now developing into the emerging concept of biorefining (Fernando et al., 2006; Clark, 2007). Biofuels offer access to a wide range of chemicals that include oxygenates (methanol, ethanol, butanol, mixed alcohols, dimethyl-ethers) and hydrocarbons (biodiesel, synthetic diesel, synthetic gasoline, and methane). Biorefining bears striking analogies with fossil-oil refining, fractionating biomass into a family of products to include transport fuels, platform chemicals, polymers, and specialty chemicals with yields and distributions that vary widely on the chemical and physical nature of the feedstock. With the production concentrated on biofuels, oil refining remains more viable as every drop of oil is used to make commodities (fuels, bulk chemicals) and specialty chemicals. Following the example of the petroleum industry, sustainable multiple product biorefineries should target a greater proportion of biomass, producing multiple streams of high volume/low value as well as low volume/high value molecules (Audsley and Annetts, 2003). Exploring possible products from biomass processing, National Renewable Energy Laboratory (NREL) and Pacific Northwest National Laboratory (PNNL) (Werpy et al., 2004; Holladay et al., 2007) produced an impressive list of potential building blocks, secondary chemicals, intermediates, and final products. Corma et al. (2007) particularly reviewed bio-based products derived from heterogeneous catalysis. Hermann and Patel (2007) and Haveren et al. (2008) discussed the substitution of mineral feedstocks by renewable biomass components, offering suggestions that include ethylene, propylene, and glycol.

The aforementioned possible biomass-derived products may be produced from various types of biomass by means of a variety of processing routes and technologies. The types of biomass that have already been processed range from sugar and starch, lignocelluloses, to oil and fat. Biomass of each of these types typically has different origins or conventional uses. Recent and growing concerns on competition with food production have gradually driven the development focus from food crop based, first-generation biofuels to the utilization of nonfood crops, forestry, and farming residues and wastes. Biomass can be converted by biological, chemical, or thermochemical routes. Biological routes include "conventional" fermentation and anaerobic digestion and have long been utilized by industry. Chemical routes mainly refer to esterification and transesterification processes for converting plant oils or animal fats into biodiesel and side products. Thermochemical routes on the other hand convert biomass at relatively high temperature; these typically include pyrolysis and gasification. It should be noted that each of these routes has a range of technologies that have been or are being developed with different efficiencies, feedstock of choice, and economics. Furthermore, in a complete biomass-based production system, biomass processing units are often to be followed by conventional processing steps; gasification offers a typical example where the resulting syngas is very seldom taken as an end product but requires to be further converted by means of, for example, chemical synthesis in order to produce more useful chemicals or fuels. For a specific intermediate chemical derived from biomass, usually a number of alternative downstream conversion options exist depending on the nature of the target chemicals.

The existence of different types of biomass feedstock, possible products, and processing technologies poses simultaneously both opportunities as well as challenges to the utilization of biomass in the process industry. One of the key challenges lies in the need of a systematic approach to selecting and integrating processing options such that production systems are developed to best suit the availability of biomass feedstock as well as the market demand of existing and potential products. In essence, this need calls for efficient, high throughput screening of a large number of options. Centered on a joint research project, Assessing Biomass to Chemicals (ABC), this contribution is intended to address this challenge particularly with a view of leveraging systems methods and tools. Section 2.2 gives a brief summary of the applications and roles of such tools in the area of biorefinery development; a detailed review of this area can be found in Kokossis and Yang (2010). In Section 2.3, the ABC project is introduced in terms of its objectives and overall methodology. The key element of the methodology, namely a process synthesis-based approach to high-throughput screening, is presented in Section 2.4. Section 2.5 reports the design and implementation of a systems platform that implements these modeling and synthesis approaches. In Section 2.6, an illustrative example is presented to demonstrate the capability of the systems platform in screening and integrating production routes, before concluding remarks are given in Section 2.7.

2.2 Applications of Systems Tools in Supporting the Development of Biorefineries

Systems technology offers powerful methods to optimize designs and systematize improvements of biomass processing systems, biorefineries in particular (Dimian, 2007; Klatt and Marquardt, 2008; Kokossis and Yang, 2010). In this section, existing and potential applications of systems tools in three areas are summarized; requirement of new applications to meet the challenge of high throughput screening is finally identified.

2.2.1 Modeling and Simulation (Flowsheeting)

This is an area where systems tools have so far been used the most. Typical applications involve the use of general flowsheeting technology for techno-economic and case-by-case analysis, using Aspen Plus for instance, as in the work of NREL on bioethanol production (Wooley et al., 1999) and that of Haas et al. (2006) for biodiesel. Cardona and Sanchez (2006) also used Aspen Plus to evaluate process configurations in the production of ethanol from lignocellulosic biomass. Configurations experimented with different arrangements for hydrolysis, fermentation and purification, and several scenarios for water recycle. Using alternative simulators, Pfeffer et al. (2007) evaluated the energy demand over several options to utilize tillage. Piccolo and Bezzo (2008) simulated the production of lignocellulosic ethanol comparing fermentation routes with enzymatic hydrolysis over gasification. Gutierrez et al. (2009) simulated scenarios to integrate biodiesel production from palm oil with ethanol produced from lignocellulosic residues of the plant. Researching at a finer level and particularly with reference to thermochemical processes, computational fluid dynamics (CFD) modeling has been performed as in the work of pyrolysis reactors (Bridgwater et al., 1999; Wurzenberger et al., 2002; Blasi, 2008) in order to gain a detailed understanding of the behavior and performance of the processing unit, hence offering guidance to the design (and possibly operation) of the unit.

2.2.2 Process Integration and Optimization

In biorefining, there is a clear need to use process integration and study strong trade-offs. In the case of sugar-based processes, the trade-offs will be between pretreatment, fermentation, and saccharification (feed conditioning, selection of pretreatment and use of acids, selection of reactors, choices and conditions of intermediates, integration of reaction and separation), all leading to challenging and interesting problems in optimization and synthesis. In biodiesel processes, similar challenges relate to the selection of pretreatment stages, the design and the selection of reactors and reaction conditions, and the potential to integrate adjacent processes (i.e., reaction and separation). In comparison to flowsheeting analysis, synthesis and process integration applications are rare but exist. Duret et al. (2005) applied energy integration to study the gasification of wood in the production of synthesis gas. Sánchez et al. (2006) provide a synthesis study for the production of lignocellulosic ethanol. More recently, Gassner and Marechal (2008) optimized wood gasification processes enabling the possible integration with electrolysis and producing promising flowsheets for validation, setting the type of screening expected from the synthesis and process integration methods.

2.2.3 Supply Chain Design and Analysis

In comparison with conventional feedstock, biomass is supplied geographically in a much less concentrated manner. Several literature studies demonstrate optimization models that combine processing and transportation costs in the supply chains of renewables-based manufacturing, using mathematical programming to illustrate techno-economic trade-offs between production and logistics. In the production of bioethanol from sugar cane and sweet sorghum, Nguyen and Prince (1996) optimized transportation costs and the scale of the production plant. Using a variety of lignocellulosic feedstocks, Kaylen et al. (2000) optimized the location and the capacity of feedstocks to maximize bioethanol and furfural production. Recent studies evaluate product portfolios, as in the work of Sammons et al. (2007) that optimized the allocation of biorefinery products accounting for sales revenues, processing, and feedstock costs. The demonstration included the production of syngas from chicken litter with syngas allocated between hydrogen and electricity stations. In reference to wheat bran and wheat-based biorefineries, Sadhukhan et al. (2008) reported a comparable value analysis approach that models marginal contributions from processing routes and products and compares different manufacturing scenarios.

Although systems tools have found applications in several areas pertaining to biorefining, most of the applications have a focus on individual processes or plants (or their parts) of known types. More recent reports, for example, that of Grossmann and coworkers (Cucek et al., 2011) do go beyond this scope and start exploring various routes and options for a specific product (e.g., bioethanol) and associated co-products. However, systematically screening a large number of options in general and exploring integration opportunities between different processing routes, although all important for optimal biomass utilization, are still largely unnoticeable in the literature.

2.3 ABC Project

In this section, the ABC project is introduced which, in essence, is aimed to address the systems challenge mentioned earlier and which has not been sufficiently tackled by previous efforts. It presents the background, scope, and objectives of the project, the general methodology adopted, as well as the development of the basis of the modeling capacities required for realizing the general methodology. A more extended view of the methodology, particularly its most novel part, will be presented separately in Section 2.4.

2.3.1 Scope and Objectives

Starting in 2009, a collaborative project supported by the Department of Environment, Food and Rural Area (DEFRA) is committed to the development of a systematic approach to screen biomass feedstocks and design options, with a reference to the developments of the Teesside industrial complex at the North East of England. The developed approach is expected to materialize in the form of a software platform which can be used to identify the best solutions for combining biomass processing routes with conventional processing facilities to produce bulk chemicals. Reflecting the industrial interests of this region, these chemicals include methanol, ethanol, butanol, ethylene, propylene glycol, acetone, styrene, and ammonia. A broad range of biomass processing technologies are considered, including both bio/chemical routes such as fermentation, anaerobic digestion, and transesterification, and thermochemical routes such as pyrolysis and gasification. These biomass processing routes are to be combined with conventional petrochemical technologies such as chemical synthesis, conversion, and reforming. This will generate candidates for viable scenarios where biomass-derived intermediate products join retrofitted conventional processing chains. The ultimate goal of this is to produce energy and chemical products in a way which reduces the dependency on nonrenewable feedstock and is both economical and environmental-friendly.

More specifically, the objective of the systems approach and the corresponding high throughput screening platform is to solve a problem that can be described as follows:

Given,

1. Types, volumes, and distribution of biomass feedstocks,
2. Sets of applicable processing routes and technologies,
3. Sets of candidate products, and
4. Sets of economical and environmental constraints.

Optimally determine the

1. Compartments of processing blocks required
2. Suitable feedstock (types, volumes, and locations) or the mix of feedstocks
3. Product portfolios and the production scales

The objective function accounts for economic and environmental targets (e.g., costs, profit, energy use, percentage consumption of nonrenewable fuels, GHG emissions).

2.3.2 General Methodology

The basis of the methodology proposed in this project is to develop mathematical modeling capacities for capturing the performance of individual compartmental processing blocks (CPBs) which may be selectively combined into simple or composite (i.e., integrative) production routes. In the following discussion, three issues are addressed, namely types of mathematical models required, levels of decomposition of production routes, and the characterization of elementary modeling units.

2.3.2.1 Types of Mathematical Models Required

Figure 2.1 shows the main types of models involved.

- *Manufacturing process models* that predict the output of each CPB from input information (material and energy streams to the block).
- *Manufacturing performance models* that evaluate, given the output from the models discussed earlier, the economical performance, the energy consumption, and the GHG.
- *Logistic performance models* addressing logistics and supply chain considerations.

High throughput evaluation shown in Figure 2.1 is a general term, which may materialize in simulation- or optimization-based analysis with predefined evaluation criteria such as economics, energy consumption, and/or GHG emission; further insights on this will be given in Section 2.4.

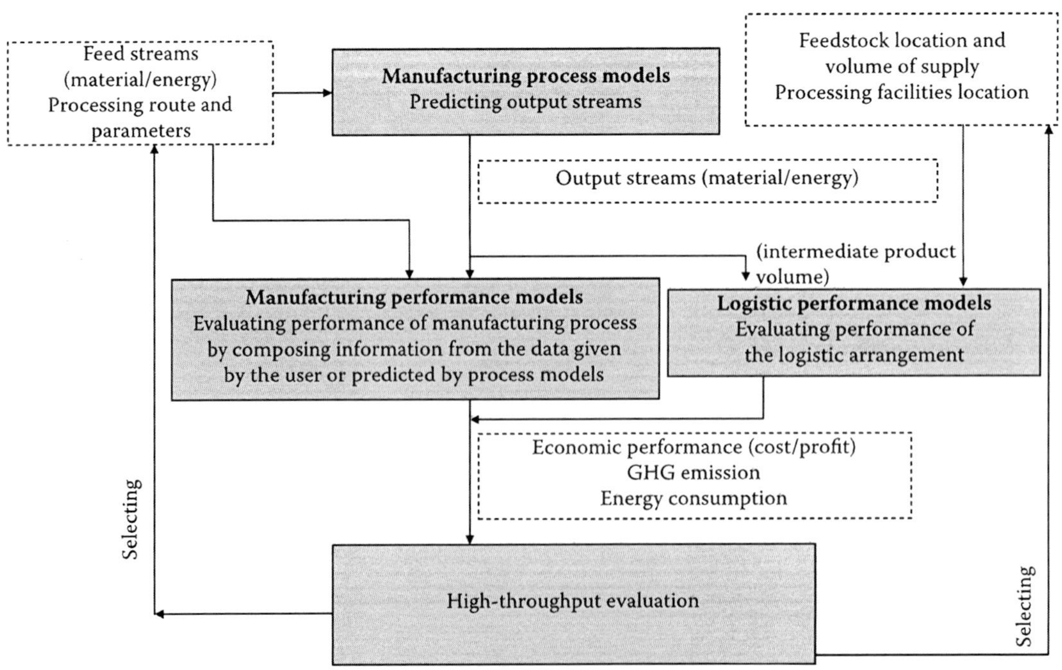

FIGURE 2.1
Types of mathematical models involved in the proposed methodology.

2.3.2.2 Levels of Decomposition

The configuration of an optimal manufacturing system essentially determines the systems components to integrate. As shown in Figure 2.2, there exist several levels to locate a particular component (Levels 2, 3, and 4) whereas, and as summarized in Table 2.1, the level of abstraction bears a significant impact on modeling and optimization.

In this ABC project, the focus of modeling has been put on Level 2 and/or Level 3. Level 2 screens a wide range of processing options whereas Level 3 is intended to distinguish between processing technologies of different efficiencies and cost. Other applications may determine other levels dictated by the intended scope of analysis and scrutiny. A mixed mode is also possible, suggesting a multiresolution approach (e.g., part of the system based on Level 2 and the rest of analysis on Level 3). Considering the current uncertainties in technologies, uncertainties in markets and products, alongside the unknown potential for process integration, the modeling and the analysis work assumed at Level 4 can be hardly justified and is probably a task to address in the future.

2.3.2.3 Elementary Units of Analysis

Depending on the analysis level (as discussed earlier), a CPB, defined as an elementary unit of analysis, may represent a whole plant (e.g., pyrolysis, gasification), a plant section (or functional unit, e.g. pretreatment, conversion, cleaning/purification), or a piece of equipment (e.g., fermentor, gasifier). A plant or a particular process to model is treated as a single CPB, or decomposed instead into a number of interconnected CPBs (each representing the plant section or individual pieces of equipment). The CPB boundaries can vary according to the intended use of the CPB. The different ways reflect on the particular scope to assess process integration benefits (i.e., emphasis on entire plant sections, namely a piece of equipment) and the intended scope remains an input for the proposed approach.

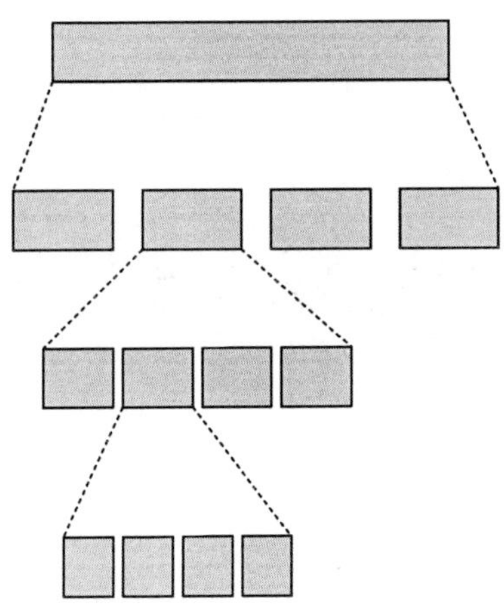

FIGURE 2.2
Different levels of analysis.

TABLE 2.1
Implication of Level Selection

Levels	Effort of Modeling and Optimization	Scope of Assessment	Resolution of Solution
Plants or processing steps (Level 2)	Low	Large	Preliminary
Plant sections or functional units (Level 3)	Medium	Medium	Medium
Equipment (Level 4)	High	Small	Fine

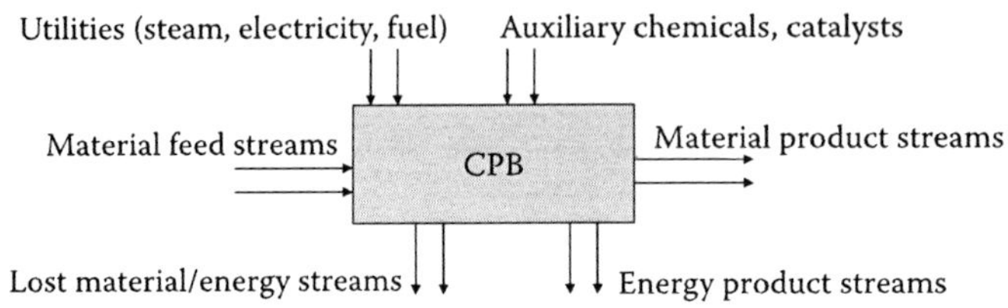

FIGURE 2.3
Representation of a CPB.

Figure 2.3 illustrates the modeling information to capture in the CPB as well as connections of CPBs toward superstructures. Reflecting on the occasional differences in boundaries, the modeling components may be preconstructed (for individual CPBs) or take a form of configurable formulations (superstructures). In principle, the former practice applies to the *manufacturing process models* (Section 2.2) and the models estimating capital and fixed operating costs (part of the models used to assess the manufacturing performance). The latter practice applies to models estimating variable operating costs, sales, revenues, energy consumption or exporting, and GHG emissions. Such models are not constructed for each individual CPB as they measure performance of the entire system and are required to account for material and energy streams exchanged between the system considered and its environment.

2.3.3 Building Modeling Capacity

In general, each CPB identified in the project is modeled such that the types of models depicted in Figure 2.1 are made ready for use in any high throughput evaluation. Figure 2.4 illustrates how this is done by starting with the modeling of the (bio)chemical process embedded in the CPB. It can be seen that the result of process modeling will drive the estimation of energy consumption, GHG emission, as well as the variable operating cost. Note that fixed costs (including capital cost and fixed operating cost) need to be estimated with additional information other than the result of process modeling. In this project, an overall protocol for cost estimation given by NREL (e.g., Ringer et al., 2006; Phillips et al., 2007) and a step- counting-based method (Taylor, 1977) for estimating capital cost have been adopted. Furthermore, the ABC project has given emphasis to production and it was decided not to address logistic modeling (indicated as part of Figure 2.1) within the project, hence it is not included in Figure 2.4.

Table 2.2 presents a list of specific CPBs for which the modeling is carried out in the ABC project with information provided by several research partners. Those marked with "N*" are "new" processing steps introduced, either directly or indirectly, for the utilization of biomass. Those marked with "C*" are existing conventional processing steps.

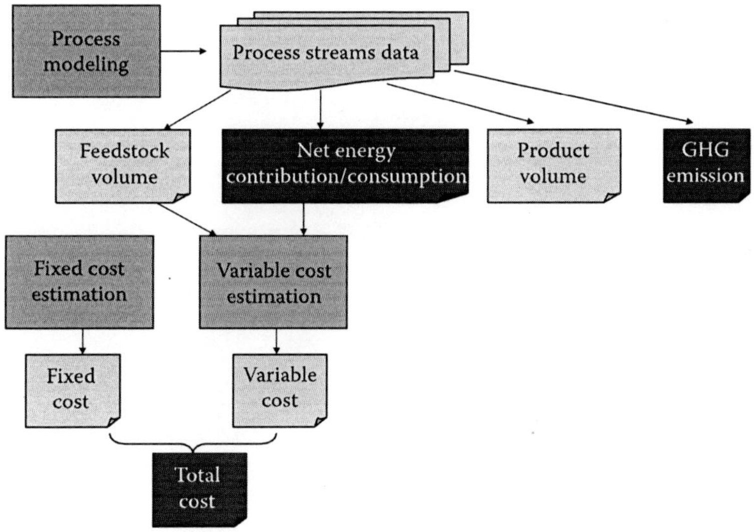

FIGURE 2.4
Relation between processing modeling and performance modeling.

TABLE 2.2
List of Compartmental Processing Blocks Investigated in the ABC Project

N1	Biomass fermentation to ethanol—grain wastes
N2	Acetone-butanol-ethanol (ABE) fermentation
N3	Syngas fermentation to ethanol
N4	Municipal solid waste (MSW) to biogas by anaerobic digestion (AD)
N5	Glycerol to biogas (CH_4) by AD
N6	Plant oil to biodiesel and glycerol
N7	Biomass gasification—indirect
N8	Biomass gasification—direct
N9	Mixed alcohol synthesis from syngas
C10	Propylene to propylene glycol
C11	Glycerol to propylene glycol
C12	Biogas (CH_4) reforming to syngas
C13	Propylene to acetone
C14	Propylene to *n*-butanol
C15	Methanol synthesis from syngas
C16	Syngas to ammonia
C17	Methanol to olefins
C18	Ethanol dehydration to ethylene

2.4 Toward a High-Throughput Platform with Process Synthesis

Using mathematical models to facilitate the evaluation of various options for biomass processing is not an entirely new idea. Focusing on biological (as opposed to thermochemical) technologies, the Bulk Chemical from Renewable Resources (BREW) project (Patel et al., 2006) developed a comprehensive spreadsheet-based tool to simulate a number of processing routes for the purpose of evaluating their performance and hence feasibility. In

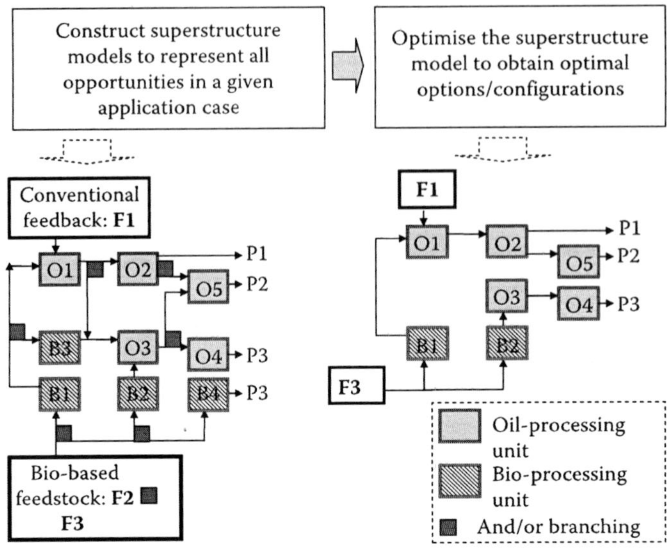

FIGURE 2.5
Methodology for the synthesis of biomass processing systems.

the ABC project, similar functionality is also developed, although the CPBs involved are not limited to biological conversions. Furthermore, a novel process synthesis approach was introduced into the area of high-throughput screening of biomass processing options with a potential to achieve the optimal combination of otherwise separate production routes as well. Process synthesis, a technique developed in process systems engineering, addresses in its classical sense the issue of identifying the best processing units and their connections within a chemical process flowsheet. It has been developed and applied to applications ranging from reaction networks to separate and heat recovery processes (e.g., Kokossis and Floudas, 1990; El-Halwagi and Srinivas, 1992; Marcoulaki and Kokossis, 2000). Introducing process synthesis to high-throughput screening and evaluation of biomass processing options, this technique is now moving from addressing typically a single process or site to a complex production chain or network.

The basic principle of the approach is illustrated by Figure 2.5. On the left is shown schematically a superstructure that represents the possible combinations of conventional and bio-based feedstock and processing units for producing a set of desired products (P1–P3). A possible result of optimization is shown on the right hand side, in the form of a configuration that involves optimal feedstock and processing options. It should be noted that each block in the superstructure is a CPB as discussed earlier; the model of which is made available in the model library mentioned in Section 2.3.3. In the ABC project, a CPB is also referred to as a unit of synthesis (UoS).

2.5 Prototyping of the Systems Platform

This section presents the design and implementation of a software platform which has been developed in the ABC project to realize the modeling and synthesis methodology outlined earlier.

2.5.1 Design of the Platform

Figure 2.6 illustrates the overall structure of the systems platform.

The platform is composed of three components:

- Central database. Shared by all computational tools in the platform, this database stores common data for individual CPBs that have been added to the platform (CPB data) as well as economic data of chemicals, fuels, and energy (Economic data). The latter is not CPB-specific and may be applied to any processes to be investigated in the platform.
- Simulation tool. Based on the data supplied by the central database, the simulation tool supports what-if studies on specific production routes selected by the user.
- Synthesis and optimization tool. Also based on the data from the central database, this tool performs systematic screening and optimization of production systems in terms of feedstock and product selection as well as production route selection (and possibly combination).

As indicated in Figure 2.6, Excel and GAMS have been chosen to implement different parts of the platform. More detailed design of the Excel-based part and the GAMS-based part is given in Figure 2.7.

2.5.2 Implementation

As indicated in Figure 2.6, the entire systems platform is composed of three components, namely the central database, the simulation tool, and the synthesis/optimization tool.

2.5.2.1 Central Database

The central database is composed of (i) 19 Excel spreadsheets each of which corresponds to one CPB (or UoS) as listed in Table 2.2 (with the exception that the N1 entry in Table 2.2 materializes in two variations for handling two different feeds, namely wood chips and corn stover) and (ii) one additional sheet for the prices of chemicals, energy streams, and services.

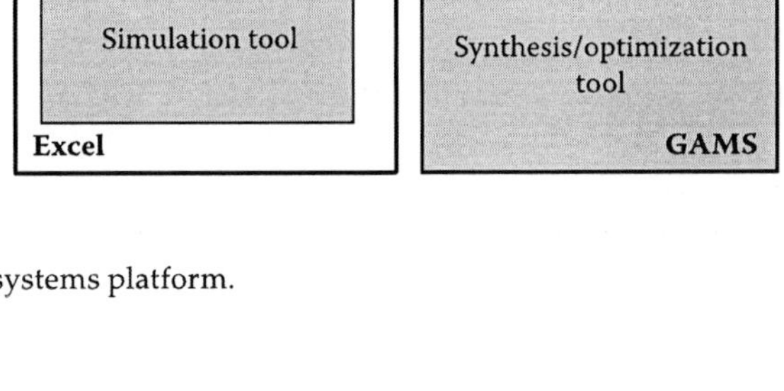

FIGURE 2.6
Conceptual design of the systems platform.

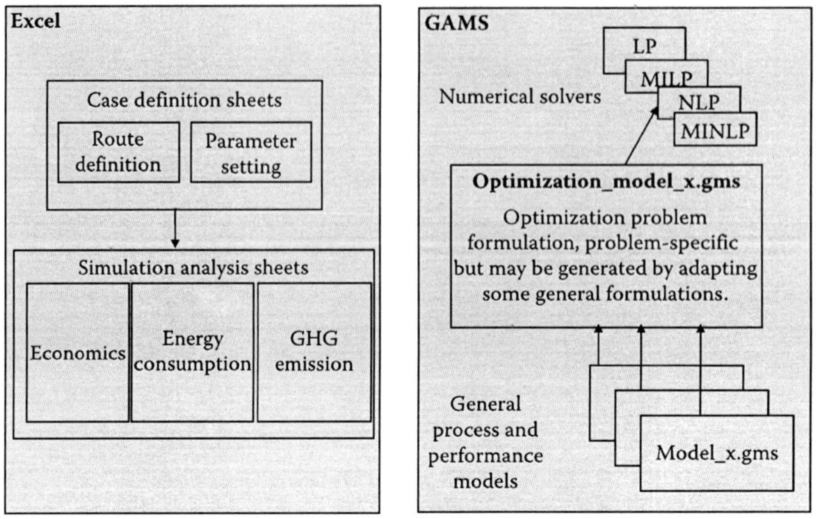

FIGURE 2.7
Design of the Excel-based and the GAMS-based parts.

2.5.2.1.1 CPB (or UoS) Data

Figure 2.8 shows an annotated screen shot of the sheet for indirect gasification (numbered as UoS1 in the platform). The sheet contains all the process and energy stream data as well as the result of cost estimation. Note that all of these data items are for a reference plant of a specific capacity as specified in the original process description document. It provides a basis for modeling a new plant by scaling.

2.5.2.1.2 Economic Data

Figure 2.9 shows a screenshot of part of the spreadsheet holding price information. This sheet includes the unit price of items of three categories, as shown in Figure 2.10. The price information was obtained from the BREW report (Patel et al., 2006) as well as sources provided by the industrial partners. It should be noted that the prices given here serves only the purpose of a default data set; it is expected that user-supplied data may be substituted into this worksheet in specific cases of analysis, which is a function supported by the central database.

2.5.2.2 Simulation Tool

The simulation tool is fully based on Excel. Figure 2.11 shows the top-level interface of the tool, where all the main functions can be seen and accessed by the users. These functions are described as follows.

2.5.2.2.1 Viewing UoSs and Production Routes

The simulation tool contains a list of UoSs included in the central database and also a brief description for each UoS on the list. Besides, the users can browse all the production routes that can be evaluated by the simulation tool; Figure 2.12 shows what the tools display for routes starting from wood chips.

UoS1_table:Biomass gasification to Raw Syngas (rng = UoS-1!A3:N19)

	A	B	C	D	E	F	G	H	I	J	K	L	M	N
3					F	T	P	H2O	CO2	Clean_syngas	methanol	steam	electricity	fuel
4					kg_hr	deg_C	atm	kg_hr	kg_hr	kg_hr	kg_hr	kg_hr	kW	kg_hr
5	UoS-1	inlet	MainS	Fbio_LignCL	166667	16	1.72	83333.4						
6	UoS-1	inlet	S-1	Air	200562	77	2.03	3913.14	98.88					
7	UoS-1	inlet	S-2	BFW1	3193.26	80	58.57	3193.26						
8	UoS-1	inlet	S-3	BFW2	54431.1	80	58.57	54431.08						
9	UoS-1	inlet	S-4	CWS	37754.5	32.2	4.15	37754.45						
10	UoS-1	inlet	S-5	Quench_water	340.19	15.6	1.01	340.19						
11	UoS-1	inlet	S-6	H_2O_scrubber	18143.7	43.3	28.67	18143.69						
12	UoS-1	inlet	S-7	Energy		235.9	31.01	33167.13				33167.1		
13	UoS-1	outlet	MainS	raw_gas	79818.7	47.9	1.01	354.68	16488.28					
14	UoS-1	outlet	S-1	Flue_gas	287943	136.2	1.62	85416.66	46961.92					
15	UoS-1	outlet	S-2	steam1	3193.26	273.9	58.57							
16	UoS-1	outlet	S-3	Wet_ash	2590.11	93.37	1.01							
17	UoS-1	outlet	S-4	Steam 2	54431.1	273.9	58.57							
18	UoS-1	outlet	S-5	CWS	37754.5	43.3	4.15							
19	UoS-1	outlet	S-6	WWT	48528.9	47.9	1.01	48469.34	3.54					

Process stream data

UoS_parameter_bt1:Biomass gasification to Raw Syngas (rng=UoS-1!A23:F26)

	A	B	C	D	E	F
23					F	F
24					kg_hr	kg_hr
25					UB	LB
26	UoS-1	inlet	MainS	Fbio_LignCL	166667	0

UoS_cost_1: Biomass Gasification to Raw Syngas (rng= UoS-1!A31:D43)

	A	B	C	D
31			M$	M$/yr
32	UoS-1	TEC	35	
33	UoS-1	WC	0.525	
34	UoS-1	SDC	3.15	
35	UoS-1	FOC		4.4682
36	UoS-1	LC		2.027
37	UoS-1	OVC		1.2162
38	UoS-1	MC	0.7	
39	UoS-1	ITC	0.525	
40	UoS-1	VOC		1.183

SHORT NAMES	
TEC	Total Equipment Cost
WC	Warehouse Cost
SDC	Site Development Cost
FOC	Fixed Operating Cost
LC	Labour Cost
OVC	Overhead Cost
MC	Maintenance Cost
ITC	Insurance and Taxes Cost
VOC	Variable Operating Cost

Costing data

FIGURE 2.8
Excel worksheet holding UoS data.

1. MATERIALS		
1.1 BIOMASS FEEDSTOCK	**UNIT**	**PRICE (£/unit)**
Corn Stover	t	21.70
Wood Chips	t	60.00
MSW	t	40.00
Plant Oil	t	711.85
1.2 CHEMICALS	**UNIT**	**PRICE (£/unit)**
Diammomium Phosphate	t	128.53
Lime (Ca(OH)2)	t	50.91
Urea	t	325.60
Acetic Acid	t	404.80
Sulpuric Acid	t	26.71
Corn Steep Liquor	t	31.72
Ammonia	t	96.82
Lo-Cat Chemicals	t	1800.00
NaOH	t	74.28
H2	t	1348.76
N2	t	39.00
HCl	t	92.85
CO2	t	30.00
CH3ONa	t	880.00
Cl2	t	272.36
KOH	t	180.00
Ethanol	t	368.91
Ethanol from corn	t	1082.51
Methanol	t	151.90

FIGURE 2.9
Excel worksheet holding price information (partially shown).

- Materials
  - Biomass feedstock
  - Chemicals
  - Catalysts
  - Water
- Energy
  - Combustion fuel
  - Steam
  - Electricity
- Services
  - Wastewater treatment
  - Solid waste disposal
  - Cooling water treatment

FIGURE 2.10
Categories of price data.

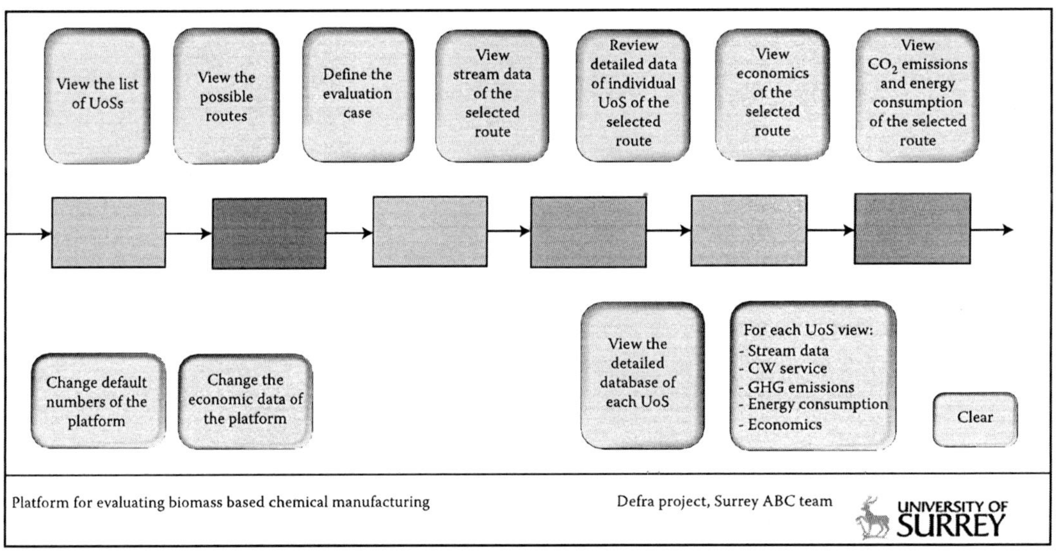

FIGURE 2.11
Top-level user interface of the simulation tool.

2.5.2.2.2 *Defining and Evaluating a Production Route*

To allow a user to study a specific production route, the tool allows the user to define the route first by indicating the UoSs involved and their order in the route. Also, the tool asks the user to specify the amount of the feedstock available to quantify the production scale. One example is shown in Figure 2.13. After the route is defined as such, the tool automatically completes all the calculations regarding the economics, energy consumption, and GHG emission of this route, producing results ready to review.

2.5.2.2.3 *Viewing Results of Production Route Evaluation*

The basis of all the analysis is the evaluation of all the inlet and outlet streams of each UoS involved in the route, with appropriate linkages between connected UoSs according

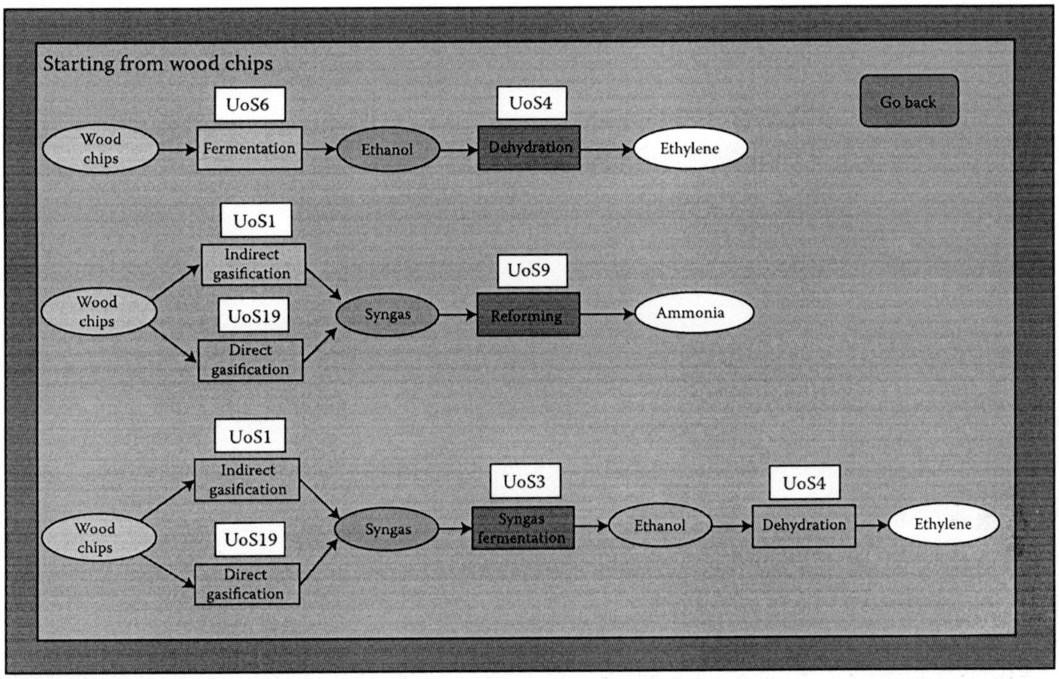

FIGURE 2.12
Production routes starting with wood chips, displayed by the simulation tool.

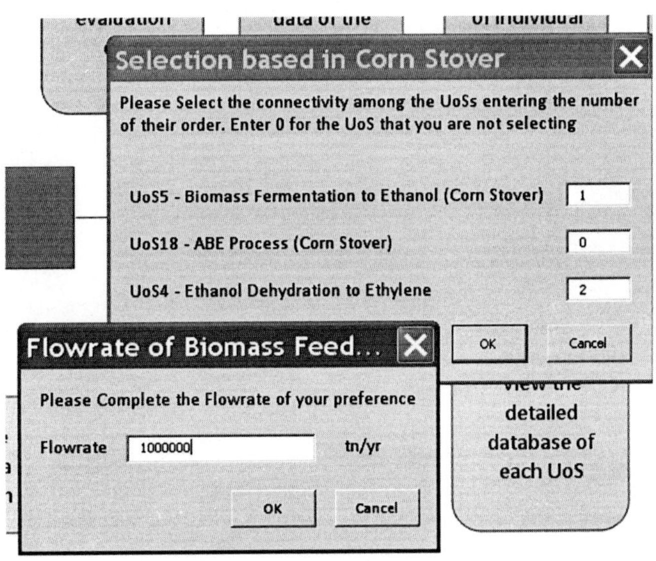

FIGURE 2.13
Entering order of UoSs and feed volume in the simulation tool.

to the route definition. To view the results, the tool is able to present to the users the following:

- Process data of selected (main) streams of the route (as illustrated in Figure 2.14) and more detailed data for individual UoSs.
- Economics of the entire route; the items calculated and displayed are shown in Figure 2.15 as appearing in a particular example.
- Energy and GHG (currently CO_2 only) figures of the entire route, as illustrated in Figure 2.16 (negative numbers for energy consumption indicate net generation for exporting).

2.5.2.2.4 Changing Default Parameters

The simulation tool also allows the user to change parameters used in the calculation. These include model parameters (number of operation hours per year; exponent for the capital cost power law; exponent for the labor cost power law) as well as the price data as described earlier in Section 2.5.2.1. It also provides the function of resuming the default data set.

2.5.2.2.5 Viewing and Evaluating Individual UoSs

An interface to the central database is provided to allow the user to select a particular UoS and then view its details as specified in the database. Furthermore, it is possible to treat the single UoS as an independent plant and evaluate it by specifying a scale of operation which is different from the reference scale.

A249

UoS 4: Ethanol Dehydration to Ethylene

	tn/yr
Ethanol	251773.90

Inlet Streams	Calculation	Outlet Streams	Calculations
Comb_air (tn/yr)	137072.427	Ethylene (tn/yr)	144997.317
Caustic (tn/yr)	1342.794	Purge_quench (tn/yr)	85970.181
Wash_water (tn/yr)	3356.985	Purge_wash_col (tn/yr)	3604.586
Steam_LP1 (tn/yr)	89295.811	Flue_gas (tn/yr)	142936.105
Steam_LP2 (tn/yr)	84931.730	Return_condesate (tn/yr)	174227.541

UoS 5: Biomass Fermenation to Ethanol (Corn Stover)

	tn/yr
Biomass	1000000.00

Inlet Streams	Calculation	Outlet Streams	Calculations
Lime (tn/yr)	24413.7102	Ethanol (tn/yr)	251773.903
Sulf_acid1 (tn/yr)	32531.5695	Water_loss (tn/yr)	116235.719
Sulf_acid2 (tn/yr)	12086.5905	Solids (tn/yr)	73601.924
Cellulase (tn/yr)	69633.1930	Vent_To_Atmosphere (tn/yr)	245941.070
Diam_Phosphate (tn/yr)	1683.7041	From_Combustor (tn/yr)	765063.139
Corn_Steep_Liq (tn/yr)	13289.2363	Purge (tn/yr)	263680.096
Water_Make_Up (tn/yr)	1924233.3133	Evaporation_Loss (tn/yr)	1383042.694

Main Menu | Stream_Data

FIGURE 2.14
Display of the main stream data.

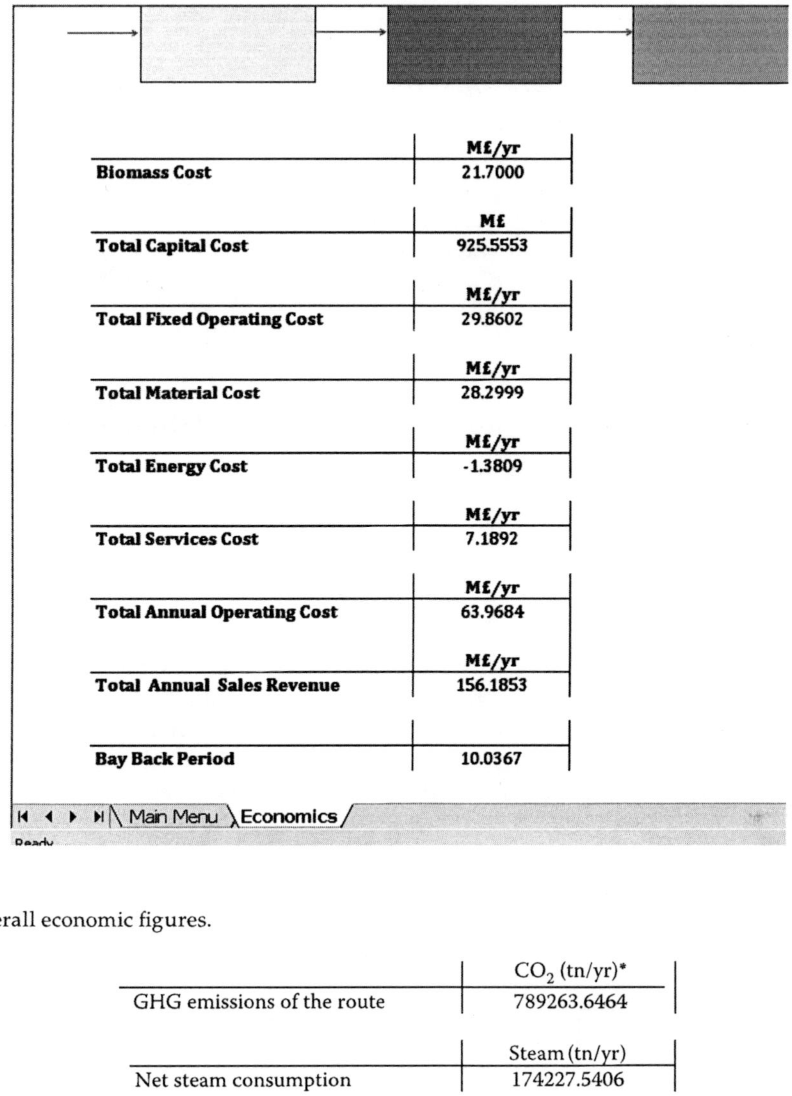

	M£/yr
Biomass Cost	21.7000
	M£
Total Capital Cost	925.5553
	M£/yr
Total Fixed Operating Cost	29.8602
	M£/yr
Total Material Cost	28.2999
	M£/yr
Total Energy Cost	-1.3809
	M£/yr
Total Services Cost	7.1892
	M£/yr
Total Annual Operating Cost	63.9684
	M£/yr
Total Annual Sales Revenue	156.1853
Bay Back Period	10.0367

FIGURE 2.15
Display of overall economic figures.

	CO_2 (tn/yr)*
GHG emissions of the route	789263.6464
	Steam (tn/yr)
Net steam consumption	174227.5406
	Electricity (MWh/yr)
Net electricity consumption	−77966.5664
	Fuel (MWh/yr)
Net fuel consumption	32227.0595

FIGURE 2.16
Display of energy consumption and GHG figures.

2.5.2.3 Synthesis/Optimization Tool

The GAMS-based prototype has implemented the following features:

- A common set of equations representing the process and performance (cost, energy, GHG) model;
- Interface with Excel to access the central database;

- Mechanism to represent a production route or a superstructure by means of a (problem-specific) connectivity matrix and a common connectivity equation;
- Adoption of a fully linearized model to be solved by a Mixed Integer (linear) solver to gain computational efficiency.

Without reporting the GAMS programming details here, the important modeling elements (particularly the representation of superstructures) and the functionality of this tool will be illustrated in the next section through a specific example.

2.6 Screening/Synthesizing Production Routes: An Illustrative Example

Consider the planning of a biomass-based chemical manufacturing complex aimed to produce ethylene and propylene glycol at a fixed rate (500,000 and 50,000 te/year, respectively). Four different types of feedstock may be used, namely corn stover, wood chips, MSW, and plant oil, each with an availability of up to 8,000,000 te/year. The objective is to identify the production route(s) with highest economic potential (i.e., net annual profit).

Based on the 19 CPBs (i.e., UoSs) available in the systems platform, a superstructure of this chemical complex can be established, as shown in Figure 2.17. Note that two by-products, namely mixed alcohol and bio-diesel may be produced as well.

In the GAMS model, this superstructure is represented by means of declaring feeds, gate (i.e., intermediate) chemicals, products/by-products, and their connections through UoSs. Specifically, the following case-specific sets are defined first:

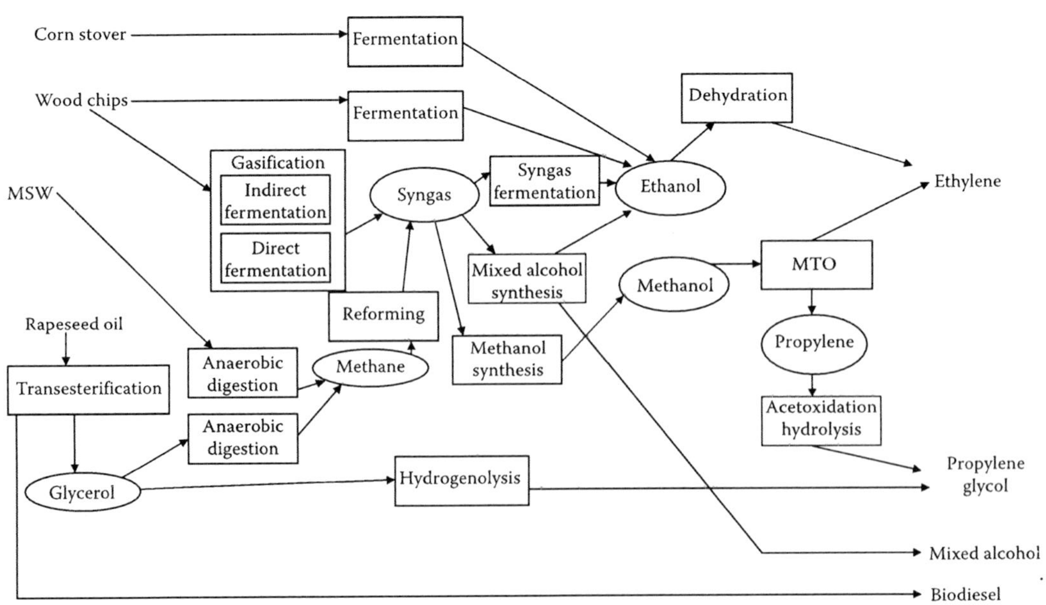

FIGURE 2.17
Superstructure of the example of biomass-based production of ethylene and propylene glycol.

```
ac "all chemicals" /inlet, cornstover, woodchips, MSW, plant_oil,
ethanol, syngas, methanol,propylene,biogas,glycerol, ethylene,acetone,
butanol,propylene_glycol,biodiesel,ammonia, mixed_alcohols/
j(ac) "raw material" /woodchips, cornstover, MSW, plant_oil/
i(ac) "gate chemical" /syngas, ethanol,methanol,biogas,glycerol,propylene/
k(ac) "all products" /ethylene, propylene_glycol,biodiesel,mixed_
alcohols, propylene,ethanol,methanol/
mp(k) "main products" /ethylene, propylene_glycol, ethanol,methanol/
bp(k) "by products" /biodiesel,mixed_alcohols, propylene/
l "processes" /corn_stover_fermentation,wood_fermentation, ind_
gasification, direct_gasification,
dehydration,syn_fermentation, meth_synthesis,alcohol_
synthesis,MTO,Ammonia_Synthesis,PG_from_Gl,Hydrogenolysis,MSW_AD,BG_
reforming_SG,
acetone_from_prl,Butanol_from_prl,transesterification,glycerol_to_biogas/
```

Tables that create connections are subsequently introduced:

```
*Connections (rm to processes, gc to processes, p to processes, bp to
processes)
*DownStream Processes
downstream_rm(j,l) "connections of raw materials with the downstream
processes"
    /cornstover.(corn_stover_fermentation)
    woodchips.(wood_fermentation,ind_gasification,direct_gasification)
    MSW.MSW_AD
    plant_oil.transesterification/
downstream_gc(i,l) "connections of the gate chemicals with the downstream
processes"
    /ethanol.dehydration
    syngas.(syn_fermentation,meth_synthesis,alcohol_synthesis)
    methanol.MTO
    biogas.BG_reforming_SG
    glycerol.(glycerol_to_biogas, PG_from_Gl)
    propylene.Hydrogenolysis/
*UpStream Processes
upstream_gc(i,l) "connections of the gate chemicals with the upstream
processes"
    /ethanol.(corn_stover_fermentation,wood_fermentation,syn_
    fermentation,alcohol_synthesis)
    syngas.(ind_gasification,direct_gasification,BG_reforming_SG)
    methanol.meth_synthesis
    biogas.(MSW_AD,glycerol_to_biogas)
    glycerol.transesterification
    propylene.MTO/
upstream_p(k,l) "connections of the products (excluding those which may
become gate chemicals) with the upstream processes"
    /propylene_glycol.(Hydrogenolysis,PG_from_Gl)
    ethylene.(dehydration,MTO)
    biodiesel.transesterification
    mixed_alcohols.alcohol_synthesis/
gate_to_products(k,i) "gate chemical to become a bio-product as an option"
    /propylene.propylene
    ethanol.ethanol
    methanol.methanol/
```

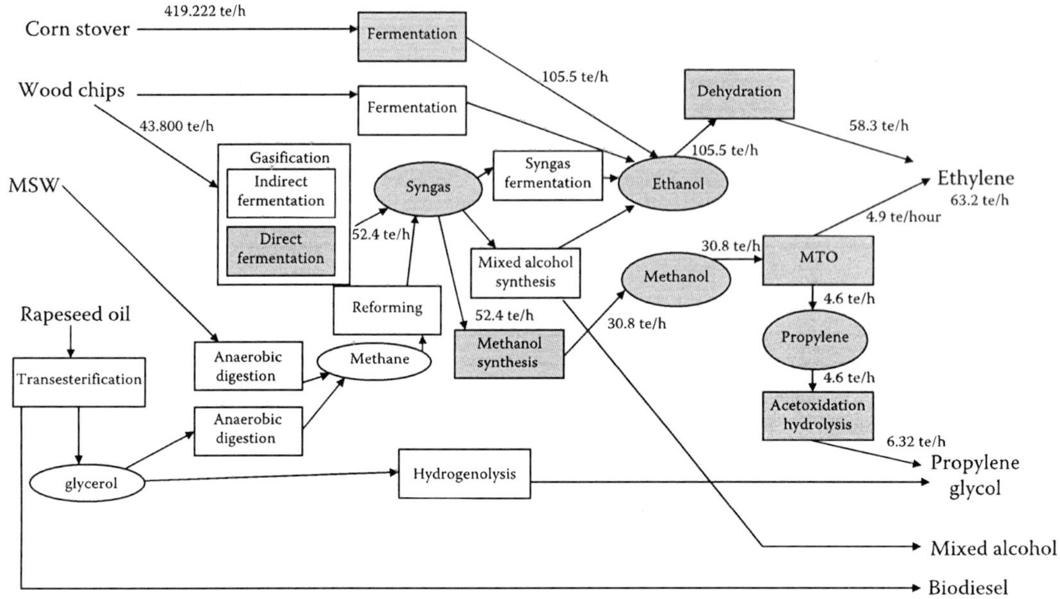

FIGURE 2.18
Selected routes for multiple feeds/products, with constraints on products.

Figure 2.18 shows the solution of this problem, with the selected routes highlighted and nonzero flowrates noted (assuming 7960 production hours per year). Corn stover and wood chips have been selected as the feedstock. Corn stover is fermented to produce ethanol which is in turn dehydrated to supply about 90% of the demanded ethylene. The remaining ethylene is produced as a co-product of the methanol to olefin (MTO) process which utilizes methanol synthesized from syngas out of the direct gasification of wood chips. The other co-product of MTO, propylene, is converted to propylene glycol by hydrolysis. No by-products are produced through the selected routes.

2.7 Concluding Remarks

Introducing renewable biomass into the manufacturing of energy and chemical products poses great challenges as well as opportunities to process systems engineering. This requires extensions of established methods and tools to address new problems in this area. This chapter reports the concepts developed in the ABC project that assesses the optimal opportunities to produce chemicals from biomass, possibly in combination with more conventional feedstock. The methodology for synthesizing complex manufacturing chains or networks is presented, and the design and implementation of a systems platform to carry out this type of synthesis described. This platform offers both a simulation tool and a synthesis/optimization tool. The simulation tool has been regarded by industrial partners as a useful tool to perform what-if studies on specific production routes. The synthesis/optimization tool, on the other hand, implements a novel approach to the fast screening of a large number of options, therefore it has the potential to support more strategic decisions.

Acknowledgment

Financial support from DEFRA (UK) via Project LK0855 and input from the project consortium are acknowledged.

References

Audsley, E., Annetts, J. E. 2003. Modeling the value of a rural biorefinery—Part I: The model description. *Agricultural Systems*, 76(1), 39–59.

Blasi, C. D. 2008. Modeling chemical and physical processes of wood and biomass pyrolysis. *Progress in Energy and Combustion Science*, 34, 47–90.

Bridgwater, T. A. V., Meier, D., Radlein, D. 1999. An overview of fast pyrolysis of biomass. *Organic Geochemistry*, 30, 1479–1493.

Cardona, C. A., Sanchez, O. J. 2006. Energy consumption analysis of integrated flowsheets for production of fuel ethanol from lignocellulosic biomass. *Energy*, 31, 2447–2459.

Clark, J. H. 2007. Green chemistry for the second generation biorefinery—Sustainable chemical manufacturing based on biomass. *Journal of Chemical Technology & Biotechnology*, 82, 603–609.

Corma, A., Iborra, S., Velty, A. 2007. Chemical routes for the transformation of biomass into chemicals. *Chemical Reviews*, 107, 2411–2502.

Cucek, L., Martin, M., Grossmann, I. E., Kravanja, Z. 2011. Energy, water and process technologies integration for the simultaneous production of ethanol and food from the entire corn plant. *Computers & Chemical Engineering*, 35(8), 1547–1557.

Demirbas, A. 2007. Progress and recent trends in biofuels. *Progress in Energy and Combustion Science*, 33, 1–18.

Dimian, A. C. 2007. Renewable raw materials: Chance and challenge for computer-aided process engineering. *Computer Aided Chemical Engineering*, 24, 309–318.

Duret, A., Maréchal, F., Friedli, C. 2005. Process design of synthetic natural gas (SNG) production using wood gasification. *Journal of Cleaner Production*, 13/15, 1434–1446.

El-Halwagi, M., Srinivas, B. K. 1992. Synthesis of reactive mass exchange networks. *Chemical Engineering Science*, 47(8), 2113–2119.

Fernando, S., Adhikari, S., Chandrapal, C., Murali, N. 2006. Biorefineries: Current status, challenges, and future direction. *Energy & Fuels*, 20, 1727–1737.

Gassner, M., Marechal, F. 2008. Methodology for the optimal thermo-economic, multi-objective design of thermochemical fuel production from biomass. *Computers & Chemical Engineering*, 33(3), 769–781.

Gutierrez, L. F., Sanchez, O. J., Cardona, C. A. 2009. Process integration possibilities for biodiesel production from palm oil using ethanol obtained from lignocellulosic residues of oil palm industry. *Bioresource Technology*, 100, 1227–1237.

Haas, M. J., Mcaloon, A. J., Yee, W. C., Foglia, T. A. 2006. A process model to estimate biodiesel production costs. *Bioresource Technology*, 97, 671–678.

Haveren, J. V., Scott, E. L., Sanders, J. 2008. Bulk chemicals from biomass. *Biofuels, Bioproducts and Biorefining*, 2, 41–57.

Holladay, J. E., Bozell, J. J., White., J. F., Johnson, D. 2007. Top Value-Added Chemicals from Biomass, Volume II: Results of Screening for Potential Candidates from Biorefinery Lignin, Pacific Northwest National Laboratory (PNNL), Richland, Washington.

Hermann, B. G., Patel, M. 2007. Today's and tomorrow's bio-based bulk chemicals from white biotechnology: A techno-economic analysis. *Applied Biochemistry and Biotechnology*, 136, 361–388.

Holladay, J. E., Bozell, J. J., White., J. F., Johnson, D. 2007 October. *Top Value-Added Chemicals from Biomass, Volume II: Results of Screening for Potential Candidates from Biorefinery Lignin*, PNNL.

Huber, G. W., Iborra, S., Corma, A. 2006. Synthesis of transportation fuels from biomass: Chemistry, catalysts, and engineering. *Chemical Reviews*, 106(9), 4044–4098.

Kaylen, M., van Dyne, D. L., Choi, Y.-S., Blasé, M. 2000. Economic feasibility of producing ethanol from lignocellulosic feedstocks. *Bioresource Technology*, 72(1), 19–32.

Klatt, K.-U., Marquardt, W. 2008. Perspectives for process systems engineering—Personal views from academia and industry. *Computers & Chemical Engineering*, 33(3), 536–550.

Kokossis, A. C., Floudas, C. A. 1990. Optimization of complex reactor networks: 1. Isothermal operation. *Chemical Engineering Science*, 45(3), 595–614.

Kokossis, A., Yang, A. 2010. On the use of systems technologies and a systematic approach for the synthesis and the design of future biorefineries. *Computers & Chemical Engineering*, 34(9), 1397–1405.

Marcoulaki, E. C., Kokossis, A. C. 2000. On the development of novel chemicals using a systematic synthesis approach. Part I: Optimisation framework. *Chemical Engineering Science*, 55(13), 2529–2546.

Nguyen, M. H., Prince, R. G. H. 1996. A simple rule for bioenergy conversion plant size optimisation: Bioethanol from sugar cane and sweet sorghum. *Biomass and Bioenergy*, 10(5–6), 361–365.

Patel, M. et al. 2006. Report of the BREW project. Available online at http://www.chem.uu.nl/brew/. Accessed on 25 January 2010.

Pfeffer, M., Wukovits, W., Beckmann, G., Friedl, A. 2007. Analysis and decrease of the energy demand of bioethanol-production by process integration. *Applied Thermal Engineering*, 27, 2657–2664.

Phillips, S. et al. 2007. Thermochemical ethanol via indirect gasification and mixed alcohol synthesis of lignocellulosic biomass. (Report produced by NREL). Available online at www.nrel.gov/docs/fy07osti/41168.pdf. Accessed on 25 January 2010.

Piccolo, C., Bezzo, F. 2008. A techno-economic comparison between two technologies for bioethanol production from lignocellulose. *Biomass and Bioenergy*, 33, 478–491.

Riese, J. 2006. Industrial biotechnology—Turning potential into profits. Plenary presentation, the third annual *World Congress on Industrial Biotechnology and Bioprocessing*, Toronto, July 11–14, 2006.

Ringer, M. et al. 2006. Large-scale pyrolysis oil production: A technology assessment and economic analysis (Report produced by NREL). Available online at http://www.forestbioproducts.umaine.edu/documents/37779.pdf. Accessed on 25 January 2010.

Sadhukhan, J., Mustafa, M. A., Misailidis, N., Mateos-Salvador, F., Du, C., Campbell, G. M. 2008. Value analysis tool for feasibility studies of biorefineries integrated with value added production. *Chemical Engineering Science*, 63, 503–519.

Sammons, N., Eden, M., Cullinan, H., Perine, L., Connor, E. 2007. A flexible framework for optimal biorefinery product allocation. *Environmental Progress*, 26, 349–354.

Sánchez, O. J., Gutiérrez, L. F., Cardona, C. A., Fraga, E. S. 2006. Analysis of extractive fermentation process for ethanol production using a rigorous model and a short-cut method, in J. Zilinskas & I. D. L. Bogle, Editors, *Computer Aided Methods for Optimal Design and Operations*, World Scientific (New Jersey), pp. 207–216.

Werpy, T., Petersen, G., Aden, A., Bozell, J., Holladay, J., White, J., Manheim, A. 2004. Top value added chemicals from biomass, Volume I: Results of screening for potential candidates from sugars and synthesis gas. NREL and PNNL, August 2004.

Wooley, R., Ruth, M., Glassner, D., Sheejan, J. 1999. Process design and costing of bioethanol technology: A tool for determining the status and direction of research and development. *Biotechnology Progress*, 15, 794–803.

Wurzenberger, J. C., Wallner, S., Raupenstrauch, H., Khinast, J. G. 2002. Thermal conversion of biomass: Comprehensive reactor and particle modeling. *AIChE Journal*, 48(10), 2398–2411.

3

Technoeconomic Assessment and Risk Analysis of Biorefinery Processes

Eemeli Hytönen and Paul R. Stuart

CONTENTS

3.1 Introduction

Traditionally, process design is defined to include process creation, synthesis, analysis, integration, and optimization activities which are carried out in a stepwise manner to arrive at an optimized solution to an engineering or societal problem that has at least satisfactory performance. In the early stages of the search for a solution and also in later steps of refining the designs, different methods are used, for example, scenario-based creation of alternatives, heuristics to screen out infeasible options, systematic mathematical formulation of all possibilities, or analysis and optimization of the superstructure generated to find the most promising set of alternatives. In particular, heuristics-based approaches, such as the onion method described by Smith [1] and the hierarchical process-flowsheet synthesis, development, evaluation, and selection method discussed by Peters et al. [2], can be applied multiple times throughout the design process and integrated with other approaches. For example, scenario development for process creation uses heuristics derived from engineering and other related perspectives such as corporate and business planning on the same design task. On the other hand, computationally heavy optimization problems can often be reduced by using heuristics.

The steps involved in a traditional process design procedure are illustrated in Figure 3.1. In the early stages of process design, low-fidelity analysis methods are applied to a large number of process-product combinations, and suitable decision-making metrics are used to screen out unpromising solutions. The methods used and the level of detail in the information often lead to low-accuracy results; moreover, especially when analyzing emerging technologies, the degree of uncertainty is high. Further on in the design process, more advanced analysis is conducted on a few most promising process-product alternatives to enhance the precision and lower the degree of uncertainty and to guarantee the selection of the best solution for current requirements.

The early stages of the design process, creation of process alternatives and concept demonstration, are tightly linked with strategic planning and management of the facility and the company. Depending on the problem statement, this linkage may also be reflected in the selection of decision-making criteria. When analyzing pure process development and improvement projects, short-term project-related performance metrics are often most useful to estimate the return on invested capital, whereas for more strategic design projects, such as analyzing the implementation of a new product-process combination or a large-scale retrofit process, longer-term performance measures combined with short-term metrics might be more suitable. Strategic long-term performance measures are commonly used in corporate finance.

The biorefinery design process follows the principles of traditional chemical process design. Some particular features of biorefining, however, play a key role, especially in the early stages of design, and need further explanation.

Many of the products under investigation and development for the biorefinery, even though attractive and perhaps valuable, have uncertain prospects. These uncertainties may arise from the market; for example, the time needed to develop markets for higher-value-added biochemical products or intermediates is often unknown. It may also be uncertain, assuming that it is possible to produce suitable quantities of these chemicals, how the market will react to the new biobased products, particularly in terms of price. On the other hand, lower-value commodity products might not provide high-enough revenues to justify

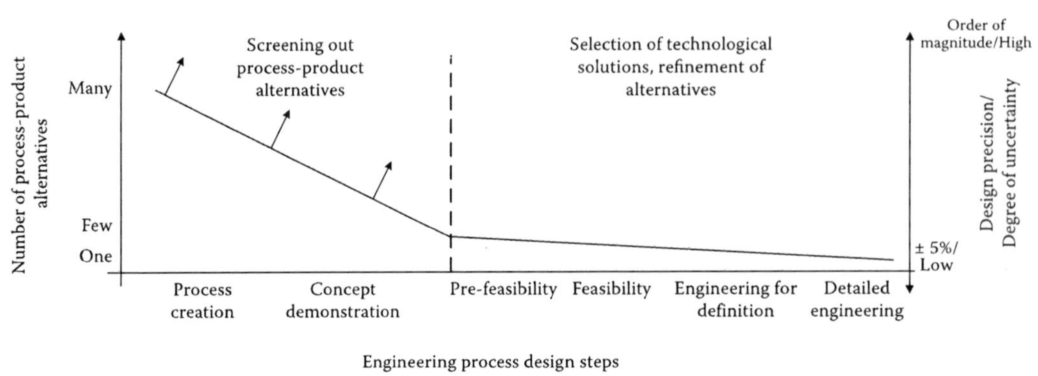

FIGURE 3.1
Process design procedure.

the investment. Hence, the selection of products and their markets from a vast range of possibilities is an extremely complex decision. These decisions are normally made separately from process design activities; however, in this case, the link between product and process development is of critical importance, and interaction and iteration between them is required for successful biorefinery implementation.

Another important specificity of biorefinery process design compared with traditional chemical process development, and also a significant source of uncertainty, is the feedstock and its preparation. Biomass exists in various forms and possesses time-varying and sometimes unwanted properties, even though the main building blocks of all biomass are the same (cellulose and lignin). The processing of this raw material is also considered to be one of the costliest process steps. Moreover, the supply of raw material is very different from that of the traditional chemical industries: point-like sources and piping are common features of the oil-based chemical industry, whereas biomass is spread over large, usually remote areas, with relatively low yield per unit area. On the other hand, the supply of a similar biomass resource, wood, is currently well managed by the forest industry. The third important factor related to feedstocks is the nature of future dedicated biomass resources and their properties and cost as biorefinery raw material.

The third specificity of biorefining is the various processing pathways; biological processes similar to those used in food manufacturing are proposed to be used in many processing steps. Many of these biological processes, such as enzymatic hydrolysis and fermentation using yeasts or bacteria, are very selective and currently established for only a few products. Because of this specificity, development of such processes for new applications such as production of value-added chemicals is still ongoing, and therefore only laboratory-level information is available. Scaling up these ideas to a commercial process introduces a new challenge and various uncertainties for process design and analysis activities.

At the industry scale, the same properties can be used to describe biorefining and traditional chemical process industries: both are capital- and energy-intensive. This leads to substantial pressure on biorefining proposals because financing large development and implementation projects in a developing industrial sector is difficult. To secure financing, biorefinery integration into existing processes such as pulp and paper mills or petrochemical plants might provide some leverage: using centralized utility systems, existing raw material supply systems, and existing knowledge and product distribution channels, as well as suitable mass and energy integration between existing and new processes, can offer substantially lower capital investment requirement and cost of production. Integration of a biorefinery into an existing business also implies modifications of the existing business model, which in turn may possibly imply a change in company strategy and a need to analyze different strategic alternatives.

In summary, specific features of the biorefinery make its future both promising and uncertain. This poses certain challenges for process design and technoeconomic evaluation, especially in early-stage design.

3.2 Technoeconomic Assessment

3.2.1 Overview of Technoeconomic Analysis

Technoeconomic analysis is normally requested by either company management, when they are developing and modifying capital spending plans, or the R&D department, when they are testing new ideas for implementation. The conventional capital appropriation process in a corporate finance context, described, for example, by Northcott or Åberg [3,4], is a stepwise process: (1) identification of potential investments; (2) project definition and screening; (3) analysis and acceptance; (4) implementation; (5) monitoring and postaudit. A comparison of this appraisal process and the process design process (Figure 3.1) reveals that they have similarities: the early stages of the process design process correspond to steps 1 and 2, and the detailed process design step 3 corresponds to planning of capital spending. This capital-spending planning process is commonly carried out as part of both operational (short-term) and strategic (long-term) capital investment decision-making. In the case of strategic decision-making, additional aspects are often also considered to account for company-level performance changes and intangible implications of strategic investments.

Technoeconomic assessment is needed at all levels of process design to provide decision-making information for the corresponding steps of the investment decision-making process or for R&D decision points. It consists of several closely integrated parts: evaluating technical feasibility, capital costs, operating costs, and revenues, and finally estimating the profitability using a set of suitable measures of technical and economic profitability. The scope and objective of the technoeconomic analysis are defined by the capital appropriation process or the R&D program of a company, which determines the required level of detail and therefore also the steps required in the overall design process and the methods needed in technoeconomic analysis.

Capital cost analysis is commonly the main focus of the engineering economic analysis literature, and because of its importance, it includes a well-established classification based on the characteristics of the various levels of the design process. For example, Christensen and Dysert have reported best practices of capital cost estimation [5] using five classes of project definition. These classes correspond to the verbal classification used in Figure 3.1: class 5 is a combination of the process creation and concept demonstration phases, and the other classes (4–1) correspond to the more detailed design phases. The skewed nominal range is a result of the level of detail of the design (a lower level of detail often leads to underestimation of the total amount of required process equipment and therefore underestimation of capital costs).

No classification is available, however, for the two other elements of the cash flow, revenue and operation and maintenance (O&M) cost estimation that will in the long term define the profitability of the business. These factors can be assumed to behave similar to the capital-cost nominal range because of the fact that more effort is made to evaluate them when fewer process-product alternatives are analyzed. Table 3.1 lists all aspects of technoeconomic analysis, including the classifications of cost analyses.

In the early stages of process design, capacity-factored methods are mainly used: capital costs are estimated based on the total actual costs of existing installations and on various factors and indices, such as scaleup factors, the Chemical Engineering Plant Cost Index, or the Marshall & Swift Equipment Cost Index. Capacity-factored methods can also be used in assembly-level or equipment-level capital cost analysis. If substantial differences between reference data and the design exist, and in the later steps of a design, equipment-factored methods can be used. These, however, require more detailed reference and design data, for example, equipment materials, process conditions, and previous project execution information.

Operating and maintenance costs are also often calculated using factors: some fraction of capital costs or a specific cost structure derived from industry averages is assumed to represent the O&M costs of the new design instead of accounting for individual material

TABLE 3.1

Characteristics of Process Design and Technoeconomic Analysis

Design Analysis Characteristics				Technoeconomic Analysis Characteristics				
Name	**Goal**	**Number of Process-Product Combinations**	**Technology Development Scale Required (Entire Process, not only Parts of the Process)**	**Analysis Basis**	**Capital and O&M Cost-Estimation Method**	**Nominal Range for Capital Cost Estimate**	**Nominal Range for O&M cost Estimate**	**Profitability Analysis Methods**
Process creation	Systematic generation of alternatives/concepts Screening of unpromising alternatives/concepts Defining production capacity	MANY (raw material-process-product combinations, capacity free)	Laboratory-commercial	Input–output process mass and energy balance	Capacity factored, parametric models	Order of magnitude	Order of magnitude	Operating profit, capital cost, gross profit, technological maturity
Concept demonstration	Systematic generation of alternatives/concepts Screening of unpromising alternatives/concepts Defining production capacity	MANY (raw material-process-product combinations, capacity free)	laboratory-commercial	Input–output process mass and energy balance	Capacity factored, parametric models	Order of magnitude	Order of magnitude	Operating profit, capital cost, gross profit, technological maturity
Prefeasibility	Screening/selection of alternatives and production capacity	FEW (fixed product, process, feedstock, capacity)	Pilot-commercial	List of main process equipment. Input-output process mass and energy balance	Equipment factored, parametric models	–30–+50%	(+/–) 20%	ROI, IRR, NPV

continued

TABLE 3.1 (continued)

Characteristics of Process Design and Technoeconomic Analysis

Design Analysis Characteristics				Technoeconomic Analysis Characteristics				
Name	**Goal**	**Number of Process-Product Combinations**	**Technology Development Scale required (Entire Process, not only Parts of the Process)**	**Analysis Basis**	**Capital and O&M Cost-Estimation Method**	**Nominal Range for Capital Cost Estimate**	**Nominal Range for O&M cost Estimate**	**Profitability Analysis Methods**
Feasibility	Screening of internal process options and technological solutions	FEW (MANY internal alternatives for process departments, technological solutions)	Demonstration-commercial	Engineering & design data: main PFDs & equipment listing, detailed mass & energy balance	Equipment factored, parametric models	−20–+30%	(+/−) 10%	ROI, IRR
Engineering for definition	Budget authorization/cost control	FEW	Demonstration-commercial	Engineering & design data: PFDs, P&IDs, utility flow drawings, equipment data sheets, motor lists, electrical diagrams, piping isometrics, equipment and piping layout drawings, plot plans, and engineering specifications	Semidetailed unit-cost estimation with assembly-level line items	−15–+20%	(+/−) 10%	ROI, IRR
Detailed engineering	Control of bid/tender	ONE	Demonstration-commercial	(same as above)	Detailed unit-cost estimation with detailed takeoff	−10–+15%	(+/−) 10%	ROI, IRR

Source: Adapted from Peters, M.S., K.D. Timmerhaus, and R.E. West, *Plant Design and Economics for Chemical Engineers*. 5th edn. 2003, McGraw-Hill; Christensen, P. and L.R. Dysert, *AACE International Recommended Practice No. 18R-97 Recommended Practice for Cost Estimate Classification–As Applied in Engineering, Procurement, and Construction for the Process Industries*. 1997, Morgantown, Va: AACE International. pp. 1–10; Dysert, L.R. Cost Engineering, 2003. **45**(6): 22–30. Sprague, J.C. and J.D. Whittaker, *Economic Analysis for Engineers and Managers: The Canadian Context*. 1st edn. 1986, Prentice-Hall Canada Inc. p. 575; Douglas, J., *Conceptual Design of Chemical Processes*. 1988, New York: McGraw-Hill. p. 601; Seider, W.D., et al., *Product and Process Design Principles: Synthesis, Analysis and Design*. 3rd edn. 2009, John Wiley and Sons Inc. p. 728.

flows and prices. These methods entail a substantial loss of accuracy; however, to be able to analyze a large number of concepts in a short time, simplifications are justified. Especially if the uncertainties arising from the method used are systematically considered and all alternatives are analyzed at the same level of detail, even this simple approach guarantees *relatively correct* results, which is the goal of the analysis at this phase.

Capacity- and equipment-factored methods are simple parametric models with capacity- or equipment-specific parameters. More sophisticated parametric models can also be developed based on the analysis objective. The models should be developed from actual costs and knowledge of the process behavior relative to the aspect that is modeled. These parametric models are very useful in the early stages of design to be able to analyze many alternatives in a short time; however, the task of developing reliable and relatively accurate models can be time consuming and data intensive.

The most detailed cost-estimation methods are based on actual equipment prices obtained from equipment manufacturers and system vendors. In addition to the equipment cost, the total capital cost is based on case-specific installation costs, labor rates, and project schedules. A critical part of the detailed cost estimation is to have the correct level of detail design information for all aspects of the process. The precision of cost results which is gained by detailed analysis of most of the process can be lost if one part or system is simply factored in or not analyzed at all.

When the scope of the design analysis becomes more detailed, the accuracy of O&M cost estimates must also increase. The basis for O&M cost estimates, a mass and energy balance with all flows, is better defined at this point, the labor requirement of all processes is established, the cost of required licenses and other annual fixed costs (operating materials, repairs, insurance, overheads) can be estimated in more detail, and facility location-specific information is more precise.

3.2.2 Technoeconomics of Biorefineries

The technoeconomics of biorefinery processes are expected to be similar to those of the petroleum-based chemical industry. Lynd et al. compared historical development of the corn wet-milling-based biorefinery industry and petroleum refining in the United States and concluded that both industries have developed into low-margin, diversified product producers, and capital-intensive industries that need to manage multiple raw materials effectively [10,11]. Because of the long time that it took to get to this point, more than 150 years, more recent lignocellulosic biorefinery business concepts need some time to develop as far, but similar future characteristics can be expected.

The key technoeconomic factors in petroleum-based as well as all biomass-based industries such as pulp and paper are feedstock, energy, and capital costs and the product portfolio. The same behavior can be observed in biorefining, as shown by public-domain studies:

- More detailed biorefinery case studies than the industry-level studies by Lynd et al. in the field of biofuel production have shown the importance of feedstock cost for overall biorefinery process economics. For example, a corn stover-based biochemical ethanol case study by Aden et al., assuming a relatively low feedstock cost of $30 US/dry US ton, found that feedstock costs contribute 31% of total ethanol production costs (including capital charges corresponding to a 10% capital recovery rate) [12]. Correspondingly, in a thermochemical ethanol production case study using woody biomass as feedstock, an even higher cost contribution from feedstock was reported (43.7% of total production costs, $35 US/dry US ton of biomass) by Phillips et al. [13].

The cost of feedstock can be expected to change in the future: competition for the same biomass is pulling the unit price up, whereas developments in biomass cultivation, harvesting, and processing could push the cost per unit of final product down. Feedstock cost is also very highly location specific: competition for the same raw material, types of raw material, growing conditions, and many other factors can vary locally. The process economics are therefore very location-sensitive, and raw material assessment is a very important part of technoeconomic assessment. Some unexpected solutions might become profitable under suitable raw material cost conditions.

- Many biorefinery technologies are energy-intensive: large amounts of heat are needed in pretreatment and purification processes, and achieving correct process conditions (pressure) may require large amounts of electricity. Many technologies are designed for energy coproduction: part of the feedstock is intentionally converted to steam and electricity to supply the process demand. This implies substantial additional capital investments in energy systems and at the same time lower product yields.
- Typically, biorefinery technologies imply relatively high-capital investment costs.
- Biochemical processes include lengthy, low-consistency, and complex biological steps such as hydrolysis or fermentation. The duration and complexity are further extended if more recalcitrant lignocellulosic feedstocks such as wood are used instead of sugar-based raw materials. These factors increase the size of equipment and therefore directly the capital cost requirement. Another costly step is the separation or purification step which is needed because of the low-consistency requirement of the biochemical processing steps.
- Thermochemical processes often use process conditions that require expensive materials to be used in process equipment. These process conditions are described by high temperature and pressure. In addition, depending on the type of thermal degradation process used, expensive oxygen separation, synthesis gas conditioning, or both may be required.
- Promising products from future biorefineries, other than and in addition to biofuels, have been studied extensively. For example, Werpy et al. and Bozell et al. conducted comprehensive studies of the most promising chemical products from biomass-based intermediates (sugars, synthesis gas, or lignin) [14,15]. This means that several potential solutions exist for a diversified product portfolio, as is the case with petroleum refining and partly for the pulp and paper industry.

The technoeconomic performance of a biorefinery, based on the key cost and revenue factors previously discussed, does not appear very attractive in general. This is also the case currently in the pulp and paper and petrochemical industries, with some exceptions. However, both these industries can possibly enhance the performance of biorefinery processes through integration of processes and businesses offering cost reductions, environmental benefits, and improved operating margins.

For example, integrating the biorefinery process into an existing facility with suitable utility systems in place could offset capital costs. For example, Hytönen and Stuart considered utility-integrated biofuel production scenarios where the host facility was a Kraft pulp mill [16]. Naturally, the existing systems have only a certain amount of excess capacity available, which in many cases might not be enough for even a small biorefinery. However, utility-system capacity expansion concurrently with a biorefinery implementation enables partial capital cost allocation to existing products, lowering the capital cost requirement for the biorefinery. In addition to the capital cost benefit, possibly a higher biomass conversion efficiency to products could be achieved.

Other prominent cost-reduction integration impacts include obtaining biomass supply from existing forest-industry supply chains or product sales and marketing using the experience of petrochemical companies. Options such as these are accessible through a correct implementation strategy.

Early-stage technoeconomic analysis of biorefinery processes often relies on publicly available information obtained from technology developers and the understanding and knowledge of the analyst. The goal of technology developers is often to attract funding for further development and demonstration of their new emerging technologies. Later on, when the technologies are developed closer to commercialization, more detailed data are not available to the public. Because of this issue of information availability and reliability, it is important to perform a critical analysis of emerging technologies as part of early-stage design analysis.

Recognizing all the above mentioned important characteristics of biorefinery technologies—key cost factors, integration possibilities, technological development stage, strategic nature of biorefinery projects, and availability and reliability of process and product information—it is obvious that accounting for uncertainties is required to be able to screen out unpromising alternatives systematically in early-stage design decision-making.

3.3 Technoeconomic Analysis under Uncertainty

In process design, several sources of uncertainty exist. Pistikopoulos classified sources of uncertainty in design as (a) model-inherent, (b) process-inherent, (c) external, and (d) discrete, based on their nature [17]. These uncertainties also try to capture the uncertainties existing at the outset of the design project, namely the uncertainties at the company, industry, or general environmental levels. This division into three levels of uncertainties to be considered by management, as listed by Miller and Waller [18], is normally called external uncertainties (category c in the above list) in process design analysis. For example, at the company level, behavioral uncertainties related to management actions are normally not included in a design analysis, whereas uncertainties in company debt, equity amounts, and required rates of return are commonly accounted for.

General risk analysis follows four main steps: (1) identification of sources of uncertainty, (2) quantification of uncertainties, (3) formulation of uncertainty for risk analysis, and (4) quantification of risk. In technoeconomic analysis, the risk analysis method should be selected depending on the process design stage (the goal of the design analysis), the sources of uncertainty, and information availability. Several methods exist for incorporating uncertainty into technoeconomic analysis; all can be classified as either qualitative or quantitative methods.

The following sections describe the different methods that are used in the early-stage process design context and show their applicability to biorefinery design.

3.3.1 Qualitative Analysis Methods

Qualitative risk analysis is best suited for investment strategy or project risk evaluation. It can be considered to be a prerequisite for any process design activities and for more detailed quantitative risk analysis. An example of a qualitative risk analysis method is the SWOT (strengths, weaknesses, opportunities, and threats) analysis approach that is commonly used in strategic planning and capital-spending planning. It subjectively "quantifies" verbally each aspect of uncertainty under a set of generic, qualitative conditions to arrive at an overall benefit–disadvantage description of each scenario under consideration.

3.3.2 Quantitative Analysis Methods

Quantitative risk analysis uses, instead of verbal "quantification" of the system's behavior, different numeric scales to categorize the input parameters or the system's behavior under certain conditions. It can be further divided into deterministic and stochastic methods. More detailed description of these two classes and of the main analysis methods used in process design is given in Sections 3.2.1 and 3.2.2.

Optimization under uncertainty is a wide class of approaches incorporating deterministic or stochastic analysis into mathematical programming-based design and operations analysis. A substantial body of knowledge has been developed during the past decades in both stochastic-based approaches (see ref. [19] for a recent review) and deterministic approaches. The main features of these methods with respect to early-stage process-design risk analysis are also given in Section 3.2.3. No attempt is made to cover optimization or different optimization methods; the focus is on incorporation of uncertainty into mathematical programming and the usefulness of these methods for early-stage process-design technoeconomic analysis.

3.3.2.1 Deterministic Analysis Methods

Deterministic risk analysis includes two types of methods: (1) methods in which the uncertain input parameters can only be given a range of possible values, all values inside the range having the same probability of occurrence. The second characteristic of these methods is that the quantified uncertainties can be propagated through the analysis model to the end results; (2) methods in which some aspect of the system is considered uncertain and can be categorized subjectively using an ordinal or verbal scale and then represented as a result. Normally, these methods quantify the uncertainty, either arbitrarily or based on knowledge of the context and heuristics of the phenomena behind the uncertain parameters, and set the parameters to certain values to create a set of scenarios for analysis.

Type I methods:

- The simplest deterministic method for evaluating the impact of uncertainty is interval analysis or sensitivity analysis. In this approach, ranges (minimum and maximum values) are used to represent uncertain model parameters; this is especially useful if no information about the likelihood or probability of parameter values is available. By solving the problem with boundary values of all uncertain model parameters, the absolutely worst/best-case scenario can be modeled; however, no indication of the likelihood of the modeled outcomes can be achieved [20]. In process-design sensitivity analysis, instead of focusing only on the worst/best-case scenarios, uncertain parameters are considered one at a time, keeping other parameters constant at their expected values (base-case values). This procedure gives the sensitivity of the system to the parameter in question. This type of analysis is especially useful in identifying risky system parameters: parameters that have substantial impact on the results if they vary over their quantified range of values.

- Another deterministic method called scenario planning or analysis uses the same principles as interval analysis. Schoemaker describes scenario planning as follows. By an iterative and interactive group decision process, including discussions among managers and other stakeholders, all different likely future scenarios, and thereby the likely values of risky parameters, will be examined, leading to a few most plausible and internally consistent future scenarios [21]. Even though this scenario planning method is intended for the strategic planning level, it also captures the main idea of risk mitigation and analysis in process design. This scenario analysis approach is often used in process design to generate input value scenarios; however the formal process of scenario definition as proposed by Schoemaker is not followed in early-stage design due to the demands of the planning procedure (time and resources are not always available). A more "streamlined" version is used instead: a group of experts in the field, such as the process design group, including main personnel from existing operations if a retrofit design problem is being considered, selects the scenarios to be analyzed. This approach has the same weakness as interval analysis: the probability of planned scenarios is not calculated systematically.

Type II methods:

- One form of scenario analysis or sensitivity analysis is scoring methods: subjective understanding and knowledge of the system being analyzed is used to quantify the magnitude of the uncertainty in some behavior of the system instead of using a formal method to convert the quantified input uncertainties into output uncertainties using a technoeconomic model. This approach uses an ordinal scale to quantify uncertainty. The uncertainty can be expressed either using the same performance measures as are used for the final results of the analysis (see the description below of discounting methods) or as an additional system behavior. In the latter case, new metrics are required for the decision-making process. Scoring methods can be useful in multicriterion decision making in early-stage design to avoid the need for complex models and time-consuming modeling and simulation.
- Discounting methods are a variation of scoring methods that are very close to qualitative risk analysis methods. Using an ordinal or verbal scale, the results are discounted in a case-specific manner to enable better comparison of different scenarios or process design alternatives. For example, if it is known that the main product of a design alternative has established markets, the profitability of the process alternative is not discounted. On the other hand, if the main product does not have an existing market, or if the market is difficult to enter, the profitability of that design alternative is discounted based on a scale. Hence, instead of changing the price of the final product, perception of market conditions is used to evaluate the impact on profitability. Other aspects of design, such as technological maturity or design data reliability, can be addressed in a similar way.

3.3.2.2 Stochastic Analysis Methods

Stochastic risk analysis is based on the idea of assigning a probability distribution to each uncertain model parameter. By selecting values randomly from these distributions, a large number of scenarios (input parameter combinations = scenarios) can be formed. When analyzing these scenarios, all uncertainties are propagated to the end results to form probability distributions, for example, the probability distribution of the profitability of the investment project. One important feature of stochastic analysis is the correlation of uncertain parameters: a strong correlation between two or more uncertain model parameters, for example, oil and gasoline prices (examples of the input and output of a chemical process) enables a simpler model because only one distribution is required to describe the behavior of all strongly correlated parameters. On the other hand, if correlations are not identified, unrealistic random parameter combinations will be included in the analysis, and these can distort the outcome distributions.

The best-known stochastic analysis method is Monte Carlo (MC) analysis. All uncertain model parameters have probability distributions; the solver randomly selects values from the distributions and calculates the outcome. By repeating this many times (often 10^5–10^7 times) many possible parameter combinations (scenarios) are calculated. Because the randomly selected input values are based on their probability, the result is correctly distributed.

As it is obvious from the above description, this method is simple, but requires much computation time. This "brute force" method of random sampling can be enhanced by other sampling methods. Often-compared methods are Monte Carlo sampling (MCS) and Latin hypercube sampling (LHS); see, for example, Diwekar and Kalagnanam [22] or Wang et al. [23]. In LHS, the probability distributions are divided into equally probable intervals, and, for example, the average value of each interval is used as the input value. All distributions are treated in this way, and the average values are organized into a matrix which is used as an input to the MC function. This "stratification" of distributions leads to faster computation and reduces variance.

Other sampling methods have also been developed for better representation of the distributions and decreased computation time. Wang et al. show the efficiency of combined Latin hypercube-Hammersley sequence sampling in several different cases, and it is clear that this combination is a much better sampling method than MCS and LHS [23]. The mathematical formulation is, however, often more complex than MCS and LHS, and therefore its application to real problems is not common.

The interpretation of the results of stochastic modeling always involves two aspects, the expected value and the variance or statistical dispersion. By presenting the result distributions as cumulative distribution functions (CDF), these two aspects can be easily combined, and mathematical methods for comparing the CDFs of various options exist. Graves and Ringuest compared six methods for comparing uncertain alternatives. They propose two, almost stochastic dominance and mean-Gini methods, to be most suitable for comparing several options [24]. However, if multiple criteria are considered in decision-making instead of only one, theoretical models for comparison are not available; in these cases, uncertainty of the criteria needs to be considered as another criterion or as additional information for subjective decisions.

Another method for estimating the uncertainty of outcomes is based on the definition of variance: variance is the second central moment of a real-valued random variable and can be estimated using a Taylor expansion of the function describing the outcome y as a function of its variables x, $y = f(x)$. The expansion is often truncated after the second-order terms, and in the case of uncorrelated variables x, the covariance between variables can be omitted, leading to a less complex formula for the variance of the outcome. This method, called the law of propagation of error (LPE), is commonly used in analysis of the impacts of measurement uncertainty on results reported using a metric calculated from the measurements (e.g., the variance of resistance is calculated from measured voltage and electric current). It is not, however, commonly applied in the process design context.

Xiao and Vien compared this method with MC analysis in for modeling a mineral processing system and concluded that, even though LPE can be used in this context, MC analysis provides more informative and accurate results. The main reason is that LPE is able to give correct answers only for linear functions, whereas Monte-Carlo analysis can also handle nonlinear and complex system models [25].

In spite of the limitations of this LPE method (inaccuracy when complex, nonlinear systems are analyzed) in a process design context, its use can, however, be justified in cases where calculation time is an important factor: the evaluation of a single equation is fast compared to the 10^5–10^7 iterations often carried out in MC analysis. Moreover, if the actual probability density function is not the objective of the uncertainty analysis and if knowing the statistical moment (variance or standard deviation calculated from variance) of the outcome is sufficient, LPE could be used in process design risk analysis.

3.3.2.3 ***Optimization-Based Analysis Methods***

Optimization under uncertainty involves a vast and rapidly expanding group of methods which are applicable to process design analysis at all design stages. Depending on the goal of the design analysis and the information available, the appropriate formulation of the uncertainty can be selected. These factors also define the optimization methods that can be used.

Optimization-based methods using finite number bounds or fixed parameter values (hence scenarios) to describe the uncertainties are deterministic approximations of the real system, and optimization of these scenarios is called deterministic optimization under uncertainty. These possible model parameter combinations (scenarios) can also be assigned probability values, for example, using expert opinions. The probability of occurrence of a scenario is then also the probability of the corresponding result. Another type of deterministic optimization under uncertainty is to conduct sensitivity analysis with a deterministic optimization model.

In stochastic programming, the uncertain model parameters can be selected either randomly from their probability distribution or based on the knowledge of the analysts (called deterministic stochastic programming). The resulting uncertainty in the decision variables can then be solved for using various methods [19]:

- Recourse-based stochastic optimization:

  Classical recourse programming divides the decision variables into two stages. The first-stage variables are fixed under some particular instantiation of the uncertain model parameters to determine the actual instantiation of the uncertainty in the uncertain second-stage variables, called also the *recourse variables*. These recourse variables represent the cost of corrective measures against the infeasibility of the solution, and because they are the result of a random parameter instantiation, the cost is random as well. The objective is to minimize the total cost (the sum of the expected first-stage cost and the random second-stage corrective cost). If the parameter probability distributions are discrete, the optimization problem can be solved using linear programming, and the problem reduces to a scenario tree of parameter instantiations. In the case of more generic distributions, sampling-based schemes can be used to solve the optimization problem.

Methods based on this classical recourse-based method are: stochastic linear programming, stochastic integer programming, stochastic nonlinear programming, and robust stochastic programming. They differ from the classical method in the types of variables (linear vs. nonlinear, continuous vs. integer/binary) that can be included in the formulation of the objective function and in the interpretation of the results. For example, robust stochastic programming includes a variability of the recourse-variable terms in the objective function to reflect the different preferences of decision-makers. Compared with other stochastic recourse-based methods, this enables the modeling of risk adversity instead of the risk neutrality that is at the basis of other methods.

In general, all recourse-based methods aim to minimize the cost of infeasibility, and therefore they allow infeasible solutions with a penalty or a cost.

- Probabilistic programming approaches the problem from another perspective: it focuses on finding solutions that are best able to meet the feasibility requirements under uncertain conditions.
- Fuzzy mathematical programming uses fuzzy numbers and fuzzy sets to describe uncertain model parameters and constraints. Membership functions are used to describe constraint violations, and hence the uncertainty in the constraints. They also describe the degree of satisfaction of the constraints and the uncertainty range of the coefficients of the objective function.
- Stochastic dynamic programming adds the dynamics of uncertainty to stochastic programming. Discrete-time systems can be solved using the stochastic programming methods described above in a recursive manner by solving all detail subproblems and then using these solutions to solve the major problem.

Algorithms and methods applying the above-mentioned methods and others have been developed in the process systems engineering (PSE) community for chemical process design and synthesis purposes. All the methods described above are applicable to early-stage design aimed at screening based on technoeconomic performance and the uncertainty of the alternatives under consideration. The decisive factor in selecting the method is the goal of the design activity because of the different capabilities of the methods. Stochastic programming methods are mostly applied in a design context (see, e.g., refs. [26,27]).

In recent publications, Svensson et al. describe a deterministic stochastic optimization method for early-stage retrofit design problems and demonstrate the method on a pulp mill retrofit investment project identification problem. After identification of a minimum number of uncorrelated uncertain model parameters (only prices and CO_2 emission charges are considered), parameter sets are defined. Each of these parameter sets is assigned a probability and a validity period. By combining these sets to define future paths and assigning a probability to each path, time scenarios up to 25 years long can be modeled. The problem is solved using a multistage mixed-binary linear programming method, where the objective function is the maximization of expected net present value (NPV) and the first-stage variables are the initial investments needed to define the outcomes of the uncertain parameters. This combination of parameter uncertainty and time dependency of decision-making in a case-study analysis leads to a robust and surprising investment solution and proves the importance of uncertainty and time considerations in strategic investment decision-making [28,29].

3.3.3 Critical Aspects of Early Design Stage Uncertainty Analysis

A generic challenge to all methods is the second main step of process design, quantification of uncertainties to be able to represent the prevailing uncertainties objectively. This is especially critical in the early phases of process design, when in general less information is available and less accurate design-analysis methods are used. When using any of the risk analysis methods discussed here, the user should try to address this challenge; however, the way of addressing it differs among the methods because the formulations are different.

Hubbard and Evans consider the possible problems of scoring methods which often rely on ordinal verbal scales [30,31]. The human aspect of risk analysis—varying perception and context of the severity of each risk factor and varying perception of the scales in use—can bring an additional uncertainty to the risk analysis results. Moreover, invisible correlations between uncertain parameters can create false outcomes in scoring-method analyses. Hubbard and Evans conclude that when recognized and carefully considered, these problems can be avoided. Another option, based on their analysis, is to use stochastic methods.

On the other hand, stochastic methods can also be considered subjective and sometimes lack the decision-makers' understanding of the underlying phenomena (the origins of uncertainty). The probability distributions are therefore only perceptions of the real distributions. Bode et al. discuss the need for a better assessment of uncertainties to be able to reduce them for decision-making in a process design context [32].

It is also important to realize whether an uncertainty is really an unknown factor that can only be estimated over time. It may not be possible to include these types of uncertainties in the analysis, but they can still be managed. One example of this type uncertainty is future products: some properties can make a product very attractive, but because no markets exist, no one will make an effort to develop a process for its production. However, a strategy to take an enabling step toward production of that specific product might provide an advantage in the future. Hence, the risk of making the wrong product decision by rushing into the unknown can be reduced and managed by taking a single step toward the new product and producing perhaps some intermediate product until the market has developed and further investment can be justified. Another example of an uncertain factor which is unknown now but might be revealed in the future are political decisions that have implications for technology development speed, market creation, prices, and costs.

In conclusion, it is critical always to evaluate the types of uncertainties to be included in a risk analysis to decide whether to address them through the risk management strategy of the company or in the actual risk analysis step of the technoeconomic analysis. In addition, an understanding of the subjectivity of the uncertainty and its formulation is important for a reliable and systematic risk analysis.

A second set of critical aspects is related to optimization-based risk analysis methods: the selection of how to formulate uncertainties and the risk analysis method. Many of the methods are computationally demanding. Therefore, in the early stages of design, simplified methods are most applicable because they can analyze a large number of design alternatives in a short time. Therefore, deterministic optimization methods and deterministic stochastic programming methods might be preferable at this stage.

When using optimization approaches, it is important to understand the type of uncertainty present: if information is missing, fuzzy programming can provide methods which account for this better than stochastic programming approaches; on the other hand, if probability distributions can be formulated, stochastic programming methods are more suitable.

3.3.4 Sources of Uncertainty in the Design of the Biorefinery

Many of the promising new biorefinery technologies are still in the laboratory or pilot-scale development phase; some biofuel technologies which are already at demonstration scale are the only exceptions. Higher levels of process-inherent uncertainty are evident for these new emerging process technologies compared to more developed processes. Many of the key aspects have not yet been addressed or even identified for an overall system such as the final commercial process must be. Therefore, design data are not very reliable and are seldom even available. Important process-related uncertainties are, for example, the conversion efficiency of biological processes, the recycling of chemicals and biochemicals in large-scale production, or the scaling up of purification and separation systems to these scales of production. Because of this nature of the information, often formulations of uncertainty which ignore the probability of uncertain parameter values can best represent process-inherent uncertainties in the early stages of design.

An equally important challenge is the understanding and quantification of external uncertainties. Biomass markets and competition are developing, and substantial changes in prices can take place in the near future. Prices of many products which are potential substitutes or replacements for petrochemical-based products are volatile, and market size can only be forecast. Project financing depends on both company level and financial market conditions. Because of the many unknown factors impacting on the uncertain external parameters, a risk management strategy becomes an important tool to account for them. For example, much market information and understanding is available already for many commodity products (biofuels), but not for many value-added biochemical products and intermediate building-block products. Therefore, it is possible to consider the external uncertainties of biofuel production scenarios systematically in a technoeconomic analysis, but this is not necessarily possible for nonexistent product scenarios. When more information becomes available for proper formulation of the uncertainties (with or without probability distributions), such analyses can be conducted. Until then, the uncertainty in value-added product scenarios can be considered through a proper risk management and implementation strategy.

At every stage of design analysis, the models that are used generate additional uncertainty:

- At the process creation level, when simple and often linear process and cost models are used, the results are as reliable as the factors that represent the aspect in question. This applies to both traditional technoeconomic analysis and optimization-based superstructure analysis. For example, when estimating the total capital investment cost of a new biorefinery process technology using the actual investment costs of an existing facility of another product as a reference cost, the capacity factor and other factors describing the differences between reference and new technologies are by definition average values. These factors can introduce bias into the total capital cost estimates. Moreover, many external economic factors are excluded from the analysis at this stage, adding uncertainty to the final profitability.
- In the concept demonstration stage, the available design data are more complete, and more detailed estimation routines can normally be used. Hence, data availability enables the use of better models to reduce the degree of uncertainty arising from the process and economic models. At this level, external factors related to financing and micro- and macroeconomics are also more often modeled. This

leads to additional uncertainty in the models that are used. For example, using NPV methods to model profitability has well-known shortfalls when comparing projects with certain properties (see, e.g., the corporate finance handbook by Ehrhardt and Brigham for a detailed discussion [33]). Internal rate of return (IRR), when used as a performance measure for mutually exclusive projects with substantially different capital investment costs, can possibly leading to different capital structures and different lifetimes. This can easily be the case in the biorefinery context, for example, when comparing a large biofuels plant with processing of an existing side-stream to a low-volume added-value product). In such a case, IRR can give results which contradict the analysis of the same concepts using NPV as the profitability measure. This model-inherent uncertainty can also be managed by selecting the models that are best suited to the particular design project and by using multicriterion decision-making (MCDM) methods.

Discrete uncertainties in the early stages of the biorefinery process design originate mainly from technology and financing availability or from political decisions. These aspects are often considered as given fixed values or included as prescreening criteria:

- If a step of the process under analysis has not been developed to the same level as the rest of the process, and its future development is not being pursued by any equipment manufacturer or technology provider, then the entire process might be screened out from the analysis that is being conducted, while not eliminating this solution from future analyses.
- If the company commissioning the design analysis has strict capital investment limits, the availability of financing might be used as a fixed-value prescreening metric.
- If a particular biobased product is a national or global priority, governmental funding for advancing technologies and developing markets is often available. Other types of subsidies are product price and feedstock cost regulations. An important feature of all subsidies is that they are available for a predefined period of time. This type of discrete uncertainty, with unknown existence, magnitude, and duration (and impacting on all other sources of uncertainty: external, process-inherent, and model-inherent), is often considered through fixed scenarios.

The origins of uncertainty and suitable methods for including them in technoeconomic analysis (resulting from the formulation of uncertainty) in a biorefinery design context are listed in Table 3.2. The overall approach in which these analysis methods are applied can be optimization based or a traditional technoeconomic analysis method.

The main sources of uncertainty given in Table 3.2 can be formulated for many analysis methods. The methods listed were selected based on the nature of the uncertainties and the objective of the design step. In the process creation phase, often many process design alternatives producing several different products are considered. Gathering and forecasting the required market data (prices and demands for all products) to define probability distributions can therefore be an excessively time consuming task even if information is available, and simpler methods to formulate the uncertainty may therefore be needed. A range of possible values is easier to establish, and therefore this might be the most suitable formulation type for considering external uncertainties. On the other hand, in the concept demonstration phase when fewer process design alternatives and also fewer products are

TABLE 3.2

Sources of Uncertainty and Suitable Analysis Methods Classified by the Type of Uncertainty in the Early Stages of Process Design Technoeconomic Analysis

		Model-Inherent	Process-Inherent	External	Discrete
Process creation	Examples of uncertainty sources	Cost-estimation models M&E balance dependencies	Product yields Energy consumption Equipment cost Process development scale	Raw material cost Product prices Market demand	Availability of overall process technology Availability of financing Availability of subsidies
	Analysis method	Scenario analysis Sensitivity analysis Scoring methods			
Concept demonstration	Examples of uncertainty sources	Cost-estimation models	Process integration impacts Cost of all required equipment Consumption of all raw materials	Feedstock price Product prices Forecasts of prices	Availability of technological solution and equipment for specific process steps Magnitude and duration of subsidies
	Analysis method	Discounting methods	Scenario analysis Sensitivity analysis Stochastic analysis		Scenario analysis

being considered, a more comprehensive market analysis can be conducted to obtain a distribution-type uncertainty formulation.

As for process-inherent uncertainties, after process creation and first screening, the concepts that are left for consideration are more constrained (feedstock options might be narrowed to only a few, or technologies for particular process steps may have been selected, leading to more detailed process parameter information). However, enough process and design data might still not be available for distribution development, and then a range-type uncertainty formulation is needed.

In further steps of process design, many of the key uncertainties analyzed in the early stages of design are reduced to a substantially lower level, and scenario or sensitivity analysis are often sufficient as a risk-analysis approach. For example, process parameter variation (a process-inherent uncertainty) can be narrowed to a more tractable form by extensive laboratory research or piloting, or more detailed process layouts and plans may enable the capital costs to be estimated based on actual vendor quotes and previous projects, leading to a very low model-inherent capital-cost uncertainty.

A group of uncertainties specific to a particular biorefinery business concept, the integrated biorefinery, is the uncertainty of integration impacts. In an integrated biorefinery, an existing facility (e.g., a pulp and paper mill, a petrochemical plant, or a corn ethanol plant) and its systems and knowledge are used to run both the new and the existing operations. To provide benefits, integration must be performed at several levels, from the process to the corporate enterprise. The impacts of this biorefinery strategy are various, for example:

- Using the same heat and power systems changes the overall energy balance, and this has impacts on the operation of the existing production systems and therefore the production costs of the existing products;
- If the existing labor force is used to operate both processes, this lowers the labor costs of existing products and also changes the facility overhead charges;
- Using the same raw material purchasing and product sales and marketing staff similarly has impacts on the production costs of existing products.

However, no experience of such integration exists in many cases, and the degree of integration possible at the various possible levels is somewhat uncertain. Quantification and formulation of these uncertainties at the process creation and concept demonstration levels can, in many cases, only be done using ranges of possible impact degrees, leading to the choice of sensitivity or scenario analysis as suitable risk analysis methods.

3.4 Selected Early Design Stage Biorefinery Technoeconomic Studies

3.4.1 Biorefinery Case Studies

The methods listed in Table 3.2 have been implemented in many types of early stage design and technoeconomic analyses of biorefineries. Table 3.3 lists some of these case studies, focusing on different characteristics of the studies. The characteristics that are analyzed include the level of design analysis, the main goal of the study, the initial and final numbers of design alternatives that were studied, the methods that were used for both capital and O&M cost calculations, the various uncertainties that were considered in uncertainty analysis, the uncertainty analysis methods used, and the criteria that were used in screening or comparison of the design alternatives. The actual results of the studies are not reviewed.

Three selected case examples are reviewed in more detail in Section 4.2, and one of the studies is further discussed as a concrete example of biorefinery technoeconomic and risk analysis in Section 4.3.

3.4.2 Selected Studies in More Detail

Three recent studies listed in Table 3.3 are reviewed in more detail to provide a better description of how key aspects of technoeconomic analysis and risk analysis in early-stage process design have been addressed. These aspects are the formulation of uncertainties and the method of risk analysis, study assumptions, and cost analysis method.

Cohen et al. used a scoring method to formulate one type of uncertainty, process-inherent uncertainty, in the process creation stage of process design analysis [35]. Emerging technologies for ethanol production in an integrated forest biorefinery were analyzed using an

TABLE 3.3

State-of-the-Art Biorefinery Process Design Studies and their Characteristics

Design Phase	Goal of the Case Study	Alternatives (In → Out)	Capital and O&M Cost Estimation Method	Considered Uncertainties	Uncertainty Analysis Method	Decision Making or Comparison Criteria Decision Making Method[a]	Ref.
Process creation	State-level roadmap development (resource-product chain screening)	650 → 8	—	Qualitative External Process-inherent	PEST (Political/legal, Economic, Social, Technological) *Process* - SWOT	Survey score Feedstock flexibility *Method:* Stepwise MCDM (MCDM using qualitative survey results)	[34]
	Identification and weighting of pertinent pathway screening criteria for integrated forest biorefinery using biofuel processes as example design alternatives	7 → 7	Capacity-factored method	Technological maturity	Scoring	Technoeconomic, environmental, feedstock flexibility, product diversification, energy, and integration *Method:* Tradeoff MCDM	[35]
	Demonstration of novel optimization-based design methodology using agricultural waste stream for fuel and energy generation	3 → 1	Capacity-factored method	—	—	Profit *Method:* Optimization	[36]
	Evaluation of alternatives for substitution of transportation fuels beyond 2010 in Europe	22 → 3	Literature values	MCDM criteria weights	Sensitivity analysis	Incremental cost of fuel substitution Total cycle GHG (green house gas) emissions Energy consumption Substitution potential *Method:* AHP (analytical hierarchy process)	[37]

Concept demonstration	Comparison of close-to-commercial gas-to-liquids design alternatives with traditional Tomlinson recovery boiler retrofit	9 → 9	*Capital costs*: equipment-factored model *O&M costs*: capacity-factored model	Energy costs Feedstock cost Product prices Monetized environmental benefits	Scenario analysis Sensitivity analysis	IRR of incremental capital investment NPV	[38]
	Economic comparison of leading pretreatment technologies for ethanol production from corn stover	5 → 5	Equipment-factored model	Ethanol yield with/without oligomer credit Solvent loading to pretreatment	Scenario analysis	MESP (Minimum ethanol selling price)	[39]
	Comparison of the economics of five pretreatment options and three downstream processing options for ethanol production from corn stover	7 → 7	Capacity-factored method	Pretreatment and saccharification conditions and conversion rates Economic assumptions	Scenario analysis Sensitivity analysis	Product value	[40]
	Comparison of biofuel production pathways Identification of future focus areas in R&D of studied pathways	14 → 14	Equipment-factored model	Energy costs Feedstock cost Capital structure	Scenario analysis Sensitivity analysis	IRR Minimum selling price Total capital cost	[41–43]
	Technoeconomic analysis of the forest biorefinery concept based on near-neutral hemicellulose pre-extraction	1 → 1	Capacity-factored parametric model	Pulp mill production capacity Availability of excess capacity and equipment	Scenario analysis	Ethanol and acetic acid production cost After-tax IRR	[44,45]
	Technoeconomic analysis of the forest biorefinery concept based on acidic hemicellulose pre-extraction	1 → 1	Capacity-factored method	Hemicellulose extraction rate Rate of cellulose degradation in kraft pulping	Scenario analysis	Production costs Capital investment cost	[46]

continued

TABLE 3.3 (continued)

State-of-the-Art Biorefinery Process Design Studies and their Characteristics

Design Phase	Goal of the Case Study	Alternatives (In → Out)	Capital and O&M Cost Estimation Method	Considered Uncertainties	Uncertainty Analysis Method	Decision Making or Comparison Criteria Decision Making Method[a]	Ref.
	Demonstration of strategy to repurpose kraft pulp mill into ethanol plant	3 → 3	Capacity-factored method	Ethanol revenue Capital cost Enzyme cost Biomass cost	Scenario analysis Sensitivity analysis	IRR Cash cost of ethanol production Total cost	[47]
	Screening of retrofit biorefinery implementation scenarios producing biofuels	42 → 6	Capacity-factored parametric model	Biofuel production capacity Raw material costs Product prices Energy costs	Sensitivity analysis Monte Carlo analysis	IRR Downside IRR Total capital cost *Method:* Single-criterion comparison	[48]
Concept demonstration/ prefeasibility	Process design and economic analysis of corn stover-based biochemical ethanol production process under investigation at NREL	1 → 1	*Capital costs:* Detailed cost estimates *O&M costs:* Equipment-factored model	Hurdle-IRR Equity fraction of financing Feedstock composition, cost, and handling Yields Enzyme loading Energy cost	Scenario analysis Sensitivity analysis Monte Carlo analysis	MESP	[12]
	Process design and economic analysis of poplar-based biochemical ethanol production process under investigation at NREL	1 → 1	*Capital costs:* Detailed cost estimates *O&M costs:* Equipment-factored model	Feedstock cost Hydrolysis and fermentation yield Capital cost	Scenario analysis	Ethanol production costs	[49]

Process design and economic analysis of corn-stover-based thermochemical ethanol production process under investigation at NREL	1 → 1	*Capital costs:* Detailed cost estimates *O&M costs:* Equipment-factored model	Hurdle IRR Financing structure Feedstock properties Plant capacity Process step efficiencies By-product credit	Scenario analysis Sensitivity analysis	Fuel yield MESP	[13,50]
Demonstrate the benefits of using risk analysis in technoeconomic analysis over traditional approaches in corn ethanol process design context	1 → 1	Capacity-factored method	Product and by-product prices Feedstock cost Energy cost Operating interest rate Inflation rate	Scenario analysis Monte Carlo analysis	PVENW (present value of end net worth) NPV ROI (Return on investment) Variable production costs Probability of economic success	[51]

[a] If actual screening/selection was done, the method is listed.

MCDM approach. The uncertain technical maturity or level of development scale of the various processing steps of ethanol production processes was considered by assessing each processing step and giving it a subjective maturity score (a value between 1 and 5). The normalized sum of the technology-specific scores was used as one decision-making criterion among others (including technoeconomic, environmental impact, feedstock flexibility, product diversification, and energy integration impact criteria). This risk criterion was weighted by the MCDM panel and given substantial importance as a decision-making criterion in the very early stages of process design.

The economic analysis in this study was a simplified capacity-factored estimation of both capital costs and O&M costs under two conditions, the current development of technologies, and the estimated future efficiencies and product yields. The profitability of each considered design alternative, measured as return on investment (ROI), under these two conditions was used as a separate decision-making variable; the estimated future ROI was given the highest importance by the MCDM panel among all criteria used. This indicates that from the two very different sets of assumptions (current efficiency of processes and future estimated efficiency), future process performance, rather than current performance, would be the preferred basis for economic evaluation.

Kazi et al. studied several bioethanol production process design alternatives using corn stover as feedstock [40]. They used published process information instead of future performance estimates as the basis for all cost analysis. Capital costs were *n*th plant base-case estimates corrected by a factor based on regression modeling of 44 processing plants (pioneering-plant analysis). This factor accounts for the uncertainty in capital cost estimates (level of design definition, e.g., process equipment not demonstrated commercially, or impurity buildup). Correspondingly, production shortfalls during the startup period were estimated using a regression model. This factor accounts for the impacts of the development scale and complexity of the process on revenue during the first years of production.

In addition to these two regression-based factors, investment cost contingency was increased to account in general for the uncertainty of estimating the *n*th plant capital costs.

Using scenario analysis, the impacts of uncertain process parameters (specific to each studied process design) on the main analysis criterion, product value, were estimated. These uncertain parameter values were taken from the open literature, and scenarios were constructed using the minimum and maximum values found. The product value at the boundary values was compared to the base-case product value. The generic economic assumptions and overall process parameters were assessed separately as a sensitivity analysis.

The second example of recent concept demonstration-level design analysis is the work performed by Laser et al. [41–43]. They made a comprehensive comparison of different biofuel and energy coproduction process design scenarios (including also biofuels as coproducts). Capital costs were estimates using the main equipment level capacity-factored method based on literature cost estimates. O&M costs were estimated using mass and energy balances (detailed simulation models) developed from raw material and energy consumption information. Literature-based values for fixed operating-cost estimates and prices (escalated to the analysis year) were used.

All design scenarios were assumed to be technologically mature, and hence the process parameters used were assumed to represent processes for which only incremental improvement in costs and benefits could be achieved with additional R&D efforts. The feedstock to all processes was switchgrass.

External uncertainties were considered using sensitivity analysis: prices of feedstock, electricity, and oil were varied, and the IRR of the designs was evaluated.

The resulting economic performance of the processes considered was aggregated based on the product mix to groups of design alternatives (bioethanol + coproducts, thermochemical fuels and power, power) so that generic conclusions could be drawn and different biofuel production strategies compared.

3.4.3 Concrete Example

Hytönen and Stuart analyzed 42 different integrated forest biorefinery design alternatives (with production capacity as a free variable) in a concept demonstration-level design analysis [48]. The following sections describe the context and the risk-analysis steps, focusing on uncertainty formulation and interpretation of the risk-analysis results.

3.4.3.1 Context and Technoeconomic Analysis Method

A North American hardwood kraft pulp mill was considered as the host process for an integrated forest biorefinery consisting of different types of biorefinery concepts. Real mill information was used as a basis for the analysis. Various feedstock options in the case mill location were considered: pulpwood, woody forest-based biomass, corn grains, corn stover, and food-processing wastes, as well as pulp mill side streams (black liquor and pre-extracted hemicelluloses). Current and possible future pulp mill process configurations with specific system capacities were analyzed. In addition, feedstock capacity was not constrained to a single value; rather, a promising production capacity and therefore a most promising feedstock capacity were examined during the study.

A detailed list of design scenarios and all economic assumptions can be found in the references. The main definitions are given below:

- Biofuel options in the study included bioethanol, Fischer–Tropsch liquids (FTL), and mixed alcohols (including ethanol).
- Fuels and acetic acid (from one design alternative) were considered as by-products or coproducts.
- The process designs assessed included both biochemical and thermochemical alternatives. Biochemical processes used either enzymatic or acidic hydrolysis, followed by fermentation and distillation to produce ethanol, and thermochemical processes used either high-temperature gasification or steam reforming combined with fuel synthesis or syngas fermentation to produce FTL or mixed alcohols from which ethanol could be separated.

Design information for a base-case design of each process was gathered from the open literature. These reference studies had assumed an *n*th plant design with estimated future performance; therefore, more optimistic future process designs were assessed in this study as well. The base-case designs were used to design other feedstock-processing alternatives based on feedstock composition (product and by-product yields, energy content of solid residue from biochemical processes), and indirectly the consumptions of energy and chemicals were modified. The reference data were used to develop parametric models of assembly-level capital costs (capacity-factored models), and the mass and energy balances (the facility-level input–output model) were used for variable cost estimation. A common basis was used for fixed operating-cost analysis; wages, overhead costs, or cost of operating supplies were constant or constant fractions of capital cost for all design scenarios.

In addition, all the economic parameters of the discounted cash flow analysis were the same for each design scenario and therefore were changed from the assumptions used in the reference analyses to a single assumption.

In this study, emphasis was placed on feedstock costs, and therefore a comprehensive raw material assessment was conducted. In the analysis, public database information on the availability of various biobased feedstocks was used. The transportation costs of the different raw materials delivered to the plant gate were calculated based on biomass availability, transportation distance, transportation method, fuel cost, and other factors. A maximum availability was set to correspond to the total available amount of feedstock in a region of approximately 150 km around the base-case pulp mill. Later in the risk analysis, the transportation cost of feedstocks was allowed to vary depending on fuel cost, and the crop cost or stumpage fee of different biomass options was also varied.

The second main point of interest in the study was the quantification of integration impacts under two pulp mill operating conditions: current operations and pulp mill configuration, and a modernized mill configuration. Two types of impacts were quantified: (1) structural process integration impacts, using the same utility and waste handling systems to the extent available and changing the investment costs, and (2) simple overhead cost impact, sharing the existing pulp mill labor force and changing the O&M costs of biofuel production. These impacts were fully allocated to the bioproduct.

A most promising production capacity for each scenario was selected using the total production costs of biofuel as the criterion. Production costs (including capital charges of 10% capital recovery) were analyzed for a wide range of production capacities, and the lowest-cost capacity was selected. No mathematical programming was used; instead, a manual comparison was conducted.

3.4.3.2 Uncertainty Analysis and Conclusions

Uncertainty analysis was conducted using two methods consecutively:

- Sensitivity analysis: evaluation of the impact of external uncertainties on the main screening criterion (after-tax IRR);
- MC analysis: evaluation of the combined impact of uncertainties having the biggest impact on the screening criterion based on the sensitivity analysis.

Because all main products were assumed to be sold as fuels, their prices were assumed to follow energy-content-corrected gasoline and diesel wholesale prices (less piping costs). Correlation analysis between other raw materials and products was not required based on sensitivity analysis results and data availability. Critical external factors were estimated to be the feedstock price (long-term reported historical price data were available only for pulpwood, and therefore correlation analysis was not possible for feedstock price). Electricity and chemicals were not substantial factors in scenario profitability, leading to a need for separate distribution of main energy-source prices. Uncertainties were formulated as probability distributions. Definitions for the uncertainties considered are presented in Figure 3.2.

Prices for ethanol and mixed alcohols were derived from gasoline price and FTL price from diesel price using the higher heating value (HHV) of the fuels. In addition, the diesel price distribution was used in the biomass transportation cost calculation. Coal price was defined as free on board (f.o.b.) price, and case-specific transportation and fixed costs were added to the base coal price. Oil-price trend forecasts from Annual Energy Outlook 2009 by DOE/EIA were used to define exponentially increasing liquid fuel prices, and a

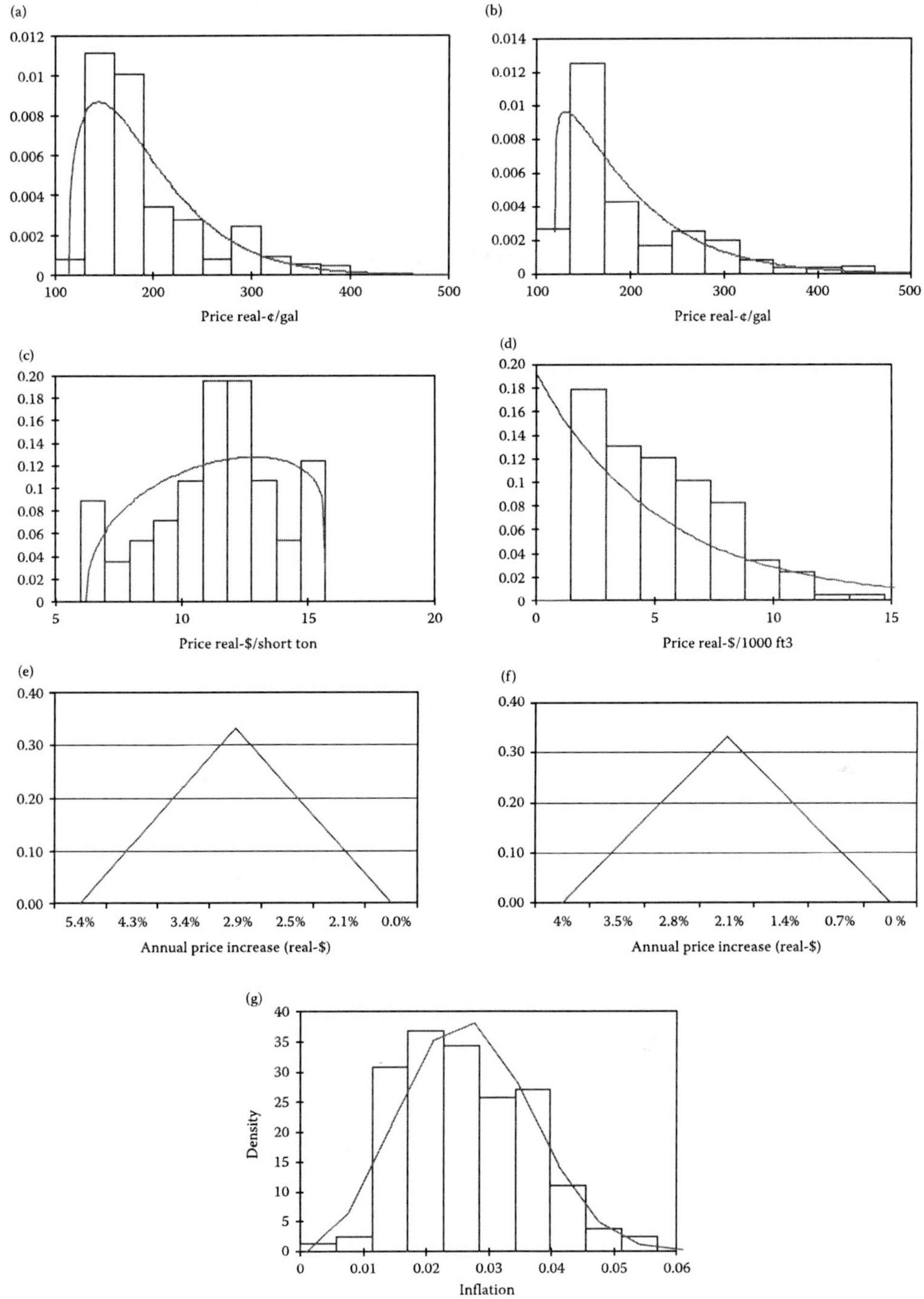

FIGURE 3.2
Monte Carlo analysis input distributions. (a) Historical retail gasoline real price (1993–2008)—Weibull distribution. (b) Historical retail diesel real price (1994–2008)—Weibull distribution. (c) Historical lignite coal real price (1949–2007)—beta distribution. (d) Historical natural-gas real price (1997–2008)—exponential distribution. (e) Annual real fuel price increase—triangular distribution. (f) Annual biomass real price increase—triangular distribution. (g) Historical inflation (%) (1997–2008)—Weibull distribution.

triangular probability distribution was assumed; the reference scenario was estimated to have the highest probability. Similarly, a price trend was defined for lignocellulosic biomass (forest-based woody biomass and agricultural waste). Here it was assumed that the real price of biomass would not exceed that of higher-quality biomass (pulpwood) during the forecast period of 20 years and that an average price increase would have the highest probability of occurrence. All other model prices were estimated to increase according to inflation throughout the analysis period.

MC analysis using 5000 iterations in Microsoft Excel-based VBA code was conducted for all process design scenarios, setting the biofuel production capacities to the previously identified most promising production capacities. In the MC analysis, after-tax IRR was calculated assuming a 20-year lifetime after a two-year construction time and a 50/75/100 production startup schedule.

An example of risk analysis results for biochemical design alternatives is given in Figure 3.3.

From these results, it is possible to draw several conclusions:

- General conclusions related to the method and its suitability for early-stage retrofit design risk analysis;
- Different design options are distinguishable based not only on their expected profitability, but also on the uncertainty of the profitability, which enables screening of alternatives considering uncertainties;
- It is possible to quantify downside profitability (worst-case profitability) for each alternative with a predetermined probability. For example, using a known measure, the standard deviation, as a criterion, the 95% confidence interval corresponds to approximately twice the standard deviation, and the lower bound (-2σ) can be used as a systematic definition of downside profitability;
- Screening conclusions;
- Processes having coproducts or by-products with revenue potential and relatively certain prices have substantially lower profitability variation (Hemicelluloses, enz. hydr. in Figure 3.3);
- More costly feedstocks are not promising for the case mill in the case of biofuel production;

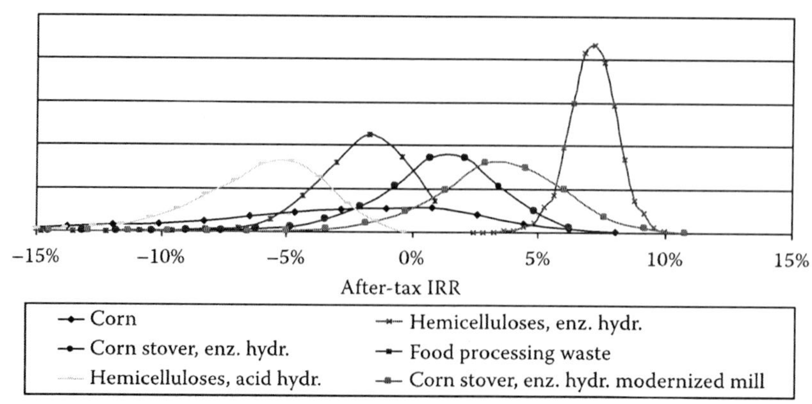

FIGURE 3.3
Probability distributions of after-tax IRR of some biochemical design alternatives.

- Thermochemical processes seem to have better economic performance in general;
- Among thermochemical design alternatives, the simplest option (mixed alcohol production without ethanol separation) seems to be the most profitable. However, this product does not have existing markets and therefore needs further consideration before a final decision is made;
- Integration impact analysis conclusions;
- Impacts of integration can be substantial, but they are most pronounced for biochemical processes that produce organic solid residues. If no capacity is available for combusting the residues, substantially higher capital investment is required, and the profitability is lower (in Figure 3.3, corn stover, enz. hydr. modernized mill represents a scenario in which excess capacity is available, whereas corn stover, enz. hydr represents a case with no excess capacity).
- Possible benefits of integration are case- and capacity- (biofuel production capacity) specific.

Moreover, because economies of scale vary among the design alternatives, substantial differences in promising production capacities and therefore also in capital costs were obtained. This raises important questions such as whether the assumed financing structure (100% equity financing) is possible for all alternatives and how important total capital cost is as a decision-making criterion compared to the profitability measure that was used in screening.

This case study conducted by Hytönen and Stuart considered only external uncertainty sources. This can be partly justified by the assumptions: references were assumed to represent *n*th plant designs, and therefore process-parameter variability (process-inherent uncertainty) is relatively low. Nevertheless, capital investment cost estimates, although having roughly similar accuracy, in early-stage process design are always uncertain. Therefore, a capital-cost uncertainty (model-inherent uncertainty) could have been included in the MC analysis. On the other hand, the possibility of capital investment subsidies can in the end offset the uncertainty in capital costs (as illustrated in Table 3.1, capital costs are often underestimated, and therefore the final project costs may be as much as 100% higher than the cost estimates at this stage of design analysis). Moreover, focusing on one type of uncertainty at a time enables better understanding of the impacts of different uncertainties on the measure that is being used.

Information about profitability uncertainty was not directly used in screening of design alternatives; however, a systematic measure of it was presented, and its implications were discussed. On the other hand, the importance of the selection of decision-making criteria, and the fact it is possible to reflect external uncertainties in commonly used decision-making criteria such as IRR, have been recognized in this chapter.

3.5 Conclusions

Methods for technoeconomic analysis under uncertainty at different stages of the design process have been reviewed in general, and particularly in the context of early-stage biorefinery design. Selected biorefinery design analysis case studies have also been presented.

Conventional technoeconomic analysis is used in early-stage biorefinery process design by applying different variants of the analysis methods. For example, capital costs are analyzed using methods ranging from simplified plant-level capacity-factored methods to detailed equipment data and vendor quotes. Similarly, different levels of accuracy are used for O&M cost estimation. This can potentially lead to additional uncertainty in the results because of the different levels of precision in the analysis methods used.

Often the goal of biorefinery design studies is to estimate the capital costs of the process being developed, and the results are reported in terms of the most common profitability measures, NPV and IRR. The simplest risk-analysis methods, sensitivity analysis and scenario analysis, are almost always used to identify and understand the important cost factors of the design(s) in question.

The analyzed process designs are commonly assumed to represent an *n*th plant implementation of the process and estimates based on engineering knowledge of the possible future development of the technologies, and sometimes on detailed simulation and modeling of the processes using advanced process simulation platforms. This leads to optimistic projections of process performance, which might be possible only in the longer term for some technologies. However, the performance of processes normally evolves to that level relatively quickly after the first implementations, and therefore this assumption of well-developed processes is justified, especially if longer-term aggregated performance measures are used in analysis.

Systematic accounting for all types of uncertainties is not common in early-stage design studies. The sensitivity of the main analysis criterion is studied, but the variability of uncertain variables is not always assessed; rather, an arbitrary range is typically selected. Often the focus is on only one type of uncertainty, giving a somewhat biased understanding of the overall risk of investing or of continuing R&D into the design. Moreover, the most studied types of uncertainty are process-inherent and external uncertainties. External uncertainties are especially important factors for biorefinery analyses because many of the products are not sold in current markets, and therefore prices and real market demand are unknown. However, in some cases, the external uncertainties may not have the biggest impact on the success of the investment, and the risk can possibly be mitigated and managed by a good implementation strategy. In addition, certain discrete phenomena might be in a decisive position for some design alternatives.

Even though some existing optimization-based methodologies have been applied in a biorefinery context, they consider uncertainties mainly in the manner described above.

Stochastic risk analysis, especially MC, is often used in process design assessments, but normally more in the detailed design phases. It is currently not commonly applied in the context of the biorefinery. In some design analyses, a systematic correlation analysis is also conducted. For example, Ince et al. concluded that correlation analysis did not have substantial benefits; the same result (IRR) could be achieved with a conventional MC approach, and when correlations are recognized, the variance is somewhat reduced [52]. This has, however, not been done systematically in any MC analyses in the biorefinery context.

Importantly, the results of uncertainty analysis are not always used to their full potential. This can be seen from the use that is made of the results: whatever may be the analyzed sources of uncertainty, how they are formulated, and how their impacts on the screening and selection are assessed, the results of uncertainty analyses are normally simply presented and their implications discussed. Hence, decisions are made using the expected values of the decision-making criteria under consideration, not based on their quantified uncertainty. Examples of using actual uncertainty information explicitly in decision-making are available, for example in the biorefinery studies by Hytonen and Stuart [48] and Cohen et al. [35].

Biorefinery design analyses in the early design stage are not always recognized to be analyses of strategic investments. This important feature of the context should perhaps be given more emphasis to be able to be of better service to investment decision-making. The criteria for screening design alternatives depend on the investment type; for example, project profitability describes only the economic performance of the new operations that are being implemented and does not reflect company-level performance changes. In addition, the need for using MCDM methods and perhaps even real weighting instead of sophisticated guesses of the possible importance of different criteria becomes more important for strategic investment decisions in the case where multiple aspects will be changed as a result of the decision, including for example environmental and social impacts. Such an inclusion of several aspects has been done, for example, by Paplexandrou et al. [37] and Cohen et al. [35], who conducted process-creation-level process-design analyses. However, these studies used different bases for the weights: educated guesses and expert-panel-based weights, respectively.

Integrated biorefinery concepts have been analyzed by several authors, mainly in the forest biorefinery context. The focus of integration impact analysis has been on structural integration (using the existing energy and utility systems), and the impacts have been fully allocated to the bioproduct as credits or additional costs. Correctly assessing the impacts and allocating them to all products, for example, using advanced costing methods, can possibly offer substantial benefits to business decision-making on various levels. When the correct production costs of all products are known, the different product margins can be used as decision-making criteria for operational-level management of production and also for assessing product and feedstock supply chains.

Nomenclature

AHP—Analytical hierarchy process
CDF—Cumulative distribution functions
CO_2—Carbon dioxide
DOE—U.S. Department of Energy
EIA—U.S. Energy Information Administration
f.o.b.—Free on board
FT—Fischer–Tropsch
FTL—Fischer–Tropsch liquids
GHG—Greenhouse gas
HHV—Higher heating value
IRR—Internal rate of return
LHS—Latin hypercube sampling
MC—Monte Carlo
MCDM—Multicriterion decision-making
MCS—Monte Carlo sampling
MESP—Minimum ethanol selling price
NPV—Net present value
O&M—Operation and maintenance

PEST—Political/legal, economic, social, technological
PFD—Process flow diagram
PNENW—Present value of end net worth
PSE—Process systems engineering
PVENW—Present value of end net worth
P&ID—Piping and instrumentation diagram
ROI—Return on investment
R&D—Research and development
SWOT—Strengths, weaknesses, opportunities, and threats
σ—Standard deviation

References

1. Smith, R., *Chemical Process Design*. 1995, New York: McGraw-Hill, Inc. p. 459.
2. Peters, M.S., K.D. Timmerhaus, and R.E. West, *Plant Design and Economics for Chemical Engineers*. 5th edn. 2003, New York: McGraw-Hill.
3. Northcott, D., *Capital Investment Decision-Making*. 1995, London: Dryden.
4. Åberg, M., *Investments and Corporate Strategy: Aligning Investment Decisions with Overall Strategy*. 2006, Karlskrona: School of Management Blekinge Institute of Technology. p. 55.
5. Christensen, P. and L.R. Dysert, *AACE International Recommended Practice No. 18R-97 Recommended Practice for Cost Estimate Classification–As Applied in Engineering, Procurement, and Construction for the Process Industries*. 1997, Morgantown, Va: AACE International. pp. 1–10.
6. Dysert, L.R., Sharpen your cost estimating skills. *Cost Engineering*, 2003. **45**(6): 22–30.
7. Sprague, J.C. and J.D. Whittaker, *Economic Analysis for Engineers and Managers: The Canadian Context*. 1st edn. 1986, Scarborough, Ontario: Prentice-Hall Canada Inc. p. 575.
8. Douglas, J., *Conceptual Design of Chemical Processes*. 1988, New York: McGraw-Hill. p. 601.
9. Seider, W.D. et al., *Product and Process Design Principles: Synthesis, Analysis and Design*. 3rd edn. 2009, Hoboken, NJ: John Wiley and Sons Inc. p. 728.
10. Lynd, L.R. et al., *Strategic Biorefinery Analysis: Analysis of Biorefineries*. 2005, Subcontract Report NREL/SR-510-35578, October 2005. p. 40, Available at http://www.nrel.gov/docs/fy06osti/35578.pdf
11. Lynd, L.R. et al., *Strategic Biorefinery Analysis: Review of Existing Biorefinery Examples*. 2005, Subcontract Report NREL/SR-510-34895, October 2005. p. 51, http://www.nrel.gov/docs/fy06osti/34895.pdf
12. Aden, A. et al., *Lignocellulosic Biomass to Ethanol Process Design and Economics Utilizing Co-Current Dilute Acid Prehydrolysis and Enzymatic Hydrolysis for Corn Stover*. 2002, Subcontract Report NREL/TP-510-32438, June 2002. p. 154, http://www1.eere.energy.gov/biomass/pdfs/32438.pdf
13. Phillips, S. et al., *Thermochemical Ethanol via Indirect Gasification and Mixed Alcohol Synthesis of Lignocellulosic Biomass*. 2007, Subcontract Report NREL/TP-510-41168, April 2007. p. 132, Available at http://www.nrel.gov/docs/fy07osti/41168.pdf
14. Werpy, T. and G. Petersen, *Top Value Added Chemicals from Biomass: Volume 1—Results of Screening for Potential Candidates from Sugars and Synthesis Gas*, Technical Report DOE/GO-102004-1992, August 2004. p. 75, Available at http://www1.eere.energy.gov/biomass/pdfs/35523.pdf
15. Bozell, J.J. et al., *Top Value Added Chemicals from Biomass: Volume II—Results of Screening for Potential Candidates from Biorefinery Lignin*. 2007, Technical Report PNNL-16983, October 2005. p. 87, Available at http://www.pnl.gov/main/publications/external/technical_reports/PNNL-16983.pdf
16. Hytönen, E. and P.R. Stuart, Integrating bioethanol production into a kraft pulp mill—Technology assessment. *Pulp & Paper Canada*, 2009. **110**(5): 58–65.
17. Pistikopoulos, E.N., Uncertainty in process design and operations. *Computers & Chemical Engineering*, 1995. **19**(Suppl. 1): 553–563.

18. Miller, K.D. and H.G. Waller, Scenarios, real options and integrated risk management. *Long Range Planning*, 2003. **36**(2003): 15.
19. Sahinidis, N.V., Optimization under uncertainty: State-of-the-art and opportunities. *Computers & Chemical Engineering*, 2004. **28**(6–7): 971–983.
20. Goh, Y.M., J. Booker, and C. McMahon, *A comparison of methods in probabilistic design based on computational and modelling issues*, in *Advances in Integrated Design and Manufacturing in Mechanical Engineering*, A. Bramley et al., Editors. 2005, Amsterdam: Springer Netherlands, p. 14.
21. Schoemaker, P.J.H., *Scenario Planning: A Tool for Strategic Thinking. Sloan Management Review*, 1995. **36**(2): 25–40.
22. Diwekar, U.M. and J.R. Kalagnanam, Robust design using an efficient sampling technique. *Computers & Chemical Engineering*, 1996. **20**: 6.
23. Wang, R., U. Diwekar, and C.E.G. Padro, Efficient sampling techniques for uncertainties in risk analysis. *Environmental Progress*, 2004. **23**(2): p. 17.
24. Graves, S.B. and J.L. Ringuest, Probabilistic dominance criteria for comparing uncertain alternatives: A tutorial. *Omega*, 2009. **37**(2009): 12.
25. Xiao, Z. and A. Vien, Limitations of variance analysis using propagation of variance. *Minerals Engineering*, 2003. **16**(5): 455–462.
26. Acevedo, J. and E.N. Pistikopoulos, Stochastic optimization based algorithms for process synthesis under uncertainty. *Computers & Chemical Engineering*, 1998. **22**(4/5): 647–671.
27. Ierapetritou, M.G., J. Acevedo, and E.N. Pistikopoulos, An optimization approach for process engineering problems under uncertainty. *Computers & Chemical Engineering*, 1996. **20**(6/7): 703–709.
28. Svensson, E. et al., An optimization methodology for identifying robust process integration investments under uncertainty. *Energy Policy*, 2009. **37**(2009): 680–685.
29. Svensson, E., T. Berntsson, and A.-B. Strömberg, Benefits of using an optimization methodology for identifying robust process integration investments under uncertainty—A pulp mill example. *Energy Policy*, 2009. **37**(2009): 813–824.
30. Hubbard, D. and D. Evans, Problems with scoring methods and ordinal scales in risk assessment. *IBM Journal of Research and Development*, 2010. **54**(3): 1–10.
31. Hubbard, D.W., *The Failure of Risk Management: Why It's Broken and How to Fix It*. 2009, Hoboken, NJ: John Wiley & Sons Inc. p. 286.
32. Bode, G. et al., Dealing with risk in development projects for chemical products and processes. *Industrial & Engineering Chemistry Research*, 2007. **46**(23): 22.
33. Ehrhardt, M.C. and E.F. Brigham, *Corporate Finance: A Focused Approach*. 3rd edn. 2008, Mason, OH: South-Western Pub. p. 680.
34. English, B. et al., *Wisconsin's Biobased Industry: Opportunities and Advantages Study—Technical Analysis Report*, Report ECW 238-1, June 2006. p. 248, http://www.ecw.org/prod/238-1.pdf
35. Cohen, J. et al., Critical analysis of emerging forest biorefinery (FBR) technologies for ethanol production. *Pulp & Paper Canada*, 2010. **111**(3): 24–30.
36. Sammons Jr., N.E. et al., Optimal biorefinery product allocation by combining process and economic modeling. *Chemical Engineering Research and Design*, 2008. **86**(7): 800–808.
37. Papalexandrou, M.A., P.A. Pilavachi, and A.I. Chatzimouratidis, Evaluation of liquid bio-fuels using the Analytic Hierarchy Process. *Process Safety and Environment Protection*, 2008. **86**(2008): 15.
38. Larson, E.D. et al., *A Cost–Benefit Assessment of Gasification-Based Biorefining in the Kraft Pulp and Paper Industry*. Final report to U.S. Department of Energy and the American Forest & Paper Association, DOE contract DE-FG26-04NT42260, December 2006. p. 365, Available at http://www.osti.gov/bridge/servlets/purl/912520-8bRNAH/
39. Eggeman, T. and R.T. Elander, Process and economic analysis of pretreatment technologies. *Bioresource Technology*, 2005. **96**(18 SPEC. ISS.): 2019–2025.
40. Kazi, F.K. et al., Techno-economic comparison of process technologies for biochemical ethanol production from corn stover. *Fuel*, 2010. **89**(Supplement 1): S20–S28.
41. Laser, M. et al., Comparative analysis of efficiency, environmental impact, and process economics for mature biomass refining scenarios. *Biofuels, Bioproducts & Biorefining*, 2009. **3**: p. 247–270.

42. Laser, M. et al., Projected mature technology scenarios for conversion of cellulosic biomass to ethanol with coproduction thermochemical fuels, power, and/or animal feed protein. *Biofuels, Bioproducts & Biorefining*, 2009. **3**: 231–246.
43. Laser, M. et al., Coproduction of ethanol and power from switchgrass. *Biofuels, Bioproducts & Biorefining*, 2009. **3**: 195–218.
44. Mao, H. et al., Technical economic evaluation of a hardwood biorefinery using the "Near-neutral" hemicellulose pre-extraction process. *Journal of Biobased Materials and Bioenergy*, 2008. **2**(2): 1–9.
45. van Heiningen, A., Converting a kraft pulp mill into an integrated forest biorefinery. *Pulp and Paper Canada*, 2006. **107**(6): 38–44.
46. Frederick, W.J., Jr. et al., Co-production of ethanol and cellulose fiber from Southern Pine: A technical and economic assessment. *Biomass and Bioenergy*, 2008. **32**(12): 10.
47. Phillips, R.B., H. Jameel, and E.C. Clark Jr. Repurposing a kraft pulp and paper mill to the production of ethanol. In *92nd Annual Meeting of the Pulp and Paper Technical Association of Canada*. 2010. Montreal, Canada.
48. Hytönen, E. and P.R. Stuart, Biofuel production in an integrated forest biorefinery—Technology identification under uncertainty. *Journal of Biobased Materials and Bioenergy*, 2010. **4**(1): 58–67.
49. Wooley, R. et al., *Lignocellulosic Biomass to Ethanol Process Design and Economics Utilizing Co-Current Dilute Acid Prehydrolysis and Enzymatic Hydrolysis current and Futuristic Scenarios*. Subcontract Report NREL/TP -580-26157, July 1999. p. 130, Available at http://www.nrel.gov/docs/fy99osti/26157.pdf
50. Phillips, S.D., Technoeconomic analysis of a lignocellulosic biomass indirect gasification process to make ethanol via mixed alcohols synthesis. *Industrial & Engineering Chemistry Research*, 2008. **46**(26): 11.
51. Richardson, J.W. et al., *Risk Assessment in Economic Feasibility Analysis: The Case of Ethanol Production in Texas*. AFPC Research Report 06-3, September 2006. p. 16 , Available at http://www.afpc.tamu.edu/pubs/2/447/RR%2006-3.pdf
52. Ince, P.J. and J. Buongiorno. *Multivariate Stochastic Simulation with Subjective Multivariate Normal Distributions*. in *The 1991 Symposium on Systems Analysis in Forest Resources*. 1991. Charleston, USA.

4

Technoeconomic Analysis of Cellulosic Ethanol Production from Hardwoods in a Repurposed Kraft Mill

Trevor H. Treasure, Hasan Jameel, Richard B. Phillips, Hou-min Chang, Yongcan Jin, and Bon Wook Koo

CONTENTS

Part I: Pretreatment Technology Development

4.1 Introduction

The North American pulp and paper industry is in decline due to falling demand for paper and board products and, in some cases, loss of market pulp share to lower-cost countries. As a result, more than 15 million tons of capacity has been permanently closed. At the same time as this decline in paper products, interest in biomass-to-biofuel has grown to unprecedented levels. Therefore, it would appear to be very attractive to repurpose kraft pulp mills to produce ethanol. In concept, repurposing features a number of significant advantages over other approaches to bioethanol production:

- A supply chain to grow, harvest, and deliver biomass is already in place, therefore avoiding new demand which might create price pressure on the raw material.
- For the reason that pulp is not produced for papermaking, chips with bark and significant levels of harvest residues may be used as raw material, offsetting some of the traditional cost of biomass from pulpwood.
- When kraft mills are permanently closed, significant equipment assets are left behind, such as a woodyard, digesters, evaporators, power plants, and waste treatment plants, which can be directly repurposed for ethanol production.
- A kraft mill closure typically involves environmental remediation and demolition costs in excess of $US 10 million.
- Much of the process equipment in a repurposed kraft mill has little technology risk since most of this equipment has been operating in place for many years.

Lignocellulosic biomass, which can be converted to glucose and other monomeric sugars for fermentation by hydrolytic enzymes,[1–7] has the potential to contribute substantially to the production of bioethanol for transportation. Enzymatic hydrolysis produces higher yields than acid-catalyzed hydrolysis and has become more attractive recently with the reduction of enzyme manufacturing cost. The enzymatic process is regarded as the most attractive way to degrade wood polysaccharides to monomeric sugars.[8–10] However, cellulose and hemicelluloses in lignocellulosic biomass are intrinsically resistant to enzyme attack because they are protected by a matrix of lignin. As a result, a pretreatment stage is necessary for lignocellulosic biomass to expose the carbohydrates by removing some of the lignin and hemicelluloses in a cost-effective manner to make the cellulose more accessible to enzymes.

Various pretreatment methods have been studied for hardwoods that would be compatible with a repurposed mill, including:

- Water prehydrolysis (autohydrolysis)
- Sodium carbonate
- Green liquor (GL) (sodium carbonate and sodium sulfide)
- Prehydrolysis-GL

These pretreatments were chosen because they could be easily implemented in a repurposed kraft mill with minimal capital. They also have the potential to decrease the labor requirement and to improve the overall energy efficiency of a repurposed kraft mill producing ethanol.

4.2 Materials and Methods

4.2.1 Materials

Mixed hardwood chips used in this study were provided by a mill in the southeastern United States. The wood chips were screened to remove the fractions greater than 38 mm and less than 6 mm in length. Bark and knots were manually removed.

4.2.2 Substrate Preparation and Pretreatment

A laboratory-scale M/K pulping digester with a capacity of 800 g of oven-dry wood was used for the pretreatment studies. Prehydrolysis was carried out with water at 160°C or 170°C for 60 min at a liquor-to-wood (oven-dry) ratio of 4 (v/w). Pretreatment with sodium carbonate was carried out at 160°C to an H-factor of 400 at a liquor-to-wood ratio of 4. The chemical charge was varied between 12% and 24% sodium carbonate as Na_2O. In the case of the GL pretreatment, the wood chips were pretreated using a mixture of sodium carbonate and sodium sulfide with a sulfidity of 25% on a total titratable alkali (TTA) basis. The ratio of pretreatment liquor to wood was 4. The TTA charge as Na_2O on OD wood chips was varied from 4% to 20% (w/w). The TTA charge takes into account both the sodium carbonate and the sodium sulfide. The wood chips were first impregnated with the pretreatment liquor at 80°C for 30 min. Immediately after impregnation, the temperature was raised to 160°C and maintained for a certain period until an H-factor of 400 was reached. At the end of all the pretreatments, the liquor was drained from the digester, the solids were collected and washed in hot water overnight to remove residual chemicals and dissolved wood compounds completely. The moisture and total weight of the cooked chips were measured to calculate the pretreatment yield. Washed wood chips were refined twice in a Bauer 148-2 disk refiner with a disk gap of 0.25 and 0.05 mm for size reduction.

4.2.3 Enzymatic Hydrolysis

Enzymatic hydrolysis of the substrates was carried out in a 250 mL Erlenmeyer flask at a consistency (insoluble solids content) of 5% (w/w) with 38 mL sodium acetate buffer (pH 4.8) at 50°C using a shaking incubator at 180 rpm. An enzyme cocktail of cellulase, xylanase, and β-glucosidase (1 FPU:1.2 FXU:1 CBU) was used for enzymatic hydrolysis. The activity loading was based on a cellulase charge between 5 and 40 FPU/g substrate. Excessive β-glucosidase was used to prevent cellobiose accumulation.[11] Commercial enzymes used for the enzymatic hydrolysis were NS50013 (cellulase), NS50014 (xylanase), and NS50010 (β-glucosidase) and were provided by Novozymes for this study. Sodium azide was charged at 0.3%, based on the total weight of the substrate slurry, as an antibiotic. Enzymatic hydrolysis residue and hydrolysate were separated by centrifugation. The residues were washed with deionized water (3 × liquid volume) to remove the enzyme protein and other dissolved substances. Frozen and vacuum-dried residues were weighed to calculate substrate weight loss. Hydrolysates were sampled for monomeric sugar content (glucose, xylose, and mannose).

4.3 Results and Discussion

The chemical composition of the wood used in this study is shown in Table 4.1. The total carbohydrate content of this wood was 66.3%. This value will be used to calculate the overall carbohydrate yield from the pretreatment and enzymatic hydrolysis processes.

4.3.1 Prehydrolysis

The process of prehydrolysis (autohydrolysis) uses only water and heat and has the advantage of requiring no inorganic chemical recovery. Moreover, the vessel used for pretreatment is compatible with the existing pulping vessel. Following prehydrolysis, the solids were separated from the soluble fraction by filtration and washing. The water-soluble fraction represents the degraded hemicelluloses. The residue was disintegrated in a refiner and then hydrolyzed using enzymes. The pretreatments were evaluated with respect to the

- Ability of the substrate to be enzymatically hydrolyzed
- Overall carbohydrate yield (monomeric sugar/total sugar in initial biomass)

After prehydrolysis at 160°C for 1 h, the pulp yield was 82.4% on wood. The composition of the prehydrolysis liquors is shown in Table 4.2 before and after acid hydrolysis. The composition before acid hydrolysis represents only the monomers produced during

TABLE 4.1

Chemical Composition of Hardwood, % on Wood

	Benzene–Ethanol Extractives	Sugar			Lignin		Total Sugar	Total Lignin	Sum
		Glucan	Xylan	Mannan	Klason	Acid Soluble			
Wood	2.0	47.8	16.3	2.2	22.7	4.0	66.3	26.7	95.0

TABLE 4.2

Sugar Analysis of Prehydrolysis Liquor, % on Wood

	Glucan	Xylan	Mannan	Total
w/o 3% H_2SO_4 hydrolysis	0.1	0.2	0.1	0.4
w/3% H_2SO_4 hydrolysis	0.9	4.2	0.5	5.6

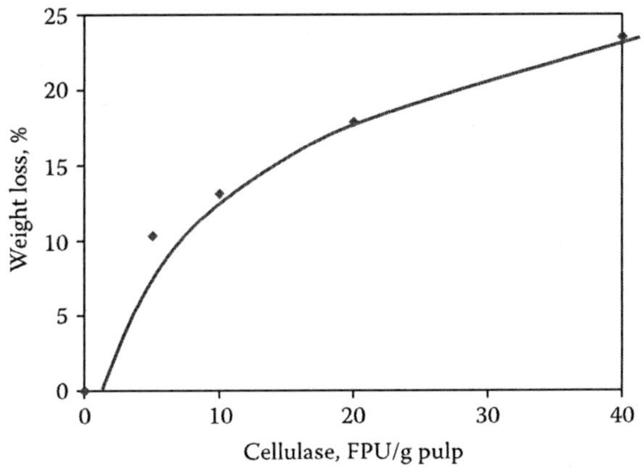

FIGURE 4.1
Effect of enzyme dosage on weight loss for prehydrolyzed pulp.

prehydrolysis, while the composition after acid hydrolysis with 3% sulfuric acid represents the total quantity of monomers and oligomers produced. The results show that 5.6% of the wood can be recovered as sugar in the prehydrolysis filtrate.

After pretreatment, the solid residues were enzymatically hydrolyzed using cellulase, xylanase, and β-glucosidase. After enzymatic hydrolysis, the weight loss was measured for different cellulase charges and the results are shown in Figure 4.1. The results can be summarized as follows:

- Total sugar in prehydrolysate (after 3% acid hydrolysis): 5.6% on wood
- Sugar from enzymatic hydrolysis (estimated according to weight loss): 24% on wood
- Total sugar: ~30% on wood
- Total sugar recovery: 45% (30% as sugar/66% as sugar in original wood)

All the processes will be compared on the basis of total sugar recovery. Sugar recovery for all options is the total amount of the original carbohydrates in the biomass that was recovered as monomeric sugars.

4.3.2 Sodium Carbonate Pretreatment

Sodium carbonate was studied as a pretreatment chemical because it can be recovered in a recovery boiler and its use does not require a caustic plant, thereby simplifying the overall operation. The results of pretreatment are shown in Table 4.3 for various sodium carbonate charges.

TABLE 4.3
Effect of Sodium Carbonate Pretreatment on Yield

Chemical charge Na_2CO_3, % Na_2O	12	16	20	24
Yield, %	86.3	85.9	84.1	83.9
Final liquor pH	8.0	9.1	9.4	9.4

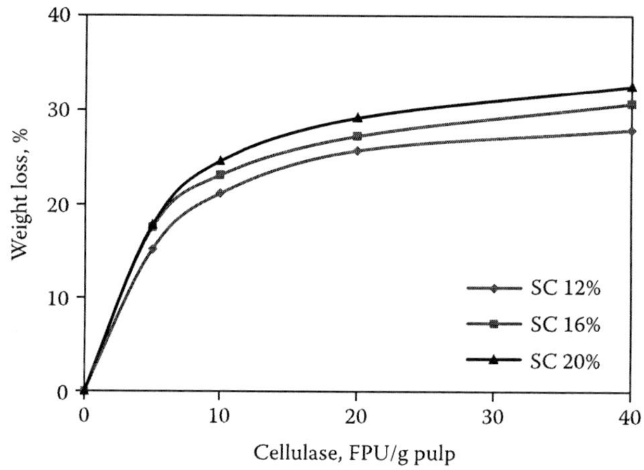

FIGURE 4.2
Effect of enzyme dosage on weight loss for sodium carbonate pretreated pulps.

The residue from the sodium carbonate was enzymatically hydrolyzed, and the results are shown in Figure 4.2. Pretreated chip yield was approximately 84% even at TTA charges greater than 20%. At a cellulase dosage of 40 FPU/g, enzymatic hydrolysis weight loss of the pulp pretreated at 20% TTA was only 33% (on wood). Because there are no recoverable sugars in the black liquor from the sodium carbonate process, the total sugar recovery is only from enzymatic hydrolysis and is approximately 50% (33% recovered sugar/66% sugar in original biomass).

4.3.3 GL Pretreatment

Green liquor, which is a mixture of sodium carbonate and sodium sulfide, was also used as a pretreatment chemical. GL is readily available in a kraft mill, and its use also does not require a caustic plant. The results of the pretreatment are shown in Table 4.4 for various

TABLE 4.4
Effect of GL Charge on Yield

GL TTA, % Na_2O	4	8	12	16	20
Yield, %	89.2	82.2	79.7	78.3	77.0
Black liquor pH	8.0	9.1	9.4	9.6	9.8

TABLE 4.5

Lignin and Sugar Content of GL-Treated Pulp, % on Wood

	Sugar, % on Wood	Lignin, % on Wood						
	Glucan	**Xylan**	**Mannan**	**Sum**	**Klason**	**Acid Soluble**	**Sum**	**Total**
Wood	47.8	16.3	2.2	66.3	22.7	4.0	26.7	93.0
GL-04	48.6	14.5	1.8	64.8	20.5	3.5	24.0	88.8
GL-08	46.2	12.3	0.9	59.5	18.5	3.0	21.4	80.9
GL-12	46.9	12.5	0.5	59.0	16.4	2.6	19.0	79.0
GL-16	47.8	11.7	0.4	60.0	15.7	2.3	18.0	78.0
GL-20	46.8	11.9	0.4	59.2	14.8	2.3	17.1	76.3

GL charges. The yields are slightly lower than those from chips pretreated with sodium carbonate at similar chemical charges. The chemical composition of the pulps pretreated with varying amounts of GL is shown in Table 4.5 and Figure 4.3. The results are expressed as a percentage of wood so that the exact retention of the different components can be easily evaluated. Most of the glucan is retained during the pretreatment, with increasing losses of hemicelluloses and lignin as the chemical charge is increased. The details of the chemical reactions were summarized by Chang.[12] As the GL charge is increased, the amount of lignin removed in the pretreatment stage becomes greater, and the lignin content decreases from the original 26.7% to approximately 17%. However, the decrease in the carbohydrate content beyond a GL charge of 8% is small; the total carbohydrate content decreases only from 66% to 60%.

The weight loss during enzymatic hydrolysis is shown in Figure 4.4 based on both pulp and wood as a function of enzyme dosage and in Figure 4.5 as a function of lignin content. The sugar recovery for the various sugars is also summarized in Table 4.6. The sugar recovery does not increase significantly as the GL charge is increased beyond 12% TTA, although the enzyme dosage may decrease with increasing GL application. The results indicate that the total sugar recovery with a GL pretreatment of 16% TTA is approximately 80% (53% recovered sugar/66% sugar in original biomass). This is significantly higher than with the prehydrolysis and sodium carbonate pretreatments.

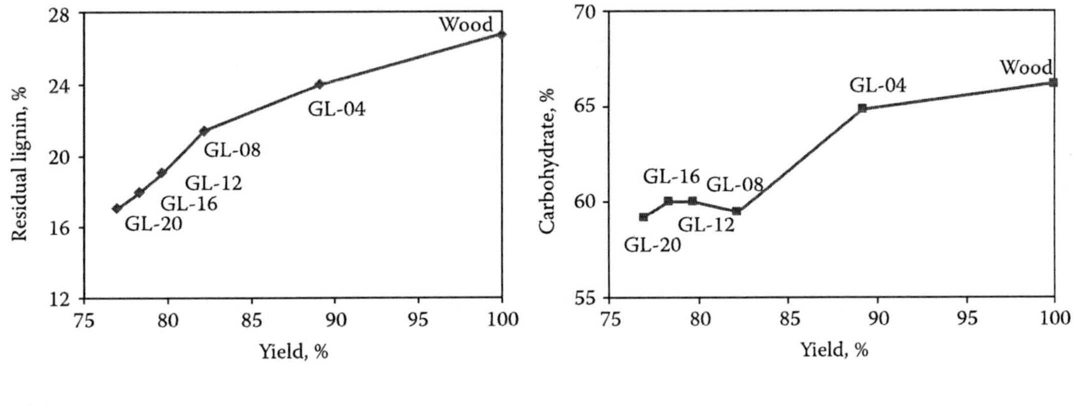

FIGURE 4.3

Lignin and carbohydrate content of green liquor-treated pulps, % on wood.

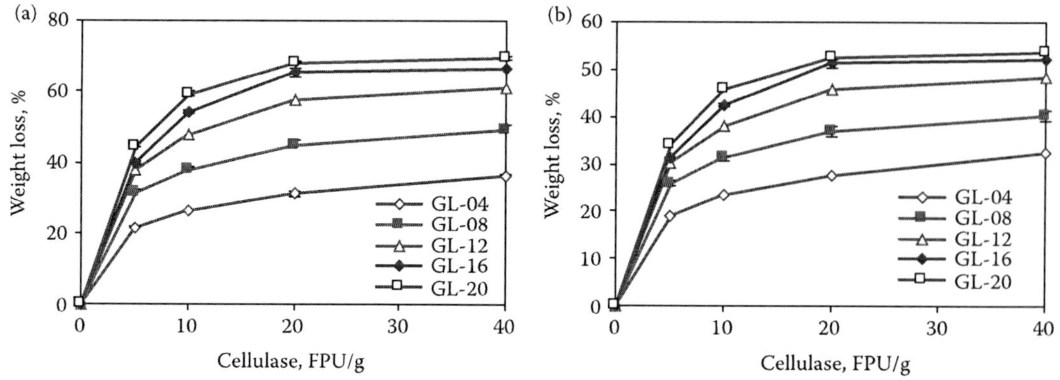

FIGURE 4.4
Effect of enzyme dosage on weight loss for green liquor-pretreated pulps. (a) Enzymatic hydrolysis weight loss, based on pulp; (b) Enzymatic hydrolysis weight loss, based on wood.

4.3.4 Prehydrolysis GL Pretreatment

Even though the GL pretreatment resulted in very high sugar recovery, a prehydrolysis step was added before the GL pretreatment to improve sugar recovery further. It was hoped that some of the hemicelluloses that were being dissolved in the black liquor could be recovered and converted to ethanol. The prehydrolysis was carried out at 160°C and 170°C for 1 h. All GL treatments were done at 8% TTA. The analyses of the prehydrolysis liquor are shown in Table 4.7 before and after acid hydrolysis to convert all the oligomers to monomers. At 160°C, 5.6% of the wood could be recovered as sugar, while at 170°C, 7.6% of the wood was recovered. The chemical composition of the pulp residue is shown in Table 4.8. Note that the overall pulp yield was comparable to that calculated from the total sugar and lignin balance. The results of the GL pretreatment for 12% and 16% TTA are also

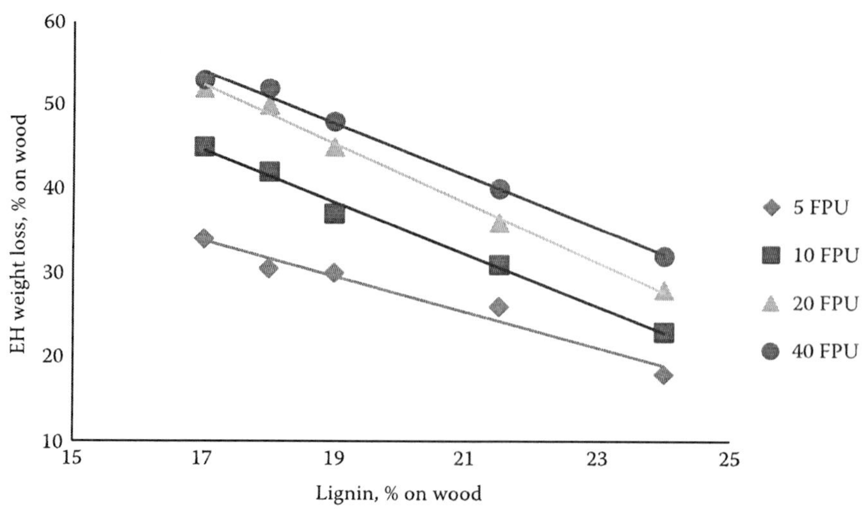

FIGURE 4.5
Enzymatic weight loss as a function of lignin content for green liquor pulps.

TABLE 4.6

Effect of GL Pretreatment and Cellulase Charge on Sugars from Enzymatic Hydrolysis

TTA, %	Cellulase FPU/g	Glucan % on Pulp	Glucan % on Wood	Xylan % on Pulp	Xylan % on Wood	Total Sugar % on Pulp	Total Sugar % on Wood
GL 4	5	14.8	13.2	6.5	5.8	21.3	19.0
	10	19.4	17.3	7.6	6.8	27.1	24.1
	20	24.1	21.5	8.5	7.6	32.6	29.1
	40	29.9	26.7	9.8	8.7	39.7	35.4
GL 8	5	22.5	18.5	8.5	7.0	30.9	25.4
	10	29.2	24.0	9.8	8.1	39.1	32.1
	20	36.1	29.6	11.0	9.0	47.0	38.6
	40	41.0	33.7	11.9	9.7	52.8	43.4
GL 12	5	28.1	22.4	9.5	7.6	37.6	29.98
	10	37.9	30.2	11.3	9.0	49.2	39.2
	20	47.0	37.5	12.7	10.1	59.7	47.5
	40	51.6	41.1	13.6	10.8	65.2	52.0
GL 16	5	30.3	23.7	9.8	7.7	40.1	31.5
	10	41.9	32.9	11.9	9.3	53.9	42.2
	20	51.3	40.2	13.3	10.5	64.7	50.7
	40	53.1	41.6	14.0	11.0	67.2	52.6
GL 20	5	32.1	24.7	10.4	8.0	42.5	32.7
	10	44.7	34.4	12.7	9.8	57.4	44.2
	20	51.8	39.9	13.9	10.7	65.8	50.6
	40	54.0	41.6	14.7	11.3	68.7	52.9

included in Table 4.8 for comparison. The results of enzymatic hydrolysis of the prehydrolysis GL pulp residues are shown in Figure 4.6. The sugar recovery for these samples can be summarized as follows:

- AH-160°C-8%GL Total sugar yield: 41.9% from enzymatic hydrolysis + 5.6% from prehydrolysis = 47.5%. Total sugar recovery: 47.5/66.3 = 72%
- AH-170°C-8%GL Total sugar yield: 41.5% from enzymatic hydrolysis + 7.6% from prehydrolysis = 49.0%. Total sugar recovery: 49.0/66.3 = 74%

The overall sugar recovery was lower than the 80% sugar recovery achieved with the GL pretreatment only, not to mention the added complexity of performing prehydrolysis

TABLE 4.7

Prehydrolysis Liquor Analysis, % on Wood

PH Temperature, °C	Glucan	Xylan	Mannan	Total
Before acid hydrolysis				
160°C	0.1	0.2	0.1	0.4
170°C	0.2	0.9	0.2	1.3
After 3% H_2SO_4 hydrolysis				
160°C	0.9	4.2	0.5	5.6
170°C	0.9	6.1	0.6	7.6

TABLE 4.8

Chemical Composition of the Pulp Residue after Prehydrolysis GL, Sugar and Lignin Expressed as % on Wood

	Sugar				Lignin				
	Glucan	Xylan	Mannan	Sum	Klason	Acid Soluble	Sum	Total	Yield %
Wood	46.7	16.3	2.2	66.2	22.7	4.0	26.7	92.9	
AH160GL	42.3	4.5	0.5	47.2	16.5	1.8	18.4	65.6	70.8
AH170GL	41.7	3.0	0.5	45.2	19.5	1.6	21.1	66.3	68.4
GL 12	46.9	12.5	0.5	59.0	16.4	2.6	19.0	79.0	79.7
GL 20	46.8	11.9	0.4	59.2	14.8	2.3	17.1	76.3	78.3

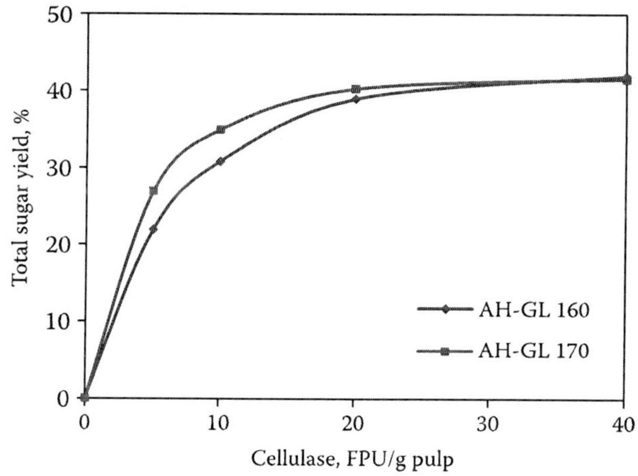

FIGURE 4.6
Sugar yield from enzymatic hydrolysis of prehydrolysis green liquor pretreatment, % on wood.

before GL pretreatment. The prehydrolysis step also brings additional water into the process, resulting in lower overall sugar concentrations, the need for larger fermenters and distillation columns, and more energy consumption for alcohol purification.

4.4 Conclusions

The results for sugar recovery from the different pretreatment studied are summarized in Table 4.9. These results show that a pretreatment based on GL (sodium carbonate and sodium sulfide) is the most effective pretreatment for a repurposed mill. The pulps produced by this process can be enzymatically hydrolyzed to monomeric sugars with a high overall sugar recovery. The use of GL for pretreatment ensures that the chemicals used during pretreatment can be recovered efficiently using proven technology and can be easily implemented in a repurposed kraft pulp mill. The yield of pulp produced by the GL

TABLE 4.9

Sugar Recovery Summary for the Different Pretreatments

Pretreatment	Sugar Recovery (%)
Autohydrolysis	45
Sodium carbonate	50
GL	80
Prehydrolysis-GL	74

pretreatment process is approximately 80%, with nearly 100% cellulose and 75% xylan retention with mixed Southern hardwoods. The relatively low pH compared to kraft pulping prevents the random hydrolysis of polysaccharides and secondary peeling reactions from occurring during the pretreatment, resulting in high retention of polysaccharides in pulp. Approximately 35% of the lignin is removed during the GL pretreatment process, which enables efficient enzymatic hydrolysis. The amount of sugar produced in enzymatic hydrolysis increased with both the GL charge and the enzyme charge. The increase in enzymatic hydrolysis efficiency was small with increases in TTA beyond 12–16%. With GL pretreatment at 16% TTA, the overall sugar recovery for hardwood was shown to be approximately 77% at a cellulase charge of 20 FPU/g of substrate. A sugar recovery of 80% could be achieved at higher enzyme charges. These levels of sugar recovery are competitive with other pretreatments for hardwood. This novel pretreatment process can be used to repurpose kraft mills, which are being closed due to decreasing demand for paper in North America, for the production of ethanol. A patent for the GL pretreatment process has been applied for by North Carolina State University.

Part II: Posttreatments to Reduce Enzyme Charge

With GL pretreatment, high sugar recovery can be achieved at an enzyme dosage of 20 FPU/g of substrate.[12,13] However, these enzyme charges are not economically viable. Posttreatments (additional treatments after pretreatment but before enzymatic hydrolysis) were applied to GL-pretreated hardwood to increase sugar-conversion efficiency and to reduce the enzyme dosage. It would be beneficial if the posttreatments applied were compatible with a repurposed mill.

The posttreatments evaluated include mechanical refining and oxygen delignification (OD), both of which are usually available in a kraft mill. Mechanical treatments such as chipping, milling, and refining improve enzyme accessibility to cellulose through the increase in surface area available to enzymes[14] and particle size reduction.[15] During refining of pulp, the primary wall of a fiber is removed, intrafiber bonds broken, and fissures generated throughout the cell wall.[16] Mechanical refining, in particular, has been used to develop paper strength in the pulp and paper industry through fibrillation.[17] During the refining process, the cellulosic fibers are mechanically treated in water, resulting in morphological and structural changes, and the fibers are fibrillated internally and externally.[18] These fibrillations, which are caused by the high shear force of the refiner, can increase the surface area significantly[19] and facilitate enzyme access to cellulose. Mechanical treatment, however, often requires considerable amounts of energy as well as increased capital cost and has been regarded as a nonviable pretreatment technology.[20] OD has also been studied as a chemical posttreatment.[10] OD has been used in the pulp and paper industry to remove lignin in both pulping and bleaching processes. OD has been shown to remove approximately 50% of lignin in a kraft pulp without significant impact on pulp yield and properties.[10,16]

4.5 Materials and Methods

4.5.1 Posttreatments (Additional Treatments Before Enzymatic Hydrolysis)

Pretreated pulps were delignified using oxygen at 110°C with 5% sodium hydroxide as a catalyst. The delignification reaction was carried out for 1 h at an oxygen pressure of 100 psig. After pretreatment and OD, the chips were thoroughly washed with tap water to remove dissolved chemicals. Following OD, the pulps were refined three times using a disk refiner at three levels of disk gap (15, 5, 1 mm). Finally, the pretreated substrates were centrifuged and stored at 4°C until use.

In a laboratory beater (PFI mill), refining was performed to improve the digestibility of the GL-pretreated substrates. Pretreated substrate samples of 30 g (oven-dried basis) were disintegrated in a blender for 5 min, and the consistency was adjusted to 10% for the refining process. Four different revolution counts were used for PFI refining in this study (2000, 4000, 6000, and 8000 revolutions).

4.6 Results and Discussion

The yields and chemical compositions obtained from OD of GL pulps are shown in Table 4.10. When 4% GL was applied, the highest yield of 89.5% was observed. As the GL charge increased, the yield decreased and then leveled off. There was no difference in yield between 8% and 12% GL charge. Jin and Wu observed that increasing the alkali charge during GL pretreatment produces a substrate with a lower residual lignin content.[12,13] This partial lignin removal during GL pretreatment provides substrates more amenable to subsequent posttreatments or enzymatic hydrolysis. OD was used as an industrially practical step to achieve a selective reduction of the residual lignin content of GL-pretreated biomass. Lignin content in hardwood pretreated with 12% GL can be reduced another 30% by use of OD.

4.6.1 Improvement of Enzymatic Hydrolysis by Posttreatments

Enzymatic conversion was improved as the GL charge increased (Figures 4.7 and 4.8). When 10 FPU of cellulase per gram of substrate was used, the enzymatic conversion rate of GL-pretreated substrates without OD increased from 29.4% to 51.5% with increasing TTA charge (Figure 4.7, based on pretreated substrates) or from 26.3% to 42.2% (Figure 4.3, based on raw materials). When OD was performed, the enzymatic conversion rate was

TABLE 4.10
Yields and Chemical Compositions with OD and Varying GL Charge

			Chemical Compositions (%)											
			Carbohydrates						Lignin					
GL Charge	Treatments	Yield (%)	Glu	Xyl	Man	Gal	Ara	Sum	KL	ASL	Sum	Ash	Ext	Total (%)
Raw material			47.8	16.3	2.2	0.0	0.0	66.3	22.7	4.0	26.7	0.5	2.6	96.1
4%	GL	89.5	45.7	14.5	1.7	0.0	0.0	61.9	20.2	2.6	22.8	ND	ND	84.7
	GL and OD	80.7	48.0	14.0	1.6	0.0	0.0	63.6	14.4	2.9	17.3	ND	ND	80.9
8%	GL	81.4	45.5	12.3	1.0	0.0	0.0	58.8	17.1	2.5	19.6	ND	ND	78.4
	GL and OD	74.2	45.2	11.9	0.7	0.0	0.0	57.8	13.5	2.3	15.8	ND	ND	73.6
12%	GL	81.9	46.6	12.5	0.6	0.0	0.0	59.7	15.9	2.3	18.2	ND	ND	77.9
	GL and OD	75.6	47.2	12.0	0.6	0.0	0.0	59.8	10.6	1.9	12.5	ND	ND	72.3

Note: All values are calculated on oven-dried weight of wood; GL: green liquor pretreatment; OD: oxygen delignification; Glu: glucan; Xyl: xylan; Man: mannan; Gal: galactan; Ara: arabinan; KL: Klason lignin; ASL: acid soluble lignin; Ext: extractives.

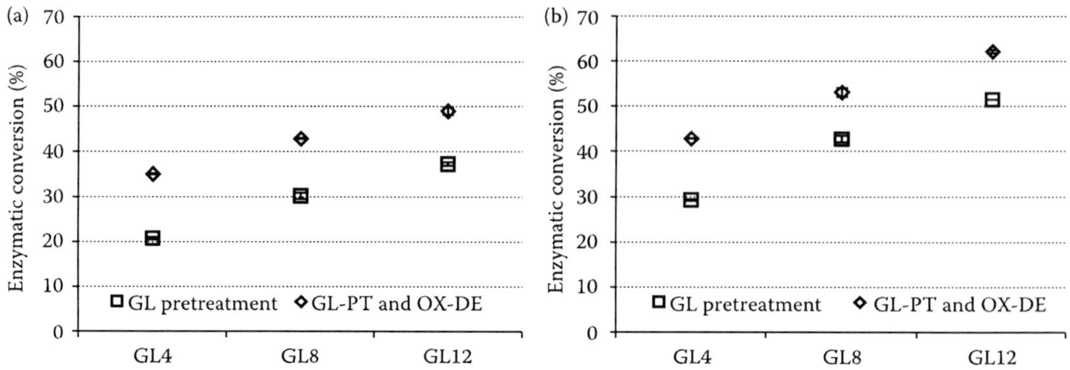

FIGURE 4.7
Enzymatic conversion depending on green liquor charge, oxygen delignification, and enzyme dosage (based on pretreated substrates, (a) 5 FPU dosages; (b) 10 FPU dosages, GL: green liquor; GL-PT: green liquor pretreatment; OX-DE: oxygen delignification).

improved from 42.8% to 62.2% (Figure 4.7, based on pretreated substrates) or from 34.5% to 47.0% (Figure 4.8, based on raw materials). This suggests that additional lignin removal during OD improves the efficiency of enzymatic hydrolysis. The correlation between lignin content and enzymatic conversion was determined as shown in Figure 4.9. It is speculated that the decreased lignin content in the pretreated substrates facilitates enzyme access to cellulose, and therefore enzymatic conversion was increased.

The refining process significantly improved enzymatic conversion under all conditions (Figures 4.10 through 4.12). The enzymatic conversion of refined substrates at 6000 PFI revolutions was 72% (based on pretreated substrates) at 5 FPU, as shown in Figure 4.10b, which was higher than the enzymatic conversion rate of 62% obtained from oxygen-delignified substrate with an enzyme dosage of 10 FPU without refining. This result clearly demonstrates the benefit of the refining process, which can enable a reduction in enzyme dosage. It was also found that the increase in enzymatic conversion was marginal at high numbers of PFI revolutions, indicating that a severe refining condition with high-energy

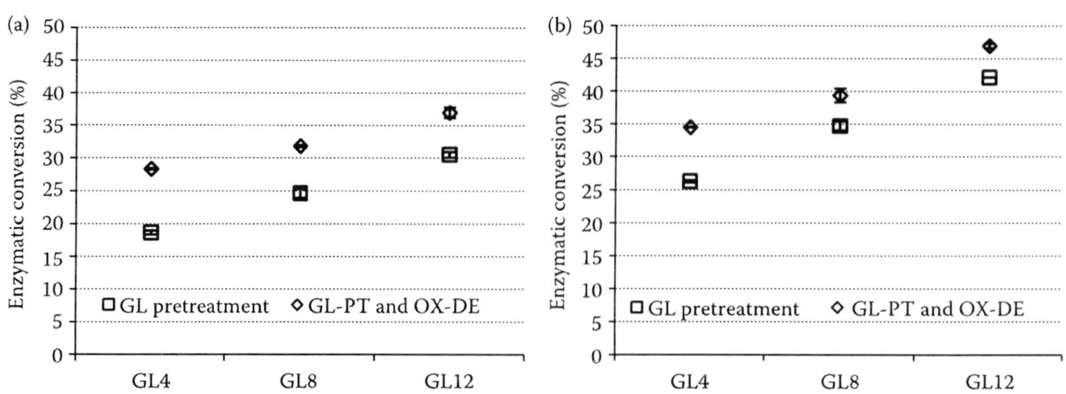

FIGURE 4.8
Enzymatic conversion depending on green liquor charge, oxygen delignification and enzyme dosage (based on raw materials, (a) 5 FPU/g dosage; (b) 10 FPU/g dosage).

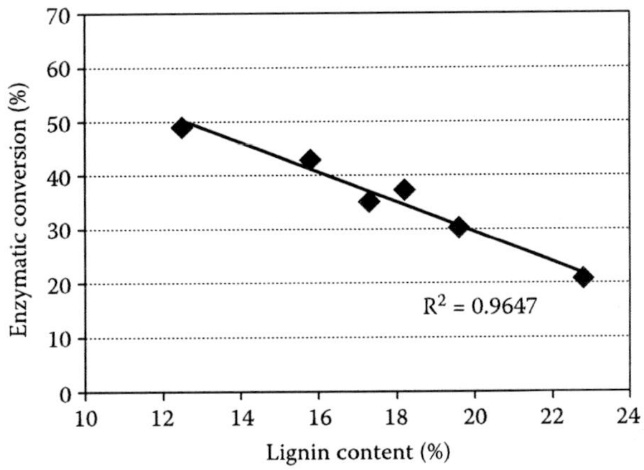

FIGURE 4.9
Correlation between lignin content and enzymatic conversion (based on pretreated substrates, with 5 FPU/g of enzyme dosage).

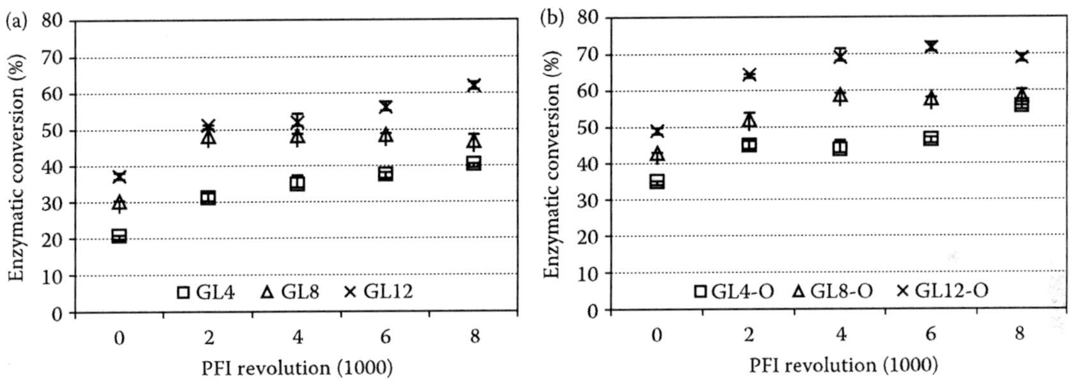

FIGURE 4.10
Enzymatic conversion by 5 FPU of cellulase depending on refining energy input (based on pretreated substrates, (a) no delignification; (b) delignification).

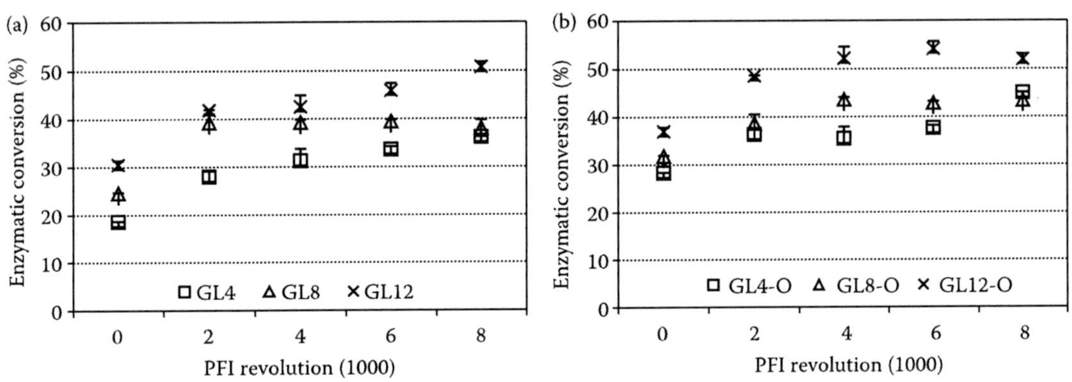

FIGURE 4.11
Enzymatic conversion by 5 FPU of cellulase depending on refining energy input (based on raw materials, (a) no delignification; (b) delignification).

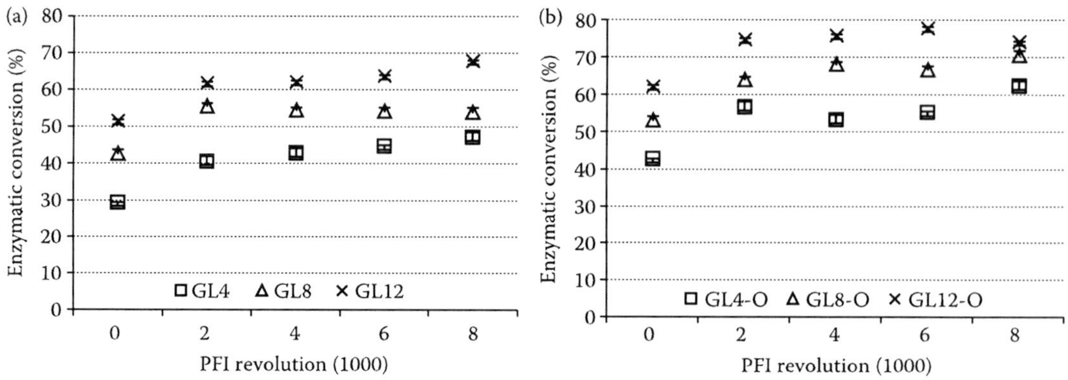

FIGURE 4.12
Enzymatic conversion by 10 FPU of cellulase depending on refining energy input (based on pretreated substrates, (a) no delignification; (b) delignification).

consumption was not necessary. The highest enzymatic conversion rate was 78% and was obtained from oxygen-delignified and refined substrates with 10 FPU of cellulase and 6000 PFI revolutions (Figure 4.12b based on pretreated substrates). The reason for the improvement of enzymatic hydrolysis by mechanical refining might be the increase in specific surface area by the development of surface fibrillation,[14] as reported in a study examining the relationship between mechanical refining and enzymatic digestibility.[15] Increase in GL charge and OD also improved enzymatic conversion. The enzymatic conversion rate of oxygen-delignified and refined (6000 PFI revolutions) substrates increased to 53% as the GL charge was increased from 4% to 12%, and OD enhanced the enzymatic conversion of refined GL 12 substrates by 29%. When enzyme dosage increased from 5 to 10 FPU, the enzymatic conversion of oxygen-delignified and refined GL12 substrates increased by 8.3% (based on pretreated substrates).

To calculate the energy consumed by PFI refining, an indirect method which uses a correlation between the energy consumption by the PFI refiner and the Canadian Standard Freeness (CSF) of bleached softwood pulp was used. First, PFI refining of the softwood pulp was performed at the same energy inputs used for the PFI refining of mixed hardwoods in this study, and the CSF of the refined pulps were measured. A reference equation between the energy consumption by the PFI mill and the CSF of refined softwood pulp was used to calculate the energy consumed by the PFI mill in this study.[21] When the PFI refining of the softwood pulp was performed at 4000 revolutions, the CSF of the refined softwood pulp was 660 mL, and the energy consumption was 188.5 kWh/ton. However, the PFI refiner is a laboratory-scale refiner which consumes more energy per unit of pulp than a commercial-scale refiner,[22] and therefore a scaleup of the refiner could reduce the specific energy. If a reference equation were used involving energy consumption by the Escher–Wyss refiner, a unit which is considered to give results indicative of a commercial-scale refining operation assuming the scaleup of the refiner, the energy consumption would be calculated as 18.6 kWh/ton for 4000 revolutions using the CSF of the refined softwood pulp.[21]

To decrease the enzyme dosage by posttreatment, OD and mechanical refining (in the PFI mill) of GL-pretreated hardwood were performed. It was determined that GL pretreatment and OD removed lignin effectively and that lignin removal increased the rate of enzymatic conversion dramatically. Carbohydrate loss was low, and therefore a combination of the GL pretreatment and OD should be considered as an effective method to improve enzymatic hydrolysis. Mechanical refining also increased the rate of enzymatic conversion dramatically, although the effect leveled off and the increase in enzymatic conversion rates was marginal at high input refining energies. A combination of OD and mechanical refining could synergistically reduce enzyme dosage, and these technologies can be incorporated into a commercial cellulosic ethanol process.

Part III: Process and Economic Analysis

4.7 Introduction

Despite many years of intensive research and development, essentially no commercial production of cellulosic ethanol exists on the market today. Simply put, the required capital investment for any of the well-researched technologies is so high that no reasonable assumptions on input costs permit an attractive level of return on investment.

Previously proposed[23] process-development targets for biochemical production which, if met, could potentially yield investment-quality economics should be used to guide development of cellulosic ethanol production:

- Biomass cost
  - Less than $70 per bone-dry ton (BDT)
- Ethanol yield
  - Greater than 70 gal of ethanol per BDT
- Capital investment
  - Less than $3.00 per annual gallon of ethanol produced
- Ethanol revenue
  - Greater than $2.50 per gallon of ethanol (market price plus subsidy)
- Enzyme cost
  - Less than $0.30 per gallon of ethanol

Despite significant progress on process simplification and enzyme efficiency, and despite potential government subsidies of $1.01/gal of ethanol, achieving these targets remains elusive.

For more than four years, the authors' group has pursued the concept of repurposing kraft pulp mills which have been closed for economic reasons away from the production of paper and toward the production of ethanol. Through two research consortia and other funding, work has continued toward industrial projects which can achieve attractive investment returns.

The technical details of the proposed concept[24]—largely based on the discovery that mild alkaline pretreatment of lignocellulosic material with GL (sodium carbonate and sodium sulfide) could remove lignin and enhance enzymatic hydrolysis—as described by

Jameel were taken as inputs to engineering and financial simulations in the present study. In addition, the analysis has been extended to include application of GL pretreatment in greenfield plants.

4.8 Technical Approach

Four completely integrated tasks were completed:

1. Construction of a model for the delivered cost of biomass
2. Construction of a model for the capital investments required to produce ethanol using the GL process (both greenfield and based in a repurposed kraft mill)
3. Process simulation using laboratory results for GL pretreatment (GL), OD, refining (R), and enzyme hydrolysis (EH), or GL–OD–R–EH
4. Integration of Tasks 1–3 into a financial investment model

4.8.1 Model for Delivered Cost of Biomass

Delivered-cost models have been prepared for a large number of potential biomass feedstocks, ranging from purpose-grown bioenergy plantations (both wood and nonwood feedstocks) to naturally regenerated hardwood forests. The following discussion relates to the methodology for estimating the delivered cost of mixed Southern hardwood.

Less than 1% of U.S. hardwood is grown in planted plantations,[25] so a cellulosic bioethanol industry must rely on hardwood forestland that is unmanaged, not replanted from either seeds or seedlings, but rather simply regenerating naturally. Much of the hardwood forest in the United States is overage and slow-growing. Table 4.11 presents the major assumptions used in the delivered cost of biomass model.

Composition of mixed Southern hardwoods varies substantially, and therefore for modeling purposes, values were selected which are typical of compositions measured in the laboratory (Table 4.12). For convenience in modeling, six-carbon hemicelluloses are categorized as *Hexans*, five-carbon hemicelluloses are lumped together as *Xylans*, and *Glucans* represent the α-cellulose portion.

Other feedstocks, including loblolly pine, eucalyptus, sweet sorghum, sweet sorghum bagasse, corn stover, and switchgrass have also been modeled as plantation investments and subsequently as feedstocks for bioethanol production, with results to be published in the future.

4.8.2 Model for the Capital Investments Required to Produce Ethanol Using the GL Process (Greenfield and Repurposed Kraft Mill Basis)

Capital investment estimates are typically carried out to a precision appropriate to the decision under consideration. To determine basic feasibility, factored estimates—based on pertinent previous detailed estimates—are frequently used to determine if a project financial return on a "reasonable but optimistic" basis is adequate to warrant additional study. Such estimates are likely to be "–0%/+50%" because of the unknowns associated with the actual project site, as well as actual plant equipment requirements. This means that the

TABLE 4.11

Framework for Analysis

Item	Values Selected
Forest density (% of land covered by hardwood forest)	5%
Average growth rate in forest area (BDT per Ha per year [BDT per Acre per year])	2.5 (1.1)
Harvesting cost (including a 6% IRR to the harvesting contractor on harvesting equipment, plus labor, taxes, insurance, and maintenance)	\$35.53 per BDT (\$32.30 per BDT)
Transportation cost including a 6% IRR to the contractor on trucking equipment, plus fuel, labor, taxes, insurance and maintenance, plus waiting time in the forest and at the conversion plant	\$20.90 per BDT (\$19.00 per BDT) for average hauling distance associated with delivering 450,000 BDT (500,000 BDT) roundwood, and \$28.06 per BDT (\$26.00 per BDT) for delivery of 900,000 BDT (1,000,000 BDT) roundwood
Stumpage (based on market price)	\$16.27 per BDT (\$14.79 per BDT)
Total delivered cost	\$39.99 per green ton (GT) (\$36.35 per GT (45% moisture) for 450,000 BDT (500,000 BDT) per year, and \$44.22 per GT (\$40.20 per GT) for 900,000 BDT (1,000,000 BDT) per year

estimates are rarely less than the actual cost and can increase by as much as 50% as a function of site specific variables. With the kraft mill basis well understood and documented, we consider our estimates "+/–25%."

Capital investment estimates were based on two major sources: (1) a confidential kraft mill study (2005) to determine capital requirements for the "front end" and "power plant" areas, and (2) a National Renewable Energy Laboratory detailed analysis of a biorefining process (2002).[26] Although the latter considered a dilute-acid process to convert corn stover to ethanol, the estimates were useful for estimating the "biorefinery" battery limits.

The approach taken was first to estimate the greenfield mill investment required to produce ethanol using the GL pretreatment–OD–mechanical refining (GL–OD–R) option, and then to deconstruct the estimate to eliminate those investment areas that would likely be transferred at no cost to the new owner of a closed kraft pulp mill operation. Two mill-specific estimates for repurposing a kraft pulp mill have been completed, and the results presented here are consistent with more detailed studies. Further details on the engineering analysis will be presented in the results section. Table 4.13 shows the factors used to scale and to escalate previous studies to a 2015 (startup year) basis.

TABLE 4.12

Compositional Analysis of Mixed Southern Hardwood

Natural Hardwood	
Moisture	45.0%
Lignin	28.3%
Glucan	42.6%
Hexan	6.7%
Xylan	19.3%
Extractives	2.5%
Ash	0.6%
Solids total	100.0%

TABLE 4.13

Basis of Capital Estimate for a Greenfield GL-OD-R–EH Process for Processing Hardwood Roundwood to Ethanol Production

		Reference Kraft Pulp Mill Study (GL Pretreatment—Oxygen—Refining)					
	Capacity Units	Reference Project	Reference Project Cost, $	Scale Factor	Reference Year	Annual Escalation	Reference
Site preparation							
Land purchase	Acres for 700,000 tons per year	1000	1,000,000	0.90	2005	1.24	Kraft Mill Study
Land preparation	Acres for 700,000 tons per year	400	12,000,000	0.90	2005	1.24	Kraft Mill Study
Raw water treatment	Gallons per day	30,000,000	10,000,000	0.7	2005	1.24	Kraft Mill Study
Waste water treatment	Gallons per day	30,000,000	15,000,000	0.7	2005	1.24	Kraft Mill Study
Biomass handling							
Roundwood receiving	GTs per year	2,000,000	25,000,000	0.6	2005	1.24	Kraft Mill Study
Chip receiving	GTs per year	1,000,000	15,000,000	0.6	2005	1.24	Kraft Mill Study
Roundwood chipping	GTs per year	2,000,000	10,000,000	0.6	2005	1.24	Kraft Mill Study
Biomass screening	GTs per year	2,000,000	5,000,000	0.6	2005	1.24	Kraft Mill Study
Pretreatment							
Green liquor pretreatment	BDTs per year Feed	700,000	30,000,000	0.6	2005	1.24	Kraft Mill Study
Posttreatment							Kraft Mill Study
Mechanical posttreatment	BDTs Pulp per year	700,000	6,000,000	0.6	2005	1.24	Kraft Mill Study
Oxygen posttreatment	BDTs per year	1,000,000	25,000,000	0.6	2005	1.24	Kraft Mill Study
Enzyme posttreatment	Hours retention per 1000000 BDT	20	15,000,000	0.5	2005	1.24	Kraft Mill Study
Lignin filter	BDTs per year	200,000	15,000,000	0.6	2005	1.24	Vendor Quote
Biorefinery							
Fermentation	Gallons per minute fermentor flow	1880	15,291,296	0.8	2000	1.48	NREL
Beer column	Gallons per minute input flow	2065	3,741,614	0.8	2000	1.48	NREL
Rectification column	Gallons per minute input flow	286	3,570,023	0.8	2000	1.48	NREL

TABLE 4.13 (continued)

Basis of Capital Estimate for a Greenfield GL-OD-R-EH Process for Processing Hardwood Roundwood to Ethanol Production

		Reference Kraft Pulp Mill Study (GL Pretreatment—Oxygen—Refining)					
	Capacity Units	Reference Project	Reference Project Cost, $	Scale Factor	Reference Year	Annual Escalation	Reference
Dehydration	Gallons per minute input flow	153	5,069,074	0.7	2000	1.48	NREL
Product storage and ship	Gallons	1,000,000	3,000,000	0.6	2005	1.24	Kraft Mill Study
Recovery and power							
Evaporation	Pounds per hour evaporation	1,000,000	40,000,000	0.6	2005	1.24	Kraft Mill Study
Recovery boiler	Tons dry solids per day	5000	140,000,000	0.6	2005	1.24	Kraft Mill Study
Power boiler	Pound per hour steam	406,496	43,576,643	0.5	2000	1.48	Kraft Mill Study
Turbine generator	MW	50	35,000,000	0.6	2005	1.24	Kraft Mill Study

4.8.3 Process Simulation of GL Pretreatment (GL)–Oxygen Delignification (OD)–Refining (R) and Enzyme Hydrolysis (EH) [GL–OD–R–EH]

Figure 4.13 depicts the current process concept, abbreviated GL–OD–R–EH. Only proven kraft pulp equipment and processes (GL rather than white liquor) are used in the greenfield scheme, which facilitates the analysis of investment requirements for repurposing. Additional equipment to accomplish fermentation and distillation are added in identical fashion for both alternatives.

The process starts with undebarked roundwood and purchased chips prepared (not shown) and screened in a conventional woodyard before cooking. Chips (A) are fed to (in this example) a continuous digester, impregnated and cooked under mild conditions with GL (B) and discharged through a first blowline refiner (C) to a blow tank (D). The pretreated chips are further refined in deshive refiners (E1), and then washed countercurrently with two vacuum-filter washers (F). Following this first washing, the high-yield GL pulp is treated with sodium hydroxide and oxygen in a medium-consistency OD reactor system (I), discharged to a second blow tank (J), further refined (E2), and washed (in this example) in a conventional wash press (L). Washing is accomplished by use of evaporator condensate (M) in the wash press and the filtrate subsequently combined with pressate (G), which then flows countercurrently to the two wash filters and is subsequently recovered as weak black liquor (H).

Wash-press discharge consistencies of 25–30% are attractive to minimize water into the biorefinery system, but laboratory work to date suggests a penalty in enzyme performance. Therefore, the model currently uses laboratory performance data with pulp at 14%

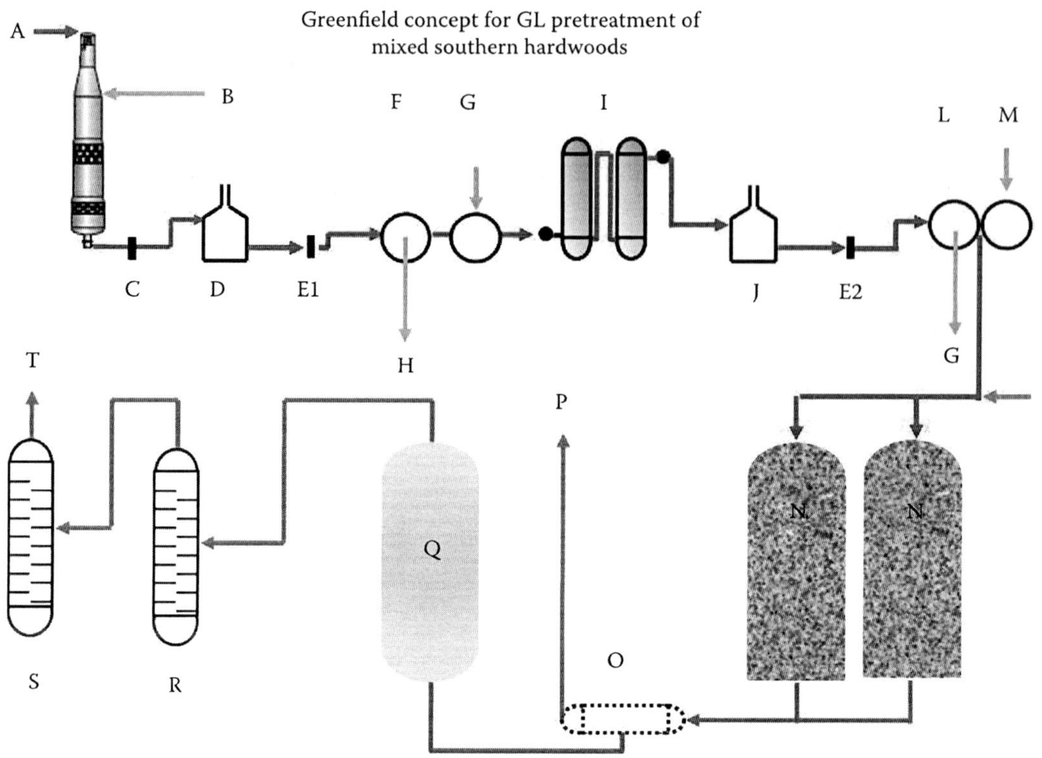

FIGURE 4.13
Unit operations of a hardwood-based green liquor (GL)–oxygen delignification (OD)–refining (R)–enzyme hydrolysis (EH) process for ethanol production.

consistency into the EH stage. Some sugar-stream recycle can be used to build up a higher equilibrium concentration of sugar that ultimately flows to fermentation.

Enzymes (K) are added in a dose sufficient to achieve 80% conversion of pulp carbohydrates to monomeric sugars. Although higher levels are achievable, the doses required are economically disadvantaged at current speculated enzyme prices.[27] The pulp impregnated with enzymes is treated for 96 h in batch towers of the same design as pulp-mill high-density storage towers. In the actual mills modeled to date, abundant storage exists to accommodate 96 h retention at an initial pulp consistency of 14%; lower consistencies favor better EH performance, but the combined economic impacts of investment in additional retention towers, plus additional water intrusion, led to the current design. Batch reaction in multiple towers is recommended because the very slow reaction of enzymes would penalize channeling to a large degree. In addition, the ability to clean out towers without shutting down the process is an important consideration.

Following EH, the hydrolyzed pulp slurry has the appearance of thin soup, with visible particles of lignin and small amounts of unhydrolyzed fibers apparent. Although the fermentation process is tolerant of lignin, the current concept uses a belt filter (O) to separate out the lignin particles as a filter cake (P), which is recovered and later mixed with weak black liquor (H), which is evaporated and burned (not shown) in a recovery boiler.

Washing efficiency	Sugar stream to fermentation	Fermentation efficiency	Sugar stream to fermentation	Fermentation stoichiometry	Ethanol stream from fermentation	
0%	—		—		—	
95%	152,125	95%	144,519	51%	73,705	
95%	5336	95%	5069	51%	2585	
95%	72,332	80%	57,866	51%	29,511	
	—		—		—	
	—		—		—	
	229,794		207,454		105,802	Tons per year ethanol

31,887,150 Gallons per year

67.2 Gallon per BDT

FIGURE 4.14

Biorefinery assumptions and simulation results.

The sugars are recovered at a concentration of 8–12%, and sent to a fermentation system (Q), where conventional fermentation with yeasts is assumed. Since hardwoods contain high levels of 5-carbon sugars, it is important to the overall economics to have a fermentation system that is efficient in converting the 5–C sugars to ethanol. Our simulation is based on laboratory verified 95% fermentation of C–6 sugars, but hypothesized C–5 sugar efficiency of 80%. Critical simulation assumptions and results are collected in Figure 4.14 for the case of 500,000 BDT per year delivered hardwood. Simulations were performed also at 1,000,000 BDT delivered hardwood roundwood (note: 5% wood loss is assumed in the handling and chipping operations, which is subsequently burned in a bark boiler).

Recovered ethanol and water are directed to a beer column (*R*), where an ethanol–water mixture is recovered at 40% concentration, and then to a rectification column (*S*), where an azeotropic mixture of 95% ethanol is recovered. Dehydration to 99.5% ethanol (not shown) is provided, along with storage for 1 million gallons of dehydrated ethanol.

4.8.4 Integration of Tasks 1–3 into a Financial Investment Model

4.8.4.1 Greenfield Assumptions

Project spending begins in 2010 with acquisition of land and site preparation activities (clearing, grading, drainage, roads), followed by subsequent installation of water supply and waste treatment facilities. Equipment is installed, and system startup occurs in 2012. A ramp-up to 100% of nameplate capacity takes place in 2015, with 30% excess material usage in Year 1 to reflect startup inefficiencies.

A project life of 10 years is assumed, along with straight-line depreciation on all new assets. Assumptions are made for working capital (10% of annual direct costs + 10% of the annual presubsidy value of ethanol product), other mill fixed costs (3% of annual revenue), sales and other overhead costs (3% of annual revenue), maintenance (including repair labor) (2% of annual replacement asset value [RAV]), and annual reinvestment of new fixed capital (1% of RAV). Each component of future direct and indirect cash costs is individually driven by a "productivity factor" and an "inflation factor," both of which is subjectively determined, but are generally set at levels that will keep net cost creep to 1% per year. As one important exception, and one of critical importance, enzyme costs in 2012 were assumed to be 50% lower than current anecdotal cost levels.

Ethanol revenues were estimated by assuming a market price ($2.00/gal today, increasing at 2% per year to track gasoline-price escalation) plus a subsidy ($1/gal, held constant).

All financial ratios were calculated for a 15 year period. Taxes were calculated at 35% of earnings before interest and tax (EBIT) when positive, and a tax loss was carried forward when EBIT was negative. Net present value (NPV) was calculated at a 12% discount rate, based on annual free cash flows from project Day 1 (free cash flow = after-tax income + depreciation – new fixed capital investment). A terminal value (TV) was used in Year 10 to reflect the enterprise value, based on a multiple of 5 × EBITDA in Year 10. Internal rate of return (IRR) was calculated on the same basis as NPV. Critical assumptions are presented in Table 4.14.

4.8.4.2 Repurposing Assumptions

As discussed in the capital section, the repurposing cases account only for the new fixed capital that would not be transferred from the existing mill to the repurposed ethanol plant, plus an assumed $10 million repiping factor. All other input assumptions were held constant at their values in the greenfield case, including the calculation of maintenance and repair expense (by using the replacement asset value basis established in the greenfield case) and the ongoing new fixed capital.

4.9 Results and Discussion

4.9.1 Biomass Cost

This report focuses only on mixed Southern hardwood as a feedstock. The delivered cost of mixed Southern hardwoods varies substantially by geography, pulpwood demand, and season. The analytical framework illustrated in Table 4.11 shows the methodology used in this study, but actual hardwood cost can easily vary by –$5.00/GT– + $20/GT in the southeastern United States.

4.9.2 Capital Investment

Using the capital-investment methodology described earlier, Table 4.15 shows a comparison of capital investments for the greenfield and repurposed mills for the 500,000 and 1,000,000 BDT per year cases. The proposed methodology penalizes the greenfield option particularly heavily in the infrastructure cost area, where very few economies of scale are possible. Both the greenfield and repurposed GL cases include additional capital for blowline refining and OD; in the repurposed cases, depending on the specific mill, these processes may already be in place. These comparisons are critical inputs to the financial model, but in reality, the overall investment returns will be shown to be equally sensitive to other inputs as well.

TABLE 4.14
Financial Model Assumptions

Value	Item	Biomass	Comments
Repurpose GL Mill	= Frontend or Process Name		Select either Process or Pretreatment from the Drop Down Menu
Fermentation	= Backend or Process Name		Select Process for converting the Biomass to Product from the Drop Down Menu. Not all combinations are legitimate
	Forest Biomass Type		
500,000	= Forest Biomass Roundwood, BDT per Year	Mixed Southern Hardwood	RESERVED FOR ROUNDWOOD—Set the input Bone Dry Tons
-	= Forest Biomass Purchased Chips, BDT perYear	None	RESERVED FOR CHIPS—Set the input Bone Dry Tons
-	= Biomass A, BDT per Year	None	RESERVED FOR OTHER BIOMASS—Set the input Bone Dry Tons
-	= Biomass B, BDT per Year	None	RESERVED FOR OTHER BIOMASS—Set the input Bone Dry Tons
2012	= Startup Year		Arbitrary input to be selected based on the expectation of commercial reality
2026	= Terminal Year		Arbitrary input to give a specific Project Life for financial purposes
29,275,899	= Ethanol Production, Gallons per Year		Calculated from Yields in "Conversion Process Summary" that are specific to the Conversion Process being evaluated
$ 92,635,744	= Total Capital Investment		Total Installed Cost from "Conversion Process Summary" of the selected Pretreatment + the selected Conversion Process
$3.16	= Capital Investment per Annual Gallon		Calculated from the two rows above
0.70	= Scaleup Exponent		Arbitrary typical value—as the number increases the value of "Economy of Scale" decreases
30%	= % of Spending in Year -2		Arbitrary typical values for a project that spends capital over a three year period
50%	= % of Spending in Year -1		
20%	= % of Spending in Year 0		
50%	= % of Nominal Capacity, Project Year 1		Arbitrary tpical value for a startup curve in the first year
80%	= % of Nominal Capacity, Project Year 2		
30%	= Excess Material Use in Project Year 1		Arbitrary typical value to recognize the inefficiencies of startup
10%	= Working Capital Per Cent of Direct Cost + Presubsidy Product Revenue		Arbitrary typical value to recognize the inefficiencies of startup
10	= Years Depreciation Schedule		Arbitrary typical value
35%	= Tax Rate, with Tax Loss Carryforward		Arbitrary typical value
12%	= Discount Rate		Arbitrary typical value, reflects the financial risk the funding entity is willing to take
5.00	= Terminal Value, Year 15 EBITDA Multiple		Arbitrary typical value, recognizes the salvage or resale value of the assets at the end of the project life
8,400	= Hours per Year		

continued

TABLE 4.14 (continued)

Financial Model Assumptions

Assumptions listed below may have three parts: (1) an estimate of the unit price or performance in 2012 (or another startup year); (2) an expected inflation increase in the cost or price of a raw material or a product; (3) an expected improvement in productivity or output over a period of time.

2012 Assumption		Annual Increase in Price	Annual Productivity Improvement	2012 Cost	
58.55	= Gallons of Ethanol per BDT (Based on Purchased Biomass)	0.00%	2.00%	0.00%	Value from the "Conversion Process Summary" should inform the calculation
$2.00	= Revenue per Gallon of Ethanol	2.00%	Arbitrary typical value		
$1.01	= Subsidy, added to Price	0.00%	Arbitrary typical value		
$0.50	= Subsidy, Tax Credit	0.00%	Arbitrary typical value (not currently used)		
$36.35	= Biomass Cost, $ per Green Ton	3.00%	1.00%	Value from the "Biomass Cost Summary" Worksheet should inform the model	
45%	= Moisture Content, % of Green Ton	3.00%	1.00%	Arbitrary typical value	
$0.68	= Enzyme Cost, $ per Gallon Ethanol	-1.00%	1.00%	Entered from "Master Conversion Model for Feedstocks"	
$0.07	= Yeast Cost, $ per Gallon Ethanol	1.00%	1.00%	Arbitrary typical value for Fermentation processes	
0.03	= Chemical Cost, $ per Gallon Ethanol	2.00%	1.00%	Value from the "Biomass Conversion Process" Summary should inform the calculation	
($1,170,248)	= Energy Cost, $ per Year	Based on Power Calculation in "Conversion Process Summary"			
$50.00	= Energy Price, $ per MWH	2.00%	1.00%	Value from the "Biomass Conversion Process" Summary should inform the calculation	
150	= Hourly and Administrative Staff (Non-Maintenance)	3.00%	1.00%	$ 70,000.00	Value from the "Biomass Conversion Process" Summary should inform the calculation
30	= Salaried Staff	3.00%	1.00%	$ 90,000.00	Value from the "Biomass Conversion Process" Summary should inform the calculation
2%	= Maintenance Expense, including Labor, % of Replacement Asset Value	3.00%	1.00%	Arbitrary typical value	
1%	= Capital Reinvestment, % of Replacement Asset Value	3.00%	1.00%	Arbitrary typical value	
3%	= Other Fixed Costs, % of Sales			Arbitrary typical value	
3%	= Sales and Other Overhead, % of Sales			Arbitrary typical value	

TABLE 4.15

Capital Investment Requirements for the Individual Greenfield and Repurpose Cases, at 500,000 and 1,000,000 BDT per Year Biomass Input Rate

	Greenfield GL Mill				Repurpose GL Mill			
	Quantity per Year	Capex	Quantity per Year	Capex	Quantity per Year	Capex	Quantity per Year	Capex
Infrastructure, acres and water volumes	1000	$19,597,279	1000	$21,777,508	1000	$0	1000	$0
Wood handling, GT per year	925,926	$31,101,949	1,851,852	$47,141,739	925,926	$0	1,851,852	$0
Alternative biomass handling, GT per year	—	$0	—	$0	—	$0	—	$0
Chemical pretreatment, BDT feed per year	1,354,630	$58,905,637	1,759,259	$0	475,000	$0	950,000	$0
Mechanical posttreatment, BDT pulp per year	405,980	$5,360,972	811,960	$8,125,714	405,980	$5,360,972	799,292	$8,049,409
Chemical posttreatment, BDT pulp per year	405,980	$18,033,962	811,960	$27,334,374	405,980	$0	799,292	$0
Enzyme treatment, hours of retention	130	$47,388,224	169	$54,003,893	46	$0	91	$0
Lignin filter, BDT Lignin per year	98,439	$12,145,680	196,878	$18,409,409	98,439	$12,145,680	196,878	$18,409,409
Biorefinery, gallons per minute feed	1790	$40,754,929	3581	$69,349,110	1727	$39,995,790	3401	$67,178,481
Evaporation, pounds water evaporated per hour	375,826	$27,548,673	751,653	$41,755,980	375,826	$0	740,863	$0
Chemical recovery, tons dry solids per day	798	$57,672,629	1596	$87,415,360	798	$0	1623	$0
Power boiler, pounds per hour steam	279,048	$53,539,678	558,095	$75,716,539	279,048	$0	566,276	$0
Power generation, MW	17	$22,779,058	34	$34,525,671	17	$0	36	$0
Other	—	$0	—	$0	—	$10,000,000	—	$10,000,000
Total capex ($2015)		$394,828,671		$485,555,297		$67,502,443		$103,637,299
Total capex per gallon ($2015)		$12.38		$7.61		$2.12		$1.66
Total biorefinery capex per gallon ($2015)		$1.28		$1.09		$1.25		$1.07

TABLE 4.16
Key Simulation Outputs

	Greenfield GL Mill		Repurpose GL Mill	
Biomass purchased, BDT per year	500,000	1,000,000	500,000	1,000,000
Biomass to pretreatment, BDT per year	475,000	950,000	475,000	950,000
Pretreatment yield, %	85.5	85.5	85.5	85.5
Mechanical posttreatment,[a] KWh per BDT pulp	200	200	200	200
Chemical posttreatment yield, %	92	92	92	92
Enzyme applied, kg per kg glucan	0.059	0.059	0.059	0.059
Carbohydrate conversion, %	70.4	70.4	70.4	70.4
Fiber consistency in EH, %	14	14	14	14
Sugar concentration, %	5.91	5.91	5.91	5.91
C-5 Fermentation yield, %	80	80	80	80
C-6 Fermentation yield, %	95	95	95	95
Gallons ethanol per BDT biomass feed	67.2	67.2	67.2	67.2

[a] Includes blowline refining and deshive refining.

4.9.3 Process Simulation

Table 4.16 shows key results from the process simulation. In general, the consumptions per gallon of product are identical in the two cases because the processes (greenfield and repurposed mills) are the same, and the unit consumptions should be independent of biomass feed rate (500,000 and 1,000,000 BDT per year).

4.9.4 Financial Modeling

The principal financial outcomes are displayed in Table 4.17. Key comparison points are:

- Repurposed-mill CAPEX per gallon is only approximately 20% than required for the greenfield project.
- Although a greenfield project could likely install process efficiencies to a higher degree than a repurposed project, most of those efficiencies would reside in the pulp mill and the power plant; the present simulation does not account for any such differences.
- Cash costs are slightly lower for the repurposed options compared to the greenfield cases; small differences in cash cost are apparent for the larger scale versus the smaller scale.
- Financial returns for the repurposed options would be attractive for a proven technology, while the greenfield concept needs substantial improvement.

Investments in bioenergy development projects are without question high risk. At the current stage of development of GL pretreatment—whether in a greenfield or a repurposed mill—investors would look for an IRR in excess of 20% to allow for the inevitable erosion of the project profitability as more details are developed. This appears possible with repurposing, but substantial progress is needed in the case of the greenfield concept.

TABLE 4.17

Key Financial Results for Greenfield and Repurpose GL Projects at 500,000 and 1,000,000 BDT per Year Feed

	Greenfield GL Mill				Repurpose GL Mill			
	Quantity per Year	Cost per Unit	Quantity per Year	Cost per Unit	Quantity per Year	Cost per Unit	Quantity per Year	Cost per Unit
Mixed Southern hardwood	500,000	$61.86	1,000,000	$67.90	500,000	$61.86	1,000,000	$67.90
None	0	$—	0	$—	0	$—	0	$—
None	0	$—	0	$—	0	$—	0	$—
Annual ethanol production, gallons	31,887,150		63,774,300		31,887,150		63,774,300	
Ethanol yield, gallons per BDT	63.8		63.8		63.8		63.8	
Capex, total per annual gallon	$394,828,671	$12.38	$485,555,297	$7.61	$67,502,443	$2.12	$104,562,495	$1.64
Total biomass cost per gallon		$1.03		$1.13		$1.03		$1.13
Total enzyme cost per gallon		$0.54		$0.54		$0.54		$0.54
Total energy credit/cost		($0.06)		($0.06)		($0.04)		($0.04)
Total direct cost per gallon		$1.62		$1.72		$1.63		$1.73
Total indirect cost per gallon		$1.92		$1.75		$1.00		$0.85
Total cash cost per gallon		$2.26		$2.20		$2.36		$2.25
Total cost per gallon		$3.53		$3.47		$2.63		$2.58
Net present value at 12%		($183,202,282)		($94,528,305)		$49,615,782		$160,709,601
IRR, %		3		9		19		26

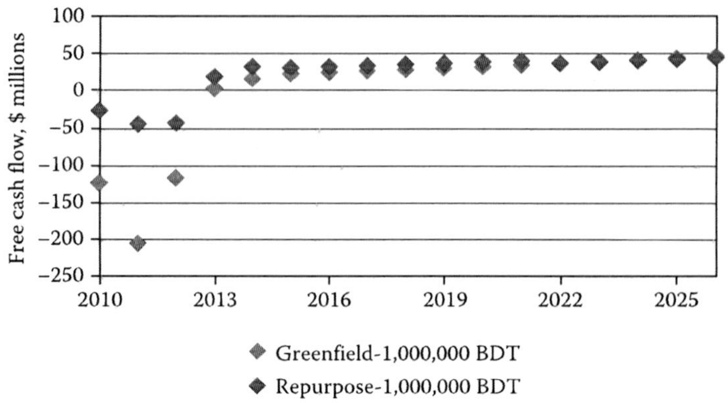

FIGURE 4.15
Free cash generated by the greenfield and repurpose options.

Figure 4.15 shows, for the 1,000,000 BDT per year option, the free cash flow (FCF) on a year-by-year basis. As is typical with large financial investments, the large negative FCF in the early investment phase before startup, plus the inefficiencies of the startup year and the three-year ramp to 100% production are all reflected in this comparison. By 2015, when both options will have achieved 100% of target production, the FCF values, which exclude depreciation, of the two options are nearly equal. The large disparity in NPV and IRR are entirely due to the initial cash outlay.

Key financial drivers derived from the laboratory work include:

- Carbohydrate conversion

Preservation of carbohydrates through GL, OD, and refining is approximately 85% of the original carbohydrates in wood, while 10% is lost due to EH inefficiency and another 5% left with the lignin on the lignin filter. The lignin-filter concept appears to be not only costly, but also an important cause of loss of monomeric sugars.

- Enzyme dose

Historically, researchers have applied enzyme doses of approximately five to ten times the levels that can be economically applied, and the enzymes have been applied at pulp consistencies of 1% or less. Optimism on the part of NREL that enzyme cost and effectiveness can be improved is reflected in the authors' own discussions and in experience that each new enzyme generation available for evaluation has been superior to the previous product. A 50% reduction in current price by 2012 and 1% per year performance improvement have been assumed. Enzyme doses have been reduced in the present laboratory evaluations through investments in OD and refining energy, and these appear to have provided a good return in reducing enzyme requirements. The doses modeled here are actual laboratory levels at practical pulp consistencies.

- Sugar concentration

A low pulp consistency of 14% into EH leads to relatively dilute sugar streams (Figure 4.16), which in turn leads to high investment in the 48 h fermentation vessel and high steam requirements for the beer column operation (Figure 4.17).

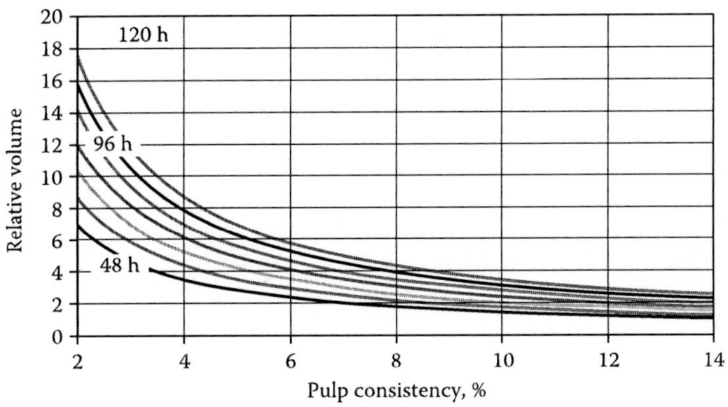

FIGURE 4.16
Impact of pulp consistency and retention time on the relative size of retention vessels required to carry out enzyme hydrolysis reactions. Most laboratory studies have been carried out at unrealistically low pulp consistencies, along with unaffordable enzyme doses.

A sensitivity analysis to the major input variables is shown in Figures 4.18 through 4.21. The sensitivity analyses are critical to understanding the financial drivers:

- Ethanol selling price and ethanol yield per BDT are essentially equal and are the primary component of financial success. Although researchers have no control over the former, the latter is only approximately 60% of theoretical and is therefore an important target for future research.
- Biomass cost is almost equal in importance to ethanol selling price, and in the case of the 1,000,000 BDT mill where demand is driving cost, is actually of more critical importance. Biomass cost can be reduced by harvesting forest residuals at the same time as roundwood cutting; seasonal use of agricultural residues can also improve profitability.
- Enzyme price, and therefore enzyme cost per gallon, is next in importance. Price has been discussed, but dose is another important research target. The authors

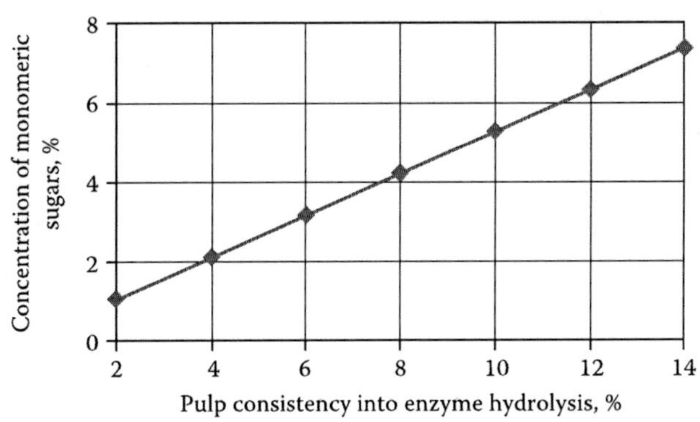

FIGURE 4.17
Impact of pulp consistency into enzyme hydrolysis stage on the concentration of monomeric sugars output. The pulp consistency has major impact on size of retention vessels for enzyme hydrolysis and the sugar concentration dictates the volume required for fermentation and subsequent concentration of ethanol.

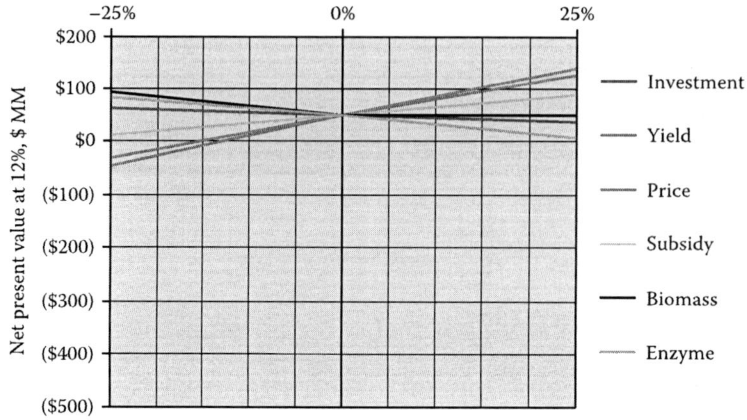

FIGURE 4.18
Repurpose GL mill—500,000 BDT per year.

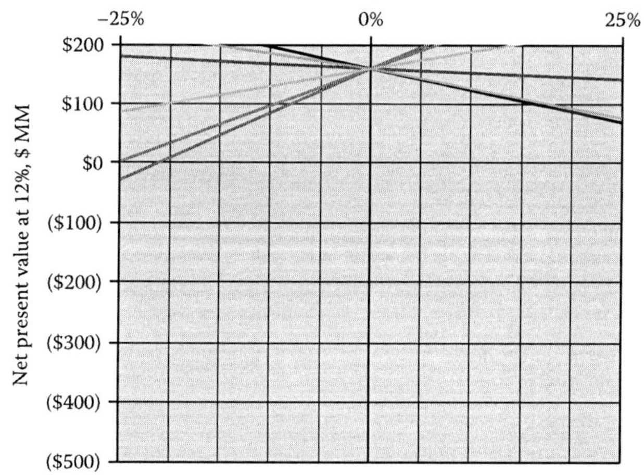

FIGURE 4.19
Repurpose GL mill—1,000,000 BDT per year.

have reduced laboratory dose levels (to achieve 80% sugar yield) by over 75% since the current research began.

- Capital investment and subsidies are least critical in importance: Capex because the impact is largely accumulated in the early years; subsidies because the center point is only half the market price and does not change over time.

4.10 Conclusions

Part I: Pretreatment Technology Development

The results show that a pretreatment based on GL (sodium carbonate and sodium sulfide) is the most effective pretreatment for a repurposed mill. The pulps produced by

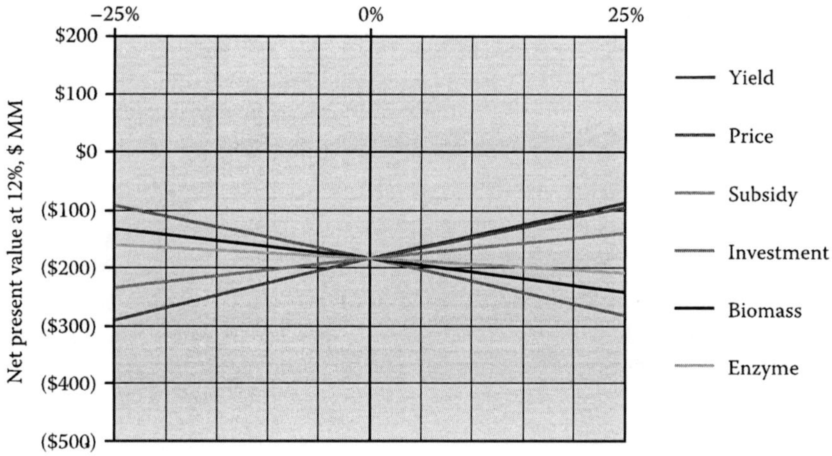

FIGURE 4.20
Greenfield GL mill—500,000 BDT per year.

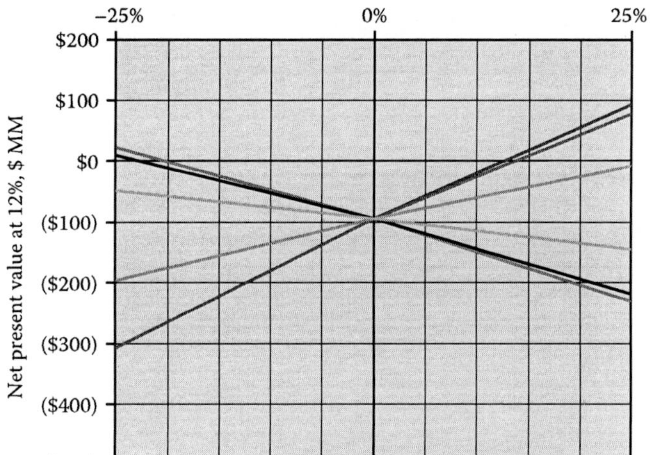

FIGURE 4.21
Greenfield GL mill—1,000,000 BDT per year.

this process can be enzymatically hydrolyzed to monomeric sugars with high overall sugar recovery. The use of GL for pretreatment ensures that the chemicals used during pretreatment can be recovered.

Part II: Posttreatments to Reduce Enzyme Charge

To decrease the required enzyme dosage by posttreatment, OD and mechanical refining (PFI mill) of GL-pretreated hardwood were performed. It was determined that GL pretreatment and OD removed lignin effectively and that lignin removal increased the enzymatic conversion rate significantly. Carbohydrate loss was low, and therefore a combination of GL pretreatment and OD is considered as the most effective method to improve enzymatic hydrolysis. Mechanical refining also increases the enzymatic conversion rate dramatically, although the effect levels off and increases in the enzymatic conversion rate were marginal at high input refining energy.

Part III: Process and Economic Analysis

Investments in bioenergy development projects are without question high risk. At the current stage of development of GL pretreatment, whether in a greenfield or repurposed mill, investors would look for an IRR in excess of 20% to allow for the inevitable erosion of project profitability as more details are developed. This appears possible with repurposing, but substantial progress is needed in the case of the greenfield concept.

Acknowledgments

We are grateful for the financial support provided by the North Carolina State University Wood-Ethanol-Research Consortium. Since 2007, we have been supported by American Process, Inc., Andritz Machinery, Arborgen, BE&K Engineering, Catchlight Energy, Evolution Resources, Genencor, Japan Pulp and Paper Research Instititute, International Paper, Metso Paper, Nippon Paper, North Carolina State Natural Resources Foundation, Novozymes, Smurfit Stone Container Corporation, Syngenta, Weyerhaeuser Company, and Xethanol. In addition, work on forest biomass cost and greenfield mill designs has been supported by a grant from the Biofuel Center of North Carolina.

References

1. Ragauskas, A.J., Williams, C.K., Davison, B.H., Britovsek, G., Cairney, J., Eckert, C.A., Frederick, W.J. et al. 2006. The path forward for biofuels and biomaterials. *Science* **311**, 484–489.
2. Kadam, K.L. and McMillan, J.D. 2003. Availability of corn stover as a sustainable feedstock for bioethanol production. *Bioresource Technology* **88**, 17–25.
3. Taherzadeh, M.J. and Karimi, K. 2007. Enzyme-based hydrolysis processes for ethanol from lignocellulosic materials: A review. *BioResources* **2**, 707–738.
4. Öhgren, K., Bura, R., Saddler, J., and Zacchi, G. 2007. Effect of hemicellulose and lignin removal on enzymatic hydrolysis of steam pretreated corn stover. *Bioresource Technology* **98**, 2503–2510.
5. Adsul, M.G., Ghule, J.E., Shaikh, H., Singh, R., Bastawde, K.B., Gokhale, D.V. and Varma, A.J. 2005. Enzymatic hydrolysis of delignified bagasse polysaccharides. *Carbohydrate Polymers* **62**, 6–10.
6. Tabka, M.G., Herpoël-Gimbert, I., Monod, F., Asther, M., and Sigoillot, J.C. 2006. Enzymatic saccharification of wheat straw for bioethanol production by a combined cellulase xylanase and feruloyl esterase treatment. *Enzyme and Microbial Technology* **39**, 897–902.
7. Zhao, Y., Wang, Y., Zhu, J.Y., Ragauskas, A., and Deng, Y. 2008. Enhanced enzymatic hydrolysis of spruce by alkaline pretreatment at low temperature. *Biotechnology and Bioengineering* **99**, 1320–1328.

8. Yu, Z. and Zhang, H. 2003. Ethanol fermentation of acid-hydrolyzed cellulosic pyrolysate with Saccharomyces cerevisiae. *Bioresource Technology* **90**, 95–100.
9. Sun, Y. and Cheng, J. 2002. Hydrolysis of lignocellulosic materials for ethanol production: A review. *Bioresource Technology* **83**, 1–11.
10. Pan, X., Arato, C., Gilkes, N., Gregg, D., Mabee, W., Pye, K., Xiao, Z., Zhang, X., and Saddler, J. 2005. Biorefining of softwoods using ethanol organosolv pulping: Preliminary evaluation of process streams for manufacture of fuel-grade ethanol and co-products. *Biotechnology and Bioengineering* **90**, 473–481.
11. Emmel, A., Mathias, A.L., Wypych, F., and Ramos, L.P. 2003. Fractionation of *Eucalyptus grandis* chips by dilute acid-catalysed steam explosion. *Bioresource Technology* **86**, 105–115.
12. Jin, Y., Jameel, H., Chang, H., and Phillips, R. 2010. Green liquor pretreatment of mixed hardwood for ethanol production in a repurposed kraft pulp mill. *Journal of Wood Chemistry and Technology* **30**, 86–104.
13. Wu, S., Chang, H., Jameel, H., and Philips, R. 2010. Novel green liquor pretreatment of loblolly pine chips to facilitate enzymatic hydrolysis into fermentable sugars for ethanol production. *Journal of Wood Chemistry and Technology* **30**, 205–218.
14. Silverstein, R.A. 2004. *A Comparison of Chemical Pretreatment Methods for Converting Cotton Stalks to Ethanol Electronic Resource*. North Carolina State University Theses Biological and Agricultural Engineering. Dissertation/Thesis.
15. Cullis, I.F., Saddler, J.N., and Mansfield, S.D. 2004. Effect of initial moisture content and chip size on the bioconversion efficiency of softwood lignocellulosics. *Biotechnology and Bioengineering* **85**, 413–421.
16. Britt, K.W., Handbook of Pulp and Paper Technology. 1964. New York, Reinhold Pub. Corp. Located in NCSU: D.H. Hill Library (Call Number: TS1105 .B59 1964); Located in NCSU: Natural Resources Library (Call Number: TS1105 .B59 1964).
17. Lecourt, M., Sigoillot, J., and Petit-Conil, M. 2010. Cellulase-assisted refining of chemical pulps: Impact of enzymatic charge and refining intensity on energy consumption and pulp quality. *Process Biochemistry* **45**, 1274–1278.
18. Gil, N., Gil, C., Amaral, M.E., Costa, A.P., and Duarte, A.P. 2009. Use of enzymes to improve the refining of a bleached Eucalyptus globulus kraft pulp. *Biochemical Engineering Journal* **46**, 89–95.
19. Zhang, J., Song, H., Lin, L., Zhuang, J., Pang, C., and Liu, S. 2010. Microfibrillated cellulose from bamboo pulp and its properties. *Biomass Bioenergy* **39**, 78–83.
20. Brown, R.C. Biorenewable Resources: Engineering New Products from Agriculture 2003. Iowa State Press. Located in Duke: Perkins/Bostock Library (Call Number: TP339 .B76 2003 c.1); Located in NCSU: D.H. Hill Library (Call Number: TP339 .B76 2003).
21. Kerekes, R.J. 2005. Characterizing refining action in PFI mills. *Tappi Journal* **4**, 9–14.
22. Lindstrom, T. *Fundamentals of Papermaking: Transactions of the Ninth Fundamental Research Symposium Held at Cambridge*, September 1989 (1989). Mechanical Engineering Publications Limited. Located in NCSU: D.H. Hill Library (Call Number: TS1080 .F87 1989 v.1).
23. Phillips, R.B. 2007. Targeting profitable wood to ethanol production by repurposing kraft pulp mill assets. *TAPPI International Conference on Renewable Energy*. Atlanta, Georgia, USA.
24. Jameel, H., Clark, E.C., and Phillips, R.B. 2008. Technical and economic analysis of repurposing a kraft pulp and paper mill to the production of ethanol. *TAPPI International Bioenergy and Bioproducts Conference*. Seattle, Washington, USA.
25. United States Department of Agriculture Forest Service, Forest Inventory and Analysis National Program, August 12, 2011, 2011, http://www.fia.fs.fed.us/tools-data/default.asp (2011).
26. Aden, A., Ruth, M., Ibsen, K., Jechura, J., Neeves, K., Sheehan, J., Wallace, B., Montague, L., Slayton A., and Lukas, J. 2002. Lignocellulosic Biomass to Ethanol Process Design and Economics Utilizing Co-Current Dilute Acid Prehydrolysis and Enzymatic Hydrolysis for Corn Stover. NREL/TP-510-32438.
27. Bryant, C. and Novozymes. Sino-US Advanced Biofuels Forum. Beijing, China. May 27, 2010.

5

Cost Integration Methodology and the Forest Biorefinery

Milan Korbel and Paul R. Stuart

CONTENTS

5.1 Introduction

In today's global environment, manufacturing companies, especially commodity manufacturers, are being challenged to define new business strategies to sustain their competitive position in the market. In recent years, the most common strategy for the North American forest industry has been the unpopular closure of unprofitable operations. This arguably straightens up enterprise-level cost curves to some extent, but will probably not sustain the company's future. Another common and yet critical strategy over the short term is to minimize production costs, for instance through continuous improvements. This exercise is typically carried out by process engineers who set up benchmarking performance parameters for individual processes or for the whole mill. Undoubtedly, well-defined process-based benchmarking is helping many mills to improve process efficiency and hence to tighten up their monthly spending. On the other hand, cost-based performance measures, generally established by mill cost accountants, assist in evaluating the difference between current production costs and their expected values, which are the budget values. The cause of this variance is difficult if not impossible to determine from only cost information. Obviously, each mill has developed its own particular cost-accounting practices, and some may even claim that a certain level of process information is involved in their cost analyses. From practical experience, it is evident that the communication link between cost accountants and process engineers is generally missing.

The ultimate goal of both cost variance analyses and process benchmarking is to achieve better control of a company's costs and hence to maximize its product margins. The task of determining true or actual profit-margin values for all products is rather difficult, especially in a multiproduct environment where this information becomes crucial. In the forest industry, most accountants are using the traditional way of determining these values, which typically provides monthly or weekly volume-based valuations. This result, however, incorporates a wide range of operating practices and production recipes, mill interruptions, and paper machine breaks. Therefore, traditional costing is a somewhat ad-hoc procedure and should be used with caution in decision-making activities. Ideally, values corresponding to each operating situation in the mill would be accessible. To capture such information, mill personnel must integrate financial and process knowledge into one costing system, for instance using an operations-driven costing approach (ODCA) (Janssen and Laflamme-Mayer, 2011). In practice, this would enable operators to avoid expensive operating regimes or accountants to understand their product margins and cost variances from a process perspective. The pillar of this approach is the principles of activity-based costing (ABC), which are generally accompanied by a certain level of complexity when setting up the system. This is one of the reasons why forest products companies do not yet recognize similar costing systems for problem-solving as an addition to their current traditional system used for financial reporting.

> One cost system isn't enough.
>
> **– Robert S. Kaplan, 1988**

The short-term values of operations-driven (ABC-like) cost-accounting practice would help companies to understand and interpret their resource consumption, hence to minimize their current core business production costs. Moreover, the understanding of individual product margins in multiproduct environment will improve planning/scheduling tasks and simultaneously will provide an input to identify the optimal product mix ratio. This new information would create a knowledge-based manufacturing opening the possibility toward margin-centric supply chain implementation for long-term company benefits. If potential business transformation is being considered, the mill personal would benefit from the most well informed decision-making information at hand.

This chapter deals briefly with the state of the art in today's cost-accounting practice, focusing on the continuous manufacturing environment. The first part of the chapter seeks to introduce the reader briefly to cost-accounting principles. With this basic knowledge, the current state of the art in cost-accounting practices in processing industries can then be presented. Next, the integration of academic knowledge about costing with process engineering knowledge will be discussed, and the first part of the chapter finishes with a discussion of how the current forest products industry would benefit from such integration. To provide the reader with a full understanding of the problem, the last part of the chapter provides a concrete case study example comparing current and leading-edge costing strategies.

5.2 Understanding Production Costs

5.2.1 Current Cost-Accounting Practices and What We Are Missing

Cost accounting is the heart of the accounting framework that provides valuable financial insight to decision makers. The information provided is confidential and is used only internally to help managers find the optimal way to maximize the company's profits. The environment and the outcome of decision-making activities is the cost-accounting system. Various companies may use several different cost-accounting systems for problem-solving. Considering that the limits of practice are entirely within the company's control, the prepared cost reports can be based on whatever rules, standards, or rational bases are chosen. Cost-accounting information is commonly used in the second pillar of an accounting framework, financial accounting. This branch of accounting deals with public corporate information used solely for a company's financial statements, and its preparation must follow generally accepted accounting principles (GAAP).

The general elements of cost accounting can be divided into three pools: material, labor, and overhead costs. Direct material and labor costs are generally variable costs and are a function of the number of units manufactured or sold. Overhead costs, on the other hand, are fixed costs that do not change with the level of production. For instance, management salaries, rent, or depreciation expenses do not vary from month to month, even though the rate of production is never the same. The ability to track these various cost elements accurately determines the value of the accounting system to final decision-making activities. In the early twentieth century, this task was not difficult because overhead costs were negligible compared to material and labor. However, it became more complex to account correctly for indirect and overhead costs once the face of manufacturing had shifted from a labor-intensive to a machine-intensive environment.

The ultimate focus in every organization is to control costs. Often a company chooses to use only one costing system, even though there are several approaches available. The most commonly used in today's industrial practice are

- Cost–volume–profit (CVP) analysis
- Standard cost accounting
- Throughput accounting
- Lean accounting
- Resource consumption accounting (RCA)
- ABC

The first two systems, cost–volume–profit and standard costing, are often referred to as traditional or normal costing and are used extensively in the pulp and paper sector. This traditional approach was created for the needs of the early industrial era when the total costs were dominated by variable elements. The overhead and other indirect costs are accounted for based on simple volume-based measures such as labor or machine hours. Therefore, a product with a low level of labor hours is allocated less overhead cost. However, the actual costs may be very different if this product requires special attention or testing. The resulting unit production cost becomes even more distorted when overhead and other indirect costs begin to dominate overall manufacturing costs. Then it is strongly recommended that other supplementary costing systems be used.

The four principles stated below are relatively new in management accounting. Throughput accounting was developed for the enterprise-wide level, to help identify factors that limit the enterprise in achieving its established goals (Goldratt and Cox 1992). In lean accounting, the essential philosophy is to preserve value with less work. This approach was developed for the car industry which was aiming to eliminate waste while simultaneously minimizing production costs and time, using techniques such as poka-yoke (Robinson 1997) or value-stream mapping (Rother and Shook 1999). RCA is a fully integrated and complex managerial approach that uses available state-of-the-art methods. The combination of the German *Grenzplankostenrechnung* (GPK) cost management system and ABC principles create a system that can be used and interpreted by nonaccountants. An extensive discussion of each approach is beyond the scope of this chapter; the interested reader is referred to, for example, Horngren et al. (2006). The pertinent ABC principles are discussed in more detail later in this chapter.

As mentioned earlier, forest industry accounting practices are dominated by traditional costing because of its simplicity and the wide understanding of this approach among accountants. An important part of standard costing is a variance analysis. By breaking down the overall variance into the three pools listed below, this analysis helps managers identify where the difference between actual and budgeted costs has occurred

- Labor-cost variation
- Material-cost variation
- Volume variation

This information helps managers to identify the source of the overall cost variance, but not the cause of it. For instance, if the variance is largely due to material-cost variation, accountants with the help of process engineers need to drill down into historical process data to interpret the variance and take appropriate action.

The problem is that traditional costing considers all costs as variable with regard to production volume. This often creates inaccuracy in fixed costs whenever the volume of production changes. Furthermore, arbitrary rather than cause-and-effect overhead allocation makes the traditional approach highly inappropriate in a multiproduct environment. Another problem in the current general accounting profession, not only in the forestry sector, is the emphasis on financial accounting. Most of the time, decision-makers must create their own cost analysis based on financial accounting reports. However, these statements contain aggregated and distorted costs with no activity data incorporated, leading to poorly informed decisions. There are a few existing advanced systems at the academic level or already being used by advanced processing industries such as the petrochemical sector. The pillar of these approaches is the principles of ABC, which is briefly discussed in the following section.

5.2.2 Cost Allocation and Activity-Based Costing

Activity-based costing (ABC) is a relatively new philosophy that emerged in the 1980s in response to overhead allocation discrepancies (Kaplan and Atkinson 1989). By simply adding an activity as a link between resource consumption and a cost object, the knowledge of costs incurred in the organization is improved significantly. The activity becomes a fundamental cost item whose value is directly assigned to the final cost objects such as

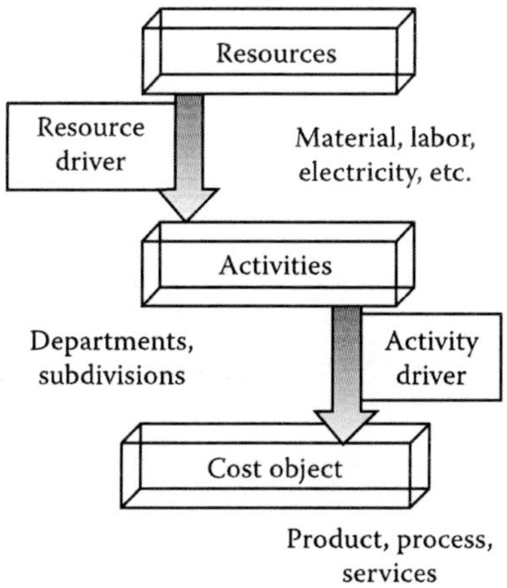

FIGURE 5.1
Resources consumed by activities and then by cost objects.

products and customers. In other words, the rate of resource spending is traced to an activity, and the activity is then traced to the product, as shown in Figure 5.1.

The ultimate advantage of using ABC is that it attempts to assign all costs to final cost objects, including marketing, engineering, and administrative costs. This added ability to trace indirect costs directly enables accountants to track overheads rationally and as closely as they track direct costs. This is done by making use of the so-called *drivers*. As shown in Figure 5.1, resources are linked to activities by resource drivers, and similarly activities are linked to cost objects by cost drivers. According to this definition, resource drivers determine the amount of a resource consumed by each activity, while activity drivers specify how different cost objects (products, customers) consume these activity costs. Labor hours, kWh, and number of shipments are examples of resource drivers, whereas number of customers and number of products are examples of the second stage, the activity driver. The difference between these drivers is that the former focuses on why things happen and the latter on what happens (Emblemsvåg and Bras, 2001). The implementation of an ABC system may be a complex and expensive task, and therefore it is important to determine the minimum number of appropriate drivers that will meet accounting objectives.

As shown in Figure 5.2, the process-oriented character of ABC means that it is implemented in two simple and logical stages, while structure-oriented traditional costing is implemented in one. This fundamental principle is the basis for increasing the accuracy of the cost data (Drucker, 1995). Traditional costing cannot encompass this critical linkage between actual causes and associated costs. Furthermore, advanced ABC has recently evolved into multistage systems where individual activities can be used by other activities before being used by final cost objects, thus enhancing even more the accuracy of cost modeling (Emblemsvåg and Bras, 2001).

In a continuous manufacturing context, the process-oriented character of an ABC system and the causal relationships between cost drivers and activities make the method highly suitable for modeling and analyzing costs. The availability of real-time cost and process data from information management systems (IMS) makes ABC easier to implement. It must be made clear that ABC is a cost-accounting system that can help managers understand

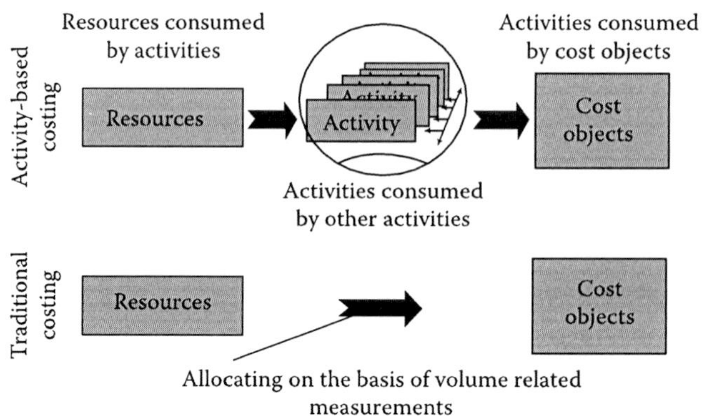

FIGURE 5.2
Activity-based costing and traditional costing. (Adapted from Korbel, M. and Stuart, P.R. 2011b. Submitted to *TAPPI Journal*.)

their actual costs and improve their profits efficiently. Traditional methods are complementary to the financial reporting prepared according to GAAP.

5.2.2.1 ABC-Like Cost Accounting

A cost-accounting system that is used by a wide spectrum of industries is the RCA, whose development has been strongly influenced by German cost accounting and ABC principles. The structure is very close to variable costing, a well-documented method discussed in cost-accounting textbooks, but rarely used by industry. RCA and its variations are extensively used by advanced processing industries such as mining, petrochemicals, and chemicals. Often their costing methods are confidential and inaccessible to the public or to researchers. In general, RCA is based on three fundamental pillars (for further details, the reader should refer to Friedl and Kupper (2005) or Keys and Merwe (1999)):

- *View of resources*: The use of a high volume of cost pools establishes a clear linkage between resource spending and a company's costs and revenues.
- *Quantity-based model*: The value of the costing system is created in this pillar by the use of operations data and models. Traditional costing uses the output of variance analysis with dollar values, creating a fixed-costs bias. By contrast, RCA exploits causal operational relationships.
- *Cost behavior*: Understanding the nature of costs is a very important aspect of the third pillar of RCA. The clear distinction between direct, indirect, variable, and fixed costs is based on aggregating pools.

There have been significant changes in recent years, although not well documented; some forestry companies are approaching now ABC-like costing for improved decision-making activities. For example, Fogelholm (2000) has discussed the difficulties of product costing in the paper making industry and its potential industrial application. This approach is now a pillar of Metso Automation's Metso DNA (Dynamic Network of Applications) that some companies are presently using for product-customer decision making as well as it helps their budgeting activities. The application seek to anticipate and determine the resource requirements for the next individual customer orders based on current raw material content, dimensions and quantities of the paper product (Fogelholm, 2004).

Some academic cost-accounting frameworks have been developed based on ABC philosophy with potential industrial applications. For example, an approach that integrates ABC principles with environmental metrics to perform analytical economic and environmental assessment for decision-making activities was developed by Emblemsvåg and Bras (2001). Their activity-based cost and environmental management (ABCEM) system is extensively discussed in their 2001 paper. The use of an uncertainty variable introduces extra complexity and versatility into the system. The ABCEM has been applied to a wide range of industries including furniture, carpets, and supply vessels, where it has provided insights and highlighted potential areas for improvement.

Finally, a sophisticated ABC-like approach that integrates process and cost information into one system, operations-driven costing (Janssen and Laflamme-Mayer, 2011), is the core of this chapter. This method will be used for cost assessment in the case study part, and its results will be compared to the outcomes from traditional costing. The basis of this approach is in making a link between costs and process operations data using principles similar to those of ABC. This approach is similar to RCA in some aspects, but is more versatile because it includes an in-depth engineering understanding of the process operation. The following section discusses this approach in more detail.

5.2.3 Operations-Driven Costing Approach

ODCA is an interdisciplinary approach developed by accountants and process engineers in the pulp and paper industry. Understanding the cost of process performance is a critical success factor for paper mills. Chemical and process engineers are concerned with developing systematic tools and methodologies for both optimal design and optimal process operation (Figure 5.3). These procedures range from nano to industrial scales. The concept can be

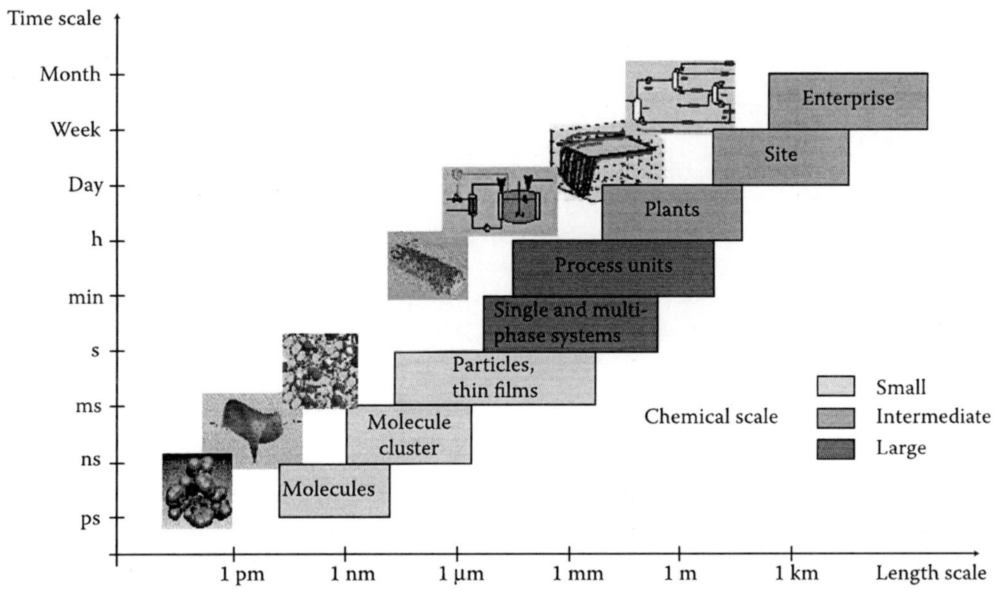

FIGURE 5.3
From microscale to macroscale complexity or the "chemical supply" chain. (Grossmann, I. E. and Westerberg, A. E. 2000. *AIChE Journal*, 46(9): 1700–1703. With permission.)

understood from the supply chain point of view, where on the one hand, product quality is determined on the nano- or microscales, and on the other hand, the desired product properties are determined by its functionality and structure.

In the pulp and paper industry, fiber microproperties influence the quality of pulp and paper products. On the macro industrial scale, a reflection of the microcomplexity of the fiber structure can be brought to light using information extracted from real-time data through IMS. The most practical way of doing this in the pulp and paper industry is to develop tools and methodologies for macro or mega scale applications that are based on real-time data and that reflect the meso and microscales according to the general chemical-engineering definition of complexity levels.

5.2.3.1 Actual Product Margins

Paper producers have multiple products satisfying numerous customers with different needs and corresponding product specifications and prices. It is therefore crucial to understand individual product margins; however, current cost practices and systems provide only approximate values that are based on time-framed (usually monthly) spending information. As discussed earlier, current practice involves the use of conventional accounting systems that aggregate costs over the manufacturing period and uses a standard recipe, for example, it is based on experience from the process operation. This overall cost information incorporates various changes in process operation due to mechanical (process) or raw material disturbances. Engineers and accountants recognize that within the manufacturing period, the generation of cost differs from one product to another as well as within the production process for the same product. However, it is not a simple task to determine these cost variances. First, current accounting practices cannot accommodate cost information from a process perspective, and second, the discrepancies in the current cost data are significant. In 2003, a survey by Ernst & Young and the Institute of Management Accountants indicate that 98% of respondents claim that cost reporting is distorted, with indirect costs and overhead allocation being the main biases reported, and almost 40% believe that the cost data they receive are significantly inaccurate.

With the current state of academic knowledge, it is possible to replace the current inaccurate margin information by the true values of product margins. However, to implement this change, actual operating knowledge must be involved. The use of IMS becomes pertinent to extracting process measurement data that provide knowledge about the underlying process. However, the lack of reliability in certain measurements as well as the lack of instrumentation on site makes this task very challenging. Many older mills have these difficulties plus a lack of process data redundancy*; however, in these cases, there are certain ways to create redundancy and proceed with ODCA (Korbel et al. 2011b).

5.2.3.2 Operations-Driven Cost Modeling Framework

Janssen and Laflamme-Mayer (2011) developed an operations-driven cost modeling framework to provide in-depth understanding of resource consumption by integrating process and cost data. The bottom-up structure (Figure 5.4) provides mutual communication between different business levels. The resulting generic framework can be used to

* Redundant measurement data points offer some level of possibility of crosschecking for potential measurement biases.

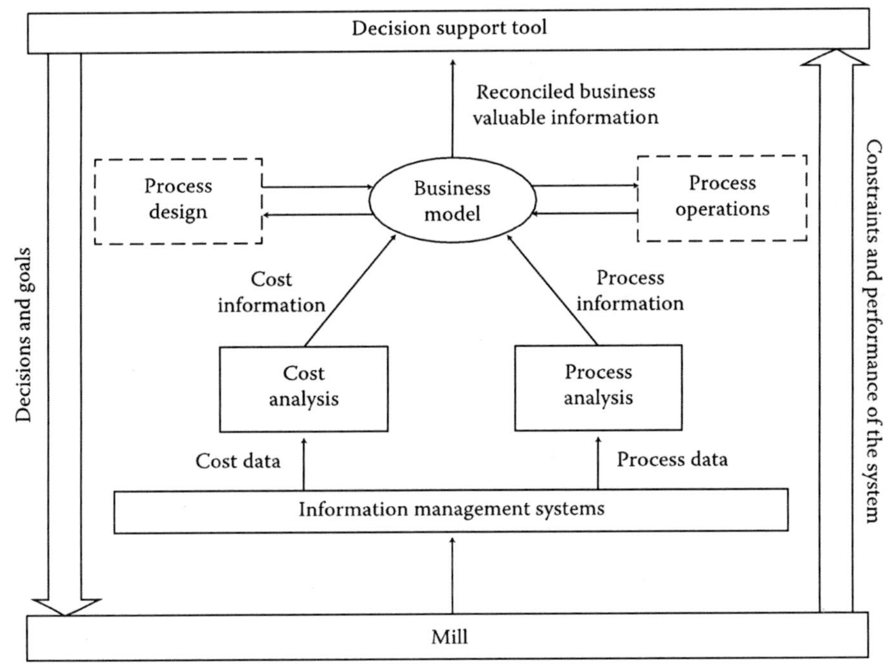

FIGURE 5.4
Overview of the bottom-up process-based approach. (Adapted from Laflamme-Mayer, M. et al., 2011. *Journal of Science & Technology for Forest Products and Processes (J-FOR)* 1(1): 32–41.)

enhance the understanding of manufacturing processes both for design and for operational decision support.

Later, Laflamme-Mayer (2011) presented in his thesis an application of operations-driven cost modeling to assess the production costs for different product campaigns. This information was then used for planning and scheduling and optimization of high-level supply-chain analysis. The understanding and differentiation of product margins for each campaign can be used to enhance the current *ad-hoc* representations of product margins. This versatile view of manufacturing costs in the paper industry is revolutionary and has tremendous value for reducing production costs.

Figure 5.5 compares standard costing with actual and operations-driven costing information. Standard costs represent how the resources should have been used to manufacture a particular grade; actual grade costs are the true resource consumptions at the end of the three-month period. The operations-driven grade-cost assessment breaks up the three-month period into segments corresponding to campaign runs. From these results, it is clear that manufacturing the same product varies significantly from one campaign to another.

5.2.3.3 Cost Assessment of Operating Regimes

With the use of advanced data processing tools and methodologies, campaign costs can be broken down further to assess the cost of different operating strategies. Every paper product is manufactured according to its production recipe. However, within this recipe, different operating strategies can be followed by operator choice or as a result of natural process-material interactions. These strategies, referred here as operating regimes, are

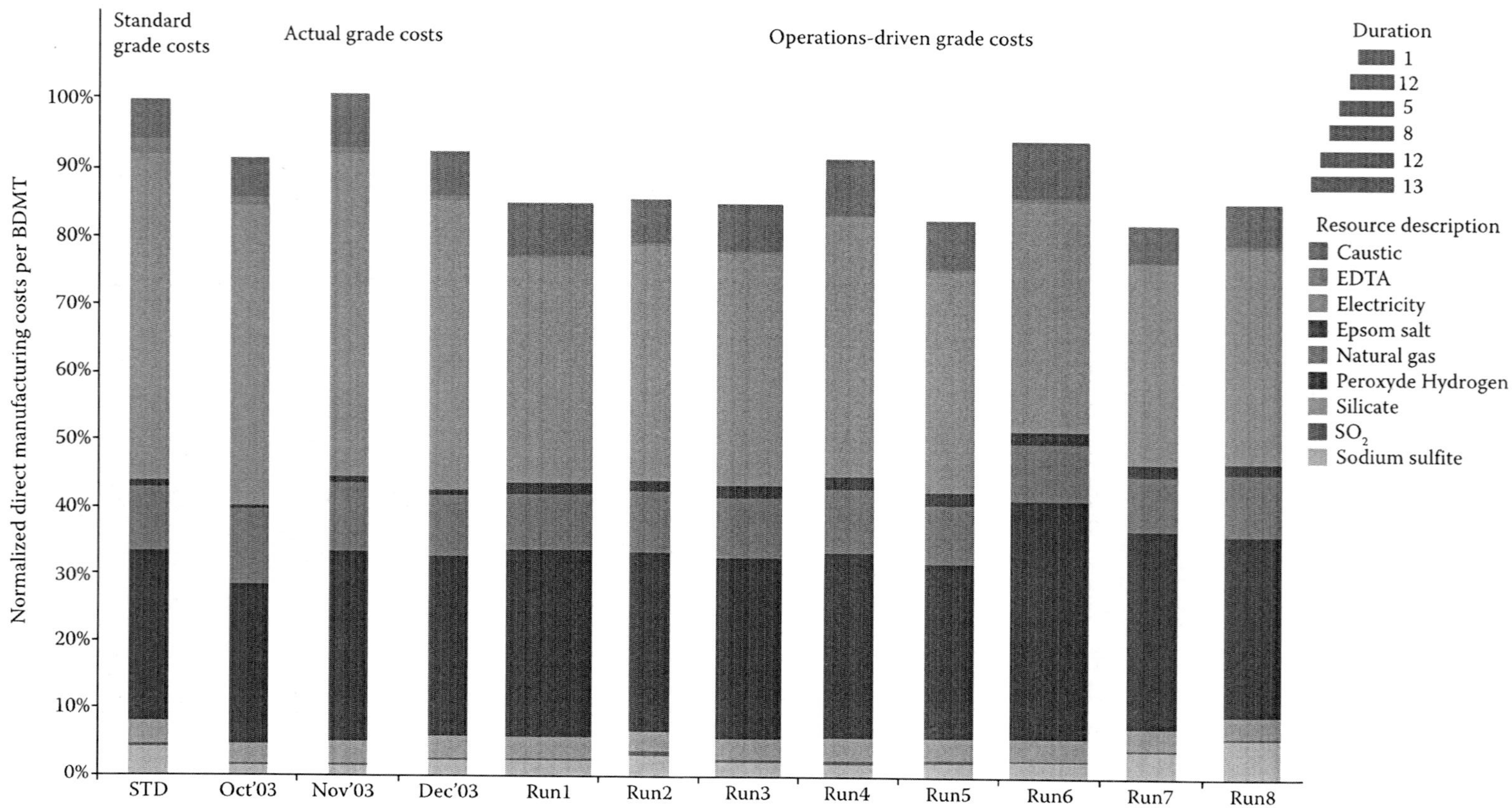

FIGURE 5.5
Comparison of standard cost, actual grade cost, and operations-driven grade costing. (Adapted from Laflamme-Mayer, M. et al., 2011. *Journal of Science & Technology for Forest Products and Processes (J-FOR)* 1(1): 32–41.)

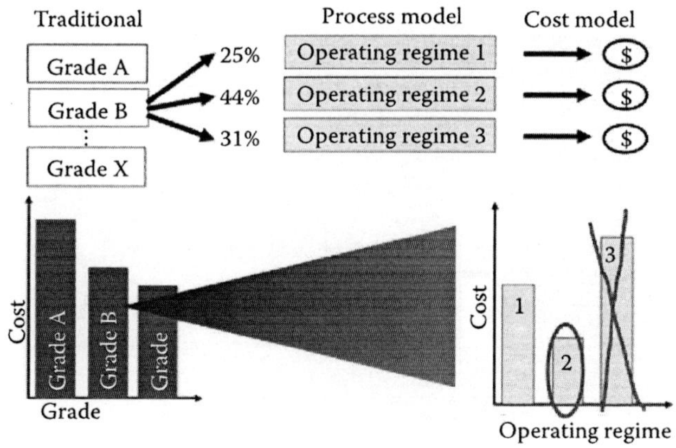

FIGURE 5.6
Smart data dissection for operations-driven cost modeling approach.

driven by the process design characteristics and operating practices. For instance, the use of different chip-refining plates, the control setpoint strategy for freeness control, and the open or closed nature of process loops and units are examples of operating regimes. Recent advances in plantwide acquisition systems, which capture real-time data from the pulp and paper operation, provide an opportunity for creative ideas to improve knowledge-based manufacturing and to support decision-making activities. In this context, it is proposed to exploit the operations-driven cost modeling framework to assess the costs of different operating regimes (Korbel, 2007).

The overall structure of this cost-modeling vision can be understood from Figure 5.6. Traditional cost-accounting procedures permit *ad-hoc* profitability analysis of different products (grades). To go further and to understand the actual costs incurred from a chemical engineering perspective, a process model should be used to assess the profitability of individual operating regimes with the probability of occurrence of each regime. At this level, the information can already be used by decision-makers to choose the most profitable operating regimes and to eliminate costly ones.

For these purposes, it is important to use a plantwide process data set that is extracted from the operation under near-steady-state conditions. Furthermore, the process operation must stay close to steady state for a satisfactory time period to guarantee reliable and accurate information. To achieve a satisfactory level of data quality, advanced cleansing techniques must be used. The mathematics of these methods is beyond the scope of this chapter; interested readers should refer to Jiang et al. (2003) and Bagajewicz (2001). Briefly, these data cleansing techniques improve the accuracy, reliability, and completeness of a given measurement data set by correcting different types of errors. First, a wavelet-based processing technique is used to identify the state of the process. When the system is at pseudo-steady-state, elimination of random white noise and abnormalities is carried out. The second step is data reconciliation, which helps to improve further the accuracy and completeness of the data and the level of compatibility of the data set with the process operation. There are many challenges in carrying out such an analysis in the pulp and paper industry, but they can be overcome using a smart and highly practical approach (Korbel, 2007).

For the reason that pulp and paper facilities operate in an item-based or order-driven manufacturing environment, the use of a regime costing system would create sustainable value for the corporation. Not only would savings in manufacturing costs be achieved, but also high-value supply-chain modeling and potential transformation of the business to a biorefinery would benefit from these valuable insights into production knowledge. What kind of information a company can access through these methods is described in a real case-study example in the next section of this chapter.

5.3 Case Study: Integrated Newsprint Mill and Its Business Transformation

5.3.1 Problem Statement

The objective of this case study is to compare the accounting practices currently in use with others awaiting industry implementation. This will make it possible to understand the value in using operations-driven cost modeling on an everyday basis to improve the cost knowledge of today's core business (short-term values) as well as to help assess the potential business transformation of the forestry sector and its margin-centric supply chain (long-term values). To clarify the differences in the two costing approaches and their consequences, a concrete example is needed. A cost analysis by conventional and operations-driven cost accounting was performed on a real single-line newsprint mill situated in Canada.

5.3.1.1 Case Study Description—Current Core Business

The current business is a simple single-line integrated thermomechanical newsprint production facility located in Canada. Figure 5.7 shows a simplified block diagram of the base-case process operation, including various manufacturing steps such as chip refining, pulp screening, and bleaching before papermaking.

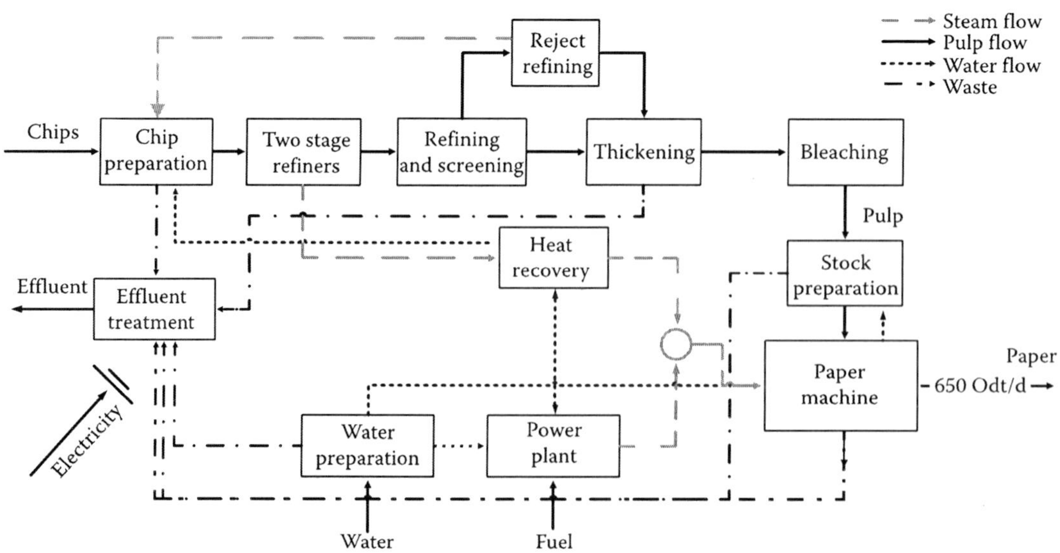

FIGURE 5.7
Block diagram of the base-case TMP pulp and paper process.

The process first heats chips by process steam in an atmospheric vessel at approximately 110°C, after which the pretreated chips are washed in hot circulating water from elsewhere in the plant. The softened warm chips are fed by a plug screw feeder into a pressurized preheater with relatively low pressure and temperature. After the preheater, the chips pass through a second plug screw feeder on their way into the first refiner, which operates at relatively high pressure and temperature. After first-stage refining, the pulp is driven to a steam separator, from which it is fed to the secondary refiner, which operates under approximately the same conditions as the first stage refiner. Then the pulp is again being separated from the steam in the second separator. After the pulp passes through the plug screw feeder at the bottom of the second separator, it falls down into a pulper for removal of latency. Screening and reject refining follow, then bleaching before storage and transport to the paper mill (Sundholm, 1999). The paper mill produces two types of products: 48 g cm^2 and 45 g cm^2 newsprint.

5.3.1.1.1 *Business Transformation: Polylactic Acid Production*

The studied company is highly competitive newsprint (in the first quartile manufacturer) and has a limited access to the biomass. Hence, they have elected to go for a biorefinery strategy that integrates into their existing processes. Thus with the relatively low amounts of hemicellulose at hand and using VPP (value prior to pulping) process, the mill could develop a robust business model by manufacturing a higher added value product, such as PLA. Therefore the potential business transformation was inspired by the combination of *VPP* and *Purac* technologies, for example, the well-established cellulosic technology from *Biopulping International* (BPI) to extract organic chains from pulp is assumed to be integrated with Purac technology to produce a specialty product, (PLA).

It is assumed that a mass corresponding to 3% of the incoming dry chips by weight is extracted before pulping, while paper production is kept unchanged with respect to the current core business. This means that the yield of paper production drops to 95% from its current 98% value. The new process design is depicted by the simplified process flow diagram in Figure 5.8. The new process flow design can be summed up* in point form as follows:

- The thermomechanical pulping and newsprint production lines are unchanged.
- Before entering the thermomechanical pulping (TMP) process, wood chips are subjected to an oxalic acid pretreatment in an impregnator, producing a solution of oligomers by the preferential extraction of hemicellulose and other wood constituents from chips.
- The extracted and separated sugars are then fed into a simultaneous saccharification and cofermentation (SSCF) unit. In this processing step, enzymatic hydrolysis of the extractives takes place together with fermentation. Acetic acid is also produced as a byproduct of the extraction process. (The efficiency of conversion is greater than 95% on carbohydrate substrate, Datta et al., 1995.)
- The flow of lactic acid is then brought to the synthesis unit where etherification and catalytic synthesis of lactic acid into cyclic ester (*Lactide*) is carried out.
- The granules of *Lactide* are then brought to the synthesis unit, where the process of catalytic and thermolytic ring-opening polymerization of lactide to polylactide takes place.

* For more details on the BPI and PURAC technologies, the reader should refer to Hunt et al., 2004 and Gruber and O'Brien, 2002, respectively.

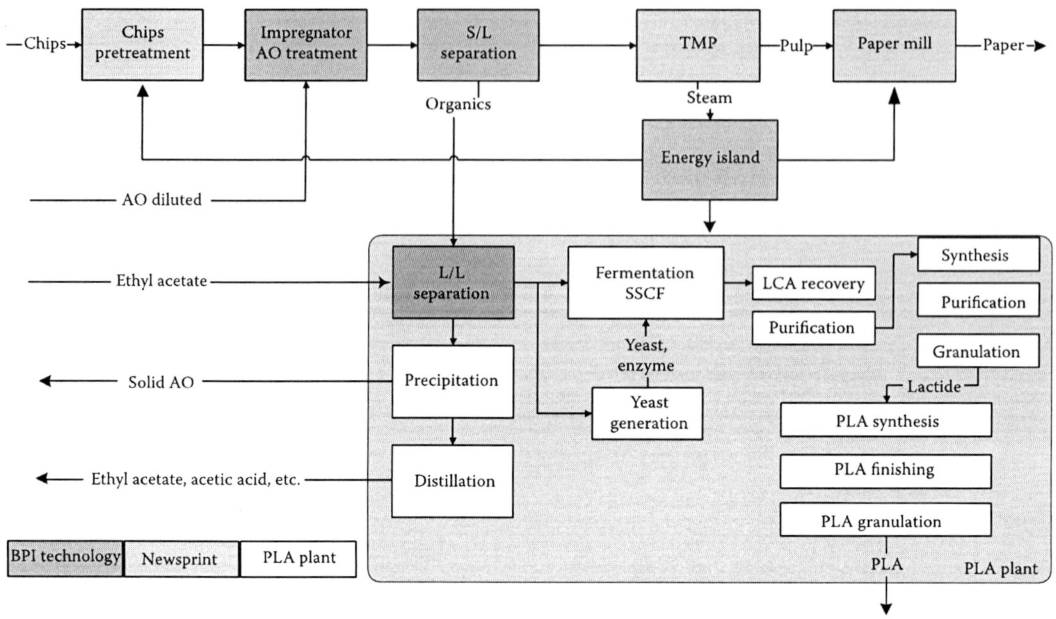

FIGURE 5.8
Block diagram for the potential business transformation of a current newsprints.

- After the finishing and granulation steps, the final product, PLA, is extracted and stored, ready for shipping to customers.
- As shown in Figure 5.8, the BPI-PURAG-TMP process model also incorporates the TMP process shown in Figure 5.7. However, with the introduction of VPP, some TMP operating parameters may require adjustment. For example, VPP chip pretreatment may result in reduced energy requirements for TMP refining (Table 5.1).

It is also recognized that different wood species yield different amount of extractives and consequently different quantities of carbon chains from the impregnating unit. Current mill furnish consists of high- and low-density chips from black or white spruce, respectively. Furthermore, a small amount of hardwood is added to the final furnish. This information must be carefully monitored since the chips ratio used can influence manufacturing costs significantly.

TABLE 5.1
List of Assumptions

Unchanged core business production volume	650 t/d
VPP	
Extractives (C5, C6, CX sugars)	865 kg/h
Energy savings assumption at refining stages (25%)	25%
PURAC	
PLA production volume	799.4 kg/h
Fermentation efficiency	95%
Purification efficiencies	98%
Assumed PLA market price	2.0 $/kg

5.3.2 Overall Methodology: From Low-Level Data to Smart Decisions

The overall methodology consists of two parallel blocks of activities aiming to assess the production cost and profit margin of the same products and to compare the outcomes. The first block represents the use of traditional accounting practices, whereas the second block represents the multidisciplinary union of process and accounting tools which make up the ODCA method. Figure 5.9 presents the methodological steps that make up each block.

5.3.2.1 The First Block: Traditional Current Practice

Traditional practices, as discussed to some extent throughout the first part of this chapter, follow a so-called top–down approach. One can simply state that the traditional cost of a product is an aggregation of liabilities and bills that the company receives with regard to the inventory state, divided by the total production within the period analyzed.

In this phase, actual costing will be used to represent the traditional way of assessing individual product margins. The following methodological steps were followed:

1. Aggregate all costs incurred during the manufacturing period to be analyzed (one month).
2. Gather information about the tonnage of production.
3. Allocate all costs to the level of production.

5.3.2.1.1 The Second Block: Operations-Driven Cost Modeling

Operations-driven cost modeling fully exploits low-level process data in a bottom-up fashion. Consistent plantwide process data sets are integrated with business data in a model whose structure was inspired by an ABC philosophy. The construction procedure

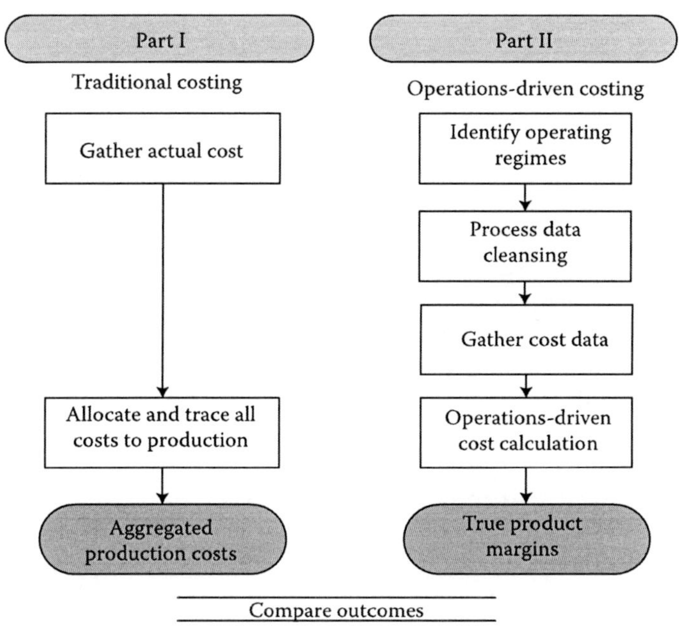

FIGURE 5.9
Overall methodology comparing traditional and operations-driven costing information.

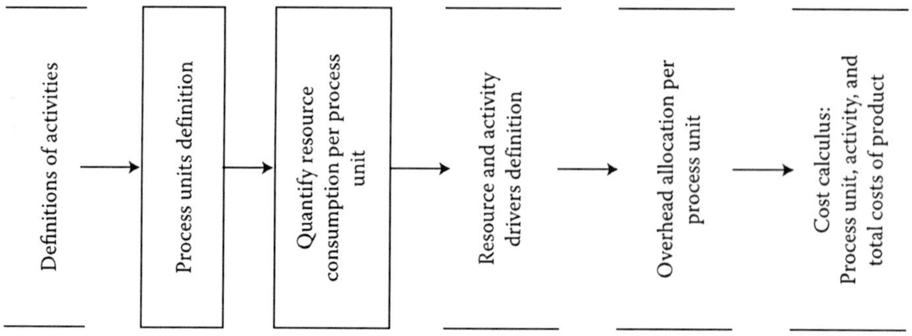

FIGURE 5.10
The ODCA modeling strategy.

for this model involves the identification of multiple parameters in the steps shown in Figure 5.10.

The definition of activities involves the division of an operation into a final number of operating activities (cost centers), each of which is an aggregation of various processing units. Each of the cost centers is characterized by two key parameters: the design layout characteristics and the operating-state knowledge. The flow and combination of information is presented in Figure 5.11. The resource, activity, and cost drivers are defined by operating knowledge corresponding to resource consumption rate, activity performance, and individual cost-center contributions respectively. The integration of cost and process data is done within the cost center and is simply defined as a step-by-step cost-calculation recipe. Similarly, overhead costs are treated in a separate cost center which accurately traces them to different operating activities and to the final cost object. Finally, the sum of the cost activities represents total millwide operating costs.

The following section describes the two main phases and their corresponding steps in the operations-driven activity costing block (Figure 5.12).

5.3.2.1.2 *Plant-Wide Consistent Manufacturing Information*

The second-block analysis starts by defining the set of operating regimes that are used to manufacture a given product. The criteria for distinguishing a regime are given by changes in process design characteristics (type/age of refining disk, opening/closing valves) as well as by variations in operating characteristics (different control setpoints or strategy, production rate). Within each defined operating regime, pseudo-steady-state data sets are identified for subsequent plantwide data reconciliation. This phase of the methodology provides profound understanding of the underlying manufacturing processes in the form of reconciled near-steady-state data sets (Korbel et al. 2012).

5.3.2.1.3 *Operations-Driven Production Costs*

The definition of relevant cost information for a given analysis is an important step to facilitate cost calculation and improve cost transparency.* The process data sets from previous phase are used to calculate the product cost simply as the sum of all production activities (cost centers) across the plant. Precision and validity of the production costs are

* All cost items identified as irrelevant at this step were excluded from the first-block activities as well.

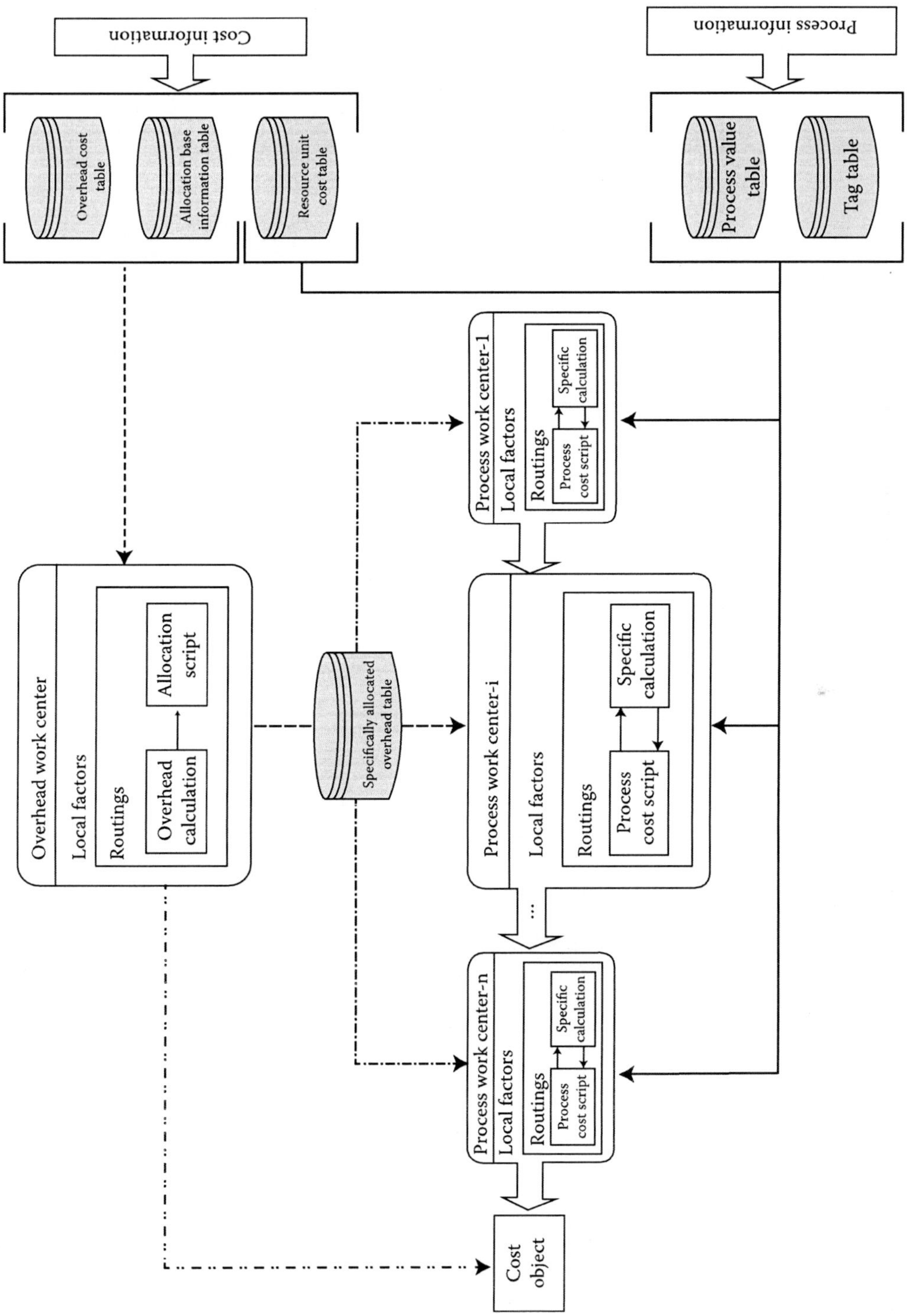

FIGURE 5.11
Work center structure within the operations-based cost model. (Adapted from Laflamme-Mayer, M. et al., 2011. *Journal of Science & Technology for Forest Products and Processes (J-FOR)* 1(1): 32–41.)

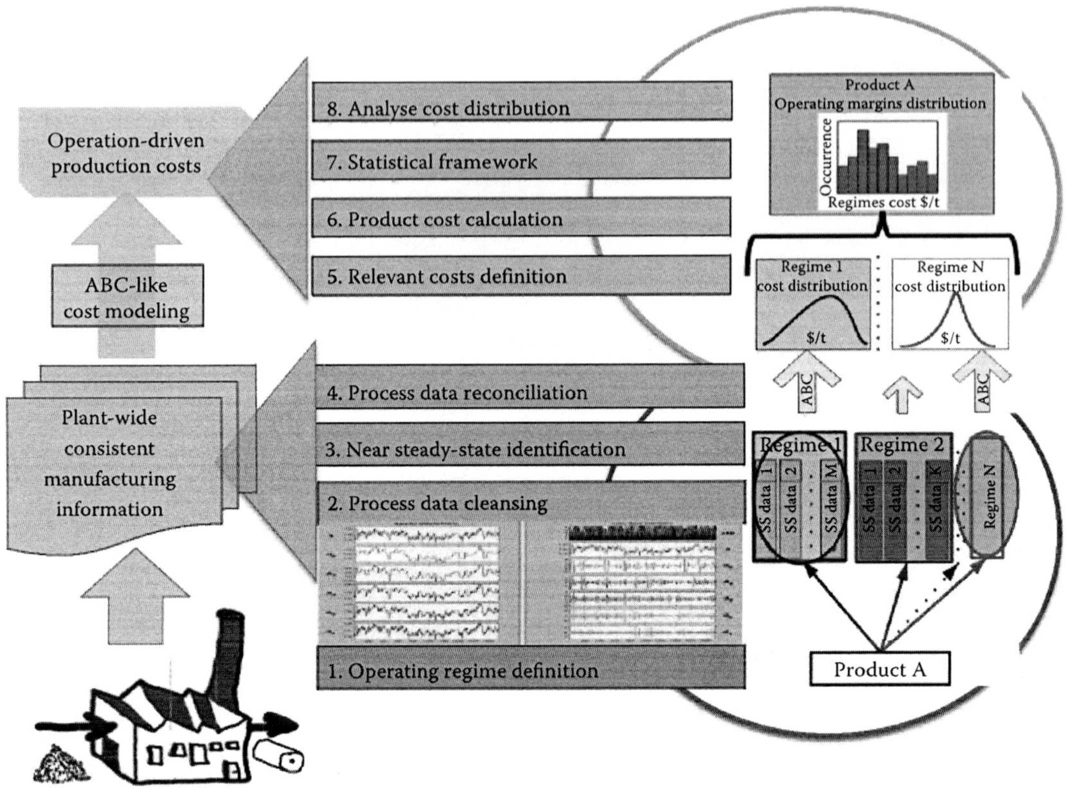

FIGURE 5.12
Step-by-step activities within the methodology to address individual product margins. (Adapted from Korbel, M. and Stuart, P.R. 2011b. Submitted to *TAPPI Journal*.)

ensured by running the same cost model with different process data sets.* The output of the analysis is a product cost distribution for a given operating regime.

5.3.3 Traditional Costing and ODCM: Economic Results and Discussion

The economic results of the case study are presented and discussed in two phases. In the first phase, the familiar cost picture based on traditional practices is compared to the operations-driven results for a current papermaking process. The second phase goes through a similar type of analysis, but with respect to a more complex production environment—the business transformation. The comparison of results, interpretation of the data, and potential improvement strategies are discussed at the end of each phase. With this in mind, let us proceed to the first phase.

5.3.3.1 Analysis of the Current Core Business

The first bar in Figure 5.13 presents the monthly aggregated resource spending as calculated using traditional costing. This enables cost accountants to understand and differentiate the resource contributors to the total product cost. A predetermined (based on standard costing) volume-based parameter is often used to multiply this overall production costs to

* These data sets were extracted during production of the same product within one operating regime.

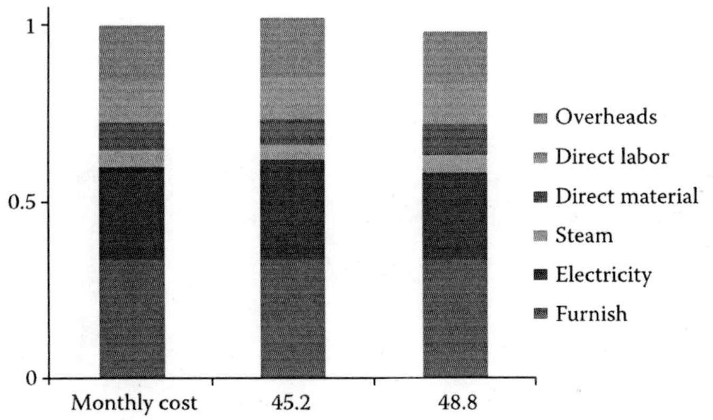

FIGURE 5.13
Traditional cost analysis results for a current core business.

differentiate individual product profit margins (second and third bars). This operation ultimately distorts the results. Another approach may be to use standard costing, although the cost variance must be properly prorated to inventories or cost of goods sold to meet GAAP requirements. This *ad-hoc* approach to distinguishing actual costs for each product in a continuous-process industry is inaccurate and can lead to wrong decisions.

The same accounting period was analyzed using operations-driven cost modeling, as shown in Figure 5.14. For the sake of simplicity, only three (the most common) operating regimes are presented per product. However, the number of different regimes that the mill operates in within a single month is significantly larger.* The outcomes of the cost model

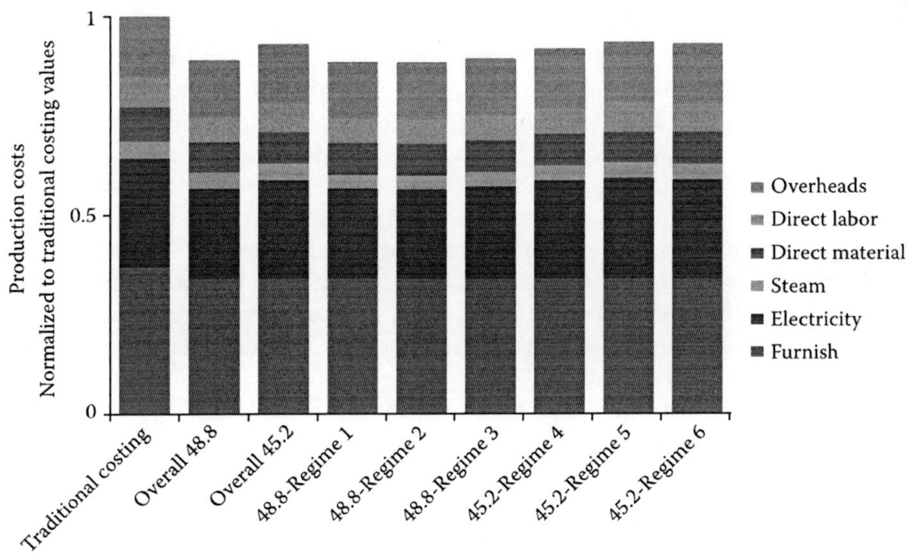

FIGURE 5.14
Operations-driven cost analysis of production activities (base-case scenario).

* For this case study, 12 and 17 operating regimes were identified for products of 48.8 and 42.5 g cm^2, respectively. However, the production time in most of these regimes was less than 1% of the total production time.

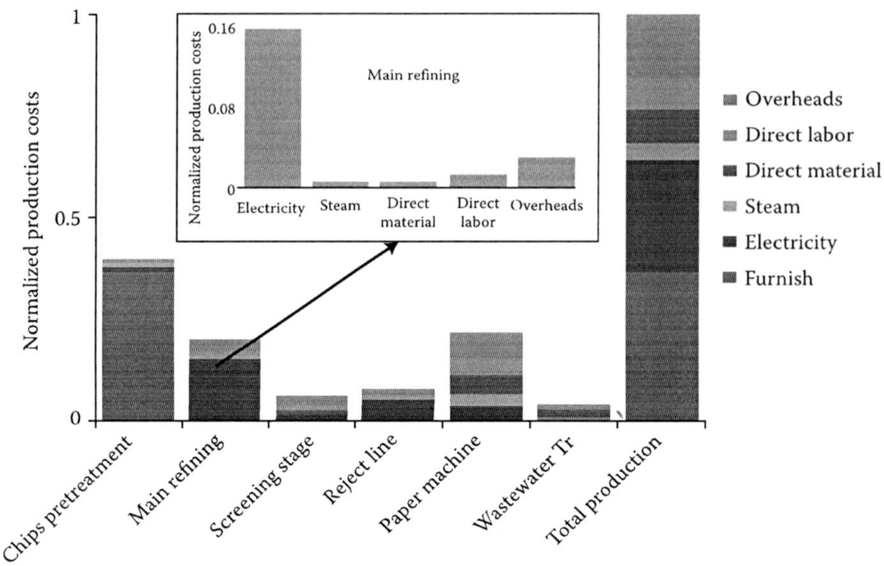

FIGURE 5.15
Operations-driven cost analysis of a current core business producing a 48.8 g cm^{-2} grade in operating regime R3.

are normalized to the values acquired by traditional costing (the first bar in Figure 5.14). The cost of each product (second and third bars) is calculated as the weighted average* of a corresponding set of operating regimes (bars 4–9). The traditional costs represent a monthly aggregation of costs, while the operating-regime costs represent the actual running costs.

It is clear from Figures 5.14 and 5.15 that the knowledge acquired from operations-driven cost analysis can have a tremendous impact on mill improvement strategies. The information about each product's actual margin within each operating regime as well as over the complete set of operating regimes will guarantee that managers can make well-informed decisions. For interpretation purposes, one can simply drill down from product costs through regime costs to the actual cost of manufacturing activities and their corresponding resource consumptions by individual processing units (Figure 5.15). This functionality helps understand and interpret the cost variances that arise from changes in resources consumption rates. The variance analysis that a cost accountant usually performs can now be well interpreted without the need to dig into data storage systems and search for answers in piles of data.

As has been briefly outlined in the context of the chapter, the apparent process improvement which can be gained from these results is to avoid operating in regimes R3 and R5 when producing grades 45.2 and 48.8 g cm^{-2} respectively. This possibility could be explored further by analyzing what would be the potential impact of this improvement on the whole business cash flow (Figure 5.16). However, it must be recognized that some operating regimes possibly cannot be bypassed due to inherent process-material characteristic interactions. Figure 5.16 presents a tree of possible future scenarios and their

* The weighting factor used is the ratio between the time that the mill operates in a given regime and the total operating hours.

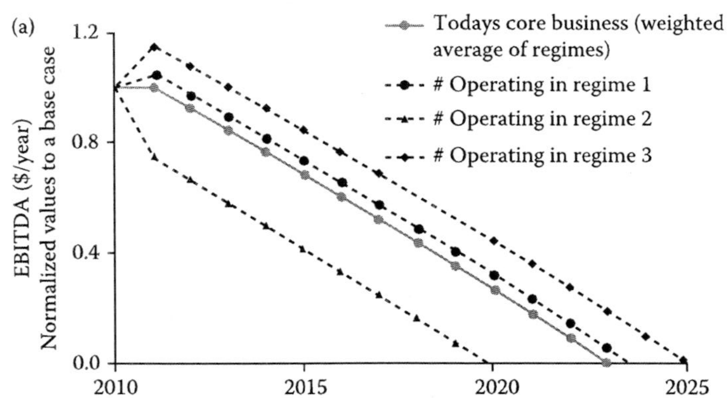

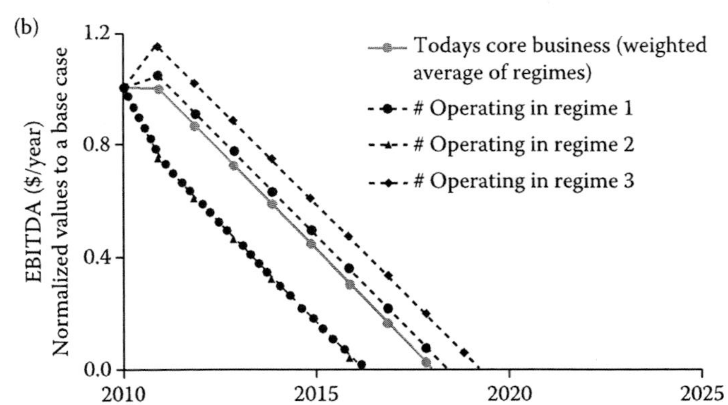

(c)

Scenario	Product margin increase (%)	Total cash increase (%)	Retirement age (years)
Base case	0.0	0.0	7.1
Scenario 1	6.7	15.2	8.9
Scenario 2	0.0	32.7	13.2

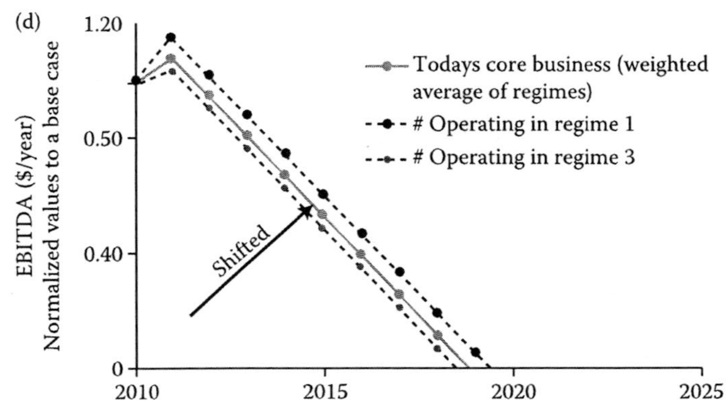

FIGURE 5.16
EBITDA forecast and the potential impact of regime costing on product margin and total cash. (a) Scenario 2: The real paper price is held constant (2001–2020), (b) Base case: The real paper price is taken from US newsprint projection by RPA (2001–2020), (c) Matrix: Scenario comparison based on three parameters, (d) Scenario 1 continuous improvements: The most costly operating regime 2 is avoided. (Adapted from Korbel, M. and Stuart, P.R. 2011b. Submitted to *TAPPI Journal*.)

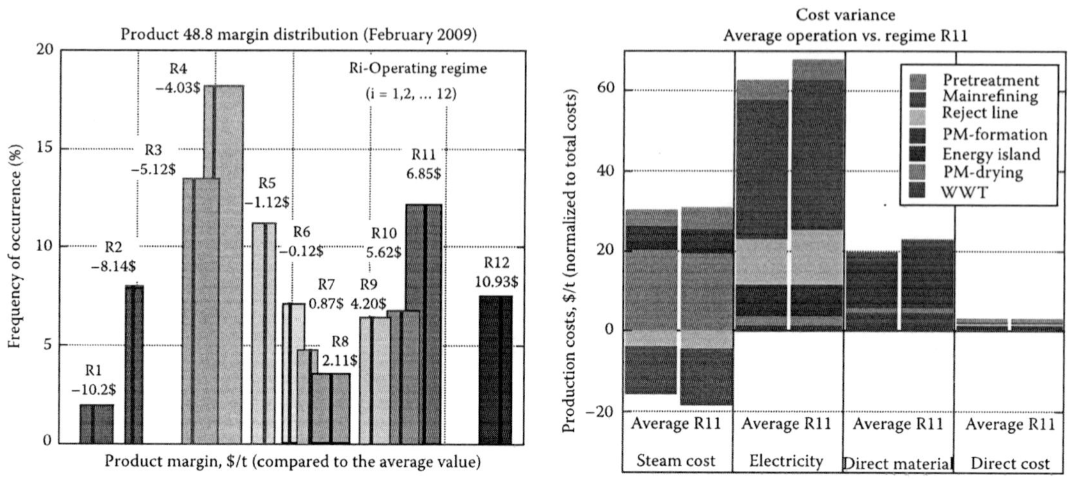

FIGURE 5.17
Wide range of product profit margins for a product 48.8 g cm^{-2} within a month of operation. (Adapted from Korbel, M. and Stuart, P.R. 2011a. Submitted to *International Journal of Production Economics*.)

interpretations in the matrix. Each of these scenarios has been analyzed under the following assumptions:

- Constant amount of paper products sold to customers,
- Yearly increase in production efficiency due to operational improvements,
- Yearly labor and raw material cost increases,
- Selling price for scenario 1 and the base case is taken from predictions by RPA (2001–2020); the selling price is held constant for scenario 2.

The results show that, solely by avoiding mill operation in regime R3 (the process from base case to scenario 1), an increase of 15.2% in total cash within the analyzed business period may be achieved. Furthermore, the product margin of the 48.8 g cm⊠2 grade is increased by nearly 7%, enabling newsprint production to continue more than 22 months longer than in the base case (under the assumptions listed above).

To provide a complete understanding of the complexity of cost modeling, Figure 5.16 shows the manufacturing information covering the whole set of operating regimes for the 48.8 g cm^{-2} product during the period analyzed. Each regime is labeled by its corresponding total production cost and its probability of occurrence. The width of the bar represents the cost range of the regime due to the use of multiple steady-state data sets for regime costing. The thick line inside each bar represents the weighted average of plantwide steady-state cost snapshots (Figure 5.17).

5.3.3.2 Analysis of a Potential Business Transformation: A Multiproduct Environment

The second phase of the case study involves assessing the individual profit margins and the potential changes in their values after retrofit design. The level of newsprint sales is assumed to stay constant. However, after the new business integration, the profit margin is significantly modified for each grade, as will be seen from the results. It is worth reminding the reader that by introducing the PLA product into the business model, the simple

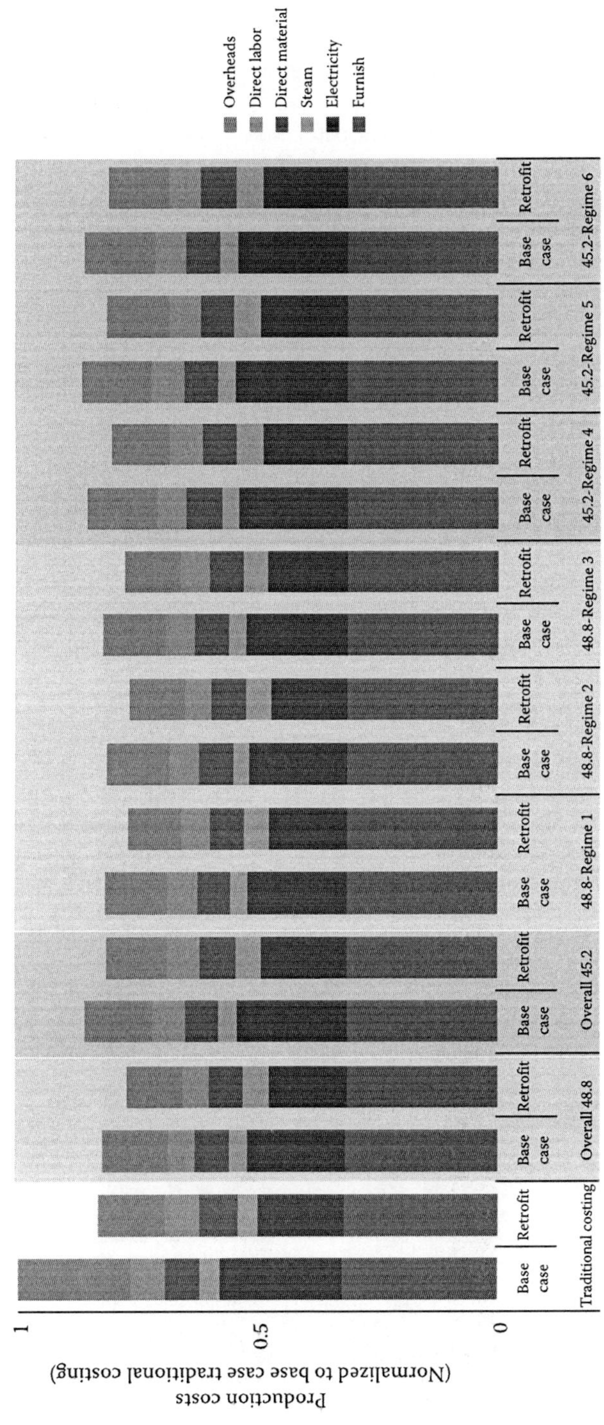

FIGURE 5.18
Operations-driven cost analysis of production activities: cost impact on a core business after business transformation.

production environment has shifted toward a more complex, simultaneous multiproduct environment. Traditional costing fails to assess actual product margins in a multiproduct industry.

An accountant using traditional costing tools would predict the cost of each product by using standard costing similar to the analysis of a current core business (Figure 5.13). This information has always been sufficient for providing investors with various financial criteria, such as return on capital employed (ROCE) or internal rate of return (IRR). With these values, more complex economic analyses can be performed to predict the potential future business cash flow. However, the internal cause-and-effect relationships due to technology integration cannot be understood with traditional thinking about aggregated production costs.

An accountant using the operations-driven cost modeling (ODCM) toolbox would try to understand the cost and process impacts of the new business integration on existing core business products. With the use of current manufacturing knowledge from a process-cost perspective and a simulation of the new process operation, this impact can be analyzed efficiently. The results shown in Figure 5.18 are normalized to the traditional cost values for easy comparison. The first two bars represent the impact of business transformation by means of a traditional cost comparison of core newsprint products. It is clear that the main impact on production arises from electricity savings and overhead sharing with the new facility. The ODCA results offer a robust and complex cost analysis of future production with an understanding of cost-process impacts on core production during different operating regimes. The interpretation of the base-case—retrofit variance can be understood by drilling down into the actual cost items and resource consumptions within operating activities. For instance, the production costs in operating regime R2 have been reduced due to significant electricity savings; however, steam costs were increased due to the reduced steam recuperation rate.

Figure 5.19 illustrates the actual production costs of each product, including PLA, for different operating regimes. The cost of PLA can vary significantly from one operating regime to another. The energy (steam costs) and material usage are the items that to some extent cause the variance in production costs. After closer analysis, it became apparent that this variance is largely due to the more expensive unit steam price when manufacturing

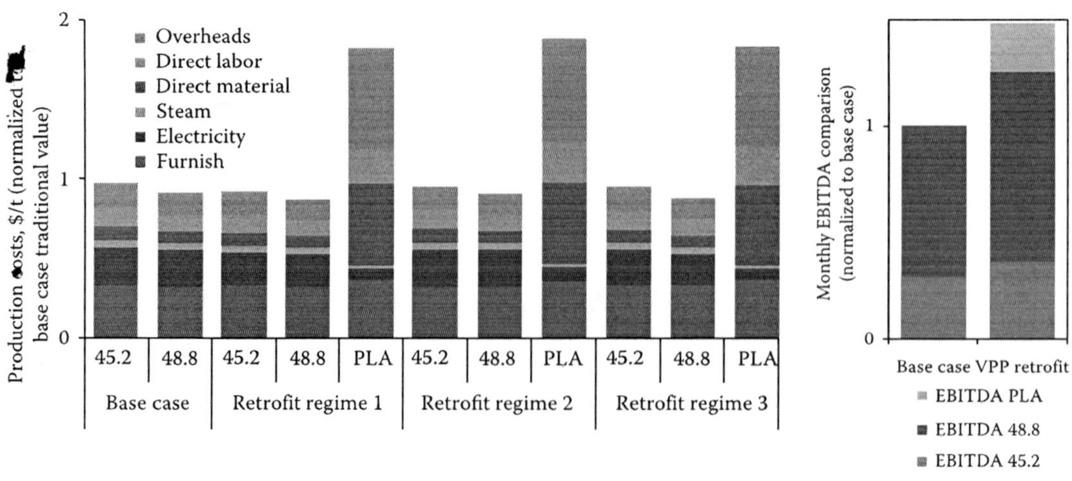

FIGURE 5.19
ODCM: The production cost of every product within a tree of operating regimes after business transformation and EBITDA comparison to a base case.

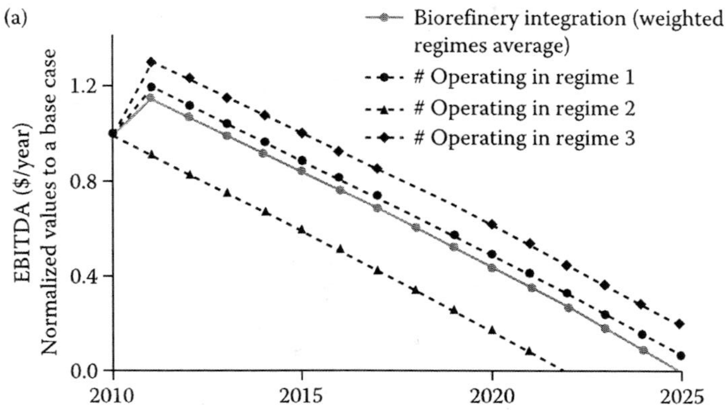

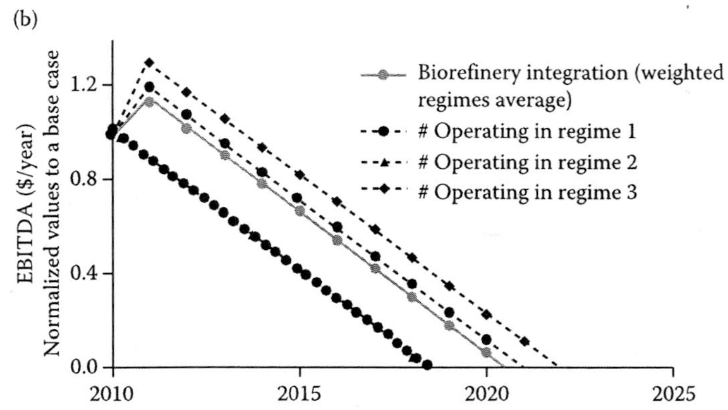

(c)

Scenario	Product margin increase (%)	Total cash increase (%)	Retirement age (years)
Base case	0	0	7.2
Scenario 1	25/31	165	15.3
Scenario 2	25/31	89	10.7
Scenario 3	33/40	106	12.3

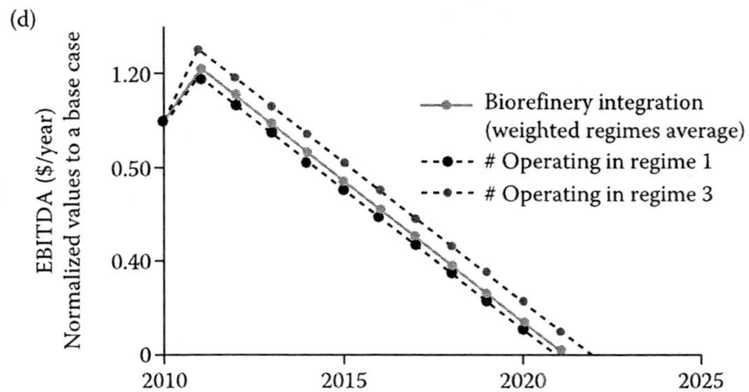

FIGURE 5.20
EBITDA forecast and the potential impact of regime costing on product margin, total cash, and retirement age of a business. (a) Scenario 1: The real paper price is held constant (2001–2020), (b) Scenario 2: The real paper price is taken from US newsprint projection by RPA (2001–2020), (c) Matrix: Scenario comparison based on three parameters, (d) Scenario 3 continuous improvements: The most costly operating regime 2 is avoided.

the 48.8 g cm^{-2} grade in operating regime R2 because of refiner plate-gap differences. Ultimately, this production state should be avoided in the future. Similarly to the first part of the case study, the EBITDA forecast can be used to assess different future scenarios and their impact on PLA production cost and total cash flow.

The assumptions for predicting EBITDA over time are identical to those used in Section 4.3.1, except that the price of newsprint is held constant in scenario 1 and is set equal to that predicted by the RPA (Figure 5.20) in scenarios 2 and 3. The assumption of constant newsprint price after biorefinery implementation shows a business cash flow increase of more than 165% compared to that of the current business. This will make the new business model break-even in more than 15.3 years from the present (8 years more than the base case).

5.3.4 Concluding Remarks

The work presented in this chapter seek to provide information to the reader about advances in cost-accounting systems and the potential benefits of using new approaches instead of the current old-fashioned systems for decision making activities. Today, the forestry sector faces difficult times and requires a systematic approach to finding an optimal path toward a more sustainable future through potential business transformation. To manage this transformation optimally, managers and decision makers need to explore the powerful and robust cost-accounting systems that are today waiting to be implemented in practice. Other industries, such as petrochemical companies or automobile manufacturers, are far more advanced in their costing systems. The use of the ABC philosophy and its variations could help managers improve forestry company profits throughout the world. One particular approach that has been developed and designed to help the forestry sector is operations-driven cost modeling. The use of lower-level process data improves the understanding of manufacturing-cost variability due to different operating regimes for current and future products. The idea has been presented throughout the chapter that forestry companies should implement ODCA systems today to enhance their current cost-savings strategies as well as to identify and sustain the best operating scenarios in the future multiple-product environment.

References

Bagajewicz, M.J. 2001. *Process Plant Instrumentation and Upgrade*. Technomic: Lancaster, PA.

Datta, G., Dong, A., Witt, J., and Tu, A.T. 1995. Biochemical characterization of basilase, a fibrinolytic protease from Crotalus basiliscus basiliscus. *Archives of Biochemistry and Biophysics* 317:365–373.

Drucker, P. 1995. Managing for business effectiveness. *Harvard Business Review* (May-June) 71: 54–55.

Emblemsvåg, J. and Bras, B. 2001. *Activity-Based Cost and Environmental Management: A Different Approach to the ISO 14000 Compliance*. Boston: Kluwer Academic Publishers, pp. 317. ISBN: 0-7923-7247-6.

Friedl, G. and Kupper, H.-U. 2005. Relevance added: ABC with German cost accounting. *Strategic Finance* June: 56–61.

Fogelholm, J. 2000. "*Cost Function Modelling in the Paper Industries.*" Report No.11. Helsinki University of Technology, Department of Industrial Engineering and Management, Helsinki. ISBN 951-22-5009-8.

Fogelholm, J. 2004. Reporting and simulation of economic steering information. *27th Annual Congress of the European Accounting Association*, Prague, Czech Republic.

Gruber, P. and O'Brien, M. 2002. Polylactides "Natureworks® PLA." In: Doi, Y., Steinbüchel, A. (eds) *Biopolymers—Polyesters III Applications and Commercial Products.* Vol. 4. Weinheim: Wiley-VCH, p. 235–249. ISBN 3-527-30225-5.

Grossmann, I.E. 2005. Challenges in the new millennium: Product discovery and design, enterprise and supply chain optimization, global life cycle assessment. *Computers and Chemical Engineering* 29: 29–39.

Grossmann, I.E. and Westerberg, A.E. 2000. Research challenges in process systems engineering. *AIChE Journal* 46(9): 1700–1703.

Goldratt, E.M. and Cox, J. 1992. The Goal: A Process of Ongoing Improvement (Second Revised Edition). Great Barrington, MA: North River Press, Inc.

Horngren, C.T., Foster, G., Datar, S.M., Teall, H.D., and Gowing, M. 2006. *Cost Accounting: A Managerial Emphasis*, 4th Canadian edition, Toronto, ON.

Hunt, C., Kenealy, W., Horn, E., and Houtman, C. 2004. A biopulping mechanism: Creation of acid groups on fiber. *Holzforschung* 58: 434–439.

Jiang, T., Chen, B., He, X., and Stuart, P.R. 2003. Application of a steady-state detection method based on wavelet transform. *Computers and Chemical Engineering* 27: 569–578.

Kaplan, R.S. and Atkinson, A.A. 1989. *Advanced Management Accounting.* Englewood Cliffs, NJ: Prentice-Hall.

Keys, D.E. and Merwe, A. 1999. German vs. United States cost management: What insights does German cost management have for U.S. companies? *Management Accounting Quarterly* 1(Fall): 1–8. ISSN 1092–8057.

Korbel, M., Garrigues, L., and Stuart, P.R. 2007. Process and Business Data Reconciliation in the Pulp & Paper Industry, *Proceedings—TAPPI Conference, Innovations in Engineering, Pulping and Environmental, Comm.* 46–3, Jacksonville, FL.

Korbel, M., Wasik, L., and Stuart, P.R. 2012. Practical methodology for plant-wide process data rectification in the pulp and paper industry. Submitted to *J-4 journal.*

Korbel, M. and Stuart, P.R. 2011a. Assessment and interpretation of advanced cost data for process improvement in an integrated newsprint mill. Submitted to *International Journal of Production Economics.*

Korbel, M. and Stuart, P.R. 2011b. Online methodology for operations-driven cost assessment of operating regimes using real-time process data. Submitted to *TAPPI Journal.*

Laflamme-Mayer, M., Janssen, M., and Stuart, P. 2011. Development of an operations-driven cost model for continuous processes—Part I: Framework for design and operations decision making. *Journal of Science & Technology for Forest Products and Processes (J-FOR)* 1(1): 32–41.

Robinson, H. 1997. Using poka-yoke techniques for early defect detection. http://facultyweb.berry.edu/jgrout/pokasoft.html. Retrieved May 4, 2009.

Rother, M. and Shook, J. 1999. *Learning to See: Value-Stream Mapping to Create Value and Eliminate Muda.* Lean Enterprise Institute, 1999. ISBN 0966784308.

Sundholm, Jan. 1999. *Mechanical Pulping, Tappi.* ISBN 952-5216-05-5 (book 5). Printed by Gummerus Printing, Jyväskylä, Finland: Fapet Oy.

6

Technoeconomic Analysis of a Lignocellulose-to-Hydrocarbons Process Using a Carboxylate Platform

Viet Pham, Mark Holtzapple, and Mahmoud M. El-Halwagi

CONTENTS

6.1 Introduction

Technologies for renewable fuels from biomass are being developed to reduce dependence on imported petroleum, decrease greenhouse-gas emissions, and increase national security. Many biochemical and thermochemical pathways have been proven technically; however, high production costs have prevented many pathways from being economically viable without government subsidies. Figure 6.1 summarizes typical pathways to produce hydrocarbon fuels from biomass via alcohols.

- The sugar-to-alcohol pathway represents industrial production of bioethanol from sugarcane in Brazil.
- The starch-to-alcohol pathway uses amylase enzymes to produce intermediate sugars, which is exemplified by the corn-to-ethanol industry in the United States.
- The lignocellulose-to-alcohol pathway uses cellulase enzymes to produce intermediate sugars, which is in the process of being commercialized.
- The thermochemical pathway gasifies biomass into syngas ($CO + H_2$), which is then catalytically converted into methanol or ethanol.
- The lignocellulose-to-alcohol pathways use mixed cultures of acid-forming microorganisms to produce intermediate carboxylates (ammonium carboxylates in Path A, calcium carboxylates in Paths B and C), which are described as carboxylate platforms. The carboxylate platform using pure-culture fermentation is described by Agler et al. (2010).

The first three biochemical platforms require aseptic fermentation, which is expensive. Furthermore, because of its recalcitrance to enzymatic hydrolysis, lignocellulose requires high cellulase loadings. The carboxylate pathways produce fuels and chemicals from biomass without encountering those problems.

In terms of intermediate chemicals, the pathways shown in Figure 6.1 use three platforms: sugar, carboxylate, and syngas. Holtzapple and Granda (2009) showed that the

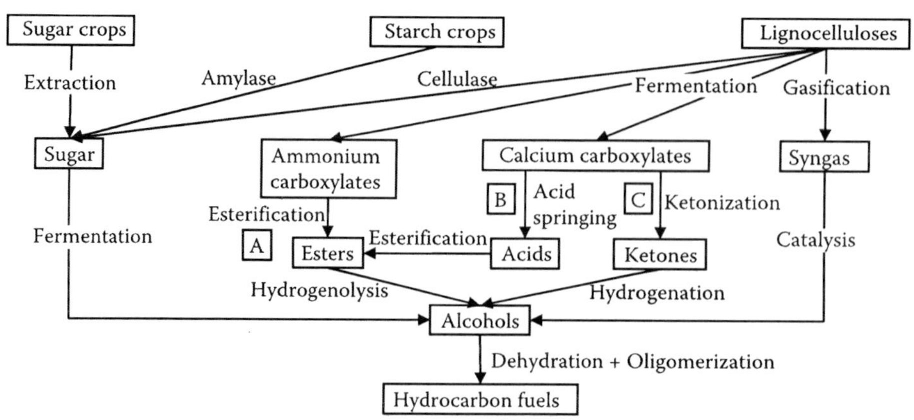

FIGURE 6.1
Pathways for converting biomass into hydrocarbon fuels. (With kind permission from Springer Science+Business Media: *J. Ind. Microbio.l Biotechnol.*, Techno-economic analysis of biomass to fuel conversion via the MixAlco process, 37(11), 2010, p. 1158, Pham, V. et al.)

carboxylate and sugar platforms give the highest theoretical yield of lignocellulosic biomass to hydrocarbon fuels when the three platforms were compared using the same ideal biomass feedstock. The biomass composition was assumed to consist of 31.7% lignin and 68.3% polysaccharides on an ash-free basis, which is similar to compositions found in hardwood biomass. The theoretical energy efficiency of the carboxylate platform is equal to that of the sugar platform and higher than that of the syngas platform.

When producing hydrocarbons from alcohols, dehydration causes a mass loss of 30% (isopropanol), 39% (ethanol), and 56% (methanol). The theoretical loss of energy is 2% (isopropanol), 5% (ethanol), and 10% (methanol). Based on these examples, the pattern is clear: higher alcohols retain more mass and energy when converted into hydrocarbons. The carboxylate platforms can produce higher alcohols (propanols and higher) than the other platforms.

A key feature of the carboxylate platform is fermentation, which uses a mixed culture of acid-forming microorganisms to convert biomass components (carbohydrates, proteins, fats) into carboxylate salts. The process does not require aseptic conditions, which reduces capital costs and improves operability. The microorganisms produce their own enzymes—a type of consolidated bioprocessing—which reduces operating costs compared to traditional enzymatic pathways. Depending on the choice of buffer, the salts are ammonium carboxylates (buffered by NH_4HCO_3) or calcium carboxylate (by $CaCO_3$). Through Pathway A (Figure 6.1), ammonium carboxylates are processed by esterification and hydrogenolysis, which produces a mixture of primary alcohols. Because there is almost no carbon loss from biomass to final products, this pathway has a high theoretical overall alcohol yield. Through Pathway B (acid springing), the calcium carboxylates are converted to the corresponding carboxylic acids. From the acids, mixed alcohols are produced by esterification and hydrogenolysis. Through Pathway C, calcium carboxylates are thermally converted into ketones, which are subsequently hydrogenated into a mixture of secondary alcohols. In the last route, the overall alcohol yield is lower than that of the previous two routes, but the process does not demand as much hydrogen.

Biomass-to-fuel pathways using a carboxylate platform have been researched and developed for 20 years in Dr. Mark Holtzapple's research group at the Department of Chemical Engineering, Texas A&M University. The technologies have been further developed and licensed by Terrabon, Inc., under the trademarked name MixAlco™.

Previous work on MixAlco™ process economics includes a study by Holtzapple et al. (1999), who estimated the economics of the calcium carboxylate platform (Pathway C) using municipal solid waste or sugarcane bagasse as feedstocks. For the dewatering process, they used water extraction with amines and multiple-effect evaporation. Lau et al. (2006) evaluated the production of ethanol from sweet sorghum through the acid springing platform (Pathway B) using various plant location, capacity, and incentives scenarios. Granda et al. (2009) analyzed the process economics of the ammonium carboxylate pathway (Pathway A) using municipal solid waste as feedstock for different scenarios of hydrogen sources and prices.

This chapter reports on a technoeconomic analysis of various lignocellulose-to-hydrocarbons pathways using the calcium carboxylate platform (Pathway C) with vapor-compression dewatering, which is a version of the MixAlco™ process. Pham et al. (2010) performed a similar technoeconomic analysis at a different plant capacity. Compared to previous efforts, this work emphasizes process synthesis, integration, and analysis with simulation in Advanced Simulator for Process Engineering (ASPEN) Plus (ASPENTech, 2010a), equipment cost estimation from ASPEN process economic analyzer (PEA) (ASPENTech, 2010b), and the most up-to-date experimental data and cost bases.

6.2 Approach

The technoeconomic analysis was performed in five key steps, as shown in Figure 6.2.

Development of process flow diagrams: From experimental data on key conversion and separation steps, designs for a plant at commercial scale were developed. All equipment necessary for conversion, separation, transportation, treatment, and storage was identified.

Simulation and calculation of mass and energy balance: The ASPEN Plus process simulation software (ASPENTech, 2010a) was used to simulate the process. Some of the biofuel components (e.g., lignin, xylan, xylose) are not available in the ASPEN Plus database. They were added to the simulator in a user-defined database, called an in-house database (INHSPCD), in ASPEN Plus, using estimated properties from National Renewable Energy Laboratory (NREL) (Wooley and Putsche et al., 1996). A few reactions (e.g., pretreatment, waste treatment) and separation processes (filtration, drying, crystallization) were difficult to simulate in ASPEN Plus and therefore were treated as "black box" simulations. Mass balances for these units were derived from experimental data or estimated by heuristics. As a result of this step, the flow rates of every stream, utility demand, and energy generation process were available in the simulation results.

Process integration: In this integration step, the process design was modified to minimize overall consumptions of resources: chemicals, fresh water, external energy demand, makeup solvent, and so on. Two types of process integration problems were solved: (1) heat integration of heat-exchanger networks using pinch analysis to target simultaneously the minimal heating and cooling utility demand (Linnhoff, 1993), and (2) recycling of water and other chemicals to minimize fresh inputs (El-Halwagi, 2006, 2012). After the targets were identified, integrated heat-exchanger networks and recycle strategies were synthesized so that the overall consumption of fresh resources was close to the target values, subject to the constraints that the processes be practical and readily controlled. The integration results were then used to update the simulation models, mass balance, and energy balance.

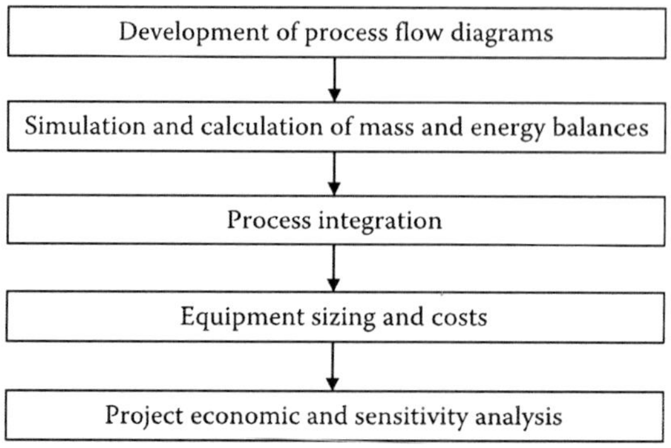

FIGURE 6.2
Major steps of technoeconomic analysis.

Equipment sizing and costs: Most of the simulation models in ASPEN Plus calculated the equipment size and were detailed enough to estimate equipment costs using the ASPEN PEA software (ASPENTech, 2010b). For unconventional equipment not available in ASPEN PEA, costs were obtained from vendor quotes or estimated from the literature. For example, some pieces of equipment common in biofuel processes, but not found in the ASPEN PEA software, are bale transport and unwrapping conveyers, truck scales, belt filters, clarifiers, and large digestion tanks. A good source of literature references for such equipment is the technoeconomic analysis reports from NREL (Aden et al., 2002; Kazi et al., 2010; Wright et al., 2010). If an equipment cost was referred to the literature, the reference cost was scaled to appropriate capacity with scaling exponents reported by Walas (1988) or Peters et al. (2003) and was updated to 2010 dollars using *Chemical Engineering*'s Plant Cost Index (2011).

Project economic and sensitivity analysis: The estimated equipment costs account for equipment only, that is, the costs of equipment materials and fabrication. Other project costs (e.g., equipment installation, instruments and piping, construction, building, contingency) were assumed to be proportional to the equipment costs by predefined factors. For biochemical processing equipment, the factors used are from the NREL method (Aden et al., 2002). For traditional chemical processing equipment, the factors used are from the Lang method (Peters et al., 2003). Cash flow analysis was performed to evaluate project economics in a base case. After that, sensitivity analysis was performed to investigate the sensitivity of the project economics to technical performance (e.g., yields, concentrations, capacities, heat transfer coefficients, the temperature approach of latent heat exchangers) and economic assumptions (e.g., raw material prices, after-tax discount rate, cost-estimation uncertainty).

The above are the five essential steps; however, a biofuel technoeconomic analysis is not limited to these steps. In addition, the following issues must be considered:

- Plant capacity, location, and logistic systems to collect biomass feedstock
- Types of feedstock
- Life-cycle assessment and greenhouse-gas emissions
- Technology alternatives for key conversion steps

These issues are outside the scope of this work, but remain interesting topics for future work.

6.3 Technical Analysis

6.3.1 Process Description

Figure 6.3 depicts biomass-to-hydrocarbons conversion using the MixAlco™ process. To make hydrocarbon fuels, the MixAlco™ process uses the following steps: (1) pretreatment with lime, (2) fermentation with a mixed culture of acid-forming microorganisms to obtain carboxylate salts, (3) dewatering with a high-efficiency vapor-compression evaporator,

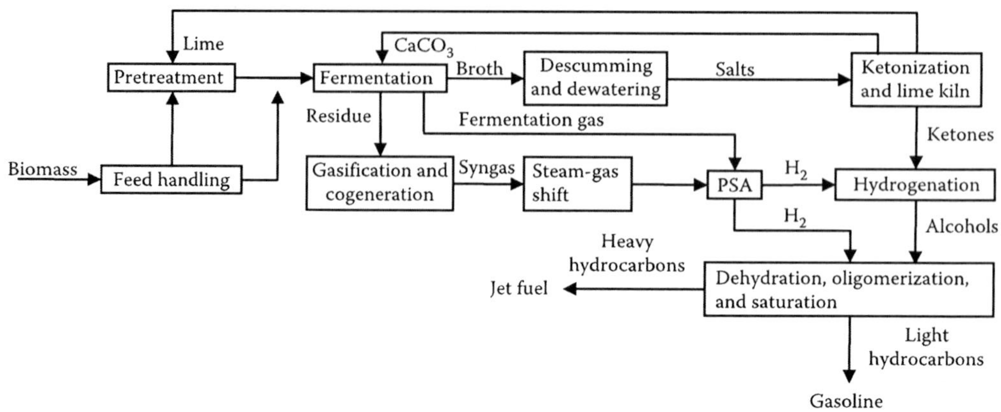

FIGURE 6.3
Simplified process block diagram of the analyzed MixAlco™ process (Pathway C). (With kind permission from Springer Science+Business Media: *J. Ind. Microbio.l Biotechnol.*, Techno-economic analysis of biomass to fuel conversion via the MixAlco process, 37(11), 2010, p. 1159, Pham, V. et al.)

(4) thermal conversion of salts to ketones, (5) hydrogenation of the ketones to mixed alcohols, and (6) oligomerization of alcohols to hydrocarbons using zeolite catalysts.

The biomass feedstock must contain a source of energy and a source of nutrients. Examples of energy sources include sorghum, bagasse, municipal solid waste, office paper, paper fines, rice straw, water hyacinths, pineapple waste, and aloe-vera pulp. Examples of nutrient sources include food scraps, sewage sludge, and manure. In addition, chemical nutrients (e.g., urea, ammonia, ammonium bicarbonate) can be added to supply essential minerals. In this technoeconomic analysis, forage sorghum and manure were used as the feedstock in a recommended ratio of 80:20 sorghum:manure (Rapier, 1995). Sorghum is an energy crop that has been well studied and developed at Texas A&M University.

If the biomass has significant lignin content, it is pretreated with lime, which can be recycled using downstream processes. In the fermentation, the nutrient source is mixed with the pretreated biomass. From the ketonization reactor, calcium carbonate is recycled to buffer the fermentation, which produces a broth of mixed calcium carboxylates. In the descumming and dewatering unit, the broth is concentrated to obtain solid salts, which are then thermally converted into ketones and calcium carbonate in the ketonization unit. Then the ketones are hydrogenated into mixed alcohols. Potential sources of hydrogen include fermentation broth, gasified fermentation residue, reformed methane, and water electrolysis. In the base case, hydrogen is recovered in the plant from fermentation gas and shifted syngas. The recovered hydrogen is sufficient to meet the demands of ketone and olefin hydrogenation. To make hydrocarbons, the mixed alcohols are dehydrated and oligomerized to produce olefins, *n*-paraffins, iso-paraffins, and aromatics with boiling ranges of gasoline and kerosene depending upon the reaction conditions in the oligomerization reactor. The light fraction can be used for gasoline and the heavy fraction as jet fuel. To improve product quality, the olefins can be saturated.

In one process option, the undigested fermentation residues (~20% of the biomass feed) are gasified and processed through cogeneration, steam-gas shift, and pressure-swing adsorption (PSA) units to generate steam, power, and hydrogen for the plant. Using Pathway C, hydrogen from gasified biomass residues and fermentation gas is sufficient to supply the hydrogen needs of the plant.

In the MixAlco™ process, the fermentation broth contains 2–6% salts, which are concentrated using vapor-compression and crystallization units. The recovered distilled water is recycled to fermentation and pretreatment. Water entering with the biomass feed is purged as distilled water, which can be sold as a by-product.

6.3.1.1 Pretreatment and Fermentation

In principle, any chemical or physical pretreatment method can be used in the MixAlco™ process; however, lime pretreatment is the best choice for purposes of mass integration, reactor design, and operation. Recycled lime from the lime kiln is assumed to contribute 65% of lime demand for pretreatment, with 35% makeup lime purchased from external vendors. Using lime, the pretreatment can be performed in simple inexpensive pile reactors (Holtzapple and Granda, 2009). Pile pretreatment integrates several fermentation piles in a round-robin system in which biomass solids are held in the same pile for both pretreatment and fermentation. Although the residence time for pretreatment (6 weeks) and fermentation (up to 8 weeks) is long, the round-robin system results in a steady flowrate and product concentration in the broth.

Using a mixed culture of microorganisms is the key feature of the MixAlco™ process. These microorganisms digest not only carbohydrates, but also proteins and fats. For food wastes, this advantage gives a significantly higher overall yield than other common fermentation methods. The MixAlco™ process does not require sterilization or external enzymes, which reduces capital and operating costs. Consequently, reactors do not require stainless steel (SS) and can be constructed using low-cost materials (e.g., concrete or plastic) that support piles or submerged fermentation. Holtzapple et al. (1999) have discussed the design of submerged fermentation ponds.

Figure 6.4 depicts a pile reactor. Shredded biomass is piled up on a one-meter-thick gravel bed which is used to filter water draining from the pile. The water is pumped back to the top. Underneath the gravel layer is a geomembrane layer to isolate the system from the surrounding soil. The pile, which can be up to 30 m high, can be used for pretreatment and fermentation. In pretreatment operation mode, air is blown through a scrubber to remove carbon dioxide and discharged to the bottom of the pile. In fermentation mode, the pile must be covered by a geomembrane, and air is excluded to maintain anaerobic conditions. To remove odors, fermentation gases are discharged through a biofilter.

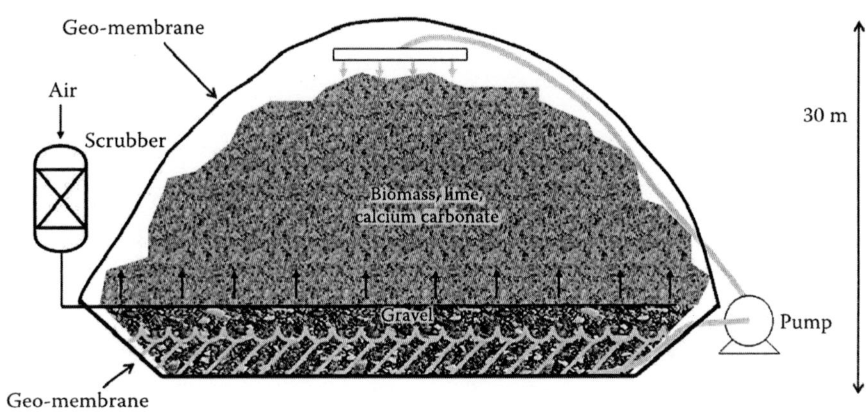

FIGURE 6.4
Pretreatment and fermentation pile reactor.

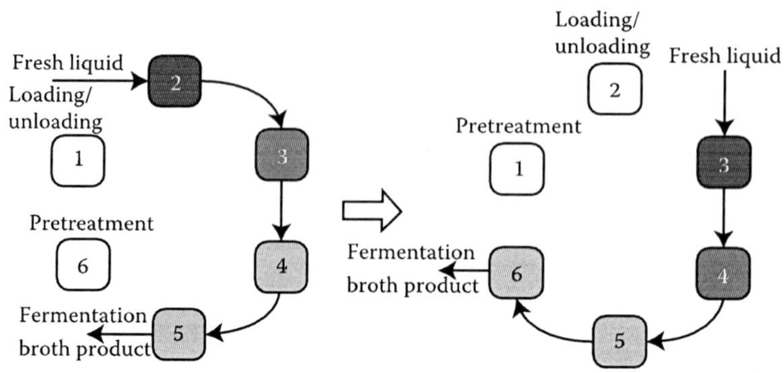

FIGURE 6.5
Round-robin operation (darker boxes represent older fermenting piles).

The pretreatment and fermentation piles operate in batch mode with a total residence time of up to 11 weeks. In contrast, the downstream sections operate in continuous mode. The pretreatment and fermentation configuration was designed to operate as a round-robin system to minimize fluctuation of outlet fermentation concentrations and flow rates. Figure 6.5 depicts a round-robin system with six pile reactors. A reactor can be in pretreatment mode and later in fermentation mode. At a given time, one of the reactors is being loaded or unloaded, one is in pretreatment mode, and the other four are in fermentation mode but at different extents of conversion. Reactor 2 is the oldest and Reactor 5 the newest fermentation. Although solid biomass does not move, water flow is countercurrent with fermentation maturity. Fresh water is pumped to the oldest fermenting pile, circulated internally, and pumped to the next newer pile. As a result, the most dilute broth contacts the most digested biomass, and the most concentrated broth contacts the freshest biomass. This countercurrent arrangement enables both high product concentrations and high conversion rates.

6.3.1.2 Dewatering

In the base-case scenario, calcium carboxylate concentration in the fermentation broth is assumed to be 5% by weight. Other components (e.g., dissolved carbon dioxide, microorganisms, undigested biomass, and other unknowns) are impurities and must be removed along with the water. To purify the carboxylate salts, the broth is degassed by stripping, descummed using flocculant, evaporated with vapor compression, and crystallized (Figure 6.6).

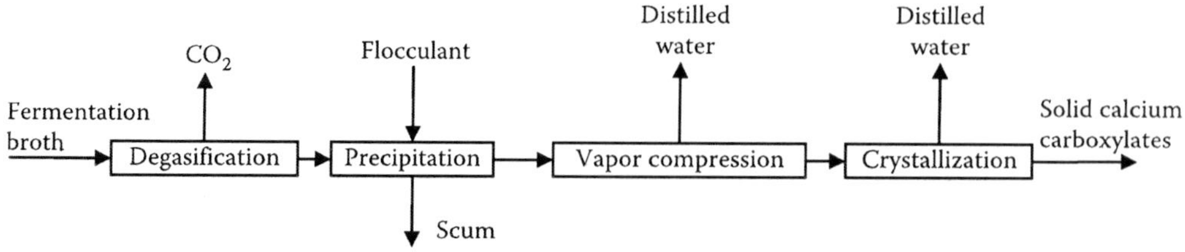

FIGURE 6.6
Simplified process block diagram of the descumming and dewatering unit. (With kind permission from Springer Science+Business Media: *J. Ind. Microbio.l Biotechnol.*, Techno-economic analysis of biomass to fuel conversion via the MixAlco process, 37(11), 2010, p. 1160, Pham, V. et al.)

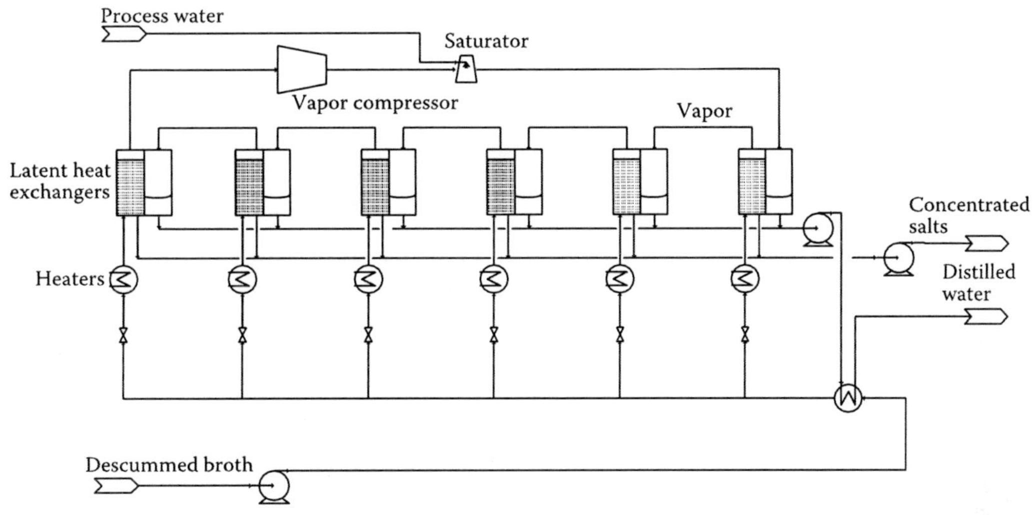

FIGURE 6.7
Parallel configuration of a multiple-effect vapor-compression evaporator.

The key to vaporizing water efficiently is the novel design and optimized operating conditions of the vapor-compression unit. Figure 6.7 is a simplified process flow diagram of the vapor-compression unit. At high pressure (8 bars), the descummed broth is preheated by countercurrent exchange of heat from product streams (which contain condensed water) and then is evenly split into many stages. The figure shows six stages, but there can be tens of stages in real-world plants. For purposes of this work, six stages were chosen. In every stage, the inlet streams pass through a valve and a heater to be adjusted to saturation conditions. After that, they enter latent heat exchangers which use heat from the condensing vapor of an adjacent stage to vaporize water from the fermentation broth. The vapor from the first stage is compressed to a higher temperature and pressure; at that point, it is saturated, and therefore it readily condenses and transfers heat to the last stage. In Figure 6.7, the pressure profile of the latent heat exchangers increases from left to right. With this profile, vapor from the right adjacent stage has a higher temperature than the salt solution in the left adjacent stage; hence, heat transfer occurs. Using copper plates with a hydrophobic coating to promote dropwise condensation, an extremely high overall heat transfer coefficient of 240 kW/(m^2 K) (42,200 Btu/(h ft^2°F)) was achieved using $\Delta T = 0.2$ K at the laboratory scale (Lara and Holtzapple, 2011a,b). This enables a very small temperature approach of only 0.20 K (0.36°F) while maintaining a high heat flux of 48 kW/m^2. As a result, the compression ratio of the compressor is small, which saves both associated capital and operating costs. The net energy consumption of this vapor-compression unit is only 1.45 kWh/m^3 (18.8 MBtu/1000 gal) of water vaporized, which is equal to 0.14% of the latent heat of vaporization of the same amount of water.

6.3.1.3 Ketonization

The solid salts are sent to a dryer to remove residual moisture (Figure 6.8). At high temperatures (430°C) in the ketonization reactor, calcium carboxylates are thermally converted into ketones and calcium carbonate:

$$\text{R–COO Ca COO–R}' \rightarrow \text{R–CO–R}' + CaCO_3.$$

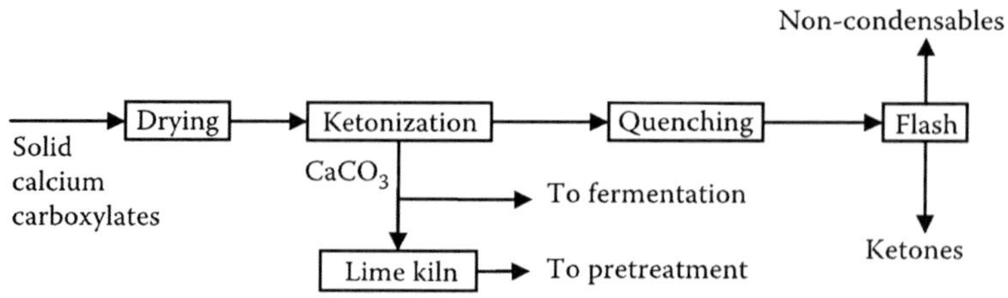

FIGURE 6.8
Simplified block diagram of the ketonization and lime kiln unit. (With kind permission from Springer Science+Business Media: *J. Ind. Microbio.l Biotechnol.*, Techno-economic analysis of biomass to fuel conversion via the MixAlco process, 37(11), 2010, p. 1160, Pham, V. et al.)

where R and R′ represent hydrocarbon groups. The reactor is kept under vacuum (30 mm Hg) which reduces residence time to avoid decomposing the produced ketones. The ketone vapor is quickly removed from the reactor, quenched, and condensed. Part of the calcium carbonate is directly recycled to the fermentors, and the remaining portion is converted into quicklime (CaO) in a kiln. The quicklime is recycled to the pretreatment reactors. Unlike conventional lime kilns that are fed coarse limestone, this kiln processes fine calcium carbonate powder, and therefore certain processing steps (grinding, drying) are not needed.

6.3.1.4 Ketone Hydrogenation

In this conversion step (Figure 6.9), the ketone carbonyl groups react with hydrogen to form alcohol groups in an exothermic reaction:

$$R{-}CO{-}R' + H_2 \rightarrow R{-}CH(OH){-}R'.$$

The reaction is performed at high pressure (55 bars) and at isothermal (130°C) conditions. The optimal design was found to be three CSTRs in series (Chang, 1994). In each CSTR, liquid ketones, solid Raney nickel catalyst, and hydrogen bubbles are well mixed. The heat of reaction is recovered by a pump-around system. Hydrogen is fed to every CSTR in 20% excess to maximize the ketone conversion. The net hydrogen demand is 0.0225 kg H_2/kg mixed alcohol (25.0 SCF/gal of mixed alcohols) or 0.00687 kg H_2/kg dry ash-free biomass (1.30 SCF per dry ash-free pound of biomass). The produced mixture of secondary alcohols can be directly used as an oxygenated transportation fuel-like bioethanol, but it has higher energy content (net heating values are 34.6 and 26.8 MJ/kg, respectively).

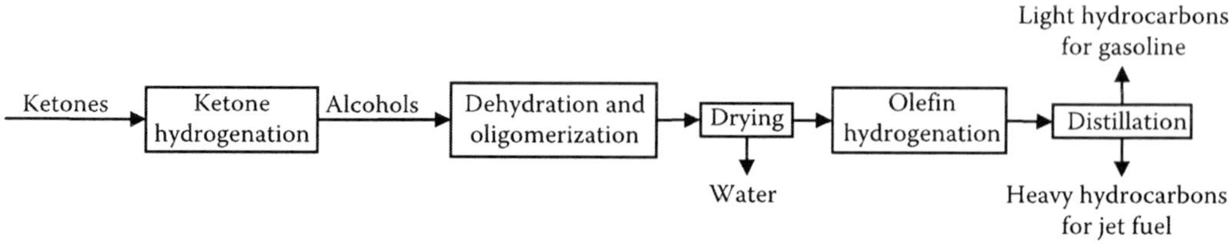

FIGURE 6.9
Simplified block diagram of ketone hydrogenation, dehydration, and oligomerization, with olefin hydrogenation units. (With kind permission from Springer Science+Business Media: *J. Ind. Microbio.l Biotechnol.*, Techno-economic analysis of biomass to fuel conversion via the MixAlco process, 37(11), 2010, p. 1161, Pham, V. et al.)

6.3.1.5 Dehydration and Oligomerization

The mixed alcohols are further processed to produce hydrocarbon fuels (Figure 6.9). Using H-ZSM-5 catalyst in a reactor at 300°C and 3 bars, the alcohols are dehydrated:

$$R{-}CH(OH){-}CH_2{-}R'' \rightarrow R{-}CH = CH{-}R'' + H_2O.$$

In the same reactor, the produced olefins are oligomerized as shown in the following simplified reaction:

$$C_mH_{2m} + C_nH_{2n} \rightarrow C_{m+n}H_{2m+2n}.$$

Depending upon the specific reaction conditions (time, pressure, temperature), the products may be very complex and include olefins, *n*-paraffins, iso-paraffins, aromatics, and cyclics. Water dissolved in the hydrocarbon products is removed in a drying unit using a salt filter.

6.3.1.6 Olefin Hydrogenation

To improve fuel quality, the olefins can be hydrogenated to make corresponding paraffins (Figure 6.9). Similarly to the design of the ketone hydrogenation unit, this conversion unit employs CSTRs in series with a Raney nickel catalyst. The carbon double-bond C=C is saturated to stabilize the hydrocarbon product:

$$C_xH_{2x} + H_2 \rightarrow C_xH_{2x+2}.$$

In this step, the net demand of hydrogen is 0.0139 kg H_2/kg hydrocarbon fuels (15.4 SCF/gal of hydrocarbon fuel) or 0.0034 kg H_2/kg dry ash-free biomass (0.64 SCF per dry ash-free pound of biomass). Once out of the reactor, the mixed hydrocarbons are distilled into C_{8-} and C_{9+} fractions. The light fraction and the heavy components can be used as blending components for gasoline and jet fuel, respectively.

6.3.1.7 Hydrogen Sources

Hydrogen is required to produce alcohols and saturate hydrocarbons. Two scenarios for hydrogen sources were analyzed.

In the first scenario, hydrogen is produced by gasifying undigested biomass from the fermentors. The investment includes not only gasification, but also cogeneration, steam-gas shift, and PSA (next section) to supply hydrogen, steam, and power for the plant. Figure 6.10 shows the gasification and cogeneration processes in this scenario. First, the small amount of biomass suspended in the pretreatment liquor is recovered by filtration. Then that filtered biomass is mixed with fermentation residue and dried in a rotary dryer using flue-gas heat. After drying, the moisture content in the biomass is ~10%. In the next step, the dried biomass is gasified to generate syngas and by-product char. Energy from the hot syngas is used to make high-pressure steam, which is expanded in a steam turbine to generate power.

Figure 6.11 shows the gasifier, a fluidized bed followed by two cyclones that effectively remove particulates (char, ash) from the syngas (Capareda, S. (2010). College Station, TX, personal communication). Atmospheric-pressure air, but no steam, is introduced to the gasifier, which requires only a simple control system and low-cost cyclones. The capital

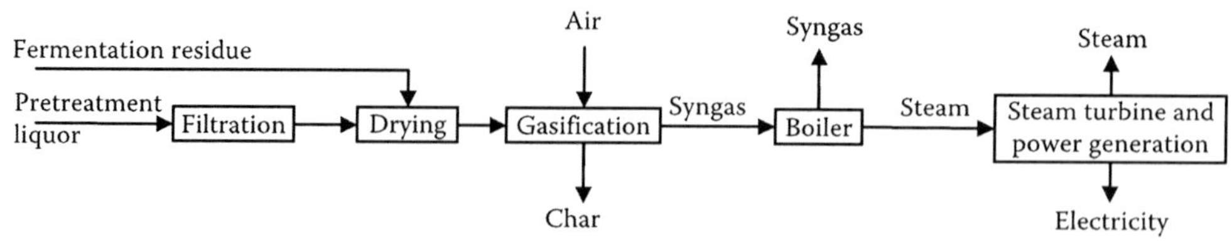

FIGURE 6.10
Simplified block diagram of gasification and cogeneration unit. (With kind permission from Springer Science+Business Media: *J. Ind. Microbio.l Biotechnol.*, Techno-economic analysis of biomass to fuel conversion via the MixAlco process, 37(11), 2010, p. 1161, Pham, V. et al.)

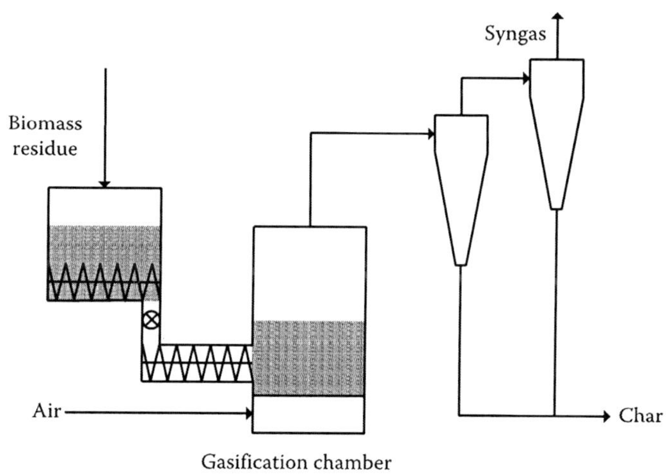

FIGURE 6.11
Schematic of the atmospheric biomass gasifier. (With kind permission from Springer Science+Business Media: *J. Ind. Microbio.l Biotechnol.*, Techno-economic analysis of biomass to fuel conversion via the MixAlco process, 37(11), 2010, p. 1162, Pham, V. et al.)

cost of this gasifier is only 25% that of commercial pressurized gasifiers that use steam (Capareda, S. (2010). College Station, TX, personal communication). There are multiple uses for the char recovered from the cyclones: (1) add it to soil to sequester carbon and improve fertility, (2) sell it to coal-fired power plants to provide "green" fuels, and (3) burn it at the plant to recover alkaline ash that can be used to replace lime in the pretreatment.

In the second scenario, hydrogen is not produced in the plant, but is purchased from external sources such as pipelines or oil refineries. In this scenario, the gasification and cogeneration unit is retained to use biomass residues to generate steam and power, but an additional combustion chamber is installed after the cyclones to burn the syngas completely to produce more steam and power.

6.3.1.8 Steam-Gas Shift and Pressure-Swing Adsorption

This section is needed only when hydrogen is produced in the plant (as described in Scenario 1). More hydrogen is made using the shift reaction between steam and carbon monoxide:

$$H_2O + CO \rightarrow H_2 + CO_2.$$

On account of the compositional characteristics of the syngas, a one-stage shift is sufficient for high conversion and low residence time. The hydrogen-rich syngas, along with the fermentation gas, is passed through molecular sieve beds in the pressure-swing adsorption unit, which purifies the hydrogen. Both of these technologies are well developed.

6.3.2 Maximum Theoretical Yields

Theoretical yields in biofuel processes can be simply calculated from the stoichiometry of the representative reactions; however, identification of representation reactions is difficult in some cases. The ensuing equations are representative reactions of the key conversion steps of the carboxylate platform. In practice, because of the mixed-culture fermentation, there are no pure intermediate chemicals. Instead, there are many chemicals with the same functional groups at each conversion step. Nonetheless, the lowest-molecular-weight chemicals can be used to represent intermediates so that maximum theoretical yields can be determined. For example, acetic acid is considered to be the only intermediate among the acid-forming fermentation products, acetone is the only intermediate ketone, and isopropanol is the only intermediate alcohol. The final product was represented by a compound that has physical properties close to those of the real final product (e.g., octane, C_8H_{18}, is used to represent gasoline).

If the feedstock is assumed to be pure cellulose ($C_6H_{10}O_5$), the ideal conversions are as follows:

- Fermentation: $2C_6H_{10}O_5 + 2H_2O + 3CaCO_3 \rightarrow 3Ca(CH_3COO)_2 + 3CO_2 + 3H_2O$
- Thermal conversion: $Ca(CH_3COO)_2 \rightarrow CH_3COCH_3 + CaCO_3$
- Ketone hydrogenation: $CH_3COCH_3 + H_2 \rightarrow CH_3CHOHCH_3$
- Dehydration: $CH_3CHOHCH_3 \rightarrow CH_3CH{=}CH_2 + H_2O$
- Oligomerization: $8CH_3CH{=}CH_2 \rightarrow 3C_8H_{16}$
- Olefin hydrogenation: $C_8H_{16} + H_2 \rightarrow C_8H_{18}$

The overall reaction is $16C_6H_{10}O_5 + 33H_2 \rightarrow 9C_8H_{18} + 24CO_2 + 32H_2O$, where the molecular weights of cellulose and gasoline (C_8H_{18}) are 162 and 114 kg/kmol, respectively. The mass yield of the final product is:

$$\frac{9 \times 114}{16 \times 162} = 0.396 \text{ ton of gasoline per ton of cellulose.}$$

Assuming that the specific gravity of gasoline is 0.74 ton/m^3, the theoretical volumetric yield of the final product is

$$0.396 \frac{\text{tonne gasoline}}{\text{tonne cellulose}} \times \frac{\text{m}^3}{0.74 \text{ tonne}} \times 264 \frac{\text{gallon}}{\text{m}^3} = 141 \text{ gal of gasoline per ton of cellulose.}$$

The overall reaction shows that ~25% of the carbon is lost as carbon dioxide in the fermentation step. Other theoretical yield losses result from oxygen removal as carbon dioxide and water produced in the fermentation and dehydration steps.

The maximum theoretical yields from specific biomass feedstocks are lower than 141 gal/ton because there is no practical biomass feedstock containing pure cellulose. (Note: Waste office paper has a composition of 87.4% cellulose, 8.4% hemicellulose, 2.3%

lignin, and 1.9% ash (Palmisano and Barlaz, 1996), which is very close to pure cellulose.) In addition, in the carboxylate platform, the optimal biomass feedstocks consist of 80% carbohydrate sources and 20% nutrient sources. In this analysis, the plant uses forage sorghum (77.1% carbohydrates, 15.2% lignin, and 7.7% ash) as a carbohydrate source and manure (48.0% carbohydrate, 16.6% lignin, and 35.4% ash) as a nutrient source in a ratio of 80:20, respectively. The maximum theoretical yield of the plant is

$$141 \times \frac{4 \times 77.1\% + 1 \times 48.0\%}{5} = 101 \text{ gal of gasoline per dry ton of biomass.}$$

Lignin and ash are not fermentable, and all carbohydrates are assumed to be cellulose.

6.3.3 Process Performance

Table 6.1 summarizes the key process performance parameters used to calculate mass and energy balances in the base case. The performance of units on the main route (i.e., excluding

TABLE 6.1

Key Process Performances

Unit	Parameter	Value
Pretreatment	Makeup lime demand	0.054 g CaO/g biomass
	Total lime loading	0.15 g CaO/g biomass
	Reaction time	6 weeks
Fermentation	Conversion	0.8 g digested/g volatile solids fed
	Selectivity	0.62 g carboxylic acids/g volatile solids digested
	VSLR	3.00 g VS/(L liquid · day)
	LRT	28 days
	Product concentration	40 g acids/L liquid
	Reaction time	32 days
	Substrate concentration	10%
Dewatering	Carboxylate recovery	95%
	Heat transfer coefficient	240 kW/(m^2 · K)
	Temperature approach	0.20 K
	Cost of latent heat exchangers	\$155/m^2
Ketonization	Conversion	99.5%
	Yield	0.583 g ketones/g carboxylic acids
Ketone hydrogenation	Conversion	98.4%
Dehydration and dimerization	Light hydrocarbon yield	0.6 g light hydrocarbon/g alcohols
	Heavy hydrocarbon yield	0.2 g heavy hydrocarbon/g alcohols
Olefin hydrogenation	Conversion	98.4%
Gasification and cogeneration	Gasification temperature	760°C
	Solid-to-air ratio	0.625
Steam-gas shift and PSA	Steam-gas shift temperature	254°C
	Hydrogen recovery	95%
Whole plant	Gasoline yield	57 gal/dry-ash-free ton biomass
	Jet-fuel yield	19 gal/dry-ash-free ton biomass

units that process fermentation residue) has been proven at the laboratory scale. A yield of more than 70 gal hydrocarbon fuel per ton of municipal solid waste was reportedly achieved at the demonstration scale (Sims, 2011). The performances of residue-processing units were typical values from the literature, as discussed above. In this technoeconomic analysis, these performances are assumed to be achieved at commercial scale for the *n*th plant.

The pretreatment unit consumes a significant amount of lime, most of which is available from recycle streams. In fermentation, the mass of carboxylic acid products is half the mass of volatile solids fed. The long residence time of pretreatment and fermentation requires large piles, which have a volume of 600,000 m^3 each. In a real-world plant, yield losses would result from formation of by-products (see previous section), but also from incomplete chemical conversions and from partial recovery of main products in the separation steps. The final products, hydrocarbon fuels, are fractions of gasoline and jet fuel produced in a ratio of 3:1.

6.3.4 Process Simulation

In ASPEN Plus, simulations of biofuel processes, physical properties of some key biomass components, and biochemical reactions are not available from standard software databases. Most of them are unconventional compounds, and their properties are difficult to estimate using available property prediction methods. Wooley and Putsche (1996) at NREL has built a database of some biofuel components that are present in lignocellulose-to-ethanol processes. The components are cellulose, glucose, xylan, xylose, lignin, zymo (bacterium), cellulase (enzyme), soluble solids, and gypsum. The authors collected their physical properties from the literature and estimated missing ones when necessary. The properties were coded in an appropriate format to be recognized as an in-house database (ASPENTech, 2001) by ASPEN Plus and are called whenever needed.

On the basis of the key performance data (Table 6.1) from experiments, the entire process was simulated in ASPEN Plus to calculate detailed mass and energy balances. Table 6.2 summarizes the types of ASPEN Plus models used to simulate key processing units. In general, the conversion steps with well-defined reaction stoichiometry and yields (hydrogenation and dehydration) were simulated using the RStoic model. The RGibbs model was used to simulate those conversions in which multiple reactants randomly react in the same types of reactions to yield multiple products with the same functional groups (ketonization and dimerization).

TABLE 6.2

ASPEN Plus Models for Key Processing Units

Processing Units	ASPEN Plus Models	Calculated Parameters
Pretreatment reactor	RStoic	Heat of dissolving lime in water
Fermentation reactors	RStoic	Heat of reactions
Latent heat exchangers	HeatX	Heat transfer area
Ketonization reactor	RGibbs	Yields and heat of reactions
Hydrogenation reactors	RStoic	Heat of reactions
Dehydration reaction	RStoic	Heat of reactions
Dimerization reaction	RGibbs	Yields and heat of reactions
Drum dryer	Flash	Heating utility consumption

The pretreatment and fermentation reactions are difficult to simulate because of their complexity. The RStoic model was used to simulate partially what happens in the reactors. For the pretreatment, only the heat of dissolving quicklime in water was estimated; lignin degradation was treated as a "black box," that is, no simulation was performed. For fermentation, the experimental yields and assumed mechanisms were used to calculate the conversions occurring in intermediate reactions in a preprocessing step without aid from ASPEN Plus; then the reactions with their calculated conversions were simulated using the RStoic model in ASPEN Plus to reproduce the yields and to estimate an energy balance. Figure 6.12 shows the simulated fermentation-section process in ASPEN Plus.

Other processing units that are difficult to simulate in ASPEN Plus (e.g., waste treatment, filtration, drying, and crystallization) were treated as "black boxes." Mass balances of these units were obtained from experimental data or estimated by heuristics.

6.3.5 Process Integration

In the simulation, the heat exchanger network was integrated by identifying opportunities to save on heating and cooling utility consumption. Pinch analysis (Linnhoff, 1993) was applied to determine the savings targets. Figure 6.13 shows a grand composite curve of the heat exchanger network. Below the pinch point, there is no curve, which indicates that if the network is ideally integrated, no cooling utility is needed and the heating utility consumption is significantly reduced. Table 6.3 shows that the reduction is 60% compared to the scenario without integration or heat recovery. However, in a practical design, some hot and cold streams should not be integrated to avoid complexity in the control system, even though it is possible. For example, process streams through condensers, reboilers, and heat exchangers which are used to fine-tune temperatures were not considered in the integration. Therefore, the expected integration needs some cooling and heating utility inputs, which are reported in the last column of Table 6.3.

Water balance is another process integration problem in biochemical processes. As in other biochemical platforms, the carboxylate platform demands a large amount of water in the fermentors. Water needs to be removed from the main fermentation products in follow-up separation units. In other biochemical platforms, the recovered water is usually treated to meet the required fermentation conditions. In contrast, water in the carboxylate platform is distilled by energy-saving vapor compression. The high quality of the vaporized water and the low-quality requirements for fermentation broth allow water recycling without treatment, which slightly reduces production costs (Table 6.4) and makes the process significantly more sustainable. In net balance, the plant generates surplus distilled water, which is approximately equal to the moisture content in the biomass feedstock.

Recycling of lime and calcium carbonate is another advantage of the carboxylate platform compared to other biochemical platforms, which enable lime pretreatment to be used very effectively. The pretreatment unit requires 1.5 tons of lime for every 10 dry ton of biomass fed (Table 6.1). The calcium flows through the pretreatment, fermentation, descumming, and dewatering units along with the main products and is finally recovered in the ketonization unit as calcium carbonate. Part of the calcium carbonate is directly recycled to the fermentors as a buffer. The remaining calcium carbonate is conveyed to the lime kiln (Figure 6.14) to produce lime, which is recycled to the pretreatment unit. As a result, a significantly smaller amount of makeup lime is needed (Table 6.1), and operating costs are significantly reduced (Table 6.4).

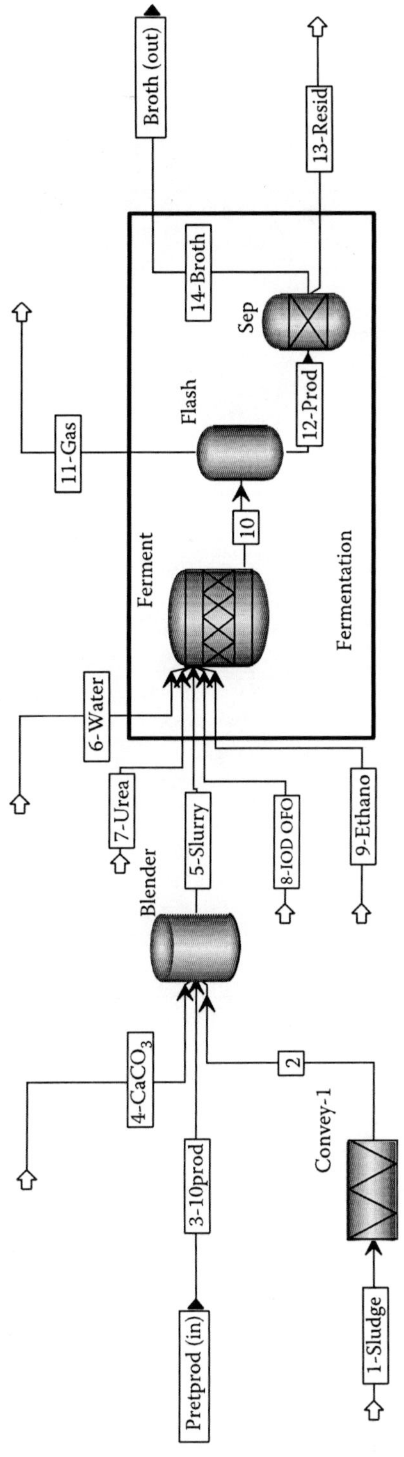

FIGURE 6.12
Simulation of the fermentation unit in ASPEN Plus.

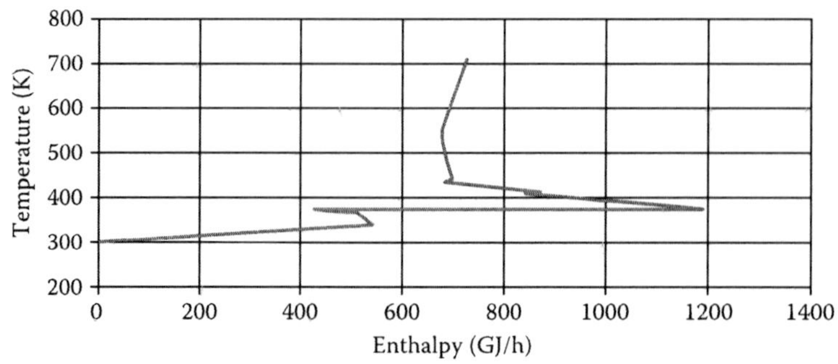

FIGURE 6.13
Grand composite curve for integration of the heat exchanger network in the plant.

TABLE 6.3
Utility Consumption in Targeted and Expected Heat Integration Scenarios

Utility	No Integration	Targeted Integration	Expected Integration
Heating utility (GJ/h)	1720	726	616
Cooling utility (GJ/h)	994	0	110

TABLE 6.4
Savings from Chemical Recycle

Chemical	Fresh Water	Fresh Calcium Carbonate ($CaCO_3$)	Fresh Lime (CaO)
Consumption without recycle (tons/h)	1,840	45.7	30.2
Consumption with recycle (tons/h)	0	0	10.7
Reduction (%)	100	100	65
Prices ($/ton)	0.13	66	70
Saved costs ($MM/year)	1.91	24.1	10.9
Saved costs ($/gal product)	0.016	0.198	0.090

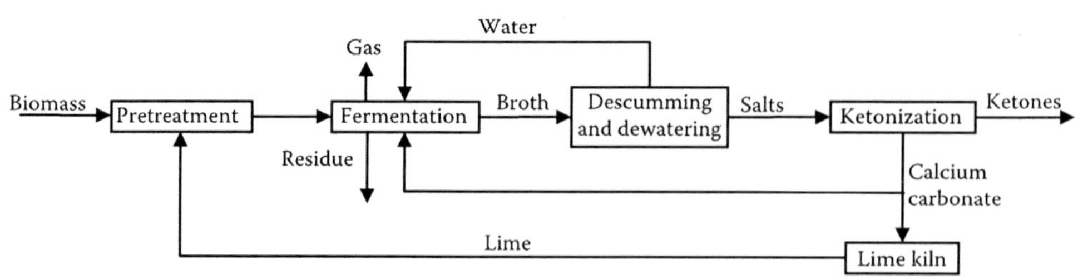

FIGURE 6.14
Recycle of water and chemicals.

6.3.6 Energy Efficiency Analysis

The maximum theoretical energy efficiency of cellulose-to-gasoline conversion through the carboxylate platform is 89.4%. This figure can be derived from the overall reaction (Section 6.3.2) given the higher heating values of cellulose (17.6 MJ/kg, Holtzapple (2003a)), hydrogen (141.8 MJ/kg, ASPENTech (2010a)), and gasoline (as C_8H_{18}, 47.9 MJ/kg, ASPENTech (2010a)).

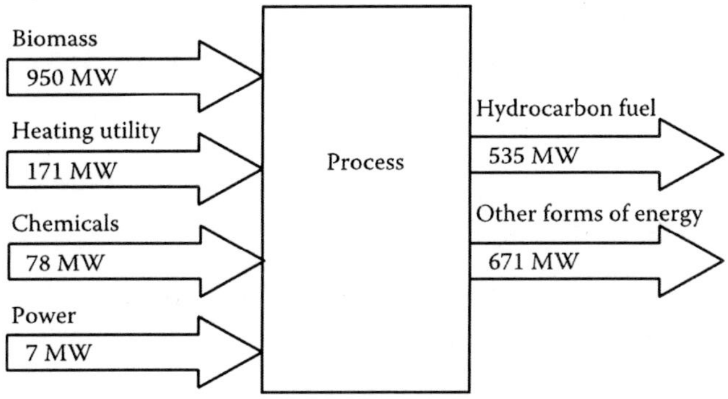

FIGURE 6.15
Energy balance of the plant.

For the plant analyzed, 80% of the biomass feedstock is forage sorghum (77.1% carbohydrates, 15.2% lignin, and 7.7% ash) and the remaining 20% is manure (48.0% carbohydrates, 16.6% lignin, and 35.4% ash). With the higher heating value of lignin (29.5 MJ/kg, Holtzapple (2003b)) and ash (assume 0 MJ/kg), the higher heating value of the feedstock is 17.1 MJ/kg. The maximum theoretical energy efficiency of the plant is calculated as 81.0%.

In the above calculations of the two maximum theoretical energy efficiencies, it was assumed that the process had no input source of energy other than biomass and hydrogen. To operate the process in a real-world plant, more energy inputs will be needed in the form of chemicals, heating utility, and electricity. All sources of energy inputs for the plant analyzed are summarized in Figure 6.15. For biomass, chemicals, and hydrocarbon fuel, the energy loads are based on higher heating values at the standard condition (i.e., sensible heat values are negligible). The total input energy and the energy contained in the products are 1206 and 535 MW, respectively. The energy efficiency of the plant is therefore 44.4%.

6.4 Economic Analysis

6.4.1 Analytical Procedure and Basis

The first step in the economic analysis is to estimate purchased equipment costs in ASPEN Icarus Process Evaluator (ASPENTech, 2010b) based on the equipment sizes. Some equipment sizes (e.g., compressor power, heat exchanger area, distillation column diameter, and height) were reported in the ASPEN Plus simulation results, whereas others (e.g., sizes of the crystallizer, drum dryer, and clarifier) were estimated using reliable heuristics and assumptions (Walas, 1988 or Peters et al., 2003). If package quotes were available from the literature or vendors, these quotes were used instead of estimating the costs from individual pieces of equipment. Scaling factors for estimating equipment costs at various capacities were obtained from NREL (Aden et al., 2002), which in turn took most of their scaling factors from Walas (1988). These scaling factors are summarized in Table 6.5 and are defined as follows:

$$\text{Equipment cost B} = (\text{Equipment cost A}) \times \left(\frac{\text{Capacity B}}{\text{Capacity A}}\right)^{\text{Scaling factor}}$$

TABLE 6.5
Scaling Factors to Estimate Equipment Costs at Various Sizes

Equipment Type	Installation Factor(s)	Scaling Factor	Scaling Base
Agitators: CS; SS	1.3; 1.2[a]	0.51[d]	Flow
Blenders	1.3[a]	0.49[b]	Flow
Blowers	1.4[a]	0.59[b]	Flow
Centrifuges, CS	1.3[a]	0.67[b]	Flow
Clarifiers, thickeners	1.51[c]	0.60[c]	Flow
Columns, distillation, CS; SS	3.0; 2.1[a]	0.62[b]	Diameter squared
Compressors, motor-driven	1.3[a]	0.69[b]	Flow
Conveyers and elevators	1.4[a]	0.60[c]	Flow
Crystallizers	1.9[a]	0.37[b]	Flow
Dryers	1.4[a]	0.40[b]	Flow
Evaporators, thin-film, CS	2.5[a]	0.54[b]	Flow
Filters, belt press	1.25[c]	0.60[c]	Solid flow
Filters, pneumapress	3.34[c]	0.60[c]	Solid flow
Heat exchangers, shell-and-tube	2.1[a]	0.44[b]	Heat-transfer area
Pumps, centrifugal, CS	2.8[a]	0.79[d]	Flow
Reactors, kettle	2.1[a]	0.54[b]	Flow
Reactors, multitubular, SS	1.6[a]	0.56[b]	Flow
Shredders	1.38[c]	0.60[c]	Flow
Tanks, field-erected, CS	1.4[c]	0.57[b]	Flow
Truck scales	2.47[c]	0.60[c]	Flow
Vessels, pressure, CS	1.7[a]	0.51[d]	Flow

[a] Walas, S.M. 1988. *Chemical Process Equipment Selection and Design.* Butterworth.
[b] Peters, M.S., Timmerhaus, K.D., and West, R.E. 2003. *Plant Design and Economics for Chemical Engineers.* McGraw-Hill Professional.
[c] Aden, A., Ruth, M., Ibsen, K., Jechura, J., Neeves, K., Sheehan, J., Wallace, B., Montague, L., Slayton, A., and Lukas, J. 2002. *Lignocellulosic Biomass to Ethanol Process Design and Economics Utilizing Co-Current Dilute Acid Prehydrolysis and Enzymatic Hydrolysis for Corn Stover.* Golden, CO: National Renewable Energy Laboratory.
[d] Garrett, D. 1989. *Chemical Engineering Economics.* New York: Van Nostrand Reinhold.

After that, other relevant costs to build the plant were estimated as factors of the purchased equipment cost. For pretreatment, fermentation, and support units (wastewater treatment, storage, and utilities), this work uses the NREL modified factor method (Aden et al., 2002), which is suitable for aqueous-based processes. This modification has a contingency factor five times greater than the original NREL method. Lang factors (Peters et al., 2003) for fluid processing were used for the other units, which are similar to chemical and petrochemical processes. The factor values for those two methods are compared in Table 6.6. The total capital investment (TCI) costs are 3.5 and 6 times the purchased equipment costs, using the modified NREL method and the Lang factor, respectively. The main difference comes from the installation factors.

Subsequently, cash flow and financial models were constructed to evaluate the project economics. The basis and assumptions of the base-case models are presented in Table 6.7. The federal income tax rate and the depreciation method and period were recommended by NREL after a review of the IRS modified accelerated cost recovery system (MARCS) (Short et al., 1995). No subsidies or incentives were applied.

TABLE 6.6

Factors in Estimation of Project Costs

Lang Method[a]		Modified NREL Method[b]	
Cost Items	**Factor**	**Cost Items**	**Factor**
Direct cost		Direct cost	
Purchased equipment	100	Purchased equipment	100
Equipment installation	47	Installation	70[c]
Instrumentation and control	36	Warehouse	2.55
Piping	68	Site development	15.3
Electrical systems	11	Total direct costs	188
Buildings	18		
Yard improvements	10		
Service facilities	70		
Total direct costs	360		
Indirect costs		Indirect costs	
Engineering and supervision	33	Prorateable costs	18.8
Construction expenses	41	Field expenses	18.8
Legal expenses	4	Office and construction	47.0
Contractor's fees	11	Contingency	28.2
Contingency	44	Other	18.8
Total indirect costs	144	Total indirect costs	132
FCI	504	FCI	320
WCI = 15% FCI	89	WCI = 10% FCI	32
TCI	593	TCI	352

[a] Peters, M.S., Timmerhaus, K.D., and West, R.E. 2003. *Plant Design and Economics for Chemical Engineers*. McGraw-Hill Professional.

[b] Aden, A., Ruth, M., Ibsen, K., Jechura, J., Neeves, K., Sheehan, J., Wallace, B., Montague, L., Slayton, A., and Lukas, J. 2002. *Lignocellulosic Biomass to Ethanol Process Design and Economics Utilizing Co-Current Dilute Acid Prehydrolysis and Enzymatic Hydrolysis for Corn Stover*. Golden, CO: National Renewable Energy Laboratory.

[c] In the NREL report, this cost factor ranges from 20 to 200 depending on the type of equipment (see Table 6.5 for details); 70 is the mean value.

After the financial model for the base case was built, the minimum product selling price (MPSP) was determined. This is the hydrocarbon fuel price that results in a net present value of zero in the cash flow calculations with a predefined after-tax discount rate. To simplify the economic analysis, the gasoline and jet-fuel prices were assumed to be identical.

Finally, to perform sensitivity analysis, the following key parameters in the simulation and financial models were varied: fermentor operating conditions, overall yield, plant capacity, sources of hydrogen supply, prices of feedstock and raw chemicals, and after-tax cash flow discount rate.

6.4.2 Base-Case Economic Analysis

For the base-case plant in which the parameter values shown in Tables 6.1 and 6.7 were used, the fixed capital investment (FCI) is \$331 MM. Figure 6.16 shows how the capital is distributed in the plant. The steam-gas shift and PSA units contribute most to the FCI,

TABLE 6.7

Basis and Assumptions of the Financial Models for the Base Case

Parameters	Values
Plant life	20 yrs
General plant depreciation	200% double declining balance (DDB) for 7 yrs
Steam generation unit depreciation	150% declining balance (DB) for 20 yrs
Financing	100% equity
After-tax discount rate	10%
Income tax rate	39%
Dollar year value	2010
Subsidy	No
Plant capacity	160 dry tons/h of forage sorghum (8.1% ash) and 40 dry tons/h of manure (35.4% ash)
Construction period	1.5 yr
Startup time	6 months
Revenues	50%
Variable costs	75%
Fixed costs	100%
Operating season	8000 h/yr
Delivered prices of raw materials:	
Forage sorghum	$60/dry ton
Manure	$10/dry ton
Hydrogen source	Processing of fermentation residue

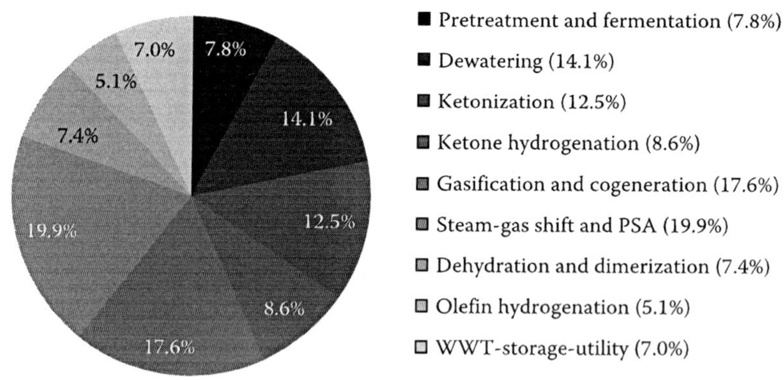

FIGURE 6.16
Breakdown of the FCI for the base case.

followed by gasification and cogeneration, the dewatering unit, and ketonization. The high capital costs of water-gas shift, PSA, and gasification indicate that producing hydrogen from fermentation residues is expensive. Most of the dewatering costs come from the expensive crystallization equipment. Because of the low-cost pile design, pretreatment and fermentation require only 7.2% of the investment.

Variable and fixed operating costs are presented in Tables 6.8 and 6.9. The variable operating cost is dominated by the main feedstock cost ($60/ton forage sorghum), followed by the cost of low-pressure steam that is primarily consumed by the crystallizer. The key

TABLE 6.8

Variable Operating Costs

Cost Items	Rate	Price	Annual Cost $1000/yr	Unit Cost $/gal product
Feedstock				
Forage sorghum	160 ton/h	$60/ton	76,800	0.630
Manure	40 ton/h	$10/ton	3200	0.026
Chemicals				
Lime	10.7 ton/h	$70/ton	6006	0.049
Ethanol	9.47 ton/h	$2.2/ton	167	0.001
Flocculant	208 kg/h	$991/ton	1650	0.014
Iodoform	7.17 kg/h	$25/kg	1434	0.012
Utility				
High-pressure steam	14.1 ton/h	$10.1/ton	1142	0.009
Low-pressure steam	476.3 ton/h	$5.50/ton	20,955	0.173
Natural gas	1.6×10^3 m^3/h	$0.113/m^3	1469	0.012
Electricity	6.58 MW	$0.062/kWh	3244	0.027
Cooling water	18,838 m^3/h	$0.013/m^3	1959	0.017
Boiler water	117 m^3/h	$0.13/m^3	121	0.001
Waste disposal	32.5 ton/h	$18/ton	4681	0.038

TABLE 6.9

Fixed Operating Costs

Cost Items	Annual Cost ($1000/yr)	Calculation[a]
Labor		
Direct wage and benefits (DW&B)	4481	For operators
Direct salary and benefits	672	15% of DW&B
Operating supplies and services	269	6% of DW&B
Technical assistance to manufacturing	747	$52,000/(operator/shift)-yr
Control laboratory	409	$57,000/(operator/shift)-yr
Maintenance		
Wages and benefits (MW&B)	11,576	4.5% of FCI
Salaries and benefits	2894	25% of MW&B
Materials and services	11,576	100% of MW&B
Maintenance overhead	579	5% of MW&B
Operating overhead		
General plant overhead	1393	7.1% of the wages, salaries, and benefits
Mechanical department services	471	2.4% of the wages, salaries, and benefits
Employee relations department	1158	5.9% of the wages, salaries, and benefits
Business services	1452	7.4% of the wages, salaries, and benefits

[a] Following the instructions of Seider et al. (2004).

contribution to the fixed operating costs is maintenance-related costs, which were estimated as a factor of the capital costs.

The MPSP of the base case was estimated to be $1.83/gal of hydrocarbon fuels at an after-tax discount rate of 10%. Table 6.10 shows the components of this selling price. The biomass

TABLE 6.10
Cost Components of MPSP in the Base Case

	Contribution	
Cost Component	**$/gallon of Product**	**Percentage (%)**
Biomass	0.656	35.8
Chemicals and waste disposal	0.114	6.3
Utility	0.237	12.9
Labor	0.054	2.9
Maintenance	0.218	11.9
Operating overhead	0.037	2.0
Average capital depreciation	0.241	13.2
Average income tax	0.140	7.7
Average return on investment (ROI)	0.134	7.3
Total	1.832	100.0

feedstock is the highest-cost component; it contributes 36% to the MPSP. This implies that using low-cost waste feedstocks will significantly reduce the product cost. Other high-cost components are capital depreciation (13.2%), utilities (12.2%), and maintenance (11.9%). This indicates the significant contributions of capital investment and low-pressure steam consumption to the selling price.

The economic viability of biofuel processes depends on how well the biofuel competes with petroleum-derived fuels. Figure 6.17 tracks the historical monthly spot prices (FOB) of conventional gasoline and kerosene-type jet fuel (at the U.S. Gulf coast) produced from crude oil (priced at Cushing, Oklahoma) during the 10 yr period from January 2001 to January 2011. Petroleum refineries sell gasoline and jet fuel at $1.83/gal when the crude oil price is ~$65–$70/barrel.

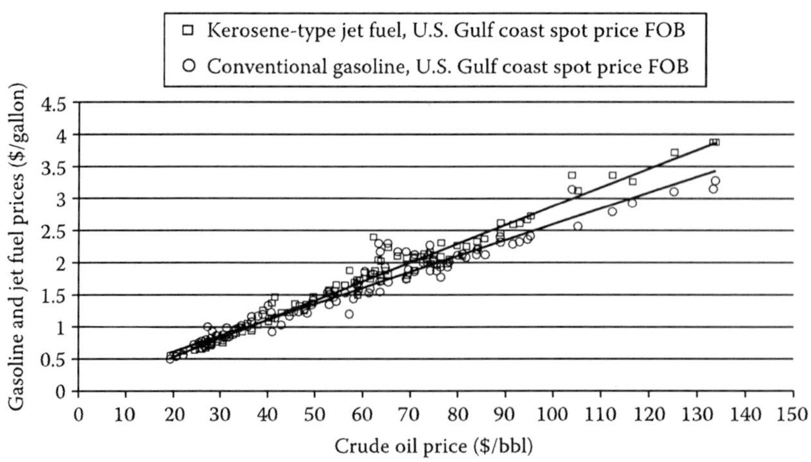

FIGURE 6.17
Historical monthly spot prices of crude oil, gasoline, and jet fuel. (Adapted from EIA. 2011. *Spot Prices for Crude Oil and Petroleum Products.* U.S. Energy Information Administration.)

6.5 Optimization and Sensitivity Analysis

6.5.1 Optimization of Yield and Fermentation Operating Conditions

Process economics are strongly affected by the practical overall yield of the plant, which is controlled by fermentor operating parameters: volatile solid loading rates (VSLR), liquid residence times (LRT), and carboxylate product concentration. These parameters along with other relevant terms are defined as follows.

$$\text{Volatile solid fed (VS)} = \text{Dry biomass} - \text{Ash in biomass} \quad (6.1)$$

$$\text{Conversion} = \frac{\text{VS digested}}{\text{VS fed}} \quad (6.2)$$

$$\text{Selectivity} = \frac{\text{Total carboxylate produced (based on weight of acids)}}{\text{VS digested}} \quad (6.3)$$

$$\text{Overall yield} = \frac{\text{Total volume of hydrocarbon fuel produced}}{\text{Dry biomass weight fed}} \quad (6.4)$$

$$\text{Volatile solids loading rate (VSLR)} = \frac{\text{Weight of VS fed}}{\text{Total liquid volume in all fermentors} \cdot \text{time}} \quad (6.5)$$

$$\text{Liquid residence time (LRT)} = \frac{\text{Total liquid in all fermentors}}{\text{Flow rate of liquid out of fermentor train}} \quad (6.6)$$

$$\text{Fermentation concentration} = \frac{\text{Total corboxylate produced (based on acid weight)}}{\text{Flow rate of liquid out of fermentor train}} \quad (6.7)$$

$$\text{Fermentation time} = \frac{\text{Dry biomass weight in all fermentors}}{\text{Dry biomass rate}} \quad (6.8)$$

$$\text{Substrate concentration} = \frac{\text{Dry biomass weight in all fermentors}}{\text{Dry biomass weight} + \text{Liquid weight in all fermentors}} \quad (6.9)$$

In this sensitivity analysis, VSLR, LRT, and fermentation product concentration were considered as independent variables. All others were calculated accordingly (yield, selectivity, fermentation time, total carboxylate produced, VS fed, VS digested, and rates of liquid) or specified as unchanged inputs (dry biomass, ash content, conversion, and substrate concentration). Table 6.1 reports their values in the base case.

The overall yield is proportional to total acids produced in the fermentation, which assumes that individual yields of all other steps are fixed. From the Equations 6.4 through 6.7, the overall yield is related to the three independent variables as follows:

$$\text{Overall yield} \sim \frac{\text{Fermentation product concentration}}{\text{VSLR} \cdot \text{LRT}} \quad (6.10)$$

TABLE 6.11

Ranges of Fermentation Operation Parameters

Parameter	Unit	Base Case	Investigated Range
Fermentation product concentration	g acids/L	40	10–60
VSLR	g/(L day)	3.0	3–10
LRT	day	28	5–30

This relation indicates that overall yield favors higher fermentation product concentration and lower multiplication of VSLR and LRT.

In fact, the three parameters (VSLR, LRT, and fermentation product concentration) interact as documented in previous work (Domke et al., 2003; Thanakoses et al., 2003; Agbogbo and Holtzapple, 2006; Fu and Holtzapple, 2010). The interactions were predicted using the continuum particle distribution model (CPDM). However, to investigate the effect of operating ranges on process economics in this work, the parameters were arbitrarily varied over predefined ranges (Table 6.11).

To perform the sensitivity analysis, the MPSP was estimated for a number of scenarios involving various combinations of the three independent parameters. Because the parameters affect fermentation outlet flow rates and compositions, the mass and energy balances of the whole plant were changed in every scenario. The capital and operating costs were updated accordingly before the MPSP was recalculated.

Figure 6.18 shows the results of these calculations over the entire parameter range. Some of their characteristics are

- For constant fermentation product concentration, the MPSP is linear with VSLR · LRT. The MPSP decreases as VSLR · LRT decreases. The reason is that VSLR · LRT is inversely proportional to yield (as discussed above), and the MPSP

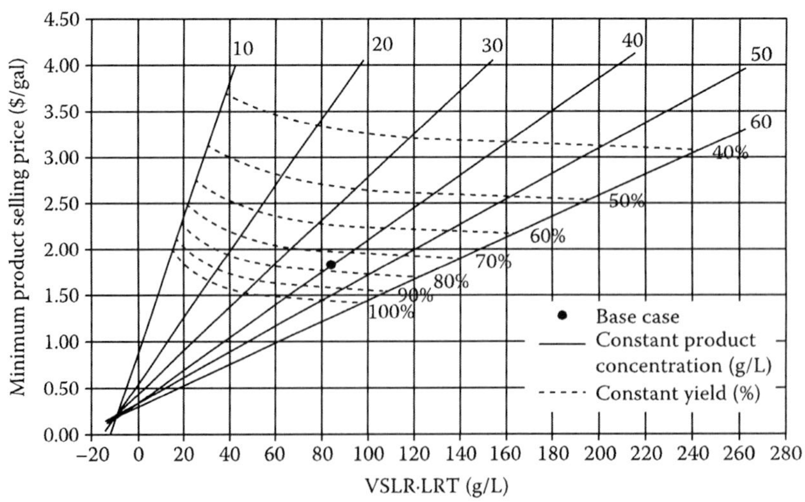

FIGURE 6.18
Minimum product selling prices with respect to multiplication of VSLR and LRT at various concentrations of carboxylic acids in fermentation broth (forage sorghum cost $60/dry ton, after-tax discount rate 10%, hydrogen produced from gasification of fermentation residue, plant capacity 200 dry tons/h, plant life 20 yr).

is generally smaller at higher yields. The value of VSLR·LRT can be reduced until the yield is equal to the theoretical yield. In Figure 6.18, the yield (as a percentage) is defined as:

$$\text{Yield (\%)} = \times 100\%.$$

- Straight lines of constant fermentation-product concentration (solid lines) converge to one point outside the operable region, which is the upper part of the 100% yield curve (dashed curve).
- For constant yield (along the dashed lines), the MPSP decreases as the fermentation-product concentration increases. In the ranges investigated, the increased concentration results in lower fermentation costs because of reduced fermentation time (with constant yield) and lower dewatering costs because less water must be vaporized. In Figure 6.18, yields of 40%, 50%, 60%, 70%, 80%, 90%, and 100% are reported.

Figure 6.18 can be used to predict quickly the overall yield and MPSP given the fermentation operating parameters (VSLR, LRT, and product concentration). In a reverse problem, it can be used to indicate what fermentation parameters are required to reach a desired yield or MPSP. The result shows that the minimum MPSP in the range investigated is $1.42/gal, which would be achieved at an acid concentration of 60 g/L and a VSLR•LRT of 98 g/L (with a yield of 100%).

6.5.2 Vapor-Compression System

The vapor-compression system (Figure 6.7) is the key dewatering unit that concentrates the calcium carboxylate solution in an energy-efficient way. The operating conditions of this system must be optimized and sensitivity analysis performed because (1) the system is highly integrated with the fermentation process, and (2) although excellent performance has been achieved at the laboratory scale, it has not yet been demonstrated at a commercial scale. This section investigates the effects of temperature approach, fouling, and fabrication costs of the latent heat exchangers on the MPSP. Table 6.12 shows the parameter ranges used.

6.5.2.1 Temperature Approach

Temperature approach is the temperature difference (ΔT) between a vaporizing salt solution and its condensing vapor in a stage of the latent heat exchangers. It is related to heat transfer by the following equation:

$$Q = q \cdot A = U \cdot A \cdot \Delta T \tag{6.11}$$

TABLE 6.12

Ranges of Latent Heat Exchanger Parameters

Parameter	Notation	Unit	Base Case	Investigated Range
Temperature approach	ΔT	K	0.20	0.20–3.0
Heat-transfer coefficient	U	kW/(m² K)	240	10–300
Equipment cost	C	\$/m²	155	0–400

where Q = heat transfer (kW), q = heat flux (kW/m^2), ΔT = temperature approach (K), U = overall heat transfer coefficient (kW/(m^2 K)), and A = heat transfer area (m^2).

For 0.203 mm thick copper plates with lead-containing Ni–P–PTFE hydrophobic coatings, saturated steam pressure of 722 kPa, and forced convective saturated liquid, Lara and Holtzapple (2011b) experimentally found that U depends on ΔT in nonfouling conditions as follows:

$$U = 61.1(\Delta T)^{-0.915}. \tag{6.12}$$

Therefore,

$$q = 61.1(\Delta T)^{0.085}, \tag{6.13}$$

$$Q = 61.1 \cdot A \cdot (\Delta T)^{0.085}. \tag{6.14}$$

In traditional heat exchangers, a higher ΔT is favored because the heat transfer area is lower, which reduces equipment costs. However, this latent heat exchanger is limited by the rate at which liquid droplets are shed from the surface. As a consequence, heat flux is nearly independent of ΔT, which is indicated by the near-zero exponent in Equation 6.3 and shown by the continuous curve in Figure 6.19.

As ΔT increases, the design outlet temperature of the vapor compressor must increase, which requires higher compressor outlet pressure, capacity, capital, and operating costs. As ΔT decreases, the design area of latent heat exchangers must increase to maintain heat transfer (see Equation 6.4), which requires higher capital costs for the heat exchangers. In the design of the vapor-compression system, the compression costs dominate at high ΔT

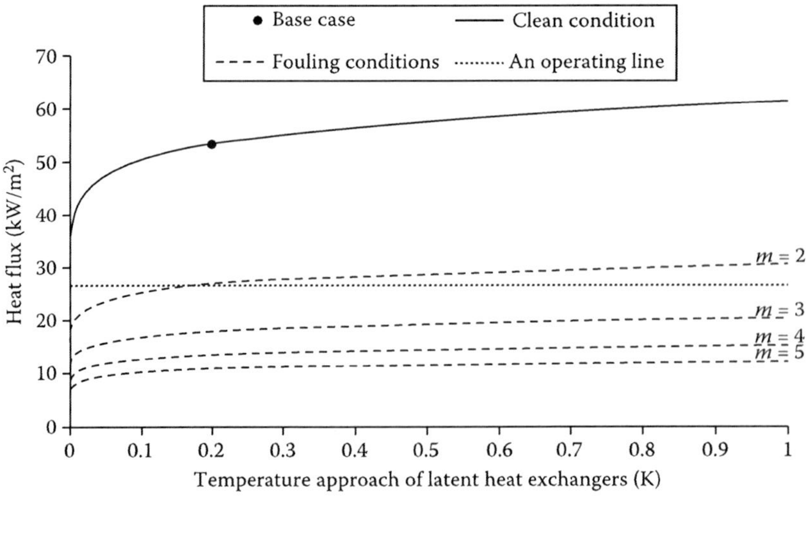

FIGURE 6.19
Heat flux of latent heat exchangers with respect to temperature approach under clean conditions and at various values of the fouling factor.

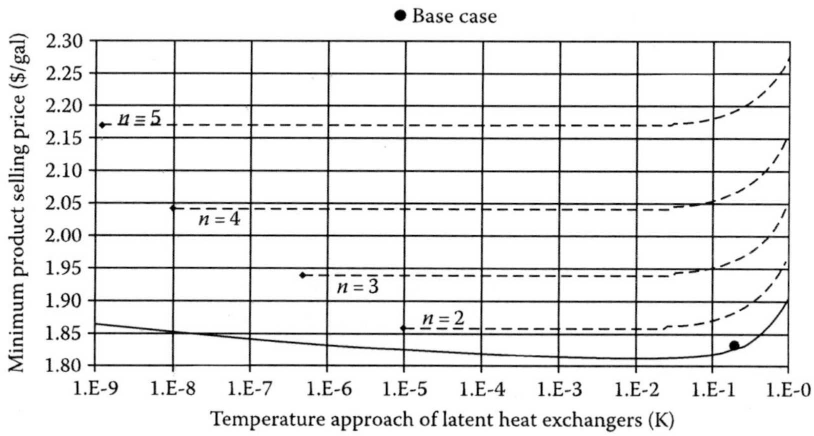

FIGURE 6.20
Minimum product selling price with respect to temperature approach in a clean condition and with various fouling expectation of the latent heat exchangers.

and the heat exchanger costs at low ΔT. A tradeoff between these costs was found at $\Delta T = 0.02$ K, as shown by the minimum of the continuous curve in Figure 6.20.

6.5.2.2 Fouling

In practical operation, the overall heat-transfer coefficient of the latent heat exchangers decreases over time between maintenance servicing. The effect of fouling on the overall heat-transfer coefficient is assumed to follow Equation 6.5:

$$U' = U/m, \tag{6.15}$$

where U' is the fouling overall heat transfer coefficient and m is the fouling factor (e.g., $m = 2$ means that overall heat transfer coefficient is reduced by a factor of two due to fouling). Figure 6.19 depicts the heat flux at some values of m.

With fouling, the heat transfer is

$$Q' = U' \cdot A' \cdot \Delta T' \tag{6.16}$$

where A' is the fouling heat-transfer area and $\Delta T'$ is the fouling temperature approach.

In an operating plant, the heat-transfer area is not changed (i.e., $A = A'$). To maintain heat transfer (i.e., $Q = Q'$) and avoid reducing the vaporization rate as the overall heat-transfer coefficient decreases from fouling, the latent heat exchangers must be designed with surplus area (i.e., they must be oversized) and lower heat flux. (Note: Increasing ΔT without oversizing does not maintain heat transfer because the heat flux is nearly independent of ΔT, as discussed in the previous section.)

In the base case, let A_0 be the ideal heat-transfer area (i.e., the design value if no fouling occurs). The heat exchangers are oversized by a factor of n:

$$A' = n \cdot A_0. \tag{6.17}$$

To maintain heat transfer right after maintenance servicing has been performed (i.e., the heat exchangers are clean), ΔT must be adjusted to a small value such that the heat flux is reduced by a factor of n compared to the ideal design:

$$q' = q_0/n,$$

where q' is the fouling heat flux and q_0 the ideal heat flux. Over time, ΔT increases during operation to keep q constant.

For example, in an ideal scenario where no fouling occurs, the system always operates at the base-case condition, where $q_0 = 53.3$ kW/m^2 and $\Delta T = 0.2$ K. To cope with fouling in a practical scenario, if heat transfer areas are oversized by a factor of two (i.e., $n = 2$), the system initially operates at $\Delta T = 10^{-5}$ K (see Figure 6.20) to keep $q = q_0/2 = 26.6$ kW/m^2. Over time as fouling develops, ΔT must increase to keep q constant at this value, which is represented by the horizontal straight operating line in Figure 6.19. When ΔT is close to 1°K, the production cost significantly increases (see Figure 6.20) because compression energy consumption is higher; therefore, at this point, the system should be shut down to clean the heat exchangers. (Note: It is not necessary to manipulate ΔT directly. It assumes whatever value is necessary to condense the vapor processed by the compressor; hence, ΔT is a dependent variable determined by independently selecting the compressor speed.)

For higher oversizing factors, the operation period of the system is longer. However, the MPSP must be higher to account for the costs of larger heat exchangers. Figure 6.20 shows the MPSP for various values of the oversizing factor n.

6.5.2.3 Cost of Latent Heat Exchangers

For the reason that this technology (dropwise condensation in a latent heat exchanger with an extremely high heat-transfer coefficient of 240 kW/(m^2 K) has not been applied at a commercial scale, the coating cost of the plates is uncertain. In the base case, the equipment cost (C) is assumed to be \$155/m^2 of heat transfer area. This sensitivity analysis investigates a wide range of costs from \$100 to \$400/m^2 (Figure 6.21) assuming $\Delta T = 0.20$ K.

At high values of U, the change of C does not increase MPSP much. For example, at $U = 200$ kW/(m^2 K), the MPSP increases by only \$0.04/gal as C increases four times, from \$100 to \$400/m^2. However, at low values of U, an increase in C significantly affects the MPSP. For example, at $U = 30$ kW/(m^2 K), the MPSP increases by \$0.24/gal for the same fourfold increase in C. The dependence of the MPSP on U and C follows this equation:

$$\text{MPSP} = C + 1.8072. \tag{9.18}$$

6.5.3 Sources of Hydrogen

The results of the base-case economic analysis shown in Section 6.4.2 reveal that producing hydrogen from fermentation residue is expensive. In a design in which hydrogen is available from an external source (e.g., a nearby petroleum refinery) and the fermentation residue is gasified only for steam and power, the MPSP will be lower if the external hydrogen is inexpensive. However, if the external hydrogen is expensive, then it is better to produce hydrogen from fermentation residues. A sensitivity analysis of the MPSP with respect to external hydrogen prices (Figure 6.22) shows that the break-even point is \$2.44/kg of external hydrogen.

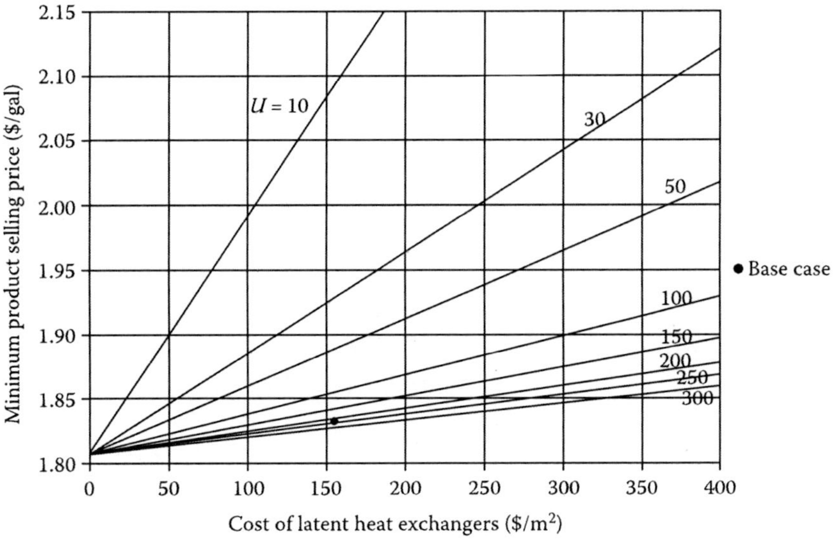

FIGURE 6.21
Minimum product selling price with respect to the purchased cost of latent heat exchangers at various values of the overall heat-transfer coefficient U (kW/(m^2 K) at $\Delta T = 0.2$ K.

In another design option in which hydrogen is internally produced from natural gas and the fermentation residue is gasified only for steam and power, the MPSP is lower than $1.89/gal if the natural-gas price is less than $14.2/GJ. When calculating this case, it was assumed that the relationship between hydrogen price and natural-gas price was linear (Holtzapple and Granda, 2009).

In other words, the approach of producing hydrogen from fermentation residues is economically favored when external hydrogen and natural-gas prices are higher than $2.44/kg and $14.2/GJ, respectively. If those prices are lower than these breakpoints, buying hydrogen or investing in a natural-gas reformer would be more economical.

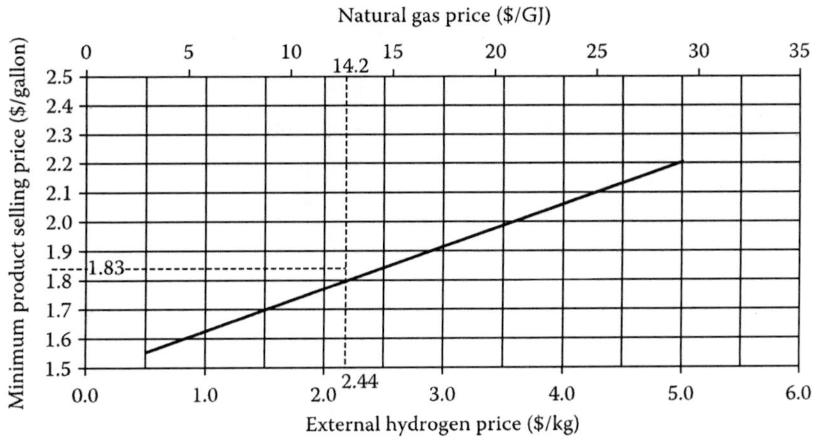

FIGURE 6.22
Plot of MPSP versus external hydrogen prices for the case of no hydrogen production.

6.5.4 Other Sensitivity Analyses

Figure 6.23 shows the sensitivity of the minimum selling price to the following key parameters: after-tax discount rate or return on investment (ROI), overall yields, FCI, external hydrogen prices, and key raw material prices. If investors expect an ROI of 20% or 25%, the product fuel must be sold for \$2.23 or \$2.47/gal hydrocarbon, respectively. For a 50% increase in FCI, the product selling price is \$2.06/gal hydrocarbon. A decrease of overall yields by 10% results in a selling price of \$2.03/gal hydrocarbon. If the lime kiln is not used and the pretreatment stage is fed fully with purchased fresh lime, the selling price increases by \$0.11/gal hydrocarbon. Investing in a lime kiln to process the calcium carbonate from ketonization and buying external lime at \$120/ton to satisfy makeup demand increases the MPSP by \$0.08/gal hydrocarbon. In the worst-case scenario of manure price, the selling price is not much affected. Unlike the main biomass (sorghum), free manure does not reduce the hydrocarbon selling price significantly.

Table 6.10 shows that feedstock costs have high impacts on product price. Figure 6.24 depicts the effect of feedstock price on hydrocarbon selling prices. This sensitivity analysis was done for energy sorghum. For other feedstocks with the same volatile solids content as sorghum (91.9% weight), the results are identical. Some feedstocks (e.g., municipal solid waste, food waste) may involve tipping fees, whereas sorghum does not. If such a feedstock is delivered for free to the plant gate (i.e., the tipping fee is assumed barely to cover collecting, sorting, and transporting costs) and the plant uses internal hydrogen, the MPSP is only \$1.20/gal hydrocarbon at the base-case capacity (Figure 6.23). In a reasonable scenario where biomass is available for free at the plant gate and external hydrogen is supplied at \$1/kg (the current price is based on natural gas costing \$5.21/GJ or \$5.50/MMBtu), the MPSP of hydrocarbon fuels is \$0.99/gal hydrocarbon, as shown in Figure 6.23.

Figure 6.24 shows the hydrocarbon selling prices and FCI with respect to capacity for a plant at various prices of biomass. The FCI curve is represented by a function of capacity to the power 0.63. At capacities of 300 ton/h or more and biomass cost of \$60/ton, the MPSP approaches \$1.65/gal hydrocarbon.

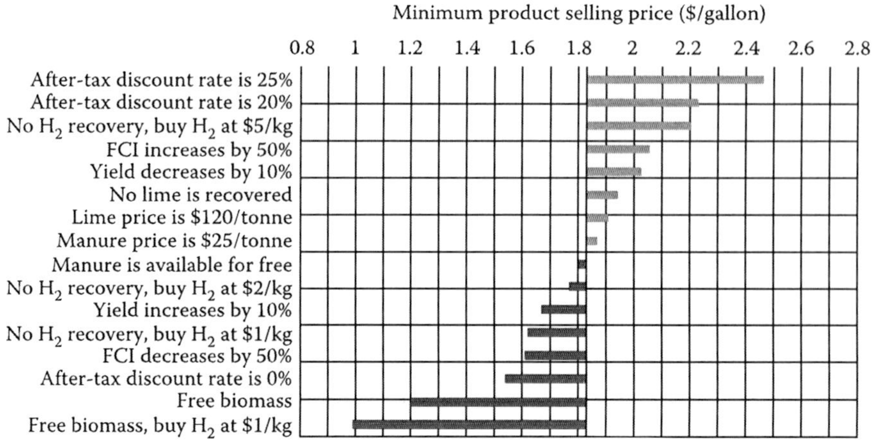

FIGURE 6.23
Sensitivity analysis of the key factors on minimum product selling price.

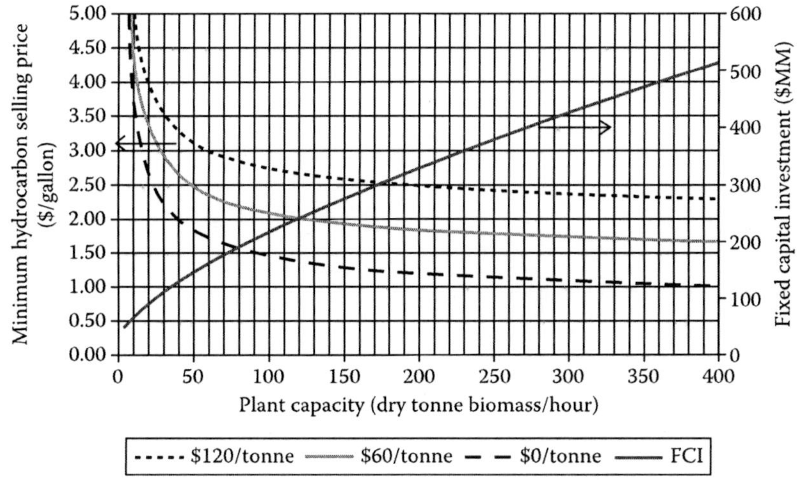

FIGURE 6.24
Minimum selling prices and FCI versus plant capacities using biomass at various prices (sorghum) and $10/dry ton (manure). Hydrogen is produced by gasification.

6.6 Conclusions

This technoeconomic analysis of producing hydrocarbon fuel from lignocellulose through the carboxylate platform was performed using extensive sources of published data and computers for simulation and cost estimation. The following technical advantages of the carboxylate platform were identified: no sterilization, no external enzymes, and low capital cost of pretreatment and fermentation. The plant is highly integrated to overcome common challenges in biorefineries such as lime consumption in pretreatment, calcium carbonate consumption for buffering, water removal for concentrating the fermentation product, and hydrogen supply.

The economic analysis shows that the effects of fermentation operating parameters on process economics can be generalized (Figure 6.18). This result can be used to estimate the MPSP quickly given any values of the operating parameters. Alternatively, it can quickly target the required parameters to achieve a desired selling price. The economic analysis also shows that the temperature approach of latent heat exchangers is preferred to be as small as 0.20°K. The analysis identified that inexpensive feedstock and replacement of the crystallizer were the keys to lowering the cost significantly.

The process does not need external hydrogen, but it is more economical if external hydrogen is available for less than $2.44/kg, which occurs when natural gas is below $14.2/GJ ($15.0/MMBtu).

The base case with reasonable expectations on technical performances and feedstock cost ($60/dry ton biomass) requires an MPSP of $1.83/gal hydrocarbon fuel ($1.25/gal equivalent ethanol) at an after-tax discount rate of 10%. In particular, the minimum selling prices of hydrocarbon fuels can be ~$1.20/gal ($0.82/gal equivalent ethanol) if municipal solid waste is available for free at the plant gate (200 ton/h plant with internal hydrogen production).

Nomenclature

Abbreviations

ASPEN	Advanced simulator for process engineering
CS	Carbon steel
SS	Stainless steel
DB	Declining balance
DDB	Double declining balance
DW&B	Direct wage and benefits
FCI	Fixed capital investment
FOB	Free on board
LRT	Liquid residence time
MW&B	Maintenance wages and benefits
NREL	National renewable energy laboratory
PEA	Process economic analyzer (a software product of ASPEN Technology, Inc.)
PSA	Pressure swing absorption
ROI	Return on investment
TCI	Total capital investment
VSLR	Volatile solid loading rate
WCI	Working capital investment

Parameters

ΔT	Temperature approach of latent heat exchangers in nonfouling conditions
$\Delta T'$	Temperature approach of latent heat exchangers in fouling conditions
A	Heat transfer area of latent heat exchanger in nonfouling conditions
A'	Heat transfer area of latent heat exchanger in fouling conditions
A_0	Base-case heat-transfer area of latent heat exchanger
m	Fouling factor
n	Oversizing factor
q	Heat flux in latent heat exchangers
q'	Fouling heat flux in latent heat exchangers
q_0	Base-case heat flux in latent heat exchangers
Q	Heat-transfer load in latent heat exchanger in nonfouling conditions
Q'	Heat-transfer load in latent heat exchanger in fouling conditions
U	Heat-transfer coefficient of latent heat exchanger in nonfouling conditions
U'	Heat-transfer coefficient of latent heat exchanger in fouling conditions

Units

gal	gallon
h	hour
kg	kilogram
m	meter
MW	megawatts
ton	a metric weight unit equal to 1000 kg.
yr	year
yrs	years

Legal Disclaimer

MixAlco™ is a registered trademark of Terrabon, Inc. Unless otherwise noted, inclusion of such trademark in this document does not imply support or endorsement by Terrabon, Inc. Except as expressly referenced in this presentation, the information, estimates, projections, calculations, and assertions expressed in this presentation have not been endorsed, approved, or reviewed by any unaffiliated third party, including Terrabon, Inc., and are based on the authors' own independent research, evaluation, and analysis. The views and opinions of the authors expressed herein do not state or reflect those of such third parties, and shall not be construed as the views and opinions of such third parties.

References

Aden, A., Ruth, M., Ibsen, K., Jechura, J., Neeves, K., Sheehan, J., Wallace, B., Montague, L., Slayton, A., and Lukas, J. 2002. *Lignocellulosic Biomass to Ethanol Process Design and Economics Utilizing Co-Current Dilute Acid Prehydrolysis and Enzymatic Hydrolysis for Corn Stover*. Golden, CO: National Renewable Energy Laboratory.

Agbogbo, F.K. and Holtzapple, M.T. 2006. Fermentation of rice straw/chicken manure to carboxylic acids using a mixed culture of marine mesophilic microorganisms. *Applied Biochemistry and Biotechnology* **129–132**: 997–1014.

Agler, M., Wrenn, B., Zinder, S., and Angenent, L. 2010. Waste to bioproduct conversion with undefined mixed cultures: The carboxylate platform. *Trends in Biotechnology* **29**(2): 70–78.

ASPENTech. 2001. *ASPEN Plus: System Management*, Version 11.1. Cambridge, MA: ASPEN Technologies, pp. 4.1–4.9.

ASPENTech. 2010a. *ASPEN Plus: ASPENOne*. Cambridge, MA: ASPEN Technologies.

ASPENTech. 2010b. *ASPEN Process Economic Analyzer. ASPENOne*. Cambridge, MA: ASPEN Technologies.

Chang, N.S. 1994. *The Kinetics Studies of Enzymatic Cellulose Hydrolysis and Catalytic Ketone Hydrogenation*. PhD dissertation, Texas A&M University, College Station, TX.

Chemical Engineering 2011. *Chemical Engineering's Plant Cost Index*. New York: McGraw Hill.

Domke, S.B., Aiello-Mazzarri, C., and Holtzapple, M.T. 2003. Mixed acid fermentation of paper fines and industrial biosludge. *Bioresource Technology* **91**(1): 41–51.

EIA. 2011. *Spot Prices for Crude Oil and Petroleum Products*. U.S. Energy Information Administration, Available at http://www.eia.gov/dnav/pet/pet_pri_spt_s1_d.htm. Accessed July 30, 2012.

El-Halwagi, M. 2006. *Process Integration*. Amsterdam: Academic Press.

El-Halwagi, M.M. 2012. *Sustainable Design through Process Integration: Fundamentals and Applications to Industrial Pollution Prevention, Resource Conservation, and Profitability Enhancement*. Oxford, UK: Butterworth-Heinemann/Elsevier.

Fu, Z. and Holtzapple, M. 2010. Fermentation of sugarcane bagasse and chicken manure to calcium carboxylates under thermophilic conditions. *Applied Biochemistry and Biotechnology* **162**(2): 561–578.

Garrett, D. 1989. *Chemical Engineering Economics*. New York: Van Nostrand Reinhold.

Granda, C., Holtzapple, M., Luce, G., Searcy, K., and Mamrosh, D. 2009. Carboxylate platform: The MixAlco process, Part 2: Process economics. *Applied Biochemistry and Biotechnology* **156**(1): 107–124.

Holtzapple, M. 2003a. Cellulose. In: Caballero, B., Trugo, L.C., and Finglas, P.M., (eds.), *Encyclopedia of Food Science, Food Technology, and Nutrition*, 2nd ed. London: Academic Press, pp. 998–1007.

Holtzapple, M. 2003b. Lignin. In: Caballero, B., Trugo, L.C., and Finglas, P.M. (eds.), *Encyclopedia of Food Science, Food Technology, and Nutrition*, 2nd ed. London: Academic Press, pp. 3535–3542.

Holtzapple, M., Davison, R., Ross, M., Aldrett-Lee, S., Nagwani, M., Lee, C.-M., Lee, C. et al. 1999. Biomass conversion to mixed alcohol fuels using the MixAlco process. *Applied Biochemistry and Biotechnology* **79**(1): 609–631.

Holtzapple, M.T. and Granda, C.B. 2009. Carboxylate platform: The MixAlco process, Part 1: Comparison of three biomass conversion platforms. *Applied Biochemistry and Biotechnology* **156**(1–3): 525–536.

Kazi, F., Fortman, J., Anex, R., Kothandaraman, G., Hsu, D., Aden, A., and Dutta, A. 2010. *Techno-Economic Analysis of Biochemical Scenarios for Production of Cellulosic Ethanol*. National Renewable Energy Laboratory (NREL), Golden, CO.

Lara, J.R. and Holtzapple, M. (2011a). Experimental investigation of dropwise condensation on hydrophobic heat exchangers. Part 1: Dimpled sheets. *Desalination* **278**(1–3): 165–172.

Lara, J.R. and Holtzapple, M. (2011b). Experimental investigation of dropwise condensation on hydrophobic heat exchangers. Part 2: Effect of coatings and surface geometry. *Desalination* **280**(1–3): 363–369.

Lau, M.H., Richardson, J.W., Outlaw, J.L., Holtzapple, M.T., and Ochoa, R.F. 2006. The economics of ethanol from sweet sorghum using the MixAlco process. http://www.afpc.tamu.edu/pubs/2/446/RR%2006–2.pdf. Accessed August 11, 2009.

Linnhoff, B. 1993. Pinch analysis: a state-of-the-art overview: Techno-economic analysis. *Chemical Engineering Research & Design* **71**(5): 503–522.

Palmisano, A. and Barlaz, M. 1996. *Microbiology of Solid Waste*. Boca Raton: CRC.

Peters, M.S., Timmerhaus, K.D., and West, R.E. 2003. *Plant Design and Economics for Chemical Engineers*. New York: McGraw-Hill Professional.

Pham, V., Holtzapple, M., and El-Halwagi, M. 2010. Techno-economic analysis of biomass to fuel conversion via the MixAlco process. *Journal of Industrial Microbiology and Biotechnology* **37**(11): 1157–1168.

Rapier, R. 1995. *Volatile Fatty Acid Fermentation of Lime-Treated Biomass by Rumen Microorganisms*. Master's Thesis, Texas A&M University, College Station, Texas.

Seider, W., Seader, J., and Lewin, D. 2004. *Product and Process Design Principles: Synthesis, Analysis, and Evaluation*. New York: Wiley.

Short, W., Packey, D.J., and Holt, T. 1995. *A Manual for the Economic Evaluation and Energy Efficiency and Renewable Energy Technologies*. Golden, CO: National Renewable Energy Laboratory, p. 19.

Sims, B. 2011. Terrabon achieves production milestone at Texas demo facility. *Biorefining* http://www.biorefiningmagazine.com/articles/5278/terrabon-achieves-production-milestone-at-texas-demo-facility. Accessed January 27, 2011.

Thanakoses, P., Black, A.S., and Holtzapple, M.T. 2003. Fermentation of corn stover to carboxylic acids. *Biotechnology and Bioengineering* **83**(2): 191–200.

Walas, S.M. 1988. *Chemical Process Equipment Selection and Design*. Newton, MA: Butterworth.

Wooley, R. and Putsche, V. 1996. *Development of an ASPEN PLUS Physical Property Database for Biofuels Components*. Golden, CO: National Renewable Energy Laboratory.

Wright, M.M., Daugaard, D.E., Satrio, J.A., and Brown, R.C. 2010. Technoeconomic analysis of biomass fast pyrolysis to transportation fuels. *Fuel* **89**(Suppl. 1): S2–S10.

Section II

Biorefinery Supply Chains

7

Value-Chain Management Considerations for the Biorefinery

Louis Patrick Dansereau, Mahmoud M. El-Halwagi, and Paul R. Stuart

CONTENTS

Recently, most of the research efforts related to the biorefinery have been focused on technology, process design, and synthesis, as well as on biomass logistics. However, research related to products and integrated planning can also provide interesting insights into how to select a product portfolio and manage its production. The objective of this chapter is to introduce value-chain and planning considerations that are of critical importance to the design of the biorefinery. In fact, different products face different market environments in terms of size, number of players, competition, level of customer service required, price volatility, and other factors. To make the biorefinery truly profitable and competitive, value-chain management (VCM) approaches should therefore be adapted to the biorefinery products that will be manufactured.

This chapter begins with a literature review on supply-chain management (SCM) (basics of planning, manufacturing environments, process characteristics, types of products, contracts, and partnerships), followed by an overview of possible biorefinery feedstocks, focusing on lignocellulosic biomass, processes, and product types (energy, biofuels, commodity biochemicals, specialty and fine chemicals, and biomaterials). The next two subsections discuss how integrated supply-chain planning from procurement to sales, that is, VCM, can be generally used in the process industries, but more specifically in biorefineries, to manage the risks of market volatility and to increase margins by revenue management (RM) and better coordination, all this depending on the type of product. Products and associated markets, location, and geographical aspects such as access to market and customers, to name a few, highlight the fact that each biorefinery configuration will be unique, and that therefore its supply chain should be carefully designed, managed, and adapted accordingly to obtain the most value.

7.1 Introduction

In the quest for sustainable development, academia and the process industry have been showing increasing interest in carbon-neutral or near-carbon-neutral technologies for producing green alternatives to fossil-based chemicals, materials, and energy. The biorefinery concept therefore appears a promising avenue for process-industry companies to enhance their environmental profile, to diversify their revenue stream, or even to survive, as in the case of the pulp and paper industry. Because no commercial biorefinery has yet been built, most research efforts have focused on technology design and biomass logistics. However, from a company's point of view, selecting the right product portfolio, as well as its associated technologies and capacities, poses a major challenge. Furthermore, it cannot be assumed that these biorefineries can compete against the petroleum industry and be profitable in a price-volatile environment. Fortunately, research related to products and integrated planning can provide interesting insights on how to select a product portfolio and manage its production to be as profitable as possible.

The objective of this chapter is to introduce value-chain and planning considerations that are of critical importance to the design of the biorefinery. In fact, different products face different market environments in terms of size, number of players, competition, level of customer service required, price volatility, and other factors. To make the biorefinery truly profitable and competitive, VCM approaches should therefore be adapted to the biorefinery products that will be manufactured.

This chapter is organized as follows. The first section is a literature review of SCM concepts as applied to the process industries in general. Aspects of planning, manufacturing environments, process characteristics, types of products, contracts, and partnerships are addressed. The second part starts with an overview of biorefinery feedstocks, focusing on lignocellulosic biomass, and continues with a brief description of biorefinery processes and their integration in pulp and paper mills. Biorefinery products are then described according to the supply-chain aspects introduced in the first section. These products have been grouped into five categories: bioenergy, biofuels, commodity biochemicals, specialty and fine chemicals, and biomaterials. In the third section, strategies for managing price volatility and market risk will be introduced and applied in the context of the biorefinery.

Finally, the last section addresses integrated supply-chain planning from procurement to sales, that is, VCM, and how it can be generally used in process industries, but more specifically in biorefineries, to increase margins by RM and better coordination.

7.2 Supply-Chain Management in Process Industry

A supply chain (SC), as defined by Christopher, consists of a network of organizations that are involved, through upstream and downstream linkages, in the different processes and activities that produce value in the form of products and services in the hands of the ultimate consumer [1].

As described by Shapiro, a "company's supply chain contains geographically dispersed facilities where raw materials, intermediate products, or finished products are acquired, transformed, stored, or sold and transportation links that connect facilities along which products flow. The facilities may be operated by the company, or they may be operated by vendors, customers, third-party providers, or other firms with which the company has business arrangements" [2].

Over the past decades, SCM concepts have become a subject of growing interest in industry and academia. Several authors have given different definitions of SCM. However, the following elements are found in the majority of these [2–6]:

- Planning and management of raw materials procurement, production, distribution, and other logistical activities (warehousing, transportation) to satisfy customer demand
- Management of facilities (plants, distribution centers, etc.) that are dispersed geographically

- Coordination and collaboration activities within the enterprise and between partners (suppliers, clients, third-party service providers, etc.)
- Integration of material, cash, and information flows between different actors, whether inside or outside the company

The objective of SCM could be summarized as follows: to deliver the right product at the right timing and at the lowest cost by the coordination of material, cash, and information flows from the supplier to the final customer in order to

1. Maximize profitability of a company (and/or all its supply-chain partners)
2. Achieve high customer satisfaction levels
3. Create a sustainable competitive advantage

There are four main planning tasks associated with SCM: procurement, production, distribution, and sales [3,6]. *Procurement* deals with all processes that provide the resources (raw materials, personnel, etc.) necessary for production. *Production* includes all the capacity-limited processes involved with the manufacturing of products. *Distribution* bridges the distance between the production site and the customers, whether they are final clients, retailers, or other enterprises that will process the products further. Finally, *sales* processes are involved with receiving and filling the customer's orders. All these logistical processes are driven by demand forecasts and order figures determined by the sales process.

Successful SCM requires many decisions about material, cash, and information flows. These decisions are usually divided into three levels or phases: strategic, tactical, and operational [2,3,6]. These levels are characterized by the decision frequency and the time horizon considered.

Strategic planning deals with decisions related to the structure of the supply chain (e.g., facility location) for the next few years. It is long-term planning (3–10 years) in which the decisions involved are related to investment or divestment activities, which are decisions that are often irreversible or very costly to alter.

Tactical (mid-term) planning establishes the parameters within which a supply chain will work over a period of time. Investment opportunities are not evaluated at this level since the SC structure has been fixed during strategic planning. Decisions concern rather the allocation of resources over a time horizon of 6–24 months to maximize the profit of the company. Short-term operating policies, such as the assignment of production targets to facilities, procurement planning, and transportation from facilities to warehouses to distribution centers, are defined, taking into account the flexibility of the SC. On the customer side, demand planning tasks such as aggregate sales forecasting and what-if scenario analyses are usually performed to support shorter-term sales decisions.

Finally, operational planning concerns short-term decisions such as the assignment of specific orders to production, the assignment of tasks to units, the sequencing of tasks in each unit, and the scheduling of raw material or product deliveries. These decisions are carried out on a daily or weekly basis and are referred to as scheduling. Demand fulfillment decisions, including order acceptance and due-date promising or capacity reservation, are also made at this level.

The general planning problems briefly discussed are presented in Figure 7.1 in the form of a supply-chain planning matrix structured according to the various planning tasks (procurement, production, distribution, and sales) and the time-dependent decision levels.

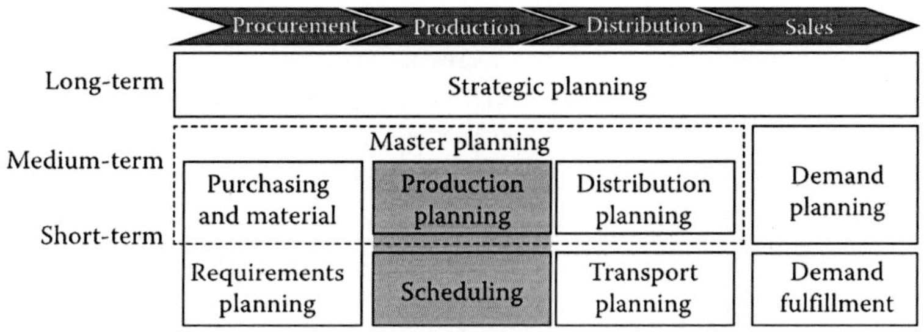

FIGURE 7.1
Supply-chain planning matrix. (With kind permission from Springer Science+Business Media: *Supply Chain Management and Advanced Planning: Concepts, Models, Software, and Case Studies*, 2008, Stadtler, H. and Kilger, C. 4th ed., Berlin: Springer-Verlag.)

As discussed by several authors, it is possible to achieve higher levels of efficiency by achieving a better coordination of material, cash, and information along the SC [2,5–7]. Shapiro defines three possible levels of integration for SCM: functional integration (or horizontal), the integration of procurement, production, and distribution activities within the same enterprise and among partners; spatial integration, the integration of these activities through the various facilities present in the SC; and intertemporal (or vertical) integration, the coherent integration of the three decision levels (strategic, tactical, and operational) [2,8]. Thus, by integrating and coordinating the different SC activities across facilities and decision levels, better planning and scheduling can be achieved, leading to higher profits.

The application of SCM and advanced planning concepts has greatly benefited from recent advances in information technologies. With the implementation of enterprise resource planning (ERP) systems and other transactional systems, from SAP [9] and Oracle [10], for example, it is now possible for managers to have rapid access to huge amounts of homogeneous information on production, logistics, sales, and procurement activities. However, as stated by several authors, competitive advantage is not gained simply through the acquisition and use of systems providing faster and cheaper communication of data [2,7,11]. Nevertheless, the use of tools to analyze this transactional data enables managers to forecast and explore possibilities while designing, planning, and operating the supply chain. These analytical tools provide comprehensive decision-making capabilities across the various functions, enabling the supply chain to run more efficiently.

One of these analytical tools that has been extensively studied is advanced planning.* Advanced planning systems (APS) do not replace, but rather supplement existing ERP systems. Although ERP systems are still required for transaction and execution purposes, planning tasks are now taken over by the APS, which uses optimization models to find better plans and schedules at the three decision levels (strategic, tactical, and operational). APS software providers typically offer their models as a set of planning modules divided more or less according to the modular structure presented in Figure 7.1.

Depending on the decision level, different objective functions might be used, such as revenue maximization, cost minimization, lead time minimization, and others. Shapiro argues that strategic and tactical planning should seek to maximize net revenues or return on investment (ROI) instead of minimizing costs [8]. According to him, minimizing costs

* See Fleischmann et al. for additional information on advanced planning [12] Fleischmann, B., Meyr, H., and Wagner, M. Advanced Planning. In: Stadtler, H. and Kilger, C. (eds.), *Supply Chain Management and Advanced Planning: Concepts, Models, Software and Case Studies*. Berlin: Springer-Verlag, 2008.

at these levels is a timid and shortsighted objective. To maximize net revenues or ROI, demand and sales should be included in medium- and long-term planning.

Rather than finding the absolute optimum plan, the power and value of APS reside in the possibility of evaluating different modes of operation. At longer-term levels, different future development scenarios can be planned to identify a robust next step for the upcoming planning interval. At a shorter-term level, the planner may evaluate the actual feasibility of the proposed plans given by the APS, compare the trade-offs between the options, and choose the best plan based on his knowledge before it is implemented.

As APS and ERP systems become more widely used, the better use of available data and the new degree of coordination involve several changes within the company. They demand better communication of information between departments that have traditionally been accustomed to working independently. In addition, managers will be increasingly called upon to use models and other tools to support their decision-making. However, managers have traditionally been reluctant to use such tools, although this reluctance is beginning to fade [2]. Human planners and decision-makers typically fear that information technology tools will replace them. In fact, APS are not meant to replace human planners, but are truly decision-support systems. Models are simplifications of the reality. Therefore, human knowledge, experience, and skill will always be needed to fill in the gap between model and reality. All these changes represent a true transformation of the company and in the way companies used to operate.

7.2.1 Decoupling Point and Manufacturing Environments

The different processes in a supply chain can be divided into two categories depending on whether they are executed in response or in anticipation to a customer order [3]. Processes that are initiated by a customer order are said to be *pull*: processes are operated in a context where the demand is known. By contrast, *push* processes are initiated and performed without knowing exactly the actual demand; they are operated using forecasts. Because no supply chain can be purely push or pull, there exists an interface that divides processes between these two views. This interface is generally called the decoupling point (DP) or the order penetration point. The position of the DP in a supply chain varies with products, markets, companies, and other factors and might also change over time as the market evolves. It should be mentioned, however, that pull processes are often constrained by inventory and capacity decision that are made in the push phase.

Manufacturing environments can be described according to the position of the DP and the lead time (average time interval between ordering and receiving a material) that the customer can accept. Three general environments are often characterized in the literature: make-to-order (MTO), assemble-to-order (ATO), and make-to-stock (MTS). These environments will be briefly discussed in the following subsections. For more information on these concepts, the reader is referred to Kilger and Meyr [13].

7.2.1.1 Make-to-Order

In MTO environments, the DP is located upstream in the supply chain. Production is triggered by incoming customer orders, and procurement of raw materials has already been done according to forecasts. These raw materials are kept in inventory at the DP and are available for manufacturing. MTO environments are appropriate primarily for high-value or customer-specific products. Customers place their orders before the start of the manufacturing process and must wait while their product is manufactured, assembled, and delivered. Because there is considerable lead time and customers must wait for their orders to be processed, one of the scheduling problems that arise in MTO is the assignment of orders to equipment and the promising of order due dates with reference to bottlenecks and available production and distribution capacity.

7.2.1.2 Assemble-to-Order

In ATO environments, basic manufacturing steps have already occurred when the order is placed. Processing steps after the DP are related to final product assembly, customization, and distribution to the client. Inventory kept at the DP consists of fairly standard products or components that are produced according to forecasts. These intermediates may be produced by the company, but are often delivered by external suppliers.

This type of manufacturing is common in the discrete industries, such as the automobile industry, where some parts are manufactured (by external suppliers or by the company) and stored. The final assembly of the car is only done once an order is placed. ATO processing is also found for certain products in some process industries, such as pulp and paper. Some types of standardized paper rolls are produced and kept in inventory at the mill, and once an order arrives, these rolls are cut and wrapped according to the customer's specifications.

7.2.1.3 Make-to-Stock

Production is based on forecasts in MTS environments. Final products are manufactured and held in inventory, ready for final distribution to the customer. Customer requests are served directly from this stock. The DP is therefore located downstream in the supply chain. MTS are especially adapted for products that require no specific customization, such as in the consumer-goods industries where standardized packaging is used. In this kind of environment, customers usually expect short lead times, and this might in fact be a condition for sale. Therefore, multiple warehouses or distribution centers may exist to reduce this lead time to the minimum. In an MTS production environment, good forecasting accuracy is important as inventory and warehousing costs can be substantial.

7.2.1.4 Position of the Decoupling Point

Deciding where the DP will be and therefore the nature of the manufacturing environment when designing a company's supply chain is an important strategic question which must be addressed because it will have numerous implications for procurement, production, distribution, and sales decisions. On one side, the closer the DP is to the market, the shorter is the customer's service time, and the higher the inventory levels will be. Because customer orders are mostly served directly from the inventory of finished products, demand forecasting accuracy becomes of critical importance. On the other side, when the DP is located further from the market, response times will be longer, and inventory levels will be lower. The promising of delivery dates and production capacity allocation decisions become important points to address.

The position of the DP in MTO, ATO, and MTS environments is shown in Figure 7.2. The type of inventory at the DP will vary depending on the production environment. For instance, in MTO, the safety stock will be constituted of raw materials. Final products will be held in inventory in MTS, while in ATO, intermediates will be kept.

In practice, manufacturing environments seldom belong purely to one of the types presented previously. They are more often a mixture of MTO, ATO, and MTS. For instance,

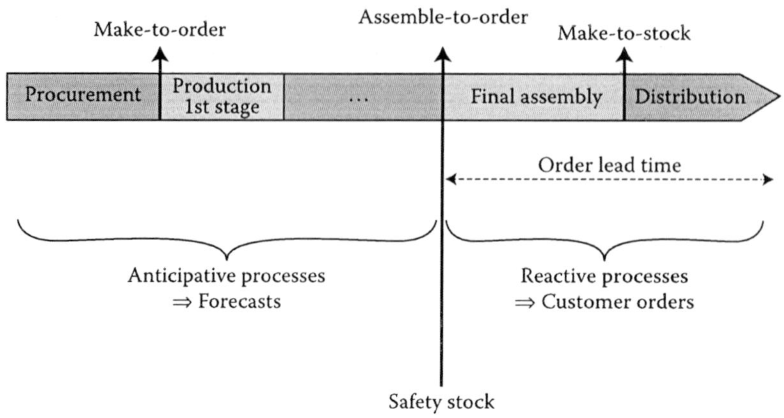

FIGURE 7.2
Decoupling point in MTO, ATO, and MTS environments. (With kind permission from Springer Science+Business Media: Demand fulfilment and ATP, *Supply Chain Management and Advanced Planning: Concepts, Models, Software and Case Studies*, 2008, Kilger, C. and Meyr, H. In: Stadtler, H. and Kilger, C. (eds.), Berlin: Springer-Verlag.)

one process line could trigger production when a specific order is received, but also at the same time keep an inventory of the same product for other customers who usually buy from this stock. In addition, it also often happens that only a fraction of the orders is known at the time of planning, and therefore the manufacturer must rely partly on forecasts. These environments are referred to as hybrid manufacturing environments and are quite common in the process industry.

7.2.2 Manufacturing Process Characteristics

Industries can be classified mainly into three groups: process, discrete, and service industries. Service industries provide intangible services that do not require physical production. Discrete industries such as the automotive industry assemble products using other discrete components. Production is therefore convergent: multiple components are assembled into the final product. In contrast, the process industries are characterized by the fact that the final product cannot be disassembled back to its components. For example, it is possible to disassemble a car relatively easily into wheels, engine, seats, radio, and so on, but it is very difficult to separate a bottle of shampoo back into its chemical components. Companies in the process industry add value to their products by mixing, separating, performing chemical reactions, or a combination of these. Production may be convergent, but it is also often divergent. Divergent processes are characterized by the fact that from one raw material (e.g., petroleum), several products are made (e.g., diesel, naphtha, aromatics, bitumen, etc.).

Subindustries of the process industry include oil and gas, chemical, steel and metals, pulp and paper, pharmaceuticals, and food products. Figure 7.3 presents the three types of industries, with examples of their subindustries.

Chemical process units (mixers, separators, or reactors) are often dedicated units that have been designed solely for their specific purpose. Their design has been optimized to maximize production yield and security, taking into account process characteristics such as the phases present, heating/cooling requirements, temperature, pressure, corrosion potential, material flow, and so on. However, some pieces of equipment are designed to

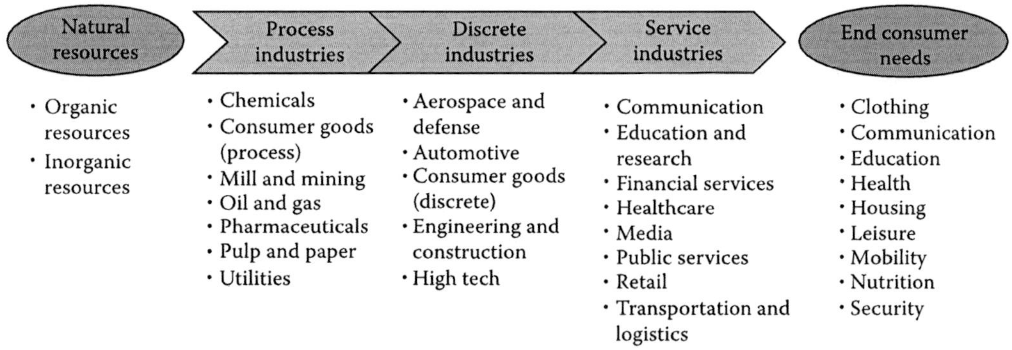

FIGURE 7.3
Types of industries. (With kind permission from Springer Science+Business Media: *Value Chain Management in the Chemical Industry—Global Value Chain Planning of Commodities*, 2008, Kannegiesser, M., Berlin: Physica-Verlag.)

withstand a range of conditions and are able to produce several products at different throughputs. Chemical process equipments may run either in continuous or in batch mode.

In continuous units, reactants are added and products removed at a constant mass flow rate. The throughput may nevertheless be varied within a certain predefined range without stopping the process. These kinds of units are adapted for large daily production rates and are mainly single-purpose assets.

Batch processes are suited for smaller production rates and when long reaction times are needed to achieve the desired selectivity. A "batch" of raw materials (reactants) is introduced into one unit which is operated at the desired conditions until the target conversion is reached. The throughput, start time, and end time for this batch of production is therefore defined. Batch processes may be multipurpose assets capable of producing several different products.

In addition to these two modes, multipurpose processes may be operated in a campaign mode. Switching from production of one product to production of another may require cleaning and other setup operations, which can be time-consuming and costly. These setup requirements are often sequence-dependent, meaning that changing from *product 1* to *product 2* might be more costly than going from *product 2* to *product 1*. Moreover, a single batch may not be sufficient to satisfy a complete order. For these reasons, processes may be run without interruption in a temporary configuration to minimize the transition costs, time, or both. A compromise must therefore be made between the ability to fulfill incoming orders which require different products on different due dates (scheduling flexibility) and longer campaigns that minimize transition costs. Usually, a minimum campaign length is calculated and used as a working rule for planning to minimize transition costs. In the case of campaign planning, one of the key decision parameters that must be determined is the duration and timing of different production runs to be able to meet accepted demand at minimum cost. Production process types are summarized in Figure 7.4 by operating mode.

An example of a continuous/campaign process is a paper machine. These multipurpose units produce different grades of paper 24 h a day. When a transition between grades occurs, off-spec paper is produced. Transitions are often planned to occur at the beginning or the end of a roll to minimize the impact on printing operations. This paper may be given to the customer as part of the contract agreement or may be repulped.

Another key characteristic of the process industry is the fabrication of *coproducts* and *by-products*. By-products are usually undesired and are produced instead of the main product

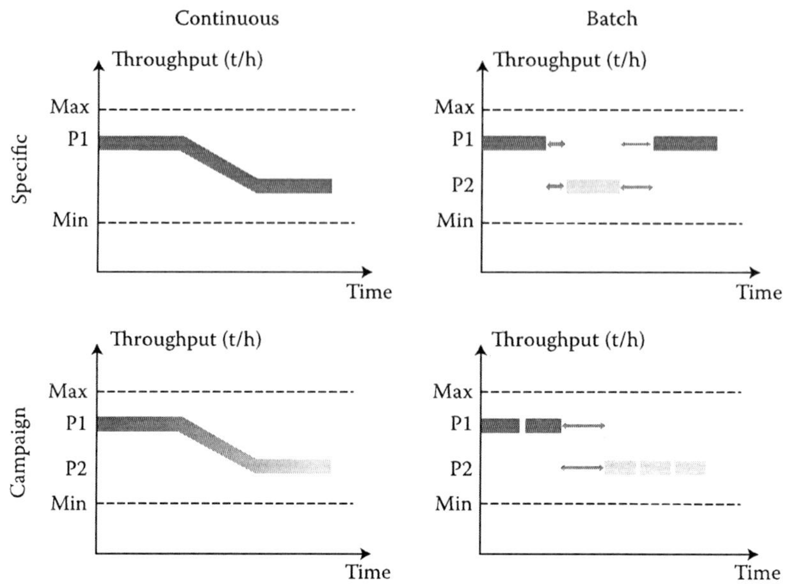

FIGURE 7.4
Types of industries. (With kind permission from Springer Science+Business Media: *Value Chain Management in the Chemical Industry—Global Value Chain Planning of Commodities*, 2008, Kannegiesser, M., Berlin: Physica-Verlag.)

because of imperfect reaction selectivity. By contrast, coproducts are produced alongside the main product and are generally necessary for further processing steps or are sold to the market. They are hence considered desirable. Valorization of these secondary products is essential to ensure the profitability of the process. However, the production of secondary materials during process operations considerably complicates the planning and scheduling of inventory, capacity, and sales, because these secondary materials must be addressed simultaneously with the manufacturing of the main products.

Moreover, as in the chemical industry, there may be more than one way of producing one material in a factory, that is, the product may be obtainable through different combinations of processes. The yields may also change depending on raw materials, time, and external conditions such as external temperature. In addition, if for certain reasons, part of the normal production does not comply with product standards (which is obviously not desirable and should be minimized), but has sufficient purity or functionality for another purpose, the off-spec products might also be reprocessed and sold to other customers at a lesser price, or might be marketed as second-quality products. These specific aspects of the process industry further complicate the planning and scheduling of production.

The different subindustries of the process industries each have specific characteristics that make its supply-chain planning problems unique. For instance, products from the food and pharmaceutical industries have expiration dates. The final product cannot be held in inventory for prolonged periods, otherwise it will be wasted. In the chemical industry, products can be intermediates and finished products at the same time, being sold or used for the manufacturing of other products (e.g., ethylene may be sold to plastic producers to produce polyethylene (PE), but may also be used by the same company to synthesize other chemicals). Hence, planning tools and heuristics that are developed in certain industries cannot always be generalized because specific constraints in each industry must be addressed.

No matter what type of process is in place, because of high capital investments, machines and plants in the process industry must be operated continuously and near capacity to realize a profit on investment [2]. Hence, production is limited by capacity. When total demand for one product cannot be entirely fulfilled by a company, which is the case most of the time, companies must make capacity-management decisions, that is, to decide which part of the market to serve from the available capacity. Unlike the discrete or service industries, strategies to enhance capacity temporarily, such as using time flexibility from the workforce (overtime work) or hiring additional seasonal workforce can be applied to a greater or lesser extent. Subcontracting can be practiced, but the customer must agree to the use of such methods. Decisions regarding which part of the demand to fulfill are therefore critical for the profitability of process enterprises.

7.2.3 Types of Products and Markets

Kline classified chemical products according to their degree of differentiation and production volume [15]. Even though this classification was made for chemicals, it could be applied to other process industries, including oil and gas, steel and metals, pulp and paper, pharmaceuticals, and food.

Materials that are produced in a substantially identical form by several suppliers are said to be *undifferentiated products*. They are usually sold according to specifications on their composition (purity) and may be used in many different applications. One example of such a product is ethylene. Ethylene produced by one company or another is the same as long as it has the same compositional purity. Ethylene is a chemical used in a wide range of processes for producing several polymers and other chemicals.

On the other hand, *differentiated products* are materials that are either produced with real differences among products from different suppliers or are at least marketed on the basis of an imputed difference. Consumer goods are generally differentiated products. For example, yogurt from two companies will have a similar, but slightly different, taste and composition. Moreover, one company could differentiate itself from its competitors by providing different packaging. Products that fall into this category are generally used in one or at most a few different applications. Rather than being sold on the basis of their compositional specifications, they are sold on the basis of their performance characteristics.

Taking into account the degree of differentiation as well as production volume, chemical products can be classified into four broad categories, as shown in Figure 7.5: true

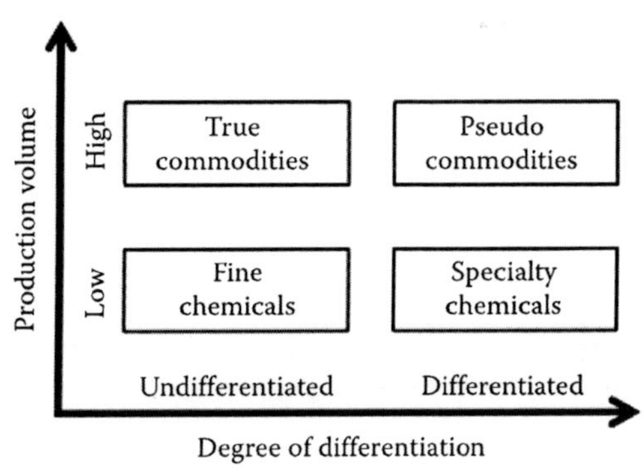

FIGURE 7.5
Chemical classification. (From Kline, C.H. 1976. Maximizing profits in chemicals. *Chemtech* 6(2): 110–117.)

commodities, fine chemicals, pseudocommodities, and specialty chemicals. *True commodities* are large-volume compounds sold to generally accepted compositional specifications. They have several possible end uses and are sold to a wide variety of customers, even if sales are usually concentrated in a small group of large buyers. *Fine chemicals* are also undifferentiated products sold to standardized specifications, but produced in much lower volumes. Most of the time, they are sold to a small number of specialized customers. Like true commodities, *pseudocommodities* are produced in large quantities. However, these compounds will be sold according to their performance specifications in their related end uses. They will often be bought by a few large customers. Ultimately, *specialty chemicals* are synthesized in small quantities and are designed according to customer-specific characteristics. These products are typically sold to several small-volume customers.

In general, goods produced upstream in the value chain are of the commodity type, while those produced toward the downstream end of the value chain are more of the specialty type. In other words, as more transformations are made, smaller volumes will be produced, but at a higher value. Process-type (continuous or batch) manufacturing environments therefore vary depending on the type of products that are manufactured.

True commodities are produced in continuous and campaign production modes 24 h a day during the whole year to take advantage of economies of scale. Equipment is highly specialized, and relatively few products and coproducts can be produced in these inflexible environments. Because their market is mature to a certain extent, production is mostly made-to-stock because demand is highly recurrent and can be forecast with reasonably good accuracy. Production is therefore planned for long periods with very few changeovers.

On the other end, specialties are generally produced on multipurpose batch equipment with smaller lot sizes and volumes. Because production must be customized to the customer's requirements, it is mostly made-to-order. In fact, efforts are focused on understanding the customer and his applications to adapt production and outputs accordingly to increase his level of satisfaction. Shorter planning periods are therefore the rule, and many transitions can be necessary and must be scheduled effectively to maintain profitability. Scheduling of specific orders on specific units is an important task.

Fine chemicals and pseudocommodities share characteristics from both true commodities and specialties. Therefore, their characterization is more difficult. Like true commodities, fine chemicals are mostly produced continuously, but may also be produced in batch campaign mode. However, they are produced on much smaller scales and may also be produced partly as coproducts in different processing steps. Because they are undifferentiated products that can be stored without additional processing or customization for different orders and customers, fine chemicals are commonly suitable for MTS environments. They may also be produced in MTO environments when produced, for example, on multipurpose batch equipment.

On the other hand, pseudocommodities may be produced partly continuously, with some customization, such as different packaging steps, performed on batch equipment. Hence, MTO, ATO, MTS environments, or hybrids of these can be found for these products. For example, even though newsprint rolls are produced in large tonnages, their production is generally MTO because different customers ask for different characteristics (roll size, quality, diameter, etc.). Raw materials are purchased in advance, but manufacturing is committed only upon order arrival because the amount of customization needed for each order is considerable. However, plastics produced in high tonnages and in a standardized quality, such as PE pellets, might be manufactured according to forecasts, that is, MTS.

These different product categories face different market environments. For instance, commodities are characterized by highly competitive businesses with substantial price volatility, while fine and specialty chemicals are less sensitive to changes in price and demand because of lower levels of competition, and because of product differentiation in the case of specialties.

As the business for a particular product matures, several forces put additional pressure on a company that wants to remain in this business segment. On one side, the arrival of competitors puts pressure on the market, lowering selling prices. Products that were traditionally fine chemicals and specialties are now confronted with commoditization, which renders their business more vulnerable to price and demand variability. On the other side, because of this increased supply, customers ask for higher service levels and customization for the same price, thus augmenting distribution and production costs and leading to even higher margin pressure. Hence, supply-chain strategies must evolve over time to adapt themselves to the market.

In addition to this product classification by Kline, products may be differentiated by their perishability and life cycle. *Perishability* refers to the amount of time during which a product can be held in inventory. Because perishable items cannot be stored for sale in the future, it is not a real possibility for the customer to wait until the price decreases. For this reason, sales in a particular period usually do not influence future sales. Most food products, like dairy products, are perishable; they cannot be stored for a long period, and customers will still purchase them even if the price is slightly higher or lower than the week before. Overstocking perishables is expensive; accurate forecasting thus plays a crucial role in the profitability of perishable products manufacturing companies. In contrast, high-tech products such as computers are durable; they last for a few years, but because customers typically buy one computer every few years, today's purchase may affect future purchases. Perishable and durable products therefore require different production, inventory, and sales planning strategies.

The *product life cycle* refers to the length of the selling season. In general, products from the process industry have especially long life cycles. For instance, chemicals, drugs, types of paper, and food are sold in a similar configuration for several years or decades. In contrast, computers have very short life cycles: after only 3 or 4 years, a computer model is considered to be outdated, and its selling price will drastically decrease because newer and more powerful models are now available. Sales and marketing strategies should be different for short and long product life cycles. In addition, product innovation and product portfolio refreshment is especially important for short life cycle products. These are also important for long life cycle products, but to a lesser extent. As a product matures, there comes a time when its market demand declines because of its replacement by better, greener, more efficient, or more acceptable products. When this situation arises, as is the case for the pulp and paper industry which faces declining markets for some of its products (e.g., people are reading fewer and fewer printed newspapers and magazines, preferring to inform themselves by other means, such as television or the Internet), product innovation becomes important for survival.

7.2.4 Contracts, Spot Markets, and Collaboration

A large part of the supply-chain literature deals with models in which demand must be met. However, in industrial practice, demand may not be completely fulfilled by a company. In fact, market demand is both price-sensitive and uncertain. Two types of demand can be distinguished: contract demand and spot-market demand. On one hand, spot-market sales can be highly variable in both price and quantity. In spot demand, no long-term agreement exists between the company and its customers; only agreements on price and quantity for this specific transaction are addressed. Therefore, this type of demand does not necessarily need to be fulfilled. The company can therefore make active sales decisions on the acceptance or rejection of spot sales requests [16].

On the other hand, contract demand is based on agreements between the company and the customer, and these agreements stipulate commitments. This type of demand is much less flexible and must be fulfilled to the level and price that were specified and agreed upon. The motivations for contracts is related to the desire to share the risks arising from various uncertainties in the business environment such as demand, supply, delivery, inventory, price, exchange rate, and so on [17]. Contracts also facilitate long-term relationships. When the demand for a specific product or raw material is predictable and constant over time, agreement on a contract will ease several repetitive supply-chain operations such as regular deliveries and payments. Service and product quality, reliability, and other nonquantifiable criteria are further reasons why a customer would consider entering a long-term commitment.

Contracts enable the various players in the SC (supplier, manufacturer, retailer) to be better coordinated and aligned with demand, leading to higher SC profits. Typically, a contract includes specifications with respect to contract length, quantity, price, delivery terms, discounts, and penalties for nonfulfillment of conditions [17,18]. These specifications should capture the material, information, and financial flows encountered in a supply chain. Several authors have characterized and classified contracts according to the clauses included in them. For example, Tsay et al. [19] classified contracts according to their specification of decision rights, pricing, minimum purchase commitments, quantity flexibility, buyback policies, allocation rules, lead time, and quality. Höhn [20] characterizes contracts according to these eight contract clauses, but adds horizon length (contract duration), periodicity of ordering, and information sharing. In the following subsections, contracts will be characterized according to contract specifications such as general terms and conditions, purchase commitments, pricing, payment terms, and other contract options. However, the intent is not to provide an exhaustive overview of all contract types and specifications.

7.2.4.1 General Terms and Conditions

Contracts are negotiated individually between the customer and the manufacturer or retailer. Because of these one-to-one negotiations, contract clauses and specifications will differ from contract to contract, from customer to customer, and over time, even if the company continues to provide the same range of products.

The *contract length* specifies how long the contract will be in force. Some contracts are transaction-specific, such as *spot contracts* and *forward contracts*, where in the latter a future exchange of goods will be involved, but at terms set today. This type of contract is very similar to a spot contract ("immediate" exchange of goods at terms set today), but it is used to hedge against the risk of future price increases. Forward contracts facilitate coordination between a buyer and a manufacturer because the latter can schedule its capacity-constrained production and transportation accordingly. Because spot sales involve transactions that need to be accomplished right now or in the near future, the final product must often be carried in inventory, ready to be shipped. On the other hand, *long-term* contracts stipulate commitments over several months or years and involve several repeated transactions. In addition to these three forms of contracts, Sykuta [21] describes *future contracts,* which combine the coordination of a forward contract with the flexibility of spot-market transactions.

The *periodicity of ordering* specifies how often the buyer can place orders during the length of the contract. In contracts with a fixed periodicity of ordering, the buyer can place orders only on predetermined dates, while with a random-periodicity clause, the buyer has the freedom to place orders when desired.

The product *quality* is an important specification of any supply relationship because it may have an influence on the product cost and on the selection of the supplier. Specific dimensions of quality, such as the purity of a chemical product or the strength and robustness of a plastic, may be specified in the contract.

The delivery terms (due date, insurance, etc.) and other logistical components such as transportation mode (truck, train, boat, plane), carrier selection, warehousing plan, and the use of a third-party logistic service provider must be agreed upon when signing a contract. These logistical decisions have an important effect on lead time (e.g., truck is faster than train, but train is better adapted to bulkier deliveries), cost, and responsiveness.

Information sharing terms state what type of information will be shared between the two parties, such as demand forecasts, inventory levels, available capacities, and so on. Sharing the right information between supply-chain partners leads to higher levels of integration and coordination, thus maximizing SC profit. However, in practice, transparent sharing of information may be hard to achieve because companies may be unwilling to provide confidential information about their manufacturing costs. Information sharing is especially important in the case of collaborative planning, as will be discussed in Section 7.2.4.4.

7.2.4.2 Purchase Commitments

From the supplier's point of view, purchase commitments guarantee orders and reduce demand and inventory uncertainty. This is an important part of process industry contracts because equipment must be continuously run at close to maximum capacity. Minimum commitments are also imposed for a discount to take effect.

Bansal et al. [17] define two types of purchase commitments: *quantity commitment* and *dollar commitment*. In a quantity-commitment contract, the buyer agrees to purchase a minimum quantity of each material under the contract, while in a dollar-commitment contract, the buyer agrees to a minimum dollar amount of total purchases under the contract.

The determination of the *due date* is a central element of a contract. It specifies the point in time (at the latest) when the manufacturer has to deliver the product to the customer [18]. Lead times in the process industry are generally long, notably because of time-consuming setups, large requested quantities, and highly utilized, limited-capacity equipments. Therefore, the customer will generally commit in advance to certain quantities based on forecasts and not necessarily on accurate and known quantities. However, to avoid bearing the risk of changing demand, the manufacturer may compel the customer to reveal his final demand at a specific point in time before the due date. This *reveal date* is particularly important in the case of *quantity-flexibility contracts,* in which the customer commits to no less than a certain percentage below the forecast with the guarantee to deliver up to a certain percentage above [22,23]. In these contracts, information becomes gradually available over time to the manufacturer, and he must keep enough inventory or adapt his available capacity to be able to produce and deliver on time. Flexibility in demand quantity and delivery date certainly increases customers' loyalty and satisfaction, but on the other side, it exposes the manufacturer to uncertainty and risk with respect to production. This uncertainty may lead to capacity shortage or excess, which requires replanning and expensive adjustments in production [24]. Chopra and Meindl [3] argued that a quantity-flexibility contract makes sense if the manufacturer has flexible capacity that can be used to produce at least the uncertain part of the order after the retailer (customer) has decided on the modification. They also mention that this type of contract can be very effective if a supplier is selling to multiple retailers with independent demand.

Some suppliers incorporate options into their contracts, such as the admission of short-term notification changes. This flexibility may give them a competitive advantage compared to other similar companies by being more responsive, but it implies additional uncertainty and risk for the manufacturer. Some contracts also contain a *cancellation* clause which stipulates that the customer has the right to withdraw his order before or on the reveal date. However, such cancellation clauses are not included by default and have to be negotiated before signing the contract.

The length of a contract may vary, and the contract may cover one or several transactions. Consequently, the purchasing commitments may cover the whole contract duration or may be periodic, that is, may consist of commitments that must be repeatedly honored several times during the contract length (e.g., a delivery every 2 weeks). However, once an order has been accepted, either as part of a long-term contract or as a spot transaction, it must be fulfilled and carried out at the same level of priority as other accepted orders.

7.2.4.3 Pricing, Payment Terms, and Other Contract Options

The *price* paid for a product is a major decision factor in a contract agreement. The price may be fixed for a period of time—this is particularly useful for hedging against future price raises—but it may also vary from order to order during the duration of the contract. Moreover, the price of some materials may be pegged to the price of another raw material. For example, liquefied natural gas in most supply contracts is pegged to the price of crude oil [17].

The process industry is characterized by highly capital-intensive equipments. As mentioned earlier, machines and equipment must be run constantly and close to their maximum capacity to minimize production costs. To push the customer to commit to high level of production to ensure long runs and to minimize production costs, the supplier may offer *quantity discounts*. Bansal et al. [17] differentiated quantity discounts into two categories: bulk and unit discounts. *Bulk discounts*, sometimes referred to as an all-units discount, apply to the total quantity of a purchase. *Unit discounts*, or incremental discounts, apply to each unit of purchase beyond a certain threshold level. These discounts may be applied to different purchasing quantity intervals, with different discounts applied for each of them. Figure 7.6 shows these two types of discount for three different purchasing levels. U represents the supplier's fixed ordering costs, and Q_1, Q_2, and Q_3 are the boundaries that mark the different quantity intervals.

Quantity discount contracts decrease overall costs but lead to higher lot sizes and therefore higher levels of inventory in the SC. This type of contract is typically justified only for commodity products for which the supplier has high fixed costs per lot [3].

To attract buyers and create additional demand, some suppliers offer discounts for buying various combinations of materials. This type of promotion is generally referred to as *product bundling*. Such contract specifications may impose various types of minimum purchase commitments.

A *customer penalty*, such as a late notification change or a cancellation, may be attached to a contract in case the buyer does not respect his engagement. There may also be a *supplier penalty* if the manufacturer cannot cover the requested demand. In some instances,

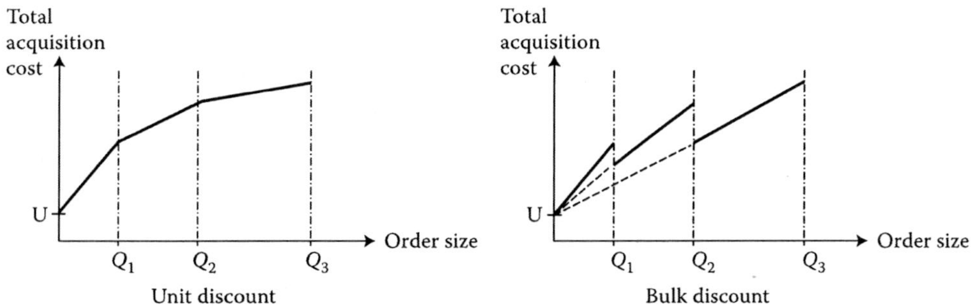

FIGURE 7.6
Bulk and unit discounts for different purchasing intervals. (With kind permission from Springer Science+Business Media: Purchasing and material requirements, *Supply Chain Management and Advanced Planning: Concepts, Models, Software and Case Studies*, 2008, Stadtler, H. In: Stadtler, H. and Kilger, C. (eds.), Berlin: Springer-Verlag.)

backlogging clauses may be included, that is, the possibility for the manufacturer to fulfill demand after the due date, with or without a penalty. This supplier penalty can be viewed either as a financial penalty for shortfall or as the cost of outsourcing production to a third company, that is, subcontracting. Subcontracting to another company must, however, be accepted by the customer.

Some contracts offer the buyer to have the right, but not the obligation, to purchase a certain quantity of a product at a specified price for an upfront premium payment. The supplier thus reserves a certain level of production capacity. This type of contract is known as an *option contract* and has been considered by several authors for planning under uncertainty [26–28]. Using the terms introduced, an option contract can be represented as contract with a cancellation clause and in which the reservation price acts as a customer penalty.

Other contracts include *buyback* options, which allow the customer to return unsold inventory up to a certain percentage at an agreed-upon price [20]. This buyback price is typically lower than or equal to the wholesale price. Manufacturers benefit the most from these contracts when they provide high-margin products, because they encourage retailers to increase the level of product availability [3].

Finally, *payment terms* specify the amount due and the requested conditions such as payment asked in advance or on delivery, the level of credit accepted, deferred payment period of x days, early-payment discounts, and so on. Other specific payment terms exist, such as revenue-sharing contracts, in which the buyer pays the supplier a wholesale price for each unit purchased plus a percentage of his revenue [29].

7.2.4.4 Collaborative Planning

With increased pressure from global competition and the shortening of product life cycles, business has become too complex and expensive for one company to go it alone and prosper [30]. It has been argued by several writers that companies must therefore collaborate to take advantage of each other's skills and expertise to compete and be more profitable. With collaboration, global supply-chain operations and decision-making among the partners should be better integrated and coordinated. However, a prerequisite for collaboration is an agreement regarding the exchange of a specific set of data [31]. Stadtler [32] defined collaborative planning as "a joint decision making process for aligning plans of individual SC members with the aim of achieving coordination in light of information asymmetry". What is meant by *information asymmetry* is the following. When a supply-chain member is about to make a decision (e.g., to buy a product), he has his own objectives, constraints, and preferences (e.g., to buy at the lowest cost possible and in small quantities to minimize inventory). However, another member might have conflicting objectives and preferences (e.g., the seller wants to sell at the highest price and maybe in bigger lot sizes to minimize production costs). Hence, these two parties do not have the necessary information to make decisions that would result in coordinating their activities and improving the profitability of the whole SC.

Without providing an overview of collaborative planning as has already been done by Kilger et al. [31] and Audy et al. [33], two types of collaborative planning approaches used in the process industry will be briefly presented: vendor-managed inventory (VMI) and continuous replenishment (CR). Implementation costs in terms of financial, material, and human resources can be very high, but can also lead to substantial benefits, as attested by several success stories available in the literature. Consequently, the potential costs and benefits of these strategies must be carefully evaluated before such agreements are entered into [33].

A producer under a VMI agreement is responsible for managing the inventory of its customer [34]. The customer provides the supplier with information about his daily consumption, and the latter is responsible for replenishing the customer's inventory to a fixed level without waiting for orders. Hence, VMIs reduce demand fluctuation and minimize safety stocks. The supplier can thus optimize his production-distribution resources to minimize costs.

CR agreements are structured on the basis of prereservation of a part of the manufacturer's capacity. The purchaser commits to buying a minimum quantity, and the supplier then reserves a part of his capacity for that customer. The customer might have the flexibility to order more than the forecasted commitment. In this case, this type of collaborative planning is equivalent to the quantity-flexibility contract with a forward sale agreement, as discussed in the previous section. CR agreements have been studied by Durango-Cohen and Yano [35].

To make collaborative planning strategies successful, it is important that collaborating companies share relevant information (demand profiles, forecasts, available capacity, etc.) and show goodwill to make these practices profitable. Too often, collaborative planning initiatives are designed by some companies as a way to pass on their inventory costs to their suppliers. However, the latter have to carry the burden of holding more products in stock or producing in different lot sizes than they are used to, thus reducing the overall supply-chain profitability or even making the supplier's operations unprofitable. Trust and goodwill between companies are therefore prerequisites for establishing these collaborative planning strategies, or else this approach might fail to yield its expected rewards.

7.2.4.5 Levels of Collaboration

Relationships between SC partners are not all equivalent. Some have been running for several years and incorporate several exchanges (material, information, and cash) to maximize the profitability of both companies, while others are transaction-specific. Lambert et al. [36] classified the different types of relationships between organizations into six categories, as shown in Figure 7.7.

The most common and standard relationship between organizations is the arm's length relationship. In this category, two organizations conduct business with each other, often over a long period of time, and involve multiple exchanges. However, there is no

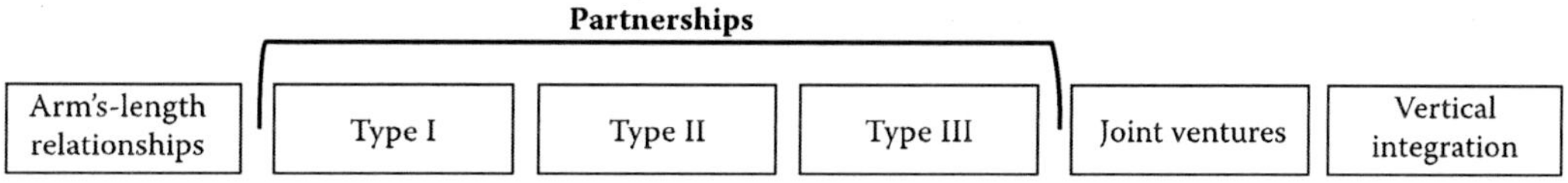

FIGURE 7.7
Types of relationships. (From Lambert, D.M., Emmelhainz, M.A., and Gardner, J.T. 1996. *International Journal of Logistics Management* 7(2): 17. With permission.)

sense of joint commitment or joint operations. Once the exchanges are finished, the relationship ends. The terms and conditions offered by a seller will typically be the same for all customers.

On the other end, when two companies (buyer and seller) merge or when one is acquired by the other, the resulting company is said to be vertically integrated. Collaboration between the two entities is usually enhanced and results in reduced costs because the business objectives of the company are better aligned.

Like vertical integration, joint ventures involve some sort of shared ownership between the two parties, but these remain independent. For example, two companies could create a joint venture that produces a new product using the technology of one and the market or raw material access of the other. The joint venture might operate independently as a third entity or might be part of both partners' operations.

Between the arm's length relationship and the joint venture, Lambert et al. identified three levels of partnerships. In a type I partnership, the organizations involved recognize each other as partners and, on a limited basis, coordinate activities and planning. The partnership usually has a short-term focus and involves only one division or functional area within each organization. Type II partnerships involve not only coordination of activities between organizations but also integration of activities. Although not expected to last "forever," the partnership has a long-term horizon. Multiple divisions and functions within the company are involved in the partnership. Finally, in type III partnerships, organizations share a significant level of operational integration. Each party views the other as an extension of its own organization. Typically, no "end date" for the partnership exists.

Several drivers are compelling companies to partner, notably the opportunity to identify asset and cost efficiencies, to improve customer service, to gain market advantage, and to stabilize profit and to grow the business. However, a company should not partner with every supplier, customer, or third-party service provider because partnerships are costly in terms of the time and effort required. In fact, type III partnerships and joint ventures, that is, strong partnerships, should be reserved only for customers or suppliers that are critical to an organization's long-term success [36]. Briefly, commodities are less well suited to strong partnerships. Because the commodity buyer is looking for a standardized item which could be easily replaced by the product from another competitor, supplier know-how becomes of limited value for downstream product or process development [37]. On the other hand, manufacturers of differentiated products, specialties in particular, can gain a lot from these collaborations because of their strong customer focus. Because these producers are often critical suppliers of a certain component, their knowledge may be crucial for further process and product development and therefore for long-term success.

Without rigorously classifying contract clauses by partnership types between manufacturers and their customers, several links can be made. Strong partnerships (type III) usually have the shortest and least specific contract agreements [36], while arm's length and type I partnerships—that is, more distant collaborations—will have many more contract specifications and will be less flexible. For instance, the contract terms for a spot sales transaction will specify quantity, price, due date, and delivery means rigidly. For a customer that buys regularly, but with whom no joint commitment is implied, that is, an arm's length relationship, a manufacturer may offer a quantity discount contract. CR contracts can loosely be characterized as type I partnerships because there is coordination of activities to gain certain advantages such as evenness of production and delivery schedules over time. VMI agreements or contracts are implemented when the collaborating companies are seeking to integrate several activities to realize further savings on global SC costs. These can thus be classified as type II partnerships. Finally, in a type III partnership, strategic planning might even be done together because one company's success depends highly on the other's. Therefore, contract clauses should not be too restrictive to leave space for flexibility.

7.3 Planning the Biorefinery Supply Chain

From the forestry point of view, biorefining implies a more complete utilization of the renewable forest biomass and the diversification of the traditional core business into the production of value-added green organic chemicals, biofuels, and energy. It offers companies an interesting opportunity to enhance their environmental profile and to become independent from petroleum derivatives, both on the energy and on the raw material side. Ragauskas et al. [38] defined the biorefinery by analogy with petroleum refining. In a conventional refinery, various fuels, chemicals, and also energy are produced from petroleum, a heterogeneous source of carbon. Similarly, biomass, a carbonaceous feedstock mainly composed of polysaccharides and lignin, can be fractionated and converted into a variety of products and also energy. Because the concept of the biorefinery relies on new ways to produce chemicals and fuels from renewable resources, it consequently implies an independence of fossil resources.

Nevertheless, one of the important differences between the biorefinery and the petroleum refinery is that chemicals and materials from biological sources contain more oxygen than their petroleum counterparts. This feature offers the opportunity to produce certain specific organic chemicals, but it also reduces the energy density of products, making them less attractive for some large-scale energy applications. Moreover, biomass raw materials have lower density and higher humidity content than fossil raw materials. Hence, the energy necessary to transport and densify biomass, as well as to convert it into products, is usually greater than for petrochemicals and fossil fuels. To be consistent with the biorefinery concept, this extra energy should come from renewable forms of energy such as solar, wind, hydro power, or bioenergy. It is apparent that two conflicting forces are opposing each other with regard to the design and size of biorefineries. On one side, the additional quantities of biomass necessary to produce energy and also to benefit from economies of scale in production units push toward biorefineries of large size. On the other side, the supply of biomass that can be brought to the plant site in a cost-efficient manner is usually limited due to transportation and preprocessing costs. The low density and high humidity content of biomass limits the quantity of valuable organic material that can be transported on one truck or railcar. Moreover, as more biomass is needed, it must be obtained from further away because yields per hectare are limited, thus increasing transportation costs on a marginal basis.

Selecting the most adequate and profitable new products and processes to add to the current business is not an easy task. The number of different possible biorefinery configurations (biomass-process-product) is very high, as attested by the large number of recent publications on the subject. On the product side, almost any organic product, be it a chemical, a plastic, or a fuel, can be produced from biomass instead of petroleum derivatives. Several production and separation alternatives using biochemical, chemical, or thermochemical pathways are under development to produce this plethora of possible products. On the raw material side, numerous types of biomass (wood, dedicated crops, agricultural and forestry residues, organic residues from food and forestry industries, or municipal solid waste) can be used exclusively or in combination to feed these processes. This complexity is illustrated in Figure 7.8 [39].

Depending on who is going to implement the biorefinery (forestry industries, agricultural companies, petrochemical companies, or independents) and where the configuration and size of the plant and the products that will be chosen will differ. Regions of the world that are suitable for biorefining differ in many ways, starting with their geography and types of biomass. For instance, softwood and bagasse, a sugarcane residual, do not come from the same types of land, and these two varieties of feedstock would therefore not represent competing biomass options in a particular region. The presence of a particular industry sector in a region will also certainly influence the type of mill and the products that will be manufactured. For example, forestry companies have a privileged access to renewable raw materials (wood) and can take advantage of their existing facilities and biomass-harvesting network. Chemical and petrochemical hubs, while being generally

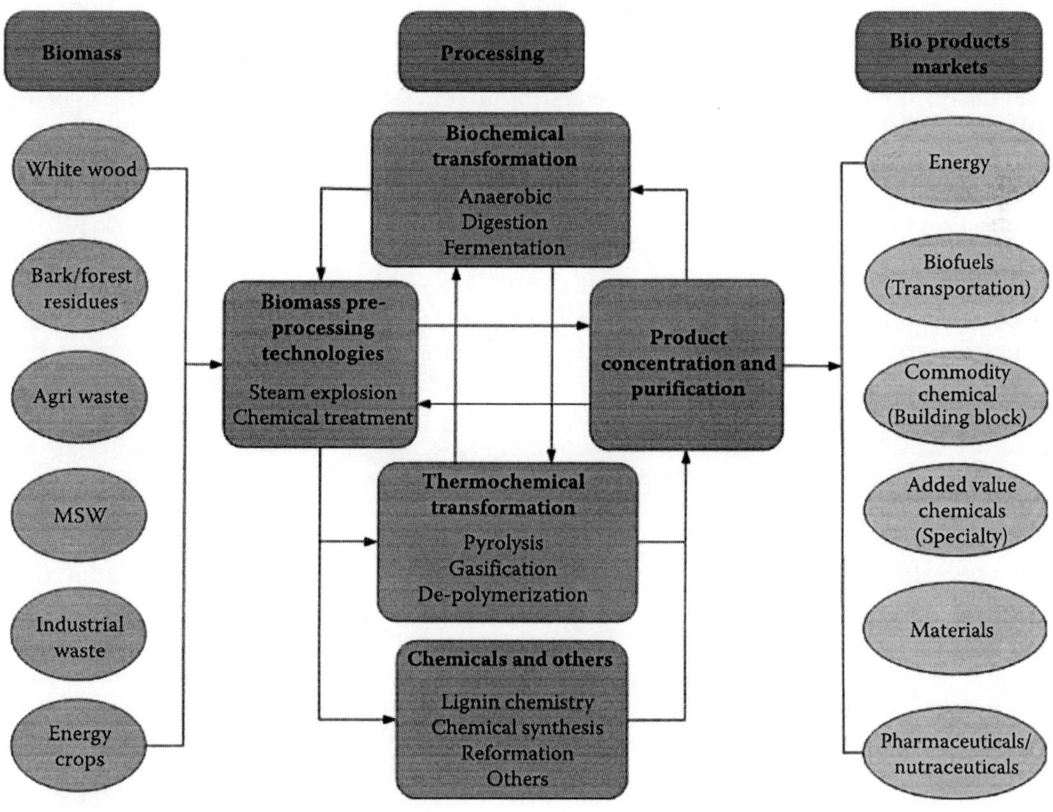

FIGURE 7.8
Biorefinery possibilities. (From Chambost, V. and Stuart, P.R. 2007. *Industrial Biotechnology* 3(2): 112–119. With permission.)

more distant from renewable raw materials, have extensive processing equipment and knowledge. Finally, policies in place in different locations will favor the production of specific products. For instance, the development of biofuels such as ethanol in the United States is highly favored by the policies in force.

Hence, because of all these differences between possible biorefining regions, it is likely that there will be no unique biorefinery solution. Rather, the biorefinery will be region-specific, industry-specific, and even mill-specific. To discuss and highlight the various supply-chain characteristics and considerations for the different biorefinery possibilities, the products, processes, and raw materials will first be described in more detail in the following subsections.

7.3.1 Biorefinery Raw Materials and Procurement

Biorefineries have many similarities with oil refineries, especially on the product side: several products are made from a heterogeneous source of carbon. However, raw material characteristics and procurement differ considerably between the two.

Crude oil is predominantly a mixture of four types of hydrocarbons: paraffins, naphthenes, aromatics, and asphaltics. Its chemical composition varies highly among different oil fields. Crude oil is first extracted from the ground or from the ocean in specific locations by oil wells or oil platforms. It is then transported by pipeline, tanker, or both to refineries, where it will be further processed and transformed into different products. Refineries typically obtain their petroleum from different sources scattered around the world. Hence, supply lead times are quite long, ranging from 2 to 8 weeks, because oil often has to travel thousands of kilometers in tankers before being refined [40]. Because crude oils differ in their composition, yields, and characteristics, depending on the type of crude that is used in a refinery, different products in different quantities will be obtained. The management of procurement operations is especially important in this industry because procurement costs may represent up to 80% of the total production costs [17] and because raw material prices have been particularly volatile recently.

Biomass is a carbonaceous feedstock that comes from living, or recently living, organisms. Hence, carbon from both vegetable and animal sources can be considered as biomass, but plants are generally preferred for industrial biorefining purposes because their production yields are much higher. Because living organisms contain significant quantities of water, the moisture content of biomass is considerable. Plants have been used and grown by humans for millennia as a source of food and energy, but also as a source of materials for construction, tools, and clothing. In general, edible plants are grown and harvested for their grains, fruits, or tubers because these contain the most valuable components in terms of nutrition (carbohydrates, fats, proteins, vitamins, and minerals). Industrial-scale biorefinery products such as bioethanol and biodiesel are currently made from grains because the components of grains, sugars and lipids, are readily accessible. However, these biorefinery uses are in direct competition with crops for food use. On the other hand, when harvested, annual crops produce residues such as straw, stover, and so on. These are used mainly for animal feed, but represent an interesting alternative raw material for the production of bioproducts. Unlike grains, they cannot be transformed into human food products, and their use would therefore avoid the "food versus fuel" debate. Similarly, bagasse, the fibrous residue of sugarcane obtained after sugar has been extracted, also represents an interesting biorefinery raw material. This nonedible part of the crop (straw, stover, and bagasse) is made of lignocellulosic material, a term which refers its main components: cellulose, hemicelluloses, and lignin. These three components also form the bulk of other plants which are not edible by humans, such as trees (conifers and angiosperm trees) and grasses (switchgrass, *Miscanthum*, etc.), which are other potential raw materials for the biorefinery. Because lignocellulosic biomass appears to be the most sustainable and publicly acceptable alternative for the biorefinery, only this type of raw material will be considered in this chapter.

The chemical composition of lignocellulosic material varies significantly by plant species, by individual plant, and even by parts of the plant. Cellulose is the main structural component of the primary cell wall of plants. It is a linear polysaccharide composed solely of glucose units. The length of this polymer has a direct impact on biomass rigidity. Hemicelluloses are also polysaccharides, but composed both of hexoses and of pentoses. Hemicellulose chains are much shorter than those of cellulose and can be found both in branched and in linear forms. They serve as support matrices around the cellulose microfibrils. Lignin is an amorphous organic polymer of high molecular weight containing multiple aromatic groups. The structure and composition of this polymer is highly complex and varies greatly depending on the plant species and even its location within the plant. Lignin is present the secondary cell wall of plants and acts as a strengthener. Because of their differences in chemical composition, these three components of lignocellulosic biomass can be used for producing different bioproducts.

As opposed to the global procurement of crudes in petrochemical sites and refineries, biomass procurement in biorefineries is most likely to be done on a local and regional basis, up to a few hundred kilometers from the mill. First of all, compared to petroleum, biomass has a much lower value in terms of mass and energy content because of its low density, its high moisture content, and its significant oxygen content. Moreover, biomass decays if not properly stored, as do most biological materials. Hence, long-distance transportation and long-term storage of bulky and low-value raw materials may not be cost-efficient unless the biomass is further treated, dried, and densified to more acceptable levels. Biomass harvesting also differs considerably from oil drilling. While oil drilling equipment is installed in one location until the well is depleted, biomass is distributed geographically and has to be harvested in different locations using the same equipment. Because the yield per unit area is limited, the more biomass is needed, the further from the mill one has to go to get the raw material. Greater distances imply additional fuel, which currently comes mostly from fossil sources, thus increasing transportation costs and reducing the green aspect of the delivered biorefinery product. All this implies that biorefineries will probably be of much smaller scale than their fossil counterparts. Moreover, it took several decades and several billion dollars of investment for refineries and the petrochemical industries to reach the point where they are right now. It would be very unlikely that biorefineries could, in just a few years, become competitive in terms of tonnage for the production of chemicals and fuels.

Biomass production, harvesting, procurement, and other related logistics activities also differ depending on whether the plant comes from the forest or is agriculturally grown. These activities also have different patterns in terms of seasonality. Annual plants and their residues are grown in specific fixed locations (farms) and usually in large quantities because farming is an intensive process designed for additional efficiency. As its name implies, harvesting can only happen once or maybe a few times a year and must be done during a short time window of a few weeks. Consequently, logistical operations are complicated because harvesting equipment must be allocated to fields and efficiently utilized.

On the other hand, forest trees may be harvested any time of the year, and even years later, although the spring thaw in northern regions limits the amounts of lumber that can be transported on some roads during that season. Once the desired quantities of wood have been cut in a specific location, the harvesting equipment is moved elsewhere.

An efficient road and transportation infrastructure is therefore required to access harvest areas and for forestry operations to be profitable. Because the different parts of the tree do not have the same value, they are not used for the same purpose. Trees are first cut, and branches are removed. Branches, treetops, and stumps, that is, forest residues, are typically left in the forest or harvested and chipped for energy purposes. The tree is then cut into logs of specific dimensions and quality, with various parts being used for veneer, lumber, or pulp. Finally, logs are transported to sawmills and pulp and paper mills, where the logs are debarked and transformed into products.

Forestry operations are conducted differently in different parts of the world. For example, bucking operations may be done in the forest, at terminals, or in mills. Branches and residues may be removed in the forest or on the side of the road, or even left in the forest, making availability highly variable. Forests may be organized into dedicated plantations to increase yield or may be left to grow naturally. Some regions practice precommercial thinning, selecting and cutting specific trees to let other surrounding trees grow bigger, thus increasing the yield of merchantable trees per acre. Intensive silviculture and genomics may also be used to increase forest yields. Hence, biomass yield, availability, and cost structure will vary by region according to geography, the existing infrastructure, and forestry and silviculture policies.

Because biomass composition and procurement vary depending on its origin, several authors have proposed various classifications for biomass. In particular, Perlack et al. [41] classified biomass into two main categories according to its origin (forest or agriculture). They further subdivided these categories into three classes, depending on the number of transformations undergone: directly harvested biomass and by-products, by-products from the industry, and postconsumer waste. Examples of biomass according to this classification are presented in Table 7.1.

TABLE 7.1

Source and Types of Biomass

Source	Secondary Classification	Biomass Examples
Forest	Primary	Trees (softwood and hardwood) Forest residuals (branches, tree tops, stumps) Precommercial thinnings Undesirable trees for forest and P&P products
	Secondary	Forestry industry residuals and coproducts (bark, sawdust, wood chips) Black liquor Wastewater sludge
	Tertiary	Construction residuals and demolition debris Recycled paper: old magazines (OMG) and old newspaper (ONP)
Agriculture	Primary	Grains Dedicated perennial crops (energy crops) Agricultural residues from annual crops
	Secondary	Food industry residuals (sludge, sugars, concentrated effluents) Bagasse Animal manure
	Tertiary	Organic fraction of municipal waste (MSW) Biogas from landfills and wastewater treatment plants

Source: From Perlack, R.D. et al. 2005. *Biomass as a Feedstock for a Bioenergy and Bioproducts Industry: The Technical Feasibility of a Billion-Ton Annual Supply.* Technical Report, Oak Ridge National Laboratory, Oak Ridge, Tennessee. With permission.

Biomass content and quality (homogeneity) also has an effect on the processing equipment that can be used. Heterogeneous biomass, like the organic fraction of municipal solid waste, needs flexible equipment that can process together the different biomass components in the feed. Processes from the thermochemical pathway would therefore generally be more suited to this type of biomass. Other biomass types have more homogeneous and specific contents, making them naturally suitable for more selective processes that may produce higher-value-added products. Furthermore, for certain products, biomass quality and freshness may have an impact on production and final product quality. For instance, pulp produced from freshly harvested wood logs or chips has better machine performance. Biomass quality and homogeneity are therefore critical aspects that must be taken into consideration when designing a biorefinery. These are especially important when selecting the size and capacity of a biorefinery: in a general way, biomass quality decreases as the quantity delivered to the plant increases.

7.3.2 Biorefinery Processes

Several processes have been developed over the years to transform lignocellulosic biomass into bioproducts and energy. Like petroleum, biomass has a complex composition; its primary separation into main groups of substances or intermediates before further transformation therefore seems appropriate. Figure 7.9 show the main biorefinery processes that produce these intermediates. It must be said that some of the intermediates shown on that figure may be used by processes other than those linked by arrows in this diagram. For instance, hemicelluloses may be hydrolyzed, but could also be gasified or combusted. These links have been omitted to simplify the figure. The goal of this subsection is not to enumerate all possible biorefinery processes and products, but rather to provide a quick overview to help formulate value-chain considerations related to the different products and product families. For a deeper discussion of biorefinery processes, the reader is referred to other chapters.

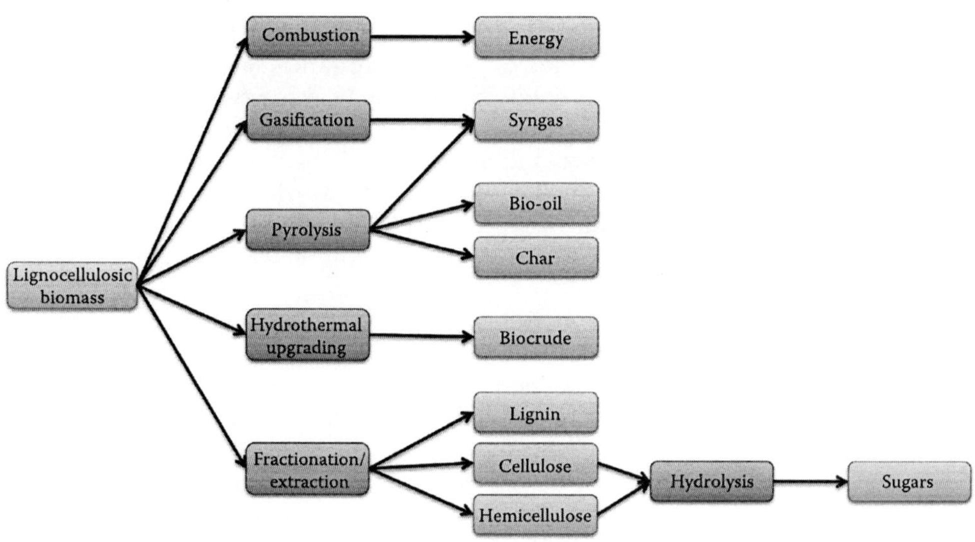

FIGURE 7.9
Biorefinery intermediates.

The first four processes (combustion, gasification, pyrolysis, and hydrothermal upgrading) in Figure 7.9 belong to the thermochemical family. These processes use heat and sometimes pressure to break organic material into smaller components. Thermochemical processes are able to process diverse and heterogeneous biomass into products and energy. Fractionation, extraction, hydrolysis, and fermentation are said to belong to the chemical or biochemical pathway, in which chemicals, enzymes, or both are used to transform lignocellulosic biomass. Being generally more selective, these processes require fairly homogeneous feedstock.

Combustion aims to extract the energy content of carbonaceous feedstock by oxidizing it into mainly carbon dioxide and water. The energy is typically recovered in the form of steam and may be further processed through turbines to produce electricity. However, apart from the energy that is recovered, no other valuable products are obtained from this process.

In gasification, steam or low quantities of oxygen are introduced into a reactor to transform organic materials into a gas phase called *syngas*. Syngas is a mixture which is composed of mostly hydrogen, carbon monoxide, carbon dioxide, and water, but contains also light hydrocarbons like methane, ethane, ethylene, and so on. Syngas can be simply burned in boilers or kilns for its energy content, but it can also be transformed by fermentation or catalytic reforming into chemicals and fuels such as hydrogen, methanol, ethanol, ammonia, mixed alcohols, dimethyl ether (DME), and Fischer–Tropsch fuels. Because of its multiple uses and multiple possible outputs, syngas is generally considered to be one of the most promising platforms for the production of chemicals and fuels.

Pyrolysis is the thermal decomposition of organic material in the absence of oxygen. It is theoretically the first stage of gasification and combustion, but in the latter two reactions, partial oxidation (gasification) or total oxidation (combustion) follows. Pyrolysis produces a charcoal-like solid, a liquid (bio-oil), and a gas similar in composition to syngas. Depending on the operating conditions (temperature, pressure, and residence time), the yield of these three products will vary: slow pyrolysis produces more solids, while fast pyrolysis produces higher quantities of liquids. Pyrolysis solids can be burned in the same way as coal or gasified to produce syngas. They can also be further processed for the production of activated carbon or even carbon nanotubes [42]. Bio-oil, the main product of pyrolysis, is a complex mixture of various oxygenated organics (acids, alcohols, ketones, aldehydes, phenols, etc.). Like pyrolysis solids, bio-oil can be burned for energy or gasified, but the diversity of organic products present in this mixture represent an interesting opportunity for the extraction and production of adhesives, resins, fertilizers, aromas, additives, and so on. However, this molecular diversity has its disadvantages because extraction becomes more complex and costly. Bio-oil has also been proposed as an intermediate densification step for transporting biomass more efficiently to centralized biorefineries.

Finally, hydrothermal upgrading is a thermochemical process in which biomass in a water medium is transformed into a liquid called bio-crude using a highly pressurized vessel. Bio-crude is a mixture of various organic chemicals with a wide molecular-weight distribution, but, unlike bio-oil, it contains lower quantities of oxygen because the critical conditions in the reactor remove much of the oxygen contained in the biomass. This mixture can be separated into two fractions, which can be either blended with fuels or further gasified.

While thermochemical processes transform lignocellulosic materials into products not present in the original raw material, fractionation and extraction transform biomass into its main constituents, cellulose, hemicelluloses, and lignin, using steam, chemicals, enzymes, or a combination of these. Kraft pulping is in fact one of the most widely known fractionation processes. It uses sodium sulfide and sodium hydroxide to convert woody biomass into a cellulose-rich stream (pulp) and black liquor, which contains lignin, a fraction of the hemicelluloses, and inorganic chemicals from pulping. Fractionation processes will separate, break, or transform biomass components differently depending on the conditions and chemicals used.

Cellulose is usually considered to be the most valuable components of biomass. In a general sense, if cellulose is recovered in its polymerized form after the fractionation, it is better suited for the production of cellulose derivatives and other biomaterials. However, if cellulose is recovered in a partially depolymerized form, it is better to hydrolyze it into sugars and other organic chemicals. Sugars can then be fermented into various oxygenated organic chemicals, including biofuels such as ethanol and butanol. In fact, along with syngas, sugars are considered as the other most promising platform for producing chemicals and fuels.

Hemicellulose streams obtained by extraction or fractionation usually come in a highly decomposed form. They are therefore better suited for conversion into sugars that can be further fermented and transformed into valuable products. Unlike cellulose, hemicelluloses typically cannot constitute a primary process because they are the least abundant of the three main components of lignocellulosic biomass.

Lignin composition differs greatly according to the biomass source and fractionation/extraction methods. Because it is a complex mixture, selective modification of lignin is currently difficult to attain. Hence, its thermochemical transformation by combustion, gasification, and pyrolysis appears to be the most promising pathway in the short term. Other usages for lignin include its chemical modification by catalysis to isolate monomers from lignin, as well as the production of macromolecules such as carbon fibers, polymer additives, resins, and adhesives.

7.3.2.1 Integrated Pulp and Paper Biorefinery

Existing pulp and paper (P&P) mills represent natural sites for biorefineries because these facilities have been processing woody biomass for decades to produce P&P products, energy, and chemicals. The process of papermaking involves first transforming wood chips into pulp, a dilute suspension of fibers. This suspension is then brought to a paper machine where it is successively drained, pressed, and dried to create a web of fibers, which is paper. The last steps consist of cutting and converting operations, depending on the application. Two main process families have been invented to transform wood chips into pulp. In mechanical pulping, wood chips are ground in refiners to produce pulp. The yield of this process may be as high as 95% because all the components of the biomass are present in the paper. However, paper made from mechanical pulp has a tendency to yellow over time because lignin reacts with light. This problem of yellowing is avoided in chemical pulping processes like the Kraft process. In these processes, the lignin is separated from cellulose, the main constituent of paper. The pulping conditions also degrade most of the hemicelluloses, which are recovered together with lignin and pulping chemicals in the black liquor. Black liquor is burned in a recovery boiler to produce the energy necessary for the pulping, chemical recovery and paper drying steps.

P&P mills possess units that could be used for both papermaking and biorefinery operations, such as the woodyard and other raw material preprocessing units, boilers, and wastewater treatment plants. However, two distinct strategies exist for biorefinery implementation in P&P mills. Biorefinery processes can be implemented either in series or in parallel with the pulping line.

The first strategy implies a better utilization of the wood chips and logs that are currently used in the P&P processes by extracting biomass components that could be better valorized than in the current configuration. Using this strategy, cellulose would still be used for P&P products, at least during the first years of implementation. New biorefinery products would instead be produced from wood lignin and hemicelluloses, which are currently mostly burned for energy. Hence, integrated P&P biorefineries like the one just described would not require the input of new types of biomass. Nonetheless, additional quantities of wood chips and logs may be necessary to counterbalance the decrease in pulp production because some components would be extracted for the biorefinery and therefore not present in the paper.

Possible processes for the integrated chemical pulping biorefinery include hemicellulose extraction before pulping (the so-called value prior to pulping process, or VPP), black liquor gasification, and lignin precipitation from black liquor. Biorefinery options for mechanical pulp mills are more limited because most of the wood is still present in the paper. However, two main biorefinery possibilities exist. First, part of the hemicellulose can be extracted before pulping (the bio-pulping process). Although this process decreases the wood-to-pulp yield, it also decreases the refining energy in addition to creating a new sugar/hemicellulose stream. The other possibility is the modification of screen retention, keeping wood fibers of a certain size for other application such as biomaterials. Figure 7.10 shows a simplified block diagram of a possible Kraft pulp integrated biorefinery.

As its name indicates, the integrated P&P biorefinery strategy implies a strong interdependence between biorefinery and P&P operations. If the pulping line is stopped for a period of time, the biorefinery lines will also be stopped, because ultimately they use the same raw material. Planning must therefore be carefully addressed because any shutdown, production increase, or production diminution for one product will have an effect on the output of the other. Thereby, the size in terms of new bioproducts tonnage for integrated P&P biorefineries remains relatively limited because of procurement, production, and product issues. On the one side, P&P processes cannot process other biomass types than wood. Any additional biomass would have to be of the wood type, but this supply to mills is generally limited to a certain extent. On the other side, the market for traditional

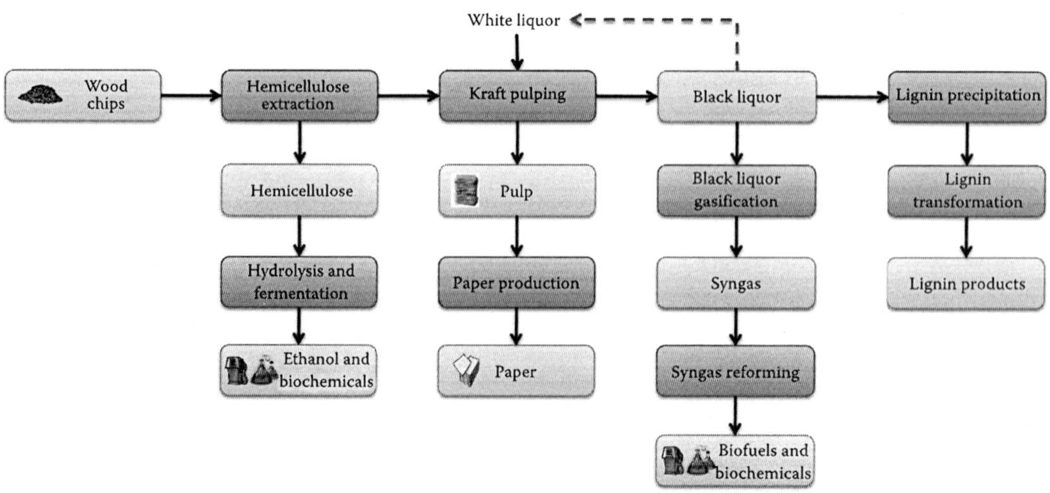

FIGURE 7.10
Block diagram of a Kraft pulp integrated biorefinery.

P&P products has recently shown limited expansion opportunities. Hence, any significant increase in the production of these products would probably be more difficult to justify in the long term, considering that this increase in production would possibly require additional investments to debottleneck the P&P processes. Therefore, the integrated P&P biorefinery strategy appears better adapted for companies that wish to remain in the P&P business.

7.3.2.2 Stand-Alone or P&P Parallel Biorefinery

In the other implementation strategy, biorefinery lines are built parallel to the existing pulp line, as shown in Figure 7.11. In this configuration, the biomass supply may include not only wood chips and logs but also additional biomass types such as forestry or agricultural residues and biomass from dedicated plantations. In fact, the typical P&P raw materials, chips and logs, would probably still be used mainly for P&P operations, while biomass residues, with their lower price, would be used for manufacturing new bioproducts. Hence, procurement operations and logistics would have to deal with additional suppliers and types of biomass, each distinguished by a specific availability, seasonality, cost, and so on. In a more distant future, wood chips and logs could even be diverted from the pulp line to the biorefinery line, if it becomes viable to have this flexibility and to use these more costly raw materials for manufacturing bioproducts.

The integration of parallel biorefinery lines into P&P mills benefits mainly from the common utilization and sharing of auxiliary units such as boilers, wastewater treatment system, and so on, as well as potential energy integration between units. Apart from these, the manufacture of P&P products would remain pretty well unaffected by and independent of the manufacture of biorefinery products. The pulp line could be stopped for, say, 1 month without impeding bioproducts production significantly because different raw materials would be used. Planning tasks for these two product families would then be less

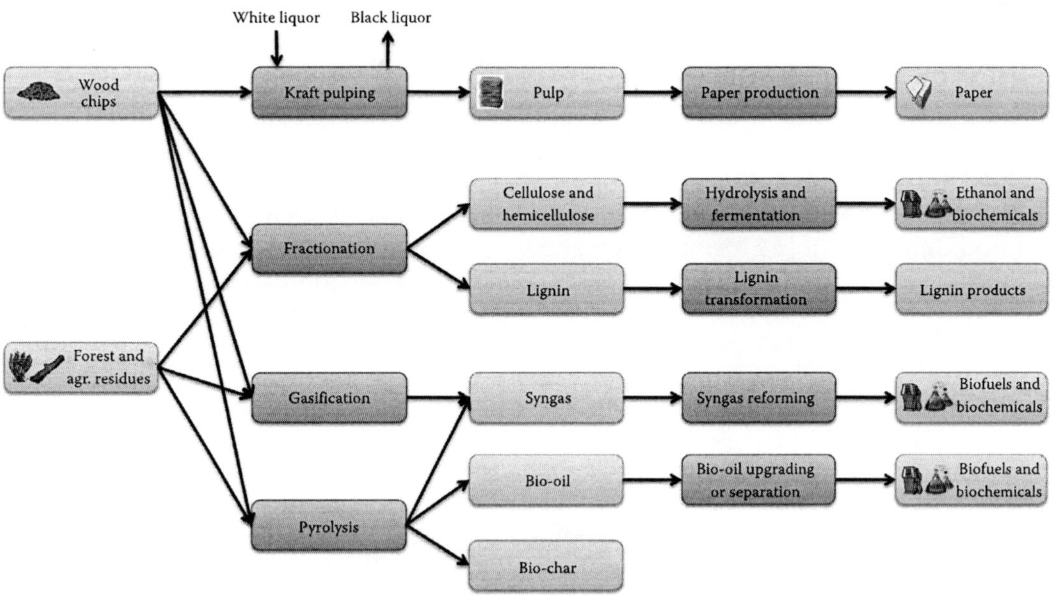

FIGURE 7.11
Block diagram of a P&P parallel biorefinery.

interrelated and could almost be performed independently. Hence, from a P&P business perspective, the parallel biorefinery strategy appears better suited for P&P companies that want to exit the P&P market and repurpose their mills to full biorefineries.

Pretty much, any biorefinery process (gasification, pyrolysis, fractionation, etc.) can be implemented in stand-alone or P&P parallel fashion to take advantage of new sources of biomass. In fact, biorefineries of larger tonnage would theoretically be possible in this configuration because larger quantities of biomass could be supplied and because the output would not be related to relatively market-limited traditional P&P products.

7.3.3 Biorefinery Products

As mentioned earlier, organic chemicals, fuels, and materials can be made out of biomass instead of from petroleum derivatives. Nevertheless, one of the main differences between biorefinery products and their fossil counterparts is their green aspect. Even if the chemical composition of the two products (green or fossil) is practically the same, the fact of being manufactured from a renewable raw material differentiates one from the other. Bioproducts might benefit from government incentives and subsidies in some regions because of this fact.

Chambost et al. classified biorefinery products into three categories [43]. *Replacement* products are identical in their chemical composition, but possess the green features mentioned above. For example, *green* ethylene is chemically identical to fossil-based ethylene, but comes from a renewable source. *Substitution* products have a different chemical composition than existing products, but similar functionality—for example, plastic bottles for water and juice currently made from polyethylene terephthalate (PET) can be replaced by polylactic acid (PLA) bottles. Finally, other products will be positioned as *novel* products. For instance, new biomaterials with enhanced functionality like nanocrystalline cellulose are likely to find a niche in new markets. While the market for replacement products is relatively secure—green products are replacing products for which a market already exists—the market for substitution and novel products is much more uncertain and typically involves more risk because new functionalities and applications must be developed.

In another vein, the green differentiation of biorefinery products provides opportunities for specific marketing and sales strategies related to this differentiation. Even for replacement products such as green ethylene, small market segments can be targeted and entirely dominated, preventing serious competitors from entering them. In that example, supposing that green ethylene is not further processed onsite, it may be sold to large PE producers who would then produce green PE. However, the demand for this material in this market would be quite high, most probably larger than the production of a biorefinery. Hence, in the short-to-medium term, competitors could enter this segment, which would put additional pressure on sale prices, reducing the chemical supplier's profit. Green ethylene could instead be sold to small specialty chemical producers that would use small quantities to manufacture products made of 100% renewable raw materials. If the production exceeds that market, the surplus could be sold on the spot market, using a different pricing scheme. This strategy would offer market segmentation and therefore RM possibilities (see Section 7.5). For substitution and novel products, additional differentiation may be obtained by specifically designing a product to perform a certain function. A constant focus on innovating to maintain differentiation will permit a supplier to dominate each of its markets by being the best supplier in each.

In addition, policies that require a certain percentage of material from renewable sources would help bioproducts producers compete against traditional petrochemical producers.

By guaranteeing market demand, and with subsidies that reduce bioproducts production costs, the impact of competition from petrochemical companies can be diminished. In the following subsections, five biorefinery products requiring different supply-chain considerations will be introduced.

7.3.3.1 Bioenergy

One of the simplest and oldest products that can be obtained from biomass is bioenergy. By burning this raw material, it is possible to retrieve its energy content. A great number of industrial processes require heat to transform products, but this heat is seldom distributed over long distances because of losses. Hence, heat and power is usually generated from fossil fuel boilers installed directly on the manufacturing plant site or in over-the-fence arrangements. In these mills and in biorefineries, this heat and power could instead be provided by biomass boilers, as long as sufficient biomass were available at a reasonable cost.

Many boilers are equipped with cogeneration capabilities. The electricity that is produced along with heat can be used directly in the process or sold to the grid. However, to be profitable, these combined heat and power plants (CHP) need heat sinks. Because pulp and paper mills need high quantities of steam, these mills represent excellent candidates for CHP facilities. Some regions around the world, such as Scandinavia, have developed district heating systems, in which the extra heat is distributed to houses and buildings within the municipality. However, the infrastructure of district heating plants is costly, which may make the development of new plants too expensive.

Biomass in many different forms can be burned for producing bioenergy. It can be burned as is (in the form of chips) or dried beforehand. Removing part of the moisture content in biomass increases the adiabatic flame temperature, thereby improving the quality of the heat provided. Biomass may also be transformed into pellets before feeding it to boilers. Pelletization can be defined as drying and pressing of biomass under high pressure to produce cylindrical pieces of compressed and extruded biomass [44]. Pellets have several advantages over wood chips for bioenergy: they have a smaller volume and a higher volumetric energy density, which makes storage, transportation, and energy conversion more efficient. They also represent a uniform and stable fuel that produces less dust [44]. Biosolids obtained by thermochemical processes (tars, chars) can also be burned for energy. Finally, biomass can be transformed into liquid or gaseous fuel by pyrolysis or gasification.

Current biomass cogeneration boilers use solid residues (bark, forest, and agricultural residues, sludge, etc.) and generate electricity using a steam turbine. More electricity could be produced with the help of gas turbines because these are more efficient. Using a combined cycle (a gas turbine coupled with a steam turbine), even more power could be generated. However, gas turbines require liquid and gaseous fuels, and further developments are still needed to use bio-oil and syngas, the pyrolysis and gasification products, directly.

The current and future market for bioenergy is and will remain highly region-specific because some grids are regulated and some not. For example, the province of Quebec in Canada already produces more than 96% of its electricity as hydroelectricity [45], a renewable source, while in its neighboring province, Ontario, nuclear- and fossil-based plants represent more than 74% of electricity production [46]. Because sustainable benefits appear easier to attain in the latter province, bioenergy efforts are naturally favored in Ontario, as attested by the recent subsidies for bioenergy given by that province's government. However, it is clear that product differentiation can be attained only with difficulty for bioenergy.

7.3.3.2 Liquid Biofuels as Transportation Fuels

Liquid biofuels are perhaps the most popular and most discussed new biorefinery products because they provide a short- and medium-term option for replacing fossil fuels while offering the potential for reducing greenhouse-gas emissions as well as ensuring energy security [47]. Recently, major oil companies have been making biofuel production a priority [48]. This is no surprise because these players have good competitive advantage and motivations: they have good knowledge of the fuel market as well as of blending operations, are used to very large-capacity production, and are willing to expend considerable efforts to green their image.

The main processes for producing lignocellulosic-based biofuels include sugar or syngas fermentation (ethanol and butanol), catalytic reforming of syngas (DME, Fischer–Tropsch (FT) fuels, ethanol, and mixed alcohols), and pyrolysis (bio-oil). Alcohols and FT fuels can be used in their pure form in combustion engines, sometimes requiring slight modifications, but they are generally blended with gasoline or diesel. DME, with characteristics similar to LPG, is envisaged for use in modified diesel engines. Because bio-oil is not miscible with hydrocarbons, it cannot be blended with diesel without modification. It must either be upgraded (deoxygenation by hydrotreatment or catalytic vapor cracking) or emulsified by the addition of surfactants [49]. Because these products have a different chemical composition from gasoline and diesel, but serve the same function, they are categorized as substitution products according to the terminology introduced earlier in this subsection.

The market for biofuels is immense: for energy security and sustainability reasons, several governments have announced policies requiring liquid transportation fuel to contain a certain percentage from renewable sources. Hence, biofuels will need to be produced in large tonnages according to specified standards for combustion engines or blending applications. They can therefore be categorized as true commodities. As a combustible, biofuel will have a selling price closely related to that of gasoline and consequently of crude oil. They will therefore face substantial price volatility and uncertainty in terms of profitability, just like typical commodities.

As commodities, biofuels should be produced in an MTS environment characterized by high production efficiency because the product does not require customization and demand is predictable (almost infinite). Hence, competitive advantage for biofuel producers would come from low-cost, effective, and high-volume procurement, production, and distribution. In a general sense, biorefineries of larger size, either stand-alone or P&P parallel, using low-cost biomass residues or high yield per year-hectare dedicated crops, would be favored for biofuel production. With a large market share, these facilities would have an efficiency advantage in this highly competitive environment. This is not to say that smaller facilities such as P&P integrated biofuel facilities could not be competitive. However, being a small player in a huge market increases exposure to fluctuations and vulnerability compared to larger producers, because a smaller producer cannot control the market. Efforts toward differentiation in the biofuel sector are therefore limited, except for supply-chain efficiency.

7.3.3.3 Biochemical Commodities

Besides energy and liquid fuel production, numerous chemicals and materials can be produced from biomass. From a sustainability perspective, the idea of replacing fossil chemicals with green equivalents seems particularly appealing. Along this line, green chemicals can be obtained in two ways: by utilizing current petrochemical routes, but using building blocks from renewable sources (by replacing commodities), or by creating new chemical derivatives through biological and chemical transformation (by substituting for commodities).

Petrochemistry is based on the use of half a dozen building-block chemicals to create all sorts of solvents, detergents, adhesives, polymers, lubricants, fertilizers, and so on. These building blocks are divided into two categories: aliphatics (mainly olefins, but especially ethylene, propylene, and butylene) and aromatics (mainly benzene, toluene, and xylene). Producing some or all of these six basic chemicals from biomass instead of from fossil sources implies different processing steps for removing oxygen from the molecules originating from biological sources. Theoretically, aromatics could be produced from lignin groups, but the process is still under development. On the aliphatic side, bio-ethylene appears to be the most promising replacement for basic petrochemical building blocks because it can be obtained relatively easily from ethanol by catalysis, and because ethylene is such a versatile and widely used chemical. Moreover, syngas reforming, notably to methanol, appears promising for the production of green versions of traditional petrochemical derivatives.

Producing these commodities from biomass offers certain advantages. To start, the market for these products is very big. Assuming that they can be produced at a competitive price compared to fossil routes, green petrochemicals would have a relatively secure market. Furthermore, these green petrochemicals could also benefit from the existing infrastructure for producing polymers, materials, and other chemicals, which has already been optimized to a certain extent. However, because these commodities are indirectly related to crude oil, their selling price would face major volatility. Just as in biofuel production, being a large producer would provide an efficiency advantage.

Instead of using the well-known petrochemical building blocks, new chemistry lines can be created from new building blocks such as, for instance, acetic acid, lactic acid, levulinic acid, succinic acid, citric acid, sorbitol, and xylitol. These building blocks are obtained by transforming sugar through fermentation, enzymatic catalysis, or chemical modification. However, the current market for these chemicals is pretty well saturated. In addition, producing these chemicals from lignocellulosic sources would not provide a green advantage because they are already manufactured mostly from other sources of sugars, which are a renewable source of carbon. To increase the demand for these chemical intermediates, new derivatives and applications need to be discovered which can substitute for existing products. Hence, in the short and medium term, producing significant quantities of these chemicals for bulk sale is not particularly appealing because there is little space left in the market. This strategy would make sense only if most of the production were dedicated to further processing to other products for which there is a market. However, with the development of new applications, the selling price of these chemicals will diminish over time as they become more and more commoditized.

Assuming that there will be demand for these "green petrochemicals" and new intermediates as commodities at some time in the future, these products would be produced and distributed in considerable quantities according to accepted composition standards. Hence, production in continuous MTS environments would be favored.

7.3.3.4 Fine and Specialty Biochemicals

Compared to commodities, fine and specialty biochemicals are characterized by much smaller production tonnages and higher added value, but generally more transformation steps. One of the main differences between specialties and fine chemicals, though, is that specialties are developed according to customer's needs for more technical applications.

Fine biochemicals, such as active medicinal ingredients, vitamins, aromas, and low-volume intermediates, and specialty biochemicals, including notably adhesives, surfactants, low-volume additives, and end-user products, can be obtained essentially either by extraction from biomass, as coproducts, or by further processing and separating biomass main components or biocommodities (biofuels, biochemicals). Products obtained by the latter pathway, that is, by additional transformation, will be referred to as *derivatives*. Producing replacement specialty chemicals, that is, green petrochemicals specialties, will require several processing steps, which represent a costly investment. Therefore, their production in the short term seems more unlikely. On the other hand, novel or substitution green specialty products appear more probable in the near term.

Solvent extraction of biochemicals from biomass is generally done before further transformation such as fractionation. However, extractibles are generally present in low concentrations in bulk lignocellulosic material. Greater, but still not very high, concentrations are found in the more specialized parts of plants such as leaves, bark, knots, seeds, and fruits. Therefore, extraction of biochemicals appears to be an interesting additional value-adding processing step, but would be hard to justify as a main process. Even though these products have high added value, their overall yield is not very high. Other biomass components should also be used to manufacture other products as part of an overall solution for transforming the entire mass of raw material. Moreover, because different plants produce different extractible compounds, only specific types of biomass should be used in an extraction process; efficient extraction requires fairly homogeneous raw materials.

Currently, fossil-based specialty chemicals are manufactured mostly from purchased raw materials. There is little to no integration back to crude oil or natural gas: companies purchase intermediates from the market to produce these derivatives. Rather than focusing on production efficiency and cost minimization, these companies focus more on the development of products which are truly adapted to their customers. Efforts toward differentiation are necessary to provide good customer service because, unlike the case of commodities, low production costs and low selling price may not be very important. In the case of biorefineries producing specialty chemicals, process efficiency will also be important to be competitive, but should not be the main focus. These chemicals are likely to be costly to produce, especially in the short term, but good customer service should remain as the main objective. As pointed out by several supply-chain authors [3,50], it is crucial to have a supply chain that is adapted to your products. There are trade-offs to be made between efficiency and responsiveness at the procurement, production, distribution, and sales levels so that the customer is served according to his needs (low cost vs. high service levels), which differ evidently depending on the type of products. As stated by Leavy, Fisher, and Lee [37,51,52], too much value gets destroyed when trying to manage supply chains for efficiency where responsiveness should be the priority and vice versa.

Because fine and specialty chemicals are produced in much smaller tonnages than commodities, the smaller production scale of biorefineries compared to their petrochemical relatives appears to be naturally better adapted to market needs. However, the market for these products would quickly become saturated; in many cases, one plant would be able to satisfy and dominate a large share of the market. This aspect would have to be carefully addressed and analyzed when designing the biorefinery to ensure that plants are built with capacity that is adapted to market needs.

Over time, as the biorefinery matures and becomes more and more fully implemented, fine and specialty chemicals will face commoditization. At this point, it will be important to reevaluate the position of these products in the market and to adapt the supply chain accordingly, for example, by shifting more toward efficiency rather than responsiveness, by standardizing and diminishing product configurations, or producing in an MTS configuration.

7.3.3.5 Biomaterials

Apart from biochemicals and biofuels, biomass components and derivatives can be used for their material properties, notably in textiles, construction, or any end-consumer applications where structure is needed. While the products described in Sections 7.3.3.3 and 7.3.3.4 are mostly chemicals in the gaseous or liquid state that are used as reactants and intermediates, materials derived from biomass, that is, biomaterials, are solids. Their handling, storage, uses, and markets differ in consequence, but many of the concerns raised in the two previous sections apply both to commodities and to specialties. In the overall supply chain, biomaterials are typically closer to the end-user. Hence, their potential for differentiation is fairly high.

Biomaterials from lignocellulose can exist in several forms, ranging from bioplastics and biocomposites, which are based on polymers, to carbon-based structures such as carbon fibers. Bioplastics can be produced either as derivatives from biochemical monomers or from the modification of naturally existing polymers. Examples of monomer-based bioplastics include green PE and PLA, which are obtained from ethylene and lactic acid, respectively. Cellulose, a naturally occurring polymer in plants, can be transformed and then regenerated into packaging plastics (e.g., cellophane) or textiles (e.g., rayon), to name a few. Biocomposites are materials formed by a polymer matrix (from a fossil or renewable source) reinforced with natural fibers. Finally, carbon fibers are produced through carbonization of polymer precursors such as rayon or lignin.

Obviously, some bioplastics would be produced in great quantities, possibly in MTS environments and according to established standards (pseudocommodities). Others, such as biocomposites and carbon fibers, would rather be engineered materials categorized as specialties and should therefore be produced in MTO mode.

7.3.4 Biorefinery Product Families

Raw material procurement in the process industry is a costly activity, notably because of harvesting/drilling operations and transportation, but also because of the high costs of investment in and maintenance of equipment and transportation infrastructure. As a consequence of this and of the divergent nature of the processes, extensive efforts are being made to use these raw materials in the most efficient way, or in other words, to produce and valorize several products at the same time. For example, petroleum is separated and cracked into fuels, chemicals, asphalt, and many other products. The different parts of the tree are used to produce lumber, panels, engineered wood products, and pulp and paper products, as well as energy. Seeds that are used for producing bioethanol and biodiesel are also transformed into several other commercial products. Distillers grain, which is traditionally used as animal feed, and syrup are also obtained from corn kernels in addition to ethanol. In the biodiesel industry, sales of glycerin, the by-product of the transformation of vegetable oil, plays a vital function in the profitability of these plants. All these examples highlight the fact that in general, one product alone cannot support the economic burden of harvesting or drilling operations. Rather, there must be a strong interdependence between the products and the various supply-chain players.

For these reasons, biorefineries would most probably have to produce more than one type of product, such as those described in the previous five subsections, to be profitable in the long run [53]. These biorefining sites, owned and operated by one or several companies just like petrochemical hubs today, would convert either forest or agricultural biomass or both in a central location to benefit from economies of scale and common operating steps. Hence, each biorefinery would have its own combination of biorefinery products, or in other words, its own *product family*, obtained by a variety of extraction processes, derivatives conversion processes, parallel lines, and so on, as shown in Figure 7.12. This example illustrates the divergent nature and complexity of managing a biorefinery product family. Because coproducts are simultaneously produced and could theoretically be sold on the market (all chemical products can be finished products), it is impossible to plan and control the production system using a single strategy. Indeed, some *core products* may be produced according to demand pull, while other *side products* will have to be pushed to the market. For example, when sawmills produce lumber, wood chips are coproduced. Because lumber provides better margins than chips (it is the core product), it may be produced in a demand-pull fashion, while the chips will be pushed to the market. Although side products may provide less marginal value than core products, they are nonetheless indispensable for a company's profitability because they provide another stream of revenues and a better utilization of raw materials.

In Figure 7.12, the excess production of *Product 1* (e.g., ethanol), the precursor of the derivative *Product 3* (e.g., ethylene), would have to be pushed to the market, while the production of the derivative could be pulled by demand. *Process 4*, an extraction process yielding a high-value chemical, removes only a small fraction of the total mass. Hence, the production of *Extractive 1* (e.g., a nutraceutical) could also be pulled by demand without much effect on downstream production. For processes that produce coproducts (e.g., *Process 3* is a gasification line with a Fischer–Tropsch reformer, producing both FT diesel, *Product 4*, and FT waxes, *Product 5*), sales of the more value-added products like *Product 5* could be pulled, while those of the other products would be pushed. For coproducts produced in processes where further derivatives (*Product 3*) are produced on a pull basis, such as *Product 2* (e.g., acetic acid produced simultaneously alongside ethanol), sales would most probably have to be pushed. Finally, parallel lines such as *Processes 1–4* and *Process 3* offer the opportunity of producing two different sets of products in a pull manner, or in other

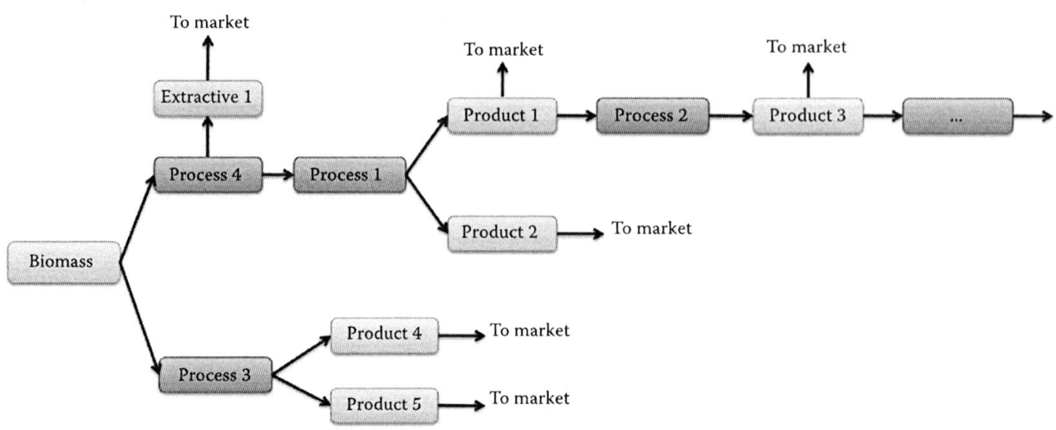

FIGURE 7.12
Example of a biorefinery product family.

words, two different core products. For instance, a P&P parallel biorefinery could have two core products: paper on one line and any other bioproducts on the other line. It must be mentioned that according to this nomenclature, core products are not necessarily those produced in the greatest quantities, but rather those with higher value for which the sales strategy should be more aligned with the market.

In brief, different biorefinery products types require different planning environments and strategies for their production, but at the same time, the process configuration (derivative chains, parallel lines, coproducts) constrains these strategies. Although the idea of having several parallel lines and a couple of derivatives chains, like petrochemical hubs, appears interesting on paper, the capital investment needed would be substantial, ranging from a few millions to several hundred millions of dollars. From a practical standpoint, companies investing in the biorefinery are more likely to focus on a few targeted products and to invest in a step-by-step fashion because access to capital is limited and the risks elevated. However, because multiproduct facilities are likely to be the endpoint of the biorefinery, it is therefore important for investors to keep the future product-portfolio strategy in mind while implementing biorefinery processes over time. The first investments might provide a lower ROI in the short term, but will have a critical impact on the flexibility and profitability of future biorefining investments.

7.4 Risk Mitigation of Market Volatility and Biorefinery

Markets in the process industry, such as oil and gas, commodity chemicals, steel and metals, and food have been facing especially volatile environments in recent years. Price volatility comes from a mismatch between supply and demand. Especially for the industries just mentioned, volatility is increased by the unpredictable aspect of raw material supply. For instance, biomass-related raw materials such as cereals show high volatility because of their seasonality and variable yields. Nonrenewable raw materials such as metals and oil are accompanied by high uncertainties in the world's supply availability, notably due to politics and natural disasters. Moreover, global consumption of and demand for commodities is constantly changing. For example, China's and India's demand for these resources has increased considerably in the last decades. All these uncontrollable reasons related to both the supply and demand sides unbalance the equilibrium between the two, thus leading to volatility.

Volatility tends to decrease downstream in the supply chain: the more a product is transformed, the less price-volatile it is [54]. Specialized product manufacturers are therefore usually favored over basic product (commodity) manufacturers from a supply-price volatility viewpoint; because they use more highly transformed raw materials, their supply costs are more predictable. In a general sense, because the process industries are upstream in the supply chain, raw material and energy price and supply volatility are important aspects to be dealt with in these industries. On the other hand, some industries, especially those manufacturing more specialized products which are sold to a small group of customers, for example, fine and specialty chemicals, might face high uncertainty on the demand side. Demand for these products is constantly evolving, but at the same time, competition between a few key suppliers might be fierce. Hence, contracts with great flexibilities in the quantities that may be purchased are traditionally negotiated to increase the customer's loyalty and satisfaction levels [18,24]. Because these contracts may be necessary

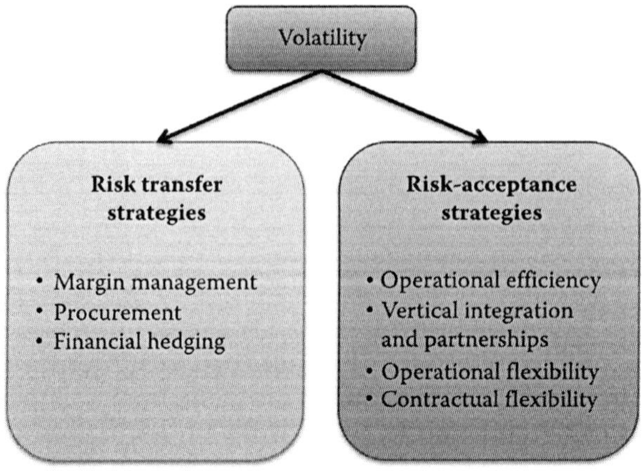

FIGURE 7.13
Risk-mitigation strategies.

to maintain a strong market presence in a particular segment, this uncertainty in demand quantities remains an important risk to address.

In the case of biorefinery products, volatility effects are likely to be rather uncertain. On one side, biorefinery products will mostly replace or substitute for traditional fossil-based products. Hence, their price should ultimately be linked to crude oil and natural gas prices because fossil products are still likely to dominate the marketplace. However, as bioproducts enter the market, the equilibrium between supply and demand will evolve. On the other side, because bioproducts come from biomass, volatility on the supply side will also be influenced by factors such as weather and government policies. Therefore, biorefineries will have to deal with significant margin pressure due to volatility.

To deal with risk related to market volatility, several strategies can be used. Standard approaches reduce short-term risk by transferring it or part of it to a third party: the customer (margin management strategies) or the supplier (procurement strategies). Financial tools, which are sometimes offered by third-party financial institutions, can also be used to reduce risk. These tools are referred to as *hedging strategies*. Process-industry companies often use these three strategies in conjunction [55]. Instead of transferring the risk, an alternative approach is to accept it and to try to reduce long-term exposure and consequences by minimizing the inputs of volatile raw materials (increasing operational efficiency), securing supply or sales (vertical integration and partnerships), or by exploiting flexibility in the supply chain. These risk-mitigation strategies are summarized in Figure 7.13.

7.4.1 Risk Transfer Strategies

7.4.1.1 Margin Management Strategies

The most common and least wide-ranging approach, *margin management*, consists of avoiding risk by passing it to the customer. The oil industry, for instance, is able to pass on part of its raw material cost volatility to the customer. When crude prices go up, the gasoline price at the pump will also go up, stabilizing the company's margins. This approach is, however, limited by the price elasticity of demand, which is a measure of how the magnitude of demand varies with price [56]. The extent of this elasticity depends notably on the strength of demand, the level of product differentiation, capacity utilization, market concentration, and the availability of substitute products [55]. Returning to the gasoline example, the market demand is very strong and global. Because there are few substitute fuels that can yet be used in a car engine, refiners have the opportunity to raise their price according to crude oil volatility without having a big impact on demand, at least in the short term. For other industries where there is excess production capacity, for example, newsprint, this strategy may not be available because the customer can switch to another cheaper supplier.

In the biorefinery, as mentioned earlier, the relationship between raw material prices and product prices will not be obvious. Unless serious efforts are made to differentiate bioproducts from the fossil resources they replace, margin management strategies will have only limited effectiveness. How can biomass volatility be transferred to a bioproduct that is pegged to crude oil prices and for which there are alternatives which are possibly cheaper? Local and regional policies enforcing a minimum content of renewable materials will, in that sense, have a significant impact on differentiation and the extent to which margin management strategies as defined above can be effective.

7.4.1.2 Procurement Strategies

Risk can also be managed by *procurement strategies*, such as entering into fixed-price contracts of various lengths with suppliers. Volatility fluctuations in raw material supply are thus eliminated to a certain extent, but often a price is paid in the form of higher costs and reduced flexibility. Indeed, in exchange for guaranteed supply and simplicity, suppliers will ask for specific volume commitments, even at higher prices. Moreover, signing long-term fixed contracts in a rapidly changing price environment may create thorny vendor management issues [55].

Because they will be part of the process industries, biorefineries will need to run their equipment to a certain capacity level to be profitable. Biomass will also represent a large fraction of the operating costs. Moreover, biomass prices will increase over time [57]. For these reasons, potential biorefiners underline the importance of signing long-term supply contracts to secure the amount of biomass that will be brought to the facility. On the other side, with several future biorefineries on the horizon, biomass suppliers may be reluctant to sign long-term fixed contracts because soon there will be more competition and more demand for their products, and therefore more opportunities for profit.

Like a good personal investment plan, raw material procurement strategies should be composed of a well-diversified portfolio of contracts plus spot opportunities adapted to the company's level of risk acceptance, with some more wide-ranging but less risky options, and some more rewarding yet riskier ones. Efficient procurement strategies should be developed to secure part of the supply, but also to benefit from the best opportunities.

7.4.1.3 Hedging Strategies

Financial derivative instruments can also be used to mitigate raw material and energy volatility. These *hedging strategies* include forward contracts, future contracts, swaps, and option contracts. The Chicago Mercantile Exchange (CME), for example, is a place where these financial derivatives are traded for several commodity materials, including crude oil, natural gas, and several agricultural crops. Several companies have been successfully using these tools to manage financial risk in energy purchases (electricity, natural gas, transportation fuels). However, financial derivatives can be highly complex and should be used carefully because improper use can be disastrous for the corporate budget.

Purchasing raw materials through commodity derivatives is often of limited effectiveness because commodity users normally have highly specific needs for product specifications, timing, and delivery terms [55]. However, using operational planning models, some authors have shown that financial derivatives can help reduce financial risk in refineries [28,58]. Therefore, using hedging as part of a raw material procurement strategy for process-industry enterprises appears to be a potential opportunity.

Because lignocellulosic materials and their biomass-based commodities are not yet exchanged globally, financial derivative instruments for these products have not been developed. However, some argue that in the near future, in 5–20 years, an international biomass trade in commodities markets will exist [59]. Nevertheless, because of the more local/regional aspect of procurement (especially for biomass residues), it is likely that these contracts will rather be one-to-one agreements. Therefore, financial derivatives seem to be limited in terms of the opportunities and effectiveness they offer to the biorefinery, at least in the short term.

7.4.2 Risk-Acceptance Strategies

Section 7.4.1 has described how risk-transfer strategies can be fairly limited in their application and effectiveness, especially for the biorefinery. Biocommodity prices will be indirectly related to crude oil and therefore dependent on its price fluctuations. For these products, it will consequently be difficult, for instance, to pass on the volatility in biomass prices to the customers without a substantial effect on future demand. Long-term fixed-supply contracts are also limited to a certain extent: as more biorefineries are built, the demand for biomass will augment, pushing costs higher. Therefore, over the long term, biomass producers will be unlikely to sell their raw materials through fixed long-term contracts because they will want to benefit from price increases. Finally, biomass procurement is likely to be local or regional, which limits the scope of financial derivative tools. For the biorefinery, risk-acceptance strategies might prove to be a more efficient approach to reducing a company's long-term risk exposure and its associated consequences. These strategies reduce risk exposure by either avoiding it or by minimizing its impact in case it occurs.

7.4.2.1 Operational Efficiency

Companies may reduce their risk exposure by increasing their *operational efficiency*, that is, making efforts to reduce waste and to use inputs more efficiently. By using fewer volatile inputs overall, exposure to risk is reduced. These efforts have been especially important in the process industry because the material and energy resources involved in these processes are enormous. In particular, techniques that analyze the process holistically to improve process design and retrofits, as well as optimizing mass and energy exchanges, namely process integration [60], have proven to be effective and are increasingly used in industry. In addition to improving process operations, improving logistical operations can also help to attain higher levels of efficiency. Better management of inventory, transportation, and distribution, or even reconfiguring the distribution network, are examples of what can be done to reduce the consumption of inputs such as fuels.

As they are implemented, biorefineries will have to compete with petroleum-based industries, which have been optimized to a great extent over the last century. In this sense, operational efficiency, both in harvesting and processing applications, is often viewed as a prerequisite for the biorefinery, even for specialty bioproducts producers, who should nonetheless strive to maintain a customer focus. Supply-chain efficiency will be especially important for reducing procurement and distribution costs at the mill gate.

On the production side, increasing the yield and selectivity of transformation and separation processes is a necessary step to reduce costs further and to increase the quality of products. In addition, using new separation technologies, such as membranes instead of distillation, would help increase production efficiency and even the flexibility of units.

On the procurement side, additional efforts should be made to increase biomass yields and to optimize transportation to reduce costs. For example, forests are currently not being used to their full potential. Of course, a fraction of the cuts should be left on the forest floor for regeneration, but much of the precommercial thinnings, branches, and stumps could also be harvested along with the logs and then further sorted by quality, homogeneity, and so on, for different uses.

On the distribution side, biorefineries are expected to be disadvantaged in terms of market access because of their remoteness: farms and forests are usually not very close to end markets, which are in cities for the most part. To compete, transport efficiency should be maximized in both directions: from the mill to the market, but also from the market to the mill, with an effort to minimize the number of one-way truck trips.

7.4.2.2 Vertical Integration and Partnerships

One of the oldest strategies for risk mitigation is *vertical integration*. By acquiring suppliers that produce raw materials, or by acquiring a downstream supply-chain producer, exposure to commodity-price volatility can be controlled. For example, several P&P companies have reduced their procurement cost variability and their dependence on external sources of biomass by performing operations ranging from harvesting, to sawmilling, to pulp and paper production themselves. Companies in this industry segment are also reducing their dependence on volatile fossil-based energy supplies by producing their own energy by building or acquiring biomass boilers, hydroelectric dams, or windmills. In addition to mitigating risk, these strategies have the benefit of reducing the company's environmental footprint. On the other side of the supply chain, downstream vertical integration ensures that part of the market for the manufactured product is secured and isolated from the competition. At the same time, the value created in subsequent process steps is captured. Although vertical integration offers several advantages for risk mitigation of market volatility, it also requires significant capital investment and can be complex to accomplish.

Like vertical integration, but without the actual acquisition of other supply-chain members, a strong collaboration between two companies, such as a type III partnership or a joint venture, could also help mitigate price volatility. In the same way, access to markets or raw materials could be secured, but at a fraction of the cost of vertical integration. However, partnerships do not come without challenges. They increase the degree of dependence on the other partner, and like human relationships, they can be costly, time-consuming, and difficult to maintain. As underlined previously, partnerships should not be concluded with everyone, nor for every type of product. They should only be made with suppliers and producers whose knowledge is critical for further process and product development. For instance, strong partnerships are typically less of a requirement for commodities than for specialties.

As for the biorefinery, there is no doubt that partnerships, vertical integration, or both will be essential for the development of this new business. Upstream, biomass costs can be volatile. Moreover, the composition, homogeneity, and quality of biomass inputs are important parameters that must be controlled for most processes and applications. Hence, to secure part of the supply in terms of quantity and to ensure a certain level of raw material quality, upstream integration and partnerships are likely to be beneficial for biocommodity and biofuel producers because they would provide an efficiency advantage. By contrast, specialty biochemicals and biomaterials would be best served by a responsiveness advantage and would therefore benefit much more from downstream integration and partnerships which could help them: (1) secure the market, (2) better serve the customer, and (3) develop new products and applications for the end user.

7.4.2.3 Operational Flexibility

Market-volatility risk can also be mitigated through the use of *operational* or *manufacturing flexibility*. For more information on this topic, the reader is referred to Chapter 9. Broadly, these flexibilities can be divided into three categories: feedstock, production, and product flexibility. Manufacturing flexibilities differ from the previously introduced risk-management strategies in the way risk is dealt with. While the previous strategies treated risk mainly by diminishing exposure to it (transferring it to third parties or reducing the quantities of volatile raw materials involved), manufacturing flexibility truly implies an acceptance of market volatility and a will to deal with it by minimizing its impact in case it occurs.

Being able to handle a wider range of input materials, either on the energy or the raw material side, is referred to as *feedstock flexibility*. Many plants obtain their energy mix using several boilers that run or are able to run on different fuel types, such as Bunker C, natural gas, biomass, sludge, electricity, and so on. On the raw material side, some metal companies have upgraded their rolling equipment to take metal or recycled material as inputs. P&P mills are able to use different mixes of tree species in their pulp lines to make products with similar properties. Likewise, refineries process different crudes to produce fuels and other chemicals. Similarly, biorefineries could take advantage of the different biomass sources available in a region to minimize procurement costs and risks, and also to obtain additional quantities and thus benefit from processing economies of scale. Because feedstocks may represent the most price-volatile products that are bought or sold by a company in the process industry, diversification of the feedstock portfolio might be an efficient strategy to minimize risk. However, a careful assessment must be performed before deciding to use different raw materials type, because a reduction in procurement costs and risks may be achieved at the expense of higher processing costs.

Two main *production flexibility* options exist. On the one hand, having the capability to slow down or speed up production rates can be referred to as *volume flexibility*. This flexibility is typically governed by one or several bottleneck units. On the other hand, many plants have several machines, units, or groups of units that are able to produce the same products. In these cases, the manufacturer has *machine flexibility* that he can take advantage of in his planning to choose the most suitable production alternative to minimize production costs. These production flexibilities enable better coordination of production with sales and thus enable planning to be adapted in case of demand volatility, in terms of both price and quantity. How to use integrated planning for better management of sales and production together will in fact be the topic of Section 7.5.

Some units can change the product mix that they manufacture. For example, some reactors may be able to produce different compounds by changing the catalyst used. These highly flexible units are generally of the batch type and are mostly found in the production of specialty products, where they serve to customize (differentiate) orders at the end of the production sequence. Not all chemical units or processes incorporate such *product flexibility* because they are often designed to maximize yield. Following this argument further, units producing commodities in capital-intensive, dedicated, and continuous processes, would not benefit much from product flexibility. However, for specialty products, flexibility permits a coordination of production with sales to manage risk related to demand, in the same way as production flexibility.

In addition to manufacturing flexibilities, external flexibilities related to distribution and purchasing may be used to minimize risk. Maintaining a diversified basket of transportation alternatives for procurement and distribution, as well as a diversified set of suppliers, might protect a company from certain price and quantity fluctuations. The idea here is not to put all one's eggs in the same basket to minimize the consequences in case of sudden fluctuations arising from a third party.

7.4.2.4 Contractual Flexibilities

As seen in Section 7.2.4, companies in the process industries traditionally negotiate contracts with a reveal-date and due-date structure because of the long lead times involved. However, this involves a certain level of uncertainty in demand, but not necessarily in price. Companies can reduce this risk by the inclusion of *contractual flexibilities*. Schiltknecht has identified several reactive components of contracts that can be included to minimize the effects of uncertainty [24].

In fact, the structure of reveal- and due-date agreements greatly influences who will bear the greatest risk. Close reveal and due dates give less time to the manufacturer to react and plan production efficiently, but decrease lead times, thus increasing customer satisfaction. Consequently, manufacturers will ask for early reveal dates and late due dates, while customers will ask for the opposite. Therefore, several clauses related to this structure can be included to minimize the manufacturer's risk. Before the reveal date, the manufacturer may ask the customer for a quantity window for his final demand. Although some contracts offer the customer the freedom to make significant changes to the quantities ordered and even to cancel the order at the reveal date, reducing the quantity window will help the producer plan production more efficiently. Some contracts have due-date postponement clauses which allow the manufacturer to postpone final product delivery within a certain time window, generally with the inclusion of a penalty. Others include the possibility of outsourcing (subcontracting). These clauses must, however, be accepted by the customer because they might entail quality and other product differences.

The possibility of using some or all of strategies described above may not always exist for a manufacturer because certain market segments may be highly competitive. In fact, contracts are often formulated to attract customers and are therefore more restrictive for the manufacturer. Giving slightly more ordering flexibility to the customer (increasing customer service) can be a differentiator from competition. This may apply even for commodities, but as mentioned before, it should not be the main focus. Responsiveness efforts should not be the center of attention when efficiency is needed. Contractual flexibilities are therefore more likely to be available in a market for differentiated products, particularly in specialties, in which the manufacturer retains a certain degree of power by making its products unique to the customer. By extension, these risk-mitigation strategies are more available to collaborating companies.

To conclude the topic of risk mitigation, when individually used, each of these strategies will be effective only to a certain extent depending on the context of the product that is manufactured. Price volatility and market uncertainty cannot be entirely controlled, but should rather be tamed. This holds especially true for the biorefinery because the price volatility and market uncertainty related to bioproducts are influenced by many different factors. As underlined by several consulting firms such as PriceWaterhouseCoopers, there is no single solution for risk mitigation. Only a robust and integrated approach using a portfolio of strategies and tactics is likely to succeed in reducing risk exposure and impacts [55].

7.5 Value-Chain Management and Biorefinery

With increasing levels of competition, more demanding customers, and more restrictive regulations related to sustainable development, doing a profitable business in the process industries is becoming more and more challenging. Process-industry companies must manage supply chains that are more complex than ever, including several locations, various raw material types, diverse processes, and many suppliers and customers, to name only a few. Decision-making related to the design of the supply chain and to operations planning must therefore be carefully addressed to make the best of available resources. Fortunately, recent advances in information technology and optimization modeling are now enabling the development of advanced planning tools that can help to manage the supply chain holistically to become more competitive. Overall, research efforts in SCM have been focused mainly on integrating procurement, production, and distribution to minimize costs and inventory, improve efficiency, and deliver better service. These are, in other words, the *costs* term of the profit equation, *Profit* = *Revenues* – *Costs*. Hence, SCM has traditionally concentrated on controlling volume and logistics while fulfilling fixed demand. However, much less research has been devoted to linking sales more efficiently with production and other logistical operations, that is, the *revenues* term. By better managing demand and by incorporating it with supply-chain activities, the business strategy can be better aligned with the plant strategy, thus helping manufacturing companies to increase their profitability, to realize additional improvements in efficiency, and to provide more value to customers. In fact, some authors, like Shapiro [2] and Chopra and Meindl [3], already consider sales and demand management as part of SCM. However, for purposes of this chapter, this holistic planning process, integrating demand with production and logistics, will be referred to as value-chain management, in accordance with VCM definitions provided by Kannegiesser [14] and Schulz et al. [61]. In this section, key VCM concepts such as RM and integrated planning concepts will be explained, as well as how they can be applied to the process industry and particularly to the biorefinery.

7.5.1 Revenue Management

The concept of *revenue management*, sometimes called yield management, was developed in the late 1970s in the airline industry. At that time, some airlines started to offer discounts to early-bird customers to optimize the capacity utilization of their flights. One decision problem that soon emerged was how many seats should be sold at the early-bird rate, while reserving enough seats for last-minute customers who would pay a higher price. These capacity-reservation decision problems have since attracted wide interest in academia and have been increasingly studied in many other industries besides airlines, such as hotels and resorts, rental cars, telecommunications systems, and cargo transportation [62–64].

As defined by Talluri and van Ryzin, RM involves demand-management decisions aimed at increasing a company's revenues [65]. The authors distinguish two RM approaches: quantity-based and price-based. The first approach relies on exploiting the differences between customers and their willingness to pay. It divides customers into multiple segments with different buying behavior, strategic importance, and/or average profitability and prioritizes these segments when allocating scarce capacity. The key idea is that giving priority to high-margin segments yields higher revenues than selling scarce capacity on a first-come-first-served basis. The second RM approach uses dynamic pricing strategies as a tool for demand management. These strategies adjust prices dynamically over time in response to nonstationary demand, during a finite selling season, or through auctions. Demand is thus actively influenced.

In brief, the basis of RM is an order acceptance and refusal process that integrates the marketing, financial, and operations functions to maximize revenue from existing capacity [66]. According to Chopra and Meindl, RM has a significant impact on supply-chain profitability when one or more of the following four conditions exist [3]:

1. The value of the product varies in different market segments.
2. The product is highly perishable, or product wastage occurs.
3. Demand has seasonal or other peaks.
4. The product is sold both in bulk and on the spot market.

Until now, revenue management concepts have been used mainly in the service and retail industries. They are, however, finding more and more acceptance and application in the process industries when some of the preceding conditions apply, as reported by some authors [67,68]. Recently, a survey done by Kolisch and Zatta with managers in this industry showed that RM concepts were considered fairly important for the future success of a company [69]. According to industry leaders, methods combining capacity management with price management seemed to be the most promising. Even so, the survey identified several hurdles that are preventing the wide development and use of RM methods in the process industry. These included the lack of a clearly defined and communicated pricing strategy, the lack of experience with RM, and the lack of appropriate RM approach.

Because the type of product that is manufactured (true or pseudocommodity, fine or specialty chemical) has an effect on the process type chosen, the manufacturing environment, and therefore the sales mechanisms that are used, RM strategies should differ depending on how production and sales are done in these particular market segments. Quante et al. [70] presented a supply-chain framework for RM and demand fulfillment because these two concepts are closely linked. Giving examples from the retail, service, and manufacturing industries, the authors identified five types of attributes that are relevant to RM and demand fulfillment and that could be used to identify applications that have the same or similar RM and demand fulfillment requirements. Explanations of these attributes and their connection to the process industries are presented in Table 7.2. They also analyzed different models and software applications supporting mainly short-term decision-making about demand fulfillment and RM, with the intention of matching applications, models, and software.

For manufacturing applications, Quante et al. [70] matched models and software according to MTO, ATO, and MTS manufacturing environments because these differ in terms of customer expectations and order fulfillment, and therefore in how RM can be applied.

TABLE 7.2

Attributes Relevant to Revenue Management and Demand Fulfillment

Attribute	Description
Replenishment	• *Buyer-driven refilling* is found when supply exceeds demand. Orders will be, in general, fulfilled as requested because there is competition. This is generally the case for commodities. Commodity-producing companies usually replenish their customers in this mode. • *Vendor-driven refilling* occurs when supply capacity is scarce. The supplier may have the "power" to ration his customers proportionally to their ordered quantities.
Decoupling point	• The different *manufacturing environments* (cf. Section 7.2.1), such as make-to-stock, assemble-to-order, and make-to-order, differ in their planning of procurement, production, and sales; by the nature of inventory kept at the decoupling point (raw materials, intermediate products, or finished products); and in the way demand is fulfilled.

Capacity	• *Flexibility of capacity* offers a lever for matching supply and demand. Short-term adaptations such as the following are possible:
	  • In case of excess capacity, shutting down unnecessary machines or production lines may add a level of flexibility. On short notice, process equipment usually cannot easily be switched off because of long and costly setups. Some units offer the possibility of intensity adjustments, i.e., producing more or less of a product. However, this flexibility depends strongly on the technology that is used (batch vs. continuous) and its specific design.
	  • In case of capacity shortages, hiring temporary workers and/or extending regular working hours can be done. This is, however, difficult in the process industry because equipment is usually run year-round almost at full capacity.

Product	• The company's ability to change product prices, i.e., *pricing flexibility*, varies by the type of product manufactured. This flexibility is characterized by the frequency and magnitude of change. • Customized products (differentiated products) will have more customized prices, while standard products will be sold at a given price over a longer time horizon. The frequency of change ranges from at the time of ordering (highly customized products), to short-term and mid-term changes (e.g., standard items with pricing catalogs issued every season or every year). • *Perishability* (cf. Section 7.2.3): For perishable products, optimal replenishment is important because overstocking is expensive. Future sales are usually not influenced by current sales. However, durable-products sales will be influenced by present sales. • *Life cycle* (cf. Section 7.2.3): Short life cycle products have few historical demand data, which complicates reliable forecasting, while the demand for long life cycle products is more predictable.

Demand	• Profitability may differ between orders. Profit heterogeneity varies over time and by customer according to three factors: revenue, cost, and the customer's strategic importance. This profit heterogeneity is at the heart of revenue management.
	  • *Revenue* earned from an order may vary in the presence of different market segments that accept different pricing (variation by customer). Also, discounts may be offered to customers reserving capacity in advance (variation over time).
	  • The *cost* for serving a customer might also differ according to logistical costs (transportation, inventory), taxes, and other variable costs downstream of production.
	  • *Strategic importance*: Some customers (partners and loyal customers) are more important for the company's long-term profitability and should be treated better than occasional customers.

Because process-industry manufacturing environments are mainly MTS or MTO (or hybrids of these), only these two environments will be discussed.

In MTS (most commodity products, undifferentiated), production is triggered according to forecasts, and orders are served from inventory. Because of general contracting practices, short-term demand fulfillment options such as order acceptance and pricing flexibility are somewhat limited. Pricing decisions and demand-management options are therefore more appropriate at the medium-term planning level. Using an integrated tactical model, quantities can be allocated to different customer classes to determine the best combination of sales, production, and replenishment quantities. The idea is to benefit from production flexibilities to link sales to production in the most effective and profitable manner. However, because sales planning at this level uses aggregated forecasts, allocation according to profitability cannot be carried down to the level of specific orders.

Still in the MTS environment, but on a short-term level, revenue management strategies, software, and applicable models can be used to deal with order promising (cf. [13,71] for more information on this subject). Simpler order-promising approaches/models take into account the current inventory and possible future replenishments, that is, the available remaining production capacity (considered to be known or fixed in time) and fulfill orders on a first-come-first-served basis. Newer and promising approaches instead process several orders together and assign the available remaining capacity by profitability classes [67,70].

Because of their high degree of customization, MTO environments are characterized by multistage and often complex production steps. Production-capacity assignment and reliable due-date promising therefore becomes critical because lead times are long. The variation in order profitability by demand attributes (revenue, logistical costs, and strategic importance) is typically greater in MTO than in MTS environments because of this customization. This gives rise to greater pricing flexibility and offers opportunities for segmentation strategies even at the order level. This difference between customer order profitability should, however, be calculated before using RM tactics. Hopefully, marginal costing methods can help to find minimum acceptable prices by accounting for the different processing and customization steps used for each order. Integrated APS, that is, optimization models, can also be used to test options and to find the best production sequence that serves more profitable orders first (revenue maximization by customer segmentation), while at the same time minimizing production costs by efficient campaign planning. This scenario analysis should, however, be run rapidly to provide a quick answer to the customers' main question: acceptance or rejection of their order. Hence, one problem that arises with this potential strategy is the development of models that yield a good balance between short response times and the quality and reliability of the solution [70].

7.5.2 Integrated Value-Chain Management for Margins-Based Planning

The process industry (including the biorefinery) is highly intensive in terms of capital spending. Assets must therefore be utilized to a certain extent to provide good return on the capital invested. However, minimizing production costs by maximizing production output does not always result in highest profitability as it is often believed, because other supply-chain costs are neglected [72]. Moreover, fulfilling demand is not necessarily the optimal way to increase profits because low-priced contracts do not always factor in variable and fixed costs [61]. To be profitable, production-capacity allocation must be performed to generate the most profitable sales (RM), while at the same time accounting for manufacturing costs and coordinating procurement and distribution accordingly.

In this context, marginal costing and integrated planning appear to be key approaches. Breaking down costs to single units or groups of units and to recipes makes it possible to evaluate trade-offs between different ways of producing and/or different capacity levels. The same marginal-costing philosophy can be applied to the procurement and sales levels by segmenting customers and suppliers into groups of different profitability or importance to the company. Finally, with the idea that optimizing the whole gives better results than optimizing one part at a time, linking the various locations (mills, distribution centers, etc.) and the four supply-chain planning tasks into a single integrated planning framework would result in a more profitable plan for adding value.

On the down side, from a mathematical optimization point of view, integrating more information and more decisions together leads to larger and larger problems that must be solved within an acceptable time for the results to be used by decision-makers. In this context, Grossmann identified four challenges related to the development of integrated planning frameworks in the process industries: (1) a modeling challenge, (2) a multiscale optimization challenge (how to coordinate long-term with medium- and short-term decisions), (3) an algorithmic and computational challenge, and (4) an uncertainty challenge [7]. These challenges mostly remain to be addressed before integrated models can be used efficiently to support decision-making activities inside a company.

Recently, several authors have incorporated certain RM concepts (without necessarily using this nomenclature) and have developed integrated planning models and strategies in accordance to a margins-based planning approach, that is, an approach that maximizes profit instead of only minimizing costs [14,16,50,73,74].

Laflamme-Mayer developed a multiscale online planning framework for a pulp mill [50]. First, an activity-based cost model was developed to show how various process conditions affect resource use and therefore costs. At the tactical level, an aggregated capacity-planning model was used to balance fiber supply with pulp demand according to forecasts. At the operational level, a campaign-planning and order-scheduling model was used to maximize short-term profitability given orders and supply-chain capabilities. Using this framework, he showed for a case study how process flexibility in terms of raw material input (chip input ratio) could be used to support more effective procurement and thus increase profit. He also demonstrated that by varying production capacity at the mill, a more profitable alignment between production and demand could be reached. Although this framework was used to prove the benefits of integrated planning, production flexibility, and capacity allocation, sales decisions such as accepting or rejecting orders, contracts, and spot sales were not considered in this framework.

Feng et al. developed a supply-chain-based sales and operations planning model for an oriented strandboard (OSB) mill running in an MTO environment [73]. Using a tactical model that integrated sales, production, distribution, and procurement, the authors showed the benefits of using such a model to align production capacity with contracted and spot demand, which is treated as a decision variable. Two of the key findings were the following: the integrated approach provides superior performance in all cases, but as market price decreases (the market environment becomes more difficult), its benefits tends to increase. Moreover, variations in cost factors (production, transportation, distribution) have a less significant impact on profitability than fluctuations in market price and demand factors.

Kannegiesser et al. [14,16,74] developed an integrated tactical model for a chemical commodity manufacturer operating in an MTS environment. The model optimizes profitability by coordinating sales quantity, price, and supply decisions throughout the value chain. In a similar way to Feng et al. [73], they studied capacity management with respect to contract and spot sales in a situation in which contract demand had to be fulfilled, but not necessarily spot demand. In addition, they considered linear recipe functions for production to represent variable production costs more accurately. Using this model, they tested two optimization approaches: a classical one-phase optimization strategy that maximizes expected profit across one or multiple price scenarios, and a robust two-phase optimization strategy that maximizes expected profit across multiple price scenarios given that a minimum profit level has been reached.

7.5.3 Value-Chain Management for Biorefinery

With new sources of raw materials, new processes, and new products and markets, the selection of products and processes, as well as the design and management of the biorefinery value chain, cannot be left to chance. In fact, several believe that technology will provide only a short-term advantage [75]. In the long run, the value-chain design (product portfolio, facility location, access to market and to raw materials, etc.) and its management will differentiate between successful and unsuccessful biorefineries. As stated by Shapiro [2] and Stadtler and Kilger [6], efficient management of the value chain can be a source of a competitive advantage that cannot be replicated by others. Hence, the use of optimization models to manage the value chain could greatly help decision makers to plan better and thus to realize this potential competitive advantage. Referring to Figure 7.12 and the earlier discussions of biorefinery products and product families, the specificities of each of these products and their associated processes require different RM strategies and a different use of operational flexibilities.

Processes that simultaneously produce coproducts usually produce undifferentiated products. For these products, MTS environments in which RM concepts are applied mainly at the tactical level, such as the Kannegiesser framework [14], seem to be better adapted. Depending on the type of undifferentiated product being produced and the scale of production (fine chemicals or true commodities), different sales strategies should be used. True commodities, such as bio-commodities and biofuels, can be sold directly to the market through contracts or spot sales, but can also be used as platform chemicals for derivatives. Hence, RM opportunities include on one side the more traditional segmentation of market customers and also the "company's internal derivative market." In addition, because these products are produced upstream in the supply chain and are therefore more vulnerable to raw material price volatility, they would benefit more from feedstock flexibility than from value-added chemicals. One the other side, fine biochemicals, for which the number of buyers is typically much smaller, would offer fewer opportunities for customer segmentation because customers should be more equally important to the company. Processes which have a certain amount of flexibility in terms of volume could better adapt to market needs and prices and their associated fluctuations. This volume flexibility could take the form of overall throughput flexibility, capacity to change the yield of output products (same throughput, but more or less of certain coproducts), or both. For these processes, distinguishing core from side products is of critical importance to identify which products would benefit from a more demand-driven sales strategy and should be managed accordingly.

Biochemicals obtained by extraction in low volumes could be produced on a demand-driven basis in an MTO environment, with the opportunity of accepting or rejecting sales at the order level. Because these products are extracted only in small volumes, the extraction steps could be bypassed or not according to demand conditions, without affecting the main production line significantly.

Specialty biochemicals and biomaterials typically involve several processing steps and are obtained in the last stages of derivative chains. Because they usually require some sort of customization and high levels of responsiveness, flexible MTO environments coupled with revenue management concepts applied at the operational level would seem to be most appropriate. Hence, batch equipment possessing both volume and product flexibility to react to customer requirements and demand conditions appears to be especially well suited to this case.

For pseudocommodity biomaterials, which are obtained by polymerization or by cellulose modification, VCM strategies should depend on the amount of customization that can be done to differentiate products between orders, as well as whether products are of the replacement or substitution type. Replacement biomaterials like "green" PE should be managed in a similar way to true commodities. However, because this "commodity" product comes from lignocellulosic raw materials, several processing steps are needed to obtain it: ethanol must first be obtained (by gasification and reforming, or by fractionation, hydrolysis, and fermentation) and then converted to ethylene and PE before being molded into a material. This is a large number of steps to manufacture a commodity product which does not bring much added value. Moreover, apart for its "green" characteristics, this material has been in existence for decades already, which makes product differentiation more difficult. On the other side, substitution biomaterials like PLA, which are relatively novel, can benefit from more differentiation opportunities and therefore more customer-segmentation opportunities. Hence, they would be able to take advantage of more flexible environments in terms of customization with a demand-pull approach, that is, MTO environments.

As for bioenergy, demand for electricity and heat is seasonal (people require more heat in winter) and varies even according to the time of the day. Hence, there seems to be opportunities for using VCM strategies, both for reducing the amount of heat and electricity consumed during more expensive periods and for maximizing extra sales of power when the grid demand for energy is high. However, these strategies depend highly on the structure and policies of the electricity grid in place in the region.

The VCM of a biorefinery product portfolio composed of both commodities and specialties will not be an easy task. Furthermore, parallel lines offer additional flexibilities and opportunities for adjusting capacity to market by switching production between different products, thus reducing the impact of price volatility and increasing revenues. For instance, a P&P parallel biorefinery could use wood chips and shift production between paper and other bioproducts. In another biorefinery configuration, part of the sugar stream could be allocated to ethanol production, while the other would produce other bioproducts like succinic acid. Because the management of these biorefinery plant configurations and their associated supply chain is not obvious, the development and use of VCM models would help understand the subtleties of fulfilling demand and operating a complex plant in different conditions. It is only by doing so that manufacturers will be able to take full advantage of the biorefinery value chain and make it more profitable overall.

7.6 Conclusions

Products and their associated markets, location, the players implementing the biorefinery, and geographical aspects such as access to markets and customers, to name only a few factors, highlight the fact that each biorefinery configuration will be unique, and its supply chain should therefore be carefully designed, managed, and adapted accordingly to yield the most value. For biorefineries, but also for any process-industry company, the value chain and its management are the elements that can provide a real competitive advantage that others cannot replicate, in both the short and the long term. The true key to profitability lies in four simple rules (inspired from [15]):

1. *Be different*: "To make unusual profits, a company must do unusual things." In this sense, the specificities and uniqueness of products and supply-chain assets such as responsiveness and efficiency can give a company an advantage. Examples of differentiators include (nonexhaustively) an access to good-quality biomass at an attractive price, an effective collaboration between partners and other supply-chain actors, distribution synergies, and product differentiation to control the marketplace.
2. *Be among the first*: "Once a new market attracts outside attention, it is too late to get in." Being first to market will undoubtedly be a key success factor for biorefinery establishment, especially for biochemical specialties. The market for these products is small, and the entire production could easily be undertaken by one plant. Any addition of production capacity would greatly affect demand patterns and thus the profitability and rationale for existence of current and entering players.
3. *Be prepared for hard times*: "Best means of entry are important, but also the best means of survival." With the current and future uncertainty about price volatility, efficient planning is necessary to minimize the impact of volatility on profitability. VCM coupled with robust planning approaches can help decision makers to test scenarios and evaluate the trade-offs between procurement, production, and sales, as well as developing better plans and making a profit even in difficult market environments.
4. *Think profit*: "Profitability frequently varies inversely with technological glamor." Technology and capacity utilization are important parameters to consider, but should not be the main focus of a company. It is by adjusting production capacity to the market and fulfilling the most profitable demand first that manufacturing companies can achieve higher levels of profitability. This profitable planning can only be done by managing the value chain in an integrated way.

References

1. Christopher, M. 1998. *Logistics and Supply Chain Management: Strategies for Reducing Cost and Improving Service*, 2nd ed. London: Financial Times Pitman Publishing.
2. Shapiro, J.F. 2007. *Modeling the Supply Chain*, 2nd ed. Pacific Grove, CA: Thomson-Brooks/Cole.
3. Chopra, S. and Meindl, P. 2007. *Supply Chain Management: Strategy, Planning, and Operation*, 3rd ed. Upper Saddle River NJ: Pearson Prentice-Hall.
4. Council of Supply Chain Management Professionals. 2009. *Supply Chain Management Definitions*. URL: http://cscmp.org/aboutcscmp/definitions.asp. [Date: 15-07-2009].
5. Maravelias, C.T. and Sung, C. 2008. Integration of production planning and scheduling: Overview, challenges, and opportunities. *Proceedings, 5th International Conference on Foundations of Computer-Aided Process Operations (FOCAPO 08)*, Cambridge MA. June 29–July 2.

6. Stadtler, H. and Kilger, C. 2008. *Supply Chain Management and Advanced Planning: Concepts, Models, Software, and Case Studies*, 4th ed. Berlin: Springer-Verlag.
7. Grossmann, I. 2005. Enterprise-wide optimization: A new frontier in process systems engineering. *AIChE Journal* 51(7): 1846–1857.
8. Shapiro, J.F. 2001. Beyond supply chain optimization to enterprise optimization. *Achieving Supply Chain Excellence through Technology (ASCET)* 3. URL: http://www.ascet.com/documents.asp?d_ID=511. [Date: 16-07-2009].
9. SAP 2010. *SAP—Business Management Software Solutions Application and Services*. No. 01-09-2010. URL: http://www.sap.com/index.epx. [Date: 01-09-2010].
10. Oracle 2010. *Oracle—Hardware and Software, Engineered to Work Together*. URL: http://www.oracle.com/index.html. [Date: 01-09-2010].
11. Shapiro, J.F. 2004. Achieving competitive supply chain management: Explore and exploit. *Achieving Supply Chain Excellence through Technology (ASCET)* 6. URL: http://www.ascet.com/documents.asp?d_ID=2515. [Date: 16-07-2009].
12. Fleischmann, B., Meyr, H., and Wagner, M. 2008. Advanced planning. In: Stadtler, H. and Kilger, C. (eds.), *Supply Chain Management and Advanced Planning: Concepts, Models, Software and Case Studies*. Berlin: Springer-Verlag.
13. Kilger, C. and Meyr, H. 2008. Demand fulfilment and ATP. In: Stadtler, H. and Kilger, C. (eds.), *Supply Chain Management and Advanced Planning: Concepts, Models, Software and Case Studies*. Berlin: Springer-Verlag.
14. Kannegiesser, M. 2008. *Value Chain Management in the Chemical Industry—Global Value Chain Planning of Commodities*. Berlin: Physica-Verlag.
15. Kline, C.H. 1976. Maximizing profits in chemicals. *Chemtech* 6(2): 110–117.
16. Kannegiesser, M., Gunther, H.O., Beek, P., Grunow, M., and Habla, C. 2009. Value chain management for commodities: A case study from the chemical industry. *OR Spectrum* 31(1): 63–93.
17. Bansal, M., Karimi, I.A., and Srinivasan, R. 2007. Optimal contract selection for the global supply and distribution of raw materials. *Industrial and Engineering Chemistry Research* 46(20): 6522–6539.
18. Reimann, M. and Schiltknecht, P. 2009. Studying the interdependence of contractual and operational flexibilities in the market of specialty chemicals. *European Journal of Operational Research* 198(3): 760–772.
19. Tsay, A.A., Nahmias, S., and Agrawal, N. 1999. Modeling supply chain contracts: A review. In: Tayur, S., Ganesham, R. and Magazine, M. (eds.), *Quantitative Models for Supply Chain Management*. Dordrecht: Kluwer.
20. Höhn, M.I. 2010. *Relational Supply Contracts—Optimal Concessions in Return Policies for Continuous Quality Improvements*. Lecture Notes in Economics and Mathematical Systems, vol. 629, Berlin: Springer.
21. Sykuta, M.E. 1996. Futures trading and supply contracting in the oil refining industry. *Journal of Corporate Finance* 2: 317–334.
22. Tsay, A.A. and Lovejoy, W.S. 1999. Quantity flexibility contracts and supply chain performance. *Manufacturing and Service Operations Management* 1(2): 89–111.
23. Barnes-Schuster, D., Bassok, Y., and Anupindi, R. 2002. Coordination and flexibility in supply contracts with options. *Manufacturing and Service Operations Management* 4(3): 171–207.
24. Schiltknecht, P. 2007. *Capacity Management and Contract Engineering*. PhD Dissertation, ETH Zurich, Zurich.
25. Stadtler, H. 2008. Purchasing and material requirements. In: Stadtler, H. and Kilger, C. (eds.), *Supply Chain Management and Advanced Planning: Concepts, Models, Software and Case Studies*. Berlin: Springer-Verlag.
26. Bonfill, A., Bagajewicz, M., Espuna, A., and Puigjaner, L. 2004. Risk management in the scheduling of batch plants under uncertain market demand. *Industrial and Engineering Chemistry Research* 43(3): 741–750.
27. Park, M., Park, S., Mele, F.D., and Grossmann, I.E. 2006. Modeling of purchase and sales contracts in supply chain optimization. *Industrial and Engineering Chemistry Research* 45(14): 5013–5026.

28. Barbaro, A. and Bagajewicz, M.J. 2004. Managing financial risk in planning under uncertainty. *AIChE Journal* 50(5): 963–989.
29. Cachon, G.P. and Lariviere, M.A. 2005. Supply chain coordination with revenue-sharing contracts: Strengths and limitations. *Management Science* 51(1): 30–44.
30. Stock, J.R. and Lambert, D.M. 2001. *Strategic Logistics Management*, 4th ed. New York: McGraw-Hill.
31. Kilger, C., Reuter, B., and Stadtler, H. 2008. Collaborative planning. In: Stadtler, H. and Kilger, C. (eds.), *Supply Chain Management and Advanced Planning: Concepts, Models, Software and Case Studies*. Berlin: Springer-Verlag.
32. Stadtler, H. 2009. A framework for collaborative planning and state-of-the-art. *OR Spectrum* 31: 5–30.
33. Audy, J.-F., D'Amours, S., Lehoux, N., and Ronnqvist, M. 2009. A review of collaborative logistics. *Proceedings, International Conference on Industrial Engineering and Systems Management (IESM 2009)*, Montreal, Canada, May 13–15.
34. Barratt, M. and Oliveira, A. Exploring the experiences of collaborative planning initiatives. *International Journal of Physical Distribution & Logistics Management* 31(4): 266–289.
35. Durango-Cohen, E.J. and Yano, C.A. 2006. Supplier commitment and production decisions under a forecast-commitment contract. *Management Science* 52(1): 54–67.
36. Lambert, D.M., Emmelhainz, M.A., and Gardner, J.T. 1996. Developing and implementing supply chain partnerships. *International Journal of Logistics Management* 7(2): 17.
37. Leavy, B. 2006. Supply chain effectiveness: Strategy and integration. *Handbook of Business Strategy* 7(1): 331–336.
38. Ragauskas, A.J., Williams, C.K., Davison, B.H., Britovsek, G., Cairney, J., Eckert, C.A., Frederick, W.J. et al. 2006. The path forward for biofuels and biomaterials. *Science* 311(5760): 484–489.
39. Chambost, V. and Stuart, P.R. 2007. Selecting the most appropriate products for the forest biorefinery. *Industrial Biotechnology* 3(2): 112–119.
40. Roitsch, M. and Meyr, H. 2008. Oil industry. In: Stadtler, H. and Kilger, C. (eds.), *Supply Chain Management and Advanced Planning: Concepts, Models, Software and Case Studies*. Berlin: Springer-Verlag.
41. Perlack, R.D., Wright, L.L., Turhollow, A.F., Graham, R.L., Stokes, B.J., and Erbach, D.C. 2005. *Biomass as a Feedstock for a Bioenergy and Bioproducts Industry: The Technical Feasibility of a Billion-Ton Annual Supply*. Technical Report, Oak Ridge National Laboratory, Oak Ridge, Tennessee.
42. Goyal, H.B., Seal, D., and Saxena, R.C. 2008. Bio-fuels from thermochemical conversion of renewable resources: A review. *Renewable and Sustainable Energy Reviews* 12(2): 504–517.
43. Chambost, V., McNutt, J., and Stuart, P.R. 2008. Guided tour: Implementing the forest biorefinery (FBR) at existing pulp and paper mills. *Pulp and Paper Canada* 109(7–8): 19–27.
44. Uslu, A., Faaij, P.C., and Bergman, P.C.A. 2008. Pre-treatment technologies and their effect on international bioenergy supply chain logistics: Techno-economic evaluation of torrefaction, fast pyrolysis and pelletisation. *Energy* 33(8): 1206–1223.
45. Wikipedia, the Free Encyclopedia. 2010. *List of Power Stations in Quebec*. URL: http://en.wikipedia.org/wiki/List_of_power_stations_in_Quebec. [Date: 17-11-2010].
46. Wikipedia, the Free Encyclopedia. 2010. *List of Power Stations in Ontario*. URL: http://en.wikipedia.org/wiki/List_of_power_stations_in_Ontario. [Date: 17-11-2010].
47. Taylor, G. 2008. Biofuels and the biorefinery concept. *Energy Policy* 36(12): 4406–4409.
48. Chemicals Technology 2010. *The Growing Biofuel Market*. http://www.chemicals-technology.com/features/feature87153/[Date: 23-09-2010].
49. Czernik, S. and Bridgwater, A.V. 2004. Overview of applications of biomass fast pyrolysis oil. *Energy and Fuels* 18(2): 590–598.
50. Laflamme-Mayer, M. 2009. Cadre de planification de la chaîne logistique basée sur la représentation des procédés pour l'amélioration de la rentabilité de l'industrie des pâtes et papiers. PhD Dissertation, École Polytechnique, Montreal (Canada).
51. Lee, H.L. 2001. Aligning supply chain strategies with product uncertainties. *California Management Review*, Spring, 105–119.

52. Fisher, M.L. 1997. What is the right supply chain for your product? *Harvard Business Review*, March–April: 105–116.
53. Lynd, L.R., Wyman, C., Laser, M., Johnson, D., and Landucci, R. 2005. *Strategic Biorefinery Analysis: Analysis of Biorefineries*. National Renewable Energy Laboratory (NREL), Golden, Colorado, USA, Technical Report NREL/SR-510-35578.
54. Regnier, E. 2007. Oil and energy price volatility. *Energy Economics* 29(3): 405–427.
55. PriceWaterhouseCooper (PWC). 2009. *Managing Commodity Risk through Market Uncertainty*. Technical Report. URL: http://www.pwc.com/gx/en/metals/commodity-risk/commodity-risk.jhtml. [Date: 17-08-2010].
56. Wikipedia, the Free Encyclopedia. 2010. *Price Elasticity of Demand*. URL: http://en.wikipedia.org/wiki/Price_elasticity_of_demand. [Date: 17-08-2010].
57. Soderholm, P. and Lundmark, R. 2009. Forest-based biorefineries: Implications for market behavior and policy. *Forest Products Journal* 59(1–2): 6–16.
58. Park, J., Park, S., Yun, C., and Kim, Y. 2010. Integrated model for financial risk management in refinery planning. *Industrial and Engineering Chemistry Research* 49(1): 374–380.
59. Sanders, J.P.M., Annevelink, B., and van der Hoeven, D. 2009. The development of biocommodities and the role of North West European ports in biomass chains. *Biofuels, Bioproducts, and Biorefining* 3(3): 395–409.
60. El-Halwagi, M.M. 2006. *Process Integration*, 1st ed. San Diego: Elsevier.
61. Schulz, O., Kannegiesser, M., and Disteldorf, H. 2007. *Supply Chain: What's Next? Managing the Entire Value Chain for Volume and Profits*. AT Kearney, Technical Report. URL: http://www.atkearney.com/index.php/Publications/supply-chain-whats-next.html. [Date: 24-06-2010].
62. Bartodziej, P., Derigs, U., and Zils, M. 2007. O&D revenue management in cargo airlines: A mathematical programming approach. *OR Spectrum* 29(1): 105–121.
63. Gosavi, A., Ozkaya, E., and Kahraman, A.R. 2007. Simulation optimization for revenue management of airlines with cancellations and overbooking. *OR Spectrum*, 29(1): 21–38.
64. Defregger, F. and Kuhn, H. 2007. Revenue management for a make-to-order company with limited inventory capacity. *OR Spectrum* 29(1): 137–156.
65. Talluri, K.T. and van Ryzin, G.J. 2004. *The Theory and Practice of Revenue Management*. Boston: Kluwer.
66. Harris, F.H. deB. and Pinder, J.P. 1995. Revenue management approach to demand management and order booking in assemble-to-order manufacturing. *Journal of Operations Management* 13(4): 299–309.
67. Meyr, H. 2009. Customer segmentation, allocation planning and order promising in make-to-stock production. *OR Spectrum* 31(1): 229–256.
68. Spengler, T., Rehkopf, S., and Volling, T. 2007. Revenue management in make-to-order manufacturing: An application to the iron and steel industry. *OR Spectrum* 29(1): 157–171.
69. Kolisch, R. and Zatte, D. 2009. Stand und Perspecktiven des Einsatzes von Revenue Management in der Prozessindustrie. *Zeitschrift für Planung und Unternehmenssteuerung* 20: 197–214.
70. Quante, R., Meyr, H., and Fleischmann, M. 2009. Revenue management and demand fulfillment: Matching applications, models, and software. *OR Spectrum* 31(1): 31–62.
71. Fleischmann, B. and Meyr, H. 2004. Customer orientation in advanced planning systems. In: Dyckhoff, H., Lackes, R., and Reese, J. (eds.), *Supply Chain Management and Reverse Logistics*. Berlin: Springer, 297–321.
72. Dansereau, L.-P., El-Halwagi, M.M., and Stuart, P. 2009. Sustainable supply chain planning for the forest biorefinery. In *Design for Energy and the Environment—7th International Conference on the Foundation of Computer-Aided Process Design*, Breckenridge, Colorado, 1101–1105, June 7–12.
73. Feng, Y., D'Amours, S., and Beauregard, R. 2008. The value of sales and operations planning in oriented strand board industry with make-to-order manufacturing system: Cross-functional integration under deterministic demand and spot market recourse. *International Journal of Production Economics* 115(1): 189–209.

74. Kannegiesser, M. and Gunther, H.O. 2010. An integrated optimization model for managing the global value chain of a chemical commodities manufacturer. *Journal of the Operational Research Society* 62(4): 711–721.
75. Chambost, V., McNutt, J., and Stuart, P.R. 2009. Partnerships for successful enterprise transformation of forest industry companies implementing the forest biorefinery. *Pulp and Paper Canada* 110(5–6): 19–24.

8

Integrated Forest Biorefinery Supply Chain Network Design Using Mathematical Programming Approach

Yan Feng, Sophie D'Amours, Luc LeBel, and Mustapha Nourelfath

CONTENTS

8.1 Introduction

Driven by the energy crisis, soaring oil prices, and environmental concerns about greenhouse-gas emissions, bioenergy and biotechnology have experienced significant advances over the last few years in their efforts to convert biomass into value-added products such as energy (heat and power), biofuels (pellets, bioethanol, and biodiesel), and chemical products. As various biotechnologies emerge, the concept of the biorefinery is evolving. Based on the concept of a petroleum refinery, the biorefinery can be defined as a facility that integrates biomass conversion processes and equipment to produce power, heat, fuels, and value-added chemical products from biomass. By converting various components of the biomass and their intermediates to produce a mix of multiple value-added products, biorefining technology can potentially generate values from traditional biomass residues, reduce nonrenewable fossil-fuel consumptions, reduce energy costs, and cut greenhouse-gas emissions.

In recent years, many individual enabling technologies have been developed to support the bioenergy and biorefining processes, including pelletization, combustion, co-combustion, gasification, pyrolysis, digestion, and fermentation (Frombo et al. 2009). All these technologies can be broadly classified into three conversion platforms: mechanical, thermochemical, and biochemical. Earlier chapters have investigated various converting platforms and technologies designed to convert biomass effectively and efficiently into various bioproducts. As organizations become eager to define strategies and to make investment decisions in this emerging business direction, they are facing many challenging decisions to find the best investment strategy that will maximize the net profitability of their organizations. Biomass can be derived from several sources, among which forest biomass will play a significant role in the biorefining industry. Consequently, the impact of biorefinery development on the existing forest products industry must be examined. It is therefore desirable to design a sustainable supply chain, leading from biomass supply through various manufacturing transformation processes and distribution networks to deliver the products to the market, which strategically maximizes the supply chain value.

In this chapter, potential opportunities for integrating the biorefinery with forest product manufacturing systems will be investigated to optimize biomass flows and utilizations. A mathematical programming model for integrated forest biorefinery (IFBR) supply chain network design will be presented. The emphasis is on demonstrating how optimization modeling techniques can be applied to the design of this integrated supply chain and how the solutions can be used to help the organizations in their investment decisions. Specifically, the following strategic questions will be addressed:

1. Should woody biomass and process residues be used for bioproducts or forest products manufacturing? What are the best tradeoffs and the optimal mix, and which market should be served?
2. What manufacturing facilities (e.g., power, pellets, biochemical/thermochemical plant) and how many of them should an organization use to satisfy its target market demand, and where should they be located?
3. Should different manufacturing facilities be built separately or partially or fully integrated?
4. Should these manufacturing facilities be built as greenfield plants or at an existing mill site, integrated with existing manufacturing activities such as a pulp and

paper mill, sawmill, or wood composite mill to produce a new product mix using mill residues without additional transportation cost?
5. What technologies and at what capacities should be installed?
6. What products and in what quantities should each manufacturing facility be capable of producing?
7. Which raw material sources (or suppliers) should be used, and what should be the raw material allocations?

This chapter is organized as follows. In the next section, the basic concepts of supply chain management and supply chain design are provided. An introduction of the modeling approach for supply chain design is presented in Section 8.3, followed by a literature review of the research carried out on various supply chain design problems. The proposed forest biorefinery supply chain design, the development of the model, and some numerical illustrations are presented in Section 8.4, with concluding remarks and future research opportunities described in Section 8.5.

8.2 The Concept of Supply Chain Management and Supply Chain Design

8.2.1 Basics of the Supply Chain

The supply chain can be broadly classified into four stages: procurement, production, distribution, and sales. The *procurement* stage describes the activities, functions, and relationships at the supply end of the business, including purchasing decisions, transactions, supply capabilities, and inbound transportation of primary raw material, parts, components, and services. Suppliers are the organizations that provide goods and services. The *production* stage describes the activities and functions associated with the manufacturing of goods using the raw materials, parts, and components procured in the first stage. The *manufacturer* is the organization that performs these production activities. In a supply chain, a manufacturer may be regarded as a customer from its suppliers' point of view and a supplier from the point of view of a downstream manufacturer or a customer. *Distribution* represents the activities and functions related to product shipments and transshipments using different transportation means and intermediate distribution facilities to transport the products effectively from production locations to customer locations. A *distribution center* is a facility that provides temporary storage as well as transshipment services to separate large loads into smaller ones or to combine smaller loads into large ones to improve transportation efficiencies and reduce costs. Distribution services may be owned by the manufacturer or subcontracted to third-party logistic service companies. The manufacturer may have its own transportation services or may use common carriers or subcontract to private companies. Last but not least, the *sales* stage refers to the general activities, functions, and relationships involving customers, such as identifying and forecasting demand, determining price, conducting promotion strategies, drawing up contracts, and developing collaborative relationships.

There are three different views of supply chain. First, there is the single-mill intraorganizational supply chain consisting of different functional units linked by flows of goods, information, and finance which fulfill the supply chain functions of procurement, production, distribution, and sales. This supply chain represents the simplest linear structure of supply chain, as shown in Figure 8.1a with a single organizational level management. Many studies have been carried out addressing the importance and benefits of supply chain management in this context. The second is the multisite intraorganizational supply chain. This supply chain is also formed within a single organization; however, it may have several supply, manufacturing, and distribution sites scattered in different locations, with material, information, and financial flows between each site forming an

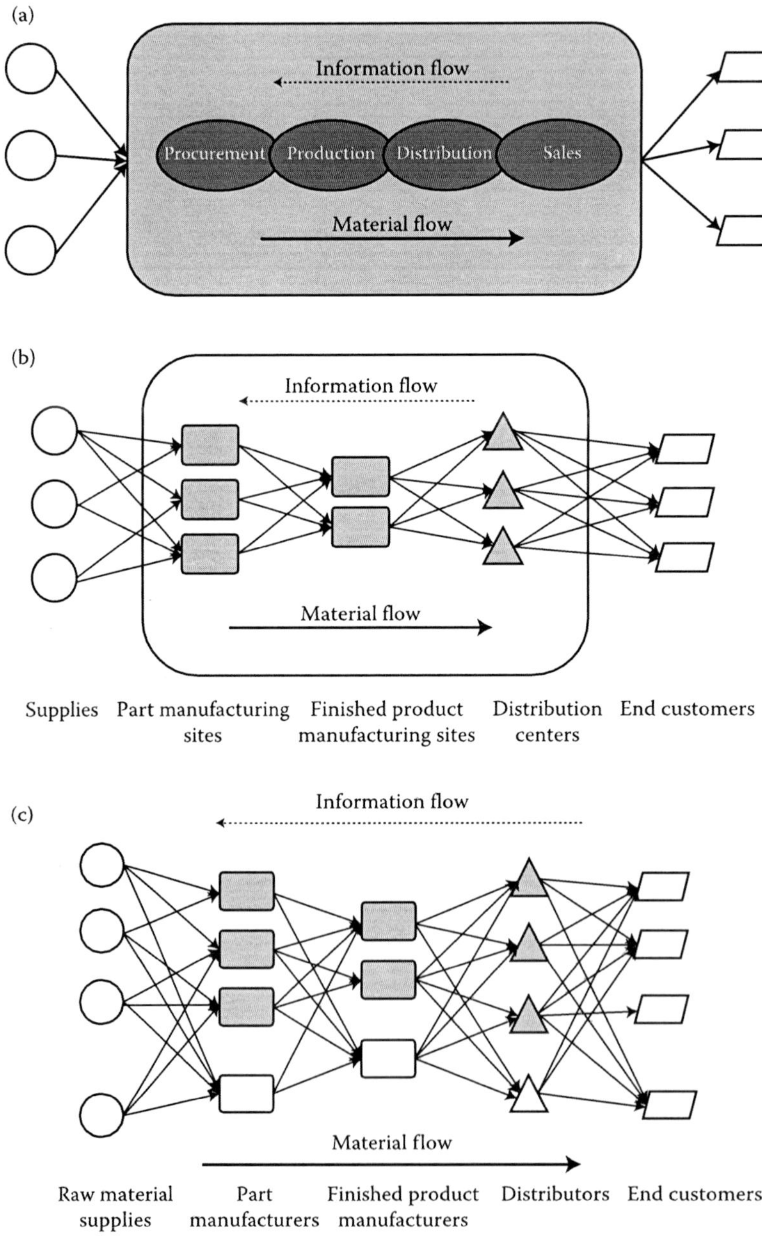

FIGURE 8.1
Supply chain structures. (a) Single-site intraorganizational supply chain. (b) Multiple-site intraorganizational supply chain. (c) Interorganizational supply chain.

intraorganizational supply chain network, as shown in Figure 8.1b. Despite the complexities of the supply chain network, supply chain management and centralized decision-making mechanism are still feasible and have been the focus of many research studies. The last is the interorganizational supply chain consisting of several suppliers, manufacturers, warehouses, distributors, and customers belonging to different organizations. These organizations are linked by material, information, and financial flows to transform raw materials into value-added parts, components, and end products for the customers (Stadtler 2002), as shown in Figure 8.1c. This large complex supply-chain network presents a complete view of a multiorganizational general supply chain network, of which the intraorganizational supply chain is thereby a subsection. Because of the complexity of the business context, the proprietary nature of information, and frequent conflicts between the interests of different organizations, integrated supply chain management, despite its significant underlying benefits, is difficult to achieve, and a game-theory approach is often used to reach at least Pareto-optimal decisions.

8.2.2 Supply Chain Management

Supply chain management (SCM) involves the tasks of integrating organizational units along a supply chain, managing, and coordinating the flows of goods, services, information, and finance in order to improve the competitiveness of the organization. The term *SCM* was first developed in the 1980s to express the need for integrating key business processes from the suppliers to the end users (Oliver and Webber 1992). The benefits of supply chain management have been well established and documented. Through coordinated and integrated supply chain planning, supply chain operations can be synchronized, inventories can be reduced, manufacturing efficiencies can be improved, and higher levels of customer satisfaction can be achieved resulting in cost reductions and high profitability. In an intraorganizational system, because the various supply, manufacturing, and distribution sites and functions are all part of one large organization, optimal multiple-site capacity allocations and resource utilizations, including manufacturing, distribution, inventory, and labor, can be achieved to satisfy customer demand in a timely manner at the lowest supply chain cost.

8.2.3 Supply Chain Design

Supply chain design, also known as strategic supply chain network planning, is a long-term planning exercise which typically covers a planning horizon of from 3 to 10 years. In strategic supply chain design, organizations set strategies and directions to make the right investment decisions for resource acquisitions and allocations to satisfy the defined market requirements. Typical decisions involve the definition of a product program and demand market locations, the establishment or closure of manufacturing and distribution facilities, and the installation of major production lines. The objectives are most often financially oriented, being either profit maximization or cost minimization, subject to customer service and budget constraints (Goetschalckx 2002). These strategic decisions are linked in a hierarchical manner to the tactical and operational planning decisions within the supply chain. One well-known framework for illustrating how these decisions are interrelated is that presented by Fleischmann et al. (2002) and illustrated in Figure 8.2. This two-dimensional supply chain framework illustrates the relationships among the various supply chain planning levels and the linkages across the four supply chain stages. In strategic supply chain design, it is particularly important that the investment decisions be

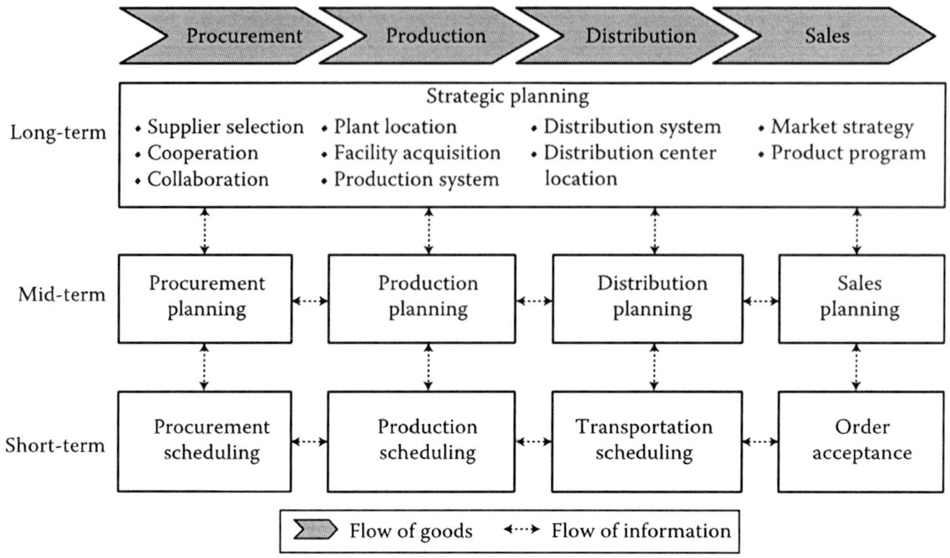

FIGURE 8.2
Supply chain planning matrix. (With kind permission from Springer Science+Business Media: Advanced planning, *Supply Chain Management and Advanced Planning—Concepts, Models, Software, and Case Studies,* Second ed., 2002, pp. 71–96, Fleischmann, B., Meyr, H., Wagner, M., In: Stadtler, H., Kilger, C. (eds.), Springer, Berlin.)

made jointly for manufacturing and distribution facilities, taking into account the product demand and market dynamics as well as long-term raw material availabilities. It is equally important that the product development and market selection decisions be made taking into consideration the manufacturing and distribution facility configurations, capabilities, costs, and efficiencies. Once the manufacturing and distribution facilities have been built, this supply chain configuration will impose constraints on the operations of the supply chain, and will have a long-term impact on the performance of the supply chain.

Designing a supply chain network is a complex task which requires comprehensive evaluations of both engineering and financial aspects, which are often company- or industry-specific. Decision support models have therefore been developed to support various supply chain design problems. These models usually involve the use of mathematical programs, predominantly mixed-integer linear programs. The next section will introduce the linear and mixed-integer linear programming techniques and refer to literature studies on how these techniques have been applied in various supply chain network design problems. The development of the forest biorefinery supply chain model will be presented thereafter.

8.3 Modeling Approach to Supply Chain Design

8.3.1 Introduction to Linear and Mixed-Integer Programming

The use of linear inequalities to solve problems by maximizing or minimizing a linear objective function subject to linear constraints can be found in the literature as early as the 1940s. The early model developed by Dantzig was originally focused on finding rapid solutions for military deployment, training, and logistical supply problems (Martin 1999).

In 1947, Dantzig formalized the concept of a linear program as we know it today. In the linear program, a planning problem was formulated with a linear objective function subject to solving a system of linear equations and inequalities. The term *program* does not refer to a computer program, but rather it is a term used by the military to describe the plan or the proposed solution to a deployment or logistical problem. The terms *linear programming* and *linear optimization* are used synonymously in many publications. The linear programming problem (*LP*) is generally formulated in the following form:

$$(LP) \quad \min c^T x \tag{8.1}$$

$$s.t. \quad Ax \geq b \tag{8.2}$$

$$x \geq 0 \tag{8.3}$$

where A is an $m \times n$ matrix of rational numbers, c is an n-component rational column vector, and b is an m-component rational column vector. Equation 8.1 is the objective function. This objective function is minimized subject to (abbreviated as *s.t.* in (8.2)) the constraint set (8.2) and the nonnegativity constraints (8.3). The unknowns, or the x-variables in the problem, are called the *decision variables*. The vector b in (8.2) is known as the *right-hand side.* The canonical structure (*LP*) is a linear program in standard form.

In an *LP*, all decision variables x are continuous variables which may take on any positive values satisfying the constraints (8.2) and (8.3). In many cases, however, a decision variable may involve "yes/no" or "0/1" decisions, such as installing a facility or not, funding a project or not, and so on, and integer variables are therefore required. Using rounded fractional values for these decisions can result in infeasibility or significant profit loss. In this case, some of the variables need to be restricted to integer values. An *LP* in which some decision variables are restricted to integer values is known as a *mixed-integer linear program* (*MILP*), or simply a mixed-integer program. A mixed-integer program (*MIP*) can be formulated as:

$$(MIP) \quad \min c^T x \tag{8.4}$$

$$s.t. \quad Ax \geq b \tag{8.5}$$

$$x \geq 0 \tag{8.6}$$

$$x_i \in Z, \; i \in I \tag{8.7}$$

where Z is a subset of integer values and certain decision variables $x_i (i \in I)$ can take on only values within this subset. When all the variables in an *MIP* are restricted to be integers, the problem becomes an *integer linear programming problem,* often called an integer programming (*IP*) problem. When all the integer variables in an *IP* are restricted to be binary variables, the problem becomes a *binary integer programming* (*BIP*) problem. Sometimes the term *integer program* is used loosely to mean a linear program with some or all the variables required to be integers.

This chapter will focus on *MIP* problems and specifically on strategic supply chain design problems. In the following sections, the various supply chain design models developed to date will be reviewed.

8.3.2 Single-Echelon Single-Period Capacitated Network Design Models

The network design models can be found as early as the 1970s which focused mainly on a subsection of the supply chain, such as manufacturing or distribution, to solve the facility location-allocation problems. The problems typically involve the selection of locations for the manufacturing or distribution facilities from a finite set of candidate facility locations that are used as potential sources to satisfy customer demand at various destinations. Examples of manufacturing facility location models can be found in Ellwein and Gray (1971), Geoffrion and McBride (1973), and Soland (1973), in which the distribution centers are regarded as demand locations with deterministic demand. Associated with each potential facility location is a capacitated facility with fixed establishment cost and capacity as well as variable operational and distribution costs. The objective is to determine the facility location that minimizes the total fixed establishment and operational cost.

Geoffrion and Graves (1974) addressed the distribution facility location problem to determine the optimal locations of intermediate distribution facilities between plants and customers. In this problem, the manufacturing plant locations are predetermined, with known capacities that produce several commodity products. The demand is known for each commodity in each of the demand zones. This demand is satisfied by shipping through regional distribution centers (DCs), with each demand zone assigned exclusively to a single DC. The candidate locations of the DCs are given, and the maximum and minimum annual allowable throughputs of each DC are known. If a DC site is selected, it will incur a DC fixed cost as well as a linear variable cost. Transportation costs between the selected DCs and demand zones are assumed to be known. The objective is to determine which DC sites to use with what DC size, serving which demand zones, and using what transportation patterns that can satisfy the demands at lowest total distribution cost.

Subsequent to these facility location-allocation models, several extensions were developed. Brown et al. (1987) presented an *MIP* multicommodity single-period model to make decisions to open and close plants, with the ability also to make facility-to-plant assignment decisions where each plant may have several facilities. The objective is to minimize total fixed investment and variable operational costs. A goal decomposition approach was introduced, yielding pure network subproblems for each commodity, which can be solved efficiently for large-scale problems. Paquet et al. (2004) developed a single-period manufacturing network design model for manufacturing facilities producing multiple products based on a nontrivial product structure and bills of material under deterministic demands. In this model, the notion of capacity options is introduced, and facility technology and capacity selection decisions are taken into account in the model. Bender's decomposition solution method is proposed. Gebennini et al. (2009) developed a mixed-integer programming model for the distribution-center location problem which can determine the number of facilities, their optimal locations, and their assignment to customer demand, taking into account anticipated tactical decisions on inventory control, production rates, and service levels in a stochastic environment.

8.3.3 Production-Distribution Network Design Models

Despite the valuable contributions of single-stage manufacturing or distribution facility location-allocation models, the weakness of the decoupled planning approach has been well recognized. One of the limitations of this approach is that because the design of an individual stage, manufacturing or distribution, seeks decisions that minimize the local fixed and operational costs of that stage, it may increase the costs of the other stage and consequently yield solutions with a higher total supply chain cost. In many business environments, the manufacturing or distribution facility location-allocation decisions are internal organizational decisions which have a significant impact on company performance and competitiveness. An integrated production-distribution network design approach would therefore present promising opportunities for further supply chain cost reductions.

Integrated production-distribution design models were mostly developed in the 1990s by several authors. Cole (1995) developed a multicommodity single-period strategic production-distribution design model where both plant and distribution center closing decisions were included. The key contribution of this study is the consideration of stochastic customer service, which is defined as the fraction of customer demand satisfied from warehouse stock and the delivery time and distance to serve each customer. Timely service is addressed by carrying safety stock, together with warehouse location, customer allocation, and channel selection. Nonlinear customer service constraints are linearized, and a tractable model is obtained which is solved using CPLEX. This model is important when safety-stock costs are significant. Vidal and Goetschalckx (1996) presented a production-distribution model incorporating supplier reliability. To represent supplier reliability, a set of linearized constraints was formulated under the assumption that the probability of supplier on-time deliveries must reach at least a specified target value. Martel and Vankatadri (1999) addressed single-period multiproduct production-warehouse location-allocation problems under economies of scale. Capacity expansion or contraction possibilities by technology types at the facilities were also considered. A nonlinear mixed-integer programming model was developed and solved using a scenario improvement approach with successive *MIP* algorithms. Thanh et al. (2008) developed a multilevel, multicommodity, and multiperiod deterministic model for the design and planning of a production-distribution system to support decisions on facility opening, closing, enlargement, and supplier selection, as well as flow decisions along the supply chain. The best period in which to open or close facilities and dynamic inventory decisions were also included.

8.3.4 Global Supply Chain Network Design Model

As companies increasingly extend their business reaching toward other countries to increase their market share while benefiting from lower labor costs, favorable currency exchange rates, and tax benefits in foreign countries, global supply chain issues and network design problems have attracted considerable attention. As discussed earlier, the supply chain design problem generally includes decisions regarding the supply chain network configuration, the number and location of production and distribution facilities, the amount of capacity at each facility, the technology to be selected, the allocation of each market to one or several facilities, and the selection of suppliers. Global supply chain design extends this definition to include the selection of facilities at international locations. Because of various international factors, the problem becomes significantly more complex (Meixell and Gargeya 2005).

One of the important international factors that are critical to global supply chain design is the transfer price. A *transfer price* is the price at which goods and services are traded across international borders between subsidiaries of a multinational company (Martel et al. 2005). It is one of the most important international tax issues facing a multinational company. In the operations management field, the transfer price has been regarded as an accounting problem rather than an important decision opportunity in global supply chain design (Goetschalckx et al. 2002). Many publications have assumed transfer price to be deterministic (Arntzen et al. 1995, Canel and Khumawala, 1996, Bhutta et al. 2003, Martel 2005, Vila et al. 2007). One of the reasons is that in most supply chain design problems, the product price is generally considered to be predetermined. Another possible reason is that making the transfer price a decision variable will result in nonlinear models which are much harder to solve. Nevertheless, according to Nieckels (1976), the impact of transfer price on taxable income, duties, and management performance is significant. Small changes in transfer prices may lead to significant differences in the after-tax profit of a company. Nieckels (1976) presented a nonlinear mathematical model in which transfer price was regarded as a decision variable, with the objective function being to maximize global after-tax profit. A systematic heuristic procedure is proposed that assigns specific values to transfer prices, enabling the model to be simplified and solved iteratively until no further improvement in the objective function is possible. Following the work of Nieckels (1976), several authors have investigated global supply chain problems with transfer prices formulated as decision variables (Cohen et al. 1989, Vidal and Goetschalckx 2001, Gjerdrum et al. 2002 , Martel et al. 2005). Cohen et al. (1989) presented a nonlinear multiperiod production-distribution supply chain design model with transfer price as decision variables. A fixed cost structure allowing economies of scale, bills of material, international tariffs, currency exchange rates, and corporate tax rates were also considered. Vidal and Goetschalckx (2001) developed a nonconvex optimization model for the global supply chain. Gjerdrum et al. (2002) proposed a nonlinear mixed-integer programming model for a two-enterprise network to examine transfer-price policies and to determine the optimal transfer prices that maximize the profits of each partner. The model is solved using a linearization technique coupled with a Nash-equilibrium-based algorithm. Martel et al. (2005) developed a single-period production-distribution supply-chain design model in which a simple formulation was proposed to make optimal transfer-price decisions. This formulation is based on the observation that transfer prices are comparatively small in practice and therefore, a finite set of varying markup-percentage multipliers can be introduced. Binary variables can then be defined corresponding to the transfer-price multipliers, enabling optimal transfer prices to be determined while preserving the linearity of the model.

Another important factor that has significant influence on global supply chain network design as well as supply chain operational performance is exchange rates. Because many of the global supply chain network design models developed to date are single-period models, static mean exchange-rate values have been used in these models (Cohen and Lee 1989, Canel and Khumawala 1996, Vidal and Goetschalckx 2001, Goetschalckx et al. 2002, Martel et al. 2005). In today's global economic environment, exchange rates may change considerably over time. A network designed under current exchange-rate and market assumptions may not achieve the expected financial returns and may experience losses in the worst-case scenario when the exchange rate becomes unfavorable. Therefore, in strategic global supply chain network design, it is important that the dynamic and stochastic characteristics of exchange rates be considered. This raises the necessity of incorporating an adequate exchange-rate forecasting model and of taking into account possible forecast errors in the supply chain design (Martel et al. 2005). However, forecasting exchange rates is a complex task which can be influenced by many factors, including random events, institutional frictions, and market distortions. Haug (1992) developed a multiperiod international location model in which dynamic exchange-rate and inflation variability are considered. Mohamed (1999) and Bhutta et al. (2003) proposed a simple linear function to estimate the exchange rate in their supply chain network design model. Other approaches such as a generalized linear autoregressive model may also be applicable (Yu et al. 2005).

Other international factors that have been widely addressed include import and export tariffs, duties, and corporate tax rates (Haug 1992, Arntzen et al. 1995, Canel and Khumawala 1996, Munson and Rosenblatt 1997, Martel et al. 2005, Martel 2005). Because these factors are typically associated with government policies in each country, they are relatively stable. Hence, these factors are generally assumed to be known parameters. In addition to the aforementioned factors, Haug (1992) addressed worker skill issues in foreign countries in the development of an international facility-location model. Arntzen et al. (1995) discussed the timing aspect of the production and distribution network that is required to produce and ship products to customer locations. In this regard, time measurements were introduced into the objective function through weighting factors, so that both cost and time are used as criteria to derive the recommended supply chain design. Customer satisfaction, balance of materials, offset trade, local content, and duty drawback restrictions are also considered. Munson and Rosenblatt (1997) investigated a global supply chain problem with local-country-content rules for supplier sourcing. These rules require that a company opening a manufacturing plant within the country must purchase a specified quantity of components from suppliers in the country. Martel et al. (2005) examined trade barriers, national labor laws, infrastructure, resource supplies, and competition in the multinational supply chain design context.

8.3.5 Supply Chain Network Design under Uncertainties

The classical method of supply chain network design modeling, as presented in the previous sections, generally assumes that all the input parameters are deterministic, and therefore the *MIP* modeling approach has been used predominantly. In real-world problems, it is almost inevitable that not all parameters are known with certainty. For instance, customer demand, market prices, and exchange rates may change unpredictably over time. Assume that a decision-maker makes facility investment and installation decisions based on the information known to date. After these strategic decisions have been made, random events may occur which will affect the actual performance of the supply chain, resulting in unfavorable financial outcomes. One way to mitigate the impact of these uncertainties is to anticipate their occurrence at the design stage to develop robust supply chain network design solutions. There are different approaches to solving optimization problems with uncertainties. One approach is stochastic programming. *Stochastic programming* (*SP*) can be regarded as an extension of *LP* or *MIP*, where some of the parameters are replaced by random variables with probability distributions that are known or can be estimated. The objective is to find policies that are feasible for all possible data instances while maximizing (minimizing) the expected value of the objective function. The concept of stochastic programming, its theoretical development, and solution methodologies can be found in Shapiro (2003) and Higle (2005). Another approach is called *robust optimization,* in which the parameters are known only within certain bounds. The objective is to optimize the worst-case performance of a system under uncertainties (Kouvelis and Yu 1997, Snyder 2006, Klibi et al. 2010) or the performance at some known degree of approximation to the worst case, according to the risk aversion of the decision-makers (Bertsimas and Sim 2004). Mulvey et al. (1995) developed a scenario-based robust optimization framework in which the probability distributions of the random parameters are assumed to be known and plausible realizations of these random variables are realized through scenario generation. The model formulation is characterized by combining goal programming with stochastic programming where resource-cost variability and model infeasibility are penalized using goal-programming weights to achieve a simultaneous tradeoff between solution optimality and model robustness.

Kogut and Kulatilaka (1994) developed a stochastic dynamic programming model for production-switching decisions under exchange-rate uncertainties. Flexibility decisions in a manufacturing system to switch locations as currency exchange rates fluctuate to benefit from the favorable exchange rates were examined, taking into account the cost of switching, shutting down, and starting up, as well as labor-related costs and management time commitments. Malcolm and Zenios (1994) presented a scenario-based robust optimization model for energy-capacity expansion decisions under power demand uncertainties. Lowe et al. (2002) developed a scenario-based two-phase multiscreening approach to incorporate exchange-rate uncertainties and risks in an international production and sourcing model. Santoso et al. (2005) proposed a two-stage stochastic programming model for a supply chain network design problem taking into account the uncertainties of demand, supply, capacity, and processing and transportation costs. Vila et al. (2007) extended the supply chain network design problem to take market opportunities into account using two-stage stochastic programming approach. Azaron et al. (2008) presented a multiobjective stochastic programming model for supply chain design, taking into account various uncertainties in demand, supply, processing, and transportation. Multiobjective functions for minimizing total investment cost and future capacity expansion cost, as well as for minimizing the variance of the total cost and the financial risk of not meeting a certain budget, are proposed. A goal-attainment technique is used to solve the model to find Pareto-optimal solutions. Franca et al. (2009) addressed a multinational supply chain design problem in which strategic tradeoff decisions regarding supply chain profitability and the quality of raw material supply are modeled. Pan and Nagi (2010) presented a robust optimization model for an integrated production and logistics supply chain design problem.

8.3.6 Supply Chain Design in Bioenergy and Biorefinery Systems

With the rapid development of bioenergy and biorefinery systems, an increasing number of publications have appeared in the area of bioenergy and biorefinery supply chain design. In bioenergy supply chain design, most of the contributions are focused on biomass supply chain optimization from harvesting areas to bioenergy plants, with or without intermediate terminals, determining plant locations, allocations, and capacities. In this vein, Kumar et al. (2003) analyzed the optimum plant size and power cost for power plants using three biomass fuels, namely agricultural residues (grain straw), whole boreal forest trees (logs, tops, and limbs), and forest harvesting residues (tops and limbs), respectively, in western Canada (Alberta). Freppaz et al. (2004) developed a decision-support system (DSS) for forest biomass systems in an effort to find suitable plant locations and sizes as well as optimal supply areas within a region. Gunnarsson et al. (2004) presented an *MIP* model for the biomass supply chain problem which determined the optimum sourcing, transportation, and allocation of forest and sawmill residues to satisfy the demand at various known heating-plant locations in Sweden. Gronalt and Rauch (2007) proposed a simple stepwise heuristic approach for a regional woody biomass supply network design problem in which biomass was supplied from several forest areas to a number of energy plants in Austria. Kanzian et al. (2009) developed a deterministic *MIP* model for a regional woody biomass supply network from the forest to heating plants in Austria, including the optional use of intermediate terminals. Frombo et al. (2009) presented an *LP*-based strategic planning model to determine the optimal annual harvesting quantities and suitable power-plant capacities with different thermochemical conversion technologies at predetermined plant locations.

In the area of biorefinery supply chain design, two contributions have been found to date. Eksioglu et al. (2009) presented an *MIP* model for designing the biorefinery supply chain by determining the number, size, and location of the biorefineries needed to produce cellulosic ethanol using corn stover and woody biomass. Parker et al. (2009) evaluated the infrastructure requirements of hydrogen production from agricultural residues. A mixed-integer nonlinear programming model was developed to find the most efficient and economical configuration for the supply chain pathway.

8.4 Integrated Forest Biorefinery Supply Chain Design

In earlier studies, supply chain design problems have focused on the design of facility and distribution networks serving primary products within a single industrial environment. Taking biorefinery supply chain design as an example, the biorefineries are generally designed as standalone facilities with potential biomass supply sources and product markets (Eksioglu et al. 2009, Parker et al. 2009). A similar approach has been taken to bioenergy supply chain design problems (Kumar et al. 2003, Freppaz et al. 2004, Gunnarsson et al. 2004, Gronalt and Rauch 2007, Kanzian et al. 2009, Frombo et al. 2009). These supply chain design problems can be represented by the supply chain illustrated on the left-hand side of Figure 8.3. Because in biorefinery supply chain systems, performance depends strongly on the cost, consistency, and efficiency of the biomass supply, coordinating biomass supply and biorefinery manufacturing offers significant opportunities for effective biomass utilization, cost reduction, and value chain optimization. In the following paragraphs, opportunities for an IFBR supply chain design, as illustrated on the right-hand side of Figure 8.3, will be discussed. The flows of energy and forest product manufacturing residues will be examined. The potentials of reallocating manufacturing residues for bioenergy and biofuel production will be investigated. The challenges faced in forest product manufacturing because of increased competition for raw material supply will also be discussed.

8.4.1 General Description of the Integrated Forest Biorefinery Supply Chain

The IFBR supply chain network can be broadly defined as a network consisting of many suppliers, manufacturers, distribution centers, and customers. Having realized the opportunities of coordinated biomass supply and biorefinery manufacturing systems, biorefineries may be built as stand-alone plants at greenfield site(s), purchasing biomass from forest suppliers and forest product manufacturers while selling bioenergy and biofuels back to the forest product manufacturers or to consumer markets, as shown on the right-hand side of Figure 8.3. Alternatively, biorefineries may be established within an existing forest product manufacturing site. The existing forest biomass supply system may thus be used to serve both biorefining and forest product manufacturing, enabling optimal biomass allocations to maximize the value of the supply chain. Traditionally unused biomass such as forest residues, as well as low-value intermediate products such as bark, sawdust, shavings, chips, and pulping liquor may be used to produce value-added bioproducts, including bioenergy, biofuels, and chemical products to satisfy internal and external demands.

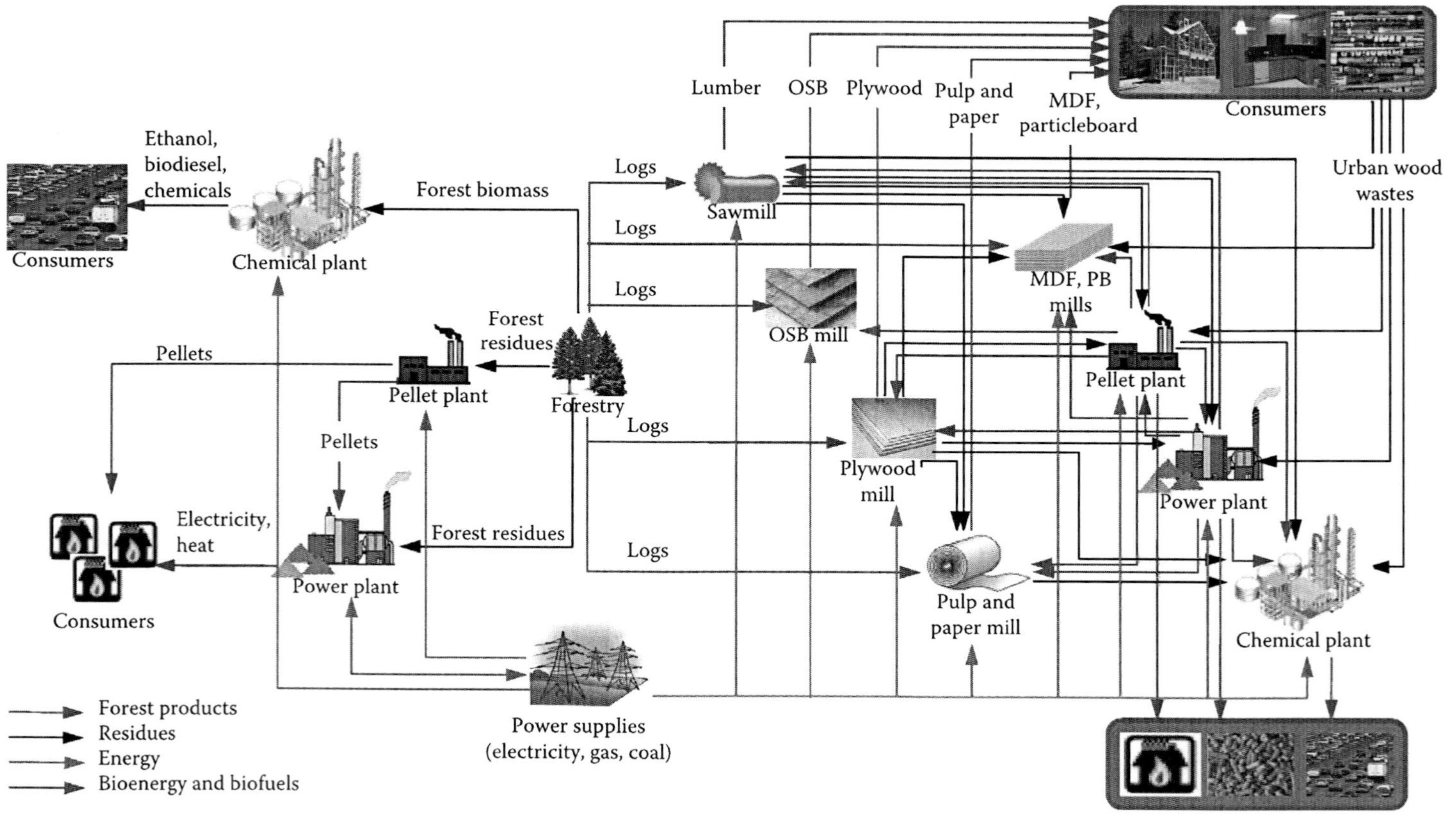

FIGURE 8.3
Integrated forest biorefinery supply chain framework.

8.4.2 Candidate Facilities and Their Characteristics

At each manufacturing site, whether existing or greenfield, one or several facilities may be established, as shown in Figure 8.4. Within each facility, a production activity can be carried out to produce a specific set of products using a defined technology under a given capacity. For example, a pulp and paper mill may have chipping, pulping, papermaking, and energy facilities. An energy facility may have a recovery boiler of a given capacity producing steam and heat to meet production requirements. New facilities may be installed, such as a CHP, pelletizing, or bioethanol production facilities, depending on economic evaluations. In this section, various candidate facilities and their characteristics which are potentially important for IFBR supply chain design will be analyzed.

8.4.2.1 Sawmill Facility

The lumber industry is at the heart of the forest industry and is a significant contributor to social and economical development in Canada. It is the driving mechanism for forest operations and the primary dominant consumer of forest biomass. Wood chips produced by the lumber industry support several other sectors of the forest products industry, including pulp and paper as well as wood composite sectors, which in turn generate significant value to the Canadian economy. The lumber industry is also a significant producer of process residues. As logs are processed to produce lumber, considerable quantities of by-products, in addition to chips, are produced simultaneously, such as bark, sawdust, and shavings. These by-products are currently used either as feedstock for the local boilers to generate heat or else sold to secondary processing mills to produce wood panels. In the development of an IFBR supply chain, existing sawmills can be one of the important biomass sources and therefore should be included as candidate locations for a biorefinery. Existing sawmill technologies and capacities can be considered as candidate technologies and capacities respectively. Unlike previous studies of supply chain design and planning

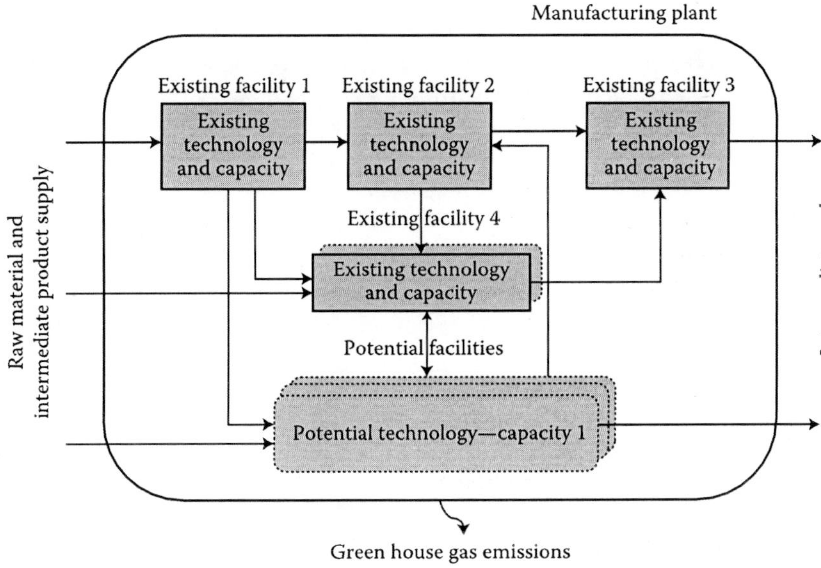

FIGURE 8.4
Facility establishments of a manufacturing site.

problems which have emphasized on lumber production, this study will focus on the co-production of lumber and its by-products to investigate the best investment opportunities for value maximization. Although a sawmill can produce various lumber products, we will focus on the product type level, assuming that the expected quantities of by-products are directly proportional to the aggregated lumber production.

8.4.2.2 Pulp and Paper Facilities

In pulp and paper production, wood chips are processed first in the pulping facility using either a mechanical or a chemical method to produce pulp. The pulp, either bleached or unbleached, is then passed in diluted form through the papermaking facility, where the randomly interwoven mat of fibers is pressed and dried to produce paper. Despite the struggles faced by many pulp and paper companies because of decreased market prices, increased wood costs, and competition from low-cost countries, the pulp and paper manufacturing system, particularly the Kraft pulping process, has a significant role to play in the development of biorefining technology. Traditionally, the pulping process has been regarded as a low-yield process in which only cellulose fibers from wood are collected for paper production. Most of the hemicellulose and lignin components of the wood are dissolved, forming spent pulping liquor which is then concentrated and burned in the recovery boiler to produce process steam and recover the inorganic cooking chemicals for reuse. Because the heating value of hemicelluloses is considerably lower than that of lignin, significant losses have been realized by burning hemicellulose. Recent technology has proposed to extract hemicelluloses from wood chips before the pulping stage to generate higher-value products without adverse impact on pulp production, yield, or properties (Wafa Al-Dajani and Tschirner 2008). Because of the unique characteristics of the Kraft pulping process, the seamless integration of biorefining technology with pulp production has been investigated to produce biofuels and biochemicals such as ethanol and acetic acid. It has been reported that this process could be economically promising if it were integrated into an existing Kraft mill and if an extraction facility were available (van Heiningen 2006). Because cellulose is an important source of fiber for pulp and paper production, which continue to be a value-recovery activity, the pulping facility will remain as a candidate facility for the existing pulp and paper mill with its existing technologies and capacity options.

Black liquor is currently used as a feedstock for the recovery boiler. This technology will remain in the technology set, but it may be upgraded with CHP technology. It is assumed here that the black liquor yield is a known function of total pulp produced (Biermann 1993). The recipes for making hemicelluloses, black liquor, and pulp, as well as the efficiencies of their yields, are known in advance.

8.4.2.3 Wood-Panel Facilities

Various wood-panel facilities and technologies are available to produce different panel products, including particleboard (PB), medium-density fiberboard (MDF), plywood, and oriented strandboard (OSB). PB and MDF are widely used for nonstructural applications such as furniture, shelves, and kitchen cabinets. PB is made of low-grade wood logs and sawmill residues such as chips, sawdust, and shavings. These raw materials are processed into small wood particles which are bonded with synthetic resin under high temperature and pressure. MDF is also made of low-grade wood logs and sawmill residues. However, instead of processing these raw materials into small particles, a mechanical refining process is used which converts the chips, sawdust, and shavings into wood fibers which are then bonded with urea-formaldehyde resin under high temperature and pressure. Because both PB and MDF require sawmill residues as their dominant raw material, the general development of biotechnology will undoubtedly have a significant impact on their raw material availabilities and costs. The sustainability of these two sectors will eventually depend on their market demand and profitability compared with emerging bioproducts.

Plywood and OSB are structural wood panels used mainly in building construction as wall, roof, and floor sheathing in North America. Plywood is made of wood logs; the logs are first debarked, then steam-treated, and peeled into veneers. The veneers are then processed through a series of drying, grading, gluing, forming, and pressing operations to produce plywood. In plywood production, peeler cores of 3–4″ in diameter are generally rejected and used to make wood chips, forming another important process biomass source. Last but not least, OSB is made of wood logs which are processed into thin wood strands that are bonded with phenol formaldehyde resin under high temperature and pressure. Because wood strands are physically ideal for hemicellulose extraction, this presents a significant opportunity for an integrated biochemical facility. Despite the current economic downturn, the projected demand for OSB products is expected to remain strong, and therefore the yield and economic return for hemicelluloses, and consequently for ethanol and acetic acid production from the OSB process, will be promising (Paredes et al. 2008).

From the analyses, to develop a comprehensive and sustainable IFBR supply chain, wood-panel facilities should be included as candidate facilities with their existing technologies and capacity options.

8.4.2.4 Biochemical Facilities

Economical conversion of woody biomass into value-added bioproducts and chemicals require efficient use of its components: hemicelluloses, cellulose, and lignin. Among these biopolymers, hemicelluloses are the easiest to extract under moderate conditions. With the rapid advancement of biorefining technology, various methods of extracting hemicelluloses have been studied.

To extract hemicelluloses and converting them into bioethanol and acetic acid, a biochemical facility is required for pretreatment, hydrolysis, and fermentation. Depending on the types of biomass used, the conditions of pretreatment and hydrolysis, the acid and enzyme used, and the cost and efficiency can vary. Despite the numerous publications found to date, the technology is still considered to be under development at the laboratory scale. For demonstration purposes, this analysis has included a biochemical facility using fermentation technology as a candidate technology to convert hemicelluloses to bioethanol. It has been assumed that the hemicelluloses, and therefore the bioethanol production rate, are known functions of the oven-dry chip quantities used (Boussaid et al. 1998, Heller et al. 2007). As an independent facility, this biochemical facility could be established at various forest product plants, such as a sawmill, a pulp and paper mill, a panel mill, or even at a greenfield site with different installation costs. The decisions on whether to select such a facility and what should be its location and its capacity are, however, decision variables to be determined by the optimization model.

8.4.2.5 Energy Facilities

The core element of an energy facility is the boiler or CHP generator. Both technologies are characterized by thermochemical reactions that convert fuel substances into energy through combustion. The difference between the two is that the boiler generates heat and steam only, whereas CHP technology, with the installation of a steam turbine, can generate heat, steam, and electricity simultaneously. After its use in the electricity generation cycle, the steam from the steam turbine can be piped to internal facilities where steam and heat are required, or to demand locations through a district heating system to provide heat energy. Traditionally, an energy facility is built to produce energy from nonrenewable fuels (natural gas, fuel oil, and coal). Recently, more and more energy facilities have used biomass, including forest residues, pellets, bark, chips, sawdust, shavings, and pulping liquor, to minimize consumption of nonrenewable fuels. This facility may be established at one or several existing mill sites or at a greenfield site to serve internal and external manufacturing or commercial energy needs. Depending on internal biomass fuel availabilities, various fuel options may be implemented. However, biomass fuel from an internal source, if available, presents the most attractive and economically advantageous option to reduce fuel outsourcing and shipping costs.

The latest development in thermochemical technologies has gone beyond the conventional combustion process. Advanced gasification and pyrolysis technologies have been developed to convert biomass into various fuel gases and pyrolysis oils for either heat and electricity generation or sale as high-value transportation fuels (Van Loo and Koppejan 2008). Within the scope of this study, the bioenergy technologies under consideration have been limited to conventional heat/steam generation and CHP generation. As gasification and pyrolysis technologies become commercially viable, they can easily be added as candidate technologies.

8.4.2.6 Pelletizing Facilities

Lignocellulosic biomass from plants usually has a low bulk density of approximately 30 kg/m^3 and varying moisture contents ranging from 10% (shavings) to 70% (bark and chips). Pelletization can transform low-density, low-heating-value biomass into high-heating-value pellets with an increased specific density of more than 1000 kg/m^3 and a reduced and constant moisture content (Mani et al. 2006). A *pelletizing facility* is the establishment housing the technology and equipment to accomplish this transformation. In pellet production, virtually any woody biomass source can be used as input material, such as wood logs, forest residues, landfill residues, and process residues like bark, chips, sawdust, and shavings. However, because the process handling and operational costs as well as product quality are affected by the raw materials used, process residues have become the preferred raw materials. Given the raw material consumptions, if the candidate pelletizing facility is installed within an existing manufacturing site, for example on a sawmill site, more effective utilization of process residues will be possible, in addition to energy generation as discussed earlier. This option could potentially reduce the chip supply to the pulp and paper mills as well as the supply of sawdust and shavings to the PB and MDF mills. If it is installed as an integration operation with pulp and paper or wood panel facilities, because it will use the same materials currently consumed by pulp and wood-panel production, the optimal allocation of raw materials among the different facility options will need to be investigated. Alternatively, the pelletizing plant may be installed on a greenfield site. In all cases, determining the most profitable option is a complex strategic decision that may be faced by many organizations. It requires a comprehensive analysis and advanced mathematical tool to support such decisions.

Table 8.1 provides a summary of the different facilities and technologies, including the input raw material they consume and their primary outputs as well as the by-products they produce.

TABLE 8.1

Summary of Facilities and Technologies with Their Inputs, Outputs, and By-Products

Facilities	Technologies	Inputs	Outputs	By-Products
Sawmill	Lumber production	Logs	Lumber	Bark, chips, sawdust shavings
Pulp mill	Pulp production	Logs, chips, chemicals	Pulp	Pulp liquor
Panel mill	Panel production	Logs, chips, sawdust, shavings, chemicals	Panels	Bark, fines
Biochemical	Fermentation	Chips	Ethanol	Extracted chips
Pellet	Pelletization	Forest residues, barks, chips, sawdust, shavings	Pellets	
Energy	Heat generation	Forest residues, landfill residues, bark, chips, sawdust, shavings, natural gas, fuel oil, electricity	Steam	Heat
Energy	CHP	Forest residues, landfill residues, barks, chips, sawdust, shavings, natural gas, fuel oil, electricity	Electricity	Steam, heat
Energy	Recovery boiler	Pulp liquor	Steam	Heat
Energy	Recovery CHP	Pulp liquor	Electricity	Steam, heat

8.4.3 Raw Materials, Intermediate Products, and Finished Products

In IFBR supply chain modeling, one of the complexities is to analyze the various types of flows of materials, products, and by-products, as well as energy, across different facilities in different industries. In addition, because of the nature of the problem, one product from one facility may serve as raw material, intermediate product, or energy for the other. For instance, chips from a sawmill can be considered as raw material for a pulp facility or as fuel for an energy facility. Electricity and heat produced from an energy facility using bark, chips, sawdust, and shavings in a sawmill can be used as energy for internal lumber production and mill energy consumption. These interactive flows of raw material, intermediates, finished products, and energy between supplier and plant, plant and plant, and plant and customer further complicate the problem. To model this integrated supply chain effectively, it is necessary to classify the biomass and define the concept and scope of raw materials, intermediates, and finished products.

8.4.3.1 Classification of Biomass

As discussed in the previous section, woody biomass can be derived from different sources in different forms. It can be purchased from publicly or privately owned forest land in the forms of forest logs and harvesting residues, from forest product manufacturers in the forms of process residues, and from landfills in the forms of bark and biowastes. For the purpose of this study, woody biomass is classified into the following nine classes:

1. *Forest saw logs*: Good-quality logs that can be used for lumber, laminated veneer lumber (LVL), or veneer production, depending on diameter and species;
2. *Forest pulp logs and thinnings*: Logs of poor quality or smaller diameter that can be used for wood composites and pulp production;
3. *Forest residues*: Treetops and branches collected in harvesting areas that can be used for thermochemical production;

4. *Bark*: Process residues from the log debarking process in sawmills, wood composite mills, and pulp and paper mills, that can be used for pellet and thermochemical production;
5. *Chips*: Sawmill process residues; can be used as raw materials for wood composites, pulp and paper, and pellets, as well as for bio- and thermochemical production;
6. *Shavings*: Sawmill process residues; can be used as raw materials for wood composites, pellets, and thermochemical production;
7. *Sawdust*: Sawmill process residues; can be used as raw materials for wood composite, pellets, and thermochemical production;
8. *Pulping liquor*: Pulp mill process residue in which the hemicellulose and lignin components may be used for bio- and thermochemical production;
9. *Landfill residues*: Woody residue deposits, containing mainly bark that have decomposed to various degrees over the years, can be used as potential biomass source for thermochemical production.

8.4.3.2 Definitions of Raw Materials, Intermediates, and Finished Products

Having classified the biomass materials, *raw materials* can be defined, within the context of this supply chain, as products that are purchased from supply sources, such as logs, forest residues, landfill residues, chemicals, coal, electricity, natural gas, and fuel oil. If a product or by-product is manufactured at a production site and can then be used to produce subsequent products either within the site or at an external production site, this product is called an *intermediate product*. A manufactured product that cannot be used further for subsequent product manufacturing within the defined supply chain scope is defined as a *finished product*. Note that the definitions of raw material, intermediates, and finished products are all relative to the defined supply chain environment. For instance, bioethanol is classified as a finished product in the current supply chain context, but it could well be an intermediate product in another. Electricity purchased from electricity suppliers is regarded as a raw material, but as an intermediate product if it is manufactured within a production site. Here, no distinction is made regarding the identity of electricity because it can be used interchangeably within the production process regardless of being purchased or produced. Table 8.2 summarizes the definitions of raw materials, intermediates, and finished products with examples.

In a general supply chain structure, each manufacturing site may provide intermediate products, such as electricity, heat, pellets, or chips to itself, to external manufacturing sites, and to demand markets. As for finished products, the manufacturer can provide them only to demand markets.

TABLE 8.2
Product Sets, Their Definitions, and Examples

Product Set	Definition	Product Examples
Raw materials	Products that are purchased from supply sources within the defined supply chain scope	Logs, forest residues, landfill residues, coal, chemicals, electricity, natural gas, fuel oil
Intermediate products	Products or by-products manufactured at a site that can be used to produce subsequent products within the defined supply chain scope	Bark, chips, sawdust, shavings, pulp, pulp liquor, pellets, electricity, heat
Finished products	Manufactured products or by-products that cannot be used for subsequent production within the defined supply chain scope	Lumber, paper, wood panels, bio-ethanol

8.4.4 Model Development

The next step is to develop the strategic design model for the IFBR supply chain. In the following paragraphs, the generic case will be described first, followed by the model formulation.

8.4.4.1 Case Description

Consider a single organizational case in which the organization has several existing forest product manufacturing mills, as shown by the solid round-cornered rectangular symbols in Figure 8.5. It would like to make an investment in one or several biorefineries to be installed either within the existing mills or at selected greenfield sites which are shown as dotted round-cornered rectangular symbols in Figure 8.5. These existing and greenfield sites form the finite list of candidate manufacturing sites, which is referred to by $\mathcal{M}$. The organization has several candidate supply sources for biomass, chemicals, electricity, and fuel. The set of supply sources is denoted as S. Downstream of the supply chain, the organization serves several demand markets, denoted by C. In its generic form, this supply chain network can be denoted by $\mathcal{T}$ and be expressed as $\mathcal{T} = \{\mathcal{A}, \mathcal{N}\}$, in which $\mathcal{N}$ represents the entire set of network nodes $\mathcal{N} = (S,\mathcal{M},C)$ and $\mathcal{A}$ is the set of arcs representing the entire set of inbound, outbound, and internal flows supporting the transportation activities between the pairs of nodes $\mathcal{A} = (S \times \mathcal{M}, \mathcal{M} \times \mathcal{M}, \mathcal{M} \times C)$. Assume that the supply chain network has a general structure, so that the products flow not only along the supply chain, as illustrated in Figure 8.1(b), but also within and across different manufacturing sites, as illustrated in Figure 8.5. Thus each plant may have several roles: a manufacturer that produces intermediate and finished products; a supplier that supplies products to internal and external production and demand locations; and a customer that receives products from suppliers and other plants. With this network structure, greater flexibility is possible, enabling the optimization model to determine the best network configuration for the supply chain.

Recall that each manufacturing site may have one or several facilities with existing or potential technologies and capacity options, and that in designing the IFBR supply chain, the facility location-allocation, technology, and capacity selection decisions must also be

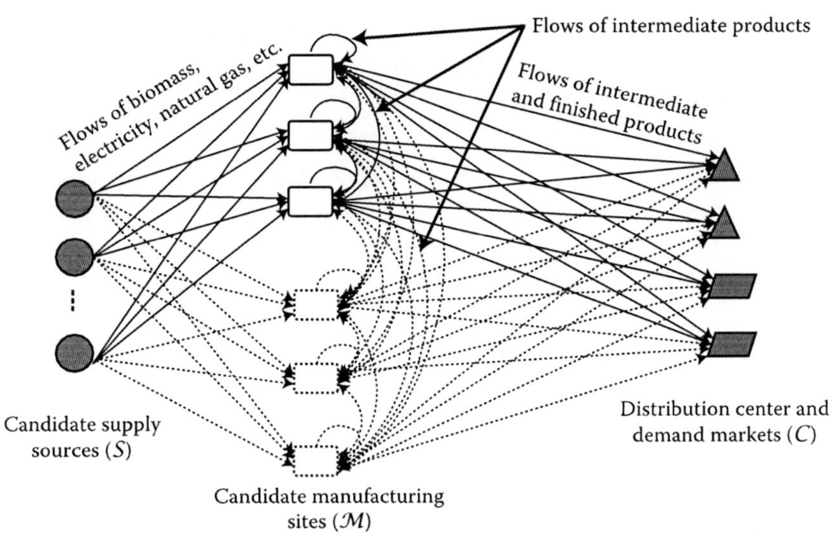

FIGURE 8.5
Generic forest biorefinery supply chain network.

made simultaneously. A technology can be defined by the products that it can produce through the combined establishment of process know-how and production equipment (Martel 2005). Once the technology has been determined, the raw materials that the technology requires and the products that it produces will also be determined. Let $\mathcal{G}$ be the set of technologies and g a particular technology, where ($g \in \mathcal{G}$). In addition, let $\mathcal{G}_n$ be the subset of technologies that can be installed at a manufacturing site $n \in \mathcal{M}$. For each candidate technology g, one or many candidate capacity options are available. A capacity option $k \in \mathcal{K}_g(\mathcal{K}_g \subset \mathcal{K})$ may correspond to the capacity that already exists or to capacity that is to be added as a new resource. In the latter case, different options can be associated with equipment of different sizes to reflect economies of scale. Because each facility will be associated only with a certain subset of technologies, the decision to establish a facility (or not to) may be determined explicitly by technology decisions. Therefore, in the model development, the facility index will be unnecessary. Once a facility with certain technology and capacity options has been chosen, the space constraints of establishing such a facility must also be considered. However, the exact facility layout is not included in the model formulation at this stage because it is a problem requiring a more detailed product flow analysis, which by itself could be another optimization problem.

Let I be the set of products, including the subsets of raw materials $I^{\mathcal{RM}}$, intermediates $I^{\mathcal{INT}}$, finished products $I^{\mathcal{FIN}}$, and energy products $I^{\mathcal{EN}}$, and let i be a particular product $i \in I = I^{\mathcal{RM}} \cup I^{\mathcal{INT}} \cup I^{\mathcal{FIN}}$ where $I^{\mathcal{EN}} \subset I$. The conversions of raw materials to intermediate and finished products are accomplished through a set of recipes $\mathcal{R}(r \in \mathcal{R})$. Assume, for a given technology, that one or several recipes can be developed according to the mixture of raw materials (and/or intermediate products) that the technology consumes and the products that it can produce. For a standard period of recipe usage, the input quantities of raw materials and intermediate products required, and the output quantities of subsequent intermediate and finished products, including by-products, produced, are assumed to be known. For a given product, its production efficiency may vary depending on the recipe used, a variation which is also assumed to be known. The planning horizon is defined as T, being the number of periods (years) the supply chain design strategy covers.

This chapter focuses on a demand-driven supply chain network design. The objectives are to provide investment decision support for organizations on the selection of manufacturing sites, facility technologies, capacity options, and material supply sources that optimize biomass utilization and maximize supply chain net present value (NPV). To develop the model, the following assumptions are necessary:

1. The existing mills proposed have potential space to build new facilities.
2. For both existing and greenfield sites, if a mill site is selected (opened), it will incur an annual fixed cost. The fixed closure cost, however, would apply only to existing mill sites.
3. If a facility is used (installed) at a mill site, it will incur an annual fixed facility cost. The fixed facility closing cost would apply only to the existing facility.
4. Products are aggregated to product types requiring the same technology.
5. Demand markets are sufficiently large to absorb all production. The decision of whether or not to serve a particular product to that market is governed by the profitability of that product in the market.

During model development, the organization's budget constraints are also built in so that investment decisions can be made within the budget limit. The next section presents the *MIP*-based IFBR supply chain design model.

8.4.4.2 Multiperiod Integrated Forest Biorefinery Supply Chain Design Model

To formulate the model, the following indexes, sets, parameters, and decision variables are defined:

Indices and Sets

$t \in \mathcal{T}$	Set of time periods (years)
$i \in I$	Set of product types, with $I^{\mathcal{RM}}$, $I^{\mathcal{INT}}$, $I^{\mathcal{FIN}}$, $I^{\mathcal{EN}}$ being the subset of raw materials, intermediate products, finished products, and energy products, where $I^{\mathcal{EN}} \subset I = I^{\mathcal{RM}} \cup I^{\mathcal{INT}} \cup I^{\mathcal{FIN}}$
$n \in \mathcal{N}$	Set of network nodes, where $\mathcal{N} = S \cup \mathcal{M} \cup C$
$g \in \mathcal{G}$	Set of candidate technologies
$k \in \mathcal{K}$	Set of candidate capacity options
$\mathcal{K}_g$	Set of candidate capacity options for technology g, $k \in \mathcal{K}_g \subset \mathcal{K}$
$r \in \mathcal{R}$	Set of production recipes
$\mathcal{R}_g$	Set of production recipes for technology g, $r \in \mathcal{R}_g \subset \mathcal{R}$

Parameters

d_{int}	Demand for product $i \in I^{\mathcal{INT}} \cup I^{\mathcal{FIN}}$ from demand market $n \in C$ in period t
p_{int}	Unit sales revenue for product $i \in I^{\mathcal{INT}} \cup I^{\mathcal{FIN}}$ in demand market $n \in C$ in period t
c_{int}	Unit cost of purchasing raw material $i \in I^{\mathcal{RM}}$ from a supply node $n \in S$ in period t
$c_{inn't}$	Unit cost for shipping product $i \in I$ from location n to location n' $(n,n' \in \mathcal{N})$ in period t
c_{irn}	Unit production cost of producing product $i \in I^{\mathcal{INT}} \cup I^{\mathcal{FIN}}$ using recipe r at plant node $n \in \mathcal{M}$
u_{ir}	Consumption quantity of product $i \in I^{\mathcal{RM}} \cup I^{\mathcal{INT}}$ using recipe r per unit usage of recipe r
q_{ir}	Production quantity of product $i \in I^{\mathcal{INT}} \cup I^{\mathcal{FIN}}$ using recipe r per unit usage of recipe r
ϕ_{ir}	Efficiency factor of producing product $i \in I^{\mathcal{INT}} \cup I^{\mathcal{FIN}}$ using recipe r
e_{in}	Estimated annual consumption of product $i \in I^{\mathcal{EN}}$ at plant node $n \in \mathcal{M}$ if it is opened
e_{ik}	Estimated annual consumption of product $i \in I^{\mathcal{EN}}$ by capacity option k with technology g $(k \in \mathcal{K}_g)$ if it is installed or used
S_k	Space required to install equipment of capacity k with technology g $(k \in \mathcal{K}_g)$
S_n	Total space available at candidate plant node $n \in \mathcal{M}$
K_{int}	Capacity availability for product $i \in I^{\mathcal{RM}}$ at supply node $n \in S$ in period t
K_{ik}	Production capacity of producing product $i \in I^{\mathcal{INT}} \cup I^{\mathcal{FIN}}$ using capacity option k with technology g $(k \in \mathcal{K}_g)$

$K_{inn'}$	Capacity of pipeline for product $i \in I^{\mathcal{EN}}$ between nodes $n,n' \in \mathcal{M}$, $n \neq n'$
a_{nt}	Annual fixed cost for selecting a candidate supply node $n \in S$
a_n	Annual fixed cost for opening candidate plant at node $n \in \mathcal{M}$
a_{kn}	Annual fixed cost for opening (installing) capacity option k with technology g ($k \in \mathcal{K}_g$) at plant node $n \in \mathcal{M}$
$a_{inn'}$	Annual fixed cost of building pipeline for product $i \in I^{\mathcal{EN}}$ between nodes n and n'
b_n	Fixed cost for closing plant node $n \in \mathcal{M}$
b_{kn}	Fixed cost for shutting down capacity option k with technology g ($k \in \mathcal{K}_g$) at plant node $n \in \mathcal{M}$
α	Discount rate
λ	Capital recovery factor
sv	Salvage value
B_t	Annual budget availability
G	Sufficiently large number

Decision Variables

Y_n	Binary variable being 1 if plant $n \in \mathcal{M}$ is opened, and 0 otherwise
Y_{kn}	variables being 1 if capacity option k ($k \in \mathcal{K}_g$) is installed at plant $n \in \mathcal{M}$, and 0 otherwise
$Y_{inn'}$	Binary variables being 1 if pipeline for product i ($i \in I^{\mathcal{EN}}$) is installed between n and n', "0" otherwise
Y_{nt}	Binary variable being 1 if supply source $n \in S$ is selected for period t, and 0 otherwise
X_{rnt}	The number of usages of recipe r at plant node $n \in \mathcal{M}$ in period t
$X_{inn't}$	Flow quantity of product i from node n to node n' ($n \neq n'$) in period t, where $n,n' \in \mathcal{N} = \{S,\mathcal{M},C\}$

The objective function is:

$$\text{Max: } \sum_{t \in T} \frac{1}{(1+\alpha)^{t-1}} \Bigg(\sum_{i \in I^{\mathcal{INT}} \cup I^{\mathcal{FIN}}} \sum_{n \in \mathcal{M}} \sum_{n' \in C} p_{in't} X_{inn't} - \sum_{i \in I^{\mathcal{INT}} \cup I^{\mathcal{FIN}}} \sum_{r \in \mathcal{R}} \sum_{n \in \mathcal{M}} c_{irn} q_{ir} X_{rnt}$$
$$- \sum_{i \in \mathrm{I}} \sum_{n \in \mathcal{N}} \sum_{n' \in \mathcal{N}\{n\}} c_{inn't} X_{inn't} - \sum_{i \in I^{\mathcal{RM}} \cup I^{\mathcal{INT}}} \sum_{n \in S} \sum_{n' \in \mathcal{M}} c_{int} X_{inn't} - \sum_{n \in S} a_{nt} Y_{nt} - \lambda(1 - sv)$$
$$\left(\sum_{n \in \mathcal{M}} a_n Y_n + \sum_{k \in \mathcal{K}} \sum_{n \in \mathcal{M}} a_{kn} Y_{kn} + \sum_{i \in I^{\mathcal{EN}}} \sum_{n \in \mathcal{M}} \sum_{n' \in \mathcal{M} \cup C} a_{inn'} Y_{inn'} \right)$$
$$- \sum_{n \in \mathcal{M}} \left(b_n (1 - Y_n) + \sum_{k \in \mathcal{K}} b_{kn} (1 - Y_{kn}) \right) \tag{8.1}$$

Subject to the following constraints:

$$\sum_{n \in \mathcal{M}} X_{inn't} \le d_{in't} \qquad \forall i \in \mathcal{I}^{INT} \cup \mathcal{I}^{FIN}, n' \in C, t \in \mathcal{T} \tag{8.2}$$

$$\sum_{n \in S \cup \mathcal{M}\{n'\}} X_{inn't} + \sum_{r \in \mathcal{R}} q_{ir}\phi_{ir}X_{rn't} - e_{in'}Y_{n'} - \sum_{k \in \mathcal{K}} e_{ik}Y_{kn'} - \sum_{r \in \mathcal{R}} u_{ir}X_{rn't} - \sum_{n'' \in \mathcal{M} \cup C\{n'\}} X_{in'n''t} = 0$$

$$\forall n' \in \mathcal{M}, i \in \mathcal{I}, t \in \mathcal{T} \tag{8.3}$$

$$\sum_{n' \in \mathcal{M} \cup C} X_{inn't} \le \sum_{r \cup \mathcal{R}} q_{ir}\phi_{ir}X_{rnt} \qquad \forall i \in \mathcal{I}^{INT} \cup \mathcal{I}^{FIN}, n \in \mathcal{M}, t \in \mathcal{T} \tag{8.4}$$

$$\sum_{r \in \mathcal{R}_g} q_{ir}X_{rnt} \le \sum_{k \in \mathcal{K}_g} K_{ik}Y_{kn} \qquad \forall i \in \mathcal{I}^{INT} \cup \mathcal{I}^{FIN}, g \in \mathcal{G}, n \in \mathcal{M}, t \in \mathcal{T} \tag{8.5}$$

$$\sum_{n' \in \mathcal{M}} X_{inn't} \le K_{\text{int}}Y_{nt} \qquad \forall i \in \mathcal{I}^{RM}, n \in S, t \in \mathcal{T} \tag{8.6}$$

$$\sum_{n \in S \cup \mathcal{M}\backslash\{n'\}} X_{inn't} \le GY_{n'} \qquad \forall i \in \mathcal{I}^{RM} \cup \mathcal{I}^{INT}, n' \in \mathcal{M}, t \in \mathcal{T} \tag{8.7}$$

$$X_{inn't} \le K_{inn'}Y_{inn'} \quad \forall i \in \mathrm{I}^{EN}, n \in \mathcal{M}, n' \in \mathcal{M} \cup C \backslash \{n\}, t \in \mathcal{T} \tag{8.8}$$

$$2Y_{inn'} \le Y_n + Y_{n'} \quad \forall i \in I^{EN}, n \in \mathcal{M}, n' \in \mathcal{M} \tag{8.9}$$

$$\sum_{k \in \mathcal{K}} S_k Y_{kn} \le S_n Y_n \qquad \forall n \in \mathcal{M} \tag{8.10}$$

$$\sum_{n \in \mathcal{M}} \Bigg(\lambda a_n Y_n + b_n(1 - Y_n) + \sum_{k \in \mathcal{K}} \big(\lambda a_{kn} Y_{kn} + b_{kn}(1 - Y_{kn})\big)$$

$$+ \sum_{i \in I^{EN}} \sum_{n' \in \mathcal{M} \cup C} \lambda a_{inn'} Y_{inn'} \Bigg) \le B_t \quad \forall t \in \mathrm{T} \tag{8.11}$$

$$X_{rnt}, X_{inn't} \ge 0, Y_n, Y_{kn}, Y_{inn'}, Y_{nt} \in \{0,1\} \;\; \forall i, r, k, n, n', t \tag{8.12}$$

The objective function is to maximize the NPV assuming that the first period is the present period. The NPV is expressed as the present value of sales revenue from selling the marketable products, minus the total variable cost for production shipping, and raw material purchasing, the annual fixed cost for contracting with suppliers, opening plants, installing (using) facilities, building pipelines, and the fixed plant or facility closing costs. The revenues for the intermediate products that are consumed internally or shipped to other manufacturing sites are not included because these are regarded as an organization's multiple-site internal flows. However, shipment costs are included in the objective function as a function of shipping distance.

Constraints (8.2) define the demand satisfaction constraints which state that for any marketable product and during any period, not all the market demand will be satisfied by the shipments from manufacturing plants. Constraints (8.3) are product flow balance constraints describing the general flow balance rule that for a product i, the total inbound shipments and internal productions at plant node $n \in \mathcal{M}$ should be equal to its internal fixed and variable consumptions and outbound shipments. For a raw material $i \in \mathcal{I}^{\mathcal{RM}} \backslash \mathcal{I}^{\mathcal{EN}}$, the internal production and outbound shipment quantities will be equal to zero, and the constraints indicate that the raw material provided by the suppliers to mill $n \in \mathcal{M}$ must be sufficient to satisfy its consumption. For intermediate products, shipments within or between plants are allowed. The model will determine whether or not such shipments are economically acceptable. For finished products, inbound shipments from external sources and internal consumptions are equal to zero, and the constraints state that the production quantity must be equal to the outbound shipments. With regard to the energy product $i \in \mathcal{I}^{\mathcal{EN}}$, a fixed energy consumption is included in constraint (8.3) in addition to its variable consumptions to take into account electricity and heating requirements upon the opening of a plant and installation or utilization of a facility. These fixed requirements generally cover the nonproduction uses of electricity and heat for lighting, space heating, cooling, and electrical appliances, as well as other miscellaneous regular uses. Constraints (8.4) define the shipment upper bound of a plant implying that for any intermediate or finished product, the total shipment from plant $n \in \mathcal{M}$ should not exceed its production in a given period. Constraints (8.5) and (8.6) provide the production and supply capacity constraints. Constraints (8.7) stress that an inbound shipment to a plant $n \in \mathcal{M}$ can be made only if the plant is open. Constraints (8.8) suggest that a pipeline must be built (used) if there is flow of product $i \in \mathcal{I}^{\mathcal{EN}}$ between nodes n and n'. Constraints (8.9) state that a pipeline can be built between the nodes n and n' only if the two nodes are both open. Constraints (8.10) provide the space availability constraints indicating that a facility of technology-capacity k can be installed at plant $n \in \mathcal{M}$ only if the plant has sufficient space. The annual budget constraints are given by constraints (8.11). Constraints (8.12) define domains for the decision variables.

8.4.5 Experimental Case Studies

To validate the model, a prototype and an experimental case are developed based on the forest industry in the Saguenay-Lac-St-Jean region, Quebec, Canada. The case includes a single organization operating two sawmills and one pulp mill that purchase raw materials from six potential suppliers and sell products to two target demand markets. The organization has a budget which enables it to invest in one or several biorefineries or bioenergy facilities with different technologies to produce value-added bioproducts. Five candidate mill locations have been identified, including the three existing mill locations and two greenfield locations. A total of 17 candidate technology-capacity options are proposed for potential investment. Given the candidate technology-capacity options and the possible recipe choices, the expected raw material and intermediate product consumptions, as well as the intermediate and finished product productions, can be defined, forming 17 products. The planning horizon is 3 years. Table 8.3 summarizes the scope of the example case.

TABLE 8.3
Scope of the Example Case

Index	Size
Candidate plants	5
Technologies	9
Technology-capacity options	17
Candidate suppliers	6
Raw materials	6
Intermediate products	9
Finished products	3
Customer zones	2
Recipes	30
Planning horizon (years)	3

8.4.5.1 Data Collection

To create a demand-driven supply chain design that maximizes the value of the biomass, end-to-end supply chain data are required. Because of the large scope of the problem, which covers several sectors of different industries, data collection becomes very challenging. In this case study, exact industrial data were not always available, and therefore data were collected from various sources, including the literature, professional and industrial magazines, official government and institutional Internet sites, and periodic market reports, as well as advice and estimates from industrial experts.

During data collection, preparation, and model validation, several challenges were encountered. The first challenge is that for different product types, ranging from forest biomass and process residues to lumbers, pulp, energy, and fuels, different units are used. Furthermore, for the same product type, different units are often used in different publications and data sources. To ensure that the model functions correctly, a standard unit must be specified for each product type, and the various units for that product type must be converted to the standard unit for that product throughout the data files. Once all instances of a product type have the same unit, cross-examination of the data from different data sources must be carried out to ensure data consistency. The second challenge is that the product types are different after mechanical, thermochemical, and biochemical conversions, and consequently the units are also different. Careful formulation and units transformation throughout production recipes must be carried out to ensure that no mixed-unit calculations are performed. The third challenge is the moisture content of the forest biomass and of process residues, such as bark, chips, sawdust, shavings, solids content of pulp liquor, and the moisture content of the finished products, such as lumber, pulp, and pellets. Because the mass quantities, prices, costs, and heating values of biomass are affected by their moisture contents, correct moisture conversions are necessary to ensure the consistency of the results.

Once all the data were collected, a Microsoft Access database was developed and populated. The *MIP* model was programmed using the Optimization Programming Language OPL 6.3 with a Microsoft Access database connection to read data inputs and write solution outputs. The model was solved using the CPLEX 11.2 optimizer. The program was run on an Intel Core 2 Duo workstation with a 2.00-GHz CPU, 4.00 GB of RAM, and Window Vista Home Edition Version 2007.

8.4.5.2 Results and Discussion

Figure 8.6 presents an example of the IFBR supply chain configuration designed using the *MIP* model. The design illustrated here is based on a single scenario with the proposed candidate plant sites, facility technologies, capacity options, and supply sources, satisfying the defined markets at the deterministic market demand and price (Table 8.3). According to this scenario, the solutions suggest that all the existing sawmills (S1 and S2) and the pulp mill (P1) should remain open and that the existing lumber and pulping facilities should continue to be used. The existing boilers at S1 and S2 should also be used. However, the existing recovery boiler at P1 should be replaced (upgraded) by a recovery CHP generator of the same capacity to produce both heat and electricity from pulp liquor. Investment should be considered for a pelletizing facility to produce pellets from forest residues, sawdust, and shavings from all the existing mills. A greenfield plant at location G1 is recommended for opening; however, it will produce only chips using pulp logs with residue bark as by-product. Three forest supply sources were selected to supply different mills. Only one electricity supplier was selected for energy supply. The suppliers of natural gas and heavy fuel oil were not selected. In this scenario, investment for ethanol production was not suggested. Despite the significant cost reduction of using process by-product chips to produce ethanol, it was found to be still comparatively less profitable than pellets and electricity. For ethanol production to become economically attractive,

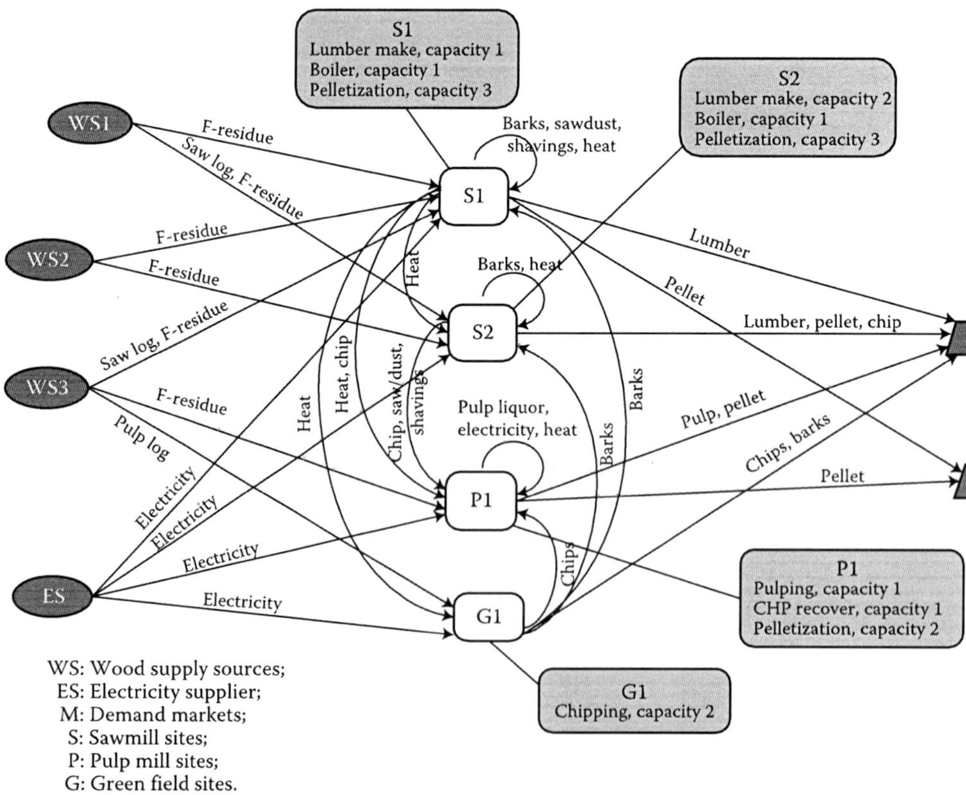

FIGURE 8.6
Example of a forest biorefinery supply chain configuration design using the MIP model.

further production cost reductions and possibly favorable tax credit and financial subsidy policies would be required.

In this demonstration case study, the problem generated approximately 2000 constraints and 2500 decision variables, including 70 binary variables. It was solved in less than 2 s with a solution gap of 0.3%, which indicating the close approximation of the best MIP objective obtained to the optimal value of the best bound upon the termination of the MIP solver. Although the test problem is relatively small, several important aspects, such as energy flows and flows of forest and process residues between suppliers and candidate plants are modeled in this general supply-chain network. The optimal investment option upon evaluating various biorefining technologies and capacity options for using biomass is revealed. The methodology provides an important basis upon which various complex models can be developed for more comprehensive full-scale problems.

It is important to note that although the model is very powerful in evaluating different design options to find the most profitable solution, the results are case- and scenario-sensitive. Each business has its own particularities. Therefore, real business data are required, and comprehensive scenario studies are essential.

8.5 Conclusions

With the rapid advancement of bioenergy and biorefinery technologies, organizations are facing the critical moment to transform from a traditional manufacturing system toward a biotechnology-enabled system to optimize biomass utilization and maximize supply chain value. However, making the right investment decisions is a complex task which requires comprehensive analysis and advanced decision-support tools. This chapter has introduced an optimization modeling approach for IFBR supply chain network design. Basic concepts of the supply chain, supply chain management, and supply chain design are provided. Mathematical programming techniques and their applications in various supply chain design problems are reviewed. Through comprehensive analysis of biomass and energy flows in the various forest product manufacturing sectors, the effective utilization of biomass are identified through IFBR supply chain optimization. In the supply chain design, a general market-driven network structure was proposed, and an *MIP*-based supply chain design model was developed which enabled optimal biomass allocations to be realized and the right facility investments to be identified. The model was validated using an experimental case. A demonstration of an optimal design based on the given market, supply, and budget scenarios was provided.

Acknowledgments

The authors would like to acknowledge the financial support provided by the Forest E-business Research Consortium (FOR@C) and would like to thank Roger Boileau, Jovani Jacques, Sebastien Lemieux, Marc Pare, Alain Perron, and Pierre Vezina for providing industrial advice and data.

References

Arntzen, B.C., Brown, G.G., Harrison, T.P., Trafton, L.L., 1995. Global supply chain management with digital equipment. *Interfaces* 25(1), 69–93.

Azaron, A., Brown, K.N., Tarim, S.A., Modarres, M., 2008. A multi-objective stochastic programming approach for supply chain design considering risk. *International Journal of Production Economics* 116, 129–138.

Bertsimas, D., Sim, M., 2004. The price of robustness. *Operations Research* 52(1), 35–53.

Biermann, C.J., 1993. *Essentials of Pulping and Papermaking*. San Diego: Academic Press. ISBN 0-12-097360-X.

Bhutta, K., Huq, F., Frazier, G., Mohamed, Z., 2003. An integrated location, production, distribution and investment model for a multinational corporation. *International Journal of Production Economics* 86, 201–216.

Boussaid, A., Robinson, J., Cai, D., Gregg, D., Saddler, J., 1998. Fermentability of the hemicellulose derived sugar from steam exploded softwood (Douglas Fir). *Biotechnology and Bioengineering* 64(3), 284–289.

Brown, G.G., Graves, G.W., Honczarenko, M.D., 1987. Design and operation of multi-commodity production/distribution system using primal goal decomposition. *Management Science* 33(11), 1469–1479.

Canel, C., Khumawala, B.M., 1996. A mixed integer programming approach for the international facilities location problem. *International Journal of Operations and Production Management* 16(4), 49–68.

Cohen, M.A., Fisher, M., Jaikumar, R., 1989. International manufacturing and distribution networks: A normative model framework. In: Goetschalckx, M., Vidal, C.J., Dogan, K., 2002. Modeling and design of global logistics systems: A review of integrated strategic and tactical models and design algorithms. *European Journal of Operational Research* 143, 1–18.

Cohen, M.A., Lee, H.L., 1989. Resource deployment analysis of global manufacturing and distribution networks. *Journal of Manufacturing Operations Management* 2, 81–104.

Cole, M.H., 1995. Service considerations and the design of strategic distribution systems. PhD Dissertation, School of Industrial and Systems Engineering, Georgia Institute of Technology, Atlanta, GA. In: Vidal, C.J., Goetschalckx, M., 1997. Strategic production-distribution models: A critical review with emphasis on global supply chain models. *European Journal of Operational Research* 98, 1–18.

Eksioglu, S.D., Acharya, A., Leightley, L.E., Arora, S., 2009. Analyzing the design and management of biomass-to-biorefinery supply chain. *Computers & Industrial Engineering* 57, 1342–1352.

Ellwein, L.B., Gray, P., 1971. Solving fixed charge location-allocation problems with capacity and configuration constraints. *AIIE Transactions* 3(4), 290–298.

Fleischmann, B., Meyr, H., Wagner, M., 2002. Advanced planning. In: Stadtler, H., Kilger, C. (eds.), *Supply Chain Management and Advanced Planning—Concepts, Models, Software, and Case Studies*, Second ed. Springer, Berlin, pp. 71–96.

Franca, R.B., Jones, E.C., Richards, C.N., Carlson, J.P., 2009. Multi-objective stochastic supply chain modeling to evaluate tradeoffs between profit and quality. *International Journal of Production Economics*, doi: 10.1016/j.ijpe.2009.09.005.

Freppaz, D., Minciardi, R., Robba, M., Rovatti, M., Sacile, R., Taramasso, A., 2004. Optimizing forest biomass exploitation for energy supply at a regional level. *Biomass and Bioenergy* 26(1), 15–25.

Frombo, F., Minciardi, R., Robba, M., Rosso, F., Sacile, R., 2009. Planning woody biomass logistics for energy production: A strategic decision model. *Biomass and Bioenergy* 33, 372–383.

Gebennini, E., Gamberini, R., Manzini, R., 2009. An integrated production-distribution model for the dynamic location and allocation problem with safety stock optimization. *International Journal of Production Economics* 122, 286–304.

Geoffrion, A.M., McBride, R.D., 1973. *The Capacitated Facility Location Problem with Additional Constraints.* Working Paper, Western Management Science Institute, UCLA.

Geoffrion, A.M, Graves, G., 1974. Multicommodity distribution system design by Benders decomposition. *Management Science* 29(5), 822–844.

Gjerdrum, J., Shah, N., Papageorgiou, L.G., 2002. Fair transfer price and inventory holding policies in two-enterprise supply chains. *European Journal of Operational Research* 143(3), 582–599.

Goetschalckx, M., 2002. Strategic network planning. In: Stadtler, H., Kilger, C. (eds.), *Supply Chain Management and Advanced Planning: Concepts, Models, Software, and Case Studies,* Second ed. Springer, Berlin, 105–121.

Goetschalckx, M., Vidal, C.J., Dogan, K., 2002. Modeling and design of global logistics systems: A review of integrated strategic and tactical models and design algorithms. *European Journal of Operational Research* 143, 1–18.

Gronalt, M., Rauch, P., 2007. Designing a regional forest fuel supply network. *Biomass and Bioenergy* 31, 393–402.

Gunnarsson, H., Ronnqvist, M., Lundgren, J.T., 2004. Supply chain modelling of forest fuel. *European Journal of Operational Research* 158, 103–123.

Haug, P., 1992. An international location and production transfer model for high-technology multinational enterprises. *International Journal of Production Research* 30(3), 559–572.

Heller, J.J.P., Shaler, S., van Heiningen, A., 2007. *Influence of Hemicellulose Extraction on Physical and Mechanical Behavior of OSB.* Advanced Engineered Wood Composites Center, University of Maine.

Higle, J.L., 2005. *Stochastic Programming: Optimization when Uncertainty Matters.* Tutorials in Operations Research, INFORMS, DOI 10.1287, Chapter 1, 1–24.

Kanzian, C., Holzleitner, F., Stampfer, K., Ashton, S., 2009. Regional energy wood logistics—optimizing local fuel supply. *Silva Fennica* 43(1), 113–128.

Klibi, W., Martel, A., Guitouni, A., 2010. The design of robust value-creating supply chain networks: A critical review. *European Journal of Operational Research* 203, 283–293.

Kogut, B., Kulatilaka, N., 1994. Operating flexibility, global manufacturing, and the option value of a multinational network. *Management Science* 40, 123–139.

Kouvelis, P., Yu, G., 1997. *Robust Discrete Optimization and its Applications.* Kluwer, Norwell MA.

Kumar, A., Cameron, J.B., Flynn, P.C., 2003. Biomass power cost and optimum plant size in western Canada. *Biomass and Bioenergy* 24, 445–464.

Lowe, T.J., Wendell, R.E., Hu, G., 2002. Screening location strategies to reduce exchange rate risk. *European Journal of Operational Research* 136, 573–590.

Malcolm, S., Zenios, S., 1994. Robust optimization for power systems capacity expansion under uncertainty. *Journal of Operational Research Society* 45(9), 1040–1049.

Mani, S., Sokhansanj, S., Bi, X., Turhollow, A., 2006. Economics of producing fuel pellets from biomass. *Applied Engineering in Agriculture* 22(3), 421–426. © 2006 American Society of Agricultural and Biological Engineers ISSN 0883-8542.

Martel, A., Vankatadri, U., 1999. Optimizing supply network structures under economies of scale. *Proceedings, International Conference on Industrial Engineering and Production Management,* Glasgow, United Kingdom.

Martel, A., 2005. The design of production-distribution networks: A mathematical programming approach. In: Geunes, J., Pardalos, P.M. (eds.), *Supply Chain Optimization.* Springer, Berlin, 265–306.

Martel, A., M'Barek, W., D'Amours, S., 2005. *International Factors in the Design of Multinational Supply Chains: The Case of Canadian Pulp and Paper Companies.* Working Paper DT-2005-AM-3, Research Consortium on e-Business in the Forest Products Industry (FOR@C), Network Organization Technology Research Centre (CENTOR), Laval University, Quebec, Canada.

Martin, R.K., 1999. *Large-Scale Linear and Integer Optimization: A Unified Approach.* Kluwer, Boston.

Meixell, M.J., Gargeya, V.B., 2005. Global supply chain design: A literature review and critique. *Transportation Research Part E* 41, 531–550.

Mohamed, Z.M., 1999. An integrated production-distribution model for a multi-national company operating under varying exchange rates. *International Journal of Production Economics* 58, 81–92.
Munson, C.L., Rosenblatt, M.J., 1997. The impact of local contents rules on global sourcing decisions. *Production and Operations Management* 6(3), 277–290.
Mulvey, J.M., Vanderbei, R.J., Zenios, S.A., 1995. Robust optimization for large-scale systems. *Operations Research* 43(2), 264–281.
Nieckels, L., 1976. *Transfer Pricing in Multinational Firms: A Heuristic Programming Approach and a Case Study*. Wiley, New York.
Oliver, R.K., Webber, M.D., 1992, Supply-Chain Management: Logistics catches up with strategy. In: Christopher, M. (ed.), *Logistics: The Strategic Issues*. Chapman Hall, London, 63–75.
Pan, F., Nagi, R., 2010. Robust supply chain design under uncertain demand in agile manufacturing. *Computers & Operations Research* 37(4), 668–683.
Paquet, M., Martel, A., Desaulniers, G., 2004. Including technology selection decisions in manufacturing network design models. *International Journal of Computer-Integrated Manufacturing* 17(2), 117–125.
Paredes, J.J., Jara, R., Shaler, S.M., van Heiningen, A., 2008. Influence of hot water extraction on the physical and mechanical behavior of OSB. *Forest Products Journal* 58(12), 56–62.
Parker, N., Fan, Y., Ogden, J., 2009. From waste to hydrogen: An optimal design of energy production and distribution network. *Transportation Research Part E*, doi:10.1016/j.tre.2009.04.002.
Santoso, T., Ahmed, S., Goetschalckx, M., Shapiro, A., 2005. A stochastic programming approach for supply chain network design under uncertainty. *European Journal of Operational Research* 167, 96–115.
Shapiro, A., 2003. Monte Carlo sampling methods. In: Roszczynski, A., Shapiro, A. (eds.), *Stochastic Programming, Handbooks in Operations Research and Management Sciences*, vol. 10. Elsevier.
Snyder, L.V., 2006. Facility location under uncertainty: A review. *IIE Transactions* 38(7), 537–554.
Soland, R., 1973. *Optimal Facility Location with Concave Costs*. Research Report CS 126, Center for Cybernetic Studies, University of Texas at Austin.
Stadtler, H., 2002. Supply chain management—an overview. In: Stadtler, H., Kilger, C. (eds.), *Supply Chain Management and Advanced Planning—Concepts, Models, Software, and Case Studies*, Second ed. Springer, Berlin, pp. 7–28.
Thanh, P.N., Bostel, N., Péton, O., 2008. A dynamic model for facility location in the design of complex supply chains. *International Journal of Production Economics* 113, 678–693.
Van Loo, S., Koppejan, J., 2008. *The Handbook of Biomass Combustion and Co-Firing*. Earthscan, UK, ISBN: 978-1-84407-249-1.
Van Heiningen, A.R.P., 2006. Converting a kraft pulp mill into an integrated forest biorefinery. *Pulp and Paper Canada* 107(6), 38–43.
Vidal, C.J., Goetschalckx, M., 1996. The role and limitations of quantitative techniques in the strategic design of global logistics systems. Georgia Institute of Technology. In: Vidal, C.J., Goetschalckx, M., 1997. Strategic production-distribution models: A critical review with emphasis on global supply chain models. *European Journal of Operational Research* 98, 1–18.
Vidal, C.J., Goetschalckx, M., 2001. A global supply chain model with transfer pricing and transportation cost allocation. *European Journal of Operational Research* 129, 134–258.
Vila, D., Martel, A., Beauregard, R., 2007. Taking market forces into account in the design of production-distribution networks: A positioning by anticipation approach. *Journal of Industrial and Management Optimization* 3(1), 29–50.
Wafa Al-Dajani, W., Tschirner, U.W., 2008. Pre-extraction of hemicelluloses and subsequent kraft pulping. Part I: Alkaline extraction. *Tappi Journal June*, 3–8.
Yu, L., Wang, S., Lai, K.K., 2005. A novel nonlinear ensemble forecasting model incorporating GLAR and ANN for foreign exchange rates. *Computers & Operations Research* 32, 2523–2541.

9

Forest Biorefinery Supply Chain Design and Process Flexibility

Behrang Mansoornejad and Paul R. Stuart

CONTENTS

9.1 Introduction

For a forestry company to improve its business model in the current market situation, it not only should diversify its revenue, but also must change its current manufacturing culture, in which the management focus is on capacity management and the profitability of the entire supply chain (SC) is generally ignored.

According to the strategic-phased approach for the forest biorefinery (FBR), revenue diversification will be achieved by means of "technology disruption" by producing building-block biorefinery chemicals, and ideally, in the longer term, by further increasing revenues by producing added-value derivatives. On the other side, manufacturing culture will be changed, in the short term, via "business disruption," through applying novel SC operating policies and exploiting production flexibility, and in the long term, by using advanced ERP and decision-making tools.

The key to success for "business disruption" is SC analysis. In the short term, to mitigate the risks of market volatility, companies should focus on improving their margins by implementing a *margins-based SC operating policy* and better exploiting the process capability for *flexible production* by using detailed knowledge of the process and its cost structure. Then advanced SC optimization techniques can be used to carry out product planning over different time horizons and to identify tradeoffs between product orders and anticipated supply and demand. This approach shows the importance of advanced cost-accounting systems that are capable of reflecting the cost of each decision at the decision-making level for short-term decision-making activities. Over the long term, companies should base their strategic SC-related decisions on a bottom-up approach, that is, designing/redesigning the SC based on the impact of the design on tactical and operational activities. These two approaches to the short- and long-term aspects of biorefinery implementation, the margins-based approach and the bottom-up approach, imply profound changes in the way forestry companies do business today, which is equivalent to business disruption.

The goal of this chapter is to show how SC-based analysis enables decision-makers to analyze various biorefinery options systematically from an SC perspective on biorefinery strategic design. These biorefinery options will be identified by the company's experts based on their experience and knowledge of the company, future forecasts, the assets of the company's existing SC, the potentials of the company's SC for biorefinery implementation given the latest developments in biorefinery technologies, and eventually the SC potential for forming partnerships with other companies. The SC-based analysis developed here analyzes these options and reveals the value that would be unleashed by the realization of each. This chapter is the evolution of the authors' recent paper published in *Computers & Chemical Engineering*, titled "Integrating product portfolio design and supply chain design for the forest biorefinery" [1]. The methodology presented in this chapter is a modified version of the methodology proposed in this paper, with more emphasis on the design aspects of the SC, that is, flexibility design and SC network design. The specific contributions of the paper and the chapter are illustrated in Figure 9.1.

At this point, the key concepts used in the chapter, margins-based SC operating policy, manufacturing flexibility, and the bottom-up approach are introduced. These concepts play key roles in the methodology, and in light of them, the goals of the methodology can be better understood.

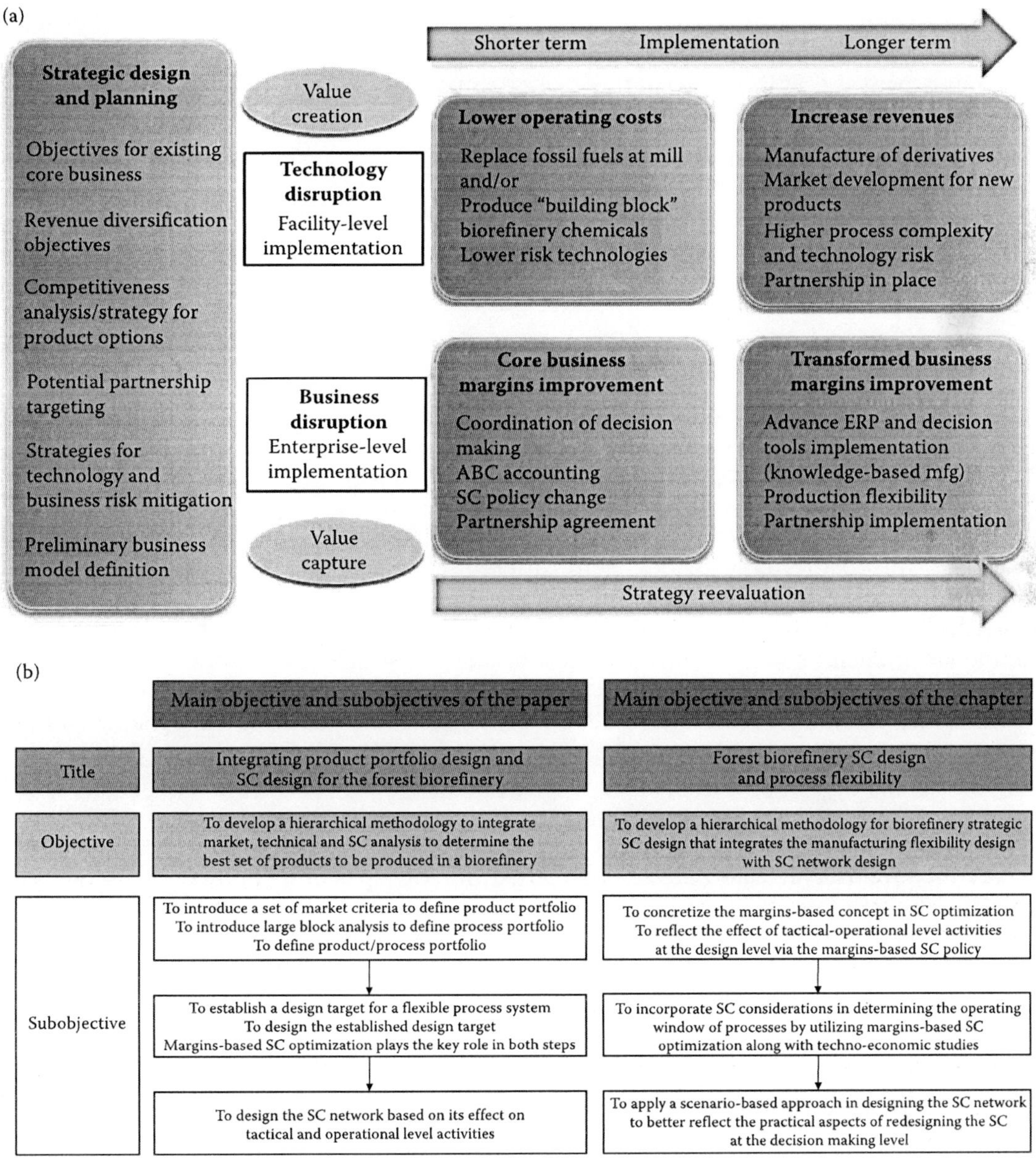

FIGURE 9.1
(a) Strategic implementation of the biorefinery by a P&P company. (b) Comparison between title, main objective, and subobjectives of the chapter and the *Computers & Chemical Engineering* paper.

9.1.1 Margins-Based Operating Policy

The operating policy in the P&P industry is said to be "manufacturing-centric." In this industrial sector, the management focus is on capacity planning, and industry participants try to achieve the efficient and effective use of machine capacity [2]. As a result, process efficiency is viewed as the key measure for profitability, and therefore it is believed that minimizing production cost will result in the highest profitability [3]. Moreover, production planning assumes a known set of orders and a fixed sequence of product grades. By treating the manufacturing process as the focal point, inventory and changeover costs are typically ignored or considered separately [2], and SC costs are often neglected, resulting in lower profitability [3].

To implement the FBR, the operating policy must shift from a manufacturing-centric approach to a margins-based one. This latter operating policy tries to maximize margins over the entire SC and to produce and select products and orders that ensure the best returns [3]. In this approach, long-term contracts and short-term order selection is made with respect, not only to process and production constraints, but to all SC constraints, including for example inventory and transportation constraints, to maximize the ultimate SC profitability.

9.1.2 Manufacturing Flexibility

Today's market is subject to huge volatilities in terms of price and demand. The price of oil, fuels, and chemicals, as well as the price of forestry products, change even on a monthly basis. The demand for some products is not always certain, and sometimes, despite strong demand, the price is too low for the production of a product to be profitable. On the feedstock side, uncertainty exists in terms of price and availability. A forestry company might be obliged to procure its feedstock from different sources over different distances and with different prices. Short product life cycles and increasing competition among companies reveal new uncertainties and risks for different industries. Specialty chemicals impose additional financial risks and uncertainties because customers are granted a very high degree of flexibility in terms of demand quantity. A reveal date is assigned to each product at which the customer specifies his final demand. Some companies even give their customers the right to cancel and withdraw the order at this point [4]. All these clauses entail more uncertainty and risk for the companies. To mitigate risks in the face of such uncertainties, it is of crucial importance to enhance adaptiveness and reactivity on one hand and proactivity on the other [4]. These capabilities are generally called *flexibility*. Based on the type of uncertainty and how it is addressed, there are different types of flexibility, which will be discussed later in this chapter.

An FBR would be exposed to this kind of volatile environment and would face these risks and uncertainties. Hence, flexibility, of any possible type, must be exploited in an FBR to mitigate risks. An FBR will be able to produce several products, including P&P products, bioproducts, and energy. Producing several products implies the opportunity to take advantage of manufacturing flexibility, that is, producing different products at different volumes in different time periods. In a volatile market, depending on feedstock and product prices as well as supply and demand, manufacturing flexibility can be exploited, and the mill can produce different products in different amounts to optimize and secure the company's margin. The company should analyze its access to feedstocks, product prices, and received as well as forecasted demands and find the best alignment between these demands and its production capacity to maximize the company's profit.

9.1.3 Bottom-Up Approach

As mentioned in the definition of the margins-based policy, the ultimate goal of this policy is to maximize profitability across the entire SC. In fact, the margins-based operating policy exploits the manufacturing system's capability for flexible production at the SC operational level to maximize the margins. Therefore, an SC-based analysis is needed to show how the flexibility capability should be managed and exploited at the SC operational level to maximize SC profit. At the design level, flexibility must be designed in a way that ensures the best performance at the operational level and the attainment of the ultimate goal, that is, maximizing the SC profit. Hence, there should be a metric representing SC profitability that can reflect it at the design stage. Moreover, from an SC design perspective, the SC network must be designed so that it enables margins-based operating policy to exploit flexibility. In other words, the SC network should be designed in such a way that it serves the maximum exploitation of flexibility for profit maximization. Therefore, an SC-based analysis is required to address both aspects: operational and design. Hence, the challenge is to develop an SC-based analysis which can be used:

- At the design stage, to reflect SC profitability as a design metric in the flexibility and SC network designs.
- At the tactical-operational level, to improve SC profitability by exploiting flexibility.

Figure 9.2 provides a schematic illustration of the linkage between SC analysis and design and operational decisions.

The bottom-up approach shows the importance of operational-level information for design-related decisions and implies that such information should be brought up to the strategic-level decision-making to obtain a system with greater flexibility in its performance.

To address the linkage between designing a flexible manufacturing system and designing its associated SC, considering the impacts of such decisions on tactical-operational SC activities, this chapter introduces a methodology which involves an SC-based analysis as the focal point. The proposed methodology incorporates SC management into designing a flexible manufacturing system using the margins-based operating policy.

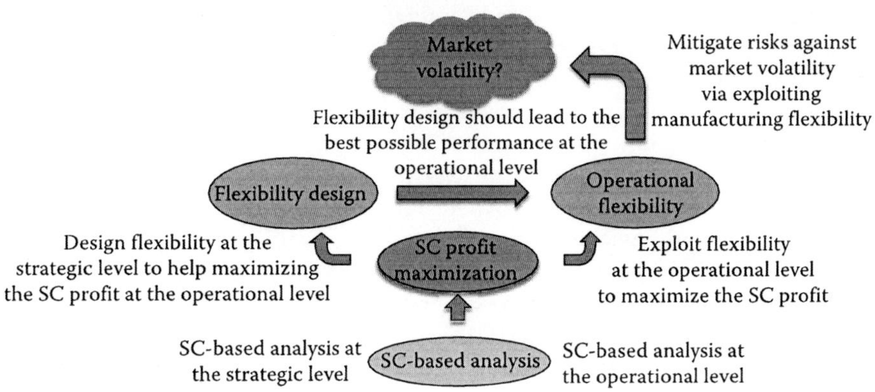

FIGURE 9.2
Linkage between SC-based analysis and design/operational decisions.

The rest of the chapter is organized as follows. First, flexibility as the major focus of this chapter is reviewed. Next, the SC optimization framework and how it is used in the margins-based approach are discussed. Afterward, the proposed methodology for SC-based analysis is presented. Each step of the methodology is described, and the way that SC modeling is used in each step is explained. An illustrative example is presented along with the methodology to highlight the importance of implementing the proposed methodology.

9.2 Biorefinery Process Flexibility: Definition, Importance, Problems, and Types

In this section, different definitions of flexibility and its importance are discussed. Then, the problems identified and studied to date related to flexibility are presented, and different types of flexibility are introduced. Finally, a concrete definition is provided of the concept of flexibility which forms the basis of the methodology presented here.

9.2.1 Definition

One of the earliest definitions of manufacturing flexibility goes back to Ropohl, who considered manufacturing flexibility as "the property of the system elements that are integrally designed and linked to each other in order to allow the adaptation of production equipments to various production tasks" [5, p. 644]. Another early definition of flexibility was proposed by Gupta and Goyal, who defined it as "the ability of a manufacturing system to cope with changing circumstances or instability caused by the environment" [6, p. 199–135]. From an operational point of view, Nagarur defined flexibility as "the ability of the system to quickly adjust to any change in relevant factors like product, process, loads and machine failure" [7, p. 799–809]. Upton provided a more comprehensive definition which addressed flexibility as "the ability to change or react with little penalty in time, effort, cost or performance" [8, p. 72–89]. Sethi and Sethi [9] did a comprehensive survey on the concept, reviewing different definitions and types of manufacturing flexibility. They defined the flexibility of a system as its adaptability to a wide range of possible environments that it may encounter. In other words, a flexible system must be capable of changing to deal with a changing environment [9]. In the chemical engineering context, Grossmann et al. defined flexibility as the ability of a manufacturing system to satisfy specifications and constraints despite variations that may occur in parameter values during operation [10].

From a hierarchical decision-making point of view, flexibility can be classified as long-term (strategic), midterm (tactical), and short-term (operational) flexibility. These levels can be defined respectively as: (i) the ability of a system to respond to changes in strategy, new product introductions, and basic design changes, (ii) the ability to operate at varying rates, to accept random, minor changes, and to convert the plant for alternative uses, and (iii) the ability to reset and readjust between known production tasks to permit a high degree of variation in sequencing and scheduling [11].

Several reasons have been mentioned for the importance of flexibility. Frazelle [12] believed that flexibility is required to maintain competitiveness in a changing business environment of which the critical features are rapidly decreasing product half-life, the influx of competitors, an increasing demand for product changes, and the introduction of new products, materials, and processes. Slack saw the incentives for flexibility in the instability and unpredictability of the manufacturers' operational environment and in developments in production technology [13].

9.2.2 Flexibility Problems

From a broad perspective, flexibility problem areas can be categorized into two groups: flexibility design and flexibility analysis.

9.2.2.1 Flexibility Design

In this type of problem, the design is unknown, and the problem is to find the optimal design of a system considering the costs incurred by that design. A design representing a higher degree of flexibility will have a lower probability of encountering an infeasible operating condition, but at a higher cost. Two major areas have been considered by Grossmann et al. optimal design with a fixed degree of flexibility, and design with an optimal degree of flexibility [10].

9.2.2.1.1 Optimal Design with a Fixed Degree of Flexibility

The flexibility of a design is optimal when the economic advantages of flexibility are balanced in relation to its cost. In this problem, a design should be identified that can operate over varying conditions. These varying conditions must be specified as a bounded range of parameter values over which the design is able to meet the specifications at minimum cost. In this type of problems the required degree of flexibility has already been specified, either by a discrete set of required operating conditions, or by requiring feasibility of operation when a set of uncertain parameters varies between fixed bounds. Therefore, this class of problem can be divided into two categories [10]:

1. Deterministic problems or problems of deterministic multiperiod design, in which the plant is designed to operate optimally under various conditions over a sequence of time periods. The goal is to ensure that the plant will be able to meet the specifications over successive periods of operation.
2. Stochastic problems or problems of design under uncertainty, which address the design of chemical plants under conditions where the values of some of the process parameters have significant uncertainty. However, a particular design problem as presented might include both these problems.

The ultimate goal in solving these types of problems is to ensure that the design, while being economic, meets the specifications under different imposed conditions.

9.2.2.1.2 Design with Optimal Degree of Flexibility

In this type of problem, the desired degree of flexibility is not known, and a design with the optimal degree of flexibility must be identified. The optimal degree of flexibility does not necessarily imply the highest degree of flexibility, because another criterion, which is the cost of the design, is important in determining optimality of a design. In fact, design with optimal degree of flexibility addresses problems which needs establishing a tradeoff between the cost of the plant and its flexibility. Therefore, the objective function can be separated into two components: minimizing capital and operating costs on the one hand,

and maximizing flexibility on the other. The result will be a tradeoff curve which relates flexibility and cost. Hence, the major task in this type of problem is to determine the degree of flexibility. In other words, a metric or a quantitative measure of flexibility in the form of a scalar index is needed that can measure the size of the region of feasible operation for the design. This metric is called the *flexibility index*. Flexibility-index problems involve designing the plant with the aim of both cost minimization and flexibility-measure maximization. Problems in this category have evolved from flexibility-index problems [14] to stochastic flexibility-index problems [15] and expected stochastic flexibility-index problems [16].

9.2.2.2 Flexibility Analysis

In flexibility analysis problems, the design of the plant is given, and the goal is to analyze the plant's capability for feasible operation. Two types of problems can be defined in this category.

9.2.2.2.1 Feasibility or Flexibility Test

In this type of problem, it is determined whether the design can operate feasibly at all uncertain points in the range. More specifically, the objective of the feasibility problem is to determine whether, for a given design, set of nominal values for the uncertain parameters, set of expected deviations in the positive and negative directions, and set of constraints, at least one set of control variables can be chosen during plant operation such that, for every possible realization of the uncertain parameters, all the constraints are satisfied [17]. Halemane and Grossmann carried out one of the earliest studies in this domain and showed how, for a given design and a fixed parameter value, the max–min–max problem provides a measure of the size of the feasible operating region [18]. Grossmann and Floudas presented mathematical formulations for the feasibility test based on the property that the number of active or limiting constraints on flexibility is equal to the number of control variables plus one, provided there is linear independence among the active constraints [19]. Bansal et al. introduced a unified theory and algorithms based on multiparametric programming techniques for the solution of feasibility test problems in linear process systems [20]. Floudas et al. presented an approach for feasibility test problems based on the principles of the αBB deterministic global optimization algorithm, which relies on a difference-of-convex-functions transformation and a branch-and-bound framework [21]. Goyal and Ierapetritou developed an algorithm for evaluating the feasibility of nonconvex processes, based on the idea of systematically determining the infeasible areas using an outer approximation procedure and a simplex approximation approach to approximate the expanded feasible space which can be constructed by the exclusion of nonconvex constraints [22].

9.2.2.2.2 Flexibility Index

The aim of flexibility-index problems is to determine how flexible a given design is. In other words, the maximum deviation that the design parameters can tolerate must be determined. The major issue in these problems is to define a quantitative measure for the degree of flexibility. A scalar metric, called the index of flexibility or the flexibility index, can be developed, for which the value characterizes the size of the region of feasible operation in the uncertain parameter space. In other words, it can be defined as the largest-scale deviation of any of the expected deviations that the design can handle and still operate feasibly [14].

Grossmann and Floudas addressed the analysis of the flexibility of a proposed design using an active constraint strategy and mixed-integer nonlinear programming (MINLP) formulations for flexibility-index problems [19]. Pistikopoulos and Grossmann [23] worked on redesigning existing process flowsheets to increase their flexibility. The major difficulty in such retrofit problems is that of deciding which parameter or structural changes are required, with the aim of increasing flexibility at the least investment cost. Their proposed approach for the retrofit design problem involves: (a) a systematic procedure for handling parametric changes of the design variables, (b) embedding a strategy for handling simultaneous structural and parametric changes, and (c) a procedure for developing tradeoff curves between cost and flexibility [23]. Bansal et al. [20] presented algorithms based on multiparametric programing techniques for the solution of flexibility analysis and design optimization problems in linear process systems which are used to solve flexibility-index problems in systems with deterministic parameters. The algorithms as developed are computationally efficient and reveal explicitly the dependence of various flexibility metrics on the values of the continuous design variables [20].

9.2.3 Flexibility Types

Many efforts have been made to categorize various types of flexibility. The common element in all types of flexibility is that they are used to mitigate the risks associated with different types of uncertainty. These uncertainties are the results of variations in the temperature, pressure, or flowrate of a stream, changes in the state of equipment, or fluctuations in the price and demand of products. Based on the type of uncertainty, specific types of flexibility can be defined. Sethi and Sethi [9] introduced 50 different terms for different types of flexibility, although their definitions were not always precise and, for identical terms, not always in agreement with one another. Swamidass [24] pointed out the difficulties of understanding and therefore categorizing flexibility to be (i) the use of flexibility terms with overlapping scopes, (ii) the use of flexibility terms with different meanings, and (iii) the use of flexibility terms which are aggregates of others. Beach et al. [11] carried out a comprehensive survey on the concept and types of flexibility and concluded that the original eight categories of flexibility defined by Browne et al. [25] represent the most comprehensive classification of flexibility. They classified manufacturing flexibility in a discrete manufacturing environment into eight categories: machine, process, product, routing, volume, expansion, operation, and production.

In the chemical engineering context, four major types of flexibility have been widely studied widely: recipe, product, volume, and process. The definition of each flexibility type is given in Table 9.1.

9.2.3.1 Recipe Flexibility

The flexible recipe concept was originally introduced as a set of adaptable recipes that can control the process output and can be modified to confront any deviation from nominal

TABLE 9.1

Types of Flexibility and Their Definition

Flexibility	Definition
Recipe	The ability to have a set of adaptable recipes that can control the process output
Product	The ability to change over to produce a new (set of) product(s) economically
Volume	The ability to operate a system profitably at different production volumes
Process	Capability of the process to operate feasibly under changing conditions

conditions. Recipes specify products and prescribe how products are to be produced. The nominal recipe for a given product represents the optimal compromise between quality and costs. According to the production scenario, recipes can be changed or modified. Verwater-Lukszo [26] developed this basic idea and introduced the concept of the flexible recipe as a way of systematically adjusting control recipes during the execution of production tasks with the aim to enable the process to perform under different operating conditions. These changing operating conditions may include different feedstock properties, changes in quality specifications, variations in process behavior, new market conditions, other real-world experiences with the process, and so on, none of which is reflected in the recipes, although it would often be profitable to be able to adapt them to the changed conditions.

One of the first attempts to do so was made by Romero et al. [27], who extended the flexible recipe approach to a plant-wide scheduling problem. Another study was carried out by Ferrer-Nadal et al. [28], who aimed to optimize production scheduling in a batch plant where flexible recipes were used. They integrated a linear flexible recipe model into a multipurpose batch-process scheduling formulation, which in turn, enabled integration between a recipe optimization procedure at the control level and a batch-plant optimization strategy. Laflamme-Mayer developed an SC planning model that exploits the capability of a market pulp mill to use different recipes in a flexible manner to provide adequate support for cost-effective fiber supply use [29].

9.2.3.2 Product/Volume Flexibility

Product flexibility, according to Browne et al. [25], is the ability to change over to produce a new product economically and quickly. This definition is consistent with the concept introduced by Sahinidis and Grossmann [30] and referred to as *flexible production,* which addresses the capability of a manufacturing system to produce different products at different times (different production modes). This type of flexibility is generally used in conjunction with volume flexibility, which is the capability of a facility to operate at different production rates. Examples of such flexible facilities include pulp and paper mills which can produce different grades of pulp and paper, or refineries that process different types of crude oil at different volumes [31]. According to Sahinidis and Grossmann [30], a flexible process network consists of dedicated and flexible production facilities that can be interconnected in an arbitrary manner. Dedicated production facilities manufacture fixed amounts of a set of high-volume products at all times, while flexible production facilities, which are normally used for producing low-volume products, manufacture different products at different times.

One of the first studies in this context was carried out by Sahinidis and Grossmann [30]. They addressed a network of existing and potential processes and chemicals. The processes can be dedicated or flexible, continuous or batch. Given a forecast of prices and demands, as well as investment and operating costs over a specific time horizon, the objective is to determine capacity expansion and shutdown policy for existing processes, selection of new processes and their capacity expansion policy, production profiles, and sales and purchases of chemicals at each time period. The objective function is the net present value which must be maximized. This work was continued by Norton and Grossmann, who considered flexibility in raw materials [32]. In this study, processes with potential flexibility on either the feedstock or the product side, as well as processes with flexibility on both sides, are considered. These two studies were dedicated to long-term planning problems. In a more recent study, Bok et al. addressed detailed operational decisions in continuous flexible process networks. The model presented in this study extends previous models by incorporating an inventory profile, changeover costs, intermittent supplies, and production shortfalls [31]. As mentioned earlier, this approach is widely used in refineries and the petrochemical industry. Petrochemical complexes are able to produce several products by means of processes which can operate over a range of production rates. Neiro and Pinto [33] and Schulz et al. [34] described SC planning in petrochemical complexes which use this strategy. Méndez et al. [35] explained the scheduling of oil-refinery operations, in which continuous processes produce a set of components at constant flowrates and then a blending process is used to transform these components into different derivatives in varying amounts.

9.2.3.3 Process Flexibility

In the chemical engineering context, process flexibility has gained the most attention. From a general point of view, process flexibility is a property of *process operability*. Grossmann et al. break down operability into a set of properties such as flexibility, controllability, reliability, and safety [10]. Flexibility is concerned with the problem of ensuring feasible operation of a plant over a whole range of conditions in both steady-state and dynamic environments, while controllability signifies the ability of a plant to move efficiently from one operating point to another as well as to deal efficiently with disturbances [36]. Reliability denotes the capability of the process to withstand mechanical and electrical failures, and safety is the prevention of major hazards given possible failures.

Grossmann et al. mentioned the need for accounting operability considerations, mainly related to flexibility and controllability, at the design stage [10]. Blanco and Bandoni named three major approaches to the design-for-operability problem [37]:

1. *Heuristics:* Heuristics rely on rules of thumb. Such recipes can be found in Douglas's famous book on conceptual design [38].
2. *Operability Measures:* Operability measures have been widely used in both open-loop and closed-loop controllability. They describe specific operability features and are used to screen or classify different designs with respect to a particular operability issue. Controllability and resiliency indices such as relative gain array (RGA) and singular value decomposition (SVD) are examples of these indices [37].
3. *Complete Integration:* This approach implies the integration between process design and process operability by including operability elements within the process design formulation. This approach takes advantage of multiobjective optimization and can be seen in the works of Grossmann and Pistikopoulos [14]. Heat exchanger networks have been used as a classical example in such studies. Hot and cold process streams are considered as uncertain parameters and, given the nominal values of the temperatures and the flowrates and assuming expected deviations of the temperatures, for example, ±10 K, the goal is to determine whether the network can tolerate changes in inlet temperatures over the specified range [19]. Pistikopoulos and Grossmann [23] addressed a stochastic flexibility problem in which the major issue is to determine the appropriate tradeoff between the investment cost for the retrofit design of a system and the expected revenue that will result from having increased flexibility. For this purpose, a number of redesign alternatives with specified degrees of flexibility were obtained from a tradeoff

curve which related retrofit cost to flexibility. Then, for these designs, the corresponding expected optimal revenue was evaluated using a modified Cartesian integration method. Pistikopoulos and Grossmann [39] extended this work for nonlinear models.

9.2.3.4 Manufacturing Flexibility in the FBR

The concept of manufacturing flexibility in the FBR implies the ability to produce several bioproducts at different volumes, that is, different production rates, in different time periods based on product price and demand. From an economic-market perspective, this type of manufacturing flexibility implies a justifiable increase in capital cost that is adequately compensated by the ability of the process to manufacture in a flexible manner so that the expected volatility in market conditions can be mitigated. The proposed definition seems to be an aggregation of product flexibility and volume flexibility. Because process flexibility is inherent in the design of each chemical process, this definition has already included process flexibility. Finally, experts believe that feedstock flexibility is a promising element in the success of the FBR. BR processes, especially thermochemical processes, can accept a wide range of feedstocks. This makes it possible to keep operations running with different types of feedstock and to have the flexibility of procuring feedstock from different sources. It will also be a competitive advantage for the company in the volatile feedstock market, where it must deal with several considerations such as feedstock price, competition from other businesses, sufficient availability, handling, proximity, seasonality, and collection. Therefore, feedstock flexibility is another dimension that must be addressed in the definition. Hence, the definition of manufacturing flexibility in the FBR can be interpreted as the aggregation of feedstock, process, product, and volume flexibility, with volume and product flexibility as the major dimensions.

The FBR presents a promising opportunity to implement manufacturing flexibility as defined here. The FBR processes are retrofit to P&P mills which are in place with a known level of flexibility. The FBR and P&P mills might be integrated in terms of feedstock, chemicals, and energy. Hence, the P&P process flexibility can be characterized first, and then the BR process and its flexibility can be designed based on the flexibility of the P&P side, as well as product price and demand. This will provide the opportunity to produce both P&P products and bioproducts, which will improve P&P companies' business model and might prevent current mill closures [40]. This strategy was used successfully in previous decades, when forestry companies were producing ethanol as a side product as well as forestry products as main products. In today's market, where the demand for P&P products is decreasing for several reasons such as the presence of global low-cost producers, this strategy can be pursued in another way, by shifting the core business from a forest-products producer to a bioproducts producer that also produces forestry products.

In the FBR context, there are two strategies with regard to the products that can be produced: large-scale commodity production, and commodity/specialty or low-value/high-value commodity production. The commodity chemicals considered in the first strategy are mainly limited to ethanol and butanol because these can be used as fuel. The idea that supports this strategy is that there is a huge market for such commodities in the fuel market, especially in the United States. However, another strategy has been gaining attention: production of specialty chemicals or a combination of small-volume value-added products and large-volume low-margin products [41–43]. Fine and specialty chemicals are said to be promising elements of an FBR product portfolio because they have bigger margins than P&P products a better market with less competition, so that the FBR does not have to compete with huge well-established commodity petrochemical products. Because fine and specialty products are produced in smaller volumes, they need less feedstock. This is a competitive advantage for a production environment like a biorefinery for which procuring biomass as feedstock is a great challenge.

A study has been done by the National Renewable Energy Laboratory (NREL) [43] on the analysis of biorefineries, in which the importance of coproducing high-margin low-volume products along with a primary product is addressed. The advantages of such a product portfolio compared to dedicated production of a single product can be classified into two levels: long-term and short-term. The long-term advantages are summarized as follows:

- Product diversity mitigates risks associated with seasonal demand cycles and market downturns.
- If selected coproducts have the potential to become platform intermediates in future, their commoditization will be fostered by taking advantage of the economies of scale provided by producing small amounts of the coproduct in a commodity-producing biorefinery. An ethanol plant producing succinic acid (SA), lactic acid (LA), and/or butanol as coproducts is an example. As mentioned in the NREL report, early-generation ethanol biorefineries can serve as incubators for chemicals that can then become high-volume products in their own right.

The short-term advantages are as follows:

- Revenues from high-value coproducts reduce the selling price of the primary product.
- The economies of scale provided by a full-size biomass refinery lower the processing costs of low-volume, high-value coproducts.
- Less fractional market displacement is required for cost-effective production of high-value coproducts as a result of the economies of scale provided by the primary product.
- Biomass refineries maximize the value generated from heterogeneous feedstock, making use of component fractions.
- Common process elements are involved in producing fermentable carbohydrates, regardless of whether one or more products are produced.
- Coproduction can provide process integration benefits (e.g., meeting process energy requirements with electricity and steam cogenerated from process residues).

For such a plant to be competitive with other commodity producers, a very important point in this strategy is to be able to produce such commodities in large volumes. Commodity buyers prefer to buy the products they need from a single seller so that they can negotiate with the product provider on price and so that they can avoid buying from several sellers to reduce their transportation costs. Therefore, it is important for a producer to be able to produce commodity products in large volumes so that it can respond to large customers. This underlines the importance of feedstock procurement for the company.

9.2.3.5 Process Considerations for a Flexible Biorefinery Option

From a generic point of view, chemical processes can be classified into two major categories, continuous and batch. Continuous processes are appropriate for large-scale production lines because they are designed to operate 24 h a day, seven days a week over the whole year at almost constant conditions. The plant is shut down only for maintenance or in emergency situations. On the other hand, batch processes are designed to be started and stopped frequently because during operation, units are filled with materials, perform their function, and are then stopped, drained, and cleaned to be ready for another cycle [38].

There are certain guidelines for when a batch process may be chosen over a continuous process [38]:

- *Production rate or capacity:* Plants with capacity of greater than 10×10^6 lb/year are usually continuous, while plants having a capacity less than that are normally batch.
- *Market forces:* Batch processes are more flexible in terms of both throughput and number of products. In fact, a large number of products can be produced in the same production line. As a result, batch processes are used to manufacture products that have seasonal demand.
- *Operational problems:* Some processes involve slow reactions, slurries with settling solids, or materials that foul the equipment rapidly. In these cases, batch processing is an ideal option, because it gives the process the time it needs, and the processing units can be stopped and cleaned after any operation involving settling or fouling materials.

On the basis of these characteristics of batch and continuous processes, if the desired strategy is to produce commodity chemicals, then a continuous process should be chosen because the production volume must be large enough to enable the company to enter the commodity market. However, it must be remembered that continuous processes have limited potential for flexibility because they are dedicated to one or a few products and their throughput is to some extent fixed. Moreover, commodity production does not have large margins needs huge amounts of biomass as feedstock.

If the desired strategy is to produce value-added products, batch processes should be chosen. Fine and specialty chemicals are produced in batch systems. Being able to respond appropriately to market changes is of great importance in today's market, and as explained earlier, flexible batch processes enable companies to produce different products in different volumes at different time periods. Therefore, based on product price and demand, production can be scheduled in such a way that in every time period, be it a week, a month, or a season, the most profitable products are produced in the right volume. Batch processes will become more important as the dedicated nature of continuous processes is addressed. Continuous processes are generally designed to produce one specific product. Therefore, even if the production rate of a continuous process is decreased to produce only the amount needed to fulfill the profitable orders, the excess capacity cannot be used to produce another product. This will have an enormous impact on the rate on return on investment (ROI). However, a batch system can be changed over to produce a more profitable product, and the whole capacity of the system can always be used.

9.3 SC Optimization with Margins-Based Operating Policy

As discussed in the introductory section, flexibility should be exploited at the operational level to maximize SC profit. Because the margins-based approach is to be used as the operating policy of the biorefinery production facility, flexibility must be exploited using the margins-based approach to maximize SC profit. Hence, an SC-based analysis must be used to reflect this approach at the SC tactical operational levels. On the other hand, according to the bottom-up approach, SC profitability must be reflected as a design metric at the strategic decision-making level, and therefore, once again, the necessity of an SC-based analysis becomes apparent. Such an analysis will use an SC optimization framework which will consider feedstock price and availability, production costs, and inventory and delivery costs, as well as product price and demand. Taking this information and these constraints into account, the SC optimization framework will exploit the potential for production flexibility and determine which orders must be fulfilled and therefore how much of which products must be produced, how they should be stored, and how they should be delivered to the market to maximize SC profit.

The focus of this chapter is on the role of SC analysis in the design problem. Therefore, in the next section, the SC framework will be explained, and in the following section, its function in a design problem will be discussed.

9.3.1 SC Framework

The SC framework aims to maximize profitability across the entire SC by first identifying the tradeoffs between demand and production capabilities and then by finding the optimal alignment of manufacturing capacity and market demand. The SC framework is formulated as an optimization problem with the objective of maximizing profit. This framework considers the management of a multiproduct, multiechelon SC, including existing production, warehousing, and distribution facilities as well as a number of customer zones, although it can also be used for design purposes, as will be discussed in Section 9.4. Production facilities can make one or several products. Warehouses can receive material, either feedstock or product, from different sources and plants and supply different distribution centers, while distribution centers can supply different markets. Each market places demand in two ways: by contract, that is, for the long term, and in the spot market, that is, for the short term. The optimization problem is formulated into a mixed-integer linear programming (MILP). Like every optimization problem, the SC framework consists of an objective function, decision variables, constraints, and parameters. In the following parts of this chapter, these components of the SC framework are explained.

9.3.1.1 Decision Variables

There are two types of decision variables in the mathematical formulation of the SC. The first type is continuous variables, which represent variables that can take on continuous values. They describe the flow of material between SC nodes, for example, the flow of feedstock from suppliers to the mill, production rates for each product showing the amount of each product to be produced, rates of product flow from plants to warehouses, to distribution centers, and to markets, and inventory levels for each type of feedstock and product. In a design problem, where the capacity of the facilities, for example, the production capacity, is unknown, a decision variable is assigned to the unknown capacities. Therefore, based on the goal of the design problem, the capacities of plants, warehouses, and distribution centers can be represented by continuous decision variables [44].

The second type of decision variable is binary variables which imply a "yes/no" type of decision, for example, which order should be taken, which product must be produced, which production line must operate, which warehouse should supply which distribution center, and which distribution center should supply which market. In a strategic design problem, decisions such as which product to produce and which market to supply for the long term, which location to choose for a facility, and which partner to incorporate with, can be represented by binary variables [44]. However, because the number of binary design variables is usually small, such decisions can also be made using a scenario-based approach, that is, generating a scenario for each design option. This approach is used in the methodology presented in this chapter and will be discussed in more detail later.

9.3.1.2 Constraints

Each node of the SC, such as suppliers, inventories, and manufacturing centers, has its own constraints which must be formulated mathematically. They can be classified as follows [44]:

- Network structure constraints determine which nodes of the SC can be linked to each other and whether material can flow between them.
- Material balance constraints relate the flow of material into and out of different SC nodes to the accumulation, production, and consumption of material in that node.
- Capacity constraints represent the minimum or maximum amount of material that can be produced or stored in a node. Production capacity is generally modeled as a linear constraint which relates the production rate of a product to the availability of feedstock. The capacities of warehouses and distribution centers are described by upper and lower bounds on their material-handling capability.
- Nonnegativity constraints ensure that all variables are greater than zero.

9.3.1.3 Objective Function

The objective function can be defined in two ways:

- *Operating costs:* Operating costs include the costs incurred in feedstock procurement and material production at plants, as well as transition and shutdown costs, material handling at warehouses and distribution centers, and transportation of material throughout the entire SC. These costs are calculated on a daily basis. If the objective function were calculated in terms of the operating cost, the goal would be to minimize the objective function. In SC design problems, the objective function can be defined as the sum of operating costs and capital costs. The capital costs are the costs associated with the establishment of the SC infrastructure, calculated on a long-term basis.
- *Profit:* Profit is the sum of revenues from different main products and byproducts minus the operating costs. In this case, the goal is to maximize the objective function. In SC design problems, the objective can also be defined in terms of SC profitability, which considers the annual revenue and operating costs as well as the

capital cost associated with the SC network. Different metrics, for example, ROI and internal rate of return (IRR), can be used to estimate the long-term profitability of the SC.

9.3.1.4 Parameters

In an optimization framework, constant values are represented by parameters. They can be classified into the following categories:

- *Cost data:* Production costs, transportation costs, material handling costs, and transition costs are among the cost data used at the tactical and operational levels. The cost of establishing different facilities and processes is used, as well as the costs just mentioned, at the strategic level.
- *Capacity data:* Production capacity or production rate, warehouse capacity, and transportation rate are important capacity data used in an optimization framework. Some of these data represent the minimum and maximum capacities of various facilities, for example, minimum and maximum production rate or minimum and maximum warehouse capacity.
- *Production-related data:* These data represent the efficiency of production lines, the quantity of materials consumed or produced, and conversion factors.
- *Price data:* These data represent the prices of feedstocks and materials as well as the prices of products in the optimization framework.
- *Demand data:* These data indicate the demand for each product in the market.

9.3.2 Executing the Margins-Based Policy

As mentioned earlier in this chapter, what drives the margins-based policy is the ultimate profitability of the entire SC. All SC activities must be executed with respect to this policy. In the forestry industry, especially in the P&P sector, some SC practices are contrary to this approach. One of the most important of these practices is treating production cost as the major driver in decision-making. In this way, operating cost is generally used as the objective function, and the costs incurred by other nodes of the SC are basically neglected. Therefore, the first point to be made in an SC optimization framework with a margins-based policy is that profit must be used as the objective function.

Another common practice in the forestry industry is that products are produced in fixed known orders and sequences and the capability of the process for flexibility in manufacturing and changeovers is not used. This can result in decreasing the profitability of the company when prices or demands change. Suppose that, based on the established sequences, the company has produced some products that in a particular period, are subject to low price or weak demand. In the case of weak demand, the company should store its products for a longer period, in which case the inventory cost rises. In such a case, the company might sell its products at a discount, which would decrease the profit. Moreover, some companies take orders based on their sequences, and they miss out on better orders just because these orders do not fit their production sequence. Hence, another point that must be respected in an SC optimization framework with a margins-based policy is to let the framework choose the best orders and to take advantage of the mill's capability of flexibility and changeover, leaving aside traditional recipes and practices.

9.4 Strategic SC Design

In the strategic design of an SC, long-term decisions should be made. Such decisions include the type of products that should be produced, the technologies that should be used, the number, location and capacity of each type of facility, for example, plants, warehouses and distribution centers, and the target markets. In a practical problem, it is difficult to address all these decision variables within a single SC optimization framework. Instead, it is preferable to pursue a systematic hierarchical methodology that addresses all these factors in a stepwise manner.

On account of the combinatorial aspect of such design problems, the hierarchical methodology might miss the global optimum. However, the methodology presented in this chapter does not seek to identify a global optimum. Rather, it seeks a set of feasible and practical biorefinery options (near-optimal solutions) that a company can strategically pursue. The decision as to what biorefinery strategy to take depends on many factors, most of which cannot be reflected in an optimization problem, for example, understanding the market and market strategies, emerging products, processes, and technologies, the capabilities of existing SC assets, and potential partners. Many of these aspects can be addressed in different scenarios instead of being modeled into an optimization formulation. In this way, a simpler model will be solved, with more practical and realistic results.

Companies seek a set of biorefinery options that would significantly improve their business model. This should include the optimum and near-optimum solutions. This set of possible strategies should be pursued by a company in parallel with potential partners to establish mutual interests and to address most effectively the competitive disadvantages of forestry companies, such as lack of capital. This methodology would end up with a set of solutions. A multicriteria decision-making framework can be used to find the best option from a specific company's point of view, considering all the complexities involved in the industrial arena.

To achieve a stepwise methodology, some of these decisions must be made by integration with other methodologies. For instance, the set of products that should be produced can be determined by a product portfolio definition and selection methodology. The processes and technologies that should be used to produce the targeted products can be chosen through a technoeconomic study. The aspects that will be determined by the hierarchical methodology include:

1. Flexibility design including the determination of the production capacity as well as the operating window as a design target, that is, a range of production rates for each process, showing the flexibility capability of the plant and designing the production lines so that they can operate over the targeted range.
2. SC network design, including determination of the number of facilities of each type, the location of each facility, and the capacity of warehouses and distribution centers, as well as partner selection. Note that network design-related decisions will be made through the generation of alternatives. These alternatives must be generated based on practical aspects of the problem that can be addressed in discussion with company experts and considering all features of the existing SC.

The methodology is illustrated in Figure 9.3. In the next section, the methodology will be explained using a hypothetical example. First, process design alternatives representing

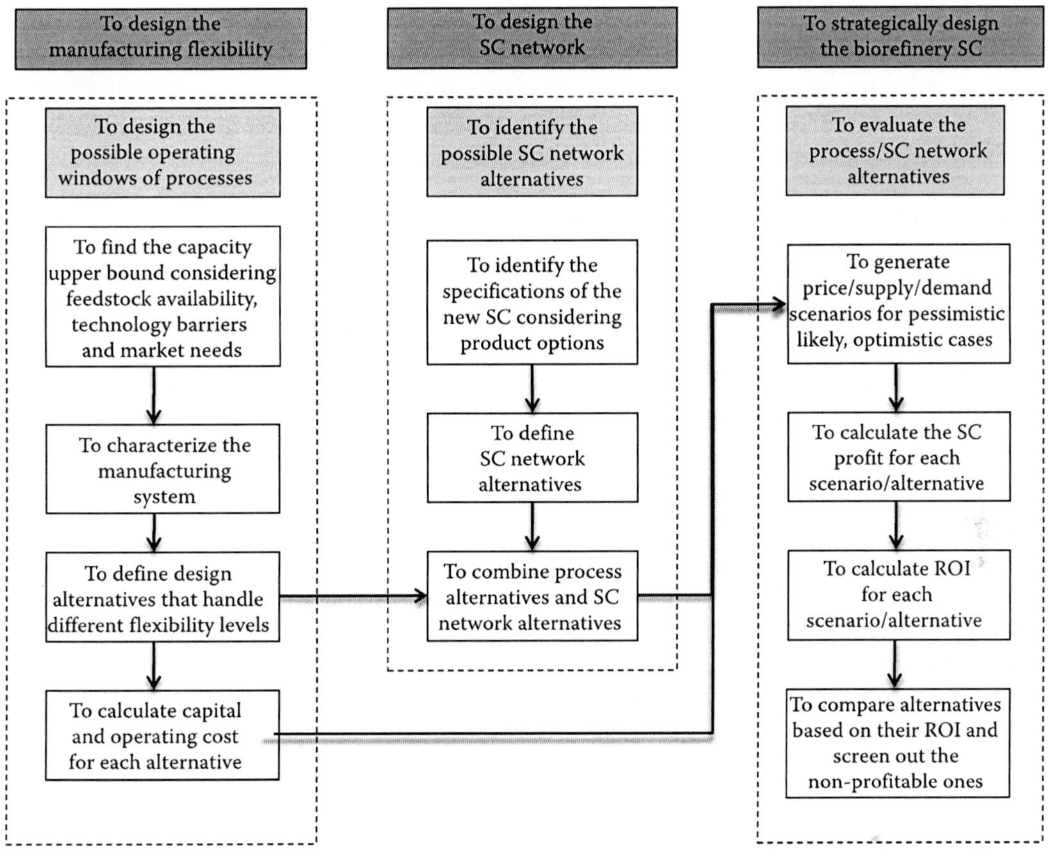

FIGURE 9.3
Hierarchical methodology for SC strategic design.

different levels of flexibility (volume flexibility) are defined. In the second step, which is independent of the first step, SC network alternatives are defined based on the assets of the existing SC and resources that are needed for new products. Then the process alternatives and the SC network alternatives are combined to create a set of process-SC network alternatives. Finally, using the SC optimization framework, the SC profitability of each combined alternative is calculated for different market scenarios. Thus, for each combined alternative, which represents a specific biorefinery with an implementation strategy, a set of SC profitability values for each market scenario is calculated. These profitability values will be used by experts to evaluate the various biorefinery options.

9.4.1 Design for Flexibility

As discussed previously, product and volume flexibility are the most important dimensions of manufacturing flexibility in the FBR. Because the products that will be produced in the FBR plant are selected using the product portfolio definition methodology, the focus of this methodology is on designing for volume flexibility. Volume flexibility has two aspects: range of variations and speed of response, with the former being useful in the long term and the latter in the short term [45]. Therefore, at the strategic design level, the production capacity and the range of production rates, that is, the operating window must be designed for the long term to serve the short-term tactical and operational activities of the SC with the ultimate goal of SC profit maximization. At the short-term tactical-operational level, the designed flexibility should be exploited in a way that ensures that this ultimate goal will be reached. Therefore, SC-based analysis can be used to establish the range of production rates and to target an operating window which, when exploited against market volatility, would ultimately maximize SC profit.

In this methodology, establishing the design target for manufacturing flexibility is viewed through the SC operating philosophy. In other words, the range of production rates for each product is designed based on the SC profitability achieved by that design. Thus, what determines the design of manufacturing flexibility is the SC profitability resulting from the design.

Designing flexibility through SC optimization has not gained much attention. The holistic approach to designing and analyzing flexibility is to examine the tradeoff between the flexibility index and the cost of having flexibility. This cost includes either the cost of modifications needed for retrofit design or the cost associated with higher flexibility for a greenfield design. In the approach presented in this work, the cost is extended to the cost incurred by the activities performed over the entire SC. This implies taking into account not only the capital cost, but also all the SC operating costs incurred by designing for flexibility. In this way, SC cost and profitability are reflected at the design stage.

In the first part of the methodology, that is, designing for flexibility, there are four steps: determining the upper bound for production capacity, characterizing the manufacturing system in terms of product and volume flexibility to recognize the modifications needed for the processes to become more flexible, generating design alternatives that can handle different production levels, and calculating capital investment and operating cost for each design alternative.

Illustrative Example To make the methodology more concrete, a hypothetical example is presented as an illustration. In this example, it is assumed that a P&P mill aims to implement FBR by producing bioproducts. Three product/process portfolios are considered and are shown in Table 9.2.

In the first portfolio, Fischer–Tropsch liquids (FTL) are produced by biomass gasification and a generic gas-to-liquid process, the products of which are separated into waxes and diesel. Finally, diesel is converted into jet fuel (JF). The second portfolio involves a series of fermenters. The majority of these are dedicated to butanol (BuOH) production, while the rest are used to produce SA and LA. All three products are produced in similar fermenters, but each needs a specific recovery system. The third portfolio involves the family of four-carbon acids, that is, SA, malic acid (MA), and fumaric acid (FA). They all can be produced and recovered on similar production lines.

TABLE 9.2

Price Change Scenarios

Product Portfolio No. 1	Product Portfolio No. 2	Product Portfolio No. 3
Fischer-Tropsch liquids (FTL)	Butanol (BuOH)	Succinic acid (SA)
Waxes and diesel	Succinic acid (SA)	Malic acid (MA)
Jet fuel (JF)	Lactic acid (LA)	Fumaric (FA)

9.4.1.1 Determining the Capacity Upper Bound

To determine the operating window of each process, two steps are required: determining the maximum capacity, and determining the turn down ratio. The maximum possible capacities for each process are identified by considering three major factors: market demand, feedstock availability, and technological or technical barriers. On the other hand, to determine process turndown ratio, process design considerations must be addressed.

To determine the capacity upper bound, a basic technoeconomic study is carried out along with a simple market study. This combination considers market demand, feedstock availability, and technical barriers. After performing a market analysis to determine the market size and market share of the targeted products based on the available amount of feedstock, available technologies, and the possible production rates from a technical point of view as provided by the technology providers, as well as P&P process constraints and the integration strategy with biorefinery processes, a number of maximum-capacity options are identified. The term *possible production rate* refers to the maximum plant size that is technically feasible. This size is basically determined by the maximum size of one or more pieces of equipment in the process. This maximum size is often fixed by technical barriers, for example, a piece of equipment cannot be built larger than a specific size, as well as restrictions on shipping the equipment to the plant site, for example, a piece equipment larger than a specific size cannot be shipped on a railroad flatcar or truck [38].

In the case of a biorefinery implementation, feedstock availability is the most important factor in calculating the capacity upper bound. After investigating market size and market share of producers through the market study and identifying the maximum-capacity options, the availability of feedstock from different sources in and around the mill region is studied, and the cost of bringing the feedstock to the mill is estimated. Various factors should be taken into consideration in calculating the amount of available feedstock, for example, price, proximity, seasonality, and transportation. Based on the results of feedstock availability and market demand studies, a number of the maximum-capacity options that were identified at the beginning of this step given the available technologies are chosen for further investigation.

Illustrative Example Figure 9.4 shows this step of the methodology for FTL. There are two maximum-capacity options identified for this product: 2000 and 3000 t/d. Figure 9.4 shows the capital, operating, and transportation costs associated with these two options. It is

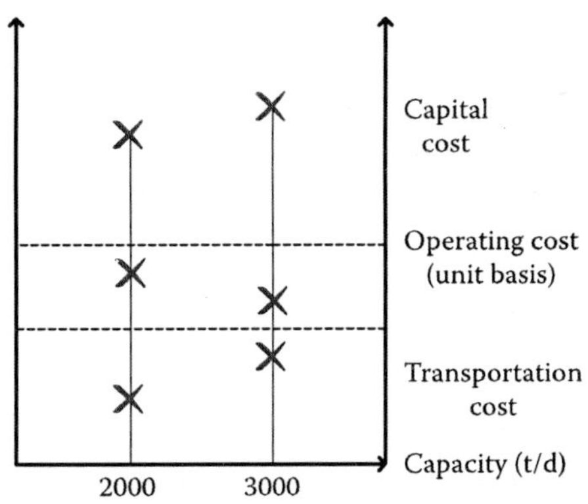

FIGURE 9.4
Capital, operating, and transportation costs for two technology options identified for FTL.

apparent that the larger the capacity, the higher are the capital costs and the transportation costs, but the lower are the operating costs, because operating costs are calculated on a unit basis.

9.4.1.2 Characterizing the Manufacturing System

It was previously stated in this chapter that manufacturing flexibility in the FBR involves four dimensions: feedstock flexibility, process flexibility, product flexibility, and volume flexibility. Feedstock and product flexibility are process-dependent attributes which are inherent in a process. In other words, a process must have the potential to accept different types of feedstock or to produce different products to be flexible. There are some processes that, by their nature, can accept only a specific type of feedstock or produce only one type of product and therefore cannot be designed to be flexible in these respects. As for product flexibility, Grossmann et al. in several studies [30–32] divided production systems into dedicated and flexible systems. Dedicated production systems operate in one mode and are used to produce one product, while flexible production systems can produce different products in different modes. Hence, processes can be either dedicated or flexible by their nature, and if a process does not have the potential for these two flexibilities, it cannot be designed to be flexible. However, process and volume flexibilities are attributes that are not process-dependent and can be designed into every process. Design for process and volume flexibility is basically part of early-stage design. Every process is designed in a way that can operate feasibly under changing conditions and over a range of production rates. The methodology presented in this chapter focuses on the volume-flexibility design for an FBR to enable it to be profitable under volatile market conditions. Note that the methodology does not deal with early-stage design, but considers the existing processes with their inherent flexibility and will retrofit the design in case the inherent flexibility is not sufficient to handle market volatility.

To design a flexible production system, this system must be characterized based on the following aspects:

- *Process configuration:* It should be verified whether the products can be produced in series, that is, they are in one product family, such as diesel and JF (Figure 9.5a), or whether they should be produced in parallel lines because they are not from one family, for example, butanol and LA (Figure 9.5b).
- *Product flexibility:* It should be verified whether the system must be dedicated in terms of products, that is, with each production line producing a specific product (Figures 9.5a and b), or whether several products can be produced in a single line, for example if a line is able to produce more than one product in different production modes (Figure 9.6). An example of such products is SA, MA, and FA. Using different reactor inputs, they can be produced in different modes of one batch system.
- *Volume flexibility:* It should be verified whether the process can handle a range of production rates. In other words, it should be determined whether the inherent flexibility of the process is enough or whether it must be made more flexible.

To clarify how a manufacturing system can be characterized according to these aspects, the system presented by Yun et al. [41] will be used as a reference. They presented a biorefinery system which produces ethanol, LA, itaconic acid, and citric acid. First, because

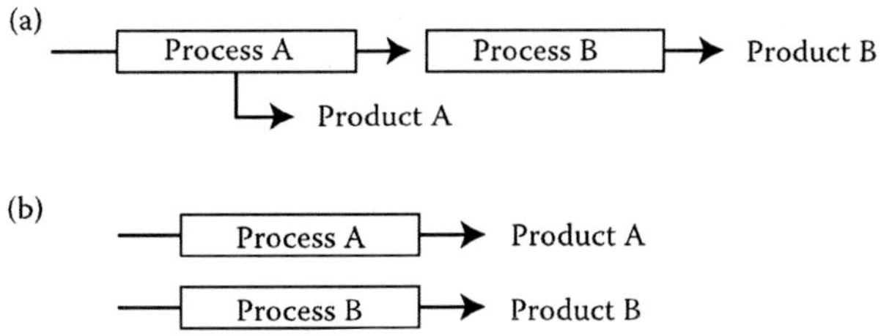

FIGURE 9.5
Separate production lines: (a) in series, (b) in parallel.

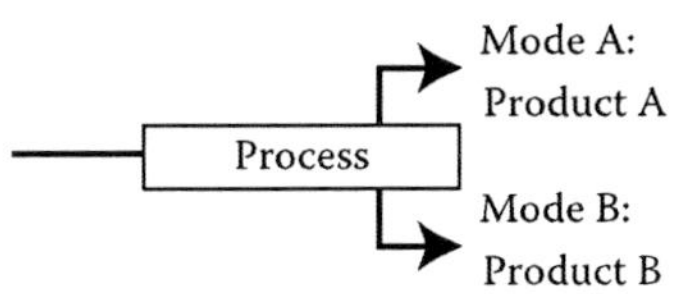

FIGURE 9.6
Flexible production line as defined by Sahinidis and Grossmann (1991).

these products do not belong to one product family, they cannot be produced in series, and separate production lines are needed. Second, all the acids can be produced in one line because the batch reactor is able to produce all of them in different modes based on the type of enzyme used in the process. A single recovery system can also be used to separate all three products from their coproducts. Therefore, only two parallel lines are needed, one for ethanol production and one for acid production. The third point is that the system can produce ethanol and lactic, itaconic, and citric acid in the ranges of 630–2100, 872–1090, 768–1100, and 0–384 kg/day, respectively. These production ranges can be used under different market conditions and, based on product price and market demand, products can be produced in different amounts. The process flow diagram of this system is shown in Figure 9.7.

Illustrative Example The manufacturing processes in all three portfolios are characterized based on their defined characteristics as shown in Table 9.3. This characterization helps to define design alternatives representing different levels of flexibility in the next step.

9.4.1.3 Defining Design Alternatives with Different Flexibility Levels

Chemical processes are designed to operate at maximum capacity, which is generally called *nominal production rate*. Under changing conditions, whether for process-related or market-related reasons, the operating rate must be reduced to some extent. The distance (as a percentage of nominal rate) between the lowest point below the nominal production rate at which the process can efficiently operate and the designed nominal production rate is called the *turndown ratio*. The turndown ratio is a key concept in volume flexibility because it measures how flexible a process is in terms of throughput with respect to changing conditions. If these changes are in market price and demand, this process capability is of crucial importance. In a volatile market, it is desirable to produce products that are more

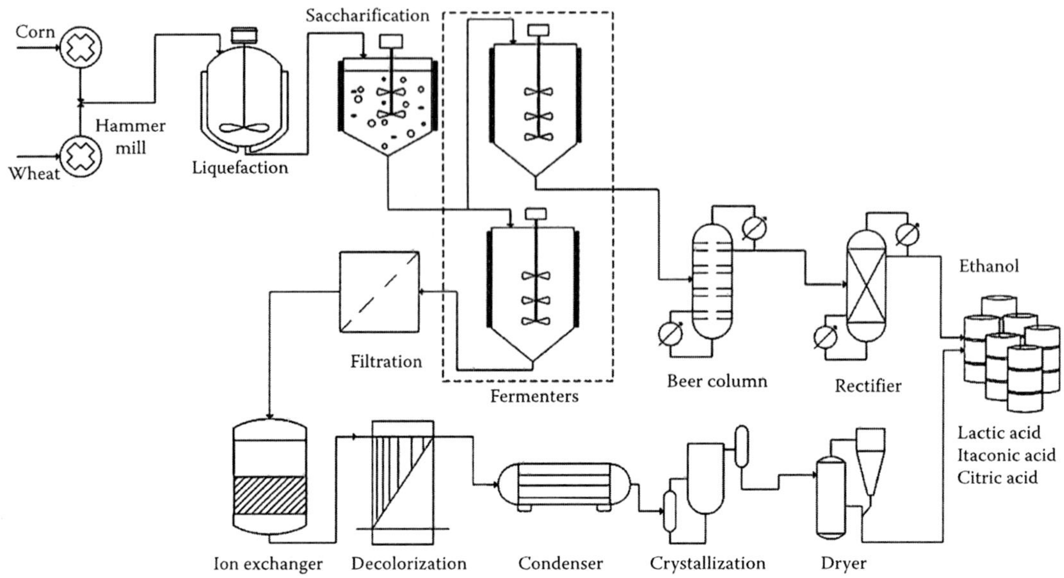

FIGURE 9.7
Process flow diagram presented by Sunwon Park, 2009. Optimal procurement and operational planning for risk management of an integrated biorefinery process. *Chemical Engineering Research and Design* 87, 1184–1190. With permission.

TABLE 9.3
Process Characteristics for Each Product/Process Portfolio

Portfolio	Characteristics
FTL to waxes and diesel + diesel to JF	Type of process: Continuous
	Process configuration: Lines in series
	Product flexibility: Each line produces only one product
	Volume flexibility: Each process has 10% turndown ratio
BuOH	Type of process: Batch
SA	Process configuration: Several lines in parallel
LA	Product flexibility: All products can be produced in similar fermenters, but in different modes. They need specific recovery systems
	Volume flexibility: Each process has 10% turndown ratio
SA	Type of process: Batch
MA	Process configuration: One line or several lines in parallel
FA	Product flexibility: All products can be produced in similar fermenters
	Volume flexibility: Each process has 10% turndown ratio

profitable. Products are sold based on either long-term contractual demand or spot demand. Therefore, when producing a particular product is not profitable or is less profitable than producing another product based on the spot price, the plant should produce the more profitable product as much as it can while fulfilling contractual demands. If this amount is less than the amount that the plant can produce according to its nominal production rate, the production rate should be reduced. It would be desirable to use the extra capacity to produce another product, as is possible in batch processes if the production is scheduled properly. Sometimes the process has been designed in a way that it has the desired turndown ratio, that is, the turndown ratio is inherent in the design of the process. But if the inherent turndown ratio of the process is not sufficient, the process must be redesigned in way that can handle the required ratio. The goal of this section is to find the potential flexibility levels that can be achieved by doing retrofit design, and then to define design alternatives that can handle the defined flexibility levels.

Generally, there is one piece of equipment that limits the maximum and minimum capacity of a plant. Different pieces of equipment have flexibility in terms of their throughput and can operate within a range of operating rates. The maximum and minimum capacity of each piece of equipment can be calculated or be provided by the manufacturer. The minimum capacity of the piece of equipment which has the lowest minimum capacity will determine the maximum inherent turndown ratio of the entire process.

Sometimes it is desirable to have a greater turndown ratio than the inherent turndown ratio of the process. A simple way is to divide the production line, or the part of it that restricts the plant to a smaller turndown ratio, into smaller lines, so that if the production rate must be decreased, some of these smaller lines can be shut down. In this regard, the most important point is to know how the production line should be divided. In other words, the division ratio should be determined. The production line can be divided into two, three, or even more lines. Moreover, it can be divided in different ratios, that is, 50–50, 70–30, 50–25–25, and so on. Therefore, both the number of lines and the capacity of each should be defined.

This task is very much case-dependent. However, some hints can be helpful in making these decisions. The first hint is that the number of lines should not be very large. According to economies of scale, certain factors lead to a reduction in the average cost per unit as the scale of output or the size of a facility is increased. Therefore, one big reactor is less expensive than two smaller reactors with an aggregate capacity equal to that of the big reactor. Therefore, as the number of divisions rises, the capital investment and the operating cost will increase as well. Hence, the number and capacity of production lines should be determined so that the increase in capital and operating costs is compensated by the flexibility increase. The second hint is that small capacities should not be chosen for production lines. Assume a division of a production line with a capacity of 100 tons/day into two smaller lines. The possible ratios are 90–10, 80–20, 70–30, 60–40, and 50–50. The first two ratios are not appropriate options because they include small-capacity lines. Such small lines will be more expensive, on a per-unit basis, than bigger lines, again due to economies of scale. This will affect the ROI. Moreover, in the case of flexible production lines which are able to produce more than one product, small lines may not be helpful because when they are changed over to produce another product, the amount produced on the small line may not be large enough to be able to respond to the customers.

Another strategy to make a system more flexible is to keep certain equipment or process sections on standby. This strategy will work for increasing capacity. The number of whatever piece of equipment restricts the process can be doubled so that when production increases, more capacity can be provided by the standby equipment. Again, the economic justification of adding a piece of equipment that will not be working all the time during plant operation is very important.

After determining the number and capacity of smaller production lines, the design alternatives must be defined in more detail. Each design alternative includes a new process configuration. The new configuration may need modifications the design of the pretreatment section, the heat exchanger network, or the separation and recovery system. In the case of flexible production lines, these modifications are of great importance. Products that are produced in a single production line might need different pretreatment and recovery processes. A batch system that is able to produce ethanol as well as acids needs different pretreatment and recovery systems for these two product families. Moreover, some products can be produced in different lines, but have some processing steps in common. For instance, biomass size processing and biomass drying areas might be similar for many processes, and therefore the important factors in their design would be their capacity and their linkage with other processing steps. Therefore, the number and capacity of all pretreatment and separation/recovery systems and the required links with all smaller lines should be considered in the design.

Illustrative Example Given the characteristics of each product/process portfolio in this illustrative example, different alternatives have been considered for each portfolio. Design alternatives representing different levels of flexibility for all portfolios are illustrated in Figure 9.8. The first portfolio, shown in Figure 9.8a, includes three design alternatives. In the first alternative, A-1, FTL is separated into waxes and diesel. By changing the process conditions, the share of waxes and diesel can change from 45–55% to 55–45%. Therefore, there is some volume flexibility in the production of waxes and diesel. The waxes are sold, and 100% of the diesel is converted to JF by a hydrotreating process. Another option would be to shut down the hydrotreating process and sell the entire diesel to market. Moreover, the inherent volume flexibility of the hydrotreating process can be used to reduce the production of JF.

In the second alternative, A-2, a smaller hydrotreating process is used to convert diesel into JF. Hence, this system would be more flexible in terms of product. Again, the hydrotreating process can be shut down or its operating rate decreased to sell more diesel. The third alternative is the most flexible and is a combination of A-1 and A-2. Two small hydrotreating processes are used in parallel. If both are in operation, alternative A-1 is selected. If one of them is shut down, alternative A-2 is selected.

For the second portfolio, two process alternatives have been considered, as can be observed in Figure 9.8b. A series of fermenters in parallel are capable of producing different products based on their inputs. The majority of these are used to produce BuOH, while the rest are dedicated to SA and LA production. BuOH is a commodity with a large market, and therefore a fixed operating rate can be assigned to BuOH production, but SA and LA are more value-added, and flexibility would be useful in the production of these two products. Because the recovery system associated with each product is unique, or in other words, each recovery system can be used to recover only one specific product, this part of the process is the process bottleneck. In the first alternative, B-1, the inherent volume flexibility of each recovery system can be used to increase or decrease the production rate of each product. In the second alternative, a spare recovery system is used for SA and LA. These recovery systems are in standby mode. When it is decided, for instance, to reduce the production of SA and increase the production of LA, the fermenter is changed over to produce LA. The SA recovery system is shut down, and the spare LA recovery system will be put in operation to recover more LA.

For the third portfolio, two process alternatives have been considered and are illustrated in Figure 9.8c. The family of four-carbon carboxylic acids, SA, MA, and FA, is produced. The production system is completely flexible, meaning that the system can produce all acids in different production modes based on the input to the fermenter. Moreover, all the acids can be recovered in a similar recovery system.

The first alternative consists of a single flexible production line which can be scheduled based on product price and demand. The production can be dedicated to the most profitable

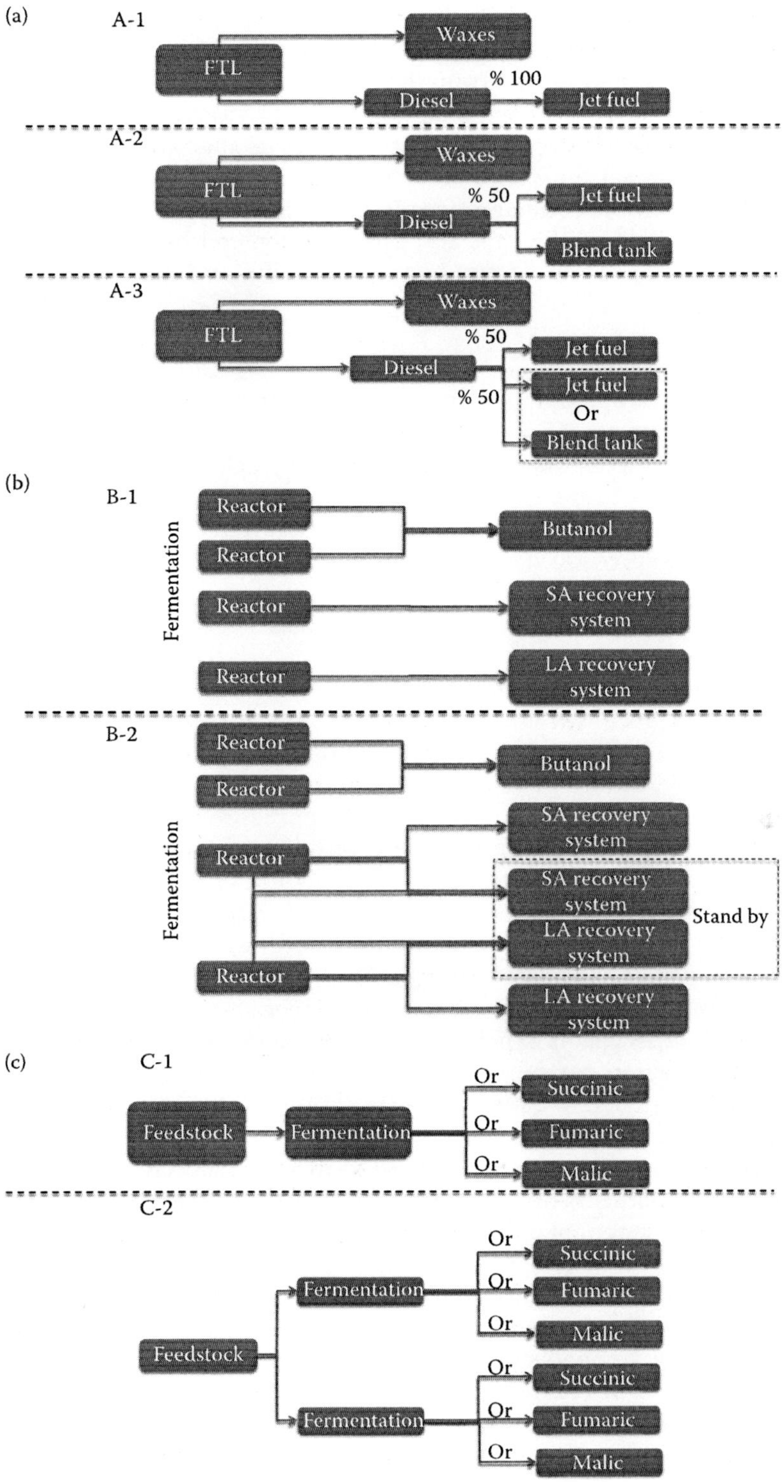

FIGURE 9.8
(a) Design alternatives for portfolio 1; (b) design alternatives for portfolio 2; (c) design alternatives for portfolio 3.

product among the three products. SC optimization can determine whether changeover to another product can be justified, that is, whether the changeover cost will be compensated and exceeded by the revenue gained from the production of the other product. This issue will be discussed in the last part of the methodology. The second alternative provides even more flexibility. Instead of having one production line, this alternative includes two smaller lines with an aggregate capacity similar to that of the first alternative. This alternative provides the opportunity of producing two out of three possible products at a time. Again, scheduling the production is complex, and optimization can help to find the best alignment between product price and demand on the one hand and production on the other.

9.4.1.4 Calculating Capital and Operating Costs for Each Design Alternative

In this step, the required capital investment for each design alternative and the major components of the operating cost associated with the nominal production rate are calculated. These costs will be used to estimate the profitability of each design alternative under different market scenarios.

The capital needed to purchase and install the required machinery and equipment, obtain the land, and to provide the service facilities, piping, and controls as well as the funds required for paying plant operational expenses before sales revenue becomes available is called the *total capital investment* (TCI). TCI consists of two components: *fixed capital investment* (FCI), which can be further divided into manufacturing FCI and nonmanufacturing FCI (also known as indirect cost), and *working capital* (WC) [45].

Manufacturing FCI includes the capital necessary for (a) purchasing equipment, (b) installing equipment, (c) instrumentation and controls, (d) piping and insulation, (e) electrical systems, (f) buildings associated with the process, for example, substructures and superstructures, auxiliary buildings such as administration and office space, maintenance buildings such as electrical, piping, and building services including heating and dust collection, (g) yard improvements, (h) service facilities including utility facilities such as steam and water, water treatment, cooling towers, nonprocess equipment such as office, safety, and medical equipment, and distribution and packaging facilities such as raw material and product storage and handling equipment, and (i) the land.

Nonmanufacturing FCI or indirect costs represent construction overhead costs, including field office and supervision expenses, home office expenses, engineering expenses, miscellaneous construction costs, contractors' fees, and contingencies, plus costs associated with all plant components that are not directly related to the process operation, such as warehouses and laboratories.

The WC is the total amount of money invested in raw material and supplies, finished products in stock, and semifinished products in the process, accounts receivable, cash kept on hand for monthly payment of operating expenses such as salaries, and finally accounts and taxes payable.

Calculating all these costs is complex. Hence, based on experience and rules of thumb, most of these costs can be estimated as a percentage of the total purchased equipment cost (TPEC). Therefore, all components of direct costs, including purchased-equipment installation, instrumentation and control, piping, electrical systems, and buildings including services and yard improvements, can be estimated as a percentage of the major component of direct costs, TPEC. The sum of all these costs will give *total installed cost* (TIC), which is equivalent to direct cost.

Indirect costs can be estimated in the same way. All components of indirect costs, such as engineering and supervision, legal and contractors' fees, construction, and project contingencies can be related to the TPEC, the direct cost or TIC, or the total project investment (TPI), which is the sum of direct and indirect costs and is equivalent to FCI.

Working capital (WC) can also be considered as a function of TCI. WC as a percentage of TCI depends on the type of plant and generally varies from 10% to 20%. It might increase to 50% in some cases, especially for plants producing products with seasonal demand, as they need large inventories [45].

There are several ways to estimate the capital investment required for a plant. Peters et al. [46] introduces seven methods for this task: detailed-item, unit cost, percentage of delivered equipment cost, Lang factors for approximation of capital investment, power factor applied to plant/capacity ratio, investment cost per unit of capacity, and turnover ratio.

Another major component of an economic analysis is the aggregation of all costs related to plant operation, selling the products, recovering the capital investment, and contributing to corporate functions such as management and research and development. This component is called *total product cost* and can be subdivided into two categories; manufacturing or production or operating costs, and general expenses. Total product costs are generally calculated on a daily basis, a unit product basis, or an annual basis. Operating costs involves all expenses that are directly connected with plant operation. These expenses can be classified into three categories: variable production costs, fixed charges, and plant overhead costs. On the other hand, general expenses can be categorized into administrative expenses, distribution and marketing expenses, and research and development expenses. The focus in this step of the methodology is on the operating-cost calculation.

Variable operating costs consist of expenses directly related to the manufacturing operation, such as expenditures for raw materials, including transportation, unloading, and so on, direct operating labor, supervisory and clerical labor directly related to the manufacturing operation, utilities, plant maintenance and repairs, operating supplies, laboratory supplies, royalties, catalysts, and solvents.

Fixed charges represent the expenses that are independent of production rate, such as expenditures for depreciation, property taxes, insurance, financing (loan interest), and rent.

Plant overhead costs cover expenses related to hospital and medical services, general plant maintenance and overhead, safety services, payroll overhead including social security and retirement plans, medical and life insurance, vacation allowances, packaging, restaurant and recreation facilities, salvage services, quality control laboratories, property protection, plant superintendence, warehouse and storage facilities, and special employee benefits.

Illustrative Example The capital investment required for each process in each portfolio is shown in Table 9.4. The more flexible an alternative is, the more capital investment it needs.

TABLE 9.4

Capital Investment Needed for Design Alternatives

Portfolio 1		Portfolio 2		Portfolio 3	
Design Alter.	**Cap. Inv. ($M)**	**Design Alter.**	**Cap. Inv. ($M)**	**Design Alter.**	**Cap. Inv. ($M)**
A-1	180	B-1	56	C-1	43
A-2	110	B-2	65	C-2	47
A-3	195				

9.4.2 SC Network Design

In the strategic design of the SC network, decisions are made to design a new SC network or to redesign an already established SC network with all its existing assets. Such decisions involve the location of plants and determination of the target markets and the location and capacity of warehouses and transportation centers. The SC of a forestry company should be redesigned so that it can be used in the FBR. In the proposed methodology, the SC network design is performed in two steps. First, the specifications of the new SC are identified based on the characteristics of the new product options. Then SC network alternatives are defined. These SC network alternatives will be combined with the process design alternatives defined in the previous part of the methodology, and in the final part, SC optimization is used to calculate the SC profitability of each alternative.

9.4.2.1 Identifying the Specifications of the New SC with Product Options

The SC networks of forest-products companies are in place with their own existing assets. Depending on the processes used in the mills, different facilities exist on the site. However, as mentioned previously, some processing steps are common among all processes in the mill, and therefore similar facilities and assets can be used or redesigned to be able to handle larger volumes.

Biomass receiving, processing, and storage areas in the mills generally include a biomass receiving and unloading station, biomass storage with a reclaimer, biomass processing involving a biomass size-reduction process, cleaning and wet storage, and finally biomass drying and dry storage. These facilities are used regardless of the fate of the biomass, that is, the final product. Therefore, the design process should identify whether the new processes need the same facilities and whether the existing facilities have enough capacity for the larger amount of biomass that will be brought to the mill. If new or additional facilities are required, there is a need to investigate how those facilities should be modified or be added to the site to enable the mill to accept more biomass.

On the product side, the characteristics of new products must be taken into account to redesign the SC network. Each product has specific properties and characteristics which imply specific facilities for transportation and storage. Some products can be stored in warehouses, while others must be stored in tanks. Moreover, some products are transferred by truck or train, while others should be transported in a tanker or by pipeline. Therefore, the specifications of each product must be identified so that they can be addressed when defining SC network alternatives.

9.4.2.2 Defining SC Network Alternatives

With the existing SC assets and the characteristics of the products, the specifications of the new SC network can be identified. Based on these specifications, several SC network alternatives can be defined, which reflect the needs of the new SC network as well as the concerns of company experts. Several issues should be addressed when generating these alternatives;

- *Partnership:* Collaborating with other companies whose expertise brings value to the company's business model must be considered in the SC network design. Partners can cooperate in producing a product, delivering the product, buying the product, and/or selling the product to the market. In this way, a part of the partner's SC assets will be used, and less capital will be needed for establishing the new SC network.

- *Location and capacity of distribution centers:* On the basis of the location of the plant, several target markets might exist in the areas around the plant. Therefore, different distribution centers with different capacities can be assigned to the target market areas. The role of partners in this issue is important. They might take the role of seller in the target markets, and they might have the required infrastructure for this purpose.
- *Transportation network:* On the basis of the characteristics of the products, different means of transportation can be used for product delivery. Again, partnerships can be used to reduce the capital costs required for establishing a transportation network. Contracts can be made with transportation companies which have a network of trucks or tankers and can simply deliver the products to the distribution centers. In addition, partners which buy the products or just deliver them to the market might have their own existing transportation network.

After defining SC network alternatives, the capital investment required to redesign the SC network based on each defined SC network alternative needs to be calculated. This capital investment should be added to the capital investment needed for the process design alternatives defined in the previous part of the methodology. This step will be discussed in the next section. The capital investment needed for the SC network alternatives which involve partnerships is smaller because a part of capital will be paid or has already been paid by the partner. However, it should be noted that the revenue will also be shared by the partner, and therefore less profit will be acquired by the company. The metric that can evaluate which strategy is better to pursue is the profitability of the entire SC, which takes into account both factors: capital investment reduction and revenue reduction. By calculating the profitability of each alternative, a tradeoff can be evaluated to unleash the value created by each strategy.

Illustrative Example Table 9.5 shows two SC network alternatives defined for the second portfolio.

9.4.2.3 Combining Process Design Alternatives and SC Network Alternatives

In this step, the process alternatives defined in the first part of the methodology are combined with the SC network alternatives defined in this part, so that the SC model can be run for each combined alternative in the next part. Each combined alternative will involve

TABLE 9.5

SC Network Alternatives for the Second Portfolio

	SC Network Alternative for B-1	SC Network Alternative for B-2
Selling	BuOH: Contract with a blender SA: Partnership with a company LA: Partnership with a company	BuOH: Contract with a blender SA: Sell on the spot market LA: Sell on the spot market
Warehousing	Expand the existing warehouse Buy a new distribution center	Expand the existing warehouse
Transportation	Buy trucks to deliver the products to the customers	Contract with a transportation company to deliver the products

a process configuration with a specific level of flexibility for each product and an SC network related to those products. A TCI is associated with each combined alternative, which is the sum of capital investment required for the process alternatives and the capital investment needed for the SC network alternatives. It is of crucial importance to note that some facilities may be considered as either a part of a design alternative or a part of a SC network alternative. For instance, product storage is considered in the capital investment calculations. On the other hand, warehouse location and capacity determination is one of the major tasks performed in SC network design. Therefore, when defining the combined design/SC network alternatives, it is necessary to ensure that these SC nodes are not considered in both alternatives and that their associated capital investment is considered only once in the TCI for each combined alternative.

9.4.3 Evaluating the Process Design/SC Network Alternatives

The goal of this part of the methodology is to evaluate all defined process design/SC network alternatives from an SC perspective. The outcome of this part would be the profitability of each alternative under different market scenarios. This part contains four steps: first, a finite number of price/supply/demand scenarios, representing price and demand volatility, are generated. Then the SC profit for each alternative is calculated by SC optimization for each scenario. Then the profitability of each alternative is estimated for each scenario. Finally, the alternatives are compared based on their profitability, and the unprofitable ones are screened out.

9.4.3.1 Generating Price/Supply/Demand Scenarios

To address the uncertainty of market conditions and to reflect market volatility in the decision-making process, a scenario-based approach is used. Each scenario represents a specific market condition with respect to price, supply, and demand. Scenarios are generated in terms of feedstock supply, that is, feedstock availability, and product demand, as well as feedstock and product prices. Scenarios must be generated to capture different market situations, that is, pessimistic, likely, and optimistic cases should be considered in scenario generation. Moreover, the relationships between supply, price, and demand should also be addressed during scenario generation. These relationships are highly complex and cannot be simplified into one holistic rule. They depend on whether the product is a commodity or specialty, price elasticity, policies, oil prices, and whether the product is a replacement product or a substitution product.

Another important factor in scenario generation is the time aspect. Scenarios can be generated for different time scales, and depending on the type of decisions to be made in the scenario analysis, scenarios can be generated for the short, medium, or long term. For strategic design-related decisions, scenarios should be generated for the long term, for example, for a period of one year. As supply, demand, and price change during the year, the values associated with them can vary on a monthly or seasonal basis. Note that buying supply and selling products can be done based either on contracts or on the spot market. Contractual prices and demands imply fixed values during specific periods, meaning that the amount of product and its price in the contract can be fixed for the whole period of the contract or can change during certain periods based on the agreements reached at the time of making the contract, while spot prices are generally subject to changes based on the market situation. Therefore, both spot and contractual prices and demands must be addressed in scenario generation.

TABLE 9.6

Price and Demand Scenarios for Contractual Demands; in the Demand Column: P = Pessimistic, L = Likely, O = Optimistic, C = Contract

	Pessimistic		Likely		Optimistic	
Product	**Price ($/lb)**	**Demand (t/y)**	**Price ($/lb)**	**Demand (t/y)**	**Price ($/lb)**	**Demand (t/y)**
BuOH	0.45	PC_B	0.50	LC_B	0.59	OC_B
LA	0.58	PC_L	0.64	LC_L	0.71	OC_L
SA	2.15	PC_S	2.68	LC_S	3.10	OC_S

TABLE 9.7

Price and Demand Scenarios for Spot Demands; in the Demand Column: P = Pessimistic, L = Likely, O = Optimistic, S = Spot

	Pessimistic		Likely		Optimistic	
Product	**Price ($/lb)**	**Demand (t/y)**	**Price ($/lb)**	**Demand (t/y)**	**Price ($/lb)**	**Demand (t/y)**
BuOH	0.40	PS_B	0.55	LS_B	0.69	OS_B
LA	0.53	PS_L	0.68	LS_L	0.83	OS_L
SA	1.95	PS_S	2.73	LS_S	3.30	OS_S

Illustrative Example In this step, scenarios are generated for one year. Scenarios are constructed for three cases: pessimistic, likely, and optimistic. For each case, only one scenario has been defined. Tables 9.6 and 9.7 show the contractual and spot prices and demands for the second portfolio. These scenarios represent price volatility and different demands in the market. In Table 9.6, the unit used for contractual demand is tons per year because the contracts are made for a period of one year. For the contracts, the price and the amount are fixed over the contract period. In Table 9.7, the unit of spot demand is tons per month. The prices can vary from month to month. For the sake of simplicity, an average monthly spot price has been shown for each product in Table 9.7. The first and third scenarios consider low price-weak demand and high price-strong demand for all products respectively, while the second scenario represents the most probable case in the market.

9.4.3.2 Calculating the SC Profit for Each Scenario/Alternative

To evaluate design/SC network alternatives, the profitability of each alternative must be estimated. Therefore, the SC profit associated with each alternative in different market situations must first be calculated, and then, using the SC profit along with the capital investment, the profitability of each alternative can be estimated. The goal is to evaluate the performance of each alternative under different market conditions. Hence, in this step, the SC profit for each process design/SC alternative is calculated for every price/supply/demand scenario. To calculate the SC profit, the SC optimization model is used. The process configurations defined in the first part of the methodology as design alternatives, along with the SC network alternatives and the market scenarios generated in the previous step, are used as inputs to the model. The model optimizes SC profit by determining which orders to fulfill and calculating the optimum value of production rate related to each product and the flows of material between SC nodes. The overall problem at this stage can be stated as follows. Given:

- Number and length of time intervals.
- Demand and price data for each feedstock, product, market, and time interval for each scenario.
- Process configuration based on what was defined in the process design alternatives.
- Configuration of the SC network based on what was defined in the SC network alternatives.
- Capacity data of the nodes of the SC.
- Direct cost parameters, that is, unit production, transport, handling, and inventory costs based on operating cost calculations.

with the aim of profit maximization, find

- Orders to fulfill: which contracts to make, which spot demand to fulfill.
- Production rates of each product for all time intervals and all market scenarios.
- Flows of materials between the plants, warehouses, distribution centers, and markets.
- SC profit.

Figure 9.9 shows this step graphically. The SC model is run for each design/SC network alternative for all market scenarios. The major output of the SC model is SC profit.

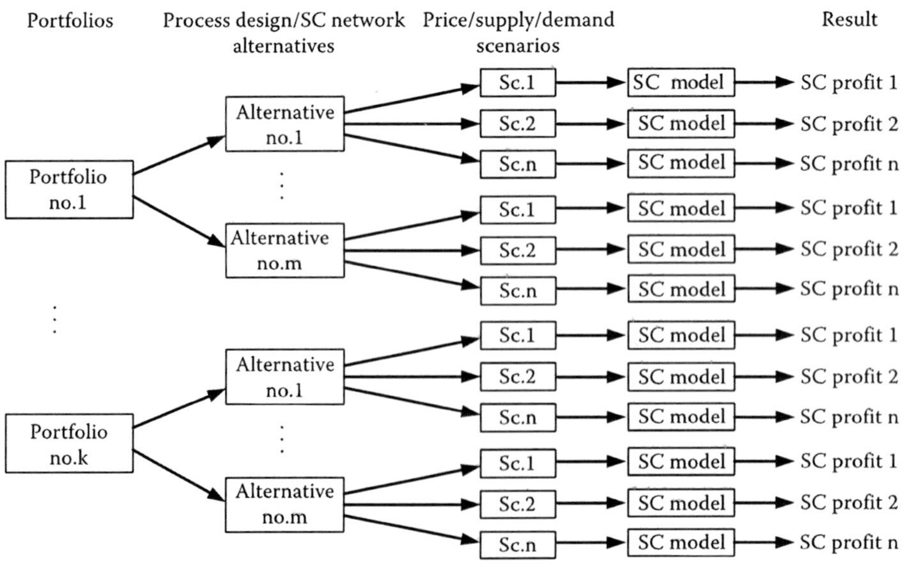

FIGURE 9.9
SC model inputs and outputs.

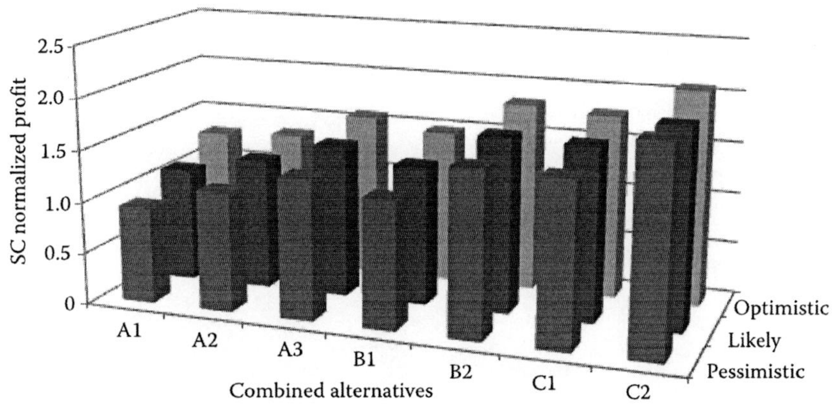

FIGURE 9.10
Normalized SC profit for each design/SC network alternative for all market scenarios.

Illustrative Example Figure 9.10 demonstrates the results of this step graphically. The SC profit has been calculated for each design/SC network alternative for all market scenarios.

9.4.3.3 Calculating the Profitability of Each Scenario/Alternative

To evaluate each process design/SC alternative, the profitability of each alternative should be estimated. There are several profitability estimation methods that can be used to estimate the profitability of a project. From a generic perspective, these methods can be divided into two main groups; methods that do not consider the time value of money, which include the rate of ROI, payback return, and net return, and methods that consider the time value of money, which include the discounted cash flow rate of return and net present worth [45].

In this methodology, ROI is used as the measure of profitability. ROI is a simple measure which is generally used for preliminary design calculations. It does not consider the time value of money, variable depreciation allowance, increasing maintenance costs over the project life, or changing sales volumes [37]. However, because the proposed methodology involves a preliminary design study, ROI can be used as the profitability measure.

ROI is defined as the ratio of profit to investment. Any measure of profit and investment can be used in this ratio, but the most common measures are annual net profit and TCI. In this way, ROI would be the annual ROI, which will be in the form of a fraction or percentage per year. Methods of estimating capital cost are introduced in Section 9.4.1.4. The net profit is calculated according to the following formula:

$$N_P = (s - c - d)\,(1 - \Phi)$$

where N_P is the net profit after taxes, s is the revenue or the money earned from sales, c is the cost for operations, d is the depreciation charge, and Φ is the percentage of the gross profit that goes to income taxes. The revenue is generated by the sales of products produced by the plant. The annual revenue is the sum of the unit price of each product multiplied by its rate of sales. Depreciation is the amount of money (a part of the revenue) that is set aside to enable the company to replace equipment which wears out at the end of its

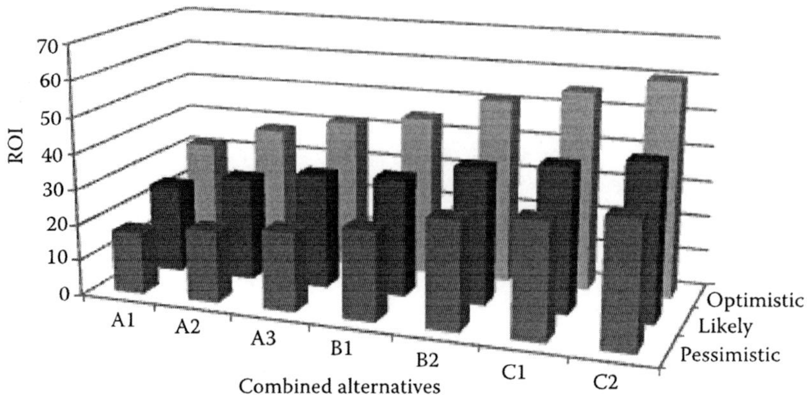

FIGURE 9.11
ROI for each design/SC network alternative for each market scenario.

lifetime. Generally, this amount of money is not used for this purpose, but instead it is invested in other ventures, and a part of the profits earned from the investment is used to replace the equipment. Governments specify the average lifetimes of different kinds of equipment. However, because there are various kinds of equipment in a plant, an average lifetime is considered for a specific plant. Depreciation can be computed by two methods: straight-line and ACRS, which are discussed briefly in Douglas [38].

For this step in the methodology, the net annual profit and capital investment are needed to calculate the ROI. Net annual profit was calculated in the previous step during SC profit maximization. The capital investment was estimated for each combined design/SC network alternative in Section 9.4.1.4. With these two components, the profitability of each combined alternative can be estimated for each market scenario.

Illustrative Example The results of this step are shown in Figure 9.11.

9.4.3.4 Comparing the Combined Alternatives Based on Their Profitability

The goal of this methodology is to evaluate the different biorefinery options that can be pursued by a company willing to implement the biorefinery. Each option involves a product/process portfolio, a process configuration with a specific level of flexibility, and an associated SC network. Each option must be analyzed from different perspectives, and the performance of the various options should be evaluated based on several metrics related to each perspective. The proposed methodology in this chapter looks at the options from an SC perspective and evaluates them based on their profitability, a concept that is then extended to SC costs and profitability. These options are analyzed by the proposed methodology, and their profitability is calculated. By comparing the profitability of alternatives and screening out the unprofitable ones, a set of biorefinery options that can be implemented by a company can be identified. The role of corporate executives in choosing the better options is a key factor. In fact, this methodology does not play the role of a decision-maker. Human knowledge and experience make the final decision; the proposed methodology simply serves this decision-making process.

Finally, it should be mentioned that an implementation strategy can also be defined for each option by defining process design/SC network alternatives. Companies often think about implementing the biorefinery through an incremental strategy. They might plan to start their biorefinery project by producing some commodity products and then, after several years, adding more processes to their portfolio to produce more value-added products from commodities. This strategy can be considered in the proposed methodology. The design/SC network alternatives can be defined in a way that reflects the incremental aspect of biorefinery implementation. Each alternative should be divided into a set of time frames, and for each time frame, a specific process configuration and a specific SC network should be defined according to the implementation strategy proposed by the company. In the first time frame of the alternatives, the process configuration and SC network related to the first set of products are defined, and in the following time frames, the necessary changes and additions which should be made to both the processes and the SC network are added. However, to apply the methodology for this purpose, some of its steps must be modified. Capital investment should be estimated by methods which consider the time value of money, design alternatives must take into account the future evolution of technology, partnerships and contracts with other companies should consider the implementation plan, and supply, demand, and price in future markets must be better forecasted.

References

1. Mansoornejad, B., Chambost, V., Stuart, P. 2010. Integrating product portfolio design and supply chain design for the forest biorefinery. *Computers & Chemical Engineering* 34(9), 1497–1506.
2. Lail, P.W. 2003. *Supply Chain Best Practices for the Pulp and Paper Industry*. Atlanta, GA: Tappi Press.
3. Dansereau, L.P., El-Halwagi, M.M., Stuart, P. 2009. Sustainable supply chain planning for the forest biorefinery. In: *Design for Energy and the Environment: 7th International Conference on the Foundation of Computer-Aided Process Design*, Breckenridge, CO, 1101.
4. Schiltknecht, P., Reimann, M. 2009. Studying the interdependence of contractual and operational flexibilities in the market of specialty chemicals. *European Journal of Operational Research* 198(3), 760–772.
5. Ropohl, G. 1967. Zum Begriff der Flexibilitaet. *Werkstattstechnik* 57, 644.
6. Gupta, Y.P., Goyal, S. 1989. Flexibility of manufacturing systems: Concepts and measurement. *European Journal of Operational Research* 43, 119–135.
7. Nagarur, N. 1992. Some performance measures for flexible manufacturing systems. *International Journal of Production Research* 30(4), 799–809.
8. Upton, D. 1994. The management of manufacturing flexibility. *California Management Review* 36(2), 72–89.
9. Sethi, A.K., Sethi, S.P. 1990. Flexibility in manufacturing: A survey. *International Journal of Flexible Manufacturing Systems* 2, 289–328.
10. Grossmann, I.E., Halemane, K.P., Swaney, R.E. 1983. Optimization strategies for flexible chemical processes. *Computers & Chemical Engineering* 7, 439.
11. Beach, R., Muhlemann, A.P., Price, D.H.R., Paterson, A., Sharp, J.A. 2000. Theory and methodology: A review of manufacturing flexibility. *European Journal of Operational Research* 122, 41–57.
12. Frazelle, E. 1986. Flexibility: A strategic response in changing times. *Industrial Engineering* 18(3), 16–20.
13. Slack, N. 1983. Flexibility as a manufacturing objective. *International Journal of Operations and Production Management* 3(3), 5–13.
14. Swaney, R.E., Grossmann, I.E. 1985. Index for operational flexibility in chemical process design. Part I: Formulation and theory. *AIChE Journal* 31(4), 621.
15. Pistikopoulos, E.N., Grossmann, I.E. 1988. Stochastic optimization of flexibility in retrofit design of linear systems. *Computers and Chemical Engineering* 12(12), 1215.

16. Straub, D.A., Grossmann, I.E. 1993. Design optimization of stochastic flexibility. *Computers and Chemical Engineering* 17(4), 339.
17. Bansal, V., Perkins, J.D., Pistikopoulos, E.N. 2002. Flexibility analysis and design using a parametric programming framework. *AIChE Journal* 48(12), 1851–1868.
18. Halemane, K.P., Grossmann, I.E. 1983. Optimal process design under uncertainty. *AIChE Journal* 29(3), 425–433.
19. Grossmann, I.E., Floudas, C.A. 1987. Active constraint strategy for flexibility analysis in chemical processes. *Computers & Chemical Engineering* 11(6), 675–693.
20. Bansal, V., Perkins, J.D., Pistikopoulos, E.N. 2000. Flexibility analysis and design of linear systems by parametric programming. *AIChE Journal* 46(2). 335–354.
21. Floudas, C.A., Gumus, Z.H., Ierapetritou, M.G. 2001. Global optimization in design under uncertainty: Feasibility test and flexibility index problems. *Industrial and Engineering Chemistry Research* 40, 4267–4282.
22. Goyal, V., Ierapetritou, M.G. 2003. Framework for evaluating the feasibility/operability of nonconvex processes. *AIChE Journal* 49(5), 1233–1240.
23. Pistikopoulos, E.N., Grossmann, I.E. 1988. Stochastic optimization of flexibility in retrofit design of linear systems. *Computers and Chemical Engineering* 12(12), 1215.
24. Swamidass, P.M. 1988. *Manufacturing Flexibility*. Monograph 2, Operations Management Association, Norman and Schneider Group, Waco, TX.
25. Browne, J., Dubois, D., Rathmill, K., Sethi, S.P., Stecke, K.E. 1984. Classification of flexible manufacturing systems. *FMS Magazine* 2(2), 114–117.
26. Verwater-Lukszo, Z. 1998. Practical approach to recipe improvement and optimization in the batch processing industry. *Computers in Industry* 36(3), 279.
27. Romero, J., Espuna, A., Friedler, F., Puigjaner, L. 2003. A new framework for batch process optimization using the flexible recipe. *Industrial and Engineering Chemistry Research* 42(2), 370.
28. Ferrer-Nadal, S., Puigjaner, L., Guillen-Gosalbez, G. 2008. Managing risk through a flexible recipe framework. *AIChE Journal* 54(3), 728.
29. Laflamme-Mayer, M. 2009. Cadre de planification de la chaîne logistique basée sur la représentation des procédés pour l'amélioration de la rentabilité de l'industrie des pâtes et papiers. PhD dissertation, École Polytechnique, Montreal, Canada.
30. Sahinidis, N.V., Grossmann, I.E. 1991. Multiperiod investment model for processing networks with dedicated and flexible plants. *Industrial & Engineering Chemistry Research* 30(6), 1165.
31. Bok, J.K., Grossmann, I.E., Park, S. 2000. Supply chain optimization in continuous flexible process networks. *Industrial and Engineering Chemistry Research* 39, 1279–1290.
32. Norton, L.C., Grossmann, I.E. 1994. Strategic planning model for complete process flexibility. *Industrial and Engineering Chemistry Research* 33, 69–76.
33. Neiro, S.M.S., Pinto, J.M. 2004. A general modeling framework for the operational planning of petroleum supply chains. *Computers & Chemical Engineering* 28(6–7), 871.
34. Schulz, E.P., Diaz, M.S., Bandoni, J.A. 2005. Supply chain optimization of large-scale continuous processes. *Computers & Chemical Engineering* 29(6), 1305.
35. Méndez, C.A., Grossmann, I.E., Harjunkoski, I., Kaboré, P. 2006. A simultaneous optimization approach for off-line blending and scheduling of oil-refinery operations. *Computers & Chemical Engineering* 30(4), 614.
36. Bahri, P.A., Bandoni, A., Romagnoli, J. 1996. Operability assessment in chemical plants. *Computers and Chemical Engineering* 20(Suppl B), S787.
37. Blanco, A.M., Bandoni, J.A. 2003. Interaction between process design and process operability of chemical processes: An eigenvalue optimization approach. *Computers & Chemical Engineering* 27(8–9), 1291–1301.
38. Douglas, J.M. 1988. *Conceptual Design of Chemical Processes*. McGraw-Hill, New York.
39. Pistikopoulos, E.N., Grossmann, I.E. 1989. Optimal retrofit design for improving process flexibility in nonlinear systems. I. Fixed degree of flexibility. *Computers & Chemical Engineering* 13(9), 1003–1016.

40. Stuart, P. 2006. The forest biorefinery: Survival strategy for Canada's pulp and paper sector? *Pulp and Paper Canada* 107(6), 13.
41. Yun, C., Kim, Y., Park, J., Park, S. 2009. Optimal procurement and operational planning for risk management of an integrated biorefinery process. *Chemical Engineering Research and Design* 87, 1184–1190.
42. Luo, L., Voet, E.V.D., Huppes, G. 2010. Biorefining of lignocellulosic feedstock—Technical, economic and environmental considerations. *Bioresource Technology* 101, 5023–5032.
43. Lynd, L.R., Wyman, C., Laser, M., Johnson, D., Landucci, R. 2002. *Strategic Biorefinery Analysis: Analysis of Biorefineries.* National Renewable Energy Laboratory.
44. Tsiakis, P., Shah, N., Pantelides, C.C. 2001. Design of multi-echelon supply chain networks under demand uncertainty. *Industrial and Engineering Chemistry Research* 40, 3585–3604.
45. Slack, N. 1987. The flexibility of manufacturing systems. *International Journal of Operations and Production Management* 7(4), 35–45.
46. Peters, M.S., Timmerhaus, K.D., West, R.E. 2003. *Plant Design and Economics for Chemical Engineers.* McGraw-Hill, New York.

Section III

Biorefinery Process Analysis

10

Optimal Synthesis of Sustainable Biorefineries

Mariano Martín and Ignacio E. Grossmann

CONTENTS

10.1 Introduction

The design of a biorefinery poses a substantial number of challenges that require the use of advanced process synthesis and optimization methods to ensure that these systems are economically viable, energy-efficient, and have minimum environmental impact.

Process synthesis involves choosing the topology of a process to convert a set of inputs into a desired set of outputs (Rudd et al., 1973). The objective is to find designs with minimum cost or maximum profit, even though objectives such as maximizing efficiency or minimizing the usage of a resource (e.g., energy or fresh water) can also be considered. Major approaches to optimize process flowsheets to meet the desired objectives include the use of heuristics, the development of physical insights (commonly based on thermodynamics), and the optimization of superstructures of alternatives. Major contributions to the first two approaches have been hierarchical decomposition (Douglas, 1988) and pinch analysis (Linnhoff, 1993), which have proved to be very successful in industrial applications. A more recent trend has been to combine some of these concepts with the mathematical-programming approach (Grossmann et al., 1999), which consists of three major steps. The first is the development of a representation of various alternatives from which the optimum solution is selected. The second is the formulation of a mathematical program, generally involving both discrete and continuous variables, to select the configuration and the operating levels respectively. The third is the solution of the optimization model (commonly a mixed-integer nonlinear programming (MINLP) model or a generalized disjunctive programming (GDP) model) to obtain the optimal solution. Although superstructures can be developed in a systematic way for subsystems, for example, for heat-exchanger networks (see, e.g., Yee and Grossmann, 1990), their development for general process flowsheets is more complex (Yeomans and Grossmann, 1999). As for the problem formulation, it is important to note that synthesis models can be formulated at three main levels of detail: (a) high-level aggregated models concentrate on major features like energy flows (e.g., the LP transshipment model for HEN [Heat exchanger Network] by Papoulias and Grossmann (1983), or the NLP heat- and mass-exchanger model by Papalexandri and Pistikopoulos (1996)); (b) short-cut models that involve cost optimization (investment and operating costs), but in which the performance of the units is predicted using relatively simple nonlinear models (e.g., the MINLP heat-exchanger networks proposed by Yee and Grossmann (1990), or the MINLP process flowsheets developed by Kocis and Grossmann (1987)); (c) rigorous models that rely on detailed superstructures and involve rigorous and complex models for predicting unit performance (e.g., the MINLP synthesis of distillation sequences by Smith and Pantelides (1995) or the GDP models proposed by Grossmann et al. (2005)).

It should be noted that at the level of process synthesis, two problems that have major significance for sustainability are optimization of energy recovery and water management. In these cases, the simultaneous optimization and heat integration of a process can be modeled using the constraints proposed by Duran and Grossmann (1986). However, this method is restricted to nonisothermal streams. Grossmann et al. (1998) have extended this model to handle isothermal streams that undergo changes of phase using a disjunctive programming approach. On the problem of synthesizing water process networks, significant work has been reported in the literature (for reviews, see Bagajewicz, 2000; Jezowski, 2008; Bagajewicz and Faria, 2009), using both pinch (Wang and Smith, 1994) and mathematical programming approaches (e.g., Karuppiah and Grossmann, 2006). The latter have emerged as the preferred choice for water networks because the pinch approach is restricted to problems with a single contaminant. These models, aside from optimizing cost, have as a major goal to minimize the consumption of fresh water. Ahmetovic and Grossmann (2010) have recently developed a model that accommodates multiple freshwater sources, recycle within units, and sources or sinks of water in a process, while enabling the user to control the complexity of the network by limiting the number of piping connections with 0–1 variables.

Finally, it should also be pointed out that systematic methods for separation systems have also played a very important role in the design of sustainable chemical processes. At this time, very few papers have reported the use of process synthesis techniques with explicit incorporation of sustainability issues (e.g., Stefanis et al., 1995; Steffens et al., 1999; Halasz et al., 2005; Carvahlo et al., 2006; Guillen-Gosalbez et al., 2008). Some of these have applied optimization methods to the molecular design of solvents and the synthesis of the associated separation processes (Pistikopoulos and Stefanis, 1998; Hostrup et al., 1999).

The design of the biorefinery must face a number of challenges: bioprocess versus chemical process, selection among alternative technologies, energy and water consumption, and waste production (Jimenez-González and Woodley, 2010), which determine the technical and economic feasibility of each design. Systematic process synthesis is a powerful tool for selecting the optimal design among a large number of alternatives. This approach has been used recently for the design of biorefineries, as will be described later in the chapter, for example by Agrawal et al. (2007); Karuppiah et al. (2008); and Martín and Grossmann (2010a, 2011a), as well for integrating the various processes for optimal operation of the biorefinery, for example by Sammons et al. (2006, 2008); Kokossis and Yang (2010); Ng (2010); and Mansoornejad et al. (2010).

10.2 History of Biofuel Production Processes

The history of biofuels has been linked to the automobile industry since its very beginning back in the nineteenth century. By the time the diesel engine was patented in 1893, R. Diesel stated, "The use of vegetable oils for engine fuels may seem insignificant today. But such oils may become in course of time as important as petroleum and the coal tar products of the present time" (Agarwal, 1997). The first car produced on an assembly line, the Ford Model T (1908–1927), can be regarded as the first flexible-fuel vehicle because it had an adjustable carburetor so that the car could be run on gasoline, ethanol, or a blend. However, the ready availability of crude-based fuels displaced the use of biofuels for decades, and the use of renewable resources was focused more on pulp and paper production from wood, saccharification of wood, nitration of cellulose for guncotton and viscose silk, production of soluble cellulose for fibers, fat curing, and the production of furfural for nylon.

Nowadays, the increasing shortage of energy resources, the increasing demand for and dependency on fossil fuels, and the growing concerns about the environment have led humans to consider alternative and renewable energy sources. Presently, the entire world is in a race to develop alternative energy resources which must be cheap, renewable, and non-polluting, or at least with "neutral" emissions. In the transportation sector, which is the most challenging one because of its need for high-density energy sources, only biomass provides an alternative that can be implemented in the short term (Cole, 2007). Therefore, bioethanol and biodiesel have become the most promising alternatives. Today's biorefinery technologies are based: (1) on the utilization of the whole plant or complex biomass, and (2) on integration of traditional and modern processes for utilization of biological raw materials. Figure 10.1 provides an overview of various processing alternatives using biomass and waste as raw material. The next section presents a number of process alternatives for the production of first- and second-generation biofuels based on systematic process synthesis.

10.3 Synthesis of Biofuel Production Processes: Energy and Water Optimization

10.3.1 Corn-Based Biofuels

Karuppiah et al. (2008) considered energy optimization of the "dry-grind" process for the corn-based bioethanol plant, which accounts for 95% of the ethanol produced in the United States. In such plants, fuel ethanol is produced using corn kernels as the feedstock. Fuel-grade ethanol must be very close to 100% pure before it can be blended with gasoline for use in automobiles. However, conventional distillation columns produce an azeotropic mixture

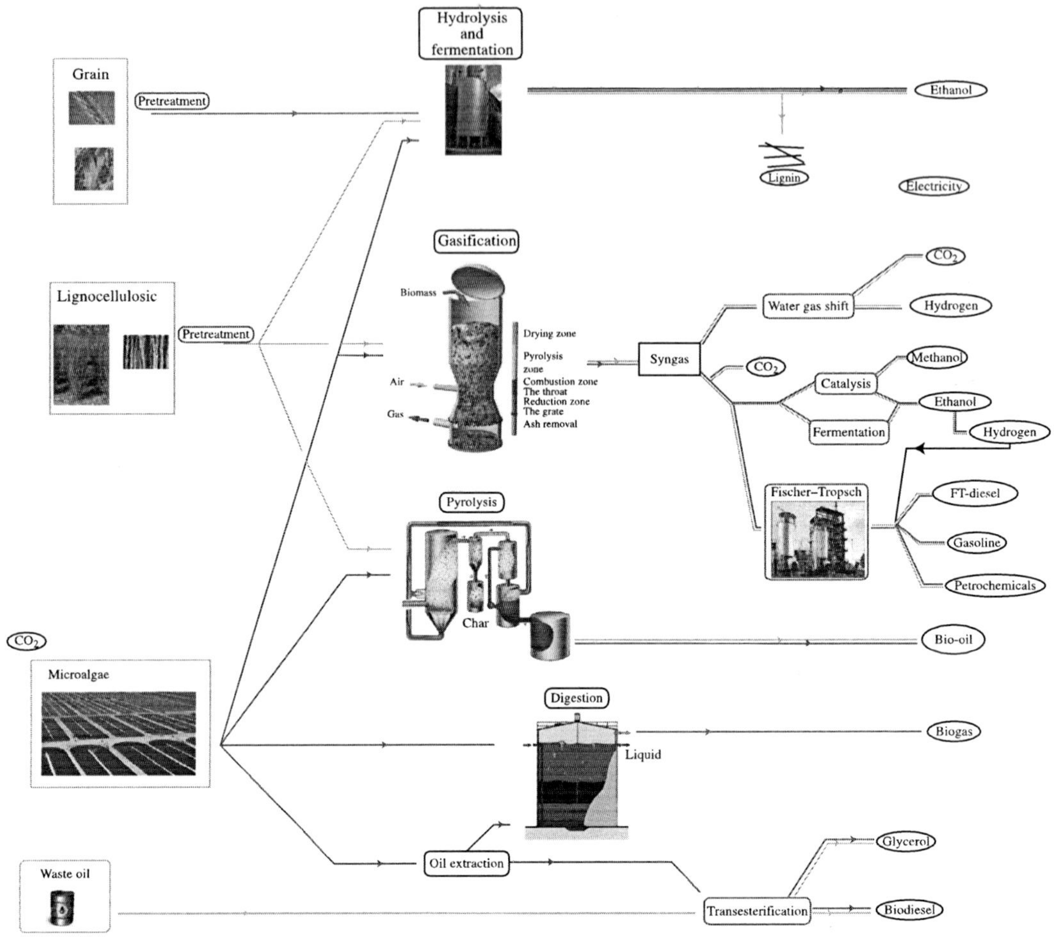

FIGURE 10.1
Raw materials, technologies, and biofuels. (Martín, M. and Grossmann, I.E. 2010d. *Industrial & Engineering Chemistry Research*. With permission.)

of ethanol and water (95% ethanol–5% water), which has to be purified further for making fuel ethanol. The main challenge impeding commercial production of fuel ethanol is that the process is very energy-intensive and requires large amounts of steam and electricity for use in the rectifiers to obtain an azeotropic mixture of ethanol and water, as well as the use of expensive molecular sieves to obtain 100% pure ethanol. Karuppiah et al. (2008) developed a simplified model to predict the performance of a bioethanol flowsheet that includes grinding, scarification, fermentation, centrifugation, and drying operations (Figure 10.2). A superstructure was also postulated in which some of the major alternatives included separation by molecular sieves, corn grits, or both, and different ways to accomplish drying of grain solids, the cattle-feed by-product. The objective was to optimize the flowsheet by determining the connections in the network and the flow in each stream in the network, such that the energy requirements of the overall plant would be minimized while the yields were maximized. Optimization without heat integration (MINLP model) led to a decrease in manufacturing cost from $1.50/gal (base case) to $1.36/gal. In the next step, heat integration was added to the optimization, which further reduced the cost to $1.28/gal. However, it

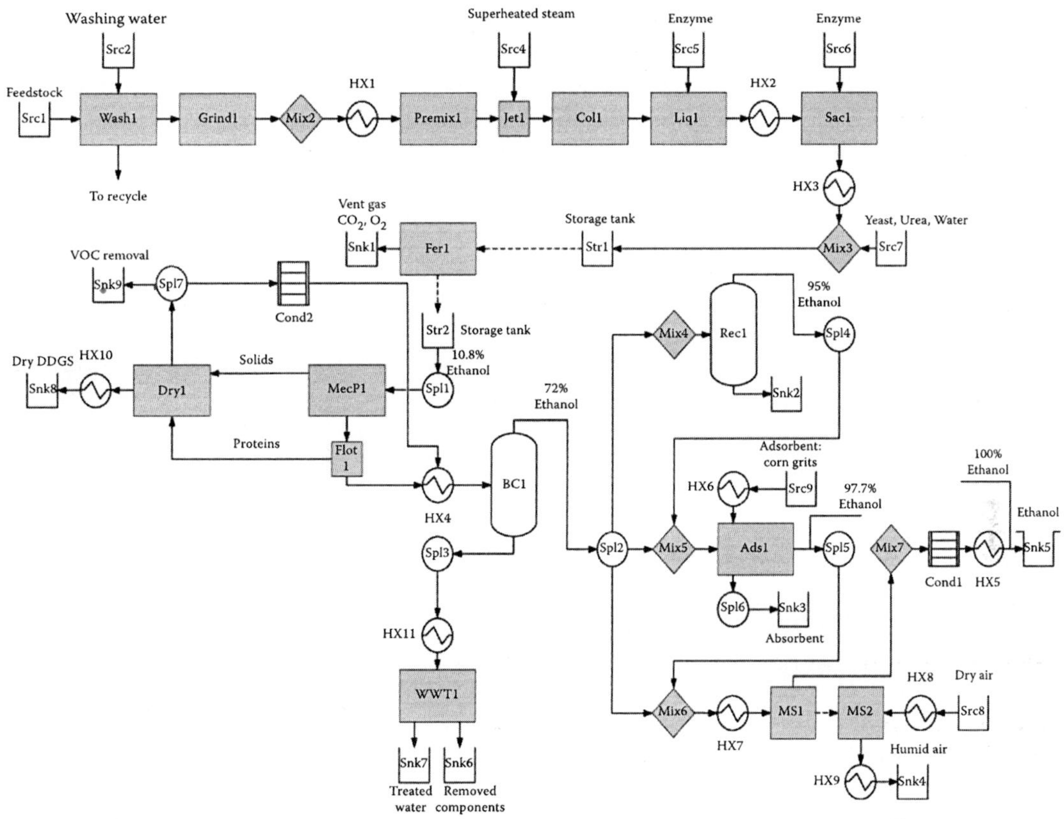

FIGURE 10.2
Corn-based ethanol production process. (Karuppiah, R. et al. 2008. *AIChE Journal* 54(6): 1499–1525. With permission.)

became clear that the scope of heat integration is limited by the relatively low temperature in the fermentor. To improve the potential for heat integration, the authors considered multiple-effect distillation in the "beer" column and in the azeotropic column as alternatives for optimization (Figure 10.2). This finally led to a 65% savings in steam and cooling-water consumption with a cost reduction to \$1.24/gal. Figure 10.5, which appears later in the chapter, shows the energy consumption and cooling needs of the process. This example illustrates the potential for cost reduction in biofuel plants with an optimization approach.

However, sustainability involves more than just low energy consumption. Corn-based ethanol has been criticized, not only for its energy balance, but also because of its high demand for water. The first data available in the literature for water consumption in ethanol plants revealed values ranging from 3 to 15 $gal_{water}/gal_{ethanol}$ (Aden 2007). Over the last decades, there has been some improvement to reach the current industrial mean value for the newest plants of 3.4 $gal_{water}/gal_{ethanol}$ (MATP, 2008). Based on this, the suggested best-practice possible water consumption for a corn-based process is 2.85 $gal_{water}/gal_{ethanol}$ (Aden 2007, Pfromm 2008). They also developed an optimum water network for the cases for which Karuppiah et al. (2008) optimized the energy consumption. It is worth pointing out that energy optimization plays a very important role in reducing water consumption because cooling needs are greatly reduced, as are the losses in the cooling tower. In this way, by coupling energy optimization and the design of optimal water networks where waste water is treated and recycled, and by using air cooling to replace water partially as a cooling agent, the water consumption of a corn-based ethanol plant can be reduced to only 1.17 $gal_{water}/gal_{ethanol}$, as shown in Figure 10.6 later in the chapter. This value is less than others recently published in the literature and also confirms the technical feasibility of the claims made by the Delta T company, which has reported consumptions of 1.5 gal/gal (Delta, 2009). Furthermore, the energy-optimization and water-network design also play an important role in the move toward zero water discharge, achieving discharge values of 0.27 $gal_{water}/gal_{ethanol}$ (Martín et al., 2010b). However, better and cheaper wastewater technologies are needed to reach the goal of zero discharge.

10.3.2 Lignocellulose-Based Biofuels

Corn ethanol has the disadvantage that it competes with the food chain. Therefore, second-generation bioethanol is based on lignocellulosic materials which offer yields (Figure 10.9) even higher than those from corn (especially switchgrass), but which have no footprint on the food supply chain. Two paths can be used to obtain ethanol from lignocellulosic raw materials: gasification or hydrolysis.

10.3.2.1 Gasification

To produce ethanol from switchgrass by gasification, syngas is produced and used to obtain ethanol. Martín and Grossmann (2011a) developed a conceptual design of a production process for ethanol from switchgrass by optimizing a superstructure of alternatives. The process consists of four steps: gasification (followed by gas cleanup), concentration adjustment, sour-gas removal, and ethanol synthesis. Two gasification alternatives were evaluated, indirect (Phillips et al., 2007) and direct (Gissy et al., 1992, Dutta and Philips, 2009). The gas obtained is treated using either steam reforming or partial oxidation to remove hydrocarbons; subsequently it must be cleaned to remove solids as well as other compounds like NH_3 using either a hot or cold cleaning process. Next, the gas composition is adjusted to a CO/H_2 ratio of one. Three alternatives (bypass, membrane-PSA, and water–gas shift) were evaluated. Then the removal of the sour gases CO_2 and H_2S is required. Three alternatives, membrane separation, absorption in ethylamines, and PSA were considered for this task. Once the syngas has been prepared, two synthetic paths were evaluated: (1) high-alcohols catalytic process with one of two possible distillation sequences (direct and indirect), and (2) a syngas fermentation followed by one of four possible dehydration processes: distillation, water adsorption in corn grits, molecular sieves, and pervaporation. The number of alternative designs is large, and therefore systematic optimization methods are a powerful tool for designing the production process. The superstructure is written as an MINLP which is solved by partial enumeration of the integer variables in terms of gasification technologies, reforming modes, and synthesis paths, generating eight subproblems. For each of the subproblems, the cleaning stages and the separation processes are optimized to minimize energy consumption. Subsequently, multiple-effect columns and the design of the optimal HEN are implemented in each of the subproblems. Finally, an economic evaluation to account for the contribution of hydrogen as a by-product, raw-material consumption, and utilities yields the flowsheet with the lowest production cost, as shown in Figure 10.3. Figure 10.5, presented later in the chapter, shows the energy consumption and cooling needs of the process (Martín and Grossmann, 2011a).

The most profitable process uses high-pressure direct gasification followed by steam reforming, which increases the production of hydrogen. Composition adjustment is carried

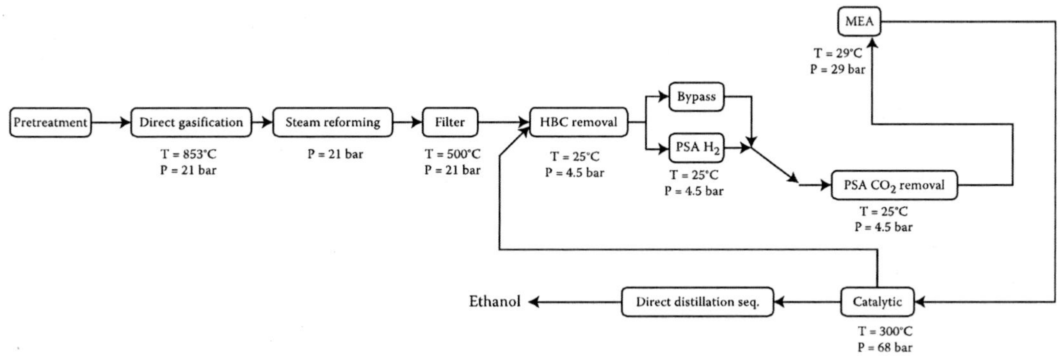

FIGURE 10.3
Thermochemical ethanol production from biomass. (Martín, M. and Grossmann, I.E. 2011a. *AIChE Journal* 57(12), 3408–3428. With permission.)

out by removing the excess of hydrogen from the stream using a hybrid system consisting of PSA and membranes. Then sour gases are removed in two steps, PSA to remove CO_2, and MEA to get rid of H_2S. Finally, the catalytic path is selected, followed by the direct distillation sequence. The process generates 18 MW of energy and requires 68 MW of cooling. The production cost of this design turned out to be \$0.41/gal due to the contribution of hydrogen as a by-product. The fermentation path is not economically attractive because of the low concentration of ethanol in the reactor, which results in high-energy consumption to dehydrate the water–ethanol mixture. It has been estimated that if the concentration of ethanol were to reach a value of 15% (Martín and Grossmann, 2011a), then this synthesis path would also become promising. Several studies have focused on increasing the concentration of ethanol in the fermentor by removing it from the liquid phase using adsorbents (Nielsen and Prather, 2009).

The production process of ethanol from switchgrass by gasification and catalytic synthesis requires a large amount of water in the form of cooling water, due to the high pressures and temperatures in the entire process, and in the form of steam to perform the gasification. Values of 5.9–9.8 $gal_{water}/gal_{ethanol}$ have been reported in the literature when using switchgrass as a raw material (Wu et al., 2009), and values of 2 $gal_{water}/gal_{ethanol}$ have been reported when using hybrid poplar (Philips et al., 2007). Because of these high cooling needs, some authors have supported the use of air-cooling technologies to replace part of the cooling water (Philips et al., 2007). By adding this technology to the process energy optimization presented earlier, it is possible to reduce cooling-water use to achieve a total freshwater consumption of 1.8 $gal_{water}/gal_{ethanol}$. This value also accounts for the production of a significant amount of hydrogen. In Figure 10.6, shown later in the chapter, the water consumption for the production of ethanol alone is presented, revealing a value of only 0.36 gal/gal when all the water sources within the process are treated and reused. (Martín et al., 2010b) There is no water discharge because the large amount of water circulating within the process enables enough dilution of the TDS to reuse the stream.

10.3.2.2 Hydrolysis

Another possible route to produce ethanol from lignocellulosic raw material is hydrolysis of the biomass. To pretreat switchgrass to expose the hemicelluloses and cellulose for further treatment, two methods have been found to be competitive for industrial ethanol production due to their high efficiency in liberating cellulose and hemicelluloses from the plant matrix: (1) dilute acid (H_2SO_4) pretreatment (Schell et al., 2003; Aden, 2009; Piccolo and Bezzo, 2009; Zhang et al., 2009) and (2) ammonia fiber explosion (AFEX) (Sun and Cheng, 2002; Alizadeh et al., 2005; Murnen et al., 2007). Martín and Grossmann (2010a) proposed an optimal conceptual design for ethanol production by hydrolysis which involved solving a superstructure that accommodates both pretreatment alternatives (AFEX and dilute acid) together with biomass hydrolysis, sugar fermentation, and four ethanol dehydration technologies (rectification, adsorption using corn grits, molecular sieves, and pervaporation). According to the MINLP solution, the optimal flowsheet includes dilute-acid pretreatment, which is cheaper and more environmentally friendly than the other options because it consumes less energy and requires less cooling water. Once the physical structure of the switchgrass is broken down to enable contact between the polymers and the enzymes, hydrolysis of the polymeric sugars takes place. This process is carried out in stirred-tank reactors at 50°C where the accessible cellulose and hemicelluloses are broken down into fermentable sugars (Gregg et al., 1995; Wooley et al., 1999; Hamelink et al., 2005; Piccolo and Bezzo, 2009; Zhang et al., 2009). The sugars, mainly glucose and xylose, are fermented in water into ethanol. The reactions are different in terms of yield and velocity. The optimal temperature is 38°C, so that both are fermented at the same time. A number of different products are obtained together with ethanol, such as different acid products of the metabolic paths of the microorganisms used (*Z. mobilis* bacterium) (Wooley et al., 1999; Zhang et al., 2009; Piccolo and Bezzo, 2009). The purification stages consist of the removal of solids, lignin, and cells from the liquid slurry coming out of the fermentor. The lignin is used to obtain energy for the process. Finally, ethanol is dehydrated by means of a beer column followed by molecular sieves to achieve fuel-grade quality. Next, heat integration is performed. Figure 10.4 shows the flowsheet as developed by Martín and Grossmann (2010a), and Figure 10.5 presents the energy consumption and cooling needs of the process. The production cost is \$0.80/gal.

The production of ethanol from hydrolysis of switchgrass is a gentler process than gasification in terms of operating pressures and temperatures resulting in cooling requirements less than half those for the thermochemical process (Figure 10.5). However, the lower temperatures prevent the extensive use of air cooling to replace water as a cooling agent. The design of the optimal water network (Martín et al., 2011b) reveals a total freshwater consumption of 1.66 $gal_{water}/gal_{ethanol}$ (see Figure 10.6) and a wastewater discharge of 0.52 $gal_{water}/gal_{ethanol}$ because the lower flowrates in the water network do not allow TDS dilution and the high cost of removing TDS results in this water being discharged instead of treated.

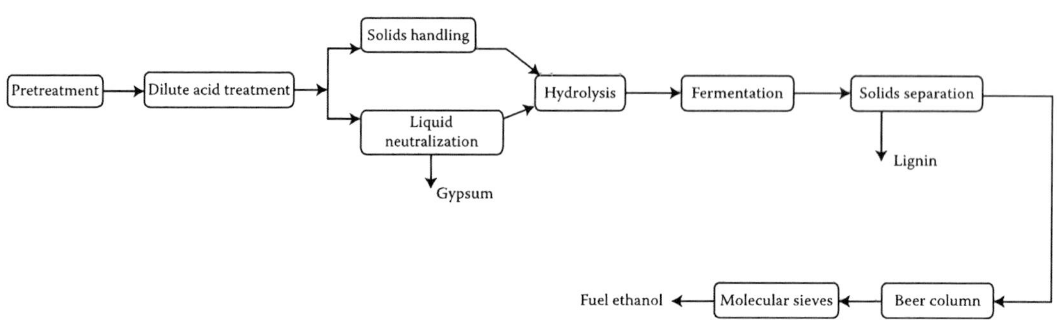

FIGURE 10.4
Hydrolysis-based production process for ethanol from switchgrass. (Martín, M. and Grossmann, I.E. 2010a. *AIChE Journal* doi: 10.1002/aic.12735. With permission.)

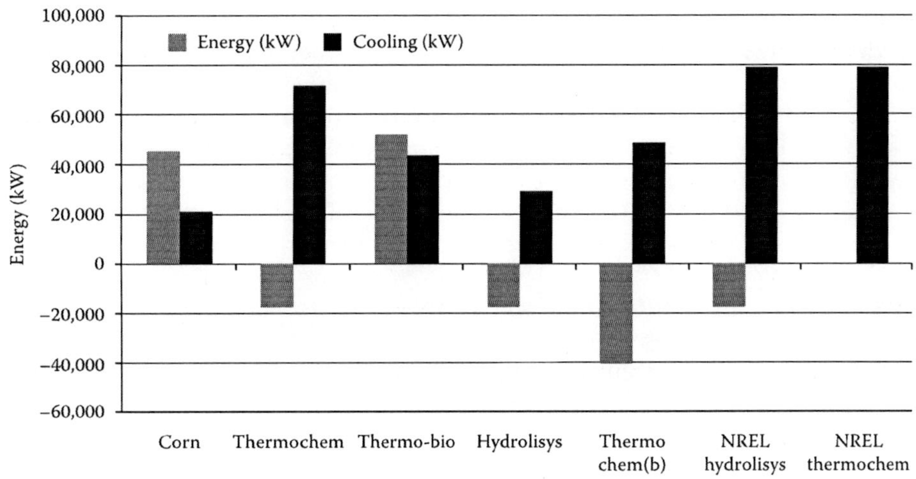

FIGURE 10.5
Energy consumption and cooling requirements for ethanol production processes. (Martín, M., Ahmetovic, E., and Grossmann, I.E. 2011c. *Industrial & Engineering Chemistry Research* 49(17): 7972–7982. With permission.)

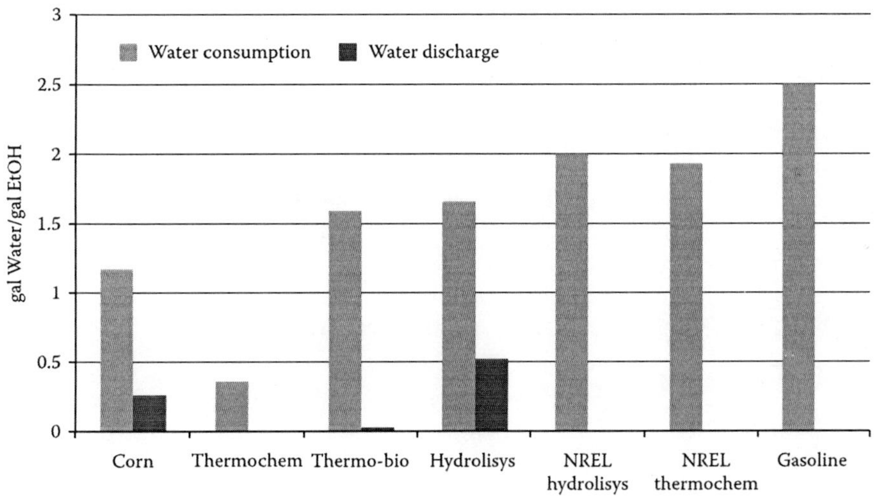

FIGURE 10.6
Water consumption of ethanol production processes. (M. Martín. E. Ahmetovic, IE. Grossmann. Ind. Eng. Chem Res. 50 (7), pp. 3705–3721, 2011. With permission.)

10.3.2.3 Fischer–Tropsch-Diesel

Not only bioethanol can be produced from lignocellulosic materials, but also diesel or hydrogen. The history of FT synthesis can be traced back to the beginning of the 20th century, when in 1902 Sabatier and Senderens discovered that CO could be hydrogenated over Co, Fe, and Ni to methane. Later, in 1925, Franz Fischer, founding director of the Kaiser-Wilhelm Institute of Coal Research in Mülheim an der Ruhr, and Hans Tropsch reported synthesis of hydrocarbon liquids and solid paraffins on Co–Fe catalyst under mild conditions of 250–300°C. FT technology was extensively developed and commercialized in Germany during World War II when that country was denied access to petroleum-rich regions of the world. Likewise, when South Africa faced a world oil embargo during their era of apartheid, it used FT technology to sustain its national economy. FT liquids can be refined to various proportions of renewable (green) gasoline, diesel fuel, and aviation fuel depending on process conditions. Depending on the types and quantities of FT products desired, either low- (200–240°C) or high-temperature (300–350°C) synthesis at pressures ranging from 10 to 40 bar is used with either cobalt- or iron-based catalysts (Brown and Wright, 2009).

Martín and Grossmann (2010c) proposed a superstructure which embeds various alternative technologies for the optimization of biodiesel production from switchgrass. The process starts with gasification of the biomass. Two different alternatives are evaluated, direct and indirect gasification. The gas obtained is reformed and cleaned up, and its composition may be adjusted in terms of the CO/H_2 ratio for the optimal production of the diesel fraction using up to three possible technologies (bypass, PSA, and water shift). Next, the removal of CO_2 and H_2S is performed by means of two cleanup processes, absorption in ethylamines and PSA. Then the FT reaction is carried out and the products separated. Hydrocracking of the heavy products is also considered to increase the yield toward diesel. Again, the number of alternative processes and tradeoffs makes it difficult to propose a flowsheet without the help of a systematic approach. The system optimization is formulated as an MINLP problem where short-cut models, reduced-order models, design equations, and mass and energy balances are used for the various units involved in the production process, with special focus on the FT reactor, the hydrocracking unit, and the CO/H_2 adjustment unit. The MINLP is solved for optimal production of the biodiesel fraction while minimizing energy consumption. Energy integration and an economic evaluation are also performed.

The optimal solution requires the use of indirect gasification followed by steam reforming. The CO/H_2 ratio is maintained at the value obtained after gas reforming because there is a reasonable excess of hydrogen, while the temperature is adjusted at the FT reactor and at the hydrocracker to increase biodiesel production, resulting in a promising production cost of \$0.71/gal. Figure 10.7 shows the optimal flowsheet.

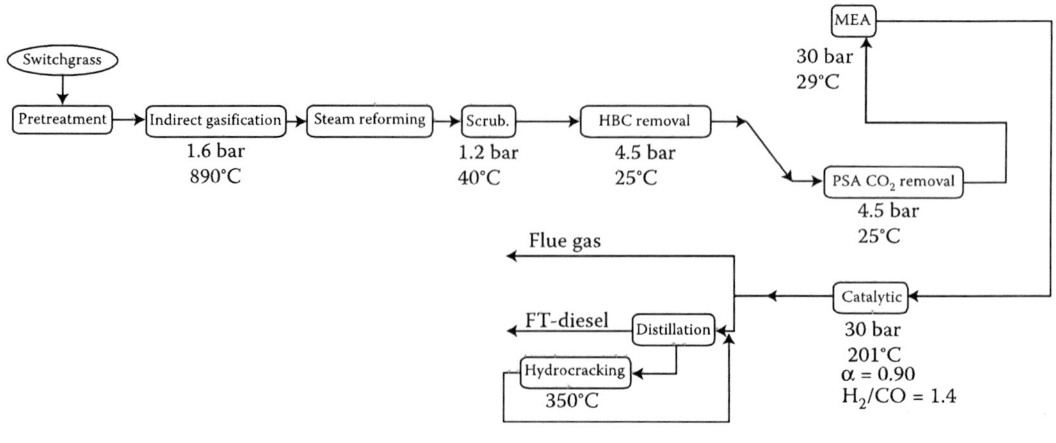

FIGURE 10.7
Optimal flowsheet for the production of FT diesel. (Martín, M. and Grossmann, I.E. 2011b. *Industrial & Engineering Chemistry Research* 50(23): 13485–13499. With permission.)

10.3.2.4 Other Bioproducts

Figure 10.8 shows a summary of the products that can also be obtained from lignocellulosic biomass. The syngas obtained from biomass is an interesting building block for a number of chemicals. This chapter does not describe the production of high-value products such as fine chemicals or polymers, but rather focuses on hydrogen, a valuable fuel for fuel-cell-powered vehicles or an intermediate needed for the production of petroleum-based chemicals.

Hydrogen can also be produced by generating syngas by gasification and then cleaning up the gas and shifting the energy value of the CO to H_2 using the water–gas shift reaction (WGSR). To reduce the production cost, a new reactor design which is capable of separating the hydrogen produced through a permeable membrane has been proposed (Doong et al., 2005; Fu and Wu, 2007; Ji et al., 2009; Adrover et al., 2009; Sa et al., 2009).

Martín and Grossmann (2011b) coupled the newly developed reactor design using membrane technology (Ji et al., 2009) with two different gasification technologies, direct and indirect gasification, and two reforming modes, partial oxidation and steam reforming, creating a superstructure in the form of an MINLP. The operating conditions in the gasifiers and at the WGSR are optimized for hydrogen production in each of the four alternative technologies. Next, energy is integrated.

Finally, an economic evaluation is performed to determine the production cost. It turns out that indirect gasification with steam reforming is the preferred technology, providing higher production yields (0.13 $kg/kg_{dryBiomass}$) than those reported in the literature for hydrogen from natural gas and at a lower and promising production cost of $0.68/kg.

10.3.3 Oil-Based Biofuels

Many researchers have concluded that vegetable oils hold promise as alternative fuels for diesel engines (Scholl and Sorenson, 1993; Wagner et al., 1984). However, using raw vegetable oils for diesel engines can cause numerous engine-related problems (Korus et al., 1982). The increased viscosity, low volatility, and poor cold-flow properties of vegetable oils lead to severe engine deposits, injector coking, and piston-ring sticking (Pestes and Stanislao, 1984; Perkins et al., 1991). Therefore, research has focused on developing

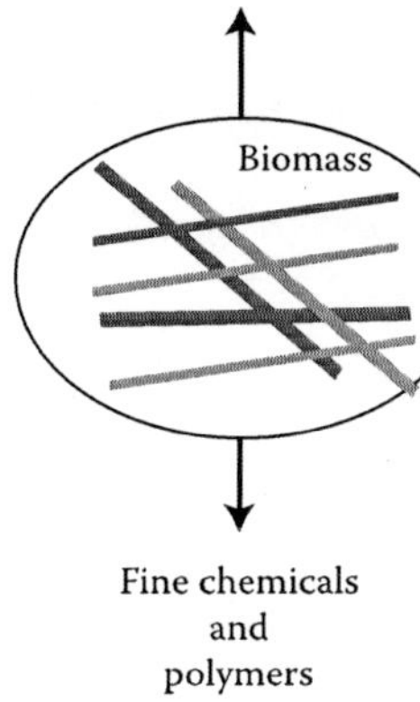

FIGURE 10.8
Biorefinery. (Martín, M. and Grossmann, I.E. 2010d. *Industrial & Engineering Chemistry Research*. With permission.)

transformation products by means such as pyrolysis, microemulsion, and transesterification. The transesterification process has been used most widely. It removes glycerol from the triglycerides and replaces it with radicals from the alcohol used for the conversion process (Kusy, 1982). This process decreases the fuel's viscosity, but maintains the cetane number and heating value. The monoesters produced by transesterifying vegetable oils with alcohol are known as *biodiesel* (Canakci and Gerpen, 1999; Fukuda et al., 2001).

10.3.3.1 First-Generation Biodiesel

Biodiesel is obtained using vegetable oils such as sunflower, cottonseed oil, rapeseed oil, soybean oil, palm oil, peanut oil, and canola oil as raw materials (Levelton Engineering Ltd, 2002; Shahid and Jamal, 2008). Their use is distributed across the globe depending on the production capabilities of different countries. For example, canola is the preferred oil in Canada, cottonseed oil is abundantly produced in Pakistan, palm oil is very frequently used in Malaysia; meanwhile, soybeans remain the largest oilseed crop produced. Biodiesel can be twice as expensive as crude-based diesel (biodiesel from soybeans is estimated to cost approximately \$2.00–\$2.50/gal), but the capital investment for a biodiesel plant is far lower (Tao and Aden, 2009). However, the competition of these technologies with the food chain and the cost of the raw material are directing the focus to the use of alternative sources of oil.

10.3.3.2 Cooking-Oil-Based Diesel

Among oils, waste cooking oil is a promising alternative to vegetable oil for biodiesel production because it is much less expensive than pure vegetable oil. So far, waste oils have been sold commercially as animal feed. However, since 2002, the European Union has enforced a ban on feeding these mixtures to animals because many harmful compounds are formed during frying. Obviously, the conditions used for frying (temperatures of 160–200°C) cause major physical and chemical changes to the oil, which differ from oil to oil depending on the composition of each. In general, three types of reactions occur during frying: thermolytic, oxidative, and hydrolytic. As a result, some of the common physical changes observed in vegetable oil are: (i) increase in viscosity, (ii) increase in specific heat, (iii) change in surface tension, (iv) change in color, and (v) increase in the tendency of fat to foam. As a combined result of all these chemical reactions, many undesirable compounds are formed, increasing the polar fraction and preventing the use of waste cooking oil as an additive to feed mixtures for domestic animals because it could result in the reintroduction of harmful compounds back into the food chain through animal meat (Kulkarni and Dalai, 2006). Hence, waste cooking oil must be disposed of safely or be used in a way that is not harmful to human beings.

The quantity of waste cooking oil generated per year is huge (Kulkarni and Dalai, 2006). The disposal of waste cooking oil is a problem because disposal methods may contaminate the water. Many developed countries have set policies that penalize the disposal of waste oil through water drainage systems. Therefore, the production of biodiesel from waste cooking oil becomes a better way to reuse the oil and to dispose of it efficiently and economically (Felizardo et al., 2006; Phan and Phan, 2008; Banerjee and Chakraborty, 2009). However, the high demand for diesel fuel and the limited availability of waste cooking oil (Wiltsee, 1998; Zhang et al., 2003; Pugazhvadivu and Jeyachandran, 2005; Kulkarni and Dalai 2006) indicate that biodiesel from cooking oil cannot replace diesel fuel, but can contribute to reducing dependency on crude-based diesel. The properties of biodiesel from waste cooking oils would be largely dependent on the physiochemical properties of these feedstocks.

10.3.3.3 Production Process of Biodiesel from Oil

The transesterification reaction depends on several parameters such as reaction temperature and pressure, reaction time, rate of agitation, type of alcohol used, molar ratio of alcohol to oil, type and concentration of catalyst used, and concentrations of moisture and free fatty acids (FFA) in the feed oil. Methanol is the alcohol most commonly used because of its low price, but ethanol has also been tested. There are four alternatives for performing the transesterification of the oil toward biodiesel depending on the catalysts used: acid-catalyzed, alkali-catalyzed, enzymatic, and no catalyst under supercritical conditions. The alkali-catalyzed reaction is limited because of the formation of soaps between the catalyst and the FFA and the negative effect of the presence of water in the oil which favors saponification, although the reaction is quick. The acid-catalyzed reaction is not limited by the purity of the reactants, but by the reaction time, which is approximately four days. The enzyme-catalyzed reaction is insensitive to water and FFA, and it is easier to purify the products. However, the catalyst is more expensive, and its conversion rate is still lower than that of the previous two processes. Finally, it is possible to transesterify oil using supercritical methanol. Although the yield of methyl ester is high, the molar ratio of methanol to oil is extremely high (1:42) and the operating conditions are extreme (350°C and 45 MPa). (Tomasevic and Siler-Marinkovic, 2003; Kulkarni and Dalai, 2006; Meher et al., 2006; Zheng et al., 2006; Sharma et al., 2008; Marchetti and Errazu, 2008; Utla and Kocak, 2008).

A number of authors have developed and evaluated production processes based on those technologies from both technical and economic points of view using a simulation approach. These studies have revealed that the acid-catalyzed process is preferable because it is less affected by impurities in the raw material, thus reducing the need for purification pretreatment stages. Recently, the use of heterogeneous catalyzed processes (enzymatic, basic, or acid-based) or supercritical-fluid-based processes is gaining support because of the reduction in the number of separation stages, resulting in a very competitive price (Zhang et al., 2003a,b; Hass, 2006; Harding et al., 2007; West et al., 2008; Marchetti, 2008; Apostolakou et al., 2009; Diaz et al., 2009; Glisic, 2009; Santana et al., 2009; Sotof et al., 2010; van Kasteren and Nisworo, 2007). Recently, Martín and Grossmann (2010d) have optimized the production of biodiesel in terms of technology (homogeneous alkali and acid-catalyzed, heterogeneous enzymatic, or basic and noncatalyzed under supercritical conditions) and operating conditions using surface-response models for the reactors. The model was formulated as a GDP in which each of the flowsheets was simultaneously optimized and heat-integrated. The result is that for cooking oil, the heterogeneous catalyzed reaction is the best process, yielding a production price of \$0.66/gal with an energy consumption of 1.94 MJ/gal (half the best estimates) and a water consumption of 0.33 gal/gal of biodiesel, almost one-third of values reported in the literature (Pate et al., 2007).

A by-product of the transesterification of oil is glycerol. Because of the expected increase in the production of biodiesel, research has focused on the use of glycerol to produce different value-added chemicals such as hydrogen (Cortright et al., 2002; Deluga et al., 2004; Davda et al., 2005; Hashaikeh et al., 2006; Dauenhauer et al., 2006; Byrd et al., 2008; Slinn, 2008; Adhikari, 2009); 1,3-propanediol which can be used in composites, adhesives, laminates, powders, UV-cured coatings, moldings, novel aliphatic polyesters, copolyesters, solvents, antifreeze, and for other end uses (Shell Chemicals, 2006); 1,2 propanediol, which is a major commodity chemical with an annual production of over 1 billion pounds in the United States (Dasari et al., 2005); dihydroxyacetone, used in the cosmetics industries as a tanning agent (Bauer et al., 2005); succinic acid, which can be used for the manufacture of synthetic resins and biodegradable polymers and as an intermediate for chemical synthesis (Zeikus, 1980); and polyglycerol esters which are used as antifogging and antistatic additives, lubricants, and plasticizers (Clacens et al., 2002), as well as polyesters and polyhydroxyalkonates (Pachauri and He, 2006).

10.3.4 Algae-Based Biofuels

Microalgae are also a biomass source with great potential. Microalgal oils are now believed to be a promising feedstock for biodiesel production because the oil they generate has a fatty-acid composition similar to that of vegetable oils (Zhang et al., 2009; Mata et al., 2010). To produce a sufficient amount of oil, microalgae must be carefully selected so that they are capable of producing large quantities of lipids while being tolerant to fluctuations in operating conditions. Algae convert sunlight, saline water, nutrients, and CO_2 into proteins, carbohydrates, and lipids, with a growth rate that doubles their biomass up to five times a day. Later, growth is stopped by nutrient limitations, and then only lipids are produced. Algae can grow not only on normal carbon sources, such as glucose, fructose, and other sugars, but also on wastes from the agriculture and food industries, and using saline water or wastewater, thus lowering the cost of microalgal oil while providing a good water treatment to remove NH_2^+, NO_3^-, and PO_4^{3-} (Mata et al., 2010). Next, the algae are harvested to eliminate the water that accompanies them, which can be recycled, using flotation, settling (helped by flocculation), centrifugation, or filtration. Finally, the oil is extracted and either transesterified (using the same processes as for any other oil), or catalytically converted into gasoline (Johnson and Sprage, 1987).

The main problem with producing biodiesel from algae is that to make it commercially viable, the algae need a concentrated and plentiful CO_2 source, which is not available in sufficient quantities, and furthermore it is necessary to reduce the high harvesting costs (Pokoo-Aikins et al., 2009). The current microalgae oil price is \$0.25/lb (http://www.soley.cn/oil-trade.html), but it could be as low as \$0.07/lb (Pokoo-Aikins et al., 2009). However, certain processes for the production of ethanol, hydrogen, or FT diesel by gasification generate significant amounts of CO_2 that could be used in this process. Furthermore, new technologies are under development to reduce the harvesting costs of microalgae (Univenture, 2009). The potential yield of biofuel from microalgae is one or two orders of magnitude higher than for any other crop and five times greater than the best yield for ethanol, as can be seen Figure 10.9 (Oak Ridge NL, 2005; Fulton, 2006; Christi, 2007; UNFAO, 2008). The use of microalgae as feedstock for the production of biodiesel has been sponsored by the U.S. Department of Energy. It is reported that microalgae can be produced using salt water (eliminating the problem of water usage), and that with 1–3 million acres, approximately 2–5% of the cropland currently used in the United States, it would be possible to meet U.S. diesel needs. Furthermore, the production of diesel from algae could be achieved at a lower cost than the current price of diesel (Pimentel and Patzek, 2005). In a study by Martín and Grossmann (2010d), it was demonstrated that oil from algae can be produced at \$0.06/lb as long as new technology released by Univenture Inc. (2009) is implemented. Next, a GDP similar to the one proposed for the production of biodiesel from cooking oil has been solved to determine the best production process among the five alternatives mentioned earlier, homogeneous acid- or alkali-catalyzed, heterogeneous enzymatic, basic, or noncatalyzed under supercritical conditions, simultaneously optimizing and with heat integration for each process. For this type of oil, the best process is the alkali-catalyzed process, with a production cost of \$0.42/gal, requiring 1.94 MJ/gal of energy and 0.60 gal/gal of water. The second-best process is fairly close, only \$0.02/gal more expensive, and is the heterogeneous basic catalyzed process.

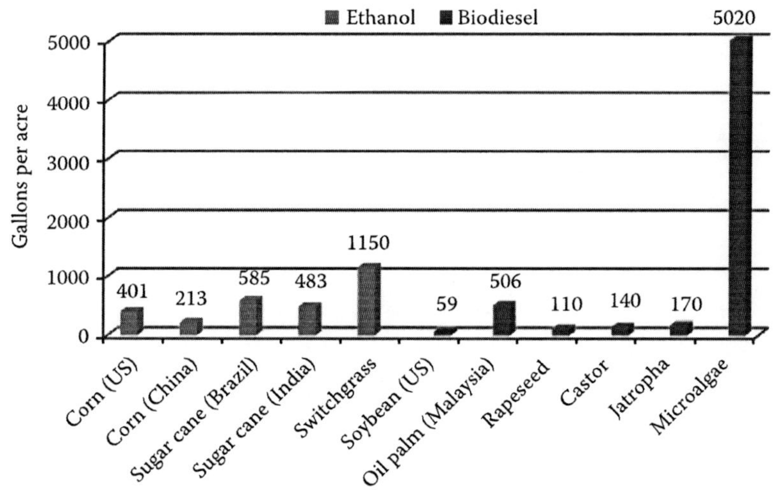

FIGURE 10.9
Yield of ethanol and diesel from different raw materials. (Martín, M. and Grossmann, I.E. 2010d. *Industrial & Engineering Chemistry Research*. With permission.)

The potential of microalgae is not limited to the oil they can generate, but they can also be used to obtain ethanol by fermenting the starch and cellulose contained in them, hydrogen by gasification or biological conversion, liquid fuels by pyrolysis or thermal liquefaction, methane by digestion, or other high-value products (Li et al., 2008; Rösch, 2009; Brennan and Owende, 2010). It is also possible to use the oil for biodiesel production while the residues are used to obtain ethanol by fermentation (Bush and Hall, 2006; Chen et al., 2009).

10.4 Concluding Remarks

Strong competition in the energy market requires alternative fuels based on biomass (e.g., bioethanol, biodiesel) to be produced in an efficient and sustainable way. Mathematical optimization techniques together with conceptual design have been traditionally used in the petrochemical industry to improve the performance and operation of processes. Therefore, as has been shown in this chapter, process systems engineering tools can be effectively used for the optimal design of biofuel processes. Sustainable designs must be optimized in terms of energy and freshwater consumption as the two most important indices for efficient performance. Mathematical programming techniques, process synthesis concepts, and newly developed technologies have been used to develop optimal processes for the production of bioethanol, biodiesel, and hydrogen. These designs must of course be validated with more detailed simulations and pilot-plant data.

As for future work, one logical step is to use the information obtained from the models reported in this chapter in the design and operation of biorefineries such as that illustrated in Figure 10.10, which are capable of producing a wide variety of products from fuels to chemicals or even energy (Bridgwater, 2003; Kamm and Kamm, 2007; Rösch et al., 2009). This line of work has begun to be addressed by several authors (Sammons et al., 2006, 2008; Dimian, 2007; Kokossis and Yang, 2010; Ng, 2010; Mansoornejad et al., 2010).

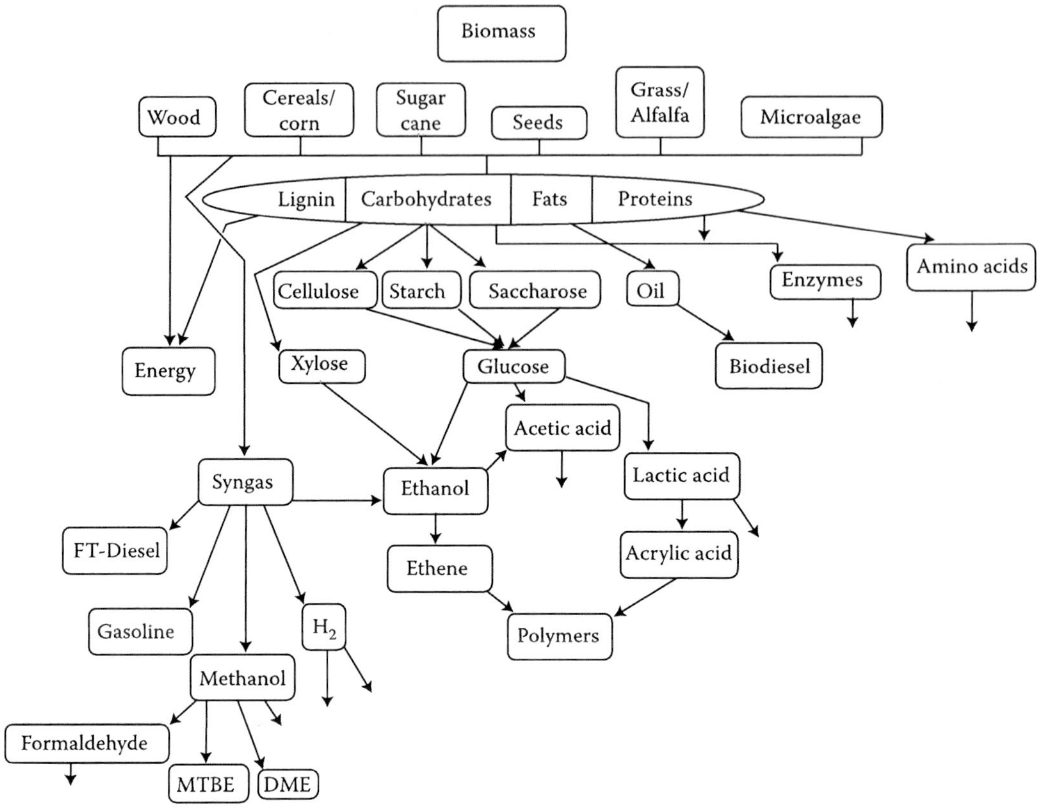

FIGURE 10.10
Network of raw materials and processes for a biorefinery.

However, there is scope for including concepts from enterprise-wise optimization (Grossmann, 2005; Varma et al., 2007) such as deterministic multiperiod mixed-integer linear design models in which biorefineries are treated as process networks (Sahinidis and Grossmann, 1989; Norton and Grossmann, 1994; Wassick, 2009). Furthermore, to address uncertainties in the supplies of and demands for raw materials and products, stochastic programming models for process network planning can also be developed (Pistikopoulos, 1995; Liu and Sahinidis, 1995; Ierapetritou et al., 1996; Cheng et al., 2003; Barbaro and Bagajewicz, 2004; Jung et al., 2008). Finally, these process network models can be coupled with inventory models to determine optimal safe inventory levels (Zipkin, 2000; You and Grossmann, 2008). This, however, requires the solution of nonconvex MINLP problems (You and Grossmann, 2010).

Acknowledgments

The authors gratefully acknowledge NSF Grant CBET0966524 and the Center for Advanced Process Decision-making at Carnegie-Mellon University. Dr. M. Martin gratefully acknowledges the financial support from the Ministry of Education and Science of

Spain and the Fulbright Commission which provided an MICINN–Fulbright postdoctoral fellowship.

References

Aden, A. 2007. Water usage for current and future ethanol production. *Southwest Hydrology* September/October, 22–23.

Adhikari, S., Fernando, S.D., and Haryanto, A. 2009. Hydrogen production from glycerol: An update. *Energy Conversion and Management* 50: 2600–2604.

Adrover, M.E., López, E., Borio, D.O., and Pedernera, M.N. 2009. Simulation of a membrane reactor for the WGS reaction: Pressure and thermal effects. *Chemical Engineering Journal* doi: 10.1016/j.cej.2009.04.057.

Agrawal, R., Singh, N., Ribeiro, F., and Delgass, W. 2007. Sustainable fuel for the transportation sector. *Proceedings of the National Academy of Sciences of the United States of America* 104(12): 4828–4833.

Alizadeh, H., Teymouri, F., Gilbert, T.I., and Dale, B.E. 2005 Pretreatment of switchgrass by ammonia fiber explosion (AFEX). *Applied Biochemistry and Biotechnology* 121–124: 1133–1141.

Apostolakou, A.A., Kookos, I.K., Marazioti, C., and Angelopoulos, K.C. 2009. Techno-economic analysis of a biodiesel production process from vegetable oils. *Fuel Processing Technology* 90: 1023–1031.

Bagajewicz, M. 2000. A review of recent design procedures for water networks in refineries and process plants. *Computers and Chemical Engineering* 24(9–10): 2093–2113.

Bagajewicz, J. and Faria, D. 2009. On the appropriate architecture of the water/wastewater allocation problem in process plants. In J. Jezowski (eds), *Computer-Aided Chemical Engineering*, Vol. 26, Elsevier Science (pp. 1–20).

Banerjee, A. and Chakraborty, R., 2009. Parametric sensitivity in transesterification of waste cooking oil for biodiesel production—A review. *Resources, Conservation, and Recycling* 53: 490–497.

Barbaro, A. and Bagajewicz, M.J. 2004. Managing financial risk in planning under uncertainty. *AIChE Journal* 50: 963–989.

Bauer, R., Katsikis, N., Varga, S., and Hekmat, D. 2005. Study of the inhibitory effect of the product dihydroxyacetone on *Gluconobacter oxydans* in a semi-continuous two-stage repeated-fed-batch process. *Bioprocess and Biosystems Engineering* 28(1): 37–43.

Brennan, L. and Owende, P. 2010. Biofuels from microalgae: A review of technologies for production, processing, and extractions of biofuels and co-products. *Renewable and Sustainable Energy Reviews* 14: 557–577.

Bridgwater, A.V. 2003. Renewable fuels and chemicals by thermal processing of biomass. *Chemical Engineering Journal* 91(2–3): 87–102.

Brown, R.C. and Wright, M. 2009. Biomass conversion to fuels and electric power. In: Howarth, R.W. and Bringezu, S. (eds.), *Biofuels: Environmental Consequences and Interactions with Changing Land Use. Proceedings, Scientific Committee on Problems of the Environment (SCOPE): International Biofuels Project Rapid Assessment*, 22–25 September 2008, Gummersbach, Germany. Cornell University, Ithaca NY, USA. (http://cip.cornell.edu/biofuels/).

Bush, R.A. and Hall, K.M. 2006. Process for the production of ethanol from algae. U.S. Patent 7135308.

Byrd, A.J., Pant, K.K., and Gupta, R.B. 2008. Hydrogen production from glycerol by reforming in supercritical water over Ru/Al2O3 catalyst. *Fuel* 87: 2956–2960.

Canakci, M. and Van Gerpen, J. 1999. Biodiesel production via acid catalysis. *Transactions of the ASAE* 42(5): 1203–1210.

Carvahlo, A., Gani, R., and Matos, H. 2006. Design of sustainable processes: Systematic generation and evaluation of alternatives. In W. Marquardt and C. Pantelides (eds.), *Computer-Aided Chemical Engineering*, Vol. 21A, Elsevier Science, pp. 817–822.

Chen, P., Min, M., Chen, Y., Wang, L., Li, Y., Chen, Q., Wang, C. et al., 2009. Review of the biological and engineering aspects of the algae-to-fuels approach. *International Journal of Agricultural and Biological Engineering* 2–4: 1–30.

Cheng, L., Subrahmanian, E., and Westerberg, A.W. 2003. Design and planning under uncertainty: Issues on problem formulation and solution. *Computers and Chemical Engineering* 27: 781–801.

Chisti, Y. 2007. Biodiesel from microalgae. *Biotechnology Advances* 25: 294–306.

Clacens, J.M., Pouilloux, Y., and Barrault, J. 2002. Selective etherification of glycerol to polyglycerols over impregnated basic MCM-41 type mesoporous catalysts. *Applied Catalysis A: General* 227(1–2): 181–190.

Cole, D.E. 2007. Issues facing the auto industry: Alternative fuels, technologies, and policies. *ACP Meeting,* Eagle Crest Conference Center, June 20, 2007.

Cortright, R.D., Davda, R.R., and Dumesic, J.A. 2002. Hydrogen from catalytic reforming of biomass-derived hydrocarbons in liquid water. *Nature* 418: 964–966.

Dasari, M.A., Kiatsimkul, P.P., Sutterlin, W.R., and Suppes, G.J. 2005. Low-pressure hydrogenolysis of glycerol to propylene glycol. *Applied Catalysis A: General* 281(1–2): 225–231.

Dauenhauer, P.J., Salge, J.R., and Schmidt, L.D. 2006. Renewable hydrogen by autothermal steam reforming of volatile carbohydrates. *Journal of Catalysis* 244: 238–247.

Davda, R.R., Shabker, J.W., Huber, G.W., Cortright, R.D., and Dumesic, J.A. 2005. A review of catalytic issues and process conditions for renewable hydrogen and alkanes by aqueous phase reforming of oxygenated hydrocarbons over supported metal catalysts. *Applied Catalysis B: Environmental* 56: 171–86.

Delta, T. 2009. http://news.thomasnet.com/companystory/525435.

Deluga, G.A., Salge, J.R., Schmidt, L.D., and Verykios, X.E. 2004. Renewable hydrogen from ethanol by autothermal reforming. *Science* 303(56–60): 993–997.

Diaz, M.S., Espinosa, S., and Brignole, E.A. 2009. Model-based cost minimization in noncatalytic biodiesel production plants. *Energy Fuels* 23: 5587–5595.

Dimian, A.C. 2007. Renewable raw materials: Chance and challenge for computer-aided process engineering. In V. Plesu and P.S. Agachi (eds.), *Computer-Aided Chemical Engineering,* Vol. 24, pp. 309–318.

Doong, S., Roberts, M., and Lau, F. 2005. Direct hydrogen production from biomass gasifier using hydrogen selective membrane. *GTI Presentation to Minnesota Renewable Hydrogen Initiative Forum*, May 27, 2005.

Douglas, J. 1988. *Conceptual Design of Chemical Processes*. McGraw-Hill.

Duran, M. and Grossmann, I.E. 1986. Simultaneous optimization and heat integration of chemical processes. *AIChE Journal* 32(1): 123–138.

Dutta, A. and Phillips, S.D. 2009. *Thermochemical Ethanol via Direct Gasification and Mixed-Alcohol Synthesis of Lignocellulosic Biomass*. NREL/TP-510-45913.

Felizardo, P., Neiva, M.J., Correia, I.R., Mendes, J.-F., Berkemeier, R., and Bordado, J.M. 2006. Production of biodiesel from waste frying oils. *Waste Management* 26: 487–494.

Fu, C.-H. and Wu, J.C.S.J. 2007. Mathematical simulation of hydrogen production via methanol steam reforming using double-jacketed membrane reactor. *International Journal of Hydrogen Energy* 32: 4830–4839.

Fukuda, H., Kondo, A., and Noda, H. 2001. Biodiesel fuel production by transesterification of oils. *Journal of Bioscience and Bioengineering* 92(5), 405–416.

Fulton, L. 2006. Biodiesel: Technology perspectives. *Proceedings, Geneva UNCTAD Conference*, 30 November 2006.

Gissy, J., Knight, R.A., Onischak, M., Carty, R.H., and Babu, S.P. 1992. Technology development and commercialization of the Renugas process. *U.S. Finland Biofuels Workshop II*, Espoo, Finland, 24–30 August.

Glisic, S. and Skala, D. 2009. The problems in design and detailed analyses of energy consumption for biodiesel synthesis at supercritical conditions. *The Journal of Supercritical Fluids* 49(2), 293–301.

Gregg, D. and Saddler, J.N. 1995. Bioconversion of lignocellulosic residue to ethanol: Process flowsheet development. *Biomass and Bioenergy* 9(1–5): 287–302.

Grossmann, I.E. 2005. Enterprise-wide optimization: A new frontier in process systems engineering. *AIChE Journal* 51: 1846–1857.

Grossmann, I., Aguirre, P., and Barttfeld, M. 2005. Optimal synthesis of complex distillation columns using rigorous models. *Computers and Chemical Engineering* 29(6): 1203–1215.

Grossmann, I., Caballero, J., and Yeomans, H. 1999. Mathematical programming approaches to the synthesis of chemical process systems. *Korean Journal of Chemical Engineering* 16(4): 407–426.

Grossmann, I.E., Yeomans, H., and Kravanja, Z. 1998. A rigorous disjunctive optimization model for simultaneous flowsheet optimization and heat integration. *Computers and Chemical Engineering* 22(Suppl. S): 157–164.

Guillen-Gosalbez, G., Caballero, J., and Jimenez, L. 2008. Application of life cycle assessment to the structural optimization of process flowsheets. *Industrial and Engineering Chemistry Research* 47(3): 777–789.

Haas, M.J., McAloon, A.J., Yee, W.C., and Foglia, T.A. 2006. A process model to estimate biodiesel production costs. *Bioresource Technology* 97: 671–678.

Halasz, L., Povoden, G., and Narodoslawsky, M. 2005. Sustainable processes synthesis for renewable resources. *Resources Conservation and Recycling* 44(3): 293–307.

Hamelinck, C.N., Hooijdonk, G.V., and Faaij, A.P.C 2005. Ethanol from lignocellulosic biomass: Techno-economic performance in short-, middle- and long-term. *Biomass and Bioenergy* 28: 384–410.

Harding, K.G., Dennis, J.S., von Blottnitz, H., and Harrison, S.T.L. 2007. A life-cycle comparison between inorganic and biological catalysis for the production of biodiesel. *Journal of Cleaner Production* 16: 1368–1378.

Hashaikeh, R., Butler, I.S., and Kozinski, J.A. 2006. Selective promotion of catalytic reactions during biomass gasification to hydrogen. *Energy Fuel* 20: 2743–2746.

Hostrup, M., Harper, P., and Gani, R. 1999. Design of environmentally benign processes: Integration of solvent design and separation process synthesis. *Computers and Chemical Engineering* 23(10): 1395–1414.

Ierapetritou, M.G., Pistikopoulos, E.N., and Floudas, C.A. 1996. Operational planning under uncertainty. *Computers and Chemical Engineering* 20: 1499–1516.

Jezowski, J. 2008. Review and analysis of approaches for designing optimum industrial water networks. *Chemical and Process Engineering* 29: 663–681.

Ji, P., Feng, W., and Chen, B. 2009. Production of ultrapure hydrogen from biomass gasification with air. *Chemical Engineering Science* 64: 582–592.

Jimenez-Gonzalez, C. and Woodley, J.M. 2010. Bioprocesses: Modelling needs for process evaluation and sustainability assessment. *Computers & Chemical Engineering* Doi: 10.1016/j.compchemeng.2010.03.010.

Johnson, D.A. and Sprague, S. 1987. Liquid fuels from microalgae. *SERI/TP*: 231-3203 US. DOE. http://www.nrel.gov/docs/legosti/old/3202.pdf.

Jung, J.Y., Blau, G., Pekny, J.F., Reklaitis, G.V., and Eversdyk, D. 2008. Integrated safety stock management for multi-stage supply chains under production capacity constraints. *Computers and Chemical Engineering* 32: 2570–2581.

Kamm, B. and Kamm, M. 2007. Biorefineries: Multi product processes. *Advances in Biochemical Engineering/Biotechnology* 105: 175–204.

Karuppiah, R. and Grossmann, I.E. 2006. Global optimization for the synthesis of integrated water systems in chemical processes. *Computers and Chemical Engineering* 30(4): 650–673.

Karuppiah, R., Peschel, A., Grossmann, I.E., Martin, M., Martinson, W., and Zullo, L. 2008. Energy optimization for the design of corn-based ethanol plants. *AIChE Journal* 54(6): 1499–1525.

Kocis, G. and Grossmann, I. 1987. Relaxation strategy for the structural optimization of process flow sheets. *Industrial and Engineering Chemistry Research* 26(9): 1869–1880.

Kokossis, A.C. and Yang, A. 2010. On the use of systems technologies and a systematic approach for the synthesis and the design of future biorefineries. *Computers & Chemical Engineering* 4: 1397–1405.

Korus, R.A., Mousetis, T.L., and Lloyd, L. 1982. Polymerization of vegetable oils. In *Vegetable Oil Fuels—Proc. Int. Conf. on Plant and Vegetable Oils as Fuels*, Fargo ND, 2–4 August, pp. 218–223.

Kulkarni, M.G. and Dalai, A.K. 2006. Waste cooking oils: An economical source for biodiesel: A review. *Industrial & Engineering Chemistry Research.* 45: 2901–2913.

Levelton Engineering Ltd. 2002. *Assessment of Biodiesel and Ethanol Diesel Blends, Greenhouse Gas Emissions, Exhaust Emissions, and Policy Issues.* September 30, 2002.

Li, Y., Horsman, M., Wu., N., Lan, C.Q., and Dubois-Calero, N. 2008. Biofuels from algae. *Biotechnology Progress* 24: 815–820.

Linnhoff, B. 1993. Pinch analysis—A state-of-the-art overview. *Chemical Engineering Research and Design* 71(5): 503–522.

Liu, M.L. and Sahinidis, N.V. 1995. Optimization in process planning under uncertainty. *Industrial and Engineering Chemistry Research* 35: 4154–4165.

Mansoornejad, B., Chambost, V., and Stuart, P. 2010. Integrating product portfolio design and supply chain design for the forest biorefinery. *Computers & Chemical Engineering* 34(9): 1497–1506.

Marchetti, J.M., Miguel, V.U., and Errazu A.F. 2008. Techno-economic study of different alternatives for biodiesel production. *Fuel Processing Technology* 89: 740–748.

Martín, M. and Grossmann, I.E. 2010a. Energy optimization of bioethanol production via hydrolysis of switchgrass. *AIChE Journal* doi: 10.1002/aic.12735.

Martín, M., Ahmetovic, E., and Grossmann, I.E. 2010b. Optimization of water consumption in second-generation bioethanol plants. *Industrial & Engineering Chemistry Research* 49(17): 7972–7982.

Martín, M. and Grossmann, I.E. 2010c. Energy optimization of hydrogen production from biomass. *Computers & Chemical Engineering* 35(9): 1798–1806.

Martín, M. and Grossmann, I.E. 2010d. Optimization and heat and water integration for biodiesel production from cooking oil and algae. *Industrial & Engineering Chemistry Research.*

Martín, M. and Grossmann, I.E. 2011a. Energy optimization of bioethanol production via gasification of switchgrass. *AIChE Journal* 57(12): 3408–3428.

Martín, M. and Grossmann, I.E. 2011b. Process optimization of FT-diesel production from biomass. *Industrial & Engineering Chemistry Research* 50(23):13485–13499.

Martín, M., Ahmetovic, E., and Grossmann, I.E. 2011. Optimization of water consumption in second generation bioethanol plants. *Industrial & Engineering Chemistry Research* 50(7): 3705–3721.

Mata, T.M., Martins, A.A., Nidia, S., and Caetano, N.S. 2010. Microalgae for biodiesel production and other applications: A review. *Renewable and Sustainable Energy Reviews* 14: 217–232.

Meher, L.C., Vidya Sagar, D., and Naik, S.N. 2006. Technical aspects of biodiesel production by transesterification—A review. *Renewable & Sustainable Energy Reviews* 10: 248–268.

Minnesota Technical Assistance Program. (MTAP) 2008. Ethanol Benchmarking and Best Practices: *The Production Process and Potential for Improvement.*

Murnen, H.K., Balan, V., Chundawat, S.P.S., Bals, B., Sousa, L. da C., and Dale, B.E. 2007. Optimization of ammonia fiber expansion (AFEX) pretreatment and enzymatic hydrolysis of *Miscanthus* × *giganteus* to fermentable sugars. *Biotechnology Progress* 23: 846–850.

Ng, D.K.S. 2010. Automated targeting for the synthesis of an integrated biorefinery with multiple products. *The Chemical Engineering Journal* 21: 421–427.

Nielsen, D.R. and Prather, K.J. 2009. Adsorption of second-generation biofuels using polymer resins with *in situ* product recovery (ISPR) applications. *Paper 564f, AIChE Annual Meeting 2009,* Nashville TN.

Nishida, N., Stephanopoulos, G., and Westerberg, A. 1981. A review of process synthesis. *AIChE Journal* 27(3): 321–351.

Norton, L.C. and Grossmann, I.E. 1994. Strategic planning model for complete process flexibility. *Industrial and Engineering Chemistry Research* 33: 69–76.

Oak Ridge National Laboratory. 2005. *Biofuels from Switchgrass: Greener Energy Pastures.*

Pachauri, N. and He, B. 2006. Value-added utilization of crude glycerol from biodiesel production: A survey of current research activities. *ASABE Annual International Meeting, Paper Number: 066223* 2006. http://www.webpages.uidaho.edu/~bhe/pdfs/asabe066223.pdf.

Papoulias, S. and Grossmann, I. 1983. A structural optimization approach in process synthesis. 2. Heat-recovery networks. *Computers and Chemical Engineering* 7(6): 707–721.

Pate, R., Hightower, M., Cameron, C., and Einfeld, W. 2007. *Overview of Energy-Water Interdependencies and the Emerging Energy Demands on Water Resources*. Report SAND 2007–1349C. Los Alamos, NM: Sandia National Laboratories.

Perkins, L.A., Peterson, C.L., and Auld, D.L. 1991. Durability testing of transesterified winter rape oil (*Brassica napus* L.) as fuel in small bore, multi-cylinder, DI, CI engines. SAE Paper No. 911764. Warrendale, Mich.: SAE.

Pestes, M.N. and Stanislao, J. 1984. Piston ring deposits when using vegetable oil as a fuel. *Journal of Testing and Evaluation* 12(2): 61–68.

Pfromm, P.H. 2008. The minimum water consumption of ethanol production via biomass fermentation. *Chemical Engineering Journal* 2: 1–5.

Phan, A.N. and Phan, T.M. 2008. Biodiesel production from waste cooking oils. *Fuel* 87: 3490–3496.

Phillips, S., Aden, A., Jechura, J., Dayton, D., and Eggeman, T. 2007. *Thermochemical Ethanol via Indirect Gasification and Mixed Alcohol Synthesis of Lignocellulosic Biomass*. Technical Report, NREL/TP-510-41168, April 2007.

Piccolo, C. and Bezzo, F. 2009. A techno-economic comparison between two technologies for bioethanol production from lignocelluloses. *Biomass and Bioenergy* 33: 478–491.

Pimentel, D. 1991. Ethanol fuels—Energy security, economics, and the environment. *Journal of Agricultural and Environmental Ethics* 4(1): 1–13.

Pimentel, D. and Patzek, T.W. 2005. Ethanol production using corn, switchgrass, and wood: Biodiesel production using soybean and sunflower oil. *Natural Resources Research* 14(1): 65–76.

Pistikopoulos, E.N. 1995. Uncertainty in process design and operations. *Computers and Chemical Engineering* 19: 553–563.

Pistikopoulos, E. and Stefanis, S. 1998. Optimal solvent design for environmental impact minimization. *Computers and Chemical Engineering* 22(6): 717–733.

Pokoo-Aikins, G., Ahmed Nadim, A., El-Halwagi, M.M., and Mahalec, V. 2010. Design and analysis of biodiesel production from algae grown through carbon sequestration. *Clean Technologies and Environmental Policy* 12(3): 239–254

Pugazhvadivu, M. and Jeyachandran, K. 2005. Investigations on the performance and exhaust emissions of a diesel engine using preheated waste cooking oil as a fuel. *Renewable Energy* 30: 2189–2202.

Rösch, C., Skarka, J., and Patyk, A. 2009. Microalgae: Opportunities and challenges of an innovative energy source. *Proceedings, 17th European Biomass Conference and Exhibition*, 29 June–3 July, Hamburg, Germany. http://www.itas.fzk.de/deu/lit/2009/roua09b.pdf

Rudd, D., Powers, G., and Siirola, J. 1973. *Process Synthesis*. Englewood Cliffs, NJ: Prentice-Hall.

Sa, S., Silva, H., Sousa, J.M., and Mendes, A. 2009. Hydrogen production by methanol steam reforming in a membrane reactor: Palladium vs. carbon molecular sieve membranes. *Journal of Membrane Science* 339: 160–170.

Sahinidis, N.V. and Grossmann, I.E. 1989. Optimization model for the long range planning in the chemical industry. *Computers and Chemical Engineering* 13: 1049–1063.

Sammons, N., Eden, M., Cullinan, H., Perine, L., and Connor, E. 2007. A flexible framework for optimal biorefinery product allocation. *Environmental Progress* 26(4): 349–354

Sammons, N.E., Yuan, W., Eden, M.R., Aksoy, B., and Cullinan, H.T. 2008. A systematic framework for biorefinery production optimization. In B. Braunschweig and X. Joulia (eds.), *Computer Aided Chemical Engineering* 25: 1077–1082.

Santana, G.C.S., Martins, P.F., de Lima da Silva, N., Batistella, C.B., Maciel Filho, R., and Wolf Maciel, M.R. 2010. Simulation and cost estimate for biodiesel production using castor oil. *Chemical Engineering Research and Design* 88(56): 626–632.

Schell, D.J., Farmer, J., Newman, M., and McMillan, J.D. 2003. Dilute sulfuric acid pretreatment of corn stover in pilot-scale reactor. *Applied Biochemistry and Biotechnology* (105–108): 69–85.

Scholl, K.W. and Sorenson, S.C. 1993. Combustion of soybean oil methyl ester in a direct injection diesel engine. SAE Paper No. 930934. Warrendale, Mich.: SAE.

Shahid, E.M. and Jamal, Y. 2008. A review of biodiesel as vehicular fuel *Renewable and Sustainable Energy Reviews* 12: 2484–2494.

Sharma, Y.C., Singh, B., and Upadhyay, S.N. 2008. Advancements in development and characterization of biodiesel: A review. *Fuel* 87: 2355–2373.

Shell Chemicals 2006. *What is 1,3-propanediol (PDO)?* http://www.shellchemicals.com/1_3_propanediol/1,1098,300,00.html. Accessed 23 April, 2006.

Slinn, M., Kendall, K., Mallon, C., and Andrews, J. 2008. Steam reforming of biodiesel by-product to make renewable hydrogen. *Bioresource Technology* 99: 5851–5858.

Smith, E.M.B. and Pantelides, C.C. 1995. Design of reaction separation networks using detailed models. *Computers and Chemical Engineering* 19(Suppl. s): 83–88.

Sotoft, L.F., Rong, B.-G., Christensen, K.V., and Norddahl, B. 2010. Process simulation and economical evaluation of enzymatic biodiesel production plant. *Bioresource Technology* 101(14): 5266–5274

Stefanis, S., Livingston, A., and Pistikopoulos, E. 1995. Minimizing the environmental impact of process plants—A process systems methodology. *Computers and Chemical Engineering* 19(Suppl. S): 39–44.

Steffens, M., Fraga, E., and Bogle, I. 1999. Multicriteria process synthesis for generating sustainable and economic bioprocesses. *Computers and Chemical Engineering* 23(10): 1455–1467.

Sun, Y. and Cheng, J. 2002. Hydrolysis of lignocellulosic materials for ethanol production: A review. *Bioresource Technology* 83: 1–11.

Tao, L. and Aden, A. 2009. The economics of current and future biofuels. *In Vitro Cellular & Developmental Biology-Plant* 45: 199–217.

Tomasevic, A.V. and Siler-Marinkovic, S.S. 2003. Methanolysis of used frying oil. *Fuel Process Technol.* 81: 1–6.

United Nations Food and Agriculture Organization 2008. *The State of Food and Agriculture.*

Univenture Inc. 2009. hppt://algaevs.com

Utla, Z. and Kocak, M.S. 2008. The effect of biodiesel fuel obtained from waste frying oil on direct injection diesel engine performance and exhaust emissions. *Renewable Energy* 35: 1936–1941.

van Kasteren, J.M.N. and Nisworo, A.P. 2007. A process model to estimate the cost of industrial scale biodiesel production from waste cooking oil by supercritical transesterification. *Resources, Conservation, and Recycling* 50: 442–458.

Varma, V.A., Reklaitis, G.V., Blau, G.E., and Pekny, J.F. 2007. Enterprise-wide modeling and optimization—An overview of emerging research challenges and opportunities. *Computers and Chemical Engineering* 31: 692–711.

Wagner, L.E., Clark, S.J., and Schrock, M.D. 1984. Effects of soybean oil esters on the performance, lubrication oil, and water of diesel engines. SAE Paper No. 841385. Warrendale, MI: SAE.

Wang, Y.P. and Smith, R. 1994. Waste water minimisation. *Chemical Engineering Science* 49: 981–1006.

Wassick, J.M. 2009. Enterprise-wide optimization in an integrated chemical complex. *Computers and Chemical Engineering* 33(12): 1950–1963.

West, A.H., Posarac, D., and Ellis, N. 2008 Assessment of four biodiesel production processes using HYSYS. *Plant Bioresource Technology* 99: 6587–6601.

Wooley, R.J. and Putsche, V. 1996. *Development of an ASPEN PLUS Physical Property Database for Biofuels Components.* NREL/MP-425-20685, http://www.p2pays.org/ref/22/21210.pdf

Wu, M., Mintz, M., Wang., M., and Arora, S. 2009. *Consumptive Water Use in the Production of Ethanol and Petroleum Gasoline.* ANL/ESD/09-1. Argonne National Lab. http://www.transportation.anl.gov/pdfs/AF/557.pdf

Yee, T. and Grossmann, I.E. 1990. Simultaneous-optimization models for heat integration. 2. Heat-exchanger network synthesis. *Computers and Chemical Engineering* 14(10): 1165–1184.

Yeomans, H. and Grossmann, I.E. 1999. A systematic modeling framework of superstructure optimization in process synthesis. *Computers and Chemical Engineering* 23(6): 709–731.

You, F. and Grossmann, I.E. 2008. Design of responsive supply chains under demand uncertainty. *Computers and Chemical Engineering* 32: 2839–3274.

You, F. and Grossmann, I.E. 2010. Stochastic inventory management for tactical process planning under uncertainties: MINLP models and algorithms. *AIChE Journal* in press. doi: 10.1002/aic.12244

Zeikus, J.G. 1980. Chemical and fuel production by anaerobic bacteria. *Annual Review of Microbiology* 34: 423–464.

Zhang, Y., Dube, M.A., McLean, D.D., and Kates, M. 2003a. Biodiesel production from waste cooking oil: 1. Process design and technological assessment. *Bioresource Technology* 89: 1–16.

Zhang, Y., Dube, M.A., McLean, D.D., and Kates, M. 2003b. Biodiesel production from waste cooking oil: 2. Economic assessment and sensitivity analysis. *Bioresource Technology* 90: 229–240.

Zhang, S., Marechal, F., Gassner, M., Perin-Levasseur, Z., Qi, W., Ren, Z., Yan, Y., and Favrat, D. 2009. Process modeling and integration of fuel ethanol production from lignocellulosic biomass based on double acid hydrolysis energy. *Fuels* 23(3): 1759–1765.

Zhang, W., Wu, H., and Zong, M. 2009. Optimal conditions for producing microalgal oil with high oleic acid content from *Chlorella vulgaris* LB 112. Technical Report, College of Light Industry and Food Sciences, South China University of Technology. http://www.paper.edu.cn/en/paper.php?serial_number=200907-153 (Accessed June 2010).

Zheng, S., Kates, M., Dube, M.A., and McLean, D.D. 2006. Acid-catalyzed production of biodiesel from waste frying oil. *Biomass Bioenergy* 30: 267–272.

Zipkin, P.H. 2000. *Foundations of Inventory Management*. Boston, MA.

11

Methodologies Developed and Applied in Process Integration Studies for the Swedish Kraft Pulp and Paper Industry: A Review

Maryam Mahmoudkhani and Thore Berntsson

CONTENTS

11.1 Introduction

Improved energy management, process integration, and water reduction are among the measures that have been proposed to improve energy efficiency in the pulp and paper industry. This industry differs from industrial chemical plants in terms of its raw material and its water consumption as well as the layout of its processes. In the pulp and paper industry, wood is the main raw material, but this material, in addition to its chemical value, has a tremendous potential to be used for its energy content. The water system in the pulp industry is highly complex, due mainly to high water demand for both utilities and the process (the so-called process water); therefore, the utility system (also called the secondary heat system or the hot- and warm-water production system) is not as straightforward as in other industries. Moreover, the piping distances in pulp and paper mills are usually longer than in typical chemical industries. Taken together, these issues make the analysis of energy systems for pulp processes a highly complex task. This fact emphasizes the importance of process integration for efficient use of energy in the process.

Process integration tools and methods have been used to analyze the use of energy and other resources in the pulp and paper industry and to identify ways to increase productivity, decrease costs, and address environmental issues. These tools have been used to perform process-integration case studies of pulp mills from 1990 onward in various locations, including the United States, Canada, Finland, and Sweden. Projects in North America have focused primarily on possibilities to improve resource use in the short and medium terms, while studies in Sweden have been more strategic in nature and have had a more long-term implementation perspective (for strategic decision-making under energy-market uncertainties in future) (IEA, 2004). Depending on technical, geographical, and legal parameters, mill specifications differ in terms of the degree of modernization of equipment, water and energy systems, fuel consumption, carbon dioxide emissions, and taxation and policy instruments. In this chapter, the interaction between various Kraft pulp specifications and the choice of an appropriate process integration method is described, because the heat and water systems in Kraft pulping are more complex than those in other mills such as thermomechanical pulp (TMP) mills.

Available data on energy use in the pulp and paper sector indicate an ~30% higher energy consumption per product unit in North America, which is mainly due to the aggressive modernization and expansion programs undertaken in Scandinavian mills and the energy optimization strategies undertaken in response to high fuel prices (Männistö and Männistö 1999). For an average Swedish nonintegrated market pulp mill, a steam demand of 15–17 GJ/ADt has been reported (KAM 2003). Männistö and Männistö (2006) stated that CO_2 emissions are lower in Sweden than in Finland and Canada due to differences in energy efficiency, degree of integration, mill size, in-house power generation, and the types of fossil fuel used for energy supply. Männistö and Smith (2006) also reported that NO_x emissions are higher from Kraft pulp mills in the United States and Canada than in Sweden, mainly because of higher energy use in the former mills. North American bleached Kraft pulp mills have a total water consumption in the range of 50–70 m^3/adt pulp and use 72–76% solids firing in the recovery boiler (Bruce 2000). For Swedish bleached Kraft pulp mills firing black liquor with 72–80% dry solids, total water consumption is reported to range from 30 to 60 m^3/adt pulp (of which process water* accounts for ~50% of the total) (Environmental Database 2008). For corresponding model mills, process water consumption has been calculated to range from 17 to 35 m^3/adt pulp (FRAM 2005). Moreover, in older mills, the pinch temperature is lower than in modern mills (well below 100°C, compared to 120–140°C). Therefore, to improve mill energy performance, different process integration methods must be considered due to differences in the specifications of each mill.

In the literature, numerous studies have described various methods and tools for process integration in the pulp and paper industry. However, a summary of the methods developed and their applications in the Kraft pulp and paper industry has not yet been published. This chapter aims to provide an overview of the available methods and tools which have been used for process integration in the Swedish Kraft pulp and paper industry. Different methods are described and discussed, and the results of applying these methods to mill energy analysis are presented.

* In general, the water intake to the mill is divided into process water and utility water.

11.2 Process Integration Studies

The term *process integration* refers to a system-oriented and integrated approach, meaning that rather than optimizing process units separately, the interaction between different parts of the process is considered, and the system as a whole is optimized. Figure 11.1 sketches a typical sequence of systematic steps taken to perform a process integration study. These steps do not necessarily need to be taken consecutively, but in general all steps are included. A fundamental step before performing any process integration study is to develop an overall mass and energy balance for the process. Process simulation can provide the data required to extract the appropriate streams in a process and to identify their potential for process integration. Various process simulation programs have been used, ranging from BALAS* to WinGEMS,† with more focus in the pulp and paper industry on broad-spectrum application software like CADSim Plus. The stream specifications can then be identified and set in the flowsheets generated by one of these simulation software packages.

Process integration studies can be based on mathematical methods, thermodynamic methods, economic methods, or a combination of these. Examples of these methods include mathematical programming, pinch analysis, and hierarchical analysis. The most widespread methods for process integration in the pulp and paper industries are pinch analysis, mathematical programming, and to a certain extent exergy analysis, among which pinch analysis methods are well proven and established and the most commonly used in commercial projects (IEA 2004).

An existing concept based on pinch technology for performing any retrofit or for integrating new energy technologies is the *grand composite curve* (GCC). Pinch analysis tools are used to construct GCCs which can then be used to analyze the mill's energy systems. Early examples such as TARGET (written at UMIST) and PROTAB (developed by ICI) have been developed into highly sophisticated programs such as SUPERTARGET (Linnhoff March) and Advent/Aspen Pinch (Aspen Tech). Later, Hyprotech developed HX-NET, which became the main Aspen specialized pinch software module. These programs use complex targeting procedures and are relatively costly to purchase or license (Kemp 2007). Tools such as Pro-Pi‡ have therefore been developed to perform the essential calculations of energy and cost targets and to generate GCCs without including the network design.

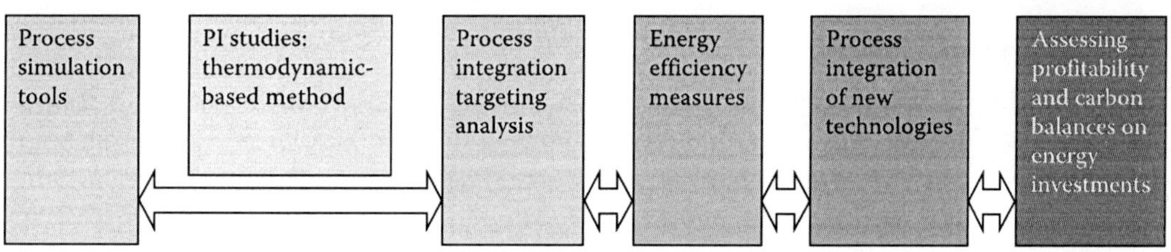

FIGURE 11.1
Example sequence of systematic steps for process integration studies based on thermodynamic methods.

* A steady-state simulation package for chemical processes with emphasis on pulp and paper, developed by VTT (Technical Research Center of Finland).

† A Microsoft Windows-based software package for studying pulp and paper production processes.

‡ Pro-Pi is an add-in module in MS Excel developed by CIT (Chalmers Industriteknik).

Pro-Pi has been used for research purposes and in applied industrial studies in Sweden and Finland. Canmet ENERGY* recently developed and tested a process integration software tool called "INTEGRATION" in Quebec. The tool aims to make process integration studies easier, faster, and cheaper and focuses on thermal energy usage (i.e., energy performance, quantity and quality of waste heat, and opportunities for heat recovery). The "INTEGRATION" tool is not yet completely documented, but as a continuation of the research work reported here, it could be interesting to study the possibilities of combining this tool with the recently developed methods presented in this chapter.

11.2.1 PI Studies: Thermodynamics-Based Methods

11.2.1.1 Pinch Analysis

Pinch analysis is a well-proven and established thermodynamics-based method with a structured approach which is used to design energy-efficient systems, that is, to maximize internal heat recovery within a process and to minimize process demands for the hot and cold duties supplied by utilities (Linnhoff 1993; Smith 1995; Kemp 2007). Pinch analysis aims to minimize the heat demand of the process. By knowing the heat content of all streams and their start and target temperatures, a heat-exchanger network (HEN) can be designed in which the external heat demand is minimized. It has, however, been reported that the traditional pinch method is not well suited to tackle certain problems. For example, pinch curves are not well suited for retrofit situations because no information on the existing heat-exchanger network (HEN) design can be extracted from these curves (Nordman and Berntsson 2001). Many efforts have therefore been made to develop new methods based on pinch analysis to extract information on existing HEN designs and to identify and evaluate opportunities for increasing energy efficiency. In addition to the lack of information on the existing HEN in traditional pinch analysis, neither the direct relation between energy and water reduction, nor the relation between the use and production of utilities, is considered. Because water handling in the pulp and paper industry is complex and closely related to both energy use and the environmental impacts of the mill, other approaches have been developed to tackle these issues. In the literature, various kinds of systematic methodologies combined with thermal pinch analysis have been presented. Examples of such methodologies are the combined method used in the "NRCan†" optimization software package used by VTT in Finland and the SDM™‡ software package (IEA 2004).

11.2.1.2 Exergy Analysis

In general, exergy analysis is used when there is a large pressure difference in a process unit or for electricity production units, particularly when electricity and heat production interact. Exergy analysis has, however, been used sporadically in the pulp and paper industry because electricity and heat are always treated separately in energy balances. If the transport and exchange of thermal energy is a dominant part of the energy system, then pinch analysis is preferable, because exergy analysis does not necessarily add further information beyond that provided by thermal pinch analysis (Grip et al. 2011). As for the energy content of the flows, in this case, temperature is suitable as a measure of energy

* http://canmetenergy-canmetenergie.nrcan-rncan.gc.ca.

† NRCan: Natural Resources Canada.

‡ SDM™: "Successive Design Methodology," developed by American Process, Inc., provides a systematic approach for simultaneous application of water and energy conservation in practical mill situations.

quality. There are, however, certain process units, for example, the recovery boiler and the electricity production units, to which exergy analysis can be applied. Although it has been shown that the recovery boiler has the lowest exergy efficiency in the mill, analysis of such process units is probably more useful for the unit manufacturer than for the pulp mill (A. Åsblad 2011, personal communication). Exergy analysis has, however, been used in combination with pinch analysis to provide a new perspective for identification of process enhancement opportunities involving energy upgrading and conversion in the pulp and paper industry. The method has, for example, been used to estimate exergy losses in the heat-exchanger network in a TMP mill (Hippinen et al. 2010) or in a black-liquor evaporation plant in a Kraft pulp mill (Marinova et al. 2009). In this method, exergy analysis is combined with pinch analysis by replacing temperature on the y-axis with Carnot efficiency.* The sequence of processes by which heat is produced and supplied to the end receptor is represented in this diagram. The area between the balanced hot and cold composite curves represents the exergy loss of the system (Marinova et al. 2009). Hippinen et al. (2010) studied the heat-exchanger network of a pulp and paper (TMP) mill using combined pinch and exergy analysis and concluded that no advantages could be obtained from this combination of techniques.

11.2.1.3 Mathematical Programming Methods

Mathematical programming is a tool by which mathematical relations can be identified in the system under analysis. The MIND (Method for analysis of INDustrial energy systems) tool developed in the Department of Management and Engineering in Linköping University is an example of a tool based on mathematical programming methods which has been applied to the analysis and optimization of industrial energy systems as well as to industries in cooperation with other players such as district heating systems. The optimization is carried out using mixed-integer linear programming (MILP) (Bengtsson et al. 2002; Karlsson 2011). To apply these mathematical methods in process integration studies, they should be combined with pinch analysis. Marechal and Kalitventzeff (1997) presented a methodology for a combined energy and environment synthesis of industrial processes in which the EMO (Effect Modeling and Optimization) mathematical formulation approach was used to combine numerical optimization involving MILP with process thermodynamic analysis. To the authors' knowledge, there exist very few studies dealing with the application of mathematical programming methods to energy analysis of Kraft pulp mills (Jönsson et al. 2008; Cakembergh-Mas et al. 2010; Karlsson 2011). No details of these methods are therefore discussed in this chapter.

11.3 Energy-Efficiency Measures Based on Newly Developed Methods

There are numerous ways to decrease energy usage in pulp and paper processes. Overviews of various measures to save both thermal and electrical energy can be found in Browne et al. (2001), Bruce (2000), and Loutfi and Ruzicka (2006). In this chapter, the focus is on measures applied for thermal energy savings, such as steam savings and excess heat utilization. The main method described in this chapter is process integration (the so-called *heat integration*)

* Carnot efficiency is defined as $1 - (T_0/T)$.

which can be achieved with or without major changes in process equipment. Several energy-efficiency measures for reducing steam consumption in pulp and paper mills can be found in the literature. The most frequently mentioned are internal heat recovery, condensate recovery, redesign of the hot- and warm-water systems (HWWS), and water reuse, as well as energy conversion and upgrading (Nordman and Berntsson 2001; Wising et al. 2002a,b, 2005; IEA 2004; Axelsson et al. 2006; Marinova et al. 2009; Mateos-Espejel et al. 2010a).

It is important to note that when performing energy analysis in a mill, the best approach is not always to eliminate or reduce the pinch violations first and then to use the remaining excess heat to reduce energy consumption. In many cases, efforts to eliminate or reduce pinch violations may encounter practical constraints, may be economically unattractive, or both. The most cost-effective solution with respect to how many, and which, pinch violations to resolve is actually an optimization problem. The target function is the net profit of the mill. For strict mathematical optimization, conditions such as the mill layout and piping distances are needed. Moreover, economic conditions such as energy prices and the annuity factor are also needed to find the best solution (Axelsson, 2008). To evaluate energy export potential from a mill in the future, both energy prices and the CO_2 emissions impact of energy exports and imports would be needed. Therefore, the need to develop new methods to provide the required information is obvious. This topic is discussed further in detail later in this chapter.

11.3.1 Pinch-Analysis-Based Advanced Composite Curves

An existing concept based on pinch technology for integrating new energy technologies is the grand composite curve (GCC). However, this curve is not well suited for retrofit situations, in which two problems occur with the GCC. First, the temperatures are not estimated correctly in existing systems having a heat demand that is large in relation to what could be achieved by enhanced heat recovery. Second, the GCC does not provide any information about the existing heat-exchanger network (HEN), which means that no indication of the changes needed in the HEN to approach theoretical levels can be given.

To cope with these problems, novel composite curves developed in the Heat and Power Department, Chalmers University of Technology, are proposed in this chapter. They are based on pinch technology and permit accurate representation of existing loads as well as possible improvements that could be obtained by retrofitting. As emphasized above, the current status and existing design of the energy systems in a plant influence considerably the available opportunities for improvement.

With the aid of four different composite curves above the pinch and four below in a temperature–heat load diagram, important information on such opportunities for a number of different measures can be extracted. The measures in question are improved heat exchange, introduction of a heat pump or a CHP plant, and the consequences of various process changes. These curves can therefore be used to assess possible heating-system improvements. The four types of curves above the pinch (see Figure 11.2) are

- Hot utility curve (HUC). This curve simply shows a composite of the hot utility as a temperature versus heat load diagram (e.g., for water, steam, or oil distribution systems).
- Actual heat load curve (AHLC). This curve shows a composite of all the cold streams or cold stream segments constituting the cold side of the heaters in the heat distribution system(s) in a temperature versus heat load diagram.

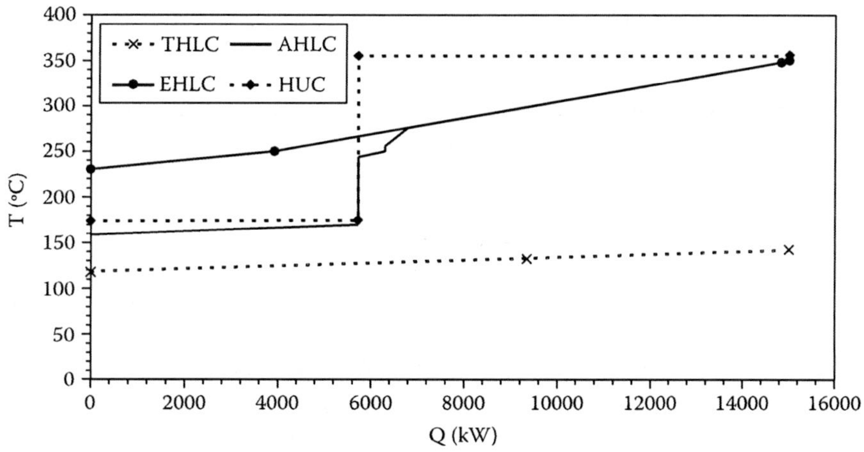

FIGURE 11.2
Temperature vs. heat load diagram above the pinch, showing the HUC, THLC, AHLC, and EHLC curves.

- Theoretical heat load curve (THLC). This curve shows the theoretically lowest temperature versus heat load at which heat can be introduced to the system when all thermodynamically correct opportunities for heat exchanger enlargement and introduction of new heat exchangers have been implemented at the *existing* external heat demand.
- Extreme heat load curve (EHLC). The EHLC curve shows the heat load vs. temperature level when the heat exchange is carried out with a minimum heat-exchanger area. This is the same as performing the heat exchange vertically in the composite curves. It also means that the externally supplied heat is transferred to the cold streams as high a temperature as theoretically possible.

In principle, the same four types of curves should also be constructed below the pinch and are called the cold utility curve (CUC), the actual cooling load curve (ACLC), the theoretical cooling load curve (TCLC), and the extreme cooling load curve (EHLC).

With the aid of these curves, three levels of complexity of changes in heating, cooling, and process systems can be identified and evaluated. These are

- Improvements involving only the HUC/CUC. Only the utility system is involved. In principle, these have no influence on the flexibility, controllability, or operability of the plant. Little or no investment cost in the heating/cooling/process system is required.
- Improvements involving AHLC/ACLC. Changes in the heating and cooling distribution systems are needed (e.g., adding one or more temperature level(s) to the heating or cooling system). No changes in the process heat-exchanger network (HEN). Relatively minor influence on flexibility, controllability, and operability. Moderate investment cost in the heating/cooling/process system.
- Improvements involving THLC/TCLC. Changes in the HEN are needed (new units or enlargements of existing units). Risk of major influence on flexibility, controllability, and operability. Risk of considerable investment cost in the heating/cooling/process system.

Generally speaking, with any improvement measure, there is a tradeoff between the possible magnitude of the energy improvement (normally influenced favorably by greater complexity of heating/cooling/process system changes) on the one hand and investment cost, flexibility, controllability, and operability on the other. Therefore, it is of high importance to identify opportunities for improvements as a function of their level of complexity.

The AHLC curve should lie between the THLC and EHLC curves. If any part of the AHLC lies below the THLC, then either that part is violating the pinch rules by heating below the pinch, or the actual $\Delta T_{min,HX}$ is less than specified. It should not be possible for the AHLC to lie above the EHLC curve because of the definition of the EHLC curve. An AHLC close to the EHLC curve means that the existing HEN has a good fundamental design with a small installed heat-exchanger area. This can, however, be a poor solution in a retrofit situation in which additional area must be installed to increase process-to-process heat exchange. The heaters that are to be freed or relieved are placed high in temperature, which leads to small ΔT's for the new or enhanced heat exchangers. A relatively large area must then be inserted for a given saving in external heating.

On the other hand, an AHLC close to the THLC means that the heaters in the existing HEN are placed low in temperature. More heat-exchanger area is then required to transfer the same amount of heat as in the previous case. This is therefore a poor design from a greenfield design viewpoint. However, this area has potential for better utilization. By increasing the internal heat exchange, heaters can be released or relieved at a lower temperature compared to the previous case. Intelligent matching using criss-cross heat exchange then leads to inserts of smaller area than in the previous case. From this discussion, it can be concluded that the cost of additional heat-exchanger area in an HEN retrofit should be relatively small when the AHLC is close to the THLC. When it lies close to the EHLC curve, the cost should be larger and should increase as the AHLC approaches the EHLC curve.

The base cost, that is, the area-independent part, for a new heat exchanger is far from negligible. If the AHLC is close to the THLC, this cost should also normally be smaller than when it is close to the EHLC. The reason is that with the original heaters placed low, the number of possible combinations between these and the hot streams is greater, and therefore the likelihood of finding matches with large streams reduces the number of matches needed. A third important cost is the cost of piping. The curves do not provide any information about how to reduce the length of piping, but fewer units generally mean less piping. Therefore, the consequences for piping cost can, at least partly, be studied, and once again, a situation with the AHCL close to the THLC is favorable.

The same discussion as above can also be held below the pinch with the four curves CUC, ACLC, TCLC, and ECLC. A more comprehensive description of the method can be found elsewhere (Nordman and Berntsson 2009a,b).

11.3.2 "Matrix Method" for Economical Retrofit of a Heat-Exchanger Network

When retrofitting an HEN using the traditional pinch method, many important parameters such as heat-exchanger area, number and type of heat exchangers, physical distance between the streams, piping, annual pressure-drop costs, and others are not considered in the analysis. An approach is desirable which would be suited for heat-exchanger networks that include many possible heat-exchanger alternatives. However, it is important that the area cost not be the only cost considered, because in practice it accounts for only a fraction of the total retrofit cost. A proposed method which systematically and rigorously solves the problem of determining retrofit solutions is the *matrix method* (Carlsson, 1996). The method was developed to make it possible to consider all the parameters that influence the final design. The main parameters taken into account in the matrix method are: area cost (size, heat-exchanger type, and material), piping cost (the distance between streams, pipe materials, and diameter), pressure-drop cost, and maintenance cost. In this method, the complex heat-exchanger network design has been simplified using certain heuristics such as temperature differences in the heat exchangers, tick-off matches, and stream splitting. The rearrangement of heat exchangers and thus the identification of profitable solutions are carried out in a simpler way in this method than in more advanced mathematics-based methods (e.g., MILP and MINLP as presented by Yee and Grossmann (1987) and Ciric and Floudas (1990)), and the final solution is found by comparing the alternatives identified. In the basic approach, certain general principles of retrofit analysis are applied, such as retaining existing heat exchangers in their original positions (matches) as much as possible, installing as few new heat exchangers as possible, maintaining a minimum heat-exchanger area, and repiping no more than necessary.

The matrix method is an interactive method in which the user decides which heat-exchanger combinations will be included in the solution. A computer program calculates the consequences of the various heat-exchanger alternatives. A heat-exchanger design consists of a combination of matches, each of which contributes an individual annual cost to the total cost of retrofitting the heat-exchanger network. The combination of matches that yields the lowest cost is the solution to the retrofit problem at each global $\Delta T_{\min}$ (Nordman and Berntsson, 2001). The method has been used successfully to find an economically feasible heat-exchanger network retrofit for the release of excess heat discovered using the curves (Bengtsson et al. 2002).

In the matrix method, a matrix is developed in which hot streams run vertically and cold streams horizontally. For each thermodynamically possible match, the cost of implementing the match is calculated at a given level of energy recovery (i.e., $\Delta T_{\min}$), using all costs discussed above. In each match, at least one stream is ticked off. For piping, the geographical distances between all streams are included as input data. The user picks one match, for example, the one having the lowest cost (except for existing ones which have zero cost), and the matrix is fully recalculated with the ticked-off stream(s) in the chosen match omitted. This goes on until all matches have been chosen and the given energy recovery has been achieved. This procedure must be repeated once or twice until the sequence with the lowest total cost has been found.

The whole procedure described above can be carried out also with other $\Delta T_{\min}$ values. This will lead to a graph of total investment cost versus energy recovery level, from which the optimal level of energy recovery for a given energy cost can be determined.

One characteristic of this method is its simple way of dealing with existing and new matches. Existing matches are regarded as new installations, but are given an appropriate value (normally zero) by means of the cost equation.

From the matrices, one can easily obtain much information about the nature of the network, such as

- Identification of which streams are difficult to match and how different streams interact with each other
- Identification of matches that are economically feasible to terminate before heat load tick-off

- Information about the need for stream splitting and the possibility of creating a network with the minimum number of units
- The influence of the global ΔT_{min} on the complexity of the network, concerning, for example, the need for stream splitting and extra units
- Identification of the parameters having the greatest impact on the solution (i.e., heat-exchanger costs and others)

A number of rules and algorithms have been developed to determine in which order the matches should be chosen. However, only a few iterations normally are necessary.

11.3.3 Combined Water and Energy Reduction Analysis

Classical water-pinch methodologies have proved difficult to use in the pulp and paper industry because there are too many contaminants in the water streams and because water is both a utility and process stream in the pulp and paper industry (Koufos and Retsina 2001; Savulescu at al. 2005a). Approaches for combining heat and water reduction techniques have therefore been developed in which modified water-pinch approaches have been used (Bedard et al. 2001; Koufos and Retsina 2001; Bengtsson et al. 2002; Savulescu et al. 2005b; Mateos-Espejel et al. 2010a). Methodologies based on the interaction of water and energy using process integration approaches such as the water-thermal composite curve and the water source-and-sink curve have therefore been developed. Some of these studies deal in addition with the complex problem of combining reductions in energy and water usage in an efficient way based on the interactions of water and energy in a mill. Heavy optimization efforts are required to consider water, not only as a source in a water balance, but as a heat source in an energy balance, and then to combine it with contaminant concentrations and temperatures (Savulescu et al. 2001; Mateos-Espejel et al. 2010a,b). Mills evaluated using these methods have had high water consumption and a low degree of backpressure electric production. Modern energy-efficient mills, however, often have low water consumption, and traditional heat-recovery projects suggested mostly for earlier mills are often not applicable to modern mills (Wising 2003).

In general, the water that is used as a hot or a cold utility in a mill is stored in tanks, normally at three to five different temperature levels. The water stream normally does not have the same temperature as the overall tank; this means that a mixing of waters at different temperatures occurs in tanks. For modern pulp mills for which the water consumption is low and the pinch temperature is already at 100°C or above, mixing of water streams at different temperatures in tanks is not considered as a pinch violation (temperature levels are below the pinch temperature), and various solutions are required principally for improving energy efficiency. The "tank curves" method was therefore developed to maximize the usable excess heat potential at high temperature in the HWWS which can be used for other processes or saved as primary heat (Nordman and Berntsson 2006). In this method, the system boundary encloses the HWWS, and making changes in the HWWS would then leave the remaining process without any changes at all. Therefore, no complex simultaneous optimization and analysis of water consumption, water contamination, number of tanks, temperature levels in tanks, and tank size is performed in this method. Instead, the number of tanks and their temperature levels are optimized for a given mill water demand. The method can be applied in mills with either high or low water consumption.

11.4 Examples of Application of New Methods

When steam is saved and excess heat is released using one or more of the energy-efficiency measures mentioned earlier, the saved steam and released heat can be used in various ways in the mill. It can be used internally to increase the level of efficiency in the mill, for example, upgrading the evaporation plant, or externally, for example by the installation of new equipment like heat pumps, or in the implementation of new biorefinery concepts, for example, precipitation of lignin from black liquor. Common uses of saved steam and released heat in Kraft pulp mills which have been studied and reported in the literature are described briefly in the following section. Examples of methodologies for integrating new concepts in Kraft pulp mills based on newly developed methods for energy-efficiency improvement will be summarized.

11.4.1 Use of Advanced Curves and Background/Foreground Analysis for Biorefinery Concepts

When analyzing the energy consequences when a subprocess, for example, a distillation plant, is changed or when a new subprocess is introduced into an existing process, the concept of background/foreground analysis is useful. It should be seen mostly as a guide for finding the integration possibilities between processes. In this analysis, the existing process and its GCC are called the *background process* and the *background curve*, respectively. Correspondingly, the subprocess and its GCC are called the *foreground process* and the *foreground curve*. If the temperature levels match, the two processes can be integrated so that the cooling demand of one process is used as heat in the other. The possibilities can then be analyzed and approximately quantified with the aid of the background and foreground curves. One example of how these advanced curves can be used is shown in Figure 11.3. It is taken from a project on how to convert a Kraft pulp mill into an ethanol plant (Fornell and Berntsson 2011).

One integration problem is how to integrate the distillation and evaporation parts into the background process, which includes the pretreatment, the ethanol line (except for distillation), and the units in the chemical recovery cycle. When assessing the use of heat in a lignocellulosic ethanol process, the most important unit is the pretreatment of the raw material. According to previous studies (Fornell and Berntsson 2011), advanced pinch curves can be used to advantage in process integration studies (shown in Figure 11.4). The background process studied here is in part a retrofit (the reused pulp-mill unit operations) and in part a greenfield design (the ethanol line). The base case for comparison in this study is a background process design in which no energy-efficiency improvements have been made in the repurposed pulp mill.

In Figure 11.4, the AHLC (actual heat load curve) and the ACLC (actual cooling load curve) show at what temperatures hot and cold utilities are needed in the base-case process. As can be seen from the ACLC, there is not much excess heat available for integration of the thermal-separation units in the original design (~3.4 MW above 90°C). If a GCC were constructed for the process at the ΔT_{min} corresponding to the same utility demand, it would appear that the excess heat available would be 13.7 MW (above 90°C). This excess heat can, however, be made available in the process only if a retrofit of the heat-exchanger network is done. The demand for low-temperature hot utility (below 110°C) is 13.5 MW, and because the THLC and the AHLC are almost completely superimposed, the supply temperature

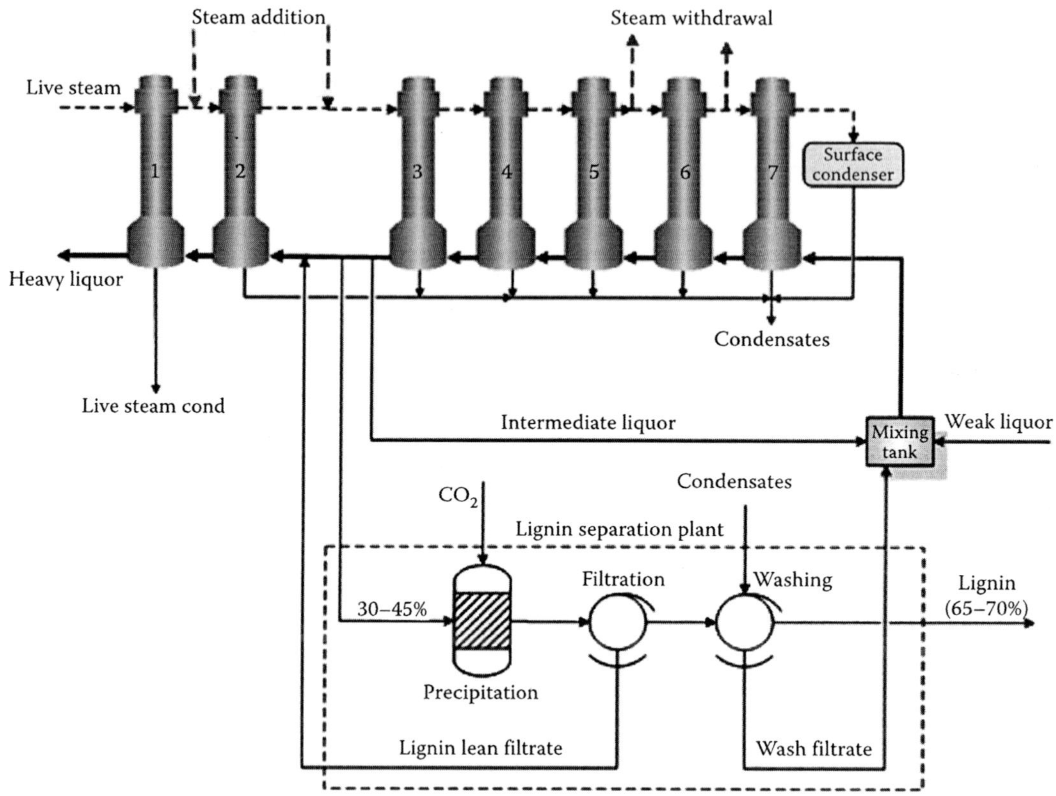

FIGURE 11.3
Evaporation plant (seven effects) with heat integration and lignin separation. (From Fornell, R. and Berntsson, T., 2011. Process integration study of a kraft pulp mill converted to an ethanol production plant—Potential for heat integration of thermal separation units. *Submitted for publication.*)

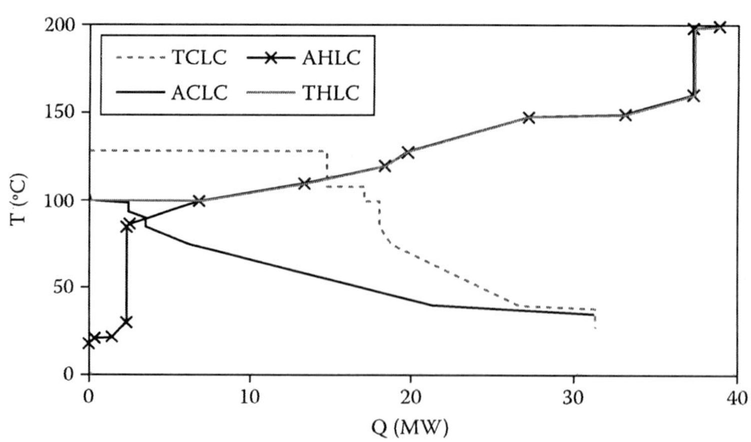

FIGURE 11.4
Advanced pinch curves for the background process. (From Fornell, R. and Berntsson, T., 2011. Process integration study of a kraft pulp mill converted to an ethanol production plant—Potential for heat integration of thermal separation units. *Submitted for publication.*)

cannot be decreased further without violating the pinch conditions. The THLC and the utility demand in the GCC are at the same temperature.

There are two ways of studying the possibilities for steam savings in this process; either the background process could be improved by increasing internal heat recovery (i.e., resolving the pinch violations), or the heat-exchanger network could be redesigned to increase the possibilities for external heat integration with distillation and evaporation. From Figure 11.4, it can be seen that resolving pinch violations will decrease the hot utility to 15 MW (where the THLC and the TCLC cross). It can also be seen that ~17 MW of excess heat at a high temperature (above 100°C) can be freed without changing the hot-utility demand (TCLC). The theoretical cooling load curve (TCLC) shows the highest temperatures at which coolers can be placed in the process for a specific ΔT_{HX} (set to 7°K in this study) while maintaining the base-case hot-utility demand. The THLC gives the same information as the TCLC, but for heaters above the pinch temperature [36]. The figure also indicates that there is a pinch violation of ~2 MW which can be resolved with the excess heat existing in the process today (to the left of where the ACLC and AHLC cross). This pinch violation actually represents the use of 4-bar steam for building heating. The total of the pinch violations that can be resolved is therefore 17 MW. From Figure 11.4, it can also be seen that the amount of excess heat that can be made available to meet the original utility demand is greater than the amount of excess heat available in the GCC (Figure 11.5) (18 MW above 90°C in TCLC compared to 13.7 MW in the GCC).

Another example of results obtained using the advanced curves is shown in Figure 11.5. The results in the figure have been calculated using the matrix method as in Nordman (2005). From this curve, it can be seen that the cost of saving energy is substantially lower in the case in which heaters are placed close to the pinch, and the more energy that is saved, the more the relative cost of the retrofit increases among the cases. The number of new or rearranged units is also lower in the case in which heaters are placed close to the pinch, indicating that the expected cost of piping is also lower in this case. Note that resolving one pinch violation can affect more than one heat exchanger. In Figure 11.5, the

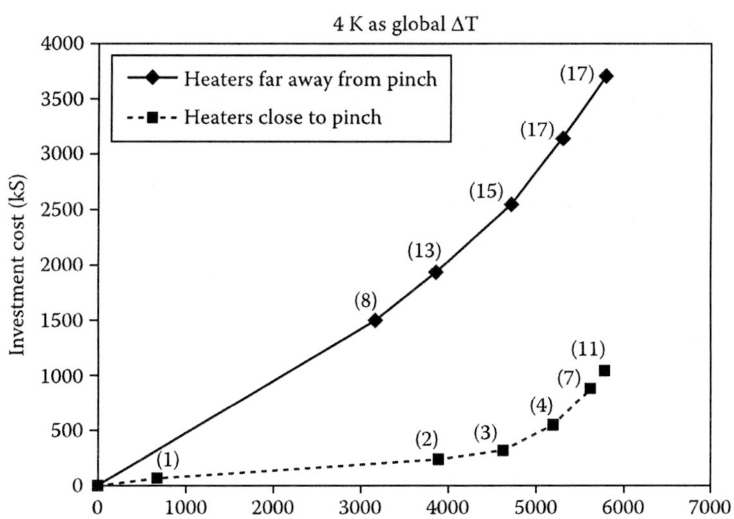

FIGURE 11.5
Cost-efficiency curve with number of new or rearranged units presented within brackets.

number of new or rearranged units is indicated within brackets for each pinch violation that is resolved. The unit cost (i.e., the area-independent part) is ~40% of the total cost in the case in which the heaters are placed close to the pinch, whereas it is only ~20% in the case in which the heaters are placed high in temperature, although there are many more units in the latter case. This can be explained by the fact that much more area needs to be introduced to perform the heat exchange when the heaters are placed high in temperature because the ΔT values serving as a driving force are low.

An example of the tank curve method is described and illustrated below.

As already mentioned, this method is based on traditional composite curves. In addition to the hot and the cold CCs, a third curve is introduced, called the *tank curve*. The steps in the method are listed below:

1. Identify the HWWS hot streams and construct the hot CC.
2. Identify the HWWS cold streams.
3. Identify the process demands and the cooling demands from the cold streams.
4. Construct the cold CC from the cold streams for the process demands.
5. Calculate the theoretical target for the maximum excess heat at a given ΔT_{min} (Figure 11.6b).
6. Replace the cold CC with the tank curve (Figure 11.7).
7. Vary the number of tanks and their temperature levels to maximize the amount of usable excess heat.

The main idea of the tank curve is that fresh water for process needs is taken at the fresh-water inlet temperature and is heated successively until it reaches the tank temperature.

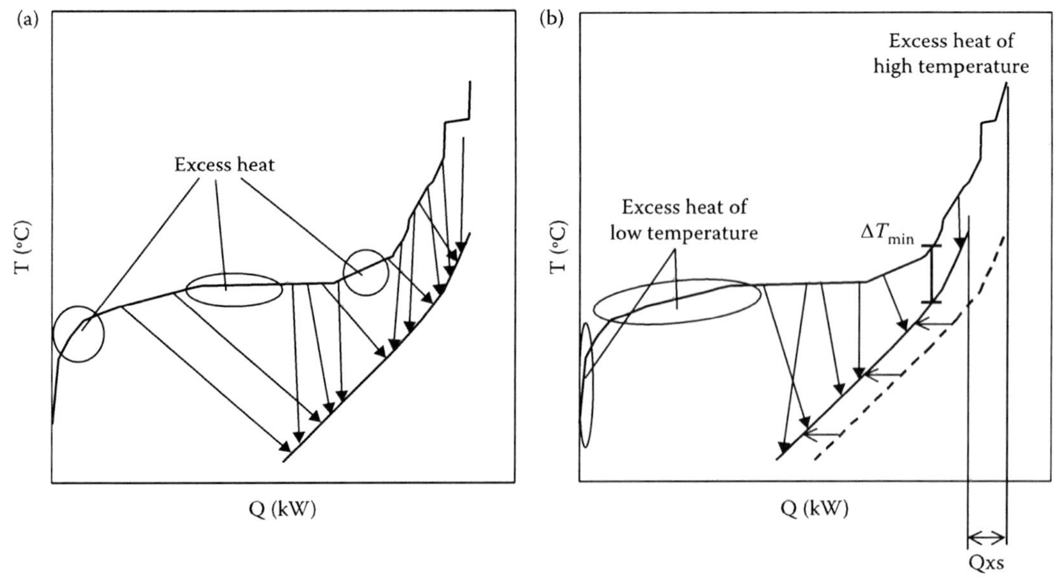

FIGURE 11.6
(a) Composite curves show how the temperature of the cooling demands, which could be used for Qxs, varies. (b) Shifting the tank curve to the left until reaching the HWWS ΔT_{min} yields the maximum usable Qxs.

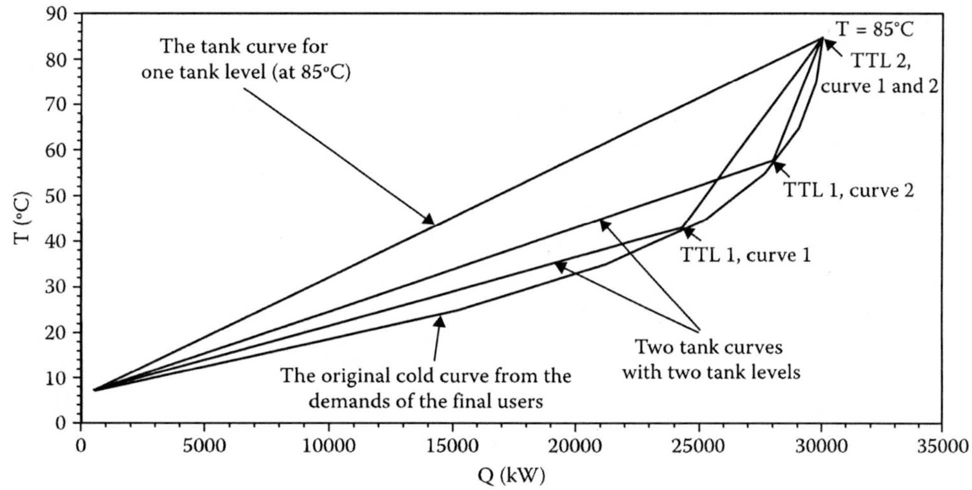

FIGURE 11.7
Tank curves are constructed from knowledge of the original cold composite curve and look different depending on the number of tanks and their placement with respect to temperature.

Process demands are then met either by direct tapping from a suitable tank (tank temperature is the same as process demand) or by mixing tank water with fresh water until the desired temperature is reached.

In a greenfield design, this method could be used to minimize the number of water storage tanks needed. In the case of a retrofit, a number of tanks are already installed. In this case, there are two options:

- Maximize the usable heat using the current number of tanks.
- Maximize the usable excess heat using fewer than the current number of tanks.

The second option makes it possible to reduce control complexity. Installing more than the current number of tanks would in most cases not be necessary because the method has shown that most systems could operate with only two tanks.

11.5 Assessing Profitability and Carbon Balances of Energy Investments

Energy systems are complex sociotechnical systems with many actors connected in an infrastructural network, including social and industrial sectors, production systems, and system operators. Besides obvious important parameters such as investment costs and performance, system culture plays a very important role. Government measures aimed at implementing new technology should therefore be characterized not only by the level of economic support provided, but also by the compatibility of that economic support with the established system cultures within the energy system (Marbe 2005). Political measures aimed at saving energy, such as green certificates, long-term agreements between the government and industry concerning energy investments, and CO_2 trading may help mills to choose the alternative which provides the largest reduction in CO_2 emissions, but could just as well have the opposite effect (Algehed 2002). The issue of CO_2 emission is important from a strategic perspective for the pulp and paper industry. Because energy use in the pulp and paper industry is high, energy savings in this sector can have a large impact on national CO_2 emission levels (Wising 2003). The profitability and the net CO_2 emissions reduction potential of energy investments in this sector must therefore be assessed by quantifying their implications within a future energy-market context. To carry out decision-making subject to uncertainty about future energy-market conditions, various strategies can be considered.

Over more than a decade, research on strategic methods for evaluating candidate investments, using different scenarios that include future fuel prices, energy carrier prices, and CO_2 emissions and policy instruments, has been conducted in the Department of Heat and Power Technology at Chalmers University of Technology (Ådahl et al. 2000; Ådahl and Harvey 2007; Axelsson et al. 2009; Axelsson and Harvey 2010). To obtain reliable benchmarks, these scenarios need to be consistent. A calculation aid or a tool can be helpful for creating scenarios adapted for this energy-intensive industry (Axelsson 2008). A methodology has therefore been developed which includes blocks with different sets of coherent energy-market prices. These blocks were initially intended to be used to construct scenarios for evaluation of industrial energy projects. These so-called *energy-market scenarios* were then upgraded by constructing a tool,* ENPAC, to enable the creation of consistent scenarios.

The ENPAC tool calculates energy prices for a large-volume customer based on forecasted world-market fossil-fuel prices and relevant policy instruments (e.g., costs associated with emitting CO_2, various subsidies favoring renewable energy sources in the electricity market or the transportation-fuel market) and on key characteristics of energy-conversion technologies in the district-heating and electric-power sectors. An overview of the procedure and purpose of the ENPAC tool for evaluation of energy-efficiency investments in energy-intensive industries is shown in Figure 11.8. For further details, the reader is referred to the report by Axelsson and Harvey (2010).

A systematic approach which trades off CO_2 emission reductions against investments in process-integration measures in industry while optimizing these investments under economic uncertainty has also been presented in the literature (Svensson et al. 2009b; Svensson and Berntsson 2010). The methodology proposed by Svensson and Berntsson (2010) is based on the assumption that investment decisions must be made before the values of uncertain parameters can be known. An example of an uncertain parameter is the availability of carbon capture and sequestration (CCS) to an extent such that coal-power plants with CCS can be assumed to be the marginal electricity producer. Many of the uncertainties affecting investment decisions are related to future CO_2 emission targets and policies. The optimization should therefore consider not only economic criteria, but also greenhouse-gas reductions. Svensson and Berntsson (2010) used a model to optimize decisions on energy-efficiency investments under uncertainty. As the decision problem became a multiobjective programming problem, the method proposed by Svensson and Berntsson attempted to provide a good framework for decision-making about energy-efficiency measures when considerations of greenhouse-gas reductions are influencing these decisions. In the Swedish case, for example, in addition to the question of future CCS availability, electricity price is also a time-dependent

* ENPAC, Energy Price and Carbon Balance Scenarios, developed by Erik Axelsson and Simon Harvey at Chalmers University of Technology and Profu (www.profu.se).

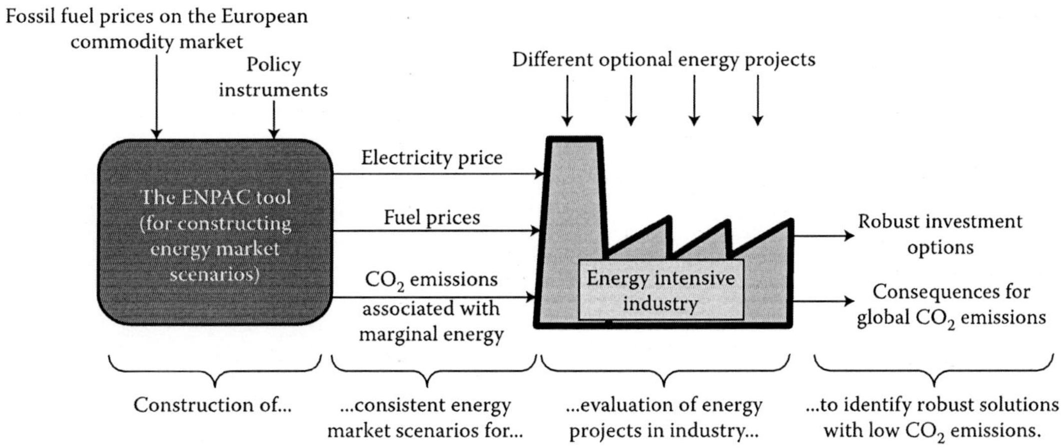

FIGURE 11.8
Overview of the purpose of energy-market scenarios for evaluation of energy-efficiency investments in energy-intensive industries when the ENPAC tools are used to construct the scenarios. (Adapted from Axelsson and Harvey 2010. *Scenarios for Assessing Profitability and Carbon Balances of Energy Investments in Industry*. AGS Pathways Report: EU1, Sweden.)

parameter. Depending on which market is setting the electricity price (e.g., a common Nordic electricity market or a North European one), different scenario building blocks will result. This method has been applied to the case of energy-efficiency investments at a chemical pulp mill (Svensson et al. 2009a). The results showed that the proposed methodology could handle both time dependence and parameter uncertainties and found a robust solution which turned out to be a combination of two opposing investment strategies. The obvious effect of policy instruments aimed at decreasing CO_2 emissions was also found to be an increased profitability for all energy-efficiency investments.

11.6 Conclusions

This chapter was intended to provide an overview of the available thermodynamics-based methods and tools that have been used to perform process integration studies for Kraft pulp and paper mills in Sweden. This overview could provide a basis for building a platform for further development of process integration tools and methods, including consideration of the implementation of new biorefinery concepts. In this context, a number of various approaches which tackle the complex issue of process integration with regards to energy and water systems, appropriate efficiency measures, and evaluation of the chosen integration measures under an uncertain future energy market have been presented. Depending on the operating history of the mill under study (modern vs. old mill) and the degree of existing integration, different methods can be used to optimize process energy consumption. In general, it can be said that energy savings can enable energy exports from the mill or can be used to integrate new technologies or concepts into the mill. To decide which measures to choose, a method for assessing profitability and CO_2 balance has been presented.

Acknowledgments

The work presented here has been performed within the "Technology Impact Assessment: System Perspectives on the Biorefinery" project in the research area of Energy Combines* within Chalmers Energy Initiatives, a strategic, governmentally funded program. The author acknowledges Anders Åsblad (CIT-Industriell Energianalys) for his valuable insights and contributions to this work.

References

Ådahl, A., Harvey, S., and Berntsson, T., 2000. Process integration as a decision-making tool for cost-effective greenhouse gas reduction in industry. *Proceedings, ECOS*, Twente, The Netherlands, pp. 1213–1224.

Ådahl, A. and Harvey, S., 2007. Energy efficiency investments in kraft pulp mills given uncertain climate policy. *Int. J. Energy Research* 31(5), 486–505.

Ådahl, A., 2004. *Process Industry Energy Projects in a Climate Change Conscious Economy*. PhD dissertation, Heat and Power Technology, Chalmers University of Technology, Sweden.

Algehed, J., 2002. *Energy Efficient Evaporation in Future Kraft Pulp Mills*. PhD dissertation, Chalmers University of Technology, Sweden.

Åsblad, A., 2011. (Personal Communication.) *Senior Engineer at CIT Industriell Energianalys*, Chalmers Research Park, Gothenburg, Sweden.

Axelsson, E., 2008. *Energy Export Opportunities from Kraft Pulp and Paper Mills and Resulting Reductions in Global CO_2* Emissions. PhD dissertation, Heat and Power Technology, Chalmers University of Technology, Sweden.

Axelsson, E. and Harvey, S., 2010. *Scenarios for Assessing Profitability and Carbon Balances of Energy Investments in Industry*. AGS Pathways Report: EU1, Sweden.

Axelsson, E., Harvey, S., and Berntsson, T., 2009. A tool for creating energy market scenarios for evaluation of investments in energy intensive industry. *Energy* 34(12), 2069–2074.

Axelsson, E., Olsson, M.R., and Berntsson, T., 2006. Heat integration opportunities in average Scandinavian kraft pulp mills: Pinch analyses of model mills. *Nordic Pulp and Paper Research Journal* 21(4), 466–475.

Bedard, S., Sorin, M., and Leroy, C., 2001. Application of process integration in water re-use projects. *Pulp and Paper Canada* 102(3), 53–56.

Bengtsson, C., Nordman, R., and Berntsson, T., 2002. Utilization of excess heat in the pulp and paper industry—A case study of technical and economic opportunities. *Applied Thermal Engineering* 22(9), 1069–1081.

Browne, T.C., Francis, D.W., and Towers, M.T., 2001. Energy cost reduction in the pulp and paper industry: An overview. *Pulp & Paper Canada* 102(2), 26–30.

Bruce, D.M., 2000. Benchmarking energy consumption to identify opportunities for conservation. *Pulp & Paper Canada*, 101(11), 35–38.

Cakembergh-Mas, A., Paris, J., and Trépanier, M., 2010. Strategic simulation of the energy management in a Kraft mill. *Energy Conversion and Management* 51, 988–997.

Carlsson, A., 1996. *Optimum Design of Heat Exchanger Networks in Retrofit Situations*. PhD dissertation, Heat and Power Technology, Chalmers University of Technology, Sweden.

* Energy Combines is one of the strategic research areas for which Chalmers University was granted an extraordinary funding from the government of Sweden.

Ciric, A.R. and Floudas, C.A., 1990. A mixed integer non-linear programming model for retrofitting heat exchanger networks. *Ind. Eng. Chem. Res.* 29, 239–251.

Environmental Database for Forest Industry in Sweden, 2008, Table 8: Complied data from all mills, http://miljodatabas.skogsindustrierna.org/si/main/reportselect.aspx?l1=report

Fornell, R. and Berntsson, T., 2011. Process integration study of a kraft pulp mill converted to an ethanol production plant—Potential for heat integration of thermal separation units. *Submitted for publication.*

FRAM (Future Resource Adapted Pulp Mills) 2005. *Final Report—Model Mills.* Stockholm, Sweden: STFI.

Grip, C.E., Elfgren, E., Soderstrom, M., Thollander, P., Berntsson, T., Asblad, A., and Wang, C., 2011. Possibilities and problems in using exergy expressions in process integration. *World Renewable Energy Congress*, Sweden.

Hippinen, I., Ruohonen, P., Sivill, L., Federley, J., Hakala, J., Manninen, J., and Ahtila, P., 2010. Methods for industry to measure and improve energy efficiency of utility systems. *Chemical Engineering Transactions* 21, 349–354.

Jönsson, J., Svensson, I.L., Berntsson, B., and Moshfegh, B., 2008. Excess heat from Kraft pulp mills: Trade-offs between internal and external use in the case of Sweden—Part 2: Results for future energy market scenarions, *Energy Policy*, 36 (11), 4186–4197.

KAM 2003. *EcoCyclic Pulp Mill Research Program, 1996–2002.* Final report, Chapter 3, STFI, Stockholm.

Karlsson, M., 2011. The MIND method: A decision support system for optimization of industrial energy systems—Principles and case studies. *Applied Energy* 88(3), 577–589.

Kemp, I.C., 2007. *Pinch Analysis and Process Integration: A User Guide to Process Integration for the Efficient Use of Energy.* 2nd ed., IChemE.

Koufos, D. and Retsina, T., 2001. Practical energy and water management through pinch analysis for the pulp and paper industry. *Water Science and Technology* 43(2), 327–332.

Linnhoff, B., 1993. Pinch analysis—A state-of-the-art overview. *Chem. Eng. Res. Des.* 71, 503–522.

Loutfi, M.H. and Ruzicka, Y., 2006. Mill-wide energy conservation for process optimization. *Proceedings, TAPPI Papermaking Conference*, Atlanta, April, 2006.

Männistö, H. and Männistö, E., 1999. Optimum energy efficiency: A comparison between North American and Nordic pulp and paper mills. *2nd Biennial Johan Gullichsen Colloquium*, Epsoo, Finland.

Männistö, H. and Männistö, E., 2006. Benchmarking CO_2 emissions in the EU and North American pulp and paper industries. *Preprints, PAPTAC 92nd Annual Meeting*, A369–A372.

Männistö, E. and Smith, S., 2006. Benchmarking the environmental performance of the pulp and paper industry: An evaluation based on statistical data. *Preprints, PAPTAC 92nd Annual Meeting*, A357–A359.

Marbe, Å., 2005. *New Opportunities and System Consequences for Biomass Integrated Gasification Technology in CHP Applications.* PhD dissertation, Heat and Power Technology, Chalmers University of Technology, Sweden.

Marechal, F. and Kalitventzeff, B., 1997. Effect modelling and optimization: A new methodology for combined energy and environment synthesis of industrial processes. *Applied Thermal Engineering* 17(8–10), 981–992.

Marinova, M., Mateos-Espejel, E., Bakhtiari, B., and Paris, J., 2009. A new methodology for the implementation of tri-generation in industry: Application to the Kraft process. *Proceedings, First European Conference on Polygeneration*, Tarragona, Spain, 16–17 October, 2007.

Mateos-Espejel, E., Marinova, M., Bararpour, S., and Paris, J., 2010a. Energy implications of water reduction strategies in kraft processes: Part I: Methodology. *Pulp and Paper Canada* 111(3), 34–37.

Mateos-Espejel, E., Marinova, M., Bararpour, S., and Paris, J., 2010b. Energy implications of water reduction strategies in kraft processes: Part II: Results. *Pulp and Paper Canada* 111(3), 38–41.

Nordman, R. and Berntsson, T., 2001. New pinch technology-based HEN analysis methodologies for cost-effective retrofitting. *Canadian Journal of Chemical Engineering* 79(4), 655–662.

Nordman, R., 2005. *New Process Integration Methods for Heat-Saving Retrofit Projects in Industrial Systems.* PhD dissertation, Chalmers University of Technology, Sweden.

Nordman, R. and Berntsson, T., 2006. Design of kraft pulp mill hot and warm water systems: A new method that maximizes excess heat. *Applied Thermal Engineering* 26(4), 363–373.

Nordman, R. and Berntsson, T., 2009a. Use of advanced composite curves for assessing cost-effective HEN retrofit: Theory and concepts. *Applied Thermal Engineering* 29(2–3), 275–281.

Nordman, R. and Berntsson, T., 2009b. Use of advanced composite curves for assessing cost-effective HEN retrofit: Case studies. *Applied Thermal Engineering* 29(2–3), 282–289.

Process integration in pulp and paper industry, International Energy Agency (IEA), 2004, Annex III within the IEA implementing agreement on process integration.

Savulescu, L., Kim, J.K., and Smith, R., 2005a. Studies on simultaneous energy and water minimization: Part I: Systems with no water re-use. *Chemical Engineering Science* 60(12), 3279–3290.

Savulescu, L., Kim, J.K., and Smith, R., 2005b. Studies on simultaneous energy and water minimization: Part II: Systems with maximum re-use of water. *Chemical Engineering Science* 60(12), 3291–3308.

Savulescu, L., Poulin, B., Hammache, A., and Bedard, S., 2001. Water and energy savings at a kraft paperboard mill using process integration. 87th Annual Meeting of the Pulp-and-Paper-Technical-Association-of-Canada, Montreal, January 2001.

Smith, R., 1995. *Chemical Process Design*. New York: McGraw-Hill.

Svensson, E. and Berntsson, T., 2010. Economy and CO_2 emissions trade-off: A systematic approach for optimizing investments in process integration measures under uncertainty. *Applied Thermal Engineering* 30(1), 23–29.

Svensson, E., Berntsson, T., and Stromberg, A.B., 2009a. Benefits of using an optimization methodology for identifying robust process integration investments under uncertainty—A pulp mill example. *Energy Policy* 37, 813–824.

Svensson, E., Berntsson, T., Stromberg, A.B., and Patriksson, M., 2009b. An optimization methodology for identifying robust process integration investments under uncertainty. *Energy Policy* 37, 680–685.

Wising, U., 2003. *Process Integration in Model Kraft Pulp Mills: Technical, Economic and Environmental Implications*. PhD dissertation, Chalmers University of Technology, Sweden.

Wising, U., Berntsson, T., and Asblad, A., 2002a. Usable excess heat in future kraft pulp mills. *TAPPI Journal* 1(9), 27–29.

Wising, U., Berntsson, T., and Asblad, A., 2002b. Energy consequences in minimum effluent market kraft pulp mills. *TAPPI Journal* 1(9), 30–32.

Wising, U., Berntsson, T., and Stuart, P., 2005. The potential for energy savings when reducing the water consumption in a kraft pulp mill. *Applied Thermal Engineering* 25(7), 1057–1066.

Yee, T.F. and Grossmann, I.E., 1987. Optimization model for structural modifications in the retrofit of heat exchanger network. *Proceedings, First International Conference on Foundations of Computer-Aided Process Operations*, Park City, Utah, July 1987.

12

Design of a Lignocellulosic Feedstock Biorefinery Based on a Biochemical Processing Platform Using Process Integration Methodologies and Exergy Analysis

Karina Ojeda, Mahmoud M. El-Halwagi, and Vyacheslav Kafarov

CONTENTS

12.1 Introduction

Lignocellulosic biomass is one of the best potential feedstocks for the bioethanol industry which does not compete with food crops. The main advantage of the production of second-generation biofuels is that it limits the direct food-versus-fuel competition associated with first-generation biofuels. However, second-generation biofuels also have technological disadvantages, for instance, high water demand and high levels of energy consumption for their production. These drawbacks generate doubts about the sustainable development of this industry. The processing of a renewable energy source usually involves the consumption of nonrenewable resources, and therefore, to design an energy-sustainable process, it must be confirmed that the energy obtained in the bioethanol is significantly greater than the energy used in the process to produce it (Ojeda and Kafarov, 2009).

Much research (Neelis et al., 2004; Tonon et al., 2006; Yang et al., 2009) has been undertaken on the exergy accounting of nonrenewable resource (NRR) consumption to measure the environmental impact of many manufacturing processes. Biomass energy crops should be grown and harvested in a way that embodies best stewardship practices to maintain or improve air, water, and soil quality. This includes the energy consumed in planting, cultivating, fertilizer or pesticide application, harvesting, and transportation to market. Consequently, second-generation biofuels represents a great alternative to reach the sustainability goals of this industry. In this sense, the overall energy consumption in ethanol production from lignocellulosic residues, like bagasse, is lower than that of nonresidual energy crops because the residual biomass from the existing sugar industry is used. In effect, the energy consumption in all stages of sugarcane farming is subsumed in the traditional sugar-production chain (Ojeda et al., 2011a).

Exergy analysis is a thermodynamic analysis technique which provides an alternative and an illuminating means of assessing and comparing processes and systems rationally and meaningfully. In particular, exergy analysis yields efficiencies which provide a true measure of how nearly actual performance approaches the ideal; it also identifies more clearly than energy analysis the causes and locations of thermodynamic losses and the impact of the built environment on the natural environment (Ojeda et al., 2011a). Consequently, exergy analysis can assist in improving and optimizing designs (Kanoglu and Bolatturk, 2008) and is associated with sustainability (Dincer and Rosen, 2007). Therefore, exergy analysis has been proposed as a tool to evaluate overall process efficiency in the second-generation bioethanol industry.

12.2 Lignocellulosic-Feedstock Biorefinery Based on Biochemical Processing Platform

Second-generation liquid biofuels are generally produced by two fundamentally different approaches, biological or thermochemical processing, from agricultural lignocellulosic biomass, which consists of either nonedible residues of food-crop production or nonedible whole-plant biomass. The feedstock involved in the process can be produced specifically for energy purposes, enabling higher production per unit land area than when using residues, and a greater proportion of the aboveground plant material can be converted and used to produce biofuels. As a result, second-generation biofuel production can be expected to increase land-use efficiency over that of first-generation biofuels (Nigam and Singh, 2010).

Chemistry, biochemistry, and biochemical engineering will contribute to developing all aspects of biorefinery operation, providing, in an analogous way to present-day petroleum refineries, fuels, lubricants, chemical solvents, intermediates, and polymer products (Koutinas et al., 2007). In addition, these plants will produce other bioproducts, such as soil amendments, food and feed components, fibers, biodegradable plastics, building materials, and various others products (Orts et al., 2008).

Second-generation biofuels can be further classified in terms of the process or method used to convert biomass into fuel. Few second-generation biofuels such as ethanol and butanol are produced through biochemical processes, while all other second-generation fuels are produced thermochemically. Many second-generation thermochemical fuels are currently being produced commercially from fossil fuels. These thermochemical fuels include methanol, refined Fischer–Tropsch liquids (FTL), and dimethyl ether (DME). Unrefined fuels (e.g., pyrolysis oils) are also produced thermochemically, but require considerable additional and refining before they can be used in engines (Nigam and Singh, 2010). In this chapter, bioethanol production from lignocellulosic biomass is described, assuming the use of enzymatic technologies.

12.2.1 Lignocellulosic Biomass to Bioethanol

The necessary stages to liberate the sugars in the lignocellulosic material are

1. Pretreatment, with the function of making material more susceptible and accessible for later stages.
2. Hydrolysis, with the function of liberating the glucose from the lignocellulosic material.
3. Fermentation of hexoses and pentoses to obtain ethanol.
4. Separation and concentration of the alcohol.

Detoxification may be necessary when strongly inhibiting hydrolyzates are fermented if high concentrations of inhibitors accumulate in the fermentation unit due to stream recycle, or when a fermenting organism with low inhibitor tolerance is used. The detoxification method should selectively remove inhibitors and be cheap and easy to integrate into the process (Palmqvist and Hahn-Hägerdal, 2000).

Plants contain three primary types of carbon-based polymers: cellulose, hemicelluloses, and lignin (Orts et al., 2008).

Cellulose, a glucose-based polymer with repeating units consisting of two β-1,4-linked glucose residues (cellobioses), can be hydrolyzed to monomeric glucose (Figure 12.1). The microfibril scaffold of the cell wall consists of both crystalline and amorphous domains within a largely cellulosic core. It cannot be readily converted into biofuels by traditional microbes in industrial-scale fermentation. Improved yeasts (Ho and Tsao, 1995; Ho and Chen, 1997) and bacteria (Ingram et al., 1991; Ingram and Clark, 1992) enable hemicellulose-derived sugars to be used, but hemicellulose conversion is not without cost. The longer fermentation times resulting from use of xylose sugars incur significantly greater capital costs (due to larger fermentors).

Hemicellulose is a heteropolymer containing sugar residues such as hexoses (D-galactose, L-galactose, D-mannose, L-rhamnose, L-fucose), pentoses (D-xylose, L-arabinose), and uronic

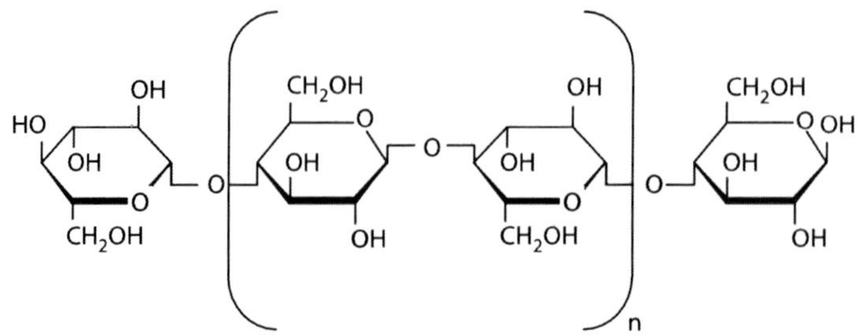

FIGURE 12.1
Cellulose.

TABLE 12.1
Composition of Common Agricultural Residues and Wastes

Lignocellulosic Materials	Cellulose (%)	Hemicelluloses (%)	Lignin (%)
Hardwood stems	40–55	24–40	18–25
Corncobs	45	35	15
Grasses	25–40	35–50	10–30
Leaves	15–20	80–85	0
Switchgrass	45	31.4	12.0
Paper	85–99	0	0–15
Nut shells	25–30	25–30	30–40
Wheat straw	30	50	15

Source: Adapted from Sun, Y., Cheng, J. 2002. *Bioresource Technology* 83: 1–11.

acids (D-glucuronic acid) (Fengel and Wegener, 1989; Zaldivar et al., 2001). The composition of a hemicellulose depends on the source of the raw material. Hemicelluloses are cross-linked by ester linkages to predominantly cinnamic (ferulic) acids, which form covalent linkages with lignin, making plant cell-wall hydrolysis even more difficult.

Lignin is a macromolecule of phenolic character and is the dehydration product of three monomeric alcohols (lignols): *trans-p*-coumaryl alcohol, *trans-p*-coniferyl alcohol, and *trans-p*-sinapyl alcohol, derived from *p*-cinnamic acid (Zaldivar et al., 2001). Lignin provides structural rigidity to the cell wall and increases its hydrophobicity, which impedes the breakdown of cell walls for easy conversion into ethanol (Orts et al., 2008).

The content of cellulose, hemicelluloses, and lignin in various agricultural residues has been reported by Sun and Cheng (2002). The main lignocellulosic materials are listed in Table 12.1.

12.2.2 Current State of Lignocellulosic Bioethanol Production

Production of bioethanol from lignocellulosic biomass can be viewed as an emerging and leading biorefinery based on biochemical technologies and should start with a sustainable development of agricultural and forestry residues which are at that time very interesting in terms of productivity and environment (Mabee et al., 2005; Lavoie et al., 2011). Currently, there is tremendous worldwide interest in the production of ethanol from biomass. In 2008, world ethanol production increased by 17 billion gallons (Figure 12.2). Major national

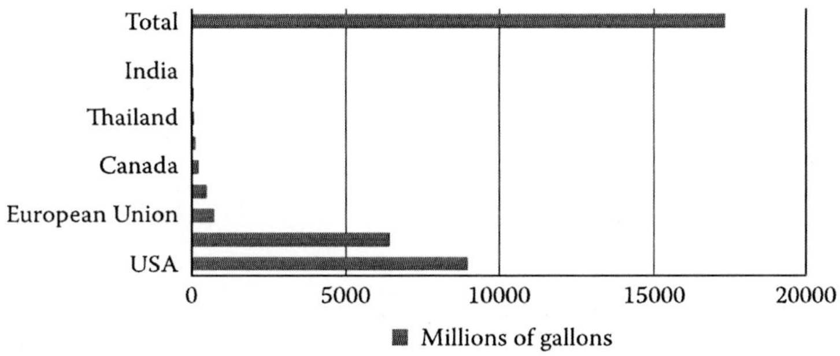

FIGURE 12.2
2008 world fuel ethanol production. (Adapted from RFA 2009. *Renewable Fuels Association: Statistics*. Available online (http://www.ethanolrfa.org/pages/statistics), RFA 2009 report. http://www.ethanolrfa.org/bio-refinery-locations/.)

biofuels programs have been initiated to produce cost-efficient ethanol and other fuels from agricultural and forest-based lignocellulosic biomass in countries like the United States and China. As for bioethanol production, fermentation technology of sugars or starch to ethanol has been done for many years. However, the cost of enzymes for hydrolysis has been considerably reduced, and other processing costs continue to improve as market economics and competition between producers evolve. This development represents a maturing of existing technologies rather than fundamental new breakthroughs in the underlying science, although major improvements in fermentation technology and other related areas are underway with renewed urgency (Orts et al., 2008).

U.S. ethanol capacity (April 2010) was 13519.4 million gallon per year (mgy) in 201 refineries, with another 1183 mgy additional capacity in refineries under construction or expansion. As a result of this kind of growth and innovation, the U.S. ethanol industry produced a record 9 billion gallons of ethanol in 2008 and ~10 billion gallons in 2009, representing almost 9% of U.S. gasoline supply (Table 12.2). U.S. ethanol producers have answered the challenge presented in the Renewable Fuels Standard (RFS) and are producing enough domestic ethanol to fill requirements. The industry has recognized new opportunities to expand the use of ethanol and to ensure the success of the RFS in the years to come (Table 12.3).

It is clear that large-scale use of ethanol will require conversion of cellulose and other underutilized biomass feedstocks (Farrell et al., 2006). Cellulosic feedstocks increase the available biomass beyond the grain supply, reducing competition with food, but not necessarily with feed. More importantly, cellulosic conversion provides a better return on input energy (i.e., the fuel required during production) and has the potential to reduce the emission of greenhouse gases per unit of energy produced (Orts et al., 2008).

12.2.3 Production Technologies

Ethanol has been known for a long time, being perhaps the oldest product obtained through traditional biotechnology. Its current applications include potable, chemical, and fuel ethanol (Zaldivar et al., 2001). The main potential feedstock for ethanol is lignocellulosic biomass such as agricultural residues (e.g., sugarcane bagasse), herbaceous crops, forestry wastes, wood, and other waste products such as municipal waste (Wyman, 2003). It has been estimated that lignocellulose accounts for ~50% of the biomass in the world (10–50 billion tons, according to Claassen et al., 1999). Lignocellulosic biomass consists of three major components (in mass fractions of dry substance): w(cellulose) = 35–50%, w(hemicellulose) = 20–35%,

TABLE 12.2

Advanced and Cellulosic Ethanol Projects

Year	Jan 2001	Jan 2002	Jan 2003	Jan 2004	Jan 2005	Jan 2006	Jan 2007	Jan 2008	Jan 2009	Jan 2010
Total ethanol plants	56	61	68	72	81	95	110	139	170[a]	187[a]
Ethanol production capacity (mgy)	1921.9	2347.3	2706.8	3100.8	3643.7	4336.4	5493.4	7888.4	12475.4[b]	13028.4[b]
Plants under construction or expansion	5	13	11	15	16	31	76	61	24	15
Capacity under construction or expansion (mgy)	64.7	390.7	483	598	754	1778	5635.5	5536	2066	1432
States with ethanol plants	18	19	20	19	18	20	21	21	26	26

Source: Adapted from RFA 2009. *Renewable Fuels Association: Statistics.* Available online (http://www.ethanolrfa.org/pages/statistics), RFA 2009 report. http://www.ethanolrfa.org/bio-refinery-locations/

[a] Operating plants.

[b] Includes idled capacity.

TABLE 12.3

New Renewable Fuels Standard Schedule (Billion Gallons per Year)

Year	2010	2012	2014	2016	2018	2020	2022
Renewable biofuels	12.0	13.2	14.4	15.0	15.0	15.0	15.0
Advanced biofuels	0.95	2.0	3.75	7.25	11.0	15.0	21.0
Cellulosic biofuels	0.1	0.5	1.75	4.25	7.0	10.5	16.0
Biomass-based diesel	0.65	1.0					
Undifferentiated advanced biofuels	0.2	0.5	2.0	3.0	4.0	4.5	5.0
Total RFS	12.95	15.2	18.15	22.25	26.0	30.0	36.0

Source: Adapted from RFA 2009. *Renewable Fuels Association: Statistics.* Available online http://www.ethanolrfa.org/pages/statistics

and w(lignin) = 10–25% (Ojeda et al., 2011b). Conversion of lignocellulosic materials into higher-value products requires fractionation of the material into its components: lignin, cellulose, and hemicelluloses. As indicated by Hayes (2009), the key to exploiting the chemical value of biomass is to depolymerize the lignocellulosic matrix to obtain smaller molecules that can be used or further converted into platform chemicals and biofuels.

Although there are similarities between the lignocellulosic and starch processes, the technoeconomic challenges facing the former are large. There are several options for a lignocellulose-to-ethanol process, but regardless of which is chosen, the following features must be assessed in comparison with established sugar- or starch-based ethanol production (Hahn-Hägerdal et al., 2006):

1. Efficient depolymerization of cellulose and hemicellulose to soluble sugars.
2. Efficient fermentation of a mixed-sugar hydrolyzate containing six-carbon (hexoses) and five-carbon (pentoses) sugars as well as fermentation-inhibitory compounds.

3. Advanced process integration to minimize process energy demand.
4. Cost-efficient use of lignin.

Research in recent years has focused on making available the glucose from lignocellulosic materials, which can act as an abundant and cheap source of fermentable sugars (Wyman, 1996; Wyman et al., 2005). In addition to the necessary pretreatment of the cellulosic material, this will involve two consecutive catalytic steps: an enzymatic conversion of the cellulose into fermentable sugars, and a fermentative conversion of these sugars to ethanol. These hydrolysis and fermentation steps may be operated sequentially (separate hydrolysis and fermentation, SHF) or concurrently (simultaneous saccharification and fermentation, SSF) (Drissen et al., 2008). Diluted ethanol is finally concentrated by, for instance, distillation or pervaporation (O'Brien et al., 2000).

Fuel ethanol production from lignocellulosic biomass includes five main steps: pretreatment, hydrolysis, hexose fermentation, purification, and effluent treatment (Cardona and Sánchez, 2007; Ojeda et al., 2011b).

12.2.3.1 Pretreatment Pathways (Main Description, Current State, Challenges)

The significant variety of pretreatment methods for biomass has led to the development of many flowsheet options for ethanol production. Pretreatment is currently one of the most expensive stages in second-generation technologies (Mosier et al., 2005). Pretreatment is of crucial importance for obtaining good yields of sugars (Hamelinck et al., 2005).

The purpose of pretreatment is to remove lignin and hemicelluloses, reduce cellulose crystallinity, and increase the porosity of the materials (Sun and Cheng, 2002). Pretreatment is mainly used to improve the yields of sugars and avoid the formation of byproducts inhibitory to the subsequent hydrolysis and fermentation processes. Physical, physiochemical, chemical, and biological processes have been used for pretreatment of lignocellulosic materials.

According to the summary information reported by Sun and Cheng (2002), the main pretreatments for biomass are listed in Table 12.4.

Many pretreatments have been proposed; however, research and development in pretreatment technologies are necessary to reduce the current cost of bioethanol production (Wyman, 2001).

12.2.3.2 Saccharification and Fermentation Pathways (SHF–SSF–SSCF, Current Status, Challenges)

After pretreatment, the next steps are hydrolysis and fermentation, which can be carried out separately or simultaneously (Tomas-Pejo et al., 2008). When enzymatic hydrolysis is used, different levels of process integration are possible: separate (or sequential) hydrolysis and fermentation (SHF), simultaneous saccharification and co-fermentation (SSCF), consolidated bioprocessing (CBP), or simultaneous saccharification and fermentation (SSF) (Ojeda et al., 2011b).

Enzymatic hydrolysis of cellulose is known to have many advantages over acidic hydrolysis, but the rate of enzymatic hydrolysis is considerably slower than that of acidic hydrolysis. This hydrolysis process is usually catalyzed by cellulase enzymes, and the fermentation is carried out by yeast or bacteria. Cellobiose and glucose are known to inhibit the activities of cellulose enzymes even at low concentrations (Ojeda et al., 2011c). In both separate hydrolysis and fermentation and simultaneous saccharification and fermentation (SSF)

TABLE 12.4

Main Pretreatment Technologies

Pretreatment		Characteristics	References
Physical pretreatment	Mechanical comminution	Reduce crystallinity Size after chipping (10–30 mm) Size after milling or grinding (0.2–2 mm)	Millet et al. (1976)
	Pyrolysis	Temperatures greater than 300°C. At lower temperatures, decomposition is slower, and less volatile products are formed	Kilzer and Broido (1965); Shafizadeh and Bradbury (1979)
Physiochemical pretreatment	Steam explosion	Biomass is treated with high-pressure saturated steam, and then the pressure is swiftly reduced. Typical conditions 160°C–260°C (0.69–4.83 MPa). Optimal hemicellulose solubilization and hydrolysis can be achieved by high temperature and short residence time (270°C, 1 min) or lower temperature and longer residence time (190°C, 10 min)	McMillan (1994); Duff and Murray (1996)
	Catalyzed steam explosion	Addition of H_2SO_4 (or SO_2) or CO_2 can improve enzymatic hydrolysis, decrease the production of inhibitory compounds, and lead to more complete removal of hemicellulose. The optimal conditions for this pretreatment of sugarcane bagasse are (220°C; 30s residence time; water-to-solids ratio, 2; and 1% H_2SO_4). Limitations of steam explosion include destruction of a portion of the xylan fraction, incomplete disruption of the lignin-carbohydrate matrix, and generation of compounds that may be inhibitory to microorganisms used in downstream processes.	Morjanoff and Gray (1987); Ballesteros (2000), Ballesteros et al. (2002), Fernández-Bolaños et al. (1998); Clark and Mackie (1987); Mackie et al. (1985); McMillan (1994).
	Ammonia fiber explosion (AFEX)	Materials are exposed to liquid ammonia at high temperature and pressure for a period of time, and then the pressure is swiftly reduced. Typical conditions (1–2 kg ammonia/kg dry biomass, 90°C, 30 min). Improves the saccharification rates of various herbaceous crops and grasses. Needs recycle to reduce the cost and protect the environment.	Mes-Hartree et al. (1987); Vlasenko et al. (1997); Holtzapple et al. (1992), (1991); Tengerdy and Nagy (1988); Dale et al. (1984)

	CO_2 explosion	4 kg CO_2/kg fiber at a pressure of 5.62 MPa. Low yields compared to steam or ammonia explosion pretreatment, but high compared to enzymatic hydrolysis without pretreatment.	Dale and Moreira (1982)
Chemical pretreatment	Ozonolysis	Effectively removes lignin; does not produce toxic residues for downstream processes; and reactions are carried out at room temperature and pressure. A large amount of ozone is required, making the process expensive.	Ben-Ghedalia and Miron (1981); Neely (1984); Vidal and Molinier (1988)
	Acid hydrolysis	Agents such as H_2SO_4 and HCl. Needs recovery and neutralization. High reactions rates significantly improve cellulose hydrolysis. High-temperature ($T > 160°C$), continuous-flow process for low solids loading (5–10%), and low-temperature ($T < 160°C$), batch process for high solids loading (10–40%).	Esteghlalian et al. (1997); Sivers and Zacchi (1995); Converse et al. (1989)
	Alkaline hydrolysis	The effect depends on the lignin content. Dilute NaOH treatment causes swelling, leading to an increase in internal surface area, a decrease in the degree of polymerization, a decrease in crystallinity, separation of structural linkages between lignin and carbohydrates, and disruption of the lignin structure.	McMillan (1994)
	Organosolv process	Organic or aqueous organic solvent mixture with inorganic acid catalysts (HCl or H_2SO_4). At high temperatures (above 185°C), the addition of catalyst was unnecessary for satisfactory delignification. Needs recycle to reduce cost.	Thring et al. (1990)
Biological pretreatment		Microorganisms such as brown-, white- and soft-rot fungi are used to degrade lignin and hemicelluloses in waste materials. Low- energy requirement and mild environmental conditions. Low rate of hydrolysis.	Fan et al. (1987)

processes, the enzyme is the most significant cost factor, and therefore enzyme loading must be kept at the lowest possible level. SSF is carried out under a sugar-limited condition, thereby reducing the inhibition effect on the enzyme. Although ethanol and other fermentation products can also inhibit enzymatic activity, this inhibition is considered to be much weaker than that caused by glucose and cellobiose (Oh et al., 2000).

The term *cellulase* refers to a group of enzymes that contribute to the degradation of cellulose to glucose. The nature of the cellulolytic enzyme system used determines the mode of action of cellulase, the activity of each enzyme component, the synergistic action among the enzyme components, and the inhibitory effect on enzyme action by reaction intermediates and products (Lee and Fan, 1983). Most cellulose systems that have been investigated were extracted from fungi such as *T. viride, T. reesei*, and *Fusarium solani* (Gan et al., 2003). It is generally agreed that complex fungal-cellulose systems may consist of four enzymes: *endo-1,4-β-glucanase, exo-1,4-β-glucanase (cellobiohydrolase), exo-1,4-β-glucosidase*, and *cellobiase* (*β-glucosidase*). *Exo-glucosidase* is not present in all cellulase preparations, and it forms glucose via a noncellobiase pathway (Gan et al., 2003). The efficiency of enzymatic hydrolysis depends on the presence of all cellulose components and the appropriate proportional ratio of the various components.

Enzymatic hydrolysis and fermentation are carried out in two different vessels (kept at optimal conditions of pH and temperature) when separate hydrolysis and fermentation (SHF) is performed (Ojeda et al., 2011c). Nevertheless, glucose and cellobiose accumulation inhibits the activity of the cellulases in the hydrolysis step. By conducting enzymatic hydrolysis and fermentation in the same vessel (SSF), the inhibition effect is minimized because glucose released by the action of cellulases is converted directly into ethanol by the fermenting microorganism. Moreover, the presence of ethanol in the culture broth helps to avoid undesired microbial contamination. These advantages increase the rate of saccharification and the ethanol productivity compared with the SHF process (Wyman, 1996).

Although processes to obtain bioethanol from lignocellulose are promising methods to produce biofuels at low cost, one of the best opportunities to reduce the cost of cellulosic bioethanol remains the enhancement of sugar yields from cellulose and hemicelluloses (Wyman et al., 2005). Therefore, microorganisms which could ferment other sugars such as xylose, mannose, arabinose, or galactose are required for an economically viable conversion from lignocellulose into ethanol (Zaldivar et al., 2001). One of the most successful microorganisms for producing ethanol is *Saccharomyces cerevisiae*, which shows high ethanol productivity, high tolerance to ethanol, and tolerance to inhibitory compounds present in the hydrolyzate from lignocellulosic biomass (Olsson and Hahn-Hägerdal, 1996).

In SSF, there is a tradeoff between the cost of cellulase production and the cost of hydrolysis/fermentation. Short hydrolysis reaction times involve higher cellulase loadings and lower hydrolysis fermentation costs than longer reaction times. The optimum is constrained by the cost of cellulase and is ~3–4 days (Hamelinck et al., 2005).

The use of lignocellulose as a raw material for a fermentation process imposes many demands on the potential microorganism (Zaldivar et al., 2001). The preferred microorganism in crop-based processes, *Saccharomyces cerevisiae*, is unable to ferment pentoses and is therefore of limited use for lignocellulose substrates with a high content of pentoses unless the necessary pathways are inserted and expressed. The same restriction applies to the ethanologenic bacterium, *Zymomonas mobilis*. Two groups of microorganisms, enteric bacteria and some yeasts, are able to ferment pentoses, but with low ethanol yields. Furthermore, in the case of xylose-fermenting yeasts (*Pachysolen tannophilus, Candida shehatae*, and *Pichia stipitis*), their large-scale use is hampered by their sensitivity to high concentrations of ethanol (≥40 g/L), the requirement for carefully monitored microaerophilic conditions, their high sensitivity to inhibitors, and their inability to ferment xylose at low pH (Hahn-Hägerdal et al., 1994).

The lack of a microorganism which is able to ferment efficiently all the sugars released by hydrolysis from lignocellulosic materials has been one of the main factors preventing use of lignocellulose (Zaldivar et al., 2001). Therefore, an obvious target in the field of metabolic engineering has been the tailoring of such a microorganism by combining advantageous qualities from different microorganisms.

Simultaneous saccharification and co-fermentation (SSCF) combines the enzymatic hydrolysis stage of cellulose-to-glucose process and the co-fermentation of pentose and hexose sugars in one reaction vessel (Lynd, 1996; Spatari et al., 2010). This reduces the number of reactors involved by eliminating the separate hydrolysis reactor and avoids the problem of product inhibition associated with enzymes (Ojeda et al., 2011c).

In consolidated bioprocessing (CBP), ethanol and all required enzymes are produced by a single microorganism community in a single reactor. CBP seems to be the logical endpoint in the evolution of biomass-conversion technology. Application of CBP involves no capital or operating costs for dedicated enzyme production (or purchase), reduced diversion of substrate for enzyme production, and compatible enzyme and fermentation systems. As yet, there are no organisms or compatible combinations of microorganisms available that both produce cellulase and other enzymes at the required high levels and also produce ethanol at the required high concentrations and yields, although various organisms already combine many of these functions (Hamelinck et al., 2005).

12.2.3.3 Purification Pathways (Description, Current State, Challenges)

The development of technologies for separation and separation integration has been linked to the development of the different involved unit operations and to new approaches for process intensification. The examples of separation and separation integration in the case of ethanol production mostly correspond to integration of the conjugated type. In this context, the possible use of saline extractive distillation and membrane technologies for the integration of several configurations for the separation and dehydration of ethanol should be highlighted (Cardona and Sánchez, 2007).

12.3 Exergy Analysis in a Lignocellulosic-Feedstock Biorefinery Based on a Biochemical Processing Platform

12.3.1 Main Concepts of Exergy

Exergy is the maximum theoretical useful work that can be obtained if a system is brought into thermodynamic equilibrium with the environment by means of a process in which the system interacts only with this environment (Sciubba and Wall, 2007). Exergy concepts have been applied to energy conversion, process optimization, diagnosis, management, analysis of complex systems, and sustainability analysis, among other problems. Exergy is a thermodynamic potential, a measure of difference, and for this reason it requires two different states for its definition. Therefore, a reference state is necessary for a definition of exergy.

Exergy is evaluated with respect to a reference environment. The reference environment acts as an infinite system and is a sink and source for heat and materials. It experiences only internally reversible processes in which its intensive state remains unaltered (i.e., its temperature T_o, its pressure p_o, and the chemical potentials μ_{ioo}, of each of the i components present remain constant) (Dincer and Rosen, 2007).

Any material, heat, and work stream can have associated with it an exergy content which is defined by the T, P, x values of the stream itself and of a reference state which normally is the environment in which the system operates (Sorin et al., 1998). The exergy content of a material stream is the amount of work which would be produced by bringing this stream into thermal, mechanical, and chemical equilibrium with a reference state by a sequence of reversible operations. This process can be achieved in two steps (Sorin et al., 1998):

1. Thermal and mechanical equilibrium is accomplished through reversible thermal exchanges only (the end temperature and pressure of the stream are the same as those of the reference state).
2. The stream is brought into chemical equilibrium with the reference state by reversible heat and mass exchange occurring at the reference temperature and pressure.

12.3.2 Exergy Applied to Production Process Analysis (Studies)

For a real process, the exergy input always exceeds the exergy output. This imbalance is due to irreversibilities (exergy destruction). An exergy balance, by definition, only exists for reversible processes. Therefore, for a real process, in which the irreversibilities are greater than zero, exergy is never in balance (Wall, 1988). It is important to distinguish between exergy destruction caused by irreversibilities and exergy waste due to unused exergy. Both represent exergy losses, but irreversibilities have, by definition, no exergy and no environment effects (Akkaya et al., 2007).

By calculating the exergy loss (i.e., destruction and waste), possible improvements can be visualized. Generally, wherever the exergy loss is high, this part of the process should be considered first for improvement; however, every part of the system depends on every other part, and therefore an improvement in one part may cause increased losses in other parts, and consequently the total losses in the modified process may be equal to or even larger than those in the original process configuration (Gong and Wall, 1997).

Exergy can also be converted into monetary units using a conversion factor which needs to be determined. In addition, restrictions may be applied to avoid toxicity and irreversible environmental effects. A higher irreversibility in a subsystem always implies a higher consumption of resources in the plant if the products remain constant. Therefore, the ratio between resources and irreversibilities defines a "cost" which is very useful for thermoeconomic analysis of a system. This kind of analysis includes costs per exergy unit of feed, product, and exergy destruction (Nafey et al., 2006).

12.3.3 Exergy to Design a Sustainable Lignocellulosic Feedstock Biorefinery

Several key indicators have been selected or developed to account for relevant aspects of performance in a quantitative way. With respect to exergy (Tonon et al., 2006), any or all of the following can be used:

- Second-law efficiency, which evaluates systems performance by converting input exergy into exergy associated with the delivered products.
- Potential second-law efficiency, which assesses the potential additional exergy efficiency deriving from exploiting the outlet flows that exist as streams but are

not considered as useful product flows and effectively used. These products are normally useful only if particular conditions occur.

- Cost of products, which determines the efficiency of using the economic resources to obtain products. To compare different products, the costs can be expressed on an exergy basis.
- Relative cost difference, which expresses the relative increase in average cost per exergy unit of the component between the feed and the product.
- Exergoeconomic factor, which compares the plant capital cost against the cost of the irreversibilities linked with the process. In fact, the latter cost is a measure of the increased amount of energy and materials (and therefore increased cost) required to obtain the same products, compared with an ideal process.
- Renewability factor, which compares the exergy in the product with the exergy consumption during waste treatment and preparation of nonrenewable inputs (restoration work) (Berthiaume et al., 2001).
- Environmental tradeoff indicator, which describes the tradeoff between the environment and the desired products. If the restoration work associated with the waste treatment process is more than the exergy content of the products, I_{et} will be less than one, indicating that the production process is not environmentally cost-effective in terms of the energy and materials exchange between the human economy and the environment.

Another point to be considered in biofuels production is "energy consumption versus energy production." The processing of a renewable energy source usually involves the consumption of NRRs (nonrenewable resources) (Ojeda et al., 2011d). However, this biomass energy should be grown or produced in a sustainable way that provides net environmental benefits (WORC, 2006). Sustainable development requires sustainable energy resources and the efficient use of residues (Dincer and Rosen, 2007). Measuring the renewability of an energy resource using traditional energy-accounting methods is also questionable because these methods are based on the first law of thermodynamics, which encompasses the principle of energy conservation. Moreover, a meaningful energy-related yield calculation, which would indicate whether an actual net gain or loss occurs during the utilization of an energy resource, should take into account the differences in all forms of energy (including the chemical energy of all materials) and the second law of thermodynamics, which recognizes changes in energy quality or usefulness (Berthiaume et al., 2001). Consequently, an exergy analysis can assist in improving and optimizing designs (Kanoglu et al., 2009). One principal advantage of exergy analysis over energy analysis is that the exergy content of a process flow provides a better valuation of the flow than the energy content, because the exergy indicates the fraction of energy that is likely useful and therefore usable. The exergy concept has been applied to processes in the petrochemical, chemical, paper, and steel industries among others (Arif et al., 2007; Zafer and Arif, 2007, 2008).

Lignocellulosic biomass, which has been extensively studied by several authors because of its contribution to reducing greenhouse gases, has also been studied in terms of exergy, as reported by Ptansinski et al. (2007), who mention the need for exergy analysis of biomass gasification because of the low efficiency of gasifiers. Martin et al. (2006) performed exergy analysis on the combined combustion of biomass and coal. Fryda et al. (2008) performed an exergy analysis of fluidized-bed gasification of biomass; Brown et al. (2007) performed an integration between the exergy analysis and economic analysis of a biomass-based energy conversion system; and Pellegrini and Oliveira (2007) worked with exergy analysis to study bagasse gasification.

As for the use of exergy analysis in biofuel production, Berthiaume et al. (2001) described a method to quantify the renewability of a biofuel (corn ethanol), considering the ideal CO_2-ethanol–glucose cycle to show that exergy can be analyzed through natural thermodynamic cycles. The amount of exergy was used to assess the deviation from ideal behavior caused by the consumption of nonrenewable resources. Yang et al. (2009) performed an exergy analysis using the cumulative energy method to determine the degree of renewability of corn-alcohol production in China, including all nonrenewable resources, transportation, agricultural production, transportation, and processing of byproducts. This study did not include the combustion of ethanol because it considered only the production process. Talens et al. (2007) analyzed biodiesel production and suggested exergetic flow analysis as a tool for evaluating the environmental contribution to account for waste and emissions, determining the exergetic efficiency, and comparing other sources of energy with economic and environmental policies defined for resource use. They determined that the biodiesel production process had lower exergy losses.

Information in the literature on exergy analysis in relation to second-generation biofuels is very limited. Velásquez-Arredondo et al. (2010) analyzed the production of ethanol from bananas and their residues using both exergetic and renewability analysis. Tan et al. (2010) compared the production of ethanol from palm fruits using exergy analysis. Recently, several articles have been published on this topic, with analyses of pretreatments and biochemical pathways (Ojeda and Kafarov, 2009; Ojeda et al., 2011a–d) for bioethanol from lignocellulosic materials.

The application of exergy-analysis methodology to processes like ethanol production from lignocellulosic material enables the identification of the steps and exergetic flows with higher exergetic consumption. The methodology also provides essential tools to compare production schedules and to select a production plan which will provide comprehensive utilization of material and which will include the most suitable design processes for second-generation biofuels production.

12.3.4 Exergy to Evaluate the Environmental Impact of Biorefinery Products and Processes

Exergy is better related to ecological effects than energy, which makes exergy flows better as ecological indicators. The exergy content of a natural resource can be interpreted as a general measure of its potential "usefulness." The exergy embodied in wastes would be one measure of the potential for causing harm to the environment (Wall and Gong, 2001).

The more exergy an emission carries, the more it deviates from the environment. An emission of substances which are common in the environment (e.g., steam or water) carries less exergy than an emission of substances that are less common. Emissions of high-exergy substances should also be regarded as a misuse of resources (Huang et al., 2007).

A higher heat exergy content means a higher temperature difference with the environment, which will result in a greater impact. Some waste materials drive the environment away from ecological equilibrium (Gong and Wall, 2001). Because of high exergy losses from irreversibilities, it becomes necessary to supply more exergy to the process to compensate for these losses, and this may have an environmental impact (an indirect effect of exergy destruction) (Wall and Gong, 2001).

Two general approaches exist to reduce the environmental impact associated with resource degradation (Rosen et al., 2008):

a. Increased efficiency preserves exergy by reducing the exergy necessary for a process and therefore reduces environmental damage. Increased efficiency also usually reduces exergy emissions.
b. Environmental damage can be reduced by taking advantage of the openness of the earth and using solar radiation.

Life-cycle analysis (LCA) is a method that evaluates all in- and outflows during the "life cycle" of a good or service with regard to its environmental impact. LCA is a poor tool for system design because it can evaluate only systems which have already been proposed (Gong and Wall, 1997). Life-cycle exergy analysis (LCEA) uses the same framework as LCA, but the only criterion is now life-cycle irreversibility, the exergy loss during the complete life cycle (Cornelissen and Hirs, 2002). In this method, a distinction has been made between renewable and nonrenewable resources. The total exergy use over time is also considered. LCEA shows in which components the losses of natural resources take place. With this information, better proposals for reducing the loss of natural resources can be generated. LCEA can also be used together with LCA. In this case, LCEA determines the depletion of natural resources, while other environmental effects are calculated with LCA. When LCEA is used separately, the intent is often to reduce the use of natural resources or the costs associated with their use (Cornelissen and Hirs, 2002).

12.4 PI and Exergy Analysis in a Lignocellulosic-Feedstock Biorefinery

Process integration is a holistic approach to process design and operation that emphasizes the unity of the process (El-Halwagi, 2006). Material recycle pinch analysis enables the identification of targets for mass recycle strategies which guarantee minimum waste discharge and fresh resource usage in the process (El-Halwagi et al., 2009). There is an exergy loss inherent in heat transfer. Minimizing exergy losses in a process will result in better use of energy inputs. Exergy losses that occur because of imperfections and inefficiencies in the process that derive from bad design, indiscriminate use of resources, leaks, and emissions can be avoided by improving the design of the process using process integration principles (Ojeda et al., 2011c).

12.4.1 PI and Exergy Concepts Applied to Process Analysis

The exergy concept has been also linked with process integration methodologies. In terms of heat integration, the exergy of a heat stream, $\dot{B}_i$, can be calculated using Equation 12.1:

$$\dot{B}_i = \dot{W}_{\max} = \left(1 - \frac{T_0}{T_i}\right)\dot{Q}_i \quad (12.1)$$

where the term within parentheses is well known as the *Carnot efficiency factor*. This implies that only a fraction of the energy content in a heat stream can be converted into work and that this fraction depends on the temperature of the source and the surroundings.

The study of heat transfer offers an interesting perspective on the inexorable production of entropy and loss of exergy. The loss of exergy that results from the transfer of heat from a *state A* to a *state B* at a lower temperature can be calculated using Equation 12.2:

$$\dot{W}_{\text{lost}} = \left(\frac{1}{T_B} - \frac{1}{T_A}\right) T_0 \dot{Q} \quad (12.2)$$

This equation implies that there is an exergy loss inherent in heat transfer. This is a striking result because it says that, no matter how efficiently heat transfer is carried out, the mere fact of having a temperature gradient is enough to incur an exergy loss.

The exergy transferred from or to a stream that goes from *state A* to *state B* can be calculated using Equation 12.3:

$$\dot{B}_i = \int_A^B \left(1 - \frac{T_0}{T_i}\right) \dot{m}_i \, Cp \, \mathrm{d}T \quad (12.3)$$

Exergy losses can be illustrated using an exergy composite curve, which can be obtained by replacing temperature by Carnot efficiency in the composite curve. The area enclosed by the two composite curves represents exergy loss. This concept makes it possible to identify areas of larger exergy loss which offer possibilities for improvement. Reducing exergy losses will result in an improved utilization of energy inputs. Nevertheless, it is not possible to eliminate all the exergy losses in a process because there is a minimum amount of exergy loss that is required as a driving force to operate the process (Ojeda et al., 2011c). Exergy losses can be decomposed into two parts: avoidable exergy loss and inevitable exergy loss. *Inevitable exergy loss* is the minimum exergy loss required to operate a process at a specific rate. *Avoidable exergy loss* is the losses that occur because of imperfections and inefficiencies in the process that derive from bad design, indiscriminate use of resources, leaks, emissions, and similar factors (Hau, 2005). Avoidable exergy loss can be avoided by improving the design and operation of the process. In summary, total exergy losses can be expressed as follows (Equation 12.4):

$$\dot{W}_{\text{lost}} = \dot{W}_{\text{lost,INE}} + \dot{W}_{\text{lost,AVO}} \quad (12.4)$$

where $W_{\text{lost,INE}}$ is inevitable exergy loss and $W_{\text{lost,AVO}}$ is avoidable exergy loss. Techniques and correlations for calculating inevitable exergy losses for different units can be found elsewhere (Szargut et al., 1988; Feng et al., 1996).

Because all real processes experience an inevitable exergy loss, they can operate only at an efficiency that is lower than the maximum achievable efficiency, η_{max}, which can be determined using Equation 12.5:

$$\eta_{\text{max}} = \frac{\sum \dot{B}_n - \dot{W}_{\text{lost,INE}}}{\sum \dot{B}_n} \quad (12.5)$$

12.5 Case Study: PI and Exergy Analysis in Bioethanol Production from Sugarcane Bagasse

In this section, three cases are compared to illustrate the concepts of exergy and process integration. Liquid hot water (LHW), catalyzed steam explosion, and organosolv processes were selected for the pretreatment stage of sugarcane bagasse based on methodologies proposed by Mosier et al. (2005). Simultaneous saccharification and fermentation (SSF) was selected based on the comparison of enzymatic process configurations described by Hamelinck et al. (2005). In this work, three technological routes for second-generation ethanol production were simulated and analyzed by use of the Aspen-Hysys^{TM} program: LHW-SSF (Case 1), catalyzed steam explosion-SSF (Case 2), and catalyzed organosolv-SSF (Case 3). The selected cases are summarized in Table 12.5. All configurations were divided into their main stages for evaluation.

Results were obtained for each technological configuration using energy and exergy analysis. The net energy ratio (NER = net energy output/net energy input) was less than one for all processes. Consequently, the fuel ethanol system has no net energy gain, meaning that these processes have high-energy consumption and that it is necessary to apply energy integration methodologies to reduce the external energy input. A renewability index (net bioenergy outputs/net fossil energy inputs) greater than one is a minimum requirement for saying that a biofuel system can help to reduce dependency on fossil energy. This index was greater than one for Cases 1 and 2. Case 3 had the highest energy consumption because of recycle in the pretreatment stage. Process integration methodologies are needed to reduce this requirement. Exergy was calculated for all material streams in the flowsheet. The reference temperature was 25°C, and the chemical exergy of ash was neglected. The main results and a comparison of exergy efficiency for the different cases are shown in Table 12.6.

TABLE 12.5

Selected Cases

Case 1	Case 2	Case 3
Bagasse (mass fractions): cellulose = 23%, hemicelluloses = 11%, lignin = 13.5%, ash = 2.25%, and moisture = 50%		
Pretreatment: LHW (461 K, 6 MPa)	Pretreatment: acid-catalyzed steam explosion (461 K, 0.6 MPa)	Pretreatment: catalyzed organosolv solution [w(ethanol) = 58.5%, w(water) = 40% and w(H_2SO_4) = 1.5%, 458 K]
Hydrolysis: SSF: Yield 85% using *S. cerevisiae*.		

TABLE 12.6

Exergy Analysis Results

				Exergy Efficiency (%)			
Case	Pretreatment	Technology	IRR[a]	Pretreatment	SSF	Purification	IR
1	LHW	SSF	23.43	76.45	38.73	67.36	1.532
2	Catalyzed steam explosion	SSF	19.88	87.53	48.01	68.22	1.180
3	Organosolv (Ethanol + water + H_2SO_4)	SSF	21.45	25.23	55.74	65.60	0.891

[a] IRR: Total irreversibility (MJ/kg ethanol), IR: Renewability index.

The highest exergy efficiency was reported for Case 2. For this case, the greatest exergy requirements were observed in the pretreatment stage because of the steam explosion process. Case 3 had the lowest exergy efficiency in the pretreatment stage due to recycle. On the other hand, the maximum irreversibilities were observed for SSF and rectification-purification. The highest improvement potential was observed in the SSF. Possible improvements included reduction of process steam requirements, adjustment of cogeneration systems, or design of a heat network for thermal integration of the process.

LCA was carried on each case using the SimaPro life cycle assessment software (Eco-indicator 99 method). The results show the greatest effect on climate change for the second bioethanol production process using the selected topologies. Case 1 (LHW) had the lowest impact in this analysis (Figure 12.3), with total environmental impacts of 5.4 mPt (milli-Point), lower than Cases 2 and 3 (7.5 and 15 mPt, respectively). However, considering both exergy and environmental results, Case 2 (catalyzed steam explosion-SSF) was selected for further improvement using process integration techniques.

The heat integration alternatives for Case 2 (catalyzed steam explosion-SSF) were analyzed by thermal pinch analysis. For this purpose, the Aspen HX-NET® software was used. Balanced composite curves were generated, and a pinch point was identified (91.5–86.5°C). Using a thermal pinch diagram and $\Delta T_{min} = 5°C$, the minimum heating and cooling utility requirements were calculated (4.84E+7 kJ/h and 6.31E+7 kJ/h, respectively) (Figure 12.4).

After performing pinch analysis using Aspen HX-NET®, the heat exchanger network (HEN) was designed. Using pinch technology rules for HEN design, the designer can make many alternatives for different combinations of connecting heat exchanger units. Figure 12.5 shows the Aspen HX-NET®-designed grid diagram for one optimal alternative for a possible HEN for this case. Case 2 can be improved by designing of a heat network for thermal integration of the process. The proposed HEN reduces cooling utilities by 63.39% and heating utilities by 69.34%. However, the HEN design requires a larger number

FIGURE 12.3
LCA for climate change; the environmental impacts are expressed in mPt (milliPoints).

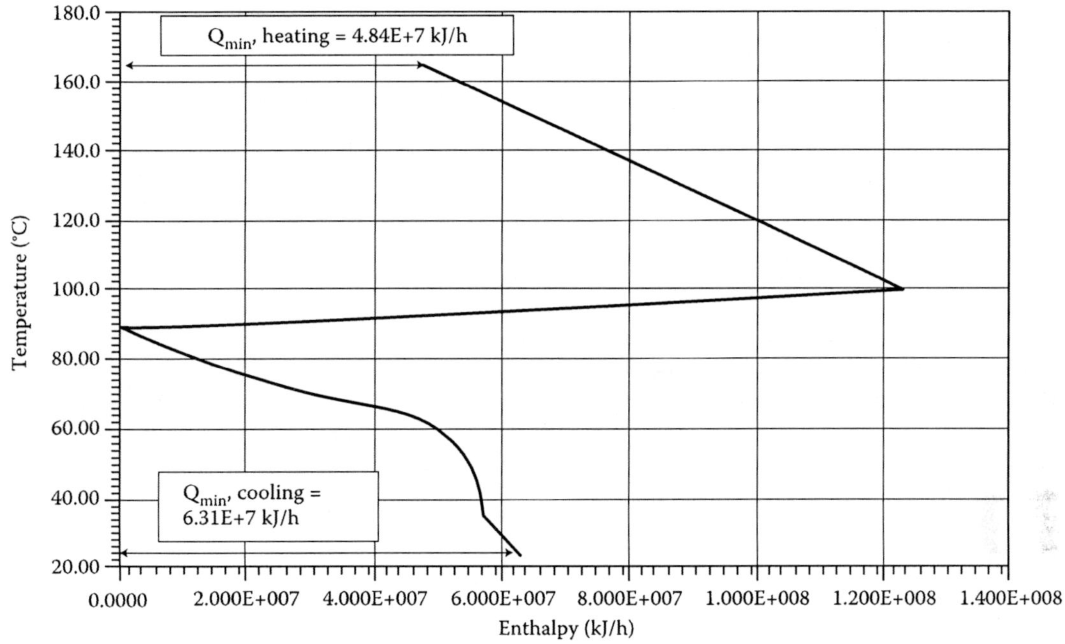

FIGURE 12.4
GCC for Case 2.

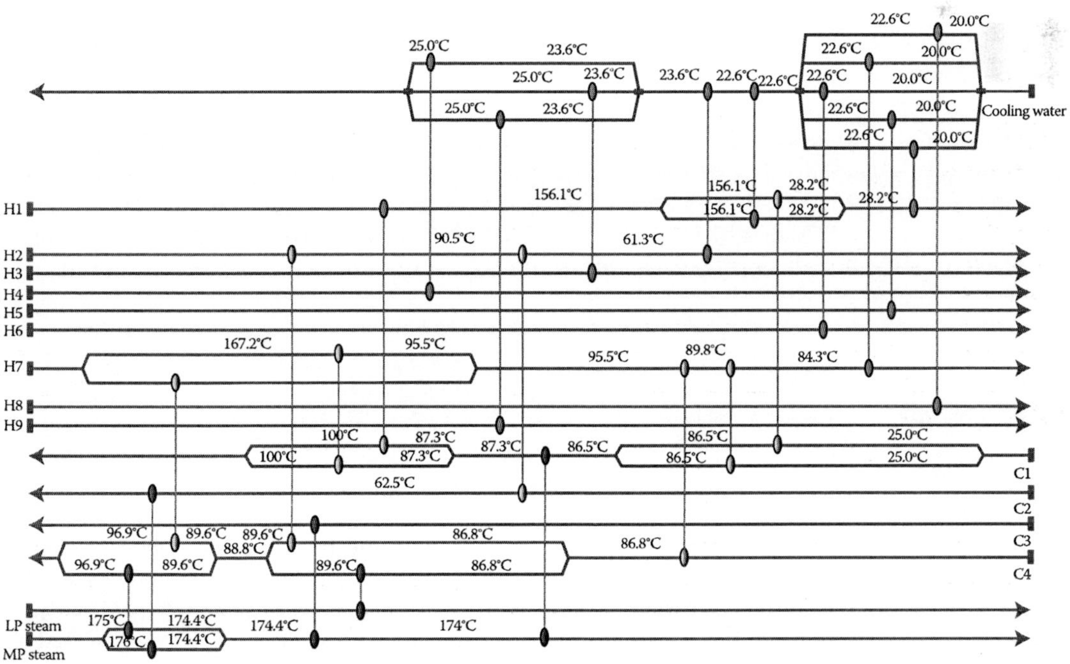

FIGURE 12.5
Grid diagram for HEN as designed.

TABLE 12.7
Utility Savings for Overall Process—Case 2 (Catalyzed Steam Explosion-SSF)

	Heating Utility	Cooling Utility
Utility without integration [MW] (13 Units)	43.44	47.53
Utility after integration [MW] (23 Units)	13.32	17.4
Savings from heat integration [MW]	30.12	30.13
Percentage of savings	69.34%	63.39%

of heat exchangers (10 new pieces of equipment), which would involve an increase in capital costs (Table 12.7). Economic analyses of this type are needed to ensure the implementation of the best HEN.

The material-recycle pinch analysis (El-Halwagi, 2006) enables the identification of targets for mass recycle strategies which guarantee the minimum waste discharge and fresh resource usage in the process. For this purpose, sinks and sources were identified based on simulation results. The main objective was to minimize the fresh water and wastewater in the process, and therefore other components in the stream were considered as impurities. In this way, if a material-recycle strategy is implemented in this process, fresh water and wastewater can be reduced by 43% and 24%, respectively. In addition to resource conservation and reduction in operating cost, the solution also offers enhanced environmental performance.

Application of exergy analysis to account both for materials use and for waste residuals is an effective way of generating and screening design alternatives for sustainable development (Ojeda et al., 2011c). The application of computer-aided process engineering tools together with energy and exergetic analysis and LCA using Aspen-HysysTM®, Aspen HX-NET®, and SimaPro life-cycle assessment software enabled the identification of exergy losses and environmental impacts in the process. This enabled the selection of a potential alternative for second-generation production.

Application of mass and heat integration to the selected process made it possible to identify minimum targets for the direct mass-recycle alternatives and thermal integration in the second-generation ethanol processes. Designs of mass-recycle strategies and a heat exchange network for thermal integration of the process were used to reduce the consumption of natural resources, reduce operating cost, and enhance environmental performance (Ojeda et al., 2011d).

12.6 Integration of a Biorefinery in Existing Industrial Facilities (with First-Generation Bioethanol)

Bioethanol from lignocellulosic materials, such as sugarcane bagasse, has been studied with great interest over the past few decades, but its production on an industrial scale has not yet become viable (Balat and Balat, 2008). Studies which include process integration, increase of fermentation yields, and integration of unit operations are still needed to make hydrolysis a competitive technology (Zaldivar et al., 2001; Cardona and Sánchez, 2007).

According to Chen (2008), the concept of the biorefinery provides opportunities for new applications of existing science and engineering principles. However, such applications require a new theoretical framework, a new integrated knowledge base, and new approaches for solving these complex problems. It is clear that the ultimate limit of the potential of a biorefinery is not the technology needed to convert the biomass materials to various products, but rather the amount of biomass materials that are available as feedstock. Other considerations for developing a successful biorefinery include the additional resources needed to provide a sustainable feedstock supply (land, water, etc.), net energy gains, and environmental impacts such as waste disposal and emissions. As stated by Du et al. (2008), selection of appropriate raw materials and development of biorefinery-based strategies to support sustainable processes are therefore essential.

Dias et al. (2009) compared two cases of biorefinery integration, in which they assumed initially that 70% of the bagasse produced in the extraction system was available for the hydrolysis process. Afterwards, an iterative procedure was used to find the maximum amount of bagasse that could in fact be delivered to the hydrolysis process, considering the steam demand of the whole plant. Two configurations were considered for the distillation columns: conventional (single-pressure) and double-effect distillation columns. It was shown that 76% of the bagasse generated in the mills was available for hydrolysis when a conventional distillation system was used, while 90% was available with double-effect distillation.

The results obtained by Dias et al. (2009) showed the importance of the process integration for the feasibility of a biorefinery made up of a conventional sugarcane mill and a hydrolysis plant. The multiple-pressure distillation columns were identified as the most important measure for process integration, enabling the simultaneous reduction of hot-utility and bagasse requirements, in spite of the reduction in the electric energy available for sale. According to this study, the use of sugarcane trash and lignin as auxiliary fuels would be of fundamental importance to fulfill the energy requirements of the biorefinery plant with hydrolysis.

On the other hand, many bioproducts that can be produced in a sugar-based biorefinery are not feedstock-specific, but rather can be derived from virtually all sugar, starch, or lignocellulosic feedstocks (Mabee et al., 2005). For instance, polyols can be derived from xylose and arabinose and used in a variety of products, including antifreeze, plastic bottles, brake fluid, synthetic fibers, resins, auto bodies, and sweeteners. Lactic acid is used as a food additive, and polylactic acid (PLA) polymers are used in biodegradable plastic bags. Glucose can be accessed microbially and chemically to produce whole family trees of industrially relevant products, including such products as lactic acid, ethanol, acetic acid, and levulinic acid. Bioethanol can be produced through the fermentation of both six- and five-carbon sugars and can be used in a gasoline blend, as a clean fuel, or as a precursor for hydrogen in fuel-cell vehicles. A specific example of an innovative glucose product is 3G (1,3-propanediol), which can be used to develop a polymer called 3GT, which has excellent stretch recovery, toughness, and dye capability.

The ethanol production process using sugarcane bagasse enables the production of valuable co-products, for which the various applications can be observed in Table 12.8.

In accordance with Huang et al. (2008), in an integrated lignocellulose biorefinery, pre-extraction of hemicelluloses followed by the production of value-added products such as ethanol, sugar-based polyesters, or other chemicals offers a tremendous economic opportunity, reducing overall production cost and improving overall profitability.

Howard et al. (2003) reported that a number of products such as organic acids, amino acids, vitamins, and a number of bacterial and fungal polysaccharides such as xanthan are produced by fermentation using glucose as the base substrate. Theoretically, these same products could be manufactured from lignocellulose waste. Boudet et al. (2003) presented

TABLE 12.8

Some Value-Added Co-Products that can be Obtained during Fuel Ethanol Production from Lignocellulosic Biomass

Co-Product	Stage Where Co-Product is Formed	Application	References
Xylitol	Xylose solutions can be converted into xylitol by chemical or biotechnological means; co-culture of *Saccharomyces cerevisiae* and *Candida tropicalis*	Anticariogenic sweetener, sugar substitute for diabetics	Converti and Del Borghi (1996); Latif and Rajoka (2001); Leathers (2003) and Saha (2003)
CMA (calcium magnesium acetate)	Xylose solutions can be converted into acetic acid by fermentation using *Clostridium thermoaceticum*	Road deicer	Bungay and Peterson (1992) and Wilkie et al. (2000)
Furfural and levulinic acid	Xylose solutions can be converted into furfural	Valuable chemical	Kaylen et al. (2000) Fitzpatrick (1990)
Lignin	Fractionation of pretreated biomass or centrifugation of stillage	Raw material for production of adsorptive materials by chemical modification	Dizhbite et al. (1999)

Source: Adapted from Cardona, C. and Sánchez, O. 2007. *Bioresources Technology* 98: 2415–2457.

a discussion of the potential value-added products which could be derived from lignin, such as vanillin and gallic acid, which are the two most frequently discussed. Vanillin is used for various purposes, including as an intermediate in the chemical and pharmaceutical industries for the production of herbicides, antifoaming agents, or drugs such as papaverine, L-dopa, and the antimicrobial agent, trimethoprim.

12.7 Conclusions

The relationship between exergy analysis and sustainability has been examined. Exergy methods can be used to improve sustainability. Exergy loss was observed in all process; the study suggests that these occurred particularly due to the use of nonrenewable forms of energy and should be minimized to achieve sustainable development. Detailed application of exergy analysis to all stages of second-generation bioethanol production using several technological routes will provide a powerful tool to respond to the "energy consumption vs. energy content in ethanol produced" debate and to verify the sustainable development of the biofuels industry using lignocellulosic biomass.

The reduction of dangerous emissions associated with biofuels production processes and the integrated use of resources constitute the main objective of the guidelines to be followed for sustainable development. For this purpose, it is of great importance to develop an exergy analysis to account both for materials use and for waste residuals. The exergetic analysis presented in this work was confirmed to be very useful for the identification of irreversibility generation in second-generation ethanol processes. Exergy methods are essential in improving efficiency, which makes it possible to maximize the benefits derived from resources while minimizing negative impacts such as environmental damage.

Acknowledgments

The authors express their acknowledgments to the Colombian Institute for Development of Science and Technology (COLCIENCIAS), Texas A&M University (USA), and the Industrial University of Santander (Colombia).

References

Akkaya, A., Sahin, B., Erdem, H. 2007. Exergetic performance coefficient analysis of a simple fuel cell system. *International Journal of Hydrogen Energy* 32(17): 4600–4609.

Arif, H., Zafer, U., Akdeniz, R.C. 2007. Energetic and exergetic aspects of cotton stalk production in establishing energy policies. *Energy Policy* 35(5): 3015–3024.

Balat, M., Balat, H. 2008. A critical review of bio-diesel as a vehicular fuel. *Energy Conversion and Management* 49(10): 2727–2741.

Ballesteros, I. 2000. *Obtención de Etanol a Partir de Biomasa Lignocelulósica Mediante un Proceso de Sacarificación y Fermentación Simultánea (SFS).* PhD dissertation, Universidad de Alcalá de Henares.

Ballesteros, I., Oliva, J.M., Negro, M.J., Manzanares, P., Ballesteros, M. 2002. Enzymatic hydrolysis of steam-exploded herbaceous agricultural waste (*Brassica carinata*) at different particle sizes. *Process Biochemistry* 38: 187–192.

Ben-Ghedalia, D., Miron, J. 1981. The effect of combined chemical and enzyme treatment on the saccharification and *in vitro* digestion rate of wheat straw. *Biotechnology and Bioengineering* 23: 823–831.

Berthiaume, R., Bouchard, C., Rosen, M. 2001. Exergetic evaluation of the renewability of a biofuel. *Exergy, An International Journal* 1: 256–268.

Boudet, A.M., Kajita, S., Grima-Pettenati, J., Goffner, D. 2003. Lignins and lignocellulosics: A better control of synthesis for new and improved uses. *Trends in Plant Science* 8(12): 576–581.

Brown, D., Gassner, M., Fuchino, T., Maréchal, F. 2007. *Thermo-Economic Analysis for the Optimal Conceptual Design of Biomass Gasification Energy Conversion Systems.* École Polytechnique Fédérale de Lausanne, LENI-ISE-STI-EPLF, Station 9, CH-1015 Lausanne, Switzerland.

Bungay, H., Peterson, J. 1992. CMA as a byproduct of the manufacture of fuel ethanol. *Resources, Conservation, and Recycling* 7: 83–94.

Cardona, C. and Sánchez, O. 2007. Fuel ethanol production: Process design trends and integration opportunities. *Bioresources Technology* 98: 2415–2457.

Chen, S. 2008. Industrial biosystems engineering and biorefinery systems. *Chinese Journal of Biotechnology* 24(6): 940–945.

Claassen, P., Van, J.B., López, A.M., Van Niel, E.W.J., Sijtsma, L., Stams, A.J.M., Vries, S.S., Weusthuis, R.A. 1999. Utilization of biomass for the supply of energy carriers. *Applied Microbiology and Biotechnology* 52: 741–755.

Clark, T.A., Mackie, K.L. 1987. Steam explosion of the softwood *Pinus radiata* with sulphur dioxide addition. I. Process optimization. *Journal of Wood Chemistry and Technology* 7: 373–403.

Converse, A.O., Kwarteng, I.K., Grethlein, H.E., Ooshima, H. 1989. Kinetics of thermochemical pretreatment of lignocellulosic materials. *Applied Biochemistry and Biotechnology* 20/21: 63–78.

Converti, A., Del Borghi, M. 1996. Selection of hemicellulosic hydrolysated pretreatments and fermentation conditions to stimulate xylitol production by ethanol-producing yeasts. *Acta Biotechnologica* 16(2–3): 133–144.

Cornelissen, R.L., Hirs, G. 2002. The value of the exergetic life cycle assessment besides the LCA. *Energy Conversion and Management* 43: 1417–1424.

Dale, B.E., Moreira, M. 1982. A freeze-explosion technique for increasing cellulose hydrolysis. *Biotechnology and Bioengineering Symposium* 12: 31–43.

Dale, B.E., Henk, L.L., Shiang, M. 1984. Fermentation of lignocellulosic materials treated by ammonia freeze-explosion. *Developments in Industrial Microbiology* 26: 223–233.

Dias, M., Ensinas, A., Nebra, S., Maciel Filho, R., Rossell, C., Wolf Maciel, M.R. 2009. Production of bioethanol and other bio-based materials from sugarcane bagasse: Integration to conventional bioethanol production process. *Chemical Engineering Research and Design* 87: 1206–1216.

Dincer, I., Rosen, M.A. 2007. *Exergy, Energy, Environment, and Sustainable Development*. Oxford, UK: Elsevier.

Dizhbite, T., Zakis, G., Kizima, A., Lazareva, E., Rossinskaya, G., Jurkjane, V., Telysheva, G., Viesturs, U. 1999. Lignin—A useful bioresource for the production of sorption-active materials. *Bioresource Technology* 67: 221–228.

Drissen, R., Maas, R., Tramper, J., Beeftink, H. 2008. Modelling ethanol production from cellulose: Separate hydrolysis and fermentation versus simultaneous saccharification and fermentation. *Biocatalysis and Biotransformation* 27(1): 27–35.

Du, C., Lin, S., Koutinas, A., Wang, R., Dorado, P., Webb, C. 2008. A wheat biorefining strategy based on solid-state fermentation for fermentative production of succinic acid. *Bioresource Technology* 99: 8310–8315.

Duff, S.J.B., Murray, W.D. 1996. Bioconversion of forest products industry waste cellulosics to fuel ethanol: A review. *Bioresource Technology* 55: 1–33.

El-Halwagi, M.M. 2006. *Process Integration*. Amsterdam: Elsevier.

El-Halwagi, M.M., Harell, D., Spriggs, H.D. 2009. Targeting cogeneration and waste utilization through process integration. *Applied Energy* 86: 880.

Esteghlalian, A., Hashimoto, A.G., Fenske, J.J., Penner, M.H. 1997. Modeling and optimization of the dilute-sulfuric-acid pretreatment of corn stover, poplar, and switchgrass. *Bioresources Technology* 59: 129–136.

Fan, L.T., Gharpuray, M.M, Lee, Y.H. 1987. *Cellulose Hydrolysis*. Biotechnology Monographs, Berlin: Springer, p. 57.

Farrell, A.E., Plevin, R.J., Turner, B.T., Jones, A.D., O'Hare, M., Kammen, D.M. 2006. Ethanol can contribute to energy and environmental goals. *Science* 311: 506–508.

Feng, X., Zhu, X.X., Zheng, J.P. 1996. A practical exergy method for system analysis. *Proceedings, 31st Inter-Society Energy Conversion Engineering Conference*, IEEE, Washington, DC, pp. 2068–2071.

Fengel, D., Wegener, G. 1989. *Wood Chemistry, Ultrastructure, Reactions*. Berlin: Walter de Gruyter, p. 613.

Fitzpatrick, S.W. 1990. *Lignocellulose Degradation to Furfural and Levulinic Acid*. U.S. Patent 4,897,497.

Fryda, L. Panopoulos, K.D., Kakaras, J.E. 2008. Exergetic analysis oxide fuel cell and biomass gasification integration with heat pipes. *Exergy* 33: 292–299.

Gan, Q., Allen, S.J., Taylor, G. 2003. Kinetic dynamics in heterogeneous enzymatic hydrolysis of cellulose: An overview, an experimental study, and mathematical modeling. *Process Biochemistry* 38: 1003–1018.

Gong, M. Wall, G. 1997. On exergetics, economics, and optimization of technical processes to meet environmental conditions. *International Conference on Thermodynamic Analysis and Improvement of Energy Systems*, Beijing, China, June, pp. 10–13.

Gong, M. Wall, G. 2001. On exergy and sustainable development—Part 2: Indicators and methods. *Exergy: An International Journal* 1: 217–233.

Hahn-Hägerdal, B., Galbe, M., Gorwa-Grauslund, M.F., Zacchi, G. 2006. Bio-ethanol: The fuel of tomorrow from the residues of today. *Trends in Biotechnology* 24: 549–556.

Hahn-Hägerdal, B., Jeppsson, H., Skoog, K., Prior, B.A. 1994. Biochemistry and physiology of xylose fermentation by yeasts. *Enzyme and Microbial Technology* 16: 933–943.

Hamelinck, C.N., Hooijdonk, G.V. Faaij, A. 2005. Ethanol from lignocellulosic biomass: Techno-economic performance in short-, middle- and long-term. *Biomass Bioenergy* 28: 384–410.

Hau, J. 2005. *Toward environmentally conscious process systems engineering via joint thermodynamic accounting of industrial and ecological systems*. PhD dissertation, Ohio State University. Available online (http://etd.ohiolink.edu/send-pdf.cgi/Hau%20Jorge%20L.pdf?osu1117650243).

Hayes, D.J. 2009. An examination of biorefining processes, catalysts, and challenges. *Catalysis Today* 145(1–2), 138–151.

Ho, N.W.Y., Chen, Z.D. 1997. *Stable Recombinant Yeasts Capable of Effective Fermentation of Both Glucose and Xylose.* Patent Appl. 60/016, 865, May 6, 1996; PCT Patent WO97/42307, November 13.

Ho, N.W.Y., Tsao, G.T. 1995. *Recombinant Yeasts for Effective Fermentation of Glucose and Xylose.* PCT Patent W095/13362.

Holtzapple, M.T., Jun, J-H., Ashok, G., Patibandla, S.L., Dale, B.E. 1991. The ammonia freeze explosion (AFEX) process: A practical lignocellulose pretreatment. *Applied Biochemistry and Biotechnology.* 28/29: 59–74.

Holtzapple, M.T., Lundeen, J.E., Sturgis, R. 1992. Pretreatment of lignocellulosic municipal solid waste by ammonia fiber explosion (AFEX). *Applied Biochemistry and Biotechnology* 34/35: 5–21.

Howard, R.L., Abotsi, E., Jansen, E.L., Howard, S. 2003. Lignocellulose biotechnology: Issues of bioconversion and enzyme production. *African Journal of Biotechnology* 2(12): 602–619.

Huang, H.-J., Ramaswamya, S., Tschirner, U.W., Ramarao, B.V. 2008. A review of separation technologies in current and future biorefineries. *Separation and Purification Technology* 62: 1–21.

Huang, L.Q., Chen, G.Q., Zang, Y., Chen, B., Luan, S.J. 2007. Exergy as a unified measure of water quality. *Communications in Nonlinear Science and Numerical Simulation* 12: 663–672.

Ingram, L.O., Alterthum, F., Conway, T. 1991. *Ethanol Production by Escherischia coli Strains Co-expressing Zymomonas PDC and ADH Genes.* U.S. Patent 5,000,000.

Ingram, L. O., Clark, D.C. 1992. *Ethanol Production Using Engineered Mutant Escherischia coli.* U.S. Patent 5,028,539.

Kanoglu, M., Bolatturk, A. 2008. Performance and parametric investigation of a binary geothermal power plant by exergy. *Renewable Energy* 33(11): 2366–2374.

Kanoglu, M., Dincer, I., Cengel, Y.A. 2009. Exergy for better environment and sustainability. *Environmental Development and Sustainability* 11: 971–988.

Kaylen, M., Van Dyne, D., Choi, Y.-S., Blasé, M. 2000. Economic feasibility of producing ethanol from lignocellulosic feedstocks. *Bioresource Technology* 72: 19–32.

Kilzer, F.J., Broido, A. 1965. Speculations on the nature of cellulose pyrolysis. *Pyrodynamics* 2: 151–163.

Koutinas, A.A., Wang, R.-H., Webb, C. 2007. The biochemurgist: Bioconversion of agricultural raw materials for chemical production. *Biofuels, Bioproducts and Biorefining* 1: 24–38.

Latif, F., Rajoka, M.I. 2001. Production of ethanol and xylitol from corn cobs by yeasts. *Bioresource Technology* 77: 57–63.

Lavoie, J.M., Beauchet, R., Berberi, V., Chormet, M. 2011. Biorefining lignocellulosic biomass via the feedstock impregnation rapid and sequential steam treatment. *Biofuel's Engineering Process Technology.* Intech: Croatia. pp. 687–714.

Leathers, T.D. 2003. Bioconversions of maize residues to value-added coproducts using yeast-like fungi. *FEMS Yeast Research* 3: 133–140.

Lee, Y., Fan L. 1983. Kinetic studies of enzymatic hydrolysis of insoluble cellulose: (II). Analysis of extended hydrolysis times. *Biotechnology and Bioengineering* 25(4): 939–966.

Lynd, L.R. 1996. Overview and evaluation of fuel ethanol from cellulosic biomass: Technology, economics, the environment, and policy. *Annual Reviews in Energy and Environment* 21: 403–465.

Mabee, W., Gregg, D., Saddler, J. 2005. Assessing the emerging biorefinery sector in Canada. *Applied Biochemistry and Biotechnology* 121–124: 765–778.

Mackie, K.L., Brownell, H.H., West, K.L., Saddler, J.N. 1985. Effect of sulphur dioxide and sulphuric acid on steam explosion of aspenwood. *Journal of Wood Chemistry Technology* 5: 405–425.

Martin, C., Villamañan, M., Chamorro, C., Otero, J., Cabanillas, A., Segovia, J. 2006. Low-grade coal and biomass co-combustion on fluidized bed: Exergy analysis. *Energy* 31: 330–344.

McMillan, J.D. 1994. Pretreatment of lignocellulosic biomass. In: Himmel, M.E., Baker, J.O., Overend, R.P. (eds.), *Enzymatic Conversion of Biomass for Fuels Production.* American Chemical Society, Washington, DC, pp. 292–324.

Mes-Hartree, M., Hogan, C.M., Saddler, J.N. 1987. Recycle of enzymes and substrate following enzymatic hydrolysis of steam-pretreated aspenwood. *Biotechnology and Bioengineering* 30: 558–564.

Millet, M.A., Baker, A.J., Scatter, L.D. 1976. Physical and chemical pretreatment for enhancing cellulose saccharification. *Biotechnology and Bioengineering Symposium* 6: 125–153.

Morjanoff, P.J., Gray, P.P. 1987. Optimization of steam explosion as a method for increasing susceptibility of sugarcane bagasse to enzymatic saccharification. *Biotechnology and Bioengineering* 29: 733–741.

Mosier, N., Wyman, C.E., Dale, B.E., Elander, R.T., Holtzapple, M., Ladisch, M.R., Lee, Y. 2005. Features of promising technologies for pretreatment of lignocellulosic biomass. *Bioresearch and Technology* 96: 673–686.

Nafey, A., Fath, H., Mabrouk, A. 2006. Exergy and thermoeconomic evaluation of MSF process using a new visual package. *Desalination* 201: 224–240.

Neelis, M.L., Van der Kooi, H.J., Geerlings, J.C. 2004. Exergetic life cycle analysis of hydrogen production and storage systems for automotive applications. *International Journal of Hydrogen Energy* 29: 537–545.

Neely, W.C. 1984. Factors affecting the pretreatment of biomass with gaseous ozone. *Biotechnology and Bioengineering* 20: 59–65.

Nigam, P.S., Singh, A. 2010. Production of liquid biofuels from renewable resources. *Progress in Energy and Combustion Science*, doi:10.1016/j.pecs.2010.01.003.

O'Brien, D., Roth, L., McAloon, A. 2000. Ethanol production by continuous fermentation and pervaporation: A preliminary economic analysis. *Journal of Membrane Science* 166: 105–111.

Oh, K.K., Kim, S.W., Jeong, Y.S., Hong, S.I. 2000. Bioconversion of cellulose into ethanol by nonisothermal simultaneous saccharification and fermentation. *Applied Biochemistry and Biotechnology* 89: 15–30.

Ojeda, K., Kafarov, V. 2009. Exergy analysis of enzymatic hydrolysis reactors for transformation of lignocellulosic biomass to bioethanol. *Chemical Engineering Journal* 154: 390–395.

Ojeda, K., Sánchez, E., Kafarov, V. 2011a. Sustainable ethanol production from lignocellulosic biomass: Application of exergy analysis. *Energy* 36(4): 2119–2128.

Ojeda, K., Suarez, J., Ávila, O., Kafarov, V. 2011b. Evaluation of technological alternatives for process integration of sugarcane bagasse for sustainable biofuels production. Part 1. *Chemical Engineering Research and Design* 89(3): 270–279.

Ojeda, K., Sánchez, E., El-Halwagi, M., Kafarov, V. 2011c. Exergy analysis and process integration of bioethanol production from acid pre-treated biomass: Comparison of SHF, SSF, and SSCF pathways. *Chemical Engineering Journal*, doi:10.1016/j.cej.2011.06.083.

Ojeda, K., Sánchez, E., Suarez, J., Ávila, O., Quintero V., El-Halwagi, M., Kafarov, V. 2011d. Application of computer-aided process engineering and exergy analysis to evaluate different routes of biofuels production from lignocellulosic biomass. *Industrial and Engineering Chemistry Research* 50(5): 2768–2772.

Olsson, L., Hahn-Hägerdal, B. 1996. Fermentation of lignocellulosic hydrolysates for ethanol production. *Enzyme Microbial Technology* 18(5): 312–331.

Orts, W.I., Holtman, K., Seiber, J. 2008. Agricultural chemistry and bioenergy. *Journal of Agricultural Food Chemistry* 56: 3892–3899.

Palmqvist, E., Hahn-Hägerdal, B. 2000. Fermentation of lignocellulosic hydrolysates. I: Inhibition and detoxification. *Bioresource Technology* 74: 17–24.

Pellegrini, L., Oliveira, S. Jr. 2007. Exergy analysis of sugarcane bagasse gasification. *Exergy* 32: 314–327.

Ptansinski, K.J., Prins, M., Pierik, A. 2007. Exergetic evaluation of biomass gasification. *Energy*, doi:10.1016/j.energy.2006.06.024.

RFA, 2009. *Renewable Fuels Association: Statistics.* Available online (http://www.ethanolrfa.org/pages/statistics), RFA 2009 report. http://www.ethanolrfa.org/bio-refinery-locations/.

Rosen, M.A, Dincer, I., Kanoglu, M. 2008. Role of exergy in increasing efficiency and sustainability and reducing environmental impact. *Energy Policy* 36, 128–137.

Saha, B.C. 2003. Hemicellulose conversion. *Journal of Industrial Microbiology and Biotechnology* 30: 279–291.

Sciubba, E., Wall, G. 2007. A brief commented history of exergy from the beginnings to 2004. *International Journal of Thermodynamics* 10: 1–26.

Shafizadeh, F., Bradbury, A.G.W. 1979. Thermal degradation of cellulose in air and nitrogen at low temperatures. *Journal of Applied Polymer Science* 23: 1431–1442.

Sivers, M.V., Zacchi, G. 1995. A techno-economical comparison of three processes for the production of ethanol from pine. *Bioresource Technology* 51: 43–52.

Sorin, M., Lambert, J., Paris, J. 1998. Exergy flows analysis in chemical reactors. *Transactions of IChemE* 76, Part A, 389–395.

Spatari, S., Bagley, D., MacLean, H. 2010. Life cycle evaluation of emerging lignocellulosic ethanol conversion technologies. *Bioresource Technology* 101: 654–667.

Sun, Y., Cheng, J. 2002. Hydrolysis of lignocellulosic materials for ethanol production: A review. *Bioresource Technology* 83: 1–11.

Szargut, J., Morris, D.R., Steward, F.R. 1988. *Exergy Analysis of Thermal, Chemical, and Metallurgical Processes*. Hemisphere: New York.

Talens, L., Villalba, G. Gabarrell, X. 2007. Exergy analysis applied to biodiesel production. *Resources, Conservation and Recycling* 51: 397–407.

Tan, H., Lee, K., Mohamed, A. 2010. Second-generation bio-ethanol (SGB) from Malaysian palm empty fruit bunch: Energy and exergy analyses. *Bioresource Technology* 101(14): 5719–5727.

Tengerdy, R.P., Nagy, J.G. 1988. Increasing the feed value of forestry waste by ammonia freeze explosion treatment. *Biological Wastes* 25: 149–153.

Thring, R.W., Chorent, E., Overend, R. 1990. Recovery of a solvolytic lignin: Effects of spent liquor/acid volume ratio, acid concentration, and temperature. *Biomass* 23: 289–305.

Tomas-Pejo, E., Oliva, J.M., Ballesteros, M., Olsson, L. 2008. Comparison of SHF and SSF processes from steam-exploded wheat straw for ethanol production by xylose-fermenting and robust glucose-fermenting *Saccharomyces cerevisiae* strains. *Biotechnology and Bioengineering* 100(6): 1122–1131.

Tonon, S., Brown, M., Luchi, F., Mirandola, A., Stoppato, A., Ulgiati, S. 2006. An integrated assessment of energy conversion processes by means of thermodynamic, economic and environmental parameters. *Energy* 31, 149–163.

Velásquez-Arredondo, H.I., Ruiz-Colorado, A.A, Oliveira, J. 2010. Ethanol production from banana fruit and its lignocellulosic residues: Exergy and renewability analysis. *International Journal of Thermodynamics* 12 (3): 155–162.

Vidal, P.F., Molinier, J. 1988. Ozonolysis of lignin: Improvement of *in vitro* digestibility of poplar sawdust. *Biomass* 16: 1–17.

Vlasenko, E.Y., Ding, H., Labavitch, J.M., Shoemaker, S.P. 1997. Enzymatic hydrolysis of pretreated rice straw. *Bioresources Technology* 59: 109–119.

Wall, G. 1988. Exergy flows in industrial processes. *Energy* 13(2): 197–208.

Wall, G., Gong, M. 2001. On exergy and sustainable development. Part 1: Conditions and concepts. *Exergy International Journal* 1(3): 128–145.

Wilkie, A.C., Riedesel, K.J., Owens, J.M. 2000. Stillage characterization and anaerobic treatment of ethanol stillage from conventional and cellulosic feedstocks. *Biomass and Bioenergy* 19: 63–102.

WORC, 2006. *Biofuels Sustainable Criteria, 2006*. Available online: http://www.worc.org/userfiles/Biofuels%20Sustainability%20Criteria.pdf.

Wyman, C.E. 1996. Ethanol production from lignocellulosic biomass: An overview. In: Wyman, C.E. (ed.), *Handbook on Bioethanol: Production and Utilization*. Taylor & Francis, Washington DC, pp. 1–18.

Wyman, C.E. 2003. Potential synergies and challenges in refining cellulosic biomass to fuels, chemicals, and power. *Biotechnology and Progress* 19: 254–262.

Wyman, C. 2001. Twenty years of trials, tribulations and research progress in biotechnology. *Applied Biochemistry and Biotechnology* 91–93, 5–21.

Wyman, C., Dale, B., Elander, R., Holtzapple, M., Ladish, M., Lee, Y. 2005. Coordinated development of leading biomass pretreatment technologies. *Bioresource Technology* 96: 1959–1966.

Yang, Q., Chen, B., Ji, X., He, Y.F., Chen, G.Q. 2009. Exergetic evaluation of corn-ethanol production in China. *Communications in Nonlinear Science and Numerical Simulation* 14(5): 2450–2461.

Zafer, U., Arif, H. 2007. A review and assessment of the energy utilization efficiency in the Turkish industrial sector using energy and exergy analysis method. *Renewable and Sustainable Energy Reviews* 11(7): 1438–1459.

Zafer, U., Arif, H. 2008. Energetic and exergetic assessment of the industrial sector at varying dead (reference) state temperatures: A review with an illustrative example. *Renewable and Sustainable Energy Reviews* 12(5): 1277–1301.

Zaldivar, J., Nielsen, J., Olsson, L. 2001. Fuel ethanol production from lignocellulose: A challenge for metabolic engineering and process integration. *Applied Microbiological Biotechnology* 56: 17–34.

13

Automated Targeting for the Synthesis of an Integrated Biorefinery

Douglas H. S. Tay and Denny K. S. Ng

CONTENTS

13.1 Introduction

According to the *Annual Energy Outlook 2010* (EIA, 2010), total world consumption of energy is projected to increase from 495 quadrillion British thermal units (Btu) in 2007 to 580 quadrillion Btu in 2015 and eventually to 739 quadrillion Btu in 2035. Throughout the projected period, fossil fuels (i.e., coal, oil, and natural gas) are expected to continue to supply a large share of energy use worldwide (EIA, 2010). Therefore, the gradual depletion and rising costs of fossil fuel supplies will soon become a major global problem. As a result, increased attention is being paid to issues of energy security, resource diversification, and efficiency enhancement.

In addition, the increase of public awareness of environmental sustainability issues has also motivated a shift from fossil fuels to renewable energy sources. Biofuels are among the most promising forms of renewable energy because they can be produced from a wide variety of feedstocks (mainly biomass), which has good potential to reduce greenhouse-gas emissions. Technologies that convert biomass to value-added products (biodiesel, bio-ethanol, charcoal, fuel pellets, etc.) through thermal, biological, and physical conversions are now well established. However, a *biorefinery* that uses biomass as a feedstock to produce multiple products, in an analogous way to a petroleum refinery, is a newly proposed concept (Fernando et al., 2006). In recent years, various concepts of the biorefinery have been proposed. For example, Kamm et al. (1998) defined a biorefinery as *a complex system of sustainable, environment- and resource-friendly technologies for the comprehensive utilization and the exploitation of biological raw materials in the form of green and residue biomass from a targeted sustainable regional land utilisation.* Meanwhile, Bridgwater (2003) defined a biorefinery as *a processing facility that is multifunctional, simultaneously producing biofuels, electricity and a considerable number of chemicals from biomass input via thermal conversion processes.* However, it should be noted that biological conversion processes (e.g., fermentation, hydrolysis) also play an important role in a biorefinery (Ritter, 2006). Therefore, the latest research and development efforts on the biorefinery are focusing on integrating the wide spectrum of biomass conversion technologies (gasification, pyrolysis, fermentation, hydrolysis, catalytic conversion, etc.) to produce a broad range of products. To enhance the overall process and economic performance, the concept of an integrated biorefinery which integrates multiple platforms into a whole has been proposed (Fernando et al., 2006).

These concepts can be summarized by saying that the function of a biorefinery is to provide a sustainable supply of biofuels and to produce both bulk and fine biochemical products (methanol, syngas, glycerol, ether, etc.) with minimum waste generation. To date, various single-product biomass-to-energy technologies such as heat and power generation (Qin et al., 2006; Mohan and El-Halwagi, 2007), biodiesel production (Elms and El-Halwagi, 2009, 2010; Myint and El-Halwagi, 2009; Pokoo-Aikins et al., 2009, 2010), and bioalcohol production (Enguidanis et al., 2002; Gnansounou and Dauriat, 2010) have been analyzed and optimized.

Because a large-scale biorefinery is a capital-intensive project, multilevel analyses of process integration should be considered. The integrated biorefinery offers a flexible pathway that can handle various types of feedstock, is potentially energy self-sufficient, reduces overall production cost, and increases process efficiency. As other biomass conversion technologies (gasification, digestion, Fischer–Tropsch synthesis, etc.) become more established, even more alternative processing pathways can be considered for an integrated biorefinery. Therefore, it is necessary to screen all potential pathways systematically and to determine the optimum pathway that produces the desired products. Bao et al. (2011) addressed this problem by using superstructure optimization to select the optimum pathways for the production of desired bioproducts. However, this work is limited to analyzing production rate (yield) and has yet to be extended to include economics and environmental performance. Recently, Ng et al. (2009a) proposed a hierarchical approach for the synthesis and analysis of an integrated biorefinery. In addition, Kokossis and Yang (2009) identified the challenges of synthesizing and designing sustainable biorefineries. They also proposed an overall systematic approach to combine multiscale formulations with multistage problem-solving capabilities for problems involving novel processes. More recently, Kokossis and Yang (2010) discussed the potential of using various systems engineering tools in the design of complex biorefineries. Tan et al. (2009) developed an extended input–output model using fuzzy linear programming to determine the optimal capacities of distinct process units given a predefined product mix and environmental goals such as carbon, land, and water footprints.

In addition to the development of synthesis tools for integrated biorefineries, a flexible framework for determining the profitability, product portfolio, and process configuration in an integrated biorefinery was proposed by Sammons et al. (2007, 2008). Sammons (2009) determined the optimum pathway of an integrated biorefinery with the consideration of both economic and environmental performance using Pareto optimum analysis. More recently, Tay et al. (2011a) proposed a fuzzy optimization approach to address the same design problem. Mansoornejad et al. (2010) have further extended the previous work by adding market aspects, supply-chain network design, and flexible manufacturing design to the framework. Tay et al. (2011b) have also presented an insight-based synthesis tool based on a carbon–hydrogen–oxygen (C–H–O) ternary diagram for synthesis of an integrated biorefinery. Based on this work, the optimum gasification parameters and the interaction of gasification with other technologies to prepare the feedstocks and deliver the required final product can be determined before the detailed design.

To achieve maximum efficiency in an integrated biorefinery, multilevel integration through process integration approaches is needed. Therefore, process integration techniques are an essential approach to develop an efficient integrated biorefinery with optimum product allocations and portfolios. By means of in-plant material and energy recovery, generation of waste (carbon dioxide, wastewater, etc.) can be minimized. Besides, the overall energy consumption of an integrated biorefinery will be lower than that of a process that operates independently (Ng, 2010). Most recently, Ng (2010) extended the pinch-based automated targeting approach that was originally developed for synthesis of resource conservation networks (Ng et al., 2009b,c,d, 2010) to determine the maximum biofuel production and revenue targets for an integrated biorefinery.

Although some research has been conducted on the synthesis of the integrated biorefinery, the latter is still considered to be in its early development stage. Therefore, research focusing on the development of systematic tools to aid in the synthesis, design, and optimization of a sustainable integrated biorefinery is needed. In this chapter, the recently proposed automated targeting approach (Ng, 2010) is first reviewed and then further extended to synthesize an integrated biorefinery with maximum revenue targets based on multiple process parameters. In addition, based on the resulting targets, the detailed allocation of raw material (biomass) and intermediate products (alcohol, syngas, etc.) to different processes to produce the final products can be determined. With the proposed approach, the biofuel production and revenue targets for a given feedstock can be determined before detailed process flowsheet synthesis of the integrated biorefinery.

13.2 Problem Statement

As described in Ng (2010), the carbon contents of raw materials and products are assumed to be the most important properties for the synthesis of an integrated biorefinery. This is because carbon is the key element in organic matter (i.e., biomass), and therefore tracking carbon content from raw material (biomass) to final products is very important. By recovering the carbon content in raw material that may be lost from conversion processes in the form of gases (e.g., carbon dioxide, carbon monoxide) and solid wastes (e.g., fermentation residues, charcoal), the overall mass efficiency of a biorefinery can be increased (Ng, 2010). On the other hand, another process parameter (syngas composition) is also considered to be a critical process constraint in the design of an integrated biorefinery. For different types of biomass, different compositions of syngas are produced by gasification. Besides, most of the conversion technologies, such as methanol synthesis, dimethyl ether synthesis, and Fischer–Tropsch synthesis, require a specific syngas composition (i.e., H_2/CO ratio). Therefore, this parameter is taken as quality measurement for the synthesis of a gasification-based integrated biorefinery, in which syngas is produced as an intermediate.

The problem of synthesizing an integrated biorefinery may be formally stated as follows. Given a set of biomass/syngas sources SR_i, which can be converted into intermediates p, $INTER_p$ or products p', $PD_{p'}$. Each source has a given flow rate F_{SRi} and is characterized by its carbon fraction (C_i) or H_2/CO ratio (R_i). A set of sinks SK_j, which are process units that can convert sources i into intermediates p and products p', is specified. Each sink is

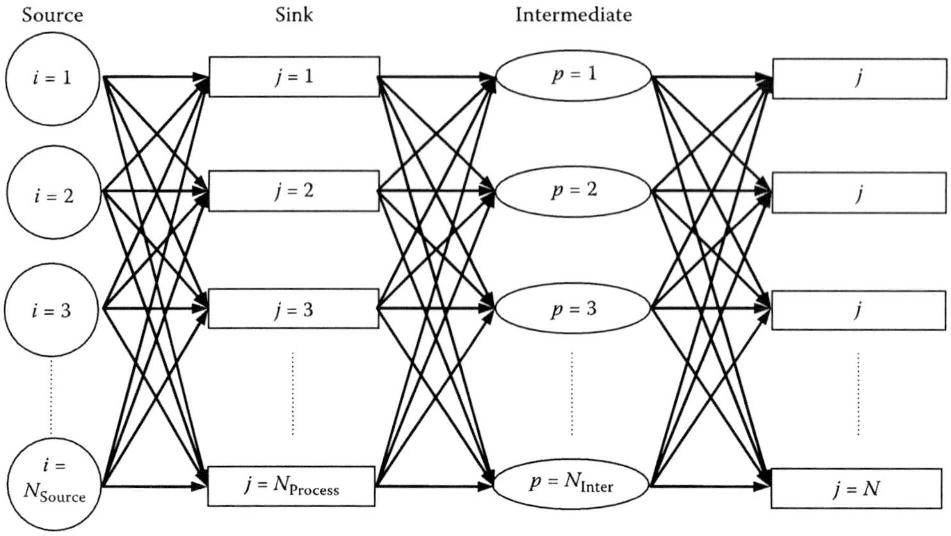

FIGURE 13.1
Graphic representation of an integrated biorefinery. (Adapted from Ng, D.K.S., 2010. *Chem. Eng. J.* 162(1), 67–74.)

characterized by predefined minimum specifications and quality requirement, C_j^{min} or R_j^{min}. In addition, the process conversion factors of sources i to intermediates p (X_{ijp}) and intermediates p to products p' ($X_{pjp'}$) via SK_j are also specified.

To reduce the complexity of an integrated biorefinery, only single-stage or two-stage processes that convert a given source i to any product p' will be considered. The problem structure can be summarized as shown in Figure 13.1. The objective of the methodology is to determine the maximum production or revenue of a single product (biofuel), or alternatively, the maximum revenue from a mix of multiple products, from given available biomass/syngas feed sources. Furthermore, alternative network designs for an integrated biorefinery that meets these targets are also determined.

13.3 Automated Targeting

The automated targeting technique was originally developed for mass exchange network (MEN) synthesis (El-Halwagi and Manousiothakis, 1990) . It was then extended to synthesis of resource conservation networks (RCN) for cases with reuse, recycle, or both (Ng et al., 2009b), interception (Ng et al., 2009c, 2009d), and waste treatment (Ng et al., 2010), based on the cascade analysis technique (Manan et al., 2004; Foo et al., 2006). Automated targeting was also used to find targets for minimizing the amount of CO_2-neutral or low-carbon energy sources for segregated energy planning problems (Lee et al., 2009). Note that in previous studies on MEN and RCN syntheses (El-Halwagi and Manousiothakis, 1990; Ng et al., 2009b, 2009c, 2009d, 2010), contaminant concentration and stream flow rate were used as the quality and quantity measures respectively of process streams. Based on the extracted data, MENs and RCNs with maximum recovery or minimum annualized operating cost can be synthesized. For carbon-constrained energy planning, carbon intensity and energy load are used as quality and quantity measures for energy systems to determine the optimum allocation of energy source(s) to the energy sector(s). Most recently, automated targeting was further extended to locate maximum production and revenue targets for an integrated biorefinery (Ng, 2010). As mentioned earlier, tracking carbon content and process parameters (H_2/CO ratio) are important in synthesizing an integrated biorefinery. Hence, carbon content or H_2/CO ratio and material flow rate are used as the dominant quality and quantity parameters respectively. Previous analogies for MEN and RCN synthesis as well as energy planning are not directly applicable to the synthesis of the integrated biorefinery. Therefore, the automated targeting approach has been further extended in this work for the synthesis of an integrated biorefinery. In addition, the model can also be used to synthesize an integrated biorefinery directly.

Based on previous work (El-Halwagi and Manousiothakis, 1990; Ng et al., 2009b, 2009c, 2009d, 2010; Ng, 2010), the technique involves arranging sources and sinks in descending order of quality, with the highest quality (lowest impurity concentration, lowest emission factor, or most desirable composition) located at the top of the cascade. In this work, the sources, sinks, intermediates, and products are arranged based on two potential quality measurements, carbon fraction level (C_k) or H_2/CO ratio level (R_k), in descending order from the highest level $k = 1$ to the lowest level $k = n$ in a *biorefinery cascade diagram* (BCD), as shown in Figure 13.2. Synthesis of a generic integrated biorefinery with multiple conversion technologies can also use carbon fraction level (C_k) as a quality measurement. However, for a more specific design such as a gasification-based integrated biorefinery, H_2/CO ratio level (R_k) provides a more accurate representation. The highest value of carbon fraction or H_2/CO ratio is added as the first level if this value does not exist among the process sinks and sources. In addition, a final fictitious level of zero ($C_n = 0$; $R_n = 0$) is added at the bottom of the cascade to enable the calculation of residue load (ε). Next, material flow rate cascading is performed across all levels. At each level k, the difference between the total available material sinks ($\Sigma_j F_{SKj}$) and sources ($\Sigma_i F_{SRi}$) is determined. Next, the *net material flow* cascaded from the earlier level k–1 (δ_{k-1}) is combined with the flow balance at level k to generate the net material flow rate at the k-th level (δ_k), as given in Equation 13.1:

$$\delta_k = \delta_{k-1} + (\Sigma_i F_{SRi} - \Sigma_j F_{SKj})_k \quad k \forall K \tag{13.1}$$

To ensure that no additional biomass or syngas flow is generated from the final level n (because that level is to be used only for the calculation of residual process load), a new constraint (as shown in Equation 13.2) is needed:

$$\delta_k = 0 \tag{13.2}$$

Process load cascading is performed next. Within each interval, the process load is determined by the product of the net material flow from level k and the difference between it and the two adjacent levels. As in the material-flow cascade, the residual of the process load at each level k (ε_k) is cascaded down to the next level. Hence, the process loads balance in the BCD, whether it is arranged based on carbon content (Figure 13.2a) or H_2/CO ratio (Figure 13.2b). Within each interval, the process load is determined by the product of the net material flow from level k and the difference between it and the two adjacent levels. As in the material-flow cascade, the residual of the process load at each level k (ε_k) is cascaded down to the next level. Hence, carbon-load balance and hydrogen-load balance at the k-th level can be determined by Equations 13.3 and 13.4, respectively:

$$\varepsilon_k = \varepsilon_{k-1} + \delta_k(C_k - C_{k+1}) \quad k \forall K \tag{13.3}$$

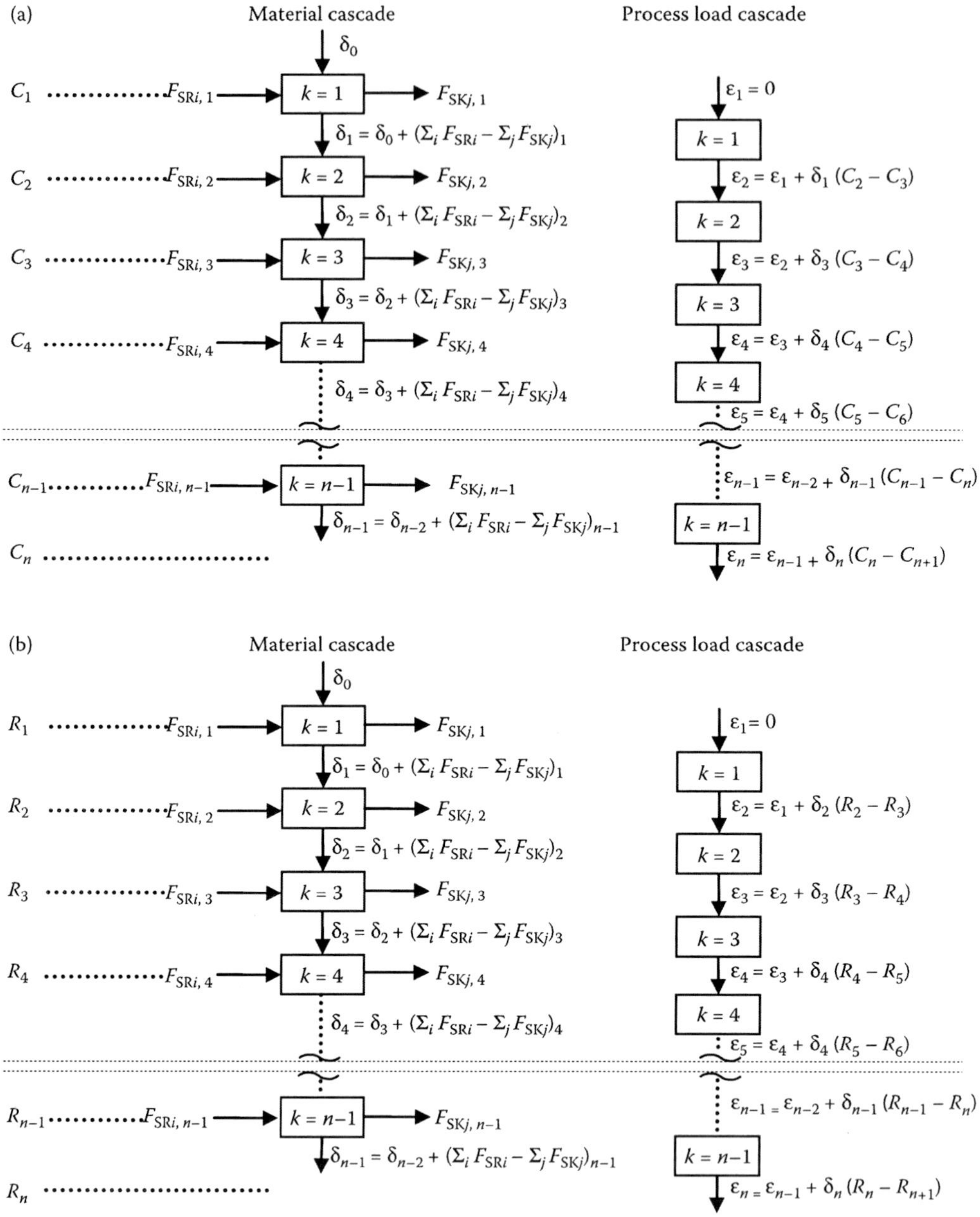

FIGURE 13.2
Generic BCD for (a) carbon content and (b) H_2/CO ratio.

$$\varepsilon_k = \varepsilon_{k-1} + \delta_k(R_k - R_{k+1}) \quad k \forall \mathrm{K} \tag{13.4}$$

where ε_{k-1} is the residue load that is cascaded from level k–1.

Conversely, the residual process load ε must have a positive value, which implies that a feasible load cascade has been achieved (Manan et al., 2004; Foo et al., 2006; Ng et al., 2009b,c,d, 2010; Ng, 2010). Therefore, the maximum allowable process load of the sink at each level is achieved. Therefore, Equation 13.5 is included as a constraint in the formulation of the model:

$$\varepsilon_k \geq 0 \quad k \forall K \tag{13.5}$$

It is interesting to note that, when the residual process load is determined as zero in the model solution at level k ($\varepsilon_k = 0$), a pinch point occurs. In physical terms, the zero process load means that, at the optimal solution, the minimum requirements of all sinks above the pinch point are completely fulfilled by the sources (Manan et al., 2004; Foo et al., 2006; Ng et al., 2009b,c,d, 2010; Ng, 2010). The identification of the pinch point provides valuable insights to decision makers. Its primary value is to identify the system bottleneck. Thus, the "golden rule" of pinch analysis can be applied to this problem to meet all the specified process requirements in the integrated biorefinery. In this case, fresh resources (i.e., biomass) must be supplied only to those process sinks below the pinch point. Allocation of this resource to process sink(s) above the pinch point will lead to a nonoptimal solution (Ng, 2010).

Note also that the above formulation is a linear programming (LP) model that can be solved easily to yield a global optimal solution if such a solution exists. In this research, the LP models were solved using LINGO v.10.0. However, in practice, the automated targeting model can be implemented using any LP solver, such as those found in common spreadsheet environments. In addition, in some cases, a nonlinear system may result from additional case-specific process constraints or objectives. Such variants can be optimized using appropriate optimization software.

To determine the maximum production of desired products (F_{PD}), the BCD (Figure 13.2) is used, with the optimization objective formulated as

$$\text{Maximize } F_{PD} \tag{13.6}$$

Alternatively, the maximum revenue solution can be obtained for cases with multiple products, in which case the optimization objective is set as

$$\text{Maximize REV} \tag{13.7}$$

where REV is the total revenue generated from selling the product. REV can be calculated as follows:

$$\text{REV} = \Sigma_{p'}(\text{MP}_{p'}F_{PDp'} - \text{TC}_{p'}F_{PDp'}) \tag{13.8}$$

where $\text{MP}_{p'}$, $F_{PDp'}$, and $\text{TC}_{p'}$ are the market price, flow rate, and total annualized fixed and variable costs of product p', respectively. In the following section, two hypothetical case studies are solved to illustrate the proposed automated targeting model for synthesis of the integrated biorefinery.

13.4 Case Study 1

Case study 1 is adapted from Ng (2010) to illustrate the proposed automated targeting approach for synthesis of an integrated biorefinery with carbon content as the dominant

TABLE 13.1

Conversion Table of Biomass to Intermediate and Final Products

Process Sink j	Raw Material	Intermediate p/Product p'	Conversion, X_{ijp} or $X_{pjp'}$(kg Product/kg Raw Material)
Digestion	Biomass	Methane (CH_4)	0.147
		Biomass residual	0.79
Fermentation	Biomass	Ethanol	0.27
		Biomass residual	0.61
Gasification	Biomass	Syngas (CO)	0.18
Pyrolysis	Biomass	Bio-oil	0.54
Dehydration	Ethanol	Biofuel	0.65
Synthetic fuel	Methane	Biofuel	0.286
Synthetic fuel	Bio-oil	Biofuel	0.1425
Fischer–Tropsch	Syngas	Biofuel	0.9

Source: Adapted from Ng, D.K.S., 2010. *Chem. Eng. J.* 162(1), 67–74.

quality parameter. Tables 13.1 and 13.2 show the conversion and other data for sinks and sources for the hypothetical case. In this case study, three scenarios are to be analyzed. In the first scenario, the optimization objective is to determine the maximum production of a single product (biofuel) from a given amount of biomass. In the second scenario, a maximum revenue target for a case with multiple products is presented. In the last scenario, the model is further used to synthesize an integrated biorefinery with multiple feedstocks and products.

In this case study, municipal solid waste (MSW) is selected as the raw material for the integrated biorefinery to produce biofuel. The MSW consists of 70%wt polysaccharides ($C_6H_{10}O_5$), 27%wt lignin ($CH_{1.12}O_{0.377}$), and 3%wt ash (Vinzant et al., 1990). Based on the above information, the carbon fraction of the biomass can be calculated as 0.477. Four processes can be used to convert raw material (biomass) to intermediates such as methane, ethanol, syngas, and bio-oil by digestion, fermentation, gasification, and pyrolysis, respectively (see Table 13.1). Note that all intermediates can also be further converted to biofuels by the four potential processes, as shown in Table 13.1. Theoretical or empirical conversions of raw material to intermediates or products for these processes are also included in Table 13.1.

Note that further recovery of biomass residues from digestion and fermentation processes enables the enhancement of the production of biofuel because these residues have high carbon content and high energy potential. Therefore, both biomass residues are taken as sources, SR2 and SR3 (see Table 13.2). Based on theoretical or experimental data reported

TABLE 13.2

Data for Case Study 1

Source		Available Source (kg)	Carbon Fraction (C_i)		Sink	Minimum Requirement of Carbon Fraction (C_j^{min})
SR1	MSW	5000	0.477	SK1	Digestion	0.477
SR2	Digested residual biomass	79% inlet biomass to digestion	0.474	SK2	Fermentation	0.477
SR3	Fermented residual biomass	61% inlet biomass to fermentation	0.170	SK3	Pyrolysis	0.250
				SK4	Gasification	0.250

in the literature, the carbon fraction of biomass residues can be estimated (Ng, 2010). As shown in Table 13.2, the carbon fractions of digested and fermented biomass residues were found to be 0.474 and 0.170, respectively.

The four processes (digestion, fermentation, pyrolysis, and gasification) that accept similar raw materials (biomass, biomass residues, or a mixture of both) to produce various intermediates are considered to be sinks (see Table 13.2). Each sink has a specific minimum value of the carbon fraction (C_j^{min}) which is required to produce intermediates or products. Note that the processes that further convert intermediates to final products are not considered as process sinks in Table 13.2 because such processes are not constrained by the carbon fractions of the intermediates. In addition, these processes require different raw materials or intermediates to produce a final product (biofuel). Examples of these processes include Fischer–Tropsch, dehydration, and synthetic fuel processes that require syngas (carbon monoxide and hydrogen), alcohol, and methane/bio-oil, respectively, to produce biofuel.

To locate the maximum production target, Equations 13.9 and 13.10 are included in the automated targeting formulation (Equations 13.1 through 13.5):

$$F_{INTERp} = X_{ijp} \, F_{SKj} \tag{13.9}$$

$$F_{PDp'} = X_{pjp'} \, F_{INTERp} \tag{13.10}$$

Based on the above equations, the flow rates of the intermediates p (F_{INTERp}) and the final product p' ($F_{PDp'}$) can be determined. Because the biomass residues are considered as SR2 and SR3, the flow rates of these sources can only be determined once the model is optimized. Therefore, the F_{SRi} of biomass residues are included as variables in the automated targeting algorithm.

13.4.1 Scenario 1: Maximum Production of a Single Product (Biofuel) from a Given Biomass

The optimization objective (Equation 13.6) is then solved, subject to the constraints in Equations 13.1 through 13.3, 13.5, 13.9, and 13.10, yielding the solution illustrated in the BCD shown in Figure 13.3. Note that for a given 5000 kg of biomass (SR1), 4115.3 kg of biomass is fed to the fermentation process (SK2), and the balance of the biomass (884.7 kg) is gasified via SK4. Based on the given conversions given in Table 13.1, 1111.1 kg, and 2510.3 kg of ethanol and biomass residue (SR3) respectively are generated when 4115.3 kg of biomass is fermented in SK2. As shown in Figure 13.3, 884.7 kg of SR1 (δ_3) and 2510.3 kg of SR3 (δ_4) are used to produce syngas via gasification (SK4). Based on Table 13.1, the conversion rate of syngas from biomass (X_{ijp}) is 0.18; hence, 611.1 kg of syngas is generated from 3395.0 kg (= 2510.3 + 884.7 kg) of biomass mixture. Because the biomass is processed via fermentation and gasification, two intermediate products (ethanol and syngas) are generated. The objective of this scenario it to find the maximum conversion of biomass to biofuel, and therefore both these intermediates are further processed via dehydration and Fischer–Tropsch processes, respectively. Based on the optimization model and the conversion data in Table 13.1, 1272.2 kg of biofuel is produced, where 722.2 and 550 kg are derived from ethanol and syngas, respectively. In line with this detailed allocation of fresh resource (biomass) and intermediates (residual biomass, ethanol, and syngas), a preliminary design of an integrated biorefinery for Scenario 1 is shown in Figure 13.4. Note that the unrecovered solid and gas wastes are not shown for simplicity.

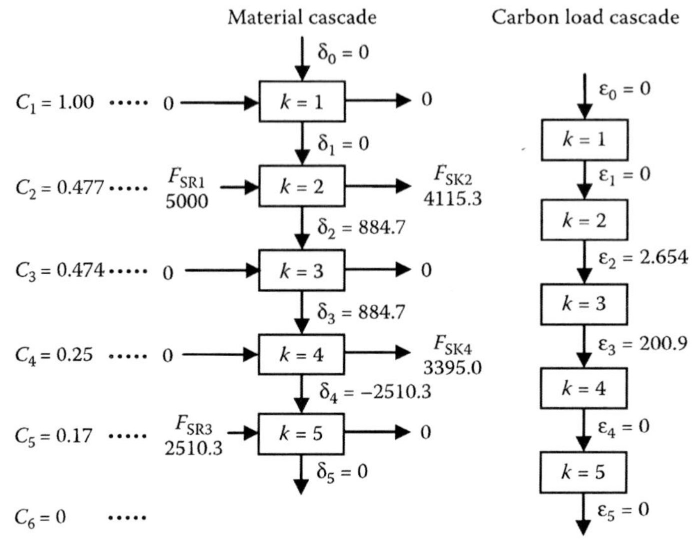

FIGURE 13.3
BCD for Scenario 1: Maximum production of biofuel.

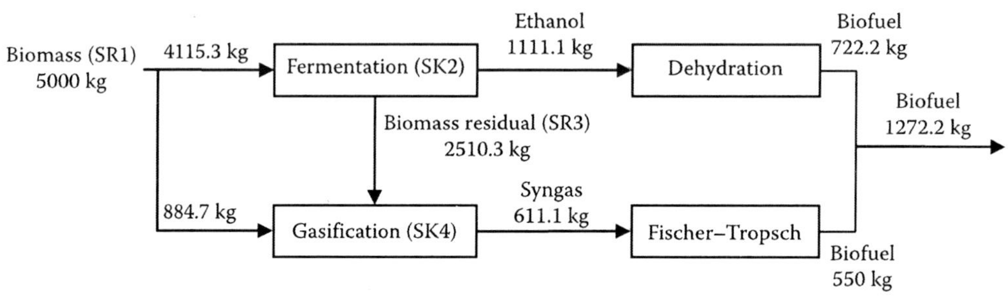

FIGURE 13.4
Preliminary integrated biorefinery design for Scenario 1.

13.4.2 Scenario 2: Maximum Revenue for an Integrated Biorefinery with Multiple Products

A second scenario with multiple products will now be examined. The optimization objective is to maximize revenue for a given fixed raw material cost by selecting the optimum product portfolio. Table 13.3 shows typical market prices for various intermediates and final products based on historical data.

Because some intermediates p have a market value ($MP_{p'}$), these intermediates can also be taken as final products. To determine the flow rates of these multiple products, a process mass balance (Equation 13.11) is included in the model:

$$F_{PDp'} = F^{I}_{PDp'} - F^{I}_{INTERp} \tag{13.11}$$

where $F^{I}_{PDp'}$ is the flow rate of product p' that can be produced from the first biomass processing step, and F^{I}_{INTERp} denotes the flow rate of product p' that requires further processing (which is also considered as an intermediate p).

TABLE 13.3

Revenues from Intermediates and Final Products

Intermediates/Final Products	Market Price, $MP_{p'}$ ($/kg)
Biofuel	0.93
Methane (CH_4)	1.02
Ethanol	0.78
Bio-oil	0.008

Source: Adapted from Ng, D.K.S., 2010. *Chem. Eng. J.* 162(1), 67–74.

Solving the model using the objective function in Equation 13.7 and subject to all constraints (Equations 13.1 through 13.3, 13.5, and 13.8 through 13.11) yields the results shown in Figure 13.5. Note that digestion (SK1), fermentation (SK2), and gasification (SK4) processes are involved in this scenario. As shown in Figure 13.5, for a given 5000 kg of biomass, 1080.8 and 3919.2 kg of biomass are digested into methane and fermented to ethanol, respectively. In this scenario, two types of residual biomass (SR2 and SR3) are generated from digestion and fermentation processes. Based on Table 13.1, the flow rates of SR2 and SR3 are determined as 853.8 kg (= 0.79 × 1080.8 kg) and 2390.7 kg (= 0.61 × 3919.2 kg), respectively (see Figure 13.5).

In addition, the maximum revenue target is found to be $1476 when 5000 kg of biomass is processed. Based on the optimized model, 1058.2 kg of ethanol and 158.9 kg of methane from fermentation and digestion processes, respectively, are considered to be final products. Meanwhile, 3244.5 kg (= 853.8 + 2390.7 kg) of biomass residues from both processes are gasified to 584 kg of syngas and then further converted to 525.6 kg of biofuel via a Fischer–Tropsch process. The preliminary design of the integrated biorefinery for Scenario 2 is shown in Figure 13.6. As in to the previous scenario, the unrecovered waste streams are neglected and excluded from Figure 13.6 for simplicity. Note that the preliminary integrated

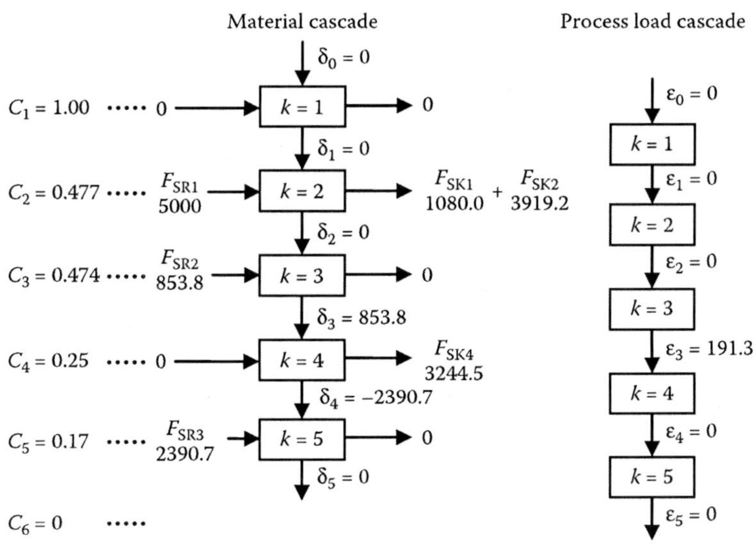

FIGURE 13.5
BCD for Scenario 2: Maximum revenue for cases with multiple products.

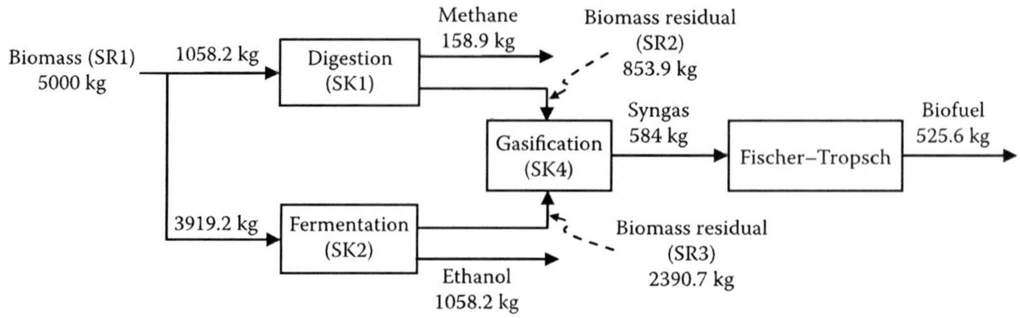

FIGURE 13.6
Preliminary integrated biorefinery design for Scenario 2.

biorefinery design of Scenario 2 is different from Scenario 1. In Figure 13.6, an additional digestion process is used to generate methane, which is considered as a product. Therefore, based on a different optimization objective, an alternative preliminary integrated design for the integrated biorefinery is obtained.

13.4.3 Scenario 3: Maximum Revenue for an Integrated Biorefinery with Multiple Feedstocks and Products

In this scenario, the proposed model is used to analyze an integrated biorefinery with multiple feedstocks and products. In this scenario, two types of available biomass (wood waste (SR1) and MSW (SR2)) are considered (see Table 13.4).

Based on the ultimate analysis, the carbon fractions of SR1 and SR2 are calculated as 0.490 and 0.477, respectively. Note that the process sink requirements and the residual biomass values shown in Table 13.4 are similar to those in previous scenarios. In addition, conversion and revenue data similar to those presented in Tables 13.1 and 13.3 are also used in this scenario. Because multiple feedstocks are available in this scenario, to ensure that the total consumption of the process sink is not greater than the available fresh resources (SR1 and SR2), Equation 13.12 is included in the model:

$$\Sigma_j F_{SKj} \leq F_{SR1} + F_{SR2} \tag{13.12}$$

TABLE 13.4
Data for Scenario 3

Source		Available Source (kg)	Carbon Fraction (C_i)		Sink	Minimum Requirement of Carbon Fraction (C_j^{min})
SR1	Wood waste	5000	0.490	SK1	Digestion	0.477
SR2	MSW	5000	0.477	SK2	Fermentation	0.477
SR3	Digested residual biomass	79% inlet biomass to digestion	0.474	SK3	Pyrolysis	0.250
SR4	Fermented residual biomass	61% inlet biomass to fermentation	0.170	SK4	Gasification	0.250

Source: Adapted from Ng, D.K.S., 2010. *Chem. Eng. J.* 162(1), 67–74.

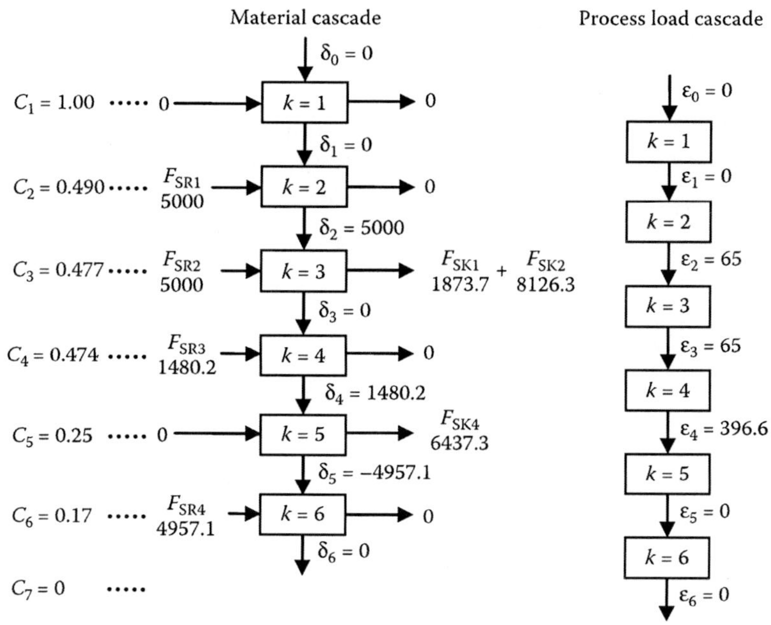

FIGURE 13.7
BCD for Scenario 3—Maximum revenue for cases with multiple feedstocks and products. (Adapted from Ng, D.K.S., 2010. *Chem. Eng. J.* 162(1), 67–74.)

Following a similar approach to Scenario 2, the model is solved using the objective function in Equation 13.7, subject to the constraints in Equations 13.1 through 13.3, 13.5, and 13.8 through 13.12, to yield the results shown in Figure 13.7.

Similarly to Scenario 2, digestion (SK1), fermentation (SK2), and gasification (SK4) are needed to produce multiple products in an integrated biorefinery to generate the maximum revenue. In this scenario (Figure 13.7), 1873.7 and 8126.3 kg of biomass (the total of SR1 and SR2) are digested into methane and fermented to ethanol via SK1 and SK2, respectively. Based on Table 13.1, 1480.2 kg (= 0.79 × 1873.7 kg) of SR3 and 4957.0 kg (= 0.61 × 8126.3 kg) of SR4 are generated, with 1873.7 and 8126.3 kg of biomass being digested and fermented, respectively. The maximum revenue target is determined to be $2962 based on the given wood waste and MSW. In addition, 2194 kg of ethanol and 275 kg of methane are produced from fermentation and digestion processes and are considered as final products. Meanwhile, all residual biomass from SK1 and SK2 (6437.3 kg) is gasified to 1168 kg of syngas and then further converted to 1042.8 kg of biofuel via a Fischer–Tropsch process. The preliminary integrated biorefinery design for this scenario is shown in Figure 13.8, which is similar to that in Scenario 2, but with different flow rates.

13.5 Case Study 2

Case study 2 illustrates the application of automated targeting to synthesis of a gasification-based integrated biorefinery. Table 13.5 shows the process data for a hypothetical case

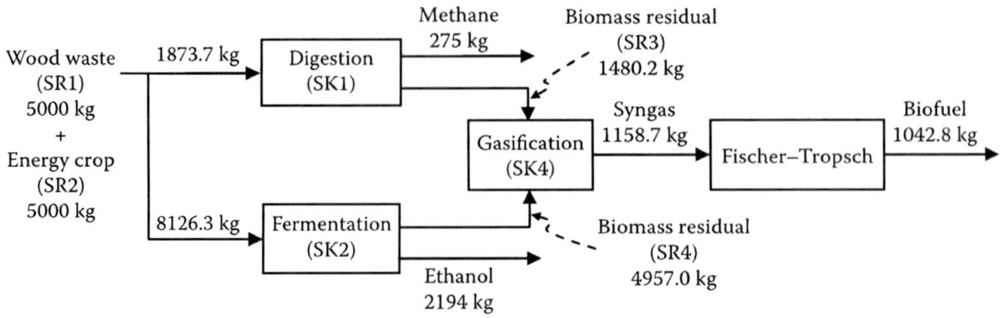

FIGURE 13.8
Preliminary integrated biorefinery design for Scenario 3. (Adapted from Ng, D.K.S., 2010. *Chem. Eng. J.* 162(1), 67–74.)

where syngas is taken as an intermediate. In this case study, various types of biomass are assumed to be first gasified to syngas. As mentioned previously, different types of biomass lead to different syngas compositions; therefore, four types of syngas (SR1–SR4) produced from oil palm empty fruit bunches (EFB), switchgrass, bagasse, and corn stover, respectively, are considered (Table 13.5). As shown, the flow rates of syngas (SR1–SR4) are assumed to have a maximum value of 100 kg/h. In this case study, four processes which convert syngas to valuable products (mixed alcohols, methanol, Fischer–Tropsch fuel, and dimethyl ether fuel) were included, as shown in Table 13.6. Note that the conversion of syngas to final products is based on CO flow rate; therefore, the corresponding CO molar flow rate of each source (100 kg/h) is taken as the quantitative measurement instead of the mass flow rate. The corresponding CO molar flow rate of syngas, which can be predicted by a gasification model (Tay et al., 2011c), is given in Table 13.5. Note also that each process requires a minimum H_2/CO ratio (C_j^{min}) to convert syngas to the desired products. To fulfill this process requirement, fresh syngas (FG) with an H_2/CO ratio (R_{FG}) of 3.0, produced from a

TABLE 13.5
Hypothetical Case Study 2

Source		Available Source (kg/h)	H_2/CO (R_i)	Corresponding CO Molar Flow Rate (kmol/h F_{SRi})	Sink		Minimum Requirement of H_2/CO (R_j^{min})
SR1	Syngas (EFB)	100	0.9	2.2567	SK1	Mixed alcohols synthesis	2.5
SR2	Syngas (switchgrass)	100	0.7	2.2583	SK2	Methanol synthesis	2.0
SR3	Syngas (bagasse)	100	0.6	2.2640	SK3	Fischer–Tropsch synthesis	2.0
SR4	Syngas (corn stover)	100	0.5	2.2600	SK4	Dimethyl ether synthesis	1.0
FG	Fresh syngas (SMR)	To be determined	3.0	To be determined			

TABLE 13.6

Conversion Table of Syngas to Final Products

Process Sink j	Raw Material	Product p'	Conversion, X_{ijp} or $X_{pjp'}$ (Gallon Product/kmol CO)
Mixed alcohols synthesis	Syngas	Mixed alcohol fuel	1.9547
Methanol synthesis	Syngas	Methanol	10.5684
Fisher–Tropsch synthesis	Syngas	Fischer–Tropsch fuel	3.7371
Dimethyl ether synthesis	Syngas	Dimethyl ether fuel	8.1620

steam methane reformer (SMR), is readily available. Theoretical and empirical conversions of syngas to the desired products for all processes are presented in Table 13.6.

The optimization objective of this case study is to determine the maximum revenue of a product portfolio from a given amount of syngas sources. In this case study, because the intermediate (syngas) is taken as a process source, Equations 13.9 and 13.10 can be combined as Equation 13.13:

$$F_{PDp'} = X_{pjp'}\ F_{SKi} \tag{13.13}$$

Based on the above equation, the flow rates of the final product p' ($F_{PDp'}$) can be determined. As shown in Table 13.5, the available CO flow rates of sources (SR1–SR4), F_{SRi}, were given as 2.2567, 2.2583, 2.2640, and 2.2600 kmol/h, respectively. Therefore, Equations 13.14 through 13.17 are included in the model based on their corresponding CO molar flow rates:

$$F_{SR1} \leq 2.2567 \text{ kmol/h} \tag{13.14}$$

$$F_{SR2} \leq 2.2583 \text{ kmol/h} \tag{13.15}$$

$$F_{SR3} \leq 2.2640 \text{ kmol/h} \tag{13.16}$$

$$F_{SR4} \leq 2.2600 \text{ kmol/h} \tag{13.17}$$

In this case study, the optimization objective was to determine the maximum revenue, and therefore Equation 13.8 was revised to yield Equation 13.18:

$$\text{REV} = \Sigma_{p'}(\text{REV}_{p'}F_{PDp'} - C_{PRODp'}\ F_{PDp'}) - \Sigma_i(C_{SRi}\ F_{SRi}) - C_{FG}F_{FG} \tag{13.18}$$

where $C_{PRODp'}$ is the annualized fixed and variable cost of producing product p', C_{SRi} is the cost of syngas SR_i, C_{FG} is the production cost of syngas by steam methane reforming, and F_{FG} is the flow rate of syngas from steam methane reforming (FG). Economic data for this case study are presented in Table 13.7.

An additional equation (Equation 13.19) is needed in the optimization model to calculate its net present value of profit/loss over the course of the plant operating life span, t. In this case study, t was assumed to be 25 years with a discount rate (r) of 15% (Sammons, 2009):

$$\text{NPV} = \sum_t \frac{(\text{REV} \times A)}{(1+r)^t} \tag{13.19}$$

where A is the annual plant operating hour (8330 h).

TABLE 13.7

Economic Data for a Gasification-Based Integrated Biorefinery

Source		C_{SRi} ($/kmol CO)	Sink		p'	$C_{PRODp'}$ ($/gallon)	$REV_{p'}$ ($/gallon)
SR1	Syngas	0.1753	SK1	Mixed alcohols synthesis	Mixed alcohol fuel	4.0080	1.77
SR2	Syngas	1.5655	SK2	Methanol synthesis	Methanol	4.1500	1.00
SR3	Syngas	0.4626	SK3	Fischer–Tropsch synthesis	Fischer–Tropsch fuel	1.1140	1.54
SR4	Syngas	0.8566	SK4	Dimethyl ether synthesis	Dimethyl ether fuel	0.6508	0.99
FG	Fresh syngas (steam reformer)	6.8000					

Source: Adapted from Sammons, N.E. Jr., 2009. *A Framework for Optimal Polygeneration Product Allocation.* PhD Dissertation (Chemical Engineering), Auburn University.

Based on the proposed approach, the optimization model (Equations 13.1 through 13.2, 13.4, 13.13 through 13.19) is solved with the optimization objective (Equation 13.7) to determine the optimum allocation of syngas for production of value-added products with maximum profit. The resulting BCD is shown in Figure 13.9. Note that all available biomasses are being gasified to produce syngas (SR1–SR4) at the corresponding CO molar flow rate, as shown in Table 13.6. Note also that 1.4694 CO kmol/h of syngas from SMR, which is equivalent to 49.96 kg/h of syngas ($1.4693 \times MW_{CO} + 3 \times 1.4693 \times MW_{H2}$, where MW_{CO} and MW_{H2} are the molecular weights of CO and H_2, respectively), is supplied to improve the syngas produced from biomass to meet the minimum SK4 requirement (R_4^{min}). As shown in Figure 13.9, all syngas from SMR and biomass gasification (10.51 kmol/h of CO, which corresponds to a total syngas flow rate of 315.30 kg/h) is sent to SK4 for production of dimethyl ether fuel (DME).

Based on the optimized model, the maximum annual profit is targeted as US$ 0.1015 million with production of 85.78 gal/h dimethyl ether fuel (= 10.51 kmol/h × 8.1620 gal DME/kmol CO). Over the course of 25 years at a discount rate of 15%, the net present value of the integrated biorefinery profit was determined to be US$ 0.7546 million. Figure 13.10 shows the preliminary integrated biorefinery design for case study 2.

It is worth mentioning that the values presented in Tables 13.1 through 13.7 may vary based on market price, state of advancement of technologies, and various operating conditions as well as the types of feedstocks. In addition, price fluctuations and feedstock quality variations will certainly be encountered in biomass supply chains. Hence, with such variations and uncertainties, the production and revenue targets as well as the network design may change. However, the proposed automated targeting can be easily revised and reformulated to locate the targets and to identify the processes involved in the integrated biorefinery.

13.6 Conclusions

In this chapter, an automated targeting approach for finding the maximum biofuel production and revenue targets of an integrated biorefinery has been presented. These targets

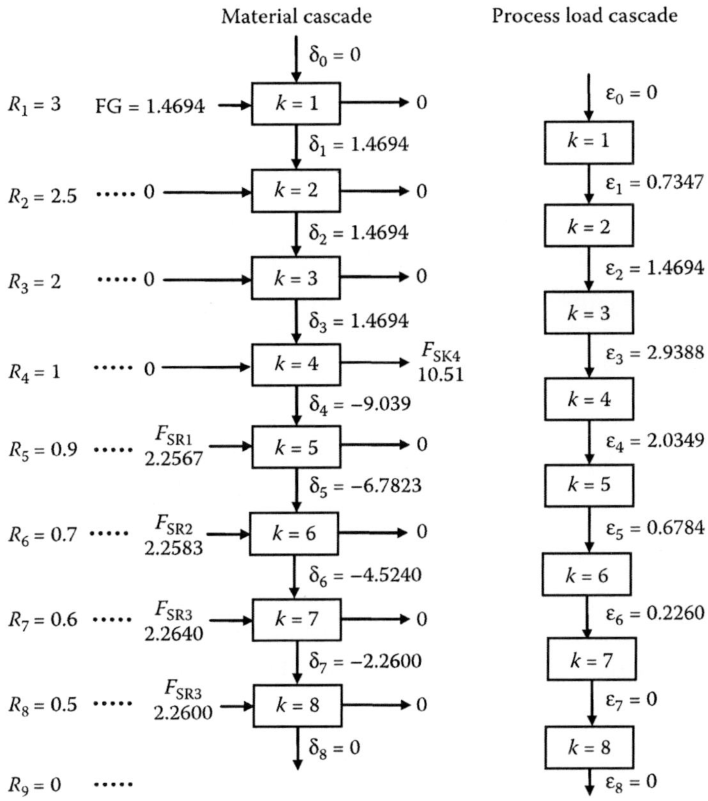

FIGURE 13.9
BCD for case study 2: Maximum revenue for a gasification-based integrated biorefinery.

can be found before detailed design of the integrated biorefinery. Then the preliminary integrated biorefinery design that achieves the targets can be determined using the proposed approach. In addition, the proposed approach can also be easily revised and reformulated to accommodate uncertainties in feedstock quality, market conditions, and process yields. Further work is still needed, however, to develop a unified approach to designing a robust and flexible integrated biorefinery that can handle different types of feedstock. In addition, extension of the current model to account for seasonality and the biomass supply chain remains to be considered.

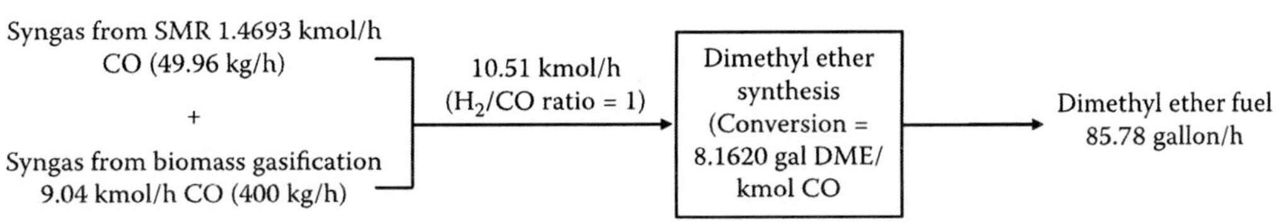

FIGURE 13.10
Preliminary integrated biorefinery design for case study 2.

Acknowledgments

Financial support from the University of Nottingham Research Committee through its New Researcher Fund (NRF 5021/A2RL32) is gratefully acknowledged. The authors would also like to express their gratitude to Professor Raymond Tan from De La Salle University, Philippines, for some helpful suggestions to improve the current work.

References

Bao, B., Ng, D.K.S., Tay, D.H.S., Jimenez-Gutierrez, A., and El-Halwagi, M.M., 2011. A shortcut method for the preliminary synthesis of process-technology pathways: An optimization approach and application for the conceptual design of integrated biorefineries. *Comp. Chem. Eng.* 35(8), 1374–1383.

Bridgwater, AV., 2003. Renewable fuels and chemicals by thermal processing of biomass. *Chem. Eng. J.* 91, 87–102.

El-Halwagi, M.M. and Manousiouthakis, V. 1990. Automatic synthesis of mass-exchange networks with single component targets. *Chem. Eng. Sci.* 45(9), 2813–2831.

Elms, R.D. and El-Halwagi, M.M., 2009. Optimal scheduling and operation of biodiesel plants with multiple feedstocks. *Int. J. Process. Syst. Eng.* 1(1), 1–28.

Elms, R.D. and El-Halwagi, M.M., 2010. The effect of greenhouse gas policy on the design and scheduling of biodiesel plants with multiple feedstocks. *Clean Technol. Environ. Policy* 12(5), 547–560.

Energy Information Administration (EIA) (2010). *International Energy Outlook 2010*. [Online] Available at: http://www.eia.doe.gov/oiaf/ieo/highlights.html [Accessed August 8, 2010].

Enguidanis, M., Soria, A., Kavalov, B., and Jensen P., 2002. *Techno-Economic Analysis of Bio-Alcohol Production in the EU: A Short Summary for Decision Makers*. [Online] Available at: http://www.senternovem.nl/mmfiles/26432_tcm24-279847.pdf [Accessed on April 10, 2010].

Fernando, S., Adhikari, S., Chandrapal, C., and Murali, N. (2006). Biorefineries: current status, challenges, and future direction. *Energy Fuels* 20(1), 1727–1737.

Foo, D.C.Y., Manan, Z.A., Tan, Y.L., 2006. Use cascade analysis to optimize water networks. *Chem. Eng. Prog.* 102(7), 45–52.

Gnansounou, E. and Dauriat, A., 2010. Techno-economic analysis of lignocellulosic ethanol: a review. *Biores. Technol.* 101, 4980–4991.

Kamm, B., Kamm, M., and Soyez, K., 1998. The green biorefinery, concept of technology. *Proceedings, 1st International Symposium on Green Biorefinery*. Neuruppin, Germany.

Kokossis, A.C. and Yang, A., 2009. Future system challenges in the design of renewable bio-energy system and the synthesis of sustainable biorefineries. *Proceedings, Seventh International Conference on Foundations of Computer-Aided Process Design 2009*, Colorado, USA.

Kokossis, A.C. and Yang, A., 2010. On the use of systems technologies and a systematic approach for the synthesis and the design of future biorefineries. *Comp. Chem. Eng.* 30(9), 1397–1405.

Lee, C.S., Ng, D.K.S., Foo, D.C.Y., Tan, R.R., 2009. Extended pinch targeting techniques for carbon-constrained energy sector planning. *Appl. Energy* 86(1), 60–67.

Manan, Z.A., Tan, Y.L., and Foo, D.C.Y., 2004. Targeting the minimum water flowrate using water cascade analysis technique. *AIChE J.* 50(12), 3169–3183.

Mansoornejad, B., Chambost, V., Stuart, P., 2010. Integrating product portfolio design and supply chain design for the forest biorefinery. *Comp. Chem. Eng.* 34(9), 1497–1506.

Mohan, T. and El-Halwagi, M.M., 2007. An algebraic targeting approach for effective utilization of biomass in cogeneration systems through process integration. *J. Clean Tech. Env. Policy* 9(1), 13–25.

Myint, L.L. and El-Halwagi, M.M., 2009. Process analysis and optimization of biodiesel production from soybean oil. *J. Clean Tech. Env. Policy* 11(3), 263–276.

Ng, D.K.S., 2010. Automated targeting for synthesis of an integrated biorefinery. *Chem. Eng. J.* 162(1), 67–74.

Ng, D.K.S., Foo, D.C.Y., and Tan, R.R., 2009b. Automated targeting technique for single-component resource conservation networks—Part 1: Direct reuse/recycle. *Ind. Eng. Chem. Res.* 48(16), 7637–7646.

Ng, D.K.S., Foo, D.C.Y., and Tan, R.R., 2009c. Automated targeting technique for single-component resource conservation networks—Part 2: Single pass and partitioning waste interception systems. *Ind. Eng. Chem. Res.* 48(16), 7647–7661.

Ng, D.K.S., Foo, D.C.Y., Tan, R.R., and El-Halwagi, M.M., 2010. Automated targeting technique for concentration- and property-based total resource conservation network. *Comp. Chem. Eng.* 34, 825–845.

Ng, D.K.S., Foo, D.C.Y., Tan, R.R., Pau, C.H., and Tan, Y.L., 2009d. Automated targeting for conventional and bilateral property-based resource conservation network. *Chem. Eng. J.* 149, 87–101.

Ng, D.K.S., Pham, V., El-Halwagi, M.M., Jiménez-Gutiérrez, A., and Spriggs, H.D., 2009a. A hierarchical approach to the synthesis and analysis of integrated biorefineries. *Proceedings, Seventh International Conference on Foundations of Computer-Aided Process Design 2009*, pp. 425–432.

Pokoo-Aikins, G., Heath, A., Mentzer, R.A., Mannan, M.S., Rogers, W.J., and El-Halwagi, M.M., 2010. A multi-criteria approach to screening alternatives for converting sewage sludge to biodiesel. *J. Loss Prev. Process Ind.* 23(3), 412–420.

Pokoo-Aikins, G., Nadim, A., Mahalec, V., and El-Halwagi, M.M., 2009. Design and analysis of biodiesel production from algae grown through carbon sequestration. *Clean Technol. Environ. Policy* 12(3), 239–254.

Qin, X., Mohan, T., El-Halwagi, M.M., Cornforth, F., and McCarl, B.A., 2006. Switchgrass as an alternate feedstock for power generation: integrated environmental, energy, and economic life cycle analysis. *J. Clean Tech. Env. Policy* 8(4), 233–249.

Ritter, S., 2006. Biorefinery gets ready to deliver the goods. *Chem. Eng. News* 84(34), 47.

Sammons, N.E. Jr., 2009. *A Framework for Optimal Polygeneration Product Allocation*. PhD Dissertation (Chemical Engineering), Auburn University.

Sammons, N.E. Jr., Mario, E., Yuan, W., Cullinan, H., and Aksoy, B., 2007. A flexible framework for optimal biorefinery product allocation. *Environ. Prog.* 26(4), 349–354.

Sammons, N.E. Jr., Yuan, W., Eden, M., Aksoy, B., and Cullinan, H., 2008. Optimal biorefinery product allocation by combining process and economic modelling. *Chem. Eng. Res. Des.* 86(7), 800–808.

Tan, R.R., Ballacillo, J.B., Aviso, K.B., and Culaba, A.B., 2009. A fuzzy multiple-objective approach to the optimization of bioenergy system footprints. *Chem. Eng. Res. Des.* 87, 1162–1170.

Tay, D.H.S., Kheireddine, H., Ng, D.K.S., and El-Halwagi, M.M., 2011b. Synthesis of an integrated biorefinery via the C–H–O ternary diagram. *J. Clean Techn. Env. Policy* 13(4), 567–579.

Tay, D.H.S., Kheireddine, H., Ng, D.K.S., and El-Halwagi, M.M., 2011c. Targeting and optimization of syngas production via thermodynamic equilibrium model of biomass gasification. *Proceedings, 13th Asia Pacific Confederation of Chemical Engineering Congress*, Paper ID: 10216.

Tay, D.H.S., Sammons, N.E. Jr., Ng, D.K.S., and Eden, M.R., 2011a. Fuzzy optimization approach for the synthesis of an integrated biorefinery. *Ind. Eng. Chem. Res.* 50(3), 1652–1665.

Vinzant, T.B., Adney, W.S., Grohmann, K., and Rivard, C.J., 1990. Aerobic and anaerobic digestion of processed municipal solid waste. *Appl. Biochem. Biotechnol.* 24–25:765–771.

14

Synthesis of Flexible Heat-Exchanger Networks for Multifeedstock Biodiesel Production Processes

Grace Pokoo-Aikins and Mahmoud M. El-Halwagi

CONTENTS

14.1 Introduction

Biodiesel is now a well-established renewable fuel source which is often used as an alternative to petroleum-derived diesel because it can be used in existing engines with comparable performance (EPA, 2002). Traditionally, biodiesel has been made from vegetable (or plant or plant-seed) oils, animal fat, and tallow. Common vegetable oils used in preparing biodiesel are soy, sunflower, safflower, canola, and palm. Palm oil has numerous positive aspects for biodiesel production. Among those attributes are a high production rate, a high oil content, and high oil production per acre (27 times more than soy). Life-cycle analysis (LCA) also reveals that production of palm-based biodiesel can reduce greenhouse gases (GHG) to a greater extent than other leading vegetable oils (Abdullah et al., 2009). Algal oil is a nonfood feedstock that is the object of much research and development and is increasingly used for biodiesel production. Like terrestrial plants, algae grow by photosynthesis through the use of CO_2, water, and sunlight. Unlike terrestrial plants, algae grow in aqueous suspensions, which enable more efficient use of H_2O, CO_2, and other nutrients. This gives algae the potential to produce more oil per unit area than other crops currently used for biodiesel production. The lipid content in algae has been known to reach 50% under appropriate conditions. An in-depth investigation into the potential use of algae to produce biodiesel was conducted by Pokoo-Aikins et al. (2010a).

The use of nonfood oil sources for biodiesel production is particularly important for the newer biodiesel plants that are being constructed. Some older biodiesel plants continue to use vegetable oils, but many are converting to utilize a variety of feedstocks in addition to vegetable oils. Government incentives to drive down the cost of vegetable oils and the cost of pretreating waste oils and triglycerides from waste animal fats and algal oil means that vegetable oils will continue to be used to produce biodiesel until the cost of using nonfood feedstocks is stable and lower than the cost of food feedstock options. In the meantime, an increasing number of biodiesel plants are considered "multi-feedstock" producers because they use a variety of feedstocks, including vegetable oils, animal fats, and recycled cooking oil or yellow grease (National Biodiesel Board, 2010). Use of a variety of feedstocks is becoming important as biodiesel continues to be an important fuel source that is helping to reduce the demand on petroleum-based fuels and shows continued promise of being one of many alternative fuel sources for a more fuel-stable future. The United States biodiesel industry has seen a dramatic increase in production since the implementation of the Biodiesel Tax Incentive. Production has increased from 25 million gallons per year in 2004 to 450 million gallons per year in 2007 and 700 million gallons per year in 2008 (National Biodiesel Board, 2009).

Synthesis of heat-exchange networks (HEN) is a topic that has attracted much interest in recent decades and in which there are still many opportunities for study, contribution, and expansion of design methods. Detailed reviews of research in the area of HEN synthesis have been reported in several literature sources. Procedures for heat integration have been described by El-Halwagi (2012), Kemp (2007), El-Halwagi (2006), Smith (2005), and Furman and Sahinidis (2002). One of the basic graphic techniques used in targeting the performance of HENs is the thermal-pinch analysis method (e.g., Linnhoff and Hindmarsh, 1983). Papoulias and Grossmann (1983) proposed an optimization approach for designing heat-recovery networks based on the transshipment model. Floudas and Grossmann (1986) extended this optimization approach to the synthesis of flexible HENs using a multiperiod formulation. Yee and Grossmann (1990) developed a mixed-integer linear program for the simultaneous optimization of utility cost and heat-exchanger areas and the selection of matches. Aaltola (2002) later applied the work of Yee and Grossmann (1990) to the multiperiod problem. Zhang (2006) then extended these two studies to the application of flexible heat-exchange networks. The synthesis of a flexible HEN generally involves considering uncertainties related to streams entering the process (e.g., number of streams, flowrate, and temperature). A systematic approach to the design of flexible HENs was introduced by Chen (2005) to consider cost while incorporating disturbances over a range of source-stream temperatures and flowrates. A strategy for synthesizing flexible networks that accounts for variations in process inlet stream flowrates, temperatures (HEN), and composition (MEN) was extended to heat- and mass-exchange networks (Chen and Hung, 2007). Xiangkun et al. (2007) also proposed a two-stage approach for the synthesis of flexible multistream HENs. An optimization approach for synthesizing HENs while accounting for scheduling was described by Al-Mutairi and El-Halwagi (2009). Scheduling was considered over a certain time horizon during which the operation was discretized into a number of periods. For each period, a HEN synthesis optimization formulation was developed.

The term *flexibility* as used in current literature refers to variations in the number of streams and in the flowrates and compositions of the streams. The use of *flexibility* in this work is broader (i.e., plant flexibility) and must in this case be achieved by incorporating a variety of feedstock scenarios in the design of the heat-exchange network for the plant. Formulations for the multiperiod heat-exchange network resemble more a scheduling problem. Multiperiod HEN formulations can be modified, extended for, and applied to the flexible multifeedstock facility by considering various time-independent scenarios in place of the time horizon of the multiperiod HEN problem.

14.2 Problem Background and Description

In considering various alternatives for feedstocks, it is important to incorporate traditional biodiesel feedstock (e.g., vegetable oils) as well as alternative feedstocks such as algal oil in the design of grassroots production facilities or in the retrofitting of existing plants. These feedstocks include soybean oil (e.g., Myint and El-Halwagi, 2009), sewage sludge (e.g., Pokoo-Aikins et al., 2010b), waste cooking oil (e.g., Elms and El-Halwagi, 2009, 2010), and algae (e.g., Pokoo-Aikins et al., 2010a). The most commonly used reaction pathway in the production of biodiesel is the transesterification reaction:

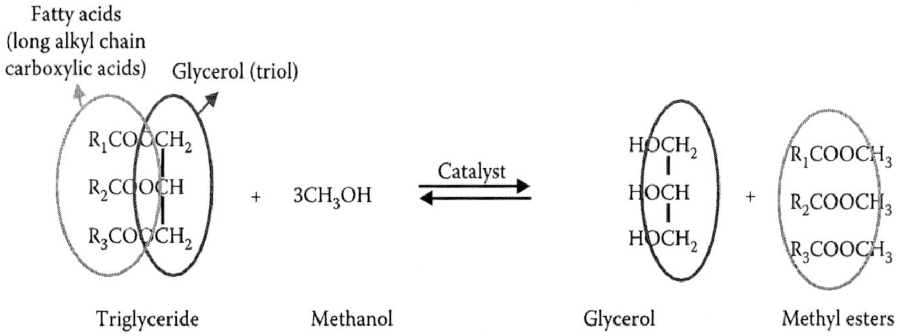

Feedstock cost makes up the bulk of the operating cost for the biodiesel production process. Flexibility in feedstock choices could provide an economic and environmental advantage to the biodiesel industry. To promote energy conservation, which results in cost savings and reduction in greenhouse-gas emissions, it is important to design HENs that are flexible enough to operate properly even when feedstocks are altered. Formally stated, the problem can be described as follows

A given biodiesel production facility has known production designs and processes a number (N_f) of feedstocks. It is desired to develop a flexible HEN that accommodates the various feedstocks that may be used while maintaining a specified production level and quality.

Question to be addressed are

- What are the targets for minimum heating and cooling utilities for different feedstock alternatives?
- What is the minimum number of heat exchangers that can be used under variations in types and flowrates of the feedstocks?
- Which heat exchangers should be used and with which feedstock? When?

The following is a detailed statement of the problem to be addressed.

Given a continuous biodiesel production process with

- A set of operational scenarios: $\{s|s = 1,2,...,N_s\}$. Each scenario corresponds to a given feedstock, throughput, and process operation.
- A set of common process units $C = \{c \mid c = 1,2,...,N_{CP}\}$. Each process unit, c, has a set of input streams $\text{INPUT}_c = \{m_c| \, m_c = 1,2,...,N_c^{\text{in}}\}$ and a set of output streams

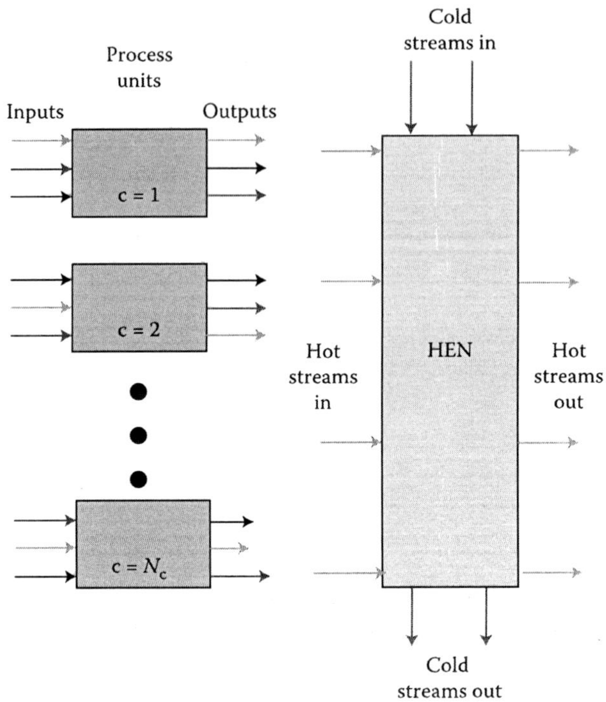

FIGURE 14.1
Diagram of the process and HEN.

$OUTPUT_c = \{n_c | n_c = 1,2,...,N_c^{out}\}$. Input stream m_c has flowrate, A_{m_c}, a composition of component x, $Y_{m_c,x}$, and a temperature, T_{m_c}. Output stream n_c has a flowrate, B_{n_c}, a composition of component x, $Z_{m_c,x}$ and a temperature, T_{n_c}.

- For each feedstock operation s, a set of hot streams $HS_s = \{h_s | h_s = 1,2,...,N_{HS,s}\}$ is defined, which are either streams that need to be cooled or external heating utilities.
- For each feedstock operation, s, a set of cold streams $CS_s = \{k_s | k_s = 1,2,...,N_{CS,s}\}$ is defined, which are either streams that need to be heated or external cooling utilities.

A systematic procedure is needed that can be used to determine an optimal process design that accounts for heat integration and that can be used to synthesize a flexible heat-exchange network (HEN) that operates under a variety of scenarios by incorporating different feedstocks.

The problem is presented in Figure 14.1 as a diagram which illustrates the common process units, process streams, subsets of hot and cold streams, and HEN.

14.3 Design Approach

The HEN must be synthesized to account for various expected scenarios. The challenge is to synthesize a single configuration which can accommodate all expected changes. A number of expected scenarios can be identified in advance. A multiscenario HEN synthesis problem is then solved to generate a single network configuration that accommodates all expected scenarios.

Simplifying assumptions were made as follows:

- The scenarios are looked at as individual processes. Each scenario is discretized into N_s components, and a set of scenarios is defined as SCENARIOS = $\{s \mid s = 1,2,\ldots,N_s\}$. The scenarios are taken into account for each feedstock and are used in determining the optimal flexible design.
- Process modifications are limited to two options. The first option is manipulation of certain design and operating variables for each unit within permissible ranges. The second option is the addition of new heat exchangers.

The approach for synthesizing the flexible multiscenario heat-exchange network is presented in Figure 14.2. First, the relevant input data and constraints are gathered, and a number of expected scenarios are identified. Next, the process model is developed with the appropriate level of detail to account for the process performance in terms of the design and operating decision variables. Then a linear program (LP) is formulated to incorporate the expected scenarios into the design, followed by a mixed-integer linear program (MILP) for network synthesis. The LP formulation is generated to determine minimum heating and cooling utilities for heat integration, using data about the hot and cold inputs and outputs for each scenario.

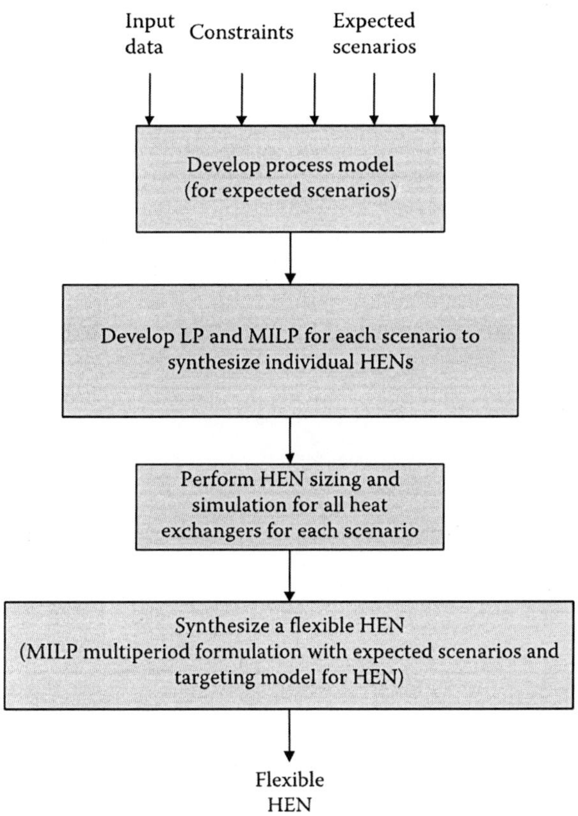

FIGURE 14.2
Flowchart of the synthesis approach.

The MILP formulation is developed to synthesize the HEN network that attains the minimum utility requirements for each scenario. Next, detailed exchanger sizing and simulation are carried out for the HEN configurations developed for the various scenarios. The result is a complete HEN design for each scenario. The MILP formulations for the various scenarios, along with the sizing information obtained from simulation, are combined to enable the simultaneous optimization of the HEN while accounting for the expected scenarios.

A transshipment model for the targeting and synthesis of HENs was developed by Papoulias and Grossmann (1983) and extended by Floudas and Grossmann (1986) for multiperiod operation. These transshipment formulations are the basis for the models used here to design the flexible HEN to handle multiple feedstocks and associated process designs. The formulation is coded using the LINGO optimization software for each individual scenario to confirm that the matching is working for all scenarios before combining the formulations and introducing a mixed-integer linear program (MILP) to include all scenarios being investigated. According to the pinch rule, no heat should be passed through the pinch. This rule is incorporated into the program formulation by decomposing the problem into two subnetworks (SN), one above the pinch and one below the pinch. Notation for the subnetworks will be as follows: index v is defined as 1 for above the pinch and 2 for below the pinch. Each interval over which heat is passed is denoted as y. The objective function is to minimize the number of heat exchangers ($E_{h,k,v}$):

$$\min \sum_{\mathrm{n}=1,2} \sum_{h \in HS} \sum_{k = CS} E_{h,k,\mathrm{n}} \tag{14.1}$$

subject to the following constraints:

Heat balance for hot stream over the temperature interval:

$$\delta_{h,y,s} - \delta_{h,y-1,s} + \sum_{k \in CS} Q_{h,k,y,s} = Q^{H}_{h,y,s} \quad h \in HS, y \in SN_{\mathrm{n}}, \mathrm{n} = 1,2 \tag{14.2}$$

Heat balance for each cold stream over the temperature interval:

$$\sum_{h \in HS} Q_{h,k,y,s} = Q^{C}_{k,y,s} \quad k \in CS, y \in SN_{\mathrm{n}}, \mathrm{n} = 1,2 \tag{14.3}$$

Matching of loads:

$$\sum_{y \in SN_{\mathrm{n}}} Q_{h,k,y,s} - U_{h,k,y,s} E_{h,k,\mathrm{n}} \leq 0 \quad h \in HS, k \in CS, y \in SN_{\mathrm{n}}, \mathrm{n} = 1,2 \tag{14.4}$$

Nonnegative residuals:

$$\delta_{h,y,s} \geq 0 \quad h \in HS, y \in SN_{\mathrm{n}}, \mathrm{n} = 1,2 \tag{14.5}$$

Nonnegative loads:

$$Q_{h,k,y,s} \geq 0 \quad h \in HS, k \in CS, y \in SN_{\mathrm{n}}, \mathrm{n} = 1,2 \tag{14.6}$$

Binary integer variables for matching streams:

$$E_{h,k,v} = 0/1 \quad h \in HS, k \in CS, v = 1,2 \tag{14.7}$$

where $Q_{h,k,y,s}$ is the heat exchanged between the hth hot stream and the kth cold stream in interval y for scenario s, $Q^{H}_{h,y,s}$ is the total heat load for the hth hot stream in interval y for scenario s, $Q^{C}_{k,y,s}$ is the total heat load for the kth cold stream in interval y for scenario s, and $U_{h,k,y,s}$ is an upper bound on the heat exchangeable between streams h and k in SN_v.

Solving the mathematical program described above will lead to matches for the minimum number of heat exchangers for the multiscenario problem. Because of the nature of the units and streams investigated for the various scenarios, careful inspection of the solution will be required to ensure that the result is indeed the minimum number.

14.4 Case Study

Three feedstocks are investigated in the case study: soy, palm, and algal oil. The biodiesel process is based on a base-case design described by Pokoo-Aikins et al. (2010a). In the base-case design, vegetable oils or algal oil can be processed to produce 44–52 MMgpy of biodiesel by transesterification (Figure 14.3). Two reactors are used to increase throughput. An 8000 h work year is assumed. Reaction temperature is 60°C in each reactor. Conversion through each reactor is assumed to be 97.7%. The alcohol to triglyceride (oil) ratio is 6:1. Oils entering the transesterification process are assumed to have a free fatty-acid (FFA) content of 0.5% or less. Glycerol is separated from the resulting methyl esters and purified. Biodiesel is washed with water. Biodiesel purity for each feedstock ranges from 99.5% to 99.7%, and glycerol purity is approximately 98% regardless of feedstock.

For each of the feedstocks, heat integration was performed for the biodiesel process without any heat exchangers to determine the pinch point and the minimum heating and minimum cooling utilities. Results from the heat integration are used to determine the minimum number of heat exchangers required for each process as well as to synthesize the heat-exchange network.

Heat integration was performed for each feedstock scenario. Integration results revealed the minimum heating and cooling utilities needed for each scenario. Integration results were used to generate a MILP for determination of the optimal HEN for the given scenario. The minimum number of heat exchangers required for each individual scenario was six for soy and seven each for palm and algal oil. The total number of heat exchangers needed to accommodate the three scenarios is nine units (Table 14.1). The heat-exchange network designs for the individual scenarios are depicted in Figures 14.4 through 14.6 for soy, palm, and algal oil, respectively. Each exchanger is represented by two circles connecting the two matched streams. The numbers inside the circles represent the rate of exchanged heat (in 10^3 Btu/h). Figure 14.7 represents the superstructure for the HEN accommodating the three scenarios. The process design configuration with the minimum number of heat exchangers for the flexible plant will have the nine heat exchangers positioned where needed. All nine heat exchangers must be present and associated with the necessary units for the flexible plant to accommodate all three of the feedstocks investigated.

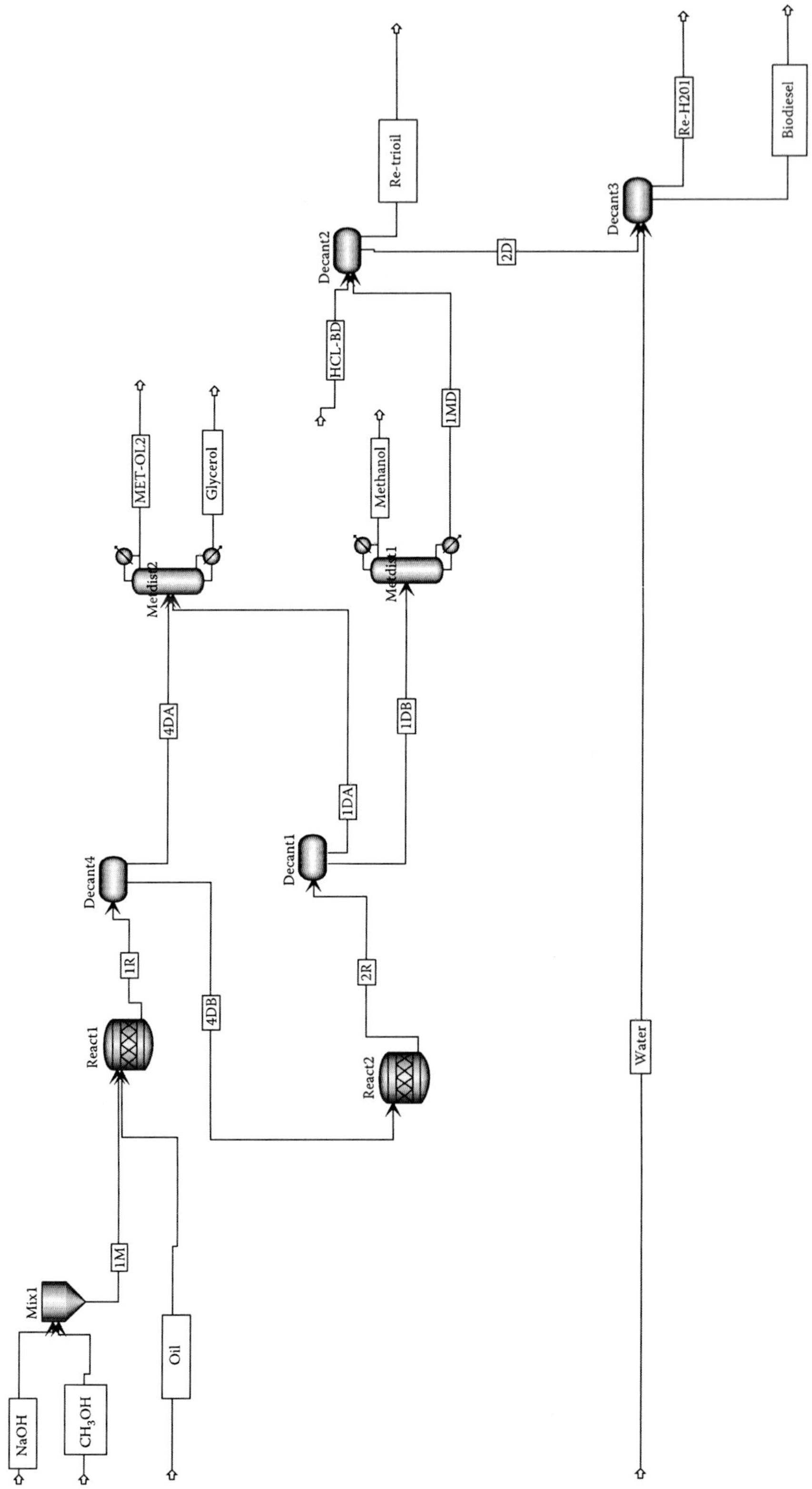

FIGURE 14.3
Biodiesel production flowsheet without heat exchangers.

TABLE 14.1

Solution for Common Heat Exchangers

Exchanger Number	From Utility or Process Unit	To Process Unit	Soy Heat Exchanger Value	Palm Heat Exchanger Value	Algae Heat Exchanger Value
E-1	Cooling Utility (CU)	MET-DIST1	E142	E142	E132
E-2	Cooling Utility (CU)	MET-DIST2	E242	E242	E232
E-3	REACT1	REACT2	E331	E332	NA
E-4	REACT1	MET-DIST1	NA	E312	E311
E-5	REACT1	MET-DIST2	E321	NA	NA
E-6	Heating Utility (HU)	MET-DIST1	E411	NA	E511
E-7	Heating Utility (HU)	MET-DIST 2	E421	E421	E521
E-8	Cooling Utility (CU)	REACT1	NA	E342	E332
E-9	Cooling Utility (CU)	REACT2	NA	NA	E432

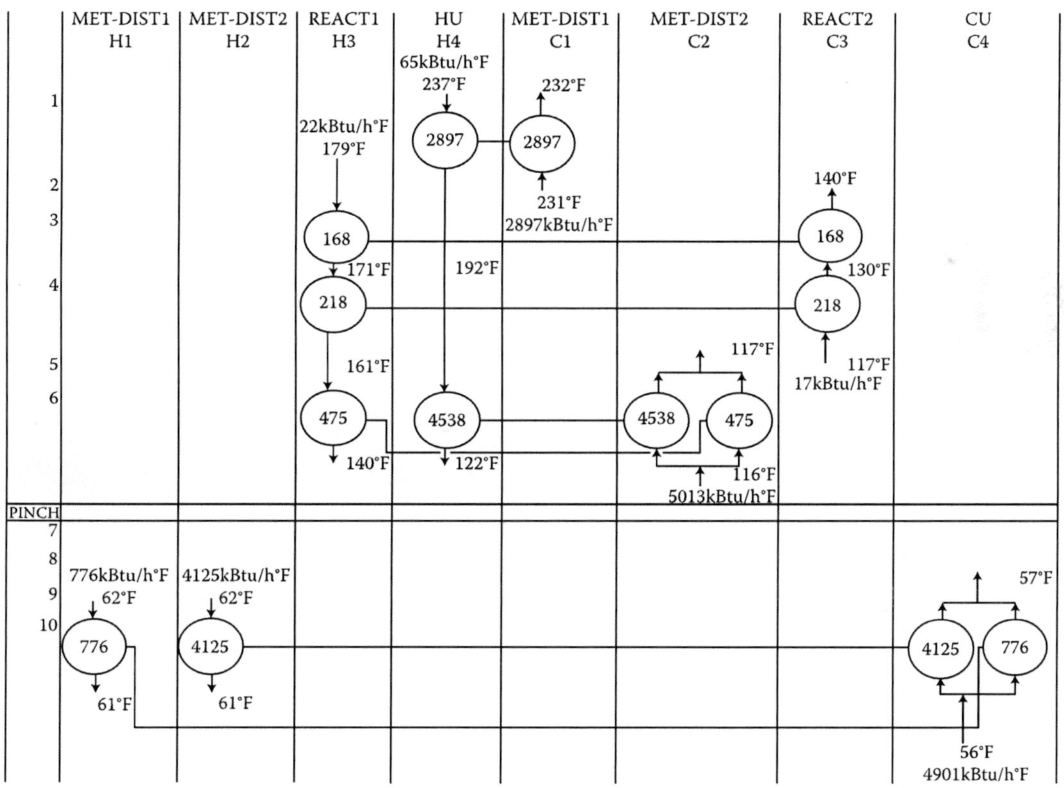

FIGURE 14.4
HEN for soy oil scenario.

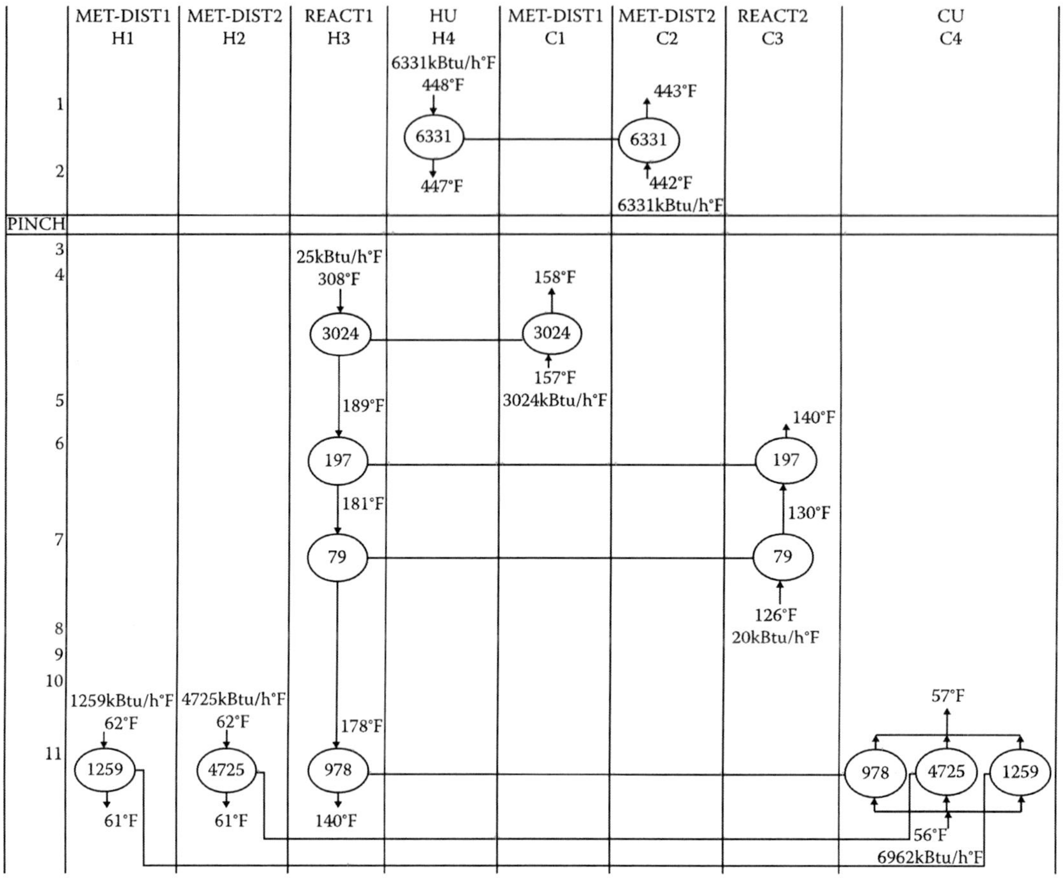

FIGURE 14.5
HEN for palm oil scenario.

14.5 Conclusions

This work has addressed the problem of designing a common and flexible HEN for a multi-feedstock biodiesel production facility. Transshipment models were used as the basis for the synthesis of individual HENs representing each scenario. For each feedstock type and throughput, a detailed simulation study was carried out to determine the process performance and the heating and cooling requirements. Minimum heating and cooling utility requirements were determined for each scenario. Then a multiperiod MILP was developed and solved to generate the common structure for all the scenarios that will minimize the number of heat exchangers required while satisfying minimum heat and cooling utility requirements for each scenario. A case study representing a three-feedstock biodiesel facility using soy bean, palm, and algal oil was solved using the proposed method.

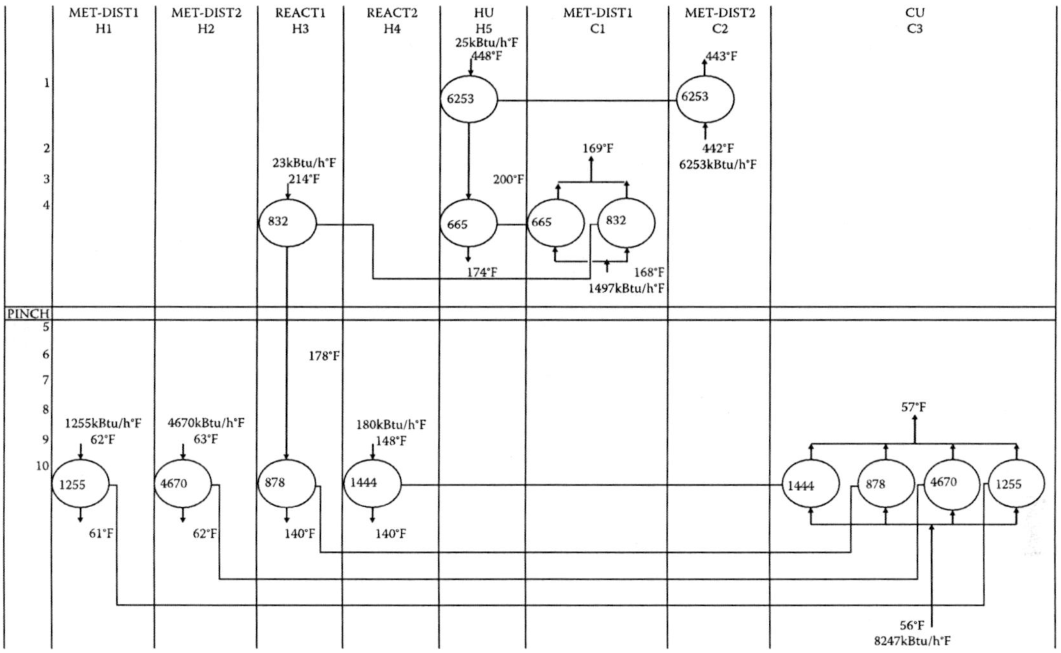

FIGURE 14.6
HEN for algal oil scenario.

Nomenclature

symbols

A_{m_c}	Flowrate of input stream m_c
B_{n_c}	Flowrate of output stream n_c
C	a set of common process units
c	index for common process units
CS	a set of cold streams
$E_{h,k,v}$	number of heat exchangers
f	index for feedstock alternatives
FFA	free fatty acid
GHG	greenhouse gas
h	index to represent hot streams
HEN	heat exchange network
HS	a set of hot stream loads
$INPUT_c$	the set of input streams for a common process unit, c
k	index to represent cold streams for CS
LCA	life-cycle analysis
LINGO	an optimization software program
LP	linear programming
m_c	index to represent input streams for a common process unit, c
MILP	mixed-integer linear program

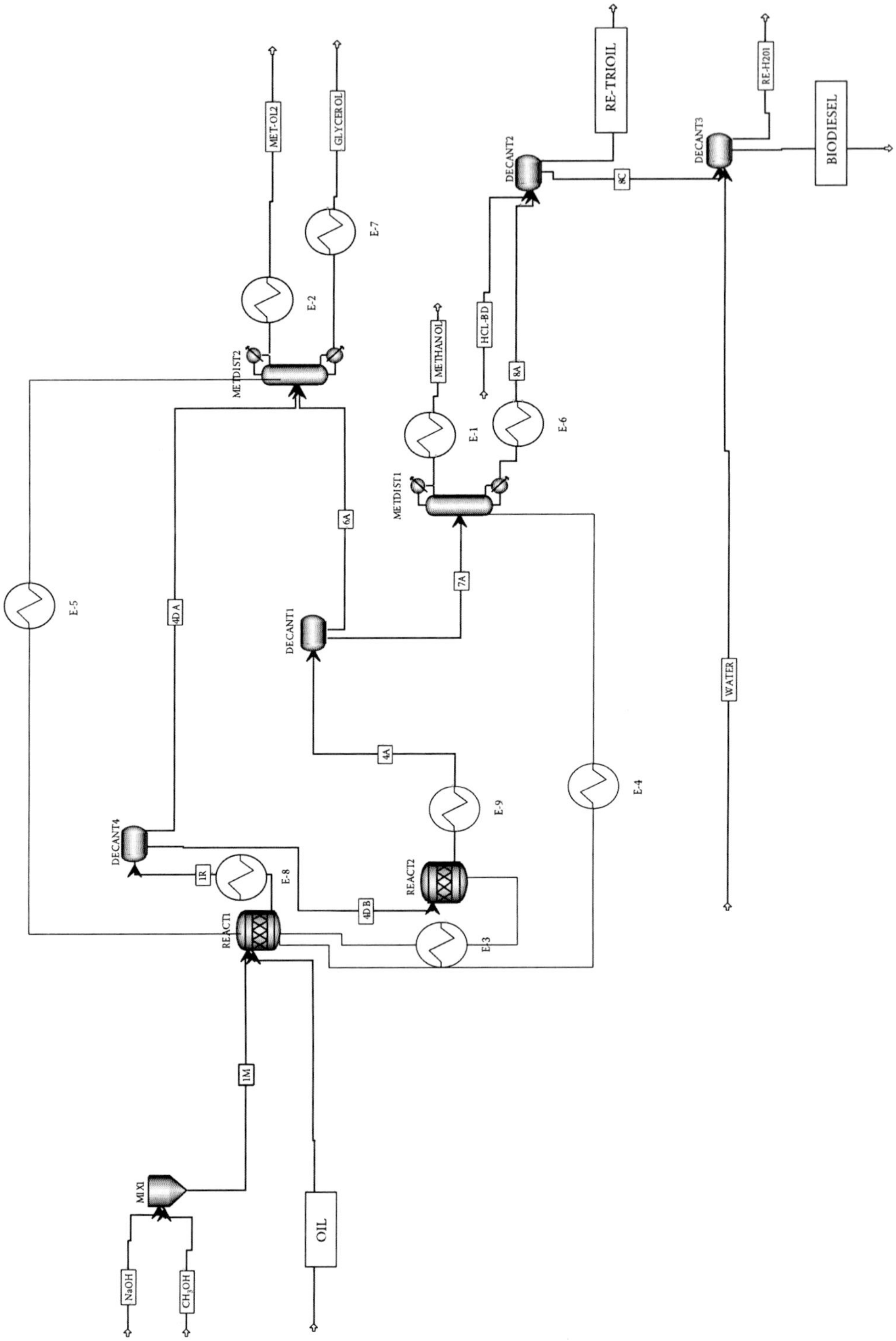

FIGURE 14.7
Flexible HEN configuration that accommodates all scenarios.

MMGPY	million gallons per year
n_c	index to represent output streams for a common process unit, c
N_c^{in}	number of input streams for a common process unit, c
N_c^{out}	number of output streams for a common process unit, c
N_{CP}	number of common process units, c
N_{CS}	number of hot streams for CS
N_f	number of feedstock alternatives, f
N_{HS}	number of hot streams for HS
N_s	number of scenarios, s
OUTPUT_c	the set of output streams for a common process unit, c
$Q^{\text{C}}_{k,y,s}$	total heat load for the kth cold stream in interval y for scenario s
$Q^{\text{H}}_{h,y,s}$	total heat load for the hth hot stream in interval y for scenario s
$Q_{h,k,y,s}$	heat exchanged between the hth hot stream and the kth cold stream in interval y for scenario s
s	a given scenario for a certain feedstock
SCENARIOS	a set of scenarios
SN	subnetworks; one above the pinch and one below the pinch
SN_v	subnetwork for index v
T_{m_c}	Temperature of process input stream m_c
T_{n_c}	temperature of process output stream n_c
$U_{h,k,y,s}$	upper bound on the exchangeable heat between streams h and k in SN_v
v	index for subnetworks, defined as 1 for above the pinch and 2 for below the pinch
y	intervals over which residual heat exchange loads are passed
$Y_{m_c,x}$	composition of component x in input stream m_c
$Z_{n_c,x}$	composition of component x in output stream n_c

Greek Symbols

$\delta_{h,y,s}$	residual heat exchange loads ($\delta_{h,y,s}$) for hot streams transferred to cold streams for each scenario
$\delta_{h,y-1,s}$	residuals from previous intervals

References

Aaltola, J. 2002. Simultaneous synthesis of flexible heat exchanger network. *Applied Thermal Engineering* 22: 907.

Abdullah, A.Z., Salamatinia, B., Mootabadi, H., and Bhatia, S. 2009. Current status and policies on biodiesel industry in Malaysia as the world's leading producer of palm oil. *Energy Policy* 37: 5440–5448.

Al-Mutairi, E. and El-Halwagi, M.M. 2009. Integration method for considering scheduling in design of heat exchange networks. *Applied Thermal Eng.* 29, 3482–3490

Chen, C.L. 2005. A novel strategy for synthesis of flexible heat-exchange networks. *Journal of the Chinese Institute of Chemical Engineers* 36(5): 421–432.

Chen, C.L. and Hung, P. 2007. Synthesis of flexible heat exchange networks and mass exchange networks. *Computers and Chemical Engineering* 31: 1619–1632.

El-Halwagi, M.M. 2006. *Process Integration*. Amsterdam: Elsevier.

El-Halwagi, M.M. 2012. *Sustainable Design through Process Integration: Fundamentals and Applications to Industrial Pollution Prevention, Resource Conservation, and Profitability Enhancement.* Oxford, UK: Butterworth-Heinemann/Elsevier.

Elms, R.D. and El-Halwagi, M.M. 2009. Optimal scheduling and operation of biodiesel plants with multiple feedstocks. *Int. J. Process Systems Engineering* 1(1): 1–28.

Elms, R.D. and El-Halwagi, M.M. 2010. The effect of greenhouse gas policy on the design and scheduling of biodiesel plants with multiple feedstocks. *Clean Technologies and Environmental Policy* 12(5): 547–560.

EPA 2002. *A Comprehensive Analysis of Biodiesel Impacts on Exhaust Emissions.* EPA420-P-02-001, October.

Floudas, C.A. and Grossmann, I.E. 1986. Synthesis of flexible heat exchanger networks for multiperiod operation. *Computers and Chemical Engineering* 10(2): 153–168.

Furman, K.C. and Sahinidis, N.V. 2002. A critical review and annotated bibliography for heat exchanger network synthesis in the 20th century. *Industrial and Engineering Chemistry Research* 41, 2335.

Kemp, I.C. 2007. *Pinch Analysis and Process Integration,* second edition. Oxford: Butterworth-Heinemann.

Linnhoff, B. and Hindmarsh, E. 1983. The pinch design method for heat exchanger networks. *Chemical Engineering Science* 38: 745.

Myint, L.L. and El-Halwagi, M.M. 2009. Process analysis and optimization of biodiesel production from soybean oil. *J. Clean Tech. Env. Policies* 11(3): 263–276.

National Biodiesel Board 2009. *Estimated U.S. Biodiesel Production by Fiscal Year (Oct. 1-Sept. 30).* Available at: http://www.biodiesel.org/pdf_files/fuelfactsheets/Production_Graph_Slide.pdf, accessed 22 December 2009.

National Biodiesel Board 2010. *NBB Member Plant Locations.* Available at: http://www.biodiesel.org/buyingbiodiesel/plants/default.aspx?AspxAutoDetectCookieSupport=1, accessed on 12 January 2010.

Papoulias, S.A. and Grossmann, I.E. 1983. A structural optimization approach in process synthesis—II: Heat recovery networks. *Computers and Chemical Engineering* 7(6): 707–721.

Pokoo-Aikins, G., Nadim, A., Mahalec, V., and El-Halwagi, M.M. 2010a. Design and analysis of biodiesel production from algae grown through carbon sequestration. *Clean Technologies and Environmental Policy* 12: 239–254.

Pokoo-Aikins, G., Heath, A., Mentzer, R.A., Mannan, M.S., Rogers, W.J., and El-Halwagi, M.M. 2010b. A multi-criteria approach to screening alternatives for converting sewage sludge to biodiesel. *J. Loss Prevention in the Process Industries* 23(3): 412–420.

Smith, R. 2005. Process integration: Current status and future potential. *Proceedings of ECOS 2005* (1–3): 35–42.

Xiangkun, M., Pingjing, Y., Xing, L., and Wilfried, R. 2007. Synthesis of flexible multi-stream heat exchanger networks based on stream pseudo-temperature with genetic/simulated annealing algorithms. *Journal of the Chinese Institute of Chemical Engineers* 38: 321–331.

Yee, T.F. and Grossmann, I.E. 1990. Simultaneous optimization models for heat integration: II. Heat exchanger network synthesis. *Computers and Chemical Engineering* 14: 1165.

Zhang, W.V. 2006. Design of flexible heat exchanger network for multi-period operation. *Chemical Engineering Science* 61(23): 7730–7753.

Section IV

Biorefinery Feedstocks

15

A Literature Review of Biomass Feedstocks for a Biorefinery

José Meléndez, Luc LeBel, and Paul R. Stuart

CONTENTS

15.1 Introduction

Ragauskas et al. [1] stated that a modern biorefinery, by definition, parallels a petroleum refinery. Raw material is brought to the refinery and processed, and a multitude of high- and low-value products, by-products, and co-products exit the plant to be distributed to a large number of users. However, unlike petroleum refineries which use only one type of raw material (petroleum), a biorefinery will most likely have multiple types of renewable feedstocks entering the process at a given time. Because of their variable characteristics (moisture, chemical composition, energy content, physical size, etc.), these feedstocks must be preprocessed before they are introduced into the main manufacturing facility. This preparation can take many forms, but two activities are most common. The first is moisture removal. For the most part, feedstock moisture does not participate in reactions within a biorefinery, but has to be removed in one way or another (e.g., evaporation or pumping out to a treatment plant). Within the system, large quantities of undesired water will increase energy consumption (e.g., in evaporation columns) or decrease process efficiency (e.g., in a boiler). In addition, moisture contained within biomass increases transportation costs because it increases biomass weight when transported. The second common preprocessing step, size reduction, is performed to increase the surface area of the feedstock which is exposed to the process, as well as to homogenize the feedstocks for the best overall outcome in the main process. Consequently, these preprocessing steps will make every biorefinery distinct because the processes will depend in part on the available feedstocks, and feedstock supply will depend on process requirements.

The best configuration for a particular biorefinery will most likely be a unique solution for each mill [2]. Like current pulp and paper mills, no two biorefineries will be exactly alike. Rather, each biorefinery will differ according to factors such as available feedstocks in the local area, final products desired (biofuels, bioenergy, and biochemicals), and existing processes and facilities. The types and quality of feedstocks will affect biorefinery processes. In turn, each feedstock may have a distinct supply-chain configuration, and different biorefinery supply-chain designs may therefore be possible for the same feedstock. Procurement decisions and supply-chain design will therefore have an impact on the facility. For example, harvesting of forestry feedstocks such as woody residues, chips, and sawdust has a different procurement supply chain than does the collection and transport of grain harvesting and agricultural crop residues. These feedstocks may all be used by a biorefinery, but different handling methods are required before they can be introduced into the main biorefinery process. Similarly, biorefineries that use forestry feedstocks will not all use the same supply chain.

Achieving a harmonious supply of locally available feedstocks will not only reduce supply costs but also help ensure that when one feedstock is not available, it can and will be replaced with others. The selection of dependable fit-for-usage feedstocks requires a detailed study of the major types of material available to supply biorefineries.

The following is a literature review of biomass feedstocks for a biorefinery. Section 15.2 first reviews the basic elements of biomass. This is followed by a characterization of different biomass sources, and a description of various classification methods for biomass. Next, Section 15.3 looks at the various sources of biomass for biorefineries, emphasizing studies of harvesting, procurement, collection, transportation, storage, and overall supply chains. Specific types of biomass sources (forestry, agriculture, and energy crops) are considered. Finally, the chapter concludes with a discussion of the key elements associated with each biomass type as a source for biorefineries, as well as possible actions required to fill certain knowledge gaps identified in the literature.

15.2 Biorefinery Feedstocks

15.2.1 Biomass Feedstocks: Generic Description and Properties

In the context of this chapter, biomass is considered to be any type of organic material derived from a living organism (plant or animal), which has the potential to be used within a biorefinery process. Although both plant and animal sources of biomass are fairly common, there is far more plant biomass in the world than animal biomass. Therefore, although biomass sources derived from animals, such as manures, lipids, processed oils, and residues are potential feedstocks for a biorefinery, the focus of this study will be on feedstocks mainly from lignocellulosic biomass, including, but not limited to, forest industry by-products, agricultural residues, energy crops, lignocellulosic residues, and waste material.

Plant biomass is created through the process of photosynthesis, in which plants use carbon dioxide (CO_2) from the air, water, and sunlight to produce carbohydrates that form the building blocks for biomass. These building blocks then serve to form the various types of biochemical structures that most plants need to survive: simple sugars, starches, cellulose, hemicellulose, and so on [3].

Typically, plants store energy in the form of sugars, and depending on the type of plant, the sugars will vary both by type (glucose, fructose, xylose, etc.) and by size (micro- or macromolecules). The use of biomass varies according to the type of sugars the plant possesses. For example, simple sugar storage is typically found in plants such as sugarcane and sugar beet, and these have commonly been used as a food source as well as a raw material for the production of ethanol through fermentation. Other plants store energy in the form of more complex sugars called starches. These plants include grains such as corn and wheat and some tubers such as potatoes, cassava, and yams. These plants are typically used for food and feed purposes, but ethanol production is also common from these sources, especially from corn [4].

A third form of plant matter, called cellulosic biomass, is made up of large sugar polymers (complex polysaccharides). Unlike simple sugars and starches that are used by the plant to store energy for later use, cellulose is a structural component of the plant that, along with other structural components (hemicellulose and lignin), gives it shape and strength. All plants, including those that produce simple sugars, typically contain significant amounts of cellulose. Plants with a large amount of structural cellulose and which are not normally used for human consumption can be attractive as a source of biomass for biorefineries. This biomass may, however, compete with other common applications such as animal feed (agricultural residues), and industrial products (timber, paper, and wood products).

In essence, photosynthesis takes solar energy absorbed by the plant and stores it in chemical bonds. Later, when biomass feedstocks are processed biologically, thermally, or chemically in a biorefinery, the bonds between carbon, hydrogen, and oxygen molecules are broken, reformed, and these substances release their stored chemical energy. If biomass is processed efficiently and the subsequent biorefinery products are then used and combined with oxygen, the carbon is oxidized to produce CO_2 and water, which are then available to reenter the photosynthesis process. This creates a cyclical process and therefore the term "renewable" can be applied to biomass feedstocks [3].

A good introduction to biomass for bioenergy has been proposed by McKendry [3]. This author describes the natural processes by which biomass is created (photosynthesis) and the different types of biomass used in bioenergy, explains the main components of lignocellulosic biomass, and describes some of the main types of biomass used for bioenergy.

Biomass feedstocks are integral to the development of biorefineries. Even before a biorefinery project becomes a reality, the financial viability of biorefinery projects will depend on the secure long-term supply of biomass feedstocks to the mill [5]. However, having a long-term supply of a feedstock does not necessarily guarantee that a viable biorefinery can be configured or that the desired product can be produced. Matching available feedstocks according to their inherent properties (moisture content, energy value, proportions of chemical compounds, and their ratios) with the best processes that fulfill required product needs will ultimately determine what type of biorefinery will be built and the level of processing difficulties that may arise [3].

In traditional biomass projects, which served as a base for the first biorefineries, grain and starch were the primary feedstocks for the production of ethanol. In these cases, the quantity of biomass was much more important than the overall quality of the feedstock. Additionally, these two components are more easily transformable than lignocellulosic compounds, as cellulose is more recalcitrant. However, the widespread use of these biomass sources in recent years to produce biofuels has created great debate. Questionable environmental benefits, land-use changes, and high production costs have reduced the attractiveness of corn and other grains as feedstocks for biofuels. Estimates indicate that replacing gasoline with corn ethanol in the United States would require more than doubling the 174 million hectares of cropland used in 2007 [6], which cannot be accomplished in the short term and is environmentally undesirable. In addition, feedstock quality has become more important in the ethanol industry; as an example, ethanol plants now prefer to buy grains that contain higher quantities of starch to reduce feedstock and production costs.

Selling grains normally used for animal and human consumption to ethanol producers has led to intense "food vs. fuel" and "feed vs. fuel" debates and poses many ethical questions. To make matters worse, the increase in the prices of corn and other grains has had a ripple effect on other commodity products, increasing the price of foods such as dairy products, meats, oils, and other cereal grains [6,7]. A 2009 report by the Food and Agriculture Organization of the United Nations (FAO) [8] on agricultural commodity markets stated that prices of all foods around the world increased significantly starting in 2006 due to higher oil prices and an increased demand for certain agricultural products as feedstocks for biofuels production.

The next generation of biorefineries will be developed with more attention to procurement alternatives. The selection of a transformation process for the biorefinery will be determined by the type of feedstocks that are best suited to it and that can provide a dependable, reliable, and sustainable supply. However, to determine the right types of feedstocks for the process, it is first important to know the process requirements. Then, available feedstocks can be characterized, and finally, the link can be made between process requirements and feedstock characteristics. This will identify which feedstocks are best suited to each process in a forest biorefinery [9].

Chambost et al. [10] have identified three categories of biorefinery processes: biochemical, thermochemical, and chemical. Process requirements for each feedstock are summarized in Table 15.1. Biorefinery processes have different biomass requirements which may be met by one or several different feedstocks. Moreover, through preprocessing of these feedstocks, some of their intrinsic properties may be modified so that they can be used in processes for which they would not be suitable otherwise. Table 15.1 exemplifies some of the generic properties that each biorefinery process must have to function.

Biochemical processes such as fermentation typically require that the feedstocks contain some type of sugar (glucose in most cases) or sugar derivatives (complex polysaccharides such as cellulose and starches) which can be broken down into simple sugars. On the other hand, thermochemical processes require feedstocks with low moisture content, and have high energy value. Of course, as mentioned earlier, properties such as moisture content can be modified through simple preprocessing operations such as drying, which can make

TABLE 15.1

Generalized Process Feedstock Requirements

Feedstock Requirements	
Biochemical Processes	
Fermentation	Sugars and starches
Digestion	Organic material
Trans-esterification	Lipids and oils
Thermochemical Processes	
Combustion, pyrolysis, and gasification	Low moisture content and high energy density
Chemical Separation	
Extraction, separation, and fractionation	Appropriate chemical composition

very high-moisture-content feedstocks suitable for thermochemical processes. However, although this is technically possible, it may not be practical if drying operations impose a high cost on the biorefinery. *It is precisely this match between biorefinery process requirements and feedstock properties which will determine how a biorefinery can best perform.*

When designing a biorefinery, it is important to understand the potential feedstocks that could be used for a specific process, be it biochemical, thermochemical, or chemical. Each process has certain feedstock requirements that must be met to produce the desired product. In addition, all feedstocks have different intrinsic properties that make them suitable for different processing methods. To design the best procurement strategy for a biorefinery, a match between these two characteristics (process requirements and feedstock properties) must be achieved.

15.2.2 Classification of Biomass Feedstocks

McKendry [3] classified biomass in a very simple manner according to one critical property: moisture. Using this property, he classified a plant biomass as of high moisture content (suited for biochemical processes) or of low moisture content (suited for thermochemical processes). This simple classification method allows a quick understanding of where a feedstock might best be used, but at the same time, limits its use by not looking at all other feedstock properties. However, other methods have also been suggested to classify biomass.

Various authors [11–18] have suggested classifying biomass sources according to supply source, physical characteristics, general usage, and conversion processes. Perlack et al. [11] classified biomass sources into two main categories: forest sources and agricultural sources. They then subclassified each category into primary, secondary, and tertiary sources. Bradley [12] classifies biomass sources into woody, agricultural, and municipal sources, but with reference only to residues in each category. Wood and Layzell [13] created a Canadian biomass feedstock inventory using three categories: forest, agricultural, and municipal. They then subdivided these categories into production of resources or residues. Chornet [18] classified biomass in a unique fashion, using the homogeneity of feedstocks as a main driver and classifying them as homogeneous, quasi-homogeneous, or heterogeneous in order to determine what best feedstocks suit the particular gasification process he works with.

The European Biomass Industry Association [14] has classified biomass according to the supply sector from which it comes, creating four main categories: forestry, agricultural, industrial, and waste. However, the emphasis in their subdivision is placed on materials which may become potential "bioproducts" and excludes all current wood products (timber resources) and industrial processing wastes (sugar waste streams, for example).

A study done by the Oak Ridge National Laboratory [15] classified biomass feedstocks according to their potential for energy production, specifying both energy density and chemical composition, both of which involved looking at biomass sources in much greater detail than had been done before. This type of classification makes it possible to determine more efficiently differences in biomass composition, which may affect process conversion efficiency, the final product. Since then, similar studies specifically looking at biomass for bioenergy have analyzed the chemical composition of feedstocks to determine more accurately those that have greater potential for energy production. Vassilev et al. [19] created a very extensive list of feedstocks and analyzed their chemical composition to obtain a better understanding of which biomass sources are better for energy production. This study, although very useful for energy production, fails to provide the biochemical makeup of the biomass sources, which is of more interest for biorefinery biochemical processes.

A different approach was taken by the National Renewable Energy Laboratory (NREL) [16], which focused more on the type of technology used in processing biomass. Six biomass platforms were presented in this study: sugar-lignin, syngas, bio-oil, biogas, carbon-rich chains, and plant products. They proposed this method of classification for use in biorefineries. However, although it provides a good way of reviewing biomass processing technologies, this approach leads to overlaps because many feedstocks can be used in several of the platforms.

The U.S. Department of Energy's (DOE) energy efficiency and renewable energy (EERE) biomass program [17] has produced a classification scheme that is similar to that created at NREL. They combined feedstock sources with processing technologies to create a classification system which includes biomass classifications as well as biorefinery pathways. The biomass feedstocks reviewed by EERE includes a greater range of sources than had been identified previously by other authors. This classification was then segregated according to the biomass-conversion platforms needed to produce specific bioproducts.

Using the classification systems described above, Melendez and Stuart [9] classified biomass using process requirements and biomass properties instead of biomass sources to classify biomass components (see Table 15.2).

Within Table 15.2, biomass feedstocks are categorized into four distinct categories [9] described below:

- *Sugar and starch feedstocks*: These feedstocks include plants that produce significant amounts of sugars and starches. Normally these are agricultural crops, specifically grains from sources such as corn, wheat, rye, and triticale; starches from potatoes and tapioca; and high-sugar-yielding juices from sugarcane and beetroot. Other such feedstocks include residual byproducts and discharges from industrial processes (e.g., the food and beverage industry [20]), which contain high concentrations of residual sugars. Processing of these feedstocks normally involves biochemical treatments to produce first-generation bioethanol.

TABLE 15.2

Biomass Characterization According to Process Requirements

		Biomass Category			
Process	**Requirements**	**Sugars and Starches**	**Vegetable Oils and Animal Fats**	**Biodegradable Waste Feedstocks**	**Lignocellulosic Materials**
Biochemical					
Fermentation	Sugars	X			X
Transesterification	Lipids and oils		X		
Digestion	Organic material			X	X
Thermochemical					
Pyrolysis	Low moisture			X	X
Gasification	and high			X	X
Combustion	energy density			X	X
Torrefaction					X
Chemical					
Extraction	Differences	X	X		X
Separation	in chemical	X			X
Fractionation	composition	X	X		X

- *Vegetable oils and animal fats*: These provide oil feedstocks for the production of first-generation biodiesel by transesterification. This category includes oils produced from crops (virgin oil feedstocks) such as rapeseed, palm, and soybean; animal fats like tallow, lard, and chicken fat; and industrially produced residual oils (waste vegetable oil).
- *Lignocellulosic feedstocks*: One of the more varied categories, this includes all types of plant material that contain three main biochemical components: cellulose, hemicellulose, and lignin. Included in this category are forest products such as white wood logs and chips as well as all forestry and agricultural residues from both harvesting and secondary processing operations (such as sawmilling, pulping, and papermaking); purpose-grown herbaceous and woody energy crops such as willows, poplars, and switchgrass; and any and all green residues from urban communities and industries (grass clippings, tree limbs, woodchips, leaves, and paper waste). Because of their lower moisture content or ease of water removal, higher energy content, and great abundance, these feedstocks are preferred for thermochemical processing, although second-generation biorefineries could process them through biochemical routes.
- *Biodegradable waste feedstocks*: Typically, this category consists of animal wastes, industrial residues (wastewater treatment sludges), and municipal solid wastes containing large amounts of organic material and high amounts of moisture that render them unsuitable (or uneconomical) for thermal treatments. In an industrial process, normally anaerobic digestion is used to produce both "biogas" and a by-product solid residue which is used as fertilizer.

Analyzing biomass in this way makes it possible to determine the specific components needed by a process and from there to work backwards to the feedstock source. The next step is to specify how much of the component is available and what other components will become available as by-products of the process selected. This back-to-front way of thinking (process, then feedstock) is an extension of the ideas proposed by Stuart et al. [21–23] for biorefinery products (product selection, then process selection). The component analysis and pairing of biomass feedstocks with processing pathways was also studied by creating a lignocellulosic biomass survey from the literature and then comparing the biochemical composition of biomass feedstocks to determine which ones were better suited for which process routes.

Another, very similar classification by Ekşioğlu et al. [24], which identified biomass sources (oil plants, corn and sugars, animal waste, and lignocellulosic) according to the characteristics necessary for the production of bioenergy using intermediate biofuels (biodiesel, bioethanol, biogas, and C-ethanol).

In recent years, other authors have also published studies which look more closely at the chemical and biochemical components of biomass feedstocks as well as their intrinsic biomass properties to combine them with process requirements. Bozell [25] stated that although many biomass feedstocks are available, three process streams will be created in a pretreatment phase (carbohydrates, aromatics, and hydrocarbons) that will supply the process with the components needed. Cherubini et al. [26] created a classification system very similar to that proposed by Melendez and Stuart [9], in which they classify by source, and then by components (Figure 15.1). Each of the feedstock components mentioned in Figure 15.1 would be used within a specified process platform within a biorefinery.

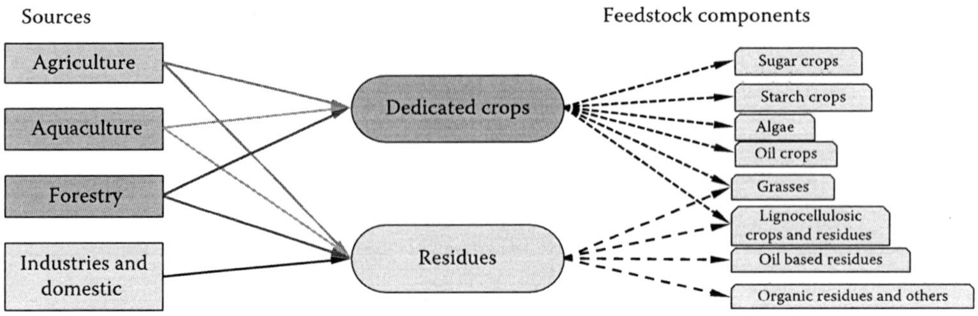

FIGURE 15.1
Feedstock features (and relative subgroups) for biorefinery systems. (From Cherubini, F. et al. 2009. *Biofuels, Bioproducts & Biorefining*, 3, 334–346. With permission.)

The recent interest in classifying biomass according to its individual components is indicative of an industry seeking to expand its current feedstock portfolio to fulfill its process requirements with the best (and lowest-cost) feedstock available. This trend also indicates that for biorefineries, the understanding of feedstocks and their component analysis will become an integral part of process design to optimize process efficiency. By classifying biomass by components, a larger biomass "basket" from which to procure feedstocks is created. No longer does a process have to be limited to a predetermined type of feedstock, as in the past. Moreover, this type of component analysis enables identification of by-product streams which will be available for production of other products in a biorefinery. This will in turn enable a process designer to integrate multiple processes into the original biorefinery design.

This section has looked at biomass in general terms, describing its properties and the way researchers have classified it. The purpose was not only to introduce the concept of biomass but also to expand the general definition and to show that a biorefinery can potentially gather its needed biomass from many sources instead of only one, as is currently the case. However, to enable the biorefinery to gather feedstocks from multiple sources, the quality of the biomass (influenced by biomass properties) must be analyzed before harvesting can take place to determine the best sources of biomass for the specific biorefinery process in question.

15.3 Biomass Procurement Logistics and Supply Chains

This section takes a closer look at the currently most common biomass sources: forestry, agriculture, and energy crops. Once a contract assures a secured supply, biomass availability for a biorefinery depends on procurement logistics. Typically, biorefinery feedstocks are residues from other industries which are sought at the lowest possible cost. However, although many sources can be purchased at low cost, procurement logistics are of critical importance because handling, transportation, and inventory management can significantly increase overall cost. Studies have shown that biorefinery production costs are highly influenced by the cost of their feedstocks [27] which account for 20–50% of total production costs [24,28,29], and of these feedstock costs, 60–90% are due to procurement logistics [24,30]. By evaluating and understanding the current methodologies involved in harvesting and processing biomass in industries such as wood products, pulp and paper, bioenergy, and agriculture, a biorefinery can set up supply chains that provide the required biomass amounts at the lowest possible costs.

Biomass harvesting involves a number of activities which not only vary according to feedstock but also depend on the desired product to be delivered to the customer (the mill). A generic division of the separate activities involved in biomass harvesting is as follows [31]:

1. Gathering or acquiring
2. Processing
3. Transport

Different types of machinery can be used within each of these activities, and these options must also be considered. The main objective of harvesting is not only to collect biomass feedstocks and transport them to the mill gate but also to alter their characteristics to fit the specific process requirements while maintaining costs as low as possible. These modifications may be of a physical nature (size, moisture, and contamination reductions). They may also include chemical changes (fractionation and separation of specific components). If harvesting activities are carefully planned and implemented, the benefits which were anticipated at the time of investment can be achieved [32].

Gathering, processing, and transportation functions are in many cases, at least in part, independent of each other and involve the use of distinct pieces of machinery. The specific operations depend on the initial state of the biomass in the field. This includes the type of biomass (herbaceous, woody, or crop residue), quality parameters, moisture content, and many other factors. End-use properties such as moisture content, particle size, contamination ratio, and homogeneity also influence the way that biomass is collected. Physical, biological, and social factors also affect biomass harvesting and, when combined, create unique process routes for each harvest area. Factors include, but are not limited to [32,33], climate, environment, topography, soils, biomass type and species, location (remoteness), labor constraints, and social acceptability.

For a successful harvesting operation, certain conditions must be met, such as [32]:

- Technical feasibility, including physical laws, engineering knowledge, and environmental relationship with the harvested area
- Economic viability, including costs, benefits, short- and long-term consequences
- Environmental feasibility, meaning that operations must be environmentally sound and must consider their impacts on the natural and social environment and favor an efficient use of natural resources, including renewable materials, nonrenewable materials, water, energy, and space
- Institutional feasibility, including the laws and regulations governing operations, landowner objectives, and social values

The process technologies and equipment used in each harvesting route have a significant influence on the way that biomass is harvested, the efficiency of harvesting, and of course the economic payback from its collection. The optimal combination of equipment and personnel will minimize harvesting costs while enabling maximum procurement of biomass. However, equipment types, capabilities, and harvesting methods vary widely and depend on the type of biomass being harvested (forest, agricultural, and energy crops).

It is important to mention that under certain circumstances, biomass harvesting operations may be determined based on sylvicultural or land-management priorities instead of on the intended end use of the biomass. For example, in many areas of the Midwest and the western United States, fire reduction is an essential activity which is carried out to maintain healthy forests and to reduce the potential for fires during the hot dry summer months [34]. The resulting thinning operations generate large quantities of woody biomass, but products and machinery are in this case dictated by other objectives. In other cases, for example, in the southeastern United States, thinning operations are needed to favor stand growth. These thinning operations make biomass continually available. Significant quantities of biomass from fires, harvesting residues, and other natural disturbances are also produced in Canada and have been quantified by Dymond et al. [35]. An additional case worth mentioning is the harvesting of wood which has been killed by insect infestations or disease. One of the most notable cases has resulted from the mountain pine beetle outbreak centered in British Columbia, Canada, which at its peak in 2004 was producing over 141 million cubic meters of killed wood annually [36].

15.3.1 Forest Biomass

Forest resource management seeks to make decisions based on balanced use and conservation of forests and related resources. It aims to provide an array of benefits, such as timber, water, wildlife, and recreation to society and to individuals [37]. It has been a part of human development and industrialization for centuries. Forest operations consist of all the technical and administrative processes needed to develop technical structures and facilities, harvest timber, prepare sites for regeneration, and maintain or improve the quality of stands and habitats [32].

Harvesting is an essential activity within forest management. When carefully planned and implemented, the benefits anticipated at the time of forest investment can be realized. Most forest management harvesting operations use similar processing methods but differ by the types of products procured. The main processing activities associated with forest management include

- *Harvesting*: Cutting of trees at the stump
- *Forwarding*: Transfer of trees from harvest area to roadside; sometimes referred to as primary transport
- *Processing*: Preparation of logs which can include operations such as delimbing, debarking, bucking, bundling, and chipping, among others. This processing may be done either at the roadside or at the stump, depending on the type of equipment used
- *Loading*: Picking up the material at the roadside and placing it on trucks
- *Transport*: Taking the material from the forest to the consuming mill. Also called secondary transport

For a biorefinery, biomass from the forest will not typically come in the form of large logs, either natural or from plantations. Rather, for the most part, biorefineries will take advantage of the large supply of residues left over from both harvesting operations and secondary industries (sawmills, pulp and paper facilities).

In the past, sawmill residues have been one of the most commonly used feedstocks in pulp and paper mills, as well as in small bioenergy projects, because of their homogeneity, ready-to-use in a process nature, and typically (although not always) low costs. However, in the wake of the recent financial crisis in the United States, the demand for Canadian lumber decreased significantly which led to a decline in the production of sawmill residues (Figure 15.2) and in many cases to the permanent shutdown of many sawmills, pulp mills, and paper mills. In a 2005 report on mill residues [38], a surplus of 2.7 million bone-dry metric tons (bdmt) of mill residues was estimated for all of Canada. In the period from 2005 to 2008, several large biomass projects were established. However with the decline in lumber production, mill-residue production dropped from an estimated 17.7 million bdmt in 2005 to just under 14.0 million bdmt in 2009 [39], wiping out the surplus. This left many companies searching for new sources of biomass.

The term *forest residues* typically refer to parts of trees such as treetops, branches, small-diameter wood, stumps, and dead wood, as well as undergrowth and low-value species [40]. Within Canada, the amounts of forest biomass available for uses other than traditional forest products has been studied by authors such as Bradley [39], Wood and Layzell [13], McCloy [38], and Lambert et al. [41] among many others. Of those mentioned, Lambert et al. [37] developed a series of biomass equations which estimate the amount of biomass produced by a certain species of tree to determine how much biomass would be available per species on a larger scale. These types of estimates make it possible to determine with reasonable accuracy the amounts (and locations) of forest residues available for harvesting.

The final aspect of the study of forest biomass is biomass costs. Currently, although a large demand for forest residues exists, there is no common market price for any type of forest-residue biomass. Many parameters influence the cost of biomass: type of residue, location, moisture content, availability, size, energy content, harvesting method, transportation distance, and many others. Therefore, defining a specific price for biomass is not easy because prices can vary significantly. That said, some authors have tried to generate estimates of biomass prices. Bradley [39] found average prices (\$/oven-dry metric ton) for bark, sawdust, and bark piles in eastern Ontario (2007 estimates) of 13.80, 23.60, and 2.00. However these

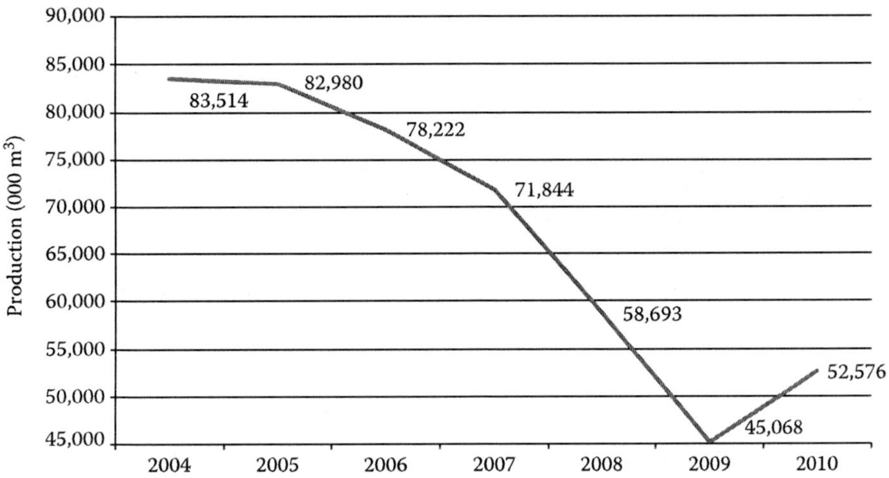

FIGURE 15.2
Canadian lumber production. (From Bradley, D. 2010. *Canada Report on Bioenergy 2010*. Climate Change Solutions, Ottawa, Ontario, p. 14. With permission.)

estimates did not include transportation, and the average value was reduced by the fact that in some cases, bark was being given away ($0 being an extremely low cost), a situation which no longer occurs. Other estimates by the authors state that standing noncommercial timber from mountain pine beetle-killed wood has a cost of approximately $80/odmt. Bradley [39] also reported on estimates performed by FPInnovations of the economic value of harvesting, collection, and transportation of roadside slash (residues left over from harvesting wood at roadside) for high-, medium-, and low-cost scenarios. In this case, the high-, medium-, and low-cost scenarios were set up according to the distance involved in collection (high: 150 km, medium: 100 km, and low: 50 km). Biomass costs in these cases were $61.54 (high), $43.60 (medium), and $24.11 (low) per bone-dry metric ton of biomass. It should be noted for these cost examples that the authors reported the costs on a dry basis. When the moisture content of biomass (approximately 50% for freshly cut wood) is factored in, transportation costs will tend to increase depending on the amount of moisture.

15.3.1.1 Forest Harvesting Equipment

Forest biomass harvesting has been reviewed by many authors in the past. Recently, emphasis has been given to the collection of forestry residues along with smaller-diameter logs and noncommercial wood species for use in other sectors such as bioenergy, biofuels, and pulp and paper.

Gringas [42] provided a literature review of potential harvesting systems for small trees and forest residues. He examined some processes which could be used to recover this material for energy purposes, using information from five different countries (Canada, Finland, Norway, United Kingdom, and Sweden). He also reviewed some potentially favorable technologies which (at the time the study was done) were still in the prototype stage, but which might result in increased benefits from biomass harvesting.

Figure 15.3 shows a number of options for processing pathways for harvesting and transporting forestry biomass. The study conducted by Dempster et al. [31] reviewed ways to lower the net cost of reducing fire damage in forested areas on the west coast of the United States, while at the same time, improving the feasibility of using the biomass removed for energy production. The objective was accomplished by first reviewing equipment currently available, under development, previously tested, or used for biomass harvesting operations in a context where the emphasis is placed on fire mitigation.

The thorough literature review provided by Dempster et al. [31] resulted in an equipment classification according to function and activities which each piece of equipment was able to carry out individually. For example, a feller-buncher gathers (by reaching out and grabbing trees), processes (by severing trees), and transports (by moving trees and creating piles on the ground). Costs, production rates, and limitations were also evaluated. Four case studies were investigated to determine the potential for cost reductions for each case, as shown in Table 15.3. All alternative paths from each case study presented in Figure 15.3 were evaluated. Finally, Dempster et al. [31] provided specific recommendations for equipment changes or modifications which could increase harvesting efficiency and help reduce fire hazards while economically using the removed biomass.

A similar study was carried out by Arnosti et al. [34]. The goals of this study were to assess the viability of fuel-reduction harvesting in conservation forests, to assess the economics of biomass harvesting, and to test the viability of different biomass-harvesting systems. The appropriate combinations of biomass-harvesting systems were determined through research, the experience of the research partners, and contractors' existing equipment inventory. In addition, nonconventional machinery was included in the study to test

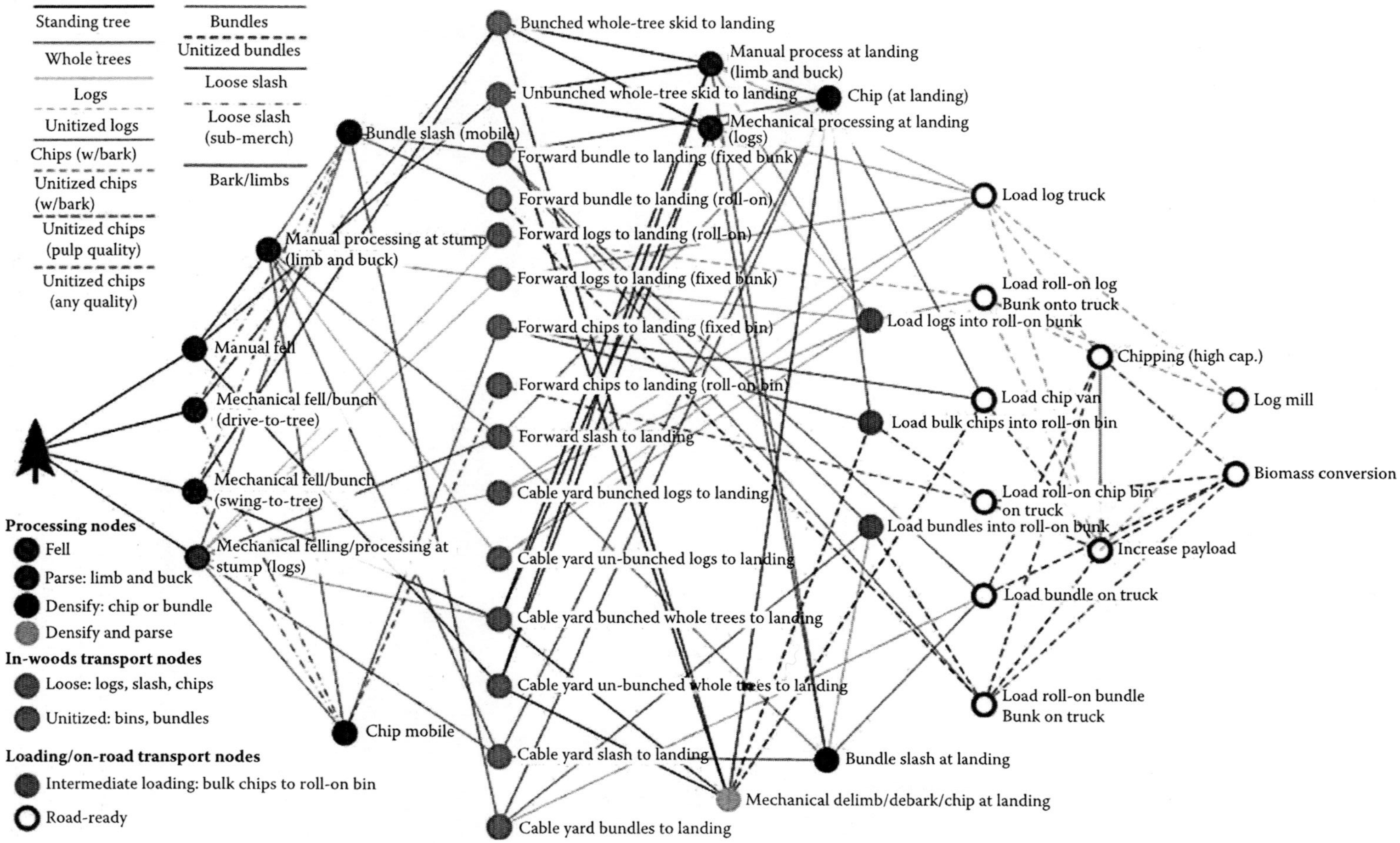

FIGURE 15.3
Network of possible processes (nodes) and forms of material (links) between standing trees and utilization facilities. (Dempster, P. et al. 2008. *Biomass Harvesting Equipment and Conversion Technology Review Final Report*. California Biomass Collaborative, UC Davis, p. 14. With permission.)

TABLE 15.3

Summary of Case Study Results from Dempster et al.

Case Study	Initial Stump-to-Plant Cost ($/Green Ton)	Cost Benefit by Making Change ($/Green Ton)	Cost Reduction (%)
Ground-based whole-tree harvesting	28	7.5	27
Ground-based cut-to-length	42	7.3	17
Cable yarding of whole trees	50	40	80
Removal of surface fuel (slash or preexisting piles)	42	20	48

Source: Dempster, P. et al. 2008. *Equipment Review: Final Report to State of California Department of Forestry and Fire Protection*. Department of Biological and Agricultural Engineering; University of California, Davis, CA, December.

Note: Case studies were carried in the central Sierra Nevada, California.

new equipment in the field. Three types of harvesters (harvester head, feller-buncher, and biomass bundler) were used on nine different sites with their accompanying forwarders, chippers, and transportation methods to evaluate the economics of biomass harvesting. The results of the study did not identify a "right way" to harvest biomass, but all three methods had varied results which proved favorable to the collection of biomass. This study also provided recommendations from both researchers and machinery operators, which will enable the learning curve of future operators to be reduced. The study was, however, limited to collection of biomass residues in a conservation forest and did not provide a scenario in which the collection of both logs and forest biomass was accomplished simultaneously.

Studies such as those conducted by Dempster et al. [31] and Arnosti et al. [39] have presented an overview of the harvesting process, the effects of interaction between equipment, and the necessity to coordinate and optimize their interaction. Although the conclusions are typically limited to the scenarios and locations observed, generic conclusions may be stated and used as guides for machinery use in biomass harvesting at other locations. These reports tend to focus on the various pieces of machinery and the efficiency at which they operate.

More specific studies, which limit their analysis to one operating unit (gathering, forwarding, or processing) within the harvesting process, have also been performed. Cuchet et al. [43] studied the Timberjack FB370 residue bundler system (the predecessor of most forest-residue bundlers available on the market today) and its performance under various working conditions in Central Europe. Analysis of this equipment showed that productivity varied directly according to the amount of biomass available on site (t/ha), its average size, and its distribution over the site. These results point toward increased productivity if a more integrated approach could be taken in harvesting operations. For example, overall bundler productivity could be improved if operators were to place residues in piles during tree harvesting.

Harvesting and processing heads were reviewed by Connell [44] with the main focus being to determine the amount and type of damage caused to logs by different mechanized harvesting and processing heads. Specific comments were made on the importance of using a size of equipment appropriate to the biomass being harvested, because harvesting larger (or smaller) diameter logs with undersized (or oversized) equipment leads to high losses (although with larger equipment, the problem may be minimized by operator training).

Anyone interested in biomass harvesting equipment and machinery should pay close attention to the specific equipment reviews available from manufacturers' websites and reports and to articles in trade magazines such as those by Burke [45], *Canadian Forest*

Industries [46,47] and *International Forest Industries* [48,49]. These sources provide the most up-to-date information about technology options and advances and a general idea of where the biomass procurement industry is headed. This last factor is exemplified by a report on the ELMIA 2009 equipment show [50], in which a clear trend was noted for manufacturers to produce harvesting heads which could handle small-diameter multistem harvesting because of the increasing importance of biomass residue and unmerchantable-stem recovery in many parts of the world.

The studies reviewed earlier contain similar elements that indicate a trend toward procurement of larger quantities of forest biomass (residuals) using specialized equipment. However, specialization must not mean that these harvesting systems can be used only for harvesting residual biomass; they must also enable the harvesting of larger logs. Combined harvesting of both logs and residues is thought to provide larger harvesting quantities at lower unit cost. An example of this situation is highlighted by a recent study on the integrated harvesting of pine stands [51]. It reviewed the potential amount of biomass produced at multiple sites under many different conditions. The results of the study compared residual biomass harvesting in clear-cut and thinning operations. They showed that when clearcutting, the amounts of biomass collected increased significantly with the use of a biomass chipper; however, during thinning operations, the use of a chipper was not as beneficial because it reduced the supply of pulpwood (lower-diameter logs) intended for pulp and paper mills.

The specific combination of equipment used for harvesting logs, pulpwood, residuals, or combinations of these is important to help reduce overall procurement costs, especially for residual harvesting. Integration of log- and biomass-harvesting equipment requires knowledge not only of the process but also of the equipment to reduce "bottlenecks" within the supply chain. Maximizing equipment productivity while at the same time minimizing the amount of equipment used is a key aspect of any forestry harvesting operation. Moreover, special attention to both processing and transportation of biomass will help minimize overall procurement costs.

15.3.1.2 Biomass Procurement and Study of Supply Chains

Supply-chain management (SCM) (which has been studied by many authors) can be viewed as the complete set of activities, resources, and information needed to plan, source, manufacture, store, sell, and deliver products to customers [52]. Supply chains are typically large, complex, and highly dynamic and involve a constant flow of information, products, and funds between different stages, with the overall objective of delivering the right product at the right time at the lowest cost. By accomplishing this, a well-functioning supply chain manages to achieve many goals which benefit all parties involved [53]:

- Maximize overall profitability (value preservation)
- Achieve high customer satisfaction levels (value growth)
- Improve product quality
- Reduce total delivered cost (procurement, manufacturing, distribution, inventory levels, and holdings costs)
- Create sustainable competitive advantage

Studies of biomass procurement and SCM within the current forestry industry have been carried out by many authors. Typically, these studies focus on the optimal integration of all processing, harvesting, and transportation operations by creating mathematical models which can then be optimized using specific modeling programs.

Weintraub and Epstein [54] reviewed the Chilean forest industry and the flow of material within its supply chain. They evaluated each activity within the supply chain, explaining in detail the complexities and interactions of each operation, from logging operations in the forest to the end of the process, depending on the products and their destination (pulp and paper mill, sawmills, plywood plants, ports, wholesalers, etc.). Next, they presented a series of mathematical models which had been successfully integrated into forestry operations to optimize various operations and the interactions between them. Their study serves as a very clear example of the different applications of SCM within the forest industry.

Jenkins and Sutherland [55] studied a U.S.-based wood procurement supply-chain system. They offered insights into the logistical (preprocessing, transportation, and storage) systems needed before and upon the arrival of forest biomass resources at the mill gate. They identified key issues (biomass composition, economical transport and storage, and processing-facility requirements) which must be addressed to ensure a viable and sustainable system of supply. They also identified, for each operation in the supply chain, certain properties which need to be carefully characterized, such as bulk density and variability of physical and chemical characteristics, because these typically impact decision-making within the supply chain.

Pulkki [56] studied how information on low-level wood characteristics (relative density, size, form, moisture content, chemical composition, etc.) is used and how procurement operations affect them to achieve a better understanding of their impact on final pulp and paper products. He explained the need to harvest more effectively by reducing biomass losses throughout the supply chain, and the importance of linking biomass characteristics with process requirements and then using these to determine the best harvesting methods to maintain these properties. In another study, he outlined the use of SCM for reviewing the overall procurement supply chain on multiple decision levels and how optimization of operations and processes can contribute to the overall goal (or at least can avoid putting the overall objectives of the organization at risk and limiting future opportunities) [57].

Using data from a Swedish case study, Gunnarsson et al. [58] studied the procurement of forest biomass to satisfy demand at heating plants. Their modeling and optimization formulation was structured as a mixed-integer linear programming (MILP) model and included the conversion (chipping) of forest biomass into forest fuel chips. Not surprisingly, their model indicated that transportation costs constitute an essential part of total costs. More interesting is their observation that, contrary to usual practice, the optimal solution often included the use of mobile chippers to generate chip flows and direct transportation to heating plants (as opposed to sending to storage first).

Another study by Carlsson and Ronnqvist [59] focused on the overall pulp and paper supply chain and its optimization. In their review, they went through five case studies on operations management at Södra Cell AB. The main purpose of the paper was to show how SCM, optimization models, and methods are applied in practice in the forest industry. Conclusions from their work included the need to conduct efficient and integrated planning for all aspects of the supply chain and to involve all parties within the supply chain to determine the best *"win-win"* scenario. They also determined that improvements to the supply chain must be made on a continual basis. They also concluded that the use of new technologies and advanced planning tools enables support of services and decisions within the organization, and that these new models and tools have the advantage that they enable testing of different scenarios within the same supply chain.

An important factor identified by Carlsson and Ronnqvist identifies a major problem in the practical use of models: although the necessary data for modeling exist within the company, obtaining them from each individual system in each machine and transferring them to a supply-chain model can be troublesome. Real-time modeling could be very useful in this case because it would make it possible to feed results back into the supply chain in a timely fashion, which would enable rapid adjustments to changing conditions.

Frombo et al. [60] used SCM tools in an effort to establish planning and management strategies for procurement of woody biomass for bioenergy. They created an optimization model which was divided into three submodels to address different kinds of decision problems on all three SCM decision levels: strategic, tactical, and operational. The results they presented focused on the strategic planning level. Decision variables were plant capacity and biomass harvested from a specific site, while the objective function sought to minimize the costs of plant installation, maintenance costs, and biomass procurement costs. A similar study was done by Gan [61], who sought to minimize the total costs of both the feedstock procurement of logging residues and electricity production in the United States. The method used involved determining the optimal power-plant size and the derivation of supply curves for biomass.

A study on biomass procurement supply chains for international transportation was carried out by Hamelinck et al. [62] to determine the feasibility of procuring biomass for the European market from South American producers. The study developed a model to compare the possible bioenergy supply chains and to assess the influence of key parameters such as distance, timing, and scale on performance. Their conclusions revealed that even with long-distance transportation, procurement of biomass from South America resulted in a lower cost than procurement in Europe, but because of truck transportation costs, the number of gathering points was restricted to a predetermined value.

This section has reviewed the use of SCM and optimization models within the forestry sector. It has shown that these tools are now in widespread use. It has further pointed out that application of SCM to biomass procurement operations can achieve promising changes that will create added value at all levels of the organization. It has also shown the importance of understanding the components of material and procurement systems. Once all this information has been obtained, it is possible to produce overall SCM models that provide results that intuition alone could not achieve.

15.3.2 Agricultural Biomass

The integration and coordination of biomass-harvesting operations (gathering, processing, and transportation) may result in significant costs savings and increases in productivity. In agriculture, these practices have been used for some time now, although little interest was paid in earlier times to the use of residues on a large commercial scale. Management of agricultural-biomass supply chains has played an important role as long as farmers have needed to remove agricultural residues to plant their fields again. Harvesting operations in agriculture follow similar processes in general, but vary according to the farm end products.

Typically, for biomass harvesting in agriculture, emphasis is placed on grain crops because of their high productivity, great abundance throughout the world, and their production of large amounts of underutilized biomass in the form of stover, straw, and other residues. Specific studies on the growth and harvesting of each type of agricultural biomass have been carried out [63]. The Food and Agriculture Organization of the United

Nations (FAO) [64] has a very complete data base of information for each crop type regarding individual harvesting and postharvest processing.

A thorough literature review has been done by Sokhansanj et al. [65] on the process and the field machinery used in the collection of corn stover. Their review revealed a surprising lack of published information on the engineering aspects of residue-collection processes (stover in this case), with little or no information available on storage conditions, densification technologies (cubing, pelletizing), and the physical characterization of corn stover throughout the postharvest process. However, using the available literature, they showed that the typical collection sequence for corn stover results in collection efficiencies of less than 40%. The collection sequence studied in this research assumed grain collection on the first machinery pass and a second pass to collect stover by shredding, windrowing, and round baling. Other factors such as weather, season, and the practice of leaving stover in the field to avoid soil depletion were also shown to influence efficiency. The same author has carried out multiple studies, both case-specific and general, on collection of agricultural feedstocks, harvesting and transportation methods, and the costs of each process [28,66–69].

Other authors have reviewed agricultural harvesting techniques and machinery with the intent to improve current collection methods. Cundiff [70] studied the effects of baling biomass by simulating five separate baling systems to determine the optimal bale size and the conditions for achieving maximum biomass collection at minimum cost. Atchison and Hettenhaus [71] studied collection, processing, storage, and transportation options for corn stover. In their report, they analyzed multiple scenarios and determined that a one-pass collection system for grain and stover has some advantages over two-pass systems and at the same time reduces overall transportation costs to a potential biorefinery. Tatsiopoulos and Tolis [72] focused on the collection of cotton stalks, and optimized biomass amounts for a combined heat and power plant (CHP) and multiple smaller power plants.

One property of biomass that is of special interest when analyzing agricultural residue collection and transport is bulk density. It is defined as the mass per unit volume of the product [65] and is of interest in agriculture because residues tend to have very low bulk densities (40–80 kg/m^3) compared to other feedstocks such as wood chips (200–400 kg/m^3) or solid wood (400–700 kg/m^3 depending on species [73]). Mani et al. [74] carried out a study on the development of densification technologies. They investigated several agricultural feedstocks (wheat and barley straw, corn stover, and switchgrass) to determine the best equipment mechanism for maximum biomass compression. Sokhansanj and Turhollow [68] also carried out a study on biomass densification using cubing operations and estimated reductions in overall delivered biomass cost. These types of studies contributed to the development of biomass densification equipment such as the machines used today for pelletization and briquetting [75].

Sokhansanj et al. [28] studied the economics, energy inputs, and carbon emissions of biomass harvesting in the agricultural sector according to harvesting methods (process), equipment used, and processing technologies. Their research was intended to serve biomass enterprises which would fill the void between the biomass producers and the biorefinery. They focused on main operations, specifically on storage, transportation, and preprocessing of biomass to a form and format which would satisfy the quantity and quality needs of a biorefinery. Reviews of a number of current and potential flow charts were carried out, and results indicated that one of the most economical ways to harvest agricultural biomass was using a single-step harvest system in which material is collected and transported to the side of the road. In addition, they looked at various operations for increasing the bulk density and flow of biomass using different densification methods.

Although the focus of this study was the harvesting of switchgrass (an energy crop), they also reviewed collection methods for wheat straw and corn stover.

Another study on the economics of corn-stover collection for production of bioethanol was carried out by Petrolia [76]. This case study took into account historical county-specific yields, transportation distances, erosion constraints, machinery specifications, and transportation, storage, and densification costs within their model. Results from this analysis were then evaluated using Monte Carlo simulation to evaluate alternative costs under different assumptions on key parameters. Overall results showed that biomass costs per ton of corn stover did not increase much when ethanol production levels were increased. Enough corn stover was available in the base area to supply the required biomass as ethanol production was increased. They also showed that modification of key parameters such as crop yields, farmer participation rates, bale moisture content, and stover collection efficiency did change biomass cost, but within a specific range of prices (±$22 per metric ton according to the assumptions made). Such an analysis helps determine what parameters must be reviewed when designing a biomass procurement supply chain, because these parameters will influence procurement costs.

Kadam et al. [77] compared two harvesting systems for rice and straw, evaluating their applicability and identifying issues that could restrict their use, such as disease and pest control, timeliness of harvesting in a two-stage harvesting operation, soil nutrient recycling, and machinery usability. They determined that a possible scenario for recuperation of rice straw was to harvest the grain first and then to remove and bale straw from the field in a second pass.

The Prairie Agriculture Machinery Institute (PAMI) has published a series of reports on farming machinery and other related issues. Their reports provide meaningful insight into advances in technology and also compare harvesting systems such as harvesting of grain, then residue (a two-step harvesting process) versus whole-grain harvesting (a one-step harvesting process) among others studies [78–80].

In addition to studies on harvesting agricultural biomass and reviews of harvesting methods, other authors have focused on biomass procurement, SCM, and optimization approaches. Sokhansanj et al. [69,81] developed a model for the SCM of biomass. The integrated biomass supply analysis and logistics model (IBSAL) is a complete modeling tool for the determination of costs for harvesting agricultural residues and energy crops and transporting them to a mill. The IBSAL model enables users to specify regional parameters such as weather conditions, crop yields, harvesting-machinery details, harvesting times, and many other case-specific details, so that the tool can take all these factors into account when calculating biomass delivered costs. It also reports back data throughout the supply chain on energy use, GHG emissions, and many other parameters of interest. Many practical papers have been published [29,82] using the IBSAL model. A separate study conducted by Sokhansanj et al. [66] focused on identifying key areas of study to address biomass supply for a biorefinery. The three areas identified were moisture control, densification, and systems modeling.

Ekşioğlu et al. [24] developed a biorefinery model based on agricultural residue use in Mississippi. The study analyzed the logistical challenges in supplying biomass to a biorefinery and centered on the integration of decision levels (strategic, tactical, and operational). The authors then created a model which focused on determining the number, size, and location of biorefineries according to the resources available within a predetermined area. Thorsell et al. [83] studied the procurement of agricultural biomass to design a supply chain and to determine the best equipment usage pattern for harvesting of lignocellulosic biomass focusing on agricultural residues (although they do mention softwoods). They reviewed the economics of harvesting from a company's (contractor's) perspective to reduce equipment capital and operational costs. Strategies for best utilization of harvesting equipment were proposed, included expanding the time during which a company uses its equipment throughout the year by following the harvesting season all the way north (from Texas to Canada). This approach, however, does not seem feasible because transportation costs to a single biorefinery would make it uneconomical.

As is the case in forestry, cost estimates for agricultural residues are many, and their values vary over a wide range depending on a number of parameters: type of residue, harvesting system used, moisture content, transportation distance, and others. However, in agriculture (specifically in the harvesting of residues), bulk density becomes of even greater importance because of its very low values. During transportation of agricultural residues, most trucks will become volume-restricted long before they become weight-restricted. Therefore, densification operations are necessary to reduce biomass transportation costs. Costs for agricultural feedstocks typically fall into two categories: one for grains and one for bales of straw or stover. Grains are a common commodity and their prices change according to market demand. Residues, on the other hand, have no real market value, so estimating their cost is difficult. A report done by IMC [84] stated that the cost of agricultural residues can be separated into two parts: a fixed cost paid to the farmer, which will typically vary between \$10 and \$20 per metric ton, and a set of variable costs depending on harvesting methods, loading and unloading, storage, and transportation costs. Within this report, values of triticale straw were estimated to be approximately \$40–\$50 per metric ton of straw, but these values can increase quickly with greater transportation distances.

This section has taken a look at potential feedstocks for a biorefinery from the agricultural sector. Most authors agree that grain crops present the greatest potential as feedstocks for a biorefinery because they produce large amounts of lignocellulosic residues (stover and straws) which have little or no current use. Similar conclusions can be drawn to those reached in the forestry supply-chain review, that single-pass systems are more economical than multipass systems. Moreover, parameters such as moisture content and bulk density as well as issues of transportation distance and crop yields per hectare, all have a significant influence on the price of these residues.

15.3.3 Energy Crops

Energy crops have two forms: herbaceous crops and woody crops. In general terms, energy crops can be considered a hybrid between forest and agricultural crops. Effectively, they are nonfood crops which are grown and cultivated on agricultural and marginal lands in a short to medium time frame (3–10 years) [85]. Because of these two characteristics (nonfood crop and use of marginal land), they are less controversial than other feedstock sources. The costs of energy crops sold to mills or plants tend to approximate those of agricultural or woody biomass, depending on how they are harvested. Herbaceous crops are baled and sold as straw bales (\$40–\$50/metric ton), and woody crops are typically chipped and sold as wood chips or forestry residues (\$50–\$60/metric ton).

Harvesting technologies for energy crops have developed from a combination of forestry and agricultural techniques. Herbaceous crops, such as miscanthus and switchgrass, are typically treated (and harvested) in the same way as stover and straw [86]. Harvesters cut and/or chop the grasses, and then balers make either round or square bales, which are then transported off the fields. Smeets et al. [87], for example, review two different harvesting methods for miscanthus and switchgrass (baled and chopped) and found baling to be the less expensive method of harvesting, specifically because of the greater bulk density of bales compared to chopped biomass (0.14 metric ton/m^3 compared to 0.11 metric ton/m^3). Similar reports have been carried out for herbaceous crops by several authors [88,89], and while their findings reflect that good adaptation of agricultural machinery has been achieved, genetic engineering and breeding programs are developing taller, stronger crops which have become more difficult for the agricultural harvester to cut and chop.

Woody energy crops (also known as "short-rotation coppice") have required the development of specialized machinery. In the case of longer-rotation woody crops such as some species of poplar which grow for 7–10 years, forest-industry harvesters have been used successfully [89]. Common varieties of woody crops include willow and poplar. In principle, two different harvesting regimens can be applied:

- The crop can be cut and chipped in one operation on the field.
- The crop is cut, and the stems are dried and chipped at a later date.

Machinery for the first type of operation has been developed by the Claas company, which has equipped a forage harvester with a chipper [90]. This machine can be used to harvest up to 4-year-old poplar plantations with stems of up to 7 cm in diameter. Machinery for the second type of operations is typically adapted from forestry operations (feller-bundlers with shears or disk saws that can handle multiple small stems at once).

Harvesting of energy crops such as willow that have multiple stems in a single plant requires its own specialized machinery, which has typically been developed by combining agricultural harvesters with specialized heads with stronger cutting blades that can handle the thicker stems. Research trials carried out by Volk [91] and coworkers combined a Case-New Holland harvester (FR-series machine), which is a standard harvester/chipper for corn and hay harvesting, with a specially designed head with larger cutting blades. The resulting machine can cut down to one-quarter of an inch all the way around a stem of approximately one-and-one-quarter inches. These types of modifications have the added benefit that they can be used on multiple types of crops simply by changing the cutting head. Willow plantations have become a common energy crop in many parts of the world. Keoleian and Volk [85] studied energy crops, specifically willow, and reviewed the energetic, environmental, and economic performance of this crop for production and conversion to electricity. Although energy and environmental considerations were favorable for willow growth and harvesting, economic performance was identified as the limiting factor.

SCM strategies have also been used to evaluate energy crops. Nienow et al. [92] used a linear programming model to determine whether cofiring of woody biomass with coal was feasible, with the intention of minimizing variable costs, including ash disposal and material procurement costs. They determined that the proposed cofiring project was suitable, not only because of its economic returns from cost reductions in feedstock but also because it provided environmental returns in the form of carbon credits.

Sims and Venturi [93] also analyzed biomass procurement strategies for energy crops, specifically for short-rotation coppice eucalyptus for use as feedstock for conversion to bioenergy. The study focused on reduction of storage costs by changing the harvesting system from one short harvesting season of 8–10 weeks to continual harvesting throughout the year. Results demonstrated that harvesting over a short period of time was more expensive than a continual harvest throughout the year.

Many of the studies mentioned have evaluated biomass procurement of single feedstocks; others examined several feedstocks, but used only one at a time. Other reviews have combined the use of multiple feedstocks within the same model to determine the optimal amount of each feedstock to procure. One such study by Rentizelas et al. [94] determined that a multiple-biomass supply chain presents significant potential for cost reductions by enabling wider distribution of capital costs and reducing warehousing requirements, especially when the seasonality of biomass is taken into consideration. Investment variables include the size and location of the facility, as well as the biomass types and quantities to procure, using a demand-driven system. A similar study by the same authors [95] reviewed different storage scenarios to determine which was best suited for a multiple-biomass system. The most important result of the case study was that the lowest-cost storage method indeed corresponded to the most efficient system-wide solution. Moreover, the multiple-biomass approach seemed to be more advantageous when combined with relatively expensive storage methods to minimize biomass losses.

Searcy et al. [96] studied the logistical costs of moving three different feedstocks (stover, straw, and woodchips) using different types of transportation (truck, rail, ship, and pipeline). Then they compared the cost of moving feedstocks to the costs of moving products in an effort to determine whether a biorefinery should be set up closer to end users or to feedstock sources. They confirmed that the costs of transporting feedstocks are much higher than the costs of shipping products (in this study the products were ethanol and energy), and therefore the conclusion was that biorefineries should be located closer to feedstock sources than to end users.

The studies reviewed in this section on energy crops present results which are similar to both the agricultural and the forestry-crop studies. However, the significant contribution that these studies have made, which will be of importance for the design of a biorefinery feedstock supply chain, is to establish that energy crops combine the supply chains for both agricultural and forestry feedstocks. It is this combination of feedstocks that will ultimately provide a biorefinery with the best mix of biomass sources for its processes.

15.4 Conclusions

Designing and implementing a biorefinery is a complex and expensive process. In addition to the engineering challenges posed by developing advanced processing technologies, decision-makers must pay close attention to the quality, cost, and dependability of their feedstock sources. The main objective of this chapter has been to present key drivers and determinants found in the literature. It provides an introduction to the many concepts and themes that can be applied to biomass procurement.

To use incoming biomass fully in a biorefinery, the complex structures and differences for each source must be understood to establish which feedstocks are better suited for which biorefinery processes and products. *It is this matching of biorefinery process requirements and feedstock properties which will determine how a biorefinery will operate best with the feedstocks available.*

To harvest biomass crops successfully and deliver them economically to the mill gate, emphasis needs to be placed on harvesting methods, equipment combinations, and the overall methodologies used according to the biomass being harvested.

In forestry as well as in agricultural operations, interest has recently increased in the procurement of previously unused biomass such as harvesting residues, thinnings, and uncommercial species for use in others industries (bioenergy, for example). Considering the low economic value of these feedstocks, procurement supply-chain costs must be maintained as low as possible to deliver these feedstocks efficiently to current plants and future biorefineries.

Another point which has been very clearly demonstrated throughout this chapter is that harvesting integration is a major component of efficient harvesting operations. Coordination between individual units operating in the field and preprocessing sites is needed to attain a smooth-running operation without major delays (or halts in operation) due to a single piece of machinery. Such integrated operations are more and more common within the forestry and agricultural sectors, but for a biorefinery, it might also be necessary to integrate operations in both areas at the same time to lower procurement costs significantly.

Many authors have reviewed different areas of biomass procurement such as equipment, harvesting, collection, transportation, and storage methods as well as the overall study of biomass supply chains and optimization. For potential biorefineries, this type of study becomes essential because it enables efficient (and low-cost) delivery of biomass feedstocks according to process requirements.

The use of SCM and optimization models within the various forestry, agricultural, and energy-crop industries reveals the wide scope of the use of these tools. It highlights the fact that use of SCM in biomass procurement operations can yield promising changes that will create added value at all levels of the organization.

References

1. Ragauskas, A.J. et al. 2006. The path forward for biofuels and biomaterials. *Science*, 311, 484–490.
2. Chambost, V. et al. 2007. Forest biorefinery: Getting on with the job. *Pulp & Paper Canada*, 108, 19–22.
3. McKendry, P. 2002. Energy production from biomass (Part 1): Overview of biomass. *Bioresource Technology*, 83, 37–46.
4. Wright, L. et al. 2006. *Biomass Energy Data Book: Edition 1*. Oak Ridge National Laboratory, Tennessee.
5. Renner, R. 2010. Where's the money. *Canadian Biomass*, May/June, 29–30.
6. Carey, J. et al. 2007. Food vs. fuel. *Business Week*, February 5, 80–83.
7. Irwin, S.H. and Good, D.L. 2009. Market instability in a new era of corn, soybean, and wheat prices. *Choices* 24, 6–11.
8. Hallam, D. et al. 2009. *The State of Agricultural Commodity Markets: High Food Prices and the Food Crisis—Experiences and Lessons Learned*. Food and Agriculture Organization, United Nations.
9. Melendez, J. and Stuart, P. 2009. Out of the box: Carbon feedstocks for the forest biorefinery. *International Biorefinery Conference*, Syracuse, NY.
10. Chambost, V. et al. 2008. Guided tour: Implementing the forest biorefinery (FBR) at existing pulp and paper mills. *Pulp & Paper Canada*, 109, 19–27.
11. Perlack, R.D. et al. 2005. *Biomass as a Feedstock for a Bioenergy and Bioproducts Industry: The Technical Feasibility of a Billion-Ton Annual Supply*. Oak Ridge National Laboratory, Oak Ridge, Tennessee.
12. Bradley, D. *Canada Biomass-Bioenergy Report*. Climate Change Solutions, Ottawa, ON.
13. Wood, S.M. and Layzell, D.B. 2003. *A Canadian Biomass Inventory: Feedstocks for a Bio-Based Economy*. BIOCAP Canada Foundation, Queen's University, Kingston, ON.
14. *European Biomass Industry Association: Biomass Characteristics*, http://www.eubia.org/115.0.html.
15. Scurlock, J. 2001. *Bioenergy Feedstock Characteristics*. Bioenergy Feedstock Development Program, Oak Ridge National Laboratory, USA.

16. *The Biomass Economy*. http://www.nrel.gov/learning/re_biomass.html.
17. *Energy Efficiency and Renewable Energy Biomass Program: Feedstock Types*. http://www1.eere.energy.gov/biomass/printable_versions/feedstocks_types.html.
18. Chornet, E. 2009. Converting non-homogeneous biomass residues into alcohols. *8th World Congress of Chemical Engineering*, Montreal, Canada.
19. Vassilev, S.V. et al. 2010. An overview of the chemical composition of biomass. *Fuel*, 89, 913–933.
20. Retka, S. 2009. Food, paper waste target for N.C. ethanol production. *Biomass Magazine*, April 15. http://www.biomassmagazine.com/article-print.jsp?article_id=2568.
21. Stuart, P.R. 2009. Biorefinery 101: Designing the forest biorefinery. *International Biorefinery Conference*, Syracuse, NY.
22. Chambost, V. and Stuart, P.R. 2007. Selecting the most appropriate products for the forest biorefinery. *Industrial Biotechnology*, 3, 112–119.
23. Chambost, V. et al. 2007. Systematic methodology for identifying promising forest biorefinery products. *Pulp & Paper Canada*, 108, 30–35.
24. Ekşioğlu, S.D. et al. 2009. Analyzing the design and management of biomass-to-biorefinery supply chain. *Computers & Industrial Engineering*, 57, 1342–1352.
25. Bozell, J.J. 2008. Feedstocks for the future—Biorefinery production of chemicals from renewable carbon. *Clean*, 36, 641–647.
26. Cherubini, F. et al. 2009. Toward a common classification approach for biorefinery systems. *Biofuels, Bioproducts & Biorefining*, 3, 534–546.
27. Towers, M. et al. 2007. Biorefinery opportunities for the Canadian pulp and paper industry. *Pulp & Paper Canada*, 108, 26–29.
28. Sokhansanj, S. and Fenton, J. 2006. *Cost Benefit of Biomass Supply and Pre-Processing*. BIOCAP, British Columbia, Canada.
29. Stephan, J.D. 2008. *Biorefinery Feedstock Availability and Price Variability: Case Study of the Peace River Region, Alberta*. M.Sc. dissertation, Chemical and Biological Engineering, University of British Columbia, Vancouver, BC.
30. Slade, R. et al. 2009. The commercial performance of cellulosic ethanol supply-chains in Europe. *Biotechnology for Biofuels*, 2.
31. Dempster, P. et al. 2008. *Equipment Review: Final Report to State of California Department of Forestry and Fire Protection*. Department of Biological and Agricultural Engineering; University of California, Davis, CA, December.
32. Sessions, J. et al. 2007. *Harvesting Operations in the Tropics*. New York: Springer-Verlag.
33. Sims, R.E. 2003. Delivering the biomass to the conversion plant. In: *Bioenergy Options for a Cleaner Environment*, New Zealand: Elsevier.
34. Arnosti, D. et al. 2008. *Harvesting Fuel: Cutting Costs and Reducing Forest Fire Hazards through Biomass Harvest*. Institute for Agriculture and Trade Policy, Minnesota.
35. Dymond, C.C. et al. 2010. Future quantities and spatial distribution of harvesting residue and dead wood from natural disturbances in Canada. *Forest Ecology and Management*, 260, 181–192.
36. Mabee, W. and Sadler, J. Bioethanol from lignocellulosics: Status and perspectives in Canada. *Bioresource Technology*, 101, 4806–4813.
37. Buongiorno, J. and Gilless, J.K. 2003. *Decision Methods for Forest Resource Management*. Oxford, UK: Elsevier Science.
38. McCloy, B. 2005. *Estimated Production, Consumption and Surplus Mill Wood Residues in Canada—2004*. Forest Products Association of Canada & National Resources Canada, Ottawa, Ontario, National Report 2005.
39. Bradley, D. *Canada Report on Bioenergy 2010*. Climate Change Solutions, Ottawa, Ontario.
40. Demirbas, A. 2010. Biomass fractionation and valorization. In: *Biorefineries for Biomass Upgrading Facilities*, Turkey: Springer-Verlag.
41. Lambert, M.C. et al. 2005. Canadian national tree aboveground biomass equations. *Canadian Journal of Forest Resources*, 35, 1996–2018.

42. Gringas, J.F. 1995. Harvesting small trees and forest residues. *Biomass and Bioenergy*, 9, 153–160.
43. Cuchet, E. et al. 2004. Performance of a logging residue bundler in the temperate forests of France. *Biomass and Bioenergy*, 27, 31–39.
44. Connell, M.J. 2003. *Log Presentation: Log Damage Arising from Mechanical Harvesting or Processing*. Forest and Wood Products Research and Development Corporation, Kingston, ACT, Australia.
45. Burke, H. 2008. Talking harvester heads: A roundup of what's available. *Forestry Journal*, 1, 28–31.
46. Burke, H. 2010. Harvesting and processing heads. *Canadian Forest Industries*. Jan–Feb, 13–18.
47. Burke, H. 2006. Getting ahead 2006: Harvesting and processing head guide. *Canadian Forest Industries*, Jan–Feb, 15–29.
48. Burke, H. 2009. Head strong. *International Forest Industries*, Aug, 12–21.
49. Burke, H. 2010. Surgical thinning: Making margins look good. *International Forest Industries*, Dec–Jan, 26–30.
50. Tice, B. 2009. Show coverage: ELMIA 09—Part 2. *Canadian Forest Industries*. Aug, 5–10.
51. Baker, S.A. et al. 2010. Evaluation of integrated harvesting systems in pine stands of the southern United States. *Biomass and Bioenergy*, 34, 720–727.
52. Grossman, I.E. 2004. Challenges in the new millennium: Product discovery and design, enterprise and supply chain optimization, global life cycle assessment. *Computers and Chemical Engineering*, 29, 29–39.
53. Chopra, S. and Meindl, P. 2008. *Supply Chain Management: Strategy, Planning & Operation*, 3rd ed. New Jersey, USA: Prentice-Hall.
54. Weintraub, A. and Epstein, R. The supply model in the forest industry: Models and linkages. In: Geunes, J. (ed.), *Supply Chain Management: Models, Applications and Research Directions*. Dordrecht: Klumer Academic Press, pp. 343–362.
55. Jenkins, T. and Sutherland, J.W. 2008. An integrated supply system for forest biomass. In: Solomon, B. and Luzadis, V.A. (eds), *Renewable Energy from Forest Resources in the United States*. New York: Routledge, pp. 92–115.
56. Pulkki, R. 1991. *A Literature Synthesis on the Effects of Wood Quality in the Manufacture of Pulp and Paper*. Forest Engineering Research Institute of Canada, Pointe Claire, QC.
57. Pulkki, R. 2001. Role of supply chain management in the wise use of wood resources. *Southern African Forestry Journal*, 191, 89–95.
58. Gunnarsson, H. et al. 2004. Supply chain modelling of forest fuel. *European Journal of Operational Research*, 158, 103–123.
59. Carlsson, D. and Ronnqvist, M. 2005. Supply chain management in forestry—Case studies at Sodra Cell AB. *European Journal of Operational Research*, 163, 589–616.
60. Frombo, F. et al. 2009. Planning woody biomass logistics for energy production: A strategic decision model. *Biomass and Bioenergy*, 33, 372–383.
61. Gan, J. 2007. Supply of biomass, bioenergy, and carbon mitigation: Method and application. *Energy Policy*, 35, 6003–6009.
62. Hamelinck, C.N. et al. 2005. International bioenergy transport costs and energy balance. *Biomass and Bioenergy*, 29, 114–134.
63. Kim, S. and Dale, B.E. 2004. Global potential bioethanol production from wasted crops and crop residues. *Biomass and Bioenergy*, 26, 361–375.
64. Mejia, D., Lewis, B., and Bothe, C. (eds) 1999. *Compendium on Post-Harvest Operations*. Food and Agriculture Organization, United Nations.
65. Sokhansanj, S. et al. 2002. Engineering aspects of collecting corn stover for bioenergy. *Biomass and Bioenergy*, 23, 347–355.
66. Sokhansanj, S. et al. 2003. Collection and delivery of biomass for fuel and power production. *Agricultural Engineering International*, 5.
67. Sokhansanj, S. et al. 2010. Techno-economic analysis of using corn stover to supply heat and power to a corn ethanol plant: Part 1: Cost of feedstock supply logistic. *Biomass and Bioenergy*, 34, 75–81.

68. Sokhansanj, S. and Turhollow, A. 2004. Biomass densification: Cubing operations and costs for corn stover. *Applied Engineering in Agriculture*, 20, 495–499.
69. Sokhansanj, S. et al. 2008. Integrated biomass supply and logistics. *Resource*, 15–18.
70. Cundiff, J.S. 1996. Simulation of five large round bale harvesting systems for biomass. *Bioresource Technology*, 56, 77–82.
71. Atchison, J.E. and Hettenhaus, J.R. 2004. *Innovative Methods for Corn Stover Collecting, Handling, Storing and Transporting*. National Renewable Energy Laboratory, Colorado, USA.
72. Tatsiopoulos, I.P. and Tolis, A.J. 2003. Economic aspects of the cotton-stalk biomass logistics and comparison of supply chain methods. *Biomass and Bioenergy*, 24, 199–214.
73. Nemestothy, K.P. 2005. *Wood Fuels: Characteristics, Standards, Production Technologies*. Australian Energy Agency.
74. Mani, S. et al. 2004. Evaluation of compaction equations applied to four biomass species. *Canadian Biosystems Engineering*, 46, 355–361.
75. Bissen, D. 2009. *Biomass Densification Document of Evaluation*. Zachry Engineering Corporation, Minneapolis, USA.
76. Petrolia, D.R. 2008. The economics of harvesting and transporting corn stover for conversion to fuel ethanol: A case study for Minnesota. *Biomass and Bioenergy*, 32, 603–612.
77. Kadam, K.L. et al. 2000. Rice straw as a lignocellulosic resource: Collection, processing, transportation and environmental aspects. *Biomass and Bioenergy*, 18, 369–389.
78. Kadam, K.L. et al. 1995. *Forage Growers' Guide to Round Baling*. Prairie Agriculture Machinery Institute, Humbolt, Canada.
79. Kadam, K.L. et al. 1998. *Modeling and Comparing Whole Crop Harvesting Systems*. Prairie Agriculture Machinery Institute, Humbolt, Canada.
80. Kadam, K.L. et al. 2001. *Does Using Grain Precleaners Make Economic Sense*? Prairie Agriculture Machinery Institute, Humbolt, Canada.
81. Sokhansanj, S. et al. 2006. Development and implementation of integrated biomass supply analysis and logistics model (ISBAL). *Biomass and Bioenergy*, 30, 838–847.
82. Kumar, A. and Sokhansanj, S. 2007. Switchgrass delivery to a biorefinery using integrated biomass supply analysis and logistics (IBSAL) model. *Bioresource Technology*, 98, 1033–1044.
83. Thorsell, S. et al. 2004. Economics of a coordinated biorefinery feedstock harvest system: Lignocellulosic biomass harvest cost. *Biomass and Bioenergy*, 27, 327–337.
84. Thorsell, S. et al. 2008. *Canadian Triticale Biorefining Initiative: Producer Value Proposition*. Ian Murray & Company, Ltd., submitted to Alberta Agriculture & Rural Development.
85. Keoleian, G.A. and Volk, T.A. 2005. Renewable energy from willow biomass crops: Life cycle energy, environmental and economic performance. *Critical Review in Plant Sciences*, 24, 385–406.
86. Austin, A. 2010. Make way for *Miscanthus*. *Biomass Magazine*, 4, 32–39.
87. Smeets, E. et al. 2009. The economical and environmental performance of *Miscanthus* and switchgrass production and supply chains in a European setting. *Renewable and Sustainable Energy Reviews*, 13, 1230–1245.
88. Lewandowski, I. et al. 2000. *Miscanthus*: European experience with a novel energy crop. *Biomass and Bioenergy*, 19, 209–227.
89. Moller, R. et al. 2007. *Crop Platforms for Cell Wall Biorefining: Lignocellulose Feedstocks*. EPOBIO, UK.
90. Spinelli, R. et al. 2009. Using modified foragers to harvest short-rotation poplar plantations. *Biomass and Bioenergy*, 33, 817–821.
91. Volk, T. 2008. Willow harvesting trials at SUNY ESF. *Biomass Conference 2008*, Mexico, New York.
92. Nienow, S. et al. 2000. Assessing plantation biomass for co-firing with coal in northern Indiana: A linear programming approach. *Biomass and Bioenergy*, 18, 125–135.
93. Sims, R.E. and Venturi, P. 2004. All-year-round harvesting of short rotation coppice eucalyptus compared with the delivered costs of biomass from more conventional short season harvesting systems. *Biomass and Bioenergy*, 26, 27–37.

94. Rentizelas, A.A. et al. 2009. An optimization model for multi-biomass tri-generation energy supply. *Biomass and Bioenergy*, 33, 223–233.
95. Rentizelas, A.A. et al. 2009. Logistics issues of biomass: The storage problem and the multi-biomass supply chain. *Renewable and Sustainable Energy Reviews*, 13, 887–894.
96. Searcy, E. et al. 2007. The relative cost of biomass energy transport. *Applied Biochemistry and Biotechnology*, 136–140, 639–652.

16

Lignocellulosic Biomass: A Raw Material for the Future

Agustin Jaime Castro-Montoya and Arturo Jiménez-Gutiérrez

CONTENTS

16.1 Introduction

Until the nineteenth century, the world economy depended basically on biomass (any organic material, either from land or from water, that was available as a renewable resource and that could be used to produce energy, raw materials, or consumer goods) and minerals. The discovery of oil and the rapid technological development of its extraction and refining provided a cheap and plentiful source of energy and raw materials that displaced biomass in the twentieth century as a basis for economic development. This change also generated a high dependence on fossil resources that could face serious depletion within the next few decades. In addition, an oil-based economy has the following drawbacks:

1. *Uncertainty*: Predictions of oil prices and supply levels have a strong impact on the strategic planning of the economy of any modern nation.
2. *Nonsustainability*: It is difficult to establish a long-term economic model based on fossil energy.
3. *Climate change*: An economy based on fossil sources has produced undesirable levels of gas emissions, mainly CO_2, that have adversely affected the environment.

It is therefore clear that there are incentives to search for alternatives that lower the dependence on fossil fuels for the production of energy and useful raw materials and that can also reduce greenhouse gas emissions. This search takes us back to the original raw material that was displaced by oil. Biomass can come from vegetal or animal origin, and it may provide the only reliable source for the future production of organic raw materials, with recent studies indicating that the transition from fossil fuels to biomass could take place over a 20-year time period (Van Haveren et al., 2008).

16.2 Lignocellulosic Materials

Lignocellulosic biomass is the most abundant material on earth. It is produced at a rate of 170×10^9 tons/year (Bozell and Patel, 2006), of which only about 3% is harvested and used as food or in other applications (Clark and Deswarte, 2008). This type of biomass is basically made up of three components, cellulose (40–50%), hemicellulose (25–30%), and lignin (20–30%), and according to its origin it can be classified as agricultural residuals, agro-industrial residuals, hardwood, softwood, cellulosic residuals, or municipal solid waste and pastureland residuals.

16.2.1 Cellulose

Cellulose is the main component in the cellular walls of plants and consists of a long linear polymeric chain $(C_6H_{10}O_5)_n$, of glucose units $(C_6H_{12}O_6)$, as shown in Figure 16.1. Due to its linear structure, it forms a crystal structure, rigid and difficult to break.

16.2.2 Hemicellulose

Hemicellulose is the second important component in the cellular walls of plants and consists of a short branched polymer chain of hexoses (glucose, mannose, galactose, etc.) and pentose (xylose and arabinose), although its main monomer is xylose. Because it is made up of a branched chain, it does not form a crystalline structure, and therefore it is easy to break. Figure 16.2 shows a typical hemicellulose structure.

16.2.3 Lignin

Lignin constitutes the third important component of lignocellulosic biomass. It is a complex phenolic polymer with phenyl-propane units, joined by C–C and C–O–C bonds that form a tridimensional amorphous structure (Lee, 1997). It is a hydrophobic material, and its main function is to join the cellulose and hemicellulose components to the cellular walls of plants, which contributes to their protection by making their degradation more difficult. The structural formula for lignin is shown in Figure 16.3.

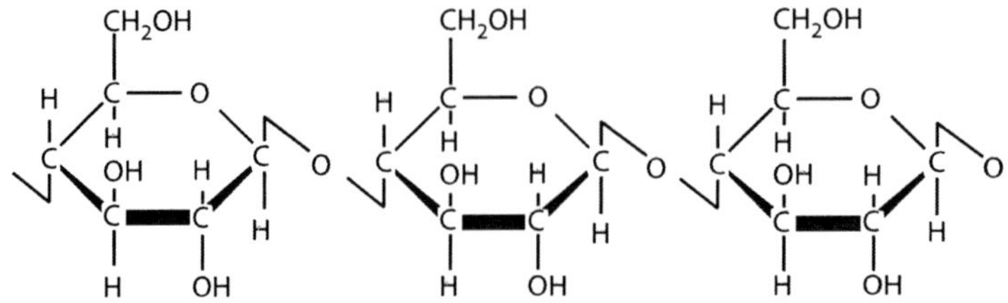

FIGURE 16.1
Cellulose structural formula.

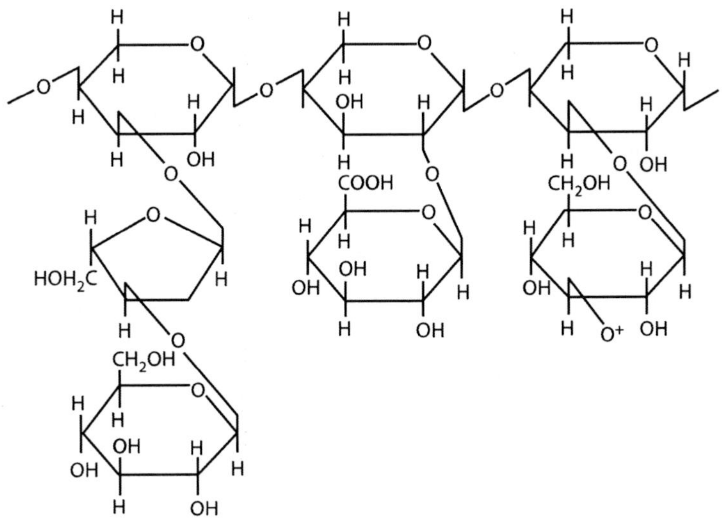

FIGURE 16.2
Hemicellulose structural formula.

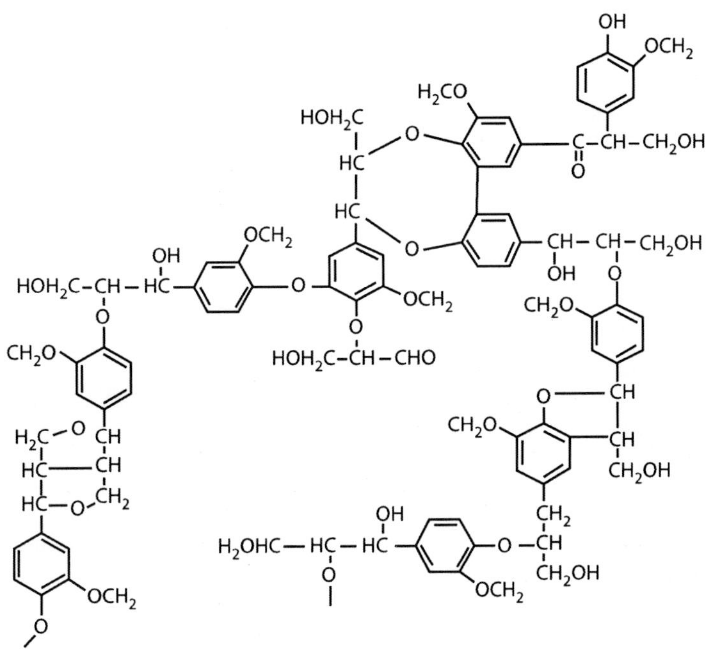

FIGURE 16.3
Lignin structural formula.

16.2.4 Chemical Composition

Because lignocellulosic biomass sources are diverse and variable, it is convenient to know their chemical composition to design possible production processes for this raw material. Table 16.1 gives the cellulose, hemicelluloses, and lignin composition of various biomass sources.

TABLE 16.1

Chemical Composition of Lignocellulosic Materials (% Weight on a Dry Basis)

Material (Variety)	Cellulose	Hemicellulose	Lignin
Hybrid poplar (DN-34)[a]	39.23	16.66	25.18
Eucalyptus (*Saligna*)[a]	48.07	12.69	26.91
Black locust (*Robinia pseudoacacia*)[a]	40.38	17.58	28.55
American sycamore (*Platanus occidentalis*)[a]	38.6	17.78	24.1
Sugarcane bagasse (*Gramineae saccharum* var. 65–7052)[a]	39.01	24.91	23.09
Corn stover (*Zea mays*)[a]	34.61	22.21	17.69
Sericea lespedeza (*Serala*)[a]	38.29	16.66	24.08
Switchgrass (Alamo)[a]	17.56	30.97	24.39
Big bluestem (Genotype, Greene County, AL)	33.41	23.74	19.35
Tall fescue (John Stone)[a]	24.83	18.81	12.63
Sweet sorghum (Cultivar M81E)[a]	22.48	13.81	11.34
Forage sorghum (sorghum–sudangrass hybrid FRR201)[a]	16.09	34.01	16.5
Eastern cottonwood (Stoneville #66)[a]	25.6	42.2	16.6
Wheat straw (Thunderbird)[a]	16.85	32.64	22.63
Nut shells[b]	25–30	25–30	30–40
Corn cobs[b]	45	35	15
Grasses[b]	25–40	35–50	10–30
Paper[b]	85–99	0	0–15
Sorted refuse[b]	60	20	20
Leaves[b]	15–20	80–85	0
Waste papers from chemical pulps[b]	60–70	10–20	5–10
Primary wastewater solids[b]	8–15	NA	24–29
Swine waste[b]	6	28	NA
Solid cattle manure[b]	1.6–4.7	1.4–3.3	2.7–5.7
Empty fruit bunches of oil palm[c]	15.47	11.73	7.14
Palm fress fiber[c]	24	14.4	12.60
Bagasse from agave (*Tequilana weber*)[d]	42	20	15

[a] U.S. Department of Energy, 2010.
[b] Sun and Cheng, 2002.
[c] Cardona et al., 2010.
[d] Saucedo-Luna et al., 2010.

Hardwoods have a high content of both cellulose and lignin, while in agricultural residuals, the lignin content is lower and the hemicellulosic content is higher. This observation illustrates how the choice of lignocellulosic material needs to be related to the products of interest.

16.3 Potential Bioproducts

In the United States alone, 100 million tons per year of organic and lubricant chemicals are produced, most of them from fossil sources (Carole et al., 2004). The dwindling supply of

TABLE 16.2

Intermediate Selection of Promising Candidates for Cellulose and Hemicellulose Reuse

Carbon Number	Potential Top 30 Candidates
1	Carbon monoxide and hydrogen (syngas)
2	None
3	Glycerol, 3-hydroxypropionic acid, lactic acid, malonic acid, propionic acid, serine
4	Acetoin, aspartic acid, fumaric acid, 3-hydroxybutyrolactone, malic acid, succinic acid, threonine
5	Arabinitol, furfural, glutamic acid, itaconic acid, levulinic acid, proline, xylitol, xylonic acid
6	Aconitic acid, citric acid, 2,5-furan dicarboxylic acid, glucaric acid, lysine, levoglucosan, sorbitol

fossil sources has directed attention once again to the production of some of these materials from renewable feedstocks such as lignocellulosic biomass. Because a large number of potential chemical products await consideration, the first step is to develop an initial screening procedure to detect the most promising compounds, from both economic and technological viewpoints.

The National Renewable Energy Laboratory (NREL), together with the Pacific Northwest National Laboratory (PNNL), have carried out a study to identify the most promising chemical products from lignocellulosic biomass (Werpy and Petersen, 2004). Starting from a database with 300 chemical compounds and considering the compatibility of substituting raw materials for petrochemical chains and their ease of synthesis, the study reported the 30 potential candidates given in Table 16.2. Note that there are no candidates with two carbon atoms, probably because bioethanol is already produced from first-generation biomass (containing natural sugars, starches, or both); however, it could be added to the list because it can also be produced from lignocellulosic materials. An additional screening of those materials can be performed if the list is restricted to the use of two out of the three components of lignocellulosic biomass, in this case cellulose and hemicellulose. The application of this criterion gives the following 12 products: Four carbon 1,4 diacids (succinic, fumaric, and malic), 2,5 furan dicarboxylic acid, 3-hydroxy propionic acid, aspartic acid, glucaric acid, glutamic acid, itaconic acid, levulinic acid, 3-hydroxybutyrolactone, glycerol, sorbitol, and xylitol. Another list of chemical products which might be produced from biomass has recently been reported (Clark and Deswarte, 2008), which in addition to the 12 products mentioned above, includes ethanol, acetic acid, oxalic acid, propanediol, acetone, butanol, butanediol, adipic acid, and citric acid.

As for lignin, the third component of lignocellulosic materials, NREL and PNNL together with the University of Tennessee have performed a similar study and have reported another list of potentially feasible bioproducts (Holladay et al., 2007). Lignin is a complex molecule that requires a greater technological effort for its reutilization than cellulose or hemicellulose, and the production opportunities were classified into three categories:

1. *Near future*: Fuels and synthetic gas, such as *n*-alcohols with one to four carbon atoms and dimethyl ether
2. *Medium-term future*: Macromolecules such as carbon fibers, modified polymers, resins, and adhesives
3. *Long-term future*: Aromatics and other chemicals, such as BTX (benzene–toluene–xylene), phenol, and quinones

16.4 Theoretical Yields

To carry out an estimation of the production potential of a bioproduct from lignocellulosic biomass so that possible production strategies can be considered, a simple and quick way to calculate theoretical yields can be developed by knowing the raw material chemical composition in terms of the percentage of cellulosic materials (cellulose and hemicellulose). The procedure is illustrated with a bioethanol case study.

16.4.1 Calculations for Bioethanol

In general, when based on a lignocellulosic biomass source, the production process for a bioproduct includes the step of breaking the polymeric chains (cellulose, $C_6H_{10}O_5$, and hemicellulose, $C_5H_8O_4$) toward fermentable sugars (glucose, $C_6H_{12}O_6$, and xylose, $C_5H_{10}O_5$, among others) followed by a fermentation stage. For the production of bioethanol, the following reactions take place:

1. Hydrolysis step

$$C_6H_{10}O_5 + H_2O \rightarrow C_6H_{12}O_6$$

$$C_5H_8O_4 + H_2O \rightarrow C_5H_{10}O_5$$

2. Fermentation step

$$C_6H_{12}O_6 \rightarrow 2\ C_2H_5OH + 2\ CO_2$$

$$3C_5H_{10}O_5 \rightarrow 5\ C_2H_5OH + 5\ CO_2$$

From the stoichiometry of the four reactions, from these two steps, and using a bioethanol density of 0.781 g/L, the following relations are obtained for the theoretical yield of bioethanol:

$$\frac{\text{liters ethanol}}{\text{ton raw material}} = 7.2715\ (\%\ \text{weight cellulose})$$

$$\frac{\text{liters ethanol}}{\text{ton raw material}} = 7.4367\ (\%\ \text{weight hemicellulose})$$

Starting with a lignocellulosic source that already contains natural sugars, the process would require only a fermentation step, in which case the yield would be

$$\frac{\text{liters ethanol}}{\text{ton raw material}} = 6.5443\ (\%\ \text{weight total sugar})$$

16.4.2 Theoretical Production Potential

Table 16.3 reports the theoretical values for the potential production of bioethanol, glycerol, xylitol, levulinic acid, and itaconic acid, chemicals that represent bioproducts of interest based on lignocellulosic materials. For these estimates, literature values for reaction stoichiometries were used, and equations similar to those shown above were developed for the ethanol case, taking as a basis the compositions given in Table 16.1. Although some reports indicate that both cellulosic components (cellulose and hemicellulose) can be used for the manufacture of bioproducts through fermentation, these calculations were based only on cellulose for glycerol, levulinic acid, and itaconic acid, and on hemicellulose for xylitol.

It can be observed in Table 16.3 how chemical composition affects the production potential of these chemical products. In general, hardwoods favor the production of cellulose-based compounds, whereas agricultural forage favors hemicellulose-based products. These values can be used as research targets for experimental programs to produce such chemicals from lignocellulosic materials.

TABLE 16.3

Theoretical Production Potential of Several Chemicals from Lignocellulosic Biomass Sources

Material (Variety)	Bioethanol (L/ton)	Glycerol (L/ton)	Xylitol (kg/ton)	Levulinic Acid (kg/ton)	Itaconic Acid (kg/ton)
Hybrid poplar (DN-34)	409.2	353.7	192.0	281.2	315.1
Eucalyptus (*Saligna*)	443.9	433.4	146.3	344.6	386.1
Black locust (*Robinia pseudoacacia*)	424.4	364.1	202.6	289.4	324.3
American sycamore (*Platanus occidentalis*)	412.9	348.0	204.9	276.7	310.0
Sugarcane bagasse (*Gramineae saccharum* var. 65–7052)	468.9	351.7	287.1	279.6	313.3
Corn stover (*Zea mays*)	416.8	312.0	256.0	248.1	278.0
Sericea lespedeza (*Serala*)	402.3	345.2	192.0	274.5	307.5
Switchgrass (*Alamo*)	358.0	158.3	357.0	125.9	141.0
Big bluestem (Genotype, Greene County, AL)	419.5	301.2	273.6	239.5	268.3
Tall fescue (John Stone)	320.4	223.9	216.8	178.0	199.4
Sweet sorghum (Cultivar M81E)	266.2	202.7	159.2	161.1	180.5
Forage sorghum (sorghum–sudangrass hybrid FRR201)	369.9	145.1	392.0	115.3	129.2
Eastern cottonwood (Stoneville #66)	500.0	230.8	486.4	183.5	205.6
Wheat straw (Thunderbird)	365.3	151.9	376.2	120.8	135.3
Nut shells	367.7	270.5	345.8	215.0	240.9
Corn cobs	587.5	405.7	403.4	322.6	361.4
Grasses	442.1	360.6	576.3	286.7	321.2
Paper	618.1	892.6	0.0	709.6	795.1
Sorted refuse	585.0	541.0	230.5	430.1	481.9
Leaves	704.0	135.2	922.1	107.5	120.5
Waste papers from chemical pulps	510.7	541.0	115.3	430.1	481.9
Primary wastewater solids	58.2	72.1	0.0	57.3	64.2
Swine waste	251.9	54.1	322.7	43.0	48.2
Solid cattle manure	22.0	14.4	16.1	11.5	12.8
Empty fruit bunches of oil palm	199.7	139.5	135.2	110.9	124.2
Palm fress fiber	281.6	216.4	166.0	172.0	192.7
Bagasse from agave (*Tequilana weber*)	454.1	378.7	230.5	301.1	337.3

16.5 Concluding Remarks

High prices for fossil fuels, together with the environmental implications of the extraction and processing steps for oil and natural gas, have spawned a new interest in renewable materials as potential sources of energy and chemicals. One such option is the use of lignocellulosic materials, which are composed mainly of cellulose, hemicellulose, and lignin. An overview of the potential for lignocellulosic materials as a source of chemicals has been presented. A list of the top candidates for industrial production has been included. Based on hydrolysis and fermentation steps, together with the chemical composition of several lignocellulosic materials, the theoretical production potential for a number of major chemical products has been calculated. These values can serve as targets for experimental research projects in the field. In addition, a systems analysis model of the lignocellulosic industry would be extremely useful to assess the impact of different development policies or economic scenarios on the production of selected chemical products.

References

Bozell, J.J. and Patel, M.K. 2006. *Feedstocks for the Future: Renewables for the Production of Chemicals and Materials*. American Chemical Society Symposium Series, United States of America.

Cardona, C.A., Sánchez, O.J., and Gutiérrez, L.F. 2010. *Process Synthesis for Fuel Ethanol Production*. CRC Press, Taylor & Francis Group, USA.

Carole, T.M., Pellegrino, J., and Paster, M.D. 2004. Opportunities in the industrial biobased products industry. *Applied Biochemistry and Biotechnology* 113–116: 871–885.

Clark, J.H. and Deswarte, F. 2008. *Introduction to Chemicals from Biomass*. Wiley, United Kingdom.

Holladay, J.E., White, J.F., Bozell, J.J., and Johnson, D. 2007. *Top Value-Added Chemicals from Biomass, Volume II—Results of Screening for Potential Candidates from Biorefinery Lignin*. U.S. Department of Energy. http://www1.eere.energy.gov/biomass/pdfs/pnnl-16983.pdf (accessed 05-22-2010).

Lee, J. 1997. Biological conversion of lignocellulosic biomass to ethanol. *Journal of Biotechnology* 56: 1–24.

Saucedo-Luna, J., Castro-Montoya, A.J., Rico, J.L., and Campos-García, J. 2010. Optimization of acid hydrolysis of bagasse from agave *Tequilana weber*. *Revista Mexicana de Ingeniería Química* 9: 91–97.

Sun, Y. and Cheng, J. 2002. Hydrolysis of lignocellulosic materials for ethanol production: A review. *Bioresource Technology* 83: 1–11.

U.S. Department of Energy, 2010. *Biomass Feedstock Composition and Property Database*. http://www.afdc.energy.gov/biomass/progs/search1.cgi (accessed 05-22-2010).

Van Haveren, J., Scott, E.L., and Sanders, J. 2008. Bulk chemicals from biomass. *Biofuels, Bioproducts, and Biorefining* 2(1): 41–57.

Werpy, T. and Petersen, G. 2004. *Top Value-Added Chemicals from Biomass, Volume I: Results of Screening for Potential Candidates from Sugars and Synthesis Gas*. U.S. Department of Energy, http://www1.eere.energy.gov/biomass/pdfs/35523.pdf (accessed 05-22-2010).

17

Food Waste as a Renewable Raw Material for the Development of Integrated Biorefineries: Current Status and Future Potential

Antonis C. Kokossis and Apostolis A. Koutinas

CONTENTS

17.1 Introduction

Potential feedstocks to future biorefineries include the food content of municipal waste and the industrial waste of the food processing industry. Potential products include energy and bioproducts (chemicals, materials, value-added ingredients) that add economic incentives to the apparent environmental benefits (reduction of fossil fuels and GHG emissions, waste disposal and treatment). Food waste is rich in organic materials (carbohydrates, proteins, oils and fats) and specialties (high-market-value products such as antioxidants). Organic components offer multiple paths to the production of chemicals but do require design studies to improve raw material and energy efficiencies. Compared to biomass-based biorefineries (e.g., biorefineries using wood chips and cereal straw), waste-based biorefineries are simpler and require a few stages (e.g., whey and molasses could be used directly as fermentation media).

The disposal of food waste is strongly associated with the potential growth of pathogens, high transportation costs (as their water content is high), the release of foul-smelling fatty acids (duc to the auto-oxidation of fats), strong odors produced as the proteins break down, and continuous substrate modifications induced by numerous embedded enzymes

(Russ and Meyer-Pittroff, 2004). Conventional disposal and treatment methods are often responsible for severe environmental burdens, as the disposal of organic waste in landfills contaminates water resources and produces greenhouse gasses (e.g., CH_4, CO_2) emitted to the atmosphere (Bingemer and Crutzen, 1987; Bogner and Matthews, 2003). Unable to properly differentiate waste, incineration, also emitting CO_2, still remains popular even though it produces energy at very low efficiencies. Bioremediation and composting produce simple products (composts), often outside the context of sustainable development as the markets for composts are limited and the actual need for useful products is instead sidelined. The selection of processing routes and product portfolios should be the combined outcome of legislation, environmental drivers, and process economics. Most commonly though, the selection is dictated by simplicity and common practices favor waste as feedstocks for fertilizers (through composting) and animal feed (e.g., bran-rich by-products from wheat milling plants, spent grains). Most promising routes have to start with the extraction of end-products before any use of waste for bioenergy (e.g., H_2 or biogas) and fertilizers.

Sustainable processing of municipal food wastes recently moved up the political and social agenda worldwide. Sources of municipal food wastes are significantly diversified including (WRAP, 2009)

- Household food waste
- Food waste generated from various "catering" services (e.g., restaurants, hotels, hospitals, schools, catering companies)
- Food waste streams generated from professional environments (e.g., offices)
- Food products that have passed a date label (e.g., use by, or best before date) from retail and wholesale markets
- Bulk quantities of gone mouldy or rotten food (mainly from third-world countries)

The components of municipal food waste are similar to those in the food industry wastes. Their use as industrial feedstock is challenged as their composition and content vary notably and depend on their origin and the sorting procedures installed. Parameters determining their potential include their content base (e.g., bread, potatoes, meat), the underlying collection and sorting procedures, and the transportation means available. Major types of carbohydrate-rich municipal waste streams are employed for electricity, H_2, and biogas production. In contrast to food processing wastes, the use of diversified biorefinery technologies is less justified for municipal waste that, without better sorting systems, will have to rely on generic approaches. This chapter discusses the potential to valorize food waste holding parallel attention to the minimization of waste and the economic benefits in producing energy and chemicals. The argument is that food wastes are valuable and renewable industrial resources promising a wide range of potential end-products.

Steps to develop biorefineries processing food wastes would require to

1. Determine geographic locations studying activities per county and region
2. Address logistics associated with waste collection and waste transportation from industrial plants (industrial waste) and sorting facilities (municipal waste)
3. Quantify the amount of each type of food processing and municipal waste produced

4. Quantify the composition of major macromolecules (e.g., starch, protein) and minor constituents (e.g., antioxidants) in each type of food waste
5. Assess the most sustainable processing route to maximize end-products at minimum cost, further minimizing greenhouse gas emissions and the utilization of nonrenewable energy

The above steps require intensive interdisciplinary and multidisciplinary collaboration between biochemical engineers, food biotechnologists, food chemists, and systems engineers. Such collaboration could quickly translate lab work into process flowsheets, further producing LCA reports and optimizing objectives in relation to economics and logistics, the flexibility and the reliability of the operation, and the integration with other available renewable resources in the area.

17.2 Food Waste as Feedstock, Types of Food Waste

In 2008, the worldwide petroleum consumption was approximately 85×10^6 barrels per day or 12.24×10^6 tonnes per day if we take into consideration the volume of a barrel (42 US gallons is equivalent approximately to 0.16 m^3) and a crude oil density of 0.9 tonnes per m^3 (Anonymous, 2010b). In addition, the chemical industry consumes only 7% of the total oil refinery output for chemical production (Wittcoff and Reuben, 1996). The European food industry generates annually approximately 2.2×10^8 tonnes of wastes and by-products (Waldron, 2004). By simply comparing the amount of oil required for chemical production and the quantity of food wastes produced annually as shown in Table 17.1, it seems that food waste could be regarded, at least in terms of capacity, as a noteworthy raw material for chemical production. However, it should be stressed that in many cases the water content is very high and the reported capacities do not reflect precisely the potential of food wastes as a substitute for fossil raw materials. For instance, whey contains more than 90% water and therefore pretreatment is necessary so as to utilize major whey components (i.e., lactose and whey proteins). For this reason, a more appropriate way to assess the potential of food waste as industrial feedstock is to take into consideration their composition in major macromolecules (e.g., carbohydrates, protein, fats, and oils) and estimate the theoretical production quantities of major chemicals based on theoretical conversion yields through either microbial bioconversion or chemical technologies. Table 17.2 presents the composition of major food processing and municipal wastes and by-products.

The food wastes presented in Table 17.2 have diversified composition and therefore could be used for the production of different chemicals and ingredients. In addition, in the case that they are used as fermentation media, various combinations of these food waste streams could lead to the production of fermentation media with no requirements for supplementation of synthetic chemicals and commercial nutrient formulations. For instance, protein-rich oilseed cakes resulted after oil extraction from oilseeds from both biodiesel plants and edible oil production plants could be converted into a nutrient-rich supplement (Wang et al., 2010) that could be mixed with sugar-rich hydrolysates or crude glycerol resulting from biodiesel production plants to produce generic fermentation media. The nutrient content of such media could be adjusted according to the fermentation selected to produce a spectrum of metabolic products ranging from platform chemicals (e.g., succinic acid) and

TABLE 17.1

Production of Major Types of Food Wastes and By-Products

Group of Food Wastes	Total Quantity Produced[a] (10^6 Tonnes)
Food Processing Wastes and By-Products	
Milling, Starch, and Confectionary Industry	
Corn, wheat, and potato starch production	4[a]
Wheat milling by-products	1.1[b]
Sugar Production from Sugar Beet and Cane	
Sugar beet pulp	5[c]
Cane and beet molasses	51[a]
Dairy Industry	
Cheese whey	145[d]
Alcoholic Beverage Production	
Spent grains from breweries	3.5[c]
White wine production	1.8[a]
Red wine production	2[a]
Oilseeds Processing Industry	
Protein-rich oilseed meals and cakes	101.5[e]
Olive Oil Production	
Olive mill wastewater	2[f]
Olive cake	0.4[g]
Nonalcoholic Beverage Production	
Orange and other citrus fruits	15.6[h]
Apple processing	1.9–4.2[h,i]
Cooking banana	9[h]
Sea Food Production	
Total fish waste (including fish farming)	18–30[c]
Fish canning, filleting, salting, and smoking	4.7[a]
Crustaceans and mollusks processing	0.5[a]
Meat and Poultry Processing Industry	
Slaughterhouse waste (beef, pig, poultry)	15.35[a]
Municipal Food Waste	
Waste cooking oil	0.7–1[j]
Bakery products	0.9[k]
Dried food (e.g., rice, pasta, flour, wheat products)	0.16[k]

The quantities were taken from

[a] Mahro and Timm (2007) and refer to EU15.
[b] Koutinas et al. (2006) and refers to UK production.
[c] Waldron et al. (2004).
[d] Affertsholt and Nielsen (2003) and refers to worldwide production.
[e] Anonymous (2010a) and refers to worldwide production.
[f] Montemurro et al. (2004) and refers to worldwide production.
[g] El Asli and Qatibi (2009) and refers to worldwide production.
[h] Djilas et al. (2009) and refers to worldwide production.
[i] Bhushan et al. (2008) and refers to worldwide production.
[j] Kulkarni and Dalai (2006) and refers to EU production.
[k] WRAP (2009) and refers to UK production.

TABLE 17.2

Composition of Major Food Processing and Municipal Wastes and By-Products

Type of Food Waste	Water Content (%)	Carbohydrate Content (%)	Protein Content (%)	Oil/Fat Content (%)	Reference
Carbohydrate-Rich Wastes					
Molasses, beet	23	65.1	6.7	—	Greasham (1993)
Spent grains from breweries	80–83	9–11.6	3.2–4.6	1.5–2.4	Russ and Meyer-Pittroff (2004)
Whey	92.7	4.9	0.9	0.9	Waldron et al. (2004)
Apple pomace	3.9–10.8	48–62	2.9–5.7	1.2–3.9	Bhushan et al. (2008)
Two-phase olive pomace	60–70	18–20	Up to 1.5	2.5–3	Borja et al. (2005)
Protein-Rich Wastes					
Soybean meal	10	29.9	42	4	Greasham (1993)
Linseed meal	8	38	36	0.5	Greasham (1993)
Yeast from breweries	5	39.5	43	1.5	Greasham (1993)
Yeast, hydrolysate	5.5	—	52.5	—	Greasham (1993)
Corn steep liquor	50	5.8	24	1	Greasham (1993)
Dried distillers solubles	8	45	26	9	Greasham (1993)
Fish meal (anchovy), 65%	8	—	65	3.8	Greasham (1993)
Blood	86	—	12	0.3	Russ and Meyer-Pittroff (2004)
Peanut meal and hulls	9.5	23	45	5	Greasham (1993)
Fat- or Oil-Rich Wastes					
Slaughterhouse waste	74	—	9	14	Russ and Meyer-Pittroff (2004)

biodegradable polymers (e.g., polyhydroxyalkanoates) to biofuels (e.g., bioethanol). Similar work has been carried out in the case of wheat which nowadays cannot be considered as a raw material for sustainable production of chemicals (Webb et al., 2004; Koutinas et al., 2004b; Arifeen et al., 2007; Du et al., 2007; Xu et al., 2010).

Classifying food waste is important but also a complex task as the food waste features notable differences in composition and physical characteristics both associated with its origin. Such differences imply different processing technologies available to select and different products worthy to produce. Based on origin, a classification of food waste is shown in Table 17.1. Industrial food waste may originate from milling, starch and confectionary industries, sugar production (sugar beet and sugarcane), dairy industries, alcoholic beverage production plants (e.g., breweries, wineries, distilleries), oilseeds-based industries for the production of edible oils, olive oil production plants, sea food production plants, fruit processing plants (e.g., nonalcoholic juices), and meat and poultry industry. In the case of municipal food waste, the classification is strongly related to the collection and the sorting technologies available. Efficient collection and/or sorting technologies could produce bulk quantities of municipal food waste, justifying viable biorefinery feedstocks and gradually diverting waste from the landfills. The challenge links to social incentives for the local authorities and commercial incentives for all parties involved. In reference to Table 17.1, for instance, as municipal waste could yield large quantities of waste cooking oil, several private companies started getting profitable and operational by collecting and distributing waste oil as raw material in biodiesel plants. To foster similar paradigms, coordinated action is required to build reliable feedstocks of desired specification, each time considering diversified resources available with different capabilities for each region.

Quite certainly, in every food industry, market, household, or catering service, a significant amount of raw material is disposed as waste. However, to build reliable feedstocks, one requires collective knowledge from the waste streams, and food industries are reluctant to share data with individuals (Mahro and Timm, 2007). National and local authorities assume a pivotal role as shown in the case of the United Kingdom and its NISP program (www.nisp.org.uk) where a national program is installed with prospects to fully valorize waste streams.

17.3 Chemical Production

Using established and emerging technologies, food industry and municipal wastes are already deployed in the production of bioenergy. Applications include the production of electricity in microbial fuel cells, the production of synthesis gas and biogas, H_2, and the production of liquid fuels such as biodiesel, bioethanol, and biobutanol (Qureshi and Maddox, 1995; Georgieva and Ahring, 2007; Rao et al., 2007; Digman and Kim, 2008; El-Mashad and Zhang R, 2010; Cercado-Quezada et al., 2010). Further to the production of energy, an exciting development remains the potential production of chemicals from waste. The section discusses this important aspect explaining the lists of products available to produce.

17.3.1 Utilization of Carbohydrate-Rich Food Wastes

Most food wastes are rich in carbohydrates, such as monosaccharides (i.e., glucose), disaccharides (i.e., sucrose, lactose), and polysaccharides (i.e., starch, cellulose, hemicellulose) that can be used as carbon sources for fermentation processes. Many reviews report long lists of potential chemicals that one could, biologically or chemically, produce from sugars contained in various sources of biomass (Werpy and Petersen, 2004; Patel et al., 2006; Koutinas et al., 2007a; Koutinas et al., 2008). The list is usually longer than what is practically possible to handle. Werpy and Petersen (2004) selected promising platform chemicals from an initial list of 300 chemicals possible to produce from sugars. Some chemicals, such as succinic acid, fumaric acid, malic acid, 3-hydroxypropionic acid, glutamic acid, itaconic acid, aspartic acid, xylitol and arabinitol, can be produced via the bioconversion of glucose, xylose, arabinose, lactose and/or sucrose. The conversion is possible using natural, mutated, or genetically modified microorganisms (Kim and Ruy, 1982; Cao et al., 1997; Yahiro et al., 1997; Zeikus et al., 1999; Tullo, 2005; Taing and Taing, 2007). Carbohydrate-rich food wastes could be used as renewable feedstocks for the microbial production of platform molecules. The rest of the platform chemicals screened by Werpy and Petersen (2004), including 2,5-furandicarboxylic acid, glucaric acid, levulinic acid, 3-hydroxybutyrolactone, and sorbitol, can be mainly produced through chemical conversion of biomass-derived saccharides. The platform molecules described above could be converted into various chemicals and polymers through bioconversion, chemical modification, or processing routes combining the previous two technologies.

Current research focuses on the microbial production of bio-based polymers, such as polylactic acid (PLA) and polyhydroxyalkanoates (PHAs), from sugar-based feedstocks (Datta and Henry, 2006; Verlinden et al., 2007). Sugars from biomass could be also used for the production of other polyesters using monomers produced entirely through microbial bioconversions, such as poly(butylene succinate) that can be produced through combination of 1,4-butanediol and succinic acid (Wolf et al., 2005). Bio-based polymers could be produced either through microbial production of bio-based monomers (e.g., lactic acid) which will be subsequently polymerized (e.g., PLA) or through direct microbial production of bio-based polymers (e.g., PHAs). Lactic acid, PHAs, and other bio-based monomers that can be used for polymer formulation have already been produced from food waste or could be produced via microbial fermentations using various monosaccharides and disaccharides that are available in most food wastes presented in Table 17.1 (Kim et al., 2003; Du et al., 2004; Datta and Henry, 2006; Solaiman et al., 2006; Verlinden et al., 2007; Koutinas et al., 2007b; Liu et al., 2010; Xu et al., 2010).

The current utilization of natural fats and oils either for food consumption or as raw material for biodiesel production creates a shortage of these raw materials for the chemical industry. In addition, in specific cases, the chemical industry may require "rare" fatty acids with industrially interesting properties that cannot be found in oilseeds. A solution to the above problem could be provided through the production of single-cell oil (or microbial oil) that does not compete with food and biodiesel production. Single-cell oil could also contain specific fatty acids in significant quantities. The fatty acids that can be produced via hydrolysis of microbial oil are of major commercial importance. The most important platform intermediates or end-products that could be derived from fatty acids generated from plant oils are fatty acid esters, soaps, fatty amines, surfactants (e.g., sorbitan ethoxylates, alkyl-polysaccharides), lubricants, dicarboxylic acids, resins, stabilizers, plasticizers, secondary alcohols, and polyols (Pollard, 2005; Meier, 2009; Carlsson, 2009). Similar or novel products could be produced from fatty acids contained in single-cell oil. Glycerol is also generated during hydrolysis of the triglyceride molecules in single-cell oil. Glycerol could be used as a versatile platform intermediate in the chemical industry (see Section 17.3.2).

Carbohydrate-rich food wastes and by-products could be employed as fermentation feedstocks for the production of single-cell oil using various oleaginous microorganisms (e.g., fungi and yeast). Oleaginous are designated the microorganisms that accumulate lipids, that is, triacylglycerol (TAG), at more than 20% of their total dry weight (Ratledge, 1991). There are yeast strains (e.g., *Rhodosporidium* sp., *Rhodotorula* sp., *Lipomyces* sp.) that accumulate intracellularly around 70% (w/w) of microbial oil (Li et al., 2007; Angerbauer et al., 2008; Meng et al., 2009). The fatty acid composition is dependent on the metabolism of the microorganism. For instance, many yeast strains produce lipids with similar fatty acid composition to common vegetable oils (e.g., palmitic acid, oleic acid, linoleic acid) (Meng et al., 2009). Various fungal strains (e.g., genera *Thamnidium, Mortierella, Mucor, Cunninghamella*) produce nutritionally important polyunsaturated fatty acids of very high added value such as arachidonic acid and γ-linolenic acid (Higashiyama et al., 2002; Ratledge and Wynn, 2002; Ward and Singh, 2005; Li et al., 2008; Fakas et al., 2009; Papanikolaou and Aggelis G, 2010). Single-cell oil could be used as raw material for the production of biodiesel, supplement in food and beverage markets (e.g., infant formulae, foods for pregnant and nursing women), and ingredient for cosmetic applications (Ratledge 1994; Ward and Singh, 2005; Wynn and Ratledge, 2006; Zhu et al., 2008; Vicente et al., 2009; Koutinas and Papanikolaou, 2011).

Apart from platform chemicals, bio-based polymers, and single-cell oil, carbohydrate-rich food wastes and by-products could be also employed as fermentation feedstocks for the production of current or future bulk and fine chemicals and food additives, such as potable ethanol, glutamic acid, citric acid, lactic acid, gluconic acid, xanthan gum, threonine, enzymes, vitamin B_{12} (Koutinas et al., 2004a).

17.3.2 Utilization of Oil-Rich Food Wastes

Some of the food wastes presented in Table 17.1 are rich in oils and fats. Waste cooking oil, animal fats, and oils extracted from fish processing wastes have all been used as feedstocks for the production of biofuels, for example, biodiesel through transesterification reactions (Chhetri et al., 2008; Lam et al., 2010; Lin and Lee, 2010; Wisniewski et al., 2010).

Plant oils are attractive raw materials for the chemical industry as they could yield fatty acids of purities appropriate to synthesize pure compounds (Metzger and Bornscheuer, 2006). As in the case of plant and microbial oils, oils extracted from food waste could be used as multipurpose or case-specific feedstocks. Multipurpose feedstocks denote oils (e.g., oils rich in oleic acid and linoleic acid) that contain common fatty acids, whereas as case-specific feedstocks denote oils containing unusual fatty acids with special properties gained from their unique molecular structure (Carlsson, 2009). As in the case of single-cell oil production from carbohydrate-rich food wastes, the oils extracted from food waste can be employed for the production of basic (e.g., free fatty acids, methyl esters, fatty alcohols, and fatty amines) or novel oleochemicals. In this way, oils and fats from food wastes could substitute plant oils in the production of oleochemicals, releasing plant oils for food consumption and the production of biodiesel. Oleochemicals could be used as environmentally friendly industrial fluids and lubricants, insulating fluid for electric utilities such as transformers and additive to asphalt (Metzger and Bornscheuer, 2006).

Oils and fats isolated from food waste could be hydrolyzed via application of subcritical water treatment to produce free fatty acids. Hydrolysis of the triglyceride molecules from oils and fats isolated from squid wastes at 240°C and 40 min reaction duration resulted in nine fatty acids, including omega-3 polyunsaturated fatty acids, eicosapentaenoic acid (EPA), and docosahexaenoic acid (DHA) (Tavakoli and Yoshida, 2006).

In the case of hydrolysis or transesterification of oils and fats isolated from food waste, crude glycerol will be produced as by-product. Worldwide research is currently focused on microbial or chemical conversion of crude glycerol into a wide range of chemicals such as 1,3-propanediol, polyhydroxybutyrate, succinic acid, 3-hydroxypropionaldehyde, single-cell oil, citric acid, propylene glycol, epichlorohydrin, dichloropropanol, and acrylic acid (Papanikolaou et al., 2000; Dubois et al., 2006; Tullo, 2007; Kraft et al., 2007; Maris and Davis, 2007; Koutinas et al., 2007a; Koutinas et al., 2008; Cavalheiro et al., 2009; Scholten et al., 2009; Papanikolaou and Aggelis, 2009). It should be stressed that glycerol has been listed by Werpy and Petersen (2004) among the most important platform molecules for the production of a wide range of chemicals and materials. Therefore, the production of glycerol during free fatty acid or biodiesel production from oils isolated from food waste could provide an additional renewable resource for chemical production.

17.3.3 Utilization of Protein-Rich Food Wastes

The protein-rich food wastes of Table 17.1 can be used in combination with carbohydrate-rich food waste to produce generic or specific fermentation media for industrial bioconversion. In this way, one could develop sustainable bioprocessing of low environmental impact that does not require supplementation of commercial nutrient formulation of inorganic or organic nature. For instance, in the case of polyhydroxybutyrate (PHB) production, the utilization of crude nutrient-rich hydrolysates would prevent the use of synthetic chemicals alone resulting in an average saving of 25,000 MJ, 600 kg oil eq. and 1750 kg CO_2 eq. per tonne of PHB produced. This concept may also improve process economics for many chemicals and polymers produced by fermentation.

This idea has been implemented successfully in the case of polyhydroxybutyrate production from wheat as the sole feedstock where the rich nutrient content of wheat was exploited so as to avoid the supplementation of expensive commercial nutrient supplements or inorganic chemicals (Xu et al., 2010). A similar process has been developed in the case of protein-rich rapeseed cakes that could be bioconverted into a nutrient-rich hydrolysate that can supplement crude glycerol from biodiesel production plants or sugar-rich hydrolysates of agro-industrial origin for the formulation of fermentation media (Wang et al., 2010). Several researchers have explored the potential of using crude protein hydrolysates of plant or food processing waste origin as fermentation supplements for the production of bio-based polymers, chemicals or even specialty products (Franek et al., 2000; Fitzpatrick and O'Keeffe, 2001; Ghorbel et al., 2005; Koller et al., 2005; Kwon et al., 2005; Farges-Haddani et al., 2006; Vazquez and Murado, 2008).

Apart from fermentation supplements, some protein-rich food wastes could be used for the production of hydrolysates with value-added applications. Whey protein hydrolysates could be used as protein source for individuals with reduced capacity of digestion, or with genetic metabolic disorders such as phenylketonuria patients (Pinto et al., 2009).

Protein-rich food wastes could be hydrolyzed via enzymatic hydrolysis that could be carried out using either commercial enzymes or crude extracts containing various enzymes from fermentations of appropriate microorganisms. In the case of protein-rich food wastes of plant origin, appropriate fungal strains (e.g., *Aspergillus oryzae*) could be cultivated in solid-state fermentations for the production of not only proteolytic enzymes but also others that could assist in the release of other nutrients (e.g., phosphorus) (Wang et al., 2010). In the case of protein-rich food wastes of animal or fish origin, commercial enzymes (such as papain and alcalase) could be employed for the production of fish protein hydrolysates (e.g., Vazquez and Murado, 2008; Pinto et al., 2009).

17.3.4 Extraction of Value-Added Constituents from Food Wastes

The utilization of food wastes for the production of only commodity products may not be economically viable. This problem could be by-passed in the case that centralized industrial plants that take advantage of "economies of scale" are developed. However, several reasons (e.g., the high water content, lignocellulosic content, or distributed availability of food wastes) hinder this option. Transportation of food wastes at centralized industrial plants should be avoided as much as possible because production costs will increase significantly. In addition, most food wastes are sensitive biological feedstocks that can be easily contaminated and, therefore, long-time storage of food wastes will also increase processing costs. For these reasons, we should develop on-site biorefining concepts that implement cascade processing routes combining extraction of high-value constituents with the fermentative or green chemical production of commodity products (e.g., biofuels and chemicals). For instance, Georgieva and Ahring (2007) suggested that bioethanol production from olive pulp waste streams generated from olive oil production plants will only be viable in the case that it is integrated with production of other value-added products. The viability of ethanol production from orange processing waste can be significantly improved via integrated processes that remove up to 90% of the limonene (with a market price of $1.1–$2.2 per kg of limonene) from the waste prior to further processing (Widmer and Grohmann, 2007).

Phytochemicals are value-added products that may be extracted from several food processing wastes (e.g., fruits processing plants, olive oil production, wineries, and oilseed processing) (Gonzalez-Paramas et al., 2004; Djilas et al., 2009). The food, pharmaceutical, and cosmetic industries are currently interested in identifying phytochemicals to substitute for synthetic substances because there are growing concerns about the potentially negative health effects of products containing synthetic chemicals. Phytochemicals contained in fruits and vegetables may have several health benefits such as reduced risk of coronary heart disease and stroke, as well as certain types of cancer (Rimm et al., 1996; Djilas et al., 2009). Fruits and vegetables contain carotenoids (e.g., β-carotene), flavonoids (e.g., catechins, flavonols, flavanones) and various phenolic compounds. Wastes from orange processing for juice production have been utilized for the extraction of flavanones (hesperidin and related compounds) and pectin (Waldron et al., 2004). Protein-rich oilseed cakes generated from oilseed processing plants are also rich in bioactive substances such as phenolic compounds and glucosinolates (Matthäus, 2002). Phytochemical-rich extracts from wastes of fruit and vegetable processing could be added to functional foods so as to increase their health-promoting properties.

Table 17.3 presents a list of high-value specialty products that may be extracted from various food processing wastes. The separation and purification of such high-value constituents could be enhanced in terms of both purity and environmental friendliness via novel green extraction processes such as extraction with supercritical solvents (e.g., CO_2) and novel membrane filtration unit operations (Novalin and Zweckmair, 2009; Temelli, 2009). Letisse and Comeau (2008) stressed that supercritical CO_2 extraction constitutes a good alternative to the use of organic solvents in classical methods for valorization of complex food industry waste as in the case of obtaining extracts enriched in EPA and DHA from fish cannery waste.

TABLE 17.3

Value-Added Constituents That Can be Extracted from Various Food Wastes

Types of Food Wastes	Value-Added Constituent	Extraction Method	Application	Reference
Winery by-products	Natural antioxidants (e.g., phenolics, flavonoids, resveratrol)	Membrane integrated with solvent extraction technologies. Supercritical CO_2 extraction	Food and medical applications due to anti-inflammatory activity, and anticarcinogenic and antimutagenic effects	Diaz-Reinoso et al. (2010); Casas et al. (2010); Gonzalez-Paramas et al. (2004)
	Grape seed oil	Supercritical extraction	High-value food product and preferred cosmetic ingredient	Passos et al. (2010); Fiori (2010)
Cheese whey	Whey protein concentrates and isolates	Membrane processing, including microfiltration and ultra/diafiltration, followed by spray or fluidized-bed drying	Protein of choice for body builders, elite athletes, and those whose health is compromised. Whey-based beverages and cheeses. Preparation of encapsulant or co-encapsulant materials for the protection and delivery of sensitive functional and bioactive components	Smithers (2008)
Fish processing waste	Extracts enriched in arachidonic acid, EPA, and DHA	Supercritical fluid (e.g., CO_2) extraction	Food, cosmetic, and pharmaceutical applications	Letisse et al. (2006); Sahena et al. (2009)
Tomato processing waste	Lycopene and β-carotene	Supercritical CO_2 extraction	Food supplements, sunscreens, and cosmetic applications	Sabio et al. (2003)
Wheat milling by-products	Arabinoxylan-rich extract	H_2O_2-based alkali extraction process	Bread manufacture, animal feeding, and medical applications	Hollmann and Lindhauer (2005)

TABLE 17.3 (continued)

Value-Added Constituents That Can be Extracted from Various Food Wastes

Types of Food Wastes	Value-Added Constituent	Extraction Method	Application	Reference
Orange peel wastes	Essential oils (e.g., D-lemonene)	Cold pressing, steam distillation or solvent extraction, usually with hexane or carbon dioxide	As flavoring agent in foods and medicines. As ingredient in the chemical industry, cosmetics, and domestic household products	Lopez et al. (2010)
	Pectin	Chemical- or enzyme-based extraction methodologies	As gelling agent in the food industry, stabilizer in fruit juices, and additive in milk drinks	
Apple pomace	Natural antioxidants (e.g., polyphenolic compounds)	Adsorption in Amberlite XAD 16 HP resin	Food, pharmaceutical, and cosmetic applications	Bhushan et al. (2008)
Olive mill wastewaters	Natural antioxidants (e.g., polyphenolic compounds)	All compounds can be recovered by physicochemical processes such as ultrafiltration, nanofiltration, and reverse osmosis.	Food processing, food products conservation, cosmetic, and pharmaceuticals	Federici et al. (2009)
Rice processing by-products	Oil enriched in antioxidants (e.g., tocochromanols and oryzanols)	Supercritical CO_2 extraction	Food applications	Perretti et al. (2003)

17.4 Biorefinery Concepts from Major Food Processing Wastes and By-Products

The biorefinery concept supports an integrated production of energy and chemicals using the paradigm of oil-based refineries. As every drop of the organic matter in the fossil oil is fully exploited, biorefineries would be viable and sustainable industries provided their organic matter is fully used. Table 17.4 illustrates the potential of specific food wastes as fermentation feedstocks that could be employed to sustain the current worldwide production capacities of ethanol, 1,3-propanediol, lactic acid, and succinic acid. Table 17.4 presents the carbon source requirements (mainly glucose equivalent) for the production of selected chemicals directly produced via fermentation.

Table 17.4 shows that 0.08 million tonnes of 1,3-propanediol could be produced from approximately 1.48 million tonnes of waste cooking oil. According to Table 17.1, the annual EU production of waste cooking oil is approximately 0.7–1 million tonnes. This means that the crude glycerol generated during biodiesel production from waste cooking oil available in EU could provide up to 67.6% of the raw material required for 1,3-propanediol worldwide production. According to Table 17.4, 0.015 million tonnes of current succinic acid production could be produced by approximately 0.03 million tonnes of waste bread. Table 17.1 shows that the annual production of household wastes from bakery products and dried food was more than 1 million tonnes only in the United Kingdom. This shows that the chemical production platform can be incorporated in future biorefineries utilizing starch-based municipal and processing wastes.

TABLE 17.4

Carbon Source Requirements for the Fermentative Production of Ethanol, 1,3-Propanediol, Lactic Acid, and Succinic Acid

Chemical Product	Current and Future Industrial Applications	Worldwide Production[a] (10^6 Tonnes)	Production Yield[a,b] (kg/kg Glucose)	Quantity of Glucose Required (10^6 Tonnes)	Quantity of Food Waste Required
Ethanol	For industrial use as solvent, disinfectant, preserving agent, and building block for chemical synthesis	31[c]	0.46	67,39	61.26 million tonnes of starch[d] that is contained in various food wastes
1,3-Propanediol	Polymeric applications (e.g., polytrimethylene terephthalate), malonic acid, polyurethans, copolyester ethers	0.08	0.54[e]	0.148	148 million kg of crude glycerol or 1480 million kg of waste cooking oil
Lactic acid	Additive in foods and beverages, industrial applications, PLA production, pharmaceuticals, and personal care products	0.15	0.95[f]	0.158	318.9 million kg of waste bread (45% starch content on average)[d]
Succinic acid	Platform molecule for the production of 1,4-butanediol, γ-butyrolactone, tetrahydrofuran, pyrrolidinones, polybutylene succinate, succinic esters, polyamides, etc.	0.015	1.01	0.015	30 million kg of waste bread (45% starch content on average)[d]

[a] Taken from Patel et al. (2006).
[b] 90% of theoretical conversion yield.
[c] Utilization of starch and sucrose as carbon sources.
[d] A starch to glucose conversion yield of 1.1 has been assumed.
[e] The carbon source is glycerol.
[f] A 95% conversion yield maybe feasible for lactic acid in the future.

The discussion underscores the strong motivation for considering food industry and municipal wastes as biorefinery feedstocks. Research studies could upgrade the installed food plants with the inclusion of new processing routes to utilize waste and by-product streams to yield products through nonfossil pathways. New processes should lead to the production of new products with sufficient value in the global market in order to offset the cost of waste disposal. Future processors should bear in mind the following aspects:

- Stabilization of waste streams to prevent microbial contamination
- Production of high-value products together with commodity products or application of "economies of scale" via the provision of adequate quantities of wastes
- Application of advanced research and development to improve process sustainability
- Minimization of waste generation and maximization of product formulation through the exploitation of the full potential of food wastes

The utilization of food processing and municipal wastes to replace petroleum as feedstock for chemical production may—in certain cases—lead to sustainable development through decentralization of chemical production. This concept has been initially introduced in the case of sustainable production of bioenergy, including biofuels, obtained from local feedstocks (e.g., agricultural and wood residues, municipal solid waste, vegetable oils, and algae residues) and adapted to the socioeconomical and environmental characteristics of the particular region where they are developed (Antizar-Ladislao and Turrion-Gomez, 2010). As in the case of bioenergy, decentralized chemical production from local food waste should be optimized following a biorefining approach where process economics and environmental benefits will be combined with local or regional societal needs.

In the following sections, three potential biorefining approaches are presented. The first approach is focused on the fractionation of components from specific types of food wastes that can be used as end-products. The second approach combines extraction with microbial bioconversion or "green" chemical conversion to produce both specialty and commodity products from specific types of food wastes. The third biorefining approach is more generic and can be applied to several types of food wastes.

17.4.1 Biorefinery Development Based on Fractionation of Components

In this case-specific processing scheme, product formulation is achieved via fractionation of components employing various combinations of physical (e.g., thermal, milling, extrusion), "green" chemical (e.g., supercritical CO_2 extraction) and enzymatic unit operations. These fractions can be used as end-products without any further biological or chemical conversion. Potential products are phytochemicals, polysaccharides, oligosaccharides, dietary fiber, protein, peptides, and oils. In this respect, orange peels could be fractionated into essential oils and pectin (Lopez et al., 2010).

Fish processing wastes could be regarded as a promising renewable resource for biorefineries as they contain proteins of high biological value, unsaturated essential fatty acids, vitamins and antioxidants, minerals or trace metals, and physiological beneficial amino acids and peptides (Waldron et al., 2004). It is important to stress that 18–30 million tonnes of fish wastes are dumped annually around the world. The utilization of high-value constituents as food for humans and as ingredients in foodstuff, health foods, pharmaceuticals, and cosmetics could increase the added value by 4–5-folds as compared to the current value achieved through conventional uses such as fishmeal, silage and animal feed (Waldron et al., 2004). Fractionation could be achieved via combinations of physical processing, supercritical fluid extraction, and enzymatic treatment.

The dairy industry has employed fractionation of whey components into three fractions enriched in its principle components, that is, proteins, lactose, and minerals (Waldron et al., 2004). The main technologies employed have been developed around crystallization, membrane, and chromatographic processes. However, the increasing volumes of these end-products have led to saturation of the respective markets, and novel processes are nowadays necessary to generate new products possibly through microbial bioconversion.

17.4.2 Biorefinery Development Based on Production of Specialties and Commodities

Another biorefinery type for case-specific food wastes combines extraction of high-value constituents with microbial bioconversion for the production of commodity products. Initial processing steps focus on the extraction of minor high-value constituents and the production of fermentation feedstocks with the desired composition of nutrients that will depend upon the subsequent microbial fermentation. To acquire the desired composition of nutrients, wastes from different food industries may be required. Downstream stages will purify the desired metabolic product which will be subsequently used as platform molecule, biodegradable polymer, bulk chemical, or fine chemical.

Hydrolysis of the macromolecules contained in food waste into fermentable monomers or oligomers could be achieved on-site. Nowadays, most bioethanol production industrial processes purchase commercial preparations of amylase, glucoamylase, cellulase, and protease to hydrolyze starch, cellulose, and protein contained in various renewable resources. Food processing and municipal wastes and by-products, especially the ones that contain relatively low water content and reasonably high carbohydrate content, have been employed in solid-state fermentations for the production of enzymes such as amylase, protease, and lipase. On-site production of hydrolytic enzymes from food wastes could be integrated in future biorefineries in order to reduce the costs currently required to purchase hydrolytic enzymes.

17.4.3 Biorefinery Development Based on a Generic Approach

This type of biorefining could be considered as generic and could be mainly employed in order to solve one of the main problems of food wastes associated with the high degree of heterogeneity that is usually encountered with food wastes. The heterogeneous and undefined nature of components within many food waste streams hinders the development of structured and defined processing routes for the valorization of food wastes in a sustainable manner. This generic approach could be applied to municipal food wastes that are collected in bulk quantities without separation of individual types of food wastes.

Renewable BioSystems LLC (RBL) has developed a process that can separate virtually any type of organic food waste (e.g., food processing and supermarket waste, offal from meat processing, DAF sludge, and fish residuals) into water, solids, and oil (Langhauser, 2009). The process initially macerates food wastes into a uniform slurry which is subsequently steam-heated in a kettle and then cooked in an inline cooker. Component separation is achieved by a highly specialized three-phase decanter centrifuge that separates the slurry into three streams, that is, oil, solid, and wastewater. This process could be applied in many food waste streams and such generic concepts may contribute toward the development of a sustainable and symbiotic relationship between food manufacturers and alternative fuel and chemical processors (Langhauser, 2009).

The oil could be converted into ASTM-certified biodiesel or combusted to generate electricity. The solid stream will mainly contain carbohydrates and protein that could be converted into fermentation media for microbial bioconversions via enzymatic hydrolysis.

17.5 Conclusion

The immense quantities of food waste generated annually could substitute for fossil raw materials in the production of a spectrum of commodities and specialties. Food waste is a complex biological entity that contains various organic components that could be either extracted as value-added ingredients or converted into various chemicals and biomaterials via microbial bioconversions. Three potential biorefining approaches have been presented based on fractionation of food waste, combination of component extraction with microbial bioconversion or "green" chemical conversion and a generic processing scheme that could be employed to various heterogeneous food wastes. Future research should focus on interdisciplinary and multidisciplinary collaboration between food biotechnologists, biochemical engineers, food chemists, and systems engineers so as to develop viable and sustainable biorefineries based on food waste.

References

Affertsholt T, Nielsen WK. December 2003. *Walk this whey. Dairy Industries International* 68(12):31.

Angerbauer C, Siebenhofer M, Mittelbach M, Guebitz GM. 2008. Conversion of sewage sludge into lipids by *Lipomyces starkeyi* for biodiesel production. *Bioresource Technology* 99:3051–3056.

Anonymous, June 2010a. Food Outlook Global Market Analysis. Food and Agriculture Organisation of the United Nations (FAO). http://www.fao.org/docrep/012/ak349e/ak349e00.pdf.

Anonymous, 2010b. International Energy Statistics. Total Consumption of Petroleum Products. US Energy Information Administration. Independent Statistics and Analysis. http://tonto.eia.doe.gov/cfapps/ipdbproject/IEDIndex3.cfm?tid=5&pid=54&aid=2.

Antizar-Ladislao B, Turrion-Gomez JL. 2010. Decentralized energy from waste systems. *Energies* 3:194–205.

Arifeen N, Wang R-H, Kookos IK, Webb C, Koutinas AA. 2007. Process design and optimization of novel wheat-based continuous bioethanol production system. *Biotechnology Progress* 23:1394–1403.

Bhushan S, Kalia K, Sharma M, Singh B, Ahuja PS. 2008. Processing of apple pomace for bioactive molecules. *Critical Reviews in Biotechnology* 28:285–296.

Bingemer HG, Crutzen PJ. 1987. The production of methane from solid wastes. *Journal of Geophysical Research* 92:2181–2187.

Bogner J, Matthews E. 2003. Global methane emissions from landfills: New methodology and annual estimates 1980–1996. *Global Biogeochemical Cycles* 17:34–41.

Borja R, Sánchez E, Rincón B, Raposo F, Martın MA, Martın A. 2005. Study and optimisation of the anaerobic acidogenic fermentation of two-phase olive pomace. *Process Biochemistry* 40:281–291.

Cao N, Du J, Chen C, Gong CS, Tsao GT. 1997. Production of fumaric acid by immobilized rhizopus using rotary biofilm contactor. *Applied Biochemistry and Biotechnology* 63–65:387–394.

Carlsson AS. 2009. Plant oils as feedstock alternatives to petroleum—A short survey of potential oil crop platforms. *Biochimie* 91:665–670.

Casas L, Mantell C, Rodriguez M, Ossa EJM, Roldan A, Ory ID, Caro I, Blandino A. 2010. Extraction of resveratrol from the pomace of Palomino fino grapes by supercritical carbon dioxide. *Journal of Food Engineering* 96:304–308.

Cavalheiro JMBT, de Almeida MCMD, Grandfils C, da Fonseca MMR. 2009. Poly(3-hydroxybutyrate) production by *Cupriavidus necator* using waste glycerol. *Process Biochemistry* 44:509–515.

Cercado-Quezada B, Delia M-L, Bergel A. 2010. Testing various food-industry wastes for electricity production in microbial fuel cell. *Bioresource Technology* 101:2748–2754.

Chhetri AB, Watts KC, Islam MR. 2008. Waste cooking oil as an alternate feedstock for biodiesel production. *Energies* 1:3–18.

Datta R, Henry M. 2006. Lactic acid: Recent advances in products, processes and technologies—A review. *Journal of Chemical Technology and Biotechnology* 81:1119–1129.

Diaz-Reinoso B, Gonzalez-Lopez N, Moure A, Dominguez H, Parajo JC. 2010. Recovery of antioxidants from industrial waste liquors using membranes and polymeric resins. *Journal of Food Engineering* 96:127–133.

Digman B, Kim D-S. 2008. Review: Alternative energy from food processing wastes. *Environmental Progress* 27:524–537.

Djilas S, Canadanovic-Brunet J, Cetkovic G. 2009. By-products of fruits processing as a source of phytochemicals. *Chemical Industry & Chemical Engineering Quarterly* 15:191–202.

Du C, Lin SKC, Koutinas A, Wang R-H, Webb C. 2007. Succinic acid production from wheat using a biorefining strategy. *Applied Microbiology and Biotechnology* 76:1263–1270.

Du G, Chen LXL, Yu J. 2004. High-efficiency production of bioplastics from biodegradable organic solids. *Journal of Polymers and the Environment* 12:89–94.

Dubois J-L, Duquenne C, Hoelderich W. 2006. *Method for Producing Acrylic Acid from Glycerol*. Patent publication number WO2006114506.

El Asli A, Qatibi AI. 2009. Ethanol production from olive cake biomass substrate. *Biotechnology and Bioprocess Engineering* 14:118–122.

El-Mashad HM, Zhang R. 2010. Biogas production from co-digestion of dairy manure and food waste. *Bioresource Technology* 101:4021–4028.

Fakas S, Papanikolaou S, Batsos A, Galiotou-Panayotou M, Mallouchos A, Aggelis G. 2009. Evaluating renewable carbon sources as substrates for single cell oil production by *Cunninghamella echinulata* and *Mortierella isabellina*. *Biomass and Bioenergy* 33:573–580.

Farges-Haddani B, Tessier B, Chenu S, Chevalot I, Harscoat C, Marc I, Goergen JL, Marc A. 2006. Peptide fractions of rapeseed hydrolysates as an alternative to animal proteins in CHO cell culture media. *Process Biochemistry* 41:2297–2304.

Federici F, Fava F, Kalogerakis N, Mantzavinos D. 2009. Valorisation of agro-industrial by-products, effluents and waste: Concept, opportunities and the case of olive mill wastewaters. *Journal of Chemical Technology and Biotechnology* 84:895–900.

Fiori L. 2010. Supercritical extraction of grape seed oil at industrial-scale: Plant and process design, modeling, economic feasibility. *Chemical Engineering and Processing: Process Intensification* 49:866–872.

Fitzpatrick JJ, O'Keeffe U. 2001. Influence of whey protein hydrolysate addition to whey permeate batch fermentations for producing lactic acid. *Process Biochemistry* 37:183–186.

Franek F, Hohenwarter O, Katinger H. 2000. Plant protein hydrolysates: Preparation of defined peptide fractions promoting growth and production in animal cells cultures. *Biotechnology Progress* 16:688–692.

Georgieva TI, Ahring BK. 2007. Potential of agroindustrial waste from olive oil industry for fuel ethanol production. *Biotechnology Journal* 2:1547–1555.

Ghorbel S, Souissi N, Triki-Ellouz Y, Dufosse L, Guerard F, Nasri M. 2005. Preparation and testing of *Sardinella* protein hydrolysates as nitrogen source for extracellular lipase production by *Rhizopus oryzae*. *World Journal of Microbiology and Biotechnology* 21:33–38.

Gonzalez-Paramas AM, Esteban-Ruano S, Santos-Buelga C, De Pascual-Teresa S, Rivas-Gonzalo JC. 2004. Flavanol content and antioxidant activity in winery byproducts. *Journal of Agricultural and Food Chemistry* 52:234–238.

Greasham RL. 1993. Media for microbial fermentations, Chapter 7, In: *Biotechnology—Bioprocessing*, Rehm H-J, Reed G, Puhler A, Stadler P, Stephanopoulos G (Editors) VCH, New York.

Higashiyama K, Fujikawa S, Park EY, Shimizu S. 2002. Production of arachidonic acid by *Mortierella* fungi. *Biotechnology and Bioprocess Engineering* 7:252–262.

Kim KI, Kim WK, Seo DK, Yoo IS, Kim EK, Yoon HH. 2003. Production of lactic acid from food wastes. *Applied Biochemistry and Biotechnology* 107:637–647.

Kim HS, Ruy DDY. 1982. Continuous glutamate production using an immobilized whole-cell system. *Biotechnology and Bioengineering* 24:2167–2174.

Koller M, Bona R, Braunegg G, Hermann C, Horvat P, Kroutil M, Martinz J, et al. 2005. Production of polyhydroxyalkanoates from agricultural waste and surplus materials. *Biomacromolecules* 6:561–565.

Koutinas AA, Du C, Wang RH, Webb C. 2008. Production of chemicals from biomass, In: Wiley series in renewable resources, Stevens CV (Series editor), *Introduction to Chemicals from Biomass*. Clark JH, Fabien EID (Editors) Wiley-VCH, Chichester, UK, pp. 77–101.

Koutinas AA, Malbranque F, Wang R-H, Campbell GM, Webb C. 2007b. Development of an oat-based biorefinery for the production of lactic acid by *Rhizopus oryzae* and various value-added co-products. *Journal of Agricultural and Food Chemistry* 55:1755–1761.

Koutinas AA, Papanikolaou S. 2011. Biodiesel production from microbial oil. In: *Handbook of Biofuels Production: Processes and Technologies*, Luque R, Campelo J, Clark J (Editors) Chapter 8, Woodhead Publishing Limited, Cambridge, UK, pp. 177–198.

Koutinas AA, Wang R-H, Campbell GM, Webb C. 2006. A whole crop biorefinery system: A closed system for the manufacture of non-food products from cereals. Chapter 8. In: *Biorefineries—Industrial Processes and Products*, Vol 1. Kamm B, Gruber PR, Kamm M (Editors) Wiley-VCH, Weinheim, Germany, pp. 165–191.

Koutinas AA, Wang R-H, Webb C. 2004a. Evaluation of wheat as generic feedstock for chemical production. *Industrial Crops and Products* 20:75–88.

Koutinas AA, Wang R-H, Webb C. 2004b. Restructuring upstream bioprocessing: Technological and economical aspects for the production of a generic microbial feedstock from wheat. *Biotechnology and Bioengineering* 85:524–538.

Koutinas AA, Wang R-H, Webb C. 2007a. The biochemurgist—Bioconversion of agricultural raw materials for chemical production. *Biofuels, Bioproducts and Biorefineries* 1:24–38.

Kraft P, Gilbeau P, Gosselin B, Claessens S. 2007. Process for producing dichloropropanol from glycerol, the glycerol coming eventually from the conversion of animal fats in the manufacture of biodiesel. Patent publication number EP1770081.

Kulkarni MG, Dalai AK. 2006. Waste cooking oils an economical source for biodiesel: A review. *Industrial and Engineering Chemistry Research* 45:2901–2913.

Kwon MS, Dojima T, Park EY. 2005. Use of plant-derived protein hydrolysates for enhancing growth of *Bombyx mori* (silkworm) insect cells in suspension culture. *Biotechnology and Applied Biochemistry* 42:1–7.

Lam MK, Lee MT, Mohamed AR. 2010. Homogeneous, heterogeneous and enzymatic catalysis for transesterification of high free fatty acid oil (waste cooking oil) to biodiesel: A review. *Biotechnology Advances* 28:500–518.

Langhauser K. January 2009. From waste stream to revenue stream. *Food Manufacturing*, 14.

Letisse M, Comeau L. 2008. Enrichment of eicosapentaenoic acid and docosahexaenoic acid from sardine by-products by supercritical fluid fractionation. *Journal of Separation Science* 31:1374–1380.

Letisse M, Rozieres M, Hiol A, Sergent M, Comeau L. 2006. Enrichment of EPA and DHA from sardine by supercritical fluid extraction without organic modifier. I. Optimization of extraction conditions. *Journal of Supercritical Fluids* 38:27–36.

Li N, Deng Z-N, Qin Y-L, Chen C-L, Liang Z-Q. 2008. Production of polyunsaturated fatty acids by *Mucor recurvus* sp. with sugarcane molasses as the carbon source. *Food Technology and Biotechnology* 46:73–79.

Li YH, Zhao ZB, Bai FW. 2007. High density cultivation of oleaginous yeast *Rhodosporidium toruloides* Y4 in fed-batch culture. *Enzyme Microbial Technology* 41:312–317.

Lin CY, Lee JC. 2010. Oxidative stability of biodiesel produced from the crude fish oil from the waste parts of marine fish. *Journal of Food Agriculture and Environment* 8:992–995.

Liu YY, Wang QH, Chen LW, Wang XQ, Wang J. 2010. Optimization of lactic acid production from food waste by the saccharification of *Bacillus subtili*. *Advanced Material Research* 113–114:1080–1083.

Lopez JAS, Qiang L, Thompson IP. 2010. Biorefinery of waste orange peel. *Critical Reviews in Biotechnology* 30:63–69.

Mahro B and Timm M. 2007. Potential of biowaste from the food industry as a biomass resource. *Engineering in Life Science* 7:457–468.

Maris EP, Davis RJ. 2007. Hydrogenolysis of glycerol over carbon-supported Ru and Pt catalysts. *Journal of Catalysis* 249:328–337.

Matthäus, B. 2002. Antioxidant activity of extracts isolated from residues of oilseeds, such as rapeseed or sunflower. *Agro Food Industry Hi-Tech* 13:22–25.

Meier MAR. 2009. Metathesis with oleochemicals: New approaches for the utilization of plant oils as renewable resources in polymer science. *Macromolecular Chemistry and Physics* 210:1073–1079.

Meng X, Yang J, Xu X, Zhang L, Nie Q, Xian M. 2009. Biodiesel production from oleaginous microorganisms. *Renewable Energy* 34:1–5.

Metzger JO, Bornscheuer U. 2006. Lipids as renewable resources: Current state of chemical and biotechnological conversion and diversification. *Applied Microbiology and Biotechnology* 71:13–22.

Montemurro F, Convertini G, Ferri D. 2004. Mill wastewater and olive pomace compost as amendments for rye-grass. *Agronomie* 24:481–486.

Novalin S, Zweckmair T. 2009. Renewable resources—green biorefinery: Separation of valuable substances from fluid-fractions by means of membrane technology. *Biofuels, Bioproducts and Biorefining* 3:20–27.

Papanikolaou S, Aggelis G. 2009. Biotechnological valorization of biodiesel derived glycerol waste through production of single cell oil and citric acid by *Yarrowia lipolytica*. *Lipid Technology* 21:83–87.

Papanikolaou S, Aggelis G. 2010. *Yarrowia lipolytica*: A model microorganism used for the production of tailor-made lipids. *European Journal of Lipid Science and Technology* 112:639–654.

Papanikolaou S, Ruiz-Sanchez P, Pariset B, Blanchard F, Fick M. 2000. High production of 1,3-propanediol from industrial glycerol by a newly isolated *Clostridium butyricum* strain. *Journal of Biotechnology* 77:191–208.

Passos CP, Silva RM, Da Silva FA, Coimbra MA, Silva CM. 2010. Supercritical fluid extraction of grape seed (Vitis vinifera L.) oil. Effect of the operating conditions upon oil composition and antioxidant capacity. *Chemical Engineering Journal* 160:634–640.

Patel M, Crank M, Dornburg V, Hermann B, Roes L, Husing B, Overbeek L et al. 2006. The BREW project report: Medium and long-term opportunities and risks of the biotechnological production of bulk chemicals from renewable resources. http://www.chem.uu.nl/brew/BREW_Final_Report_September_2006.pdf.

Perretti G, Miniati E, Montanari L, Fantozzi P. 2003. Improving the value of rice by-products by SFE. *Journal of Supercritical Fluids* 26:63–71.

Pinto GA, Giordano RLC, Giordano RC. 2009. Remote engineering for a cheese whey biorefinery: An Internet-based application for process design, economic analysis, monitoring, and control of multiple plant sites. *Bioprocess and Biosystems Engineering* 32:69–78.

Pollard G. 2005. Catalysis in renewable feedstocks—A technology roadmap. Report prepared on behalf of The Department of Trade and Industry, Report number CR7656. BHRSolutions Project No 180 2421. http://www.bhrgroup.co.uk/extras/renewcatfull.pdf.

Qureshi N, Maddox IS. 1995. Continuous production of acetone-butanol-ethanol using immobilized cells of *Clostridium acetobutylicum* and integration with product removal by liquid- liquid extraction. *Journal of Fermentation and Bioengineering* 80:185–189.

Rao K, Chaudhari V, Varanasi S, Kim DS. 2007. Enhanced ethanol fermentation of brewery wastewater using genetically modified strain *E. coli* KO11. *Applied Microbiology and Biotechnology* 74:50–60.

Ratledge C, Wynn JP. 2002. The biochemistry and molecular biology of lipid accumulation in oleaginous microorganisms. *Advances in Applied Microbiology* 51:1–51.

Ratledge C. 1991. Microorganisms for lipids. *Acta Biotechnologica* 11:429–438.

Ratledge C. 1994. Yeasts, moulds, algae and bacteria as sources of lipids. In: *Technological Advances in Improved and Alternative Sources of Lipids*. Kamel BS, Kakuda Y (Editors) Blackie Academic and Professional, New York, pp. 235–291.

Rimm EB, Katan MB, Ascherio A, Stampfer MJ, Willett WC. 1996. Relation between intake and risk of coronary heart disease in male health professionals. *Annals of Internal Medicine* 125:384–389.

Russ W and Meyer-Pittroff R. 2004. Utilizing waste products from the food production and processing industries. *Critical Reviews in Food Science and Nutrition* 44:57–62.

Sabio E, Lozano M, Montero de Espinosa V, Mendes RL, Pereira AP, Palavra AF, Coelho JA. 2003. Lycopene and β-carotene extraction from tomato processing waste using supercritical CO_2. *Industrial Engineering and Chemistry Research* 42:6641–6646.

Sahena F, Zaidul ISM, Jinap S, Saari N, Jahurul HA, Abbas KA, Norulaini NA. 2009. PUFAs in fish: Extraction, fractionation, importance in health. *Comprehensive Reviews in Food Science and Food Safety* 8:59–74.

Scholten E, Renz T, Thomas J. 2009. Continuous cultivation approach for fermentative succinic acid production from crude glycerol by *Basfia succiniciproducens* DD1. *Biotechnology Letters* 31:1947–1951.

Smithers GW. 2008. Whey and whey proteins—From "gutter-to-gold". *International Dairy Journal* 18:695–704.

Solaiman DKY, Ashby RD, Foglia TA, Marmer WN. 2006. Conversion of agricultural feedstock and coproducts into poly(hydroxyalkanoates). *Applied Microbiology and Biotechnology* 71:783–789.

Taing O, Taing K. 2007. Production of malic and succinic acids by sugar-tolerant yeast *Zygosaccharomyces rouxii*. *European Food Research and Technology* 224:343–347.

Tavakoli O, Yoshida H. 2006. Squid oil and fat production from squid wastes using subcritical water hydrolysis: Free fatty acids and transesterification. *Industrial & Engineering Chemistry Research* 45:5675–5680.

Temelli F. 2009. Perspectives on supercritical fluid processing of fats and oils. *Journal of Supercritical Fluids* 47:583–590.

Tullo A. 2005. Industrial biotech: Codexis, Cargill claim chemical breakthrough. *Chemical Engineering News* 83(26):11.

Tullo A. 2007. News of the week: Chemicals from renewables. *Chemical Engineering News* 85(19):14.

Vazquez JA, Murado MA. 2008. Enzymatic hydrolysates from food wastewater as a source of peptones for lactic acid bacteria productions. *Enzyme and Microbial Technology* 43:66–72.

Verlinden RAJ, Hill DJ, Kenward MA, Williams CD, Radecka I. 2007. Bacterial synthesis of biodegradable polyhydroxyalkanoates. *Journal of Applied Microbiology* 102:1437–1449.

Vicente G, Bautista LF, Rodriguez R, Gutierrez FJ, Sadaba I, Ruiz-Vazquez RM, Torres-Martínez S, Garreb V. 2009. Biodiesel production from biomass of an oleaginous fungus. *Biochemical Engineering Journal* 48:22–27.

Waldron K, Faulds C, Smith A. 2004. Total food: Exploiting co-products—minimizing waste. Total Food Proceedings, Institute of Food Research, Norwich Research Park, http://www.ifr.ac.uk/sfc/links/ProceedingsTotalFood2004.pdf.

Wang R-H, MD Shaarani S, Godoy LC, Melikoglu M, Vergara CS, Koutinas A, Webb C. 2010. Bioconversion of rapeseed meal for the production of a generic microbial feedstock. *Enzyme and Microbial Technology* 47:77–83.

Ward OP, Singh A. 2005. Omega-3/6 fatty acids: Alternative sources of production. *Process Biochemistry* 40:3627–3652.

Webb C, Koutinas AA, Wang R-H. 2004. Developing a sustainable bioprocessing strategy based on a generic feedstock. *Advances in Biochemical Engineering/Biotechnology* 87:196–268.

Werpy T, Petersen G. 2004. Top value added chemicals from biomass. Volume I—Results of screening for potential candidates from sugars and synthesis gas. Produced by staff at the Pacific Northwest National Laboratory (PNNL) and the National Renewable Energy Laboratory (NREL). http://www1.eere.energy.gov/biomass/pdfs/35523.pdf.

Widmer W, Grohmann K. 2007. Ethanol from orange processing waste. National Meeting of Institute of Food Technologists/Food Expo. Presentation No. 140–09, Agricultural Research Service, http://www.ars.usda.gov/research/publications/publications.htm?SEQ_NO_115=207114.

Wisniewski A, Wiggers VR, Simionatto EL, Meier HF, Barros AAC, Madureira LAS. 2010. Biofuels from waste fish oil pyrolysis: Chemical composition. *Fuel* 89:563–568.

Wittcoff HA, Reuben BG. 1996. *Industrial Organic Chemicals*, John Wiley & Sons, New York, pp.13–84.

Wolf O, Crank M, Patel M, Marscheider-Weidemann F, Schleich J, Husing B, Angerer G. 2005. Techno-economic feasibility of largescale production of bio-based polymers in Europe, Institute for Prospective Technological Studies, Technical Report Series, EUR 22103 EN.

WRAP, 2009, *Household Food and Drink Waste in the UK*. Report prepared by WRAP. Banbury.

Wynn JP, Ratledge C. 2006. Microbial production of oils and fats, In: Food Biotechnology, Sheetty K, Paliyath G, Pometto A, Levin R (Eds), Boca Raton, London, Taylor & Francis Group LLC, pp. 443–472.

Xu Y, Wang R-H, Koutinas AA, Webb C. 2010. Microbial biodegradable plastic production from a wheat-based biorefining strategy. *Process Biochemistry* 45:153–163.

Yahiro K, Shibata S, Jia S-R, Park Y, Okabe M. 1997. Efficient itaconic acid production from raw corn starch. *Journal of Fermentation and Bioengineering* 84:375–377.

Zeikus JG, Jain MK, Elankovan P. 1999. Biotechnology of succinic acid production and markets for derived industrial products. *Applied Microbiology and Biotechnology* 51:545–552.

Zhu LY, Zong MH, Wu H. 2008. Efficient lipid production with *Trichosporon fermentans* and its use for biodiesel preparation. *Bioresource Technology* 99:7881–7885.

Section V

Biorefinery Technologies

18

Forest Biorefinery: Current State of the Art

Tom Browne, Naceur Jemaa, Ibrahim Karidio, Talat Mahmood, Mike Paice, Mike Paleologou, Doug Singbeil, and Xiao Zhang

CONTENTS

18.1 Introduction

The forest biorefinery concept has attracted widespread interest recently as the forest sector looks for new products and processes to improve the returns on wood consumed and capital invested. Much work has been focused on kraft mills, but opportunities are also needed for mechanical pulp mills. In the last few years, a huge number of potential approaches have been proposed, and there is now a need for some structure in the search for solutions. A large number of presentations and articles have described high-level [1–5] or detailed potential projects [6]. This chapter tries to summarize briefly the first 18 months of effort, from April 2007 to September 2008, in the so-called Transformative Technologies program carried out by FPInnovations with funding from Natural Resources Canada and the Canadian forest sector. Questions remain, and while the detailed scientific and engineering results are still to be confirmed, the high-level objectives presented here are not likely to change much as the work progresses.

18.2 A Structured Approach

From the perspective of the pulp and paper industry, there are two large-scale approaches to the forest biorefinery: first, to modify an existing (pulp) mill by creating new, higher-value uses of the existing biomass supplied to that mill, or by improving the use of existing streams; second, to build a new plant (possibly next to an existing one, to take advantage of potential synergies) which makes use of new biomass sources while the existing plant continues as before.

The first option requires identifying underutilized streams in an existing mill: what are these streams, what are their properties and chemical compositions, what are they currently used for, and are there higher-value uses for them, bearing in mind the technical and economic implications of no longer using them as before? Many of these streams are currently burned for heat, and frequently power, for use onsite. What other changes are necessary in the mill if these streams are diverted to new uses?

Existing streams found in pulp mills, each with its own floor value based on its current use, include:

- Forest residues
- Bark
- Sawdust, planer shavings, other sawmill residues
- Fibers of various types
- Lignin
- Hemicellulose
- Cellulose
- Sludge
- Effluent treatment streams
- Condensates, pressates, extractives

The second pathway requires identifying new biomass sources: what are they, what are their properties, and what will they cost delivered to the mill gate? Once delivered, what can be done with them?

New biomass sources include:

- New chip supplies
- New bark supplies
- New forestry residue supplies
- New agricultural supplies (corn, soy, canola)
- New agricultural residues (corn stover, wheat straw)
- Sorted municipal residues such as construction and demolition waste
- Other recovered wood or biomass (railway ties, telephone poles, etc.)

These sources of biomass are all very different in composition, properties, and availability. The end products that can be made from them will depend on these and other characteristics.

At the same time, new products and markets must be identified. As with feedstocks, a range of product types can be considered, each with its own market advantages and limitations. Generally, they can be divided into the following categories:

- Energy
  - Stationary heat and power (combined heat and power, CHP), using solid, liquid, or gaseous fuels derived from biomass
  - Liquid transportation fuels (ethanol, butanol, bio-diesel, synthetic gasoline or diesel)
- Products used within the forest industry
  - Methanol for ClO_2 generators
  - CO_2 for pH control in pulp mills
  - Wet-end strength additives
  - Resins for panels and boards; and so on
- Chemicals used outside the forest industry
  - Resins
  - Plastics polymer
  - Xylose, xylitol, furfural, levulinic acid, acetic acid, and so on [7,8]
  - Lignosulfonates and other products from lignin
  - Carbon fiber and other carbon-based products; and so on

It is necessary to consider which products can be made from which biomass sources, in what quantities, and at what cost. Not all biomass sources will be suitable for all products, for technical or economic reasons; not all pathways will make sense. There is a key difficulty in identifying some of these pathways: they require detailed market knowledge. In some cases, the knowledge is quite simple: it should be possible to sell every drop of fuel-grade ethanol or biodiesel that one can produce, at fair market prices, without affecting the market in any significant way given the enormous size of the market; the only requirement is to meet a set of prescribed quality specifications. Other products, whether on the United States Department of Energy (DOE) lists [7,8] or not, will require more intimate knowledge of a specific chemicals market, and in this case, it will be wise to build relationships with players in these fields.

Typical pathways include:

- Fermentation routes
  - Ethanol or other biofuels from hydrolyzed cellulose or hemicellulose
  - Lactic acid, propanediol, polyhydroxybutyrate, and other polymer precursors
- Biodiesel from classic processes such as esterification
- Thermochemical routes
  - Heat, power, or synthetic fuels and chemicals from syngas
  - Heat, power, fuels, or chemicals from pyrolysis oils
- Chemical routes
  - Synthetic gasoline or diesel from chemical processes (other than thermochemical)
  - Chemicals and other products from lignin
  - Chemicals from hemicellulose
  - Chemicals from cellulose

Geographically, biomass resources are distributed thinly and unevenly, with concentrations of one type or another in different areas. Biomass also tends to be very bulky. These two factors limit the maximum economic transportation distance to the primary transformation site. Geographic considerations beyond biomass availability will also play into the decision on what to build where and for what feedstock. Opportunities will therefore depend on a regional combination of biomass sources, existing forestry or other infrastructure, transportation logistics, local availability of specialized R&D or engineering support, and possibly local end-users for any novel products developed. The specific pathway or pathways that make sense in one geographic area may not make sense elsewhere.

Regional considerations include the following:

Western Canada

- Large quantities of mountain pine beetle-killed wood
- Large oil and gas industries; unknown synergies with these industries
- Chemical industry focused on large amounts of a limited range of products (e.g., intermediates such as ethylene)
- Potential for government R&D support
- Availability of a mix of forestry and agricultural residues

Ontario

- Forest sector based in the north and northwest
- Large urban residue streams available in southwestern Ontario

- Chemical sector based in Sarnia, focused on a larger number of smaller-volume products than in western Canada, and in some instances dependent on pipelined raw materials from the west (e.g., ethylene). Note that the imminent closure of the Cochin ethylene pipeline from Alberta has already led to plant closures in Sarnia
- Automotive sector based in southern Ontario; potential large demand for novel, biobased materials for structural and nonstructural applications
- Great Lakes and rail lines joining south and north
- University of Western Ontario Research Park, Sarnia, offers pilot facilities focused on biomass sources for petrochemicals substitution
- CRIBE, Thunder Bay offers R&D and other support for modifications to existing pulp mill infrastructure

Quebec

- Pulp and paper dominated by mechanical pulping (more limited mill streams)
- Smaller but relatively more diverse petrochemical industry
- Transportation facilitated by the St. Lawrence Seaway
- Critical need to make better use of hardwood resources in mixed forests
- Critical need to secure a local source of ethanol given limited corn supply

Atlantic Canada

- Largest concentration of privately held forest lands in Canada
- Proximity to deepwater ports for overseas shipping
- Barge, rail, and road links to markets in New England and New York

Urban areas

- Concentrated volumes of sorted municipal solid waste
- Construction and demolition residues

18.3 Ongoing Semicommercial Activities

The importance of partnerships cannot be underestimated. In several cases, the products which can be made are already in use in our industry; resins for use in panel-making are one example. Product volumes, performance levels, and pricing are therefore well known. In other cases, the market, while external, is also large and well known; transportation fuels, where the market is essentially bottomless and where there are relatively well-known quality standards, is an example. There are two areas where partnerships will be necessary: with companies having access to novel, specialized technologies, and with companies (frequently outside the forest sector) with existing supply chains, market knowledge, and access to capital.

Companies with specialized technology are numerous and range in size from small startups to large engineering firms. No list can be exhaustive because new companies and joint ventures are formed daily. Canadian companies like Enerkem, Nexterra, Iogen,

Lignol, Ensyn, and Dynamotive are developing markets and technologies. Outside Canada, the list includes companies with expertise in everything from fermentation to gasification. A significant recent player in the field is UOP, a company that supplies the bulk of refinery technology to the oil and petrochemical industry. The list of companies with established supply chains and market knowledge begins with the oil and gas industry and continues with various petrochemical or fine-chemical firms.

Given all this, where are the opportunities? In the biofuels market, a number of international pulp and paper companies have announced links to the traditional fossil fuel-based industries. There are also many smaller companies, with a range of nontraditional technologies for making new products from biomass, looking for partners to help build pilot or commercial demonstration plants. To illustrate the scope of activities underway around the world, a selection of recent announcements and ongoing pilot plant work is discussed next.

It is worth noting that with a very few exceptions, these projects currently exist only on paper, and that all are (or almost certainly will be) receiving substantial subsidies and grants from multiple levels of government to move their projects forward. Partnerships with all levels of government are clearly still key to managing the financial risks posed by these projects. For example, in 2007–2008, the DOE announced a series of grants totaling \$385 million over four years (\$30–\$60 million each) to kick-start construction of six commercial-scale lignocellulosic biorefineries, as well as an additional \$240 million to fund nine smaller demonstration plants [9]. Another \$600 million was allocated out of the American Recovery and Reinvestment Act funds to support construction of 19 additional pilot, demonstration and commercial-scale projects beginning in 2010. Similar financial aid has been provided by other governments, including that of Canada, to projects within their boundaries. Even with government support, not all the announced projects will be built; several of the companies selected to receive funding for commercial-scale projects in the United States have already dropped out, and others are likely to follow as they struggle to meet milestones.

18.3.1 North American Initiatives

18.3.1.1 Chevron and Weyerhaeuser

These two companies have formed a joint venture called Catchlight Energy LLC, based at least partly on Weyerhaeuser's extensive private timberland holdings. Land stewardship, crop management and collection, and delivery and transformation of biomass are listed as key strengths brought to the table by Weyerhaeuser [10]. Recent presentations have outlined a scheme of silviculture for the Weyerhaeuser lands called intercropping—rows of plants grown specifically for energy purposes will be placed between rows of trees grown for lumber and pulp production. No details of specific technical pathways have been published, although it has been stated that they will buy technology rather than develop their own.

18.3.1.2 Enerkem and the City of Edmonton

Edmonton has signed an arrangement with Enerkem to build a \$85–90 million facility to produce 36 million liters of ethanol per year, using Enerkem's gasification and synthesis process running on sorted municipal waste [11]. The funding is largely from government agencies and the City of Edmonton and will help divert up to 90% of Edmonton's municipal garbage away from landfill. At the same time, Enerkem is operating a small demonstration gasification plant making ethanol from recovered telephone poles in Quebec. The common thread here is feedstock costs, which are zero or possibly negative because of avoided landfill costs.

18.3.1.3 UOP and Ensyn

UOP is a major engineering and technology supplier to petrochemical refineries, similar to companies like Metso and Andritz in the pulp and paper industry. UOP has identified waste lignocellulosic materials and algae as preferred precursors for biofuels [12]. In particular, the value of the aromatic content of lignin was highlighted as making it the perfect feedstock for fungible biofuels: those that can be "dropped into" the existing petroleum-based fuel production network and delivered to the consumer without modification to infrastructure. To this end, a joint venture between UOP and Ensyn has been formed to offer technology and equipment to convert second-generation biomass into oil for power generation, heating fuel, and transportation fuel [13]. A key piece of technology in the joint venture is the RTP™ pyrolysis process provided by Ensyn, which generates an aromatic-rich bio-oil that can be stabilized and used as a "green" feedstock in existing refineries. Unlike most companies in this market, Ensyn has commercial experience with their process. They have already built several small plants to produce artificial flavorings, resins, and other products from waste wood and have had their technology adapted for conversion of heavy oil sands in large-scale production. The entrance of UOP into the market is significant, given their background and expertise. The focus of this joint venture on production of hydrocarbon-based biofuels, rather than oxygenated biofuels like ethanol, is an interesting statement on their view of the future of the biofuel market.

18.3.1.4 DOE Awards to Forest-Based Companies

The DOE has offered $30 million each to Flambeau River Biofuels and NewPage Wisconsin Rapids for small-scale, gasification-based synthesis plants designed to make between 12 and 23 million liters of synthetic diesel per year. The lack of dependence on food crops or agricultural land was stressed in the press release regarding the Flambeau River project [14]. The fluidized-bed steam reformer provided to these projects by TRI will be similar to the one used by Norampac to process OCC black liquor in their Trenton mill, although substantial new ground will be broken because the gasifiers will be operating at much higher bed temperatures than those used for black liquor. Three other DOE grants, totaling up to $86 million, were also announced around the same time [15]. One is focused on forestry operations: Red Shield Environmental Pulp and Chemicals LLC was selected to receive a DOE grant toward construction of a plant at the existing pulp mill in Old Town, Maine. The process, developed at the University of Maine, involves preextraction of hemicellulose before pulping, followed by fermentation. There is no certainty that any of these pilot projects will proceed to completion—Red Shield almost immediately entered bankruptcy, and has been reconstituted as Old Town Fuel and Fiber, with a new plan (and new technology partners) to produce bio-butanol. Progress on the Flambeau River Biofuels and New Page projects has been intermittent.

Range Fuels is notable for being one of the very few companies on the DOE list of supported projects to have actually built a commercial-scale demonstration plant. Their proposed process involves gasification of forest residue and wood waste, followed by a catalytic stage generating mixed alcohols. Along with ethanol, Range Fuels claims that methanol, propanol, and butanol are also products of their process. However, recent news regarding the closure of this plant illustrates the challenges facing start-up companies in this field.

Lignol is working to commercialize an updated version of the Repap AlCell process. Chips are exposed to an acidic ethanol solution at high temperature to separate the lignin from the hemicellulose and cellulose. Although the original ethanol pulping process failed in part because of pulp quality problems, the new approach is to ferment the cellulose portion of the feed to ethanol while looking for value-added chemicals from the lignin portion. Lignol has a number of government grants to build pilot and demonstration plants, both in the United States and in Canada. The Lignol process also generates a lignin product which has significantly different properties from either kraft lignin or lignosulfonates.

Virent is paired with Shell Global, but is still an early-phase technology company that grew out of work at the University of Wisconsin. Relative to some of the other players in the market, it is relatively small and has raised only about $30 million to fund ongoing process development. Its technology is unique and aims to circumvent the high gas-phase pressure required in Fischer–Tropsch plants. Like UOP, Virent is aimed at producing fungible biofuels that can directly substitute for current petrochemically based gasoline. Their process generates alkanes, the building blocks of gasoline, from sugars in an aqueous phase. Pressures required are still high, but the cost of gas compression is avoided, and the result is straight hydrocarbon chains up to about C6. A pilot facility in Madison, Wisconsin, is being used to develop the process. A key issue will be accessing the sugars; it does not appear that the lignin fraction will be converted, and some pretreatment to extract sugars will be needed. As well, use of the lignin fraction will be critical.

Many *fermentation plants* are also being funded by the DOE, of which a number address the food-versus-fuel debate by focusing on lignocellulosic material such as agricultural and forestry residues. Most focus on production of ethanol, but butanol is an alternative product that is being pursued by some companies, including Butamax, Cobalt Biofuels, and Gevo which claim to be able to process both hexose and pentose sugars into butanol. If feasible, this would be a significant advance in technology. All these companies face the same hurdle: pretreatment of the fiber to release the cellulose and expose the sugar chains to enzymatic hydrolysis is technically challenging and expensive. Although these companies do not necessarily intersect directly with forestry activities, there may be interesting synergies if such a plant can be located next to a pulp mill as described below. It is also likely that some of the results may be applicable to fermenting cellulose from forestry sources. Iogen, Mascoma, Verenium, Poet, Pure Energy Fuels, and many others are active in this field.

18.3.2 European Initiatives

18.3.2.1 Güssing and the Repotec Gasifier Design

This gasifier plant, developed by researchers at the Technical University of Vienna, reaches high efficiency levels because waste heat is sold to the district heating system of the town of Güssing, Austria. Economics are distorted by power prices of €0.16 per kWh (about $0.25 per kWh at current exchange rates), while heat to the district heating system is sold for €0.02 per kWh (€5.56 per GJ) [16]. Prices paid for wood were not given. As well, a substantial portion of the capital came from Austrian or EU government sources. Technically, the process works, and the equipment had logged well over 30,000 hours by late 2008, averaging over 7000 hours per year since 2005. See, among others, Ref. [17] for more information.

18.3.2.2 Choren

This German firm, with support from Shell, Daimler, and Volkswagen, is in the process of commissioning a multistage gasification and synthesis plant designed to make up to 16 million liters per year of synthetic diesel fuel from 75,000 tonnes of biomass, largely white wood and bark [18]. Significant levels of government funding are supporting this initiative, both in the form of outright grants and as loan guarantees. This project builds on previous pilot-plant work at the 4 million-liter-per year scale. Choren envisages a next-generation plant making up to 250 million liters per year [19], although securing the necessary biomass supply (well over a million dry tonnes a year) might be problematic. The proposed location of this larger plant, in Schwedt, Germany, is next to Germany's largest existing petrochemical refinery. Partners include Linde Engineering, Shell, Pöyry, and the refinery operated by PCK. The project is expected to cost in excess of €800 million, which Choren expects to finance with the help of a government loan guarantee. The company also considers a mandatory blending quota for second-generation biofuels as critical to the success of the project. Choren's bankruptcy announcement in mid-2011 illustrates the challenges facing even the best-financed startup companies.

18.3.2.3 UPM and Andritz/Carbona

This agreement begins with testing of Carbona's gasification technology at the Gas Technology Institute's pilot plant. The eventual goal, like Choren, is to obtain synthetic diesel by means of a Fischer–Tropsch process [20]. The pilot work is expected to cost €5–€10 million. In an eventual commercial process located next to a UPM mill, UPM would supply biomass to a plant assembled by Andritz/Carbona.

18.3.2.4 Neste Oil and Stora Enso

This partnership, called NSE Biofuels OY, is also focused on thermochemical routes to synthetic diesel and will be closely tied to researchers at VTT in Helsinki, where significant gasification expertise is available [21]. The initial step is to be a €14 million demonstration plant located at Stora Enso's Varkaus mill. Although startup was initially announced for sometime in 2008, a contract for construction of the circulating fluid-bed gasifier designed by Foster Wheeler Energia OY was only awarded in mid-2008. Plans for a full-scale commercial plant have been announced, but are not imminent.

18.3.2.5 Chemrec

This Swedish company operates a high-temperature pressurized black-liquor pilot plant at a kraft mill in Piteå, Sweden, and has also built an atmospheric black-liquor gasifier operated by Weyerhaeuser at their mill in New Bern, NC [22,23]. Chemrec has constructed a small-scale dimethyl ether (DME) plant that will use syngas from the pilot gasifier at Piteå as feedstock, as part of the €28 million EU-funded BioDME project to produce green transportation fuels. There is a large amount of literature on this topic (see, for example, [22,23]), which will not be discussed here.

18.3.2.6 Lignoboost

Lignoboost is a commercial company, spun off from STFI (now Innventia) and now owned by Metso, which offers a commercial lignin precipitation process. The pilot unit in operation at the Backhammer mill in Sweden is capable of producing 10,000 tonnes per year of dry, powdered lignin, some of which has been burned in a district heating plant in Stockholm and in extended lime kiln trials. The technology works and is similar to processes developed at Paprican (now FPInnovations) in the 1980s [24].

Overall, large amounts of government money and venture capital are being spent in the United States and the European Union to develop second-generation biofuels. The drivers are the same in both jurisdictions: security of energy supply first, and carbon emissions second. It is also clear that the technical and financial risks are still high and that partnerships of various sorts, including with governments, are key to bringing the right skills and resources to the table. The projected size of the announced commercial-scale plants to date, typically aimed at producing approximately 40–60 million liters of liquid fuels per year, may be too small to be commercially successful in the long term. Most companies, including Catchlight, envision full-scale plants that produce 200 million liters or more annually.

Finally, the biofuels market, even with mandated usage levels and various other incentives and subsidies, remains a commodity market with thin profit margins. With kraft pulp at $800 per tonne, and with ethanol revenues at the plant gate of $0.50 per liter, it may make little sense to convert the cellulose component of wood to ethanol because revenues and profit margins will be significantly lower. Ethanol prices and yields will have to rise significantly before the revenues from ethanol are the equivalent of current pulp prices of $800 per tonne. Alternatively, the cost of feedstock will have to be significantly lower than it is for wood or bark delivered to a typical Canadian mill. Processes relying on expensive white wood or bark will have a hard time competing with those using zero- or negative-cost feedstocks such as sorted municipal waste or recovered telephone poles. To be economically viable, processes will be needed to convert existing mill streams to fuels and chemicals more effectively, to generate products with higher market values, or both.

The remainder of this chapter will address not only ethanol or biodiesel, but also potential pathways to converting existing mill streams or forest residues to higher-value products, thus maximizing the spread between feed costs and product revenues.

18.3.3 Opportunities Specific to Canadian Operations

A short list of opportunities specifically tailored to the Canadian forest industry is presented next. For lack of space, not every combination of feed, process, product, and geographic area will be described; broad approaches with reasonably wide applicability across the Canadian industry will be described.

18.3.3.1 Fermentation Pathways from Hemicellulose

Hemicellulose can be extracted from wood before pulping. The process has been evaluated by FPInnovations and others for kraft and mechanical pulps and for both paper and nonpaper processes such as dissolving pulp. The process generally has a negative impact on the papermaking properties of pulp, which needs to be addressed. The hemicellulose removed before chemical pulping processes would otherwise go to the recovery system; because the heating value of hemicellulose is significantly lower than that of lignin, removing it from the black liquor is one way of removing material and water from evaporation equipment and recovery furnaces where these unit operations are a bottleneck to increased pulp production. A large economic advantage is obtained if the hemicellulose stream can be fully used for by-products.

When prehydrolysis is applied to kraft pulps, either hardwood or softwood, the hemicellulose content in the pulp will be lower. There may also be some change in the degree of polymerization of cellulose. These chemical changes increase the tear and decrease the tensile strength of the resulting pulps. The extent of the changes in pulp strength will depend on the prehydrolysis conditions and are most critical for softwood pulps which tend to be sold for their strength properties. For hardwood pulps, prehydrolysis also has some positive features, including faster cooking and increased pulp drainage. Again, prehydrolysis conditions are critical. Van Heiningen has proposed a "near-neutral" preextraction process for hardwood kraft pulp integration, which has been selected by Red Shield Environmental Ltd. (now Old Town Fuel and Fiber) for application. The process uses green liquor (sodium carbonate and sodium sulfide) for prehydrolysis [25]. The process results in the same quantity and quality of hardwood pulp as from conventional cooking. However, less hemicellulose is produced than from autohydrolysis and other prehydrolysis technologies, and the extract contains inorganic salts. These factors result in a rather low return on investment (between 1% and 13% according to the authors). Perhaps the best application of prehydrolysis will initially be to dissolving-pulp kraft mills, where the hemicellulose needs to be removed anyway, and where the strength losses are unimportant when the cellulose is to be dissolved.

FPInnovations and others have also investigated prehydrolysis of chips before mechanical pulping. A study of mild acid hydrolysis of aspen chips before BCTMP manufacturing [26] found that more xylan than glucan hemicelluloses were solubilized compared to the usual steaming of chips. The acid pretreatment resulted in reduced energy requirements to reach the target freeness, accompanied by losses in bulk, tensile, and bleachability [27]. In conventional TMP manufacturing, much of the solubilized hemicellulose is in the form of medium-chain-length galactoglucomannans. If these could be efficiently concentrated, there might be a market for these hemicelluloses as chemical additives or nutraceuticals.

Prehydrolysis of wood chips, either with steam or with dilute acids, results in mixtures of monosaccharide wood sugars, oligomeric wood sugars, and other products, predominantly furfural, hydroxyl-methylfurfural, and levulinic acid (from dehydration of sugars), acetic acid (from deacetylation of acetylated hemicelluloses), and phenols (from lignin). For hardwoods, the predominant sugars are pentoses such as xylose and xylooligomers. For softwoods, hexoses such as mannose, galactose, and glucose are found.

Further chemical processing of the prehydrolysis streams is discussed later; fermentation routes to transportation fuels and other chemicals will be discussed here. Conventional fermentation of sugars with the yeast *Saccharomyces cerevisiae* to yield ethanol is effective only for hexose sugars. Therefore, other fermentation routes must be found for the predominantly pentose streams from hardwoods. Several groups have developed genetically engineered yeasts, either *S. cerevisiae* or others such as *Pichia* or *Pachysolen*, which are claimed to be capable of fermenting both pentose and hexose sugars [28]. Others have worked with bacteria such as *Zymomonas* and *Escherichia coli*, both native and genetically modified, to transform xylose into ethanol. Other factors which influence the success of fermentation are the concentration of sugars and the amount of inhibitors present in the prehydrolysate.

Assuming that all hemicellulose can be removed from wood chips and converted, the production of ethanol from hemicellulose could be as high as 158 liters per tonne of wood supplied to the mill, or up to 100 million liters per year for a pulp mill producing 1000

TABLE 18.1

Fermentation Routes to Biofuels and Industrial Chemicals

Fermentation Route	Microorganism	Status
Sugars → butanol	*Clostridium acetobutylicum*	Traditional, but low yield, also gives ethanol and acetone
Sugars → volatile organic acids (VOA) (e.g., butyric acid) → (reduction) butanol	Acetogens	BP and DuPont have sponsored development
Sugars → succinic acid	*E. coli*	Both pentoses and hexoses fermented
		DSM, Bioamber
Sugars → lactic acid → polylactide	*Lactobacillus*	Commercial scale production by Natureworks and others
Sugars or VOAs → polyhydroxyalkanoates (PHA)	Various	Commercial
Cellulose → methane	Rumen bacteria	Ruminants do it
Mixed substrates → H_2	*Clostridium* sp.	Research stage
Synthesis gas (H_2 and CO) → ethanol + other alcohols	Mixed cultures	Under development by Coskata and others

tonnes/day of pulp (and consuming 2300 tonnes of wood per day). (It is theoretically possible to generate an additional 278 liters of ethanol per tonne of wood if cellulose is also converted to ethanol, but this comes at the cost of pulp production.) Realistic production rates are probably half this figure. Although this is at the low end of the 100–400 million liters per year produced by large corn-based ethanol plants, the economics may be driven not by scale, but by feedstock cost: if the hemicellulose in a dissolving pulp mill is not converted to ethanol or some other value-added product, it will need to be evaporated and burned in the recovery boiler or elsewhere, adding a load to these unit operations and potentially contributing to a production bottleneck.

Overall, the technology is available now, with the following caveats. It will work best in hardwood dissolving-pulp mills, where the fiber properties are not critical, provided that the pentose sugars can be efficiently fermented. It will also work in softwood mills where hexoses will be produced, if the papermaking properties of the resulting pulps can be maintained.

In principle, many biofuels and industrial chemicals can be made by fermentation routes, as summarized in Table 18.1. These include biofuels such as butanol, chemical feedstocks such as lactic acid, succinic acid, and propanediol, and polymers such as PHA. In general, a fermentation stage from glucose or other wood sugars requires a selected microorganism which is capable of producing the required product. Alternatively, hybrid gasification/fermentation routes have been proposed, as have completely microbial processes such as anaerobic digestion of cellulose to yield methane or hydrogen, or combined hydrolysis and fermentation by cellulolytic organisms.

18.3.3.2 Fermentation Pathways from Cellulosic Residues

One of the major obstacles to bioconversion of cellulose to ethanol is the need to pretreat cellulose before enzyme hydrolysis and fermentation. The FPInnovations approach to this has been to identify delignified cellulosic residues available at a mill site which are much more amenable to enzymatic hydrolysis. These substrates have been combined with enzymes in a high-consistency hydrolysis which results in high concentrations of wood sugars suitable for efficient fermentation to ethanol [29]. The results showed that knot rejects are easily hydrolyzed and that the optimum enzyme dosage of about 10 FPU (0.08 mL) per gram of knots can be lowered to 2 FPU by incorporating an enzyme recycle process developed at FPInnovations. High-consistency hydrolysis, combined with enzyme recycling, leads to a significant reduction in capital cost for the hydrolysis tank and lower operating costs for cellulase enzymes.

18.3.3.3 Chemical Pathways from Hemicellulose

Chemical conversion of hemicellulose to chemicals is an alternative to fermentation pathways. In hardwood mills, extracted hemicellulose in the form of xylan can be hydrolyzed to xylose and sold to market as is. Alternatively, the xylose could be converted to such high-value products as xylitol (a sweetener) or furfural (a base chemical for furfuryl alcohol, tetrahydrofuran, and levulinic acid). In softwood mills, the extracted hemicellulose (a mixture of xylan and galactoglucomannan) could be hydrolyzed to a mixture of five- and six-carbon sugar monomers. Five-carbon sugars can be converted to the same products as in hardwood mills, while six-carbon sugars can be chemically converted to hydroxymethylfurfural and other products.

18.3.3.4 Chemical Pathways from Lignin

The U.S. Department of Energy has commissioned a series of reports on potential products from biomass. The first [7] described sugar platforms and identified a dozen pathways leading to large markets with reasonably low technical risk. The second report [8] outlined potential products from so-called biorefinery lignin, defined as lignin left over after the carbohydrates (cellulose and hemicellulose) have been removed and converted in a biorefinery. The definition is broad enough to include lignin from softwood, hardwood, and agricultural residues, as well as from a range of processing steps such as kraft pulping, steam explosion pulping, organosolv pulping, and others. The report starts with the idea that once lignocellulosic ethanol plants are generating the 230 billion liters of ethanol that the United States estimates it will need in 2030, there will be as much as 200 million tonnes of lignin available every year as a by-product. New outlets for this material will be needed to minimize disposal issues and to enhance the economics of ethanol plants.

The second DOE report [8] summarized the opportunities for lignin in three major types of products, which also appear to coincide with short-, medium-, and long-term opportunities. The first category is to make power, green fuels, or syngas using combustion, gasification, pyrolysis, or hydroliquefaction. Potential products, other than from simple heat and power from combustion, include olefins such as ethylene, synthetic motor fuels, and methanol. With the exception of black liquor gasification and hydroliquefaction, these technologies are available now, or soon will be, although capital cost may still be prohibitive. The issue of the sulfur content of kraft lignins was mentioned in the DOE report, although the precipitation of lignin from kraft black liquors was not discussed.

The second category is larger molecules or polymers, such as carbon fibers, polymer modifiers, and various resins, adhesives, and binders. Many of these technologies are close to commercial reality. Resin applications will be market-driven because the use of formaldehyde is likely to be increasingly restricted by legislative or regulatory action. A significant level of R&D work is still required to tailor lignin to the needs of the customer; given the variability of lignin with biomass source and type of processing stage, the technology required may be relatively site-specific. These processes are identified as medium-term solutions, with more significant R&D needs.

The third category is aromatic chemicals that take advantage of lignin's inherent molecular structure. Aggressive depolymerization processes can lead to the BTX family (benzene, toluene, xylene) of chemicals, which serve as a basis for a range of chemicals and plastics. A less aggressive depolymerization might retain more of the structure of lignin and could lead to a generation of entirely new processes and products. The authors acknowledge that this approach will require significant levels of R&D and that results are likely to reach commercial maturity only in the long term, but point out that the opportunities are large because lignin is the main biobased source of aromatics.

Overall, the authors make the point that processing steps generally have been optimized to protect the cellulose or sugar structures at the expense of lignin properties. This contributes to the variable nature of lignin, which is in addition to its inherent variability as a function of feedstock. They recommend that R&D work should address methods of removing lignin as early as possible in the process, and as gently as possible, to retain its potentially valuable aromatic structure.

Lignin precipitation is a process for removing lignin from black liquor by acidification. A number of processes exist, including a process developed by Paprican (now FPInnovations) and commercialized by NORAM, and one developed by STFI (now Innventia) in Sweden and now licensed to Metso. Lignin was removed commercially from the kraft mill in Cornwall, ON, in the 1940s through to the 1960s. Removal of lignin in a recovery boiler-limited chemical pulp mill will enable added pulp production. Depending on the initial heating value of the black liquor, as much as 25% of the lignin could be removed without any major impact on recovery boiler operation. In a typical mill, this figure would be closer to 10%, but even this figure leads to large numbers: in a 1000 tonnes/day kraft mill, this represents up to 50 tonnes/day of lignin, which is associated with up to 50 tonnes/day of added pulp production if no other production bottlenecks limit this potential production increase.

Lignin can be used as a fuel in the lime kiln or power boiler; based on the price of fossil fuel displaced, this represents a value of about $200 per tonne of lignin, or $10 in savings per tonne of pulp production. This represents the floor value of lignin. At this value, preliminary estimates suggest that a 30 tonnes/day lignin plant would require an installed capital cost of $10 million and would have a payback period of 3–4 years, assuming first that the lignin is used as a fuel in the lime kiln or the power boiler, and second that the removal of lignin from black liquor enables production of an additional 25–30 tonnes/day of pulp in a recovery-boiler-limited mill.

Two potentially high-volume, high-value applications of lignin are its use as a replacement for carbon black in tires and for phenol in phenol-formaldehyde (PF) resins. Another promising but technically more challenging application involves use of lignin as a feedstock in the manufacture of carbon fibers in place of polyacrylonitrile (PAN) [30]. These three applications are discussed in more detail below.

Because *carbon black* is made from petroleum-based feedstocks, its price has increased substantially over the last few years. Recent reports suggest that list prices are as high as $1700–$2700, depending on the grade. Given current petroleum prices, tire manufacturers are likely to be very interested in a less expensive product that provides the same functions as carbon black.

In research dating to the 1950s, lignin was shown to be an acceptable substitute for carbon black in tires, providing similar properties as a reinforcing filler and as a curing agent without being an exact duplicate at the molecular level. Lignin also has the potential to increase tire life through increased wear and abrasion resistance and reduced rolling resistance. In addition, the potential exists to use lignin in place of the expensive antioxidants and antiozonants currently used in tires. Previous research has shown that it is technically feasible to incorporate kraft lignin into tires in place of carbon black without any major penalty with respect to the physical or mechanical properties of the tire. However, this work is dated, and tire design requirements have changed substantially since. Further experimentation is required to determine if, how, and to what extent lignin can replace carbon black in modern tires.

A stepwise integration strategy for lignin is the one that is most likely to succeed. Because different grades of carbon black are used depending on the tire component, it would make sense to target initially the replacement of small fractions of lower grades of carbon black, such as that used in tire inner liners. Building on acquired knowledge, lignin can then be modified to meet the specifications of higher grades of carbon black.

Lignin as a substitute for carbon black in tires or other rubber products is therefore one good example of a potentially high-value application for lignin. If successful, the size of the carbon black market (1.1 million tonnes/year in North America for use in tires) will provide a large outlet for the lignin which could be removed from the black liquor of recovery-boiler-limited mills (estimated to be approximately 1.5 million tonnes/year in North America). The low price and availability of raw lignin are great advantages that increase the chances of success of making a lignin reinforcer for tires.

Because *phenols* are also made from petroleum-based feedstocks, prices have also increased substantially over the last few years. Current list prices are reported to be as high as \$1500–\$1600 per tonne. Therefore, a second potential high-value application for lignin is as a replacement for phenol in PF resins used as an adhesive in various solid wood products (e.g., plywood and OSB). An extensive amount of R&D has already been conducted in this area, showing that the manufacture of lignin-phenol formaldehyde (LPF) resins with properties comparable to PF resins is technically feasible. For example, plywood has been made in a mill trial with lignin replacing 50% of the phenol in an LPF resin. The physical properties of the product compared quite well with the control as long as the press time was extended by 15%. To address this and other related issues, lignin size and heterogeneity will probably need to be carefully controlled by fractionating or chemically modifying the lignin using approaches such as selective precipitation from black liquor through pH control, ultrafiltration of black liquor or lignin solutions, and thermal or chemical depolymerization. These steps will be necessary because lignin will have to be refined to the point where it can be delivered in sufficient quantity at consistent quality.

If successful, the size of the PF resin market (600,000 tonnes/year in wood products for North America) will provide another outlet for lignin. Lignin-based resins can also penetrate other markets, and in the specific case of urea formaldehyde (UF) resins, they can replace a product with known formaldehyde emission problems. This market is currently estimated at 800,000 tonnes/year.

As with substitution for carbon black in tires, a stepwise integration strategy for lignin is most likely to succeed. The market for PF resins is divided into multiple applications, each with its own formulation. The largest sector is the wood products area, and it is this market that should be initially targeted. One option is to make and market LPF resins in partnership with existing resin manufacturers. In this case, lignin would be added as a starting ingredient and would therefore be copolymerized with other chemicals such as phenol or formaldehyde.

A third promising application for lignin is its use as a feedstock in the production of *carbon fiber* in place of PAN which currently sells for over \$3000 per tonne. Today, the carbon fiber market is quite strong, with demand far exceeding supply of 75,000 tonnes/year. The market is growing rapidly and will grow faster if prices drop fast enough. Volume production of lower-cost carbon fibers for use in car body panels and transportation and aerospace structural components represents a significant and large market opportunity, given the anticipated pressure on manufacturers to build lightweight, fuel-efficient vehicles. Key manufacturing steps and quality issues have been the focus of a DOE-funded research program at Oak Ridge National Laboratories [30]. Carbon fiber currently sells for approximately $14 per kg or $14,000 per tonne, and although this will drop as production increases to meet demand, these prices are still very high. The limiting economic factors for faster growth in the current market relate mostly to infrastructure and energy costs.

On the technical side, there are still outstanding questions about how to make lignin a reliable and cheap precursor to carbon fiber. Sustained effort toward addressing the issues of lignin purity, suitability, and modification is required. Contaminants must be removed, the most appropriate form of lignin identified, and, if necessary, the lignin must be modified to change it into the best form for carbon fiber production. Further technical issues relate to the scale-up of carbon fiber production from lignin. Lignin-based carbon fibers may have to be modified to withstand the temperature of the thermo-stabilization stage, or the process itself may have to be changed to account for the relatively low glass transition and decomposition temperatures of lignin and lignin/polymer blends as compared to PAN.

Overall, lignin precipitation is a technically simple process which could offload the recovery boiler with respect to calorific load and thus free up some material for other uses. The technology works and could be implemented today. There exist uses for lignin beyond the small lignosulfonates market which can best be exploited by sulfite mills. The initial use of the lignin extracted could be in the lime kiln, to displace fossil fuel, while R&D efforts continue to identify and develop higher-value uses.

18.3.3.5 Thermochemical Pathways: Gasification

Gasification is a process for converting organic material to a so-called syngas made up largely of CO and H_2 [31]. Depending on gasifier design, operation, and feedstock, there may be quantities of CO_2, N_2, oxides of nitrogen, CH_4, tars, and small levels of other contaminants. The gas is flammable and can have heating values as high as 50–60% the value of natural gas. Approximately one-third of the incoming biomass must be burned, or else added fuel must be supplied, to provide heat for the reaction to proceed.

The discussion here will exclude gasification of black liquor because there is already a large international R&D effort on this topic and because significant technical challenges remain, especially if the added chemical recovery function of a kraft black liquor recovery boiler is considered. Conventional biomass sources such as wood, bark, and residues are most likely to be used in the short term.

The simplest use of the gas, and the one requiring the least complex technology, is in straightforward heating applications. Nexterra, a Canadian company based in Vancouver, provides a simple gasifier which is well suited to heat applications, and installations at Tolko's Heffley Creek sawmill and Kruger Products' New Westminster tissue mill are effective and meet expectations around natural gas substitution. Nexterra also claims that the syngas can be burned in industrial natural-gas-fired internal combustion engines, providing a pathway to small, distributed heat- and power-generating stations without the capital cost and operating expenses of a steam plant. These applications require very little in the way of gas cleanup, and it is claimed that emissions of NO_x and particulate matter are lower than with conventional biomass combustion.

Use of syngas in gas turbines for power generation was tried in Varnamo, Sweden in the late 1990s, but was not economical. Gas cleanup requirements are higher for these applications than for heat or piston engines. The Swedish gasifier was a high-capital, pressurized unit, and while the process is said to have worked, the costs were prohibitive.

Syngas from biomass has been used in lime kilns in six European mills, starting in the mid-1980s. Of the six gasifiers, one is still running; most have been shut down because of age or because the dramatic decrease in fossil-fuel prices through the 1990s made them uneconomical. New, higher fuel prices are presumably changing that calculation. The lime-kiln application requires a high-heating-value syngas with high adiabatic flame temperature and low levels of contaminants, and the type of gasifier used in Europe appears to have been technically successful at fossil-fuel substitution rates up to 95%. However, there are reports of low gasifier availability, mainly due to fuel handling and dryer issues; at one mill, gasifier availability is reported to average between 60% and 90% due to problems with auxiliary equipment such as conveyors, fans, and pumps. As well, the gasifier design required for this application is not currently economical in Canada given current natural gas prices, because of which payback periods will likely exceed 5 years.

Finally, the use of syngas for the synthesis of fuels or chemicals is a well-known technology where the feed is coal. The South African company SASOL developed this technology in the 1950s, based on work in Germany in the 1930s; in North America, Eastman Chemicals operates a large coal gasification and chemical synthesis plant in Kingsport, TN. SASOL makes synthetic gasoline and diesel fuels, while Eastman Chemicals makes methanol and a range of other chemical intermediates for internal use.

The use of syngas for synthesis, where the feed is biomass, is being developed and commercialized by a range of European companies. Several were described above. These follow from the EU focus on diesel fuels (unlike the United States where the focus is on ethanol) and a desire to avoid food crops, or crops requiring high-grade agricultural land, as a feedstock.

Gasification is likely to be capital-intensive. The technology is reasonably well proven for heat applications, with lime kiln applications in Europe in the 1980s and the steam-blown, circulating fluidized-bed unit at Güssing, Austria, all operating at reasonably high operating rates for multiple years in a row. However, the cost of a new gasification unit for lime-kiln use will not likely have a payback period of less than 5 years with natural gas at $10 per GJ and bark delivered to the mill at $40 per odt. Higher-revenue applications, such as methanol generation or synthesis of motor fuels, remain technically risky and require further R&D, and the synthesis equipment will not be inexpensive.

Assuming that wood and bark are approximately 55% carbon by weight, and assuming that approximately one-third of the carbon in wood is converted to CO_2 (i.e., fully combusted) to provide heat to drive the gasification reaction, there remains about 360 kg of carbon available from the initial tonne of wood as CO for synthesis. Methanol is $CH_3(OH)$, and assuming no losses in the synthesis stage and sufficient O and H for the reaction to proceed until all carbon is consumed, this 360 kg of carbon will make 960 kg of methanol. At current prices, and subtracting transportation costs, this should be worth at least $600 (assuming a sale price, including freight, of $650 per tonne) at the mill gate.

When gasified, the same tonne of wood will generate 10–12 GJ of heat, worth $100–$150 at current natural gas prices. Therefore, the potential revenues from methanol are approximately six times higher per tonne of wood consumed. Obviously, this does not include the fact that the methanol synthesis plant will require additional capital beyond the gasifier, but it does point to the potential for obtaining higher value from chemicals than from fuels.

The application of biomass gasification technology in Canada will therefore require serious investment, first from government agencies to help build a commercial-scale demonstration gasification plant supplying, for example, fuel to a lime kiln, and then by partners in the chemical or petroleum industries to design the cleanup and synthesis stages to follow it and to scale up the process. Quebec-based Enerkem is pursuing this approach, building a pilot plant to make methanol and ethanol from recovered telephone poles; financing is largely from governments and venture-capital investors. It is likely that economies of scale will demand large amounts of biomass, and this may lead to competition between traditional uses of wood for lumber or paper and novel uses for fuels and chemicals. If the use of the syngas is for fuel substitution, for example, in a lime kiln or flash dryer, the demand will compete with existing demands for biomass and may impact its availability for use in power boilers. However, it is possible to recover low-pressure steam from a range of synthesis processes because these are generally exothermic, which might compensate to a certain degree for the loss of fuel in a power boiler.

18.3.3.6 Thermochemical Pathways: Pyrolysis

Pyrolysis is a high-temperature process which gasifies biomass, followed rapidly by a condensation step to generate a liquid commonly called a bio-oil. A certain amount of gas and char is also generated, and these are usually burned to provide the heat to drive the reaction. As with gasification, combustion of a portion of the carbon in the feed is therefore required to drive the process, and yields will reflect this fact.

Bio-oil is not strictly speaking an oil because it does not mix with petroleum oils. Typically, bio-oils are viscous, very acidic, and contain significant levels of water. They are often two-phase mixtures when water content is high and contain a wide range of different compounds, mainly phenolics.

Two Canadian companies operate commercial-scale pyrolysis plants. Ensyn operates a plant in Renfrew, ON, and makes a range of chemicals and food flavorings from the oil. Their technology is also being used on a large scale to process heavy oil sands in Alberta. At their Renfrew plant, one key product is a substitute for resins used in panel and board manufacture. A large portion of the feedstock comes from a hardwood flooring plant across the road, but specialized feedstocks, such as mesquite wood for charcoal flavorings, are purchased for specific products. Dynamotive operated plants in West Lorne and Guelph, both in Ontario; the feedstocks were generally municipal construction and demolition waste, which can be obtained at little or no cost, and the product was sold as a substitute for heavy fuel oils. The biggest of these plants consumes about 200 odt/day of material. At the time of writing, it appears that neither Dynamotive plant is running at full capacity.

Other potential uses for pyrolysis oils could be as a feed to a gasification system, which would reduce the capital cost of a pressurized gasifier because the feed is simply a pressurized nozzle rather than the complex systems required to feed solid biomass into a pressure vessel. The aromatic nature of the oil could also be an opportunity to generate more specialized chemicals.

Some companies are proposing mobile pyrolysis units as a way of extracting more material from the forest. The basic concept is that of a trailer-mounted unit that would follow harvesting crews and convert roadside slash, a bulky and difficult material to transport long distances, into a dense, pumpable liquid. Volumetric and energy densities are said to be significantly higher than with slash chipped or ground at roadside, and one company claims that the heat value, in GJ, that can be transported in a truck is increased by a factor of up to 10 when slash is pyrolyzed. The potential problems with this approach include designing and building a fleet of tankers capable of handling the highly acidic material safely, as well as concerns around moving a refractory-lined vessel repeatedly over logging roads. Finally, the water content of the bio-oil is high when the feed material is wet, and the added challenges of designing, building, and operating a mobile dryer, or dealing with high water contents, need to be addressed.

18.3.4 Sludges for Biomethane Production

Biotreatment of pulp and paper mill effluent generates large quantities of sludge which is mainly organic in nature with high energy content. Despite these properties, which make it a good feedstock for energy recovery, sludge is typically managed as waste, consuming up to half the total effluent treatment plant operating budget [32]. It is generally produced at a solids content of 1–2% and therefore requires dewatering before final disposal. A large portion of the cost incurred in processing and management of sludge as waste (approximately 70%) is attributed to dewatering chemicals [33], a majority of which are petroleum-derived products. Clearly, the practice of managing sludge as waste is increasingly difficult to justify, whether economically or environmentally. In the light of energy price increases and Canada's commitment to reduce greenhouse gas emissions, it is extremely important to recover bioenergy from this negative-cost feedstock rather than dispose of it as waste. Recognizing this, the Federal government has encouraged biomethane recovery from sludge through tax incentives in the 2007 Federal Budget. Along the same lines, the Canmet Energy Technology Centre, Natural Resources Canada, completed a state-of-the-art review of pulp and paper sludge to energy conversion technologies and provided directives for optimization and further development of promising technologies [34].

Anaerobic digestion of carbon-rich feedstocks such as sludge to produce biomethane is a well-known process. There are numerous installations of this technology worldwide for municipal sludge digestion and stabilization, many of which recover the biomethane for its energy value. However, there are no full-scale anaerobic sludge-digestion facilities currently operating within the Canadian pulp and paper sector [34]. In a few cases, anaerobic treatment technology has been used to treat pulp and paper liquid waste streams [35]. One reason for the low number of anaerobic treatment plants within the pulp and paper sector is that it is inherently a slow process requiring large reactor size. Conventional municipal anaerobic digesters are typically designed for a solids retention time of 20–30 days [36]. Building digesters with such long holding times is capital-intensive and represents the main hurdle limiting biomethane generation from pulp and paper sludge.

Sludge conversion to biomethane is a multistep process where the rate of reaction is generally limited by biochemically mediated hydrolysis of particulate matter to soluble species. However, recent advances in sludge preconditioning technologies using physiochemical means have shown promise in efficiently disrupting sludge solids to produce solubilized species, which in turn enhances the rate of biogas production. FPInnovations has investigated the use of several sludge pretreatment technologies, including commercially available ones, to assess their potential for increasing the rate of biomethane formation. Research results show that digester size can be reduced substantially through sludge preconditioning without compromising biomethane production [37].

Sludge as produced from an effluent treatment plant is odorous, potentially pathogenic, and decomposable. An added benefit of sludge conversion to biomethane is that the digested residue has reduced odor and pathogenic potential and is mineralized by generation of ammonia and phosphates. Moreover, it is a rich source of organic carbon, potassium, calcium, and other constituents of high agronomic value. Such improved properties enable beneficial uses of the digested residue, such as land application for soil amendment as a replacement for commercial fertilizers, the prices of which have increased recently. In addition, a significant portion of the sludge is converted to biomethane in the digestion process, leaving a relatively small amount of residue. Savings in petroleum-derived dewatering polymers can also be realized when a smaller amount of digested residue is blended with the primary clarifier rejects before combined dewatering.

18.3.5 Condensates, Extractives, and Other Smaller Streams

One product that can be extracted from kraft mill streams is *methanol*. At a chemical pulp mill, condensates are generated during cooking and evaporation of the residual pulping liquor. In kraft pulp mill condensates, hydrogen sulfide, methyl mercaptan, dimethyl sulfide, dimethyl disulfide, and methanol are the major contaminants. In general, hardwood pulping generates approximately 10–15 kg of methanol per tonne of pulp, while softwood pulping generates approximately 7–10 kg/tonne [38]. The most prevalent treatments for kraft mill evaporator and digester foul condensates are air and steam stripping, which is efficient in removing total reduced sulfur (TRS) compounds and methanol, but the use of live steam, which is widespread, increases the operating cost of the stripper. Condensate segregation is an important approach for reducing effluent treatment costs [39]. After the steam stripper, a stream rich in methanol is obtained and is generally burned in the lime kiln or in a dedicated incinerator because of its high concentration of TRS compounds. Samples taken from several Canadian kraft mills indicated that the methanol concentration in the stripper off-condensate ranges from 12% to 70% by weight.

Kraft mills purchase methanol to use in the production of chlorine dioxide. The use of the stripper off-condensate from both hardwood and softwood kraft mills was investigated in FPInnovations' chlorine dioxide generator pilot plant. Condensates were characterized with respect to potential contaminants in this application. The total TRS compounds represent only 1% to 5% of the methanol on a molar basis. The majority of terpenes are alcohols which act (along with hydrogen sulfide) as reducing agents. Results from pilot trials were satisfactory and did not show any difference between the use of pure methanol and stripper off-condensate for chlorine dioxide generation. A typical 1000 tonnes/day kraft mill requires about 4–5 tonnes/day of methanol, which can be obtained from the steam stripper. This enables the pulp mill to save approximately $1.0M per year (assuming a cost of $600 per tonne of methanol) and to use methanol generated onsite from the biomass used to make pulp.

FPInnovations is developing a technology that removes TRS compounds from condensate. This technology is based on the use of hollow-fiber contactors. Laboratory tests revealed that the removal efficiency of TRS compounds can be as high as 99%. The removal of TRS compounds enables kraft pulp mills with steam strippers to sell excess methanol on the open market. Pulp mills with no steam stripper will benefit from this technology by addressing odor problems associated with the TRS compounds. A pilot plant for testing the concept at kraft pulp mills is under construction, and two U.S. patent applications have been filed [40,41].

It is also possible to make a *biodiesel* fuel from tall oil. This process has the potential to generate 3–5 liters of biodiesel per tonne of kraft pulp production by esterification of the fatty acids in tall oil. Methanol from condensates can be used for this purpose, implying that all the raw materials are available onsite. Although production rates are low, capital and feedstock costs are also low, so that the overall economics may be favorable. Technically, the challenge remains to remove enough sulfur from the tall oil to meet ASTM standards for biodiesel; the standard has recently been reduced from 500 to 15 ppm, and this new low standard is proving to be a challenge. Tests have shown that the sulfur comes from the kraft process itself and not from sulfuric acid used for pH control, because trials with different acids showed the same level of sulfur. Work is underway to identify the chemical form of sulfur in the biodiesel produced and to evaluate novel sulfur-removal techniques.

A Swedish company, Sun Pine, is planning to make biodiesel from tall oil at the Smurfit Kappa plant in Piteå, Sweden. The biodiesel contains 600 ppm sulfur, which is well above EU limits of 10 ppm. Their partner is Preem, a Swedish oil company, which will take the high-sulfur, tall-oil-based crude and further refine it, including desulfurization, in an existing petroleum refinery.

18.3.6 Synergies with Crop-Based Biofuel Plants (First- or Second-Generation)

Ethanol from corn or lignocellulosics is a large and growing industry. The process generates CO_2, which in at least one case is sold to a neighboring chemicals plant. It also generates a solid residue which can be burned for heat or sold to farmers as animal feed. Potential synergies with a pulp mill include sharing transportation logistics and effluent treatment plants and using mill steam for the distillation step.

Oil-based biodiesel plants are based on converting fatty acids to fatty acid methyl esters. The plants require methanol and caustic, both of which are available at a pulp mill site. The process generates waste glycerol, and although glycerol prices have slumped as a result of oversupply, new processes using glycerol are being developed, and prices are recovering. In particular, it is possible to use glycerol in its unpurified form in pulp mills as an additive to the recovery cycle, which may provide added synergies with the pulp mill.

Other novel processes for first- or second-generation biofuels plants will each have their own needs and residue streams and should be looked at on a case-by-case basis.

18.3.7 District Heating and Other Combined Heat and Power Opportunities

Stationary combustion is perhaps the best-known technology in the area of biomass transformation. Large fluidized-bed boilers with modern feed, ash handling, and flue gas treatment equipment are cleaner than oil-fired plants due to their low sulfur emissions. With high-pressure steam-generating banks and associated steam turbines driving generators, a range of heat and power options are available. Although power can be sold to the grid, heat must be used locally, and this is the biggest impediment to greater efficiency.

A condensing turbine will maximize power generation, but the low-grade waste heat will be in the form of hot water, and the overall efficiency will be lower. Moreover, the plant may not qualify as a cogeneration plant under Canadian accelerated depreciation rules.

Backpressure turbines generate less power, but the low-grade heat is in the form of low-pressure steam, and the overall efficiency is therefore higher. This approach makes sense where there is a steam host nearby.

The key here is to identify a steam host. Too many mills have excess steam available due to newly reduced demand (for instance, because one or more paper machines has been shut down since the cogeneration plant was built). It is of critical importance to match the cogeneration plant to a steam host, because otherwise steam will have to be vented if power production is not to suffer. Many European pulp and paper mills manage to sell heat to local district heating corporations; in the Canadian context, this makes sense only where a small number of large users is located a short distance from the steam plant [42]. A typical list of users might include the pulp mill and its pulp or paper machines; sawmills or panel mills; and local infrastructure such as hospitals, light industry or large educational institutions. Although the steam demand may well be seasonal, it should ideally be full time (i.e., 24 h per day) during the winter months.

18.3.7.1 Opportunities for Mechanical Pulp Mills

Most of the opportunities discussed so far are applicable mainly to the various types of chemical pulp mill. The principal reason for this is low pulp yields in chemical mills, which mean that a significant portion of the wood consumed is burned for heat and power. Alternate uses for these streams have the potential to provide new products with no need for new wood supply, as long as the energy implications (of burning less lignin, for instance) can be addressed. Mechanical pulp mills, with pulp yields generally well in excess of 90%, will have to identify new biomass sources for new processes. In turn, these new processes will need to be integrated into the existing mill to achieve the best synergies.

Most Canadian mechanical pulp mills use some form of chip refining process, such as the thermo-mechanical pulping (TMP) process; very few groundwood mills are still operating in Canada, and these will not be discussed here. There are two general uses of mechanical pulps: as a papermaking pulp (usually supplied to an integrated paper machine onsite) and as a market pulp (frequently hardwood, generally sold for use in tissue grades or board grades where bulk is a key property). Opportunities will be somewhat different for these two types of mill.

18.3.7.2 Papermaking Grades of Mechanical Pulp

Most mills making these grades are integrated with one or more paper machines. Wood supply is almost exclusively softwood because fiber length is critical for paper properties. Pulp yields are generally 90–95%, with the balance going to sewer as condensates or as anionic trash from peroxide bleaching plants. Changes in pulp production levels are likely to impact paper production, depending on where the production bottleneck is located, and therefore it will generally not be possible to redirect substantial amounts of the existing wood supply to new uses unless one or more paper machines are to be shut down, leading to large levels of excess wood supply and pulping capacity.

At the same time, the energy profile of a modern TMP mill with an integrated paper machine is significantly different from that of a chemical pulp mill. The TMP plant requires substantial electrical power, in the order of 80–85 MW for a 1000 tonnes/day, newsprint-grade pulp mill, and increasing to 130–140 MW for a similar amount of a value-added grade such as coated mechanical or supercalendered paper. On the other hand, a well-operated TMP plant with modern heat-recovery systems can recover sufficient low-grade steam to match the steam demand in the paper machines; very little additional steam capacity is then needed. What steam capacity is required can usually be met by a small, natural-gas-fired package boiler, where small capacity and quick response to steam demand changes are critical. Obviously the installed mechanical pulping equipment in Canada is not all at this level, with many mills operating with little or no heat recovery, but this remains the target as mills are upgraded. Plans for large combined heat and power plants next to mechanical pulp mills need to take potential upgrades to the TMP plant into account.

Opportunities for these mills therefore revolve around identifying new power generation opportunities from new sources of biomass. Clever use of the residual heat from a combined heat and power plant, such as sales to neighboring lumber mills, will be of key importance because a modern TMP mill with an associated paper machine should not need large amounts of external steam.

18.3.7.3 Market Mechanical Pulp Mills

These mills make a product which is significantly different from the standard papermaking grade of TMP. Bulk and brightness are key attributes, with fiber length and opacity being less critical; wood supply is frequently hardwood. The process usually involves a chemical pretreatment stage before refining and a peroxide bleach plant afterwards and is therefore typically called a BCTMP plant. Yields are lower due to the chemical pretreatment stage and aggressive bleaching and can be in the range from 85% to 90%. Finally, few Canadian BCTMP mills are integrated with a paper machine; finishing lines typically involve a flash dryer (fired with Bunker C or natural gas) and a pulp baling line. Given that the flash dryer requires a liquid or gaseous fuel (unlike a paper machine that runs with steam), a modern BCTMP mill will usually have an excess of steam from the pulping area, in addition to the high power requirements typical of mechanical pulps.

Opportunities here begin with the chemical pretreatment stage. Ongoing work [26] has identified some potential use of BCTMP effluent streams, although optimizing for sugar production in pressates has led to problems with bulk and brightness in the pulp. Although the flows are small and dilute, there may still be opportunities because the capital cost to remove the streams is small. As an added benefit, removing sugars here will frequently reduce loading on the effluent treatment system, a common bottleneck in these mills. Added work to identify new pathways, while minimizing the impact on pulp properties, would be useful.

Several hardwood mills purchase whole logs and operate their own wood rooms, and therefore added bark and chip fines should be available onsite. These or other additional biomass supplies, if available, should target the production of biomass-based fuels to replace the fossil fuel demand in the flash dryer; gasification or pyrolysis systems appear to be well suited to this purpose. Although generating power onsite would be beneficial given the power demand, combined heat and power plants are not the best approach because there is no large steam demand in the mill which cannot be met by steam recovered from the refiners. Steam sales to a neighbor, such as a lumber kiln, might make this approach feasible.

18.4 Conclusions

Many opportunities are available in the forest sector today to generate new products from biomass. Several are ready for implementation today, while others require more R&D work. Several will require partnerships with companies offering specialized knowledge (Enerkem, Nexterra, Ensyn, Mascoma, etc.), access to markets (oil and chemical companies, tire companies), capital, or some combination of these. Many will require that integration issues within the existing pulp mill be addressed. Although it is always risky to predict the future, a short list of opportunities, broken out by their state of technology readiness and economic feasibility, follows:

Available now

- Lignin precipitation to generate a fuel for the kiln
- Hemicellulose extraction to generate ethanol through fermentation (HW or dissolving pulp)
- Methanol generation from stripper off-gas condensates
- Synergies with first-generation biofuels plants (corn to ethanol, soy or canola to biodiesel)
- District heating approaches where a small number of large steam users can be identified
- Stationary pyrolysis plants for bio-oil production
- Small-scale gasification of bark or other residues for heat applications

Available in 1–3 years

- Lignin precipitation to generate new chemicals and products
- Hemicellulose extraction to generate value-added chemicals
- Small-scale gasification to produce fuel for the lime kiln
- Anaerobic digestion of sludges to generate methane
- Biodiesel from tall oil
- Mobile pyrolysis units for in-woods use
- Large-scale gasification to produce fuels for stationary heat and power

Available in 3–5 years

- Large-scale gasification to produce synthetic liquid transportation fuels and other synthetic chemicals
- Large-scale district heating plants

Acknowledgments

This work has been made possible through integrated research along the entire forest sector value chain at FPInnovations in partnership with Natural Resources Canada. A large number of FPInnovations staff members were involved, either directly or peripherally, in generating the results; the short list of coauthors represents only a small fraction of the people who have contributed to this work. All contributions are gratefully acknowledged. Doug Singbeil's very thorough review contributed to the content and organization and is very much appreciated.

References

1. Thorp, B.A., Thorp, B.A. IV, and Murdock-Thorp, L.D. 2008a. A compelling case for integrated biorefineries, Part I. *Paper360*, 3(3): 14–15.
2. Thorp, B.A., Thorp, B.A. IV, and Murdock-Thorp, L.D. 2008b. A compelling case for integrated biorefineries, Part II. *Paper360*, 3(4): 20–22.
3. Thorp, B.A., Thorp, B.A. IV, and Murdock-Thorp, L.D. 2008c. A compelling case for integrated biorefineries, Part III. *Paper360*, 3(5): 12–14.
4. Thorp, B.A., Thorp, B.A. IV, and Murdock-Thorp, L.D. 2008d. A compelling case for integrated biorefineries, Part IV. *Paper360*, 3(6): 16–17.
5. Ragauskas, A.J., Nagy, M., Kim, D.H., Eckert, C.A., Hallett, J.P., and Liotta, C.L. 2006. From wood to fuels: Integrating biofuels and pulp production. *Industrial Biotechnology* 2(1): 55–65.
6. Tunc, M.S. and van Heiningen, A.R.P. 2008. Lignin carbohydrate complexes (LCC) in the autohydrolysis extract of hardwoods. *Preprints, 94th Annual Meeting, Pulp and Paper Technical Association of Canada*, Montreal, 6–7 February.
7. Werpy, T. and Petersen, G. 2004. *Top Value Added Chemicals from Biomass. Volume I—Results of Screening for Potential Candidates from Sugars and Synthesis Gas.* U.S. Department of Energy, August. Available online at: http://www1.eere.energy.gov/biomass/pdfs/35523.pdf.
8. Bozell, J.J., Holladay, J.E., Johnson, D., and White, J.F. 2007. *Top Value Added Chemicals from Biomass. Volume II—Results of Screening for Potential Candidates from Biorefinery Lignin.* U.S. Department of Energy, October. Available online at: http://www.pnl.gov/main/publications/external/technical_reports/PNNL-16983.pdf.
9. http://www.eesi.org/publications/Fact%20Sheets/eesi_cellethanol_factsheet_072308.pdf.
10. Chevron press release, 12 April 2007. Available online at: http://www.chevron.com/news/press/Release/?id=2007-04-12.
11. Edmonton Biofuels Factsheet, November 22, 2011. Available online at: http://www.edmontonbiofuels.ca/assets/files/Ressources/Edmonton%20Waste-to-Biofuels%20Project_Factsheet.pdf.
12. Holmgren, J. 2008. Refining of bio-oil into green hydrocarbons. *BIO Pacific Rim Summit*, Vancouver, BC, September 10–12.
13. Ensyn press release, October 2008. Available online at: http://www.ensyn.com/news/UOP_Ensyn_Joint_Venture.pdf .
14. DOE press release, 14 July 2008. Available online at: http://www.doe.gov/news/6413.htm.
15. DOE press release, 18 April 2008. Available online at: http://www.doe.gov/6164.htm.
16. Simander, G.R. 2004. *Case Study: 2 MW_e biomass gasification plant in Güssing (Austria).* Austrian Energy Agency, April. Available online at http://www.opet-chp.net/download/wp3/g%FCssingaustria.pdf.
17. http://www.renet.at/english/sites/guessing/technology.php.
18. Rudloff, M. 2007. Operation experiences of Carbo-V process for FTD production. Presented at *Synbios II: International Conference on the Introduction of Second Generation Automotive Biofuels*, Stockholm, Sweden, 23–24 May.
19. Choren press release, 11 October 2007. Available online at: http://www.choren.com/en/choren_industries/information_press/press_releases/?nid=177.
20. UPM-Kymmene press release, 23 May 2007. Available online at: http://w3.upmkymmene.com/upm/internet/cms/upmcms.nsf/($all)/3ed4d077ffc471b1c22572e4002de186?OpenDocument&qm=menu,0,0,9.
21. Neste Oil press release, 16 March 2007. Available online at: http://www.nesteoil.com/default.asp?path = 1;41;540;1259;1261;7440;7846.
22. Brown, C.A., Leary, R., Gorog, J.P., and Abdullah, Z. 2004. The Chemrec black liquor gasifier at New Bern—a status report. *Preprints, 2004 International Chemical Recovery Conference*, Charleston SC, 6–11 June.

23. Landälv, I. 2007. The status of the Chemrec black liquor gasification concept. *2nd European Summer School on Renewable Motor Fuels*, Warsaw, Poland, 29–31 August. Available online at: http://www.baumgroup.de/Renew/download/8%20-%20Landalv%20-%20slides.pdf.
24. Davy, M.F., Uloth, V.C., and Cloutier, J.-N. 1994. Economic evaluation of black liquor treatment processes for incremental kraft pulp production. *Pulp and Paper Canada* 99(2): T51–T55 (1998).
25. Mao, H., Genco, J.M., van Heiningen, A.R.P., and Pendse, H.J. 2008. Towards economic evaluation of a hardwood biorefinery using the "near-neutral" hemicellulose pre-extraction process. *Biobased Materials Bioenergy* 2: 1–9.
26. Boluk, Y., Yuan, Z., Tosto, F., and Browne, T.C. 2008. Acid prehydrolysis and extraction of hemicellulose prior to aspen chemi-thermomechanical pulping, *AICHE Annual Spring Meeting*, New Orleans, LA, April, 2008.
27. Willfor, S., Rehn, P., Sundberg, A., Sundberg, K., and Holmbom, B. 2003. Recovery of water-soluble acetylgalactoglucomannans from mechanical pulp of spruce. *Tappi J.* 2(11): 27–32.
28. Jeffries, T.W. and Jin, Y.S. 2004. Metabolic engineering for improved fermentation of pentoses by yeasts. *Applied Microbiology Biotechnology* 63: 495–509.
29. Zhang, X., Qin, W. Paice M., and Saddler, J.N. 2009. High consistency hydrolysis of pretreated hardwood substrate. *Bioresource Technology*, 100(23): 5890–5897
30. Krause, C. 2007. ORNL's carbon fiber pilot plant. *Oak Ridge National Laboratory Review* 40(2) (available on-line at: http://www.ornl.gov/info/ornlreview/v40_2_07/index.shtml).
31. Spath, P.L. and Dayton, D.C. 2003. *Preliminary Screening—Technical and Economic Assessment of Synthesis Gas to Fuels and Chemicals with Emphasis on the Potential for Biomass-Derived Syngas.* U.S. Department of Energy, December. Available online at: http://www.nrel.gov/docs/fy04osti/34929.pdf
32. Springer, A.M., Dietrich-Velazquez, G., Higby, C.M., and DiGiacomo, D. 1996. Feasibility study of sludge lysis and recycle in the activated sludge process. *Tappi J.* 79(5): 162–170.
33. Kantardjieff, A. and Jones, J. 2000. Pulp and paper bio-solids dewatering: Why we can win the war with water. A freeze/thaw technology increased final dryness to 35% from 13%. *Pulp and Paper Canada* 101(10): 56–59.
34. Kantardjieff, A. and Jones, J. 2005. *Pulp and Paper Sludge to Energy: Preliminary Assessment of Technologies.* A Canmet Energy Technology Centre Report prepared by ADI Limited; Canmet report #34–0173–479.1.
35. Smith, M., Fournier, P., and de Vegt, A. 1994. Operating experience at Lake Utopia Paper Co., St. George, New Brunswick, increases confidence in UASB process. *Proceedings, TAPPI International Environmental Conference*, Portland, OR, 1, 153–156.
36. Tchobanoglous, G., Burton, F.L., and Stensel, H.D. (eds.) 2003. *Wastewater Engineering: Treatment and Reuse*, 4th ed. (originally by Metcalf and Eddy). McGraw-Hill, New York.
37. Elliott, A. and Mahmood, T. 2012. Comparison of mechanical pretreatment methods for the enhancement of anaerobic digestion of pulp and paper waste activated sludge. *Water Environment Research* 84(6): 497–505.
38. Zhu, J.Y., Yoon, S.H., Liu, P.H., and Chai, X.S. 2000. Methanol formation during alkaline wood pulping. *Tappi J.* 83(7): 65–78.
39. Burgess, T.L. 1990. The basics of foul condensate stripping, *Tappi Engineering Conference Proceedings*, TAPPI Press, Atlanta, pp. 835–839.
40. Jemaa, N., Paleologou, M., and O'Connor, B. 2010. US Patent Application US20100122784, Process for treating pulp mill condensates using a hollow fiber contactor, May 20, 2010.
41. Jemaa, N., Paleologou, M., Sacciadis, G., and Berry, R. 2010. US Patent Application US20100263816, Use of chemical pulp mill steam stripper off gas condensate as a reducing agent in chlorine dioxide production, October 21, 2010.
42. Marinova, M., Baudry, C., Taoussi, A., Trépanier, M., and Paris, J. 2008. Economic assessment of rural district heating by bio-steam supplied by a paper mill in Canada. *Bulletin of Science, Technology & Society*. 28(2): 159–173.

19

Thermochemical Conversion of Biomass to Liquid Fuels and Chemicals

W. J. Frederick, Jr.

CONTENTS

19.1 Gasification Chemistry

Biomass gasification has been used commercially since the mid-1800s. The products of gasification are combustible, containing CO, H_2, and methane. Gasification was first used to generate a fuel gas for heating or to power combustion engines, and this is still the most prevalent application of biomass gasification. Examples of fuel-gas production from biomass and its use for power and heat generation include the 40–70 MW Kymijärvi power plant in Lahti, Finland, Carbona's 5.5 MW_e/11.5 MW_{th} district heating plant, the gasifier-driven CHP plant in Skive, Denmark, and the Norampac 18 MW_{th} black liquor gasifier-driven steam plant in Trenton, Ontario. Today, many chemicals, fuels, and fertilizers are produced from synthetic gas (syngas), as illustrated in Figure 19.1. As sustainable use of renewable biomass feedstocks for these products increases in importance, biomass-derived syngas will assume an increasingly important role in meeting society's needs for energy, materials, and food production.

Gasification deconstructs biomass and converts it to light gases, primarily CO, H_2, CO_2, water vapor, and methane. Other organic materials, including tar and light hydrocarbon species, are also produced. Because gasification breaks down biomass completely to light gases and a few other volatile compounds, it can be used to convert any biomass source.

The important chemical processes for converting biomass to syngas are shown in Equations 19.1 through 19.4:

$$\text{Biomass} + \text{heat} => \text{pyrolysis products } (CO, H_2, CO_2, \text{water vapor}, CH_4, \text{light hydrocarbons, tar, C}) \tag{19.1}$$

$$\text{Pyrolysis products} + O_2 => CO, CO_2, H_2O(v) \tag{19.2}$$

$$C + H_2O => CO + H_2 \tag{19.3}$$

$$C + CO_2 => 2CO \tag{19.4}$$

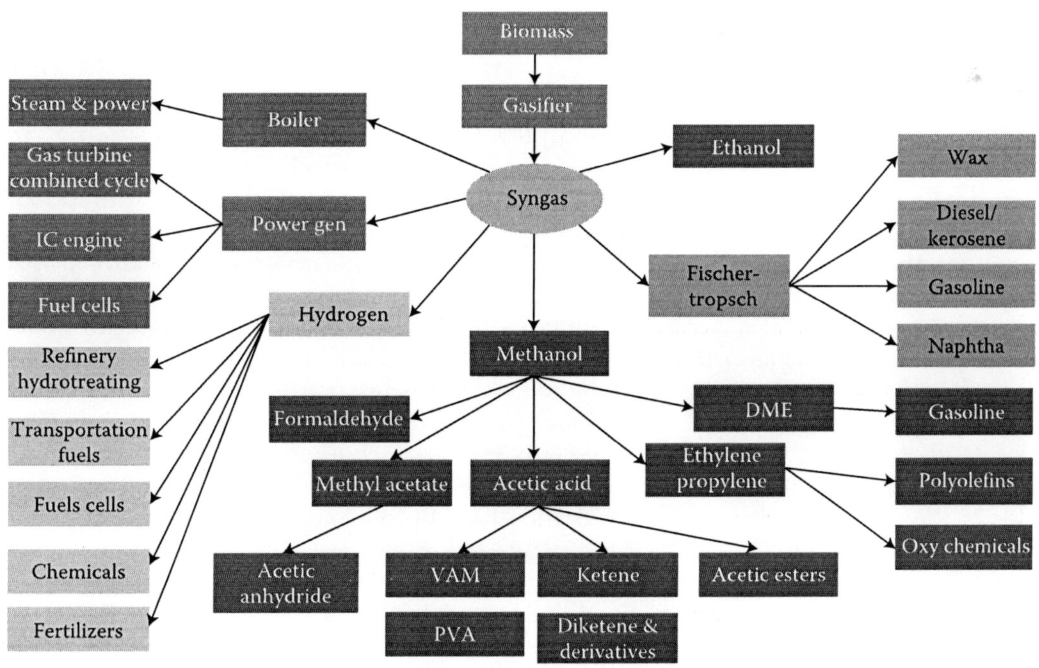

FIGURE 19.1
Some of the biobased products that can be produced from syngas. (Adapted from Elbert, J. 2008. *Biomass Magazine*, January. (http://www.biomassmagazine.com/article.jsp?article_id=1399&q=&page=all).)

As Equations 19.1 through 19.4 suggest, biomass gasification proceeds in several stages. In the first, biomass decomposes to a number of light gases, volatile organic compounds, and fixed carbon (reaction 19.1). At typical gasifier temperatures, the time frame for devolatilization is in the order of 1 s. The second stage occurs only if O_2 is supplied to the gasifier, either as air or as nearly pure O_2. In this stage, the combustible volatile gases are oxidized rapidly (in seconds) to produce CO, CO_2, and water vapor (reaction 19.2). In the third stage, fixed carbon is converted to CO by reaction with water vapor (reaction 19.3) and CO_2 (reaction 19.4). The time frame for the third stage is typically minutes to hours, depending on the type of gasifier and the gasification temperature and pressure.

Biomass is inherently oxygen-rich, containing about 43–46 wt% O on a dry, ash-free basis, or 0.64–0.71 atoms of oxygen per atom of carbon. However, conversion of the C and H in biomass to CO, CO_2, H_2, and H_2O(v) requires twice the amount of oxygen that is available in biomass. The additional oxygen required for gasification is supplied as elemental oxygen, either in air, as nearly pure O_2, or as steam.

The char carbon produced during devolatilization of biomass can be gasified completely with air, O_2, or mixtures that contain both air or oxygen and steam. However, complete gasification of char carbon is difficult to obtain with steam alone, and carbon conversions with steam are more typically 86–92% (Kurkela and Kurkela, 2009).

At gasification temperatures, the water gas shift (WGS) reaction, Equation 19.5, plays an important role in determining the distribution of carbon, hydrogen, and oxygen, the main chemical species in syngas. This reaction acts as an equilibrium constraint at temperatures above 700°C when alkali or alkaline earth metals are present to catalyze it (Huttinger et al., 1986; Meijer et al., 1991a):

$$CO + H_2O(v) <=> CO_2 + H_2 \tag{19.5}$$

Methane, other light hydrocarbons, and polyaromatic hydrocarbons (PAHs or tar) are undesired products from gasification of biomass. Methane can be present in substantial quantities in raw syngas, as much as 10–15 mol% (dry basis) (Evans and Milne, 1987; Sricharoenchaikul, 2002a; Kurkela and Kurkela, 2009). It derives primarily from decomposition of methoxy groups on lignin. Aromatic hydrocarbon structures, originating as primary devolatilization products from lignin, provide one starting point for PAH formation. Aromatic precursors of PAHs are also obtained from cellulose and xylan where aromatic structures do not exist as primary products. At gasification temperatures of 1100°C or below, tar is always produced. It is present in raw syngas in quantities equivalent to 3–10% of carbon input as biomass, with the amount produced decreasing with increasing gasification temperature (van der Drift et al., 2001; Sricharoenchaikul et al., 2002b; Kuo et al., 2010; Wolfesberger et al., 2009). The PAHs that make up the tar must be reformed for higher yields of the desired syngas species (CO and H_2), and to avoid coking and poisoning of syngas conversion catalysts.

Some inorganic components in biomass, particularly alkali metals, catalyze carbon gasification by creating oxygen-exchange sites within the carbon matrix (Meijer et al., 1991b). They create sites onto which CO_2 and water vapor adsorb and give up oxygen atoms, making the oxygen atoms available to react with the solid carbon matrix to produce CO.

Gasification of black liquor falls within the realm of biomass gasification. Black liquor is a biomass-rich residue from cellulose fiber production, containing about 60 wt% biomass on a dry basis. It is a major industrial energy source in many parts of the world, including North America, the Nordic countries, parts of South America, and some Asian countries.

The biomass in black liquor is derived from delignification of woody biomass to produce cellulose fibers. It consists of approximately half lignin degradation products and half carbohydrate-derived sugars, oligomers, acetate, formate, and other organic matter. Black liquor also contains far more inorganic matter than virgin biomass, including sodium (<20 wt%), sulfur (3–7 wt%), and smaller amounts of potassium and chloride (Frederick, 1997). The inorganic products of gasification of black liquor are typically 90 wt% Na_2CO_3, Na_2S and NaCl and 10 wt% of the equivalent potassium salts. Black liquor gasification has the same chemistry as gasification of other biomass feedstock. However, it gasifies three to four orders of magnitude faster than conventional biomass at the same temperature. Its higher rate of gasification is attributable to catalysis by alkali metal salts (Cerfontain et al., 1988; Li and van Heiningen, 1991).

References

Baxter, L.L. 2005. Biomass-coal co-combustion: Opportunity for affordable renewable energy. *Fuel* 84(10): 1295–1302.

Cerfontain, M.B., Kapteijn, F., and Moulijn J.A. 1988. Characterization of alkali carbonate catalysts for carbon gasification with ^{18}O labeled CO_2. *Carbon* 26(1): 41–48.

Elbert, J. 2008. Syngas 101. *Biomass Magazine*, January. (http://www.biomassmagazine.com/article.jsp?article_id=1399&q=&page=all)

Frederick, W.J. 1997. Black Liquor Properties. Chapter 3 In: Adams, T.N. (ed.), *Kraft Recovery Boilers*, TAPPI Press, Atlanta, pp. 63–65.

Huttinger, K.J., Masling, B., and Minges, R. 1986. Catalytic activity of carbon-supported potassium in the carbon monoxide shift reaction. *Fuel* 65(7): 932–936.

Kuo, J-H., Lian, Y-H., Rau, J-Y., Wey, M-Y., and Lin, C-L. 2010. Effect of operating conditions on emission concentration of PAHs during fluidized bed air gasification of biomass. *Chemistry and Chemical Engineering (ICCCE), 2010 International Conference*, Kyoto, 1–3 August, 76–79.

Li, J. and van Heiningen, A.R.P. 1991. Kinetics of gasification of black liquor char by steam. *Ind. Eng. Chem. Res.* 30(7): 1594–1601.

Llorente, M.J.F., Laplaza, J.M.M., Cuadrado, R.E., and García, J.E.C. 2006. Ash behaviour of lignocellulosic biomass in bubbling fluidised bed combustion. *Fuel* 85(9): 1157–1165.

Meijer, R., Sibeijn, M., van Dillen, M.R.B., Kapteijn, F., and Moulijn, J.A. 1991a. Kinetics of the alkali-metal-carbonate-catalyzed gasification of carbon: The water-gas-shift reaction. *Ind. Eng. Chem. Res.* 30: 1760–1770.

Meijer, R., van der Linden, B., Kapteijn, F., and Moulijn, J.A. 1991b. The interaction of H_2O, CO_2, H_2, and CO with the alkali-carbonate/carbon system: a thermogravimetric study. *Fuel* 70: 205–214.

Senneca, O. 2007. Kinetics of pyrolysis, combustion, and gasification of three biomass fuels. *Fuel Process. Technol.* 88: 87–97.

Sricharoenchaikul, V., Agrawal, P., and Frederick, W.J. Jr. 2002a. Black liquor gasification characteristics. 1. Formation and conversion of carbon-containing product gases during gasification of black liquor. *Ind. Eng. Chem. Res.* 41(23): 5640–5649.

Sricharoenchaikul, V., Agrawal, P., and Frederick, W.J. Jr. 2002b. Black liquor gasification characteristics. 2. Measurement of condensable organic matter (tar) at rapid heating conditions. *Ind. Eng. Chem. Res.* 41(23): 5650–5658.

Thy, P., Jenkins, B.M., Grundvig, S., Shiraki, R., and Lesher, C.E. 2005. High-temperature elemental losses and mineralogical changes in common biomass ashes. *Fuel* 85: 783–795.

van der Drift, A., van Doorn, J., and Vermeulen, J.W. 2001. Ten residual biomass fuels for circulating fluidized-bed gasification. *Biomass Bioenergy* 20(1): 45–56.
Wolfesberger, U., Aigner, I., and Hofbauer, H. 2009. Tar content and composition in producer gas of fluidized bed gasification of wood: Influence of temperature and pressure. *TCbiomass2009*, Chicago, 16–18 September.

19.2 Gasification Process Technology

The development of biomass gasification process technology has followed the coal gasification technology development, although the process technologies differ to accommodate several important differences in the character and composition of the two feedstocks. For one, biomass, with its higher oxygen content, gasifies one to two orders of magnitude more rapidly than high-rank coals at the same temperature and behaves similarly to lignite when gasified (Higman and van der Burgt, 2003). As a result, biomass is normally gasified at temperatures between 750 and 900°C, while high-rank coals are gasified at temperatures of 1100–1500°C (Higman and van der Burgt, 2003).

Another important difference is inorganic content. Woody biomass normally contains 1% or less inorganic matter, with potassium and calcium as two of the major inorganic components. Agricultural residues and energy crops contain more, from 2% to 20%, inorganic matter, with silicon and potassium as the main components. By comparison, higher-rank coals contain 5–9% inorganic matter, while lignite coals contain even higher amounts (Smoot, 1991). Silica and calcium are the predominant ash components of high-rank coals (Smoot, 1991). Biomass contains approximately two orders of magnitude less sulfur than coals, but it contains chloride, which is not a significant inorganic component of coals. Biomass ash has a lower melting point than coal ash (<600°C versus >1200°C) (Baxter, 2005; Thy et al., 2005; Llorente et al., 2006).

A third important difference is that biomass is a fibrous material while coal is not. As a result, biomass is more difficult to feed because of its tendency to bridge and plug feeders.

The development path for biomass gasifiers has almost always been first to produce a combustible fuel gas. Combustion of hot, raw syngas in a steam power boiler does not require a fuel gas with low levels of tar, methane, and other contaminants. This is a simpler application than producing clean syngas for synthesis of fuels and chemicals because of the rigorous impurity removal requirements for catalytic synthesis of organic compounds. Once a gasifier design has been proven in CHP applications, production of clean syngas for synthesis of fuels and chemicals is a logical next stage of development.

19.2.1 Types of Biomass Gasifiers

A number of biomass gasifiers are available commercially, varying in design, capacity, operating pressure and temperature, and raw product-gas quality.

19.2.1.1 Single Fluidized-Bed Gasifiers

Single fluidized-bed gasifiers are the most prevalent types of biomass gasifiers. They are designed either as stationary-bed or as circulating-bed gasifiers as illustrated in Figure 19.2. Stationary-bed gasifiers consist of a hot bubbling fluidized bed of inorganic particles, such as 1-mm quartz sand or olivine particles, into which biomass is fed and converted mainly

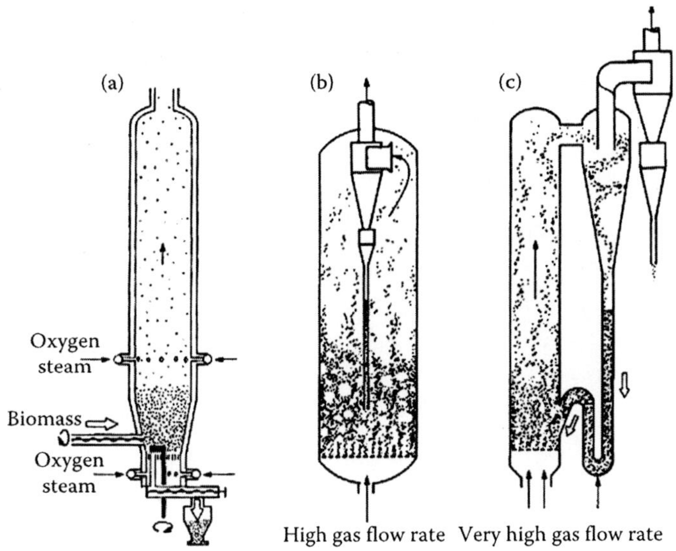

FIGURE 19.2
Types of fluidized-bed gasifiers: (a) stationary bed, (b) turbulent bed, and (c) circulating bed. (Adapted from Kunii, D. and Levenspiel, O. 1991. *Fluidization Engineering*, 2nd ed. Butterworth-Heinemann, Boston.)

to CO, CO_2, H_2, H_2O(v), and methane. The term *stationary-bed* implies that the bed material in this type of gasifier remains in the gasifier and is not transported out with the product gas. In circulating-bed reactors, the bed material is finer, typically 0.2–0.4 mm particles. Gas velocities are high enough to transport bed material out of the gasifier where it is collected along with unconverted carbon and ash particles and returned to the bed. The unconverted carbon may be burned off to preheat the bed material before it is returned to the gasifier.

Turbulent flow in both stationary and circulating fluidized-bed gasifiers provides rapid mixing of biomass particles with the hot bed material and rapid heat transfer to them. As a result, devolatilization is rapid, but gasification of the remaining fixed carbon is slower and strongly temperature-dependent (Senneca, 2007).

Both stationary and circulating fluidized-bed gasifiers operate at temperatures from 750°C to 900°C. The advantages of these gasifiers are rapid gasification of biomass, uniformity of conditions within the gasifier, and flexibility to process a range of biomass materials. The disadvantages include incomplete carbon conversion, especially at lower temperatures and with steam gasification (Kurkela, 2009), significant tar production (Sricharenchaikul, 2002b), and agglomeration of bed material by low-melting-point ash. At temperatures below 750°C, conversion of fixed carbon occurs so slowly that carbon may accumulate in the reactor.

The gasifiers listed in Table 19.1 are single fluidized-bed gasifiers of both stationary- and circulating-bed types. Most operate at atmospheric pressure, although one operates at elevated pressure (<20 bar).

19.2.1.2 Indirectly Heated Twin Circulating Fluidized-Bed Gasifiers

Indirectly heated twin circulating fluidized-bed gasifiers (Table 19.2), illustrated in Figure 19.3, recirculate bed material between them as a heat carrier. With this type of gasifier, biomass is heated and gasified in one of the fluidized beds by mixing it with hot sand at

TABLE 19.1

Single Fluidized-Bed Biomass Gasifiers

Type of Gasifier	Stationary Fluidized Bed					Circulating Fluidized Bed	Stationary Fluidized Bed with Internal Heat Exchangers
Provider	Carbona-Andritz (pressurized)	Carbona-Andritz (near-atmospheric)	Energy Products of Idaho (EPI)[a]	Enerkem	Frontline BioEnergy	Foster Wheeler	TRI
Gasifier type	Pressurized stationary fluidized bed	Stationary fluidized bed	Staged stationary fluidized bed	Stationary fluidized bed	Stationary fluidized-bed gasifier	Atmospheric circulating fluidized bed	Stationary fluidized-bed steam reformer, indirectly heated with in-bed pulse combustors
Operating temperature	850–900°C	850°C	540–980°C	750°C	Unknown	750–850°C	780°C
Operating pressure	21 bar	2 bar	1 bar	2.1 bar	1 bar	1 bar	5.5 bar
Fluidizing medium	Air	Air	Air	Air- or O_2-enriched air and steam	Air	Air	Steam
Heating method	Partial combustion	Partial combustion	Partial combustion	Partial combustion	Partial combustion	Partial combustion	Indirectly heated with in-bed pulse combustors
Product gas heating value	Low	Low	Low	Low to medium	Low	Low	Medium

TABLE 19.2

Indirectly Heated Twin Circulating Fluid-Bed Gasifiers

Provider	SilvaGas/RenTech	Taylor Energy	Repotech GmbH
Gasifier type	Indirectly heated twin circulating fluidized bed	Indirectly heated twin circulating fluidized bed	Indirectly heated two-stage fluidized bed (CFB gasifier and BB combustor)
Operating temperature	Gasifier 815°C Combustor 925°C	Gasifier 815°C Combustor 980°C	Gasifier 850°C
Operating pressure	Slightly above atmospheric	Slightly above atmospheric	Slightly above atmospheric
Fluidizing medium	Gasifier: steam Combustor: air	Gasifier: steam Combustor: air	Gasifier: steam Combustor: air
Product gas heating value	Medium	Medium	Medium

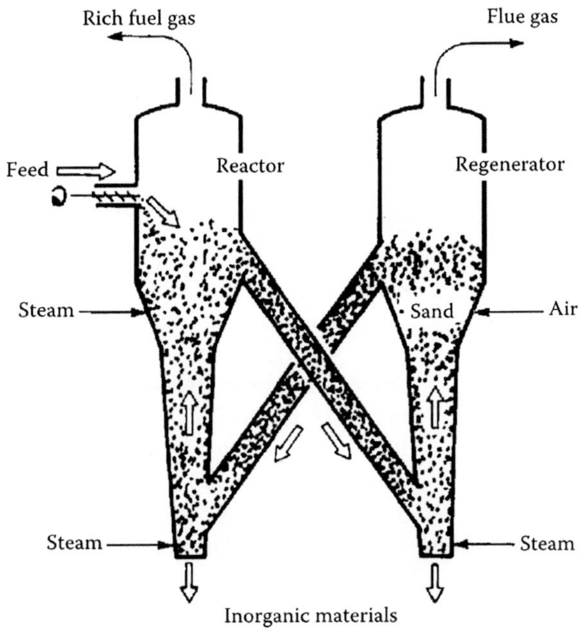

FIGURE 19.3
Twin-bed fluidized-bed gasifier for syngas or fuel gas production. (Adapted from Kunii, D. and Levenspiel, O. 1991. *Fluidization Engineering*, 2nd ed. Butterworth-Heinemann, Boston.)

820–850°C. The hot sand provides heat to drive the net endothermic gasification reactions. The bed is fluidized with steam, and the gasifier operates as a circulating fluidized bed in which the fluidizing gas carries the bed contents out of the top of the reactor. Carbon is not gasified completely. The raw product gas is separated from residual solid carbon and sand in cyclones at the exit near the top of the gasifier.

The residual carbon and sand exiting the gasifier are fed to a second circulating fluidized bed where the residual carbon is burned to reheat the sand. The combustor is fluidized with air. The fluidizing combustion gas and bed contents exit at the top of the combustor. Hot sand at 925–980°C is separated from the combustion gas by cyclones. The sand is returned to the bottom of the gasifier to complete the heat cycle between the gasifier and the combustor.

TABLE 19.3

Entrained-Flow Biomass Gasifiers

Provider	Lurgi-Air Liquide-KIT	CHEMREC
Gasifier type	Pyrolysis followed by oxygen-blown entrained-flow gasifier with cooling shield	Oxygen-blown entrained-flow gasifier
Operating temperature	Pyrolysis: 500°C Gasification of pyrolysis oil: 1200°C	950–1000°C
Operating pressure	30–80 bar	30 bar
Intended use	Production of liquid fuels	Syngas from gasification of black liquor for liquid fuels production
Product gas heating value	Medium	Medium

These gasifiers produce a medium-heating-value gas because residual carbon rather than combustible product gases is burned to provide heat for the gasifier (Green Tech Media, 2010). Air is used for char combustion, but not for gasification so that nitrogen from combustion air will not dilute the gasifier product gas.

Characteristics of three of these types of gasifier are presented in Table 19.2. The design of the twin circulating fluidized-bed gasifiers is similar to that of the twin-bed catalytic crackers used in petroleum refining (Gary et al., 2007).

19.2.1.3 Entrained-Flow Gasifiers

Entrained-flow gasifiers form a third category of biomass gasifiers; some of their characteristics are included in Table 19.3. These gasifiers are oxygen-blown. They operate at elevated pressure (30–80 bar) and temperatures between 1000°C and 1200°C, depending upon the reactivity of the char carbon generated. Residence times are short (a few seconds), and therefore carbon gasification must proceed rapidly. At the higher temperatures used, tar and methane are not produced, and ash is molten at gasification conditions. The clean product gas is a medium-heating-value gas that can be used for synthesis of fuels and chemicals or as a combustion fuel.

The CHEMREC gasifier, illustrated in Figure 19.4, combines the features of pressurized entrained-flow gasification with a unique gas cleaning and chemical recovery capability. The gasifier is designed to convert black liquor to syngas and regenerated pulping chemicals. It operates at 950–1000°C and 30 bar. Black liquor is introduced to the gasifier as a fine droplet spray, and the residence time of the droplets is a few seconds. The products of gasification are quenched, and inorganic matter, primarily sodium and potassium carbonates, sulfides, and chlorides, is removed with the quench solution. The hot product gas is cooled further to generate process steam. H_2S is scrubbed from the cooled product gas. The H_2S solution and the inorganic matter are recycled for reuse as pulping chemicals.

19.2.1.4 Other Types of Gasifiers

Other types of gasifiers have been developed that are combinations of pyrolysis and gasification reactors (Table 19.4). One is a multistage gasifier in which biomass is pyrolyzed, the pyrolysis gas and char are separated, the pyrolysis gas is burned, and the hot combustion gas is used to gasify the carbon produced in the char during pyrolysis (Choren design; Bienert, 2007, 2010). Its advantages are high carbon conversion to syngas and absence of tar or methane production. Another type of biomass gasifier is the ClearFuels hydrothermal reformer (2010). A third combines a fixed-bed gasifier with a plasma assist (Alter, 2008).

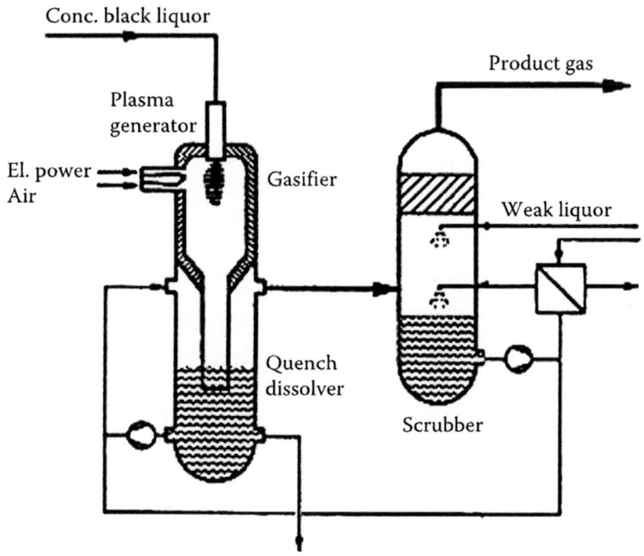

FIGURE 19.4
CHEMREC gasifier, quench dissolver for green liquor production, and gas scrubber for hydrogen sulfide absorption.

TABLE 19.4
Other Gasifiers

Provider	Alter NRG	CHOREN	ClearFuels
Gasifier type	Fixed bed with plasma-assisted cupola	Pyrolyzer, oxygen-enriched bio-char combustor, and entrained-flow gasifier	Hydrothermal reformer (HEHTR). Based on technology developed by Pearson (2008) (see US7375142)
Operating temperature	1100°C	1400–1600°C	956°C
Operating pressure	Near atmospheric	5 bar	Pressurized

Fixed-bed gasifiers of both updraft and downdraft designs are used for biomass gasification, but have smaller capacities than the gasifiers discussed above. These gasifiers are not discussed further in this chapter.

19.2.2 Direct and Indirect Heating of Biomass Gasifiers

The overall process of biomass gasification is net endothermic, and therefore heat must be provided to gasifiers to drive the gasification reactions. When air or O_2 are the gasification agents, biomass is partially combusted, and the heat released provides the heat needed for gasification. These gasifiers are called *autothermal*. With an autothermal gasifier, less total CO and hydrogen are produced per unit of biomass, decreasing the heating value of the gasifier product gas. Using air as the gasifying medium also reduces the heating value of the product gas by diluting it with nitrogen. Gasifiers that use partial combustion of CO and H_2 to heat themselves produce a low-heating-value gas.

TABLE 19.5

Product Gas Composition and Heating Value Ranges (Dry Gas Basis) for Different Classes of Biomass Gasifiers and Gasification Agents

Heat Provision	Autothermal		Allothermal
Gasifying agent	Air	$O_2/H_2O(v)$	$H_2O(v)$
H_2 (vol%)	11–16	23–28	35–40
CO (vol%)	13–18	45–55	22–25
CO_2 (vol%)	12–16	10–15	20–25
CH_4 (vol%)	2–6	<1	9–11
N_2 (vol%)	45–60	<5	<1
LHV (MJ/Nm³)	4–6	10–12	12–14

Source: Adapted from Bacovsky D. et al. 2010. *Status of Second-Generation Biofuels Demonstration Facilities in June 2010.* IEA Bioenergy Task 39, Report T39-P1b (27 July).

Indirect heating of a fluidized-bed gasifier eliminates in-bed partial combustion of CO and H_2 and avoids dilution of the product gas with nitrogen. Gasifiers that used indirect heating are called *allothermal*. An allothermal gasifier produces a medium-heating-value gas. This concept is used in twin-circulating-bed gasifiers, where carbon is burned separately from the gasification reactor, and in the TRI gasifier, where combustible gas is burned in pulse-combustor heat exchangers located within the fluidized bed. Overall, the fuel combusted in either type of gasifier decreases the net energy yield from the gasification plant, and careful heat integration is required to maximize energy product yield. The heating value of the product gas from indirectly heated fluidized-bed gasifiers is higher than that from directly heated fluidized-bed gasifiers.

Ranges of product gas composition and lower heating value for different categories of gasifiers and gasifying agents are shown in Table 19.5.

A gasification concept proposed by Lurgi-Air Liquide-KIT (Lane, 2009) uses vacuum pyrolysis at 500°C to generate a bio-oil from wood and agricultural residues. The bio-oil is subsequently gasified and heated to 1400°C at 30–80 bar. The resulting syngas is catalytically converted into synthetic diesel, hydrogen, or methanol.

A concept for producing biofuels by combining pyrolysis and gasification is under development in Germany (Heinrich et al., 2009; Oasmaa et al., 2010): Bio-oil/char slurries are produced by gasification of straw in distributed pyrolysis plants. The bio-oil slurry is transported to a centralized gasification plant coupled to a Fischer–Tropsch plant that produces hydrocarbon biofuel. Currently, 40 small pyrolysis plants process enough straw to produce 1 Mt/annum of hydrocarbon biofuel.

References

Alter, N.R.G. 2008. *The Plasma Gasification Process.* Calgary & Region Waste Reduction Partnership, 13 February. http://www.calgaryregion.ca/crp/media/32152/the_alter_nrg_plasma_gasification_process_-_turner_valley_presentation.ppt.pdf.

Bacovsky, D., Dallos, M., and Wörgetter, M. 2010. *Status of Second-Generation Biofuels Demonstration Facilities in June 2010.* IEA Bioenergy Task 39, Report T39-P1b (27 July).

Bienert, K. 2007. The status of the Choren Carbo V gasification plant. *2nd European Summer School on Renewable Motor Fuels*, Warsaw, Poland, 29–31 August.

Bienert, K. 2010. *ClearFuels Hydrothermal Reformer*. http://www.clearfuels.com/technology.php

Gary, J.H., Handwerk, G.E., and Kaiser, M.J. 2007. *Petroleum Refining. Technology and Economics*, 5th ed. CRC Press, Boca Raton, FL, pp. 122–133.

Gary, J.H., Handwerk, G.E., and Kaiser, M.J. 2010. Rentech buys Silvagas gasification technologies. Green Tech Media, http://www.greentechmedia.com/green-light/post/green-patent-acquisitions-rentech-buys-silvagas-gasification-technologies/, August.

Henrich, E., Dahmen, N., and Dinjus, E. 2009. Cost estimate for biosynfuel production via biosyncrude gasification. *Biofuels, Bioprod. Bioref.* 3: 28–41.

Higman, C. and van der Burgt, M. 2003. *Gasification*. Elsevier, Amsterdam.

Jenkins, B.M., Baxter, L.L., Miles, T.R. Jr., and Miles, T.R. 1998. Combustion properties of biomass. *Fuel Process. Technol.* 54: 17–46.

Kunii, D. and Levenspiel, O. 1991. *Fluidization Engineering*, 2nd ed. Butterworth-Heinemann, Boston.

Kurkela, E. and Kurkela, M. 2009. Fluidized-bed gasification of biomass for syngas applications: Comparison of gasification process alternatives. *TCbiomass2009*, Chicago, 16–18 September.

Lane, J. 2009. KIT, Lurgi to develop pyrolysis, gasification pilot plant, with goal of $2.49 renewable diesel. *BioFuels Digest*, 30 January. www.biofuelsdigest.com/blog2/2009/01/30/kit-lurgi-to-develop-pyrolysis-gasification-pilot-plant-with-goal-of-249-renewable-diesel/.

Oasmaa, A., Solantausta, Y., Arpiainen, V., Kuoppala, E., and Kai Sipilä, K. 2010. Fast pyrolysis bio-oils from wood and agricultural residues. *Energy Fuels* 24: 1380–1388.

Pearson, S.R. 2008. *Process and Apparatus for the Production of Useful Products from Carbonaceous Feedstock*. Patent No. US7373142, 20 May, assigned to Pearson Technologies, Inc.

Smoot, L.D. 1991. Coal and char combustion. Chapter 10 In: Bartok, W. and Sarofim, A.F. (eds.), *Fossil Fuel Combustion*. Wiley, New York.

19.3 Syngas Cleaning and Conditioning

19.3.1 Syngas Cleanliness Requirements

Raw synthesis gas contains CO and H_2, the desired components for production of synthetic fuels and chemicals, but also other species that are not wanted. CO_2 and water vapor are major products of biomass gasification that must be removed prior to catalytic processing of syngas (Ciferno and Marano, 2002). For synthesis of chemical and fuels, organic matter, including tar species, methane, and other light hydrocarbons, represent both incomplete conversion of biomass to the desired energetic products (CO and H_2) and contaminants that poison syngas conversion catalysts. However, for power generation, methane and other light hydrocarbons are not problematic. Sulfur- and nitrogen-containing gases, alkali metal salt vapors, and chlorides are not tolerable in gas turbines or in catalytic syngas reactors at their concentrations in raw syngas. The range concentrations of these contaminants found in raw biomass-derived syngas are shown in Table 19.6. Table 19.7 shows the maximum acceptable levels of these contaminants in clean synthesis gas for catalytic conversion to synthetic fuels and chemicals.

Overall, gas cleaning trains that provide sufficiently clean syngas for catalytic conversion to synthetic fuels and chemicals are not in widespread use industrially. Cleanup of raw syngas from coal is practiced in South Africa for syngas used as a fuel and not for chemical synthesis. Raw syngas cleanup trains for synthesis-compatible syngas is an active area of development in the United States and Europe (Kurkela, 2009; Gibson, 2011).

The challenges, status, and limitations of the stages of syngas cleaning are discussed in the following subsections.

TABLE 19.6

Syngas Contaminants, Problems they Cause in Biomass Gasifiers, and Cleanup Methods

Contaminant	Examples	Problems	Cleanup Method
Particulates	Ash, char, bed material	Erosion, plugging, catalyst damage	Filtering, scrubbing
Alkali metals	Sodium, potassium	Catalyst poisons	Condensation, filtering, guard beds
Fuel-bound nitrogen	Mainly NH_3 and HCN	NO_x formation in combustion; catalyst poisons	Scrubbing, selective catalytic reduction
Tar	Reactive aromatics	Filter plugging, internal condensation and deposition, catalyst poisons	Tar reforming, absorption
Sulfur, chlorine	H_2S, COS, HCl	Catalyst poisons, corrosion, gaseous sulfur emissions	Limestone or dolomite, zinc-based guard beds, absorption

Source: Adapted from Konttinen, J. 2010. Evaluation of gasification-based energy opportunities for forest industries. Presented at the AEBIOM Conference, Brussels, 28–29 April; Bridgewater, A. et al., 1998. An Assessment of the Possibilities for Transfer of Biomass Gasification Technology to China. Part 1. Report of the Mission to China. 65 p.

TABLE 19.7

Maximum Acceptable Levels of Various Contaminants in Clean Synthesis Gas for Catalytic Conversion to Synthetic Fuels and Chemicals

Process	Contaminant	Maximum Level
Methanol synthesis	Sulfur gases excluding COS	<0.1 ppmv
	Halides	1 ppbv
	Fe and Ni	5 ppbv
Low-pressure methanol synthesis	Sulfur gases including COS	0.1 ppmv
	Halides	10 ppbv
	Acetylene	5 ppmv
	Total unsaturates	300 ppmv
	NH_3	10 ppmv
	HCN	10 ppbv
	Fe and Ni	10 ppbv
Ethanol synthesis	Very little work published	
Higher alcohol synthesis	Same as for methanol synthesis	
Fischer–Tropsch synthesis	Sulfur gases excluding COS	0.2 ppmv
	Halides	10 ppb
	NH_3	10 ppmv
	NO_x	0.2 ppmv
	HCN	10 ppb
Steam methane reforming		
Ammonia synthesis	H_2O	200 ppm
	CO	200 ppm
	CO_2	100 ppm
	O_2	100 ppm
	H_2S	0.1 ppm
	Cl	0.1 ppm

Source: Adapted from Spath, P.L. and Dayton, D.C. 2003. *Preliminary Screening: Technical and Economic Assessment of Synthesis Gas to Fuels and Chemicals with Emphasis on the Potential for Biomass-Derived Syngas.* NREL/TP-510–34929, December.

19.3.2 Gas Cleaning Technology

Syngas contaminants: contaminants fall into five categories: particulates, tar, alkali metals, and sulfur gases and other acid gases (H_2S, NH_3, HCl, etc.).

Typical gas cleaning techniques include sorbents, scrubbers, filters, and reforming catalysts to crack tars into smaller molecules, ideally to CO and H_2. Guard beds are used to protect both catalytic tar removal units and catalytic synthesis reactors from small but harmful amounts of contaminants that may pass through upstream gas cleaning units, along with guard columns that can be used to scrub syngas prior to reforming.

The key questions around the design of gas cleaning systems are (a) the order of removal of contaminants and (b) the temperature limit(s) needed for each of the gas cleaning operations. For example, with catalytic tar reforming, the raw syngas must enter the tar reforming reactor above 250°C to avoid tar condensation in transport lines, and hot enough for catalytic reforming of the various tar components, typically 700–900°C, the range at which syngas would exit most gasification reactors. But most tar reforming catalysts are deactivated by sulfur gases, especially H_2S, and sulfur is removed at lower temperatures, around 400°C, and sulfur sorbents may be contaminated by hydrocarbons. Similar problems are encountered with nitrogen-containing gases, HCl, trace metals, and other fine particulates. Temperature is important for both the performance of cleaning operations and avoidance of energy losses if gas streams must be cooled and reheated. All together, this poses challenging problems for designers of gas cleaning systems for raw syngas.

19.3.3 Particulate Removal

Particulates in raw syngas include char fragments from partially gasified biomass (Scahill et al., 1996) and soot produced by cracking of hydrocarbons upstream of tar reforming reactors. They may also include alkali metal condensation aerosols from potassium and sodium that are volatilized during gasification.

Several separation methods for particle removal are shown in Table 19.8. Hot gas filtration has been developed for removing particles from raw syngas (Purolator-facet, 2011; Tri-mer, 2011). The filters may be ceramic or sintered metals. Metals of construction include 316L stainless steel, Hastelloy, Inconel, and Carpenter 20, a high Ni–Cr stainless-steel alloy. Filters of both materials can operate up to 900°C, and some have been imbedded with catalysts for gas conversion reactions. They can be designed with particle

TABLE 19.8

Separation Methods for Particle Removal

Separation Method	Particle Removal Limits	Limitations
Cyclones	>10 microns	
Ceramic or sintered metal filters	>0.5 microns	
Bag filters	>1 micron	$T < 400°C$
Electrostatic precipitators	>0.2 microns	$T < 400°C$
Wet or oil scrubbers	>0.2 microns	Syngas cooling required

Source: Adapted from Coulson, J.M. and Richardson, J.F. 1991. *Chemical Engineering—Volume 2. Particle Technology and Separation Processes.* 4th Edition. Butterworth-Heinemann, Oxford, Figure 8-1; Graham, G. 2011. Controlling stack emissions in the wood products industry. http://www.ppcesp.com/ppcart.html (11 September); Waste-to-Energy Research and Technology Council. 2011. http://www.wtert.eu/default.asp?Menue=12&ShowDok=23 (11 September).

size cutoff from <0.5 micron to >100 microns, for inlet particulate loadings to 12,000 mg/Nm^3 (5.2 gr/dscf) and outlet loadings <1 mg/Nm^3.

Cyclones have too high a particle size cutoff limit for effective removal of the fine particles in raw syngas.

Electrostatic precipitators and bag filters are limited to temperatures below 400°C. They are not used for hot gas filtration in the temperature range required for hot raw syngas cleaning.

Wet or oil scrubbers require that the raw syngas be cooled to temperatures well below the hot gas temperature required for downstream tar reforming.

19.3.4 Reforming or Removal of Tar and Other Organic Contaminants

Tar and other organic matter are ubiquitous in raw syngas, and they contain a significant percentage of the energy in biomass. Tar is condensable, it is soluble in various oils, and it can be reformed to smaller, more desirable molecules. These characteristics point to three different methods to deal with tar produced during gasification. Catalytic reforming in the presence of water vapor is an effective way to convert tar to CO and H_2.

19.3.5 Water Scrubbing

Aqueous scrubbing of raw syngas has been used but is no longer considered a viable option for tar removal. Scrubbers can be difficult to operate and are prone to tar deposition and plugging. They create a challenging tar–water separation problem once tar is scrubbed from syngas, and they create a wastewater stream that must be cleaned up (Dayton, 2002).

19.3.6 Solvent Absorption

Absorption of tar in a tar-compatible organic solvent has been developed and demonstrated at the Energy Research Centre of the Netherlands (ECN). The ECN process, named OLGA, consists of three stages. The first is a collector in which heavy tars are condensed by contacting the raw syngas with a cool scrubbing oil. The second stage is a demister to separate entrained oil droplets from the syngas. The scrubbing oil from the two stages are combined, cooled, filtered to remove solid particles, and returned to the first scrubber. The combined oil collected from the demister is passed through a separator in which solid particles are filtered from it and returned to the first-stage scrubber.

The third stage, absorber, separates light tar components by absorption in oil. The absorber operates above the dew point of the gas to avoid condensation of water in the absorber oil. At this point, the syngas is designated as tar-free. The scrubbing oil is stripped of light tar species and returned to the process. The light tar compounds stripped from the oil are combusted to provide process heating.

In a 675-h pilot-scale test with syngas from biomass, the tar content of the product syngas was decreased from 16,900 to 200 mg/Nm^3.

At the time of this writing, one industrial-scale OLGA unit is operating in France, and a second had been scheduled to start up in Portugal in 2010.

19.3.7 Tar Reforming

Tar reforming is performed catalytically at moderate to high temperatures. Tens of catalysts have been proposed and evaluated for tar reforming. Excellent reviews of various aspects of biomass tar reforming have been published by Dayton (2002), Gerber (2007), and Yung et al. (2009). The content of this section is based largely on these three reviews. Readers are referred to them for more detail and a wealth of references.

Tar reforming involves decomposition of tar compounds to light gases, ideally CO and H_2. Water vapor or CO_2/H_2 are essential ingredients in tar reforming because tar components are rich in heterocyclic hydrocarbons with H/C atom ratios near 1.0 and they contain little oxygen.

Reforming reactions usually occur on metal catalysts. The most effective tar reforming catalysts contain transition metals, particularly Fe, Co, Ni, Cu, Zn, Ru, Rh, Pd, Ir, Pt, and Au. The last six elements in this list are noble metals and are very expensive. Ni has been established as an effective catalyst for methane and naphtha reforming. As a result, much development work for metallic biomass tar reforming catalysts has focused on Ni.

19.3.7.1 Limestone and Dolomitic Catalysts

Calcined dolomites and limestones were used early as tar reforming catalysts, and are still the most commonly used for that purpose. Their advantages are that they are effective, low cost, available, and cheap to dispose of. They become active upon calcining as CO_2 is driven off, creating particles with internal surface areas of <10 m^2/g. They contain small amounts of potassium and iron oxides which act as promoters for cracking reactions and reforming with CO_2 and water vapor (Delgado et al., 1996).

In processing raw syngas, calcined dolomites and limestones must be used at temperatures above the calcination range (>800°C) to avoid recarbonization that can result in nearly complete loss of activity. Carbonization limits the use of dolomites at elevated pressures, and even higher temperatures are required.

Calcined dolomites are generally more active catalysts than calcined limestones.

19.3.7.2 Nickel-Based Catalysts

Nickel-based catalysts supported on alumina (Ni/Al_2O_3) have shown promise for reforming of tar in biomass-derived syngas. They are very effective at tar reforming in the absence of H_2S, performing well in the temperature range 550–700°C. They are also active catalysts for water gas shift and ammonia decomposition (Tamaru et al., 1965; Simell et al., 1997; Nagai et al., 2006; Haryanto et al., 2009).

One limitation with nickel-based catalysts is that they are poisoned by H_2S which adsorbs strongly onto active sites. To minimize sulfur poisoning, it is important to remove sulfur from raw syngas before catalytic tar reforming with Ni-based catalysts, or, for very low sulfur-containing syngas, to place a guard bed immediately upstream of the tar-reforming reactor. However, even small amounts of sulfur in syngas will eventually decrease the performance of nickel catalysts, making regeneration of these catalysts necessary (Bartholomew et al., 1982; Lombard et al., 2006; Magrini-Bair et al., 2007).

During regeneration, sulfided nickel catalysts are first treated with steam to drive off H_2S. This forms a nickel oxide that is subsequently reduced with hydrogen to regenerate metallic nickel.

19.3.7.3 Ruthenium

In his review of metal catalysts other than Ni for biomass tar reforming, Gerber (2007) noted that ruthenium stood out as a metal catalyst for both tar cracking and methane reforming, with minimal carbon buildup on the catalyst (Tomishige et al., 2003; Yoon et al., 2011).

In one study, Tomishige et al. (2003) reported results for tar reforming over Ni and noble metal catalysts (Ru, Rh, Pd, and Pt) on a CeO_2/SiO_2 support. The syngas used in the tests was generated by gasification of cedar wood containing 10 wt% moisture. Measurements were made over a 550–650°C temperature range. In these tests, the Rh catalyst performed best as a tar cracking catalyst. All of the tar species in the product gas were eliminated over the entire temperature range. Coke deposits were 1% of the catalyst weight over the temperature range. None of the other catalysts were as effective in tar removal at temperatures up to 600°C. The platinum catalyst was the only other catalyst to eliminate tar completely at 650°C. The Rh catalyst became deactivated above 600°C. More coke was deposited on all of the other catalysts than on ruthenium and platinum.

19.3.7.4 In-Gasifier Catalytic Tar Reduction

Another option for catalytic tar reduction is catalytic conversion of tar inside the gasifier, with an objective of minimizing downstream tar treatment. Typically, gasifiers are optimized to produce a fuel gas with minimum tar concentration. Bed additives or catalyst are used to convert tar species to CO, H_2, and light hydrocarbons. Dolomite, olivine, or char have been evaluated to catalyze tar reformation. In-reactor Ni-based catalysts have been found to be very effective for tar reforming and also for ammonia reduction (Devi et al., 2005).

19.3.8 Sulfur Capture

Biomass contains low concentrations of sulfur primarily as proteins. During gasification, sulfur is converted primarily to H_2S, but COS and other sulfur-containing gases may also be produced (Sricharoenchaikul et al., 1997).

Commonly used methods for sulfur capture in industrial processes include scrubbing with aqueous solutions of Na_2CO_3 or NaOH, absorption in amines or other acid gas absorbents, and adsorption on solid sorbents. Sulfur gas concentrations in biomass-derived syngas are low, from about 20 ppm to 600 ppm (Sato et al., 2007). At these low concentrations, sulfur sorbents are the preferred capture agent.

Various materials will adsorb H_2S, but the concentrations to which H_2S can be removed depend on the temperature and water vapor content of the gas being processed. A review of available sulfur gas sorbents, their performance characteristics, and their limitations was published by Cheah et al. (2009). In this section, we present a summary of the available information about sulfur sorbents. Readers are referred to Cheah's excellent review for more details about each sorbent.

Metal oxides of zinc, manganese, copper, iron, rare earth, and calcium have been found to be good H_2S sorbents. At high water vapor concentrations, H_2S desorbs to regenerate these sorbents according to the equilibrium sulfidation reaction:

$$M_xO_y(s) + yH_2S(g) <=> M_xS_y + yH_2O(g) \tag{19.6}$$

The magnitude of the equilibrium constant for this reaction and the water vapor partial pressure determine the extent to which H_2S is adsorbed, and thus the effectiveness of the sorbent. Cheah and coworkers summarized published results for various sulfidatable materials, some of which are included in Table 19.9.

The criteria for a good H_2S adsorbent are (a) a high adsorption capacity; (b) rapid adsorption; (c) mechanical strength with low attrition and tolerant of high temperatures; (d) stable in a steam and reducing environment; and (e) regenerable over repeated cycles while retaining its sulfur sorption capacity and sorption rate.

TABLE 19.9
H_2S Adsorbents

Material	Sulfidation Conditions	Outlet [H_2S]
ZnO–TiO_2	1% H_2S in 13% H_2, 19% H_2O, 67% N_2	<10 ppm in all cases
Limestone, dolomite	800–1000°C, 0.25–1.0% H_2S in mix, 10 bar	<250 ppm
Cu–Cr–O and Cu–Ce–O	750–850°C, 0.5–2% H_2S, 10–20% H_2, 0–10% H_2O, balance N_2	<5 ppm
CuO + CrO_3, Al_2O_3	550–650°C, 2% H_2S in mix	5–10 ppm
MnO or FeO on γ-Al_2O_3	400–800°C, 0–1% H_2S, 0–1% COS	0–100 ppm
Cu/Mo/Mn on zeolite	871°C, 0.2% H_2S, 1% CH_4, 19% H_2O	<10–100 ppm
γ-Al_2O_3 + MnO	400–1000°C, 1.0% H_2S, in H_2/Ar	<5 ppm
Mn ore (β-MnO_2)	550–850°C, 1.0% H_2S in mix	<1 ppm
Zn-doped manganese oxide	700°C, 1.0% H_2S in mix	<15 ppm
CeO_2 + La, Cu	600–850°C, 1000 ppm H_2S in 50% H_2/10% H_2O/balance helium, ambient pressure	<1 ppm

19.3.8.1 Zinc Oxide

ZnO is an effective H_2S sorbent in the mid-temperature range, to about 400°C. H_2S adsorption is limited by the presence of water vapor (Kim et al., 2007). H_2S removal from syngas to concentrations <10 ppm can be achieved with ZnO.

One problem with ZnO is that Zn vaporizes from it at syngas operating temperatures. However, ZnO is stabilized when combined with TiO_2. Zinc titanate is physically stable up to 700°C.

ZnO is also stabilized when combined with iron oxide as a zinc ferrite ($Zn_xFe_{1-x}O_4$). These compounds have a high desulfurization efficiency up to 500°C but they disintegrate at higher temperatures. Zinc ferrites can produce soot when hydrocarbons and water vapor are present.

19.3.8.2 Cu-Based Sorbents

Copper-based sorbents, either CuO or as mixed oxides, are effective sorbents but lose most of their effectiveness if CuO is reduced to Cu. $CuCr_2O_4$ has a high H_2S removal efficiency and is very stable because Cu in this form is difficult to reduce.

Copper supported on an alumina and manganese oxide matrix is an effective H_2S sorbent over the temperature range 350–600°C. It is able to remove H_2S to concentrations of <1 ppmv. These sorbents are highly attrition-resistant sorbents and are regenerable at 650–725°C using a 6% O_2–N_2 gas mixture.

19.3.8.3 Calcium-Based Sorbents

Calcined dolomites and limestone are inexpensive sorbents that are effective at higher temperatures, above $CaCO_3$ calcining temperatures (<800°C). Calcining creates porous particles that are effective as sorbents to high conversions to CaS. Sulfidation rates are faster and sorbent conversions are higher than with uncalcined materials (Yrjas et al., 1996) where conversions are limited to about 10% (Fenouil and Lynn, 1995). Calcined limestone and dolomites behave the same way at elevated pressure. However, higher CO_2

partial pressures in pressurized gas cleaning require higher sorbent temperatures to avoid recarbonation of the calcined sorbents.

19.3.8.4 Limitations with Sulfur Sorbents

Water vapor: Equilibrium constraints determine how low a concentration of H_2S can be achieved with a given sorbent when water vapor is present. Water vapor may also adsorb in competition with H_2S on sorbents, limiting the amount of H_2S that can be adsorbed.

Temperature: Calcined limestone and dolomite sorbents require temperatures greater than 800°C to avoid recarbonation. ZnO is effective only at lower temperatures, to about 400°C, although zinc titanates and ferrites can be used at higher temperatures, to at least 500°C. Copper-based sorbents function up to 600°C.

Other sulfur species: Under some conditions, H_2S can be transformed to COS, SO_2, and CS_2 (Gupta et al., 2001). H_2S, for example, reacts with CO and CO_2 to produce COS. H_2S sorbents are not as effective for the other sulfur gases.

19.3.8.5 Regeneration of Sulfur Sorbents

Sorbents are regenerated with steam according to the equilibrium illustrated in Equation 19.1. However, the H_2S removed from the sorbent must be recovered and either reused or disposed of. Some SO_2 may be produced during catalyst regeneration.

Removal of H_2S by oxidation is the same chemical process as roasting sulfide-containing ores:

$$2ZnS(s) + 3O_2(g) <=> 2ZnO(s) + 2SO_2(g) \tag{19.7}$$

The SO_2 generated by oxidation is more easily captured than H_2S. The oxidation reaction is exothermic, so it must be carried out at low O_2 concentrations. Low O_2 concentrations are also necessary to avoid formation of $ZnSO_4(s)$ that consumes the sorbent.

19.3.9 Alkali Metal Capture

Alkali metal salt particles in biomass syngas result from volatilization of K and Na in the strongly reducing environment within biomass particles during gasification (Reis et al., 1995; Dayton et al., 1995; Dayton and Frederick, 1996; Wåg et al., 1997; Salo and Mojtahedi, 1998). These vapors condense and react with gas components, ultimately converting to carbonates and sulfates. The resulting particles generated are alkali metal salt condensation aerosols with diameters falling between 0.1 and 1 micron (Mikkanen et al., 1994). Particles in this size range are easily trapped by larger particles (Frederick and Vakkilainen, 2003).

Aerosol particles containing K–Na–CO_3–SO_4 are condensed and solidified at temperatures below 600°C. At these conditions, the particles can be removed with fine particle filters. There is, however, an energy penalty if raw syngas has to be cooled and then reheated prior to a subsequent cleaning operation such as tar reforming.

A second particle removal option is to use alkali getters to chemisorb alkali vapors before they condense to aerosol droplets. Sorbents that have been evaluated include diatomaceous earth, kaolinite and various other clays, bauxite, alumina, alundum, silica gel, dolomite, coal char, and coal ash (Turn et al., 2000).

References

Bartholomew, C.H., Agrawal, P.K., and Katzer, J.R. 1982. Sulfur poisoning of metals. *Adv. Catal.* 31, 173.

Bridgewater, A. et al.. 1998. *An Assessment of the Possibilities for Transfer of Biomass Gasification Technology to China. Part 1.* Report of the Mission to China. 65 p.

Cheah, S., Carpenter, D.L., and Magrini-Bair, K.A. 2009. Review of mid- to high-temperature sulfur sorbents for desulfurization of biomass- and coal-derived syngas. *Energy Fuels* 23: 5291–5307.

Ciferno, J.P. and Marano, J.J. 2002. *Benchmarking Biomass Gasification Technologies for Fuels, Chemicals and Hydrogen Production.* Report to USDOE, National Energy Technology Laboratory, Pittsburgh, PA, June, 2002.

Coulson, J.M, Richardson, J.F. 1991. *Coulson, and Richardson's Chemical Engineering. Volume 2—Particle Technology & Separation Processes.* 4th Edition. Butterworth-Heinemann, Oxford, Figure 8-1.

Dayton, D. 2002. *A Review of the Literature on Catalytic Biomass Tar Destruction.* NREL Report TP-510-32815, December, pp. 12–18.

Dayton, D.C. and Frederick, W.J. Jr. 1996. Direct observation of alkali vapor release during biomass combustion and gasification. 2. Black liquor combustion at 1100°C. *Energy Fuels* 10(2): 284–292.

Dayton, D.C., French, R.J., and Milne, T.A. 1995. Direct observation of alkali vapor release during biomass combustion and gasification. 1. Application of molecular beam/ass spectrometry to switchgrass combustion. *Energy Fuels* 9(5): 855–865.

Delgado, J., Aznar, M., and Corella, J. 1996. Calcined dolomite, magnesite, and Calcite for cleaning hot gas from a fluidized bed biomass gasifier with steam: Life and usefulness. *Ind. Eng. Chem. Res.* 35: 3637–3643.

Devi, L., Ptasinski, K.J., Janssen, F.J.J.G., van Paasen, S.V.B., Bergman, P.C.A., and Kiel, J.H.A. 2005. Catalytic decomposition of biomass tars: Use of dolomite and untreated olivine. *Renewable Energy* 30(4): 565–587.

Fenouil, L. A. and Lynn, S. 1995. *Ind. Eng. Chem. Res.* 34(7): 2324–2333.

Frederick, W.J. Jr. and Vakkilainen, E.K. 2003. Sintering and structure development in alkali metal salt deposits formed in kraft recovery boilers. *Energy Fuels* 17(6): 1501–1509.

Gerber, M.A. 2007. *Review of Novel Catalysts for Biomass Tar Cracking and Reforming.* PNNL-16950, October, http://www.pnl.gov/main/publications/external/technical_reports/PNNL-16950.pdf (September 11, 2011).

Gibson, L. 2011. Syngas tasks, biomass power & thermal, http://biomassmagazine.com/articles/3451/syngas-tasks (9 August).

Graham, G. 2011. Controlling stack emissions in the wood products industry. http://www.ppcesp.com/ppcart.html (11 September).

Gupta, R.P., Turk, B.S., Portzer, J.W., and Cicero, D.C. 2001. *Environ. Prog.* 20(3): 187–195.

Haryanto, A., Fernando, S.D., Filip To, S.D., Steele, P.H., Lester Pordesimo, L., and Adhikari, S. 2009. Energy hydrogen production through the water–gas shift reaction: Thermodynamic equilibrium versus experimental results over supported Ni catalysts. *Energy Fuels* 23(6): 3097–3102.

http://www.purolator-facet.com/gasfilter.htm, September 9, 2011.

http://www.tri-mer.com/hot-gas-filtration.html, September 9, 2011.

Kim, K., Jeon, S.K., Vo, C., Park, C.S., and Norbeck, J.M. 2007. Removal of hydrogen sulfide from a steam-hydrogasifier product gas by zinc oxide sorbent. *Ind. Eng. Chem. Res.* 46(18): 5848–5854.

Konttinen, J. 2010. Evaluation of gasification-based energy opportunities for forest industries. Presented at the AEBIOM Conference, Brussels, 28–29 April.

Kurkela, E. 2009. Fluidized-bed gasification of biomass for syngas—BTL in forest industry. Presented at Biomass NoE, Brussels, November.

Lombard, C., Le Doze, S., Marencak, E., Marquaire, P.-M., Le Noc, D., Bertrand, G., and Lapicque, F. 2006. *In situ* regeneration of the Ni-based catalytic reformer of a 5 kW PEMFC system. *Int. J. Hydrogen Energy* 31(3): 437–440.

Magrini-Bair, K.A., Czernik, S., French, R., Parent, Y.O., Chornet, E., Dayton, D.C., Feik, C., and Bain, R. 2007. Fluidizable reforming catalyst development for conditioning biomass-derived syngas. *Appl. Catal., A* 318, 199–206.

Mikkanen, P., Kauppinen, E.I., Jokiniemi, J.K., Sinquefield, S.A., Frederick, W.J., and Mäkinen, M. 1994. The particle size and chemical species distributions of aerosols generated in kraft black liquor pyrolysis and combustion. *AIChE Symp. Ser.* 90(302): 46–54.

Nagai, M., Zahidul, A.M., and Matsuda, K. 2006. Nano-structured nickel–molybdenum carbide catalyst for low-temperature water-gas shift reaction. *Appl. Catal. A: General* 313(2): 137–145.

Reis, V.V., Frederick, W.J., Wåg, K. J., Iisa, K., and Sinquefield, S.A. 1995. The effects of temperature and oxygen concentration on potassium and chloride enrichment during black liquor combustion. *Tappi Jl.* 78(12): 67–76.

Salo, K., and Mojtahedi, W. 1998. Fate of alkali and trace metals in biomass gasification. *Biomass Bioenergy* 15(3): 263–267.

Sato, K., Shinoda, T., and Fujimoto, K. 2007. New nickel-based catalyst for tar reforming, with superior resistance to coking and sulfur poisoning in biomass gasification processes. *J. Chem. Eng. Jpn.* 40(10): 860–868.

Scahill, J., Diebold, J.P., and Feik, C. 1996. Removal of residual char fines from pyrolysis vapors by hot gas filtration, In: *Developments in Thermochemical Biomass Conversion, Volume 1*, A. V. Bridgwater, Ed. London: Blackie Academic & Professional, pp. 253–264.

Simell, P.A., Hepola, J.O., Outi, A., and Krause, I. 1997. Effects of gasification gas components on tar and ammonia decomposition over hot gas cleanup catalysts. *Fuel* 76(12): 1117–1127.

Sricharoenchaikul, V., Frederick, W.J., and Grace, T.M. 1997. Sulfur species transformations during pyrolysis of kraft black liquor. *J. Pulp Paper Sci.* 23(8): J394–J400.

Tamaru, K., Tanaka, K-i., Fukasaku, S., and Ishida, S. 1965. Decomposition of ammonia on a nickel catalyst *Trans. Faraday Soc.* 61: 765–772.

Tomishige, K., Miyazawa, T., Asadullah, M., Ito, S., and Kunimori, K. 2003. Catalyst performance in reforming of tar derived from biomass over noble metal catalysts. *Green Chem.* 5: 399.

Turn, S., Kinoshita, C., Ishimura, D., Zhou, J., Hiraki, T., and Masutani, S. 2000. *Control of Alkali Species in Gasification Systems—Final Report*. NREL/SR-570-26160, July, 2000.

Waste-to-Energy Research and Technology Council. 2011. http://www.wtert.eu/default.asp?Menue=12&ShowDok=23 (11 September).

Wåg, K.J., Reis, V.V., Frederick, W.J., and Grace, T.M. 1997. Mathematical model for the release of inorganic emissions during black liquor char combustion. *Tappi J.* 80(5): 135–145.

Yoon, S.J., Kim, Y.K., and Lee, J.G. 2011. Catalytic oxidation of biomass tar over platinum and ruthenium catalysts. *Ind. Eng. Chem. Res.* 50(4): 2445–2451.

Yrjas, P., Iisa, K., and Hupa, M. 1996. Limestone and dolomite as sulfur absorbents under pressurized gasification conditions. *Fuel* 75(1): 89–95.

Yung, M.M., Jablonski, W.S., and Magrini-Bair, K.A. 2009. Review of catalytic conditioning of biomass-derived syngas. *Energy Fuels* 23: 1874–1887.

19.4 Conversion of Synthesis Gas to Products

As Figure 19.1 shows, a wide range of synthetic chemicals and fuels are made today from synthetic gas derived from coal and natural gas. As thermochemical-based biorefineries with biomass feedstocks are developed, specific fuels or chemicals must be targeted and catalytic conversion technologies adapted or further developed for producing them. In this section, five products from syngas and six technical routes to produce them are reviewed.

19.4.1 Hydrocarbons by Fischer–Tropsch Synthesis

A process for the production of hydrocarbons by Fischer–Tropsch synthesis was invented in the 1920s. It was commercialized in Germany in 1936 to produce synthetic transportation fuels. Since the 1980s, SASOL in South Africa has produced most of that country's diesel fuel from coal and natural gas.

Conversion of syngas to hydrocarbons by Fischer–Tropsch synthesis produces a distribution of mostly linear hydrocarbons according to the stoichiometry shown below:

$$(2n+1)H_2 + nCO \rightarrow C_nH_{(2n+2)} + nH_2O \tag{19.8}$$

Cobalt-, iron-, and ruthenium-based catalysts promoted with alkali metal oxides and copper and supported on silica, alumina, or zeolites are used in this synthesis (Khodakov et al., 2007). Process conditions are 150–300°C and 25 bar operating pressure. Per-pass carbon monoxide conversion is approximately 40%, while overall conversion is approximately 67%. Contaminant limits for syngas entering the reactor are 200 ppb sulfur and 10 ppm ammonia (Spath and Dayton, 2003).

The molecular-weight distribution of the hydrocarbons produced is broad, ranging from methane to waxes, and follows the Anderson–Schulz–Flory distribution (Song et al., 2004). The desired products are liquids with chain lengths in the right range for gasoline (C_5–C_{12}), jet fuel (C_{11}–C_{13}), or diesel (C_{13}–C_{17}). When unconstrained, the product distribution from Fischer–Tropsch synthesis favors methane and light hydrocarbons that must be recycled to a steam methane reformer and reconverted to synthesis gas. Waxes must be cracked for the same reason.

The use of zeolites or other catalyst substrates with fixed-size pores has been proposed to restrict formation of hydrocarbon chains longer than about 10 carbons. This would shift the chain length of hydrocarbons produced away from methane while constraining production of long-chained hydrocarbons. This concept has not been commercialized.

After liquid hydrocarbons and water are separated from the reactor product, part of the unconverted syngas is recycled to the FT reactor, part to the acid–gas removal system, and the rest is used for power generation.

19.4.2 Methanol Synthesis

Methanol is used both as a transportation fuel or as a blending stock and also as an intermediate for the production of synthetic gasoline, dimethyl ether, and ethanol (Cheng and Kung 1994). The technology for producing methanol from syngas is well established. Syngas from methane and coal is the main feedstock for methanol synthesis, although other carbon sources including biomass can be used.

The overall reaction stoichiometry for methanol production is

$$CO + 2H_2 \rightarrow CH_3OH \tag{19.9}$$

Methanol is produced catalytically, using a catalyst that is a mixture of copper, zinc oxide, and alumina. Process conditions are 50–100 bar and 220–275°C. Overall selectivity for methanol is greater than 99% with recycle. Limits for contaminants are 0.5 ppm_v sulfur, 1 ppb_v halides, and 5 ppb_v Fe and Ni (Spath and Dayton, 2003).

19.4.3 Dimethyl Ether Synthesis

Dimethyl ether (DME) is of interest as an alternative to diesel as a transportation fuel in Sweden and China. DME is produced from syngas through methanol as an intermediate, with dehydration methanol yielding DME. The stoichiometry for methanol to DME proceeds according to Equation 19.10:

$$2CH_3OH \Leftrightarrow CH_3OCH_3H_2O \tag{19.10}$$

Methanol is dehydrated using catalysts such as silica–alumina. Both two-step (indirect synthesis) and single-step (direct synthesis) conversion technologies are commercially available. The direct synthesis route uses a dual-catalyst system where both methanol synthesis and dehydration occur in the same reactor. This eliminates the need for an intermediate methanol synthesis stage. Both the one-step and two-step processes are commercially available. The two-step process is used more extensively because it is relatively simple and startup costs are low.

19.4.4 Methanol-to-Gasoline Conversion

A synthesis route to convert methanol to gasoline (MTG) and the process technology for it were developed by Mobil in the 1970s (Chang, 1992). The overall reaction stoichiometry is

$$\frac{n}{2}[2CH_3OH \Leftrightarrow CH_3OCH_3 + H_2O] \xrightarrow{-nH_2O} C_nH_{2n} \tag{19.11}$$

The MTG process consists of two reaction stages. In the first, methanol is dehydrated to produce dimethyl ether. Dimethyl ether is then dehydrated to form olefins, and the olefins are hydrogenated to aromatic and paraffinic hydrocarbons using a zeolite catalyst (ZSM-5). The Mobil MTG process produces hydrocarbons that are predominantly in the gasoline boiling range. Processes based on both adiabatic fixed-bed and circulating fluidized-bed reactor configurations have been developed (Jones and Zhu, 2009).

The Mobil MTG technology was deployed as part of a commercial gas-to-gasoline plant in New Zealand in the 1980s. In this plant, natural gas was steam-reformed to produce syngas, and the syngas was converted to synthetic gasoline using the Mobil technology.

19.4.5 Direct Synthesis of Ethanol

Ethanol can be synthesized directly from syngas using silica-supported cobalt catalysts prepared from dicobalt octacarbonyl [$Co_2(CO)_8$]. Takeuchi et al. (1989) measured 20% selectivity for alcohols and 80% for hydrocarbons, but modification with alkaline-earth metal oxides increased selectivity for oxygenates to 60%. Barium and strontium were more effective than Mg and Ca.

More recently, Rathke and coworkers reported "a carbonium ion mechanism that produces ethanol in a selective SN2 process without the coproduction of significant amounts of either water or higher alcohols" (Rathke et al. (2006). The reaction is catalyzed by $HFe(CO)_4$ and occurs at 180–220°C and 300 bar pressure in an atmosphere of CO and H_2 at pressures ($3CO + H_2$) close to 300 atm. The overall stoichiometry is as follows:

$$CH_3OH + 2CO + H_2 = CH_3CH_2OH + CO_2 \tag{19.12}$$

Direct synthesis of ethanol is still at the laboratory exploration level.

19.4.6 Ethanol by Synthesis of Mixed Alcohols

Mixed alcohols containing linear-chain alcohols from methanol through octanol have been produced from synthesis gas with various catalysts.

The overall stoichiometric reaction for alcohol synthesis can be summarized as

$$nCO + 2nH_2 \Rightarrow C_nH_{2n+1}OH + (n-1)H_2O \quad (19.13)$$

Equation 19.11 suggests an optimum H_2:CO ratio of 2.0. However, the mixed-alcohol synthesis catalysts are also active as water gas shift catalysts that generate H_2 from CO and H_2O:

$$CO + H_2O = H_2 + CO_2 \quad (19.14)$$

As a result, the optimal H_2:CO ratio delivered to the reactor is close to 1.0.

Catalysts that have been evaluated for mixed alcohol production include modified Fischer–Tropsch catalysts, modified methanol catalysts, and alkali-enhanced rhodium (Rh).

The main components of the mixed alcohols produced are methanol, ethanol, and propanol. Methanol is normally separated and recycled to extinction to the synthesis reactor to increase ethanol yield. Propanol and higher alcohols have greater value than ethanol. They can be separated from ethanol and sold for other uses.

Studies have been conducted at the bench PDU scale, but no commercial plants are currently producing mixed alcohols.

19.4.6.1 Modified Fischer–Tropsch Catalysts

Modified Fischer–Tropsch catalysts include alkali-promoted iron or cobalt, or alkali- and cobalt-promoted molybdenum sulfide (MoS_2). The promoted MoS_2 catalysts increase selectivity for ethanol and higher alcohols. Inclusion of these promoters shifts production away from linear hydrocarbons and toward alcohols. Reaction conditions are typically 220–350°C and 50–200 bar.

In their patent on mixed-alcohol production from syngas, Stevens and Conway (1989) showed that on a single-pass basis, either 40% ethanol selectivity in the mixed alcohols product or 40+ % CO conversion could be achieved, but not both simultaneously (Table 19.10). These results were attainable with Fe- and Co-modified MoS_2 catalysts. Without modifiers, hydrocarbon and CO_2 production increased and selectivity for higher alcohols increased, both at the expense of ethanol productivity. Productivity for mixed alcohols produced with MoS_2-based catalysts is low to moderate, 0.15–0.40 kg/kg catalyst/h; by comparison, productivity of methanol synthesis is approximately 1 kg/kg catalyst/h (Phillips et al., 2007).

19.4.6.2 Modified Methanol Synthesis Catalysts

Modified methanol catalysts for mixed-alcohol synthesis are promoted with alkali and other active elements. Methanol catalysts fall into two categories: those used at higher pressure (125–250 bar, 300–425°C) and at lower pressure (50–100 bar, 250–400°C). Zinc-chromium oxides and manganese-chromium oxides are used as catalysts for high-pressure methanol production, while copper is used as a lower-pressure methanol catalyst. Mixed alcohols produced with these catalysts are primary branched alcohols, with isobutanol as a main component. CO_2 production is significant. As with modified Fischer–Tropsch

TABLE 19.10
Single-Pass Products from Mixed-Alcohol Synthesis Using Alkali-Promoted Modified Fischer–Tropsch Catalyst

Catalyst	66% Mo/Fe Sulfide, 20% Clay, 10% K_2CO_3, 4% Sterotex			48.8% MoS_2, 38.0% CoO_2, 13.2% K_2CO_3		78.7 MoS_2, 21.3% K_2CO_3
T (°C)	305	295	300	307	309	307
P (bar)				98.9	99.1	98.6
H_2/CO mole ratio	0.98	0.98	1.03	1.06	1.06	1.06
GHSV (h)	1300	1050	1330	651	2529	572
CO conversion (%)	23	13	26	42.6	14.4	46.5
Gas-Phase Products						
CO_2 (%)	33.5	31.3	32.9	31.4	23.2	39.8
CH_2 (%)	12.6	11.3	18	14.2	7.6	28.4
C_2+ HCs (%)	5.7	3.2	2.7	1.5	0.0	34.9
Liquid-Phase Products						
Methanol (%)	16.1	22.7	15.2	27.0	45.6	11.5
Ethanol (%)	39.9	40.7	41.8	27.3	26.3	11.8
Propanols (%)	14.9	12.7	11.5	11.8	6.3	19.6
Butanols (%)	4.3	3.5	1.4	6.2	3.4	9.3
Pentanols (%)	0.5	1.2	1.5	0.1	0.0	1.9

Source: Adapted from Stevens, R.R. and Conway, M.M. 1989. *Mixed Alcohols Production from Syngas*. U.S. Patent 4,831,060, 16 May; Assignee: The Dow Chemical Company.

catalysts, there is a trade-off between achieving high CO conversion on the one hand versus higher ethanol selectivity and yield on the other.

19.4.6.3 Rhodium-Based Catalysts

Catalysts composed of rhodium, manganese, and other promotors, including transition metals and noble metals on SiO_2 supports,* were evaluated by Gerber et al. (2008). Reaction conditions for most of the measurements were 255–350°C, 80 bar pressure, and space velocities of 7500–15,000 L/L_{cat}/h. Gerber et al. (2008) summarized their results as follows:

> Of all of the catalysts evaluated, the Li-promoted catalysts had the highest carbon selectivity to C2+ oxygenates (47%) under the conditions at which the maximum C2+ -oxygenate STYs† were obtained. The catalysts promoted with Ir, Pt, and Au had carbon selectivities to C2+ oxygenates at maximum C2+ -oxygenate STYs, of 39, 39, and 37%, respectively, which were better than the unpromoted catalyst (carbon selectivity of 32%). None of the catalysts with higher STYs than the baseline catalyst had high carbon selectivities to C2+ alcohols relative to the total oxygenates. Only the gallium (Ga)- and indium(In)-promoted catalysts had very high ratios of C2+ alcohols to total C2+ oxygenates ratios, with ratios of 0.67 and 0.71, respectively. These same catalysts had total alcohol:oxygenates ratios of 0.87 and 0.89, respectively.

* Rh:Mn:M (M = promoter) atomic ratios were 1.00:0.57:0.10, except for Li promotion, which was 1.00:0.57:0.30.
† Space time conversions (STYs) in units of g/L_{cat}/h^2.

TABLE 19.11

Performance of Selected Rhodium-Based Catalysts for Conversion of Syngas to Mixed Alcohols at 80-Bar Pressure

Catalyst	T, C	GHSV ($L/L_{cat}/h$)	STY of C_2+ Oxygenates ($g/L_{cat}/h$)	Selectivity to C_2+ Oxygenates (%)	Ratio of C_2+ Alcohols to C_2+ Oxygenates
$Rh/Mn/Ir/SiO_2$	256–328	7500–15,000	290–810	23–58	0.10–0.37
$RhMnPt/SiO_2$	255–315	7500–15,000	200–660	38–51	0.11–0.17
$Rh/Mn/Au/SiO_2$	256–325	7500–19,000	110–500	28–55	0.15–0.44
$Rh/Mn/Li/SiO_2$	256–350	7500–11,000	90–480	34–62	0.19–0.56
$Rh/Mn/Ga/SiO_2$	256–345	7500–11,000	10–190	26–46	0.64–0.69
$Rh/Mn/In/SiO_2$	256–345	7500–11,000	20–140	33–66	0.56–0.93

Source: Adapted from Gerber, M.A., White, J.F., Gray, M., and Stevens, D.J. 2008. *Evaluation of Promoters for Rhodium-Based Catalysts for Mixed Alcohol Synthesis*. PNNL-17857 technical report, December.

The results described by Gerber et al. (2008) are included in Table 19.11. These results indicate that several of the rhodium-based catalysts evaluated show technical promise for conversion of syngas to mixed alcohols.

Two limitations on large-scale use of rhodium catalysts are the cost and availability of rhodium. Rhodium prices have varied from $2125/oz to $2975/oz during 2010, after peaking at $10,000/oz in mid-2008 and dropping to $1500/oz in late 2008. The annual world production of rhodium is only approximately 25 tonnes (Johnson Mathey, 2011).

References

Chang, C.D. 1992. The New Zealand gas-to-gasoline plant: An engineering tour de force. *Catalysis Today* 13: 103–111.

Cheng, W. and Kung, H. 1994. *Methanol Production and Use*. New York: Marcel Dekker.

Gerber, M.A., White, J.F., Gray, M., and Stevens, D.J. 2008. *Evaluation of Promoters for Rhodium-Based Catalysts for Mixed Alcohol Synthesis*. PNNL-17857 technical report, December.

Johnson Matthey, 2011. *Platinum* 2011. Johnson Matthey Public Limited Company, Royston, Hertfordshire, SG8 5HE, England.

Jones, S.B., Zhu, Y. 2009.Techno-economic Analysis for the Conversion of Lignocellulosic Biomass to Gasoline via the Methanol-to-Gasoline (MTG) Process. Report PNNL-18481 Prepared for the U.S. Department of Energy under Contract DE-AC05-76RL01830 Pacific Northwest National Laboratory, Richland, Washington 99352.

Khodakov, A.Y., Chu, W., and Fongarland, P. 2007. Advances in the development of novel cobalt Fischer-Tropsch catalysts for synthesis of long-chain hydrocarbons and clean fuels. *Chemical Review* 107: 1692–1744.

Phillips, S., Aden, A., Jechura, J., Dayton, D., and Eggeman, T. 2007. *Thermochemical Ethanol* via *Indirect Gasification and Mixed Alcohol Synthesis of Lignocellulosic Biomass*. NREL/TP-510–41168, April.

Rathke, J.W., Chen, M.W., Klingler, R.J., Gerald, R.E., Marshall, C.L., and Rodgers, J.L. 2006. Selective catalytic ethanol synthesis. *Proceedings, 2006 Meeting of the DOE/BES Catalysis and Chemical Transformations Program*, Cambridge, Maryland, 21–24 May.

Rathke, J.W., Chen, M.W., Klingler, R.J., Gerald, R.E., Marshall, C.L., and Rodgers, J.L. 2010. *Rhodium Uses, Supply, and Prices*. www.kitco.com/charts/rhodium.html, 17 September.

Song, H.-S., Ramkrishna, D., Trinh, S., and Wright, H. 2004. Operating strategies for Fischer Tropsch reactors: a model-directed study. *Korean J. Chem. Eng*. 21(2): 308–317.

Spath, P.L. and Dayton, D.C. 2003. *Preliminary Screening: Technical and Economic Assessment of Synthesis Gas to Fuels and Chemicals with Emphasis on the Potential for Biomass-Derived Syngas.* NREL/TP-510-34929, December.

Stevens, R.R. and Conway, M.M. 1989. *Mixed Alcohols Production from Syngas.* U.S. Patent 4,831,060, 16 May; Assignee: The Dow Chemical Company.

Takeuchi, K., Matsuzaki, T., Hanaoka, T.-A., Arakawa, H., Sugi, Y., and Wei, K. 1989. Alcohol synthesis from syngas over cobalt catalysts prepared from $Co_2(CO)_8$. *J. Mol. Catal.* 55(1): 361–370.

19.5 Pyrolysis Chemistry

Pyrolysis is the thermal decomposition of biomass in the absence of oxygen. Pyrolysis produces three products: light gases, oil, and char. The distribution of pyrolysis products among these three depends both on the type of biomass pyrolyzed and on the pyrolysis conditions used.

Slow pyrolysis of biomass has been used since ancient times to produce solid carbonaceous fuels (char) for cooking and heating. During slow pyrolysis, the components of biomass undergo complete deconstruction by the loss of pendant groups on rings and chains and by depolymerization and fragmentation of its three main component polymers (Czernik and Bridgewater, 2004). Slow pyrolysis at approximately 450°C produces relatively high yields of char, 35–40% of the dry biomass, and approximately equal amounts of oil and light gases.

Fast pyrolysis of biomass is necessary to produce high yields of pyrolysis liquids. Fast pyrolysis decomposes biomass thermally at moderate temperatures (<500°C) and high heat transfer rates (>1000°C/s). Pyrolysis oil yields can exceed 60wt%; light gas and char yields each account for about half of the remaining biomass. Short vapor residence times (less than 1s) and rapid quenching of vapors are necessary to prevent undesired vapor-phase cracking of the volatile organic compounds produced, because cracking decreases oil yield (Czernik and Bridgewater, 2004).

The chemistry of pyrolysis is complex, due in part to the different characteristics of the three main organic components of biomass (cellulose, hemicellulose, and lignin), and in part on the complex chemical structure of each of these components. Cellulose is a polysaccharide, a linear polymer that consists of 1–4 linked D-glucose units as shown in Figure 19.5. Cellulose molecules contain between 7000 and 15,000 glucose units, corresponding to molecular weights between 1 million and 2 million AMU.

Hemicellulose is a class of polysaccharides, which are branched polymers each constructed of five or six linked carbon–sugar monomers. The molecular structure of arabinoglucuronoxylan, shown in Figure 19.6, is a representative of hemicelluloses. Sugar monomers in different hemicelluloses include xylose, mannose, galactose, rhamnose, arabinose, or glucose. Hemicellulose molecules contain between 500 and 3000 sugar units, which is an order of magnitude less than cellulose. Some hemicelluloses contain acetyl side groups. Birch wood contains six or seven acetyl groups per 10 xylose units in xylan (Maloney et al., 1985), the most prevalent hemicellulose in hardwoods. Softwoods contain less xylan and fewer acetyl groups.

Lignin is the most complex of the three polymers. It is a three-dimensional network, illustrated in Figure 19.7, constructed of phenyl propane units connected by ether and carbon–carbon linkages. The aromatic structural units contain methoxyl and hydroxyl side groups.

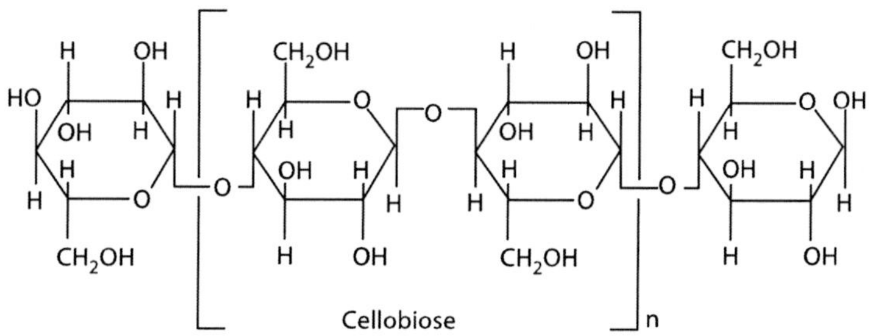

FIGURE 19.5
Molecular structure of the cellobiose unit in cellulose.

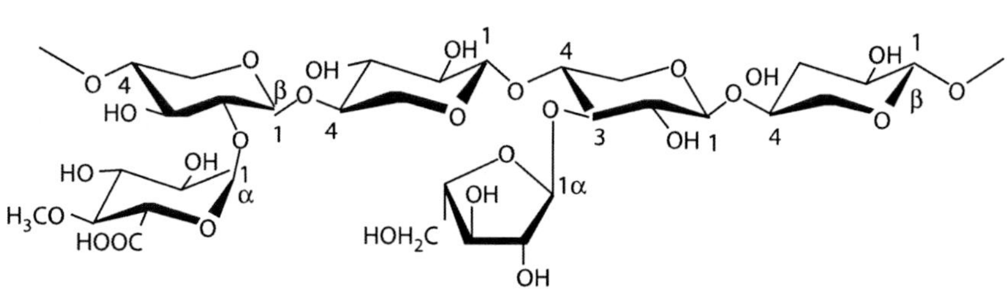

FIGURE 19.6
Molecular structure of arabinoglucuronoxylan, a hemicellulose found in softwoods.

The short residence times and other conditions used for fast pyrolysis yield the primary pyrolysis products. Vapors from primary pyrolysis include light gases (CO, CO_2, H_2, methane, and light hydrocarbons) and over a hundred condensable organic compounds that are of relatively low molecular weight (50–300 AMU) and consist largely of monomers and fragments of monomers from the polymeric constituents of biomass (Evans and Milne, 1987). Lignin undergoes primary pyrolysis by structurally controlled depolymerization (Caballero et al., 1997), which is enough to liberate molecules that are volatile at pyrolysis temperatures (Grant et al., 1989; Fletcher et al., 1990). Cellulose and hemicellulose are decomposed by two main mechanisms (Liao et al., 2004; Satrio et al., 2009). One is the cracking of glycosidic bonds between glucose monomer units to produce levoglucosan and isomers. The second is a set of ring-opening and reforming reactions that produce low-molecular-weight oxygenates (acids, aldehydes, ketones), furfural and hydroxymethylfurfural, heterocyclic compounds, and light gases. Levoglucosan produced during pyrolysis is also subject to ring-opening reactions. Char residues from the pyrolysis of cellulose and hemicellulose are low (Frederick and Mentzer, 1975; Yang et al., 2007).

Heavier-molecular-weight compounds present in pyrolysis oil originate from the lignin present in biomass. Chemical interactions among the organic components of biomass during pyrolysis are minimal (Evans and Milne, 1987; Raveendran et al., 1996).

Alkali metal salts have a greater detrimental impact than any other process or compositional variables on pyrolysis oil yields. The pyrolysis reactions of biomass are catalyzed by alkali metal salts (Di Blasi et al., 2009). Addition of relatively small amounts of these salts, <0.4 wt% on a dry basis, lowers the temperature at which pyrolytic decomposition proceeds and accelerates conversion rates by approximately a factor of 2. Pyrolysis oil yield decreases by 20–30% with this level of addition of alkali metal salts. Data from a wider range of biomass feedstocks showed a 3–6% decrease in pyrolysis oil yield per

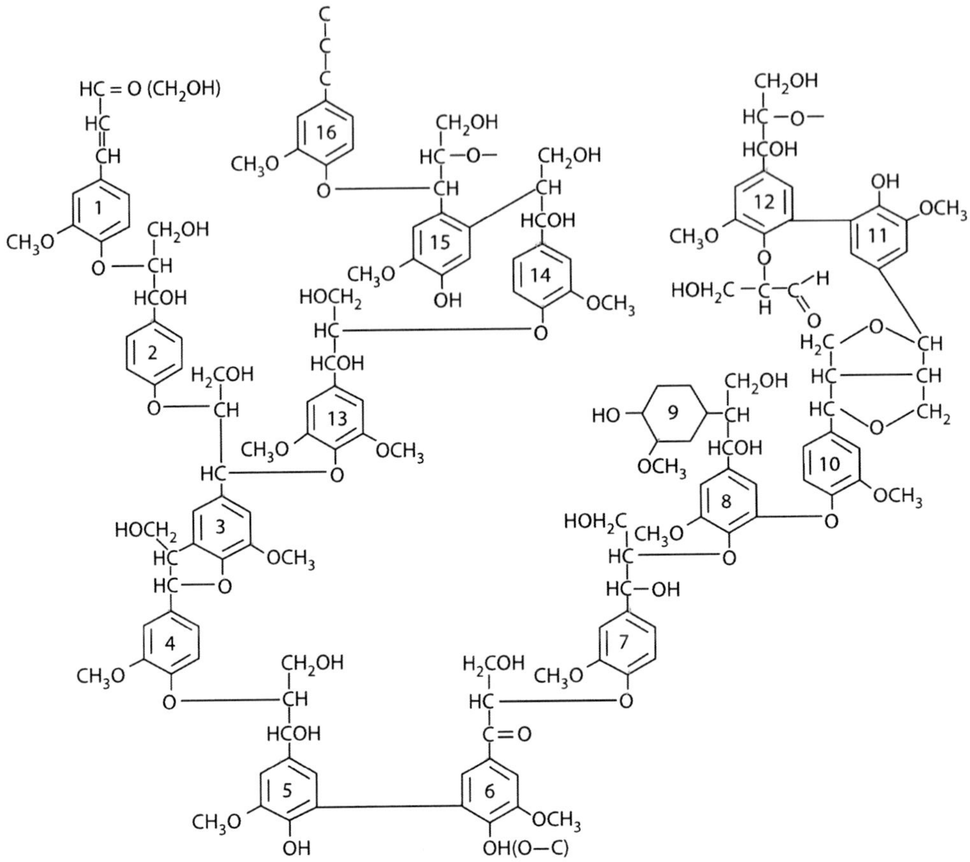

FIGURE 19.7
A representation of the molecular structure of softwood lignin. (Adapted from Alder, E. 1977. *Wood Sci. Technol.* 11: 169.)

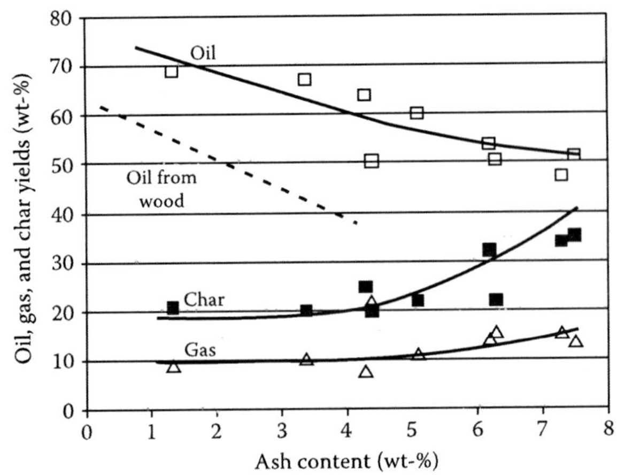

FIGURE 19.8
Yields of oil, char, and gas from pyrolysis versus ash content of biomass. (Adapted from Venderbosch, R.H. and Prins, W. 2010. *Biofuels, Bioprod. Bioref.* 4: 178–208.)

TABLE 19.12

Organic Acids in Pyrolysis Oil from Three Biomass Sources and their Contribution to Carboxylic Acid Number

	Straw oil (wt%)	Pine oil (wt%)	Hardwood Oil (wt%)	Straw oil (mg KOH/g)	Pine Oil (mg KOH/g)	Hardwood Oil (mg KOH/g)
Formic acid	1.85	2.47	1.10	22.5	30.1	13.4
Acetic acid	7.41	3.45	4.59	90.3	42.0	55.9
Glycolic acid	0.58		0.78	0.11	9.5	1.3
Other organic acids	0.96	0.60	0.40	11.8	7.4	5.4
Total organic acids	10.80	7.30	6.20			
Carboxylic acid number (CAN)				131.7	89.1	76.1

Source: Adapted from Sipilä, K. et al. 1998. *Biomass Bioenergy* 14(2): 103–113.

percentage-point increase in total inorganic content in the feedstock (Figure 19.8) (Fahmi et al., Fuel 2008; Oasmaa et al., 2010; Venderbosch et al., 2010). Agricultural biomass has an order of magnitude higher content of alkali metals and total inorganics than woody biomass and energy crops: 1–4wt% alkali metals versus 0.1–0.3wt% alkali metals (Jenkins et al., 1998; Fahmi et al., 2008). This contributes to the poorer yields and higher water content of pyrolysis oil from agricultural biomass (Fahmi et al., 2007; Oasmaa et al., 2010).

Alkali metals also impact the average molecular weight of pyrolysis oils. Data from Fahmi et al. (2008) show that the average molecular weight decreases linearly at a rate of 4.5 molecular weight units per 1000 ppm alkali metals (K + Na) for alkali metal contents between 800 and 30,000 ppm.

During fast pyrolysis, many of the functional groups on lignin are not removed, but remain intact. It is these functional groups, as well as the reactive end-groups formed when ether and carbon–carbon linkages between phenylpropane structures are broken, that account for the high reactivity of lignin-derived organic matter in crude pyrolysis oil, largely through condensation, esterification, and other polymerization reactions (Evans and Milne, 1987; Czernik et al., 1994; Diebold, 2000). Methoxy groups on lignin fragments are dissociated, yielding methane and dihydroxybenzene (Evans and Milne, 1987; Sada et al., 1992); this is the main source of methane production during pyrolysis. Pyrolysis at higher temperatures and over longer times produces polycyclic aromatic hydrocarbons from primary pyrolysis products (Williams and Horne, 1995; Sricharoenchaikul et al., 2002; Britt et al., 2004).

Crude pyrolysis oil produced during fast pyrolysis has a high oxygen content. This oil is reactive and unstable during processing or storage.

Crude pyrolysis oils are acidic, with pH values between 3 and 4 (Sipilä et al., 1998). The main acids are formic, acetic, and glycolic acids. They account for over 90 mol% of the organic acids in pyrolysis oil, as shown in Table 19.12. Acetic acid accounts for half to three-quarters of the organic acids in pyrolysis oil. It comes from deacetylization of hemicelluloses with pendant acetyl groups and from ring opening and reforming reactions in carbohydrates.

The sum of the organic acids in crude pyrolysis oil constitute the *carboxylic acid number* (CAN), which is one measure of the acidity of pyrolysis oils. The *total acid number* (TAN) is commonly used as a measurement of acidity in petroleum and has been used with pyrolysis oils as well. TAN values are higher than CAN values because the acid-base titrations performed to measure TAN values can neutralize phenolic hydroxyls as well as carboxylic acids, while CAN values are determined by quantifying individually each of the carboxylic acids in crude pyrolysis oil.

19.5.1 Characteristics of Pyrolysis Oil

The characteristics of crude pyrolysis oil as reviewed by Diebold et al. (1995) are summarized here.

Composition: The elemental composition of crude pyrolysis oils is typically 54–58 wt% C, 5.5–7.0 wt% H, and 35–40% O. The most important classes of oxygen-containing compounds are water, hydroxyaldehydes, hydroxyketones, sugars, carboxylic acids, and phenolics (Czernik and Bridgewater, 2004).

Water content: Water in pyrolysis oils originates from two sources: moisture in biomass feedstocks and as a product of dehydration reactions that occur during pyrolysis. The water formed during pyrolysis is formed by dehydration reactions, which for the most part are related to char-forming reactions. If water vapor is not condensed with the biocrude oil, large losses in biocrude oil yield and quality would be expected due to the loss of volatile organic matter. Crude pyrolysis oils have moisture contents of 15–30% by weight. Too much water causes the biocrude to separate into two phases, a thin aqueous phase and a thick tar phase.

Alkali metal content: The total alkali content of pyrolysis oil from woody biomass differs with feedstock and depends upon the method used to remove fine particles from the oil (Table 19.9). Neither cyclones nor liquid filters are effective in removing particles below about 10 microns, and the alkali content of oils processed this way is in the order of 100 ppm. Hot-gas filtration of pyrolysis oil vapors reduces the total alkali metal content more effectively, to 1–10 ppm. Alkali, alkaline earth, and other metals accumulate in biochar, which contains much more alkali than does oil, in the range of 6000–20,000 ppm K, Na, Ca, and other metals (Diebold, 2000). The high alkali content of these chars is likely the reason why they catalyze polymerization of organic matter in pyrolysis oils.

Viscosity: The viscosity of freshly produced pyrolysis oil varies by more than an order of magnitude at moderate temperatures (10–400 cP at 40°C). It depends on the feedstock, processing conditions, amount of cracking that occurs, and collection efficiency of lower-molecular-weight volatile compounds. Loss of low-viscosity volatile compounds results in a higher-viscosity oil. Temperature has a strong impact on viscosity in pyrolysis, more so than in petroleum refining. Viscosity can increase with time in storage for weeks to months, depending on the amount of char fines remaining in the oil. It increases with time even at room temperature and 100 times faster at 90°C (Diebold, 2000). The presence of water reduces the viscosity of pyrolysis oil.

Corrosiveness: The pH of crude pyrolysis oil is low, typically between 2 and 4. The total acid number can exceed 100. Pyrolysis oil is very corrosive to carbon steel and aluminum, but not to stainless steels. Increasing the temperature or water content of pyrolysis oil increases its corrosivity.

Combustibility: Pyrolysis oil is combustible, but not flammable. Loss of volatile matter with aging decreases combustibility.

Stability with aging (from Czernik et al., 1994): Three observable and related changes take place as pyrolysis oil ages. One is that ether and ester (C–O–C) linkages increase with time, while alcohol groups (C–O–H) decrease. Together, these changes indicate polymerization by condensation reactions that increases the average molecular weight of the constituent molecules in pyrolysis oil. The increase in viscosity with aging is a result of the increase in average molecular weight over time. The pH of pyrolysis oil does not change with time upon aging. These changes were observed for oil stored at temperatures from

TABLE 19.13

Alkali Metals in Pyrolysis Oils after Char Removal and in Bio-Char

Char Separation Method		Cyclone and Oil Filter		Hot-Gas Filtration
Biomass	**Alkali in Biomass (ppm[a])**	**Alkali in Pyrolysis Oil (ppm)**	**Alkali in Pyrolysis Char (ppm[b])**	**Alkali in Pyrolysis Oil (ppm)**
Hybrid poplar wood[c,f,g]	1227	12–42		1–9
Oak wood[d]		82	1360	
Pine wood[d]		17	1149	
Switchgrass	8531[d]	165[d]	5900[d]	
		192[e]	9190[e]	

[a] Alkali is the sum of K + Na + Li in ppm.
[b] Char also contains 4000–8000 ppm calcium.
[c] Diebold et al., 1995.
[d] Elliott, 1994.
[e] Agblevor et al., 1994.
[f] Diebold et al., 1996.
[g] Scahill et al., 1997.

37°C to 90°C, with a more rapid increase as temperature increased. The viscosity change during aging is very small below 20°C (Diebold, 2000).

The presence of char fines in pyrolysis oil has been shown to be a major factor in its stability. Char fines contain alkali and other metals (see Table 19.13) that catalyze acetal formation and polymerization reactions in pyrolysis oil. The viscosity of oils with higher lignin content increases more rapidly on aging, as shown in Table 19.14. Both removal of the catalytic char fines by hot vapor filtration and deoxygenation of pyrolysis oil by mild hydrotreating are effective in stabilizing pyrolysis oil (Diebold et al., 1995; Scahill et al., 1997; Diebold, 2000).

The effects of both storage temperature and method of char fines removal on the molecular weight and viscosity of pyrolysis oil are shown in Table 19.14. Cyclones have been used to remove char particles from hot pyrolysis oil vapors, but the efficiency of cyclones in removing solids from gases decreases as the particle size decreases to less than 10 microns. Both the average molecular weight and viscosity of pyrolysis oil increase rapidly with time for oils processed with cyclones to remove char fines. Hot gas filtering removes char particles to submicron size and as well removes adsorbed alkali metals to levels less than 10 ppm. Hot filtration decreases the rate of viscosity increase upon storage significantly more than does cyclonic char removal (Oasmaa and Czernik, 1999).

TABLE 19.14

Stability of Molecular Weight and Viscosity for Pyrolysis Oils during Storage from 37°C to 90°C with Cyclonic Removal of Fine Char Particles

	Cyclonic Char Removal	
Treatment Temperature (°C)	**Time for 50% Increase in MW_w (h)**	**Time for Viscosity to Double (h)**
37	250	2500
60	115	160
90	8	8

Source: Adapted from Czernik, S., Johnson, D.K., and Black, S. 1994. *Biomass Bioenergy* 7(1–6): 187–192.

References

Agblevor, F.A., Besler, S., and Evans, R.J. 1994. Inorganic compounds in biomass feedstocks: Their role in char formation and effect on the quality of fast pyrolysis oils. In: Milne, T.A. (ed.), *Proceedings, Biomass Pyrolysis Oil Properties and Combustion Meeting*, 18–26 September, Estes Park, CO, NREL-CP-430-7215, pp. 77–89.

Agblevor, F.A., Beis, S., Mante, O., and Abdoulmoumine, N. 2010. Fractional catalytic pyrolysis of hybrid poplar wood. *Ind. Eng. Chem. Res.* 49: 3533–3538.

Alder, E. 1977. Lignin chemistry: Past, present, and future. *Wood Sci. Technol.* 11: 169.

Britt, P.F., Buchanan, A.C., Owens, C.V. Jr., and Skeen, J.T. 2004. Does glucose enhance the formation of nitrogen containing polycyclic aromatic compounds and polycyclic aromatic hydrocarbons in the pyrolysis of proline? *Fuel* 83 (11–12): 1417–1432.

Caballero, J.A., Font, R., and Marcella, A.J. 1997. Pyrolysis of Kraft lignin: Yields and correlations. *Anal. Appl. Pyrolysis* 39(2): 161–183.

Czernik, S., Johnson, D.K., and Black, S. 1994. Stability of wood fast pyrolysis oil. *Biomass Bioenergy* 7(1–6): 187–192.

Czernik, S. and Bridgewater, A.V. 2004. Overview of applications of biomass fast pyrolysis oil. *Energy Fuels* 18: 590–598.

Czernik, S. 2010. *Review of Fast Pyrolysis of Biomass*. www.nh.gov/oep/programs/energy/documents/biooil-nrel.pdf, 2 September.

Di Blasi C., Galgano A., Branca C. 2009. Influences of the chemical state of alkaline compounds and the nature of alkali metal on wood pyrolysis. *Industrial & Engineering Chemistry Research*, 48: 3359–3369.

Diebold, J.P. 2000. *A Review of the Chemical and Physical Mechanisms of the Storage Stability of Fast Pyrolysis Bio-Oils*. NREL/SR-570-27613, January.

Diebold, J.P., Scahill, J.W., Czernik, S., Phillips, S.D., and Feik, C.J. 1995. *Progress in the Production of Hot-Gas Filtered Biocrude Oil at NREL*. NREL/TP-431-7971, May.

Evans, R.J. and Milne, T.A. 1987. Molecular characterization of the pyrolysis of biomass. 2. Applications. *Energy Fuels* 1(4): 311–319.

Fahmi, R., Bridgwater, A.V., Darvell, L., Yates, N., and Thain, S. 2007. The effect of alkali metals on combustion and pyrolysis of *Lolium* and *Festuca* grasses, switchgrass, and willow. *Fuel* 86: 1560–1569.

Fahmi, R., Bridgwater, A.V., Donnison, I., Yates, N., and Jones, J.M. 2008. The effect of lignin and inorganic species in biomass on pyrolysis oil yields, quality, and stability. *Fuel* 87: 1230–1240.

Fletcher, T.H., Kerstein, A.R., Pugmire, R.J., and Grant, D.M. 1990. Chemical percolation model for devolatilization: II. Temperature and heating rate effects on product yields. *Energy Fuels* 4: 54.

Frederick, W.J. and Mentzer, C.C. 1975. Heats of volatilization by differential scanning calorimetry. *J. Applied Polymer Sci.* 19(7): 1799 C.C. 1804.

Grant, D.M., Pugmire, R.J., Fletcher, T.H., and Kerstein, A.R. 1989. A chemical model of coal devolatilization using percolation lattice statistics. *Energy Fuels* 3: 175–186.

Jenkins, B.M., Baxter, L.L., Miles, T.R. Jr., and Miles, T.R. 1998. Combustion properties of biomass. *Fuel Processing Technology* 54: 17–46.

Liao, Y.-F., Wang, S.-R., and Ma, X.-Q. 2004. Study of reaction mechanisms in cellulose pyrolysis. *Am. Chem. Soc., Div. Fuel Chem.* 49(1): 407.

Maloney, M.T., Chapman, T.W., and Baker, A.J. 1985. Dilute acid hydrolysis of paper birch: kinetic studies of xylan and acetyl-group hydrolysis. *Biotechnol. Bioeng.* 27: 355–361.

Oasmaa, A. and Czernik, S. 1999. Fuel quality of biomass pyrolysis oils. *Energy Fuels* 13: 914–921.

Oasmaa, A., Solantausta, Y., Arpiainen, V., Kuoppala, E., and Sipilä, K. 2010. Fast pyrolysis bio-oils from wood and agricultural residues. *Energy Fuels* 24: 1380–1388.

Raveendran, K., Ganesh, A., and Khilar, K.C. 1996. Pyrolysis characteristics of biomass and biomass components. *Fuel* 75(8): 987–998.

Sada, E., Kumazawa, H., and Kudsy, M. 1992. Pyrolysis of lignins in molten salt media. *Ind. Eng. Chem. Res.* 31: 612–616.

Satrio, J.A., Brown, R.C., and Shanks, B.R. 2009. Product distribution from fast pyrolysis of glucose-based carbohydrates. *J. Anal. Appl. Pyrolysis* 86(2): 323–330.

Scahill, J., Diebold, J.P., and Feik, C. 1997. Removal of residual char fines from pyrolysis vapors by hot gas filtration. In: Bridgwater, A.V. (ed.), *Developments in Thermochemical Biomass Conversion*, Volume 1, Kluwer, London, pp. 253–266.

Sipilä, K., Kuoppala, E., Fagernäs, L., and Oasmaa, A. 1998. Characterization of biomass-based flash pyrolysis oils. *Biomass Bioenergy* 14(2): 103–113.

Sricharoenchaikul, V., Agrawal, P., and Frederick, W.J. Jr. 2002. Black liquor gasification characteristics. 2. Measurement of condensable organic matter (tar) at rapid heating conditions. *Ind. Eng. Chem. Res.* 41(23): 5650–5658.

Venderbosch, R.H. and Prins, W. 2010. Fast pyrolysis technology development. *Biofuels, Bioprod. Bioref.* 4: 178–208.

Venderbosch, R.H., Ardiyanti, A.R., Wildschut, J., Oasmaa, A., and Heeres, H.J. 2010. Stabilization of biomass-derived pyrolysis oils. *J. Chem. Technol. Biotechnol.* Published online in Wiley Interscience: http://www.btgworld.com/uploads/documents/Stabilization%20of%20biomass-derived%20pyrolysis%20oils_Venderbosch%20et%20al.,%202010.pdf (26 October).

Williams, P.T. and Horne, P.A. 1995. Analysis of aromatic hydrocarbons in pyrolytic oil derived from biomass. *J. Anal. Appl. Pyrolysis* 31: 15–37.

Yang, H., Yan, R., Chen, H., Lee, D.H., and Zheng, C. 2007. Characteristics of hemicellulose, cellulose and lignin pyrolysis. *Fuel* 86(12–13): 1781–1788.

Zhang, H., Xiao, R., Wang, D., Zhong, Z., Song, M., Pan, Q., and He, Q. 2009. Catalytic fast pyrolysis of biomass in a fluidized bed with fresh and spent fluidized catalytic cracking (FCC) catalysts. *Energy Fuels* 23 (12): 6199–6206.

19.6 Catalytic Pyrolysis and Vapor Upgrading

With the recent attention focused on liquid transportation fuels from biomass, fast pyrolysis has emerged as a promising alternative. However, rapid pyrolysis per se produces a raw product that requires considerable upgrading to stabilize it, reduce its acidity, increase its heating value, and make it compatible with conventional petroleum fuels (French and Czernik, 2010). Catalytic pyrolysis of biomass is one route to high-grade pyrolysis oils that, if successful, requires substantially less downstream upgrading of crude pyrolysis oils for compatibility in processing with conventional transportation fuels.

Catalytic pyrolysis is conversion of biopolymers into pyrolysis oils that contain less oxygen and have lower acidity than pyrolysis oils produced thermally and in the absence of catalysts. It encompasses two alternative pathways. One is *in situ* catalytic pyrolysis, in which cracking catalysts are placed within the pyrolysis reactor to contact pyrolysis vapors as they are produced. The other is catalytic cracking of pyrolysis vapors immediately downstream of the pyrolysis reactor.

19.6.1 *In Situ* Catalytic Pyrolysis

In situ catalytic pyrolysis uses cracking catalysts *in situ* during pyrolysis to deoxygenate pyrolysis oils as they are being produced. It increases the yields of olefins by transforming oxygenates into liquid hydrocarbons and olefins. It also reduces the temperature required before thermal cracking occurs, although some thermal cracking proceeds in parallel with catalytic cracking. Zeolites, silica–alumina, and molecular sieves are some of the cracking catalysts that have been evaluated at atmospheric pressure conditions and without hydrogen (Williams and Horne, 1995; Bridgwater, 1996; Horne and Williams, 1996; Lappas et al., 2002; Vitolo et al., 1999).

Synthetic zeolites such as ZSM-5 have been used to convert organic oxygenates to hydrocarbons since the late 1970s. At 350–450°C, zeolite catalysts deoxygenate alcohols, aldehydes, ketones, acids, and esters by dehydration, decarboxylation, cracking, aromatization, alkylation, condensation, and polymerization reactions (French and Czernik, 2010). Products of these reactions are two-phase liquids with both aqueous and organic phases present, gaseous products, and coke deposited on catalysts. According to French and Czernik (2010), "high conversions (<90%) were obtained for alcohols, aldehydes, ketones, acids, and esters, while phenols and ethers remained mostly unchanged. Alcohols and ketones reacted to produce high yields of aromatic hydrocarbons, while acids and esters were mostly converted to gas, water, and coke with low yield of hydrocarbons."

Currently, catalytic pyrolysis produces oils of reduced oxygen content and acidity, with oil yields in the 10s to mid-40s as a percentage of dry biomass pyrolyzed (Table 19.15). These are not hydrocarbon yields because the oils still contain 10–30% oxygen. With improvements in catalytic pyrolysis performance, yields of oxygen-free hydrocarbon-containing oil would be expected to be 30–35% of dry biomass.

Other catalysts have also been explored in recent studies of catalytic pyrolysis. French and Czernik (2010) listed 40 different catalysts and evaluated them in their study. They included commercial and synthetic modified zeolites and silica and alumina catalysts. Other catalysts evaluated by others include FCC catalysts and silica sand (Lappas, 2002), activated alumina (Demiral and Sensöz, 2008; Putun, 2008), mesoporous aluminosilicates (Al-MCM-41; Iliopoulou et al., 2007), and the commercial cracking catalysts BP 3189 and Criterion-424 (Onay, 2007). These catalysts were evaluated in a number of different reactor configurations, including batch, fixed bed, and circulating fluidized bed. A wide range of pyrolysis oil yields (13–44%) were reported, but the oxygen content of the oils produced were often not reported. The pyrolysis oil yields reported cannot be compared directly because of their differing oxygen contents and the different reactors and reaction conditions used to produce them.

19.6.2 Vapor Upgrading of Pyrolysis Oil

In conventional biomass pyrolysis, vapors exiting the pyrolyzer may be filtered to remove fine char particles and then are cooled rapidly and condensed. In vapor upgrading, hot filtered oil vapors are deoxygenated and cracked immediately, after hot filtration. Acidic, shape-selective catalysts such as HZSM-5 zeolite have been used successfully for this purpose. Process conditions are moderate, 350–500°C and atmospheric pressure. Hydrogen is not required.

During vapor upgrading, oxygen is rejected as H_2O, CO_2, and CO through simultaneous dehydration and decarboxylation over acidic zeolite catalysts. The resulting upgraded oil has a low H/C ratio, which limits hydrocarbon yield. As a result, the upgraded oil product contains primarily aromatics. Rejection of oxygen as CO_2 and CO rather than as water vapor is preferable to produce oil with a higher H/C ratio.

Bridgwater (1994) estimated that typical yields of aromatics suitable for gasoline blending from biomass would be approximately 20 wt%, or 45% in energy terms, both

TABLE 19.15

Results from Catalytic Pyrolysis of Biomass in Fluidized-Bed Gasifiers

Reference	Biomass	Catalyst	System	Temperature	Total Liquid (%)	Oil (%)	Water (%)	Char + Coke (%)	Gas (%)	Oil O Content
Agblevor (2010)	Hybrid poplar	HZSM5 (ExxonMobil)	50 mm ID, N_2, 1 s (vapor), semibatch, 200 g catalyst, wood fed at 100 g/h for 2–3 h (200–300 g)	500°C	33.5	18–21	12.5–5.5	Char + coke 12.6	53.6	22%
Agblevor (2010)	Hybrid poplar	HZSM5 (ExxonMobil)	100 mm ID, N_2 + producer gas (2:13), 1–2 s, semibatch, 1000 g catalyst, 4200 g wood fed at 2 kg/h	450°C	34.2			Char 14.2	51.6	
Zhang et al. (2009)	Corncob	HZSM-5 Si:Al = 24 (Nankai University)	30 mm ID, N_2, batch, 30 g catalyst, 6 g conrcob fed at once	550°C	38.8	13.2	25.6	Char 20.1 coke 8.4	26	15% in 2nd condenser
	Pine	HZSM5–23 (Zeolyst)	45 mm ID, N_2, 5.9 s (vapor), semibatch, 12 g catalyst, 30 g wood fed at 20 g/h	450°C	33.7	20.7	12.0	Char 9.0, coke 5.2	52.1	
Lappas et al. (2002)	Lignocell HBS (Rettenmeier)	ZSM-5-based FCC additive (10% zeolite)	CFB, 7.1 mm ID (riser), N_2, 0.66 s (biomass) Catalyst:biomass = 18	400°C	65.2	30.6	34.6	Char 22.5	13.9	
Lappas et al. (2002)	Lignocell HBS (Rettenmeier)	ZSM-5-based FCC additive (10% zeolite)	CFB, 7.1 mm ID, N_2, 0.66 s (biomass) Catalyst:biomass = 4.3	400°C	70.5	44.4	26.1	Char 18.4	13.9	
Lappas et al. (2002)	Lignocell HBS (Rettenmeier)	ZSM-5-based FCC additive (10% zeolite)	CFB, 7.1 mm ID, N_2, 0.66 s (biomass) Catalyst:biomass = 2.9	400°C	60.3	36.4	23.9	Char 11.5	13.6	

TABLE 19.16
Distillate Product Yield and Composition from Vapor Upgrading of Pyrolysis Oil from Maple Wood

Catalyst	**None**	**HZSM-5**				
Temperature	**425**	340	340	370	370	390
Steam/Bio-Oil wt Ratio	0	0	0.4	0	0.4	0
Organic liquid yield (wt% of biomass)	15.1	10.5	10.4	19.1	17.2	16.8
Product distribution (wt% of crude bio-oil)						
Aliphatic hydrocarbons	5.3	3.2	36.9	2.7	12.7	
Aromatic hydrocarbons	3.2	62.2	49.1	80.7	71.0	83.4
Phenols	9.6	7.9	2.2	3.3	2.4	3.1
Acids	19.2	1.0	4.8		0.8	0.4
Ethers	10.5	0.7				
Furans	3.2	5.0				
Aldehydes	12.9	7.8	0.6	1.1	1.2	
Ketones	16.1	12.2	1.6	2.6	3.5	5.1
Alcohols	13.8	100	0.7	1.2	0.8	2.2
Unaccounted for	6.2	8.4	4.1	8.4	7.6	5.8
Total	100	100	100	100	100	100

Source: Adapted from Sharma, R.K. and Bakhshi, N.N. 1993. *Energy Fuels*. 7: 306–314.

based on biomass feedstock. In one study, the distillates produced by vapor upgrading of pyrolysis oil from maple wood are predominantly aromatic hydrocarbons from catalytic vapor which were upgraded at 340–400°C from 49% to 83% of crude pyrolysis oil (Table 19.16). For comparison, the aromatics product yield from uncatalyzed thermal treatment at 425°C was only 3.2%. Vapor upgrading in the presence of steam reduced aromatics yield by approximately 10%, but increased aliphatic hydrocarbons by a factor of 5–10.

A number of recent studies have been carried out to evaluate processes and catalysts for vapor upgrading of pyrolysis oil (Sharma and Bakhshi, 1993; Scholze, 2002; Park et al., 2010). Although it is often difficult to compare directly the results from studies made under different reactor conditions and with different crude pyrolysis oils, some of the common findings are

- Acidic zeolite catalysts produce predominantly aromatic organic compounds, both single- and multiring.
- HZSM-5, with its strong acidic sites, is more efficient in bio-oil upgrading than less acidic catalysts.
- The optimal catalyst bed temperature for producing aromatic hydrocarbons differs with process configurations and conditions, ranging from 350°C to 500°C.
- Water vapor shifts hydrocarbon production from aromatics to aliphatics.

The major compounds in bio-oil were found to be phenolics, including cresol, guaiacol, eugenol, benzendiol, and their derivatives, ketones, and aldehydes. Similar compounds have been found in other bio-oils produced by fast pyrolysis.

In particular, because ketones and aldehydes are highly reactive, crude pyrolysis oil is unsuitable for use as a fuel. Therefore, it is necessary to convert these compounds into more stable compounds.

As might be expected, different classes of oxygenates in pyrolysis oil behave quite differently during catalytic deoxygenation. In their broad study of this topic, Gayubo et al. (2004a, 2004b, 2005) investigated the transformations of various classes of species found in pyrolysis oils when processed over a HZSM-5 catalyst. They found that aliphatic alcohols were transformed to hydrocarbons by a route similar to that for methanol and ethanol. Aromatic hydrocarbons (phenol and 2-methoxy-phenol), on the other hand, have low reactivities to hydrocarbons.

The reactivities of acetaldehyde, ketones (acetone and butanone), and acetic acid varied widely. Acetaldehyde had a low reactivity to hydrocarbons and deposited coke that deactivated the catalyst significantly. Ketones were less reactive than alcohols. Acetic acid was transformed to acetone as the primary product. Their transformations occur mainly through decarboxylation and dehydration. Above 400°C, olefins and aromatics were produced through a reaction scheme similar to that for the reaction of alcohols. Coke generation was greater than with alcohols; it shifts the product mix toward olefins and away from aromatics.

The reactivities of mixtures of these five classes of compounds differed from the reactivity of the individual classes of components. Coking was worse when mixtures were processed than with individual classes of compounds. To minimize coking, components such as aldehydes, oxyphenols, and furfural that undergo severe thermal degradation and form carbonaceous deposits should be separated before catalytic processing.

References

Bridgwater, A.V. 1994. Catalysis in thermal biomass conversion. *Appl. Catal. A* 116: 5–47.

Bridgwater, A.V. 1996. Production of high-grade fuels and chemicals from catalytic pyrolysis of biomass. *Catal. Today* 29: 285–295.

Demiral, I. and Sensöz, S. 2008. The effects of different catalysts on the pyrolysis of industrial wastes (olive and hazelnut bagasse). *Biores. Technol.* 99: 8002–8007.

French, R. and Czernik, S. 2010. Catalytic pyrolysis of biomass for biofuels production. Fuel Process. Technol. 91: 25–32.

Gayubo, A.G., Aguayo, A.T., Atutxa, A., Aguado, R., and Bilbao, J. (2004a). Transformation of oxygenated components of biomass pyrolysis oil on an HZSM-5 zeolite. I. Alcohols and phenols. *Ind. Eng. Chem. Res.* 43: 2610–2618.

Gayubo, A.G., Aguayo, A.T., Atutxa, A., Aguado, R., Olazar, M., and Bilbao, J. (2004b). Transformation of oxygenated components of biomass pyrolysis oil on an HZSM-5 zeolite. II. Aldehydes, ketones, and acids. *Ind. Eng. Chem. Res.* 43: 2619–2626.

Gayubo, A.G., Aguayo, A.T., Atutxa, A., Valle, B., and Bilbao, J. 2005. Undesired components in the transformation of biomass pyrolysis oil into hydrocarbons on an HZSM-5 zeolite catalyst. *J. Chem. Technol. Biotechnol.* 80: 1244–1251.

Horne, P.A. and Williams, P.T. 1996. Upgrading of biomass-derived pyrolytic vapours over zeolite ZSM-5 catalyst: effect of catalyst dilution on product yields. *Fuel* 75: 1043–1050.

Iliopoulou, E.F., Antonakou, E.V., Karakoulia, S.A., Vasalo, I.A., Lappas, A.A., and Triantafyllidis, K.S. 2007. Catalytic conversion of biomass pyrolysis products by mesoporous materials: Effect of steam stability and acidity of Al-MCM-41 catalysts. *Chem. Eng. J.* 134: 51–57.

Lappas, A.A., Samolada, M.C., Iatridis, D.K., Voutetakis, S.S., and Vasalos, I.A. 2002. Biomass pyrolysis in a circulating fluid bed reactor for the production of fuels and chemicals. *Fuel* 81: 2087–2095.

Onay, O. 2007. Fast and catalytic pyrolysis of *Pistacia khinjuk* seed in a well-swept fixed bed reactor. *Fuel* 86: 1452–1460.

Park, H.J., Heo, H.S., Yim, J.-H., Jeon, J.-K., Ko, Y.-S., Kim, S.-S., and Park, Y.-K. 2010. Catalytic pyrolysis of Japanese larch using spent HZSM-5. *Korean J. Chem. Eng*. 27(1), 73–75.

Putun, E., Ateş, F., and Putun, A.E. 2008. Catalytic pyrolysis of biomass in inert and steam atmospheres. *Fuel* 87: 815–824.

Scholze, B. 2002. Long-term Stability, Catalytic Upgrading, and Application of Pyrolysis Oils—Improving the Properties of a Potential Substitute for Fossil Fuels. PhD Dissertation, University of Hamburg, Germany.

Sharma, R.K. and Bakhshi, N.N. 1993. Catalytic upgrading of pyrolysis oil. *Energy Fuels* 7: 306–314.

Vitolo, S., Petarca, L., and Bresci, B. 1999. Treatment of olive oil industry wastes. *Biores. Technol.* 67(2): 129–137.

Williams, P.T. and Horne, P.A. 1995. The influence of catalyst type on the composition of upgraded biomass pyrolysis oils. *J. Anal. Appl. Pyrol.* 31: 39–61.

19.7 Upgrading Crude Pyrolysis Oil

Upgrading pyrolysis oils from biomass entails removal of char particles to stabilize the oils, removal of water, removal of oxygen moieties from the organic matter, and removal of acidic compounds such as formic and acetic acids to increase pH and reduce corrosivity. Commercial processes for upgrading biomass-derived pyrolysis oils were not established as of the date this section was written, although active research and development programs on upgrading were underway (Elliott, 1994; Elliott et al., 2009; French et al., 2010; Oasmaa et al., 2010; Venderbosch, 2010).

The upgrading requirements for pyrolysis oils depend on the intended end use of the oil, and at least two options need to be considered. One is upgrading and refining of pyrolysis oil by itself to end products. The other is partial upgrading of pyrolysis oil to produce a blendstock suitable for further processing in oil refineries. Standards and requirements for upgrading and subsequent refining of pyrolysis oils for either application remain to be defined.

19.7.1 Removal of Char Particles to Stabilize Oils

The char particles in crude pyrolysis oil are submicron in size and contain alkali metals at concentrations of 1000–10,000 ppm (Elliott, 1994). They catalyze polymerization of pyrolysis oils, thereby increasing oil viscosity during aging. Removal of char particles has been shown to stabilize pyrolysis oil by reducing or eliminating viscosity increase over time.

Hot gas filtration is effective in removing fine char particles because oil vapors pass easily through filters (Diebold et al., 1995). Filtration of char particles from liquid pyrolysis oil is difficult because the oil-char sludge clogs liquid-oil filters. Removal of the submicron char particles with cyclones is not efficient enough to eliminate their impact on pyrolysis-oil stability.

19.7.2 Removal of Water

About half the water in crude pyrolysis oil comes from the water in the biomass feedstock, while the rest comes from dehydration reactions that occur during pyrolysis. Drying the feedstock before pyrolysis reduces the water content of crude pyrolysis oil, but water is formed during pyrolysis and remains in the crude pyrolysis oil.

One way to remove water from pyrolysis oil is by partial condensation of the pyrolysis oil vapor at a temperature near 100°C. This method can eliminate nearly all the water in crude pyrolysis oil, but with a yield loss from organic components that are not yet condensed.

Water can also be eliminated by phase splitting, which separates pyrolysis oil into a water-rich phase and an oil-rich phase. Phase splitting is accomplished by adding water to crude pyrolysis oil. The oil-rich phase contains almost no water, while the water-rich phase contains some of the water-soluble organic matter in crude pyrolysis oil.

Another option for removing water from crude pyrolysis oil is first to stabilize the oil and then to evaporate or distill it to remove water and light organic acids. Although this method shows promise, it is still in the research stage and remains to be demonstrated at a larger scale (French, 2010). One problem with removing water and its associated organics is that the residue may be highly viscous and not easily handled.

19.7.3 Removal of Oxygen Moieties from Organic Matter

Oxygen in pyrolysis oil can be removed by hydrotreating. Hydrotreating is a process used in oil refineries to stabilize petroleum by removing contaminants such as oxygen-containing moieties and to convert olefins to paraffins in petroleum before hydrocracking. Hydrotreating converts chemically bound oxygen to water by reacting it with hydrogen. The catalyst used may be tungsten- or molybdenum-based, possibly modified with cobalt or nickel sulfides. Nickel sulfide by itself is also active as a hydrotreating catalyst. Process conditions are typically 270–340°C and 100–300 bar.

One challenge in applying hydrotreating technology to pyrolysis oil is the cost of the hydrogen consumed. Crude pyrolysis oil contains from 44 to 60 wt% oxygen (Oasmaa and Czernik, 1999), while petroleum contains 0.1–1.5 wt% (Speight, 1999). This means that hydrotreating of pyrolysis oil would consume from 30 to 1000 times as much hydrogen for deoxygenation than would hydrotreating of petroleum for the same yield from both oil sources. These estimates are based on nearly complete removal of oxygen. It is not yet known to what extent oxygen needs to be removed from pyrolysis oil before further processing, but the cost of hydrogen for hydrotreating could be reduced substantially if higher oxygen levels in hydrotreated pyrolysis oil could be tolerated.

The definition of the tolerable level of oxygen in pyrolysis oil as a refinery feedstock is a critical question that is currently being addressed. Research shows that with two-stage hydrotreating, the carboxylic acid number (CAN) in pyrolysis oil can be reduced by prestabilizing the oil at 150°C and then hydrotreating the oil at 340–400°C in the presence of a molybdenum-promoted nickel catalyst on an alumina support and a hydrogen pressure of 137–171 bar (Elliott, 2007; Elliott et al., 2009; French et al., 2010). With this approach, oxygen contents of 2–7% and CAN numbers of 14–35 can be obtained. Both of these values decrease with increasing treatment temperature. This treatment produces low-viscosity oils that range from slightly to highly soluble in hydrocarbons. Work is currently underway to determine the compatibility of these hydrotreated, low- to moderate-oxygen-content pyrolysis oils with various petroleum streams in oil refineries.

19.7.4 Removal of Acidic Compounds Such as Formic and Acetic Acids to Increase pH and Reduce Corrosivity

Removal of acidic compounds from crude pyrolysis oil to reduce acidity is essential for oil stability and to minimize corrosion of processing equipment. As shown in Table 19.17, the acidity of these oils, expressed as carboxylic acid number (CAN), can exceed 100 mg

TABLE 19.17
Properties of Flash Pyrolysis Oils

Property	Pyrolysis Liquids from Different Feedstocks			
	Birch	Pine	Poplar	Various
Solids (wt%)	0.06	0.03	0.045	0.01–1
pH	2.5	2.4	2.8	2.0–3.7
Water (wt%)	18.9	17.0	18.9	15–30
Density (kg/m³)	1.25	1.24	1.20	1.1–1.3
Viscosity, cSt (50°C)	28	28	13.5	13–80
LHV (MJ/kg)	16.5	17.2	17.4	13–18
Ash (wt%)	0.004	0.03	0.01–1	0.004–0.3
C (wt%)	44.0	45.7	46.5	32–49
H (wt%)	6.9	7.0	7.2	6.9–8.6
N (wt%)	<0.1	<0.1	0.15	0–0.2
S (wt%)	0.00	0.02	0.02	0.00–0.05
O (wt%)	49.0	47.0	46.1	44–60
Na + K (ppm)	29	22	6.9–8.6	5–500
Ca (ppm)	50	23	4	4–600
Mg (ppm)	12	5	3	
Flash point (°C)	62	95	64	50–100
Pour point (°C)	–24	–19		–36 to –9

Source: Adapted from Oasmaa, A. and Czernik, S. 1999. *Energy Fuels* 13: 914–921.

KOH/g. Organic acids derived from biomass are the main, if not the predominant, contributors to the acidity of pyrolysis oils.

The methods for removal of organic acids from pyrolysis oil are established or being developed at the laboratory scale, but none are yet practiced at the production scale. One method uses dilute acid or alkali at 100°C or higher to cleave acetyl groups and extract them as acetic acid (Maloney et al., 1985). This method can remove essentially all the acetic acid precursors before pyrolysis.

Two methods for water removal also separate organic acids from pyrolysis oil. In phase splitting of pyrolysis oil, water-soluble components, including light organic acids are enriched in the aqueous phase and can be recovered from it. Evaporation or distillation of stabilized crude pyrolysis oil removes both water and volatile organic acids. Both of these methods offer promise, but both are still in the research stage and remains to be demonstrated at a larger scale (French, 2010). One problem with removing water and the associated organics is that the residue may be highly viscous and not easily handled.

Hydrotreating of pyrolysis oil results in a decrease in overall yield to approximately 30% because the oxygen content of the oil is reduced from almost 50% to 7% or less. However, the energy value of the hydrotreated oil increases as oxygen is removed, and therefore the loss in energy content of the oil is much less than the loss in mass yield.

References

Diebold, J.P., Scahill, J.W., Czernik, S., Phillips, S.D., and Feik, C.J. 1995. *Progress in the Production of Hot-Gas Filtered Biocrude Oil at NREL*. NREL/TP-431-7971, May.

Elliott, D.C. 1994. Water, alkali, and char in flash pyrolysis oils. *Biomass Bioenergy* 7(1–6): 179–185.
Elliott, D.C. 2007. Historical developments in hydroprocessing bio-oils. *Energy Fuels* 21(3): 1792–1815.
Elliott, D.C., Hart, G.G., Neuenschwander, L.J., Rotness, T.R. Jr, and Zacher, A.H. 2009. Catalytic hydroprocessing of biomass fast pyrolysis bio-oil to produce hydrocarbon products. *Environ. Prog. Sustainable Energy* 28(3): 441–449.
French, R. 2010. Bio-oil stabilization by mild hydrotreatment. *Symposium for Thermal and Catalytic Sciences for Biofuels and Biobased Products*, Ames, IA, 21–23 September.
French, R.J., Hrdlicka, J., and Baldwin, R. 2010. Mild hydrotreating of biomass pyrolysis oils to produce a suitable refinery feedstock. *Environ. Prog. Sustainable Energy* 29(2): 142–150.
Maloney, M.T., Chapman, T.W., and Baker, A.J. 1985. Kinetics studies of xylan and acetyl-group hydrolysis. *Biotechnol. Bioeng.* 27: 355–361.
Oasmaa, A. and Czernik, S. 1999. Fuel quality of biomass pyrolysis oils. *Energy Fuels* 13: 914–921.
Oasmaa, A., Solantausta, Y., Arpiainen, V., Kuoppala, E., and Sipilä, K. 2010. Fast pyrolysis bio-oils from wood and agricultural residues. *Energy Fuels* 24: 1380–1388.
Speight, J.G. 1999. The chemical and physical structure of petroleum: Effects on recovery operations. *J. Petroleum Sci. Eng.* 22(1–3): 3–15.
Venderbosch, R.H., Ardiyanti, A.R., Wildschut, J., Oasmaa, A., and Heeres, H.J. 2010. Stabilization of biomass-derived pyrolysis oils. *J. Chem. Technol. Biotechnol.* 85: 674–686.

19.8 Pyrolysis Technology

Technology development for converting lignocellulosic biomass to liquid transportation fuels began in earnest in the mid-1970s. One outcome was the recognition of fast pyrolysis as the leading contender for new pyrolysis processes. This led to the development of several fast pyrolysis technologies (Bridgewater, 2004; Czernik and Bridgewater, 2004; Ringer et al., 2006; Oasmaa, 2010; Venderbosch, 2010). Five of the pyrolysis reactors which were developed to the scale of process demonstration units are shown in Table 19.18. Yields of crude liquid product as high as 70–80 wt% dry biomass feedstock were reported. Today, two of these, the stationary fluidized bed and the circulating fluidized bed, have become leaders in pyrolysis process technology development.

TABLE 19.18
Pyrolysis Processes Developed as of 2006

Reactor Design	Capacity (Dry Biomass Feed)	Organization or Company
Fluidized bed	400 kg/h (11 tons/day)	DynaMotive, Canada
	250 kg/h (6.6 tons/day)	Wellman, UK
	20 kg/h (0.5 tons/day)	RTI, Canada
Circulating fluidized bed	1500 kg/h (40 tons/day)	Red Arrow, WI Ensyn design
	1700 kg/h (45 tons/day)	Red Arrow, WI Ensyn design
	20 kg/h (0.5 tons/day)	VTT, Finland Ensyn design
Rotating cone	200 kg/h (5.3 tons/day)	BTG, Netherlands
Vacuum	3500 kg/h (93 tons/day)	Pyrovac, Canada
Other types	350 kg/h (9.3 tons/day)	Fortum, Finland

Source: Adapted from Ringer, M., Putsche, V., and Scahill, J. 2006. *Large-Scale Pyrolysis Oil Production: A Technology Assessment and Economic Analysis.* NREL Technical Report TP-510-37779, November.

From the 1990s onward, process development moved toward technology commercialization. The rest of this section reviews briefly the currently commercial fast pyrolysis technologies and some that are near commercialization.

19.8.1 Circulating-Bed Fast Pyrolysis

Ensyn Technologies (Ottawa, Canada) has developed a fast pyrolysis reactor that uses a circulating fluidized-bed, thermal pyrolyzer design. In the Ensyn process, biomass is introduced into a vessel where it is mixed with hot sand, heated rapidly to 500°C, and then rapidly cooled within seconds. The yield of crude pyrolysis oil can exceed 70 wt% based on dry biomass feedstock. The process also produces by-product char and noncondensable gas, both of which are used to provide process energy. The company claims that its plants can break even on process energy when the moisture content of the biomass is 45%.

Ensyn has designed, built, and commissioned seven commercial fast pyrolysis plants in the United States and Canada. The largest of these is located in Renfrew, Ontario, and processes 100 tonnes of dry residual wood per day. Ensyn expects to design next-generation plants of "five to 10 times the size of the Renfrew plant."

19.8.2 DynaMotive (Vancouver, Canada)

DynaMotive (Vancouver, Canada) uses a bubbling fluidized-bed design to achieve rapid heating of biomass with short pyrolysis oil-vapor residence times. Feedstock consisting of 1–2 mm particles dried to less than 10% moisture content is fed into a bubbling fluid-bed reactor that operates at 450–500°C. Char is separated from the pyrolysis oil vapors, and the vapors are quenched rapidly. Gases exiting the quench system are burned for process heating. The process also produces a bio-char. The DynaMotive process has been demonstrated in a 15 t/day pilot plant and in two commercial plants; a 130 t/day plant in West Lorne, ON and a 200 t/day plant in Guelph, ON are currently operating.

A companion two-stage pyrolysis-oil upgrading process is under development by Dynamotive. The first stage is intended to improve oil stability and miscibility with hydrocarbon liquids as well as to reduce viscosity, corrosivity, and oxygen content. The objective is hydroreforming of bio-oil to a first improved liquid that can either be directly used in blends with hydrocarbon fuels as an industrial fuel or be further upgraded in a second step to transportation-grade liquid hydrocarbon fuels. The second stage is conventional hydrotreatment over a commercial catalyst to convert the pyrolysis oil to diesel, gasoline, and jet fuel.

19.8.3 Metso Corporation—UPM Corporation—VTT (Helsinki, Finland)

Metso and UPM have developed a concept for the production of bio-oil. The concept is based on a technology developed at the Technical Research Centre of Finland (VTT).

References

Bridgewater, A.V. 2004. Biomass Fast Pyrolysis. *Thermal Sci.* 8(2): 21–49.

Czernik, S. and Bridgewater, A.V. 2004. Overview of applications of biomass fast pyrolysis oil. *Energy Fuels* 18: 590–598.

Ringer, M., Putsche, V., and Scahill, J. 2006. *Large-Scale Pyrolysis Oil Production: A Technology Assessment and Economic Analysis*. NREL Technical Report TP-510-37779, November.

Oasmaa, A., Solantausta, Y., Arpiainen, V., Kuoppala, E., and Kai Sipilä, K. 2010. Fast pyrolysis bio-oils from wood and agricultural residues. *Energy Fuels* 24, 1380–1388.

Venderbosch, R.H. and Prins, W. 2010. Fast pyrolysis technology development. *Biofuels, Bioprod. Bioref.* 4: 178–208.

19.9 Application to Integrated Biorefineries

Although the concept of an integrated biorefinery is relatively new in the biomass arena, forest products companies have traditionally coproduced wood products, cellulose fiber, paper, turpentine, tall oil, and power at a single site or sites located near each other. Older sulfite pulp mills have produced ethanol by fermentation of dissolved wood carbohydrates. More recently, with constraints on the release of environmentally harmful chemicals, methanol, a coproduct of wood pulping, is being recovered and used as fuel.

Studies have been carried out by a number of research teams to evaluate the technical and economic feasibility of integrated biorefineries. In one of the first published studies, Larson et al. (2006, 2008, 2009) evaluated options for the production of liquid fuels from biomass in a fuel plant fully integrated with a pulp and paper mill. In their study, "full integration" meant complete mass and energy integration so that residual energy from process cooling in the fuel synthesis plant, cooling of syngas, and other sources would be used for evaporation and drying in the pulp and paper mill, offsetting the need for some of the fuel purchased by the mill. Pulp fiber production was 1580 dry t/day.

Larson et al. considered four cases in which either Fischer–Tropsch hydrocarbons or mixed alcohols were the primary product from the fuel synthesis plant. A base-case scenario consisting of only the pulp and paper mill with no synthetic fuel production was evaluated for comparison. Electrical power for export was generated in each case. In two of the three cases where Fischer–Tropsch hydrocarbons were produced, the overall complex was configured in one case for maximum power production and in the other case for maximum liquid fuel production. In all cases, additional capital was required for the liquid fuel plant, and additional purchased biomass was required to generate additional syngas for the fuel synthesis plant. A summary of the energy inputs and outputs for the overall complex is shown in Table 19.19.

The capacity to coproduce liquid fuels requires additional capital investment. The incremental capital cost for an integrated biorefinery producing both liquid fuels and pulp fiber relative to a plant producing pulp fiber only is shown in Table 19.20.

An additional benefit from the integration of synthetic fuels and pulp fiber production is a reduction in CO_2 emissions. Achievable reductions for the four biorefinery configurations reviewed here are shown in Figure 19.9. The reductions are "forest to wheels" reductions per biorefinery relative to conventional-fuel-chain fuels and pulp fiber production. The capacity of the biorefineries considered here corresponds to a pulp fiber production rate of 1580 dry t/day.

One of the findings of the Larson et al. studies is that a significant increase in the annual output of energy products (F-T liquids and power) can be achieved with an integrated biorefinery. Furthermore, the conversion efficiency of biomass energy to energy products increases by a factor of 2–4 depending on which fuel is produced, the relative production rates of

TABLE 19.19

Overall Energy Inputs and Outputs for an Integrated Biorefinery Consisting of a Fischer–Tropsch Fuel Synthesis Plant and a Kraft Pulp and Paper Mill Producing 1580 Dry t/Day of Pulp Fiber

Case	Base	FTa	FTb	FTc	MA
Energy Inputs (MW)					
Wood	722.6	675.1	675.1	675.1	675.1
Hog fuel	71.4	66.8	66.8	66.8	66.8
Purchased wood residues	0.0	125.4	454.9	557.2	110.1
Lime kiln fuel oil	32.7	37.8	37.8	37.8	35.9
Total	826.7	905.1	1234.6	1336.8	887.9
Energy Outputs (MW)					
Fiber	315.8	315.8	315.8	315.8	315.8
Liquid fuel	0.0	111.6	111.6	342.7	59.6
Power	64.3	87.7	228.9	77.3	91.9
Losses	446.6	390.0	578.3	601.0	420.6
Total	826.7	905.1	1234.6	1336.8	887.9
Overall Energy Efficiency					
Energy products	7.8%	22.0%	27.6%	31.4%	17.1%
All products	46.0%	56.9%	53.2%	55.0%	52.6%

Source: Adapted from Larson, E.D., Consonni, S., Katofsky, R.E., Iisa, K., and Frederick, W.J. Jr. 2006. *A Cost-Benefit Assessment of Gasification-Based Biorefining in the Kraft Pulp and Paper Industry, Volume 1–4*. Final report under US Department of Energy contract DE-FC26-04NT42260, 21 December. Downloadable from www.osti.gov/servlets/purl/912520-8bRNAH/.

TABLE 19.20

Incremental Capital Cost for an Integrated Biorefinery Producing Both Liquid Fuels and Pulp Fiber Relative to a Plant Producing Pulp Fiber Only

Case	Incremental Capital Cost, Million US$ (1995)
FTa	181
FTb	337
FTc	308
MA	249

energy products versus pulp fiber, and the relative distribution of liquid fuel versus power as products. Over the range evaluated, the complex with maximized F-T liquids production had the highest efficiency for conversion to energy products. Perhaps more than anything else, the Larson et al. studies showed the value and importance of a detailed analysis of technical and economic feasibility when considering options for integrated biorefineries.

19.9.1 Integrated Biorefineries under Development

19.9.1.1 New Page Corporation, Wisconsin Rapids, WI

New Page Corporation plans to construct and operate a demonstration-scale thermal gasification and gas-to-liquids plant integrated into their Wisconsin Rapids pulp and

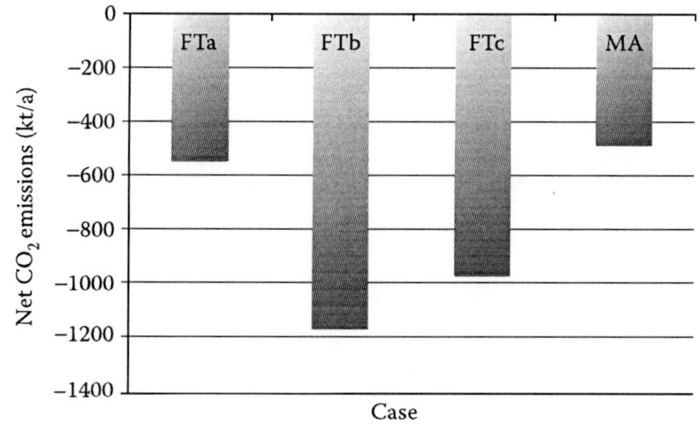

FIGURE 19.9
Net decrease in CO_2 emissions from producing synthetic fuels and fiber in an integrated biorefinery for the conditions shown in Table 19.19. (Adapted from Larson, E.D., Consonni, S., Katofsky, R.E., Iisa, K., Frederick, W.J. Jr. 2006. A Cost-Benefit Assessment of Gasification-Based Biorefining in the Kraft Pulp and Paper Industry, Volume 1–4," Final report under US Dept. of Energy contract DE-FC26-04NT42260, 21 December. Downloadable from www.osti.gov/servlets/purl/912520-8bRNAH/; Larson, E.D., Consonni, S., Katofsky, R.E., Iisa, K., and Frederick, W.J. Jr. 2009. *Tappi J.* 8(1): 27–45.)

paper mill. The plan calls for a 15% scale plant that will produce 5.5 million gal/year of Fischer–Tropsch hydrocarbons from synthesis gas generated by gasification of 500 t/day of dry biomass. The gasification plant will use ThermoChem Recovery International's biomass gasification technology. The F-T synthesis plant will use the EFT, Inc., catalytic F-T synthesis process. Detailed engineering for the plant began in the second half of 2010. The U.S. Department of Energy supports this project financially.

19.9.1.2 Flambeau River Biofuels (Park Falls, Wisconsin)

Flambeau River Papers plans to construct and operate a demonstration-scale thermal gasification and gas-to-liquids plant integrated into their Wisconsin Rapids pulp and paper mill. The production-scale plant will produce 18 million gal/year of Fischer–Tropsch hydrocarbons from synthesis gas generated by gasification of 900 t/day of dry biomass. The gasification plant will use ThermoChem Recovery International's biomass gasification technology. The F-T synthesis plant will use the EFT, Inc., catalytic F-T synthesis process. Detailed engineering for the plant is scheduled to be completed by the end of 2010. The U.S. Department of Energy supports this project financially.

19.9.1.3 Metso and UPM (Helsinki, Finland)

Metso and UPM have developed a process concept for the production of biomass-based fuel oil for heating and power generation. The process consists of a pyrolyzer colocated with a power boiler. Heat from the boiler is used to pyrolyze biomass. Pyrolysis oil, char, and gas are separated, and the char and gas are returned to the boiler as fuel. The pyrolysis liquid generated can be used for fuel elsewhere. Wood harvesting residues and sawdust are pyrolyzed. Testing of the concept began in June 2009.

TABLE 19.21

Other Nonintegrated Thermal Conversion Plants under Development in the United States as of 2010

Company	Technology	Capacity	Feedstock	Status
Gasification-Based Plants				
Clear-Fuels-Rentech, Commerce City, CO	Biomass gasification, F-T hydrocarbons synthesis, upgrading to diesel and jet fuel	18 t/day biomass, 140,000 gal/year	Wood residues and bagasse	Pilot plant operational; target for commercial facility to be operational by 2013
Enerchem, Pontotoc, MS	Biomass gasification, ethanol synthesis through methanol intermediate	10 million gal/year	Wood residues and post-recycling municipal solid waste	DOE award in March, 2010
Haldor Topsoe, Des Plaines, IL	Biomass gasification with Carbona gasifier	19 dry t/day biomass, 345,000 gal/year of biomass-derived gasoline	Wood residues and unmerchantable wood	Startup planned for 2012
Range Fuels, Soperton, GA	Biomass gasification, mixed alcohols synthesis	775 dry t/day; 20 million gal/year methanol and ethanol	Woody biomass, forest residues, and thinnings	Plant under construction
Renewable Energy Institute International, Toledo, OH	Biomass gasification and diesel synthesis	25 dry t/day biomass	Agricultural and forest residues	Awaiting DOE award
Pyrolysis-Based Plants				
Gas Technology Institute (GTI), Des Plaines, IL	Integrated hydropyrolysis and hydroconversion	1 dry t/day biomass	Wood, agricultural residues, and algae	DOE award date: early 2010; pilot facility scheduled for startup in 2013
KiOR, Houston, TX	CFB catalytic pyrolysis with an *in situ* FCC catalyst and catalyst regeneration	200,000 gal/year crude oil substitute for oil refinery feedstock	Wood, wood residues, and urban wood; agricultural residues; energy crops; algae	Operating pilot plant
UOP, Kapolei, Oahu, HI	Biomass pyrolysis with Ensyn pyrolysis and upgrading process	Pilot scale: 1 t/day biomass, 55,000 gal/year synthetic hydrocarbon fuels	Agricultural residues, wood, energy crops, algae	DOE award: early 2010

19.9.1.4 Nonintegrated Thermal Conversion Plants under Development in the United States

Several nonintegrated thermal conversion plants under development in the United States are listed in Table 19.21. They range from small pilot plants to full commercial plants and include both gasification-based and pyrolysis-based processes. All but one have been awarded financial support by the U.S. Department of Energy.

19.10 Future Direction for Integrated Biorefineries

In the longer term, the concept of the integrated biorefinery may need to await the more advanced development of industrial parks based on the model of "industrial symbiosis" on which the Kalundborg, DK park is based (Figure 19.10). Kalundborg is the site of pharmaceutical and enzyme manufacturing facilities, an oil refinery, a power station, a gypsum board plant, concrete and cement manufacturing, and a fertilizer plant. Residues from one manufacturing facility are used by another within a carefully constructed overall utilization plan. Water discharged from each of the plants and from the municipality of Kalundborg is treated, purified, and recycled. Heat discharged from the power

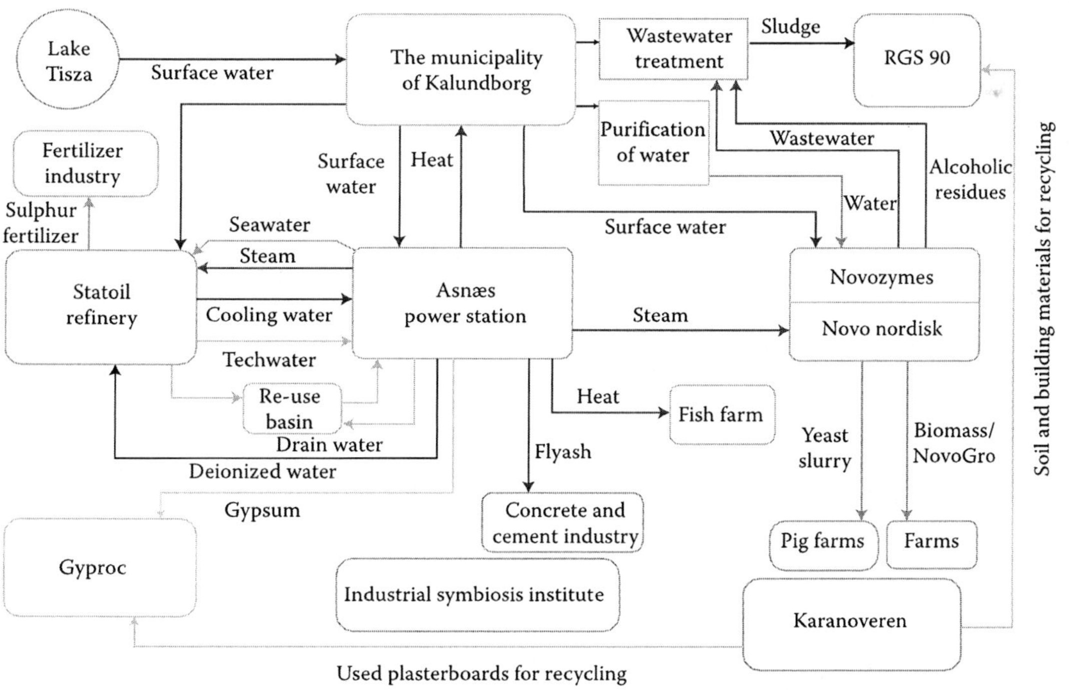

FIGURE 19.10
Diagram of industrial facilities and their interconnection for mass and energy integration at the industrial park in Kalundborg, DK. (Adapted from *Industrial Symbiosis*, 2010, Kalundborg, DK. http://ie.tudelft.nl/index.php/File:Diagram_flow_kalundborg1.jpg (October 6, 2012).)

plant is used for process operation and for municipal heating. A second-generation bioethanol plant is being built adjacent to the power plant and will use its residual heat for process energy.

The integration of municipal and industrial facilities, along with the cooperation between them, generates better results and provides opportunities for companies to increase production while consuming less energy, water, and raw materials per unit of product.

The Kalundborg model and its application may well serve as a blueprint for biorefinery development together with other industrial facilities and municipalities.

Acknowledgment

I would like to acknowledge the intellectual contributions and support of Dr. Kristiina Iisa, without which the *Pyrolysis* section of this chapter would not have been completed.

References

Industrial Symbiosis, 2010, Kalundborg, DK. http://www.symbiosis.dk/industrial-symbiosis.aspx (October 30, 2010).

Larson, E.D., Consonni, S., Katofsky, R.E., Iisa, K., and Frederick, W.J. Jr. 2006. *A Cost-Benefit Assessment of Gasification-Based Biorefining in the Kraft Pulp and Paper Industry, Volume 1–4*. Final report under US Department of Energy contract DE-FC26-04NT42260, 21 December. Downloadable from www.osti.gov/servlets/purl/912520-8bRNAH/.

Larson, E.D., Consonni, S., Katofsky, R.E., Iisa, K., and Frederick, W.J. Jr. 2008. An assessment of gasification-based biorefineries at Kraft pulp and paper mills in the United States. Part 1: Background and assumptions. *Tappi J*. 7(11): 8–14.

Larson, E.D., Consonni, S., Katofsky, R.E., Iisa, K., and Frederick, W.J. Jr. 2009. An assessment of gasification-based biorefineries at kraft pulp and paper mills in the United States. Part 2: Results. *Tappi J*. 8(1): 27–45.

20

Enzyme Applications in the Forest-Based Biorefinery

Alexey Vishtal and Andrzej Kraslawski

CONTENTS

20.1 Introduction

Refining of lignocellulosic feedstock (biorefining) is a real alternative to oil-derived products and fuels. Biorefining can be defined as "a sustainable processing of biomass into a spectrum of marketable products and energy" (IEA 2005). The main distinction between the biorefinery and the oil refinery is the feedstock. Conventional oil refining is based on various hydrocarbons, while the biomass biorefinery uses mainly carbohydrates and lignin.

There are various biorefinery concepts which use different feedstocks. For countries with a developed pulp and paper industry, the forest-based biorefinery is seen as the most attractive option because of relatively easy conversion of existing facilities. There are two main approaches to the forest-based biorefinery: thermochemical and biochemical. The first is based on high-temperature transformation of wood biomass by means of gasification or pyrolysis. The main product of the gasification process is syngas, which can be used for production of different organic substances. In pyrolysis, there are three types of products: the gaseous fraction, pyrolysis oil, and solid residues (resins and coal). The ratio between solid, liquid, and gaseous products varies according to process conditions (Demibras 2009). In general, the products obtained by the thermochemical path are highly diversified and require sophisticated separation processes.

In the second case, the biochemical pathway is associated with use of various microorganisms and enzymes to produce various chemical products and fuels. The biochemical pathway is considered to be more complex than the thermochemical. Nevertheless, its high yields and greater product purity along with higher added value make the biochemical pathway a very interesting option. In comparison to thermochemical processes, biochemical reactions require higher retention times. However, use of microorganisms makes it possible to perform a multistep synthesis in a single processing stage. Use of natural biocatalysts is one of the most promising tools in the biotechnology portfolio because of the possibility of running the reactions at moderate pressure and temperature. Several review papers on the industrial applications of enzymes have been published recently (Kirk and Jeffries 1996, Kashyap et al. 2001, Houde et al. 2003, Kenealy and Jeffries 2003, Sukumaran et al. 2005, Couto and Herrera 2006, Hasan et al. 2006, Erb et al. 2009, Dhiman et al. 2010). However, these papers do not analyze developments related to enzyme applications in the forest-based biorefinery.

The objective of this review is to present both existing and potential applications of enzymatic processes, concentrating on forest-based biorefineries.

20.2 Concept of the Biorefinery

The concept of the biorefinery originated relatively recently when the question of effective biomass transformation into a variety of products, energy, and fuels become critical (Clark and Deswarte 2008, pp. 1–20).

The first generation of biorefineries was targeted at a single product using food-based feedstocks. Classical examples of first-generation biorefineries are biodiesel production from rapeseed or sunflower oil and bioethanol production from sugar cane, corn, and other sugar-containing crops (Kokossis and Yang 2010). The residues from these production sites were mainly used for the generation of energy. First-generation biorefineries were established in the United States and Western Europe (France, Germany) (GBEP 2007). The major issues are ethical (food vs. fuel) and concern the rationality of land use (decrease of soil fertility by biofuel-designed species) (Cherubini 2010).

Second-generation biorefineries are intended to overcome these issues; they use lignocellulosic biomass as their feedstock. Examples of such feedstocks are agricultural crops, wood and wood residues, and algae. The overall organization of a biorefinery process is shown in Figure 20.1.

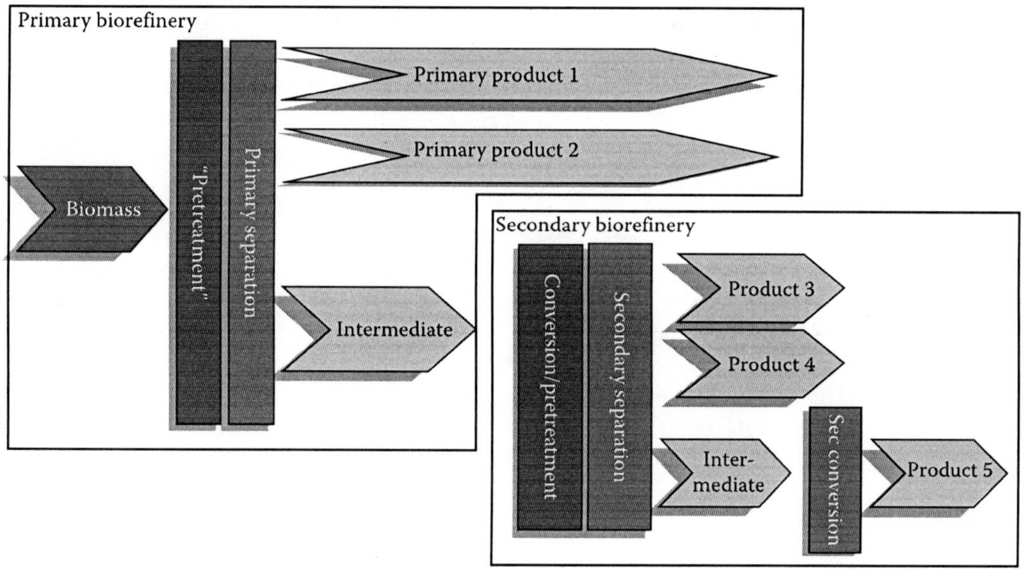

FIGURE 20.1
Overall organization of the biorefinery process. (Adapted from De Jong et al. 2010. Biorefineries for the chemical industry—a Dutch point of view. Available at: http://www.biorefinery.nl/fileadmin/biorefinery/docs/biorefineries_for_the_chemical_industry_a_dutch_point_of_view.pdf.)

The biomass is subjected to pretreatment to separate it into two or more fractions with different chemical compositions. This could include grinding, steaming, chemical treatment, or other processes (Alvira et al. 2009). The major objective of pretreatment is to increase the selectivity of the subsequent separation processes. The feasibility of a biorefinery is determined by the effectiveness of its pretreatment and separation processes. Note that one of the objectives of biorefining is elimination of waste generation; every compound should be converted into useful products.

Different designs exist for biorefining processes, with different pathways for biomass conversion and different final products. A biorefinery can be built as a greenfield mill, or it can be integrated into existing production facilities such as pulp and paper mills, agricultural complexes, or other biomass-processing sites (Clark and Deswarte 2008). The product portfolio can also vary: the biorefinery can be a single-product-oriented plant, a multiple-product plant, or primarily an energy production site. According to Cherubini et al. (2010), biorefineries can be classified on the basis of their platforms, products, feedstocks, and processes. The proposed classification is shown in Table 20.1.

20.3 Forest-Based Biorefinery

The forest-based biorefinery is the most promising biorefinery concept for countries in the northern hemisphere (Canada, Finland, Sweden, Russia) where a well-developed forest sector, and especially a pulp and paper industry, exist. The big advantage here is the fact that pulp and paper mills have the needed infrastructure for biomass transformation into the desired products, fuels, and energy. However, the pulp and paper companies still face

TABLE 20.1

General Classification of Biorefineries

Platforms	Products	Feedstocks	Processes
1. C5 sugars	1. Energy products	1. Dedicated crops	1. Thermochemical
2. C6 sugars	1.1. Biodiesel	1.1. Oil crops	1.1. Combustion
3. Oils	1.2. Bioethanol	1.2. Sugar crops	1.2. Gasification
4. Biogas	1.3. Biomethane	1.3. Starch crops	1.3. Hydrothermal
5. Syngas	1.4. Synthetic biofuels	1.4. Lignocell. crops	1.4. Pyrolysis
6. Hydrogen	1.5. Electricity and heat	1.5. Grasses	1.5. Supercritical
7. Organic juice	2. Material products	1.6. Algae	2. Biochemical
8. Pyrolytic liquid	2.1. Food	2. Residues	2.1. Fermentation
9. Lignin	2.2. Animal feed	2.1. Lignocell. residues	2.2. Anaerobic digestion
10. Electricity and heat	2.3. Fertilizers	2.2. Oil-based residues	2.3. Aerobic conversion
	2.4. Glycerin	2.3. Organic residues	2.4. Enzymatic conversion
	2.5. Biomaterials		3. Chemical processes
	2.6. Bio-based chemicals		3.1. Catalysis
	2.7. Polymers and resins		3.2. Pulping
	2.8. Biohydrogen		3.3. Esterification
			3.4. Hydrogenation
			3.5. Hydrolysis
			3.6. Methanization
			3.7. Steam reforming
			3.8. Other
			4. Mechanical/Physical
			4.1. Extraction
			4.2. Grinding
			4.3. Pressing
			4.4. Other

Source: Adapted from Cherubini, F. et al. 2010. *Biofuels, Bioproducts and Biorefining* 3: 534–546.

a major challenge because the needed technologies require significant research and development efforts. The economic uncertainty of the market for biorefinery products is an additional risk factor.

Noncommercial wood (forest and sawmill residues, sawdust, stumps, etc.) and black liquor is major feedstocks for the integrated biorefinery. At the moment, the forest residues from wood cutting are burned to produce heat or, in the worst case, are simply left in the forest. During a conventional pulping process, 20–40% of wood mass is dissolved in the black liquor and is consequently burned for energy production (Van Heiningen 2006). Forest residues and organic matter dissolved in the black liquor could be a source of new value-added products for the forest biorefinery (Closset et al. 2005).

Efficient use of wood is based on a deep knowledge of its chemical composition and the processes used for separation of its components. The chemical composition of wood is shown in Figure 20.2.

Wood is composed of three main fractions: carbohydrates, lignin, and extractive compounds. Carbohydrates and lignin represent ~95% of the wood mass, while the share of other compounds is insignificant. Extractives have a highly distinct chemical origin, and the separation of these compounds is very complicated. Some extractives are bioactive and could be objects of great interest to the pharmaceutical industry. The components of

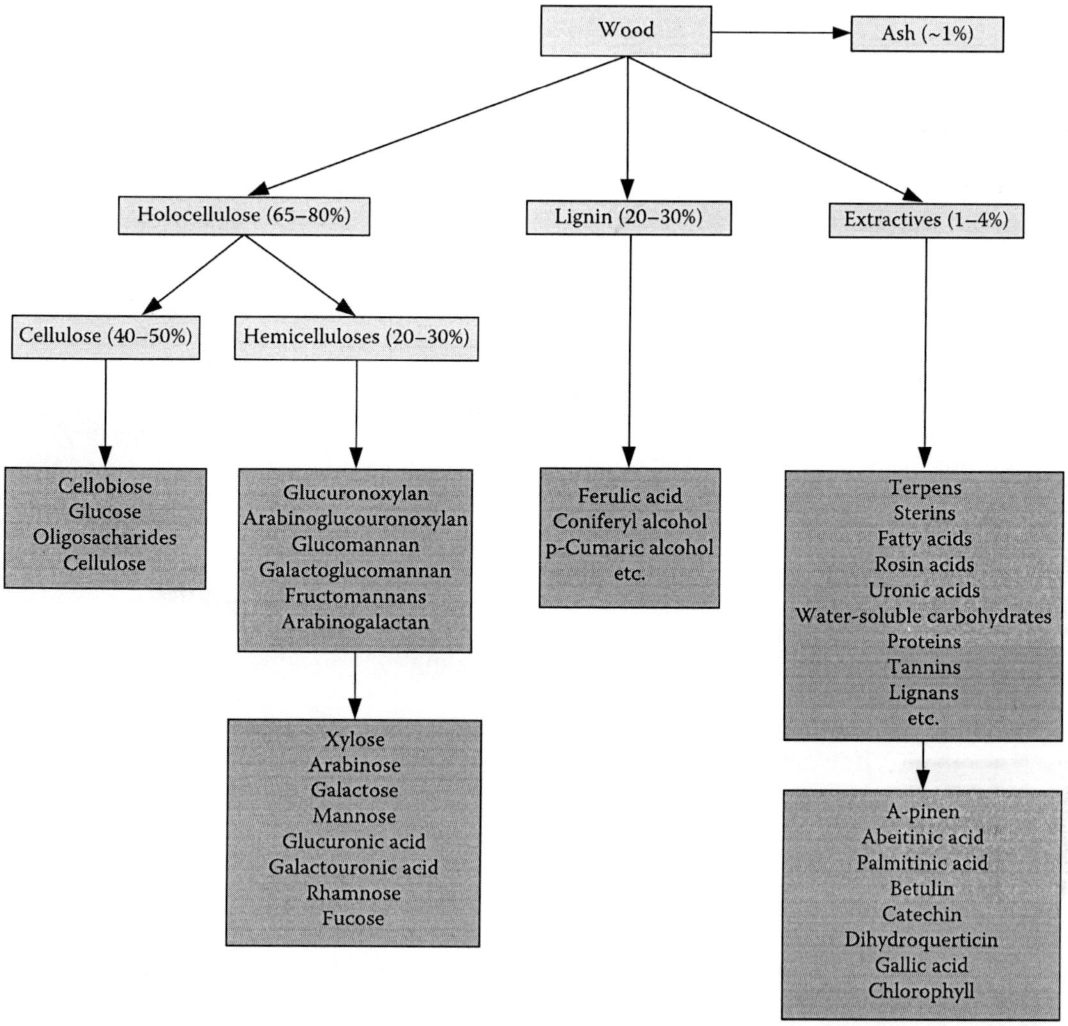

FIGURE 20.2
Chemical composition of wood. (Adapted from Azarov, V., Burov, A., and Obolenskaya, A. 1999. *The Chemistry of Wood and Synthetic Polymers*. pp. 183–191, SPbLTA, St. Petersburg (Book in Russian).)

wood can be considered as the starting platforms for further chemical or biochemical synthesis of various products.

20.3.1 Cellulose

Cellulose is the most abundant polymer in the cell walls of plants; it is composed of the repeating linear chains of beta 1–4 anhydroglucopyranose. Each of the pyranose units is rotated 180° in relation to the preceding. The degree of polymerization (DP) of wood cellulose varies in the 3000–5000 range, while the DP of cotton cellulose is ~10,000. Cellulose in wood exists in both amorphous and crystalline forms. In the crystalline form, cellulose fibrils have a strictly ordered structure which implies low chemical activity of the cellulose. Amorphous regions are less ordered, and therefore amorphous cellulose can be more easily decomposed than crystalline cellulose (Azarov et al. 1999, pp. 225–239).

Cellulose is used in the production of paper, board, and other composite materials. There are several opportunities for enzyme applications in the modification and transformation of cellulose. For example, enzymatic catalysis can be used for hydrolysis, esterification, and production of nanocellulose and oligosaccharides from cellulose.

20.3.2 Hemicelluloses

Hemicelluloses are a group of natural polymers of linear or branched structure composed of different monosaccharides of nonglucosic origin. The main chain of a hemicellulose is composed of 1–4 linked anhydrosugars, and the side chains are linked by 1–6 and 1–3 bonds. The main chain is shorter than in cellulose, and usually the DP of hemicelluloses does not exceed 200–250. The side branches are always shorter than the main branch. Hemicelluloses can contain chains which have been partially esterified with acetic acid. Hemicelluloses also form additional covalent and noncovalent linkages between cell-wall components. Because of their noncrystalline structure and short chain length, hemicelluloses can be easily hydrolyzed by dilute acids. However, part of the hemicelluloses is tightly bonded with cellulose in crystalline regions, and in consequence, this part can be hydrolyzed only together with cellulose. The main noncellulosic polysaccharides in wood are glucuronoxylan in hardwoods and galactoglucomannan in softwoods (Azarov et al. 1999, pp. 269–278).

Potentially, hemicelluloses could be used not only for production of commodity products such as ethanol, butanol, or furfural, but also to produce specialty chemicals for various applications. Backlund and Axegard (2006) proposed the use of modified hemicelluloses as a barrier for fiberboard packaging; for example, it could replace conventional aluminum cans in beer packaging. Enzymes can be used in the separation of hemicelluloses and in the esterification and production of value-added products such as xylitol.

20.3.3 Lignin

Lignin is the second most abundant natural polymer after cellulose. It is built from propanophenolic units which are divided into three groups: syringyl, guaiacyl, and *p*-hydroxyphenol. The structure of native lignin has still not been completely identified because of its complexity. Some researchers claim that it is a net polymer (Derkacheva and Sukhov 2008), while others argue that it is a highly branched three-dimensional polymer (Schoemaker and Piontek 1996). This complex structure has a significant influence on the chemical reactivity of lignin, giving it, for example, superior resistivity to hydrolytic agents. However, it can be subjected to oxidative destruction more readily than carbohydrates; the action of enzymes, laccases, and ligninases is based on the oxidative destruction of lignin. In general, lignolytic enzymes are effective tools for biodelignification and modification of low-molecular phenolic compounds originating from lignin (Riva 2006).

Lignin and other compounds contained in black liquor could be gasified to produce synthetic gas. The synthetic gas could then be used to synthesize various organic compounds, including hydrocarbons. Kraft lignin can be separated and used for the production of various chemicals such as vanillin and dimethylsulfoxide (Kulshreshtha 2007). Lignin can also be used for production of carbon fibers and synthesis of anthraquinone (Johnson et al. 1975, Dimmel and Bozel 1991). Conventional chemical methods are more applicable to lignin transformation because of its low accessibility to enzymes.

20.3.4 Extractives

The term "extractive compounds" is simply a common label for the different classes of organic compounds which can be extracted from wood by alcohol, water, and nonpolar solvents. Proteins, fatty acids, tannins, rosin acids, and terpenes are the main classes of extractive compounds. The diversity of their chemical composition and the relatively low content of extractives in wood limit their industrial use. In the pulp and paper industry, tall oil is the valuable byproduct produced from a mixture of fatty and rosin acids. It is obtained from black liquor before evaporation by separating the light fraction of the fatty acids by simple sedimentation. Tall oil is available in fairly large quantities and is used as a component of paints and lacquers as well as in the production of biodiesel (Keskin et al. 2007).

Extractives are the source of the various wood products which are applicable in the pharmaceutical industry. The bioactive substances are found mainly in the bark, leaves, and resin channels of the wood. For instance, suberin and betulinol are found in birch bark and can be used as antioxidants (Azarov et al. 1999, pp. 207–211). Softwood needles contain unique polyphenols which are used to cure hepatitis (Bespalov and Alexandrov 2006). Sterols, proteins, and vitamins found in the living cells of the wood can be used as a vitaminized additive to animal feed or cosmetics (Clark and Deswarte 2008). These substances can be obtained by alkaline extraction from leaves and needles (Gordon et al. 1988).

To be cost-effective, the forest biorefinery should combine the most comprehensive approach to the treatment of wood compounds, using chemical, thermochemical, and biochemical methods as appropriate. In this context, enzymatic bioconversion seems to be very attractive as a way to obtain value-added products from wood.

20.4 Wood Structure

The morphological structure of wood cells is very complex, and their compounds are tightly bound (Alen 2000, pp. 11–38). The structure of the cell wall and the fibrillar packing of cellulose is shown in Figure 20.3.

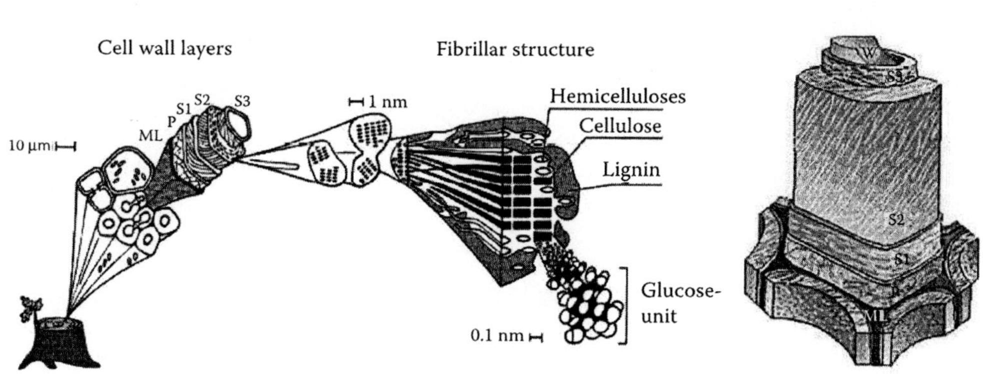

FIGURE 20.3
Morphological structure of the cell-wall layer and fibrillar structure of wood. (Koch, G. In: Sixta, H. (ed.): *Handbook of Pulp*, Vol. 1. pp. 21–68. 2006. Weinheim, Germany: Wiley-VCH Verlag. Copyright Wiley-VCH Verlag GmbH & Co. KGaA. Reproduced with permission.) W: Warty layer, S3(T): inner layer of the secondary cell wall, S2: middle layer of the secondary cell wall, S1: outer layer of the secondary cell wall, P: primary cell wall, ML: middle lamella.

The cell wall, a main component of wood fiber, is composed of several layers which are differentiated by their function and chemical composition, as shown in Figure 20.3. The S1 layer is the first layer that affects fiber strength. It is composed of crystalline microfibrils and has a thickness of ~0.3 μm. The P layer is the primary cell-wall layer and contains mainly lignin with pectinic substances and some extractives. It is quite dense and thin (0.1–0.2 μm), but can easily be destroyed during pulping. The ML layer, the middle lamella, contains mainly lignin. Its main function is to bind the fibers together. The S2 layer represents ~90% of the entire cell wall and is ~1–5 μm thick. The cellulose microfibrils in this layer are oriented spirally at a specific angle to fiber axis (5–10° in earlywood and 20–30° in latewood) (Azarov et al. 1999, p. 231). The chemical composition of the cell layers is shown in Figure 20.4.

The fibrillar structure is the determining factor which is responsible for the poor reactivity of wood with respect to chemical and biochemical transformations. There are three types of fibrils: elementary, micro-, and macrofibrils. The thickness of elementary fibrils, including crystalline and amorphous regions with regular hydrogen bonds, ranges from 2 to 5 nm, and their length varies significantly from 50 to 200 nm. The microfibrils are composed of aggregates of elementary fibrils which are shrouded by hemicelluloses and are 10–30 nm thick. These aggregates formed of lignin microfibrils and hemicelluloses are called macrofibrils. The cellulose chains contained in the fibrils are encapsulated into a matrix structure and surrounded by lignin and hemicelluloses, as shown in Figure 20.3. Lignin plays a role of a binder gluing the fibrils together and is bonded to cellulose fibrils by means of hemicellulosic bridges (Azarov et al. 1999, p. 234; Alen 2000, pp. 11–38).

The sophisticated packing of wood components in cell walls and their tight bonds obstruct any action by enzymes. The activity of enzymes depends on the affinity between

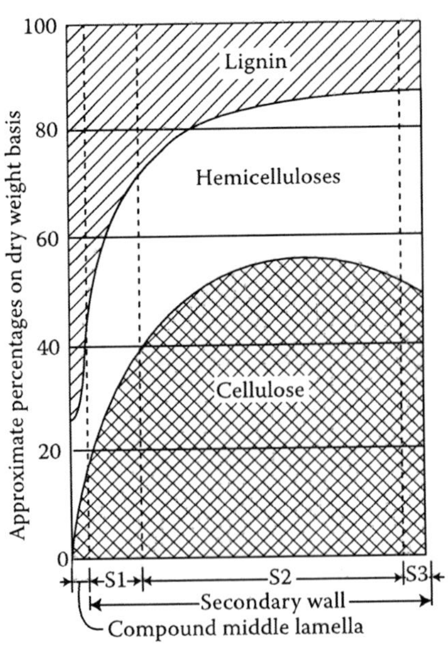

FIGURE 20.4
Chemical composition of cell-wall layers in wood. (Adapted from Salmen, L. et al. 1999. In: Sundholm, J. (ed.), *Papermaking Science and Technology 5. Mechanical Pulping*. Fapet Oy, Jyvaskyla, Finland, pp. 34–65.)

the active center of the enzyme and a substrate, which, on the other hand, depends on the processing conditions. The enzymes cannot attack the target molecule in the cell wall because the substrates of the enzymatic reactions are enclosed in the lignocellulosic matrix. Therefore, wood pretreatment has a crucial importance in enzyme-based processing (Alvira et al. 2009). During pretreatment, the cell walls are partially damaged, and enzymes can penetrate into the disrupted parts of the cell wall and initiate chemical reactions. The penetration of enzymes is also restricted by their size. The selection of an appropriate pretreatment method tailored to the particular enzymatic reaction used could significantly improve product yield of the products and recued the retention time required.

20.5 Enzymes Applications in the Forest-Based Biorefinery

The use of enzymes is an important option for the forest-based biorefinery because it offers the possibility of conducting reactions under milder conditions than those needed for conventional chemical or thermomechanical methods. Another factor supporting the use of enzymes is the significantly higher purity of the end products. A vast spectrum of enzymes can be used to process lignocellulosic materials. Therefore, a systematic classification of enzymes could considerably facilitate the analysis of existing options for the treatment of lignocellulosic materials.

The numbers of enzymes isolated is growing constantly. Modern analytical techniques make it possible to identify the structure of enzymes and their catalytic functions. The naming of enzymes is based on the name of the substrate (to be used by a given enzyme) plus the suffix -ase (e.g., xylanase). Enzymes are classified into six groups according to the nature of the chemical reaction catalyzed. The classification of enzymes into groups is shown in Table 20.2.

The oxireductases catalyze oxidation/reduction reactions. Enzymes from this group can be divided into two types: reductases and oxidases, and the systematic name for this group of enzymes is donor:acceptor oxireductase. The transferases are enzymes which catalyze the transfer of functional groups, for instance, methyl or glycosyl groups. The systematic name for this group is donor:acceptor group transferase. The enzymes from group EC3 catalyze the hydrolytic breakage of C–O, C–N, C–C, S–N, and peptidic bonds. The hydrolytic

TABLE 20.2

Enzyme Classification

Type of Enzyme	Main Function	Example
Oxireductases (EC1)	Catalysis of RedOx reactions	Glucose oxidase
Transferases (EC2)	Catalysis of group transfer	Transaminases
Hydrolases (EC3)	Catalysis of bond breakage in water medium	Cellulases
Lyases (EC4)	Cleavage of groups from substrates	Aldolases
Isomerases (EC5)	Catalyzing intramolecular transformations	Maleylacetoacetate isomerase
Ligases (Synthetases) (EC6)	Catalyzing the joining of two molecules	Glutamine synthetase

Source: Adapted from Leisola et al. 2010. *Industrial Use of Enzymes.* Available at: www.tkk.fi/Units/BioprocessEngineering/Kem-70.415/INDUSTRIAL_USE_OF_ENZYMES.DOC; Lu, L. et al. *Computational Biology and Chemistry* 31: 226–232.

enzymes could be regarded as transferases, where hydrolysis is the process of transferring the hydrolyzed substances into water. The lyases act on C–O, C–N, C–C, C–S, P–O, and C–H bonds through an elimination reaction which leads to the creation of double bonds and rings in the place of the cleaved bond. The names of lyases are formed by addition of the suffix *lyase* to the name of the substrate, for instance, decarboxylase. The isomerases catalyze geometric or structural modifications within a single molecule. A distinction can be made among racemases, epimerases, *cis-trans*-isomerases, isomerases, tautomerases, mutases, and cycloisomerases, depending on the type of isomerism generated. The enzymes belonging to the ligase class catalyze reactions which join two molecules together, usually combined with the hydrolysis of a diphosphate bond in ATP or a similar molecule (Enzyme Nomenclature 2010).

The number of enzymes in each group varies. The size of the groups is shown in Figure 20.5.

As can been seen in Figure 20.5, the enzymes belonging to groups EC2 and EC3 make up ~50% of the entire number of enzymes isolated. This can be explained by the fact that the processes catalyzed by the enzymes in groups EC2 and EC3 are the most common phenomena in living organisms. Strict classification and nomenclature are extremely important for enzyme science because new enzymes are constantly being discovered (Lu et al. 2007).

Hydrolases, oxireductases, isomerases, and transferases are the main groups of enzymes that can be used in the forest-based biorefinery. Wood is composed of natural polymers which need to be decomposed into simpler molecules to be used in the subsequent steps of biorefining. Different enzymes are used to decompose the various types of polymers; for example, hydrolases are finding great use for cleaving glucosidic bonds in carbohydrates. Hydrolases offer the possibility of controlling the hydrolysis process and obtaining products with certain chain lengths. To obtain high added value, the substances obtained from wood must be modified and converted. Oxireductases transferases and isomerases are mainly used for this purpose. Oxireductases catalyze oxidation reactions involving wood compounds. Transferases are used to graft wood polymers onto other compounds to obtain new properties of the composite materials. Isomerases initiate intramolecular transformations of carbohydrates. Ligases and lyases do not readily use wood as a substrate, and therefore their

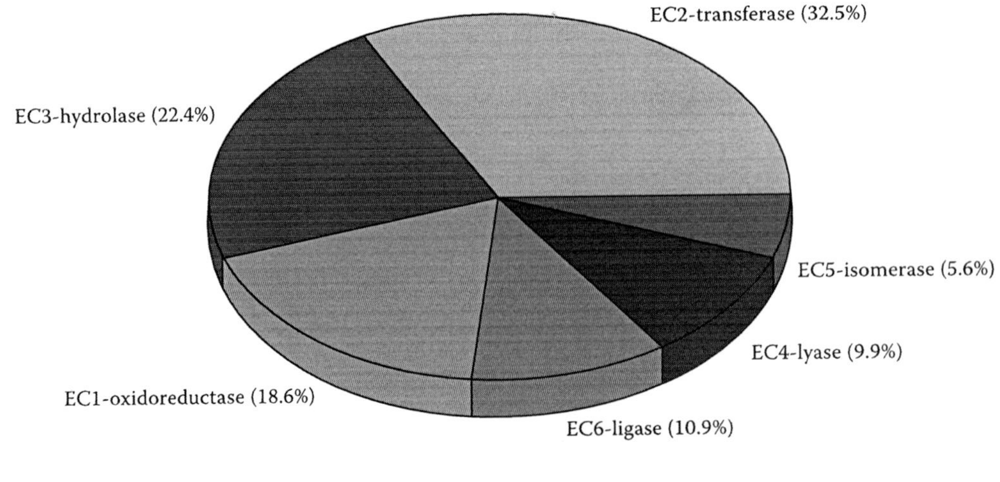

FIGURE 20.5
Shares of the enzyme types. (Adapted from Lu, L. et al. 2007. *Computational Biology and Chemistry* 31: 226–232.)

use in the forest-based biorefinery is rather limited. The enzyme groups which can be used in the treatment of lignocellulosic materials are shown in Table 20.3.

At the moment, the hydrolases are the major group of enzymes used in industry. Other enzymes have been applied only in the laboratory or at the bench-scale level.

It is worth underlining that many of the enzymatic reactions which can be used for the treatment of wood biomass have not yet been systematically studied. One of the obstacles to this is the highly complex structure of wood. Moreover, even well-described reactions still require research effort to optimize process conditions and to identify the most effective enzyme cocktails. A radical change in this situation could be achieved by research into the mechanisms of enzymatic reactions.

Enzymes can be successfully incorporated into emerging processes for separation of wood components. For instance, they can be combined with ionic liquids. Despite the risk of possible denaturization of the enzymes due to the high salinity of these liquids, their activity can be preserved by addition of polyethyleneglycol (Turner et al. 2003).

TABLE 20.3

Enzymes Which Can be Used in the Treatment of Lignocellulosic Materials

Group	Enzymes	Substrate	Utilization	Reference
Hydrolases	Cellulases	Cellulose	Enzymatic hydrolysis, production of nanocellulose, dissolving pulp refinement, effluent treatment	Teter et al. (2005), Henriksson et al. (2007), Rahkamo et al. (1996), Lakshimedevi and Muthukumar (2010)
	Hemicellulases	Hemicelluloses	Enzymatic hydrolysis, Separation of hemicelluloses, production of dissolving pulp	Helmerius et al. (2010), Willfor et al. (2007), Christov et al. (2000)
Oxireductases	Glucose oxidase	Sacharides	Synthesis of aldonic acids from sugars	Pezzotti and Therisod (2006), Gajda et al. (1998)
	Xylose Reductase	Xylose	Synthesis of xylitol from xylose	Makinen (2000), Jeun et al. (2003), Martinez et al. (2007)
	Peroxidases	Lignin	Synthesis of lignin-phenolic resins	Liu et al. (2001)
Transferases	Various esterases	Cellulose, Hemicelluloses	Esterification of cellulose and hemicelluloses with various compounds (acetic acid, xylo-oligosacharides, tannins etc.	Kumar et al. (2003), Kremlinicky et al. (2004), Seteri et al. (1998), Zhou et al. (2006), Chang and Shaw (2009)
	Glycosyltransfe rase	Cellulose, Xylan, Starch	Synthesis of oligosacharides	Srisomsap et al. (1999), Perugino et al. (2004), Cheng and Gross (2005), Erb et al. (2007)
	Ferulic acid esterase	Ferulic acid	Release of ferulic acid from pectinic compounds	Bonnin et al. (2002), Gottschalk et al. (2010)
	Transferase complex	Carbohydrates and levulinic acid	Synthesis of polyhydroxyalkanoates	Keenan et al. (2006), FitzPatrick et al. (2010)
Isomerases	L-arabinose isomerase	Arabinose, Galactose	Synthesis of ribulose and tagatose	Roh et al. (2000), Ryttersgaard et al. (2004)

Another important aspect of enzymatic reactions is the end use of the final product. Some chemicals are used as commodities, while others are needed amounts of several kilograms per year. Cellulose, hemicelluloses, and lignin can be considered as the main platforms for enzymatic conversion; together, these compounds account for ~95% of the total biomass of wood. This means that the major production line of the potential biorefinery will be concentrated on conversion of one or more of these three compounds. Extractives are mainly considered to be used in production of low-tonnage, high-value-added products, because the amount of these substances in wood is fairly low. The classification approach based on substrates enables easy search for and identification of appropriate enzymatic reactions.

20.6 Enzymes Acting on Cellulose and Derived Substances

20.6.1 Enzymatic Hydrolysis

Probably the most important enzyme application in the lignocellulosic biorefinery is enzymatic hydrolysis. By this process, cellulose and hemicellulose can be separated from lignin. Compared to conventional acid-catalyzed hydrolysis, enzymatic hydrolysis does not produce large quantities of byproducts, thereby increasing sugar yield close to theoretical levels (Holkin 1989, p. 197).

Efficiency of enzyme action is often determined by the accessibility of the substrate. In the case of lignocellulosic materials, this is especially important because cellulose is tightly packed into fibrils which prevent enzyme penetration into the cell-wall structure. Moreover, the fiber matrix is surrounded by lignin and hemicelluloses. Pretreatment is of crucial importance in this case (Alvira 2009). A schematic mechanism for enzymatic hydrolysis is shown in Figure 20.6.

As can be seen in Figure 20.6, endoglucanases play a crucial role in the hydrolysis of the crystalline part of cellulose. They attach themselves to the fibril surface and initiate

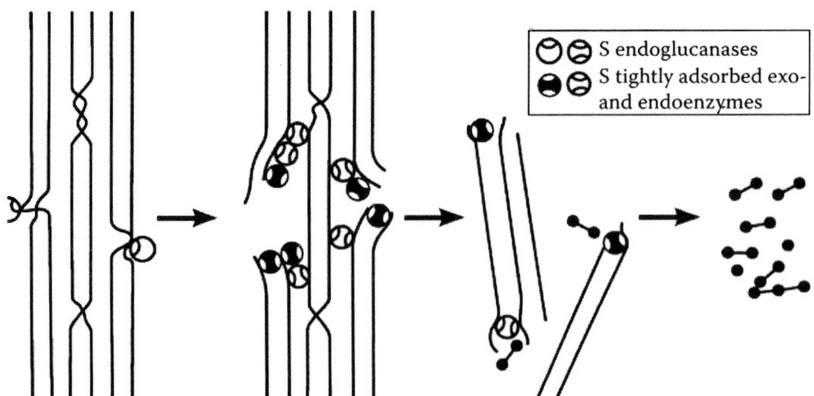

FIGURE 20.6
Graphical representation of the mechanism of enzymatic hydrolysis by cellulases.

cellulose decomposition. Endoglucanases have an additional cellulose binding site, or adsorption center. Enzymes with poor adsorption ability act mainly on the amorphous parts of cellulose. The adsorption ability of the endoglucanases is therefore the determining factor in the hydrolysis of cellulose (Rabinovich and Melnik 2000). Exoglucanases and other enzymes of the cellulose complex then settle in the places where the crystalline structure of the cellulose has been partially destroyed and mechanically draw out the cellulose microfibrils. Hydrolysis of the crystalline part of cellulose always starts from either the reducing or the nonreducing end of the chain; it cannot start from the middle of the molecule (Rabinovich 1984).

Cellulases are composed of a mixture of different enzymes. Each kind of enzyme has its specific role in hydrolysis, for example, endoglucanases act on the amorphous part of the cellulose and facilitate cleavage of glycosidic bonds. Another important group of enzymes is the cellobiohydrolases. They act on the crystalline regions of cellulose by degrading the glycosidic bond either from the reducing or the nonreducing end (Teter et al. 2005). The structure of the cellulose, its origin, and its location in the plant tissues are the decisive factors for the selection of the appropriate enzymes (Rabinovich et al. 2002). The main factors affecting the performance of hemicellulase and cellulase enzymes are the degrees of crystallinity and polymerization and the size of the particles (chips, sawdust, etc.). These factors in general determine the availability of the wood polysaccharides to enzymatic action (Holkin 1989, p. 200).

Cellulases are used to hydrolyze the cellulose, which is contained in the wood, into glucose. To hydrolyze as much of the carbohydrates as possible, cellulases are used together with xylanases and other hemicellulases. This has a positive influence on product yield and reaction speed through synergetic action. However, xylanases can also impede the enzymatic hydrolysis of cellulose if the raw material has already been pretreated to remove lignin (Varnai et al. 2010). Glucose and other released sugars can be used in various fermentation processes, for example, for ethanol or butanol production. Enzymatic hydrolysis is becoming a more and more important process because of the increased demand for biofuels. Compared to acid hydrolysis, enzymatic hydrolysis provides hydrolysates of better purity which are free from harmful byproducts (Taherzadeh and Karimi 2007). Unfortunately, because of their structure, lignocellulosic materials are not easily hydrolyzable, and pretreatment is still needed (Lee et al. 2008). Accessibility of the cellulose can be improved in numerous ways by thermal, mechanical, or chemical methods (Alvira et al. 2009).

20.6.2 Production of Nanocellulose

Cellulolytic enzymes can be used to produce nanocellulose from microcrystalline cellulose (MCC). Nanofibers are used to make durable high-strength polymer nanocomposites. Enzymes from *Trichoderma reesei* decrease the crystallinity of the MCC and provide ~22% yield of nanocellulose. Compared to the acid treatment, partial enzymatic hydrolysis provides a more uniform structure and decreases environmental impact (Satyamurthy et al. 2011).

MCC is usually produced by acid hydrolysis of cellulose; however, enzymes can be used for this purpose as well (Ioelovich and Leykin 2006). For better results, enzymatic treatment can be combined with mechanical refining (Pääkkö et al. 2007). Enzymatic treatment enables significant decrease in energy consumption in the production of microfibrillated cellulose (MFC) (Ankerfors and Lindstrom 2007). Endoglucanase pretreatment facilitates disintegration of pulp fibers into MFC and ensures a favorable nanofiber structure after the disintegration process. Nanocomposites prepared from enzymatically treated fibers have good mechanical properties (Henriksson et al. 2007). Endoglucanase treatment is also favorable for dissolving pulp because it increases the reactivity of the pulp and does not substantially lower its viscosity (Engstrom et al. 2006).

20.6.3 Conversion of Xylose into Bacterial Cellulose

Bacterial cellulose can also be considered as nanocellulose because of its high tensile strength and its ability to form durable films (Ishihara et al. 2002). Conventionally, bacterial cellulose is produced by the biotechnological conversion of glucose; xylose and other pentoses are not converted into cellulose. Use of the xylose isomerase during synthesis of bacterial cellulose makes it possible to convert xylose into bacterial cellulose. The xylose isomerase converts xylose into xylulose, which can be used in the synthesizing bacterial cycle. Use of this enzyme makes it possible to a 33% conversion of xylose, which is almost the same as the conversion rate of glucose to cellulose (Ishihara et al. 2002).

20.6.4 Enzymatic Esterification of Cellulose

Esterification of celluloses is a process which is used to improve the functionalities of cellulose to obtain new properties. Sereti et al. (1998) have used commercial lipase preparation for the production of cellulose acetates; they reported that it was possible to increase the percentage of ester bonds in cellulose to 2%. Cheng and Gross 2005, Chang and Shaw (2009) have identified many enzymatic reactions which can be used for the esterification of cellulose and other carbohydrates. Enzymatic esterification can be applied to monomeric sugars, polymers, and oligosaccharides. Yields vary significantly different over a range of 2–96% (Chang and Shaw 2009, Cheng and Gross 2005). Partially pretreated agricultural residues such as straw can be esterified as well (Biswas et al. 2006).

Xyloglucan endotransglucosylase is used to modify cellulose using xyloglucan oligosaccharides. Such modification leads to irreversible changes in the cellulose surface, providing new opportunities for its usage as a novel biocomposite (Zhou et al. 2006a,b).

A bifunctional enzyme isolated from the cellulose complex of *Trichoderma viride* can be used for the selective degradation of chitosan-containing compounds. This enzyme contains chitosinase and carboxymethylcellulase. Chitosan oligomers can be used in tablet preparation as an agent to improve the absorbability of drugs (Liu and Xia 2006).

Kraft pulp can be grafted onto tannins using laccases. This procedure provides good antimicrobial properties, including resistance to Gram-positive and Gram-negative bacteria such as *Escherichia coli*. Grafting is performed by means of ester linkages between the glucose and the gallic acid contained in the tannins. The modified pulp has good resistance to biodegradation and can be used in pharmaceutical and medical applications (Widsten et al. 2010).

20.6.5 Production of Dissolving Pulp

Dissolving pulp is used for the preparation of various composite materials and for further chemical conversion to plastics. The chemical conversion is performed quickly due to the high reactivity of dissolving pulp. Paice and Jurasek (1984) have shown that xylanases are effective for removal of xylan compounds from dissolving pulps. Unlike conventional methods, xylanase treatment does not lead to significant losses in cellulose yield because of its high selectivity (Christov and Prior 1993, Christov et al. 2000, Ibarra et al. 2010).

Cellulases can also be used for production of dissolving pulps. Rahkamo et al. (1996) have tested endoglucanases and cellobiohydrolases on hardwood dissolving pulp. Endoglucanases were found to lower the viscosity and thus to increase the alkali solubility of pulp. Cellobiohydrolases have a less evident effect than endoglucanases.

Christov et al. (1998) have studied the applicability of a white-rot fungal treatment before sulfite cooking and called this process "biosulfite" pulping. A two-week treatment of wood with fungi provided better selectivity in lignin removal during the cooking and bleaching processes. The yield of dissolving pulp was increased by 1% in average. An interesting observation was that white-rot fungal treatment can interfere with the xylanase biobleaching effect. In these experiments, xylanase treatment did not yield a sufficient brightness gain. It was suggested that fungal treatment before pulping might modify the lignin structure and thus restrict xylanase action.

20.6.6 Treatment of Fiber-Containing Effluents

Mill effluents containing lignocellulosic materials can be used for biohydrogen production (Lakshmidevi and Muthukumar 2010). Biohydrogen is one of the fuels that can be used in fuel cells. These researchers first treated effluents from an Indian pulp mill with cellulases, to hydrolyze the sugars, and then subjected the effluents to fermentation with *Enterobacter aerogenes*. The main properties of the effluents studied are shown in Table 20.4.

Cellulase treatment of the effluents yielded ~22 g/L of fermentable sugars, or 2.3 mol H_2/mol sugar. Note that acetic acid is produced as a byproduct during fermentation (Lakshmidevi and Muthukumar 2010).

20.6.7 Synthesis of Oligosaccharides

Oligosaccharides play important roles in the digestive cycle and other processes in living organisms and are used as additives in food, cosmetics, and pharmaceutical products.

TABLE 20.4

Main Properties of the Effluents Studied

Property	Value (g/L)
COD	85.5
BOD	26.5
TS	88.0
TDS	64.0
Total volatile solids	42.0
Total sugars	23.7
Volatile fatty acids	3.2
Total nitrogen	3.5
Total protein	72.5
pH	8.2

Source: Adapted from Lakshmidevi, R. and Muthukumar, K. 2010. *International Journal of Hydrogen Energy* 35(8): 3389–3400.

Oligosaccharides could find a potential application in the textile and packaging industries and in environmental protection (Mussatto and Mancilha 2007). Oligosaccharides are low-molecular-weight carbohydrate compounds with chain lengths between 2 and 20 and are formed from various sugar residues (Azarov et al. 1999, p. 342). They occupy an intermediate position between simple sugars and polysaccharides. Building blocks for the synthesis of oligosaccharides can be obtained from the polymers in wood, mainly by acid or enzymatic hydrolysis.

Oligosaccharides can be synthesized from sugars by enzyme-catalyzed reactions (Srisomsap et al. 1999, Perugino et al. 2004, Erb et al. 2007, Avci and Donmez 2009, Wongsangwattana et al. 2010). One way to produce oligosaccharides is controlled acid hydrolysis of various polymeric carbohydrates such as xylan, starch, and cellulose (Mussatto and Mancilha 2007, de Menezes et al. 2010). However, this method does not yield oligosaccharides with a well-defined structure because their formation is a result of random bond breakage in a polymer. Bacterial glycosyltransferase can catalyze direct synthesis of oligosaccharides from sugars such as sucrose, cellobiose, and lactose. These enzymes catalyze the reaction of the sugar as it attaches to the anglycon (acceptor) (Palcic 1999). Glycolsyltransferases are classified according to the donor position and the transfer stereochemistry of the acceptor.

20.7 Enzymes Acting on Hemicelluloses and Derived Substances

Several biorefinery concepts are proposing partial removal of the hemicelluloses from wood before or after pulping (Marinova et al. 2009, Mendes et al. 2009, Helmerius et al. 2010, Huang et al. 2010). Removal of the hemicelluloses can be performed by various methods: acid hydrolysis, auto-hydrolysis, alkali extraction, or enzymatic hydrolysis.

Xylanases and other hemicellulases are commonly used for hydrolysis of already extracted hemicelluloses. Hemicelluloses in this case are usually extracted by alkali, which leads to the solubilization of hemicelluloses, but does not break them down into elementary sugars (Helmerius et al. 2010). Water extraction of the hemicelluloses is performed as well.

20.7.1 Enzymatic Separation and Modification of the Galactoglucomannan

Galactoglucomannan is a major fraction of the softwood hemicelluloses. Recently, it was reported by Willfor et al. (2007) that galactoglucomannan (GGM) could serve as a valuable substrate for production of novel materials and products. Possible applications might be found in the papermaking, textile, and pharmaceutical industries. It is suitable for production of various films in a mixture with other wood-derived polymers (Mikkonen et al. 2008). The authors also mentioned the possible bioactivity of this compound. Enzymes can be used for separation and modification of GGM, enhancing its applicability. One possible source of GGM is the effluents from TMP production from Norway spruce, in which case 50% of dry matter is composed of solubilized GGM (Lundqvist et al. 2002, Willfor et al. 2007). Mannose obtained by enzymatic hydrolysis from galactoglucomannan can be used in the production of fast-dissolving tablets. Mannanases can be also used in detergent production (Van Zyl et al. 2010).

20.7.2 Enzymatic Synthesis of Aldonic Acids

Pezzotti and Therisod (2006) have synthesized several aldonic acids (D-mannonic, D-galactonic, D-xylonic, 2-deoxy-D-arabinohexonic) by an oxidation reaction catalyzed by glucose oxidase. Aldonic acids are used as a binding agent for metal ions such as Fe, Cu, and Ca. They are used as an acidulant in foods, as a protein coagulant, and in the pharmaceutical industry as a calcium carrier (Chotani et al. 2007). Conventionally, aldonic acids are produced by chemical oxidation using a silver or copper catalyst (Gajda et al. 1998, Pezzotti and Therisod 2006). Pezzotti and Therisod (2006) found that the yield of aldonic acid obtained from enzymatic conversion varied from 70% to 92%. However, demand for aldonic acid is relatively low, and therefore its production cannot be considered as a full-scale industrial process.

20.7.3 Functionalization of Hemicelluloses by Esterification

Compounds obtained from wood, such as xylose, can be further modified by lipases. For instance, Kumar et al. (2003) have shown that commercially available lipase can catalyze direct polycondensation of sorbitol or glycerol with diacids. The substances obtained are further used to produce various biobased materials. This approach has considerable advantages over conventional methods, namely reduced energy consumption, milder processing conditions, and control of polymer branching during the process. These advantages are the main factors in the competitiveness of this process compared to the oil-based processing methods. Acetyl esterase and acetylxylan esterase from *T. reseei* can be used for esterification of xylan-containing hemicelluloses into mono- and diacetates (Kremlinicky et al. 2004).

20.7.4 Production of Xylitol

Conventionally, xylitol is produced by reduction of xylose obtained from birch chips or sugarcane bagasse, with further multistep chromatographic separation of xylitol from the other polyols (Makinen 2000). However, it is possible to convert xylose into xylitol by means of enzymes. Xylose reductase can convert xylose into xylitol in the presence of cofactor NADH and NADPH (Jeun et al. 2003). Because the xylitol is a product of the reduction of five-atomic sugar xylose, raw materials rich in pentosans are preferred. Martinez et al. (2007) used hemicellulosic hydrolyzate from sugar cane to produce xylitol using xylose reductase from *Candida guilliermondii*. The yield of xylitol was ~75% based on xylose, and the purity of the xylitol was 94%. Hydrolysis of the hemicelluloses from hardwood tree species is a very interesting process for the forest-based biorefinery because hydrolysates can be used for further xylitol production (Saukkonen 2009). Hemicelluloses can be isolated by various methods (Girio et al. 2010). Enzymatic separation of xylose provides superior purity and decreased separation costs compared to other methods. Unfortunately, most of the xylose reductases catalyze, at the same time, a reaction of arabinose to arabinitol. Arabinose is the epimer of xylose, and therefore its separation from xylose is complex. Nair and Zhao (2010) have isolated a new xylose reductase with a decreased affinity to produce arabinitol from arabinose. This enzyme could make it possible to increase considerably the purity of xylitol because is almost free of arabinitol.

20.7.5 Production of Tagatose

Galactose-containing carbohydrates can be used to produce the alternative sweetener, tagatose. D-Tagatose is a sugar substitute suitable for diabetics. Its structure and organoleptic properties are similar to those of sucrose (Levin 2002). A conventional way to produce D-tagatose is by hydrolysis of lactose with subsequent isomerization under alkaline conditions (Ryu et al. 2003). Arabinogalactan is a very attractive wood compound, contained in many plant species, for the production of D-tagatose. For instance, Siberian larch (*Larix sibirica*) can contain up to 20% arabinogalactan as a mass fraction of wood (Azarov et al. 1999, p. 313). Arabinogalactan is easily removed from wood by water extraction (Azarov et al. 1999, p. 313). Consequently, acid or enzymatic hydrolysis can be used to break down polymer chains into arabinose and galactose, or in the case of enzymatic hydrolysis, arabinogalactan endo-1,4-beta-galactosidase (Ryttersgaard et al. 2004).

L-Arabinose isomerases can catalyze the transformation of L-arabinose into L-ribulose and are also capable of transforming D-galactose into D-tagatose (Roh et al. 2000).

20.8 Enzymes Acting on Lignin and Other Wood-Derived Substances

20.8.1 Enzymatic Transformation

The complex of enzymes synthesized from microbial cells is able to polymerize various carbohydrates and levulinic acid into polyhydroxyalkanoates (PHA) (Keenan et al. 2006). PHA compounds have various structures and exhibit a huge variety of properties, which makes them useful for production of various plastics (Fitzpatrick et al. 2010). It is important to mention that PHAs are totally biodegradable and biocompatible (Reddy et al. 2003). Tall oil and lignin can also be used as substrates for enzymatic conversion into PHAs (Keenan et al. 2006).

Peroxidases are an environmentally friendly alternative to the synthesis of phenol-formaldehyde resins. It is possible to polymerize monomeric phenols with lignin. Lignin can react with phenol in an HRP-catalyzed copolymerization reaction to produce lignin phenolic resins (Liu et al. 2001).

Tyrasinases can be used to polymerize soluble lignin fragments in effluents to turn them into phenolic poly mers with a predefined structure; polymerization takes place through cross-linking reactions (Guerra et al., 2000).

20.8.2 Ferulic Acid Modification

Ferulic acid can be used as a precursor for vanillin production (Bonnin et al. 2002). This process has been realized at an industrial scale by Givaudan (Switzerland) (Ghisalba et al. 2010). Ferulic acid is found in the pectinic compounds of plants. It can be released from plant raw material by feruloyl esterase. It has been shown that the enzyme efficiently releases ferulic acid (Bonnin et al. 2002). Feruloyl esterase acts in synergy with hydrolytic enzymes and can be produced by various *Trichoderma* and *Aspergillus* species (Gottschalk et al. 2010). It is also worth mentioning that ferulic acid itself has important pharmaceutical and food applications (Kikuzaki et al. 2002). Esterification of ferulic acid with polyols could produce compounds with new useful properties (Tsuchiyama et al. 2006).

20.9 Problems in Using Enzymes for Treatment of Lignocellulosic Raw Materials

The preceding sections have described how enzymes might be applied in many processes in the forest-based biorefinery. However, the overall scope of their usefulness has still not been determined.

Enzymes are essential elements in the chemistry of living organisms, facilitating numerous metabolic processes. It is very common that enzymes catalyze a series of complex reactions. In consequence, it is extremely difficult to identify an exact match between an enzyme and a particular reaction in a very complex chain of chemical transformations. At this moment, lack of understanding of the mechanisms of the complex enzymatic reactions occurring in multiphase systems is one of the major obstacles to the propagation of enzyme use in biorefineries.

Usually, implementation of enzyme applications requires some modifications to existing processing facilities: bigger storage volumes, changes in processing conditions, and additional pretreatment, for example. All such changes add financial and technical risk to the mill operation.

Enzymatic reactions usually require considerably longer retention times than conventional chemical reactions, which has a negative impact on operating costs. The sensitivity of enzymatic reactions to the presence of impurities and to changes in processing conditions makes their industrial application a challenging task. In multiphase reactions, mixing and proper dosage strategy significantly affect product yield. It seems that a key issue in this problem is the development of new, pH- and thermo-tolerant enzymes.

Moreover, it should be remembered that there are many limiting factors governing the use of enzymes in the forest-based biorefinery when the products are to be used in the pharmaceutical and food industries. For instance, it is necessary to eliminate residual bioactivity and to avoid enzymes originating from pathogenic microorganisms or fungi.

The existing tradeoffs between the efficiency of enzymatic processes, their ecological advantages, and the opportunities for production of new materials on the one hand and the costs and risks of doing so on the other do not always work in favor of the industrial use of enzymes.

20.10 Conclusions

The existing and foreseeable applications of enzymes in the forest-based biorefinery have been reviewed and classified according to the substrates used in the enzymatic reactions. The major conclusions are

- Wood is a difficult material from the viewpoint of enzymatic conversion. Pretreatment of wood is required and is of crucial importance.
- Enzymes can be used for high-tonnage production, for example, enzymatic hydrolysis, as well as for low-tonnage, highly specialized products such as phosphorylated oligosaccharides for the pharmaceutical industry.
- Many enzymatic reactions have not yet been studied as potential industrial processes.

- Serious scientific and technical problems remain to be solved to optimize the existing applications of enzymes in the treatment of lignocellulosic feedstocks.
- The major problems related to the broad application of enzymes in forest-based biorefining are the cost of enzymes and the difficulty of controlling enzymatic processes (process inertia and low tolerance for changes in process parameters).

References

Alen, R. 2000. Structure and chemical composition of wood. In: Stenius, P. (ed.). *Papermaking Science and Technology 3. Forest Products Chemistry*. Fapet Oy, Jyvaskyla, Finland, pp. 11–54.

Alvira, P. et al. 2010. Pretreatment technologies for an efficient bioethanol production process based on enzymatic hydrolysis: A review. *Bioresource Technology* 101: 4851–4861.

Ankerfors, M. and Lindstrom, T. 2007. On the manufacture and uses of nanocellulose. *Proceedings, 9th International Conference on Wood and Biofiber Plastic Composites*, 21–23 May, Madison, WI.

Avci, A. and Donmez, S. 2009. A novel thermophilic anaerobic bacterium producing cyclodextrin glycosyltransferase. *Process Biochemistry* 44: 36–42.

Azarov, V., Burov, A., and Obolenskaya, A. 1999. The Chemistry of Wood and Synthetic Polymers. ISBN-5-230-10569-0.626 p. SPbLTA, St. Petersburg (Book in Russian).

Backlund, B. and Axegard, P. 2006. Skogsravaran: ger mangder av nya produkter. *SkogsVarden* (3).

Bespalov, V.G. and Alexandrov, V.A. 2006. *Clinical Use of Conifer Green Needle Complex: A Review of the Medical Applications*. N.N. Petrov State Scientific Research Institute of Oncology, Federal Agency for Healthcare and Social Development of the Russian Federation.

Biswas, A., Saha, B., Lawton, J., Shogren, R., and Willett, J. 2006. Process for obtaining cellulose acetate from agricultural by-products. *Carbohydrate Polymers*, 64(1): 134–137.

Bonnin, E., et al. 2002. Release of ferulic acid from agroindustrial by-products by the cell wall-degrading enzymes produced by *Aspergillus niger* I-1472. *Enzyme and Microbial Technology* 31: 1000–1005.

Chang, S. and Shaw, J. 2009. Biocatalysis for the production of carbohydrate esters. *New Biotechnology* 26(3/4): 109–116.

Cheng, H. and Gross, R. 2005. Polymer biocatalysis and biomaterials. In: *Polymer Biocatalysis and Biomaterials*. ACS, ISBN13: 9780841239173, pp. 1–12.

Cherubini, F., et al. 2010. Toward a common classification approach for biorefinery systems. *Biofuels, Bioproducts and Biorefining* 3: 534–546.

Cherubini, F. 2010. The biorefinery concept: Using biomass instead of oil for producing energy and chemicals. *Energy Conversion and Management* 51: 1412–1421.

Chotani, G.K., et al. 2007. Industrial biotechnology: Discovery to delivery. In: Kent, J.A. (ed.). *Handbook of Industrial Chemistry and Biotechnology*. Springer, New York, pp. 1311–1374.

Christov, L. and Prior, B. 1993. Xylan removal from dissolving pulps using enzymes of *Aureobasidium pullulans*. *Biotechnology Letters* 15(12): 1269–1274.

Christov, L., Akhtar, M., and Prior, B. 1998. The potential of biosulfite pulping in dissolving pulp production. *Enzyme and Microbial Technology* 23: 70–74.

Christov L., Biely, B., Kalogeris, E., Christakopoulos, P., Prior, B.A., and Bhat, M.K. 2000. Effects of purified endo-2-1,4-xylanases of family 10 and 11 and acetyl xylan esterases on eucalypt sulfite dissolving pulp. *Journal of Biotechnology* 83(3): 231–244.

Clark, J,H. and Deswarte, F.I. 2008. *Introduction to Chemicals from Biomass*. Chichester, UK: Wiley.

Closset, G., Raymond, D., and Thorp, B. 2005. The integrated forest products biorefinery. *Agenda 2020 Program: Coordination Chemistry of Polyhydroxy Acids: Role of the Hydroxyl Groups* (Brochure). Available at: http://www.agenda2020.org.

Couto, R. and Herrera, T. 2006. Industrial and biotechnological applications of laccases: A review. *Biotechnology Advances* 24(5): 500–513.

Johnson, D.J., Tomizuka, I., and Watanabe, O. 1975. The fine structure of pitch-based carbon fibres. *Carbon* 13: 321.

De Jong, E. et al. 2010. Biorefineries for the chemical industry—a Dutch point of view. Available at: http://www.biorefinery.nl/fileadmin/biorefinery/docs/biorefineries_for_the_chemical_industry_a_dutch_point_of_view.pdf.

de Menezes, C., Silva, I., Pavarina, E., de Faria, A., Franciscon, E., and Durrant, L. 2010. Production of xylooligosaccharides from enzymatic hydrolysis of xylan by white-rot fungi *Pleurotus. Maringa* 32(91): 37–42.

Demibras, A. 2009. Biorefineries: Current activities and future developments. *Energy Conversion and Management* 50: 2782–2801.

Derkacheva, S. and Sukhov, D. 2008. Investigation of lignins by FTIR spectroscopy. *Macromolecular Symposia* 265: 61–68.

Dhiman, S., Sharma, J., and Battan, B. 2010. Industrial applications and further prospects of microbial xylanases: A review. *Bioresources* 3(4): 1377–1402.

Dimmel, D. and Bozell, J. 1991. Pulping catalysts from lignin. *Tappi Journal*, 74(5): 239–241.

Engstrom, A., Ek, M., and Henriksson, G. 2006. Improved accessibility and reactivity of dissolving pulp for the viscose process: Pretreatment with monocomponent endoglucanase. *Biomacromolecules* 7: 2027–2031.

Engstrom, A., Ek, M., and Henriksson, G. 2010. *Enzyme Nomenclature.* Available at: http://www.chem.qmul.ac.uk/iubmb/enzyme/index.html.

Erb, T.J., Berg, I.A., Brecht, V., Müller., M, Fuchs, G., and Alber, B.E. 2007. Synthesis of C5-dicarboxylic acids from C2-units involving crotonyl-CoA carboxylase/reductase: The ethylmalonyl-CoA pathway. *Proceedings of the National Academy of Sciences USA* 104: 10631–10636.

Erb, A., Wei, H., Harle, J., and Bechthold, A. 2009. A bacterial glycosyltransferase gene toolbox: Generation and applications. *Phytochemistry* 70: 1812–1821.

FitzPatrick, M., et al. 2010. A biorefinery processing perspective: Treatment of lignocellulosic materials for the production of value-added products. *Bioresource Technology* 101: 8915–8922.

Gajda, T., Gyurcsik, B., Jakusch, T., Burger, K., Henry, B., and Delpuech, J. 1998. Coordination chemistry of polyhydroxy acids: role of hydroxy groups. *Inorganica Chimica Acta, 275–276*: 130–140.

GBEP 2007. *A Review of the Current State of Bioenergy Development in G8+5 Countries.* GBEP (Global Bioenergy Partnership) Secretariat, Food and Agriculture Organization of the United Nations (FAO). Available at: ftp://ftp.fao.org/docrep/fao/010/a1348e/a1348e00.pdf [Referenced 25.10.10].

Ghisalba, O., Meyer, H., and Wohlgemuth, R. 2010. Industrial biotransformation. In: Flickinger, M.C. (ed.), *Encyclopedia of Industrial Biotechnology: Bioprocess, Bioseparation, and Cell Technology.* Wiley, Hoboken, NJ.

Girio, F., Fonseca, C., Carvalheiro, F., Duarte, L., and Bogel-Łukasik, M. 2010. Hemicelluloses for fuel ethanol: A review. *Bioresource Technology* 101: 4775–4800.

Gordon, L.V., Skvortsov, S.O., and Lisov, V.I. 1988.*The Technology and Equipment of the Wood-Chemical Industry.* Wood Industry, Moscow, USSR. [In Russian].

Gottschalk, L., Oliveira, R., and da Silva, E. 2010. Cellulases, xylanases, glucosidase, and ferulic acid esterase produced by *Trichoderma* and *Aspergillus* act synergistically in the hydrolysis of sugarcane bagasse. *Biochemical Engineering Journal* 51: 72–78.

Guerra, A., Ferraz, A., Cotrim, A.R., and da Silva, F.T. 2000. Polymerization of lignin fragments contained in a model effluent bipolyphenoloxidases and horseradish peroxidase/hydrogen peroxide system. *Enzyme Microb. Technol.* 26(5–6): 315–323.

Hasan, F., Ali Shah, A., and Hameed, A. 2006. Industrial applications of microbial lipases. *Enzyme and Microbial Technology* 39(2): 235–251.

Helmerius, J., et al. 2010. Impact of hemicellulose pre-extraction for bioconversion on birch Kraft pulp properties. *Bioresource Technology* 101: 5996–6005.

Henriksson, M., et al. 2007. An environmentally friendly method for enzyme-assisted preparation of microfibrillated cellulose. *European Polymer Journal* 43: 3434–3441.

Holkin Y. 1989. *Tehnologiya Gidroliznih Proizvodstv* (*The technology of wood hydrolysis*). Lesnaya Promishlennost'. ISBN-5-7120-0158-6. 496p (Book in Russian).

Houde, A., Kademi, A., and Leblanc, D. 2003. Lipases and their industrial applications: An overview. *Applied Biochemistry and Biotechnology* 118(3): 155–170.

Huang, H.J., et al. 2010. Process modeling and analysis of pulp mill-based integrated biorefinery with hemicellulose pre-extraction for ethanol production: A comparative study. *Bioresource Technology* 101: 624–631.

Ibarra et al. 2010. Combination of alkaline and enzymatic treatments as a process for upgrading sisal paper-grade pulp to dissolving-grade pulp. *Bioresource Technology* 101(19): 7416–7423.

IEA 2005. Brochure. Available at: http://www.globalbioenergy.org%2Fuploads%2Fmedia%2F0909_IEA_Bioenergy_-_Biorefineries_adding_value_to_the_sustainable_utilisation_of_biomass.pdf&ei=kUnpT4TmOfDO4QThqLiQDg&usg=AFQjCNHRVGIjb-QeQhcFctvJsH92NNEGqQ&sig2=0kMhIfShPGLay0uZCeSdrA

Ioelovich, M. and Leykin, A. 2006. Microcrystalline cellulose: Nano-structure formation. *Cellulose Chemical Technology* 40(5): 313–317.

Ishihara, M., Matsunaga, M., Hayashi, N., and Tišler, V. 2002. Utilization of d-xylose as carbon source for production of bacterial cellulose. *Enzyme and Microbial Technology* 31: 986–991.

Jeun, Y.S, et al. 2003. Expression of *Azotobacter vinelandii* soluble transhydrogenase perturbs xylose reductase-mediated conversion of xylose to xylitol by recombinant *Saccharomyces cerevisiae. Journal of Molecular Catalysis B: Enzymatic* 26: 251–256.

Kashyap, D., Vohra, P., Chopra, S., and Tewari, R. 2001. Applications of pectinases in the commercial sector: A review. *Bioresource Technology* 77(3): 215–227.

Keenan, T., Nakas, J., and Tannenbaum, S. 2006. Polyhydroxyalkanoate copolymers from forest biomass. *The Journal of Industrial Microbiology and Biotechnology.* 33: 616–626.

Kenealy, W. and Jeffries, T. 2003. Enzyme processes for pulp and paper: A review of recent developments—increasing strength with cellulases. *American Chemical Society, ACS Symposium Series* 845: 210–239.

Keskin, A., Gürü, M., and Altuparmak, D. 2007. Biodiesel production from tall oil with synthesized Mn and Ni based additives: Effects of the additives on fuel consumption and emissions. *Fuel*, 86(7–8): 1139–1143.

Kikuzaki, H., Hisamoto, M., Hirose, K., Akiyama, K., and Taniguchi, K. 2002. Antioxidant properties of ferulic acid and its related compounds. *Journal of Agricultural and Food Chemistry* 50: 2161–2168.

Kirk, K. and Jeffries, T. 1996. *Roles for Microbial Enzymes in Pulp and Paper Processing.* Institute of Microbial and Biochemical Technology, USDA, Madison WI. *Book of Abstracts, 211th ACS National Meeting*, New Orleans, LA, March, pp. 24–28.

Koch, G. 2006. Raw material for pulp. In: Sixta, H. (ed.), *Handbook of Pulp*, Vol. 1. Weinheim, Germany: Wiley-VCH Verlag, pp. 21–68.

Kokossis, A. and Yang, A. 2010. On the use of systems technologies and a systematic approach for the synthesis and the design of future biorefineries. *Computers and Chemical Engineering* 34: 1397–1405.

Kremlinicky, L., Mastihuba, V., and Cote, G.L. 2004. *Trichoderma reesei* esterase catalyzes transesterification in water. *Journal of Molecular Catalysis B: Enzymatic* 30: 229–239.

Kulshreshtha, A.K. 2007. Chemicals come out of Mother Nature's womb. *Chemical Business*, June 21(6): 29–38.

Kumar, A., Kulshrestha, A.S., Gao, W., and Gross, R.A. 2003. Versatile route to polyol polyesters by lipase catalysis. *Macromolecules* 36: 8219–8221.

Lakshmidevi, R. and Muthukumar, K. 2010. Enzymatic saccharification and fermentation of paper and pulp industry effluent for biohydrogen production. *International Journal of Hydrogen Energy* 35(8): 3389–3400.

Lee, S.-H., Teramoto, Y., and Endo, T. 2008. Enzymatic saccharification of woody biomass micro/nanofibrillated by continuous extrusion process: I—Effect of additives with cellulose affinity. *Bioresource Technology* 100: 275–279.

Leisola, et al. 2010. *Industrial Use of Enzymes.* Available at: www.tkk.fi/Units/BioprocessEngineering/Kem-70.415/INDUSTRIAL_USE_OF_ENZYMES.DOC

Levin, G.V. 2002. Tagatose, the new GRAS sweetener and health product. *Journal of Medicinal Food* 5: 23–36.
Liu, J. and Xia, W.S. 2006. Purification and characterization of a bifunctional enzyme with chitosanase and cellulase activity from commercial cellulase. *Biochemical Engineering Journal* 30(1): 82–87.
Liu, J., Yang, F., Xian, M., Qiu, L., and Ye, L. 2001. Molecular weight and distribution of the copolymer of lignin/phenol in the copolymerization catalyzed by peroxidase. *Macromolecular Chemistry and Physics* 202: 840.
Liu, J. and Xia, W. 2006. Purification and characterization of a bifunctional enzyme with chitosanase and cellulase activity from commercial cellulose. *Biochemical Engineering Journal* 30: 82–87.
Lu, L., Qian, Z., Cai, Y.-D., and Li, Y. 2007. ECS: An automatic enzyme classifier based on functional domain composition. *Computational Biology and Chemistry* 31: 226–232.
Lundqvist, J., Teleman, A., Junel, L., Zacchi, G., Dahlman, O., Tjerneld, F., and Stålbrand, H. 2002. Isolation and characterization of galactoglucomannan from spruce (*Picea abies*). *Carbohydrate Polymers* 48(1): 29–39.
Makinen, K.K. 2000. The rocky road of xylitol to its clinical application. *Journal of Dental Research* 79: 1352–1355.
Marinova, M., et al. 2009. Addressing the increased energy demand of a Kraft mill biorefinery: The hemicellulose extraction case. *Chemical Engineering Research and Design* 8(7): 1269–1275.
Martinez, E., Silva, J., Giulietti, M., and Solenzal, A. 2007. Downstream process for xylitol produced from fermented hydrolysate. *Enzyme and Microbial Technology* 40: 1193–1198.
Mendes, C.V.T., et al. 2009. Valorisation of hardwood hemicelluloses in the kraft pulping process by using an integrated biorefinery concept. *Food and Bioproducts Processing* 8(7): 197–207
Mikkonen, K.S., et al. 2008. Blended films: Galactoglucomannan. *Bioresources* 3(1): 178–191.
Mussatto, S. and Mancilha, I. 2007. Non-digestible oligosaccharides: A review. *Carbohydrate Polymers* 68: 587–597.
Nair, U. and Zhao, H. 2010. Selective reduction of xylose to xylitol from a mixture of hemicellulosic sugars. *Metabolic Engineering* 12: 462–468.
Pääkkö, M. et al. 2007. Enzymatic hydrolysis combined with mechanical shearing and high-pressure homogenization for nanoscale cellulose fibrils and strong gels. *Biomacromolecules* 8(6): 1934–1941.
Paice, M.G. and Jurasek, L. 1984. Peroxidase-catalyzed color removal from bleach plant effluent. *Biotechnology and Bioengineering* 26: 477–480.
Palcic, M. 1999. Biocatalytic synthesis of oligosaccharides. *Current Opinion in Biotechnology* 10: 616–624.
Perugino, G., Trincone, A., Rossi, M., and Moracci, M. 2004. Oligosaccharide synthesis by glycosynthases. *Trends in Biotechnology* 22: 31–37.
Pezzotti, F. and Therisod, M. 2006. Enzymatic synthesis of aldonic acids. *Carbohydrate Research* 341: 2290–2292.
Rabinovich, M. 1984. *Materials of Soviet–Finland Seminar on Bioconversion of Plant Raw Materials by Microorganisms*. Institute of Biochemistry and Physiology of Microorganisms, Pushchino, pp. 31–48.
Rabinovich, M. and Melnik, M. 2000. Progress in the studying of cellulolitical enzymes and mechanisms of enzymatic hydrolysis. *The Advances in Biological Chemistry* 40: 205–266.
Rabinovich, M.L., Melnick, M.S., and Bolobova, A.V. 2002. The structure and mechanism of action of cellulolytic enzymes. *Biochemistry (Moscow)* 67(8): 850–871.
Rahkamo, L., et al. 1996. Modification of hardwood dissolving pulp with purified *Trichoderma reesei* cellulases. *Cellulose* 3: 153–163.
Reddy, C., Ghai, R, Rashmi, V., and Kalia, V.C. 2003. Polyhydroxyalkanoates: An overview. *Bioresour. Technol.* 87: 137–146.
Riva, S. 2006. Laccases: Blue enzymes for the green chemistry. *Trends in Biotechnology* 24(5): 219–225.
Roh, H.J., Kim, P., Park, Y.C., and Choi, J.H. 2000. Bioconversion of D-galactose into D-tagatose by expression of L-arabinose isomerase. *Biotechnology and Applied Biochemistry* 31: 1–4.

Ryu, S.-A., et al. 2003. Continuous D-tagatose production by immobilized thermostable L-arabinose isomerase in a packed-bed bioreactor. *Biotechnology Progress* 19(6): 1643–1647.

Ryttersgaard, C., Nours, J. L., Leggio, L. L., Jørgensen, C. T., Christensen, L. L. H., Bjørnvad, M., and Larsen, M. 2004. The structure os endo-beta-1,4-galactanse from Bacillus licheniformis in complex with two oligosaccharide products. *Journal of Molecular Biology* 341: 107–117.

Salmen, L., Lucander, M., Harkonen, E., and Sundholm, J. 1999. Fundamentals of mechanical pulping. In: Sundholm, J. (ed.), *Papermaking Science and Technology 5. Mechanical Pulping*. Fapet Oy, 1999, Jyvaskyla, Finland, pp. 34–65.

Satyamurthy, P. et al. January 2011. Preparation and characterization of cellulose nanowhiskers from cotton fibres by controlled microbial hydrolysis. *Carbohydrate Polymers* 83(1): 122–129.

Saukkonen, E., 2009. Hemicellulose extracted Kraft pulp as a raw material of paper, M.Sc. Thesis, Lappeenranta University of Technology, Faculty of Technology, Chemical Technology, Paper and Fiber Technology. 98p.

Schoemaker, H. E. and Piontek, K. 1996. On the interaction of lignin peroxidase with lignin. *Pure and Applied Chemistry* 68: 2089–2096.

Sereti, V., Stamatis, H., Koukios, E., and Kolisis, F.N. 1998. Enzymatic acylation of cellulose acetate in organic media. *Journal of Biotechnology* 66: 219–223.

Srisomsap, C., Subhasitanont, P., Techasakul, S., Surarit, R., and Svasti, J. 1999. Synthesis of homo- and hetero-oligosacharises by Thai rosewood-glucosidase. *Biotechnology Letters* 21: 947–951.

Sukumaran, R., Singhania, R., and Pandey, A. 2005. Microbial cellulases—production, applications, and challenges. *Journal of Scientific and Industrial Research* 64: 832–844.

Taherzadeh, M.J., and Karimi, K. 2007. Process for ethanol from lignocellulosic materials I: Acid-based hydrolysis processes. *Bioresources* 2(3): 472–499.

Teter, S.A., Xu, F., Nedwin, G.E., and Cherry, J.R. 2005. Chapter 16. Enzymes for biorefineries. In: Kamm, B., Gruber, V.R., and Kamm, M. (eds.), *Biorefineries: Industrial Processes and Products*, vol. 1, Wiley: Weinheim, Germany, p. 357.

Tsuchiyama, M., Sakamoto, T., Fujita, T., Murata, S., and Kawasaki, H. 2006. Esterification of ferulic acid with polyols using a ferulic acid esterase from *Aspergillus niger. Biochimica et Biophysica Acta* 176: 1071–1079.

Turner, M.B., Spear, S.K., Huddleston, J.G., Holbrey, J.D., and Rogers, R.D. 2003. Ionic liquid salt-induced inactivation and unfolding of cellulase from *Trichoderma reesei. Green Chem.* 5: 443–447.

Van Heiningen, A. 2006. Converting a kraft pulp mill into an integrated forest biorefinery. *Pulp & Paper Canada* 107(6): 38–43.

van Zyl, W., Rosea, S., Trollopeb, K., and Gorgens, J. 2010. Fungal mannanases: Mannan hydrolysis, heterologous production, and biotechnological applications. *Process Biochemistry* 45: 1203–1213.

Varnai, A., Siika-Aho, M., and Viikari, L. 2010. Restriction of the enzymatic hydrolysis of steam-pretreated spruce by lignin and hemicelluloses. *Enzyme and Microbial Technology* 46: 185–193.

Widsten, P., et al. 2010. Enzymatic surface functionalisation of lignocellulosic materials with tannins for enhancing antibacterial properties. *Process Biochemistry* 45: 1072–1081.

Willfor, S., et al. 2007. Spruce-derived mannans—A potential raw material for hydrocolloids and novel advanced natural materials. *Carbohydrate Polymers* 72: 197–210.

Wongsangwattana, W., Kaulpiboon, J., Ito, K., and Pongsawasdi, P. 2010. Synthesis of cellobiose-containing oligosaccharides by intermolecular transglucosylation of cyclodextrin glycosyltransferase from *Paenibacillus* sp. A11. *Process Biochemistry* 45: 947–953

Zhou, Q., Baumann, M., Brumer, H., and Teeri, T. 2006a. The Influence of surface composition on the adsorption of xyloglucan to chemical and mechanical pulps. *Carbohydrate Polymers* 62: 449–458.

Zhou, Q., Baumann, M., Piispanen, P.S., Teeri, T., and Brumer, H. 2006b. Xyloglucan and xyloglutransglycosylases (XET): Tools for *ex vivo* cellulose surface modification. *Biocatalysis Biotransformation* 24: 107–120.

21

Biochemical Conversion

Shijie Liu, Yang Wang, John P. Buyondo, Yuanzhen Wang, and Michael Garver

CONTENTS

21.1 Introduction

Increased environmental awareness has promoted the development of alternate forms of chemicals, materials, and energy to traditional fossil sources, which produce approximately 73% of greenhouse gas emissions, particularly CO_2 (Lokhorst and Wildenborg, 2005). For example, ethanol has attracted attention worldwide for use as an alternative automotive fuel. Unlike fossil fuels, ethanol is a renewable energy source which can be produced by fermentation of sugars. It is widely used as a partial gasoline replacement in Brazil and the United States. During combustion, ethanol acts as an oxidizing agent, and ethanol/gasoline blends therefore produce fewer harmful emissions, such as carbon monoxide (CO) and nitrogen oxides (NO_x), compared with gasoline (Takahashi et al., 2000; Mielenz, 2001).

Brazil and the United States are the world's two largest producers of bioethanol. Ethanol is currently blended with gasoline. In the United States, a 15% (v/v) blend is used, and in Brazil pure ethanol is presented as an option to gasoline. Bioethanol today is derived from sugarcane (Brazil) and corn (United States) feedstock. These and other traditional feedstocks, such as molasses, have been widely used for bioethanol production, but encounter social and economic barriers because these materials are also used for animal feeds and human consumption (Kumar et al., 2009).

Woody biomass (or in short, biomass) is the most abundant renewable resource in the world, with an annual production estimated at 170 billion tons (Klass, 1998), and is an alternative feedstock for bioethanol production. This biomass includes forest residues such as wood (aspen, poplar, pine, spruce, etc.); agricultural residues such as sugarcane bagasse, corncobs, corn stover, wheat straw, and rice straw; industrial residues such as pulp and paper processing waste and municipal solid wastes; and energy crops such as switchgrass. Woody biomass is a potential feedstock for fuel ethanol and is abundant enough to produce approximately 491 billion liters of bioethanol per year, which is approximately 16 times higher than the current annual production rate (Balata et al., 2008).

Woody biomass is composed of three major components: cellulose (20–50%), hemicellulose (20–35%), and lignin (5–35%), with other components present in small quantities. These components are tightly bound via hydrogen and covalent bonds which are major factors in the recalcitrance of biomass to enzymatic degradation (Himmel et al., 2007).

Cellulose is a structural compound and a polymer of glucose linked by β-1,4-glycosidic bonds. This bonding pattern enables extensive, ordered hydrogen bonding between cellulose molecules, which enables the formation of crystalline cellulose, which is compact and highly resistant to chemical and biological attack (Mielenz, 2001). For hardwood and herbaceous plants, hemicellulose is composed of a main-chain backbone made of xylan (β-1,4-linkages) with various branches of mannose, arabinose, galactose, and glucuronic acid in many plants. The degree of branching and the nature of the minor sugars present in the hemicellulose vary depending on the origin of the biomass. In comparison to cellulose, which is crystalline, strong, and resistant to hydrolysis, hemicellulose has a random and amorphous structure with little strength. It is easily hydrolyzed by dilute acid or base during regular pretreatment (Amidon et al., 2008). Lignin is a large biopolymer consisting of phenolic compounds, with a molecular mass exceeding 10,000 Da. It is hydrophobic and covalently links to hemicellulose, cross-linking different plant polysaccharides to confer mechanical strength to the cell wall (Charbannes et al., 2001).

The strong, tight interconnections of heterogeneous substances in woody biomass make biomass conversion challenging (Mielenz, 2001). Figure 21.1 shows four typical biochemical routes of biomass conversion to materials, chemicals, and fuel. These four pathways are defined by the number of steps involved and how the woody biomass is deconstructed. The closest approach to thermochemical conversion is route (4). Unlike thermochemical processes, biochemical conversion proceeds at lower temperatures with lower reaction rates and offers higher selectivity for products. However, the thermodynamic barrier remains the same. Using milder conversion conditions, a slightly better efficiency can be expected. The energetics are no worse than for thermochemical processes because the thermodynamic barrier remains as the true controlling parameter. One can infer that processing lignocellulosics to bioproducts such as chemicals and biofuel consists of four

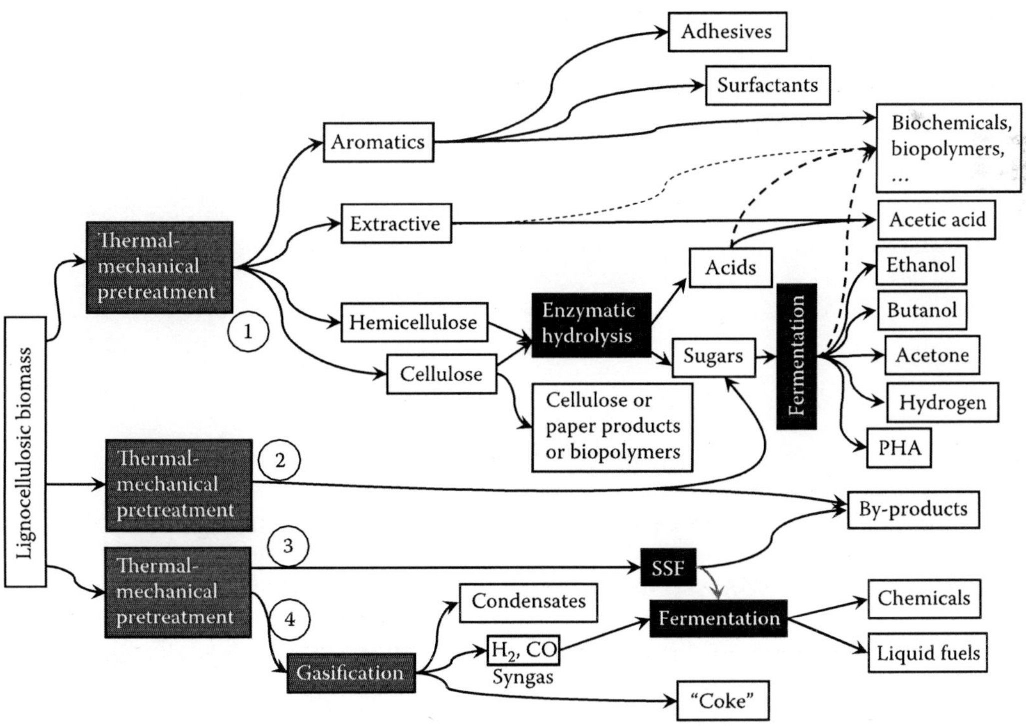

FIGURE 21.1
Schemes of biochemical conversion to materials, chemicals, and fuels: (1) sequential incremental deconstruction; (2) two-step saccharification and fermentation; (3) simultaneous saccharification and fermentation; (4) gasification and fermentation.

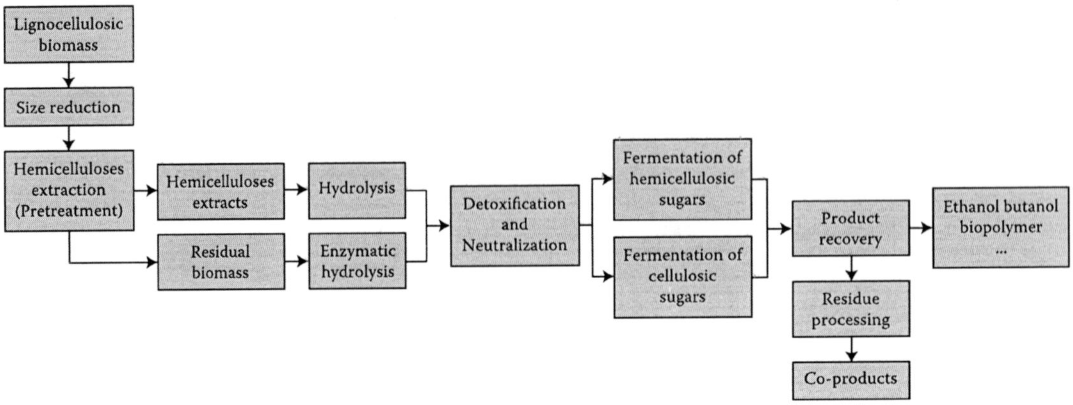

FIGURE 21.2
Schematic flowsheet for biomass conversion to ethanol. (U.S. Department of Energy, 2006. Breaking the biological barriers to cellulosic ethanol: A joint research agenda. p. 14. With permission.)

major unit operations: pretreatment, hydrolysis, fermentation, and product separation and distillation. The strength of biochemical conversion is that incremental deconstruction can be implemented as shown in Figure 21.1, route (1). A popular approach used in biochemical conversion is schematically shown in Figure 21.2. The overall scheme for many proposed bioethanol facilities reflects a strategy similar to that shown in Figure 21.2. One of the major drawbacks is detoxification, because various waste by-products may be generated mainly depending on the pretreatment step(s) chosen. One of the advantages of bioconversion with lignocellulosics is the opportunity to create a biorefinery, producing value-added coproducts plus biofuel. For instance, sugars (or syngas) may be subjected to bacterial fermentation under aerobic or anaerobic conditions. This produces a variety of products, including ethanol, butanol, lactic acid, and so on. These products in turn can be processed into jet fuel, chemicals, or plastics. The noncarbohydrate components or lignin also have potential for use in value-added applications (Mabee et al., 2006).

21.2 Pretreatment Technologies

The first steps in biochemical conversion of lignocellulosics to biofuels are size reduction and pretreatment (Graf and Koehler, 2000). The development of an effective pretreatment method for lignocellulosic biomass is a very important aspect of the saccharification process. The goal of any pretreatment technology is to alter or remove structural and compositional impediments to hydrolysis in order to improve the rate of hydrolysis and to increase the yields of fermentable sugars from cellulose or hemicelluloses (Mosier et al., 2005).

There are several options for a pretreatment process, but regardless of which one is chosen, the following overall downstream features must be assessed to establish biofuel production:

1. Efficient depolymerization of cellulose and hemicellulose to soluble sugars
2. Efficient fermentation of a mixed-sugar hydrolysate containing six-carbon (hexoses) and five-carbon (pentoses) sugars, as well as dealing with compounds that inhibit fermentation

3. Advanced process integration to minimize energy demand
4. Lignin content: lower lignin content in the feedstock decreases the cost of biofuel (Hahn-Hägerdal et al., 2006)

The pretreatment stage promotes the physical disruption of the lignocellulosic matrix to facilitate enzyme-catalyzed or chemical hydrolyses. The choice of pretreatment can have significant implications for the configuration and efficiency of the rest of the process and ultimately its economics (Mosier et al., 2005). Several techno-economic analyses have been performed recently that assess the cost and performance of various pretreatment technologies. Studies have shown that pretreatment is the most significant determinant of success of a biofuel technology because it defines the extent of conversion and the cost of converting the carbohydrates of cellulose and hemicellulose into biofuel. This presents a huge opportunity and a challenge to lower the cost of pretreatment processes through extensive research and development (Chandel et al., 2007). Cost-effective pretreatment of biomass is therefore a major challenge in biofuel technology research and development (Hamelinck et al., 2005).

Various pretreatment methods have been developed, including biological pretreatment, physical or mechanical pretreatment, chemical pretreatment, and physicochemical pretreatment.

21.2.1 Biological Pretreatment

Biological pretreatment can be performed by applying lignin-degrading microorganisms, such as white, brown, or soft-rot fungi, to lignocellulose materials (Lee, 1997; McMillan et al., 1999). Xu et al. (2010) investigated the effect of high-yield concentrated recombinant MnP (rMnP) produced from the yeast *Pichia pastoris* on the biobleaching of Kraft pulps. It was found that rMnP applied at 30 U/g pulp for 24 h followed by alkali extraction caused significant lignin removal in both hardwood and softwood unbleached Kraft pulps. Compared with the control pulp, the rMnP-treated pulp was more susceptible to subsequent peroxide bleaching. More than 60% of the kappa number was reduced by sequential rMnP treatments combined with alkaline extraction. Barber (2007) reported that by using white-rot fungi (*Ceriporiopsis subvermispora*) to treat sugar maple chips, the extraction of hemicelluloses can be enhanced. This was because the noncarbohydrate mass had been removed and the physical and chemical structures of the chips had been changed during biotreatment.

This method of biological pretreatment is considered to be safe, environmentally friendly, and energy-saving. Biological pretreatment is increasingly being advocated because it requires little energy, it is performed at a low temperature, and it needs no chemicals to remove lignin from the lignocellulosic biomass (Okano et al., 2005). However, biological pretreatment processes are too slow for current industrial use, some material is lost because the microorganisms used also consume, to some extent, hemicelluloses, cellulose, and lignin, and the microorganisms can also be poisoned by lignin derivatives (Bohlmann, 2006; Hamelinck et al., 2003). Nevertheless, this method could be useful as a first step when followed by other pretreatment methods.

21.2.2 Physical or Mechanical Pretreatment

Natural cellulose is a crystalline polymer which is generally associated with hemicellulose and lignin in a matrix and is highly resistant to enzymatic attack (Mielenz, 2001). Therefore, feedstock pretreatment has been recognized as a necessary upstream process. Woody biomass can be comminuted by a combination of chipping, grinding, and milling. Shearing and impact processes yield a fine substrate with a low crystallinity index and high specific surface area, enhancing its susceptibility to further pretreatment actions. Similarly to wood-fiber production, two steps are involved in reducing the size of woody biomass. The first step is a coarse mechanical size reduction, such as wood chipping, that produces wood chips of approximately 10–50 mm in two dimensions and 5–15 mm in the third dimension. The second step is to reduce the size of the wood further from chips to particles, fibers, fiber bundles, or a combination of these (Zhu and Pan, 2010). Ball milling gives the most promising results in terms of hydrolysis rate and sugar yield, and this pretreatment stage is clean and easy to operate. Vibratory ball milling has been found to be more effective in breaking down cellulose crystallinity and improving the digestibility of the biomass than ordinary ball milling (Millet et al., 1976). However, mechanical milling requires long operating times and large energy inputs and is therefore a costly option (Lynd et al., 1996). The energy requirement for the mechanical comminution of agricultural materials depends on the final particle size desired and the biomass characteristics (Cadoche and Lopez, 1989). A comparison of various pretreatment methods is shown in Table 21.1.

Another method that has been suggested for physical pretreatment is irradiation of lignocellulosic biomass by high-energy radiation. High-energy radiation such as gamma rays or an accelerated electron beam could exert a great influence on the physical and chemical properties of biomass, depending on the dosage. Many studies have been carried out on the degradation of lignocellulose using high-energy radiation (Wasikiewicz et al., 2005). These studies show a decrease in the degree of polymerization (Bouchard et al., 2006) and microstructural changes of the irradiated cellulose pulp (Dubey et al., 2004). These changes lead to an increase in carbonyl content and an improvement in the accessibility of the cellulose crystalline regions to reagents (Stepanik et al., 1998). For example, wheat straw could be pretreated by γ-radiation, followed by enzymatic hydrolysis. The irradiation of wheat straw by γ-radiation can cause significant structural breakdown and increase the rate of subsequent enzymatic hydrolysis. As irradiation doses are increased, the weight loss of the irradiated samples increases, and the size distribution after crushing shifts toward fine particles. Glucose yield was found to increase with increasing irradiation dose to reach a maximum value of 13.40% at 500 kGy, after which it decreased with excessive irradiation doses. Crushing of irradiated wheat straw can also accelerate the rate of enzymatic hydrolysis. In addition, a synergistic effect between irradiation and crushing was observed. After irradiation, the postdegradation of wheat straw continued over time. It has been observed that irradiation leads to a large improvement in sugar yield via enzymatic hydrolysis of wheat straw (Yang et al., 2008).

TABLE 21.1

Cost Analysis of Mechanical Pretreatment Methods and of γ-Irradiation Pretreatment

Type of Pretreatment	Yield of Sugar from Wheat Straw (g/kg)	Pretreatment Cost Based on Sugar ($/kg)
Ball milling, 8 h	260	5.82
Roller milling, 0.25 h	160	14.0
Fitz milling, fine	100	0.10
Extrusion with pressure	64	0.16
γ-Irradiation	210	0.48

Source: Fan, L.T., Gharpuray, M.M., and Lee, Y.H. 1981. *Biotechnol. Bioeng. Symp.* 11: 29–45. With permission.

Electron-beam irradiation has also been used as a pretreatment to improve the enzymatic hydrolysis of biomass. Electron-beam irradiation can lead to chain scission of biopolymers such as cellulose, hemicellulose, and lignin. This has been indirectly demonstrated by the increased number of carbonyl groups, as shown using Diffuse Reflectance Infrared Fourier Transform Spectroscopy (DRIFTS) on irradiated samples. Electron-beam irradiation had several benefits compared to irradiation using a radioisotope source. One was that, without power, no radiation is produced, which makes the method safer. Moreover, electron-beam irradiation can produce a much higher power level per dose, which could enable effective application in continuous treatment of biomass (Auslender et al., 2002). However, all these size-reduction methods are currently still far too expensive to be used in a full-scale process. It is also doubtful that irradiation can be efficiently used in combination with other supposedly environmentally friendly technologies because of its high energy requirements (Table 21.2).

21.2.3 Chemical Pretreatment

21.2.3.1 Ozonolysis

Ozone can be used to degrade lignin and hemicellulose in many lignocellulosic materials such as bagasse, green hay, peanut, pine, cotton straw, wheat straw, and poplar sawdust (Ben-Ghedalia and Miron, 1981; Vidal and Molinier, 1988). The degradation is essentially limited to lignin, with hemicellulose only slightly attacked and cellulose hardly affected at all. The rate of enzymatic hydrolysis increased by a factor of five following 60% lignin removal from wheat straw using ozone pretreatment (Ben-Ghedalia and Miron, 1981). Enzymatic hydrolysis yield increased from 0% to 57% as the percentage of lignin decreased from 29% to 8% after ozonolysis pretreatment of poplar sawdust.

Ozonolysis pretreatment has the following advantages: (1) it effectively removes lignin; (2) it does not produce toxic residues for downstream processes; and (3) the reactions are

TABLE 21.2

Energy Requirements of Mechanical Comminution of Agricultural Lignocellulosic Material for Different Size Reductions

		Energy Consumption (kWh/ton)	
Lignocellulosic Material	**Final Size (mm)**	**Knife Mill**	**Hammer Mill**
Hardwood	1.6	130	130
	2.54	80	120
	3.2	50	115
Straw	6.35	25	95
	1.6	7.5	42
	2.54	6.4	29
Corn stover	1.6	NA	14
	3.2	20	9.6
	6.35	15	NA
	9.5	3.2	NA

Source: From *Radiat. Phys. Chem.*, 73, Wasikiewicz, J.M. et al., Degradation of chitosan and sodium alginate by gamma radiation, sonochemical and ultraviolet methods, 287–295, Copyright 2005, with permission from Elsevier.

Note: NA, not available.

carried out at room temperature and pressure (Ben-Ghedalia and Miron, 1981). However, a large amount of ozone is required, making the process expensive (Sun and Cheng, 2002).

21.2.3.2 Alkaline Pretreatment

During alkaline pretreatment, the first reactions to take place are solvation and saponification. The mechanism of alkaline hydrolysis is believed to be saponification of intermolecular ester bonds cross-linking xylan hemicelluloses and other components, for example, lignin and other hemicelluloses. The porosity of the lignocellulosic materials increases with removal of the cross-links. This causes the biomass to enter a swollen state and makes it more accessible to enzymes and bacteria.

At "strong" alkali concentrations, dissolution can take place. This is known as "peeling" of end-groups or the alkaline hydrolysis, degradation, and decomposition of dissolved polysaccharides. Loss of polysaccharides is caused mainly by peeling and hydrolytic reactions (Fengel and Wegener, 1984). This peeling can be an advantage for later conversion because it leads to an increase in internal surface area, a decrease in the degree of polymerization, a decrease in crystallinity, a separation of the structural linkages between lignin and carbohydrates, and a disruption of the lignin structure. As a result, lower-molecular-weight compounds are formed, and the risk of degradation and loss of carbon in the form of carbon dioxide also increases.

Xylan can be selectively removed with aqueous potassium hydroxide. The temperature is kept low during extraction (room temperature or lower) to prevent peeling (Hon and Shiraishi, 2001). Glucomannans and xylans can both be subjects of the peeling reaction. This is not a problem, but the higher the monomeric hemicellulose fraction, the lower is the total recovery of the hemicelluloses. For example, Zhao et al. (2008) found that when using 3% NaOH/12% urea in –15°C alkaline pretreatment, a glucose conversion of over 60% was achieved. The monomeric forms of hemicelluloses are probably easily degradable to other (volatile) compounds. One example is furfural, which leads to losses of digestible substrate for the ethanol process (Bobleter, 1994). An important aspect of alkali pretreatment is that the biomass itself consumes some of the alkali. The residual alkali concentration after alkali consumption by the biomass is the alkali concentration left over for the reaction. Lime can work remarkably better than sodium hydroxide as an alkali source (Gossett et al., 1982).

Alkaline pretreatment can also cause solubilization, redistribution of lignin, and modifications in the crystalline state of the cellulose. These effects can reduce or counteract the positive effects of lignin removal and cellulose swelling. The digestibility of NaOH-treated hardwood increased from 14% to 55% with the decrease of lignin content from 24–33% to 20%. However, no effect of dilute NaOH pretreatment was observed for softwoods with lignin content greater than 26%. Dilute NaOH pretreatment was also effective for the hydrolysis of straws with a relatively low lignin content of 10–18% (Bjerre et al., 1996). Chosdu et al. (1993) used a combination of irradiation and 2% NaOH for pretreatment of corn stalk, cassava bark, and peanut husk. The glucose yield of corn stalk was 20% in untreated samples, compared to 43% after treatment with electron-beam irradiation at a dose of 500 kGy and 2% NaOH. The glucose yields of cassava bark and peanut husk were only 3.5% and 2.5%, respectively. Ammonia has also been used for pretreatment to remove lignin. Iyer et al. (1996) described an ammonia recycle percolation process (temperature, 170°C; ammonia concentration, 2.5–20%; reaction time, 1 h) for the pretreatment of a corncob/stover mixture and switchgrass. The efficiency of delignification was 60–80% for corncobs and 65–85% for switchgrass.

Another important aspect of alkaline pretreatment is the change in the cellulose structure to a form that is denser and thermodynamically more stable than native cellulose. Generally, alkaline pretreatment is more effective on agricultural residues and herbaceous crops than on wood materials (Liu and Wyman, 2003; Hendriks and Zeeman, 2009).

21.2.3.3 ***Acid Pretreatment***

Pretreatment of lignocellulose with acids at ambient temperature is performed to enhance anaerobic digestibility. The objective is to solubilize the hemicellulose and make the cellulose more accessible. Dilute or strong acids have been used to treat lignocellulosic biomass. During acid pretreatment, the main reaction that occurs is the hydrolysis of hemicellulose, especially xylan, because glucomannan is relatively acid-stable. Solubilized hemicelluloses can be subjected to hydrolytic reactions producing monosaccharides, furfural, hydroxymethylfurfural (HMF), and other volatile products in acidic environments. Solubilized lignin will quickly condense and precipitate in acidic environments (Esteghlalian et al., 1997; Shevchenko et al., 1999; Liu and Wyman, 2003).

Concentrated acids such as H_2SO_4 and HCl have been introduced to lignocellulosic biomass pretreatment. Although they are powerful agents for lignocellulose hydrolysis, concentrated acids are toxic, corrosive, and hazardous, and the reactors must be resistant to corrosion. In addition, the concentrated acid must be recovered after hydrolysis to achieve economic feasibility and to avoid serious environmental issues.

Dilute acid hydrolysis has been successfully developed for pretreatment of lignocellulosic biomass. A dilute sulfuric acid pretreatment can achieve high reaction rates and significantly improve lignocellulose hydrolysis (Hinman et al., 1992). Because of sugar decomposition, direct saccharification suffers from low yields at moderate temperatures. High temperature is efficient for cellulose hydrolysis in dilute acid treatment. Using less severe conditions to achieve high xylan-to-xylose conversion yields has been investigated in dilute acid hydrolysis processes. Because xylan, the major component in hemicellulose, accounts for up to one-third of the total carbohydrate in many lignocellulosic materials, achieving high xylan-to-xylose conversion yields is necessary to achieve favorable overall process economics (Chum et al., 1988).

There are two main types of dilute acid pretreatment processes: high-temperature (T greater than 160°C) continuous-flow processes for low solids loadings (5–10% [weight of substrate/weight of reaction mixture]), and low-temperature (T less than 160°C) batch processes for high solids loadings (10–40%). Although dilute acid pretreatment can significantly improve lignocellulose hydrolysis, its cost is usually higher than other physicochemical pretreatment processes such as steam explosion or ammonia fiber explosion. Neutralization of the acid is necessary for any downstream enzymatic hydrolysis or fermentation processes (Sun and Cheng, 2002).

21.2.3.4 ***Organosolv Process***

In the organosolv process, an organic or aqueous organic solvent mixture with an inorganic acid catalyst (HCl or H_2SO_4) is used to eliminate the internal lignin and hemicellulose bonds. Organic solvents such as formic acid, methanol, ethanol, acetone, ethylene glycol, triethylene glycol, and tetrahydrofurfuryl alcohol have been used in this process. Organic acids, including oxalic, acetylsalicylic, and salicylic acid, can also be used as catalysts in the organosolv process (Sarkanen, 1980; Thring et al., 1990; Pan et al., 2006b).

Pan et al. (2006a) investigated total recovery of xylose (monomeric and oligomeric), the most abundant pentose sugar. They found that recovery was equivalent to ~72% of the xylose present in the untreated wood. They used the organosolv process to evaluate the bioconversion of hybrid poplar to ethanol at center-point conditions (180°C, 60 min, 1.25% H_2SO_4, and 60% ethanol). Approximately 74% of the lignin was recovered as a precipitate following ethanol extraction (Pan et al., 2006a).

At high temperatures (greater than 185°C), the addition of the catalyst was unnecessary for satisfactory delignification. Generally, a high yield of xylose can be achieved by adding more acid. Solvents used in the process need to be drained from the reactor, evaporated, condensed, and completely recycled to reduce costs. Removal of solvents from the system is necessary because the solvents may inhibit organism growth, enzymatic hydrolysis, and fermentation. Together with the problem of residues in downstream processes, loss of solvents to the atmosphere is another major hurdle. Therefore, the organosolv process is not ideal for large-scale operations.

21.2.3.5 Oxidative Pretreatment

An oxidative pretreatment consists of the addition of an oxidizing compound, such as hydrogen peroxide or peracetic acid, to the biomass while it is suspended in water. The purpose of this pretreatment is to remove hemicellulose and lignin to increase the accessibility of the cellulose. During oxidative pretreatment, several reactions can take place, including electrophilic substitution, displacement of side chains, cleavage of alkyl aryl ether linkages, or the oxidative cleavage of aromatic nuclei (Hon and Shiraishi, 2001). In many cases, the oxidant used is not selective, and losses of hemicellulose and cellulose can occur. As lignin is oxidized and soluble aromatic compounds are formed, a high risk of formation of inhibitors exists.

Azzam (1989) investigated the use of hydrogen peroxide as a pretreatment method for cane bagasse and reported that susceptibility to enzymatic hydrolysis can be greatly enhanced. Hemicelluloses and approximately 50% of the lignin were solubilized by 2% H_2O_2 at 30°C for 8 h, and then a 95% efficiency of glucose production from cellulose was achieved in a subsequent saccharification by cellulase at 45°C within 24 h.

Peracetic acid is very lignin-selective, and under some conditions, no significant carbohydrate losses occur. The use of peracetic acid at ambient temperatures as a pretreatment method for hybrid poplar and sugarcane bagasse has been investigated by Teixeira et al. (2000). Enzymatic hydrolysis of the cellulose increased from 6.8% (untreated) to a maximum of approximately 98% (pretreated) after a 21% peracetic acid pretreatment.

The disadvantage of oxidative pretreatment is that often large amounts of sugars can be lost because of nonselective oxidation. Moreover, soluble lignin compounds are formed, which can inhibit the subsequent conversion of hemicelluloses and cellulose to ethanol or methane.

21.2.4 Physicochemical Pretreatment

21.2.4.1 Steam Explosion

High-pressure steaming with rapid decompression (explosion) has been the most commonly proposed pretreatment method for fractionating lignocellulosic biomass into its three major components and enhancing the susceptibility of cellulose to enzymatic attack. As early as 1928, Mason was granted a patent for a process in which wood chips or shavings were treated with high-pressure saturated steam and then the pressure was swiftly reduced, which made the materials undergo explosive decompression (Mason, 1928). Since then, steam explosion has become a well-known method to separate lignocellulosic materials or to disrupt different biomasses into their main components: cellulose, lignin, and hemicelluloses (Nguyen, 1989).

Generally, it is agreed that the steam process is basically an autohydrolysis because of the naturally acid conditions and the influence of the water and steam. It has been recognized as a cost-effective method for pretreating lignocellulosic biomass because no addition of an external catalyst is necessary. The chemical changes which take place during steam explosion in the three main constituents of lignocellulosics (hemicelluloses, cellulose, and lignin) are very much dependent on temperature, time at temperature, chip size, and moisture content. Steam explosion is typically initiated at a temperature of 160–260°C (corresponding pressure 0.69–4.83 MPa) for several seconds to a few minutes before the material is exposed to atmospheric pressure.

The process causes hemicellulose degradation and lignin transformation due to the high temperature, thus increasing the potential yield of subsequent cellulose hydrolysis. The hemicelluloses are hydrolyzed to soluble sugars by the organic acids, mainly acetic acid derived from the acetylated hemicelluloses. In other words, the acetyl groups in the hemicelluloses are cleaved to form acetic acid. The pH during steam explosion is kept low (pH 3–4), and most of the hemicelluloses are degraded (Sun et al., 2005). However, the solubilized hemicelluloses undergo a series of secondary reactions to form furfural and hydroxymethyl furfural under more severe conditions of higher temperature or longer incubation time. Cellulose domains can be accessed by cellulolytic enzymes after treatment, while lignin is depolymerized and partially removed from the fibers. The reactive precursor, besides undergoing self-polymerization, can participate in condensation reactions with lignin. The high reaction severity improves sugar solubility and enzyme digestibility, but sugar degradation can also be increased.

Compared to only 15% with untreated poplar chips, the hydrolysis efficiency of residual cellulose treated by steam explosion can reach 90% after enzymatic hydrolysis for 24 h (Grous et al., 1986). Optimal hemicellulose solubilization and hydrolysis can be achieved by either high temperature and short residence time (270°C, 1 min) or lower temperature and longer residence time (190°C, 10 min) (Duff and Murray, 1996). Recent studies indicate that lower temperature and longer residence time are more favorable (Wright, 1998).

Steam explosion pretreatment can be improved by using an acid catalyst such as H_2SO_4 (or SO_2) or CO_2. The acid can achieve a more complete removal of hemicellulose, effectively improves enzymatic hydrolysis of the solid residue, and decreases the production of inhibitory compounds (Morjanoff and Gray, 1987). The three most important pretreatment parameters that affect total sugar yields are the H_2SO_4 (or SO_2) or CO_2 level, the residence time, and the temperature. The optimal conditions for steam-explosion pretreatment of sugarcane bagasse have been found to be 1% H_2SO_4; 220°C; 30 s residence time; and 2:1 water-to-solids ratio (Holtzapple et al., 1989). Sugar production from bagasse was 65.1 g sugar/100 g after steam-explosion pretreatment. The use of an acid catalyst in steam pretreatment results in an action similar to that of dilute acid hydrolysis, but with less liquid involved. It is especially important to use an acid catalyst for softwood because softwood in general is more difficult to degrade.

Compared to mechanical comminution, the advantages of steam-explosion pretreatment include its relatively low energy requirement. Conventional mechanical methods require 70% more energy than steam explosion to achieve the same size reduction

(Dale and Moreira, 1982). However, steam pretreatment includes a risk of production of compounds like furfural, HMF, and soluble phenolic compounds. These compounds inhibit fermentation and some chemical production processes. Destruction of a portion of the xylan and incomplete disruption of the lignin–carbohydrate matrix are also limitations of steam explosion. Because of the formation of degradation products that are inhibitory to microbial growth, enzymatic hydrolysis, and fermentation, treated biomass needs to be washed with water to remove these inhibitory materials along with water-soluble hemicelluloses. This decreases overall saccharification yields due to the removal of soluble sugars such as those generated by hydrolysis of hemicellulose. Typically, 20–25% of the initial dry matter is removed by a water wash (Sun and Cheng, 2002).

21.2.4.2 Ammonia Fiber Explosion

The AFEX (ammonia fiber explosion) process is another type of physicochemical pretreatment, similar to the steam pretreatment process, which may offer both an effective and an economically attractive means of increasing the yields of fermentable sugars from lignocellulosic biomass (Holtzapple et al., 1991, 1992). In the AFEX process, lignocellulosic materials are treated with liquid ammonia (not ammonium hydroxide) at moderate temperatures and elevated pressures for a period of time. The pressure is then rapidly discharged, literally exploding the fibrous biomass. Typically, the dosage of liquid ammonia is 1–2 kg ammonia/kg dry biomass at moderate temperatures (below 100°C) and high pressures (above 3 MPa), with a residence time between 10 and 60 min. AFEX pretreatment can significantly improve the saccharification rates of various herbaceous crops and grasses. It has been tested for pretreatment of many lignocellulosic materials including aspen chips, softwood newspapers, kenaf newspapers, alfalfa, wheat straw, wheat chaff, barley straw, corn stover, rice straw, municipal solid waste, coastal Bermuda grass, switchgrass, and bagasse. During the AFEX process, only an insignificant amount of the solid material is solubilized. Compared to acid pretreatment and acid-catalyzed steam explosion, almost no hemicellulose or lignin is removed. However, the structure of the material is changed, resulting in increased water-holding capacity and higher digestibility. The composition of the materials after AFEX pretreatment was essentially the same as that of the original materials. Over 90% hydrolysis of cellulose and hemicelluloses has been obtained after AFEX pretreatment of Bermuda grass (approximately 5% lignin) and bagasse (15% lignin) (Holtzapple et al., 1992). However, this method has not been proven efficient on wood, with its higher lignin content. The resulting AFEX-treated low-lignin-content biomass is readily hydrolyzed at near-theoretical yields of sugars, and the resulting sugars can be fermented rapidly and at high yield to yield various products.

To reduce the cost and to protect the environment, ammonia must be recycled after pretreatment. In an ammonia recovery process, a superheated ammonia vapor with a temperature up to 200°C is used to vaporize and strip the residual ammonia from the pretreated biomass, and the evaporated ammonia is then withdrawn from the system by a pressure controller for recovery (Holtzapple et al., 1990). Over 99% of the ammonia can be recycled and recovered by this process. According to Dale et al. (1984) and Mes-Hartree et al. (1988), the AFEX process does not produce inhibitors that may affect downstream biological processes, and therefore a water wash is not necessary. Furthermore, AFEX pretreatment does not require a small particle size for efficacy (Larson and King, 1986). Because pure ammonia is used in the AFEX process, the stringent environmental and recovery requirements are similar to those of an organosolv process. The capital and operating cost are therefore high.

21.2.4.3 Supercritical Carbon Dioxide Explosion

Supercritical carbon dioxide explosion has been recently developed as a potentially more cost-effective process than ammonia explosion. This process operates at lower temperatures and produces fewer inhibitory compounds than steam explosion. The term *supercritical fluid* refers to a fluid that at standard temperature and pressure would be in a gaseous state. However, when compressed to higher pressures at temperatures above its critical point, the fluid changes into a liquid-like dense state. The fluid then possesses mass transfer characteristics that are "gas-like" with a solvating power that is "liquid-like" (Kim and Hong, 2001).

These unique features of supercritical fluids are applicable in many industrial processes. For example, in food and pharmaceutical applications, carbon dioxide used as a supercritical fluid has advantages over many other fluids. Carbon dioxide (critical temperature, 31.1°C; critical pressure, 73 atm) is nontoxic, physiologically safe, and inexpensive. Carbon dioxide molecules are comparable in size to those of water and ammonia and should be able to penetrate the small pores that are accessible to water and ammonia molecules. It was hypothesized that CO_2 would form carbonic acid and thus increase the hydrolysis rate. Moreover, low temperature might prevent any appreciable decomposition of monosaccharides by the weak acid. Upon an explosive release of the carbon dioxide pressure, the disruption of the biomass structure should increase the accessible surface area of the substrate to enzymatic hydrolysis (Zheng et al., 1998; Conner and Lorenz, 1986).

Using carbon dioxide as a solvent is more advantageous than using the organosolv process or NH_3, and emissions of carbon dioxide cause less harm to the atmosphere. However, the operating and capital costs are still on the high side.

21.2.4.4 Liquid Hot-Water Extraction (Hydrothermal Pretreatment)

Hydrothermal pretreatment is of interest because water is the only reagent used, making it an environmentally friendly and inexpensive process compared to other pretreatment methods (Mosier et al., 2005). This process uses pressure to maintain water in the liquid state at elevated temperatures (Hendriks and Zeeman, 2009). Water under pressure can penetrate the cell structure of biomass, hydrate cellulose, and remove hemicelluloses. Water also has an unusually high dielectric constant that enables ionic substances to dissociate and dissolve the hemicelluloses and a fraction of the lignin.

When the temperature increases above 150°C, hemicelluloses start to solubilize. The composition of the hemicellulose backbone and the branching groups determines the thermal, acid, and alkali stability of the hemicelluloses. Hydrothermal treatment of sugar maple wood selectively solubilized hemicelluloses. A maximum xylan solubilization of approximately 75% of the initial amount was achieved at 175°C after 2 h, whereas only about 30% of the initial xylan was removed over the same duration at 152°C (Mittal et al., 2009). An exothermal reaction (probably solubilization) of the hemicelluloses starts above 180°C. Because the thermal reactivity of lignocellulosic biomass depends largely on its composition, this temperature of 180°C is probably just an indication of the temperature at which an exothermal reaction of the hemicelluloses starts (Brasch and Free, 1965).

During thermal processes, portions of the hemicelluloses are hydrolyzed and form acids, such as acetic acid, which are released from the acetylated polysaccharides in wood, thus lowering the pH of the extract into the 3–4 range (Gregg and Saddler, 1996). These acids are assumed to catalyze the further hydrolysis of hemicelluloses (Liu, 2003; Zhu et al., 2005; Liu, 2008; Tunc and van Heiningen, 2008).

Depending on the intensity of hot-water extraction, sugar dehydration reactions may also take place, leading to hydroxyfurfural (5-(hydroxymethyl)-2-furaldehyde or HMF) and furfural formation from hexose and pentose sugars, respectively. In addition to causing solubilization of hemicelluloses, thermal pretreatment at temperatures of 160°C and higher causes solubilization of portions of lignin (Ramos, 2003). The compounds produced are almost always phenolic heterocyclic compounds, like vanillin, vanillin alcohol, furfural, or HMF, especially in acidic environments. In many cases, these compounds have an inhibitory or toxic effect on bacteria, yeasts, and methanogens or archaea, which is a disadvantage of using hot-water extraction as a pretreatment method (Brownell et al., 1986).

When the solid concentration was 3% or higher and the temperature exceeded 220°C during a 2-min pretreatment time, the resulting hydrolysate was strongly inhibitory to fermentation. This was due to the formation of furfural and other compounds (probably soluble lignin compounds). Temperatures of 250°C and higher should be avoided during pretreatment because unwanted pyrolysis reactions start to take place at such temperatures (Laser et al., 2002). According to Weimer et al. thermal pretreatment can also cause an increase in the crystallinity index (CrI) of cellulose, although no increase was observed when the CrI was already high (Weimer et al., 1995). Soluble lignin compounds are very reactive and will, if not removed quickly, recondense and precipitate on the biomass (Liu and Wyman, 2003). Severe pretreatment conditions promote the condensation and precipitation of soluble lignin compounds, sometimes even with soluble hemicellulosic compounds, such as furfural and HMF (Mittal et al., 2009).

Compared to other biomass pretreatment methods, liquid hot water has the major advantage that the solubilized hemicelluloses and lignin products are present in lower concentrations because of the higher water input. As a result of these lower concentrations, the risk of degradation products like furfural and the condensation and precipitation of lignin compounds are reduced.

There are three types of liquid hot water reactors, which are differentiated by their configurations as cocurrent, countercurrent, and flow-through (illustrated in Figure 21.3). In cocurrent pretreatments, biomass and water are heated to the desired temperature and held at the pretreatment conditions for a controlled residence time before being cooled. Countercurrent pretreatment is designed to move water and lignocellulosic biomass in opposite directions through the pretreatment reactor. In a flow-through reactor, hot water passes over a stationary bed of lignocellulose, hydrolyzes and dissolves lignocellulose components, and carries them out of the reactor (Hendriks and Zeeman, 2009).

21.2.4.5 Ionic Liquids

Ionic liquids have been proposed to separate woody biomass by dissolving the biomass components in liquid at room to moderately high temperatures. Ionic liquids have tunable properties depending on the selection of the anion and cation. Typically, they exhibit very low vapor pressures, and therefore they can be considered to be green solvents. Properties reported to be very important in wood dissolution are viscosity, melting point, dipolarity, and hydrogen-bond basicity. In addition, the corrosivity and toxicity of ionic liquids are important to keep in mind when considering their practical applications in industrial conditions (Maki-Arvela et al., 2010).

The most important factors affecting ionic liquid use are price, physical properties, toxicity, corrosivity, availability, water tolerance, and biodegradability. These issues have

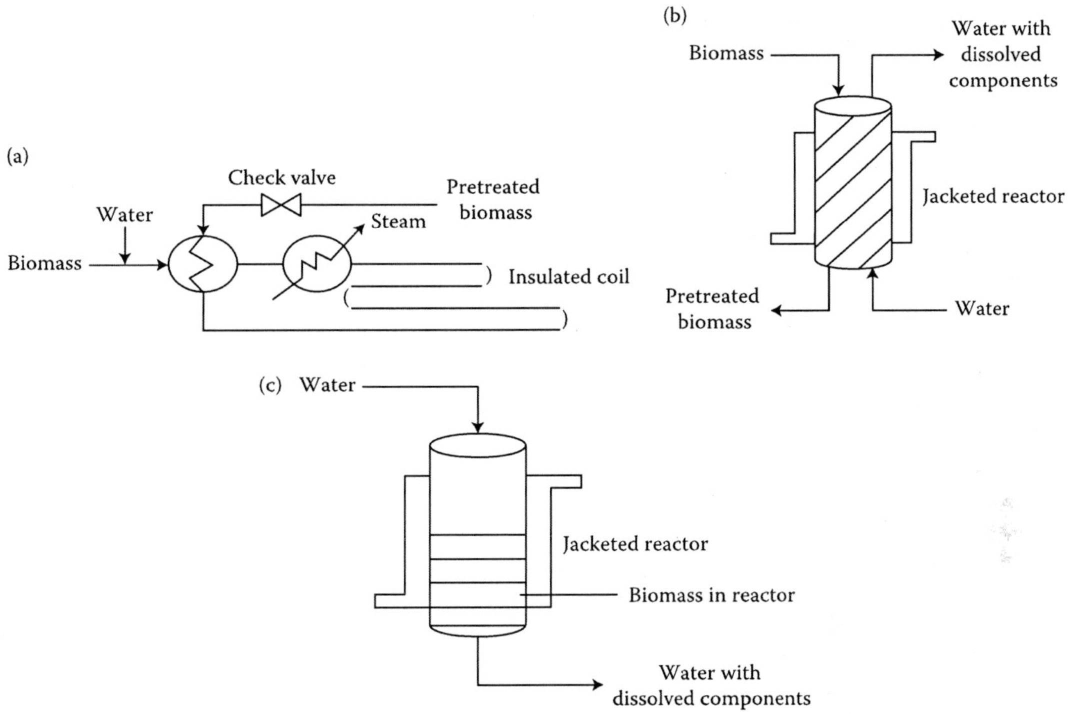

FIGURE 21.3
Schematic illustrations of (a) cocurrent liquid hot water pretreatment, (b) countercurrent reactor, and (c) flow-through reactor. (From *Bioresour. Technol.*, 96, Mosier, N. et al., Features of promising technologies for pre-treatment of lignocellulosic biomass, 673–686, Copyright 2005, with permission from Elsevier.)

until now been rarely investigated, which limits the scope of investigation into potential industrial applications. Wood dissolution is possible if both the ionic liquid and the dissolution conditions are properly selected.

The toxicity of ionic liquids is one important criterion when considering their practical application on a large scale. 1-Butyl-3-methylimidazolium chloride ([BMIM][Cl]), which is a good solvent for cellulose, exhibits relatively moderate toxicity (Swatloski et al., 2004) compared to that of 1-ethyl-3-methylimidazolium chloride ([EMIM][Cl]) (Wu et al., 2004). A better alternative, instead of using chloride-containing ionic liquids, could be the use of 1-ethyl-3-methylimidazolium acetate ([EMIM][OAc]), which is considered biodegradable and is reasonably nontoxic and noncorrosive (Liebert, 2010).

The corrosivity of ionic liquids raises very important economic questions when selecting materials for the apparatus to be used in large-scale applications for wood pretreatment. Good candidates for suitable, noncorrosive ionic liquid solvents are halogen-free ionic liquids such as [EMIM][OAc] and 1,3-dimethylimidazolium-dimethylphosphate ([MMIM][$(MeO)_2PO_2$]), which are good solvents for cellulose (Zavrel et al., 2009).

Viscosity is important from a practical point of view because if the viscosity of an ionic liquid is very high, a high dissolution temperature is also required. This in turn can potentially give rise to other side effects such as stability problems in the ionic liquids or the emergence of unwanted side reactions of wood compounds. One method to reduce the viscosity of an ionic liquid system is to use a cosolvent. It has been stated that the use of dimethyl sulfoxide (DMSO) should be avoided and that instead polyethylene glycol could be used as a viscosity-reducing cosolvent (Willauer et al., 2000).

The water content and particle size of wood chips have substantial effects on dissolution efficiency. The dissolution rate depends on the wood-particle size as follows: ball-milled wood powder > sawdust ≥ TMP fibers >> wood chips (Kilpelainen et al., 2007). For example, Zavrel et al. (2009) carried out high-throughput screening of ionic liquids for dissolution of wood in the particle size range of 1–2 mm. The effect of wood-chip particle size on the dissolution power of [EMIM][OAc] and [BMIM][Cl] was systematically investigated by Sun et al. (2009a). The presence of water significantly hampered the dissolution efficiency of ionic liquids. Therefore, before the dissolution experiments, the tested wood chips were kept in ionic liquids at 50°C (Kilpelainen et al., 2007) or at 90°C (Sun et al., 2009a) to decrease the water content in the wood.

The dissolution power of an ionic liquid for wood also depends on the type of wood, dissolution time, temperature, and wood-to-ionic-liquid ratio. Moreover, for comparable results, the humidity (water content in the sample) and the size of the wood chips should be similar, regardless of the type of wood under investigation. This is a challenge because different species from various parts of the globe have different levels of natural humidity. Moreover, if wood is extensively dried, the nature of its structure is subject to changes. Consequently, the comparison of published data from various sources can be very challenging. It is, however, obvious that those ionic liquids, which are relatively powerful in dissolving both cellulose and lignin, are also quite good at dissolving wood.

One of the best solvents reported for wood chips is 1-allyl-3-methylimidazolium chloride ([AMIM][Cl]) (Zavrel et al., 2009) together with [EMIM][OAc]. In addition, ionic liquids derived from polycyclic amidine bases, such as a 1,8-diazabicyclo[5,4,0] undec-7-enium salt (or DBU), for example, 1,8-diazabicyclo[5,4,0] undec-7-enium chloride [HDBU][Cl], dissolved aspen wood chips completely (D'Andola et al., 2008). [AMIM][Cl] was capable of dissolving both hardwood and softwood wood chips. Only partial dissolution of Norway spruce was achieved in [AMIM][Cl] (Kilpelainen et al., 2007). The efficiency of [AMIM][Cl] for dissolution of wood was stated to originate from the presence of π-electrons both in the alkenyl chain and in the imidazolium ring. Possible π–π interactions can also occur between the aromatic part of lignin and the ionic liquid (Kilpelainen et al., 2007; Hunter and Sanders, 1990). The highest solubilities of maple wood powder were achieved in [AMIM][Cl] and in [BMIM][Cl] (Lee et al., 2009).

[EMIM][OAc] could dissolve three different types of wood chips completely, those from spruce, beech, and chestnut, but only partial dissolution was achieved for silver fir (*Abies alba*) chips (Zavrel et al., 2009). The most probable reason for the poor dissolution of silver fir is that it contains more cellulose (50.3%) and lignin (27.7%) than other common wood species (Kuznetsov et al., 2002). A direct comparison of the dissolution capability of [EMIM][OAc] and [BMIM][Cl] was performed for the same size of Southern yellow pine chips at 110°C; the results indicated a 3.6-fold greater dissolution of the chips in [EMIM][OAc] compared to [BMIM][Cl]. This result can be explained by the fact that the basicity of the acetate anion is higher than that of the chloride ion, and therefore the former is more powerful in breaking intramolecular hydrogen bonds (Sun et al., 2009a). Moreover, the capability of [BMIM][Cl] to dissolve different wood species (oak, eucalyptus, poplar, and pine) was investigated at 100°C in a [BMIM][Cl/DMSO-d6] solvent (Fort et al., 2007). After a dissolution period of 24 h, an approximately 67 wt% of cellulosic material and lignin had been dissolved from poplar and pine, whereas slightly lower dissolution levels were achieved for eucalyptus and oak.

Almost complete dissolution of wood chips was also achieved in DBU-formate (D'Andola et al., 2008), whereas only partial dissolution was achieved using [EMIM][OAc], [BMIM]

[Cl], [MMIM][$(MeO)_2PO_2$] (Zavrel et al., 2009), and DBU-acetate, as well as 8-butyl- and 8-octyl-DBU chlorides (D'Andola et al., 2008).

One challenge in wood dissolution is the recovery of all the biomass released from wood, because during the process proposed by Sun et al. (2009a), the losses of hemicelluloses during the washing steps were very pronounced. This indicates that total utilization of biomass-derived compounds is unlikely to be possible with this procedure.

After dissolution of wood, cellulose can be precipitated and reconstituted. Pretreatment of wood with ionic liquids followed by enzymatic hydrolysis to synthesize nonreducing sugars is a promising technology, because it has been confirmed that in the presence of 40% of the original lignin in wood, 90% of the cellulose can be hydrolyzed by enzymatic hydrolysis (Sun et al., 2009a). To use the wood material completely, another challenge is to recover the hemicelluloses, because these are usually lost during the washing steps. The recovery and reuse of ionic liquids in wood dissolution is possible, but the price, availability, and toxicity issues of ionic liquids are very important when designing new fractionation and pretreatment technologies for wood. These factors are the main hurdle today which is impeding the application of ionic liquids in industrial developments.

21.3 Hydrolysis

Successful pretreatment can alter or remove structural and compositional impediments to full wood utilization. Next, the polysaccharides in lignocellulosic materials can be further broken down to monosaccharides before fermentation through a process called hydrolysis. Various methods have been developed for the hydrolysis of lignocellulosic materials. The most common methods can be classified into two groups: chemical hydrolysis and enzymatic hydrolysis. Another class of hydrolysis methods do not use either chemicals or enzymes. Instead, these methods use radiation, such as γ-ray, microwave, or electron-beam irradiation.

21.3.1 Chemical Hydrolysis

Chemical hydrolysis involves exposure of lignocellulosic materials to a chemical for a period of time at a specific temperature. This results in the formation of monosaccharides from cellulose and hemicelluloses (Taherzadeh and Karimi, 2007). Acid hydrolysis is widely used in chemical hydrolysis processes because it is a fast, low-energy-consumption method that produces a high yield of reducing sugars. Acid hydrolysis normally uses a catalyst such as sulfuric acid, hydrochloric acid, nitric acid, or phosphoric acid. Acid hydrolysis processes can be divided into two general approaches: concentrated acid at low temperature or diluted acid at high temperature (Girio et al., 2010). Table 21.3 shows a comparison between concentrated acid and dilute acid hydrolysis methods.

21.3.1.1 Concentrated Acid Hydrolysis

In 1819, Braconnot first demonstrated that cellulose dissolved in concentrated and heated H_2SO_4 can be transformed into a fermentable sugar (Braconnot, 1819). Figure 21.4 shows a schematic of concentrated H_2SO_4-catalyzed hydrolysis. The concentrated acid plays a dual role in biomass hydrolysis. The concentrated acids first disrupt the network of intra- and

TABLE 21.3

Comparison between Concentrated Acid and Dilute Acid Hydrolysis Methods

Hydrolysis Method	Advantages	Disadvantages
Concentrated acid process	Operated at low temperature	High acid consumption
	High sugar yield	Equipment corrosion
		High energy consumption for acid recovery
		Longer reaction time (e.g., 2–6 h)
Dilute acid process	Low acid consumption	Operated at high temperature
	Short residence time	Low sugar yield
		Equipment corrosion
		Formation of undesirable by-products

Source: Taherzadeh, M.J. and Karimi, K. 2007. *BioResources* 2: 472–499. With permission.

interchain hydrogen bonds in the biomass, decrystallizing the cellulose and thus making it accessible to reagents; then it catalyzes the hydrolysis of glycosidic bonds and cleaves cellulose and hemicellulose into sugars (Binder and Raines, 2010).

Different acid concentrations have been used to hydrolyze polysaccharides, such as 72% H_2SO_4, 41% HCl, or 100% TFA (Fengel and Wegener, 1984). These provide a complete and rapid conversion of cellulose to glucose and hemicelluloses to five-carbon sugars with little degradation (Balata et al., 2008). Compared to dilute acid processes, the sugar yield from a concentrated acid hydrolysis process is much higher, reaching approximately 90% of the theoretical glucose yield (Taherzadeh and Karimi, 2007). Concentrated acid processes have the further advantage that they can be operated at low or medium temperatures, leading

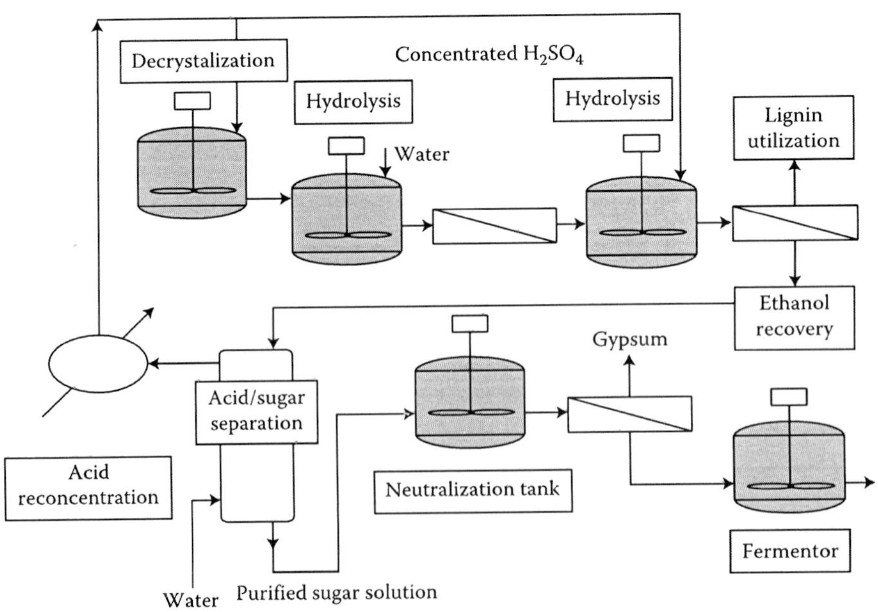

FIGURE 21.4
Concentrated-acid hydrolysis. (From *Renew. Sustain. Energ.* Rev., 13, Saxena, R.C. et al., Biomass-based energy fuel through biochemical routes: A review, 167–178, Copyright 2009, with permission from Elsevier.)

to a reduction in operational costs. However, due to the high acid concentrations and to heating during the hydrolysis process, it requires either expensive alloys or specialized nonmetallic construction to prevent corrosion. Furthermore, concentrated acid is difficult to work with and needs to be recovered and reconcentrated to be economically and environmentally friendly (Jeffries and Jin, 2000).

21.3.1.2 Dilute Acid Hydrolysis

Among the chemical hydrolysis methods, dilute acid hydrolysis is probably the most commonly used. The first established dilute acid hydrolysis process was probably the Scholl process, published in the 1940s (Faith, 1945). The acid hydrolysis process normally uses sulfuric acid and hydrochloric acid at concentrations of 0.5–10% and moderate temperatures in the range of 100–150°C (Wingren et al., 2003). In dilute acid hydrolysis, hemicellulose hydrolyzes at lower temperatures than cellulose.

As shown in Figure 21.5, when some links of the xylan hemicellulose or cellulose are attacked by protons (H^+) provided by the acids, the backbones can break down (Liu, 2008). Hydrolysis breaks down long-chain sugar oligomers into shorter-chain oligomers and monosaccharides (mainly xylose), liberates acetic acid, and produces furfurals and other by-products. Meanwhile, cellulose can be hydrolyzed to glucose and hydroxymethylfurfural (HMF), a product of the dehydration of hexoses. However, hydrolysis is less effective at forming hexoses under these relatively moderate operating conditions (Gámez et al., 2006). In addition, phosphoric acid was used in this dilute acid hydrolysis process. Using phosphoric acid may be more advantageous than using sulfuric or hydrochloric acid because phosphoric acid is less aggressive than these other acids. The harsher acids yield solutions

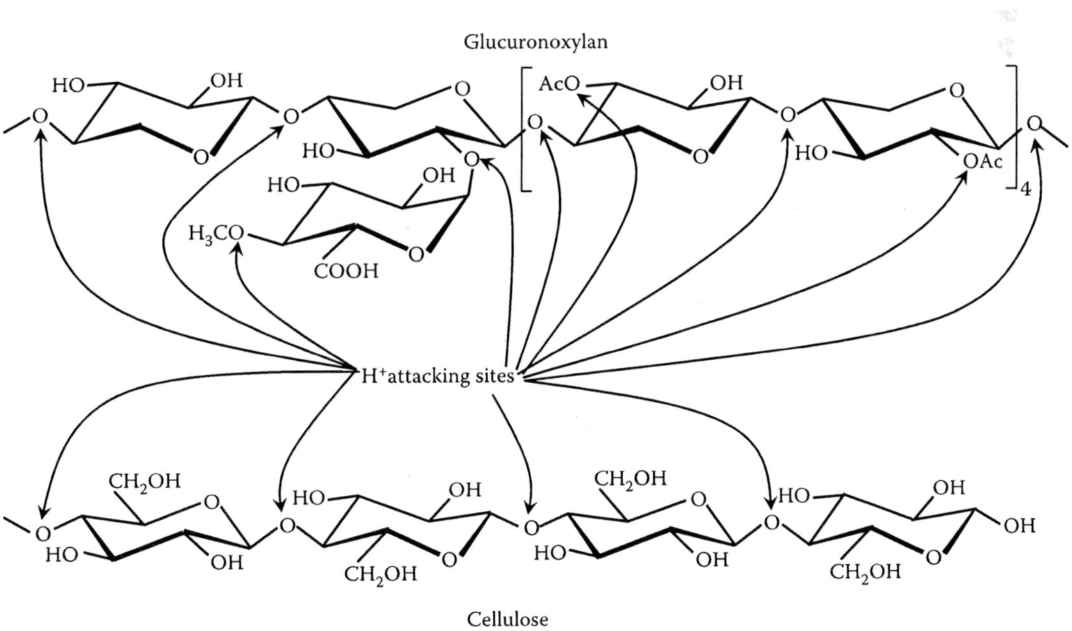

FIGURE 21.5
Sites and structure of *O*-acetyl-4-*O*-methylglucuronoxylan from hardwood and cellulose. Ac = $CH_3C(O)$–.

with higher concentrations of furfural or acetic acid, which inhibit the growth of the microorganisms used in fermentation (Romero et al., 2007). The biggest advantage of dilute acid processes is their fast rate of reaction, which facilitates continuous processing. Their biggest disadvantage is their low sugar yield. Most dilute acid processes are limited to a sugar recovery efficiency of approximately 50% (Badger, 2002). However, compared to concentrated acid hydrolysis processes, hydrolysis with a dilute mineral acid results in much less corrosion than with a concentrated acid, and recycling of the acid is not necessary.

21.3.2 Enzymatic Hydrolysis

Enzymatic hydrolysis is, along with acid hydrolysis, the most common method of saccharification (Harun et al., 2010). The goal of enzymatic hydrolysis is to depolymerize the polysaccharide solids that remain insoluble in the water after pretreatment into monosaccharides. After most pretreatment processes, the bulk of the remaining polysaccharides are cellulose. A class of enzymes known as *cellulase* is used as a catalyst in the hydrolysis of cellulose. These enzymes can be produced by fungi such as *Trichoderma reesei* and *Aspergillus niger* (Table 21.4) or by bacteria such as *Clostridium cellulovorans* (Arai et al., 2006). These microorganisms can be aerobic or anaerobic, and they can also be mesophilic or thermophilic.

Most research into commercial cellulase production has focused on fungi because the majority of the relevant bacteria are anaerobes with very low growth rates (Duff and Murray, 1996). At least three classes of enzymes act synergistically to hydrolyze cellulose: endo-β-1,4-glucanases (EG, EC 3.1.2.4), cellobiohydrolases or exo-glucanases (CBH, EC 3.2.1.91), and β-glucosidases (BG, EC 3.2.1.21). Endo-glucanases attack regions of low crystallinity in the cellulose fiber and create free chain ends. Exo-glucanases degrade the molecule further by removing cellobiose units from the free chain ends, which are then cleaved to glucose by the action of β-glucosidase. The enzymatic hydrolysis can be influenced by substrate and end-product concentrations, enzyme activity, and reaction conditions. β-Glucosidase plays a significant role in the hydrolysis process because cellobiose is an

TABLE 21.4
Sugar Yields from the Enzymatic Hydrolysis of Wheat Straw after Various Pretreatments

Pretreatment	Enzyme Mixture	Source of Enzyme	Hydrolysis Condition	Sugar Yield (g/g DM)	% Maximum Theoretical
Dilute H_2SO_4 impregnation + SE	Cellulase	*T. reesei*	40°C; pH 5.0	0.612	99.6
	α-Glucosidase	*A. niger*	96 h		
0.75% (v/v)	Cellulase	*T. reesei*	45°C; pH 5.0	0.565	74
H_2SO_4	α-Glucosidase	*A. niger*	72 h		
121°C	Xylanase	*T. longibrochiatum*			
2.15% (v/v)	Cellulase	*T. reesei*	45°C; pH 5.0	0.672	96.7
H_2O_2	α-Glucosidase	*A. niger*	120 h		
35°C	Xylanase	*T. longibrochiatum*			
Fine grinding + wet oxidation	Cellulase	*T. reesei*	50°C; pH 5.0	0.638	92
	α-Glucosidase	*A. niger*	24 h		

Source: From *Bioresour. Technol.*, 101, Talebnia, F. et al., Production of bioethanol from wheat straw: An overview of pretreatment, hydrolysis and fermentation, 4744–4753, Copyright 2010, with permission from Elsevier.

end-product inhibitor of many cellulases, including both exo- and endo-glucanases (Coughlan and Ljungdahl, 1988; Rabinovich et al., 2002; Galbe and Zacchi, 2002; Zhang et al., 2006). Cellulase enzymes are also highly specific, affecting only cellulose, and do not treat the hemicelluloses contained in the biomass. These factors mean that any hemicelluloses and lignin must have either been removed during pretreatment or hydrolyzed by accessory or "helper" enzymes. These "helper" enzymes include hemicellulases such as glucuronidase, acetylesterase, xylanase, β-xylosidase, galactomannanase, and glucomannanase and ligninases (Duff and Murray, 1996). Pretreatment methods and hydrolyzing conditions such as temperature and pH are among the factors influencing the efficiency of enzymatic hydrolysis of cellulosic materials. Most cellulose enzymes show optimum activity at temperatures in the range of 45–55°C and at pH values between 4 and 5 (Galbe and Zacchi, 2002). Due to these mild conditions, the utility costs are low compared to chemical hydrolysis.

The action of cellulase consists of three steps: adsorption of cellulase enzymes onto the surface of the cellulose, biodegradation of cellulose to fermentable sugars, and desorption of cellulase. The rate of enzymatic hydrolysis of the cellulosic materials always decreases fairly quickly. The irreversible adsorption of cellulase on cellulose is partially responsible for this deactivation (Converse et al., 1988). Generally, enzymatic cellulose degradation is characterized by a rapid initial phase followed by a slower secondary phase that may last until all the substrate is consumed. This is most often explained by the rapid hydrolysis of the readily accessible fraction of cellulose, strong product inhibition, and slow inactivation of absorbed enzyme molecules (Nutt, 2006).

Several factors influence the enzymatic hydrolysis of cellulose, including the substrate, cellulase activity, and reaction conditions (temperature, pH, and other parameters), as well as the presence or absence of strong product inhibition. To improve the yield and rate of enzymatic hydrolysis, research has focused on optimizing the hydrolysis process and enhancing cellulase activity (Sun and Cheng, 2002). The rate of enzymatic hydrolysis of cellulose depends on several structural features of cellulose. The features of cellulose which are known to affect hydrolysis rate include (1) molecular structure, (2) crystallinity, (3) surface area of fibers, (4) degree of swelling of the fibers, (5) degree of polymerization, and (6) associated lignin or other materials (Detroy and Julian, 1982). Lignin interferes with hydrolysis by blocking access of cellulases to cellulose and by irreversibly binding hydrolytic enzymes. Therefore, removal of lignin can dramatically increase the hydrolysis rate (McMillan, 1994).

Use of cellulase enzymes supplemented with other enzymes has been extensively studied. It is well known that the conjugated action of cellulases and hemicellulases can increase the rate of enzymatic hydrolysis and achieve higher sugar production. Hemicellulases facilitate cellulose hydrolysis by exposing the cellulose fibers, thus making them more accessible to cellulases (Shallom and Shoham, 2003). In contrast to cellulose, which is a homopolysaccharide, hemicelluloses are heteropolysaccharides. The major component of hemicelluloses in lignocellulosic materials is xylan. The backbone chain of a hemicellulose consists of xylose units with different side groups, including L-arabinose, D-galactose, acetyl, feruloyl, *p*-coumaroyl, and glucuronic acid residues. Therefore, enzymatic hydrolysis of hemicelluloses requires a multiple-enzyme system.

The main enzymes involved are endo-1,4-β-xylanase (1,4-β-D-xylan xylanohydrolase; EC 3.2.1.8) and β-xylosidase (1,4-β-D-xylan xylohydrolase; EC 3.2.1.37), which can cleave the glycosidic bonds in the xylan backbone. The activities of several accessory enzymes to cleave the various side chains are also essential (Polizeli et al., 2005; Gray et al., 2006; Ryabova et al., 2009). α-D-Glucuronidase (EC 3.2.1.139) hydrolyzes the α-1,2 bonds between the glucuronic acid residues and the β-D-xylopyranosyl backbone units found in glucuronoxylan. α-L-Arabinofuranosidase (EC 3.2.1.55) removes the L-arabinose residues which are substituted at positions 2 and 3 of the β-D-xylopyranosil. Acetylxylan esterase (EC 3.1.1.72) removes the *O*-acetyl groups from positions 2 and/or 3 on the β-D-xylopyranosyl residues of acetyl xylan (Polizeli et al., 2005; Chavez et al., 2006). Ferulic acid and *p*-coumaric acid esterases (EC 3.1.1.73) hydrolyze the ester bonds on xylan, liberating the respective phenolic acids linked to the arabinofuranoside residues (Girio et al., 2010). These enzymes are produced by a variety of fungi, bacteria, yeast, marine algae, protozoans, snails, crustaceans, insects, and seeds, such as *Trichodrema* spp., *Penicillium* spp., *Talaromyces* spp., *Aspergillus* spp., and *Bacillus* spp. (Sunna and Antranikian, 1997; Balat, 2011).

21.4 Fermentation and Major Product Choice

After pretreatment and conversion of cellulose and hemicelluloses to monosaccharides, fermentation is the next step. Fermentation is defined as the process of converting a fermentable substrate into valuable products using microorganisms, enzymes, or both. These products are primary or secondary metabolites of microbial origin, such as ethanol, butanol, acetone, organic acids, antibiotics, amino acids, vitamins, glycols, aldehydes, flavorings, enzymes, lipids, alkaloids, surfactants, hormones, esters, ethers, and ketones.

21.4.1 Bioethanol

21.4.1.1 Historical Background

The process of converting biorenewable materials into useful higher-value compounds is not new. For instance, sugarcane has been used in the production of bioethanol since 6000 BC. The ancient Egyptians produced alcohol by naturally fermenting vegetative materials. Also in ancient times, the Chinese discovered the art of distillation, which increases the concentration of alcohol in fermented solutions (Demirbas, 2009).

Ethanol and ethanol/gasoline blends have a long history as alternative transportation fuels. The history of using ethanol as a fuel extender in gasoline dates back to the development of the internal combustion engine by Nicholas Otto in 1876. The first car produced by Henry Ford in 1896 could use pure ethanol as fuel. In 1908, the Ford Model T, the first mass-produced car, was a flexible vehicle which could use ethanol as a fuel in the same way as gasoline or any mixture of the two (Solomon et al., 2007; Mussatto et al., 2010). Bioethanol as a fuel was in widespread use in Europe and the United States until the early 1900s. However, after the First World War, there was a decrease in demand for ethanol because gasoline prices were significantly reduced. In the 1970s and early 1980s, the fuel ethanol market was revived for economic reasons. At this time, there was a global oil crisis along with overproduction in the international sugar market and an oversupply of agricultural products, which all contributed to renewed interest in bioethanol. At least two major programs were created in response to heightened demand: The National Alcohol Program (ProAlcohol) in Brazil, and the U.S. Fuel Ethanol Program (Gasohol). These programs were intended to target the large-scale use of ethanol to reduce dependency on fossil fuel. Nowadays, the United States and Brazil are still the world leaders in the production and usage of bioethanol. However, as history shows, the price of gasoline has a significant effect of the production of bioethanol and will continue to do so until the technologies are developed to a point at which the cost of bioethanol is similar to or lower than that of gasoline.

21.4.1.2 Bioethanol Fermentation

Hexoses and pentoses can be obtained from the hydrolyzates of lignocellulosic materials. Depending on the lignocellulosic source, the hydrolyzate typically consists of xylose, arabinose, glucose, galactose, mannose, and rhamnose.

Pentosan to pentose:

$$(C_5H_8O_4)_n + nH_2O \rightarrow nC_5H_{10}O_5 \tag{21.1}$$

Hexosan to hexose:

$$(C_6H_{10}O_5)_n + nH_2O \rightarrow nC_6H_{12}O_6 \tag{21.2}$$

The maximum theoretical yields of pentose and hexose depend on whether aerobic or anaerobic conditions are used.

The anaerobic fermentation of pentose and hexose to ethanol is based on the following overall stoichiometry:

$$3C_5H_{10}O_5 \rightleftharpoons 5C_2H_5OH + 5CO_2 \tag{21.3}$$

$$C_6H_{12}O_6 \rightleftharpoons 2C_2H_5OH + 2CO_2 \tag{21.4}$$

Therefore, the theoretical ethanol yield is 0.511 kg/kg pentose or hexose. For xylan, the theoretical ethanol yield on a mass basis is 58.12%:

$$3C_5H_9O_4(C_5H_8O_4)_{n-2}C_5H_9O_5 + 3(n-1)\,H_2O \rightleftharpoons 5nC_2H_5OH + 5nCO_2 \tag{21.5}$$

However, in practice, monosaccharides are consumed during microorganism growth; therefore, the maximum ethanol yield is decreased (Liu, 2008).

Microorganisms for ethanol fermentation can best be described in terms of their performance parameters and other requirements such as compatibility with existing products, processes, and equipment. The performance parameters for fermentation are temperature range, pH range, alcohol tolerance, growth rate, productivity, osmotic tolerance, substrate utilization, specificity, yield, genetic stability, and inhibitor tolerance (Demirbas, 2005). The characteristics required for an industrially suitable microorganism are summarized in Table 21.5 (Dien et al., 2003).

Many types of microorganisms are capable of bioethanol fermentation, including bacteria, yeasts, and fungi.

The best-known microorganisms for ethanol production from hexoses are the yeast *Saccharomyces cerevisiae* and the bacterium *Zymomonas mobilis* (Claassen et al., 1999).

S. cerevisiae, also known as baker's yeast, is generally recognized as safe (GRAS) and the most commonly used type of yeast. It has proved to be very robust and well suited to the

TABLE 21.5

Important Traits for Bioethanol Fermentation Organisms

Trait	Requirement
Bioethanol yield	>90% of theoretical
Bioethanol tolerance	>40 g/L
Bioethanol productivity	>1 g/L h
Robust grower and simple growth requirements	Inexpensive medium formulation
Able to grow in undiluted hydrolysates	Resistance to inhibitors
Culture growth conditions retard contaminants	Acidic pH or higher temperature

Source: With kind permission from Springer Science+Business Media: *Appl. Microbiol. Biotechnol.*, Bacteria engineered for fuel ethanol production: Current status, 63, 2003, 258–266, Dien, B.S. et al.

fermentation of lignocellulosic hydrolysates. *S. cerevisiae* can efficiently ferment simple hexoses such as D-glucose, D-mannose, and D-galactose and disaccharides like sucrose and maltose and can yield bioethanol concentrations of up to 20% (v/v) in the fermentation broth (Figure 21.6). Moreover, it can tolerate a wide range of inhibitors and elevated osmotic pressure (Banerjee et al., 2010). In addition, the fermentation rate is not significantly reduced at ethanol concentrations below 10% (v/v) (Casey and Ingledew, 1986). However, the main disadvantage of the native strains of *S. cerevisiae* is their inability to process pentoses such as D-xylose and L-arabinose. This is because *S. cerevisiae* lacks the enzymes that convert xylose to xylulose. However, this yeast can ferment xylulose. For xylose-using *S. cerevisiae*, high bioethanol yields from xylose also require metabolic-engineering strategies to enhance the xylose flux (Balat, 2011).

Ethanologenic bacteria, such as *Escherichia coli*, *Klebsiella oxytoca*, and *Zymomonas mobilis*, have recently attached particular interest for industrial exploitation. *Z. mobilis*, a Gram-negative bacterium, is well recognized for its ability to produce efficiently a high yield of bioethanol as a consequence of the anaerobic utilization of D-glucose using the Entner–Doudoroff (ED) pathway. The ED pathway yields only 1 ATP per molecule of glucose, in contrast to 2 ATP per molecule of glucose when using the Embden–Meyerhoff–Parnas (EMP) pathway, the common glycolytic pathway which is present in both *E. coli* and *S. cerevisiae*. As a consequence, *Z. mobilis* produces less cell mass (Girio et al., 2010). Therefore, higher specific ethanol productivity has been observed for *Z. mobilis* when *Z. mobilis* and *S. cerevisiae* were compared for their efficiency in producing bioethanol from glucose hydrolyzate. Comparative performance trials on glucose have shown that *Z. mobilis* can achieve 5% higher bioethanol yields and up to fivefold higher bioethanol volumetric productivity compared to traditional *S. cerevisiae* yeast (Saez-Miranda et al., 2008). Its theoretical yield is 97%. However, there are some disadvantages for *Z. mobilis* in the fermentation process. Although a xylose-fermenting *Z. mobilis* has been generated by introducing a xylose-metabolizing pathway from *E. coli*, *Z. mobilis* still has a narrow substrate range, lacking in particular the ability to process pentoses (Hahn-Hägerdal et al., 2006). Moreover, *Z. mobilis* requires a neutral pH range and has a low tolerance to lignocellulose-derived inhibitors. Its overall tolerance to ethanol allows fermentation to proceed up to concentrations of 12–13.5% (v/v), but ethanol causes significant inhibition of fermentation at concentrations as low as 2% (v/v) (Osman and Ingram, 1985; Girio et al., 2010). However, a modified *Z. mobilis* strain was investigated recently and has the advantages of requiring a minimum of nutrients, growing at low pH and high temperatures, and being considered generally recognized as safe (GRAS).

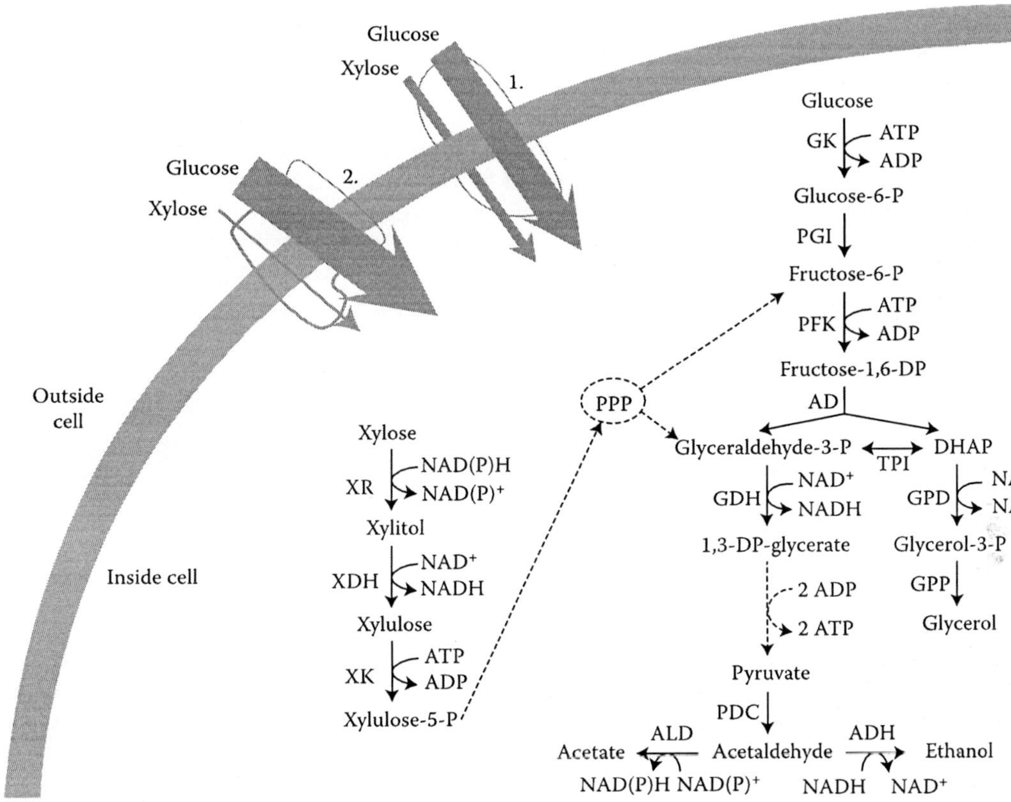

FIGURE 21.6
Simplified scheme of sugar transport and metabolism in *S. cerevisiae*. 1. Low- and intermediate-affinity hexose transporters. 2. High-affinity hexose transporters. PPP, pentose-phosphate pathway; XR, xylose reductase; XDH, xylitol dehydrogenase; XK, xylulokinase; GK, glucokinase; PGI, phosphoglucose isomerase; PFK, phosphofructokinase; AD, aldolase; TPI, triose phosphate isomerase; GDH, glyceraldehyde-3-P dehydrogenase; GPD, glycerol-3-P dehydrogenase; GPP, glycerol-3-phosphatase; PDC, pyruvate decarboxylase; ALD, acetaldehyde dehydrogenase; ADH, alcohol dehydrogenase. (Olofsson, K. et al. 2008. *Biotechnol. Biofuels* 1: 7. With permission.)

E. coli and *K. oxytoca* naturally metabolize arabinose, and therefore the ethanologenic strains can ferment all lignocellulose-derived sugars (Hahn-Hägerdal et al., 2007). The construction of *E. coli* strains to produce bioethanol selectively was one of the first successful applications of metabolic engineering (Balat, 2011). *E. coli* presents several advantages as a biocatalyst for ethanol production, such as the ability to ferment a wide range of sugars, including D-xylose and L-arabinose, no requirements for complex growth factors, and previous industrial use (e.g., for the production of recombinant proteins). However, several major disadvantages associated with *E. coli* cultures reduce ethanol yield. These disadvantages include its need for a narrow and neutral pH range (6–8), which makes bacterial fermentation susceptible to contamination, a low tolerance to lignocellulose-derived inhibitors, a low ethanol tolerance, and a tendency to form mixed products (ethanol, acetic acid, lactic acid, and others). *K. oxytoca* is an enteric bacterium which has been found growing in paper and pulp streams as well as around other sources of wood. It is capable of growing at a pH at least as low as 5.0 and temperatures as warm as 35°C. It can grow on a wide variety of sugars, including hexoses and pentoses, as well as on cellobiose and cellotriose (Dien et al., 2003).

Contrary to *S. cerevisiae*, the yeast *P. stipitis* is able to metabolize xylose (pentoses) by the action of xylose reductase (XR), which converts xylose to xylitol, and by the action of xylitol dehydrogenase (XDH), which converts xylitol to xylulose. Therefore, bioethanol fermentation from xylose can be successfully performed by recombinant *S. cerevisiae* carrying heterologous XR and XDH from *P. stipitis* and xylulokinase (XK) from *S. cerevisiae* (Figure 21.7) (Katahira et al., 2006; Margeot et al., 2009).

When considering hemicellulose conversion to ethanol, *P. stipitis* has received special interest. *P. stipitis*, its anamorph *Candida shehatae*, *Candida parapsilosis*, and *Pachysolen tanophilus* efficiently ferment xylose, but only under oxygen-limited conditions. Moreover, these yeasts are less tolerant to pH, ethanol, and hydrolysate inhibitors than *S. cerevisiae* (Hahn-Hägerdal et al., 2007). In addition, *P. stipitis* is able to produce hemicellulolytic enzymes.

Some naturally occurring microorganisms, such as bacteria, fungi, and yeasts, can convert xylose (the primary pentose from hemicellulose) into bioethanol. However, each has its own disadvantages which make it unsuitable for industrial use, as outlined in Table 21.6.

Although xylose is the dominant hemicellulose sugar, other five- and six-carbon sugars are part of hemicellulose as well. Besides xylose, arabinose, galactose, and mannose are present in noticeable abundances. The fermentation of six-carbon hemicellulose sugars (galactose and arabinose) has not yet been demonstrated. They can be fermented by, for example, *E. coli*, *Z. mobilis*, *S. cerevisiae*, and *P. stipitis*. The metabolic pathway for uptake of these sugars is shown in Figure 21.8.

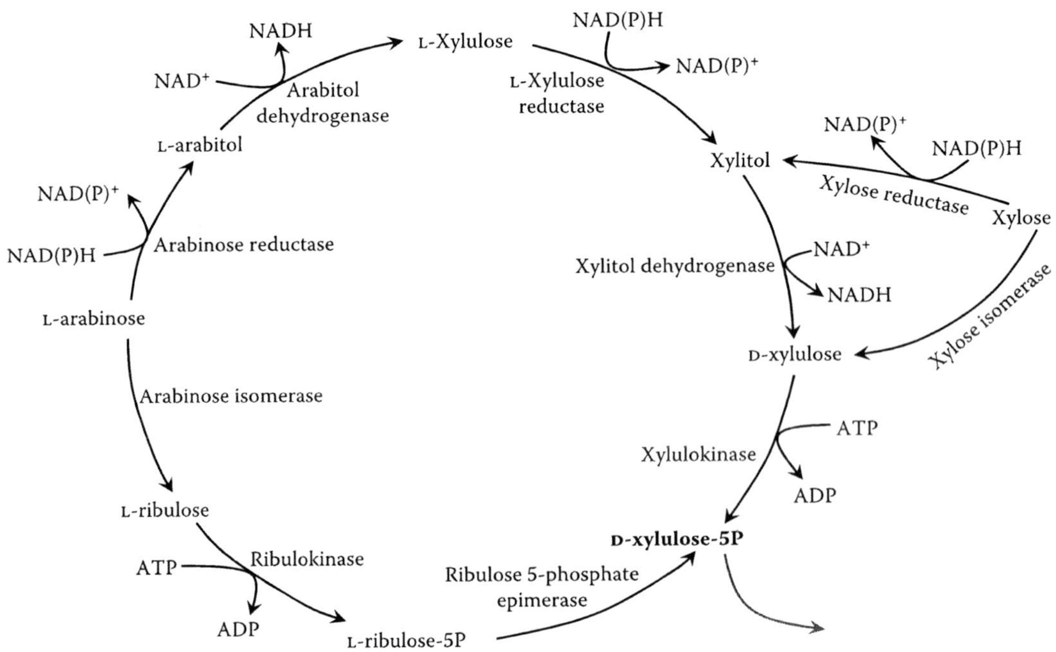

FIGURE 21.7
Pentose metabolism. (From *Curr. Opin. Biotechnol.*, 20, Margeot, A. et al., New improvements for lignocellulosic ethanol, 372–380, Copyright 2009, with permission from Elsevier.

TABLE 21.6

Advantages and Disadvantages of Various Natural Microorganisms for Industrial Ethanol Production

	Natural Sugar Utilization Pathways					Major Products		Tolerance			O_2	
Organism	Glu	Man	Gal	Xyl	Ara	EtOH	Other	Alcohols	Acids	Hydrolyzate	Needed	pH
Anaerobic bacteria	+	+	+	+	+	+	+	−	−	−	−	Neutral
E.coli	+	+	+	+	+	−	+	−	−	−	−	Neutral
Z.mobilis	+	−	−	−	−	+	−	+	−	−	−	Neutral
S.cerevisiae	+	+	+	−	−	+	−	++	++	++	−	Acidic
P.stipitis	+	+	+	+	+	+	−	−	−	−	+	Acidic
Filamentous fungi	+	+	+	+	+	+	−	++	++	++	−	Acidic

Source: With kind permission from Springer Science+Business Media: *Appl. Microbiol. Biotechnol.*, Towards industrial pentose-fermenting yeast strains, 74, 2007, 937–953, Hahn-Hägerdal, B. et al.

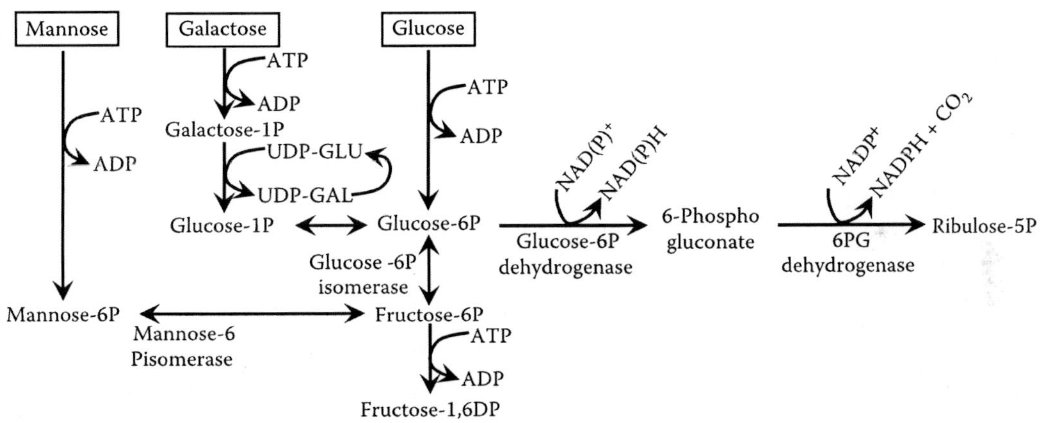

FIGURE 21.8
Fermentation pathways of hemicellulosic six-carbon sugars.

21.4.2 Butanol

Industrial production of butanol was first achieved in 1916. ABE fermentation (ABE is an acronym for acetone, butanol, and ethanol) was one of the main methods used at that time. Fermentation of mixed alcohols from sugars by *Clostridium* can be represented by

$$7CH_2O \rightarrow C_4H_9OH(\text{butanol}) + 3CO_2 + 2H_2 \tag{21.6}$$

$$5CH_2O \rightarrow C_3H_7OH(\text{propanol}) + 2CO_2 + H_2 \tag{21.7}$$

$$3CH_2O \rightarrow C_2H_5OH(\text{ethanol}) + CO_2 \tag{21.8}$$

$$5CH_2O \rightarrow CH_3COCH_3(\text{acetone}) + 2CO_2 + 2H_2 \tag{21.9}$$

Hydrogen is usually accompanied by mixed alcohol production because the metabolism of strictly anaerobic microorganisms tends to reject excess O_2 via CO_2.

During the Second World War, butanol was used as a chemical to produce synthetic rubber. However, producing butanol by fermentation began to fall into disuse after 1954 and was at a complete standstill by the 1960s except in Russia and in South Africa, where costs of substrate and labor were low. Half the butanol currently produced is used in the form of butyl acrylate and methacrylate esters in latex surface coatings, enamels, and lacquers (Kirschner, 2006). Other important derivatives of butanol are butyl glycol ether, butyl acetate, and plasticizers. Butanol is also an excellent diluent for brake-fluid formulations and a solvent for the manufacturing of antibiotics, vitamins, and hormones (Lee et al., 2008). Using butanol as a direct replacement for gasoline or as a fuel additive has only been reported recently. Biobutanol is an attractive liquid fuel that can be blended with gasoline. It can be blended with almost no limitations, and it can be used without modifying the engines of existing motor vehicles.

21.4.2.1 Properties of Butanol

Butanol (C_4H_9OH) is a higher-chain alcohol with a four-carbon structure. It is a colorless liquid with a distinct odor, and its vapor has an irritant effect on mucous membranes and a narcotic effect in higher concentrations. Butanol is completely miscible with organic solvents and partly miscible with water. Some of its physical properties are listed in Table 21.7.

Butanol exists in different isomers based on the location of the hydroxyl (OH) group and the carbon-chain structure. These structures have a direct impact on the physical properties of butanol. 1-Butanol, which has a straight-chain structure with the alcohol at the terminal carbon, has physical properties similar to those of gasoline and is therefore

TABLE 21.7

Properties of Butanol

Property	Value
Molecular weight	74.12
Melting point	−88.9°C
Boiling point	117.5°C
Relative density (water = 1)	0.81
Relative vapor density (air = 1)	2.55
Saturated vapor pressure (at 25°C)	0.82 kPa
Heat of combustion	2673.2 kJ/mol
Critical temperature	287°C
Critical pressure	4.9 MPa
Flash point	35°C
Autoignition temperature	340°C
Explosive limit (V/V)	1.4–11.2%

considered a potential candidate for replacing ethanol as an oxygenate or for replacing gasoline entirely. *sec*-Butanol is not considered a potential alternative engine fuel due to its low motor octane number. *tert*-Butanol at low volumetric fractions is used as a denaturant for ethanol or to improve the knock behavior of gasoline. It is not considered to be effective as an alternative fuel at higher volumetric concentrations because of its high melting point (Wallner et al., 2009).

21.4.2.2 Advantages and Disadvantages as Fuel

A comparison of different alcohols and gasoline is shown in Table 21.8. It is clear that butanol has several advantages: butanol contains 30% more energy than ethanol and only 10% less than gasoline. Butanol can be mixed with gasoline and can even replace gasoline without limitations. Butanol is preferable to a mixture of ethanol and gasoline because it improves the fuel-utilization efficiency of cars and increases the running distance per unit of consumed fuel. Butanol is also less evaporative than either ethanol or gasoline. This makes butanol safer when used as an oxygenate, and there is no significant need to change the mixture proportions with the seasons. In addition, butanol is less aggressive than ethanol, making it transportable by currently used fuel pipelines. Ethanol, on the other hand, needs to be transported by rail. In addition, producing butanol simultaneously solves certain problems connected with the infrastructure to supply hydrogen. Therefore, increasing the production and use of butanol would provide a solution to a number of problems regarding the rapid adoption of biofuels all around the world.

The major disadvantage of butanol is its high flashpoint (butanol 35°C, ethanol 13°C, gasoline <–40°C). Butanol has a lower research octane number (butanol 96, ethanol 129, and gasoline 91–99) and a higher viscosity (varies by temperature) than competitive fuels. Butanol emits a foul odor as well. Ethanol has a faster biodegradation rate than butanol, particularly in water.

21.4.2.3 Butanol Biosynthesis Pathway with *C. acetobutylicum*

It is well known that the fermentation substrate is an important factor influencing the cost of biobutanol production. Clostridia can secrete numerous enzymes that facilitate the breakdown of polymeric carbohydrates into monomers, which makes these good substrates for biobutanol production (Ezeji et al., 2007). Starch hydrolysis involves α-amylase, β-amylase, pullulanase, glucoamylase, and α-glucosidase, while cellulose hydrolysis needs cellulases and β-glucosidase. Hydrolysis yields glucose which is then harvested. Xylose and arabinose converted from hemicellulose are subsequently broken down through the transketolase–transaldolase sequence. This produces fructose 6-phosphate and glyceraldehyde 3-phosphate, which are subsequently metabolized by the EMP pathway (Ounine

TABLE 21.8

Properties of Butanol Compared with Those of other Fuels

Properties	Methanol	Ethanol	Butanol	Gasoline
Energy density (MJ/L)	16	21.2	29.2	32.5
Air–fuel ratio	6.5	9	11.2	14.6
Research octane number	136	129	96	91–99
Heat of vaporization (MJ/kg)	1.2	0.92	0.43	0.36
Flash point (°C)	79	13	35	≤ −40

et al., 1983). Finally, glucose passes through the cell membrane with the aid of the glucose phosphotransferase system (PTS) and is then consumed by ABE fermentation.

The various cellulosic and hemicellulosic monosaccharides produced as described above are transported into the cell by specific membrane-bound transport systems, and the carbohydrates are subsequently metabolized by glycolysis or the pentose-phosphate pathway. *C. acetobutylicum* was first isolated in 1916 by Weizmann, who wanted to discover how it produced acetone, butanol, and ethanol from starch using the so-called ABE process. A typical ABE fermentation is characterized by two distinct phases (Figure 21.9), which are called acidogenesis and solventogenesis (Ennis et al., 1986). Bacteria in the first phase, acidogenesis, experience an exponential growth phase, with active hydrogen evolution and acetic and butyric acid production. The pH of the medium decreases to as low as 4.5 because of acid formation. As the accumulation of organic acids increases, the fermentation switches to solventogenesis, which is accompanied by steady cell growth and decreased

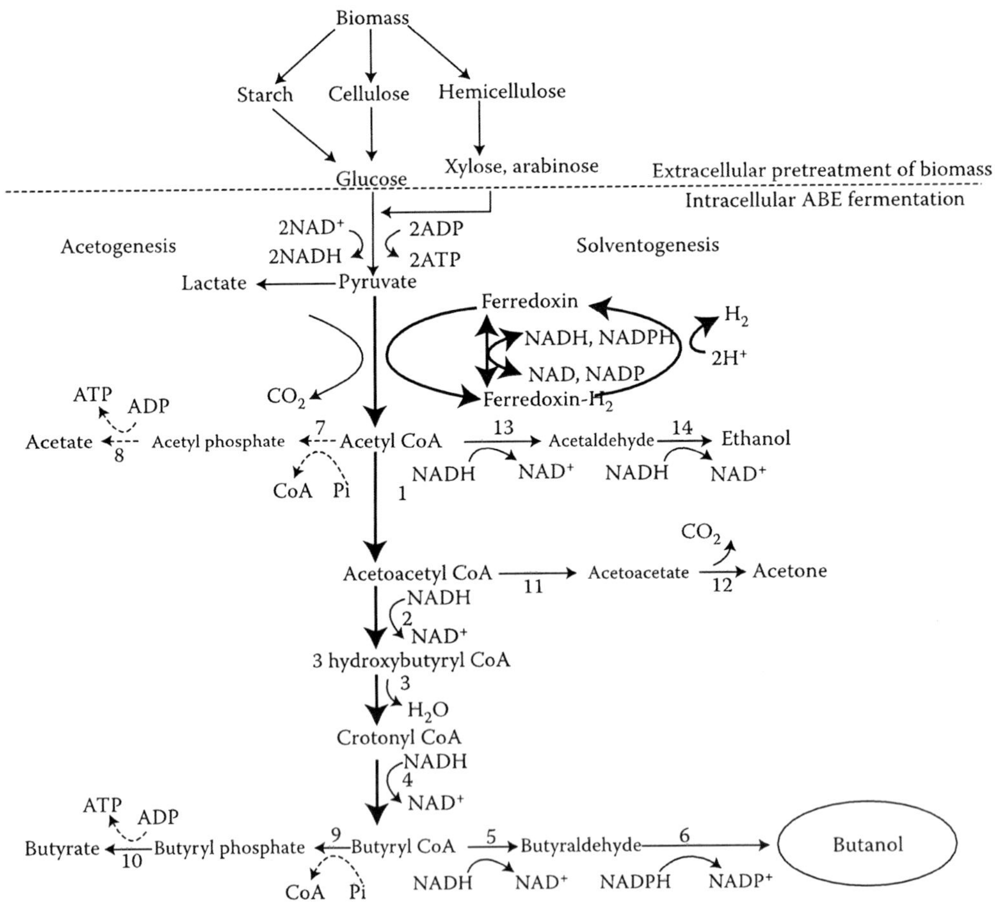

FIGURE 21.9
Butanol biosynthesis pathway in *C. acetobutylicum*. Circled numbers refer to the enzymes used: 1. acetyl-CoA acetyltransferase (thiolase); 2. β-hydroxybutyryl-CoA dehydrogenase; 3. 3-hydroxybutyryl-CoA dehydratase (crotonase); 4. butyryl-CoA dehydrogenase; 5. butyraldehyde dehydrogenase; 6. butanol dehydrogenase; 7. phosphate acetyltransferase; 8. acetate kinase; 9. phosphate butyryltransferase; 10. butyrate kinase; 11. acetoacetate decarboxylase; 12. CoA transferase; 13. coenzyme A (CoA)-acylating; 14. NAD(P)H alcohol dehydrogenase.

hydrogen evolution. Acetone, butanol, and ethanol are produced, with partial uptake of the acids formed during the first growth phase. Then, as *C. acetobutylicum* ages, it gradually loses its original bioactivity and enters the cell autolysis phase, forming spores until the end point is reached.

21.4.2.4 Enzymes Involved in Biobutanol Synthesis by Clostridium

In nature, bacterial butanol production is carried out exclusively by members of the genus *Clostridia*, including *C. acetobutylicum*, *C. beijerinckii*, *C. saccharoperbutylacetonicum*, *C. saccharoacetobutylicum*, *C. aurantibutyricum*, *C. pasteurianum*, *C. sporogenes*, *C. cadaveris*, and *C. tetanomorphum* (Keis et al., 1995). Acetone–butanol–ethanol (ABE) fermentation has been widely studied, and microorganisms that can perform ABE fermentation can be viewed as a model system for understanding the regulation and genetics of complex and primary metabolism. Although a number of different *Clostridia* are continually being isolated and their abilities in butanol production being verified, *Clostridium acetobutylicum* is still the species which is most often used for the industrial production of butanol.

As can be seen in Table 21.9, two enzymes are responsible for conversion of acetyl-CoA during the synthesis of lactate, phosphotransacetylase (PTA), and acetate kinase (AK). The acetyl-CoA (A-CoA) node serves as a connecting point where several metabolic pathways intersect. It is also the switch point between the acid- and solvent-producing pathways. During acidogenesis, the effects of AK and PTA increase, and ATP is produced as an important energy source for cell growth. During solventogenesis, the activity levels of AK and PTA decrease, and they have unclear roles in this stage (Hartmanis et al., 1986). The activities of PTA and AK in batch fermentation of *C. acetobutylicum* have been observed to be at a maximum in the early stage of the exponential phase and to drop off significantly with the increase of acetate concentration in the medium (Ballongue et al., 1986). This reference also stated that AK was synthesized inside the organism until acetate accumulated to a concentration of 3 g/L in the fermentation broth. Then the higher concentration of acetate caused a rapid drop in acetate kinase activity.

For butyrate formation, phosphate butyryltransferase (PTB) and butyrate kinase (BK) are essential. PTB catalyzes the interconversion of butyryl-CoA and butyryl-P and, together with BK, enables the conversion of butyryl-CoA to butyrate (Wiesenborn et al., 1989b). Butyryl-P is formed in the physiological direction when the pH value is specifically 8.1. In the reverse direction, butyryl-CoA is formed. In this direction, the enzyme is less sensitive and shows a broad pH tolerance, ranging from 7.5 to 8.7. In the forward direction, BK catalyzes the reaction by which butyryl-P is converted to butyrate, with concomitant ADP phosphorylation. The reverse reaction converts butyrate to butyryl-P using ATP as the phosphate donor (Sullivan et al., 2010).

In the butanol synthetic pathway, six enzymes are needed to complete the conversion of acetyl-CoA to butanol, including thiolase, also called acetyl-CoA acetyltransferase (THL), β-hydroxybutyryl-CoA dehydrogenase (BHBD), 3-hydroxybutyryl-CoA dehydratase (CRT, also called crotonase), butyryl-CoA dehydrogenase (BCD), butyraldehyde dehydrogenase (BAD), and butanol dehydrogenase (BDH). Thiolase plays a key role in the production of both acids and solvents. It effectively controls the flow of carbon into these pathways by catalyzing the condensation of two molecules of acetyl CoA into one of acetoacetyl-CoA. Acetoacetyl-CoA is then converted to acetone, butyrate, or butanol. Regulation of thiolase activity is important for maintaining the C2:C4 acid ratio, which has a profound effect on the energy management of the cell. THL may play an indirect role in acid uptake. BAD is NADH-dependent and catalyzes the conversion of

TABLE 21.9

Key Enzymes in the Butanol Synthetic Pathway by *C. acetobutylicum*

Pathway	Key Enzyme	Ab.	Gene	EC No.	M (kDa)	Reference
Lactate synthetic pathway	Phosphate acetyltransferase	PTA	*pta*	2.3.1.8	36.2	Boynton et al. (1996b)
	Acetate kinase	AK	*ack*	2.7.2.1	44.3	Boynton et al. (1996b)
Butyrate synthetic pathway	Phosphate butyryltransferase	PTB	*ptb*	2.3.1.19	264	Hartmanis. (1987)
	Butyrate kinase	BK	*buk*	2.7.2.7	85	Hartmanis. (1987)
Butanol synthetic pathway	Acetyl-CoA acetyltransferase	THL	*thiL*	2.3.1.9	41	Stim-Herndon et al. (1995)
	β-Hydroxybutyryl-CoA dehydrogenase	BHBD	*hbd*	1.1.1.35 (157)	30.5	Boynton et al. (1996a)
	Enoyl-CoA hydratase (crotonase)	CRT	*crt*	4.2.1.17	158	Waterson et al. (1972)
	Butyryl-CoA dehydrogenase	BCD	*bcd-etfB etfA*	1.3.99.2	33	Boynton et al. (1996a)
	Butyraldehyde dehydrogenase	BAD	*aad*	1.2.1.57	56	Palosaari and Rogers (1988)
	Butanol dehydrogenase	BDH	*bdhAB*	1.1.1	42	Petersen et al. (1991)
Acetone synthetic pathway	Acetoacetate decarboxylase	AADC	*adc*	4.1.1.4	28	Petersen and Bennett (1990)
	CoA transferase	CoAT	*CtfA-B*	2.8.3.9	93	Palosaari and Rogers (1988)
Ethanol synthetic pathway	Acetaldehyde dehydrogenase	ALDH	*aad*	1.2.1.10	96	Palosaari and Rogers (1988)
	NAD(P)H alcohol dehydrogenase	ADH	*adh*	1.1.1.2	44	Youngleson et al. (1988)

butyryl-CoA to butyraldehyde. Butyraldehyde is a key intermediate in the formation of butanol and is accompanied by the oxidation of NAD(P)H and the release of CoA. BDH catalyzes the final step in butanol synthesis with the reduction of butyraldehyde to butanol at the expense of a reduced NAD(P). An assay performed with butanol dehydrogenase from *C. acetobutylicum ATCC 824* (Welch et al., 1989) clearly revealed that the enzyme has two different coenzyme requirements: the NADH-dependent butanol dehydrogenase (NADH-BDH), with higher activity at lower pHs, and a NADPH-dependent one, with higher activity at basic pHs. BCD catalyzes the α- or β-desaturation of acyl-CoA substrates.

CoA transferase and acetoacetate decarboxylase are two important enzymes involved in the acetone synthetic pathway. The former enzyme seems to be a heterotetramer with two α- and two β-subunits, which code for the genes *ctfA* and *ctfB*, respectively (Wiesenborn et al., 1989a). The formation of acetoacetate is catalyzed by a CoA-transferase which transfers CoA from acetoacetyl-CoA to either acetate or butyrate. Acetate and butyrate harvested during the growth phase reduce the pH value of the cell culture, and the resulting acidic condition in the stationary phase is widely regarded as a trigger to shift metabolism into solvent production (Jones and Woods, 1986). The uptake of acetate and butyrate by CoA-transferase from *C. acetobutylicum* serves as a detoxification mechanism which reduces the inhibitory effects of these acids on cell growth. Acetoacetate decarboxylase (AADC) is another critical enzyme in this pathway, which is responsible for catalyzing the conversion of acetone to acetoacetyl-CoA. The acetoacetate formed by CoA-transferase is subsequently decarboxylated by AADC to yield acetone and carbon dioxide. Thus, the formation of acetone is coupled with the uptake of acetate and butyrate. The activity of this decarboxylase was found to be almost 40 times higher in solvent-producing cells (4.93 U/mg) than in acid-producing cells (0.13 U/mg) (Andersch et al., 1983).

There are two enzymes in the ethanol synthetic pathway, coenzyme A-acylating aldehyde dehydrogenase and NAD(P)H alcohol dehydrogenase. The former enzyme is also known as an important enzyme at the junction in the pathway leading to the production of either acids (acetate or butyrate) or solvents (acetone, butanol, or ethanol) during the growth of *C. acetobutylicum ATCC 824*. The latter enzyme catalyzes the reduction of acetaldehyde to ethanol. EDH activity is five to eight times higher in the solvent-producing phase than in the acid-producing phase (Vasconcelos et al., 1994).

21.4.2.5 Butanol Fermentation

Recent advances have occurred in butanol fermentation, including upstream and downstream processing. Using excess carbon under nitrogen limitation is required to achieve high levels of solvent production (Madihah et al., 2001). Iron is an important mineral supplement since the conversion involves a ferredoxin oxidoreductase iron–sulfur protein. Classical fed-batch and continuous cultivation do not seem to be economically feasible because of solvent toxicity and the biphasic nature of acetone–butanol fermentation, respectively. To overcome this problem, fed-batch cultures are coupled with an *in situ* recovery process (Ezeji et al., 2004).

Butanol has been produced from various raw materials, including molasses, whey permeate, and corn. Qureshi et al. (2001) used spray-dried soy molasses (SDSM) to produce ABE using *C. beijerinckii* BA101 in batch cultures. Higher concentrations of SDSM resulted in poor solvent production because of the presence of excessive salt and inhibitory components. However, when supplemented with 25.3 g/L glucose, a 113% (10.7–22.8 g/L) increase in ABE production was achieved with 80 g/L SDSM. These studies suggest that SDSM can be a useful substrate for producing solvents if it is used with an additional sugar such as glucose or sucrose.

Butanol has also been produced from low-cost waste products such as starch-based waste from packing peanuts and agricultural waste. Low-grade glycerol has also been used to produce butanol in a chemostat culture of *C. acetobutylicum*, resulting in high yield and productivity of 0.34 mol/mol (0.42 g/L/h) (Vasconcelos and Andrade, 2003).

Another low-cost substrate explored is corn-fiber xylan (Qureshi et al., 2006). Hydrolysis of corn-fiber xylan, fermentation, and recovery of ABE were performed in a single, integrated system with a high yield of 0.44 g ABE per g substrate and a high productivity of 0.47 g/L/h. Recently, a 214% increase in butanol productivity was achieved in a batch culture of *C. beijerinckii P260* using wheat straw hydrolysate as a carbon source (Ezeji et al., 2007).

A recent report describes the impact of the components of agricultural hydrolysates on butanol production. Furfural and hydroxymethyl furfural were found to have a stimulatory effect on the growth of *C. beijerinckii* BA101 and on butanol production (Ezeji et al.,

2007). Mono- and disaccharide uptake occurs mainly through the phosphotransferase (PTS) system, with galactose being the only sugar reported to be transported by a non-PTS mechanism (Zverlov et al., 2006). In addition, solvent-producing *Clostridia* are capable of utilizing pentose sugars efficiently, including xylose and arabinose.

21.4.2.6 Solvent Toxicity

One of the most critical problems in ABE fermentation is solvent toxicity. Cellular metabolism in *Clostridium* may cease when the concentration of solvents exceeds 20 g/L. This limits the concentration of carbon substrate that can be used for fermentation, resulting in low final solvent concentration and low productivity. The lipophilic solvent butanol is more toxic than others because it disrupts the phospholipid components of the cell membrane, causing an increase in membrane fluidity. Moreira found that 0.1–0.15 M butanol caused 50% inhibition of both cell growth and sugar-uptake rate by negatively affecting ATPase activity. Increased membrane fluidity causes destabilization of the membrane and disruption of membrane-associated functions such as various transport processes, glucose uptake, and membrane-bound ATPase activity. Moreover, butanol is the only solvent that is produced during fermentation to a level that becomes toxic to *Clostridia* cells. It is known that the addition of 7–13 g/L butanol to a culture medium results in a 50% inhibition of growth, while the addition of acetone and ethanol up to 40 g/L reduced growth by 50% (Jones and Woods, 1986).

Biotechnological production problems such as solvent toxicity have traditionally been addressed by random mutagenesis and process optimization. Antisense RNA has also been used to improve butanol tolerance. A butanol-tolerant mutant strain of *C. beijerinckii NCIMB 8052* was constructed using antisense RNA and targeting the gldA gene which encodes glycerol dehydrogenase (Liyanage et al., 2000). It is expected that a more solvent-tolerant strain of Clostridia will be developed by metabolic and cellular engineering. So far, metabolic engineering approaches have been explored more extensively for *C. acetobutylicum ATCC 824*, and these are reviewed below.

21.4.2.7 Recovery and Separation

High product recovery cost is another problem in biological butanol production. Besides the traditional distillation process, several other processes, including pervaporation, adsorption, liquid–liquid extraction, gas stripping, and reverse osmosis have been developed to improve recovery performance and reduce costs.

The traditional recovery process that uses distillation suffers from high operating cost due to the low concentration of butanol in the fermentation broth. To solve this problem and the solvent toxicity problem at the same time, *in situ* recovery systems have been used. From an economic point of view, reverse osmosis is preferable. However, it has the disadvantages of membrane clogging and fouling. In contrast, liquid–liquid extraction has high capacity and selectivity, although it can be expensive to operate. Therefore, there are advantages and disadvantages of using each recovery system, which need to be thoroughly examined.

Gas stripping is a simple but efficient way to recover butanol from fermentation broth. The fermentation gas is bubbled through the fermentation broth, then passed through a condenser for solvent recovery. The stripped gas is then recycled back to the fermentor, and the process continues until all the sugar in the fermentor has been utilized. Gas stripping enables the use of a concentrated sugar solution in the fermentor (Ezeji et al., 2005), reduces butanol inhibition, and enables high rates of sugar utilization. Gas stripping was applied to a batch reactor to recover solvents from fermentation broth of *C. beijerinckii* BA101 (Evans and Wang, 1988). A 161.7 g/L sugar solution was successfully fermented, and 75.9 g/L total solvent was produced by the integrated process. Fed-batch fermentation was also integrated with gas stripping to reduce substrate inhibition and increase cell mass. In this system, 500 g glucose was consumed, and 233 g solvent was produced with a productivity of 1.16 g/L/h and a yield of 0.47 g/g. In the case of continuous fermentation integrated with gas stripping, 460 g of total solvent was produced from 1163 g glucose with a productivity of 0.91 g/L/h.

Liquid–liquid extraction is another efficient technique used to remove solvents from fermentation broth. This approach takes advantage of the differences in the distribution coefficients of the chemicals. Because butanol is more soluble in the extractant (organic phase) than in the fermentation broth (aqueous phase), it is selectively concentrated in the extractant. Common extractants used include decanol and oleyl alcohol (Evans and Wang, 1988). Liquid–liquid extraction has critical problems such as the toxicity of the extractant to the cell and emulsion formation. These problems can be overcome if the fermentation broth and the extractant are separated by a membrane that provides a surface area for butanol to exchange between the two immiscible phases. *C. acetobutylicum* has been cultivated in a continuously operated membrane bioreactor connected to a four-stage mixer-settler cascade. In this system, butanol was extracted with *n*-decanol (extractant) from the cell-free fermentation broth and was fed back into the fermentor. This system yielded a high solvent productivity of 3.08 g/L/h. Among the several extractants reported, oleyl alcohol, as a good extractant, was the most promising for *C. beijerinckii BA 101* because it is relatively nontoxic.

Pervaporation is a membrane-based process that enables the selective removal of volatile compounds from a fermentation broth. The membrane is placed in contact with a fermentation broth, and the volatile liquids or solvents diffuse through the membrane as a vapor which is recovered by condensation. Both liquid and solid pervaporation membranes have been used. A liquid membrane containing oleyl alcohol was used in pervaporation of dilute aqueous butanol solution (Matsumura et al., 1988). The selectivity of this liquid membrane was better than that of a silicon rubber membrane. Pervaporation using an oleyl alcohol liquid membrane for the pretreatment of butanol purification requires just one-tenth the energy of conventional distillation. To develop a stable membrane with a high degree of selectivity, Qureshi and Blaschek (1999) synthesized a siliconsilicalite-1 composite membrane which showed a 2.2-fold improvement in selectivity. Using this membrane in an integrated batch-pervaporation process with *C. beijerinckii* BA101 achieved a twofold increase in total solvent concentration (from 24.2 g/L in the control batch to 51.5 g/L in the pervaporation batch). The pervaporation conditions did not affect the growth of *C. beijerinckii* BA101. Because the membrane permeate contains acetone, butanol, and ethanol, distillation is still required for further purification. Pervaporation was also used in a fed-batch reactor, resulting in increased solvent productivity from 0.35 g/L/h (batch reactor) to 0.98 g/L/h (fed-batch reactor) due to the reduction in product inhibition. Recently, the overall solvent productivity for continuous fermentation of *C. acetobutylicum* was increased to 2.34 g/L/h by integrating it with a pervaporation system using an ionic liquid polydimethylsiloxane ultrafiltration membrane (Izák et al., 2008).

It should be emphasized that the recovery and purification processes are directly affected by fermentation and that performance is determined by the strain characteristics. For example, when a strain is metabolically engineered to produce butanol without, or with much less, acetone and ethanol, the purification process will be considerably simplified.

When the butanol tolerance of a strain is increased by metabolic engineering, this will also facilitate the recovery process because higher butanol concentrations can be achieved during fermentation. Therefore, the overall process must be optimized, from strain development to fermentation to downstream processes. This will lead to the reduction of overall production costs. For example, an engineered strain capable of producing butanol to a high concentration with high productivity will result in a significant reduction in the direct fixed capital costs (e.g., the costs of the fermentor and recovery units) and associated depreciation costs because a smaller fermentor can be used to produce the desired amount of butanol (Zhao et al., 2003).

21.4.2.8 Biological Processes

In principle, there are two metabolic engineering options for altering the fermenting organisms: (1) the genes required for butanol formation can be cloned into a desired production organism, or (2) a butanol producer such as *C. acetobutylicum* can be mutated in a way that it produces only butanol (becomes a homobutanol producer) plus some CO_2 and H_2.

The first option would be suitable for biotechnology workhorses such as *Escherichia coli*. Because this organism does not possess the ability to synthesize C_4 compounds, the respective genes would have to come from *C. acetobutylicum* as well. A problem that might develop is that the essential *Clostridial* enzymes, such as *adhe*, might become oxygen-sensitive. Another problem with this approach is that it does not take into account the tolerance of the envisaged production strain against butanol. As already mentioned, this alcohol damages membranes and certain proteins.

Another approach is to overexpress cyclopropane fatty acid synthase. Tolerance improves, but butanol production decreases (Harris et al., 2001). However, even without such targeted constructs, a number of other mutants can produce butanol in concentrations that a few years ago were considered impossible. The current best performer (238 mM butanol) is an *orf5*-negative strain, overexpressing *adhE* (Festel, 2008). Therefore, it may be wise to start with *C. acetobutylicum* to construct a tailor-made production strain. Acetate and butyrate production could be inactivated by mutations in the phosphotransacetylase and phosphotransbutyrylase genes or the corresponding kinase genes respectively. Lactate formation can be inhibited by inactivating the lactate dehydrogenase gene, by knocking out *adc* and thus increasing acetone synthesis, and by mutating the 2-acetolactate synthase gene to stimulate acetoin production. A gene knockout procedure has been developed for *Clostridia* (Dürre, 2007).

Yet another approach is possible. A patent from the Fraunhofer Institute in Stuttgart, Germany, describes butanol formation from butane by means of a specific monooxygenase. Overexpression of the respective gene might result in a different, yet suitable, recombinant production strain.

21.4.2.9 Economic Analysis

Biofuels must meet the following criteria: competitive production costs, no additional infrastructure costs, problem-free, blending with existing fuel types, and similar chemical/physical properties. Fuel cost, selling price, demand, and customer acceptance are important fuel-buying criteria.

Usually, most of the total costs of a bioproduct are for substrate delivery and processing. In addition, substrate cost is a major factor impacting the economics of butanol production.

Biobutanol shows advantages over bioethanol with regard to blending costs because bioethanol has to undergo an expensive drying process before being added to normal fuel.

Theoretical calculations show that the cost of butanol from corn will be 1.20 U.S. dollars per gallon, excluding side hydrogen. This is comparable with the cost of producing ethanol (approximately 1.28USD per gallon). For comparison, the cost of producing butanol from petroleum is approximately 1.35USD per gallon. Currently, the average retail price of gasoline is about 2.80USD in the United States, but with a tendency to rise continually. Ethanol sells for 3.30USD per gallon while butanol currently has a higher price of approximately 3.70USD per gallon. An economic assessment of butanol production from corn using the hyper-butanol-producing strain *C. beijerinckii* BA101 has been conducted by Qureshi et al. (2006). It showed that an improvement of butanol yield by 19.0% (0.42–0.50 g butanol per g glucose) would reduce the butanol price by 14.7%.

Based on the production data of an average industrial ABE plant in 2008 in China, for each ton of ABE solvent produced, the consumptions of materials and energy typically are corn 4–4.5 tons, steam 13–25 tons, water 20–30 tons, and electricity 700–1000 kWh. From 100 kg corn starch, the solvent products consist of 11 kg acetone, 22.5 kg butanol, and 2.7 kg ethanol. The by-products include 36 m^3 CO_2 and 24 m^3 H_2 as well as corn kernels, fuel oil, and waste mash. The CO_2 and H_2 evolved could also be separated and then compressed as valuable by-products of the fermentation process. The distillation residue, that is, waste mash, containing nutrient components such as proteins, organic acids, and vitamins (particularly vitamins B_2 and B_{12}) as well as saccharides, is an excellent animal feed. In industrial fermentations, other by-products such as isopropanol, isopentanol, acetic acid, and butyric acid can be formed; the product components can be slightly different depending on the substrate and strain used. Competitive production costs compared to conventional fuels are imperative to gain market share because any current tax advantages are only temporary.

21.4.3 Lactic Acid

Lactic acid (2-hydroxy propionic acid) has been known since it was first isolated from sour milk by Scheele in 1780. Lactic acid is nontoxic, naturally occurring (Narayanan et al., 2004; Jin et al., 2005), and renewable raw material. Lactic acid is the simplest hydroxy acid with an asymmetric carbon atom (Tuominen, 2003). It exists in two optically active configurations; D- and L-enantiomers (Garlotta, 2001; Naveena et al., 2005) (Figure 21.10).

L-Lactic acid rotates polarized light to the left, whereas D-lactic acid rotates polarized light to the right. The L-isomer is an intermediate of carbohydrate metabolism in humans and other mammals, whereas both the D- and L-enantiomers are produced in bacterial systems. In general, lactic acid can be obtained either by chemical synthesis or by microbial fermentation. Chemical synthesis by hydroxypropanoic acid gives a racemic mixture of L(+) and D(−) lactic acid isomers, whereas microbial fermentation has the advantage of selectively producing either of the single lactic acid enantiomers (Garlotta, 2001; Rao et al., 2008). However, some lactic acid bacteria (LAB) possess lactate racemase(s) that can catalyze the conversion of L(+)-lactate to D(−)-lactate until a racemic mixture is formed (Liu, 2003). DL-Lactic acid is

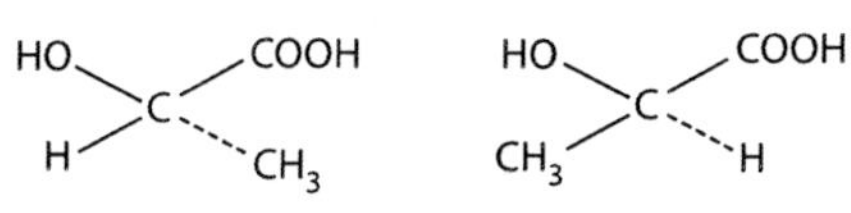

FIGURE 21.10
L- and D-enantiomers of lactic acid.

optically inactive (Tuominen, 2003). The production of lactate isomers L(+)- and D(–)-lactate varies with the genus and species within a genus of LAB. L(+)-Lactate is produced by *Aerococcus, Carnobacterium, Enterococcus, Lactococcus, Tetragenococcus, Streptococcus,* and *Vagococcus.* D(–)-Lactate is produced by *Leuconostoc* and *Oenococcus.* The genera of *Lactobacillus, Pediococcus,* and *Weissella* produce L(+)-, D(–)-, and DL-isomers, where the specific isomer formed is dependent on the species (Liu, 2003).

The organisms that predominantly yield the L-isomer are *Lactobacilli amylophilus, L. bavaricus, L. casei, L. maltaromicus,* and *L. salivarius.* Strains such as *L. delbrueckii, L. jensenii,* and *L. acidophilus* yield the D-isomer or mixtures of both.

21.4.3.1 Applications of Lactic Acid

Lactic acid is an industrially important product with a large market because of its attractive properties. The global marketplace produces about 100,000 tons of lactic acid and lactate (polymers excluded) per year, with an annual growth rate of 15% (Jin et al., 2005). Considering global market trends, agro-based chemicals have a promising future, and the demand for lactic acid is expected to shoot up to approximately 200,000 MT by the end of 2011 (Naveena et al., 2005).

Lactic acid is used primarily in food-related applications in the United States and accounts for approximately 85% of the commercially produced product (Garlotta, 2001). Lactic acid is used as a buffering agent, an acidic flavoring agent, and an acidulant, and as a bacterial inhibitor in many processed foods. Calcium lactate is a good dough conditioner, whereas sodium lactate acts both as a conditioner and as an emulsifier (John et al., 2007). The acid and salts are preferred to other acids in the food industry because they do not dominate other flavors and because they also act as preservatives (Monteagudo et al., 1997; Schmidt and Padukone, 1997).

In the chemical industry, lactic acid is used for metal etching and as a raw material for acrylic acid production (Monteagudo et al., 1997; Schmidt and Padukone, 1997). Technical-grade lactic acid is extensively used as an acidulant in vegetable and leather-tanning industry for deliming hides. Lactic acid is also used as a descaling agent, solvent, cleaning agent, and low-acid-releasing agent, and as a humectant in a variety of technical processes (John et al., 2007).

Lactic acid has long been used in pharmaceutical formulations, mainly in topical ointments, lotions, and parenteral solutions. The natural occurrence of lactic acid in the human body makes it very useful as an active ingredient in cosmetics. The water-retaining capacity of lactic acid makes it suitable for use as a moisturizer in cosmetic formulations. Lactic acid suppresses the formation of tyrosinase, which permits its use for skin lightening and rejuvenation.

Currently, the most promising application of lactic acid is as a biodegradable and biocompatible polylactic acid polymer, an environmentally friendly alternative to nonbiodegradable plastics derived from petrochemicals. The physical properties of polylactic acid (PLA) polymers depend on the isomeric composition of the lactic acid used. Therefore, it has become much more important to be able to synthesize high-grade polymers to produce optically pure lactic acid. PLA is one of the few polymers in which the stereochemical structure can easily be modified by polymerizing a controlled mixture of L and D isomers to yield high-molecular-weight amorphous or semicrystalline polymers. PLA has wide applications not only as a biodegradable plastic, but also as a biomedical material because of its excellent mechanical properties and its adjustable hydrolyzability (Moon et al., 2001). These polymers have potential markets in commodity packaging, fabrication of prosthetic devices, and controlled delivery of drugs in humans (Schmidt and Padukone, 1997; Garlotta, 2001). PLA can also be used to enhance the strength properties of paper sheets made from bleached or unbleached Kraft pulp produced from hot-water-extracted and unextracted sugar maple woodchips (Hasan et al., 2010).

21.4.3.2 Lactic Acid Fermentation

Most of the world's commercially produced lactic acid is made by the bacterial fermentation of carbohydrates (Jin et al., 2005) using homolactic organisms such as various strains of the genus *Lactobacilli* which have been optimized or modified to form lactic acid exclusively. The various types of carbohydrates that can be used in the fermentation depend on the particular strain of *Lactobacillus* used. In general, most of the simple sugars obtained from agricultural by-products can be used. These sugars include (1) glucose, maltose, and dextrose from corn or potato starch; (2) sucrose from cane or beet sugar (Monteagudo et al., 1997), (3) lactose from cheese whey (Moon et al., 2001), and (4) hemicellulosic sugars from wood and agricultural residues (Garde et al., 2002; Saha, 2003; Bustos et al., 2005b).

The use of refined sugars in the production of lactic acid is not economically favorable because pure sugars are expensive, while lactic acid is a cheap product (Hofvendahl and Hahn-Hägerdal, 2000). Lignocellulosic biomass offers a favorable alternative as a feedstock for the biological production of lactic acid because it is readily available, has no competing food value, and is less expensive than either cornstarch or sugar. Hemicellulose is the second most abundant renewable biomass, accounting for 25–35% of lignocellulosic material. Hemicelluloses are heterogeneous polymers made up of pentoses (D-xylose, D-arabinose), hexoses (D-mannose, D-glucose, D-galactose), and sugar acids (Kumar et al., 2008; Liu, 2010). Hemicelluloses in hardwood contain mainly xylans, while in softwood, glucomannans are the most common (Saha, 2003). However, several obstacles preclude efficient use of lignocellulosic materials in lactic acid production.

Treatment of hemicellulosic material with dilute acid at an elevated temperature leads to liberation of fermentable sugars and generation of microbial-growth-inhibitory compounds such as furfural, 5-hydroxymethyl furfural (HMF), and weak acids (Sreenath et al., 2001; Liu et al., 2010). Weak acids inhibit cell proliferation through uncoupling and intracellular anion accumulation, and either furfural or HMF inactivates cell replication by inhibiting specific enzymes related to glycolysis or inhibitor reduction (Palmqvist and Hahn-Hägerdal, 2000). To improve fermentation yields of wood hydrolysates, many methods for detoxification of inhibitory compounds have been investigated. Some of the methods studied include neutralization, overliming, evaporation, the use of ion-exchange resins, and activated charcoal adsorption (Liu et al., 2010). Although several microorganisms can efficiently ferment hexose sugars, conversion of the pentose sugars contained in the hemicellulose has proven more difficult. Fermentation of xylose, the most abundant pentose sugar in hemicellulose, remains a primary challenge for the economical conversion of biomass. The few microorganisms that can grow in both pentose and hexose sugars in lignocellulosic feedstock typically grow slowly and demonstrate marginal yields and productivities (Picataggio et al., 1998). Therefore, much research has focused on isolating naturally occurring bacterial strains which can use xylose to produce lactic acid. Such microorganisms include *Lactobacillus xylosus*, *L. pentosus*, *L. lactis* IO-1, *L. coprophilus*, *L. hilgardii*, and *L. casei* subsp. *Rhamnous* (Iyer et al., 2000). In addition, genetically engineered microbial strains have been developed to use both pentose and hexose sugars to produce lactic acid (Picataggio et al., 1998; Dien et al., 2002).

Along with carbohydrates, organisms require proteins and other complex nutrients such as B-vitamins, amino acids, and nucleotides. These supplies can come from corn steep liquor, yeast extract, cottonseed flour, or soy flour (Garlotta, 2001; Jin et al., 2005). These requirements are very species-dependent, and it is typical to develop the strains around the available nutrients because these can add considerable cost to the process. LAB have complex nutrient requirements because of their limited ability to synthesize B-vitamins and amino acids (Hofvendahl and Hahn-Hägerdal, 2000).

Commercial fermentation of lactic acid is usually conducted in a batch process which takes 3–6 days to complete. Sugar concentrations of 5–10% are used, giving production rates of 2 g of acid per liter of broth per hour. A high final lactic acid concentration is desired to obtain the greatest process efficiency, but high concentrations lead to lactic acid toxicity and growth inhibition. To obtain high efficiencies while maintaining cell growth and production, various methods to neutralize or extract the produced acid have been developed (Garlotta, 2001).

21.4.3.3 Metabolism of Sugars in Lactic Acid Bacteria

Conversion of hexoses and pentoses by *Lactobacillus* is described by the following overall stoichiometric equations (Lampen and Peterjohn, 1951; Iyer et al., 2000):

$$C_6H_{12}O_6 \rightarrow 2C_3H_6O_3 \tag{21.10a}$$

$$3C_5H_{10}O_5 \rightarrow 5C_3H_6O_3 \tag{21.10b}$$

$$C_5H_{10}O_5 \rightarrow C_3H_6O_3 + CH_3COOH \tag{21.11}$$

During this process, carbon dioxide, lactate, and either acetate or ethanol are produced in variable proportions depending on several factors such as aeration, initial sugar concentration, and even the presence of other proton and electron acceptors (Martínez-Anaya et al., 1994; Bustos et al., 2005a). Fermentation of 1 mol of a hexose sugar like glucose yields 2 mol of lactic acid, as shown in Equation 21.10. Equation 21.11 predicts the yield of acetic acid to be in excess of 65% of that of lactic acid. Two mol of adenosine triphosphate (ATP) are formed per mole of xylose consumed, the same amount of energy as produced per mole of glucose. Alternatively, ethanol may be produced in addition to acetate, but 1 mol less of ATP per mole of xylose is formed by this route (Iyer et al., 2000).

Two main sugar metabolism pathways are used to classify LAB genera: homofermentative LAB and heterofermentative LAB.

21.4.3.3.1 Homofermentative Lactic Acid Bacteria

During anaerobic growth and in the presence of excess substrate, obligatory homofermentative LAB catabolizes 1 mol of glucose to yield 2 mol of pyruvate by the EMP pathway. Pyruvate is further metabolized to lactate (Bailey and Ollis, 1986; Bustos et al., 2005a). The homofermentative pathway includes a first phase in which the reactions of glycolysis lead from hexose to pyruvate. The terminal electron acceptor in this pathway is pyruvate, which reduces to lactic acid. The intracellular redox balance is maintained through the oxidation of nicotinamide adenine dinucleotide (NADH) and the concomitant reduction of pyruvate

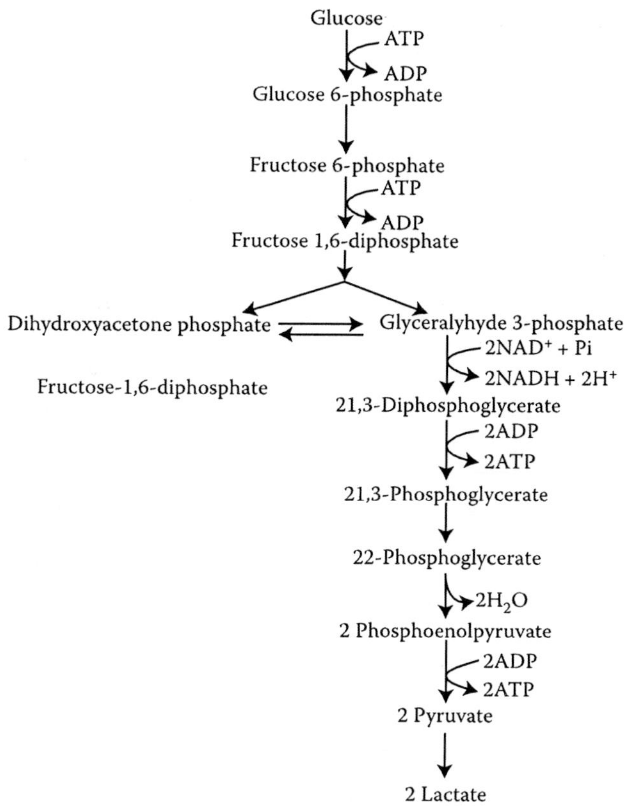

FIGURE 21.11
Biochemical steps during the homolactic fermentation of glucose.

to lactic acid. This process yields 2 mol of adenosine triphosphate (ATP) per glucose consumed (Figure 21.11). Representative homolactic LAB genera include *Lactococcus, Enterococcus, Streptococcus, Pediococcus,* and group I lactobacilli.

Although 100% yield can be predicted for the glucose-to-lactic-acid conversion, a small part of the sugar is used for biomass maintenance and generation. These strains yield high carbon conversions from feedstocks under standard fermentation conditions, such as relatively low to neutral pH, temperatures of approximately 40°C, and low oxygen concentrations.

21.4.3.3.2 Heterofermentative Lactic Acid Bacteria

Heterofermentative LAB use the pentose-phosphate pathway, otherwise referred to as the pentose-phosphoketolase pathway. After being transported into the cell, a glucokinase phosphorylates the glucose into glucose 6-P (glucose 6-phosphate).

Next, two oxidation reactions occur: the first leads to gluconate 6-P, and the second, accompanied by a decarboxylation, forms ribulose 5-P (see Figure 21.12). In each of these reactions, a molecule of NADP+ is reduced. Ribulose 5-P can then be epimerized either to ribose 5-P or to xylulose 5-P. The resulting xylulose-5-phosphate is cleaved into 1 mol glyceraldehyde 3-phosphate (GAP) and 1 mol acetyl phosphate (Figure 21.12). GAP is metabolized into lactic acid following a pathway similar to the homofermentative pathway (Bailey and Ollis, 1986). The acetylphosphate has two possible destinations, depending on environmental conditions.

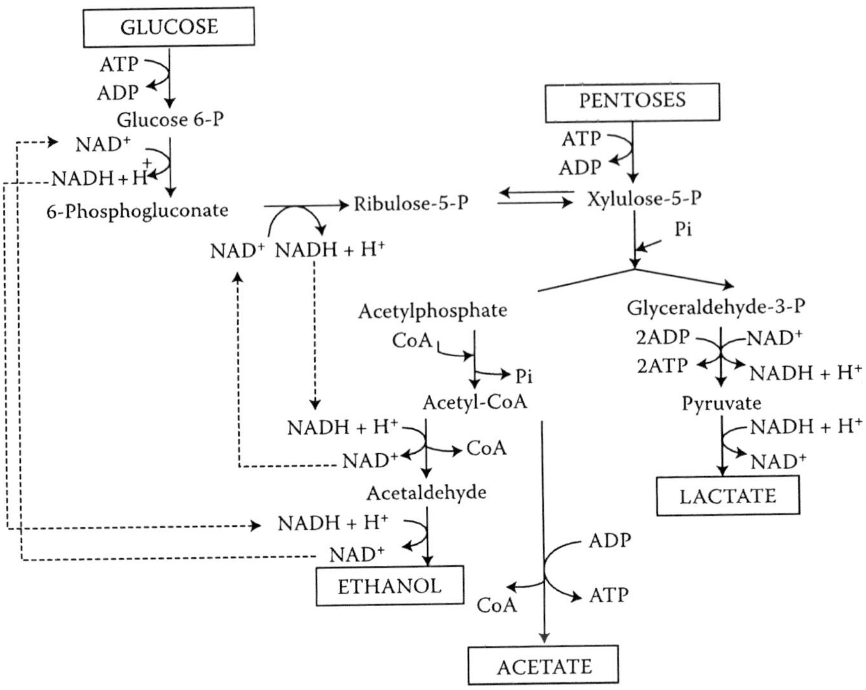

FIGURE 21.12
Biochemical steps of the heterolactic fermentation of glucose and pentoses (xylose and arabinose). (From Bustos, G. et al. 2005a. *Biotechnol. Prog.* 21: 793–798. With permission.)

Under one condition, this molecule successively reduces to ethanol via acetyl-CoA and acetaldehyde intermediates. In this case, the molecules of the coenzyme NADPH formed during the two oxidation reactions of glucose at the beginning of the heterofermentative pathway reoxidize. Under the second condition, acetyl phosphate can produce acetate (acetic acid) through the enzyme acetate kinase. This reaction also yields a molecule of ATP. The final products of this pathway are then lactate and acetate. In theory, the end-products (including ATP) are produced in equimolar quantities from the catabolism of 1 mol of glucose (Bailey and Ollis, 1986).

Heterolactic LAB can be divided into obligatory heterofermentative species, in which both hexoses and pentoses are fermented through the PK-P, and facultative heterofermentative organisms, which degrade hexoses through the EMP-P and pentoses through the PK-P following the overall stoichiometry for both pathways (Bustos et al., 2005a).

21.4.3.3.3 *Pyruvate Catabolism*

Although lactic acid bacteria are named after their ability to form lactic acid, many are able to degrade lactic acid as well, especially if oxygen is available as an electron acceptor (Oude Elferink et al., 2001). In some LAB such as *Lb. curvatus, Lb. sake, Lb. casei,* and *Lb. plantarum,* lactic acid catabolizes under aerobic conditions by lactate oxidase or NAD+-independent LDH to produce pyruvate, which is further catabolized (Liu, 2003):

$$\text{Lactate} + O_2 + \text{lactate oxidase} \rightarrow \text{pyruvate} + \text{pyruvate oxidase} \rightarrow \text{acetate} + CO_2 \quad (21.12)$$

$$\text{Lactate} + O_2 + \text{LDH} \rightarrow \text{pyruvate} + \text{pyruvate oxidase} \rightarrow \text{acetate} + CO_2 \quad (21.13)$$

Under anaerobic conditions, some lactobacilli such as *Lb. brevis, Lb. buchneri,* and *Lb. plantarum* catabolize lactate through NAD+-independent lactate dehydrogenase (LDH). These lactate-degrading lactobacilli use other electron acceptors (e.g., shikimate or oxaloacetate derived from citrate and 3-hydroxypropionaldehyde derived from glycerol) for anaerobic catabolism of lactate as shown below (Oude Elferink et al., 2001; Liu, 2003):

$$\text{Lactate} \rightarrow \text{pyruvate} + \text{acetate} + \text{carbon dioxide} \quad (21.14)$$

or

$$\text{Lactate} \rightarrow \text{pyruvate} + \text{acetate} + \text{formate} \quad (21.15)$$

The products formed from pyruvate degradation depend on the presence of specific enzyme(s) in a particular LAB: pyruvate oxidase, pyruvate formate lyase, or both. So far, the only LAB that have been shown to degrade lactic acid under anoxic conditions without an external electron acceptor are L. *bifermentans*, L. *buchneri*, and L. *parabuchne*. L. *bifermentans* produces hydrogen gas to get rid of its excess of reducing equivalents, whereas L. *buchneri* and L. *parabuchneri* do not produce hydrogen gas during lactic acid degradation, but instead produce large quantities of 1,2-propanediol (Oude Elferink et al., 2001).

21.4.4 Polyhydroxyalkanoate

Polyhydroxyalkanoates (PHAs) are a family of polyesters synthesized by numerous bacteria as an intracellular carbon- and energy-storage compound. PHAs have at least 150 monomers. PHAs can be used in many fields as biodegradable plastics or biomedical materials. Many new potential applications are still emerging. Currently, the major problem inhibiting the development and wide use of PHAs is their high cost compared to petroleum-derived plastics. One of the factors leading to this high cost is the high cost of the substrate. The substrate cost may be as high as 50% of the PHA production cost. A cheaper substrate that is a sufficient carbon resource must be developed to solve this problem. Wood biomass represents a renewable carbon resource and an excellent substrate for the production of PHAs (Mai et al., 2004).

The most common PHA produced from sugars is PHB or PH3B, $H[OCH(CH_3)CH_2C(O)]_nOH$. Each H3B ($C_4H_6O_2$) unit is produced from a sugar carbon source. The stoichiometry for producing one unit of PHB is given by

$$9CH_2O \rightarrow 2\ \text{-}OCH(CH_3)CH_2C(O)\text{-} + CO_2 + 3H_2O \quad (21.16)$$

Therefore, the theoretical yield of PHA from sugar, excluding cell biomass production, is 0.6371 g PHA/g sugar.

21.4.4.1 PHA Production from Glucose

As the most abundant sugar obtainable from nature, the hexose glucose can be used as an efficient carbon substrate for the growth of numerous microorganisms. Much research has been done on the use of glucose to produce PHAs. Sun et al. (2009b) reported feeding a fed-batch culture of *Pseudomonas putida* KT2440 by cofeeding glucose and nonanoic acid to produce PHAs at a yield of 56%. Increasing glucose in the feed resulted in a slightly higher yield, but lower PHA content and productivity. Valentin and Dennis (1997) developed a recombinant *E. coli* strain to produce P(3HB-3HV), a copolymer, using glucose in the culture medium. The yield of PHA reached 50%. Jang and Rogers (1996) used *Alcaligenes* sp. SH-69 to produce PHA with glucose as a sole carbon source. Shang et al. (2004) studied sequential and simultaneous feeding of glucose and valerate to produce P(3HB-co-3HV) in a fed-batch culture of *Ralstonia eutropha*. Results showed that the composition of the copolymer varied with different feeding strategies.

21.4.4.2 PHA Production from Xylose and Minor Sugars

Xylose is the second most abundant sugar obtainable from nature. Although regarded as a waste material by a number of cellulosic biofuel and chemical proponents, xylose is a pentose with great potential. Research has demonstrated that numerous microorganisms can produce PHAs using xylose as the carbon resource. Several recombinant *E. coli* strains have been used to produce PHAs from xylose with yields as high as 0.226 g PHB/g xylose (Lee, 1998). Ramsay et al. (1995) demonstrated the ability of *Burkholderia cepacia* to accumulate PHAs from xylose. Young et al. (1994) reported producing poly-f-hydroxybutyric acid (PHB) from xylose and lactose using *Pseudomonas cepacia*. Approximately 50% PHB (grams of PHB total/grams of biomass total) was produced.

Other sugars that can be produced from woody biomass include mannose, galactose, arabinose, rhamnose, and glucuronic acid. Some of them have been investigated as a cofeeding carbon resource for PHA production, and the influence of growth rate, yield, and polymer composition have been examined (Young et al., 1994).

Microorganism growth rate and PHA production rate both vary with different sugar substrates. Young et al. (1994) studied the production of PHA from glucose, xylose, and lactose by *Pseudononas cepacia*. The maximum specific growth rates of cells grown on xylose and glucose were almost the same, while that on lactose was much lower. The biomass yields were approximately the same on all three sugars. Bertrand et al. (1990) studied the potential of *Pseudomonas pseudoflava* to produce PHAs using glucose, xylose, or arabinose as the sole carbon source. Their results showed that the final PHB content varied from 17% (wt/wt) of the biomass dry weight on arabinose to 22% (wt/wt) of the biomass dry weight on glucose and xylose. Specific growth rates were 0.58 h^{-1} on glucose, 0.13 h^{-1} on xylose, and 0.10 h^{-1} on arabinose, while the specific PHB production rates, based on total biomass, ranged from 0.02 g/(g*h) on arabinose to 0.11 g/(g*h) on glucose. The weight-average molecular weights of PHB were 640,000 on arabinose and 1,100,000 on glucose and xylose.

21.4.5 Simultaneous Saccharification and Fermentation (SSF)

One-step consolidated bioprocessing (conversion) integrates cellulase production, cellulose hydrolysis, and product fermentation into a single step. This decreases the cost of using cellulase, increases volumetric productivity, and reduces capital investment. Therefore, it is a widely accepted method for the low-cost production of biofuels or other value-added biochemicals (Lynd et al., 2002, 2008; van Zyl et al., 2007; Zhang and Lynd, 2008; Tsai et al., 2009; Xu et al., 2009; Wen et al., 2010; Zhang and Zhang, 2010). Earlier approaches to SSF used a potpourri of yeast, enzymes, or both in a single reactor. The activities of these mixtures and the reuse of the enzymes were problematic. The preference commonly found in the current literature is for single microbes that produce the needed enzymes for conversion. Considering the narrow margins between the high cost of biomass (especially for a dedicated biomass crop) and the low value of biocommodities, it is crucial that industrial microbes demonstrate good physiological performance for economically viable production (Zhang and Zhang, 2010). The ideal microorganism for biocommodity production from nonfood biomass should possess multiple capabilities. These characteristics include the abilities to hydrolyze cellulose, process substrate quickly, transport sugar along fast and deregulated pathways, tolerate inhibitors and high-titer products, and produce a high yield of a sole product with a high and efficient metabolic flux. Now, more than ever before, it may be possible to find or modify organisms that will meet these demanding criteria. Consider that there is an industrial imperative demanding these organisms, that well-developed genetic engineering tools are available, and that a tremendous amount of information on the metabolism, biochemistry, and physiology of a huge number of organisms has been generated. Now consider recombinant cellulolytic *B. subtilis* strains as an important platform on which to engineer and produce cellulases, as well as a platform for producing diverse biocommodities from nonfood biomass or waste in a single step. Therefore, the development of these microbes is very active.

It is unclear whether an ideal biocommodity-producing organism will be engineered using a native isolated strain or a recombinant model organism as the starting point (Alper and Stephanopoulos, 2009). *E. coli* has been engineered to produce ethanol, butanol, PHA, and other compounds in high yields and to secrete some active cellulases (Wood and Ingram, 1992; Zhou and Ingram, 2001). However, when considering the industrial need for biocommodity production, recombinant *E. coli* has several weaknesses. These include a relatively high nutrient requirement, a weak ability to secrete active cellulases, a high sensitivity to end-product inhibition, and the need to treat residual cell waste. *B. subtilis* has recently been presented as an alternative recombinant cellulolytic industrial-quality microorganism that can produce the desired biocommodities at high yields and that exhibits a strong physiological robustness and fitness for industrial bioprocesses (Table 21.10).

TABLE 21.10

Comparison of Potential One-step-Bioconversion Microorganisms to Produce Low-value Biocommodities

Key Feature	*S. cerevisiae*	*E. coli*	*C. thermocellum*	*B. subtilis*
C5 sugar utilization	D	A	F	A
Oligosaccharide utilization	F	D	A	A
Protein secretion capacity	C	D	A	A
Ease of genetic modification	A	A	D	A
Medium cost benefits	B	F	B	A
Resistant to product inhibition	A	F	D	A
Resistant to salt/toxic inhibition	C	F	F	A
Value of cell residues	B	F	F	A
Growth rate	B	A	C	A
Anaerobic fermentation	A	A	A	B
Optimum culture temperature (°C)	30	37	60	30–45

Source: Adapted from Zhang, X.Z. and Zhang, Y.H.P. 2010. *Eng. Life Sci.* 10: 398–406.
Note: A: excellent; B: good; C: fair; D: maybe; F: poor.

It can be expected that other *Bacillus* species may have similar properties, such as thermophilic *Geobacillus*, alkaliphilic or acidophilic *Bacillus*, and *Brevibacillus*. They could be developed into one-step consolidated bioprocessing microorganisms or platforms for improved cellulase performance by directing evolution to acquire the desired properties for industrial applications (such as thermostability, alkaliphilic, or acidophilic properties). It is noteworthy that *Geobacillus* spp. can be either thermophilic and aerobic or facultative and anaerobic.

Members of this genus are capable of growth between 40°C and 70°C and can ferment lignocellulose-derived hexose and pentose sugars and oligosaccharides to generate lactate, formate, acetate, and ethanol as products (Nazina et al., 2001). Thermophilic *Geobacillus* spp. have a number of potential advantages for development. These developments may improve the fermentation process. High growth temperatures promote higher rates of feedstock conversion, reduce cooling costs, and lower the risk of contamination in the fermentation. However, development of other *Bacillus* species, for example, *B. subtilis*, may be slow due to the nonavailability of genomic information, inefficient genetic engineering tools, and undefined expression systems. Gene cloning and expression systems have been established for *Geobacillus* (Imanaka et al. 1982; Aiba et al. 1983; Wu and Welker, 1989; Noguchi et al., 1994; Studholme et al., 1999). Metabolic engineering has been successfully conducted in *G. thermoglucosidasius* to produce high yields of ethanol (Cripps et al., 2009). Other potential consolidated bioprocessing microorganisms include *Clostridium* spp., *Escherichia coli*, and the yeast *Saccharomyces cerevisiae*.

One-step bioconversion, or SSF, is a marvelous concept in which conversion of pretreated woody biomass to either biochemicals or fuels is biologically accomplished in a single unit operation. It has the potential to lower capital cost, although at the expense of overall efficiency. Suboptimal operating conditions can be used, at best, for all the steps involved in the single unit operation (Chen et al., 2002, 2008; Chen and Liu, 2007; Chen and Sun, 2007; Chen and Qiu, 2007).

One example of an integrated SSF demonstration project was described by Chen and Qiu (2010). They built a straw-to-ethanol production facility with an annual output of 3000 tons in Dongping, Shandong Province, China. This demonstration facility achieved a reasonably high efficiency and an environmental-friendly production of ethanol by integrating the following technologies: steam explosion coupled with value-added by-product recovery, solid-state fermentation to produce cellulase, solid-state enzymatic hydrolysis coupled with liquid ethanol fermentation and CO_2-gas-stripping ethanol production technology, and organic fertilizer production from fermented residue (Figure 21.13).

As shown in Figure 21.13, ethanol, organic fertilizer (a by-product of the cellulose production and the SSF processes), and xylo-oligosaccharides (a by-product of the steam explosion process) are produced simultaneously. In this process, hemicellulose was extracted after either corncobs or straw were pretreated by steam explosion, and the hemicellulose was used for the production of xylo-oligosaccharide. A small portion of straw residues were used for cellulase production, and a majority of the residues were hydrolyzed by the crude enzyme prepared during the solid-state fermentation process. Then the hydrolysate was converted into ethanol (in the same reactor that produced the hydrolysate), and the unhydrolyzed residues, enriched in lignin, were used for the preparation of organic fertilizers.

The major pieces of equipment used at this demonstration facility include a 5-m^3 steam explosion tank, two 100-m^3 solid-state fermentation reactors, a 110-m^3 solid-state simultaneous enzymatic hydrolysis unit, an ethanol fermentation and ethanol separation plant, and four ethanol distillation towers (Chen and Qiu, 2007). The results indicate that the

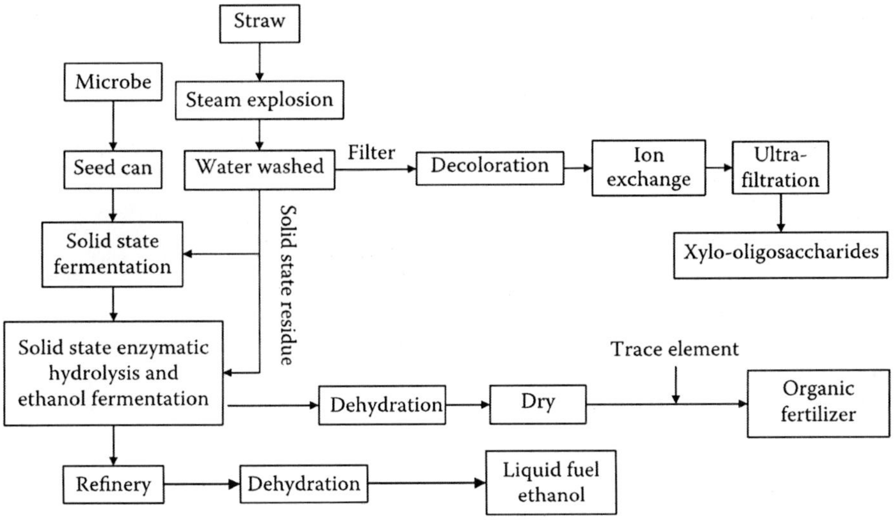

FIGURE 21.13
Schematic of a fuel-ethanol demonstration facility integrating solid-state fermentation with simultaneous saccharification to produce cellulase and ethanol utilizing agricultural residual straw. (Adapted from Chen, H.Z. and Qiu, W.H. 2010. *J. Biotech. Adv.* 28: 556–562.)

average yield of ethanol was more than 0.15 g/g dry straw and that the final ethanol concentration, as desorbed from activated carbon, was higher than 69.8% (Chen and Qiu, 2007). Bioethanol was produced in this plant for a cost of approximately 5900 RMB/ton and at an annual volume of 3000 tons. Simultaneously, the plant produced 200 tons of xylo-oligosaccharides and 3000 tons of organic fertilizers.

SSF can be applied to produce virtually any of the fermentation products as listed in Figure 21.13. Steam explosion, or steam explosion with prior chemical impregnation, is commonly proposed in an SSF process. It is essential to wash the biomass thoroughly with water after steam explosion to achieve sufficient efficiency with SSF. Washing removes furfural, phenolics, and organic acids from the fine powders resulting from the woody biomass.

21.5 Conclusions

The increasing trend toward biomass use is one way of reducing global greenhouse gas emissions. Woody biomass in its various forms presents a viable alternative to traditional food-based crops such as corn and sugarcane for the production of ethanol. However, many technical challenges must be overcome before woody-biomass-based processes can be fully commercialized.

Biochemical conversion holds tremendous promise to relieve the pressures on petroleum fuel consumption. This chapter has reviewed pretreatment methods, hydrolysis, and the variety of products produced by biochemical conversion.

Pretreatment prepares woody biomass by weakening the strong lignocellulosic bonds and networks and by increasing the surface area on which enzymes and other hydrolyzing agents can act to convert the macromolecules fully into five- or six-carbon monomeric sugars. Several pretreatment methods have been presented in this chapter, including biological, physical, and chemical methods, as well methods that combine physical and chemical approaches, such as ionic liquids. Of these, the physicochemical methods, for example, hydrothermal pretreatment and supercritical carbon dioxide, have the best potential to prepare woody biomass for hydrolysis.

The process of hydrolysis disassembles celluloses and hemicelluloses into sugars. Hydrolysis can be divided into two approaches: chemical hydrolysis and enzymatic hydrolysis. The specific method chosen depends on the desired product. Barriers and inhibitors to efficient and cost-effective solutions were discussed.

Ethanol and butanol were compared as valuable end-products, and it was demonstrated that butanol is more energy-dense than ethanol and easier to produce. The properties and production methods of butanol were reviewed in detail. Methods of fermentation and the advantages and disadvantages of a variety of organisms were described.

Finally, lactic acid and PHA were shown to be valuable by-products of fermentation or SSF, and further research in these areas is strongly encouraged.

References

Aiba, S., Imanaka, T., and Koizumi, J. 1983. Transformation of *Bacillus stearothermophilus* with plasmid DNA and its application to molecular cloning. *Ann. NY Acad. Sci.* 413: 57–70.

Alper, H. and Stephanopoulos, G. 2009. Engineering for biofuels: Exploiting innate microbial capacity or importing biosynthetic potential? *Nat. Rev. Microbiol.* 7: 715–723.

Amidon, T.E., Wood, C.D., Shupe, A.M., Wang, Y., Graves, M., and Liu, S.J. 2008. Biorefinery: Conversion of woody biomass to chemicals, energy and materials. *J. Biobased Mater. Bioenergy* 2: 100–120.

Andersch, W., Bahl, H., and Gottschalk, G. 1983. Level of enzymes involved in acetate, butyrate, acetone and butanol formation by *Clostridium acetobutylicum*. *Appl. Microbiol. Biotechnol.* 18: 327–332.

Arai, T., Kosugi, A., Chan, H., Koukiekolo, R., Yukawa, H., Inui, M., and Doi, R.H. 2006. Properties of cellulosomal family 9 cellulases from *Clostridium cellulovorans*. *Appl. Microbiol. Biotechnol.* 71: 654–660.

Auslender, V.L., Ryazantsev, A.A., and Spiridonov, G.A. 2002. The use of electron beams for solution of some ecological problems in pulp and paper industry. *Radiat. Phys. Chem.* 63: 641–645.

Azzam, A.M. 1989. Pre-treatment of cane bagasse with alkaline hydrogen peroxide for enzymatic hydrolysis of cellulose and ethanol fermentation. *J. Environ. Sci. Health B* 24(4): 421–433.

Badger, P.C. 2002. Ethanol from cellulose: A general review. In: Janick, J. and Whipkey, A. (eds.), *Trends in New Crops and New Uses*. ASHS Press, Alexandria, VA, 17–21.

Bailey, J.E. and Ollis, F.D. 1986. *Biochemical Engineering Fundamentals*. 2nd ed. McGraw-Hil, New York.

Balat, M. 2011. Production of bioethanol from lignocellulosic materials via the biochemical pathway: A review. *Energy Convers. Manage.* 52: 858–875.

Balata, M., Balata, H., and Oz, C. 2008. Progress in bioethanol processing. *Prog. Energ. Combust.* 34: 551–573.

Ballongue, J., Amine, J., Petitdemange, H., and Gay, R. 1986. Regulation of acetate kinase and butyrate kinase by acids in *Clostridium acetobutylicum*. *FEMS Microbiol. Lett.* 35: 295–301.

Banerjee, S., Mudliar, S., Sen, R., Giri, B., Satpute, D., Chakrabarti, T., and Pandey, R.A. 2010. Commercializing lignocellulosic bioethanol: Technology bottlenecks and possible remedies. *Biofuels Bioprod. Biorefin.* 4: 77–93.

Barber, V.A. 2007. *Extraction of Hemicelluloses from Sugar Maple Chips after Biotreatment with Ceriporiopsis subvermispora.* PhD dissertation, SUNY-ESF.

Ben-Ghedalia, D. and Miron, J. 1981. The effect of combined chemical and enzyme treatment on the saccharification and *in vitro* digestion rate of wheat straw. *Biotechnol. Bioeng.* 23: 823–831.

Bertrand, J.L., Ramsay, B.A., Ramsay, J.A., and Chavarie, C. 1990. Biosynthesis of poly-*p*-hydroxyalkanoates from pentoses by pseudomonas pseudoflava. *Appl. Environ. Microbiol.* 56: 3133–3138.

Binder, J.B. and Raines, R.T. 2010. Fermentable sugars by chemical hydrolysis of biomass. *Proc. Natl. Acad. Sci.* 107: 4516–4521.

Bjerre, A.B., Olesen, A.B., Fernqvist, T., Ploger, A., and Schmidt, A.S. 1996. Pre-treatment of wheat straw using combined wet oxidation and alkaline hydrolysis resulting in convertible cellulose and hemicellulose. *Biotechnol. Bioeng.* 49: 568–577.

Bobleter, O. 1994. Hydrothermal degradation of polymers derived from plants. *Prog. Polym. Sci.* 19: 797–841.

Bohlmann, G.M. 2006. Process economic considerations for production of ethanol from biomass feedstocks. *Ind. Biotechnol.* 2: 14–20.

Bouchard, J., Methot, M., and Jordan, B. 2006. The effects of ionizing radiation on the cellulose of woodfree paper. *Cellulose* 13: 601–610.

Boynton, Z.L., Bennett, G.N., and Rudolph, F.B. 1996a. Cloning, sequencing, and expression of clustered genes encoding β-hydroxybutyryl-Coenzyme A (CoA) dehydrogenase, crotonase, and butyryl-CoA dehydrogenase from *Clostridium acetobutylicum ATCC 824. Bacteriology* 6: 3015–3024.

Boynton, Z.L., Bennett, G.N., and Rudolph, F.B. 1996b. Cloning, sequencing, and expression of genes encoding phosphotransacetylase and acetate kinase from *Clostridium acetobutylicum ATCC 824. Appl. Environ. Microbiol.* 8: 2758–2766.

Braconnot, H. 1819. Memoir on the conversion of wood particles in rubber, in sugar, and in a special natural acid, by means of sulfuric acid; conversion of the same woody substance in ulmin by potash (Translated from French). *Ann. Chim. Phys.* 12: 172–195.

Brasch, D.J. and Free, K.W. 1965. Prehydrolysis-Kraft pulping of *Pinus radiata* grown in New Zealand. *Tappi* 48: 245–248.

Brownell, H.H., Yu, E.K.C., and Saddler, J.N. 1986. Steam explosion pre-treatment of wood: Effect of chip size, acid, moisture content, and pressure drop. *Biotechnol. Bioeng.* 28: 792–801.

Bustos, G., Moldes, A.B., Cruz, J.M., and Domínguez, J.M. 2005a. Influence of the metabolism pathway on lactic acid production from hemicellulosic trimming vine shoots hydrolyzates using *Lactobacillus pentosus. Biotechnol. Prog.* 21: 793–798.

Bustos, G., Moldes, A.B., Cruz, J.M., and Domínguez, J.M. 2005b. Production of lactic acid from vine-trimming wastes and viticulture lees using a simultaneous saccharification fermentation method. *J. Sci. Food Agric.* 85: 466–472.

Cadoche, L. and Lopez, G.D. 1989. Assessment of size reduction as a preliminary step in the production of ethanol from lignocellulosic wastes. *Biol. Wastes* 30: 153–157.

Casey, G.P. and Ingledew, W.M. 1986. Ethanol tolerance in yeasts. *Crit. Rev. Microbiol.* 13: 219–280.

Chandel, A.K., Es, C., Rudravaram, R., Narasu, M.L., Rao, L.V., and Ravindra, P. 2007. Economics and environmental impact of bioethanol production technologies: An appraisal. *Biotechnol. Molec. Biol. Rev.* 2:14–32.

Charbannes, M., Ruel, K., Yoshinaga, A., Chabbert, B., Jauneau, A., Joseleau, J.-P., and Boudet, A.-M. 2001. *In-situ* analysis of lignins in transgenic tobacco reveals a differential impact of individual transformations on the spatial patterns of lignin deposition at the cellular and subcellular levels. *Plant J.* 28: 271–282.

Chavez, R., Bull, P., and Eyzaguirre, J. 2006. The xylanolytic enzyme system from the genus *Penicillium. J. Biotechnol.* 123: 413–433.

Chen, H.Z. and Liu, L.Y. 2007. Unpolluted fractionation of wheat straw by steam explosion and ethanol extraction. *Bioresour. Technol.* 98: 666–676.

Chen, H.Z. and Qiu, W.H. 2007. The crucial problems and recent advance on producing fuel alcohol by fermentation of straw. *Prog. Chem.* 19: 1116–1121.

Chen, H.Z. and Qiu, W.H. 2010. Key technologies for bio-ethanol production from lignocellulose. *J. Biotech. Adv.* 28: 556–562.
Chen, H.Z. and Sun, F.B. 2007. Novel bioconversion of wheat straw to bio-organic fertilizer in a solid-state bioreactor. *Bioproc. Biosyst. Eng.* 30: 99–105.
Chen, H.Z., Han, Y.J., and Xu, J. 2008. Simultaneous saccharification and fermentation of steam exploded wheat straw pretreated with alkaline peroxide. *Process Biochem.* 43: 1462–1466.
Chen, H.Z., Xu, F.J., Tian, Z.H., and Li, Z.H. 2002. A novel industrial-level reactor with two dynamic changes of air for solid-state fermentation. *J. Biosci. Bioeng.* 93: 211–214.
Chosdu, R., Hilmy, N., Erizal, Erlinda, T.B., and Abbas, B. 1993. Radiation and chemical pretreatment of cellulosic waste. *Radiat. Phys. Chem.* 42: 695–698.
Chum, H.L., Johnson, D.K., and Black, S. 1988. Organosolv pre-treatment for enzymatic hydrolysis of poplars: 1. Enzyme hydrolysis of cellulosic residues. *Biotechnol. Bioeng.* 31: 643–649.
Claassen, P.A.M., Lopez Contreras, A.M., Sijtsma, L., Weusthuis, R.A., Van Lier, J.B., Van Niel, E.W.J., Stams, A.J.M., and De Vries, S.S. 1999. Utilisation of biomass for the supply of energy carriers. *Appl. Microbiol. Biotechnol.* 52: 741–755.
Conner, A.H. and Lorenz, L.F. 1986. Kinetic modeling of hardwood prehydrolysis. Part III: Water and dilute acetic acid prehydrolysis of southern red oak. *Wood Fiber Sci.* 18: 248–263.
Converse, A.O., Matsuno, R., Tanaka, M., and Taniguchi, M. 1988. A model for enzyme adsorption and hydrolysis of microcrystalline cellulose with slow deactivation of the adsorbed enzyme. *Biotechnol. Bioeng.* 32: 38–45.
Coughlan, M.P. and Ljungdahl, L.G. 1988. Comparative biochemistry of fungal and bacterial cellulolytic enzyme system. In: Aubert, J.-P., Beguin, P., and Millet, J. (eds.), *Biochemistry and Genetics of Cellulose Degradation*. Academic Press, London, 11–30.
Cripps, R.E., Eley, K., Leak, D.J., Rudd, B., Taylor, M., Todd, M., Boakes, S., Martin, S., and Atkinson, T. 2009. Metabolic engineering of *Geobacillus thermoglucosidasius* for high yield ethanol production. *Metab. Eng.* 11: 398–408.
D'Andola, G., Szarvas, L., Massonne, K., and Stegmann, V. 2008. *Ionic Liquids for Solubilizing Polymers.* WO Pat 2008/043837.
Dale, B.E. and Moreira, M.J. 1982. A freeze-explosion technique for increasing cellulose hydrolysis. *Biotechnol. Bioeng. Symp., Biotechnology in Energy Production and Conservation* 12: 3143.
Dale, B.E., Henk, L.L., and Shiang, M., 1984. Fermentation of lignocellulosic materials treated by ammonia freeze-explosion. *Dev. Ind. Microbiol.* 26, 223–233.
Demirbas, A. 2005. Bioethanol from cellulosic materials: A renewable motor fuel from biomass. *Energy Sources* 21: 327–337.
Demirbas, M.F. 2009. Biorefineries for biofuel upgrading: A critical review. *Appl. Energ.* 86: S151–S161.
Detroy, R.W. and Julian, G.S. 1982. Biomass conversion: Fermentation chemicals and fuels. *Crit. Rev. Microbiol.* 10: 203–228.
Dien, B.S., Cotta, M.A., and Jeffries, T.W. 2003. Bacteria engineered for fuel ethanol production: Current status. *Appl. Microbiol. Biotechnol.* 63: 258–266.
Dien, B.S., Nichols, N.N., and Bothast, R.J. 2002. Fermentation of sugar mixtures using *Escherichia coli* catabolite repression mutants engineered for production of L-lactic acid. *J. Ind. Microbiol. Biotechnol.* 29: 221–227.
Dubey, K.A., Pujari, P.K., Rammani, S.P., Kadam, R.M., and Sabharwal, S. 2004. Microstructural studies of electron beam-irradiated cellulose pulp. *Radiat. Phys. Chem.* 69: 395–400.
Duff, S.J.B. and Murray, W.D. 1996. Bioconversion of forest products industry waste cellulosics to fuel ethanol: A review. *Bioresour. Technol.* 55: 1–33.
Dürre, P. 2007. Biobutanol: An attractive biofuel. *Biotechnol. J.* 2: 1525–1534.
Ennis, B.M., Gutierrez, N.A., and Maddox, I.S. 1986. The acetone-butanol-ethanol fermentation: A current assessment. *Process Biochem.* 21: 131–147.
Esteghlalian, A., Hashimoto, A.G., Fenske, J.J., Penner, and M.H. 1997. Modeling and optimization of the dilute-sulfuric-acid pre-treatment of corn stover, poplar and switchgrass. *Bioresour. Technol.* 59: 129–136.

Evans, P.J. and Wang, H.Y. 1988. Enhancement of butanol formation by *Clostridium acetobutylicum* in the presence of decanol-oleyl alcohol mixed extractants. *Appl. Environ. Microbiol.* 54(7): 1662–1667.

Ezeji, T.C., Qureshi, N., and Blaschek, H.P. 2004. Butanol fermentation research: Upstream and downstream manipulations. *Chem. Rec.* 4: 305–314.

Ezeji, T.C., Qureshi, N., and Blaschek, H.P. 2005. Continuous butanol fermentation and feed starch retrogradation: Butanol sustainability using *Clostridium beijrinckii BA* 101. *J. Biotechnol.* 115: 179–187.

Ezeji, T.C., Qureshi, N., and Blaschek, H.P. 2007. Bioproduction of butanol from biomass: From genes to bioreactors. *Curr. Opin. Biotechnol.* 18: 220–227.

Faith, W. 1945. Development of the Scholler process in the United States. *Ind. Eng. Chem.* 37: 11.

Fan, L.T., Gharpuray, M.M., and Lee, Y.H. 1981. Evaluation of pre-treatments for enzymatic conversion of agricultural residues. *Biotechnol. Bioeng. Symp.* 11: 29–45.

Fengel, D. and Wegener, G. 1984. *Wood: Chemistry, Ultrastructure, Reactions.* Walter de Gruyter, Berlin.

Festel, G.W. 2008. Biofuels—Economic aspects. *Chem. Eng. Technol.* 31: 715–720.

Fort, D.A., Remsing, R.C., Swatloski, R.P., Moyna, P., Moyna, G., and Rogers, R.D. 2007, Can ionic liquids dissolve wood? Processing and analysis of lignocellulosic materials with 1-n-butyl-3-methyl-imidazolium chloride. *Green Chem.* 9(1): 63–69.

Galbe, M. and Zacchi, G. 2002. A review of the production of ethanol from softwood. *Appl. Microbiol. Biotechnol.* 59: 618–628.

Gámez, S., González-Cabriales, J.J., Ramírez, J.A., Garrote, G., and Vázquez, M. 2006. Study of the hydrolysis of sugar cane bagasse using phosphoric acid. *J. Food Eng.* 74: 78–88.

Garde, A., Jonsson, G., Schmidt, A.S., and Ahring, B.K. 2002. Lactic acid production from wheat straw hemicellulose hydrolysate by *Lactobacillus pentosus* and *Lactobacillus brevis. Bioresour. Technol.* 81: 217–223.

Garlotta, D. 2001. A literature review of poly(lactic acid). *J. Polym. Environ.* 9: 63–84.

Girio, F.M., Fonseca, C., Carvalheiro, F., Duarte, L.C., Marques, S., and Bogel-Lukasik, R. 2010. Hemicelluloses for fuel ethanol: A review. *Bioresour. Technol.* 101: 4775–4800.

Gossett, J.M., Stuckey, D.C., Owen, W.F., and McCarty, P.L. 1982. Heat treatment and anaerobic digestion of refuse. *J. Environ. Eng. Div.* 108: 437–454.

Graf, A. and Koehler, T. 2000. *Oregon Cellulose-Ethanol Study: An Evaluation of the Potential for Ethanol Production in Oregon using Cellulose-Based Feedstocks.* Oregon Department of Energy. Available at: http://secureinfo.oregon.gov/ENERGY/RENEW/Biomass/docs/OCES/OCES.PDF?ga=t

Gray, K.A., Zhao, L.S., and Emptage, M. 2006. Bioethanol. *Curr. Opin. Chem. Biol.* 10: 141–146.

Gregg, D. and Saddler, J.N. 1996. A techno-economic assessment of the pre-treatment and fractionation steps of a biomass-to-ethanol process. *Appl. Biochem. Biotechnol.* 57–58: 711–727.

Grous, W.R., Converse, A.O., and Grethlein, H.E. 1986. Effect of steam explosion pre-treatment on pore size and enzymatic hydrolysis of poplar. *Enzyme Microb. Technol.* 8: 274–280.

Hahn-Hägerdal, B., Galbe, M., Gorwa-Grauslund, M.F., Liden, G., and Zacchi, G. 2006. Bioethanol: The fuel of tomorrow from the residues of today. *Trends Biotechnol.* 24: 549–556.

Hahn-Hägerdal, B., Karhumaa, K., Fonseca, C., Spencer-Martins, I., and Gorwa-Grauslund, M.F. 2007. Towards industrial pentose-fermenting yeast strains. *Appl. Microbiol. Biotechnol.* 74: 937–953.

Hamelinck, C.N., van Hooijdonk, G., and Faaij, A.P.C. 2003. *Prospects for Ethanol from Lignocellulosic Biomass: Techno-economic Performance as Development Progresses.* Report NWS-E-2003-55, Utrecht University, Copernicus Institute for Science, Technology, and Society, Utrecht, The Netherlands, 35.

Hamelinck, C.N., van Hooijdonk, G., and Faaij, A.P.C. 2005. Ethanol from lignocellulosic biomass: Techno-economic performance in short-, middle- and long-term. *Biomass Bioenergy* 28: 384–410.

Harris, L.M., Blank, L., Desai, R.P., Welker, N.E., and Papoutsakis, E.T. 2001. Fermentation characterization and flux analysis of recombinant strains of *Clostridium acetobutylicum* with an inactivated *solR* gene. *J. Ind. Microbiol. Biotechnol.* 27: 322–328.

Hartmanis, M.G.N. 1987. Butyrate kinase from *Clostridium acetobutylicum. J. Biol. Chem.* 26: 617–621.

Hartmanis, M.G.N., Klason, T., and Gatenbeck, S. 1986. Uptake and activation of acetate and butyrate in *Clostridium acetobutylicum. Appl. Microbiol. Biot.* 24: 159–167.

Harun, R., Jason, W.S.Y., Cherrington, T., and Danquah, M.K. 2010. Biomass as a cellulosic fermentation feedstock for, bioethanol production. *Ren. Sust. Energ. Rev.*, doi:10.1016/j.rser.2010.07.071.

Hasan, A., Bujanovic, B., and Amidon, T. 2010. Strength properties of kraft pulp produced from hot-water extracted woodchips within the biorefinery. *J. Biobased Mater. Bioeng.* 4: 46–52.

Hendriks, A.T.W.M. and Zeeman, G. 2009. Pre-treatments to enhance the digestibility of lignocellulosic biomass. *Bioresour. Technol.* 100: 10–18.

Himmel, M.E., Ding, S.-Y., Johnson, D.K., Adney, W.S., Nimlos, M.R., Brady, J.W., and Foust, T.D. 2007. Biomass recalcitrance: Engineering plants and enzymes for biofuels production. *Science* 315: 804–807.

Hinman, N.D., Schell, D.J., Riley, C.J., Bergeron, P.W., and Walter, P.J. 1992. Preliminary estimate of the cost of ethanol production for SSF technology. *Appl. Biochem. Biotechnol.* 34: 639–649.

Hofvendahl, K. and Hahn-Hägerdal, B. 2000. Factors affecting fermentative lactic acid production from renewable resources. *Enzyme Microb. Technol.* 26: 87–107.

Holtzapple, M.T., Davison, R.R., and Stuart, E.D. 1992. *Biomass Refining Process.* U.S. Patent 5, 171, 592.

Holtzapple, M.T., Humphrey, A.E., and Taylor, J.D. 1989. Energy requirements for the size reduction of poplar and aspen wood. *Biotechnol. Bioeng.* 33: 207–210.

Holtzapple, M.T., Jun, J.H., Ashok, G., Patibandla, S.L., and Dale, B.E. 1990. Ammonia fiber explosion (AFEX) pretreatment of lignocellulosic wastes. *Proceedings, American Institute of Chemical Engineers National Meeting*, Chicago, IL.

Holtzapple, M.T., Jun, J.H., Ashok, G., Patibandla, S.L., and Dale, B.E. 1991. The ammonia freeze explosion (AFEX) process: A practical lignocellulose pre-treatment. *Appl. Biochem. Biotechnol.* 28: 59–74.

Hon, D.N.S. and Shiraishi, N. 2001. *Wood and Cellulosic Chemistry*, 2nd ed. Marcel Dekker, New York.

Hunter, C.A. and Sanders, J.K.M. 1990. The nature of PI–PI interactions. *J. Am. Chem. Soc.* 112: 5525–5534.

Imanaka, T., Fujii, M., Aramori, I., and Aiba, S. 1982. Transformation of *Bacillus stearothermophilus* with plasmid DNA and characterization of shuttle vector plasmids between *Bacillus stearothermophilus* and *Bacillus subtilis*. *J. Bacteriol.* 149: 824–830.

Iyer, P.V., Thomas, S., and Lee, Y.Y. 2000. High-yield fermentation of pentoses into lactic acid. *Appl. Biochem. Biotechnol. Enzym. Eng. Biotechnol.* 84–86: 665–677.

Iyer, P.V., Wu, Z.-W., Kim, S.B., and Lee, Y.Y. 1996. Ammonia recycled percolation process for pretreatment of herbaceous biomass. *Appl. Biochem. Biotechnol.* 57–58: 121–132.

Izák, P., Schwarz, K., Ruth, W., Bahl, H., and Kragl, U. 2008. Increased productivity of *Clostridium acetobutylicum* fermentation of acetone, butanol, and ethanol by pervaporation through supported ionic liquid membrane. *Appl. Microbiol. Biotechnol.* 78: 597–602.

Jang, J.H. and Rogers, P.L. 1996. Effect of levulinic acid on cell growth and poly-β-hydroxyalkanoate production by *Alcaligenes sp. SH-69*. *Biotechnol. Lett.* 18: 219–224.

Jeffries, T.W. and Jin, Y.S. 2000. Ethanol and thermotolerance in the bioconversion of xylose by yeasts. *Adv. Appl. Microbiol.* 47: 221–268.

Jin, B., Yin, P., Ma, Y., and Zhao, L. 2005. Production of lactic acid and fungal biomass by *Rhizopus* fungi from food processing waste streams. *J. Ind. Microbiol. Biotechnol.* 32: 678–686.

John, R.P., Nampoothiri, K.M., and Pandey, A. 2007. Fermentative production of lactic acid from biomass: An overview on process developments and future perspectives. *Appl. Microbiol. Biotechnol.* 74: 524–534.

Jones, D.T. and Woods, D.R. 1986. Acetone-butanol fermentation revisited. *Microbiol. Rev.* 50: 484–524.

Katahira, S., Mizuike, A., Fukuda, H., and Kondo, A. 2006. Ethanol fermentation from lignocellulosic hydrolysate by a recombinant xylose- and cellooligosaccharide-assimilating yeast strain. *Appl. Microbiol. Biotechnol.* 72: 1136–1143.

Keis, S., Bennett, C.F., Ward, V.K., and Jones, D.T. 1995. Taxonomy and phylogeny of industrial solvent-producing. *Clostridia. Int. J. Syst. Bacteriol.* 45: 693–705.

Kilpelainen, I., Xie, H., King, A., Granstrom, M., Heikkinen, S., and Argyropoulos, D.S. 2007. Dissolution of wood in ionic liquids. *J. Agr. Food Chem.* 55: 9142–9148.

Kim, K.H. and Hong, J. 2001. Supercritical CO_2 pre-treatment of lignocellulose enhances enzymatic cellulose hydrolysis. *Bioresour. Technol.* 77: 139–144.

Kirschner, M. 2006. *n*-Butanol. *Chemical Market Reporter*, January 30–February 5, ABI/INFORM Global; 42.

Klass, D.L. 1998. *Biomass for Renewable Energy, Fuels, and Chemicals.* Academic Press, New York.

Kumar, R., Singh, S., and Singh, O.V. 2008. Bioconversion of lignocellulosic biomass: Biochemical and molecular perspectives. *J. Ind. Microbiol. Biotechnol.* 35: 377–391.

Kumar, S., Singh, S.P., Mishra, I.M., and Adhikari, D.K. 2009. Recent advances in production of bioethanol from lignocellulosic biomass. *Chem Eng. Technol.* 32: 517–526.

Kuznetsov, B.N., Kuznetsova, S.A., Danilov, V.G., Kozlov, I.A., Taraban'ko, V.E., Ivanchenko, N.M., and Alexandrova, N.B. 2002. New catalytic processes for a sustainable chemistry of cellulose production from wood biomass. *Catal. Today* 75: 211–217.

Lampen, J.O. and Peterjohn, H.R. 1951. Studies on the specificity of the fermentation of pentoses by *Lactobacillus pentosus. J. Bacteriol.* 62: 281–292.

Larson, K.A. and King, M.L. 1986. Evaluation of supercritical fluid extraction in the pharmaceutical industry. *Biotechnol. Prog.* 2: 73–82.

Laser, M., Schulman, D., Allen, S.G., Lichwa, J., Antal, M.J. Jr, and Lynd, L.R. 2002. A comparison of liquid hot water and steam pre-treatments of sugar cane bagasse for bioconversion to ethanol. *Bioresour. Technol.* 81: 33–44.

Lee, J. 1997. Biological conversion of lignocellulosic biomass to ethanol. *J. Biotechnol.* 56: 1–24.

Lee, S.Y. 1998. Poly(3-hydroxybutyrate) production from xylose by recombinant *Escherichia coli. Bioproc. Biosyst. Eng.* 18: 397–399.

Lee, S.H., Doherty, T.V., and Linhardt, J.S. 2009. Ionic liquid-mediated selective extraction of lignin from wood leading to enhanced enzymatic cellulose hydrolysis. *Biotechnol. Bioeng.* 102: 1368–1376.

Lee, S.Y., Park, J.H., Jang, S.H., Nielsen, L.K., Kim, J., and Jung, K.S. 2008. Fermentative butanol production by *Clostridia. Biotechnol. Bioeng.* 101: 209–228.

Liebert, T. 2010. Cellulose solvents—remarkable history, bright future for analysis, shaping and chemical modification. In: Liebert, T.F., Heinze, T.J., and Edgar, K.J. (eds.), *ACS Symposium Series 1033*. American Chemical Society, Washington, DC, 3–54.

Liu, S.Q. 2003. Practical implications of lactate and pyruvate metabolism by lactic acid bacteria in food and beverage fermentations. *Int. J. Food Microbiol.* 83: 115–131.

Liu, S. 2008. A kinetic model of autocatalytic reactions in woody biomass hydrolysis. *J. Biobased Mater. Bioenergy* 2: 135–147.

Liu, S. 2010. Woody biomass: Niche position as a source of sustainable renewable chemicals and energy and kinetics of hot-water extraction/hydrolysis. *Biotechnol. Adv.* 28: 563–582.

Liu, C. and Wyman, C.E. 2003. The effect of flow rate of compressed hot water on xylan, lignin and total mass removal from corn stover. *Ind. Eng. Chem. Res.* 42: 5409–5416.

Liu, T., Lin, L., Sun, Z., Hu, R., and Liu, S. 2010. Bioethanol fermentation by recombinant *E. coli* FBR5 and its robust mutant FBHW using hot-water wood extract hydrolyzate as substrate. *Biotechnol. Adv.* 28: 602–608.

Liyanage, H., Young, M., and Kashket, E.R. 2000. Butanol tolerance of *Clostridium beijerinckii* NCIMB 8052 associated with down-regulation of *gldA* by antisense RNA. *J. Mol. Microbiol. Biotechnol.* 2: 87–93.

Lokhorst, A. and Wildenborg, I. 2005. Introduction to CO_2 geological storage: Classification of storage options. *Oil Gas Sci. Technol.* 60: 513–515.

Lynd, L.R., Elander, R.T., and Wyman, C.E. 1996. Likely features and costs of mature biomass ethanol technology. *Appl. Biochem. Biotechnol.* 57: 741–761.

Lynd, L.R., Laser, M.S., Bransby, D., Dale, B.E., Davison, B., Hamilton, R., Himmel, M., et al. 2008. How biotech can transform biofuels. *Nat. Biotechnol.* 26: 169–172.

Lynd, L.R., Weimer, P.J., van Zyl, W.H., and Pretorius, I.S. 2002. Microbial cellulose utilization: Fundamentals and biotechnology. *Microbiol. Mol. Biol. Rev.* 66: 506–577.

Mabee, W.E., Saddler, J.N., Nielsen, C., Nielsen, L.H., and Jensen, E.S. 2006. Renewable-based fuels for transport. In: *Renewable Energy for Power and Transport*. Risø Energy Report 5, 47–50.

Madihah, M.S., Ariff, A.B., Khalil, M.S., Suraini, A.A., and Karim, M.I.A. 2001. Anaerobic fermentation of gelatinized sago starch-derived sugars to acetone-I-butanol-ethanol solvent by *Clostridium acetobutylicum. Folia Microbiol.* 46: 197–204.

Mai, C., Kues, U., and Militz, H. 2004. Biotechnology in the wood industry. *Appl. Microbiol. Biot.* 63: 477–494.

Maki-Arvela, P., Anugwom, I., Virtanen, P., Sjoholm, R., and Mikkola, J.P. 2010. Dissolution of lignocellulosic material and its constituents using ionic liquids: A review. *Ind. Crops Prod.* 32: 175–201.

Margeot, A., Hahn-Hägerdal, B., Edlund, M., Slade, R., and Monot F. 2009. New improvements for lignocellulosic ethanol. *Curr. Opin. Biotechnol.* 20: 372–380.

Martínez-Anaya, M.A., Llin, M.L., Macías, M.P., and Collar, C. 1994. Regulation of acetic acid production by homo-and heterofermentative lactobacilli in whole-wheat sour-doughs. *Zeitschrift für Lebensmittel-Untersuchung und -Forschung* 199: 186–190.

Mason, W.H. 1928. *Apparatus for the Process of Explosion Defibration of Lignocellulose Material.* U.S. Patent 1,655,618.

Matsumura, M., Kataoka, H., Sueki, M., and Araki, K. 1988. Energy saving effect of pervaporation using oleyl alcohol liquid membrane in butanol purification. *Bioproc. Biosyst. Eng.* 3: 93–100.

McMillan, J.D. 1994. Pretreatment of lignocellulosic biomass. In: Himmel, M.E., Baker, J.O., and Overend, R.P. (eds.), *Enzymatic Conversion of Biomass for Fuels Production.* American Chemical Society, Washington, DC, 292–324.

McMillan, J.D., Newman, M., Templeton, D., and Mohagheghi, A. 1999. Simultaneous saccharification and cofermentation of dilute-acid pretreated yellow poplar hardwood to ethanol using xylose-fermenting *Zymomonas mobilis. Appl. Biochem. Biotechnol.* 79: 649–665.

Mes-Hartree, M., Dale, B.E., and Craig, W.K. 1988. Comparison of steam and ammonia pretreatment for enzymatic hydrolysis of cellulose. *Appl. Microbiol. Biotechnol.* 29: 462–468.

Mielenz, J.R. 2001. Ethanol production from biomass: Technology and commercialization status. *Curr. Opin. Microbiol.* 4: 324–329.

Millet, M.A., Baker, A.J., and Scatter, L.D. 1976. Physical and chemical pre-treatment for enhancing cellulose saccharification. *Biotech. Bioeng.* 6: 125–153.

Mittal, A., Chatterjee, S.G., Scott, G.M., and Amidon, T.E. 2009. Modeling xylan solubilization during autohydrolysis of sugar maple and aspen wood chips: Reaction kinetics and mass transfer. *Chem. Eng. Sci.* 64:3031–3041.

Monteagudo, J.M., Rodríguez, L., Rincón, J., and Fuertes, J. 1997. Kinetics of lactic acid fermentation by *Lactobacillus delbrueckii* grown on beet molasses. *J. Chem. Technol. Biotechnol.* 68: 271–276.

Moon, S., Ikuo, T., Masatoshi, M., Yoshiharu, K., and Chan-Woo, L. 2001. Synthesis and properties of high-molecular-weight poly(L-lactic acid) by melt/solid polycondensation under different reaction conditions. *High Perform. Polym.* 13: S189–S196.

Morjanoff, P.J. and Gray, P.P. 1987. Optimization of steam explosion as a method for increasing susceptibility of sugarcane bagasse to enzymatic saccharification. *Biotechnol. Bioeng.* 29: 733–741.

Mosier, N., Wyman, C., Dale, B., Elander, R., Lee, Y.Y., Holtzapple, M., and Ladisch, M. 2005. Features of promising technologies for pre-treatment of lignocellulosic biomass. *Bioresour. Technol.* 96: 673–686.

Mussatto, S.I., Dragone, G., Guimarães, P.M.R., João Silva, P.A., Carneiro, L.M., Roberto, I.C., Vicente, A., Domingues, L., and Teixeira, J.A. 2010. Technological trends, global market, and challenges of bio-ethanol production. *Biotechnol. Adv.* 28: 817–830.

Narayanan, N., Roychoudhury, P.K., and Srivastava, A. 2004. L(+) lactic acid fermentation and its product polymerization. *Electron. J. Biotechnol.* 7: 167–179.

Naveena, B.J., Altaf, M., Bhadriah, K., and Reddy, G. 2005. Selection of medium components by Plackett-Burman design for production of L(+) lactic acid by *Lactobacillus amylophilus GV6* in SSF using wheat bran. *Bioresour. Technol.* 96: 485–490.

Nazina, T.N., Tourova, T.P., Poltaraus, A.B., Novikova, E.V., Grigoryan, A.A., Ivanova, A.E., Lysenko, A.M., et al. 2001. Taxonomic study of aerobic thermophilic bacilli: Descriptions of *Geobacillus subterraneus gen. nov., sp. nov.*, and *Geobacillus uzenensis sp. nov.* from petroleum reservoirs and transfer of *Bacillus stearothermophilus, Bacillus thermocatenulatus, Bacillus thermoleovorans, Bacillus kaustophilus, Bacillus thermodenitrificans* to *Geobacillus* as the new combinations *G. stearothermophilus, G. thermocatenulatus, G. thermoleovorans, G. kaustophilus, G. thermoglucosidasius*, and *G. thermodenitrificans. Int. J. Syst. Evol. Microbiol.* 51: 433–446.

Nguyen, Q.A. 1989. *Continuous Leaching of Lignin or Hemicellulose and Lignin from Steam Pretreated Lignocellulosic Particulate Material.* Canada Patent 1,322,366.

Noguchi, S., Yamazaki, T., Yaginuma, A., Sakamoto, J., and Sone, N. 1994. Over-expression of membrane-bound cytochrome *c-551* from the thermophilic bacillus *PS3* in *Bacillus stearothermophilus K1041. Biochim. Biophys. Acta* 1188: 302–310.

Nutt, A. 2006. *Hydrolytic and Oxidative Mechanisms Involved in Cellulose Degradation.* PhD Dissertation, Acta Universitatis Upsaliensis.

Okano, K., Kitagaw, M., Sasaki, Y., and Watanabe, T. 2005. Conversion of Japanese red cedar (*Cryptomeria japonica*) into a feed for ruminants by white-rot basidiomycetes. *Animal Feed Sci. Technol.* 120: 235–243.

Olofsson, K., Bertilsson, M., and Liden, G. 2008. A short review on SSF—An interesting process option for ethanol production from lignocellulosic feedstocks. *Biotechnol. Biofuels* 1: 7.

Osman, Y.A. and Ingram, L.O. 1985. Mechanism of ethanol inhibition of fermentation in *Zymomonas mobilis CP4. J. Bacteriol.* 164: 173–180.

Oude Elferink, S.J.W.H., Krooneman, E.J., Gottschal, J.C., Spoelstra, S.F., Faber, F., and Driehuis, F. 2001. Anaerobic conversion of lactic acid to acetic acid and 1,2-propanediol by *Lactobacillus buchneri. Appl. Environ. Microbiol.* 67: 125–132.

Ounine, K., Petitdemange, H., Raval, G., and Gay, R. 1983. Acetone-butanol production from pentoses by *Clostridium acetobutylicum. Biotechnol. Lett.* 5: 605–610.

Palmqvist, E. and Hahn-Hägerdal, B. 2000. Fermentation of lignocellulosic hydrolysates. I: Inhibition and detoxification. *Bioresour. Technol.* 74: 17–24.

Palosaari, N.R. and Rogers, P. 1988. Purification and properties of the inducible coenzyme A-linked. *J. Bacteriol.* 170: 2971–2976.

Pan, X., Gilkes, N., Kadla, J., Pye, K., Saka, S., Gregg, D., Ehara, K., Xie, D., Lam, D., and Saddler, J. 2006a. Bioconversion of hybrid poplar to ethanol and co-products using an organosolv fractionation process: Optimization of process yields. *Biotechnol. Bioeng.* 94: 5.

Pan, X., Kadla, J.F., Ehara, K., Gilkes, N., and Saddler, J.N. 2006b. Organosolv ethanol lignin from hybrid poplar as a radical scavenger: Relationship between lignin structure, extraction conditions, and antioxidant activity. *J. Agric. Food Chem.* 54: 5806–5813.

Petersen, D.J. and Bennett, G.N. 1990. Purification of acetoacetate decarboxylase from *Clostridium acetobutylicum ATCC 824* and cloning of the acetoacetate decarboxylase gene in *Escherichia coli. Appl. Environ. Microbiol.* 11: 3491–3498.

Petersen, D.J., Welch, R.W., Rudolph, F.B., and Bennett, G.N. 1991. Molecular cloning of an alcohol (butanol) dehydrogenase gene cluster from *Clostridium acetobutylicum ATCC 824. J. Bacteriol.* 173: 1831–1834.

Picataggio, S.K., Zhang, M., Franden, M.A., McMillan, J.D., and Finkelstein, M. 1998. *Recombinant Lactobacillus for Fermentation of Xylose to Lactic Acid and Lactate.* United States Patent: 5,798,237.

Polizeli, M.L.T.M, Rizzatti, A.C.S., Monti, R., Terenzi, H.F., Jorge, J.A., and Amorim, D.S. 2005. Xylanases from fungi: Properties and industrial applications. *Appl. Microbiol. Biotechnol.* 67: 577–591.

Qureshi, N. and Blaschek, H.P. 1999. Butanol recovery from model solution/fermentation broth by pervaporation: Evaluation of membrane performance, *Biomass Bioenergy*, 17: 175–184.

Qureshi, N., Li, X.-L., Hughes, S., Saha, B.C., and Cotta, M.A. 2006. Butanol production from corn fiber xylan using *Clostridium acetobutylicum. Biotechnol. Prog.* 22: 673–680.

Qureshi, N., Lolas, A., and Blaschek, H.P. 2001. Soy molasses as fermentation substrate for production of butanol using *Clostridium beijerinckii* BA101. *J. Ind. Microbiol. Biotechnol.* 26(5): 290–295.

Rabinovich, M.L., Melnik, M.S., and Bolobova, A.V. 2002. Microbial cellulases (review). *Appl. Biochem. Microbiol.* 38: 305–321.

Ramos, L.P. 2003. The chemistry involved in the steam treatment of lignocellulosic materials. *Quim. Nova* 26: 863–871.

Ramsay, J.A., Alyhassan, M.C., and Ramsay, B.A. 1995. Hemicellulose as a potential substrate for production of poly(β-hydroxyalkanoates). *Can. J. Microbiol.* 41: 262–266.

Rao, C.S., Prakasham, R.S., Rao, A.B., and Yadav, J.S. 2008. Production of L(+) lactic acid by *Lactobacillus delbrueckii* immobilized in functionalized alginate matrices. *World J. Microbiol. Biotechnol.* 24: 1411–1415.

Romero, I., Moya, M., Sánchez, S., Ruiz, E., Castro, E., and Bravo, V. 2007. Ethanolic fermentation of phosphoric acid hydrolysates from olive tree pruning. *Ind. Crop Prod.* 25: 160–168.

Ryabova, O., Vrsanska, M., Kaneko, S., van Zyl, W.H., and Biely, P. 2009. A novel family of hemicellulolytic α-glucuronidase. *FEBS Lett.* 583: 1457–1462.

Saez-Miranda, J.S., Saliceti-Piazza, L., and McMillan, J.D. 2008. Measurement and analysis of intracellular ATP levels in metabolically engineered *Zymomonas mobilis* fermenting glucose and xylose mixtures. *Biotechnol. Prog.* 22: 359–368.

Saha, B.C. 2003. Hemicellulose bioconversion. *J. Ind. Microbiol. Biotechnol.* 30: 279–291.

Sarkanen, K.V. 1980. Acid-catalyzed delignification of lignocellulosics in organic solvents. *Prog. Biomass Convers.* 2: 127–144.

Saxena, R.C., Adhikari, D.K., and Goyal, H.B. 2009. Biomass-based energy fuel through biochemical routes: A review. *Renew. Sustain. Energ. Rev.* 13: 167–178.

Schmidt, S. and Padukone, N. 1997. Production of lactic acid from wastepaper as a cellulosic feedstock. *J. Ind. Microbiol. Biotechnol.* 18: 10–14.

Shallom, D. and Shoham, Y. 2003. Microbial hemicellulases. *Curr. Opin. Microbiol.* 6: 219–228.

Shang, L., Yim, S.C., Park, H.G., and Chang, H.N. 2004. Sequential feeding of glucose and valerate in a fed-batch culture of *Ralstonia eutropha* for production of poly(hydroxybutyrate-co-hydroxyvalerate) with high 3-hydroxyvalerate fraction. *Biotechnol. Prog.* 20: 140–144.

Shevchenko, S.M., Beatson, R.P., and Saddler, J.N. 1999. The nature of lignin from steam explosion/enzymatic hydrolysis of softwood. *Appl. Biochem. Biotechnol.* 79: 867–876.

Solomon, B.D., Barnes, J.R., and Halvorsen, K.E. 2007. Grain and cellulosic ethanol: History, economics, and energy policy. *Biomass Bioenerg.* 31: 416–425.

Sreenath, H.K., Moldes, A.B., Koegel, R.G., and Straub, R.J. 2001. Lactic acid production from agriculture residues. *Biotechnol. Lett.* 23: 179–184.

Stepanik, T.M., Rajagopal, S., Ewing, D., and Whitehouse, R. 1998. Electron-processing technology: A promising application for the viscose industry. *Radiat. Phys. Chem.* 52: 505–509.

Stim-Herndon, K.P., Petersen, D.J., and Bennett, G.N. 1995. Characterization of an acetyl-CoA C-acetyltransferase (thiolase) gene from *Clostridium acetobutylicum ATCC 824*. *Gene* 154: 81–85.

Studholme, D.J., Jackson, R.A., and Leak, D.J. 1999. Phylogenetic analysis of transformable strains of thermophilic *Bacillus* species. *FEMS Microbiol. Lett.* 172: 85–90.

Sullivan, L., Cates, M.S., and Bennett, G.N. 2010. Structural correlations of activity of *Clostridium acetobutylicum ATCC 824* butyrate kinase isozymes. *Enzyme Microb. Technol.* 46: 118–124.

Sun, Y. and Cheng, J.Y. 2002. Hydrolysis of lignocellulosic materials for ethanol production: A review. *Bioresour. Technol.* 83: 1–11.

Sun, N., Rahman, M., Qin, Y., Maxim, M.L., Rodriguez, H., and Rogers, R.D. 2009a. Complete dissolution and partial delignification of wood in the ionic liquid 1-ethyl-3-methylimidazolium acetate. *Green Chem.* 11: 646–655.

Sun, Z.Y., Ramsay, J., Guay, M., and Ramsay, B. 2009b. Enhanced yield of medium-chain-length polyhydroxyalkanoates from nonanoic acid by co-feeding glucose in carbon-limited, fed-batch culture. *J. Biotechnol.* 143: 262–267.

Sun, X.F., Xu, F., Sun, R.C., Geng, Z.C., Fowler, P., and Baird, M.S. 2005. Characteristics of degraded hemicellulosic polymers obtained from steam exploded wheat straw. *Carbohydr. Polym.* 60: 15–26.

Sunna, A. and Antranikian, G. 1997. Xylanolytic enzymes from fungi and bacteria. *Crit. Rev. Biotechnol.* 17: 39–67.

Swatloski, R.P., Holbrey, J.D., Memon, S.B., Caldwell, G.A., Caldwell, K.A., and Rogers, R.D. 2004. Using *Caenorhabditis elegans* to probe toxicity of 1-alkyl-3-methylimidazolium chloride based ionic liquids. *Chem. Commun.* 6: 668–669.

Taherzadeh, M.J. and Karimi, K. 2007. Acid-based hydrolysis processes for ethanol from lignocellulosic materials: A review. *BioResources* 2: 472–499.

Takahashi, C.M., Lima, K.G.D, Takahashi, D.F, and Alterthum, F. 2000. Fermentation of sugar cane bagasse hemicellulosic hydrolysate and sugar mixtures to ethanol by recombinant *Escherichia coli KO11. World J. Microbiol. Biotechnol.* 16: 829–834.

Talebnia, F., Karakashev, D., and Angelidaki, I. 2010. Production of bioethanol from wheat straw: An overview of pretreatment, hydrolysis and fermentation. *Bioresour. Technol.* 101: 4744–4753.

Teixeira, L.C., Linden, J.C., and Schroeder, H.A., 2000. Simultaneous saccharification and cofermentation of peracetic acid-pretreated biomass. *Appl. Biochem. Biotechnol.* 84: 111–127.

Thring, R.W., Chorent, E., and Overend, R. 1990. Recovery of a solvolytic lignin: Effects of spent liquor/acid volume ratio, acid concentration and temperature. *Biomass* 23: 289–305.

Tsai, S.-L., Oh, J., Singh, S., Chen, R., and Chen, W. 2009. Functional assembly of mini-cellulosomes on the yeast surface for cellulose hydrolysis and ethanol production. *Appl. Environ. Microbiol.* 75: 6087–6093.

Tunc, M.S. and van Heiningen, A.R.P. 2008. Hemicellulose extraction of mixed southern hardwood with water at 150°C: Effect of time. *Ind. Eng. Chem. Res.* 47: 7031–7037.

Tuominen, J. 2003. *Chain Linked Lactic Acid Polymers: Polymerization and Biodegradation Studies.* Polymer Technology Publication Series No.25, Polymer Technology, Department of Chemical Technology, Helsinki University of Technology.

U.S. Department of Energy, 2006. Breaking the biological barriers to cellulosic ethanol: A joint research agenda.

Valentin, H.E. and Dennis, D. 1997. Production of poly(3-hydroxybutyrate-co-4-hydroxybutyrate) in recombinant *Escherichia coli* grown on glucose. *J. Biotechnol.* 58: 33–38.

van Zyl, W.H., Lynd, L.R., den Haan, R., and McBride, J.E. 2007. Consolidated bioprocessing for bioethanol production using *Saccharomyces cerevisiae. Adv. Biochem. Eng. Biotechnol.* 108: 205–235.

Vasconcelos, I., Girbal, L., and Soucaille, P. 1994. Regulation of carbon and electron flow in *Clostridium acetobutylicum* grown in chemostat culture at neutral pH on mixtures of glucose and glycerol. *J. Bacteriol.* 176: 1443–1450.

Vasconcelos, J.C. and Andrade, I. 2003. Continuous cultures of *Clostridium acetobutylicum*: Culture stability and low-grade glycerol utilization. *Biotechnol. Lett.* 25: 121–125.

Vidal, P.F. and Molinier, J. 1988. Ozonolysis of lignin-improvement of *in-vitro* digestibility of poplar sawdust. *Biomass* 16: 1–17.

Wallner, T., Miers, S.A., and McConnell, S. 2009. A comparison of ethanol and butanol as oxygenates using a direct-injection, spark-ignition engine. *J. Eng. Gas Turbines Power* 131: 1–9.

Wasikiewicz, J.M., Yoshii, F., Nagasawa, N., Wach, R.A., and Mitomo, H. 2005. Degradation of chitosan and sodium alginate by gamma radiation, sonochemical and ultraviolet methods. *Radiat. Phys. Chem.* 73: 287–295.

Waterson, R.M., Castellino, F.J., Hass, G.M., and Hill, R.L. 1972. Purfication and characterization of crotonase from *Clostridium acetobutylicum. J. Biol. Chem.* 247: 5266–5271.

Weimer, P.J., Hackney, J.M., and French, A.D. 1995. Effects of chemical treatments and heating on the crystallinity of celluloses and their implications for evaluating the effect of crystallinity on cellulose biodegradation. *Biotechnol. Bioeng.* 48: 169–178.

Welch, R.W., Rudolph, F.B., and Papoutsakis, E.T. 1989. Purification and characterization of the NADH-dependent butanol dehydrogenase from *Clostridium acetobutylicum (ATCC 824). Arch. Biochem. Biophys.* 273: 309–318.

Wen, F., Sun, J., and Zhao, H.M. 2010. Yeast surface display of trifunctional minicellulosomes for simultaneous and synergistic saccharification and fermentation of cellulose to ethanol. *Appl. Environ. Microbiol.* 76: 1251–1260.

Wiesenborn, D.P., Rudolph, F.B., and Papoutsakis, E.T. 1989a. Coenzyme A transferase from *Clostridium acetobutylicum ATCC 824* and its role in the uptake of acids. *Appl. Environ. Microbiol.* 55: 323–329.

Wiesenborn, D.P., Rudolph, F.B., and Papoutsakis, E.T. 1989b. Phosphotransbutyrylase from *Clostridium acetobutylicum ATCC 824* and its role in acidogenesis. *Appl. Environ. Microbiol.* 55: 317–322.

Willauer, H.D., Huddleston, J.G., Li, M., and Rogers, R.D. 2000. Investigation of aqueous biphasic systems for the separation of lignins from cellulose in the paper pulping process. *J. Chromatogr. B Biomed. Sci. Appl.* 743: 127–135.

Wingren, A., Galbe, M., and Zacchi, G. 2003. Techno-economic evaluation of producing ethanol from softwoodComparison of SSF and SHF and identification of bottlenecks. *Biotechnol. Progr.* 19(4): 1109–1117.

Wood, B.E. and Ingram, L.O. 1992. Ethanol production from cellobiose, amorphous cellulose, and crystalline cellulose by *Klebsiella oxytoca* integrated *Zymomonas mobilis* genes for ethanol production and plasmids expressing thermostable cellulase genes from *Clostridium thermocellum*. *Appl. Environ. Microbiol.* 58: 2103–2110.

Wright, J.D. 1998. Ethanol from biomass by enzymatic hydrolysis. *Chem. Eng. Prog.* 84: 62–74.

Wu, J., Zhang, J., Zhang, H., He, J., Ren, Q., and Guo, M. 2004. Homogeneous acetylation of cellulose in a new ionic liquid. *Biomacromolecules* 5: 266–268.

Wu, L.J. and Welker, N.E. 1989. Protoplast transformation of *Bacillus stearothermophilus NUB36* by plasmid DNA. *J. Gen. Microbiol.* 135: 1315–1324.

Xu, H.W., Scott, G.M., Jiang, F. and Kelly, C. 2010. Recombinant manganese peroxidase (rMnP) from *Pichia pastoris*. Part 1: Kraft pulp delignification. *Holzforschung* 64: 137–143.

Xu, Q., Singh, A., and Himmel, M.E. 2009. Perspectives and new directions for the production of bioethanol using consolidated bioprocessing of lignocellulose. *Curr. Opin. Biotechnol.* 20: 364–371.

Yang, C.P., Shen, Z.Q., Yu, G.C., and Wang, J.L. 2008. Effect and aftereffect of γ radiation pre-treatment on enzymatic hydrolysis of wheat straw. *Bioresour. Technol.* 99: 6240–6245.

Young, F.K., Kastner, J.R., and Sheldon, W.M. 1994. Microbial production of poly-β-hydroxybutyric acid from D-xylose and lactose by *Pseudomonas cepacia*. *Appl. Environ. Microbiol.* 60: 4195–4198.

Youngleson, J.S., Santangelo, J.D., Jones, D.T., and Woods, D.R. 1988. Cloning and expression of a *Clostridium acetobutylicum* alcohol dehydrogenase gene in *Escherichia coli*. *Appl. Environ. Microbiol.* 54: 676–682.

Zavrel, M., Bross, D., Funke, M., Buchs, J., and Spiess, A.C. 2009. High-throughput screening for ionic liquids dissolving (ligno)-cellulose. *Bioresour. Technol.* 100: 2580–2587.

Zhang, Y.H.P. and Lynd, L.R. 2008. Chapter 16. New generation biomass conversion: Consolidated bioprocessing. In: Himmel, M.E. (ed.), *Biomass Recalcitrance: Deconstructing the Plant Cell Wall for Bioenergy*. Blackwell, Chichester, UK, 480–494.

Zhang, X.Z. and Zhang, Y.H.P. 2010. One-step production of biocommodities from lignocellulosic biomass by recombinant cellulolytic *Bacillus subtilis*: Opportunities and challenges. *Eng. Life Sci.* 10: 398–406.

Zhang, Y.H.P., Himmel, M.E., and Mielenz, J.R. 2006. Outlook for cellulase improvement: Screening and selection strategies. *Biotechnol. Adv.* 24: 452–481.

Zhao, Y., Hindorff, L.A., Chuang, A., Monroe-Augustus, M., Lyristis, M., Harrison, M.L., Rudolph, F.B., and Benett, G.N. 2003. Expression of a cloned cyclopropane fatty acid synthase gene reduces solvent formation in *Clostridium acetobutylicum ATCC 824*. *Appl. Environ. Microbiol.* 69: 2831–2841.

Zhao, Y.L., Wang, Y., Zhu, J.Y., Ragauskas, A., and Deng, Y.L. 2008. Enhanced enzymatic hydrolysis of spruce by alkaline pre-treatment at low temperature. *Biotechnol. Bioeng.* 99: 6.

Zheng, Y.Z., Lin, H.M., and Tsao, G.T. 1998. Pre-treatment for cellulose hydrolysis by carbon dioxide explosion. *Biotechnol. Prog.* 14: 890–896.

Zhou, S. and Ingram, L.O. 2001. Simultaneous saccharification and fermentation of amorphous cellulose to ethanol by recombinant *Klebsiella oxytoca SZ21* without supplemental cellulase. *Biotechnol. Lett.* 23: 1455–1462.

Zhu, J.Y. and Pan, X.J. 2010. Woody biomass pretreatment for cellulosic ethanol production: Technology and energy consumption evaluation. *Bioresour. Technol.* 101: 4992–5002.

Zhu, Y., Lee, Y.Y., and Elander, R.T. 2005. Optimization of dilute-acid pre-treatment of corn stover using a high solids percolation reactor. *App. Biochem. Biotechnol.* 121: 1045–1054.

Zverlov, V.V., Berezina, O., Velikodvorskaya, G.A., and Schwarz, W.H. 2006. Bacterial acetone and butanol production by industrial fermentation in the Soviet Union: Use of hydrolyzed agricultural waste for biorefinery. *Appl. Microbiol. Biotechnol.* 71: 587–597.

22

Separation of Hemicellulose by Hot-Water Extraction from Woody Biomass

Shijie Liu, Houfang Lu, Yichao Lei, Christopher D. Wood, Thomas E. Amidon, Bin Liang, Runcang Sun, Gary M. Scott, Daryl Nichol, and Al Ward

CONTENTS

22.1 Introduction

Although humans first learned to use biomass as their almost exclusive source of chemicals and energy, fossil fuels have become the dominant energy and chemical source since the Industrial Revolution. Petroleum, natural gas, and coal were regarded as waste materials from ancient times. More recently, they have become the cheapest sources of energy and chemicals and have retained this status for over half a century. Full utilization of these "waste" materials has been the focus of much research and development, and the achievements have been remarkable. Compared with biomass, fossil energy sources have a higher energy density with much lower carbon dioxide output per unit energy output than biomass (not considering the other known pollutant gases that are emitted due to the composition of biomass, as described by Walker et al. (2010)). This situation has confused many members of the public and even policy-makers, leading them to label biomass as the worst or dirtiest fuel (worse than the generally accepted coal). However, the quantity of fossil-fuel deposits on the Earth is finite. The time when fossil fuels will be completely depleted is now predictable and lies in the foreseeable future despite discoveries of new deposits. Therefore, it is imperative to find alternatives to fossil sources because the time scale for fossil fuels to be recycled or to replenish naturally is, at best, on the order of 280 million years (Amidon and Liu, 2009), as shown in Table 22.1. Fossil fuels are replenished when a carbonaceous age occurs on the Earth, preserving a large amount of organic matter. Given the advanced forms of life existing now, the chance of such an event occurring is small. In any case, the 280-million-year cycle is far too long and the amount of possible reserves negligible compared to the recharge duration from a human-use standpoint, even if humankind could survive a fossil-replenishment era. Therefore, fossil resources in general are deemed to be nonrenewable and cannot possibly be managed sustainably to compete with biomass. Using fossil energy sources is like sending CO_2 on a one-way trip to the atmosphere. Societal awareness of environmental impact as well as problems in the stability and sustainability of the energy supply have lent greater urgency to the development and implementation of biobased chemicals and energy. Domestic and regional energy security and rural economies would all benefit from a plant biomass-derived chemical and energy economic base. Reverting to reliance on biomass as materials, chemicals, and energy sources is therefore our destiny.

Woody biomass is the most abundant organic source on Earth, with an annual production in the biosphere of approximately 1.7×10^{14} billion metric tons (Klass, 1998). Tapping into the chemical energy of biomass and reclaiming the historically important position of biomass in chemicals, energy, and transportation is imperative to the sustainability of the world economy. Forests cover approximately 9.5% of the Earth's surface or approximately 32% of the land area, but account for 89.3% of the total standing biomass, producing *73 billion metric tons per year*, or 42.9%, of the total annual biomass production of the Earth. Savanna and grasses are a distant second on the list, accounting for 11% of total biomass production. When measured in energy terms, the amount synthesized by the forest alone is 1030 quadrillion Btu/year (Klass, 1998), which is equivalent to more than double the world's total primary energy consumption of approximately 460 quadrillion Btu in 2005 (Amidon et al., 2008). Research and development are required to boost the energy conversion efficiency of forest biomass to meet industrial and residential energy and commodity chemical requirements. Plants synthesize chemicals from solar energy. The chemical energy stored by the biomass can be converted to energy and chemicals that humankind can use (Amidon et al., 2008; Mao et al., 2008; Tunc and van Heiningen, 2008; Boussarsar et al., 2009). The flow

TABLE 22.1

Recycle Times and Productivities of Chemical and Energy Feedstocks

Feedstock	Recycle Time	Standing Biomass (tons/ha)	Biomass Production (tons/(ha·year))
Algae	1 month	9.0	11.25
Agricultural crops	3 months–1 year	4.5	2.93
Temperate grasses	1 year	7.2	2.70
Savanna	1 year	18.0	4.05
Shrubs	1–5 years	27.0	3.15
Tropical forest	5–25 years	202.5	9.90
Tropical seasonal forest	5–25 years	157.5	7.20
Boreal forest	25–80 years	90.0	3.60
Temperate deciduous	10–50 years	135.0	5.40
Temperate evergreen	10–80 years	157.5	5.85
Oil, gas, and coal	280 million years	(38.4×10^{27} J)	(0)

Source: Liu, S.J. 2010. *J. Biotech. Adv.* 28: 565. With permission.

of energy for a plant-biomass-based economy is illustrated in Figure 22.1. Plants grow naturally "even without human intervention." Dedicated or managed energy crops (agriforest or agricultural biomass) can further increase biomass availability, as described, for example, in Keoleian and Volk (2005), Ragauskas et al. (2006), and Boussarsar et al. (2009). A sustainable plant-biomass-based economy is achievable.

CO_2 is simply a catalyst or carrier for the energy, chemicals, and materials we use. Renewable forest material is carbon-neutral, which means that using forest material will not create a carbon imbalance over the life cycle of the forest, which extends for 5–80 years for managed forests, as shown in Table 22.1. *"Catch and Release"* is the key in biomass utilization (Amidon and Liu, 2009). Carbon dioxide is drawn from the atmosphere into plants to enable growth, while planting, management, conversion of biomass to bioproducts, and use and decomposition of bioproducts will all produce carbon dioxide. In an optimally balanced operation, carbon dioxide is simply recycled during the life span of plant growth and of bioproducts (Liu et al., 2006). The net effect is that solar energy and atmospheric carbon dioxide are converted to energy and materials that are used by humans.

A manufacturing concept for converting woody biomass to energy and chemicals is the *biorefinery*. Woody biomass is fractionated into a number of compounds, and energy is recovered. Table 22.2 shows the chemical composition of various wood species. Use of these heterogeneous materials which are bonded together in wood is key to a future sustainable supply of materials, chemicals, and energy. Using renewable carbon eliminates an additional fossil-fuel-derived carbon dioxide burden on the environment and reduces the greenhouse-gas (GHG) driver for global climate change. Therefore, managed forests have significant potential to reduce GHG emissions through the conversion of forest material into liquid fuels, electricity, and other products now being derived from nonrenewable carbon.

Using woody biomass does not mean that we should consume all the biomass on the planet. However, it does mean that trees will be harvested for conversion to usable energy and chemicals and to free space for new trees to grow. Figure 22.2 illustrates the time evolution of total aboveground biomass in a given area when left untouched and when used by humans. Trees, shrubs, grasses, and algae grow with the CO_2 available in the atmosphere and the H_2O in or on the ground. However, the density of available biomass, expressed as a mass per unit surface area, is in the order of highest to lowest: trees, shrubs, then grasses and algae, as shown in Table 22.1. The biomass production rate, however, is

FIGURE 22.1
Energy flow from the Sun to power the needs of humans. (From Liu, S.J. 2010. *J. Biotech. Adv.* 28: 564. With permission.)

TABLE 22.2

Chemical Composition (%wt Based on Dry Wood Mass) of Various Wood Species

Species	Common Name	Total Extractives (CH_2Cl_2 Followed by C_2H_5OH)	Lignin	Cellulose	Glucomannan (Including Galactose and Acetyl in Softwood)	Glucuronoxylan (Including Arabinose in Softwood and Acetyl in Hardwood)	Other Polysaccharides	Residual Constituents
Softwoods								
Abies balsamea	Balsam fir	2.7	29.1	38.8	17.4	8.4	2.7	0.9
Pseudotsuga menziesii	Douglas fir	5.3	29.3	38.8	17.5	5.4	3.4	0.0
Tsuga canadensis	Eastern hemlock	3.4	30.5	37.7	18.5	6.5	2.9	0.5
Juniperus communis	Common juniper	3.2	32.1	33.0	16.4	10.7	3.2	1.4
Pinus radiata	Monterey pine	1.8	27.2	37.4	20.4	8.5	4.3	0.4
Pinus sylvestris	Scots pine	3.5	27.7	40.0	16.0	8.9	3.6	0.3
Picea abies	Norway spruce	1.7	27.4	41.7	16.3	8.6	3.4	0.9
Picea glauca	White spruce	2.1	27.5	39.5	17.2	10.4	3.0	0.3
Larix sibirica	Siberian larch	1.8	26.8	41.4	14.1	6.8	8.7	0.4
Hardwoods								
Acer rubrum	Red maple	3.2	25.4	42.0	3.1	22.1	3.7	0.5
Acer saccharum	Sugar maple	2.5	25.2	40.7	3.7	23.6	3.5	0.8
Populus tremuloides	Trembling aspen	3.8	18.1	49.4	3.6	23.0	1.7	0.4
Fagus sylvatica	Common beech	1.2	24.8	39.4	1.3	27.8	4.2	1.3
Betula pendula	Silver birch	3.2	22.0	41.0	2.3	27.5	2.6	1.4
Betula papyrifera	Paper birch	2.6	21.4	39.4	1.4	29.7	3.4	2.1
Alnus incana	Gray alder	4.6	24.8	38.3	2.8	25.8	2.3	1.4
Eucalyptus	River red gum	2.8	31.3	45.0	3.1	14.1	2.0	1.7
camaldulensis	Blue gum	1.3	21.9	51.3	1.4	19.9	3.9	0.3
Eucalyptus globulus	Yemane	4.6	26.1	47.3	3.2	15.4	2.5	0.9
Gmelina arborea	Black wattle	1.8	20.8	42.9	2.6	28.2	2.8	0.9
Acadia mollissima	Balsa	2.0	21.5	47.7	3.0	21.7	2.9	1.2
Ochroma lagopus								

Source: Adapted from Amidon, T.E., Liu, S.J. 2009. *Biotechnol. Adv.* 27: 542–550.

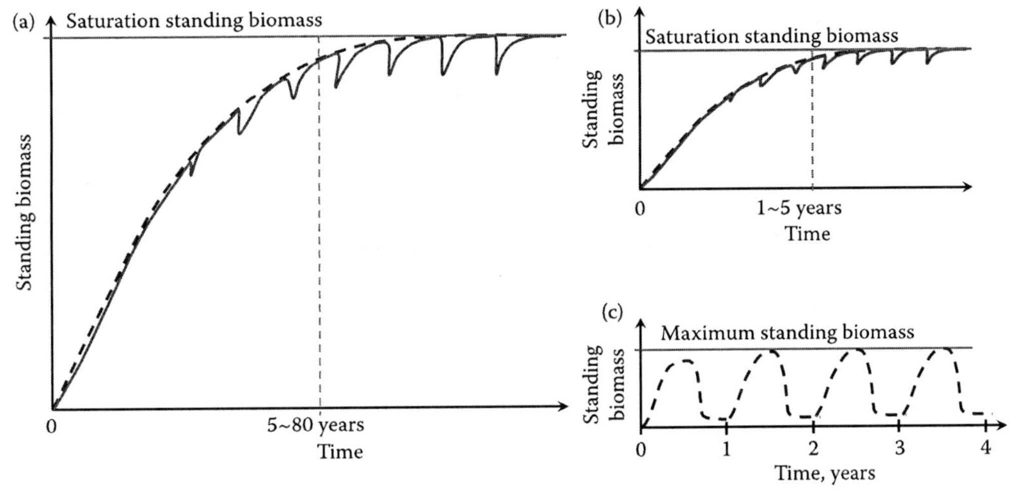

FIGURE 22.2
Schematic of aboveground standing biomass accumulation as a function of time in a given area. The dotted line indicates the situation in which no disturbances occur, whereas the solid line shows the impact of periodic minor disturbances. (a) Forest, trees, (b) shrubs, bamboo, and so on, and (c) grasses. (From Liu, S.J. 2010. *J. Biotech. Adv.* 28: 565. With permission.)

highest for algae, followed by trees, shrubs, and grasses. This ordering of biomass production rate could be changed if intensive cultivation and harvesting techniques were used at various levels. Intensive cultivation can shorten recycle time and increase yield. The combined effects could make the shorter-rotation plants produce much more incremental biomass than the longer-rotation plants. Agricultural crops can be even more beneficial, potentially producing the same amount of woody biomass while also producing food and animal feed. Therefore, when labor is unlimited, but land area is limited, agricultural crops can play a significant role in a renewable economy.

There is an upper limit on how much biomass the Earth can "host." If the Earth were saturated with plant biomass, no more net additional biomass could be added or grown. Natural disturbances promote plant biomass renewal while temporarily changing the level of standing biomass. Wildfires sweep the ground from time to time, reducing the standing biomass to ashes and releasing CO_2 and H_2O back to nature. This cycle of natural growth and replacement keeps the plant biomass fluctuating below the saturation level. Therefore, not using plant biomass, in particular forest biomass, does not lead to an increase in the total amount of standing biomass on Earth beyond its saturation level. Constantly harvesting plant biomass at or below the rate at which it can be replenished has the same effect as if the plant biomass were left untouched. Sustaining a similar level of plant biomass on the Earth is achievable, as nature has taught us through natural disturbances. Therefore, harvesting and using plant biomass, whether it be trees, forests, or grasses, does not automatically lead to any net change in the biomass coverage on Earth. However, intensive cultivation, which we will increasingly need to practice to satisfy our needs as we prepare for the eventual depletion of fossil energy sources, will reduce the level of total carbon storage compared to that of an undisturbed forest. Nevertheless, forests and trees have the highest potential net energy benefit and stand quantity for sustainable renewable materials, chemical products, and energy development.

Greenhouse gas emissions are a complex issue in energy utilization. So far, the focus has been "high above the ground," looking down on the effects of energy sources. However, let us not forget about the short-term effects. CO_2 is CO_2, whether it comes from fossil or biogenic sources. Fossil energy sources are dense in energy content, and our technology for fossil-energy utilization has peaked. Table 22.3 shows a number of the environmental factors that affect energy production from fossil sources and woody biomass. Both the higher efficiency and the denser energy content of the fossil raw materials have contributed to much lower CO_2 emissions from fossil fuels than from wood. This gives the illusion of much higher carbon dioxide emissions from woody-biomass use. In addition to the clear impact of woody-biomass use on CO_2 emissions, there are hidden effects as well. Trees need to be harvested for energy. Harvesting trees, in particular aboveground biomass, affects the forest ecosystem. Old-growth forest continues to sequester CO_2 from the atmosphere, decaying and sending mineralized carbon into the soil. For forests between 15 and 800 years of age, the net ecosystem productivity (the net carbon balance of the forest, including soils) is usually positive. The mineralization of carbon and its storage in soil is very slow, and the carbon saturation limit takes a long time to reach. Old-growth forests accumulate carbon for centuries and contain large quantities. However, much of this carbon, even the soil carbon, will move back into the atmosphere if these forests are disturbed (Luyssaert et al., 2008). The world's soils store more carbon than is present in biomass and in the atmosphere. Little is known, however, about the factors controlling the stability of soil organic-carbon stocks, and the response of the soil carbon pool to climate change remains uncertain. The supply of fresh plant-derived carbon to the subsoil (0.6–0.8 m depth) stimulates the microbial mineralization of 2567 ± 226-year-old carbon. In the absence of fresh organic carbon, an essential source of energy for soil microbes, the stability of organic carbon in deep soil layers is maintained and may prevent the decomposition of the organic carbon pool in deep soil layers in response to future changes in temperature. Any change in land use and agricultural practices that increases the distribution of fresh carbon along the soil profile could, however, stimulate the loss of ancient buried carbon (Fontaine et al., 2007). Managed forest stands on 80-year rotations store only half the carbon of old-growth forests (Janisch and Harmon, 2002). Therefore, if all the forests were converted to productive woody-biomass fields, there would be a net CO_2 input into the atmosphere (if the ocean and open land did not take in the newly available excess CO_2). A well-managed woody-biomass forest will have a net CO_2 contribution to the atmosphere if it were converted from natural forest. Still, the net CO_2 contribution to the atmosphere would be lower than other crop

TABLE 22.3

Emission Specifications of Some Power Plants in the State of Massachusetts, USA

Power Plant	Fuel Type	CO_2/MW (tpy)	NO_x/MW (tpy)	Particulate Matter/MW (tpy)
Boardman (PGE)	Coal	9067	3.38	0.59
Mount Tom Power Plant	Coal	7167	5.97	0.80
Pioneer Valley Energy Center	Natural gas	3130	0.23	0.12
Russell	Biomass wood	12,644	3.9	1.69
Pioneer Renewable Energy	Biomass wood	11,312	3.46	1.08
Palmer Renewable Energy (planned, but dropped for greenfield instead)	Construction and demolition debris	12,415	3.53	0.71

operations once steady state were reached (biomass harvesting rate equal to growth rate) simply because of its unparalleled high standing mass. When steady state is reached (80 years into the future for an 80-year-rotation forest management scheme), woody biomass becomes truly carbon-neutral, although then the CO_2 level in the atmosphere will be higher than what we know today. The consequence of not starting to use woody biomass today would simply be to increase the CO_2 in the atmosphere to an even higher level.

Evolution in values has energized the progress of humankind. Fossil fuels were once seen as waste and an obstacle to land development. Economic utilization of these wastes has fueled significant progress in our way of living. Since roughly 1900, people have started to use fossil energy to process and even to generate food and fiber. Both of these tend to be much more desirable and valuable than the fuel itself. Therefore, the idea of turning food and fiber into fuel is somewhat contrary to what the market has "recommended" through prices and what humans have in fact done over the past century. For example, as of 2010, cellulose for paper sold for roughly 50–80 cents per kg, while sugar (the substrate for alcohol fermentation) sold for only about 30 cents per kg. Food cereal grains and liquid biofuels such as ethanol were at approximately the same price level, \$1.00/kg. The price structure today does not favor a renewable economy because fossil sources are still treated as waste materials (in terms of price). Even doubling the price of liquid fuel while maintaining the food price leaves no room for renewable liquid-fuel production. Therefore, the arrival of a truly renewable economy will be accompanied by a fundamental change in commodity prices.

Trees (Table 22.2) and herbaceous material (Table 22.4) are made up of three basic components plus many trace materials. The most abundant components are cellulose, hemicellulose, and lignin, all based on carbon backbones. Cellulose is the long, flexible, natural, linear homopolymer in fibers that is predominantly responsible for giving trees and wood their strength. Cellulose fibers are held together by lignin, a natural phenolic glue. Hemicellulose consists of the lower-molecular-weight heterosugar polymers which are found encrusting the cellulose and lignin. Structural wood is simply the tree with all these components cut and shaped to the desired form. White papers are made from relatively pure cellulose in the original fiber shape.

Cellulose is the most abundant constituent in woody biomass (Tables 22.2 and 22.4). It is a linear polymer of glucose residues bound together by β 1-*O*-4 glycosidic bonds, as

TABLE 22.4

Chemical Compositions of Common Fiber/Oil Seed Crops, Cereal Straws, and Grasses

Straw Species	Cellulose	Lignin	Pentosans	Ash	Extractives
Hemp bast	53.0–62.0	2.0–4.0	4.2–6.6	5.2–7.1	17.1–23.3
Hemp core	38.9–48.6	17.5–21.7	16.2–22.8	2.8–4.4	8.2–17.4
Flax bast	52.9–67.2	2.6–6.9	3.9–6.4	3.2–7.0	11.0–23.0
Flax core	44.6–53.1	19.1–24.6	19.0–25.8	1.0–4.1	4.1–11.7
Triticale	27.9–38.6	15.1–17.6	21.0–25.4	4.5–7.7	15.2–22.9
Wheat	27.5–40.3	17.0–21.2	22.0–26.2	4.8–6.7	12.9–19.3
Barley	35.5–41.6	16.4–19.7	20.2–26.9	3.6–5.7	10.5–18.9
Oats	35.3–40.9	18.2–18.8	22.8–25.8	4.4–6.3	9.9–14.0
Lupine	35.0–43.3	15.3–18.0	16.8–18.1	4.4–7.2	14.5–16.81
Switchgrass	27.6–30.7	13.1–14.9	23.2–26.8	6.1–6.9	16.2–22.8
Reeds	21.4–29.4	12.3–17.1	17.5–22.5	6.8–8.8	20.9–35.8
Fescue	13.7–16.7	7.7–7.9	14.9–18.1	9.4–10.0	39.7–40.1

Source: Adapted from Liu, S.J. et al. 2006. *Indus. Biotechnol.* 2: 113–120.

depicted in Figure 22.3. Breaking those connections would appear to be simple. It would appear that a long fiber of cellulose could become thousands of simple glucose molecules, but there is a reason why this is very difficult. Cereal grains (starch) and fructans are food storage materials, whereas cellulose is a structural material. When structure is needed, cellulose is a natural choice because it does not break apart easily. In fact, easy breakage would be rapidly selected against because trees would topple, and more generally, woody plants could not grow very tall. An economical way of imitating these naturally occurring processes has not yet been discovered. Still, cellulose fibers with glucose in the polymeric form are more valuable than their derived sugars because paper and particleboard products can be made from them and both have a higher value per pound than sugars. Cellulose is a valuable source of materials. The sheer amount of sustainably available cellulose creates an urgency for the development of chemical and fuel conversion technologies as well, so that we can begin to establish the foreseeable pattern of sustainable (equilibrium or steady state) utilization of these resources to curtail the level of CO_2 in the atmosphere.

The amount of hemicellulose in woody biomass is significant, being approximately one-quarter of the total dry mass (Tables 22.2 and 22.4). Hemicellulose is a collection of heteropolymers of five- and six-carbon sugars with short branched side connections. Figure 22.4 shows a typical hardwood hemicellulose molecule: glucuronic xylan. In contrast to cellulose, which is linear or one-dimensional, hemicellulose is a two-dimensional polymer, as shown in Figure 22.4. Table 22.5 shows common hemicellulose sugars from various woody-biomass sources (Mittal, 2006). Xylose is the dominant hemicellulose sugar found in angiosperm woody biomass. Hemicellulose oligomers can be partially extracted from woody biomass such as hardwood chips through solution in water at high temperatures (Garrote et al., 1999, 2001a,b; Liu et al., 2006; Liu and Amidon, 2007; Liu, 2010), and xylose or hemicellulose sugars can be fermented to ethanol by microorganisms (Jeffries and Jin, 2004; Qureshi et al., 2006; Liu et al., 2010). Hemicellulosic sugars may also be used as building blocks for biodegradable plastics (Keenan et al., 2004) or other products and chemicals that are currently made from petroleum (Liu et al., 2006).

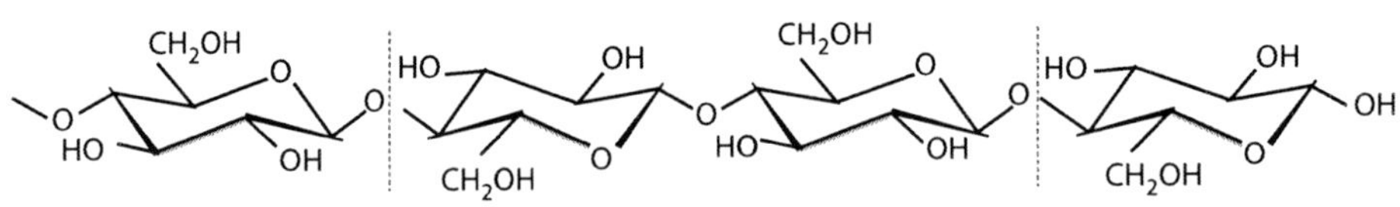

FIGURE 22.3
Stereochemical structure of cellulose. The repeating units are between the gray dotted lines.

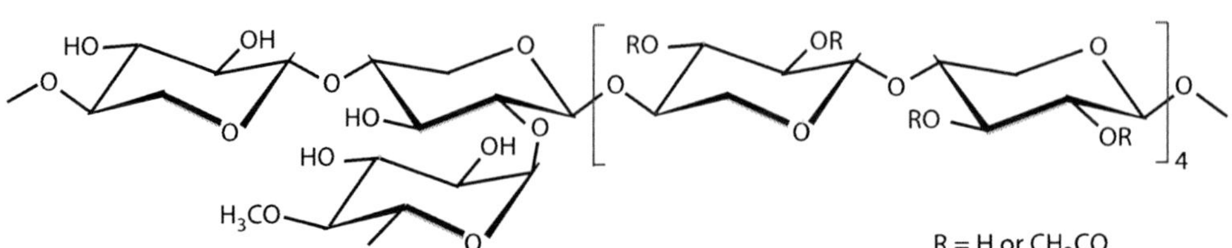

FIGURE 22.4
Stereochemical structure of a common hardwood hemicellulose: glucuronoxylan. The OH groups in the mannose units have been partially substituted by *O*-acetyl groups on the C-2 or C-3 positions (approximately 7 acetyls in 10 xylose units), that is, $R = CH_3CO$ or H.

TABLE 22.5

Sugar Compositions of the Hemicellulose Fraction of Selected Woody-Biomass Sources

Raw material	Mannose	Galactose	Xylose	Arabinose	Total Sugars	Reference
a. Hardwoods						
Aspen	2.3	0.8	16.0	0.4	19.5	Holtzapple and Benjamin (2003)
White birch	1.5	0.6	24.6	0.5	27.2	Holtzapple and Benjamin (2003)
Red maple	3.5	0.6	17.3	0.5	21.9	Holtzapple and Benjamin (2003)
b. Softwoods						
Balsam fir	12.4	1.0	4.8	0.5	18.7	Holtzapple and Benjamin (2003)
Jack pine	10.6	1.4	7.1	1.4	18.5	Holtzapple and Benjamin (2003)
White spruce	11.6	1.2	6.8	1.6	21.2	Holtzapple and Benjamin (2003)
Pinus pinaster	5.6	2.9	5.9	2.0	16.4	Ballesteros et al. (2000)
Douglas fir	13.6	3.9	2.7	1.2	21.4	Boussaid et al. (2000)
c. Non-Forest-Based Materials						
Corn stover	0.8	1.6	20.8	2.7	25.9	Kim and Lee (2006)
Arundo Donax L. (giant reed)	—	—	19.4	1.5	20.9	Nabarlatz et al. (2005)
Almond shells	—	—	26.1	2.4	28.5	Nabarlatz et al. (2004)
Corncobs	—	—	29.5	3.3	32.8	Kim and Lee (2006)
Barley husks	—	—	26.8	5.7	32.5	Garrote et al. (2004)
Barley straw	1.26	1.71	15.0	4.0	22.0	Holtzapple and Benjamin (2003)
Rice straw	1.6	0.4	13.0	4.0	19.0	Holtzapple and Benjamin (2003)
Wheat straw	0.72	2.16	16.9	2.1	20.0	Holtzapple and Benjamin (2003)
Sorghum straw	0.8	0.2	15.0	3.0	19.0	Allen et al. (2001)
Corn fiber		4.0	19.0	10.0	33.0	

The most energy-dense constituent of woody biomass is lignin, which makes up more than half of woody biomass. Lignin is an amorphous heteropolymer which consists mainly of methoxylated phenylpropylene alcohol. Figure 22.5 shows the three primary precursors of lignin (Alén, 2000). Lignin is a three-dimensional macromolecule with a carbon-to-oxygen atomic ratio greater than 2:1, as can be inferred from Figure 22.5. This is at least double that of hemicellulose or cellulose, in which the carbon-to-oxygen ratio is approximately 1:1. Therefore, lignin is a more energy-dense substance than sugars. Its aromatic ring structure also provides excellent functional chemical sources when depolymerized. However, there is no efficient depolymerization method currently available that acts directly and specifically on lignin. In fact, lignin acts as a natural barrier to biological degradation of cellulose and hemicellulose in woody biomass. Development of conversion technologies other than direct burning on-site holds the key to the viability of a sustainable biorefinery.

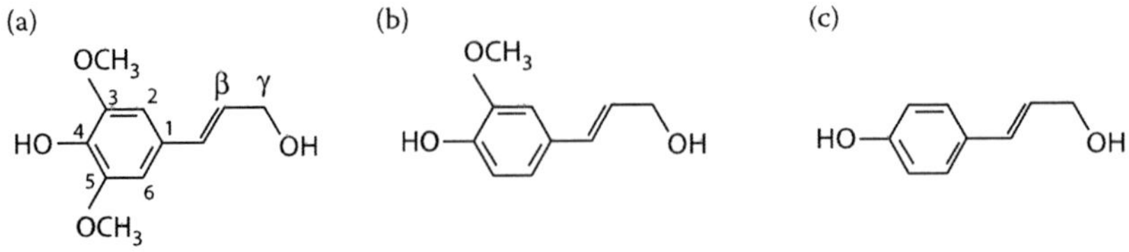

FIGURE 22.5
Precursors of lignin. (a) *Trans*-snapyl alcohol, syringyl unit (S), (b) *trans*-coniferyl alcohol, guaiacyl unit (G), and (c) *trans*-p-coumaryl alcohol, *p*-hydroxyphenyl unit (H).

The structure of lignin is more complex than that of carbohydrates (cellulose and hemicelluloses). The methoxylated phenylpropylene alcohol monomeric units are interconnected by a variety of stable carbon–carbon and carbon–oxygen–carbon (ether and ester) linkages (Conner and Lorenz, 1986), as shown in Figure 22.6. The linkages are identified based on the carbon positions indicated in Figure 22.5a. The β-*O*-4 interunit linkages are the most abundant in lignin and are estimated to be as high as 50% in softwoods and almost 60% in hardwoods (Sjöstrom, 1993). More than two-thirds of the phenylpropane units are linked by ether bonds and the rest by carbon–carbon bonds. Structurally, lignin is a three-dimensional macromolecule. While the lignin of gymnosperms (also called softwoods, conifers, needle trees, or evergreens) is primarily an

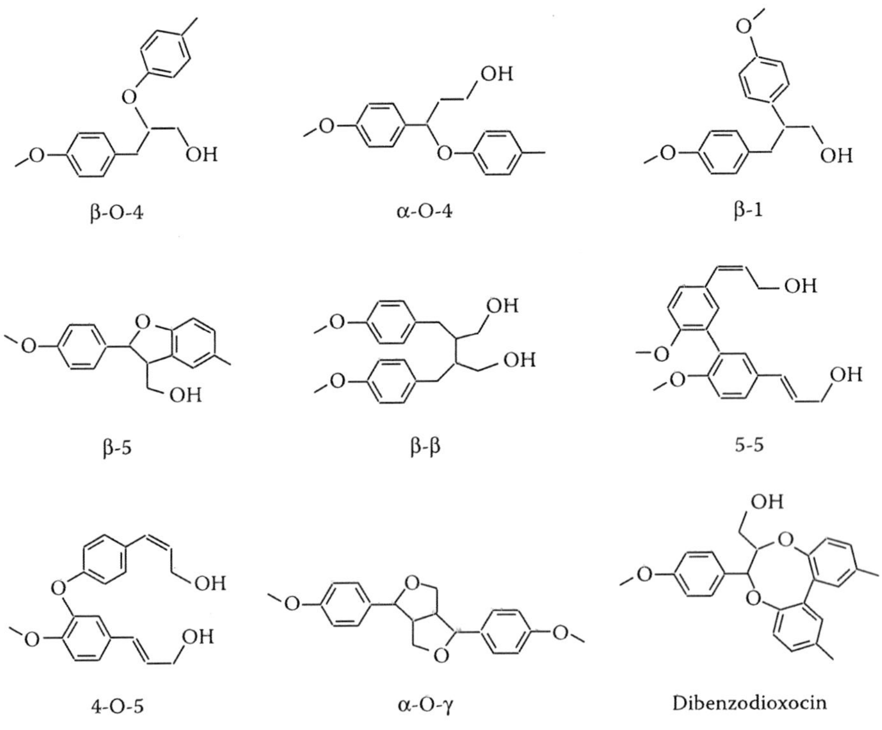

FIGURE 22.6
Common linkages between phenylpropane units in lignin.

enzyme-mediated dehydrogenative polymerization product of coniferyl alcohol (Fengel and Wegener, 1989), the lignin of angiosperms (also called hardwoods, deciduous, or broad-leaved trees, as well as of grasses) is derived primarily from a mixture of coniferyl and sinapyl alcohols.

Lignins can be divided into several classes according to their structural elements. The guaiacyl-syringyl lignin (GS-lignin) found in hardwoods is a copolymer of coniferyl and sinapyl alcohols (the precursors of G- and S-units, respectively), with the G/S ratio varying from 4:1 to 1:2. Guaiacyl lignin (G-lignin), which occurs in almost all softwoods, is largely a polymerization product of coniferyl alcohol. The different contribution of *p*-hydroxycinnamyl alcohols to the biosynthesis of hardwood and softwood lignins causes significant differences in their structures, including the contents of different types of bonds and the main functional groups, for example, methoxyl (OMe), phenolic hydroxyl (PhOH), aliphatic hydroxyl, carbonyl, and carboxyl groups. Softwood lignin is more condensed than hardwood lignin because of the difference in substitution degree of the lignin precursors: substituted C_3 and unsubstituted C_5 in coniferyl alcohol, versus substituted C_3 and C_5 in sinapyl alcohol. Significant differences observed between the hardwood and softwood lignin structure indicate the different physical and chemical properties of these two lignin types.

A mature industry which uses woody biomass to produce pulp, chemicals, or both is the pulp and paper manufacturing industry. The throughput of a single pulp and paper mill is in the order of thousands of tons per day. However, in the paper industry, the product focus is on cellulose, which is primarily used to make paper. Before the 1940s, lignin was considered a waste product and a severe environmental problem. This lignin was in an aqueous mixture with degraded hemicelluloses and cooking chemicals, which are collectively called *black liquor*. Black liquor is a gooey substance that looks like molasses. In the 1930s and 1940s, paper engineers learned how to burn lignin and hemicelluloses while recovering the cooking chemicals, and therefore that our rivers became much cleaner, and much less fossil energy was used to make paper—at least in the developed world. In a modern chemical pulp mill, net electrical energy is exported to the power grid. This makes many wood-based production facilities good examples of industries that can run predominately on renewable energy. What was once a pollutant is now a useful fuel that has displaced fossil fuels.

Still, current technologies for the chemical utilization of wood use a "destructive strategy" to obtain a single, relatively pure component (cellulose). Other fractions are either wasted or severely degraded and used for power generation that provides a low added value. The "biorefining" philosophy provides an alternative approach to utilization of lignocellulosic biomass which enables higher value recovery. Lignocellulosic biomass can be "fractionated" into its main components by sequential treatments to yield separate streams that can be used for different product applications, thus maximizing the benefits of a renewable and complex resource that will become increasingly important as fossil sources become more constrained and more expensive.

Therefore, woody biomass occupies a niche position in the development of a sustainable renewable economy. It exists in the quantity and quality needed for serious consideration and has a current technology base for its utilization. In other words, an industrial base already exists to use woody biomass. However, woody biomass remains the most difficult source of chemicals and energy to convert efficiently because of its highly stable structure (Liu, 2010). This chapter will focus on the separation and recovery of hemicelluloses from woody biomass.

22.2 Hemicelluloses and Wood-Based Biorefinery

The major constituents of woody biomass (cellulose, hemicelluloses, and lignin) cannot be simultaneously isolated as polymers, and therefore several processes must be used in sequence, each involving the degradation of one or more of the polymers. One approach is to treat the woody biomass initially with a hemicellulose-degrading step. Hemicellulose separation is a preferable approach to the more commonly proposed pretreatment techniques. For example, one could use extraction or autohydrolysis in the absence of mineral acids or caustics, leaving both cellulose and lignin as essentially undegraded polymers (Conner and Lorenz, 1986; Garrote et al., 1999, 2001a,b; Liu and Wyman, 2005; Mosier et al., 2005a,b; Amidon et al., 2008). Hot-water extraction uses water and lignocellulosic biomass as the only reagents. Hydronium ions from both water and *in situ*-generated compounds (e.g., acetic, formic, uronic, and phenolic acids) catalyze the hemicellulose depolymerization (Overend and Chornet, 1989; Heitz et al., 1991; Aoyama et al., 1995; Saska and Ozer, 1995; Weil et al., 1997; Garrote et al., 2001a,b; Liu, 2008).

Hot-water extraction has a wide range of applications (Heitz et al., 1991; Amidon and Liu, 2009). First, there are fractionation or pulping processes, in which hemicelluloses are removed with selectivity toward cellulose degradation and toward splitting the α- and β-aryl ether bonds of lignin (Lora and Wayman, 1978; Schultz et al., 1984). A second application is defibering for fiberboard production, which is carried out in processes using high-pressure steam. A third application is as a pretreatment for the enzymatic hydrolysis of cellulose (Dekker and Wallis, 1983; Grethlein and Converse, 1991). That study (Heitz et al., 1991) focused on the degradation kinetics of *Eucalyptus globulus* wood hemicellulose to assess which operating conditions produce the maximum concentration of sugar oligomers. Because xylan is the main component of hardwood and grass hemicelluloses, xylooligomers and xylose are the main products obtained in hydrothermal treatments of this raw material. Sugar-degradation products (such as furfural or hydroxymethylfurfural) and acetic acid (generated from acetyl groups) can also appear in the reaction media.

There are many applications for this hemicellulose stream. Xylooligomers can be used as novel sweeteners in food additives (Aoyama, 1996; Ichikawa and Mitsumura, 1996). Health applications of indigestible oligosaccharides have also been reported. For example, Imaizumi et al. (1991) found that a xylooligosaccharide-based diet reduced the blood concentrations of sugars and lipids in diabetic rats, and Toyoda et al. (1993) reported that xylooligosaccharides improved calcium absorption by rats. The phagocytic activity of neutrophils in mice was enhanced by either oral or intraperitoneal administration of xylooligosaccharides (Aoyama, 1996). Moreover, *Clostridium difficile* increased resistance to infection in mice. May et al. (1995) reported that this phenomenon was caused by xylo- and fructo-oligosaccharides. Improvements in the gastrointestinal health of rats caused by a xylooligosaccharide-containing diet have also been reported by Campbell et al. (1997). In relation to human health, xylooligosaccharides selectively enhance the growth of bifidobacteria, thus promoting a favorable intestinal environment (Okazaki et al., 1990, 1991). The largest potential demand for hemicelluloses would be conversion to liquid fuels in a world thirsty for transportation fuel. Therefore, an integrated biomass-conversion facility needs a hemicellulose separation step to extract value and provide much-needed feedstocks for materials, chemicals, and energy.

Figure 22.7 shows a schematic of a biorefinery with woody biomass as the feedstock. Wood chips can be replaced or supplemented by other woody biomass such as agricultural

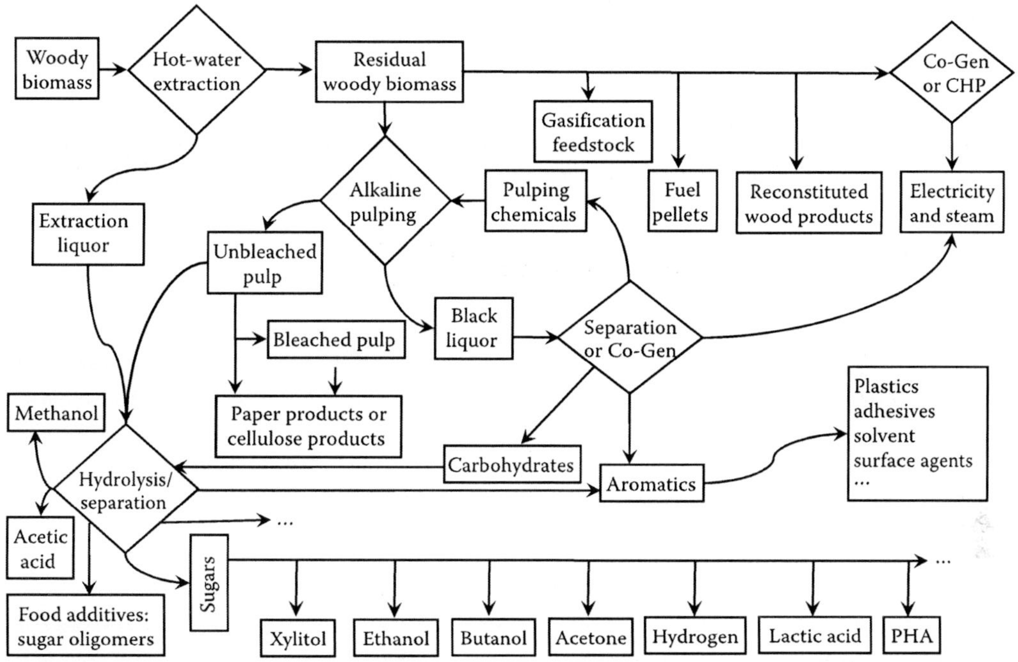

FIGURE 22.7
Schematic of a biorefinery. (Adapted from Amidon, T.E., Liu, S.J. 2009. *Biotechnol. Adv.* 27: 542–550.)

residues. Although woody-biomass extracts can be turned into chemicals and energy, the residual materials from hot-water extraction can have many applications as well, including chemicals, energy, paper products, and particleboard. For example, stronger and lighter board materials can be produced from hemicellulose-extracted woody biomass because hemicellulose contributes more to the weight than to the strength of reconstituted materials and products. This approach to the biorefinery is an incremental sequential deconstruction, harvesting valuable products along the path of woody-biomass deconstruction.

There are 5–9 main component stages in a wood-based biorefinery (Figure 22.7): (1) hot-water extraction (Liu, 2008, 2010); (2) hydrolysis of hot-water extracts (Hu et al., 2010); (3) separation of xylan, sugars, and acetic acid (Liu et al., 2008); (4) fermentation of sugars to ethanol or bioplastics (Liu et al., 2010); (5a) pulping of hot-water-extracted wood chips followed by bleaching and papermaking (Liu et al., 2009); (5b) burning of the hot-water-extracted residual woody biomass and organics for CHP (combined heat and power) use; (5c) use of extracted wood chips or woody biomass for reconstituted wood products such as particleboard; (5d) use of the extracted woody biomass to produce wood-fuel pellets; or (5e) further separation and hydrolysis of the extracted woody biomass to platform chemicals such as aromatics and wood sugars. These components are designed so that the process can be readily adapted to fit into existing industrial operations.

Therefore, hemicellulose separation and utilization play an important role in the use of woody biomass for sustainable chemical and energy developments. This is the first step in a sequential incremental deconstruction of woody biomass for efficient utilization.

22.3 Hot-Water Extraction: Kinetics and Process Characterization

Hemicellulose and extractives are the two constituents of woody biomass which can be readily extracted by hot water. Extractives, although low in quantity (Tables 22.2 and 22.6), are extractable compounds in the woody biomass that can be readily dissolved with organic solvents or water under room temperature and atmospheric conditions. In addition, there are inorganic compounds present in the woody biomass. Over 70 metals, earth elements, and inorganic compounds are found in woody biomass. Potassium and calcium are the most abundant elements, followed by magnesium and phosphorus (Gullichsen and Paulapuro, 2000). In hardwood, for example, calcium contributes 0.08–0.2% of the dry stemwood mass and 0.85–3.05% of the stem bark mass. Magnesium contributes 0.02–0.04% of the dry hardwood stem mass and 0.07–0.11% of the dry hardwood stem bark mass. Table 22.6 shows the major components in the extractives and hemicellulose fractions of wood. These components are the first ones that can be extracted from wood. Extraction of the readily hydrolyzable carbohydrates with hot water, in the absence of added mineral acids or bases, is desirable to facilitate the recovery and use of the hydrolyzate components. When the heating values of the residual chips are recovered after extraction, there is no reduction in value due to caustic or metal hydroxides and no increase in corrosion from mineral acids. Furthermore, the environmental and recovery side effects are avoided because caustic or sodium is not added to the process streams and no by-products from mineral acid neutralization are produced.

22.3.1 Extraction Kinetics

As shown in Table 22.6, hemicelluloses are a group of 2-D branched polymers that are heterogeneous in nature and can be composed of multiple five- and six-carbon sugars, including glucose, xylose, mannose, rhamnose, and arabinose. The hemicelluloses have an average degree of polymerization of 200–500, which is much less than that of cellulose (Glaudemans and Timell, 1958; Goring and Timell, 1960; Timell, 1960; Koshijima et al., 1965) (Table 22.7).

The xylose molecules in the backbone of xylan are linked by β (1-*O*-4) bonds as shown in Figure 22.4. Residues on the backbone include acetyl groups at the C-2 or C-3 positions with a substitution rate of approximately 7 per 10 xylose residues. In addition, (1 → 2)-linked 4-methoxyl-α-D-glucuronic acid residues occur at an average rate of approximately 1 per 10 xylose residues (Fengel and Wegener, 1989). Figure 22.4 shows the structure of hardwood xylan with its associated side groups. It can be observed that the ring or chair structures (5-carbon + 1-O backbones of sugar units) are interconnected through –O– bonds and also connected to acetyl (Ac) and methyl groups through –O– bonds. All the interresidue –O– bonds could be broken by hydrolysis, that is

$$\text{–O–} + \text{H–OH} \rightarrow \text{–OH} + \text{HO–} \tag{22.1}$$

The two residues are split apart by ending each of them with a hydroxyl (–OH) group. Thus, hydrolysis of xylan forms sugars, acetic acid, and methanol.

Although efforts have been directed toward modeling chemical reactions involving wood chips or wood fibers (e.g., Kraft pulping, sulfite pulping, pulp bleaching, extraction, and hydrolysis), most investigators have either implicitly assumed or clearly stated that

TABLE 22.6

Major Components of Hemicellulose and Extractives

	Type	Distribution (wt%)	
		Softwoods	Hardwoods
Hemicellulose		**25–30**	**25–35**
Galactoglucomannan (1:1:3)	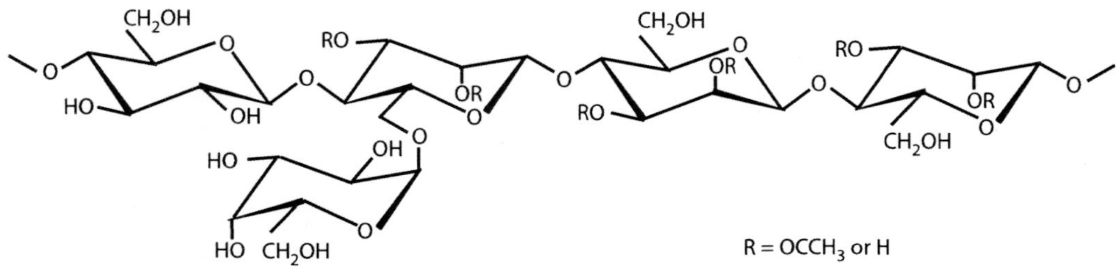 R = $OCCH_3$ or H	5–8	0
(Galacto) glucomannan (0.1:1:4)	The OH groups in the xylose units were partially substituted by *O*-acetyl groups on the C-2 or C-3 positions, that is, R = CH_3CO or H	10–15	0
Glucomannan (1:2-1:1)	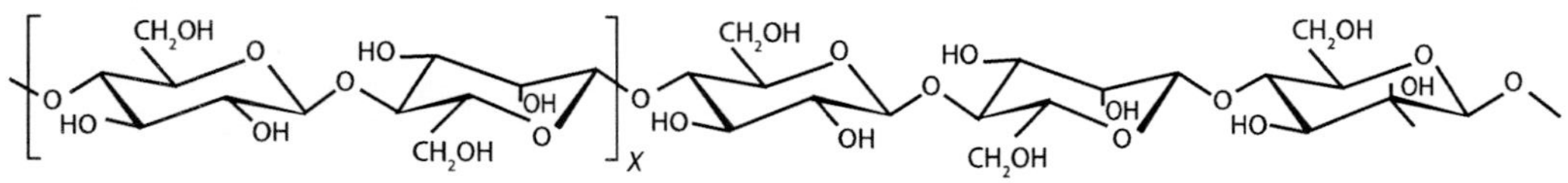	0	2–5
Arabinoglucurono-xylan	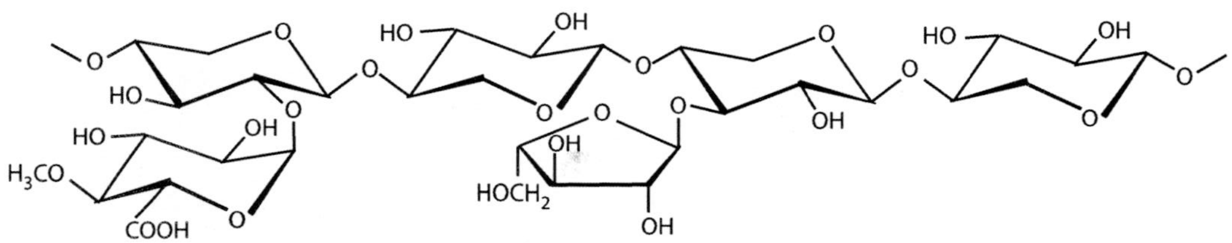	7–10	Trace

continued

TABLE 22.6 (continued)

	Type	Distribution (wt%)	
		Softwoods	Hardwoods
Glucuronoxylan	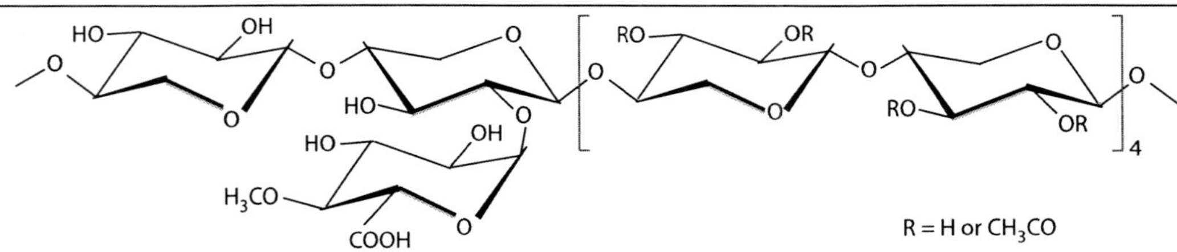 The OH groups in the xylose units were partially substituted by O-acetyl groups on the C-2 or C-3 positions (about 7 in 10 xylose units), that is, R = CH_3CO or H	Trace	15–30
Extractives		**5–8**	**2–4**
Aliphatic and alicyclic	Terpenes; terpenoids; esters; fatty acids; alcohols; . . .		
Phenolics	Phenols; stilbenes; lignans; isoflavones; . . .		
Carbohydrates	Arabinose; galactose; glucose, xylose; raffinose; starch; pectinic material		
Inorganics	Ca, K, Mg, Na, Fe, SO_4^{2-}, Cl^-, . . .		
Others	Cyclitols; tropolones; amino acids, protein, alkaloids, . . .		
Ash		0.2–0.5	0.2–0.8

Source: From Liu, S.J. 2010. *J. Biotech. Adv.* 28: 566. With permission.

TABLE 22.7

Reported Molecular Weights of Xylan from Various Sources

Species	MW	MN	DP	Source
Sugar maple (*Acer saccharum*)		51,000		Glaudemans and Timell (1958)
White birch (*Betula papyrifera*)			190	Goring and Timell (1960)
	76,000	51,000		Glaudemans and Timell (1958)
Yellow birch	*75,000*	46,000		Glaudemans and Timell (1958)
White elm	*70,000*	46,000		Glaudemans and Timell (1958)
Trembling aspen (*Populus tremuloides*)	31,500		200	Koshijima et al. (1965)

Note: The data were measured based on osmometry or light scattering (with numbers *italicized* in the table).

their underlying reactions are pseudo-homogeneous processes and have neglected the heterogeneous nature of the process in the literature (Saeman, 1945; Vroom, 1957; Conner, 1984; Maloney et al., 1985; Kim and Lee, 1987; Overend and Chornet, 1987; Abatzoglou et al., 1992; Carrasco and Roy, 1992; Montane et al., 1994; Zhuang and Vidal, 1996; Montane et al., 1998; Garrote et al., 1999, 2001a,b, 2002; Nabarlatz et al., 2004, 2005; Kabel et al., 2007; Mittal et al., 2009a,b). In fact, pseudohomogeneous or even homogeneous reaction models (with model chemical compounds) for reactions involving wood, fibers, or both are commonly seen in the pulp and paper literature. The reactive components of the wood or fiber are implicitly assumed to be in the same phase as the reactive agents in the liquor or the liquid phase. Although considering reactions involving wood or fibers as homogeneous reduces the complexity of the kinetic model, the discussions and conclusions based on this model become less reliable.

Based on surface reaction theory, Liu (2003, 2004) and Yang and Liu (2005) inserted an LWWH type of kinetics into bleaching and pulping reactions involving wood and fibers. This introduction did explain why the order of reaction with regard to the reactants in the liquid phase changes with concentration. There are multitudes of chemical components in wood that participate in the reactions. Liu (2008) introduced the concept of surface reaction to the kinetic study of hot-water extraction and hydrolysis of wood. In the case of acid hydrolysis, hydrogen ions act as a catalyst to soften the glycosidic bonds and induce their depolymerization. The glycosidic bond breakage is enhanced when the breakaway oligomers are attracted away from the solid surface. High temperature and high electrolyte content make it easier for the oligomers to be in the liquor phase.

The hydrolysis mechanism can become very complex when mass transfer is coupled with chemistry. However, by viewing one component at a time, one can deduce that

1. Hydrogen ions or small strong hydrogen-bond-forming molecules from the bulk liquor migrate to the surface of the solid particle
2. Chemisorption (hydrogen-bond formation, for example) occurs on the solid surface
3. Hydrogen ions (protons) and xylan react at the surface
4. Dissoluble xylooligomer desorbs or cleaves
5. Dissolved xylooligomer diffuses and is transported into the bulk liquor

This is typical of LHHW kinetics. However, the solid phase is not acting as a catalyst. The solid phase contains both the reactant and the desired product.

Under normal conditions, transport steps 1 and 5 are probably unimportant, especially when the effective thickness of the wood chips is less than the critical thickness, that is, 2–3 mm (Wilder and Daleski, 1964). These transport steps are negligible compared to the overall observed extraction rate. However, recognizing the fact that the extraction reactions occur on the woody-biomass surface, the extent of extraction can be expected to increase with decreasing size of woody-biomass particles as more surface area becomes exposed and available for extraction. In wood-chip form, the hemicelluloses and extractives that are not exposed to an external surface are very difficult to dissolve in the extraction liquor, because this would require diffusion of the molecules through the cellulose matrix. Step 3, the surface reaction step, can be complex in nature. However, there exists a controlling substep in the chemistry of xylan hydrolysis: breakage of a glycosidic bond. Therefore, step 3 can be simplified to an elementary step representing the glycosidic bond breakage. The chemical reaction involved is said to be the rate-controlling step. In this case, the overall reaction rate is dependent on steps 3–5 directly. We address the intrinsic kinetics, that is, steps 3–5, only.

Under hot-water extraction conditions, the hydrogen ion concentration is initially very low. However, acetyl groups are present in woody biomass because they are associated both with lignin and with hemicellulose. Hydration of the acetyl groups leads to the acidification of the liquor and thus to formation of hydrogen ions.

According to these considerations, the extraction–hydrolysis process for solid wood chips can be represented by the following simplistic model (Liu, 2008):

Reaction steps forming proton and/or hydronium ion:

$$H_2O(aq) \rightleftarrows H^+(aq) + OH^-(aq) \tag{22.2}$$

$$R\text{–}POAc(s) + H^+(aq) \rightleftarrows R\text{–}POAc \cdot H^+(s) \tag{22.3}$$

$$R\text{–}POAc \cdot H^+(s) + H_2O \rightleftarrows R\text{–}POH \cdot H^+(s) + HOAc(aq) \tag{22.4}$$

$$R\text{–}POAc \cdot H^+(s) + H_2O \rightleftarrows R\text{–}OH \cdot H^+(s) + HPOAc(aq) \tag{22.5}$$

$$HOAc(aq) \rightleftarrows H^+(aq) + OAc^-(aq) \tag{22.6}$$

$$R\text{–}POH \cdot H^+(s) \rightleftarrows R\text{–}POH(s) + H^+(aq) \tag{22.7}$$

$$R\text{–}OH \cdot H^+(s) \rightleftarrows R\text{–}OH(s) + H^+(aq) \tag{22.8}$$

Reaction steps to solubilize hemicellulose:

$$R\text{–}X_n OH(s) + H^+(aq) \rightleftarrows R\text{–}X_n OH \cdot H^+(s) \tag{22.9}$$

$$R\text{–}X_n OH \cdot H^+(s) + H_2O(aq) \rightarrow R\text{–}X_m OH \cdot H^+(s) + HX_s OH(aq) \tag{22.10}$$

Reaction steps to reduce the chain length inside the liquor (hydrolysis):

$$HX_nOH(aq) + H^+(aq) \rightleftarrows HX_nOH \cdot H^+(aq) \tag{22.11}$$

$$HX_nOH \cdot H^+(aq) + H_2O(aq) \rightarrow HX_mOH \cdot H^+(aq) + HX_sOH(aq) \tag{22.12}$$

where $m + s = n$, R– denotes the cellulose, lignin, or both bonded to woody biomass, and P represents a segment or subunit of hemicellulose or lignin. X_n represents an n-xylooligomer middle group, that is, $(\text{–O–}C_5H_8O_3\text{–})_n$,

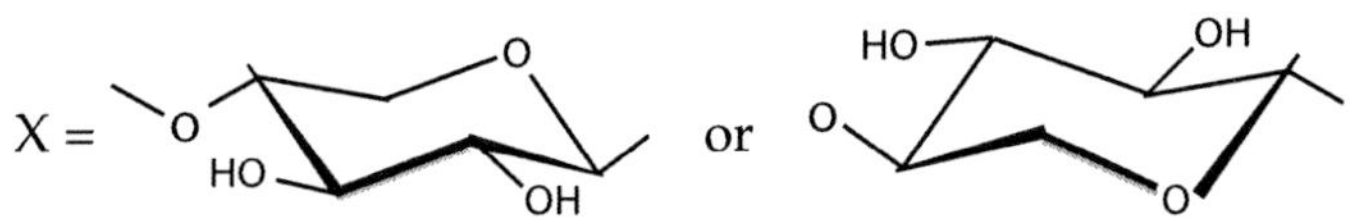

whereas HX_nOH is an n-xylooligomer. HOAc represents the acetic acid molecule, where $Ac = CH_3CO$.

Equations 22.3 and 22.9 represent the adsorption of hydrogen ions onto the woody-biomass surface. Equations 22.4, 22.5, and 22.10 represent the surface reaction in which one acetyl group, one acetylated group, or one xylooligomer is cleaved from the woody biomass. Note that the actual entity participating in the extraction and hydrolysis reactions can be either H^+ or H_3O^+. Kinetically, the steps as shown in Equations 22.2 through 22.12 can be taken to represent the extraction and hydrolysis reactions.

Monomeric sugars can further dehydrate in the presence of protons to dehydrated products such as furfural, humic acid, levulinic acid, and others:

$$HX_1OH \cdot H^+(aq) \rightarrow H^+ + H_2O + \text{furfural, and other dehydration products} \tag{22.13}$$

Besides the extraction reaction, the polysaccharides can condense back onto the woody biomass:

$$R\text{–}X_nOH \cdot H^+(s) + HX_sOH(aq) \rightarrow R\text{–}X_{n+s}OH \cdot H^+(s) + H_2O(aq) \tag{22.14}$$

or

$$R\text{–}X_nOH(s) + HX_sOH \cdot H^+(aq) \rightarrow R\text{–}X_{n+s}OH \cdot H^+(s) + H_2O(aq) \tag{22.15}$$

In addition, the dehydration polymerization of the xylooligomers can be written as:

$$HX_nOH \cdot H^+(aq) + HX_mOH(aq) \rightarrow HX_{n+m}OH \cdot H^+(aq) + H_2O(aq) \tag{22.16}$$

The polymerization reactions (22.14) through (22.16) can have a significant effect on the extraction and hydrolysis processes, especially when woody-biomass extracts are recycled.

Reactions (22.4), (22.5), (22.10), and (22.14) through (22.16) are pseudoelementary steps. Numerous studies have been performed with the intent of revealing the chemistry of these steps. However, for the present purposes, they will be considered as elementary steps. The reaction rate so defined represents that of the rate-controlling substeps based on their detailed chemistry.

With the reaction mechanism given by Equations 22.2 through 22.16, it is possible to develop a reaction rate expression for any given component of interest. However, the resulting reaction rate expression could potentially involve a large number of kinetic parameters. These kinetic parameters are not easily determined theoretically based on the chemistry and physics of matter. Therefore, further simplification is needed. The following assumptions are therefore made:

1. Condensation reactions, that is, reactions (22.14) through (22.16), are neglected
2. Breakage of glycosidic bonds has an equal rate or reactivity as long as they are in the same phase
3. Affinities of glycosidic bonds to hydrogen ions are the same as long as they are in the same phase

These three assumptions form the basis for the discussions in the rest of this section.

It is assumed here that adsorption and desorption reactions occur at much faster rates than those of surface reactions on the woody-biomass surface. Therefore, one can assume that the adsorption steps are always in equilibrium. The adsorption process can be represented by the elementary reversible reactions shown in Equation 22.9. Therefore, at the time scale of the surface reactions, the equilibrium of adsorption of hydrogen ions is given by

$$K_S = \frac{\theta_{n-H^+}}{\theta_{nV} C_{H^+}} \tag{22.17}$$

where K_S is the adsorption equilibrium constant of H^+ onto X_n on the biomass solid surface (or $R\text{–}X_nOH$); θ_{n-H^+} is the concentration or fraction of X_n on the biomass solid surface that is associated with H^+; θ_{nV} is the concentration or fraction of X_n on the biomass solid surface that is not associated with H^+; and C_{H^+} is the concentration of H^+ in the extract liquor. The total concentration of xylan on the biomass solid surface is given by

$$\theta_{n-H^+} + \theta_{nV} = \theta_n. \tag{22.18}$$

Substituting Equation 22.17 into Equation 22.18, the result is

$$\theta_{n-H^+} = \frac{K_S C_{H^+} \theta_n}{1 + K_S C_{H^+}}. \tag{22.19}$$

Ideally, θ_n is the surface concentration of component n on the woody-biomass surface. θ_n can be conveniently expressed as moles per g of inert (or water-insoluble/insoluble) biomass. The total surface area can be assumed to be constant during the extraction process.

The following is a description of the kinetic reaction for the intermediate formation in the liquid phase, as shown in Equation 22.9:

$$K_H = \frac{C_{n-H^+}}{C_{nV} C_{H^+}} \tag{22.20}$$

where K_H is the equilibrium constant of reaction (22.9) for H^+ to be chemically associated with HX_nOH in the extract liquor; C_{n-H^+} is the concentration of HX_nOH in the extraction liquor that is chemically associated with H^+; and C_{nV} is the concentration of HX_nOH in the extraction liquor that is not associated with H^+. The concentration of HX_nOH in the extraction liquor is then given by

$$C_{n-H^+} + C_{nV} = C_n. \tag{22.21}$$

Therefore,

$$C_{n-H^+} = \frac{K_H C_{H^+} C_n}{1 + K_H C_{H^+}}. \tag{22.22}$$

The reaction rates, based on reactions (22.10) and (22.12) respectively, are given by

$$r_{\theta n \to s} = k_{en}\theta_{n-H^+} = k_{en}\frac{K_S C_{H^+}}{1 + K_S C_{H^+}}\theta_n, \tag{22.23}$$

$$r_{n \to s} = k_{hn}C_{n-H^+} = k_{hn}\frac{K_H C_{H^+}}{1 + K_H C_{H^+}}C_n. \tag{22.24}$$

Neglecting condensation reactions, the formation of s-xylooligomers in the liquor can be represented as

$$\begin{aligned} r_s = \sum_{n=s+1}^{N} r_{n \to s} + \sum_{n=s}^{N} r_{\theta n \to s} - \sum_{l=1}^{[s/2]} r_{s \to l} &= \sum_{n=s+1}^{N} k_{hn}\frac{K_H C_{H^+}}{1 + K_H C_{H^+}}C_n \\ + \sum_{n=s}^{N} k_{en}\frac{K_S C_{H^+}}{1 + K_S C_{H^+}}\theta_n - \sum_{n=1}^{s-1} k_{hn}\frac{K_H C_{H^+}}{1 + K_H C_{H^+}}C_s \end{aligned} \tag{22.25}$$

Cleavage of xylooligomers from the woody biomass when the condensation reactions are neglected can be expressed as

$$r_{\theta m} = -\sum_{s=1}^{m} k_{em}\frac{K_S C_{H^+}}{1 + K_S C_{H^+}}\theta_m + \sum_{n=m+1}^{N} k_{en}\frac{K_S C_{H^+}}{1 + K_S C_{H^+}}\theta_n. \tag{22.26}$$

The reaction is initially of nearly zeroth order with respect to xylan content and becomes nearly first-order close to the complete dissolution of xylan from woody biomass.

Because the reaction sites on the woody biomass are of equal reactivity, the effective extraction rate constant is given by

$$k_E = mk_{em}\frac{K_S C_{H^+}}{1 + K_S C_{H^+}} \tag{22.27}$$

that is, the depolymerization or extraction occurs equally at any xylo-*O*-xylo bond, and each "xylan" molecule contributes to one equal reactive site each time:

$$r_{\theta m} = -k_E\theta_m + \sum_{n=m+1}^{N} \frac{k_E}{n}\theta_n. \tag{22.28}$$

The solution for the xylooligomer distribution in the extract liquor can be determined using the same simplification. Let

$$k_H = (n-1)k_{hn}\frac{K_n C_{H^+}}{1+K_n C_{H^+}}, \tag{22.29}$$

that is, the depolymerization or hydrolysis occurs equally at any interxylose residue –O– bond.

The formation rate of the s-xylooligomer in the liquor, for s > 1, is given by

$$r_s = \sum_{n=s+1}^{N} \frac{2k_H C_n}{n-1} + R_{SL}\sum_{n=s}^{N}\frac{k_E\theta_n}{n} - k_H C_s, \tag{22.30}$$

where $R_{SL} = W_S/V_L$ is the solid mass to liquid volume ratio, W_s is the mass of the inert solid (not dissoluble), and V_L is the volume of the water. For monosaccharide units,

$$r_1 = \sum_{n=2}^{N} \frac{2k_H C_n}{n-1} + R_{SL}\sum_{n=1}^{N}\frac{k_E\theta_n}{n} - k_d C_1, \tag{22.31}$$

where k_d is the degradation rate constant. The degradation reaction is illustrated by Equation 22.13.

The extraction of acetyl units (or acetic acid residues) from the woody biomass is described by Equations 22.2 through 22.4. There are three types of acetyl units, which are distinguished in reactions (22.2) through (22.4): acetyl groups on woody biomass, bonded acetyl groups in extraction liquor (HPOAc), and free acetyl groups in extraction liquor (HOAc or OAc^-). Because H_2O molecules are abundant in an aqueous medium, one can effectively neglect the H_2O "adsorption" step. Neglecting the reverse reactions depicted in Equations 22.2 and 22.3 or in the absence of acetic acid dehydration, polymerization, and condensation, the reaction rates can be obtained as

$$r_{\theta Ac} = -k_E\theta_{Ac}, \tag{22.32}$$

$$r_{BAc} = \alpha k_E\theta_{Ac} - k_{Ac}C_{Bac}, \tag{22.33}$$

$$r_{Ac} = (1-\alpha)k_E\theta_{Ac} + k_{Ac}C_{Bac}, \tag{22.34}$$

where r_{BAc} is the rate of formation of bonded acetyl groups in the extraction liquor, r_{Ac} is the rate of formation of "free" acetyl groups in the extraction liquor, $r_{\theta Ac}$ is the rate of formation of acetyl groups in the woody biomass, θ_{Ac} is the concentration of acetyl groups in the woody biomass, C_{BAc} is the concentration of "bound" acetyl in the extraction liquor, and α is the ratio of acetyl groups to the sum of acetyl and glycosidic groups in POAc.

The extraction can be examined starting with a mole balance on the m-th xylooligomer on the woody-biomass surface:

$$-\frac{d\theta_m}{dt} = -r_{\theta m} = k_E\theta_m - \sum_{n=m+1}^{N} \frac{k_E}{n}\theta_n. \quad (22.35)$$

Mass balance of the s-xylooligomer in the liquor leads to

$$\frac{dC_s}{dt} = r_s = \sum_{n=s+1}^{N} \frac{2k_H C_n}{n-1} + R_{SL}\sum_{n=s}^{N} \frac{k_E\theta_n}{n} - k_H C_s, \text{ for } s > 1. \quad (22.36)$$

and

$$\frac{dC_1}{dt} = r_1 = \sum_{n=2}^{N} \frac{2k_H C_n}{n-1} + R_{SL}\sum_{n=1}^{N} \frac{k_E\theta_n}{n} - k_d C_1. \quad (22.37)$$

The dynamic behavior of batch extraction can be simulated by solving Equations 22.36 and 22.37 simultaneously. This kinetic model yields detailed information on the extract composition as it changes over time. Figures 22.8 through 22.10 show one set of solutions at a time scale before the dehydration and degradation of sugars have become important.

Figure 22.8 shows that the monosaccharide concentration in the extract liquor increases with extraction time, very slowly at first and more sharply as time goes on. On the other hand, the polysaccharide concentration increases quickly over short extraction times and slows down at longer durations. Note that at extremely long extraction times, there will be no increase in extract concentration because the extractable carbohydrates in the solid phase have been exhausted.

Figure 22.9 shows the change in degree of polymerization (DP) with extraction time. Observe that the degree of polymerization decreases during the extraction. The DP of the

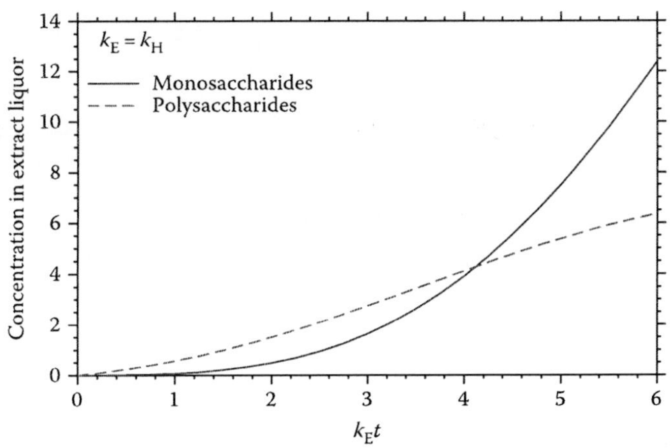

FIGURE 22.8
Model simulation of batch extraction: variation of monosaccharides and polysaccharides in the extract liquor as a function of time, all as relative values.

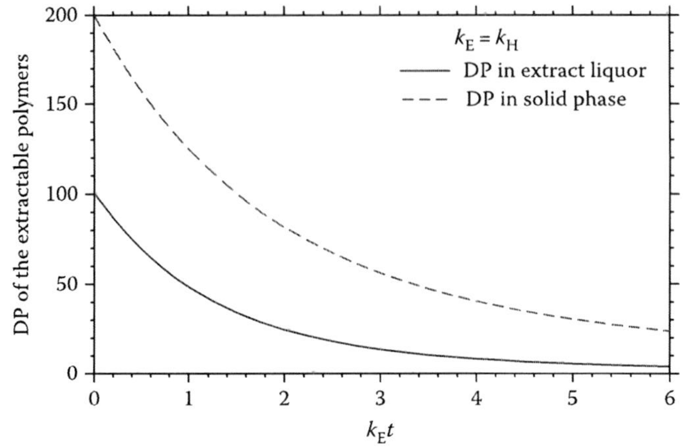

FIGURE 22.9
Degree of polymerization as a function of time, all as relative values. (Amidon, T.E., Liu, S.J. 2009. *Biotechnol. Adv.* 27: 544. With permission.)

extractable carbohydrates in the solid phase (with wood) is higher than that in the extract liquor. The decrease in DP is slower at long extraction times.

Figure 22.10 shows the acetyl concentrations in the extract liquor. It can be seen that the "free" acetic acid in the extract liquor increases sharply at short extraction times and then largely levels off. The acetyl groups associated with the polysaccharides in the extract liquor also increase sharply over short time periods and then decrease slightly over longer extraction times.

Most of the experimental data in the literature were collected without knowledge of the detailed distribution of substances either in the woody biomass or in the extraction liquor. A lumped-kinetics approach was commonly used. The total dissolution rate can be obtained by adding up all the extraction rates. Summing Equation 22.35 on a monosaccharide unit (residue) basis, the result is

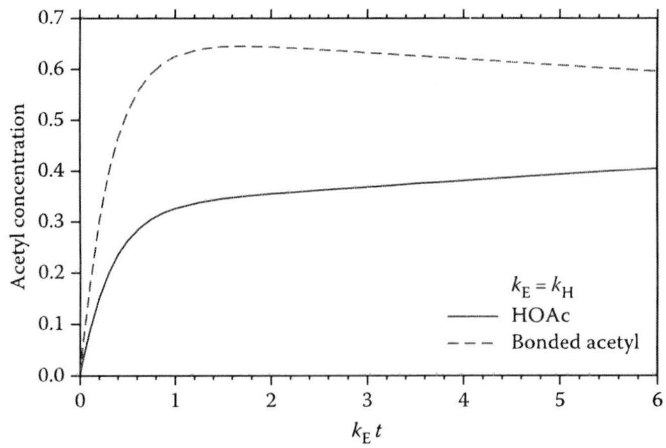

FIGURE 22.10
Variation of acetyl concentrations in the extract liquor with extraction time, all as relative values.

$$r_E = \sum_{m=1}^{N} m(-r_{\theta m}) = \sum_{m=1}^{N} mk_E\theta_m - \sum_{m=1}^{N} m \sum_{n=m+1}^{N} \frac{k_E}{n}\theta_n, \tag{22.38}$$

which yields

$$r_E = \frac{1}{2}k_E\left(\sum_{m=1}^{N} m\theta_m + \sum_{m=1}^{N}\theta_m\right) = \frac{1}{2}k_E P\left(1 + \frac{M_1}{M_n}\right), \tag{22.39}$$

where P is the total extractable substance concentration in terms of monosaccharide units in the woody biomass, M_1 is the molecular weight of the monosaccharide, and M_n is the number-averaged molecular weight of the extractable molecules in the woody biomass. Equation 22.39 can also be written as

$$r_E = \frac{1}{2}k_E\left(1 + \frac{1}{DP}\right)P, \tag{22.40}$$

where DP is the degree of polymerization of the extractable constituents remaining in the woody biomass. Therefore, it is reasonable to approximate the extraction process as a first-order process in terms of the overall mass concentration remaining in the woody biomass. The overall concentrations based on mass or monomer units are convenient measures when detailed distributions are not available. However, it should be noted that the first-order rate is apparently slower at the beginning of the extraction process where DP is high (as shown in Figure 22.9) and that the proton concentration is lower, as shown in Equation 22.27. The rate is higher at the end (or toward the completion) of the reaction because the DP value is lower and the proton concentration has reached a maximum.

Increasing temperature will cause the number of extractable constituents to increase as larger macromolecules become soluble. Therefore, hot-water extraction reactions are more complex than catalytic reactions.

22.3.2 Lumping of Temperature Effects into Residence Time

The extraction and hydrolysis processes are normally not isothermal. They include periods during which temperature is ramping up or cooling down. From a quality control and performance analysis point of view, it would be more convenient if the time and temperature effects could be lumped together. For example, each time the reactor is loaded or discharged, the temperature profile or temperature change with time during the intended reaction period,

$$T = f(t), \tag{22.41}$$

is different. The concept of H-factor, which is the effective reaction time equivalent to that at a constant temperature T_R, was developed by Vroom (1957) for chemical pulping for the case where the constituent intended to be removed is lignin. It will be demonstrated that this concept can be used to approximate extraction and hydrolysis of wood as well.

Examining the kinetics in the light of the above discussions, it becomes clear that lumping the time and temperature together directly would be difficult because there are too many components and reaction rate constants involved. However, with the assumption that all the extraction rate constants have the same effective activation energy, that is,

$$k_{en}\frac{K_S C_{H^+}}{1+K_S C_{H^+}} = k_{en0}\exp\left(-\frac{E_a}{RT}\right), \tag{22.42}$$

then the temperature dependence of the extraction can be described by a single activation energy E_a. Because the proton concentration changes with contacting time and temperature and the saturation constant K_s and even the maximum extractable constituent concentration are functions of the extraction temperature, Equation 22.42 is approximate. The validity of Equation 22.42 will be improved if (1) the temperature range is narrower and (2) the pH variation is small.

For any component of the woody biomass, a mass balance yields the following:

$$r_{\theta m}W = \frac{dn_m}{dt}. \tag{22.43}$$

Because the inert mass of the woody biomass is constant (a constant-volume batch reactor), it can be further determined that

$$-k_{em0}\exp\left(-\frac{E_a}{RT}\right)g(\theta_j) = \frac{d\theta_m}{dt}, \tag{22.44}$$

where

$$g(\theta_j) = g(\theta_1,\theta_2,\ldots,\theta_N) \tag{22.45}$$

Rearranging Equation 22.44 yields

$$\frac{d\theta_m}{g(\theta)} = -k_{em0}\exp\left(-\frac{E_a}{RT}\right)dt. \tag{22.46}$$

For a desired reaction end condition or conversion, the reaction time can be obtained by integrating Equation 22.46:

$$-\int_{\theta_{m0}}^{\theta_m}\frac{d\theta_m}{g(\theta)} = \int_0^t k_{em0}\exp\left(-\frac{E_a}{RT}\right)dt. \tag{22.47}$$

Although the true functional form of $g(\theta)$ is complex, it is not necessary to carry out the integration for now. Equation 22.47 is applicable to reactions at a constant reference temperature, T_R:

$$-\int_{\theta_{m0}}^{\theta_m}\frac{d\theta_m}{g(\theta)} = \int_0^t k_{em0}\exp\left(-\frac{E_a}{RT_R}\right)dt = k_{em0}\exp\left(-\frac{E_a}{RT_R}\right)t_R, \tag{22.48}$$

where t_R is the time required to reach the reaction end condition (concentration θ_m from θ_{m0}) at a constant reaction temperature T_R. Substituting Equation 22.48 into Equation 22.47 leads to

$$k_{em0}\exp\left(-\frac{E_a}{RT_R}\right)t_R = -\int_{\theta_{m0}}^{\theta_m}\frac{d\theta_m}{g(\theta)} = \int_0^t k_{em0}\exp\left(-\frac{E_a}{RT}\right)dt, \tag{22.49}$$

which can be reduced to

$$t_R = \int_0^t \exp\left(\frac{E_a/R}{T_R} - \frac{E_a/R}{T}\right)dt. \tag{22.50}$$

Therefore, for a given temperature and time progression $T(t)$, an equivalent time t_R (based on a constant temperature T_R) can be determined. The change in extraction yield with time and temperature is captured by its variation with the equivalent time t_R.

In chemical pulping, this temperature T_R is usually taken as 100°C (or 373.15 K), and the resulting t_R is called the H-factor. Once the activation energy is known, the effects of temperature and time can be easily lumped together into this single parameter, t_R, or H-factor. This has turned out to be a convenient parameter for use by operators to control the reaction effectively even if the temperature varies during the process. Frequently quoted values of the constants are those proposed Vroom for Kraft pulping of wood chips:

$$\mathrm{HF} = \int_0^t \exp\left(43.186 - \frac{16115\mathrm{K}}{T}\right)dt. \tag{22.51}$$

The H-factor treatment is only an approximation of the reaction system because the proton concentration changes with time and the activation energy may not be equal in all the kinetic constants. One variant of the H-factor is the severity factor. Equation 22.50 can also be written as

$$t_R = \int_0^t \exp\left(\frac{T - T_R}{RT_RT / E_a}\right)dt. \tag{22.52}$$

If one approximates:

$$\frac{RT_RT}{E_a} \approx T_E \tag{22.53}$$

as a constant, then Equation 22.52 can be written as:

$$t_R \approx \int_0^t \exp\frac{T - T_R}{T_E}dt = t\exp\frac{T - T_R}{T_E}. \tag{22.54}$$

The severity factor is defined as the log of t_R, that is,

$$SF = \log\left(t\exp\frac{T - T_R}{T_E}\right) = \frac{T - T_R}{T_E\ln 10} + \log(t). \tag{22.55}$$

For hot-water extraction of woody biomass, the most frequently used T_E value is 14.75°C, or 14.75 K to be precise (Overend and Chornet, 1987; Montane et al., 1994; Kabel et al., 2007). As one would expect, the severity factor is not a universally applicable parameter to use because by definition it ignores the temperature ramping-up and cooling-down periods, as defined in Equation 22.55. However, if the ramping-up and cooling-down times were short, neglecting them would not cause much error.

It can be inferred from the above discussion that the H-factor or the severity factor can be applied to determine the extraction efficiency, that is, mass removal efficiency, with time and temperature. When more parameters are considered, such as the compositions in the extraction liquor, the H-factor or severity factor can be a good indicator if they are small, that is, over short time periods.

To use the relative time t_R to lump temperature and residence time together to characterize hot-water extraction, the correct value for the activation energy E_a is needed. When $T_E = 14.75$ K, this is equivalent to saying that $E_a/R \approx 10{,}800$ K. This value is much lower than the value of $E_a/R = 16{,}115$ K in the H-factor evaluation. This is understandable because the activation energy for hemicelluloses is lower than that for lignin. It confirms that hemicellulose is easier to degrade than lignin. However, Garrote et al. (1999, 2001a,b) found that $E_a = 139–146$ kJ/mol. Even at the lower end, $E_a/R \approx 16{,}840$ K, which is a value even higher than that of lignin. The high value of E_a may be an artifact of a change in the extractable quantity of macromolecules with temperature. For sugar-maple wood meals, Mittal (2006) and Mittal et al. (2009a) found that $E_a = 117.4$ kJ/mol. The value of $E_a/R \approx 10{,}800$ K may be more plausible.

The computation of the H-factor or the relative time t_R can be demanding because the integration cannot be carried out exactly when the temperature varies over time. If the temperature changes linearly over time, the following equation can be used to estimate the t_R factor:

$$t_R = \int_0^t \exp\left(\frac{E_a}{RT_R} - \frac{E_a}{RT}\right)dt = ET\left(\frac{E_a}{RT}\right) - ET\left(\frac{E_a}{RT_0}\right), \tag{22.56}$$

where the exponential integral function can be computed using the following asymptotic expression:

$$ET(x) \approx -\frac{E_a}{R\dfrac{dT}{dt}}\exp\left(\frac{E_a}{RT_R} - x\right)x^{-2}\left[1 - 2x^{-1} + \cdots + (-1)^n(n+1)!x^{-n}\right] \tag{22.57}$$

with $n \le x$. Note that $E_a/R = 16{,}115$ K and $E_a/RT_R \approx 43.186$ when calculating the H-factor. For hot-water extraction, the activation energy may change with species and with temperature. Here, the value $E_a/R \approx 10{,}800$ K will be used when no other information is available. The asymptotic series converges very fast, especially at low temperatures. Even for $T = 250°C = 523.15$ K, Equation 22.57 estimates the function accurately to a relative error of 1.53×10^{-6} with only eight terms ($n = 7$). When 10 terms are used ($n = 9$), the relative error is reduced to 1.45×10^{-7}. If the temperature does not change much, for example by one degree or less, it is then better to estimate the H-factor or the relative time t_R using numerical integration. For example, by Simpson's rule,

$$t_R = \int_0^t \exp\left(\frac{E_a}{RT_R} - \frac{E_a}{RT}\right) dt$$

$$\approx \frac{t}{6}\left[\exp\left(\frac{E_a}{RT_R} - \frac{E_a}{RT_0}\right) + 4\exp\left(\frac{E_a}{RT_R} - 2\frac{E_a/R}{T + T_0}\right) + \exp\left(\frac{E_a}{RT_R} - \frac{E_a}{RT}\right)\right] \quad (22.58)$$

The concept of the relative time t_R simplifies extractor design and performance analysis. Once the species and the extraction temperature range are fixed, E_a can be determined, and t_R become a reliable parameter for characterizing the extraction efficiency or mass removal efficiency. However, it still cannot be used to characterize the extraction liquor composition, especially for long residence times. In general, lower temperature and shorter residence time favor the recovery of hemicelluloses from hot-water extraction because further degradation of the hemicelluloses in the extraction liquor is limited at low temperature and over short times. The following sections will examine the effects of various parameters on hot-water extraction.

22.4 Mass Removal and Dissolved Solids in Extraction Liquor

The extent of extraction is best quantified by the amount of mass removed from the woody biomass. Dissolved solids represent the extracted substances that are not volatile at 105°C. The amount of dissolved solids is both a measure of the extent of extraction (for short residence times) and a measure of the extracted substances in the extraction liquor with minimum degradation. These two quantities are key to extraction performance analysis and reactor design.

Figure 22.11 shows the mass removal by hot-water extraction of sugar-maple wood chips in a 4-L M/K digester. The residence time and temperature effects were collapsed together into the relative time t_R for $E_a/R = 10{,}800$ K and $T_R = 100°C$. The time to ramp up to temperature, the residence time at temperature (1, 2, and 4 h), and the quenching time required to top off the extraction reactions are all lumped together into the relative-time parameter. It is apparent that the experimental data are reasonably well connected in a smooth curve, despite the differences in residence times and extraction temperatures. Therefore, it is possible to conclude that the relative time can be used to lump time and temperature effects together.

One can observe from Figure 22.11 that the mass removal, or the amount of substances extracted from the sugar-maple hardwood chips, increases rapidly with t_R during hot-water extraction between $t_R = 20$ and 140 h. It is evident that the mass removal initially increases very slowly with t_R, as would be expected from Equation 22.40. The increase in mass removal slows down as t_R is increases beyond 140 h. At long residence (or relative) times, the extractable constituents are nearly depleted from the residual woody biomass, and therefore mass removal gradually levels off as the relative time is further increased. Higher temperature results in higher maximum mass removal even for the same value of t_R.

Figure 22.12 shows the total dissolved solids in the hot-water extracts as a function of the relative time t_R corresponding to the extractions shown in Figure 22.11. Although high temperature and long extraction time (high values of t_R) result in higher mass removal (Figure 22.11), the actual total dissolved solids content in the hot-water extracts decreases with t_R at high t_R, indicating lower levels of recoverable carbohydrates in the hot-water

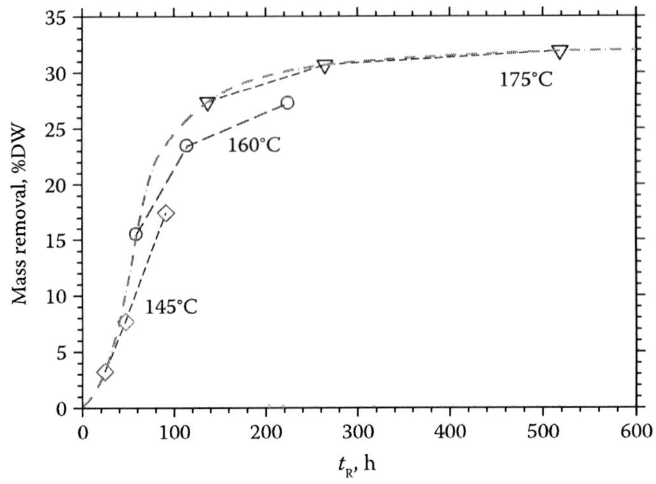

FIGURE 22.11
Mass removal (% dry wood mass) as a function of t_R, with $E_a/R = 10800$ K and $T_R = 100°C$, for the hot-water extraction of sugar-maple wood chips. The experimental data were collected at different extraction time durations (1, 2, and 4 h) at the desired temperatures (145°C, 160°C, and 175°C). The liquor-to-wood ratio was 5.92.

extracts at high temperature and long extraction durations. A maximum dissolved solids content can be expected at a t_R value of approximately 120 h with high temperatures. In one of the authors' experiments, high temperature caused the complete dissolution of the woody biomass, but no recoverable materials were found in the extraction liquor during extraction at 250°C for 60 min (results not shown here). The total dissolved solids concentration increases with increasing temperature at the same and shorter t_R values.

Figure 22.13 shows the concentrations of dissolved solids in the extraction liquor (or extract) as a function of t_R for hot-water extraction sugar-maple wood chips in a 1.84-m^3

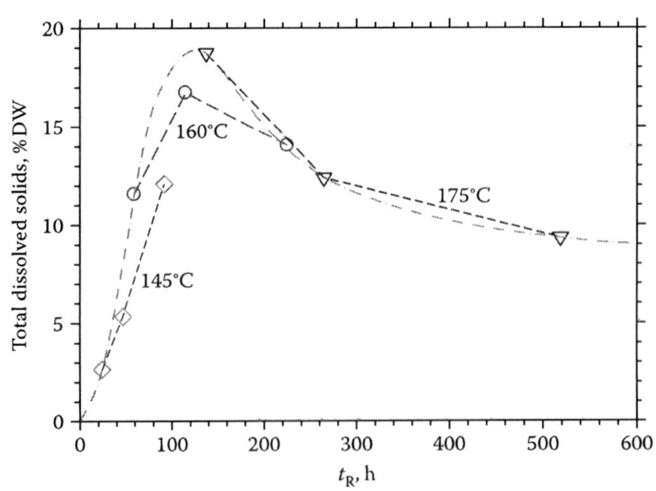

FIGURE 22.12
Total dissolved solids (% dry wood mass) in extraction liquor as a function of t_R for the hot-water extraction of sugar-maple wood chips. The experimental data were collected at different extraction time durations (1, 2, and 4 h) at desired temperatures (145°C, 160°C, and 175°C). The liquor-to-wood ratio was 5.92.

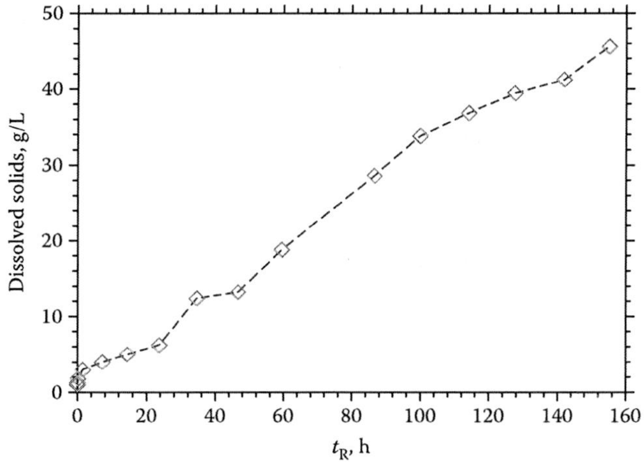

FIGURE 22.13
Variation of dissolved solids with t_R during hot-water extraction of sugar-maple wood chips in a 1.84-m^3 digester. (Data from Amidon, T.E. et al. 2008. *J. Biobased Mater. Bioenergy* 2: 100–120.)

reactor with a water-to-wood ratio of 4 and a maximum extraction temperature of 160°C. It is clear that the concentration of dissolved solids increases with increasing t_R or increasing duration of extraction. The increase in dissolved solids is fast initially due to the solubilization of small hydrophilic materials in wood. After a short initial t_R period, the concentration of dissolved solids increases rather slowly. After $t_R = 25$ h, the dissolved solids concentration increases more rapidly. As shown in Figure 22.11, the total mass removal from woodchips increases with increasing t_R. The hot-water extraction proceeds faster and more components are dissolved under higher extraction temperatures. As the extraction time is extended, a higher degree of extraction is observed.

Figures 22.14 and 22.15 show the mass removal based on the dry mass of original aspen wood chips (from Alberta, western Canada) as a function of temperature, duration, and

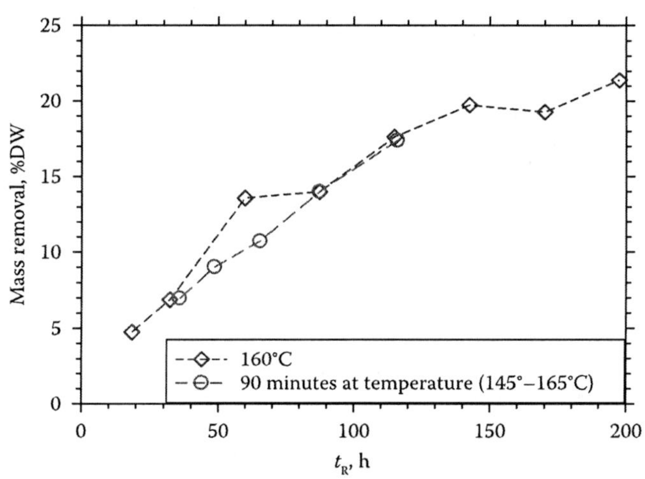

FIGURE 22.14
Mass removal (% dry wood mass) in extraction liquor as a function of t_R for the hot-water extraction of aspen wood chips from northern Alberta, Canada. The liquor-to-wood ratio was 4.

water-to-wood ratio. As the extraction temperature is increased, mass removal is also increased. With constant temperature, mass removal increases with duration of extraction. It is also apparent that mass removal from the hot-water extraction increased sharply at higher temperatures. The mass removal increased more sharply with extraction time before $t_R = 140$ h. Beyond that point, the increasing trend slows down. The mass removal increased markedly with increasing liquor-to-wood ratio below 4.5:1 (Figure 22.15), and remained nearly unchanged with further increases in liquor-to-wood ratio. The small differences in mass removal at the same temperature and duration indicated that the water-to-wood ratio had little influence on mass removal during hot-water extraction beyond a ratio of 4.5:1.

Figure 22.16 shows mass removal as a function of extraction temperature for hot-water extraction of fast-rotation shrub willow cuts. It can be seen that the mass-removal level is similar to that of aspen wood chips, but lower than that of sugar-maple wood chips. The trend is identical to the other cases.

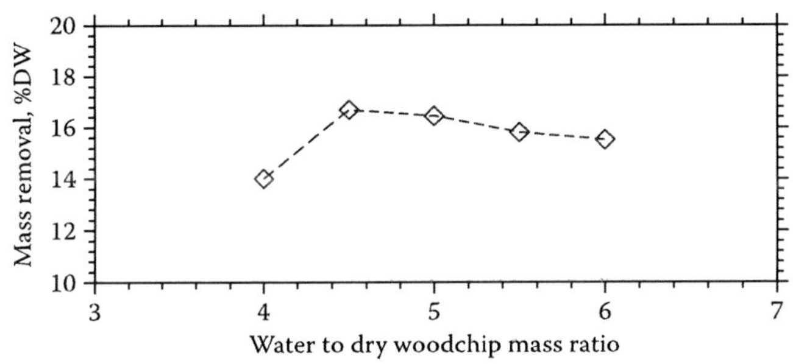

FIGURE 22.15
Mass removal (% dry wood mass) in extraction liquor as a function of water to dry wood chip mass ratio for the hot-water extraction of aspen wood chips from northern Alberta, Canada. The extraction was conducted at 160°C for 90 min.

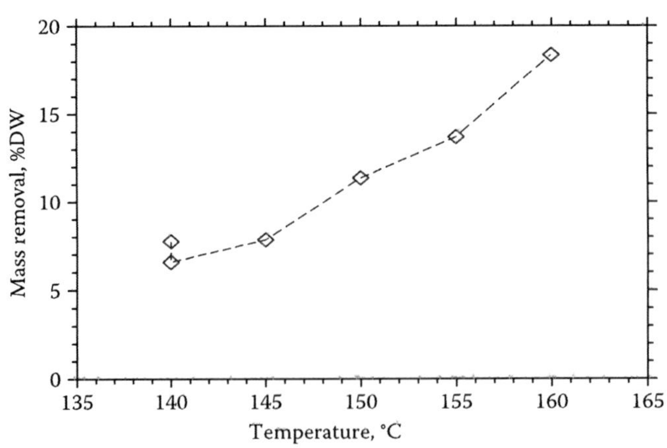

FIGURE 22.16
Mass removal by hot-water extraction as a function of temperature for willow cuts. Solids-to-water ratio of 1:4 and 2 h at temperature. (Data from Blower, M. 2003. Xylan extraction from short rotation willow biomass. MSc Thesis. SUNY ESF, Syracuse, NY.)

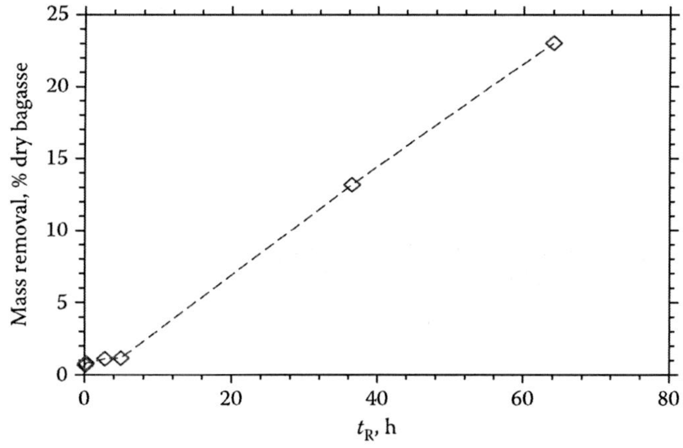

FIGURE 22.17
Mass removal by hot-water extraction as a function of temperature for depithed bagasse, with a solids-to-water ratio of 1:8 and an extraction time of 30 or 60 min at temperatures ranging from 80°C to 160°C (Data from Lei, Y. et al. 2010. *J. Biotech. Adv.* 28: 609–612.)

Figure 22.17 shows mass removal by hot-water extraction of depithed bagasse as a function of relative time. It is clear that the extraction of bagasse behaves very similarly to the extraction of wood chips and shrub willow cuttings. Soluble hydrophilic constituents are quickly dissolved from the woody biomass. Mass removal changes with relative time very slowly initially (other than the dissolution of hydrophilic constituents) and then increases more rapidly. The extraction rate from bagasse is much faster than from shrub willow cuts or from sugar-maple wood chips.

To show the effect of temperature on mass removal due to hot-water extraction at high temperatures, Figure 22.18 shows the residual biomass of rice bran as a function of temperature for a constant residence time of 10 min. The residue after hot-water extraction consisted mainly of unreacted rice bran, carbonized rice bran, and hydrolyzed but still insoluble parts of rice bran, as well as inorganic compounds. The amount of residual solids decreased slightly from 100°C to 140°C and then decreased sharply to a minimum of 8% at 320°C. This sharp

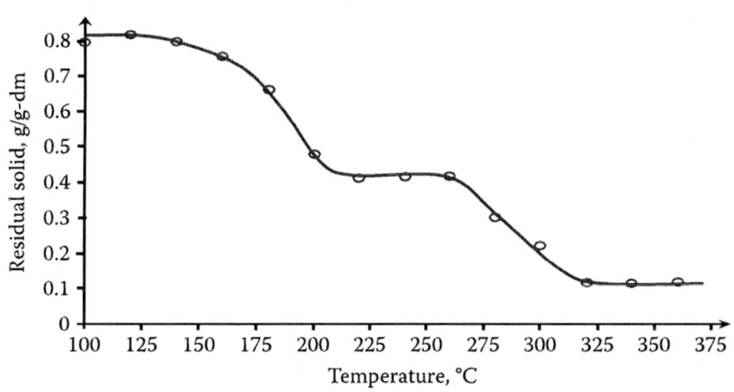

FIGURE 22.18
Effect of extraction temperature on residual solid amounts of rice bran at a residence time of 10 min. (Adapted from Pourali, O., Asghari, F.S., Yoshida, H. 2010. *Chem. Eng. J.* 160: 259–266.)

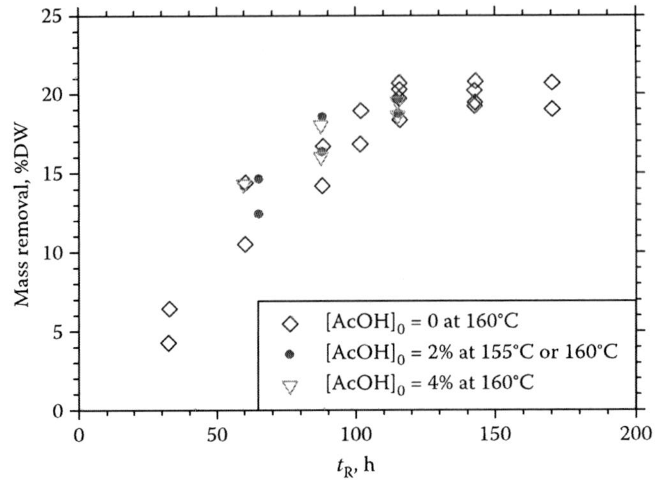

FIGURE 22.19
Effect of acetic acid addition on mass removal for the hot-water extraction of *Eucalyptus globulus* woodchips. The extractions were conducted in two reactors, and therefore the actual temperatures in the reactors might have been slightly different because of instrumentation limitations.

decrease proved that hot water could effectively decompose insoluble macromolecules of rice bran into smaller soluble compounds in a short residence time at elevated temperatures, which agrees with the earlier observation on hot-water extraction of wood chips.

Figure 22.19 shows that the addition of acetic acid to the extraction liquor has a positive effect on mass removal. However, the increase in mass removal due to acetic acid addition is small, especially when t_R is high.

The mass removal from woody biomass and the dissolved solids content in the extraction liquor both exhibit a plateau at short t_R times. This plateau at short times is a useful design point for applications where minimum mass removal is an objective. For bagasse, this relative time is found to be $t_R < 5$ h; for aspen wood chips, $t_R < 20$ h; and for sugar-maple wood chips, $t_R \leq 25$ h, all based on $E_a/R = 10{,}800$ K. However, there is an upper limit on t_R times for maximum recoverable dissolved solids content. A combination of short residence time and high temperature is most desirable if maximum dissolved solids content is the objective. For eucalyptus, Figure 22.19 shows that t_R should be approximately 115 h for maximum mass removal. However, this plateau was observed within a narrow temperature range. More plausible results were obtained for sugar-maple wood chips, as shown in Figure 22.12. The maximum dissolved solids content was obtained at $t_R \approx 120$ h. Higher temperature (at the same t_R) is beneficial to recoverable dissolved solids content. The optimum water-to-wood ratio for maximum extraction is approximately 4.5.

22.5 Carbohydrates in Extraction Liquor

The carbohydrates content in the extraction liquor is an important parameter for hot-water extraction applications. It measures both hemicellulose removal and the amount of sugars that can be recovered.

Figure 22.20 shows the concentration of xylooligomer and xylose in the wood extract as a function of t_R. It is apparent that the concentration of xylooligomers increases with t_R. Hemicellulose is extracted from the wood chips in the form of xylooligomers with various degree of polymerization. Data on this were first reported in Amidon et al. (2008) as xylose. Because the data were obtained by ^{1}H NMR, xylooligomers and xylose overlap, and therefore the concentration reading is actually the sum of xylose and xylooligomers, or more precisely, polysaccharides with a xylose residue end group. Initially, the temperature and the proton concentration were both low, and the amount of xylooligomer dissolved in the extraction liquor increased very slowly with relative time. After 40 h of t_R, the concentration of dissolved xylooligomers in the extraction liquor began to increase more rapidly.

Figure 22.21 shows the concentrations of minor sugar oligomers of glucose, mannose, galactose, rhamnose, and arabinose as functions of t_R. Compared with Figure 22.20, the

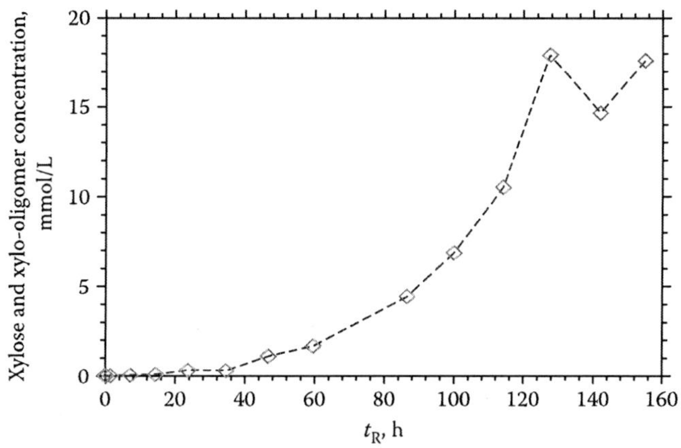

FIGURE 22.20
Variation of xylose concentration with t_R during hot-water extraction of sugar-maple wood chips in a 1.84-m^3 digester. (Pourali, O., Asghari, F.S., Yoshida, H. 2010. *Chem. Eng. J.* 160: 263. With permission.)

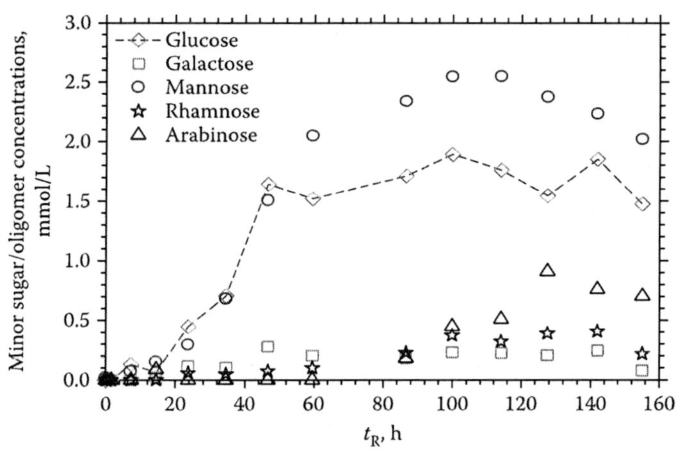

FIGURE 22.21
Variation of minor sugar concentrations with t_R during hot-water extraction of sugar-maple wood chips in a 1.84-m^3 digester. (Data from Amidon, T.E. et al. 2008. *J. Biobased Mater. Bioenergy* 2: 100–120.)

amounts of these six-carbon and five-carbon sugars are only small fractions of the amount of xylooligomers. Among these minor sugars, glucose- and mannose-ended oligomers lead the group, with a nearly one-to-one ratio. This ratio suggests that glucomannan was the main origin of the glucose and mannose and their oligomers, which is consistent with the structures shown in Table 22.5.

Figures 22.2 and 22.3 show the xylose and polysaccharides with a xylose residue end group for hot-water extraction of aspen woodchips. Figure 22.22 shows that the trend is rather similar to that shown in Figure 22.20; the concentration of xylooligomers increases rather slowly when $t_R < 20$ h. Figure 22.23 shows that the concentration of xylooligomers is at a maximum when the water-to-wood ratio is 5.

Figure 22.24 shows the effect of hot-water extraction time on xylose unit removal from an eastern Canada hardwood mixture (7:2:1 ratio of maple, poplar, and birch) (Li et al., 2010). The total xylose saccharides content in the mixed wood species was 20.47% (expressed as xylose). At the maximum of 10.54% obtained at 90 min, approximately 51.49% of the xylose

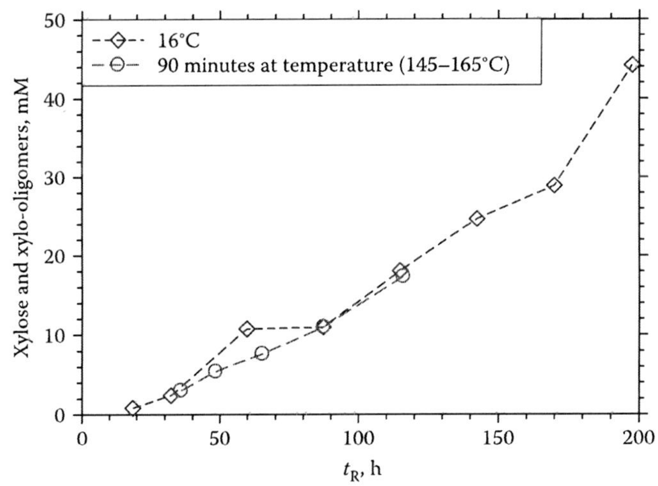

FIGURE 22.22
Xylo units in extraction liquor as a function of t_R for the hot-water extraction of aspen wood chips from northern Alberta, Canada. The liquor-to-wood ratio was 4.

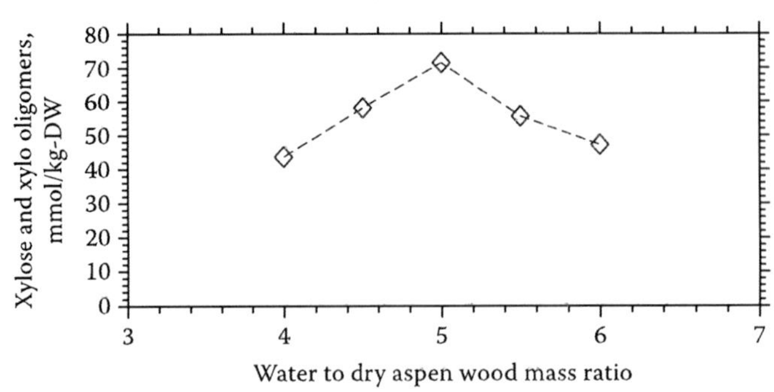

FIGURE 22.23
Xylo units in extraction liquor as a function of water-to-dry-wood mass ratio for the hot-water extraction of aspen woodchips from northern Alberta, Canada. The extraction was conducted at 160°C for 90 min.

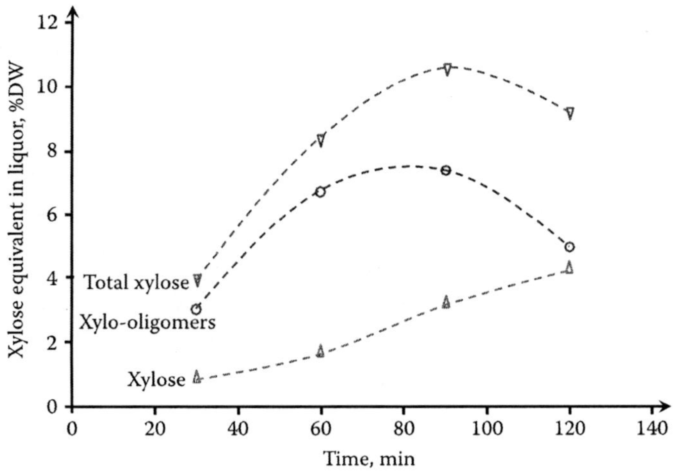

FIGURE 22.24
Xylo units in the hot-water extraction liquor of an eastern Canada hardwood-chip mixture obtained at 170°C. (Adapted from Li, H. et al. 2010. *J. Wood Chem. Techn.* 30: 1, 48–60.)

saccharides were removed from the wood chips, 30.55% of which was in the monomeric form and the remainder in the oligomeric form.

The effect of extraction temperature on hemicellulose removal is shown in Table 22.8 (Li et al., 2010). It can be seen that maximum temperature is the key parameter that determines the sugar composition in the extraction liquor. The temperature, rather than the acid concentration, significantly affects the extraction composition (Esteghlalian et al., 1997). Only a small amount of hemicelluloses was extracted at 150°C, and the total xylose saccharides was only 0.98%. At 170°C, the total xylose saccharides produced was 8.34%, with the majority (6.66%) in the oligomeric form. At 190°C, the total xylose saccharides decreased significantly (3.20% in the monomeric form and 0.72% in the oligomeric form). At 210°C, the presence of xylose saccharides was negligible. The decrease in xylose saccharides was certainly due to their further conversion to other compounds, such as furfural. For the purpose of utilization of xylose saccharides from extraction liquor, a 170°C temperature appears to be ideal. When the maximum temperature was increased to 190°C, a significant amount of furfural was produced in the extraction liquor (2.99%). The mass balance indicated that the amount of furfural generated could not account for the loss of five-carbon sugars, indicating that other chemicals, including furfural-based polymers (Root et al., 1959; Shin and Cho, 2008) could have been formed during the course of the reactions. The total of acetyl groups was further increased to 4.41%, 1.08% of which were still bound to oligomers. The amounts of glucose and mannose were almost unchanged, but at higher temperatures, more were in the monomeric form. When the maximum temperature was increased to 210°C, almost all the five-carbon sugars were consumed, which caused a large increase in the amount of furfural in the extraction liquor (3.65%). The amount of monomeric galactose was also increased. The total of the acetyl groups was 5.04%, which was higher than the original acetyl content of 4.89% in the raw material. Takeuchi et al. (2008) proposed as an explanation the decomposition of six-carbon sugars to levulinic acid or HMF and their further degradation into acetic acid. Even at 210°C, the sum of monomeric and oligomeric glucose remained at a low level, which is consistent with the results reported by Garrote et al. (2004).

TABLE 22.8

Chemical Composition of Original Wood and the Extraction Liquors Obtained at Different Temperatures for 60 min (Based on Wood, %)

Extraction Temperature (°C)	Wood	150	170	190	210
Arabinose	0.89				
Monomeric		0.15 (0.36)	0.28 (0.29)	0.09 (0.27)	0.00 (0.21)
Oligomeric		0.15 (0.00)	0.02 (0.02)	0.02 (0.00)	0.00 (0.01)
Rhamnose					
Monomeric		0.02 (0.10)	0.13 (0.15)	0.10 (0.18)	0.01 (0.19)
Oligomeric		0.07 (0.11)	0.14 (0.13)	0.04 (0.11)	— (0.04)
Galactose	0.85				
Monomeric		0.02 (0.13)	0.16 (0.20)	0.23 (0.24)	0.52 (0.28)
Oligomeric		0.11 (0.26)	0.30 (0.32)	0.06 (0.27)	— (0.16)
Glucose	56.90				
Monomeric		0.02 (0.07)	0.11 (0.11)	0.38 (0.18)	0.03 (0.23)
Oligomeric		0.22 (0.56)	0.63 (0.70)	0.35 (0.65)	0.66 (0.52)
Xylose	20.47				
Monomeric		0.05 (0.72)	1.68 (2.02)	3.20 (3.35)	0.03 (4.03)
Oligomeric		0.93 (6.59)	6.66 (7.78)	0.72 (6.25)	0.02 (3.94)
Mannose	2.06				
Monomeric		0.00 (0.04)	0.08 (0.11)	0.42 (0.20)	0.06 (0.28)
Oligomeric		0.25 (0.77)	0.71 (0.86)	0.34 (0.85)	— (0.65)
Acetyl	4.89				
AcOH form		0.32 (2.67)	1.10 (3.22)	3.33 (3.70)	4.97 (4.43)
Bound to oligomers		0.28 (2.04)	1.72 (1.78)	1.08 (1.75)	0.07 (1.38)
Furfural		0.01 (0.11)	0.37 (0.50)	2.99 (0.80)	3.65 (1.49)

Source: Data from Li, H. et al. 2010. *J. Wood Chem. Techn.* 30: 1, 48–60.
Note: The numbers in the brackets are those obtained when the initial water contained 6 g/L of acetic acid.

In a similar operation, rice bran was subjected to hot-water extraction by Pourali et al. (2010). Figure 22.25 shows the influence of extraction temperature on the yield of total produced sugars in the aqueous solution. It illustrates that total soluble sugar yield increased with temperature, reaching a peak at 190°C, and then decreased drastically to zero at temperatures above 300°C for a residence time of 10 min. The results proved that the water-insoluble carbohydrate part of rice bran could be effectively hydrolyzed into water-soluble oligomers and monomers at elevated temperatures. In addition, the decrease in yield at high temperatures could be interpreted as a sign of the conversion of soluble sugars into other constituents, mainly to HMF and soluble polymers. The effect of residence time on the yield of total produced sugars at an extraction temperature of 220°C is shown in Figure 22.26. The yield profile showed a peak at 3 min and then decreased rapidly with residence time.

For hot-water extraction of sugar-maple wood chips, similar responses have been observed with temperature. When the extraction temperature was raised to 250°C for 60 min, all the solid substances were dissolved. However, neither monomeric sugars nor polysaccharides were detected in the extraction liquor.

Below a t_R value of 30 h, very few carbohydrates can be found in the hot-water extraction liquor. This minimal level of carbohydrates removal could be a design objective for

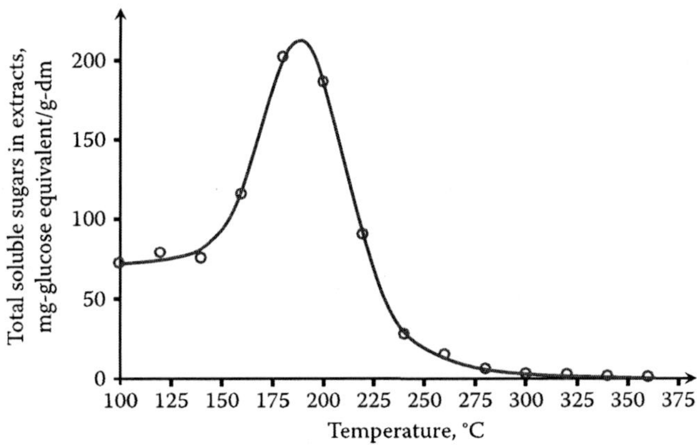

FIGURE 22.25
Effect of hot-water extraction temperature on the yield of total soluble sugars from rice bran at a residence time of 10 min. (Adapted from Pourali, O., Asghari, F.S., Yoshida, H. 2010. *Chem. Eng. J.* 160: 259–266.)

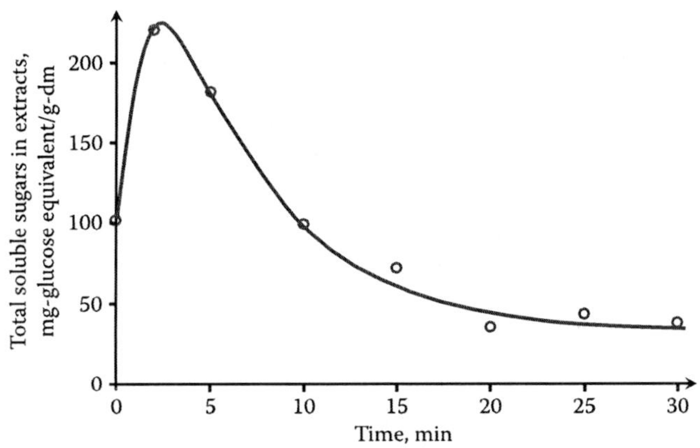

FIGURE 22.26
Effect of residence time on the yield of total soluble sugars from rice bran at an extraction temperature of 220°C. (Adapted from Pourali, O., Asghari, F.S., Yoshida, H. 2010. *Chem. Eng. J.* 160: 259–266.)

maximizing pulp yield by inserting a hot-water extraction step (Liu et al., 2009; Lei et al., 2010). There is an optimum t_R value for maximum carbohydrates retention in the hot-water extraction liquor. Based on Figures 22.20 and 22.21, this value of t_R is approximately 115 h for sugar-maple wood chips. For mixed northeastern North American hardwoods, the optimal value should be $120 < t_R < 150$ h, based on Table 22.8 and Figure 22.24. The optimum water-to-wood ratio is approximately 5, as shown in Figure 22.23.

22.6 Acetyl and Acetic Acid in Extraction Liquor

Acetyl and acetic acid content in extraction liquor is a measure of the proton source available for catalyzing hot-water extraction. It is also an indication of how much acetic acid can be recovered from hot-water extraction.

Figure 22.27 shows the variation of pH during hot-water extraction of sugar-maple wood chips in a 1.84-m^3 digester. It is apparent that the pH profile in the extraction liquor drops from initial near-neutral conditions (~6.3) to acidic conditions. The nominal extraction temperature was 160°C. This acidic condition during the hot-water extraction process further catalyzes the extraction and hydrolysis reactions. Therefore, the hot-water extraction reaction is referred to as autocatalytic, and sometimes as autohydrolysis. The extraction proceeded slowly initially and then faster as the pH decreased during the extraction. As indicated by Equation 22.27, a decrease in pH or an increase in the proton concentration can increase the extraction rates marginally. This increase in rate was nearly first order (or proportional), at least when the pH was high. It can be inferred that the extraction rate increases more sharply at small t_R times. Recycle of extracts or extract components can increase the extraction rates, as shown in Figure 22.19. Efficient reactors could be designed to perform hot-water extraction operations using existing wood-chip processing units as model systems.

Figure 22.28 shows the concentrations of acetic acid (AcOH) and acetyl groups associated with polysaccharides (AcOX) in the wood extract as functions of t_R in a 1.84-m^3 digester. It is clear from Figure 22.28 that the acetic acid concentration increases with extraction time from the start of the hot-water extraction. However, there were no measurable acetylated polysaccharides until almost 20 h of t_R had elapsed, at which time the extraction liquor pH was close to the final pH, as shown in Figure 22.27. The concentration of acetyl groups rapidly increased over that of the acetic acid in the wood extract. From Figure 22.20, there were approximately 0.018 mol/L of xylooligomers in the wood extract near the end of the extraction. The acetic acid concentration was much higher, as shown in Figure 22.28, at 0.053 mol/L. For every xylooligomer molecule dissolved in the wood extract, there were nearly three acetic acid molecules dissolved in the liquor. Still, the acetyl dissolved, and the amount of xylooligomers still associated with polysaccharides was more than the amount of acetic acid. The main hemicellulose macromolecule in sugar-maple wood is shown in Figure 22.4. Figure 22.4 shows that there are only about 7 units of acetyl

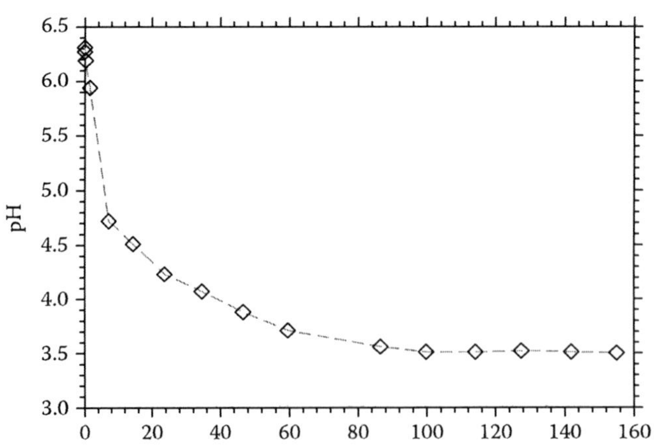

FIGURE 22.27
pH variation with t_R during hot-water extraction of sugar-maple wood chips in a 1.84-m^3 digester with a wood-to-liquor ratio of approximately 1:4. The digester was heated by direct steam injection. The pH values were measured in liquor samples taken from the digester and cooled to 30°C. (Pourali, O., Asghari, F.S., Yoshida, H. 2010. *Chem. Eng. J.* 160: 263. With permission.)

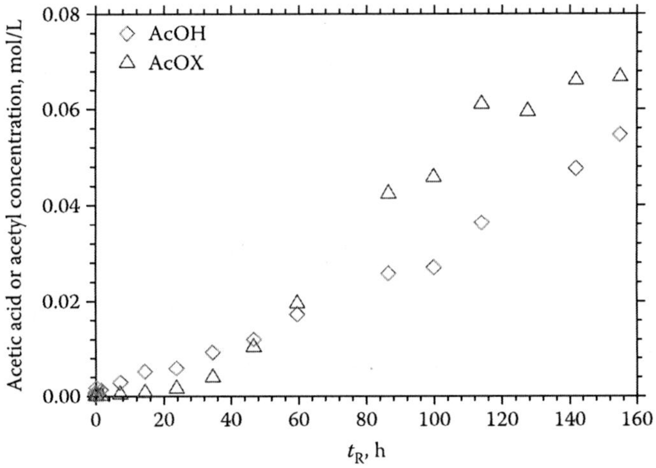

FIGURE 22.28
Variations of acetic acid (AcOH) and polymer acetyl group (AcOX) concentrations in the wood extract with t_R during hot-water extraction of sugar-maple wood chips in a 1.84-m^3 digester. (Pourali, O., Asghari, F.S., Yoshida, H. 2010. *Chem. Eng. J.* 160: 263. With permission.)

groups per 10 xylose residues. Therefore, acetic acid is more easily soluble than either xylose or sugar units. This is consistent with the expectation that acetic acid acts as a catalyst in the extraction process.

Figures 22.29 and 22.30 show the variation of acetic acid and bonded acetyl concentrations in the extraction liquor for hot-water extraction of aspen wood chips. It is clear that the extent of acetyl dissolution in the extraction liquor increases with increasing t_R (Figure 22.29). The maximum amount of acetyls in the extraction liquor occurs when the water-to-wood ratio is between 4.5 and 5.

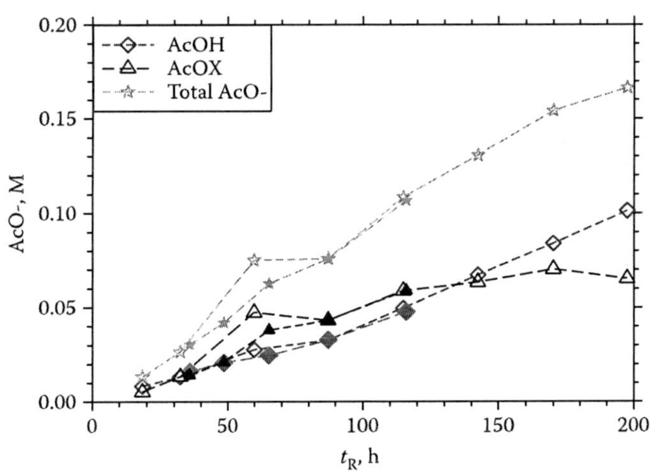

FIGURE 22.29
Acetyl units in extraction liquor as a function of t_R for hot-water extraction of aspen wood chips from northern Alberta, Canada. The open symbols represent extraction conducted at 160°C, while the filled symbols represent 90 min of extraction at various temperatures (145–165°C). The liquor-to-wood ratio was 4.

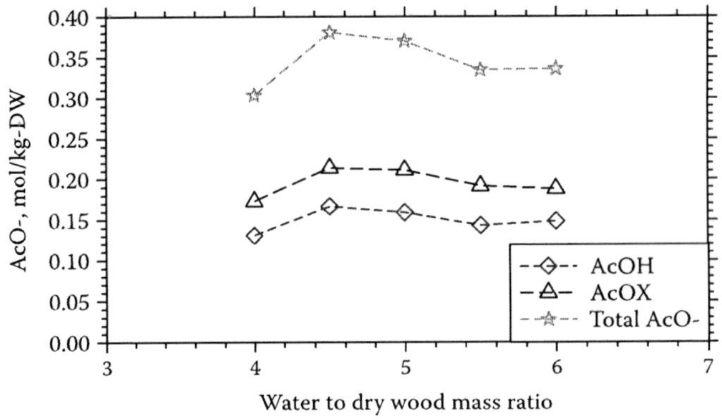

FIGURE 22.30
Acetyl units in extraction liquor as a function of water to dry wood mass ratio for the hot-water extraction of aspen woodchips from northern Alberta, Canada. The extraction was conducted at 160°C for 90 min.

Figure 22.31 shows the variation of acetyl in the extraction liquor. The amount of acetic acid increased with the duration of the extraction, while the concentration of acetyl groups bound to oligomers increased to a maximum at 90 min and then decreased at 120 min. The total concentration of acetyl groups, expressed as acetic acid, in the extraction liquor reached a maximum of 3.76% at 90 min. The ratio of total acetyl groups to total xylose saccharides was 0.615, 0.338, 0.357, and 0.411 at extraction times of 30, 60, 90, and 120 min, respectively. These results support the conclusion that the removal rate of acetyl groups from wood is higher than that of xylose saccharides in the initial phase of hydrolysis, while the opposite is true in the latter phase of the process.

Acetic acid is generated in the extraction process, and it can catalyze extraction and hydrolysis. In industrial applications, due to the recycling or reuse of weak extraction liquor, acetic acid may be present at the beginning of the extraction process. Table 22.8

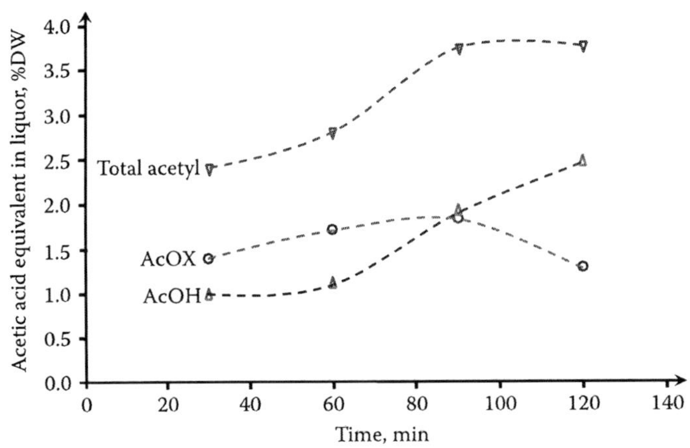

FIGURE 22.31
Acetyl solubilized in the hot-water wood extract of an eastern Canada hardwood chip mixture obtained at 170°C. (Adapted from Li, H. et al. 2010. *J. Wood Chem. Techn.* 30: 1, 48–60.)

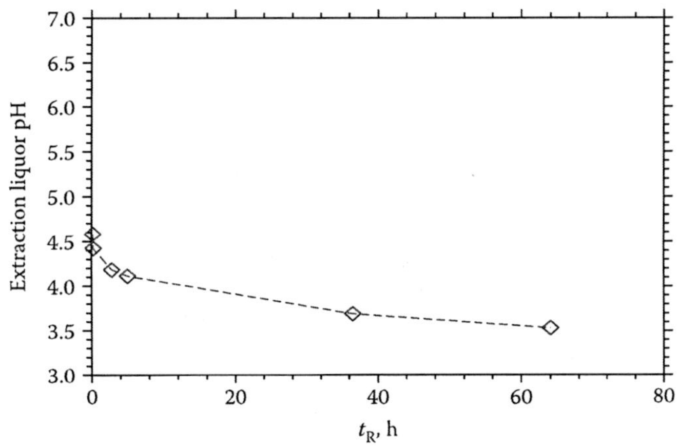

FIGURE 22.32
pH values of extraction liquor as a function of temperature for depithed bagasse. Hot-water extractions were conducted with a solids-to-water ratio of 1:8 and extraction times of 30 or 60 min at temperatures ranging from 80 to 160°C. (Data from Lei, Y. et al. 2010. *J. Biotech. Adv.* 28: 609–612.)

shows that the time needed to reach the maximum total xylose-saccharide yield was shortened by the addition of acetic acid, while the amount of xylose in monomeric form at extraction times of 60–90 min was also increased. These results support the conclusion that the addition of acetic acid can speed up: (i) the degradation of hemicelluloses into oligomers; (ii) the conversion of oligomers into monomeric sugars; and (iii) the degradation of monomeric sugars into other products.

Figure 22.32 shows the pH value change with relative time for hot-water extraction of depithed bagasse. The trend is very much the same as for extraction of wood chips, as shown in Figure 22.27. The decrease in pH is fairly rapid at short extraction times. The initial rapid decrease in pH during hot-water extraction is the reason why acetic acid addition into the extraction liquor has only a minor effect on the extraction process.

22.7 Aromatic Constituents in Extraction Liquor

Aromatic substance dissolution in hot-water extraction liquor is a measure of extractive and lignin removal from woody biomass. In a hot-water extraction process, the exact percentage of the aromatics that is lignin is of less importance or interest, but the total amount of aromatics is an important process characteristic which affects product recovery and sugar stream purification needs. The content of aromatic constituents in extraction liquor is presently quantified using modified TAPPI standards for lignin measurements without extractives removal.

Figure 22.33 shows the variation of the total aromatic materials in woody biomass as a function of the relative extraction time t_R for aspen wood chips. Temperature and time individually can be seen to have slightly different effects on aromatic materials removal than the single t_R parameter. Nevertheless, it is clear that the concentration of aromatic substances in the residual woody biomass decreases with increasing relative extraction time t_R. As t_R approaches 120 h, the decrease in aromatic substances in residual woody

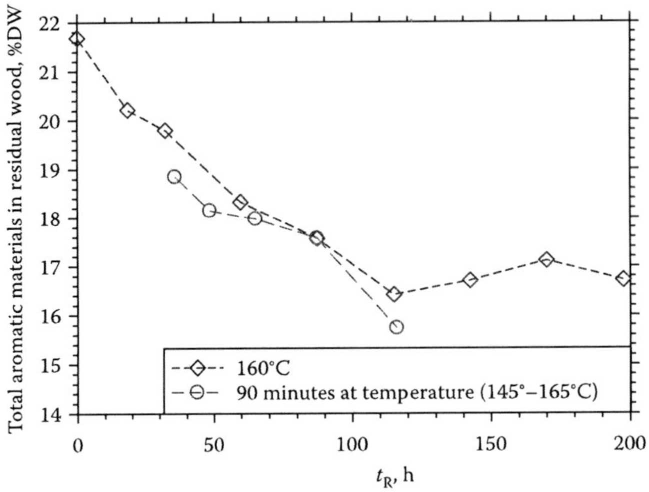

FIGURE 22.33
Aromatic materials in the wood residue as a function of t_R for the hot-water extraction of aspen wood chips from northern Alberta, Canada. The liquor-to-wood ratio was 4.

biomass levels off, although total mass removal and hemicellulose removal continue to increase (Figures 22.14 and 22.22). This indicates that hot-water extraction removes very few aromatics in lignin form. Still, Figure 22.34 shows that approximately 5% of the dry wood mass (DW) removed into the hot-water extracts consists of aromatic materials. Figure 22.35 shows the concentration of total aromatic materials in extracts as a function of the water-to-wood ratio. Clearly, there is no observable influence of the water-to-wood ratio on aromatic substance removal during hot-water extraction.

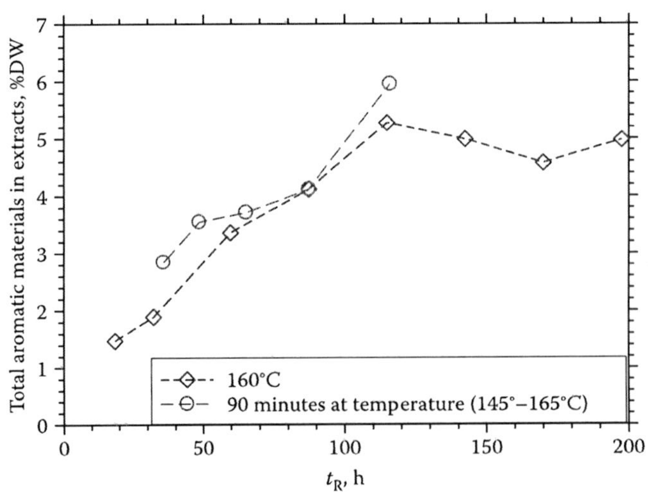

FIGURE 22.34
Aromatic materials dissolved (as a percentage of dry wood mass) in the extraction liquor as a function of t_R for the hot-water extraction of aspen wood chips from northern Alberta, Canada. The liquor-to-wood ratio was 4.

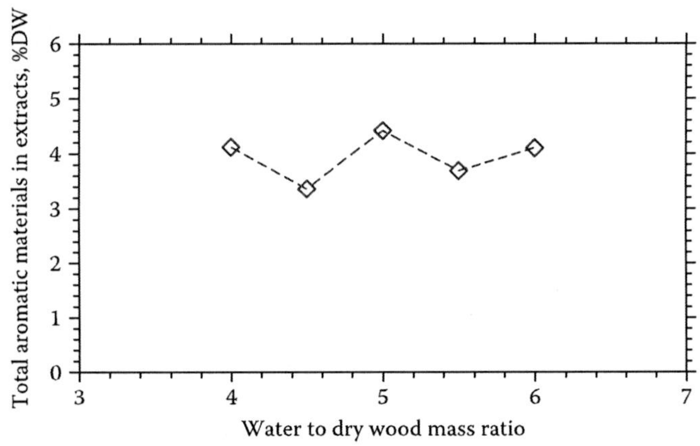

FIGURE 22.35
Aromatic materials in the extraction liquor as a function of water-to-dry-wood mass ratio for the hot-water extraction of aspen wood chips from northern Alberta, Canada. The extraction was conducted at 160°C for 90 min.

Figure 22.36 shows the rate of aromatic substance removal as a function of temperature for hot-water extraction of shrub willow cuts. It is apparent that aromatic material dissolution in hot water increases with increasing water temperature. Figure 22.37 shows the rate of aromatic substance removal as a function of the relative extraction time t_R for hot-water extraction of depithed bagasse. It is clear that aromatic substance removal increases with increasing t_R.

Based on the observations in this section, hot-water extracts contain significant amount of aromatics. Up to 5% of the dry wood mass from aspen wood can be dissolved into hot-water extracts as aromatic materials. The amount of aromatic material removal increases with increasing temperature and extraction time, although aromatic material removal from woody biomass levels off at long residence times.

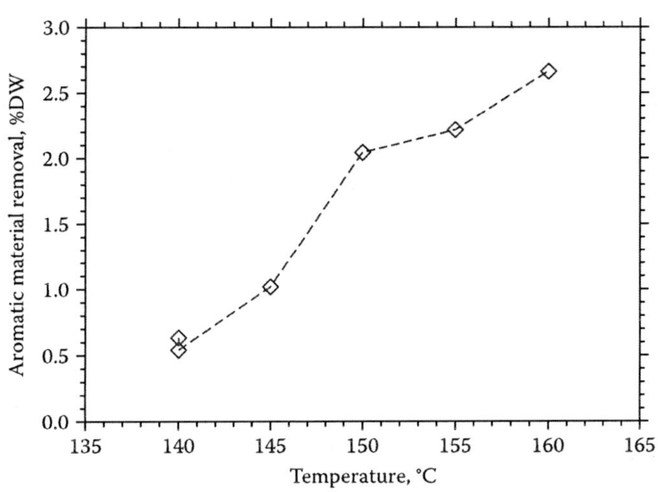

FIGURE 22.36
Aromatic materials (based on lignin measurement) dissolved in the wood extract as a function of temperature for willow cuts. Solids-to-water ratio of 1:4 and 2 h at temperature. (Data from Blower, M. 2003. Xylan extraction from short rotation willow biomass. MSc Thesis. SUNY ESF, Syracuse, NY.)

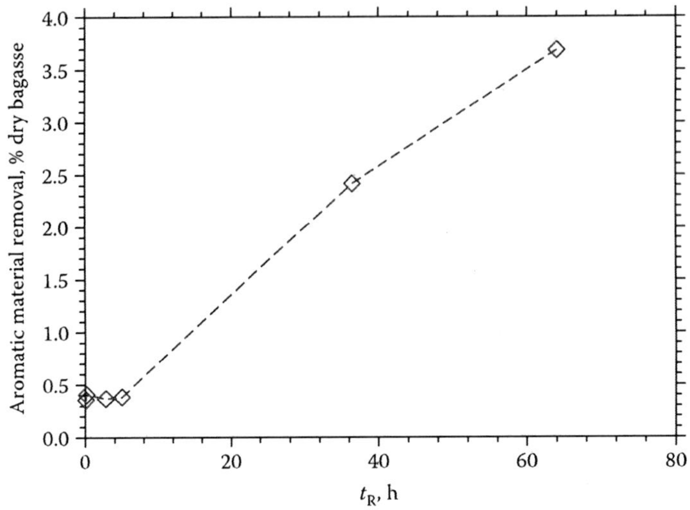

FIGURE 22.37
Aromatic material (based on lignin measurement) dissolved in hot-water extract as a function of temperature for depithed bagasse. Hot-water extractions were conducted with a solids-to-water ratio of 1:8, extraction times of 30 or 60 minutes, and temperatures ranging from 80°C to 160°C. (Data from Lei, Y. et al. 2010. *J. Biotech. Adv.* 28: 609–612.)

22.8 Minor Constituents in Extraction Liquor

Minor constituents such as methanol, formic acid, and degradation products such as furfural and hydroxylmethylfurfural can be detected in hot-water extraction liquor. Methanol can be cleaved off xylan or formed from extractives or lignin. Formic acid can be cleaved off extractives as well as produced from sugar degradation. Formic acid can act as a catalyst for hot-water extraction as well.

Figure 22.38 shows concentrations of the sugar dehydration products furfural and hydroxylmethylfurfural (HMF), as well as of minor extractives, methanol, and formic acid, as functions of extraction time during the hot-water extraction process. It is evident that the concentrations of formic acid and methanol are nearly equal, with more formic acid being added to the extract initially and more methanol forming near the end of the extraction. Both furfural and HMF concentrations increase with extraction time as well. Furfural is a dehydration product of the five-carbon sugars xylose and arabinose. Therefore, the relative time t_R is not a good measure of dehydration because the dehydration reaction may not have an activation energy similar to that of the extraction and hydrolysis reactions. For a single extraction run, the relative time t_R is equivalent to time and therefore will not cause a problem in interpretation. Figure 22.38 shows that the increase in furfural concentration is slow initially during the first 100 h of t_R. However, the concentration of furfural increases more sharply after 100 h t_R. Figures 22.20 and 22.22 shows that more xylose is available starting at approximately this time. Therefore, xylose and arabinose monomers are available later in the extraction, which favors the formation of furfural. HMF is a dehydration product of glucose or six-carbon sugars. Figure 22.38 shows that HMF increases with time more or less gradually, unlike the furfural concentration. This

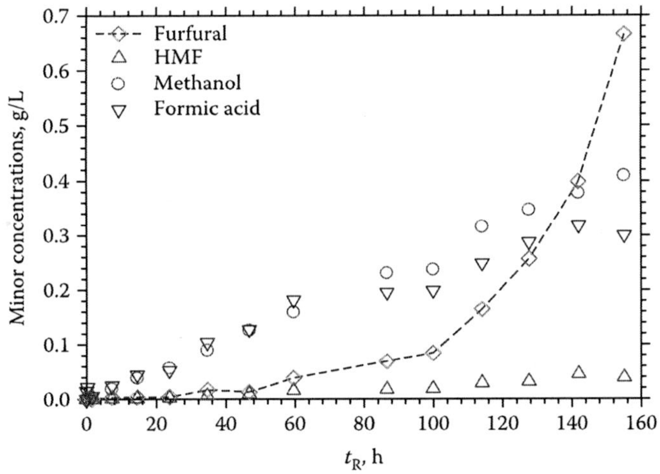

FIGURE 22.38
Variations of minor chemicals in the wood extract with t_R during hot-water extraction of sugar-maple wood chips in a 1.84-m^3 digester.

can be expected because the level of six-carbon sugars in the extract did not change much after 50 h of t_R, as shown in Figure 22.21.

The appearance of formic acid in the extraction liquor early in the extraction process is an indication that the autocatalytic process starts fairly rapidly. The addition of formic or acetic acid to hot water may not have the significant effect that would be expected if acetic acid (Figure 22.28) and formic acid (Figure 22.38) were formed more slowly in the early stages.

22.9 Hot-Water Extraction Process Design Considerations

Hot-water extraction is an important woody-biomass processing technology for pulp and paper, food processing, and the woody-biomass biorefinery. In pulp and paper, hot-water extraction can be used to increase pulp yield and extractives value (Liu et al., 2009). In the woody-biomass biorefinery, hot-water extraction can replace the pretreatment step in hemicellulose recovery to expand the product portfolio (Amidon et al., 2008; Amidon and Liu 2009; Liu 2010).

The relative time t_R as defined by Equation 22.50 is an important parameter for extraction vessel design and extraction process control and performance analysis. It is recommended that extract processing (hydrolysis) be conducted in separate reactors to optimize the extraction yield and product value. The design of the extraction reactor is strongly dependent on the design objectives. If the application is intended to increase pulp yield, minimum removal of carbohydrates is desired, and therefore the relative time should be short. Below $t_R = 30$ h, there are very few carbohydrates in the extraction liquor. For maximum carbohydrates recovery, a combination of high temperature and short residence time is recommended. In this case, $t_R \approx 120$ h has been found to be appropriate.

The residence time for hot-water extraction is limited by the reactor operations. For batch extractions, the temperature ramp-up stage (and a ramp-down stage if explosive decompression is not available) and extraction-process economics play an important role. For continuous extractions, the residence time requirement for the woody biomass counter-currently free-flowing down the extractor and the extraction contents heating and flowing up, together with the process economics, form the basis to compute the residence time and extraction temperature. The overall guidelines for hot-water extraction are high temperature and short residence time. These two parameters are linked by a desirable relative extraction time t_R:

$$t_R = \int_0^t \exp\left(\frac{E_a/R}{T_R} - \frac{E_a/R}{T}\right) dt. \tag{22.50}$$

If the temperature can be assumed constant, Equation 22.50 reduces to

$$t_R = t\exp\left(\frac{E_a}{RT_R} - \frac{E_a}{RT}\right) \tag{22.59}$$

which can also be written as

$$t = t_R \exp\left(-\frac{E_a}{RT_R} + \frac{E_a}{RT}\right), \tag{22.60a}$$

TABLE 22.9

Residence Time and Temperature

t_R (h)	t (min)	T (°C)	t_R (h)	t (min)	T (°C)
5	9.814514	150	110	215.9193	150
5	5.444598	160	110	119.7811	160
5	3.101787	170	110	68.23932	170
10	19.62903	150	110	39.85343	180
10	10.8892	160	110	23.82225	190
10	6.203575	170	110	14.55279	200
10	3.623039	180	110	9.073399	210
30	58.88708	150	110	5.766532	220
30	32.66759	160	120	235.5483	150
30	18.61072	170	120	130.6703	160
30	10.86912	180	120	74.4429	170
30	6.496976	190	120	43.47646	180
30	3.968943	200	120	25.9879	190
30	2.474563	210	120	15.87577	200
30	1.572691	220	120	9.898254	210
			120	6.290763	220

or

$$T = \left[\frac{1}{T_R} + \frac{R}{E_a}\ln\frac{t}{t_R}\right]^{-1}. \tag{22.60b}$$

Table 22.9 shows the relationship between temperature and residence time. Although higher temperature is favorable for the extraction of hemicellulose, the residence time requirement is short and can cause operational difficulties. The residence time limitation places a restriction on the highest temperature at which the reactor can be operated.

22.10 Conclusions and Discussions

Woody-biomass use is imperative to the future economy and a key to conversion to a sustainable society. Although fossilized carbons (petroleum, natural gas, and coal) may be regarded as waste materials from the distant past, their use has caused a carbon imbalance. More carbon dioxide is emitted to the atmosphere from these buried waste through their utilization and release. This imbalance is not to be confused with the high energy density of fossil fuels compared with woody biomass, which results in a lower CO_2 release from fossil sources per unit of energy use than from woody biomass. The use of fossil energy is like sending CO_2 on a one-way trip to atmosphere, which is a different scenario from the use of woody biomass. Fossil energy and chemical sources are not a sustainable way to meet the needs of humans because their deposits on Earth are finite. Their availabilities are limited, and they will run out. Fossil energy sources will become expensive as they become less available. On the other hand, biomass can be regenerated at a steady rate as required by human needs. Although using woody biomass will result in a rise in the CO_2 in the atmosphere because of the change in the carbon storage scheme as a result of human intervention, there is a steady state (sustainable state) which is reachable within the cycle time of woody-biomass growth. The net release of CO_2 into the atmosphere can be as high as half of that available in the untouched standing forest for an 80-year rotation in boreal forest. An 80-year rotation is a long time; however, it is imperative to start using woody biomass now because by doing so, atmospheric CO_2 can be controlled at a lower level than if we continue to send CO_2 into the atmosphere without any means to recapture it. Therefore, biomass is sustainable as a materials, chemical, and energy source. Over a time scale equal to or less than its recycle time (1–80 years), biomass does not contribute to any net change in the carbon dioxide content of the atmosphere once steady state is reached. Of all the biomass types, algae are potentially the highest unit biomass producer, while trees or forests have the highest standing biomass available and are therefore the most readily available biomass source for industrial development. Woody biomass is renewable and serves as a catalyst in that energy from the Sun is effectively converted into energy we can use. Woody biomass represents a major source of renewable chemicals and energy.

Woody biomass is composed of four components: extractives, hemicelluloses, lignin, and cellulose. Each component has a different degree of degradability by chemical, thermal, or biological means. The sequential conversion of woody biomass promotes higher conversion efficiency and an expanded product portfolio. Easily degradable extractives and hemicelluloses are extracted, separated, and converted to valuable chemicals or energy. Hot-water extraction is preferred over other pretreatment techniques because both value and products can be generated.

Woody biomass is heterogeneous and contains many kinds of macromolecular carbohydrate and aromatic components, including hemicelluloses, lignin, and cellulose. Efficient use of woody biomass requires selective removal and hydrolysis of these components as a means of fractionation. One step in the hydrolysis of cellulose can result in the degradation of more easily separable components, for example, extractives, into undesired by-products. Hot-water extraction is designed for the removal of the more easily extractable components.

Besides extractives, hemicelluloses are the most easily separable main component of woody biomass and therefore form the bulk of the wood extracts obtained from hot-water extraction. Xylan is the main polymeric molecule in hardwood hemicelluloses. Xylan is a polymer made up of β-xylopyranose units linked through (1 → 4)-glycosidic bonds, where arabinose, acetyl groups, and uronic acids are also present as lateral chains through inter-residue –O– bonds. Hydrolysis of hemicelluloses or xylan breaks the interconnecting β(1 → 4)-glycosidic –O– bonds as well as the side-chain interresidue –O– bonds, replacing them with –OH and HO–, and produces five-carbon sugars (mainly xylose), six-carbon sugars (mainly glucose), acetic acid, methanol, and other compounds.

Hydrolysis, extraction, and dissolution are similar in that protons are effective catalysts for the breaking of β(1 → 4)-glycosidic bonds. Under high temperatures, acetyl groups in woody biomass hydrate and dissociate, creating acetic acid and causing the extraction liquor pH to drop. The protons formed further catalyze deacetylation and hydrolysis. The extraction reactions are therefore autocatalytic. All the β(1 → 4)-glycosidic bonds are susceptible to breakage, leading to the presence of polydispersed polysaccharides in the extraction liquor. In the case of extraction and dissolution, proton and water molecules need to be adsorbed onto a solid surface at the β(1 → 4)-glycosidic bond sites for the reaction to begin. This reaction occurs on the solid (biomass) surface, whereas hydrolysis occurs in the liquid phase. Therefore, particle size is an important factor in determining the total amount of extractable constituents. The smaller the particles and the greater the surface area, the greater is the extractable mass obtainable in the extraction liquor. The formation reactions for methanol, acetic acid, and uronic acid are similar to the formation of sugars in the hydrolysis of polysaccharides, where the C–O–C linkage breaks because of the insertion of a water molecule catalyzed by protons. The resulting liquor contains a host of polysaccharide or oligosaccharide species, methanol, acetic acid, and other compounds. The monosaccharide concentration increases as the residence time increases, with polysaccharides reaching a maximum during the extraction and dissolution process. However, there are no uronic acid being detected in the extraction liquor. Formic acid and methanol are the closest derivatives of uronic acid found in the hot-water extract.

A simplistic mechanistic model was presented by Liu (2008) to describe the extraction and hydrolysis of hemicelluloses and extractives. Extraction and hydrolysis reaction rates have been obtained from the kinetic model as a function of proton concentration and oligosaccharide concentration distribution. Based on the kinetic model, the distribution of oligosaccharides, the polydispersivity, the degree of polymerization, and the bound and free acetate concentrations can be determined as functions of time, solids-to-water ratio, and temperature. At short extraction times, more oligosaccharides than monosaccharides can be expected in the extraction liquor. At longer extraction times, the average degree of polymerization decreases, while more sugar dehydration degradation products are formed. When the hydrolytic reactivities of all the interresidue –O– bonds are assumed equal, the extraction kinetics may be simplified to a nearly first-order extraction rate expression in terms of the number of extractable monomer units. The extraction duration and temperature can be lumped together into a relative time measure, t_R, reducing the complexity of describing the control and performance of the extraction process. Lumping together the effects of temperature and residence time also enables optimization of the extraction process to be carried out more easily. Furthermore, a first-order kinetic expression based on mass concentration, or more specifically, on equivalent monomer unit concentration, can be used to approximate the dynamic behavior of the extraction process. In this approach, the actual reaction (or extraction) rate is lower than that predicted by the first-order kinetic model because of the change in the molecular distribution of the woody biomass during extraction.

The effect of water-to-wood ratio on the efficiency of hot-water extraction is small. However, a water-to-wood ratio between 4.5 and 5 does have some advantages: higher mass removal, higher carbohydrate content, and the presence of acetyls in the extraction liquor. These effects should not be underestimated.

The mass removal from woody biomass and the dissolved solids content in the extraction liquor each show a plateau at short t_R times. This plateau at short times is a useful design point for applications where minimum mass removal, especially minimum carbohydrate removal, is an objective. For bagasse, this relative time has been found to be $t_R < 5$ h; for aspen wood chips, $t_R < 20$ h; and for sugar-maple wood chips, $t_R \leq 25$ h, all based on $E_a/R = 10{,}800$ K. There is an upper limit on the t_R time for maximum recoverable dissolved solids content. A combination of short residence time and high temperature is most desirable if maximum dissolved solids content is the objective. For eucalyptus, Figure 22.19 shows that t_R is approximately 115 h for maximum mass removal. However, this maximum was obtained within a limited temperature range. A more plausible range can be obtained from Figure 22.14 for sugar-maple wood chips. The maximum dissolved solids content is obtained at $t_R \approx 120$ h, with higher temperatures gives rise to lower dissolved solids concentrations.

Below a t_R of 30 h, there are very few carbohydrates in the hot-water extraction liquor. This minimal carbohydrate removal could be an design objective to maximize pulp yield by inserting a hot-water extraction step (Liu et al., 2009; Lei et al., 2010). There is an optimum t_R for maximum carbohydrate retention in the hot-water extraction liquor. Based on Figures 22.20 and 22.21, this value of t_R is approximately 115 h for sugar-maple wood chips. For mixed northeastern North American hardwoods, the desirable range is $120 < t_R < 150$ h, based on Table 22.8 and Figure 22.24. Typical desired temperature and residence time relations are shown in Table 22.9.

The concentration of acetyl in the extraction liquor increases with increasing extraction time and temperature. More acetic acid is produced initially or at short t_R times, while more bound acetyl is found in the extraction liquor after the initial period. Addition of acetic acid to the water can speed up the extraction slightly.

Significant amounts of aromatic substances are found in the extraction liquor in amounts second only to carbohydrates. For example, aromatic substances up to 5% of dry wood mass can be extracted into the extraction liquor from aspen wood chips, for which the overall mass removal is only approximately 20% of the dry wood mass.

Acknowledgments

The authors are indebted to NYSERDA, DOE and Alberta Pacific Forest Product Industries Inc. for financial support. Research support from the Biorefinery Research Institute (BRI) is also greatly appreciated. The authors are indebted to Drs. Cornelius B. Murphy Jr., Neil H. Ringler, and Bruce C. Bongarten, whose support from SUNY ESF was invaluable to the success of this research project.

References

Abatzoglou, N., Chornet, E., Belkacemi, K., Overend, R.P. 1992. Phenomenological kinetics of complex systems: Development of a generalized severity parameter and its application to lignocellulosics fractionation. *Chem. Eng. Sci.* 47: 1109–1122.

Alén, R. 2000. Structure and chemical composition of wood. In: Gullichsen, J. and Paulapuro, H. (eds.), *Forest Products Chemistry*. Jyväskylä, Finland, Fapet Oy.

Allen, S.G., Schulman, D., Lichwa, J., Antal, M.J., Laser, M., Lynd, L.R. 2001. A comparison between hot liquid water and steam fractionation of corn fiber. *Ind. Eng. Chem. Res.* 40: 2934–2941.

Amidon, T.E., Liu, S.J. 2009. Water-based woody biorefinery. *Biotechnol. Adv.* 27: 542–550.

Amidon, T.E., Wood, C.D., Shupe, A.M., Wang, Y., Graves, M., Liu, S.J. 2008. Biorefinery: Conversion of woody biomass to chemicals, energy and materials. *J. Biobased Mater. Bioenergy* 2: 100–120.

Aoyama, M. 1996. Steaming treatment of bamboo grass. II. Characterization of solubilized hemicellulose and enzymatic digestibility of water-extracted residue. *Cellul. Chem. Technol.* 30: 385–393.

Aoyama, M., Seki, K., Saito, N. 1995. Solubilization of bamboo grass xylan by steaming treatment. *Holzforschung* 49: 193–196.

Ballesteros, I., Oliva, J.M., Navarro, A.A., Gonzalez, A., Carrasco, J., Ballesteros, M. 2000. Effect of chip size on steam explosion pretreatment of softwood. *Appl. Biochem. Biotech.* 84–86: 97–110.

Blower, M. 2003. Xylan extraction from short rotation willow biomass. MSc Thesis. SUNY ESF, Syracuse, NY.

Boussaid, A.L., Esteghlalian, A.R., Gregg, D.J., Lee, K.H., Saddler, J.N. 2000. Steam pretreatment of Douglas-fir wood chips—Can conditions for optimum hemicellulose recovery still provide adequate access for efficient enzymatic hydrolysis? *Appl. Biochem. Biotech.* 84–86: 693–705.

Boussarsar, H., Roge, B., Mathlouthi, M. 2009. Optimization of sugarcane bagasse conversion by hydrothermal treatment for the recovery of xylose. *Bioresource Technol.* 100: 6537–6542.

Campbell, J., Fahey, J., Wolf, B. 1997. Selected indigestible oligosaccharides affect large bowel mass, cecal and fecal shortchain fatty acids, pH and microflora in rats. *J. Nutr.* 127: 130–136.

Carrasco, F., Roy, C. 1992. Kinetic study of dilute-acid prehydrolysis of xylan-containing biomass. *Wood Sci. Technol.* 26: 189–208.

Conner, A.H. 1984. Kinetic modeling of hardwood prehydrolysis, Part 1: Xylan removal by water prehydrolysis. *Wood Fiber Sci.* 16: 268–277.

Conner, A.H., Lorenz, L.F. 1986. Kinetic modeling of hardwood prehydrolysis, Part 33: Water and dilute acetic-acid prehydrolysis of Southern red oak. *Wood Fiber Sci.* 18: 248–263.

Dekker, R., Wallis, A. 1983. Autohydrolysis-explosion as pretreatment for the enzymic saccharification of sunflower seed hulls. *Biotechnol. Lett.* 5: 311–316.

Esteghlalian, A., Hashimoto, A.G., Fenske, J.J., Penner, M.H. 1997. Modeling and optimization of the dilute-sulfuric-acid pretreatment of corn stover, poplar, and switchgrass. *Bioresource Technol.* 59: 129–136.

Fengel, D., Wegener, G. 1989. *Wood: Chemistry, Ultrastructure, Reactions*. New York, Walter de Gruyter.

Fontaine, S., Barot, S., Barre, P., Bdioui, N., Mary, B., Rumpel, C. 2007. Stability of organic carbon in deep soil layers controlled by fresh carbon supply. *Nature* 450: 277–281.

Garrote, G., Dominguez, H., Parajo, J.C. 1999. Mild autohydrolysis: An environmentally friendly technology for xylooligosaccharide production from wood. *J. Chem. Technol. Biot.* 74: 1101–1109.

Garrote, G., Dominguez, H., Parajo, J. 2001a. Kinetic modeling of corncob autohydrolysis (Oxford, United Kingdom). *Process Biochem.* 36: 571–578.

Garrote, G., Dominguez, H., Parajo, J.C. 2001b. Generation of xylose solutions from *Eucalyptus globulus* wood by autohydrolysis-posthydrolysis processes: Posthydrolysis kinetics. *Bioresource Technol.* 79: 155–164.

Garrote, G., Dominguez, H., Parajo, J.C. 2002. Interpretation of deacetylation and hemicellulose hydrolysis during hydrothermal treatments on the basis of the severity factor. *Process Biochem.* 37: 1067–1073.

Garrote, G., Dominguez, H., Parajo, J.C. 2004. Production of substituted oligosaccharides by hydrolytic processing of barley husks. *Ind. Eng. Chem. Res.* 43: 1608–1614.

Glaudemans, C., Timell, T.E. 1958. The polysaccharides of white birch (*Betula papyrifera*). IV. The constitution of the hemicellulose. *J. Am. Chem. Soc.* 80: 1209–1213.

Goring, D., Timell, T. 1960. Molecular properties of six 4-O-methylglucuronoxylans. *J. Phys. Chem.* 64: 1426–1430.

Grethlein, H., Converse, A. 1991. Common aspects of acid prehydrolysis and steam explosion for pretreating wood. *Bioresource Technol.* 36: 77–82.

Gullichsen, J., Paulapuro, H. 2000. *Forest Resources and Sustainable Management*. Helsinki, Finland: Fapet Oy.

Heitz, M., Capekmenard, E., Koeberle, P.G., Gagne, J., Chornet, E., Overend, R.P., Taylor, J.D., Yu, E. 1991. Fractionation of *Populus tremuloides* at the pilot-plant scale—Optimization of steam pretreatment conditions using the stake-II technology. *Bioresource Technol.* 35: 23–32.

Holtzapple, M., Benjamin, C. 2003. Hemicelluloses. In: *Encyclopedia of Food Sciences and Nutrition*, Oxford, Academic Press.

Hu, R.F., Lin, L., Liu, T.J., Liu, S.J. 2010. Dilute sulfuric-acid hydrolysis of sugar-maple wood extract at atmospheric pressure. *Bioresource Technol.*, 101(10): 3586–3594. (doi:10.1016/j.biortech.2010.01.005).

Ichikawa, T., Mitsumura, Y. 1996. Effects of the addition of new saccharide sweeteners on the properties of sponge cakes. *Nippon Kasei Gakkaishi* 47: 445–452.

Imaizumi, K., Nakatsu, Y., Sato, M., Sedarnawati, Y., Sugano, M. 1991. Effects of xylooligosaccharides on blood glucose, serum, and liver lipids and cecum short-chain fatty acids in diabetic rats. *Agric. Biol. Chem.* 55: 199–205.

Janisch, J.E., Harmon, M.E. 2002. Successional changes in live and dead wood carbon stores: Implications for net ecosystem productivity. *Tree Physiol.* 22(2–3):77–89.

Jeffries, T.W., Jin, Y.S. 2004. Metabolic engineering for improved fermentation of pentoses by yeasts. *Appl. Microbiol. Biot.* 63: 495–509.

Kabel, M.A., Bos, G., Zeevalking, J., Voragen, A.G.J., Schols, H.A. 2007. Effect of pretreatment severity on xylan solubility and enzymatic breakdown of the remaining cellulose from wheat straw. *Bioresource Technol.* 98: 2034–2042.

Keenan, T.M., Tanenbaum, S.W., Stipanovic, A.J., Nakas, J.P. 2004. Production and characterization of poly-beta-hydroxyalkanoate copolymers from *Burkholderia cepacia* utilizing xylose and levulinic acid. *Biotechnol. Progr.* 20: 1697–1704.

Keoleian, G.A., Volk, T.A. 2005. Renewable energy from willow biomass crops: Life cycle energy, environmental and economic performance. *Crit. Rev. Plant Sci.* 24: 385–406.

Kim, S., Lee, Y. 1987. Kinetics in acid-catalyzed hydrolysis of hardwood hemicelluloses. *Biotechnol. Bioeng. Symposium* 17: 71–84.

Kim, T.H., Lee, Y.Y. 2006. Fractionation of corn stover by hot-water and aqueous ammonia treatment. *Bioresource Technol.* 97: 224–232.

Klass, D. 1998. *Biomass for Renewable Energy, Fuels, and Chemicals*. New York, Academic Press.

Koshijima, T., Timell, T., Zinbo, M. 1965. The number-average molecular weight of native hardwood xylans. *J. Polymer Sci., Part C: Polymer Symposia* 11: 265–279.

Lei, Y., Liu, S.J., Li, J., Sun, R.C. 2010. Effect of hot-water extraction on alkaline pulping of bagasse J. Biotech. Adv 28: 609–612.

Li, H., Saeed, A., Jahan, M.S., Ni, Y., van Heiningen, A. 2010. Hemicellulose removal from hardwood chips in the extraction step of the Kraft-based dissolving pulp production process. *J. Wood Chem. Techn.* 30: 1, 48–60.

Liu, C.G., Wyman, C.E. 2005. Partial flow of compressed hot water through corn stover to enhance hemicellulose sugar recovery and enzymatic digestibility of cellulose. *Bioresource Technol.* 96: 1978–1985.

Liu, S.J. 2003. Chemical kinetics of alkaline peroxide brightening of mechanical pulps. *Chem. Eng. Sci.* 58: 2229–2244.

Liu, S.J. 2004. A simplistic mechanistic model and effect of consistency on alkaline peroxide brightening of mechanical pulps. *Chem. Eng. Sci.* 59: 4377–4383.

Liu, S.J. 2008. A kinetic model on autocatalytic reactions in woody biomass hydrolysis. *J. Biobased Mater. Bioenergy* 2: 135–147.

Liu, S.J. 2010. Woody biomass: Niche position as a source of sustainable renewable chemicals and energy and kinetics of hot-water extraction/hydrolysis. *J. Biotech. Adv.* 28: 563–582.

Liu, S.J., Amidon, T.E., Francis, R.C., Ramarao, B.V., Lai, Y.-Z., Scott, G.M. 2006. From forest biomass to chemicals and energy: Biorefinery initiative in New York. *Indus. Biotechnol.* 2: 113–120.

Liu, S.J., Amidon, T.E. 2007. Essential components of the wood-based biorefinery. *O Papel* 68: 84–105.

Liu, S.J., Amidon, T.E., Wood, C.E. 2008. Membrane filtration: Concentration and purification of hydrolyzates from biomass. *J. Biobased Mater. Bioenergy* 2(2): 121–134.

Liu, S.J., Mishra, G., Amidon, T.E., Gratien, K. 2009. Effect of hot-water extraction of woodchips on the Kraft pulping of eucalyptus woodchips. *J. Biobased Mater. Bioenergy* 3(3): 363–372.

Liu, T.J., Lin, L., Sun, Z., Hu, R.F., Liu, S.J. 2010. Bioethanol fermentation by robust recombinant *E. coli* FBHW using hot-water wood extract hydrolyzate as substrate J. Biotech. Adv. 28: 602–608

Liu, S.J., Luo, Q., Ampong-Nyarko, K., Kloeck, T. 2006. AgFiber properties vs. species/varieties and locations in Alberta. *Preprints, PAPTAC 92nd Annual Meeting*, Montreal, Canada, February.

Lora, J., Wayman, M. 1978. Delignification of hardwoods by autohydrolysis and extraction. *Tappi J.* 61: 47–50.

Luyssaert, S., Schulze, E.-D., Börner, A., Knohl, A., Hessenmöller, D., Law, B.E., Ciais, P., Grace, J. 2008. Old-growth forests as global carbon sinks. *Nature* 455: 213–215.

Maloney, M.T., Chapman, T.W., Baker, A.J. 1985. Dilute acid hydrolysis of paper birch: Kinetics studies of xylan and acetyl-group hydrolysis. *Biotechnol. Bioeng.* 27: 355–361.

Mao, H., Genco, J.M., Yoon, S.H, van Heiningen, A., Pendse, H. 2008. Technical economic evaluation of a hardwood biorefinery using the "near-Neutral" hemicellulose pre-extraction process. *J. Biobased Mater. Bioenergy* 2: 177–185.

May, T., Mackie, R., Garleb, K. 1995. Effect of dietary oligosaccharides in the intestinal growth of and tissue damage by *Clostridium difficile*. *Mikrooekol. Ther.* 23: 158–170.

Mittal, A. 2006. *Kinetics of Hemicellulose Extraction During Autohydrolysis of Sugar Maple Wood.* PhD dissertation, State University of New York, College of Environmental Science and Forestry, Syracuse, New York,.

Mittal, A., Chatterjee, S.G., Scott, G.M., Amidon, T.E. 2009a. Modeling xylan solubilization during autohydrolysis of sugar maple wood meal: Reaction kinetics. *Holzforschung* 63: 307–314.

Mittal, A., Scott, G.M., Amidon, T.E., Kiemle, D.J., Stipanovic, A.J. 2009b. Quantitative analysis of sugars in wood hydrolyzates with H-1 NMR during the autohydrolysis of hardwoods. *Bioresource Technol.* 100: 6398–6406.

Montane, D., Overend, R.P., Chornet, E. 1998. Kinetic models for non-homogeneous complex systems with a time-dependent rate constant. *Can. J. Chem. Eng.* 76: 58–68.

Montane, D., Salvado, J., Farriol, X., Jollez, P., Chornet, E. 1994. Phenomenological kinetics of wood delignification: Application of a time-dependent rate constant and a generalized severity parameter to pulping and correlation of pulp properties. *Wood Sci. Technol.* 28: 387–402.

Mosier, N., Hendrickson, R., Ho, N., Sedlak, M., Ladisch, M.R. 2005a. Optimization of pH-controlled liquid hot water pretreatment of corn stover. *Bioresource Technol.* 96: 1986–1993.

Mosier, N., Wyman, C., Dale, B., Elander, R., Lee, Y.Y., Holtzapple, M., Ladisch, M. 2005b. Features of promising technologies for pretreatment of lignocellulosic biomass. *Bioresource Technol.* 96: 673–686.

Nabarlatz, D., Farriol, X., Montane, D. 2004. Kinetic modeling of the autohydrolysis of lignocellulosic biomass for the production of hemicellulose-derived lignosaccharides. *Ind. Eng. Chem. Res.* 43: 4124–4131.

Nabarlatz, D., Farriol, X., Montane, D. 2005. Autohydrolysis of almond shells for the production of xylo-oligosaccharides: Product characteristics and reaction kinetics. *Ind. Eng. Chem. Res.* 44: 7746–7755.

Okazaki, M., Fujikawa, S., Matsumoto, N. 1990. Effect of xylooligosaccharide on the growth of bifidobacteria. *Bifidobacteria Microflora* 9: 77–86.

Okazaki, M., Koda, H., Izumi, R., Fujikawa, S., Matsumoto, N. 1991. *In-vitro* digestibility and *in-vitro* utilization of xylobiose. *J. Jpn. Soc. Nutr. Food Sci.* 44: 41–44.

Overend, R.P., Chornet, E. 1987. Fractionation of lignocellulosics by steam-aqueous pretreatments. *Philos. T. Roy. Soc. A* 321: 523–536.

Overend, R., Chornet, E. 1989. Steam and aqueous pretreatments: Are they prehydrolyses? In: *Wood Processing and Utilization*. New York, Elsevier. P. 395–400.

Pourali, O., Asghari, F.S., Yoshida, H. 2010. Production of phenolic compounds from rice bran biomass under subcritical water conditions. *Chem. Eng. J.* 160: 259–266.

Qureshi, N., Dien, B.S., Nichols, N.N., Saha, B.C., Cotta, M.A. 2006. Genetically engineered *Escherischia coli* for ethanol production from xylose: Substrate and product inhibition and kinetic parameters. *Food Bioprod. Process.* 84: 114–122.

Ragauskas, A.J., Williams, C.K., Davison, B.H., Britovsek, G., Cairney, J., Eckert, C.A., Frederick, W.J., et al. 2006. The path forward for biofuels and biomaterials. *Science* 311: 484–489.

Root, D.F., Saeman, J.F., Harris, J.F. 1959. Kinetics of the acid-catalyzed conversion of xylose to furfural. *Forest Products J.* 9: 158–165.

Saeman, J. 1945. Kinetics of wood hydrolysis: Decomposition of sugars in dilute acid at high temperature. *J. Indus. Eng. Chem.* 37: 43–52.

Saska, M., Ozer, E. 1995. Aqueous extraction of sugarcane bagasse hemicellulose and production of xylose syrup. *Biotechnol. Bioeng.* 45: 517–523.

Schultz, T., McGinnins, G., Biermann, C. 1984. Similarities and differences in pretreating woody biomass by steam explosion, wet oxidation, autohydrolysis, and rapid steam hydrolysis/continuous extraction. *Energy Biomass Wastes* 8: 1171–1198.

Shin, S., Cho, N. 2008. Conversion factors for carbohydrate analysis by hydrolysis and 1H-NMR spectroscopy. *Cellulose* 15: 255–260.

Sjöstrom, E. 1993. *Wood Chemistry: Fundamentals and Applications*. New York, Academic Press.

Takeuchi, Y., Jin, F.M., Tohji, K., Enomoto, H. 2008. Acid catalytic hydrothermal conversion of carbohydrate biomass into useful substances. *J. Mater. Sci.* 43: 2472–2475.

Timell, T. 1960. Isolation and properties of an O-Acetyl-4-O-methylglucurono-xyloglycan from the wood of white birch (*Betula papyrifera*). *J. Am. Chem. Soc.* 82: 5211–5215.

Toyoda, Y., Hatanaka, Y., Suwa, Y. 1993. Effect of xylooligosaccharides on calcium absorption. *Proc., 47th Ann. Meeting Jpn. Soc. Nutr. Food Sci.*, Tokyo.

Tunc, M.S., van Heiningen, A.R.P. 2008. Hydrothermal dissolution of mixed Southern hardwoods. *Holzforschung* 62(5): 539–545.

Vroom, K. 1957. The H factor: A means of expressing cooking times and temperatures as a single variable. *Pulp Paper Mag. Can.* 58: 228–231.

Walker, T., Cardellichio, P., Colnes, A., Gunn, J., Kittler, B., Perschel, B., Recchia, C., Saah, D. 2010. *Biomass Sustainability and Carbon Policy Study*. Commonwealth of Massachusetts Department of Energy Resources and Manomet Center for Conservation Sciences, June.

Weil, J., Sarikaya, A., Rau, S.L., Goetz, J., Ladisch, C.M., Brewer, M., Hendrickson, R., Ladisch M.R. 1997. Pretreatment of yellow poplar sawdust by pressure cooking in water. *Appl. Biochem. Biotech.* 68: 21–40.

Wilder, H.D., Daleski, E.J. Jr. 1964. Kraft pulping kinetics. I. Literature review and research program. *Tappi J.* 47(5): 270.

Yang, L., Liu, S.J. 2005. Kinetic model for Kraft pulping process. *Ind. Eng. Chem. Res.* 44: 7078–7085.

Zhuang, Q.Y., Vidal, P.F. 1996. Hemicellulose solubilization from *Populus tremuloides* via steam explosion and characterization of the water-soluble fraction: 1. Autocatalytic processes. *Cellul. Chem. Technol.* 30: 371–384.

23

Integrated Utilization of Algae Biomass in a Biorefinery Based on a Biochemical Processing Platform

Eduardo Sánchez-Tuirán, Mahmoud M. El-Halwagi, and Vyacheslav Kafarov

CONTENTS

23.1 Benefits of Algae Biofuels

Algae crops have been used through time for various purposes. Fish, vitamins, and other products have been produced on a large scale using algae. However, in recent years, the world's energy demand has grown dramatically, leading the scientific community to undertake research and development of new and efficient sources of energy to provide cleaner and greener performance. In this scenario, algae-to-biofuels has emerged as a very promising process option because algae have most of the desirable characteristics of a biomass feedstock, such as

1. Efficiency: Algae can transform sunlight efficiently into biomass (lipids and other products).
2. Lipid production: Some species can produce up to 80% of their weight in dry biomass (Metting, 1996; Spolaore et al., 2006). Oil production levels of 20–50% are fairly common (Chisti, 2007a).

3. Fast growth: Algae have exponential growth rates and can double their biomass in periods as short as 3.5 h (Chisti, 2007a).
4. Carbon dioxide: Microalgae biomass production can provide biofixation of waste CO_2 (1 kg of dry algal biomass utilizes approximately 1.83 kg of CO_2) (Chisti, 2007a).
5. Avoidance of the food versus fuel debate: Unlike many other crops, use of microalgae avoids the food versus fuel debate because algae are not used as a primary source of human food. Furthermore, algae cultivation could even use arid land that is not suitable for other traditional crops.

As pointed out by Valderrama et al. (2002), microalgae have emerged as one of the most promising biorefinery feedstocks because of their widespread availability and high oil yields. Unlike traditional oilseed crops, microalgae can grow in places away from farmlands and forests, such as ponds, fermentation units, and even wastewater (Converti et al., 2009; Pokoo-Aikins et al., 2010; Doušková et al., 2010; Huang et al., 2010; Liu et al., 2010; Najafabady et al., 2010; Teresa et al., 2010; Ahmad et al., 2011; Amaro et al., 2011; Chen et al., 2011; Demirbas, 2011; Harun et al., 2011; Levine et al., 2011; Moazami et al., 2011; Tabatabaei et al., 2011; Tang et al., 2011; Yang et al., 2011).

There are several ways to cultivate microalgae. They can be cultivated on a small scale inside a laboratory photobioreactor to ensure the controllability of the process to obtain a very specific product. The most important variables that affect the growth and composition of microalgae on a laboratory scale are temperature, pH, CO_2 concentration in the broth, nutrients concentration, mixing, geometry, and of course light intensity.

For the reason that microalgae are plants, they need sunlight, CO_2, and nutrients. It is well known that either an excess or a lack of any of these factors may have a negative impact on microalgae growth and oil accumulation. Optimum conditions must be found by searching the experimental space to maximize both growth rate and lipid accumulation.

On the other hand, algae cultivation in open ponds requires an initial capital investment for construction and adjustment, but then the control, maintenance, and operating costs are fairly low (Borowitzka, 1999). Very low-quality soil areas that are unsuitable for traditional energy crops can be used to construct the ponds. Large-scale microalgae production is used today to obtain vitamins and other components for the food supplements industry. However, if biorefineries are implemented, most of the worldwide microalgae production will be used in the production of biofuels and other high-value coproducts such as pharmaceuticals and precursors for other industries.

23.2 Basic Algae Biology

Algae are a group of oxygenic photosynthetic organisms made up of macro- and microalgae. They represent an extremely diverse, yet highly specialized, group of microorganisms that live in diverse ecological habitats such as freshwater, brackish, marine, and hypersaline environments, with a range of temperatures and pH and unique nutrient availabilities (Falkowski and Raven, 1997).

These organisms use energy from the sun to combine water with carbon dioxide to create biomass. They include macroalgae, microalgae, and emergents. Macroalgae, more commonly known as "seaweed," are fast-growing marine and freshwater plants that can grow to a considerable size (up to 60 m in length). Emergents are plants that grow partially submerged in bogs and marshes. Microalgae are, as the name suggests, microscopic photosynthetic organisms. Like macroalgae, these organisms are found in both marine and freshwater environments.

Biologists have categorized microalgae into a variety of classes, distinguished mainly by their pigmentation, life cycle, and basic cellular structure. The four most important (at least in terms of abundance) are

- Diatoms (*Bacillariophyceae*). These algae dominate the phytoplankton of the oceans, but are also found in fresh and brackish water. Approximately 100,000 species are known to exist. Diatoms contain polymerized silica (Si) in their cell walls. All cells store carbon in a variety of forms. Diatoms store carbon in the form of natural oils or as a polymer of carbohydrates known as chrysolaminarin.
- Green algae (*Chlorophyceae*). These are also quite abundant, especially in fresh water. They can occur as single cells or as colonies. Green algae are the evolutionary progenitors of modern plants. The main energy storage compound in green algae is starch, although oils can be produced under certain conditions.
- Blue-green algae (*Cyanophyceae*). Much closer to bacteria in structure and organization, these algae play an important role in fixing nitrogen from the atmosphere. There are approximately 2000 known species found in a variety of habitats.
- Golden algae (*Chrysophyceae*). This group of algae is similar to the diatoms. They have complex pigment systems and can appear yellow, brown, or orange in color. Approximately 1000 species are known to exist, primarily in freshwater systems. They are similar to diatoms in pigmentation and biochemical composition. The golden algae produce natural oils and carbohydrates as energy storage compounds.

Microalgae are the most primitive form of plants. Although the mechanism of photosynthesis in microalgae is similar to that of higher plants, they are generally more efficient converters of solar energy because of their simple cellular structure. In addition, because their cells grow in aqueous suspension, they have more efficient access to water, CO_2, and other nutrients than other plants. For these reasons, microalgae are capable of producing 30 times more oil per unit area of land than terrestrial oilseed crops.

23.3 Emerging and Leading Algae-Feedstock Biorefinery Technologies Based on a Biochemical Processing Platform (Technological Alternatives and Challenges)

Biochemical processes are an emerging choice among biorefinery technologies because they depend highly on process conditions and their speed is usually less than that of physical or thermochemical processes. The most commonly used biochemical processes involve fermentation and anaerobic digestion (Cherubini, 2010). Through fermentation, the sugars obtained from the cellulosic residues of microalgae are converted into other desirable products like alcohols that are removed from the mixture by distillation and then by molecular sieves. However, other products are also obtained and can represent real possibilities for the biorefinery concept to maximize benefits by using the whole biomass feedstock, for example *Botryococcus braunii* is known to contain large fractions lipids and hydrocarbons as well as some pigments like carotenoids (Dayananda et al., 2005, 2006, 2007; Li and Qin, 2005; Berberoglu et al., 2009; Choi et al., 2010; Eroglu et al., 2011; Eroglu and Melis, 2010; Gardner et al., 2011; Kita et al., 2010; Linh et al., 2010; Rivas et al 2010; Samorì et al., 2010; Tanoi et al., 2011; Weiss et al., 2011; Baba et al., 2012; Zhang et al., 2011; Zhila et al., 2011), also *Chlorella* sp., is also known for its great content of carbohydrates and proteins (González et al., 1997; Valderrama et al., 2002; Hernández et al., 2006, 2009; Travieso et al., 2006; Kim et al., 2010; Lim et al., 2010; Li et al., 2011), even the algae *Haematococcus pluvialis* may work due to its range of lipids (Dong and Zhao, 2004; Domínguez-Bocanegra et al., 2004; Jeon et al., 2006; Kang et al., 2006; Suh et al., 2006; Ranjbar et al., 2008; Issarapayup et al., 2011; Razon and Tan, 2011). In this way, the biorefinery not only takes advantage of the resources available, but also minimizes the amount of residues sent to the environment.

One of the most important challenges for biochemical processes is effective control of sudden changes in process variables such as temperature and concentration. Biochemical kinetics are extremely sensitive, and therefore they can be severely affected by variations. For this reason, strict control of process variables will improve the selectivity and quality control of the products.

23.3.1 Algae Mass Cultivation Systems

On account of the microalgae production at a large scale for biofuels or other purposes provides technical and economic benefits, several companies have developed their own systems of cultivation. However, many of these share a number of characteristics that make a more general description possible. Large-scale production of microalgae biomass uses continuous culture during daylight, achieving steady-state conditions of feed of fresh culture and withdrawal of microalgae broth (Chisti, 2007a; Molina Grima et al., 1999).

Some losses of biomass will occur due to respiration in the night period. The amount of biomass that is lost will depend on the temperature during the day and the night and on the light intensity during the growth period (Chisti, 2007a) also, due to different light/night cycles, which strongly affect biomass production. Huppe and Turpin (1994) reported that up to 55% of net plant carbon is committed to nitrogen assimilation and metabolism in some tissues. Therefore, it is important to consider those pathways that consume the largest amounts of carbon during photosynthesis to produce lipids and starch.

Two methods have been studied and applied successfully for large-scale production of microalgae: raceway ponds (Terry and Raymond, 1985; Molina Grima et al., 1999) and tubular photobioreactors (Chini Zittelli et al., 1999; Molina Grima et al., 1999; Sánchez et al., 1999).

A biorefinery is a factory consisting of a collection of processes which takes agricultural inputs from the surrounding area and produces a wide range of products which are specifically targeted at different market uses and are refined to market specifications (Audsley and Annetts, 2003).

Nowadays, the biorefinery appears to be a good alternative with great social significance because it connects four cornerstones of our modern society: (1) energy security, because it offers an alternative to the limited fossil-fuel supply; (2) environmental protection, specifically for combating climate change by reducing greenhouse gas emissions; (3) economic development, because it creates new opportunities for agricultural and energy industries as well as bringing about fundamental change in the future economy; and (4) allocation of natural resources such as land and water, because bioenergy and bioproducts may compete with the human food supply for the production of feedstock (Chen, 2008).

In this context, a raw material such as microalgae biomass is a novel and promising option for biorefineries. Microalgae have the capacity to produce a vast array of high-value bioactive compounds that can be used as pharmaceutical material, health foods, and natural pigments (Jiang, 2000; Oh et al., 2003). Some well-studied examples include acetylic acids, β-carotene (Huang et al., 2006; Del Campo et al., 2007), vitamin B (He et al., 2005), ketocarotenoid astaxanthin (Ip and Chen, 2005), polyunsaturated fatty acids (Wen and Chen, 2003; Jiang et al., 2004), and lutein (Shi et al., 2002; Blanco et al., 2007).

According to Wijffels (2007), to develop products from microalgae in an industrial process, several requirements must be met. First, a large amount of biomass must be produced, preferably in a compact installation such as an efficient photobioreactor; second, the biomass should contain high concentrations of the desired product, which can be achieved with a thorough understanding of the metabolic pathways involved and with effective control of these pathways. Finally, the entire process should be applicable on an industrial scale, which would involve cultures of several thousands of cubic meters.

23.3.2 Harvesting Technology

Microalgae cultures require, in addition to nutrients, sunlight, and CO_2, large amounts of water to dilute the broth. According to Brennan and Owende (2010), there are several possible ways of harvesting microalgae: flocculation and ultrasonic aggregation, flotation, gravity or centrifugal sedimentation, and biomass filtration. The choice of harvesting technique depends on the physical characteristics of the culture. The final extraction of the microalgae concentrate generally requires a two-step procedure. First, bulk harvesting is done by means of flocculation, flotation, or gravity sedimentation, which achieves approximately 5% total solids in the broth. Then centrifugation, filtration, and ultrasonic aggregation techniques, which are generally energy-intensive, are used to increase the concentration.

These techniques take advantage of the natural properties of the microalgae to improve the separation factor. Flocculation and ultrasonic aggregation make use of the fact that microalgae cells carry a negative charge that prevents natural aggregation of cells in suspension. Therefore, the addition of multivalent cations and cationic polymers neutralizes or reduces the negative charge, making separation less difficult (Brennan and Owende, 2010). Another alternative is to use very simple harvesting techniques like flotation, where the idea is to form clusters of algae cells by means of very small air bubbles.

In gravity and centrifugal sedimentation, separation is performed using either a natural gravitational field or an artificial one. In the first case, generally the sedimentation time is longer, and therefore this process is used only when the microalgae cells are large enough to make the process viable. In an artificial gravitational field, an external force is applied by means of continuous rotation around an axis. This technique is faster, yet energy-intensive. However, it is appropriate when use of a natural gravitational field use makes large-scale operation infeasible.

Biomass filtration makes use of the difference in pressure before and after the filter. This technique is useful for microalgae which are considered large, that is, >70 μm, like *Coelastrum* and *Spirulina*, but it is unsuitable for the very popular *Chlorella* species and for *Scenedesmus* and *Dunaliella* (Mohn, 1980).

23.3.3 Oil-Extraction Technologies

After the harvesting stage, drying is needed to remove excess water. Several strategies can be used for this. Solar drying is the cheapest, but inefficient because it needs long drying times and large drying surfaces and risks material loss (Prakash et al., 1997). After the biomass has been dried, the next step is to extract the lipids by means of several techniques such as hexane and cyclohexane solvents or the *soxhlet* method. With these techniques, the high affinity between lipids and solvents makes possible the extraction of lipids from the microalgae. After extraction, separation from the solvent is achieved by means of repeated evaporations and condensations. Recovery of solvent is mandatory to obtain a cost-effective and safe process.

23.3.4 Biodiesel Production from Algae

Various oils are used worldwide to obtain biodiesel. The choice of oil depends strongly on geographical location and therefore on raw material availability. Generally, the transesterification reaction is used to transform oils in the presence of an alcohol and a catalyst because of its performance and ease of operation. Several alcohols can be used to react with the oil to obtain biodiesel (from methanol going through ethanol and then propanol, butanol, etc.). The selection of the alcohol depends mainly on technical, economic, and safety criteria.

In the transesterification reaction, one mole of triglyceride reacts with three moles of alcohol to obtain three moles of alkyl esters and one mole of glycerol. The composition of the alkyl esters obtained will depend on the oil composition as well as the alcohol used. To ensure good performance of the fuel obtained in diesel engines, the European Committee for Standardization (CEN) and the American Society for Testing and Materials (ASTM) have established criteria that must be satisfied to meet their technical standards. Some differences exist between the EN and ASTM criteria because European biodiesel is obtained primarily from rapeseed and American from soybeans. The principal differences are the specified flash point (101.0°C vs. 130.0°C min., EN and ASTM, respectively), and that the European criteria include stricter control of the content of certain substances such as unreacted tri-, di-, and monoglycerides. Table 23.1 shows the properties, the limits, and in most cases, the test used to ensure good biodiesel performance.

From these criteria, it is clear that the oil source must have certain compositional characteristics to satisfy both standards. For example, the linolenic-acid methyl ester content must remain smaller than 12% m/m. Petkov and García (2007) have reported the oil composition produced by several species of *Chlorella* microalgae. They showed that the composition of linolenic acid in three different species of *Chlorella* (*C.* sp., *C. vulgaris*, and *C. pyrenoidosa*) under various culture conditions (indoor, outdoor, nitrogen starvation, photoautotrophic, and heteroautotrophic) was between 17% and 27% m/m. These percentages are higher than those found in other traditional feedstocks for biodiesel production. For example, African palm oil has a linolenic acid content of less than 1% m/m, and soy oil has approximately 7% m/m. Therefore, it is important to check whether a microalgal species is suitable for use as an oil source for biodiesel production.

If the microalgal species is a candidate oil source for biodiesel production, then a traditional transesterification reaction can be performed. Freedman et al. (1984) reported the principal variables that were found to affect the yield of fatty esters from transesterified vegetable oils. They demonstrated that the molar ratio of alcohol to vegetable oil, the type of catalyst (alkaline vs. acidic), the temperature, and the degree of refining of the vegetable oils were the most important factors. Although they worked with cottonseed, peanut, soybean, and sunflower oils, the general conclusions that they reported for these oils can be extended to algae-derived oil. Some of their conclusions were that to obtain maximum ester formation by transesterification of vegetable oils, refined oils which are substantially

TABLE 23.1

European Union Normative (EN) and American Standards for Biodiesel Production

		Limits		Test Method	
Property	**Unit**	**ASTM D6751**	**EN14214**	**ASTM D6751**	**EN 14214**
Flash point	°C	130.0 min.	101.0 min.	D93	ISO CD3679e
Kinematic viscosity at 40°C	mm^2/s	1.9–6.0	3.5–5.0	D445	EN ISO 3104
Cetane number	—	47 min.	51 min.	D613	EN ISO 5165
Sulfated ash content	% (m/m)	0.020 max.		D874	ISO 2987
Copper-strip corrosion	—	No. 3 max.	Class 1	D130	EN ISO 2160
Acid value	mg KOH/g	0.80 max.	0.5 max.	D664	Pr EN 14104
Free glycerol	% (m/m)	0.020 max.		D6584	Pr EN 14105m Pr EN 14106
Total glycerol	% (m/m)	0.240 max.	0.25 max.	D6584	Pr EN 14105m
Phosphorus content	% (m/m)	0.001 max.	0.01 max.	D4951	Pr EN 141101
Carbon residue					
ASTM D6751 (100% sample)	% (m/m)	0.050 max.	—	D4530	—
EN 14214 (10% bottoms)		—	0.3 max.	—	EN ISO 10370
Cloud point	°C	Report costumer	—	D2500	—
Density at 15°C	kg/m^3	—	860–900	—	EN ISO 3675 EN ISO 12185
Distillation T90 AET	°C	360 max.	—	D1160	—
Sulfur (S 15 grade)	ppm	0.0015 max.	—	D5453	—
Sulfur (S 500 grade)	ppm	0.05 max.	—	D5453	—
Sulfur content	mg/kg	—	10 max.	—	—
Water and sediment	% vol.	0.050 max.	—	D2709	—
Water content	mg/kg	—	500 max.	—	EN ISO 12937
Total contamination	mg/kg	—	24 max.	—	EN 12662
Oxidation stability at 110°C	H	—	6 min.	—	Pr EN 14112
Iodine value	—	—	120 max.	—	Pr EN 14111
Linolenic acid methyl ester	% (m/m)	—	12 max.	—	Pr EN 14103d
Polyunsaturated (four double bonds) methyl esters	% (m/m)	—	1 max.	—	Pr EN 14103
Ester content	% (m/m)	—	96.5 min.	—	Pr EN 14103d
Methanol content	% (m/m)	—	0.2 max.	—	Pr EN 14101
Monoglycerides content	% (m/m)	—	0.8 max.	—	Pr EN 14105m
Diglycerides content	% (m/m)	—	0.2 max.	—	Pr EN 14105m
Triglycerides content	% (m/m)	—	0.2 max.	—	Pr EN 14105m
Alkaline metals (Na + K)	mg/kg	—	5 max.	—	Pr EN 14108 Pr EN 14109

anhydrous with a free fatty-acid content of less than 0.5% (acid number less than one) should be used. The alcohol must be moisture-free. A molar ratio of alcohol to oil of 6:1 gives optimum conversion to the ester. Either 0.5% sodium methoxide (laboratory scale) or 1% sodium hydroxide (for larger-scale reactions) are effective catalysts. Finally, ester conversions of 96–98% can be obtained by transesterifying the refined oils with methanol, ethanol, or butanol at 60°C, 75°C, and 114°C, respectively, for 1 h with 0.5% sodium methoxide.

23.3.5 Bioethanol Production from Algae Bagasse

After extraction of the microalgae oil, the residues generally represent approximately 75% of the initial biomass. Given this quantity of potential residues, it is appropriate to transform them into other products using well-known techniques which have been applied successfully in other biomass residue industries. In this case, the sugar industry can act as a model, and their techniques could be applied to microalgae residues.

However, certain variations in the traditional pretreatment steps for lignocellulosic material (the organosolv process, steam explosion, dilute-acid treatment, and alkaline processes among others) must be implemented because microalgae residues have experimentally shown a lack of lignin polymers. These residues could therefore be converted into bioethanol with a lower resistance in the hydrolysis step. Experimental observation has revealed that microalgae residues have little or no lignin content, making the pretreatment stage unnecessary and the hydrolysis stage easier. Therefore, pretreatments such as the organosolv process, alkaline hydrolysis, and acid-diluted hydrolysis as well as acid and enzymatic hydrolysis can be avoided, making the process more straightforward and cost-effective.

Finally, ethanol production is already known to make the process more profitable (more high-value products with the same feedstock) and greener (fewer residues). Now, with microalgae residues, it is possible to add a new characteristic: safer (because the pretreatment stage, with its many process chemicals, is no longer needed).

23.3.6 Coproducts

Biodiesel production by means of the transesterification reaction has glycerol as a coproduct. According to Rahmat et al. (2010), biodiesel coproduction of glycerol might create a glycerol surplus which might destroy the economic viability of glycerol production as well as of the biodiesel/biofuel refinery and of the petrochemical industries (because glycerol synthesis and coproduction are closely linked to these industries).

Therefore, several alternative uses for glycerol, including fuel additives (Rahmat et al., 2010), pharmaceutical-industry precursors, the food industry, and the cosmetic industry, must be considered to transform the glycerol into other precursors for other industries and in this way avoid glutting the market.

23.4 Process Integration Applied to the Algae Biorefinery

Process integration is a holistic approach to process design, retrofitting, and operation which emphasizes the unity of the process (El-Halwagi, 1997). An algae biorefinery can be treated similarly to a petroleum refinery. In a petroleum refinery, a feedstock which is treated using several techniques under different conditions and is thereby transformed into high-value products in different process stages. When working with an algae biorefinery, the same process integration concepts can be applied to reduce the consumption of fresh water and other resources, the discharge of wastes, and the requirements for industrial services. However, some differences can be determined based on operating conditions, because these will be less aggressive in the algae biorefinery to avoid material decomposition. Nevertheless, a large number of high-value products can be obtained under the gentle conditions of the algae biorefinery, and the whole process will probably achieve a higher profit because of its low requirements for industrial services.

23.4.1 Design Objectives for the Sustainable Microalgae Biorefinery

According to Adams (2006), the concept of sustainability suggests the possibility of achieving economic growth and industrialization without environmental damage.

Figure 23.1 shows a triangular representation in which economic, social, and environmental aspects interact to create four zones:

1. Bearable
2. Equitable
3. Viable
4. Sustainable

If the concept of sustainability is applied to the microalgae biorefinery, then this biorefinery must be economically profitable, socially acceptable, and environmentally friendly.

In accordance with these three pillars, the following are the design objectives that must be achieved in a microalgae biorefinery:

1. Use as much as possible of the microalgae biomass in the process.
2. Perform mass integration analysis to reduce consumption of fresh raw materials and reduce waste effluents.
3. Perform heat integration analysis to reduce the industrial services of cold water and steam.
4. Perform an economic analysis after mass and heat integration to check the profitability of the process.

If these design objectives are achieved, the following benefits will be obtained:

Economic: Lower cost of external materials because of recycling strategies and lower waste treatment costs.

Social: Less dependency on oil prices and a more equitable distribution of the fuel industry.

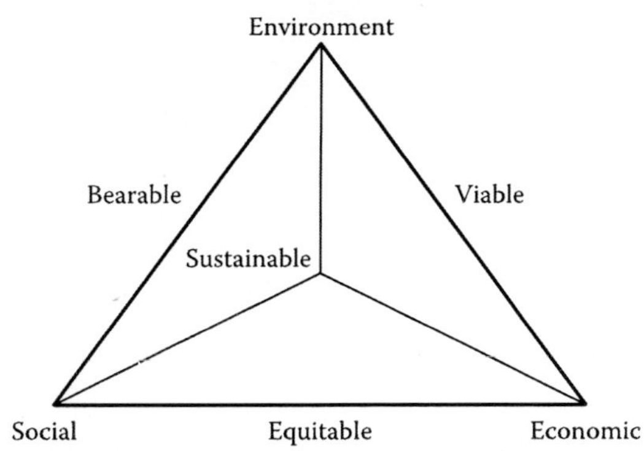

FIGURE 23.1
Graphical representation of economic, social, and environmental aspects. (Adapted from Adams, W. 2006. *The Future of Sustainability: Rethinking Environment and Development in the Twenty-First Century*. Report of the IUCN Renowned Thinkers Meeting, Zurich, January 29–31.)

Environmental: Lower levels of greenhouse gases because more carbon dioxide is sequestered by the microalgae during their growth. Less waste is disposed because of material recycling. Less consumption of industrial services and therefore less consumption of fossil fuels (heating) and cold water from rivers.

23.4.2 Integration Alternatives in Microalgae Biorefineries

Figure 23.2 illustrates a nonintegrated strategy to obtain biodiesel by the transesterification of microalgae oil with methanol.

As can be seen in Figure 23.2, the only element of process integration is the recycle of water after harvesting the microalgae. The decision to do this is very straightforward because the water contains high levels of nutrients that are then available for the growth of new microalgae.

To analyze whether various alternatives are profitable, it is necessary to simulate the process as realistically as possible and then to apply economic analysis.

23.4.3 Case Study: Process Integration in Biofuels Production from Microalgae

To perform an accurate calculation of the mass and heat integration possibilities, an approximately 200,000 L/day biodiesel–bioethanol process was designed. Several scenarios and biorefinery topologies were analyzed for integral microalgae use using Aspen Plus

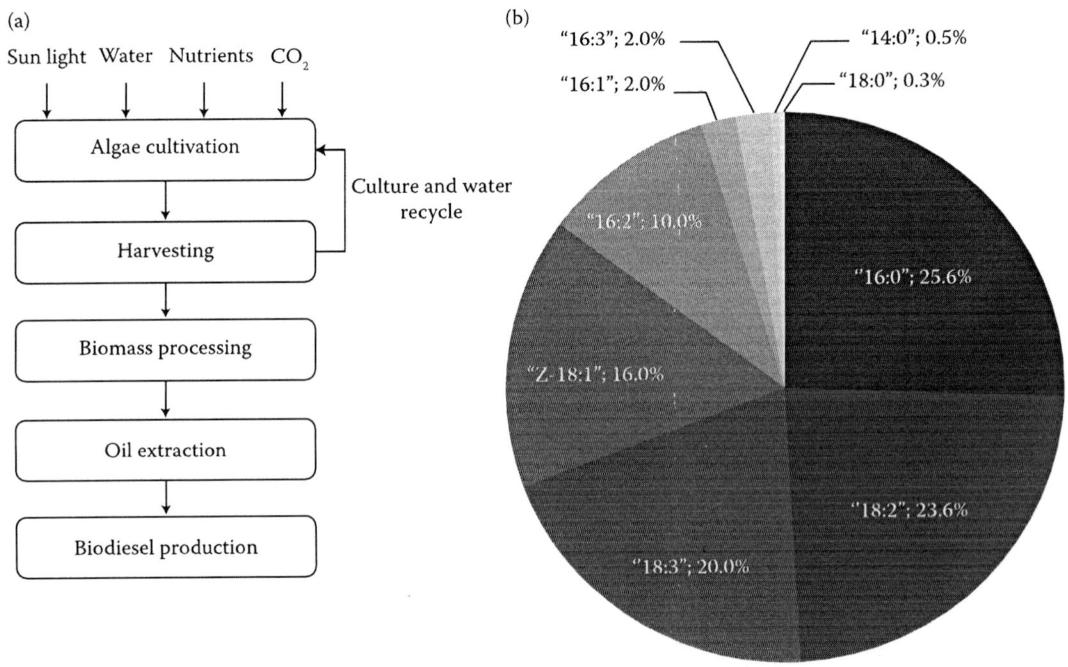

FIGURE 23.2
(a) Nonintegrated strategy to obtain biodiesel by the transesterification of microalgae oil with methanol. (Adapted from Chisti, Y. 2007b. *Trends in Biotechnology* 26: 3.) (b) Percentage fatty-acid composition of *Chlorella vulgaris* under outdoor conditions. (Adapted from Petkov, G. and García, G. 2007. *Biochemical Systematics and Ecology* 35: 281–285.)

2006.5™ software. The alcohol and catalyst used for the transesterification stage were selected based on a physical and chemical properties analysis.

Methanol, ethanol, iso-propanol, butanol, and iso-amyl alcohol (Demirbas, 2007) can be used in the transesterification process in the presence of a catalyst to break down the molecules of the oil into alkyl esters and glycerol as a by-product. Salis et al. (2005) have shown that C_1–C_5 alcohols are the most useful in chemically catalyzed biodiesel production. Methanol and ethanol are used most often. Methanol is more frequently used because of its low cost and the fact that it offers a faster conversion of triglycerides into biodiesel (Demirbas, 2007). According to Meher et al. (2006), when basic catalysts are used in methanolysis, emulsions form that quickly and easily break down to form a lower glycerol-rich layer and an upper methylester-rich layer. The ethyl esters are more difficult to separate than methyl esters because of the formation of this stable emulsion during ethanolysis. Therefore, in this case, the ethanol obtained from microalgae residues will not be used inside the process, but will be sold as a raw material for other industries.

According to Meher et al. (2006), the catalysts used for the transesterification of triglycerides can be classified as alkalis, acids, or enzymes. Alkali-catalyzed transesterification is currently the most widely used technology for biodiesel production. It needs moderate temperatures and pressures and a low or zero water content to avoid soap formation. Several compounds such as sodium hydroxide (NaOH) or potassium hydroxide (KOH) are effectively used as catalysts (Demirbas, 2007). Although chemical transesterification using an alkaline catalysis process gives high conversion levels of triglycerides to their corresponding alkyl esters in short reaction times, the reaction has several drawbacks: it is energy-intensive, the acidic or alkaline catalyst has to be removed from the product, and free fatty acids (FFA) and water decrease reaction yield. However, according to Rashid and Anwar (2008), higher yield of esters can be reached using KOH as the alkali catalyst. Several types of catalysts were analyzed, and KOH was chosen for both processes based on reaction yields and kinetic data reported by several authors (Dorado et al., 2004; Colucci et al., 2005; Xie et al., 2007).

The oil content in microalgae can exceed 80% by weight of dry biomass, and oil levels of 20–50% are fairly common (Metting, 1996; Spolaore et al., 2006). Depending on species, microalgae produce many different kinds of lipids, hydrocarbons, and other complex oils (Banerjee et al., 2002; Metzger and Largeau, 2005; Guschina and Harwood, 2006). Microalgae like *Botryococcus braunii* have high lipid levels of 55% of their organic mass for N-deficient cells (Sheehan et al., 1998). Most of these lipids are in the form of hydrocarbons, including C_{29}–C_{34} aliphatic hydrocarbons and a variety of branched and unsaturated isoprenoids. Glycerolipids are less abundant than the hydrocarbons and are composed primarily of 16:0 and various C_{18} fatty acids. These data, coupled with the fact that this species grows very slowly (one doubling per 72 h), suggested that *Botryococcus* would not function well as a feedstock for lipid-based fuel production. Instead, the microalga *C. vulgaris* was selected for this study on the basis of experimental data reported in the literature (Shi et al., 2002; Petkov and García, 2007).

Figure 23.2 shows Petkov and García's (2007) representation of *Chlorella vulgaris* composition as a circular diagram. Note that 16:0, 18:2, 18:3, and 16:0 acids represents more than 85% of the fatty acids in the microalgae oil.

The percentage of oil contained in microalgae was assumed to be 30% in the report by Chisti (2007a). According to Pienkos (2008) of the National Renewable Energy Laboratory (NREL), the dry algae mass yield for an open pond is 20 g/m^2·day. A standard depth of 30 cm was used in the raceway ponds.

Previous work on biodiesel production used up to five triglyceride components to simulate the molecule (García et al., 2010). Further improvements simulation modeling of the molecule using up to nine components has been reported by Sánchez et al. (2011). Each of the nine components of triglycerides and fatty-acid ethyl ester molecules was created using the ISIS/DRAW™ software and then exported to Aspen Plus 2006.5™. Binary interaction coefficients were unavailable in the Aspen Plus 2006.5™ library. Therefore, the UNIFAC (Universal Functional Group Activity Coefficient) method for liquid–liquid equilibrium was used to calculate them. The NRTL (nonrandom two-liquid) thermodynamic model was chosen for liquid-phase equilibrium calculations because of the existence of polar compounds such as methanol and glycerol in the reaction mixture. The Peng–Robinson equation of state was used to estimate the physical and chemical properties of the vapor phase. Other physical and chemical characteristics were taken from Perry's *Chemical Engineers' Handbook* (1997) or the *International Critical Tables* (Washburn, 1926–1930; 2003). At the cultivation stage, the raceway pond was simulated using user-developed models. To model the separations of water and nutrients from microalgae and of water from the ethanol–water azeotrophic mixture, a filter and a molecular sieve were simulated using a component splitter. Oil extraction from microalgae was simulated in a three-phase separator in which the heavy-liquid phase consists of residues + solvent, the light-liquid phase of oil + solvent, and the vapor phase of water and other minor components. Process operating parameters such as temperature and pressure conditions, as well as equipment dimensions, are the same as for biodiesel from vegetable oils. The separation stages for biodiesel from glycerol are highly similar or identical to those for biodiesel from traditional vegetable oils.

A process of enzymatic hydrolysis of microalgae bagasse obtained in the oil-extraction stage was simulated, taking into account the chemical composition (w/w) of 30% cellulose and 35% hemicellulose reported for microalgae by Leemhuis and de Jong (1997). The NRTL model was chosen to calculate the activity coefficients for the liquid phase and the SRK equations of state for the vapor phase. No pretreatment stage was needed because of the absence of lignin. Xylose and glucose were used in this simulation to represent both pentoses and hexoses. Properties such as critical temperature, volume and pressure, acentric factor, heat of formation at 298.15 K, vapor pressure, and heat of vaporization among others for cellulose, xylose, and glucose components were taken from Wooley and Putsche (1996).

Simulations in Aspen 2006.5™ were constructed for biodiesel production (see Figure 23.3) and bioethanol production (see Figure 23.4) from microalgae biomass.

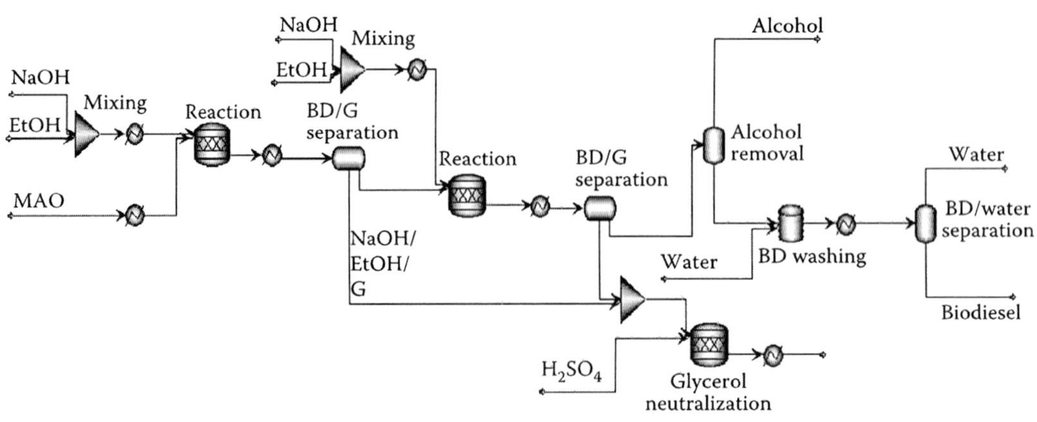

FIGURE 23.3
Biodiesel simulation in Aspen Plus 2006.5™ software.

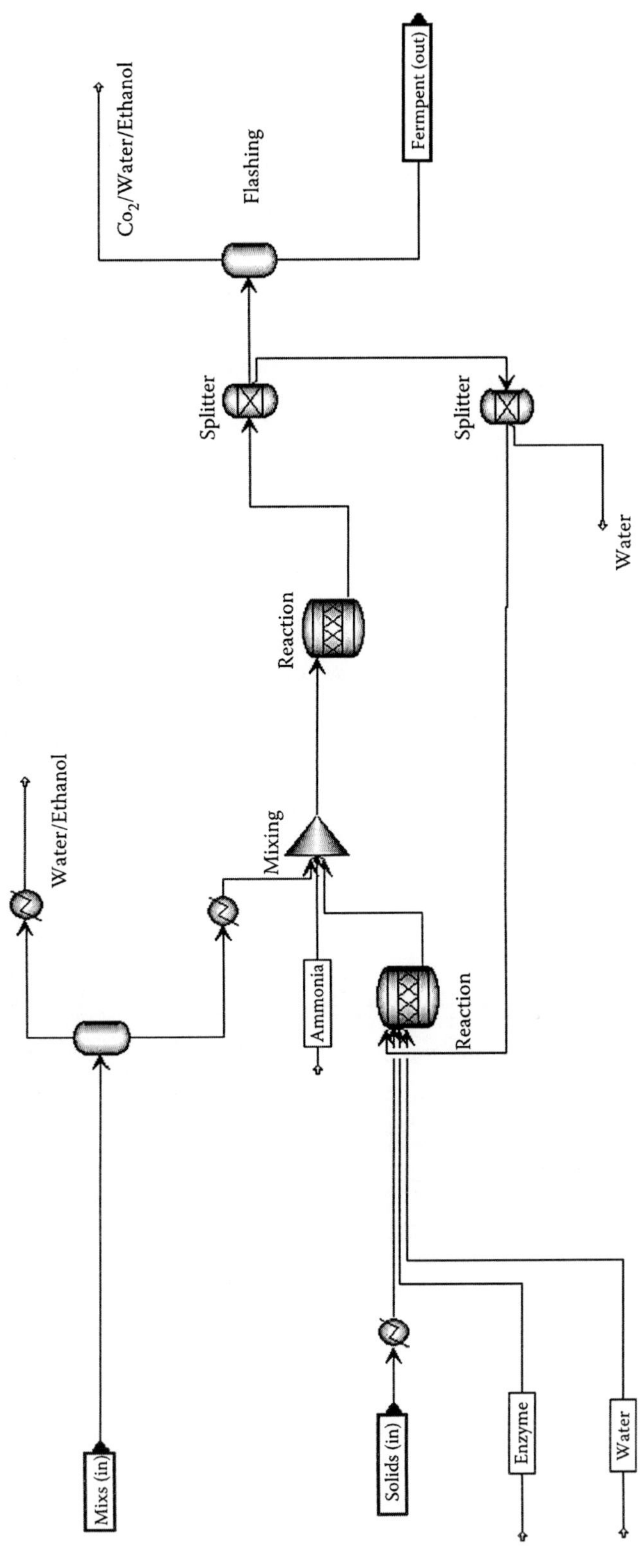

FIGURE 23.4
Bioethanol simulation in Aspen Plus 2006.5™ software.

TABLE 23.2

Source Data for Biodiesel and Bioethanol Production from Microalgae

Source	Flowrate (kg/day)	Mass Fraction of Impurities	Load (kg/day)	Cumulative Load (kg/day)
Water after washing	20,604.98	0.1	2060	2060
Water after drying	19,986.83	0.1	1998	4059
Water from molecular sieves	641.59	0.1	64	4123

TABLE 23.3

Source Data for Biodiesel and Bioethanol Production from Microalgae

Sink	Flowrate (kg/day)	Maximum Inlet Mass Fraction of Impurities	Inlet Load (kg/day)	Cumulative Load (kg/day)
Culture	2077.27	0.05	103	103
Pretreatment	848,330.52	0.05	42,416	42,520
SSCF	32,721.32	0.05	1636	44,156

The data summarized in Tables 23.2 and 23.3 were obtained from the previously discussed simulations using the Aspen Plus 2006.5™ software.

Table 23.2 shows that more than 98% of the water and the impurities load are shared between two streams: water after washing, and water after drying. To improve process performance, the mass fraction of impurities in these two streams must be reduced by interception, or the flowrate must increase while keeping the mass fraction of impurities constant, or both. Of these two options, the first can be achieved by installing a filter or another device to retain and separate the impurities. This option will not change the process design in terms of flows of microalgae obtained, but will represent a change in terms

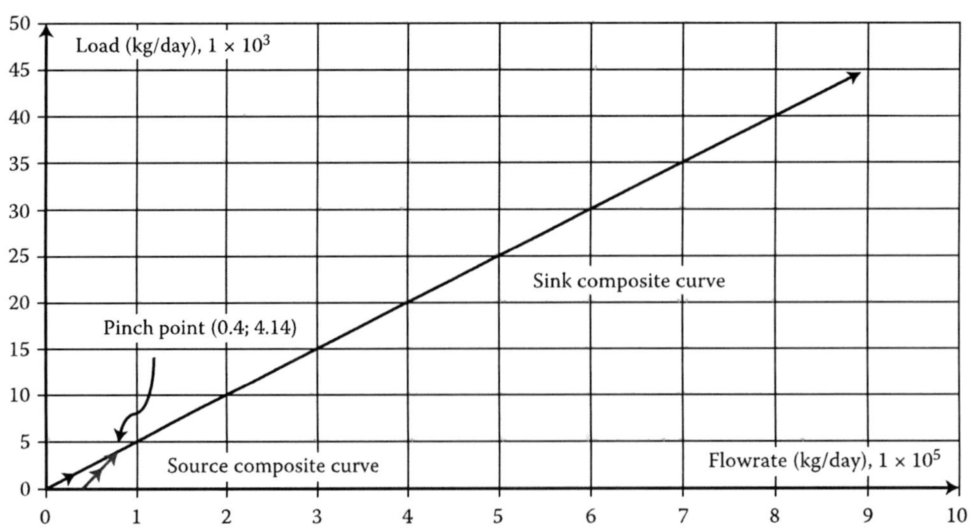

FIGURE 23.5
Material recycle pinch diagram.

of external water usage. The other option can be applied only if this will not cause problems with process equipment capacity.

In Table 23.3, it can be seen that more than 96% of the impurities load can be accommodated in the pretreatment stage. The other 4% can be treated in the simultaneous saccharification and cofermentation (SSCF) and culture stages. To improve process performance, it would be necessary to increase either the maximum inlet mass fraction of impurities that the pretreatment stage can accept, or the flowrate in the stage, or both. For instance, a pretreatment option could be selected that is capable of treating biomass with a higher load of impurities while maintaining bioethanol yield. Again, the other option can be applied only if this will not cause problems with process equipment capacity.

Using the data from Tables 23.2 and 23.3, a material recycle pinch diagram was constructed to find the pinch point (Figure 23.5).

23.5 Conclusions

Microalgae offer an excellent opportunity to develop biorefineries because their biomass can be completely used to produce biofuels and other products by means of process integration analysis. Microalgae oil and residues are used together to obtain biodiesel by transesterification and bioethanol by means of SSCF. Pinch analysis reduced the external water consumption and provided a general overview of the entire process. It pointed out those stages that have the most potential to improve overall performance. The pinch analysis methodology enabled a reduction of 41,243 kg/day in freshwater requirements by use of direct recycle, which represents a 4.67% reduction. More improvements would be possible by analyzing the sink/source data. It was shown that more than 98% of the water and the impurities load came from two streams: water after washing and water after drying. To improve overall process performance, the mass fraction of impurities in these two sources must be reduced, or the flowrate must increase keeping the mass fraction of impurities constant, or both. The decision on which of these two options is selected will depend on a technical and economic analysis.

References

Adams, W. 2006. *The Future of Sustainability: Rethinking Environment and Development in the Twenty-First Century*. Report of the IUCN Renowned Thinkers Meeting, Zurich, January 29–31.

Ahmad, A., Yasin, N., Derek, C., Lim, J. 2011. Microalgae as a sustainable energy source for biodiesel production: A review. *Renewable and Sustainable Energy Reviews* 15: 584–593.

Amaro, H., Guedes, A., Malcata, A. 2011. Advances and perspectives in using microalgae to produce biodiesel. *Applied Energy* 88: 3402–3410.

Audsley, E. and Annetts, J.E. 2003. Modeling the value of a rural biorefinery—Part I: The model description. *Agricultural Systems* 76(1): 39–59.

Baba, M., Kikuta, F., Suzuki, I., Watanabe, M., Shiraiwa, Y. 2012. Wavelength specificity of growth, photosynthesis, and hydrocarbon production in the oil-producing green alga *Botryococcus raunii*. *Bioresource Technology* 169: 266–270.

Banerjee, A., Sharma, R., Chisti, Y., Banerjee, U.C. 2002. Botryococcus braunii: A renewable source of hydrocarbons and other chemicals. *Critical Reviews in Biotechnology* 22: 245–279.

Berberoglu, H., Gomez, P., Pilon, L. 2009. Radiation characteristics of *Botryococcus braunii*, *Chlorococcum littorale*, and *Chlorella* sp. used for CO_2 fixation and biofuel production. *Journal of Quantitative Spectroscopy & Radiative Transfer* 110(17): 1879–1893.

Blanco, A., Moreno, J., Del Campo, J., Rivas, J., Guerrero, M. 2007. Outdoor cultivation of lutein-rich cells of *Muriellopsis* sp. in open ponds. *Applied Microbiology and Biotechnology* 73(6): 1259–1266.

Borowitzka, M. 1999. Commercial production of microalgae: Ponds, tanks, and fermenters. *Progress in Industrial Microbiology* 35: 313–321.

Brennan, L. and Owende, P. 2010. Biofuels from microalgae—a review of technologies for production, processing, and extractions of biofuels and co-products. *Renewable and Sustainable Energy Reviews* 14: 557–577.

Chen, S. 2008. Industrial biosystems engineering and biorefinery systems. *Chinese Journal of Biotechnology* 24(6): 940–945.

Chen, C., Yeh, K., Aisyah, R., Lee, D., Chang, J. 2011. Cultivation, photobioreactor design and harvesting of microalgae for biodiesel production. *A Critical Review Bioresource Technology*, 102: 71–81.

Cherubini, F. 2010. The biorefinery concept: Using biomass instead of oil for producing energy and chemicals. *Energy Conversion and Management* 51(7): 1412–1421.

Chini Zittelli, G., Lavista, F.M., Bastianini, A., Rodolfi, L., Vincenzini, M., Tredici, M.R. 1999. Production of eicosapentaenoic acid by *Nannochloropsis* sp. cultures in outdoor tubular photobioreactors. *Journal of Biotechnology* 70(1–3, 30): 299–312.

Chisti, Y. 2007a. Biodiesel from microalgae. *Biotechnology Advances* 25: 294–306.

Chisti, Y. 2007b. Biodiesel from microalgae beats bioethanol. *Trends in Biotechnology* 26: 3.

Choi, G., Kim, B., Ahn, C., Oh, H. 2010. Effect of nitrogen limitation on oleic acid biosynthesis in *Botryococcus braunii*. *Journal of Applied Phycology* 23(6): 1031–1037.

Colucci, J.A., Borrero, E.E., Alape, F. 2005. Biodiesel from an alkaline transesterification reaction of soybean oil using ultrasonic mixing. *Journal of the American Oil Chemists' Society* 82: 525–530.

Converti, A., Casazza, A., Ortiz, E., Perego, P., Borghi, M. 2009. Effect of temperature and nitrogen concentration on the growth and lipid content of Nannochloropsis oculata and *Chlorella vulgaris* for biodiesel production. *Chemical Engineering and Processing* 48: 1146–1151.

Dayananda, C., Sarada, R., Bhattacharya, S., Ravishankar, G. 2005. Effect of media and culture conditions on growth and hydrocarbon production by Botryococcus braunii. *Process Biochemistry* 40: 3125–3131.

Dayananda, C., Sarada, R., Ranib, U., Shamala, T., Ravishankar, G. 2007. Autotrophic cultivation of Botryococcus braunii for the production of hydrocarbons and exopolysaccharides in various media. *Biomass and Bioenergy* 31: 87–93.

Dayananda, C., Sarada, R., Srinivas, B., Shamala, T., Ravishankar, G. 2006. Presence of methyl branched fatty acids and saturated hydrocarbons in botryococcene producing strain of *Botryococcus braunii*. *Acta Physiologiae Plantarum* 28(3): 251–256.

Del Campo, J., García-González, M., Guerrero, M. 2007. Outdoor cultivation of microalgae for carotenoid production: Current state and perspectives. *Applied Microbiology and Biotechnology* 74(6): 1163–1174.

Demirbas, A. 2007. Progress and recent trends in biofuels. *Progress in Energy and Combustion Science* 33: 1–18.

Demirbas, A. 2011. Biodiesel from oilgae, biofixation of carbon dioxide by microalgae: A solution to pollution problems. *Applied Energy* 88: 3541–3547.

Domínguez-Bocanegra, A., Legarreta, I., Jeronimo, F., Campocosio, A. 2004. Influence of environmental and nutritional factors in the production of astaxanthin from *Haematococcus pluvialis*. *Bioresource Technology* 92: 209–214.

Dong, Q. and Zhao. X. 2004. In situ carbon dioxide fixation in the process of natural astaxanthin production by a mixed culture of Haematococcus pluvialis and *Phaffia rhodozyma*. *Catalysis Today* 98: 537–544.

Dorado, M.P., Ballesteros, E., Mittelbach, M., Lopez, F.J. 2004. Kinetic parameters affecting alkali-catalyzed transesterification process of used olive oil. *Energy Fuels* 18: 1457–1462.

Doušková, I., Kaštánek, F., Maléterová, Y., Kaštánek, P., Doucha, J., Zachleder, V. 2010. Utilization of distillery stillage for energy generation and concurrent production of valuable microalgal biomass in the sequence: Biogas-cogeneration-microalgae-products. *Energy Conversion and Management* 51(3): 606–611.

El-Halwagi, M.M. 1997. *Pollution Prevention through Process Integration*. Academic Press, San Diego.

Eroglu, E. and Melis, A. 2010. Extracellular terpenoid hydrocarbon extraction and quantitation from the green microalgae Botryococcus braunii var. *Showa*. *Bioresource Technology* 101: 2359–2366.

Eroglu, E., Okada, S., Melis, A. 2011. Hydrocarbon productivities in different Botryococcus strains: Comparative methods in product quantification. *Journal of Applied Phycology* 23(4): 763–775.

Falkowski, P. and Raven, J. 1997. *Aquatic Photosynthesis*. Malden, MA: Blackwell Science.

Freedman, B., Pryde, E., Mounts, T. 1984. Variables affecting the yields of fatty esters from transesterified vegetable oils. *JAOCS* 61(10): 1638–1643.

García, M., Gonzalo, A., Sánchez, J., Arauzo, J., Peña, J. 2010. Prediction of normalized biodiesel properties by simulation of multiple feedstock blends. *Bioresource Technology* 101: 4431–4439.

Gardner, R., Peters, P., Peyton, B., Cooksey, K. 2011. Medium pH and nitrate concentration effects on accumulation of triacylglycerol in two members of the chlorophyta. *Journal of Applied Phycology* 23(6): 1005–1016.

González, L., Cañizares, R., Baena, S. 1997. Efficiency of ammonia and phosphorus removal from a Colombian agroindustrial wastewater by the microalgae *Chlorella vulgaris* and *Scenedesmus dimorphus*. *Bioresource Technology* 60: 259–262.

Guschina, I.A. and Harwood, J.L. 2006. Lipids and lipid metabolism in eukaryotic algae. *Progress in Lipid Research* 45: 160–186.

Harun, R., Davidson, M., Doyle, M., Gopiraj, R., Danquah, M., Forde, G. 2011. Technoeconomic analysis of an integrated microalgae photobioreactor, biodiesel and biogas production facility. *Biomass and Bioenergy* 35: 741–747.

He, H., Li, H., Chen, F. 2005. Determination of vitamin B1 in seawater and microalgal fermentation media by high-performance liquid chromatography with fluorescence detection. *Analytical and Bioanalytical Chemistry* 383(5): 875–879.

Hernández, J., de-Bashan, L., Bashan, Y. 2006. Starvation enhances phosphorus removal from wastewater by the microalga *Chlorella* spp. co-immobilized with *Azospirillum brasilense*. *Enzyme and Microbial Technology* 38: 190–198.

Hernández, J., de-Bashan, L., Rodriguez, J., Rodriguez, Y., Bashan, Y. 2009. Growth promotion of the freshwater microalga *Chlorella vulgaris* by the nitrogen-fixing, plant growth-promoting bacterium *Bacillus pumilus* from arid zone soils. *European Journal of Soil Biology* 45: 88–93.

Huang, J., Chen, F., Sandmann, G. 2006. Stress-related differential expression of multiple β-carotene ketolase genes in the unicellular green alga Haematococcus pluvialis. *Journal of Biotechnology* 122(2): 176–185.

Huang, G., Chen, F., Wei, D., Zhang, X., Chen, G. 2010. Biodiesel production by microalgal biotechnology. *Applied Energy* 87: 38–46.

Huppe, H. and Turpin, D. 1994. Integration of carbon and nitrogen metabolism in plant and algal cells. *Annual Review of Plant Physiology and Plant Molecular Biology* 45: 577–507.

Ip, P. and Chen, F. 2005. Employment of reactive oxygen species to enhance astaxanthin formation in *Chlorella zofingiensis* in heterotrophic culture. *Process Biochemistry* 40(11): 3491–3496.

Issarapayup, K., Powtongsook, S., Pavasant, P. 2011. Economical review of *Haematococcus pluvialis* culture in flat-panel airlift photobioreactors. *Aquacultural Engineering* 44: 65–71.

Jeon, Y. Cho, C. Yun, Y. 2006. Combined effects of light intensity and acetate concentration on the growth of unicellular microalga Haematococcus pluvialis. *Enzyme and Microbial Technology* 39: 490–495.

Jiang, F. 2000. *Algae and Their Biotechnological Potential*. Kluwer Academic Publishers: Dordrecht/Boston/London.

Jiang, Y., Fan, K., Wong, R. Chen, F. 2004. Fatty acid composition and squalene content of the marine microalga Schizochytrium mangrovei. *Journal of Agricultural and Food Chemistry* 52(5): 1196–1200.

Kang, C., An, Y., Park, T., Sim, S. 2006. Astaxanthin biosynthesis from simultaneous N and P uptake by the green alga Haematococcus pluvialis in primary-treated wastewater. *Biochemical Engineering Journal* 31: 234–238.
Kim, k., Lingaraju, B., Rheaume, R., Lee, J., Siddiqui, K. 2010. Removal of ammonia from wastewater effluent by *Chlorella vulgaris. Tsinghua Science and Technology* 15: 391–396.
Kita, k., Okada, S., Sekino, A., Imou, K., Yokoyama, S., Amano, T. 2010. Thermal pre-treatment of wet microalgae harvest forefficient hydrocarbon recovery. *Applied Energy* 87: 2420–2423.
Leemhuis, R.J. and de Jong, R.M. 1997. Biomassa: Biochemische samenstelling en conversiemethoden (confidential report, in Dutch).
Levine, R., Costanza-Robinson, M., Spatafora, G. 2011. *Neochloris oleoabundans* grown on anaerobically digested dairy manure for concomitant nutrient removal and biodiesel feedstock production. *Biomass and Bioenergy* 35: 40–49.
Li, Y., Chen, Y., Chen, P., Zhou, W., Martinez, B., Zhu, J., Ruan, R. 2011. Characterization of a microalga *Chlorella* sp. well adapted to highly concentrated municipal wastewater for nutrient removal and biodiesel production. *Bioresource Technology* 8: 5138–5144.
Li, Y. and Qin, J. 2005. Comparison of growth and lipid content in three Botryococcus braunii strains. *Journal of Applied Phycology* 17: 551–556.
Lim, S., Chu, W., Phang, S. 2010. Use of *Chlorella vulgaris* for bioremediation of textile wastewater. *Bioresource Technology* 19: 7314–7322.
Linh, H., Kwon, J., Kim, Z., Oh, Y., Lee, C. 2010. Statistical optimization of culture media for growth and lipid production of *Botryococcus braunii* LB572. *Biotechnology and Bioprocess Engineering* 15: 277–284.
Liu, J., Huang, J., Fan, K., Jiang, Y., Zhong, Y., Sun, Z., Chen, F. 2010. Production potential of *Chlorella zofingienesis* as a feedstock for biodiesel. *Bioresource Technology* 101: 8658–8663.
Meher, L.C., Sagar, D.V., Naik, S.N. 2006. Technical aspects of biodiesel production by transesterification—A review. *Renewable & Sustainable Energy* 10: 248–268.
Metting, F. 1996. Biodiversity and application of microalgae. *Journal of Industrial Microbiology and Biotechnology* 17: 477–489.
Metzger, P. and Largeau, C. 2005. *Botryococcus braunii*: A rich source for hydrocarbons and related ether lipids. *Applied Microbiology and Biotechnology* 66: 486–496.
Moazami, N., Ranjbar, R., Ashori, A., Tangestani, M., Nejad, A. 2011. Biomass and lipid productivities of marine microalgae isolated from the Persian Gulf and the Qeshm Island. *Biomass and Bioenergy* 35: 1935–1939.
Mohn, F.H. 1980. Experiences and strategies in the recovery of biomass in mass culture of microalgae. In: Shelef, G., Soeder, C.J. (eds.), *Algal Biomass.* Amsterdam: Elsevier, pp. 547–571.
Molina Grima, E., Acién Fernández, F., García Camacho, F., Chisti, Y. 1999. Photobioreactors: Light regime, mass transfer, and scaleup. *Journal of Biotechnology* 70: 231–247.
Najafabady, N., Rasoul-Amini, S., Ghasemi, Y. 2010. Production of biodiesel by a naturally isolated strain of *Chlorella vulgaris* in bubble-column photobioreactor. *Journal of Biotechnology* 150 (Suppl1): 371.
Oh, H., Choi, A., Mheen, T. 2003. High-value materials from microalgae. *Korean Journal of Microbiology and Biotechnology.* 31(2): 95–102.
Perry, R.H., Green, D.W., Maloner, J.O. 1997. *Perry's Chemical Engineers' Handbook*, 7th ed., 0-07-049841-5, Australia: McGraw-Hill.
Petkov, G. and García, G. 2007. Which are fatty acids of the green alga *Chlorella*? *Biochemical Systematics and Ecology* 35: 281–285.
Pienkos, P. 2008. *Historical Overview of Algae Biofuels Technoeconomic Analyses. NREL National Renewable Energy Laboratory*, U.S. Department of Energy, Office of Energy Efficiency and Renewable Energy.
Pokoo-Aikins, G., Heath, A., Mentzer, R., Mannan, S., Rogers, W., El-Halwagi, M., 2010. A multi-criteria approach to screening alternatives for converting sewage sludge to biodiesel. *Journal of Loss Prevention in the Process Industries* 23: 412–420.
Prakash, J., Pushparaj, B., Carlozzi, P., Torzillo, G., Montaini, E., Materassi, R. 1997. Microalgae drying by a simple solar device. *International Journal of Solar Energy* 18(4): 303–311.

Rahmat, N., Zuhairi, A., Rahman, A. 2010. Recent progress on innovative and potential technologies for glycerol transformation into fuel additives: A critical review. *Renewable and Sustainable Energy Reviews* 14: 987–1000.

Ranjbar, R., Inoue, R., Shiraishi, H., Katsuda, T., Katoh, S. 2008. High efficiency production of astaxanthin by autotrophic cultivation of Haematococcus pluvialis in a bubble column photobioreactor. *Biochemical Engineering Journal* 39: 575–580.

Rashid, U. and Anwar, F. 2008. Production of biodiesel through optimized alkaline catalyzed transesterification of rapeseed oil. *Fuel* 87: 265–273.

Razon, L. and Tan, R. 2011. Net energy analysis of the production of biodiesel and biogas from the microalgae: Haematococcus pluvialis and Nannochloropsis. *Applied Energy* 88: 3507–3514.

Rivas, M., Vargas, P., Riquelme, C. 2010. Interactions of Botryococcus braunii Cultures with bacterial biofilms. *Microbial Ecology* 60: 628–635.

Salis, A., Pinna, M., Monduzzi, M., Solinas, V. 2005. Biodiesel production from triolein and short chain alcohols through biocatalysis. *Journal of Biotechnology* 119: 291–299.

Samorì, C., Torri, C., Samorì, G., Fabbri, D., Galletti, P., Guerrini, F., Pistocchi, R., Tagliavini, E. 2010. Extraction of hydrocarbons from microalga *Botryococcus braunii* with switchable solvents. *Bioresource Technology* 101: 3274–3279.

Sánchez, A., Contreras, A., García, F., Molina, E., Chisti, Y. 1999. Comparative evaluation of compact photobioreactors for large-scale monoculture of microalgae. *Journal of Biotechnology* 70: 249–270.

Sánchez, E., Ojeda, K., El-Halwagi, M., Kafarov, V. 2011. Biodiesel from microalgae oil production in two sequential esterification/transesterification reactors: Pinch analysis of heat integration. *Chemical Engineering Journal* 176–177: 211–216.

Sheehan, J., Dunahay, T., Benemann, J., Roessler, P. 1998. A look back at the U.S. Department of energy's aquatic species program—Biodiesel from Algae. Prepared by the National Renewable Energy Laboratory, 1617 Cole Boulevard. Goden, Colorado 80401-3393.

Shi, X., Jiang, Y., Chen, F. 2002. High-yield production of lutein by the green microalga *Chlorella protothecoides* in heterotrophic fed-batch culture. *Biotechnology Progress* 18(4): 723–727.

Spolaore, P., Joannis-Cassan, C., Duran, E., Isambert, A. 2006. Commercial applications of microalgae. *Journal of Bioscience and Bioengineering* 101: 87.

Suh, I. Joo, H. Lee, H. 2006. A novel double-layered photobioreactor for simultaneous Haematococcus pluvialis cell growth and astaxanthin accumulation. *Journal of Biotechnology* 125: 540–546.

Tabatabaei, M., Tohidfar, M., Jouzani, G., Safarnejad, M., Pazouki, M. 2011. Biodiesel production from genetically engineered microalgae: Future of bioenergy in Iran. *Renewable and Sustainable Energy Reviews* 15: 1918–1927.

Tang, H. Abunasser, N. Garcia, M. Chen, M. Simon, K. Salley, S. 2011. Potential of microalgae oil from Dunaliella tertiolecta as a feedstock for biodiesel. *Applied Energy* 88: 3324–3330.

Tanoi, T., Kawachi, M., Watanabe, M. 2011. Effects of carbon source on growth and morphology of Botryococcus braunii. *Journal of Applied Phycology* 23(1): 25–33.

Teresa, M., Martins, A., Caetano, N. 2010. Microalgae for biodiesel production and other applications: A review. *Renewable and Sustainable Energy Reviews* 14: 217–232.

Terry, K. and Raymond, L. 1985. System design for the autotrophic production of microalgae. *Enzyme and Microbial Technology* 7: 474–487.

Travieso, L., Benítez, F., Sánchez, E., Borja, R., Martín, A., Colmenarejo, M.F. 2006. Batch mixed culture of *Chlorella vulgaris* using settled and diluted piggery waste. *Ecological Engineering* 28: 158–165.

Valderrama, L., Campo, C., Rodriguez, C., Luz, E., de-Bashana, B., Bashan, Y. 2002. Treatment of recalcitrant wastewater from ethanol and citric acid production using the microalga *Chlorella vulgaris* and the macrophyte Lemna minuscula. *Water Research* 36: 4185–4192.

Washburn, E.W. 1926–1930; 2003. *International Critical Tables of Numerical Data*, Physics, Chemistry and Technology (1st Electronic Edition). Knovel.

Weiss, T., Johnston, W., Fujisawa, K., Okada, S., Devarenne, T. 2011. Genome size and phylogenetic analysis of the A and L races of Botryococcus braunii. *Journal of Applied Phycology* 23(5): 833–839.

Wen, Z. and Chen, F. 2003. Heterotrophic production of eicosapentaenoic acid by microalgae. *Biotechnology Advances* 21(4): 273–294.

Wijffels, R.H. 2007. Potential of sponges and microalgae for marine biotechnology. *Trends in Biotechnology* 26(1): 1–6.

Wooley, R.J. and Putsche, V. 1996. *Development of an ASPEN PLUS Physical Property Database for Biofuels Components*. Technical Report NREL/MP-425-20685, National Renewable Energy Laboratory, Colorado, USA.

Xie, W., Huang, X., Li, H. 2007. Soybean oil methyl esters preparation using NaX zeolites loaded with KOH as a heterogeneous catalyst. *Bioresource Technology* 98: 936–939.

Yang, J., Xu, M., Zhang, X., Hu, Q., Sommerfeld, M., Chen, Y. 2011. Life-cycle analysis on biodiesel production from microalgae: Water footprint and nutrients balance. *Bioresource Technology* 102: 159–165.

Zhang, F., Cheng, L., Gao, W., Xu, X., Zhang, L., Chen, H. 2011. Mechanism of lipid extraction from Botryococcus braunii FACHB 357 in a biphasic bioreactor. *Journal of Biotechnology* 154(4): 281–284.

Zhila, N., Kalacheva, G., Volova, T. 2011. Effect of salinity on the biochemical composition of the alga Botryococcus braunii Kütz IPPAS H-252. *Journal of Applied Phycology* 23: 47–52.

Section VI

Biorefinery Policy and Environmental Analysis

24

Policy Options to Support Biorefining

Warren E. Mabee

CONTENTS

24.1 Introduction

This chapter will explore the use of policy to support development of the biorefinery. Biorefining introduces (or reintroduces) the concept of generating energy and fuel products from forest feedstocks, in addition to solid wood and other value-added materials including biomaterials and biochemicals; this is somewhat analogous to a petroleum refinery (Chum and Overend 2001). Development of biorefineries could add significant value to natural resource sectors, including forestry and agriculture, by providing access to new markets, particularly in the energy sector, where demand is often described as "bottomless." Biorefineries could have a positive environmental impact, reducing net carbon dioxide emissions by substituting "green" carbon for fossil carbon. Economic returns from biorefining will include rural employment, increased economic diversity, and increased tax returns for local governments. Ultimately, the development of biorefineries worldwide will reduce dependence on uncertain fossil energy reserves and increase energy security.

The first likely step for biorefining is the development of energy products. Off-the-shelf, commercial options exist for both thermal and electricity production, although these tend to be relatively small compared to fossil generation facilities. As an example, the largest biomass power facility in Canada is the plant in Williams Lake BC, a 60-MW electricity-generating plant that has been operational since 1993. This facility consumes approximately 600,000 tonnes of wood residues annually, including bark, chips, and sawdust. The wood-residue fuel is provided by five surrounding sawmill operations; the electricity generated at the facility is sold under a 25-year electricity purchase agreement to the provincial utility. This facility uses standard boilers and a high-pressure steam turbine for electricity production. Although this facility is an electricity-only plant, it is possible with similar facilities to capture process steam to supply other industrial processes or to support district heating grids for residential heating; this is known as cogeneration or combined heat and power production.

Liquid biofuels suitable for use in the transport sector are likely to be of higher value than electricity or heat and therefore may be more attractive to industry. Existing, commercially viable technologies for biofuel production are often referred to as "first-generation" plants. Most biofuels available today are produced from agricultural crops traditionally grown for human or livestock consumption. Today, biofuels provide just over 1% of global demand for transport fuels (34 million tonnes of oil equivalent in 2007 on an energy basis), while crops grown for biofuel feedstock occupy approximately 2% of the world's arable land (Sims et al. 2010). Emerging biorefining platforms will use nonfood feedstocks, including wood from forests, agricultural residues, and municipal solid wastes. Significant advances must be achieved, however, before forest-based biorefining operations are considered economically viable. Projections for when second-generation fuels will become commercial are wide-ranging, but this is often considered unlikely to occur before 2015 (IEA 2008). Some scenarios suggest that the total share of biofuels by 2030 may be as high as 5–10% of all liquid-fuel use (IEA 2006, IPCC 2007). Such scenarios depend heavily on deployment of second-generation technologies because first-generation biofuels based on agricultural systems may see steady or declining production due to land and water constraints and issues of system sustainability (Sims et al. 2010). Considerable investment in pilot and demonstration plants using second-generation technologies has been made worldwide, but how and when commercial scale-up can be realized is the key question.

Two main biorefining platforms are being explored by researchers and industry. These use biochemical and thermochemical pathways respectively to produce a variety of fuel and chemical products. Both the biochemical enzyme hydrolysis process and the thermochemical biomass-to-liquid (BTL) process have reached the demonstration stage, and several plants in North America and Europe are operating, planned, or under construction following recent substantial government grants to help reduce the commercial risks. As these international demonstration plants come online over the next 2 or 3 years, significant data on the performance of different conversion routes will become available, enabling governments to inform strategic policy decisions for further second-generation technology development.

It is clear that successful deployment of biorefining has the potential to support resource-dependent sectors, including forestry and agriculture. Over the longer term, these technologies will likely be applied to processing the biological components of municipal solid waste as well. Countries that take the initiative in developing biorefinery technologies and approaches may also be able to foster lucrative business opportunities in technology transfer. Between 2002 and 2008, China's exports of renewable energy technologies rose by 40%, while Germany's exports grew by 15%; by comparison, Canada's exports of such technologies have dropped by 2% (Goldfarb 2010). Finally, with rapid rises in global energy use and expectations of continued growth well toward the mid-twenty-first century, the markets for biorefining products can be continued to increase. International Energy Agency (IEA) scenarios predict an increase in liquid-fuel production to meet global transport demand to 2050.

Developing biorefinery capacity, using forest and agricultural feedstocks, could help meet a growing global demand for energy and fuel products.

Developing a biorefinery with an initial energy focus can be done with the aid of certain policy tools which can be used to increase fiber supply, create biorefining capacity, and develop a market for biorefining products. Each of these areas will be considered in turn.

24.2 Increasing Fiber Supply for Biorefining

24.2.1 Forest Feedstocks

In considering the development of the biorefinery, it is useful to consider the pathways leading to bioproducts and the likely places where new technologies might be introduced. In the forest sector, for example, a differentiation may be made between primary products (high-value products which remain the cornerstone of the forest industry, including solid wood and engineered wood products), intermediate coproducts (relatively low-cost products that might be used for energy production or secondary forest product manufacture, including pellets, syngas, and pulp) and biorefining coproducts (representing a wide range of fuels, chemicals, and energy, in addition to primary products). In temperate and northern climates, where forestry is practiced in an extensive rather than an intensive fashion, it is likely that multiple upstream processing stations, producing a mix of primary and intermediate products, would feed a single downstream biorefinery. In such scenarios, upstream processing lends itself to small- to medium-scale operations, while downstream processing is likely to trend toward large-scale facilities. This is very similar to the sawmill–pulpmill differentiation seen in the forest sector today. The importance of linking traditional and biorefining products has been emphasized in recent papers (e.g., Soderholm and Lundmark 2009).

From a logistical perspective, forest biomass is an ideal feedstock for biorefining because it can be recovered year-round from forests, thus reducing the need for storage. In many countries, forest operations are under severe financial pressure. A long-term structural shift in the industry has seen great reductions in demand for newsprint; the global economic slowdown that began in 2008 has exacerbated this problem by impacting both solid wood and paper markets, particularly in the United States. As an example, in the 5 years since 2005, 8 of the 16 pulp and paper mills that then operated in the Canadian province of Ontario have been taken out of service, and only five of the remaining facilities continue to operate on regular schedules (various sources including CIBC 2005, CPAWS 2007). As the integrated forest products industry requires economic uptake of a range of products, introducing alternative uses through the biorefining model is one means of revitalizing the sector. However, identifying wood supplies suitable for biorefining presents a challenge. Available statistics describe only two categories of forest residues: chips and particles that are used in the manufacture of value-added products, including particle board, oriented strandboard, and paper products; and unused wood residues generated in forest-products processing (FAOStat 2010). Additional feedstocks not covered by existing inventories include harvest residues and small-diameter wood that is currently left in the forest. An analysis of global data comparing growth and yield figures with removals indicates that the amount of forest-harvest residues varies greatly; expressed as a percentage (w:w) of total roundwood removals, these residues range from a high of almost 100% to a low of only 0.08% (Bull et al. 1998, FAOStat 2010). The average forest-harvest-residue generation rate for temperate northern countries is estimated to be approximately 12% (using data from Bull et al. 1998), which translates into approximately 35 million tonnes of wood across North America in 2009 alone (FAOStat 2010). This estimate does not include small-diameter or nonmerchantable wood species, which are not included in most inventories and which may exist in very large amounts depending on the forest type.

24.2.1.1 Policy Implications

It is clear that significant amounts of forest feedstocks exist that might be suitable for biorefining activities. To make these feedstocks available for an emerging biorefining sector, however, sustainability concerns must be addressed. Current indicators for sustainable forest management have been geared toward traditional forestry, in which larger-dimension wood is removed for primary and secondary product production. Some have expressed concerned that development of biorefining, and particularly bioenergy options, might lead to nonsustainable removal of harvest residues and small-diameter material which play an important part in ecosystem functioning as currently understood. Some go so far as to predict significant deforestation in environmentally sensitive areas if climate policy creates a strong enough incentive for bioenergy products (Persson and Azar 2010).

Recently, Thiffault et al. (2010) addressed this issue, noting that jurisdictions such as Sweden have developed information on the impacts of forest biomass removal which has supported development of specific guidelines for the extraction of forest fuels. On private forest lands, the enforcement of sustainable operations might fall to third-party certification systems such as the Forest Stewardship Council. In certified forests, as long as operations are carried out in a fashion acceptable to the certifying agency, the choice of biomass for harvesting and its use downstream is up to the landowner. It may be necessary, however, to revisit FSC indicators of sustainability to ensure that they do not prevent the harvest of small-diameter wood for energy purposes.

In contrast to private forest operations, the regulation of forestry operations on public land is generally conducted through forest tenure arrangements, wherein local governments award licenses to various industrial or community concerns. In these landscapes, the amount of wood that can be removed is often regulated on a volume or area basis; again, it might be possible to remove additional biomass sustainably, but tenure reform would need to take place to allow such removals to happen. In Canada, various provinces, including British Columbia, Ontario, and Quebec are currently undergoing forest-tenure review in which one of the driving features is the development of biorefining products as a complement to existing forest products. Ontario's government released a proposed framework for a new forest tenure and pricing system in May 2010, the objectives of which include meeting the needs of the emerging forest sector, supporting forest-reliant communities, and safeguarding the province's world-class forest management system (Ontario 2010). It has been suggested that sustainable forest management frameworks, such as adaptive forest management, can help to link the growing body of research on forest bioenergy production to reviews of these policies, ensuring the long-term sustainability of the rapidly evolving forest bioenergy sector (Lattimore et al. 2010).

In Canada and elsewhere, there is a fear that the current emphasis on bioenergy products may damage existing forestry companies. The concern is that forest allocations will be taken away from large mills which are already struggling with deficits and even in bankruptcy protection (McKinley 2010). Fiber supply is seen to be one of the key assets that remain with these companies, and therefore any attempt to reallocate fiber under new tenure arrangements could be interpreted as an attack on the existing forest industry.

In the recent tenure review process in Ontario, approximately 200 forest industry companies representing over 8600 employees signed an open letter asking the provincial government to defer any action on the proposed tenure reform (OFIA 2010).

One of the main purposes of changing forest policy may be to achieve a pricing model for timber that reflects both social and market needs. Demand for green energy may require subsidies for forest biomass to be used in energy production. Determining the value of such a subsidy is extremely difficult and is made more complex by the intricate nature of forest operations, in which transportation costs, diversity of forest ownership, and externalities associated with timber harvesting ensure that a perfectly competitive market will never exist (Kant 2009). The challenge for policymakers is to design a pricing mechanism which has the highest potential given forest, social, and economic conditions. New forms of tenure may also consider nontimber values, cultural values, and ecosystem services such as carbon storage to reflect local and regional interests (NOSCP 2010).

Changes to policies that govern the sustainable management of forests need to be both politically and economically feasible if these changes are to be successful (Haley and Nelson 2007). This means designing a system that will achieve government objectives related to forestry and the forest environment while considering the interests of key stakeholders, including community members, industry, and others. A comprehensive communication strategy should be integrated into policy reviews to ensure that stakeholders are well informed at all stages.

24.2.2 Agricultural Feedstocks

In the agricultural sector, a differentiation can be made between food crops (of highest value and for human consumption), feed crops (grown for starch and protein content and used in downstream food production or increasingly for energy products such as corn-to-ethanol), and agricultural residues and energy crops (recovered or grown for lignocellulosic biomass content and potentially used in thermal, electrical, or biofuel production). Energy crops may take the form of annual crops (e.g., switchgrass, *Miscanthus*, hemp, and *Jatropha*) as well as fast-growing forest species that might be harvested on an accelerated schedule of 3–7 years (e.g., willow and poplar). The estimated yields associated with energy crops can also vary significantly. Robinson (1987) assumes annual yields of 10 dry t/ha, while Mabee et al. (2006a) assumes more conservative yields of 3–6.5 dry tonnes/ha. Conservative estimates for energy crop outputs in Canada range from a low of zero (Yemshanov and McKenney 2008) to 17.3 million dry tonnes per annum (Mabee et al. 2006b; Mabee and Saddler 2010).

Development of dedicated energy crops will bring a number of other concerns to the attention of land managers, including (but not limited to) the impacts of indirect land-use change, such as potential increases in deforestation as marginal forests are turned over to perennial energy crops, or alternatively increases in afforestation as marginal farmland is put under short-rotation forest plantations. Few statistics exist on the production of energy crops at this time, although some governments (notably in the United States) have instituted aggressive programs to develop these alternatives. Where potential energy crops can be grown on marginal and degraded land, these crops would not compete directly with food and fiber crops on better-quality arable land. Note that energy crops are not widely grown in current agricultural practice, although trial operations have begun in North America and Europe. Today, the intensive nature of agricultural practices makes large-scale food- and feed-processing facilities the norm in Western countries; under crop rotation practices, the same land might be used for food production one year and feed production the next.

One issue with agricultural-based biorefining in the temperate regions of the world is that the production of residues or energy crops is associated with the agricultural cycle, meaning that harvesting in late summer (with the potential for spring cropping in some regions) is the norm. Therefore, storage and delivery of these residues must be addressed. There is also concern about recovery of these residues from the field. It has been noted that the crop under consideration will dictate the total amount of residue left behind. In Canada, average residues for corn stover, wheat straw, barley straw, and oat straw are reported to be on the order of 1.0, 1.3, 1.0, and 1.2 tonnes per hectare, respectively (Wedlin and Klopfenstein 1985, Bowyer and Stockmann 2001). Further study of wheat straw residues has shown that variations of between 1.37 and 8.46 tonnes per hectare can be expected, depending on the harvesting method used (Bowyer and Stockmann 2001). In North America, approximately 29 million tonnes of wheat were produced in 2007 (FAOStat 2010), and therefore wheat-straw production can be estimated at between 40 and 240 million tonnes across the continent.

24.2.2.1 Policy Implications

Significant amounts of biomass, in the form of agricultural residues and energy crops, can be expected to be available for biorefining activities. Many have questioned, however, the sustainability of using the land base for energy production. One of the most incendiary debates has been about indirect land-use change (ILUC), or the concept that converting a hectare of land to bioenergy production will lead to land-use change elsewhere (Mathews and Tan 2009). Indirect land-use change effects can be categorized and investigated under environmental, economic, and social impacts (Ros et al. 2010). The biofuel-production value chain interacts with the global economic system and with natural ecosystems and climatic systems, thereby resulting in indirect effects. The compound and complex interactions in these dynamic systems are constantly evolving, which makes attainment of a final equilibrium a challenge (Ros et al. 2010). The present study will review only the environmental aspects of indirect land-use change associated with biofuels.

The first paper that urged scientists to extend the LCA system boundary to incorporate indirect land-use GHG emissions associated with biofuels was published in 2008 (Searchinger et al. 2008). Although a number of methodological challenges and uncertainties were associated with the key variables used, scientists recognized the importance of accounting for indirect impacts. Indirect land-use change occurs when pressure from market forces results in land conversion from prevailing feed or food crop production to biofuel cultivation, which consequently can lead to land-use change in other regions of the world to make up for the loss in feed or food production.

Searchinger et al. (2008) estimated the carbon payback periods resulting from ILUC for biofuels associated with production levels of 56 billion liters by 2016. The authors assumed that this consumption growth would occur as a direct result of the biofuel mandate approved by the U.S. Congress and would lead to the diversion of 12.8 million hectares of U.S. cropland from corn-for-feed production to corn-for-ethanol production. Consequently, this diversion would result in the cultivation of an additional 10.8 million hectares globally, which in turn would result in soil and vegetative carbon emissions of 351 tonnes of CO_2-eq. per converted hectare, or a total of 3.8 billion tonnes of CO_2-eq. (Searchinger et al. 2008; Mathews and Tan 2009).

Estimates on the ILUC of biofuels are sensitive to several variables, including biofuel type and the geographic location of production (Searchinger 2010). In a recent study, challenges in determining the optimum balance between cropland expansion and cropland intensification (i.e., increased production on current farmlands) were raised. Most recently, Searchinger (2010) stressed that biofuel production must generate and use "additional carbon," or carbon that plants would not otherwise absorb or that would be emitted to the atmosphere anyway.

Currently, energy crops are not expected to be profitable compared to corn production or other conventional agricultural systems. A price analysis carried out by James et al. (2010) found that cellulose would need to be priced between US$110 and US$130 per tonne for poplar (*Populus* spp.), switchgrass (*Panicum virgatum* L.), and mixed grasses. For *Miscanthus*, breakeven cellulose prices were found to be US$200 per tonne given current costs, with the potential for significant reductions if European costs for rhizomes were to be introduced. At US$45 per tonne under a modified price structure, *Miscanthus* would be well within the range of U.S. cost levels for biomass harvesting, transportation, and storage. Without targeted subsidies, however, energy crops were not found to be competitive at current yields and foreseeable prices (James et al. 2010).

24.2.3 Production of Intermediate Feedstocks

Developing a feedstock supply chain is essential to supporting widespread biorefining operations. One way to do this is through the production of "intermediates." To reduce feedstock transportation costs and to enable conversion-efficiency economies of scale, many studies have investigated means of increasing the bulk density of lignocellulose from wood and from energy crops (e.g., Sokhansanj and Turhollow 2004; Mani et al. 2006). An important goal of the production of intermediates is to achieve some degree of homogeneity in the physical and chemical quality of downstream feedstocks (Stephen et al. 2010). This is important because large-scale production of second-generation biofuels will most likely require a variety of lignocellulosic feedstocks, including agricultural residues, wood and wood residues from forestry and forest products, dedicated energy crops, and ultimately municipal solid waste.

Examples of existing, viable intermediates include solid products such as chips, pellets, and pulp (e.g., Mani et al. 2006), and liquid products such as bio-oil (Holbein et al. 2005; Holmgren et al. 2008). Pellets are typically produced by the compression of dried biomass particles, for the most part currently derived from sawmill industry by-products such as sawdust and shavings. The pellet production process, consisting of particle-size reduction, drying, pelletizing, and cooling, is well established, and a number of manufacturers exist for each type of processing equipment. Global pellet production was close to 10 Mt in 2008, with Europe being the largest consumer of these products. It is likely that wood pellets would be more suitable as an intermediate for thermochemical biorefining platforms. Pulp products include chemical pulps (Kraft) and mechanical pulps. Yields associated with mechanical processes tend to be relatively high (>85% on a w/w basis), while chemical yields tend to be relatively low (40–50%) (Biermann 1996). Both chemical and mechanical pulp could be used as an intermediate feedstock and would be best suited for the biochemical platform (Stephen et al. 2010).

Bio-oil, more properly known as *pyrolysis oil*, is the dominant product from fast pyrolysis (Pootakham and Kumar 2010). Approximately 1.7 t of wood or agricultural biomass at 20% moisture content is required to produce 1000 liters of bio-oil. The heterogeneous nature of this product can make its transportation, storage, and utilization by downstream processors challenging (Badger and Fransham 2006). World bio-oil production is currently severely limited, and bio-oil is currently a specialty product rather than a tradeable commodity (Holbein et al. 2005). Bio-oil could be used as an intermediate feedstock for both biochemical and thermochemical processing, but is more likely suited for the thermochemical platform (Stephen et al. 2010).

Intermediates that could be developed in the near future to support biorefineries include synthetic gas, which is produced by gasification at temperatures between 700°C and 1500°C. Syngas consists of a mix of mainly CO and H_2 with some CO_2, methane, and higher-carbon compounds. The gases generated must be cleaned (by removing the tars), filtered, and the clean gas collected ready for use (Stephen et al. 2010). This process releases a large amount of heat from the highly exothermic synthesis reactions, and reactor design improvements tend to focus on heat removal and temperature control. Syngas may be an ideal feedstock for thermochemical biorefining, but the gas needs to meet a number of criteria for the production of liquid fuels using the FT process. The number of inhibitory substances in the syngas, as well as its overall composition, varies with biomass feedstock and gasifier design, and therefore it is more problematic to process than other bioproducts.

From a policy perspective, it may be useful to explore the creation of a value chain in which intermediate feedstocks are generated in a supply sufficient to support large-scale downstream processing. The advantage of most intermediate products is that they can be used on their own for thermal heat production or (in the case of pulp) for secondary forest-products manufacture, which means that they already have intrinsic value in their current form. In the future, however, higher-value products might be achieved through a biorefining approach.

24.3 Creating Biorefinery Capacity

24.3.1 Current Status of Biorefinery Development

It is important to consider that different biorefinery technologies are likely to operate at different scales and that larger facilities will be more costly and more difficult for industry to develop. Bacovsky et al. (2010) has suggested that the scaling of these facilities is based largely on feedstock characteristics; for example, agricultural residues may be dispersed over large areas, leading to high collection and transport costs and constraining commercial plant sizes to smaller outputs (less than 200 million liters per year), while intensively produced forests or energy crops or trees might be more easily accessed, facilitating larger facilities (greater than 150 million liters per year). At the same time, the rates at which different technological platforms are developed may not be the same. Taken together, these facts suggest that different policy approaches might be appropriate for different technology platforms.

Although no commercial lignocellulosic biorefineries are currently operating, a database detailing demonstration facilities has been prepared by IEA Bioenergy Task 39; at the time of writing, it has detailed information on 74 projects (IEA Bioenergy 2009; Bacovsky et al. 2010). Production capacity for each facility has been normalized to metric tonnes per year using standard densities and energy contents. Both biochemical and thermochemical facilities are tracked in the database, as are "hybrid" facilities that use elements of each biorefining platform. A summary of the scale of these facilities and their known level of investment is provided in Table 24.1. Note that this figure does not represent the total investment in these facilities, nor does it accurately summarize public investment. Rather, the figure included is the sum of all public investment reported by the project owners; some owners may have chosen not to report all investments, and some owners declined to provide specific details.

TABLE 24.1

Summary of Lignocellulose-Based Biorefinery Facilities

	Total Number of Plants	Agricultural	Forest	Municipal/ Industrial	Energy Crops	Ethanol	FT Liquids	Synthetic Natural Gas	Mixed Alcohols	Green Hydrocarbons	Planned and Constructed Production Capacity (t/year)	Known Investment[a] (Million USD)
Bioconversion	32	19	3	4	6	32					405,000	757
Thermochemical	22	3	14	5	0	5	9	3	4	1	748,000	525
Hybrid	5	1	2	1	1	4	1				305,000	418
Other	15	0	0	11	4	1			4	10	3,370,000	933
Total	138	23	19	21	11	42	10	3	8	11	4,828,438	2633

[a] Known investments may not reflect total investment of public funds and do not represent total project costs.

Although significant technological challenges still exist for each platform under consideration, the results of this review can inform future investment. From a public-policy perspective, investment in bioconversion facilities might be viewed as the safest alternative because there are more projects underway worldwide using this technology, and therefore data detailing process performance and yields are likely to be fairly robust. Biochemical processes tend to be smaller in scale and to work efficiently with green, never-dried feedstocks (Sims et al. 2008). These processes are therefore likely less costly to implement compared to thermochemical alternatives. The majority of these projects use agricultural feedstocks, and therefore this technology platform might be most appropriate for these types of biomass.

At the same time, a significant number of thermochemical projects are underway, focused on larger scales and largely using forest biomass. Countries like Canada with significant forest feedstocks might consider these technologies in developing biorefinery policy. As reported by Stephen et al. (2010), thermochemical conversion, which is more similar to traditional fossil-fuel utilization than biochemical conversion, benefits more from economies of scale due to its technological complexity (and hence large capital costs) and its ability to accept a dry, compacted feedstock (Sims et al. 2008). It is worth noting that four of the thermochemical projects on the books today are classed as very large (>50,000 tonnes per year production capacity) and that these projects are planned to be commercial facilities, indicating confidence in the technology and its ability to scale up (Bacovsky et al. 2010).

To date, most public investment has been made in either bioconversion or thermochemical conversion platforms. Of the 74 projects for which detailed data are available, only 15 are not classified as bioconversion, thermochemical, or a hybrid of the two systems. The companies included in this category are largely focused on using industrial feedstocks (including synthetic gases) for the production of "green" or biobased hydrocarbons, thus avoiding the conversion stage altogether. This emerging pathway for biofuel production may lead to plants of very large capacity because these facilities are not bound by the same constraints as those working with raw feedstocks from agriculture or forests; however, the relatively small number of projects means that progress will be slow on this front.

For policymakers, the biochemical platform is the one with the most industrial experience, primarily focused on smaller-scale operations fed by agricultural feedstocks. The thermochemical platform tends toward larger demonstration projects, with most operations today choosing forest feedstocks. In any jurisdiction, feedstock type and availability are the first aspects to consider in selecting technologies to support through public policy.

24.3.2 Scales of Operation, Current and Future

Of the currently known pilot and demonstration biorefinery projects planned, under construction, in commissioning, or operational, the most common facility size is <5000 tonnes per year of production capacity. Few very large facilities are planned, and even fewer of these are bioconversion facilities, indicating again the importance of scale and technology type. Only six very large facilities are still in planning or under construction, and five of these will be commercial facilities (IEA Bioenergy 2009; Bacovsky et al. 2010).

A common theme in these results is the relative immaturity of all available biorefining technologies. This emphasizes the importance of research and development funding in developing each biorefinery platform. At this time, integration of the various process steps and overall process efficiency are being improved by integrated research programs which combine process development units with pilot- or demonstration-scale facilities around the world. Process development units are operating in Canada at the University of British Columbia, at Lund University in Sweden, at RISO/DTU in Denmark, and at the U.S. National Renewable Energy Laboratory (NREL). These facilities are providing important information to help guide policymakers in selecting larger projects for investment.

A factor that will greatly influence the ability of biorefining projects to meet a series of policy goals, including economic, social, and environmental performance, is the final scale of operations. The results of the IEA survey indicate that bioconversion technologies might work best in the small to medium portion of the size spectrum, while thermochemical technologies, which are largely derived from experience in the fossil energy sector, tend to work best in the large-scale range (Bacovsky et al. 2009). One way to understand this variation in scale is to consider the relative size of the world's largest ethanol, pulp, power (coal), and oil facilities. In a recent study by Stephen et al. (2010), these facilities have been ranked by the relative size of their feedstock inputs in terms of energy content. Although the average plant size for all four industries studied is significantly smaller than these largest-of-type facilities, industrial trends toward larger plants that provide economies of scale make it worthwhile to consider these capacities for logistical comparison. For each facility, the delivery mode was also considered. It seems likely that in the short term, the realistic modes of feedstock delivery will be truck or rail, while ship and pipeline remain less likely.

Some authors have speculated that the optimal size for a lignocellulose-based biorefining facility, particularly in the case of thermochemical conversion, is greater than 1 billion litres per year in capacity (e.g., see Wright and Brown 2007). Stephen et al. (2010) use logistics associated with transportation mode to dictate the maximum scale at which a facility can operate, with coast-based facilities that can receive feedstock by ship having an advantage over landlocked facilities that require delivery by truck or rail. Very large facilities tend to have multimodal opportunities for feedstock delivery, accepting biomass by road, rail, and ship. This type of strategy enables mitigation of feedstock supply risk.

Stephen et al. (2010) considered a number of intermediate products, including chips, pellets, and bio-oil, as feedstocks for biochemical and thermochemical biorefineries. As shown in Table 24.2, the potential of thermochemical biorefineries to deliver a given volume of biofuels is greater than that of the biochemical platform because thermochemical facilities can use dense bio-oil as a feedstock more effectively. By comparison, biochemical facilities work best with wood chips (or pulps), largely because these feedstocks have never been dried and are more suitable for biochemical conversion.

It seems likely that biochemical facilities will be similar in scale to a pulp and paper mill or a first-generation ethanol plant, using approximately 1 million tonnes of feedstock to produce approximately 200 million liters of product per year (Gallagher et al. 2005; Zamboni et al. 2009). By contrast, thermochemical facilities are likely to be much larger, demanding significantly more feedstock delivered through a more costly intermediate product. Stephen et al. (2010) pointed out that if facilities must be located inland, as is the case for most facilities, biochemical conversion may have an advantage due to its ability to

TABLE 24.2

Potential Scale of Commercial Facilities by Delivery Mechanism

Biochemical Platform		Thermochemical Platform	
Truck, chips	~4 million tonnes/800 M liters	Truck, bio-oil	~4 million tonnes/1.1 B liters
Rail, chips	~11 million tonnes/2.3 B liters	Rail, bio-oil	~14 million tonnes/4.0 B liters
Ship, pellets	~16 million tonnes/2.3 B liters	Ship, bio-oil	~21 million tonnes/6.0 B liters

operate economically at a smaller scale, which reduces the risk associated with securing feedstock supply (Wooley et al. 1999; Wright and Brown 2007).

Once the technology associated with either or both platforms has been successfully demonstrated at scale, the industry should see predictable but accelerated growth, provided that the drivers of biofuel production remain constant. This point is important; a decline in energy prices, for example, could greatly reduce the interest in biofuel industry expansion (as was observed in the 1990s). If individual projects in the current round of demonstration projects should prove successful in the next 24 months, it is estimated that the production of significant volumes from these plants might be seen as soon as 2013. Commercial uptake would follow within 1 or 2 years. Bacovsky et al. (2010) suggest that commercial biorefining remains at least 5 years away on any platform, which creates a timeline for investment in research, development, and demonstration.

For policymakers, the choice of technology for investment is crucial. No technology is a clear winner today. Many small pilot and demonstration facilities are using biochemical technologies, and therefore certainty about process performance is likely to be greater with these technologies. The expected industrial scale of operations for biochemical facilities is smaller, requiring less feedstock and making them open to sourcing using road or rail options. The expected industrial scale of operations for thermochemical facilities tends toward large operations, and favors locations where feedstock availability is greater or where delivery by marine transport is an option. In any given jurisdiction, a combination of geography (access to feedstock) and risk acceptance (willingness to invest in large-scale operations) will likely dictate policy decisions.

24.3.3 Funding Infrastructure and Production

A common policy instrument used to support development of biorefining capacity is direct government program funding in the form of contracts, grants, or loan guarantees. These instruments can help a company overcome the capital-cost hurdle involved in creating capacity for thermal, electrical, or fuel production. It is difficult to evaluate the effectiveness of direct funding by comparing projects in different countries because synergistic policies (such as renewable fuel mandates, excise tax exemptions, etc.) or simply more favorable market conditions may play a role in determining capacity. However, within a single country, it may be easier to see the impact of direct funding on the establishment of biofuel capacity. The bioethanol industry in the United States has been chosen for an analysis of the effectiveness of direct funding in establishing biofuel production capacity. Direct funds are considered to be funds earmarked for all aspects of research, development, and demonstration, including all biofuel production as well as biomass production for general energy purposes.

A study by Mabee (2007) found that bioethanol production was strongly correlated with the availability of funding for infrastructure development, biomass procurement, and plant operation, particularly in the early years. Key U.S. biofuel-producing states, including Illinois, Iowa, Minnesota, South Dakota, and Nebraska, have offered significant incentives over the last decade to establish corn-based bioethanol facilities (DSIRE 2010; RFA 2010; USDOE 2010). Each of these states has followed a different approach in creating these incentives. The tools of tax exemptions, direct loans, and cost-share schemes are shown to be effective in attracting capacity to individual jurisdictions, and the tools are shown to be flexible in achieving different results. The Minnesota example, in particular, shows the potential impacts of small changes to policy. By limiting the capacity to which the incentive applied, the state government was able to spur the creation of many smaller facilities, which in turn spurred iterative development of corn-to-ethanol technology while having an impact on the total number of jobs created and the benefits to the local economy (Mabee 2007). It was noted that each of these state-based strategies complemented an existing commitment to research and development funded by the U.S. federal government.

Although direct funding of infrastructure can be powerful in creating initial capacity, it may not be the greatest influence on the ultimate distribution of biorefining facilities. Mabee (2007) observed that the amount of direct funding available to capital projects has become less important in relation to other factors, particularly feedstock availability and access to market. Follow-up analyses using corn production data indicated that the availability of corn feedstocks has slowly become the dominant factor in determining where corn-based biorefining facilities are located (Mabee 2007).

For policymakers, it is clear from these results that availability of funds for the establishment of infrastructure and the development of capacity within each jurisdiction is a powerful tool in the early stages of biorefinery growth. Policy tools that offer funding for infrastructure have been shown to be very successful in delivering on broader goals of employment. One important finding was that a balance between research funding and funding for the creation of facilities is necessary to support the emerging biorefining sector. Mabee (2007) noted that the United States has devoted a significant amount of funds to research as well as to supporting facility creation.

24.3.4 Production Incentives

New biorefining capacity can be developed by offering a producer incentive, which might be delivered as a payment per unit produced, or through a tax exemption. An example of the former mechanism, widely used in this fashion, is a feed-in tariff. This tool can be used to increase public participation in the generation of renewable energy. It works by paying a premium price to the producer for the energy, in the form of electricity or fuel; this product can then be distributed through normal channels to consumers. In principle, the feed-in tariff should cover the additional cost of producing a green energy alternative over conventional fossil products (Fell 2009). Feed-in tariffs have been most widely used in the electricity sector in Europe, with Germany playing a lead role in refining this policy tool (Runci 2005).

In a study carried out by the Forest Products Association of Canada, the potential for forest-based biorefining products was measured by a number of economic tools, particularly return on capital investment (ROCE) and earnings before interest, taxes, depreciation, and amortization (EBITDA) (Roberts 2009). These products included traditional primary and secondary forest products as well as energy outputs. Based on the ROCE measure, the Province of Ontario seemed particularly well suited to build capacity in bioenergy products. This is due to the Green Energy and Green Economies Act (Ontario 2009) which was recently passed in Ontario. Part of the Act creates a feed-in tariff for renewable electricity; the presence of the feed-in tariff dramatically changes ROCE for energy products compared to other jurisdictions in Canada.

At the time of writing, no known jurisdiction has attempted to use feed-in tariffs or similar policy incentives to support biorefining operations. The necessary values of such tariffs are therefore unknown. A simplified analysis by the International Energy Agency, using second-generation biofuel cost ranges extracted from the literature at the time, indicates that biofuel production costs (excluding any subsidies) need to reach US$0.80/liter of gasoline equivalent to be competitive with oil at US$100/barrel, based on a medium-scale plant producing approximately 50–150 million liters per year and costing in the range of US$125–250 million (Sims et al. 2008).

The future projected costs of lignocellulosic biofuel vary over a wide range, partly depending on the feedstock costs chosen for the assessment. The estimated production costs of second-generation bioethanol in the literature currently range from US$0.60 to US$1.30 per liter. The potential for cost reductions is estimated to drive production costs down, possibly to as low as US$0.25 per liter (Riese 2007). Other predictions cluster around US$0.30–0.40 per liter by 2020; these predictions are the basis for biofuel policy in the United States, where the most recent target is approximately US$0.28 per liter (IEA 2005; USDOE 2010; Perlack et al. 2005). The actual rate of cost reduction will depend on feedstock prices, economies of scale from large plants, integration of new technologies, and the benefits of experience and learning. Based on these estimates, feed-in tariffs for biofuel produced from the biorefinery might be in the range of US$0.30 per liter in the short term. By comparison, the Canadian province of Ontario currently offers US$0.13 per kWh of biomass-based electricity.

Exemption from excise taxes or mineral spirits taxes is a commonly used tax-based mechanism to incentivize production of biofuels and could be used in support of the biorefinery. Excise taxes are commonly used in the transportation sector and are designed to fill the gap between property and income taxes. These types of taxes can be imposed on the sale or use of certain articles, including fuels, and on certain transactions and occupations. In many cases, these taxes are not itemized in sales receipts and cannot be easily detected, therefore resulting in a hidden cost to the consumer (USDOE 2001). Excise tax rates in different countries range considerably, but tend to be a fixed charge per liter of fuel rather than a percentage of selling prices.

In North America, excise tax exemptions have been used for some time as a tool to support renewable biofuels. In a review of excise tax exemptions, Mabee (2007) found that this incentive could provide a benefit for producers, but that providing these incentives did not correlate well with the development of new production capacity. In Europe, high excise taxes mean that exemptions for bioethanol (and other biofuels) are orders of magnitude larger than those found in North America; in these jurisdictions, excise tax exemptions seem to be a valuable incentive for developing biorefining capacity (Bacovsky et al. 2009). For example, until 2006, German biodiesel production enjoyed a complete tax exemption, which drove significant development of capacity in that country. Since August 1, 2006, the former complete tax exemption on biofuels has been reduced, allowing pure biodiesel to be taxed at an increasing rate. Introducing a tax on domestic biofuel production has had two effects. One effect has been a reduction in biofuel production. German biofuel production peaked in 2007 at 3560 million liters of biodiesel, 580 million liters of bioethanol, and 790 million liters of pure plant oil. By 2008, production capacity for biodiesel (which felt the most immediate impacts of the tax resumption) fell to 3180 million liters—a drop of approximately 12%. By March 2008, 85% of the existing biodiesel production capacity had been idled. Approximately 70% of the companies engaged in the business had stopped production or gone bankrupt; 14% of filling stations stopped selling biodiesel. This devastating effect on the industry has been slightly ameliorated by increases in fuel prices over the summer of 2008, and again into the summer of 2009, but the industry remains well below its 2007 peak.

From the perspective of a policymaker, production-based incentives or tax exemptions targeted at energy products could be effective in driving development of lignocellulose-based biorefining capacity. The review indicates that producer incentives, such as feed-in tariffs, would have to be quite significant to overcome the cost differential related to renewable fuels. Incentives for renewable electricity tend to be significantly lower. Tax exemptions might also be an effective means for creating capacity, but only in jurisdictions where the base tax rate is high enough to make the exemption attractive. The German example shows the danger of providing short-term support to an industry and the risks associated with changing support mechanisms mid-stream.

24.4 Creating Markets for Biorefinery Products

Many products might be generated from the lignocellulose-based biorefinery. As stated previously, however, the most likely short-term outputs of the biorefinery are likely to be energy products, which may be thermal, electrical, or fuel for transport.

Combined heat and electrical power (CHP) can be generated in several different ways within a biorefinery. A portion of the incoming biomass can be diverted to a power boiler, which generates steam (heat); the steam can be then used to drive a turbine to generate electricity. A more advanced version of this approach is to use a gasifier to reduce biomass (chips) to synthetic gas (see the next section), which can then be used to drive a gas turbine. In existing Kraft pulp mills, for example, the recovery boiler associated with black liquor recovery could be modified to produce gas for heat and power generation. Gasification tends to be more energy-efficient than steam power boilers; if the energy in flue gas were recovered, it could provide up to 80% energy recovery (Karlsson and Gustavsson 2001). CHP is not widely used with agricultural materials, largely because so little agricultural biomass has been used for energy in the past. Multiple different CHP configurations might be developed within the forest sector, including heat and power associated with BCTMP (bleached chemithermomechanical pulp) mills and with NBSK (northern bleached softwood Kraft pulp) mills (in the latter case, most likely through black liquor gasification). Stand-alone installations producing heat and power might be developed with both traditional and gasification approaches and could be used with both forest and agricultural feedstocks. Although combined heat and power is not an intermediate product per se, a gasification process generating syngas could be used for CHP in the short to medium term and could eventually be seen as providing an intermediate product for thermochemical biorefining, with downstream processing of syngas implemented at a later date.

Wetterlund and Soderstrom (2010) have considered policy tools to create combined heat and power capacity in Sweden. Their study anticipated the profitability of biomass gasification for a Swedish district heating system in 2025, using two technological approaches. Their study indicated that investment in a thermochemical biorefinery facility with synthetic natural gas as an output would require support in the range in the range of US$0. 32–US$0.55 per kWh of production. They concluded that combined heat and power for district heating systems would require a large capital commitment and that the necessarily strong dependency on policy instruments to make these projects work would require policies to be enacted over long periods (Wetterlund and Soderstrom 2010).

As for biofuel for transportation, a number of jurisdictions have considered renewable fuel standards as a means of driving consumer demand. In the United States, the most significant policy development in recent years was the legislation of a Renewable Fuel Standard (RFS) as part of the Energy Independence and Security Act (EISA) of 2007. The RFS requires 136.3 billion liters of biofuel in the transportation fuel mix by 2022 and includes specific provisions for advanced biofuels, including cellulosic ethanol and biomass-based diesel that pave the way for advanced technologies. Classifications for biofuels include conventional (first-generation) biofuels, as well as advanced biofuels (including cellulosic ethanol and "green" gasoline, "green" diesel, and other synthetic fuels, all of which can be classified as second-generation biofuels). Under the RFS, ethanol production from conventional corn-based systems will be capped at 56.8 million liters per year as of 2015, which represents approximately 22 million liters of additional capacity over 2008 figures. New conventional bioethanol facilities will be held to tight environmental performance regulations. At the same time, advanced biofuel production (including second-generation cellulose-based biofuels) is expected to rise dramatically. In 2009, 2.3 million liters of biofuel produced in the United States must be of types classified as advanced biofuels. The ability of the industry to meet the RFS for advanced biofuels is limited, with only one or two projects under construction for production of these fuels and with little active capacity to meet the 2010 requirement. It should be noted that the requirements for advanced biofuels would exceed those for conventional biofuels by 2021, but that rollout of these fuels may take longer than originally proposed in the Act (Bacovsky et al. 2009).

Brazil is one of the world's largest bioethanol producers and remains the largest exporter of bioethanol to other markets. Brazil produces bioethanol from sugar- or starch-based material in the form of sugarcane and sugarcane residues. Because of Brazil's optimal climate, two seasons of sugarcane growth can be achieved, adding greatly to the potential production of both sugar and bioethanol products. In response to the first oil crisis of the 1970s, Brazil invested heavily in fuel alcohol, primarily as a means of increasing fuel security and avoiding having to use foreign currency for petroleum purchases. Today, Brazil controls more than 75% of the world's export market, with most exports going to the United States and Europe. Biofuel use is mandated within the country, with most vehicles using a blend of up to 25% ethanol in gasoline. Many countries, including Japan and members of the European Union, have made Brazilian bioethanol a part of their renewable-fuel strategies. Unlike the United States and Germany, Brazil has not made dramatic policy shifts in the past few years and continues to develop a strong domestic and export market for bioethanol. The introduction of flexible-fuel cars capable of running on either Brazilian "gasoline" (containing 20–25% ethanol) or hydrous ethanol (~93% ethanol) has spurred internal use of bioethanol in Brazil in the past couple of years. Brazil is also rapidly growing its biodiesel sector, primarily for domestic use at this time.

In the European Union, the primary policy tool behind the development of a bioethanol industry has been the European Commission Directives on the promotion of the use of biofuels for transport (Directive 2003/30/EC) and later on the use of renewable energy (Directive 2009/28/EC). The motivations behind these Directives include improving the security of energy supply and reducing the environmental impact of the transportation sector. The most recent targets are 20% renewable energy and 10% biofuels by 2020. Most member states have passed these Directives into national law, including Belgium, the Czech Republic, France, Germany, Greece, Latvia, Lithuania, Sweden, and the Netherlands (Neeft et al. 2007). A parallel Directive was created to restructure the community framework for the taxation of energy products and electricity (Directive 2003/96/EC), allowing excise tax exemptions for biofuels produced or blended within European countries (Neeft et al. 2007). This mandate at the EC level has led various member countries to develop their own policy responses.

For policymakers, it is clear that mandates can drive significant development of the market for specific biorefinery products. These mandates have rarely been used in countries without other means of support, including tax exemptions and other funding mechanisms that promote the creation of infrastructure or the import of bioenergy or biofuel products. Therefore, mandates should be seen as a component of a more wide-ranging biorefining strategy.

24.5 Conclusions

Policy options to support biorefining can take a number of forms, but generally support three major endeavors: (1) increasing biomass supply for biorefining activities, (2) creating biorefining capacity in terms of infrastructure and production, and (3) developing a market for biorefining products.

To increase biomass supply, two major sources of biomass have been considered. In the area of forest biomass, it is expected that major policy initiatives will be related to increasing access to harvest residues and currently nonmerchantable, small-diameter wood. Similar initiatives might be put into place to increase the supply of agricultural residues. The sustainability of such operations, in terms of nutrient losses and other environmental impacts, needs to be informed by research, and therefore policy should include funding for these activities in both forestry and agricultural management schemes. An adaptive management framework, in which research constantly supports improved management practices, should be put into place.

Significant volumes of biomass may be produced through the establishment of energy crop systems. It is expected that the fundamental policy question pertaining to these systems is the issue of indirect land-use change, meaning that the loss of agricultural or forest land to energy systems will require conversion of other land elsewhere in the world to make up for the lost productivity. There is an ongoing scientific debate over the scale of the impacts of indirect land-use change, but policy that favors management techniques that focus on "additional" carbon uptake might prove beneficial. Energy crop systems would require significant subsidies because current market prices are less than half the cost of producing biomass through these systems.

Policymakers may wish to explore the creation of intermediate feedstocks, which would enable reduced feedstock transportation costs and conversion-efficiency economies of scale in large downstream facilities. A number of different intermediates exist, but research and development is still required to determine the optimal business plan for structuring the biorefinery.

Selection of technology within policy programs is an important question that needs to be addressed. This review indicates that the biochemical platform is the one with the most industrial experience, primarily focused on smaller-scale operations fed by agricultural feedstocks. The thermochemical platform tends toward larger demonstration projects, with most operations today choosing forest biomass. Feedstock type and availability are the first aspects to consider in selecting technologies to support through public policy. Although no technology is a clear winner today, there may be more certainty about process performance with biochemical technologies because of their wider deployment in demonstration facilities. The expected industrial scale of operations for biochemical facilities is smaller, requiring less feedstock and open to sourcing from road or rail options. The expected industrial scale of operations for thermochemical facilities tends toward large operations and favors locations where feedstock availability is greater or where delivery by marine transport is an option. A combination of geography and willingness to assume risk needs to be considered in selecting technologies for support through public policy.

To create biorefining capacity, the most effective means is the availability of funding through loan guarantees, public–private partnerships, or similar mechanisms. Production-based incentives, such as feed-in tariffs, or tax exemptions targeted at energy products can be effective in driving development of lignocellulose-based biorefining capacity, but need to be substantial to overcome the cost differential related to renewable fuels. Furthermore, if policy support is removed suddenly, as has happened in Germany, there is a risk of derailing the entire biorefining sector.

It is clear that successful policy options to support biorefineries, and in turn biorefining, may take a number of forms, including targets and mandates, exemption of biofuels from national excise taxation schemes, direct government funding of capital projects to increase capacity or upgrade distribution networks, or consumption mandates for government or corporate vehicle fleets. These policies can be differentiated by their relative emphasis on government, industry, or consumer actions. Development of lignocellulose-based biorefining is likely to be of major interest to most of the major producers of biofuels; these nations are beginning to rethink their biofuel strategies, largely with an eye toward moving away from food-based feedstocks and toward residues, energy crops, and other cellulosic material, as well as algae and other innovative pathways to biofuels. At this point in time, very little cellulosic biofuel is being produced, with less than 0.1% of total production across 21 countries attributed to cellulosic feedstocks. An innovative and comprehensive biorefinery strategy could greatly increase this figure in the future.

References

Bacovsky, D., Barclay, J., Bockey, D., Dornelles, R., Edye, L., Foust, T., Grabowski et al. 2009. *Update on Implementation Agendas 2009*. Mabee, W.E., Neeft, J., van Keulen, B. (eds.), IEA Task 39 Report T39-P5, 75 pp.

Bacovsky, D., Mabee, W.E., Wörgetter, M. 2010. How close are second-generation biofuels? *Biofuels, Bioproducts & Biorefining* 4(3): 249–252.

Badger, P.C., Fransham, P. 2006. Use of mobile fast pyrolysis to densify biomass and reduce biomass handling costs: A preliminary assessment. *Biomass and Bioenergy* 30: 321–325.

Biermann, C.J. 1996. *Handbook of Pulping and Papermaking*. Academic Press, Toronto ON.

Bowyer, J.L., Stockmann, V.E. 2001. Agricultural residues: An exciting bio-based raw material for the global panels industry. *Forest Products Journal* 51: 10–21.

Bull, G., Mabee, W.E., Scharpenberg, R. 1998. *Global Fibre Supply Model*. Food and Agriculture Organization of the United Nations, Rome, Italy.

Chum, H.L., Overend, R.P. 2001. Biomass and renewable fuels. *Fuel Processing Technology* 71(1–3): 187–195.

CIBC 2005. *Ontario Government's Package for the Forest Products Industry*. Equity Research Industry Update, September 29, 2005. CIBC World Markets, Toronto, Canada.

CPAWS 2007. *Ontario's Forest Economy...Adrift in a Global Marketplace*. Special Report, June 2007. CPAWS Wildlands League, Toronto, Canada. Available online at: http://www.wildlandsleague.org/attachments/Adrift.Ontario.mill.closures.2007.pdf.

DSIRE 2010. *Database of State Incentives for Renewables and Efficiency*. North Carolina State University, Raleigh NC.

FAOStat 2010. *FAOStat Forestry Data*. Food and Agriculture Organization of the United Nations, Rome, Italy. Available online at www.faostat.fao.org.

Fell, H.J. 2009. *Feed-In Tariff for Renewable Energies*. Berlin: BMU.

Gallagher, P.W., Brubaker, H., Shapouri, H. 2005. Plant size: Capital cost relationships in the dry mill ethanol industry. *Biomass and Bioenergy* 28: 565–571.

Goldfarb, D. 2010. Canada isn't cleaning up on green technology exports. Special to *Globe and Mail*. Update Published on Monday, March 22, 2010, 7:50 PM EDT Last updated on Tuesday, Mar. 23, 2010, 8:52AM EDT. www.theglobeandmail.com.

Haley, D., Nelson, H.W. 2007. Has the time come to rethink Canada's Crown forest tenure systems? *Forestry Chronicle* 83(5): 630–641.

Holbein, B.E., Stephen, J.D., Layzell, D.B. 2005. *Canadian Pyrolysis Initiative: New Directions in Biorefining*. BIOCAP Canada Foundation, Kingston, ON.

Holmgren, J., Marinangeli, R., Nair, P., Elliott, D., Bain, R. 2008. Consider upgrading pyrolysis oils into renewable fuels. *Hydrocarbon Process* 87: 95–103.

IEA. 2005. *Biofuels for Transport: An International Perspective*. International Energy Agency, OECD/IEA, Paris. www.iea.org.

IEA. 2006. *World Energy Outlook, 2006*. International Energy Agency, OECD/IEA, Paris. www.iea.org.

IEA. 2008. *Energy Technology Perspectives to 2050*. International Energy Agency, OECD/IEA, Paris, www.iea.org.

IEA Bioenergy. 2009. *Update on Demonstration Initiatives*. Report to IEA Bioenergy ExCo64, Liege, Belgium. www.task39.org.

IPCC. 2007. *Climate Change 4th Assessment Report—Mitigation*. Intergovernmental Panel on Climate Change, Working Group III, www.ipcc.ch.

James, L.K., Swinton, S.M., Thelen, K.D. 2010. Profitability analysis of cellulosic energy crops compared with corn. *Agronomy Journal*, DOI: 10.2134/agronj2009.0289.

Kant, S. 2009. Sale of Canada's public forests: An economically non-viable option. *Forestry Chronicle* 85(6): 841–848.

Karlsson, A., Gustavsson, L. 2001. External costs and taxes in heat supply systems. *Energy Policy* 31(14): 1541–1560.

Lattimore, B., Smith, T., Richardson, J. 2010. Coping with complexity: Designing low-impact forest bioenergy systems using an adaptive forest management framework and other sustainable forest management tools. *Forestry Chronicle* 86(1): 20–27.

Mabee, W.E., Fraser, E.D.G., McFarlane, P.N., Saddler, J.N. 2006a. Canadian biomass reserves for biorefining. *Applied Biochemistry and Biotechnology* 129: 22–40.

Mabee, W.E., Gregg, D.J., Arato, C., Berlin, A., Bura, R., Gilkes, N., Mirochnik, O. et al. 2006b. Update on softwood-to-ethanol process development. *Applied Biochemistry and Biotechnology* 129: 55–70.

Mabee, W.E. 2007. Policy options to support biofuels. *Advances in Biochemical Engineering/Biotechnology* 108: 329–357.

Mabee, W.E., Saddler, J.N. 2010. Bioethanol from lignocellulosics: Status and perspectives in Canada. *Bioresource Technology* 101(13): 4806–4813.

Mani, S., Tabil, L.G., Sokhansanj, S. 2006. Effects of compressive force, particle size and moisture content on mechanical properties of biomass pellets from grasses. *Biomass and Bioenergy* 30: 648–654.

Mathews, J.A., Tan, H. 2009. Biofuels and indirect land use change effects: The debate continues. *Biofuels, Bioproducts, and Biorefining* 3: 305–317.

McKinley, K. 2010. Talks on forest rules one-sided: Critics. *The Chronicle Journal*, Thunder Bay ON, May 20, 2010, p. A14.

Neeft, J., van Thuijl, E., Wismeijer, R., Mabee, W.E. 2007. *Biofuel Implementation Agendas*. IEA Task 39 Report T39-P5, 52 pp.

NOSCP 2010. *Response to Ontario's Proposed Framework to Modernize Ontario's Forest Tenure and Pricing System*. Northern Ontario Sustainable Communities Partnership.

OFIA 2010. *Coalition for Putting Ontario's Wood Back to Work*. Ontario Forest Industries Association. Available online at: http://www.ofia.com/tenure_coaliton/index.html.

Perlack, R.D., Wright, L.L., Turhollow, A.F., Graham, R.L., Stokes, B.J., Erbach, D.C. 2005. *Biomass as Feedstock for a Bioenergy and Bioproducts Industry: The Technical Feasibility of a Billion-Ton Annual Supply*. Department of Energy/GO-102005-2135, April 2005.

Persson, U.M., Azar, C. 2010. Preserving the world's tropical forests: A price on carbon may not do. *Environmental Science and Technology* 44(1): 210–215.

Pootakham, T., Kumar, A. 2010. A comparison of pipeline versus truck transport of bio-oil. *Bioresource Technology* 101: 414–421.

Province of Ontario. 2010. *Ontario's Forests, Ontario's Future: A Proposed Framework to Modernize Ontario's Forest Tenure and Pricing System*. Ontario Ministry of Northern Development, Mines, and Forestry, Toronto, ON.

Province of Ontario. 2009. *Ontario's Green Energy and Green Economy Act.* Ontario Ministry of Energy, Toronto, ON. Available online at http://ontariogreenenergyact.ca/.

Riese, J. 2007. *Personal Communication.* Associate Principal, McKinsey & Company, Frankfurt, Germany.

RFA. 2010. *U.S. Fuel Ethanol Production Capacity.* Renewable Fuels Association, Washington DC.

Roberts, D. 2009. *Future Bio-Pathways.* Report to the Forest Products Association of Canada.

Robinson, J. 1987. An embarrassment of riches: Canada's energy supply resources. *Energy* 12: 379–402.

Ros, J.P.M., Overmars, K.P., Stehfest, E., Prins, A.G., Notenboom, J., van Oorschot, M. 2010. Identifying the indirect effects of bio-energy production. Netherlands Environmental Assessment Agency, Publication Number 5001439003, 10 pp.

Runci, P. 2005. *Renewable Energy Policy in Germany: An Overview and Assessment.* Maryland: Joint Global Change Research Institute.

Searchinger, T.D. 2010. Biofuels and the need for additional carbon. *Environmental Research Letters* 5(2): 1–10.

Searchinger, T.D., Heimlich, R., Houghton, R.A., Dong, F.X., Dlobeid, A., Faviosa, J., Tokgoz, S., Hayes, D., Yu, T.H. 2008. Use of U.S. croplands for biofuels increases greenhouse gases through emissions from land-use change. *Science* 319(58–67): 1238–1240.

Sims, R.E.H., Mabee, W.E., Saddler, J.N., Taylor, M. 2010. An overview of second-generation biofuel technologies. *Bioresource Technology* 101(6): 1570–1580.

Sims, R.E.H., Taylor, M., Saddler, J.N., Mabee, W.E. 2008. *From 1st- to 2nd-Generation Biofuel Technologies: An Overview of Current Industry and RD&D Activities.* International Energy Agency, Paris, France, 124 pp.

Soderholm, P., Lundmark, R. 2009. The development of forest-based biorefineries: Implications for market behaviour and policy. *Forest Products Journal* 59(1–2): 6–16.

Sokhansanj, S., Turhollow, A.F. 2004. Biomass densification—Cubing operations and costs for corn stover. *Applied Engineering in Agriculture* 20(4): 495–499.

Stephen, J.D., Mabee, W.E., Saddler, J.N. 2010. Biomass logistics as a determinant of second-generation biofuel facility scale, location and technology selection. *Biofuels, Bioproducts and Biorefining* 4(5): 503–518.

Stephen, J., Mabee, W.E., Saddler, J.N. 2010. Biomass logistics as a determinant of second generation biofuel scale, location, and technology selection. *Biofuels, Bioproducts, and Biorefining* (in press).

Thiffault, E., Pare, D., Brais, S., Titus, B.D. 2010. Intensive biomass removals and site productivity in Canada: A review of relevant issues. *Forestry Chronicle* 86(1): 36–42.

USDOE. 2010. *Biomass Program Biofuels Initiative.* Available online at http://www1.eere.energy.gov/biomass/biofuels_initiative.html.

USDOE. 2001. *A Primer on Gasoline Prices.* U.S. Department of Energy, Office of Energy Efficiency and Renewable Energy. Rockville MD.

Wedlin, W.F., Klopfenstein, T.J. 1985. Cropland Pastures and Crop Residues. In: Heath, M.E., Barnes, R.F., Metcalfe, D.S. (eds.), *Forages: The Science of Grassland Agriculture.* Iowa State University Press, Ames IA, pp. 496–506.

Wetterlund, E., Soderstrom, M. 2010. Biomass gasification in district heating systems: The effect of economic energy policies. *Applied Energy* 87(9): 2914–2922.

Wright, M.M., Brown, R.C. 2007. Establishing the optimal sizes of different kinds of biorefineries. *Biofuels, Bioproducts, and Biorefining* 1: 191–200.

Wooley, R., Ruth, M., Glassner, D., Sheehan, J. 1999. Process design and costing of bioethanol technology: A tool for determining the status and direction of research and development. *Biotechnology Progress* 15: 794–803.

Yemshanov, D., McKenney, D. 2008. Fast-growing poplar plantations as a bioenergy supply source for Canada. *Biomass & Bioenergy* 32: 185–197.

Zamboni, A., Shah, N., Bezzo, F. 2009. Spatially explicit static model for the strategic design of bioethanol production systems. 1. Cost minimization. *Energy & Fuels* 23: 5121–5133.

25

Biopower in the U.S. South: Barriers, Drivers, and Potential for Expansion

Youngsun Baek and Marilyn A. Brown

CONTENTS

25.1 Introduction

Biomass as a renewable energy resource has received increased attention in the search for clean, renewable energy alternatives. Worldwide, biomass combustion (including cogeneration) accounts for approximately 62 GW of electric power capacity, and both large- and small-scale systems have been expanding, with 2–4 GW of power capacity added in 2010 (REN21, 2011, Table R1). In the United States, biopower (including municipal solid waste) is the third largest form of renewable electricity generation after hydropower and wind energy (EIA, 2011).

Biomass can be (1) used as fuel for direct combustion or cofired with coal, (2) gasified, or (3) used in biochemical conversions. Because of the wide range of feedstocks, biomass has a broad geographic distribution. If a national renewable electricity standard (RES) target were to be set, some analysts estimate that a majority of the growth in renewable electricity would come from electricity generated from wood and other biomass (Brown and Baek, 2010; EIA, 2009). However, other analysts show very little biopower growth, relative to wind (NREL, 2010). The possible dominance of biomass is due to its dispatchability and the relatively low capital and operating costs it requires to generate electricity. In addition, compared to other renewable resources, the feedstock is readily provided in terms of gross supply and ease of delivery. Regionally, the South has a potential to supply over 35% of U.S. biomass energy resources. However, while biopower provided 1.1% total U.S. electricity generation in 2008, the South produced only 0.6% of its total electricity from biomass (Brown et al., 2012).

25.2 Biopower in the South

To estimate the potential of biopower in the U.S. South, this study defines the South in two different ways. The definition of the South provided by the U.S. Census Bureau for the purposes of data analysis that relies principally on census statistics, state-based data from the Energy Information Administration (EIA), and energy end-use statistics from the EIA's National Energy Modeling System (NEMS). This definition of the South includes the District of Columbia and 16 states (Figure 25.1), and it divides the region into three census divisions. The *South Atlantic* division is the largest in both population and geography, with eight states and the District of Columbia; all but West Virginia lie along the eastern

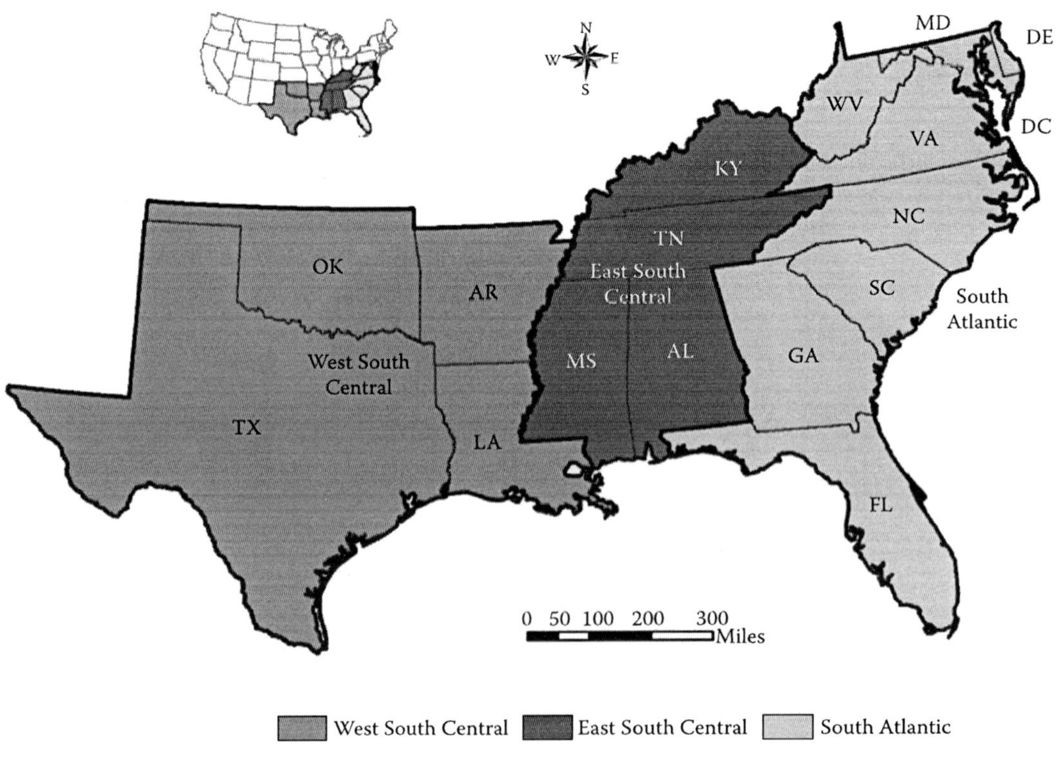

FIGURE 25.1
The South census region and its three divisions.

seaboard. The *East South Central* division includes Alabama and three states with western borders that touch the Mississippi River. The *West South Central* division also includes four states, which all lie west of the Mississippi River. In 2009, the South accounted for 42% of U.S. energy consumption and 45% of U.S. electricity consumption, but only 37% of U.S. population. In the future, its electricity consumption is expected to grow more rapidly than in the rest of the country reflecting the region's relatively stronger economy; despite this growth, electricity rates in the South are expected to remain below the national average (Energy Information Administration, 2011).

The South is also defined as a subset of four of the 13 regions used by the North American Electric Reliability Corporation (NERC) covering the continental United States (Figure 25.2). The four NERC regions that are used to define the South are

- Electric Reliability Council of Texas (ERCOT)
- Florida Coordinating Council (FRCC)
- Southeast Electricity Reliability Council (SERC)
- Southwest Power Pool (SPP)

NERC's regions are the basis for managing U.S. electricity generation and are used in the NEMS electricity market module. Coal accounts for the largest share of power generation in the South, as it does nationwide. In contrast, the South depends less on renewable sources of electricity than any other region, with only 4.9% of its electricity coming from renewables compared with 10.4% nationwide. As is true nationwide, biopower is the third largest generator of renewable electricity in the South, following hydroelectricity and windpower (Energy Information Administration, 2011).

Biomass resources are generally classified into five major categories: urban wood wastes, mill residues, forest residues, agricultural residues, and dedicated energy crops. Urban wood wastes are woody materials such as yard and tree trimmings, site-clearing wastes, pallets, packing materials, and construction and demolition debris. Mill residues include

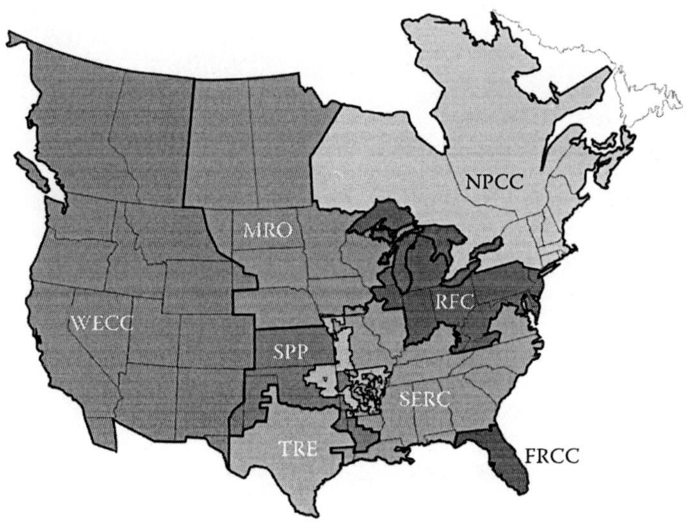

FIGURE 25.2
The NERC regions. (From Electricity Market Module. 2010. EIA. Report #:DOE/EIA-0554(2010). http://www.eia.gov/oiaf/aeo/assumption/pdf/electricity.pdf.)

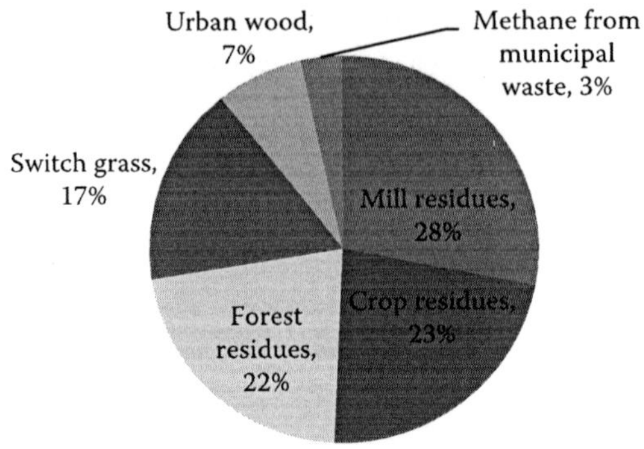

FIGURE 25.3
Biomass availabilities in the South. (Adapted from Milbrandt, A. 2005. *A Geographic Perspective on the Current Biomass Resource Availability in the United States.* NREL, TP-560-39181. Retrieved from: http://www.nrel.gov/docs/ fy06osti/39181.pdf.)

residues from the processing of lumber, pulp, veneers, and composite wood fiber materials. Forest residues are logging residues, including small branches, limbs, tops, and leaves. Agricultural residues are the stalks and leaves of crops that remain after the grains have been harvested. Dedicated energy crops include short-rotation woody crops such as hybrid poplar and hybrid willow and herbaceous crops such as switch grass. Among the listed biomass resources, urban wastes are the least expensive biomass resource, followed by mill residues, forest residues, agricultural residues, and energy crops (NREL, 2011).

The current availability of biomass resources in the South is shown in Figure 25.3. Clearly, solid waste from mill, forest, and agricultural sources is dominant. Mill and forest residues account for 50% of biomass resources, and supply biomass stably with fewer seasonal variations than energy crops and agricultural residues. Some industries such as the pulp and paper industry operate their own electricity generators to recycle their waste and produce electricity on site.

Using heat content values from best engineering estimates of heat rates and a 70% capacity factor, the maximum achievable potential of biopower in the South has been estimated by McConnell et al. (2010) using data from Milbrandt (2005). Figure 25.4 shows that the maxi-

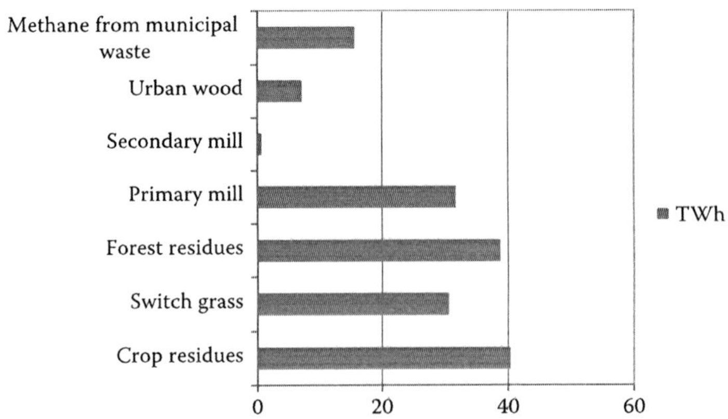

FIGURE 25.4
Approximation of biopower potential by source in the South.

mum achievable potential of biopower in the South is 165 TWh. Clearly, not all available biomass would be used for power generation, but in keeping with the national goals set by the Biomass R&D Technical Advisory Committee, 5% of electricity generation in the South is approximated to be met using biomass as a primary fuel.

25.3 Technology Characterization

Biopower technologies are generally categorized into direct combustion, cofiring, and gasification. Existing coal-fired power plants can significantly reduce sulfur emissions by involving the biopower technologies because biomass has a relatively low sulfur content compared to fossil fuels. Among these three categories, direct firing has been widely used for biomass- and waste-fired power plants in the United States. Biomass gasification is an emerging technology that can be used in advanced power cycles such as an integrated gasification combined cycle (IGCC).

Cofiring involves using biomass as a supplementary energy source in high-efficiency coal boilers. Cofiring with coal in existing boilers is the lowest-cost biopower option among the three. Extensive demonstrations and commercialization have improved the effective substitution rate of biomass up to approximately 15% of the total energy input; the rate is approximately 5% for cofeed systems and 15% for separate injection systems (McGowin, 2007). In addition, cofiring operations suggest an SO_x and NO_x reduction potential of up to 20% with woody biomass (NREL, 2011).

Direct combustion involves the oxidation of biomass with excess air to yield hot flue gases, which produce steam in the heat-exchange sections of a boiler. The steam is used to produce electricity in a Rankine cycle. Direct-combustion systems include multipressure, reheat, and regenerative steam turbine cycles as well as supercritical steam turbines. The common boiler designs used for steam generation with biomass are stationary- and traveling-grate stokers and atmospheric fluid-bed stokers. The addition of dryers and the incorporation of more rigorous steam cycles are expected to raise the efficiency of direct combustion systems by approximately 20% over today's efficiency (DeMeo and Galdo, 1997).

Gasification involves the conversion of biomass in an atmosphere of steam or air/oxygen to a medium- or low-calorific gas. A medium-calorific-value gas has a heating value of 30–50% that of natural gas, and a low-calorific-value gas has a heating value of 10–15% that of natural gas. Gasifiers are typically referred to as *direct* (pyrolysis, gasification, and partial combustion take place in one vessel) or *indirect* (pyrolysis and gasification occur in one vessel, combustion in a separate vessel). For direct gasification, air and sometimes steam are directly introduced to the single gasifier vessel. For indirect gasification, an inert heat-transfer medium such as sand carries heat generated in the combustor to the gasifier to drive the pyrolysis and char-gasification reactions. While direct gasification systems have been demonstrated at both elevated and atmospheric pressures, indirect gasification systems operate close to atmospheric pressure (NREL, 2011).

25.4 Barriers, Drivers, and Policies

Despite advances in technologies, biopower makes up less than 1% of the South's electricity supply. The following barriers illustrate significant challenges that currently impede the full deployment of biopower technologies in the South.

A major limitation of agricultural residues is the limited collection season. Agricultural residues are usually collected over the course of a few months after the grain harvest. For that reason, storage of up to 10 months is generally required for year-round use. In addition to the storage issue, loading and transportation costs affect the market prices of feedstock. Compared to the total amount of resources available; the amount of resources available for power generation is limited by the economical transportation range surrounding the power plant.

It is well known that one of the advantages of using biomass is the relatively low capital and operational costs for biopower generation. However, there are still technical issues associated with cofiring, such as limits on the percentage of biomass that can be cofired. The current biomass integrated-gasification combined cycle (BIGCC) technology has high costs for installation and maintenance, but its performance is better than conventional options. Therefore, the BIGCC option still has potential to be improved technologically and economically by active R&D and demonstration. In addition, relative to wind, the level of production tax credits (PTCs) for biopower is low (as discussed later in this chapter).

Unlike other renewable resources, biomass is regulated by the Environmental Protection Agency (EPA) tailoring rule. The rule tailors the applicability criteria that determine which stationary sources and modification projects become subject to permitting requirements for greenhouse-gas (GHG) emissions under the Prevention of Significant Deterioration (PSD) and Title V programs of the Clean Air Act (CAA) (EPA, 2010). The EPA's final Tailoring Rule does not exempt biomass power producers from GHG permitting requirements, despite past EPA affirmations that biomass is carbon neutral. Instead, biomass power producers are required to satisfy the same GHG reporting obligations as fossil-fuel consumers (Nelson, 2010). Moreover, there are controversies about defining a "sustainable" harvest of biomass, and conflicts over feedstock use with other applications, such as cellulosic ethanol, wood products, paper, and chemicals, as well as wood pellets for export to Europe. Finally, the relatively small scale of viable biopower plants prevents them from enjoying the economies of scale that large solid-fuel (coal) plants enjoy.

In addition, biopower is a thermoelectric generating technology and has consumptive water use requirements. Consumptive water use is the amount of water withdrawn from the source and not returned to the source. Significant water consumption would also result from dedicated crops in areas requiring irrigation (NRC, 2008).

To develop realistic and feasible scenarios for biopower in the South, this study reviews various policies promulgated in the Southern states. Georgia has enacted legislation (HB 1018) creating an exemption for biomass materials from the state's sales and use taxes. To qualify for the exemption, biomass material must be used in the production of energy, including electricity, steam, and cogeneration. In 2007, Kentucky established the Incentives for Energy Independence Act to promote the development of renewable energy and alternative fuel facilities, as well as energy efficiency. Especially for renewable energy facilities, the bill provides incentives to companies that build or renovate facilities that use renewable energy. The maximum recovery for a single project may not exceed 50% of the capital investment. In Alabama, the Biomass Energy Program assists businesses in installing biomass energy systems. Program participants receive up to $75,000 in interest subsidy payments to help discharge the interest expense on loans to install approved biomass projects. Technical assistance is also available through the program. Bioenergy-supportive policies in Southern states are summarized in Table 25.1.

Box 25.1 describes one of the larger biopower projects currently underway in the South. It illustrates that stable financial support from diverse sources and purchase agreements for sale of the renewable electricity can be key to the success of biopower projects.

TABLE 25.1

Summary of Bioenergy-Supportive State Policies in the South

Type of Policy	State	Applicability and Amount	Requirements and Limits
Biomass Energy Production Incentive	SC	Amount: $0.01/kWh for electricity produced Duration: 2008–2018	$100,000 per fiscal year per taxpayer
Clean/Renewable Energy Tax Credit	GA, NC, KY	35% of Corporate/Personal Tax Credit (GA, NC) 30% of Corporate/Personal Tax Credit (KY)	
Green Jobs Tax Credit	VA	Amount: $500 per each job created Maximum incentive: $175,000	Must create a new job in the alternative energy/renewable energy field
TVA-Generation Partners Program	GA, AL, MS, TN, NC, VA, KY	Amount: $1000 per project plus $0.03/kWh above the retail rate (biomass) Performance-Based Incentive (PBI) payments continue for 10 years	
Biomass Sales and Use Tax Exemption	GA	100% exemption of sales and use tax Biomass, CHP/cogeneration Duration: 2006–2012	Must be utilized in production of electricity, steam, or cogeneration
Sales Tax Exemption for Large-Scale Renewable Energy Projects	KY	100% exemption from sales and use tax	Maximum incentive: 50% of capital investment Equipment requirements: >1 MW for biomass
Sales Tax Credit for Clean Energy Technology	TN	100% of sales and use tax	Certified green energy production facility as defined in TN Code Annotated (Section 67-4-2004)
Sales and Use Tax Credit for Emerging Clean Energy Industry	TN	Amount: Up to 99.5% credit	Taxpayer must make $100 million investment (minimum) and create 50 full-time jobs at 150% of TN's average occupational wage
Biomass Energy Tax Credit (Corporate)	SC	Amount: 25% of eligible costs Maximum incentive: $650,000 per year; credit may not exceed 50% of tax liability Carryover provisions: Excess credit may be carried forward for 15 years	
Green Power Production Incentive	NC	Amount: Varies Terms: Payments contingent on program success	

continued

TABLE 25.1 (continued)

Summary of Bioenergy-Supportive State Policies in the South

Type of Policy	State	Applicability and Amount	Requirements and Limits
Renewable Energy Systems Property Tax Exemption	TX, KS	Amount: 100%	Eligible system size: None specified, but system must be used primarily for on-site energy needs (TX)
		Applicable sectors: commercial, industrial, and residential	
Loan Program	KY, MS, NC, MO	Amount: $15,000–$300,000 (MS)/minimum $50,000 (KY)/not specified (NC and MO)	Payback: 15 years (KY)/up to 5 years ($4999 and less); up to 6 years ($5000 and greater) (MS)
State Grant Program	AL	Maximum incentive: $75,000	

Data source: Database of State Incentives for Renewables & Efficiency (DSIRE). Retrieved on September 15, 2011 from: http://www.dsireusa.org/.

BOX 25.1 GAINESVILLE RENEWABLE ENERGY CENTER (GREC)

Nameplate capacity: 100 MW
Location: Gainesville, Florida
Estimated operation: 2013

American Renewables is developing a 100-MW wood-fired biomass electricity generating project in Gainesville, Florida. Construction began in March 2011, and the Gainesville Renewable Energy Center (GREC) is expected to be operational by late 2013. The biopower project being constructed secured approximately $400 million of project financing in June 2011. The ongoing involvement of capital partners and financiers allows the project to remain on schedule with operations. The GREC is currently undergoing the bank loan syndication process, which has been successful to date. Chris Smith, financial advisor to project sponsors, said in a press release, "the Bank of Tokyo-Mitsubishi UFJ acted as coordinating lead arranger and, along with Natixis and Rabobank, as initial mandated lead arrangers." He also said, "ING Capital, SG Americas Securities, and Credit Agricole came in as mandated lead arrangers. CIT and DZ Bank have now also joined the bank group."

In addition, the GREC has entered into a 30-year power purchase agreement with Gainesville Regional Utilities (GRU) for sale of the biopower.

The GREC will be fueled by woody biomass, including tops and limbs from harvested trees, urban wood waste from landscaping contractors and power line/road contractors, and other lumber mill residues. One million tons of woody biomass will be necessary to fuel the plant annually, which will be obtained locally within a 75-mile radius of Gainesville. The GREC estimated that it will provide enough electricity to power 70,000 homes.

Gainesville Renewable Energy Center. (Image source: American Renewables, http://www.amrenewables.com/our-projects/gainesville-renewable-energy-center.php)

Source: Biomass Power & Thermal, Biomassmagazine.com (http://biomassmagazine.com/articles/5780/florida-biomass-project-loan-financing-successful-to-date).

25.5 Expanded Biopower

25.5.1 Methodology

This assessment of biopower potential in the South uses a version of the National Energy Modeling System (NEMS). NEMS models U.S. energy markets and is the principal modeling tool used by EIA and DOE. It consists of four supply-side modules, four demand-side modules, two conversion modules, two exogenous modules, and one integrating module (Figure 25.5). NEMS is one of the most credible national modeling systems used to forecast the impacts of energy, economic, and environmental policies on the supply and demand of energy sources and end-use sectors. Its "reference case" forecasts are based on federal, state, and local laws and regulations in effect at the time of the prediction. The baseline projections developed by NEMS are published annually in the *Annual Energy Outlook*, which is regarded as a reliable reference in the field of energy and climate policy. It is also widely used to conduct the sensitivity analyses of alternative energy policies and to validate research findings conducted by other government agencies and laboratories, including the U.S. Environmental Protection Agency, Lawrence Berkeley National Laboratory, Oak Ridge National Laboratory, and the Pacific Northwest National Laboratory.

The version of NEMS used for this modeling is GT-NEMS (GT is short for Georgia Institute of Technology). It is calibrated to the "stimulus" version of NEMS, released by EIA in March 2009. Any references to "NEMS" in this report indicate generic attributes of EIA's model. The distinctive feature of GT-NEMS is that while it uses all the same initial data as

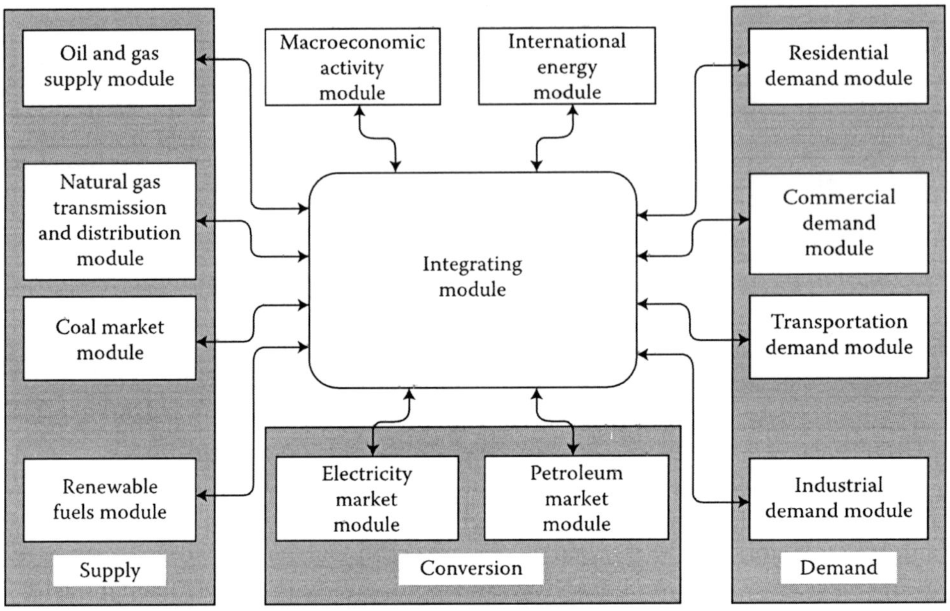

FIGURE 25.5
National Energy Modeling System (NEMS). (Adapted from U.S. Energy Information Administration (EIA). 2003. *National Energy Modeling System: An Overview*. Retrieved from: http://www.eia.doe.gov/oiaf/aeo/overview/index.html.)

NEMS, GT-NEMS incorporates changes specific to this study and does not run on EIA's system.

25.5.2 Case for Expanded Biopower

This study examines the potential for expanded biopower independent of changes that might occur with other renewable sources. It characterizes the biopower generation that would occur in an expanded biopower scenario as the result of (1) increased R&D and demonstration on biopower technologies; (2) extended production tax credits; and (3) improved feedstock supply.

These three assumptions underlying our expanded biopower scenario address the key barriers described above with supporting policies. Detailed explanations of the three policies are presented below with results in Section 6.

25.5.3 Modeling Scenario Assumptions

Based on capital and operating costs and capacity factors, as well as fuel costs, generation by the electricity sector is modeled using the NEMS Electricity Market Module (EMM). The fuel costs are provided in sets of regional supply schedules and are passed to the EMM, where biomass competes with other sources. Among the seven submodules of the EMM, the biomass electric power submodule (BEPS) handles biopower.

25.5.3.1 Description of Biomass Supply Curves

EIA's biomass feedstock prices for electricity generation are estimated from regional supply curves which are inputs to the BEPS. The raw data for the supply schedules are collected at the state or county level. These resource availabilities are aggregated to form the regional supply schedule by NERC region. Biomass resources are generally classified into five categories: urban wood waste, mill residues, forestry residues, agricultural residues, and energy crops.

Merging urban wood waste and mill residues in one category and agricultural residues and energy crops in another, the BEPS uses three different biomass-resource supply curves. The annual supply curves for agricultural residues, energy crops, and forestry residues have recently been updated based on the biomass supply data from the POLYSIS model developed by the University of Tennessee. To estimate the supply curves, the USDA annual projection forecasts are used to determine the yield rates of energy crops and agricultural residues. The supply plans of urban wood waste are provided by Oak Ridge National Laboratory (Perlack et al., 2005).

Unlike other renewable resources, biomass is traded in the feedstock market. For that reason, the growth of biopower production depends highly on feedstock prices and supply. Figures 25.6 through 25.8 show the variation in the resource availability as a function of price in 2020 and 2030. The supply curves of urban wood waste and mill residues are anticipated to remain the same until 2030. Figure 25.3 shows that ERCOT (TX) has the greatest potential supply of urban wood waste and mill residues among the four southern NERC regions.

The SERC region (MO, AR, MS, TN, AL, GA, FL, VA, NC, SC, and some parts of LA) has a far larger supply of agricultural residues and energy crops than other regions in the South. At the same price point, the supply from the agricultural sector is expected to increase by approximately 15% from 2020 to 2030. The supply in the SERC region at $20/MMBtu in 2030 is expected to reach 2795 trillion Btu.

EIA's supply curves for forestry residues remain the same from 2010 until 2030. The SPP region (KS, OK, and a part of LA) and the FRCC (FL) region have a larger potential for forestry residues than the SERC and ERCOT regions.

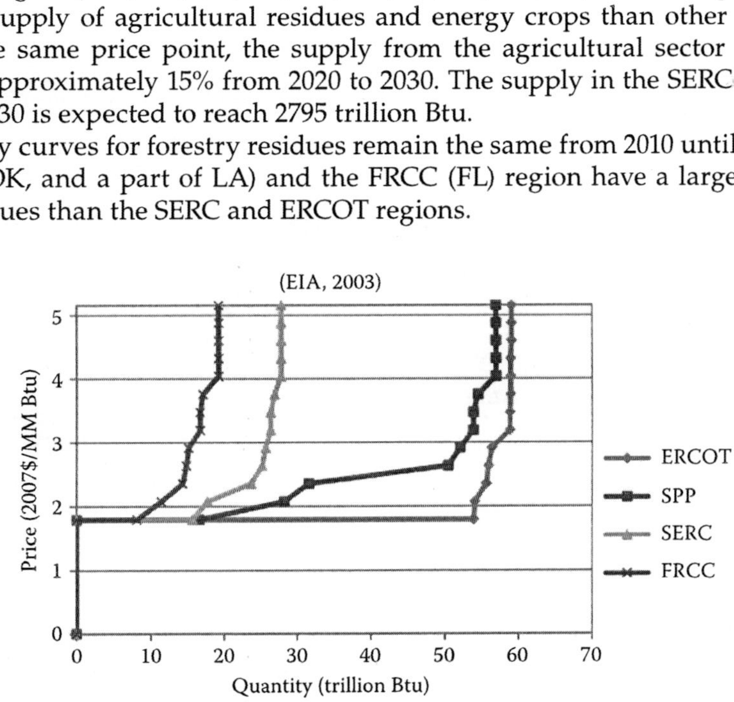

FIGURE 25.6
Supply curve of urban wood waste and mill residues by NERC region.

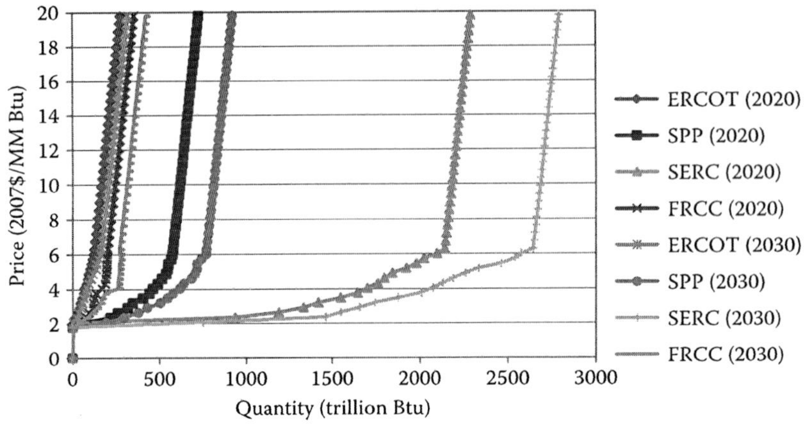

FIGURE 25.7
Feedstock Supply from the agricultural sector in 2020 and 2030 by NERC region (agricultural residues and energy crops).

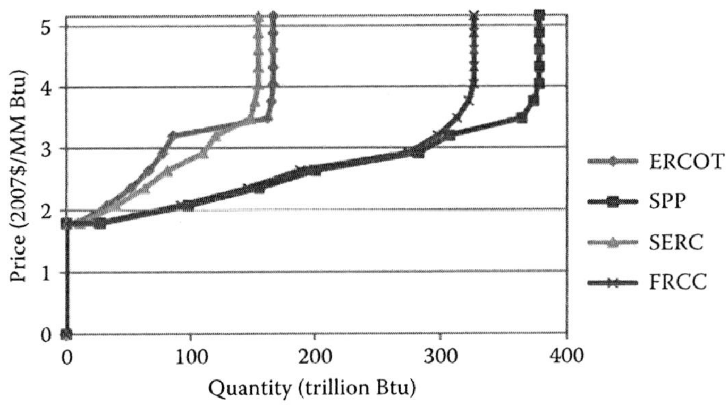

FIGURE 25.8
Supply curve of forestry residues by NERC region.

25.5.3.2 Technological Characteristics

In addition to biomass supply, technology-specific inputs are used to predict the magnitude of biopower potential. GT-NEMS represents both dedicated biomass and biomass cofiring plants to estimate the capacity of biomass for electricity generation. NEMS assumes that biomass cofiring can account for up to a maximum of 15% of the fuel used in coal-fired generating plants. The BEPS considers both dedicated biomass and biomass cofiring plants to forecast the capacity of biomass for electricity generation. The cofiring levels are assumed to vary by region as determined by the availability of biomass and coal-fired capacity in each region. Most states in the region have biomass-cofiring opportunities because of the widespread distribution of coal plants.

GT-NEMS models dedicated biomass plants in the same way as other generation options which use a single kind of fuel, such as coal, petroleum, and nuclear generation. The main inputs for the dedicated biomass generators are capital, operating, and maintenance costs, project life, production tax credit, and heat rate. Biomass cofiring plants are modeled in GT-NEMS by assuming that plant owners can retrofit their coal-fired plants and transform them into biomass cofiring plants. In addition, GT-NEMS assumes that no additional operating and maintenance costs would be required after the retrofitting because the biomass would be commingled with coal, and the mixture would be fed into the boiler through the existing coal feed system. However, if the cofiring system operated at higher levels, it would incur an additional capital cost to enhance the capacity and performance (Haq, 2002; EIA, 2003).

25.6 Expanded Biopower Scenario Results

25.6.1 Potential from Financial Incentive Policy

The federal renewable production tax credit (PTC) is a per-kWh tax credit for electricity generated by qualified renewable resources and sold by the taxpayer to an unrelated consumer during the taxable year. The PTC was originally enacted in 1992 and has been renewed multiple times. Although the tax credits for an open-loop biomass project are half of those for a wind project with the same production, closed-loop biomass projects are eligible to receive the same level of PTC as wind. The higher tax credits provide a motivation for building closed-loop biopower generating plants, which are relatively less widely used because of their higher costs. Indeed, there are no closed loop biopower projects in operation in the United States.

This study modeled a scenario in which the current PTC continues until 2030 and the rate stays at 0.9 cents per kWh. The extended PTC is forecast to lead to a dramatic increase in electricity generation from biomass in the South by 2030. It would result in an 8% increase in biopower in 2020 and approximately a threefold increase in 2030 in the South (Figure 25.9). The SPP region is anticipated to respond to the policy most actively.

25.6.2 Supportive R&D

Among the three technological options of cofiring, direct combustion, and BIGCC, the last is the most advanced technology and has the most room for improved performance.

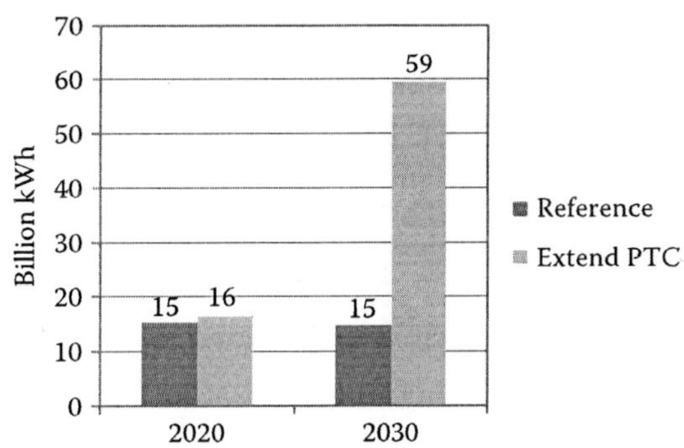

FIGURE 25.9
Increase in utility-scale biopower generation by expanding the PTC in the South.

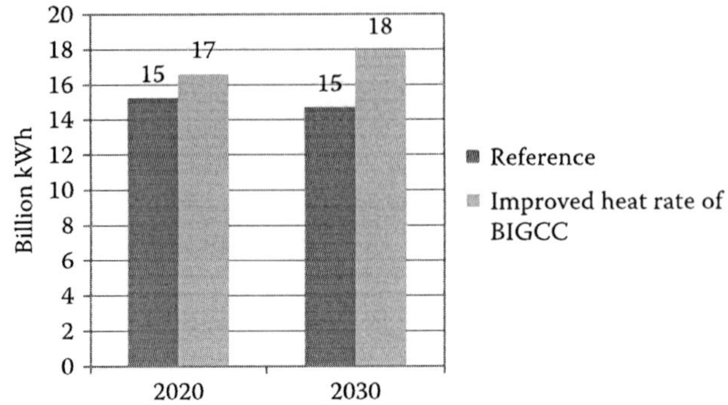

FIGURE 25.10
Increase in utility-scale biopower generation in the South through supportive R&D.

Historical capital costs do not show the cost reduction achieved by many of the biomass electricity technologies because direct combustion technologies are a mature commercial technology, and the industry has been static for the past 15 years. BIGCC systems are at the demonstration stage, while smaller-scale gasification internal combustion systems are in a more pioneering commercial stage. Using the Regional Energy Development System (ReEDS), NREL (2011) modeled the potential of biopower, assuming that BIGCC will displace direct-combustion systems gradually over 40 years (2010–2050), and the average system heat rate will improve by 14% over the same 40 years. The heat rate of the GT-NEMS reference scenario is assumed to be 9450 Btu/kWh in 2010, to decrease by 1.76% annually, to reach 7766 Btu/kWh in 2021, and then to stay at the same level through 2030.

Instead of a constant heat rate from 2022 to 2030, for this policy, the GT-NEMS model assumes that the heat rate will continue to improve beyond 2021 until 2030 at the same rate (1.76%) and will finally reaches 6620 Btu/kWh.

This policy, active R&D in the BIGCC technology, when modeled by itself, increases the biopower generation in the South. The ERCOT region in particular would respond to this scenario most sensitively of the four NERC regions and could produce three times more electricity than the reference case due to the technological advances. The improved BIGCC performance is expected to lead to a 9% rise in biopower generation in 2020 and a 22% increase in 2030 (Figure 25.10).

25.6.3 Improved Feedstock Supply

Sales tax incentives typically provide an exemption from the state sales tax (or use tax) for the purchase of a renewable energy system. Several states have established tax incentives by allowing an exemption from the state sales tax. The sales tax among the states in the South ranges between 0% and 7%. Whereas Georgia, Kentucky, and Tennessee enacted legislation creating an exemption for biomass materials from the states' sales and use taxes, many states in the South do not have such sales tax incentives for biomass purchased for electricity generation.

Therefore, the third expanded biopower policy is a sales tax exemption program involving all states in the South, with an improvement in loading and transportation systems. This study assumes that these measures would increase the biomass supply by 10%. The South would generate more biopower under an improved feedstock supply-only scenario;

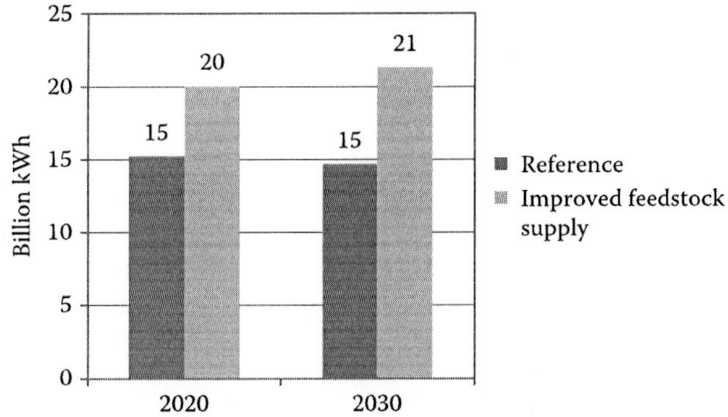

FIGURE 25.11
Increase of utility-scale biopower generation in the South through improved feedstock supply.

with a 32% rise in 2020 and a 45% increase in 2030 projected (see Figure 25.11). ERCOT is the NERC region with the greatest potential for increasing biopower generation as a result of a sales tax exemption.

25.6.4 Expanded Biopower Scenario

The expanded biopower scenario is defined as the combination of the three preceding policies. To reflect the second-order effect from the electricity demand side and other fuel markets, all of the modules in GT-NEMS, including the macroeconomic activity module, are involved in running this combined scenario. The utilities in the South could generate about four times more biopower under the supportive policy, market, and technological environment than in the reference scenario.

In addition, the end-use sectors (especially the industrial sector) would generate a substantial amount of biopower on site. For instance, the pulp and paper industry has its own electricity generation system using black liquor extracted from the mill residues. The ERCOT region in particular would be the greatest contributor to this trend because Texas has a large potential in urban wood waste and mill residues, which are a relatively low-cost feedstock. Unlike other regions, in the SERC region, the amount of biopower generated from the end-use sectors has been greater than that from the utilities in the SERC region in the past. However, if the combined policy suggested in this study is implemented, the utility-scale generation will outstrip the customer-owned generation in the future. Customer-owned biopower resources include industrial facilities that produce electricity from wastes coming from their own manufacturing process. On-site electricity generation reduces their requirement to consume electricity from utilities (Figure 25.12).

Table 25.2 shows the percentages of total power generation in the South that could be met by biopower in 2030. With biomass, the electric power sector and the end-use sector could provide respectively 4.0% and 1.9% of the total power supply of the South in 2030 compared with the reference case forecasts of 0.9% (utility-scale) and 1.8% (customer-owned). Thus, the biggest increase is for utility-scale biopower. Utilities in the FRCC (FL) region could produce approximately 13% of that region's electricity supply using biomass. Florida currently has the largest biopower of any state in the South, and it would likely strengthen this leadership position in the expanded biopower scenario (Brown et al., 2011). Although the ERCOT region is anticipated to increase the absolute amount of its biopower generation

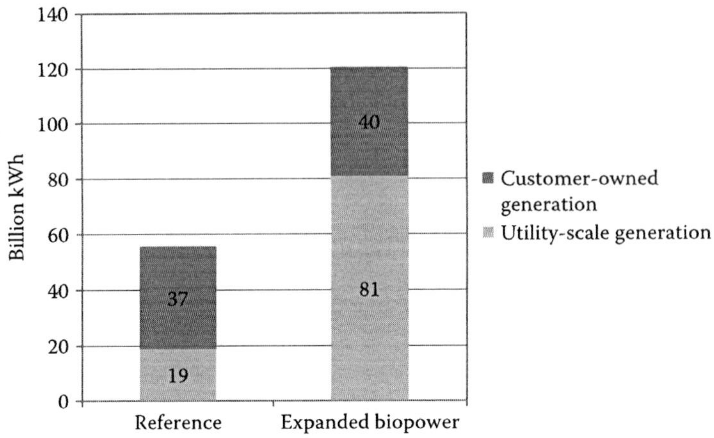

FIGURE 25.12
Total biopower potential in the South in 2030.

TABLE 25.2
Share of Biopower to the Total Electricity Generation by NERC Region in 2030

		Utility-Scale Biopower			
Region	ERCOT	SPP	SERC	FRCC	South Total
Reference	0.1%	0.5%	0.7%	3.3%	0.9%
Expanded biopower	0.0%	2.6%	3.3%	12.8%	4.0%
		Customer-Owned Biopower			
Region	ERCOT	SPP	SERC	FRCC	South Total
Reference	0.0%	1.8%	2.8%	0.7%	1.8%
Expanded biopower	0.0%	1.9%	3.0%	0.7%	1.9%

under the influence of the three policies, its share of biopower in the total electricity generation would not increase significantly because of the dominance of fossil-fuel-based electricity in the region.

The majority of the increase in customer-owned biopower, on the other hand, is expected to occur in the SERC region where it grows by 5.0% in the expanded biopower scenario, but only by 2.8% in the reference case. In SPP, the share of customer-owned biopower generation would increase from 1.8% to 1.9% under the expanded biopower scenario. In the remaining NERC regions (ERCOT and FRCC), the extended PTC underpinning the expanded biopower scenario does not motivate more customer-owned biopower, and the sales tax and R&D policies have only a minor stimulating effect.

25.6.5 Wind- and Biopower Trade-Offs

Renewable resources, fossil fuels, and nuclear compete with one another in the electricity generation market. Even among the renewable resources, the competition still exists. GT-NEMS enables trade-offs between different renewable resources and between utility-scale and customer-owned resources to be examined. It is noteworthy that wind would be the strongest rival in the competition. To clarify the trade-off relationship between wind and biomass in the future renewable electricity market, additional scenarios were run, including renewable electricity standards (RES) and a carbon-constrained future (CCF) scenario. Many policy analysts believe that an energy bill (with an RES) or a climate bill (with some type of CCF) will be passed sometime over the next several years. If these are promulgated, the electricity market would have to be reorganized. Three scenarios involving expanded use of renewables in the South are compared with the reference forecast reflecting EIA's analysis of the Stimulus Bill and the 2008 economic downturn:

- *Expanded renewables*: Uses updated estimates of renewable resources in the South. In addition, it assumes a number of renewable policies such as an extension of R&D and tax subsidies, but no new state or Federal carbon pricing or renewable energy portfolio policies are enacted (see Brown et al., 2011, for further details).
- *Expanded renewables + renewable electricity standard (RES)*: Uses all of the renewable policies and updated estimates of renewable resources from the *expanded renewables scenario*, along with a Federal requirement of 25% renewable electricity production by 2025. The scenario exempts small retailers from the RES mandate and excludes hydroelectric power and municipal solid waste from the sales baseline. An RES-only scenario was also created for purposes of comparison.
- *Expanded renewables + carbon-constrained future (CCF)*: Uses all of the renewable policies and updated estimates of renewable resources from the *expanded renewables scenario*, along with a carbon price of $15 (in $2005) per metric ton of carbon dioxide in 2012, growing annually at 7%. Allowances are redistributed to load serving entities as described above, and there are no carbon offsets. A CCF-only scenario was also created for purposes of comparison.

Given aggressive investment in R&D and installation of advanced technologies, and its attractive price-competitiveness in renewable electricity generation, wind would substitute for a sizable portion of the market share of biomass in the South (Figure 25.13).

The marginal resource of substitution depends on the policy, technology and regional characteristics of the change that causes the displacement. The trade-off relationship

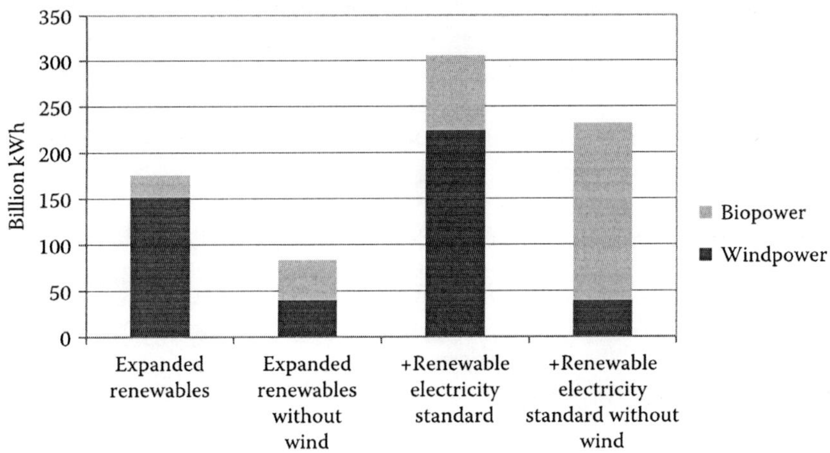

FIGURE 25.13
Comparison of windpower and biopower generation in the South in 2030.

TABLE 25.3

Comparison of Windpower and Biopower Generation in the South in 2030

	Windpower (Billion kWh)	% of Total Electricity Generation	Biopower (Billion kWh)	% of Total Electricity Generation
Expanded renewables (ER)	151	8	24	1
ER without wind	40	2	44	2
ER + RES	224	12	82	4
ER without wind + RES	39	2	192	11

between wind and biomass is most obvious under the RES scenarios (Table 25.3). However, biopower could not be considered as a perfect substitute for windpower because there still exists a portion of wind generation that could not be replaced with biopower. If an RES is implemented, the share of biopower will be even larger because biomass cofiring in existing coal plants is a low-cost and low-risk option for supporting compliance with the RES.

25.6.6 Comparison with Other Studies

When comparing the results of this analysis with the findings of other recent assessments of the potential for renewable electricity in this region, an array of similarities and differences can be noted. Findings vary, as expected, because of the variety of assumptions used in each study, the definition of the term "potential," alternative modeling methodologies, smaller or larger geographic coverage, and shorter or longer time frames.

The present and the EIA (2009) report on *Impacts of a 25-Percent Renewable Electricity Standard as Proposed in the American Clean Energy and Security Act* cover four NERC regions: ERCOT, SPP, SERC, and FRCC. In contrast, the report by the Southern Alliance for Clean Energy (SACE) published in 2009 (*Yes We Can: Southern Solutions for a National Renewable Energy Standard)* and *Powering the South: A Clean Affordable Energy Plan for the Southern United States* by Beck et al. (2002) cover much smaller regions. This analysis as well as the SACE (2009) and EIA (2009) studies provide estimates for 2030, while Beck et al. (2002) projects only to 2020. These distinctions help explain the lower estimates of renewable energy potential by Beck et al. (2002).

The SACE (2009) and EIA (2009) reports were intended to describe how the South could meet a federal renewable electricity standard, and were not intended simply to model the market-based adoption of biopower under a continuation of today's policies. Therefore, it is useful to compare these reports to the "expanded renewables + RES" case in this study. In both cases, the present analysis provides lower estimates for biopower and solar potential and a higher estimate for on-shore wind potential.

Biopower featured prominently in the EIA (2009) analysis of the Waxman–Markey Bill. It is especially dominant as a source of renewable electricity in the SPP and SERC regions, with an RES. Biopower is also seen as becoming the largest resource in the FRCC region, although wind becomes a significant contributor, primarily as a result of imports. Although wind is projected to dominate in the ERCOT region under an RES, biopower also contributes a significant amount in the EIA assessment.

Beck et al. (2002) characterizes the potential for efficiency and renewable resources in a six-state region of the Southeast (including the Carolinas, Georgia, Tennessee, Alabama, and Florida) in 2020. After accounting for a 23% reduction in electricity demand from investments in energy efficiency, it estimates that 84 billion kWh of renewable electricity generation could meet 10% of the region's electricity demand in 2020. Biomass cofiring

TABLE 25.4
Levelized Cost of Electricity by Generating Option in 2020 and 2030 (2010 Cents per kWh)

	Gasification	Cofiring	Direct Combustion
2020	8.0	4.4	8.5
2030	6.2	4.2	6.9

accounts for four percentage points of the 10%; all six states have biomass cofiring opportunities because of the distribution of coal plants. Biomass CHP accounts for 2.1 percentage points of the 10%, with slightly more variation across states because of the distribution of pulp and paper mills.

These comparisons support the conclusion of this chapter that wind- and biopower are close trade-offs, with comparable costs foreseen for the Southern U.S. This suggests that biopower could fill much of the gap left by windpower if it is unable to grow rapidly. Of course, these two sources of renewable electricity have very different operational characteristics. Windpower is an intermittent resource that tends to be greatest at night when load requirements are lowest. As a result, it poses challenges for large-scale integration into grid systems. Biopower, on the other hand, operates more as a base-load resource that can meet a much larger share of a power system's requirements without needing the addition of reserve requirements to manage rapid fluctuations in available power supplies. Its future role in the United States appears to be more dependent on the management of its air quality and biodiversity impacts, which are issues that do not challenge windpower.

25.6.7 Cost Effectiveness

The levelized cost of biopower reflects the cost to generate a particular amount of electricity with biomass through supportive policies and environments. The levelized cost of electricity (LCOE) calculated for biomass generation in the South ranges from 4.4 to 8.5 cents per kWh in 2020 and 4.2 to 6.9 cents per kWh in 2030. Cofiring biomass with coal has the lowest estimated LCOE, with costs that are considerably lower than the other two biopower options (Table 25.4).

25.7 Conclusions

The expanded biopower scenario modeled in this study suggests that the potential of biopower generation in 2030 could reach 120 billion kWh, which would account for about 6% of the total electricity generation in the South. (Additional electricity could be generated from municipal solid waste, which is not examined in this study.) This economic potential reflects the competition among renewable resources in the electric power market. The combined scenario of the production tax credit, the supportive R&D environment, and the improved feedstock supply is expected to have a great impact on utility-scale biopower and to increase the market share of biopower to 4%. On the other hand, only the third scenario with a sales tax exemption would be influential on the customer-owned electricity generation.

The production tax credit is expected to be the most effective driver to enhance biopower potential. The ERCOT (TX) region is anticipated to increase the absolute amount of biopower generation significantly, but the proportion of biopower in its total electricity generation mix would remain small, because fossil fuels are cost-competitive in the region. The FRCC (FL) region is expected to generate 40 billion kWh of biopower, which would amount to approximately 14% of the electricity generated in the region in 2030. The SERC region would continue to be the greatest producer of customer-owned biopower in the future.

Several issues must be resolved to realize the maximum economic biopower potential presented here. Unless biopower incentives and mandates are carefully managed, they could negatively influence other manufacturing industries in terms of jobs and economic activities. There are some arguments that mill residues already have a beneficial use in other industries, and the economic impact of the use of the residues for producing secondary woody products is greater than that of power generation (Thorp and Akhtar, 2009). In addition, the lack of incentives for closed-loop crop production and use has been pointed out as a problem. Although energy crops (including short-rotation woody crops) are often thought to be a part of the solution, there are few incentives for utilities or farmers and foresters to start producing these crops.

Other than the known air- and water-quality environmental issues associated with permitting and operation of biopower plants, the primary environmental issues that must be addressed for biopower are overall sustainability and indirect impacts caused by land-use changes. These issues are the same as those associated with biofuels processes, and biopower is now subject to the same EPA reporting requirements (EPA, 2009; EPA, 2010; NAS, 2010). In addition, the optimal zones for growing energy crops such as switch grass are close to or overlapping metropolitan areas, which are continuing to experience population growth and urban expansion. Competition for housing and other urban development may result in higher land prices and zoning challenges for biomass growers in these areas (Howell et al. 2010). On balance, the biopower potential is anticipated to depend on the supply of feedstock materials, policy environments, and technological advancements. For supporting compliance with renewable electricity standards, biomass is a low-cost and low-risk option in the U.S. South.

Acknowledgments

The authors gratefully acknowledge the support received by the Georgia Institute of Technology's Institute of Paper Science and Technology. Valuable insights were provided by Norman Marsolan and Marko Hakovirta (IPST) and Paul Stuart (École Polytechnique de Montréal). Any errors in the manuscript are the sole responsibility of the authors.

References

Beck, F., Kostiuk, D., Woolf, T., and Singh, V. 2002. *Powering the South: A Clean Affordable Energy Plan for the Southern United States.* Renewable Energy Policy Project. Retrieved from: http://www.repp.org/articles/static/1/binaries/pts_repp_book.pdf.

Brown, M.A. and Baek, Y. 2010. The forest products industry at an energy/climate crossroads. *Energy Policy* 38(2010): 7665–7675.

Brown, M.A., Gumerman, E., Baek, Y., Wang, J., Morris, C., and Wang, Y. 2011. *Renewable Energy in the South.* Atlanta, GA: Southeast Energy Efficiency Alliance. December (http://www.seealliance.org/PDFs/RENEWABLE%20ENERGY%20IN%20THE%20SOUTH.pdf).

Brown, M. A., Gumerman, E., Sun, X., Sercy, K., and Kim, G. 2012. Myths and facts about clean electricity in the U.S. South. *Energy Policy* 40: 231–241.

DeMeo, E.A. and Galdo, J.F. 1997. *Renewable Energy Technology Characterizations.* TR-109496. Washington, DC: U.S. Department of Energy. Palo Alto, CA: Electric Power Research Institute.

Haq, Z. 2002. *Biomass for Electricity Generation.* Energy Information Administration (EIA). Retrieved from: http://www.eia.doe.gov/oiaf/analysispaper/biomass/.

Howell, F.M., Porter, J.R., Mason, P.B., and Blanchard, T.C. 2010. Spatial Contours of Potential Biomass Crop Production: An Extension and Measures of Change. *Journal of Rural Social Sciences* 25(2): 1–32.

McGowin, C. 2007. *Renewable Energy Technological Assessment Guide.* TAG-RE:2007. Palo Alto, CA: EPRI.

McConnell, B.W., Hadley, S.W., and Xu, Y. (2010, July). *A Review of Barriers and Opportunities for the Integration of Renewable Energy,* Final Draft. Oak Ridge, TN: Oak Ridge National Laboratory.

Milbrandt, A. 2005. *A Geographic Perspective on the Current Biomass Resource Availability in the United States.* NREL, TP-560-39181. Retrieved from: http://www.nrel.gov/docs/fy06osti/39181.pdf.

National Academy of Science (NAS). 2010. *Electricity from Renewable Resources: Status, Prospects, and Impediments.* IBN No. 13: 978-0-13708-0. Washington, DC. Retrieved from: www.eenews.net/greenwire/climate_courts/2010/10/12/1.

National Renewable Energy Laboratory (NREL). 2010. Evaluating Renewable Portfolio Standards and Carbon Cap Scenarios in the U.S. Electric Sector. Technical Report NREL/TP-6A2-48258. May 2010 (http://www.nrel.gov/docs/fy10osti/48258.pdf).

National Renewable Energy Laboratory (NREL). 2011. Renewable Electricity Futures Study Report. Volume 1: Exploration of High-Penetration Renewable Electricity Fugures (http://www.nrel.gov/docs/fy12osti/52409-1.pdf).

National Research Council of the National Academies (NRC). 2008. *Water Implication of Biofuels Production in the United States.* ISBN:0-309-11362. Washington, DC.

Nelson, G. 2010. It's red states vs. blue in legal war over EPA rules. *ClimateWire,* E&E Publishing, LLC. October, 12, 2010 (http://www.eenews.net/public/Greenwire/2010/10/12/1).

Perlack, R.D., Wright, L.L., Turhollow, A.F., Graham, R., Stokes, B.J., and Erbach, D.C. 2005. *Biomass as Feedstock for a Bioenergy and Bioproducts Industry: Billion-Ton Annual Supply.* ORNL/TM-2005/66 & DOE/GO-102995-2135. Oak Ridge, TN: Oak Ridge National Laboratory.

Renewable Energy Policy Network for the 21st Century (REN21). 2011. *Renewables Global Status Report.* Retrieved from: http://www.ren21.net/Portals/97/documents/GSR/GSR2011_Master18.pdf.

Southern Alliance for Clean Energy (SACE). 2009. *Yes We Can: Southern Solutions for a National Renewable Energy Standard.* Knoxville, TN: SACE.

Thorp, B. A. and Akhtar, M. 2009. The best use of wood. *Paper360°,* January/February, TAPPI. 2009 (http://www.epoverviews.com/oca/The%20best%20use%20of%20wood.pdf).

U.S. Energy Information Administration (EIA). 2011. *Annual Energy Outlook 2011.* DOE/EIA-0383. Washington, DC: Energy Information Agency.

U.S. Energy Information Administration (EIA). 2009. *Impacts of a 25-Percent Renewable Electricity Standard as Proposed in the American Clean Energy and Security Act.* Discussion Draft. SR/OIAF/2009-04.

U.S. Energy Information Administration (EIA). 2003. *National Energy Modeling System: An Overview.* Retrieved from: http://www.eia.doe.gov/oiaf/aeo/overview/index.html.

U.S. Environmental Protection Agency (EPA). 2010. *Prevention of Significant Deterioration and Title V Greenhouse Gas Tailoring Rule.* Final Rule. 40 CFR Parts 51, 52, 70, and 71 RIN 2060–AP86. 3 June. Vol. 75, No. 106.

U.S. Environmental Protection Agency (EPA). 2009. *Clearinghouse for Inventories and Emissions Factors/AP42.* http://www.epa.gov/ttn/chief/.

26

Biorefinery Design from an Earth Systems Perspective

Damon Honnery, Gil Garnier, and Patrick Moriarty

CONTENTS

26.1 Introduction

A biorefinery (BRF) uses biomass as a feedstock or a raw material to produce multiple products, much like a petroleum refinery. Because biomass has various components such as lignin, cellulose, and hemicellulose, BRF products can include various fibers, fuels, chemicals, and plastics that are today often produced from petroleum (Ragauskas et al. 2006). Advantages of the BRF concept over more conventional uses of biomass include greater utilization of the entire biomass feedstock, greater product income, and less waste generated. Wider advantages claimed include greater energy security because of replacement of fossil fuels, lower greenhouse-gas emissions, economic and social benefits to primary producers, and even the creation of a new industry.

However, it is necessary to examine whether BRFs will in fact produce the benefits claimed, and if so, under what conditions. This chapter assesses some of these claims using an approach which attempts to integrate all aspects of BRF operation into the Earth's environmental systems (e.g., atmosphere, land, water, flora, and fauna) in combination with the human economic system. For example, if a BRF is intended to provide energy as a major output, then it must provide net energy to the system in which it operates. In providing this energy, the BRF must operate in a manner which is economically sustainable, and it must have a lower environmental impact than the processes it seeks to replace.

This approach taken here is an Earth Systems Science (ESS) approach. ESS seeks to understand the planetary life-support system as an integrated whole. This approach extends life cycle assessment (LCA), which is limited to quantifying energy and materials flows, by examining the interactions of these flows with the major components of the Earth's systems. To illustrate this, when evaluating greenhouse-gas emission benefits from use of a biomass feedstock, it is necessary to evaluate emissions emanating from land-use practices as well as those from fossil fuel use and nitrogen fertilizers. Extending this approach could lead to examining what alternative uses of land there might be and whether one use has better outcomes than another. Determining whether land should be used for food, fuel, or fiber or set aside to service the Earth's ecosystems (Moriarty and Honnery 2010a) is an example of where this type of approach has value.

Undertaking an ESS-type analysis of a complex industrial system like a BRF requires, at a minimum, the following:

- Detailed definitions of the processes that make up the BRF
- Quantification of the BRF material and energy flows
- Knowledge of the interactions of the BRF material and energy flows with the Earth's environmental systems
- Economic and resource data to enable projection of BRF growth
- Quantification of the link between BRF impact and scale

Given the relative novelty of most BRF concepts, little, if any, of this type of data is available, and what is available is often inconsistent. Moreover, given the ongoing development of the foundation technology, the point of reference for much of the data is continually shifting. This is also true for BRFs which might be classified as being traditional, such as those based on pulp and paper, because these are undergoing changes of sufficient magnitude that the lessons learned over many years of operation might no longer be valid. These problems give rise to the central paradox of new technologies: we will not fully understand the impact of a technology until it exists in the form and at the scale that we intend to use it.

Despite this, and noted already, a number of general rules can be used to assess whether a BRF warrants further development. Examples of these are

1. If energy is the major output of the BRF, it must supply net energy to the system.
2. The environmental impact of the BRF must at least be better than that of other proposed, but competing systems.
3. The cost of production must at least be equal to or better than other proposed, but competing systems.
4. BRF cost and environmental impact per unit output must reduce or at least be constant with scale if output is to increase.

The term *environmental impact* here includes the interaction of the BRF with the Earth's environmental systems.

While it is preferred to consider the complete set of rules, they need not all apply to each individual BRF. Taking production of aviation fuels as an example, it is likely, in the medium term at least, that these fuels will need to be liquids with particular thermochemical and physical properties. Producing these fuels in the future might require more energy than is available from the fuel produced, thus contravening the first rule. Balancing this energy shortfall is required at the larger system level rather than at the smaller individual BRF level. Under such circumstances, scale becomes an important consideration. Total aviation fuel production would be limited by the system's capacity to make up the energy shortfall. Extending this example, if production at the scale necessary were only possible through use of fossil fuels, this might be allowed to occur given sufficiently low environment impact and carbon-free energy to drive air-capture systems to remove aviation carbon emissions from the atmosphere (Moriarty and Honnery 2010a,b). The economic cost of producing aviation fuels under either circumstance would have to include the externalities of energy supply or else air capture and carbon storage.

In this chapter, three BRFs have been selected for analysis from an Earth Systems Science (ESS) perspective. The first, which is the most speculative of the three, examines the potential for using algae to produce fuels and chemicals. The second is based on the marriage of existing pyrolysis technology to newly developed upgrading processes to produce liquid fuels and chemicals, while the third examines the potential for an existing industry, pulp and paper, to alter its production to generate a range of new products.

26.2 Microalgae for Energy, Food, and Chemicals

Biorefineries based on microalgae are very much a future concept only. The idea is not recent; the use of algae for fuels was proposed half a century ago. Presently, only about 5000 tonnes (dry weight) of microalgal products, chiefly for cosmetics and human and animal nutrition, are produced annually worldwide (Wijffels and Barbosa 2010). These have product values (PVs) between 200 and 2000 euros per kg (~US$ 255–2550) (Rosch et al. 2009). As another estimate, costs for producing basic algal biomass in 100-tonne annual capacity production plants have been given as US$3000 per tonne, but these costs should decrease with increasing plant capacity (Chisti 2008). Clearly, there is a vast gap between this price and that for energy feedstock for power stations, for example, which must be available at a few tens of dollars per tonne.

Nevertheless, supporters claim that microalgae have many potential advantages over conventional biomass production (Chisti 2008, Marsh 2009, Rosch et al. 2009, Mata et al. 2010, Scott et al. 2010, Stephenson et al. 2010). Microalgae

- Have higher yield/ha/year than land-based plants and can use land not suitable for land-based plants
- Can be grown in brackish or sea water
- Can be used to remediate waste water
- Can use CO_2-rich flue gases from fossil fuel plants for enhanced algae growth
- Can be harvested continuously, even daily, given the appropriate climate

- Can be designed to produce different feedstocks such as lipids, carbohydrates, hydrocarbons, and hydrogen
- Can have up to 80% oil content on a total dry biomass basis, compared with 5% for agricultural oil crops like soybeans or oil palms
- Can potentially use advances in biotechnology to increase yield and cut costs
- Can potentially revolutionize many biotechnology sectors such as biofuels, pharmaceuticals, and nutritional additives

However, microalgae production also has a number of critics (e.g., Clarens et al. 2010, Walker 2009), who point out several serious problems that must be overcome if microalgae are ever to become a major energy source either for the transport sector (including air travel) or as a power-plant fuel, perhaps cofired with fossil fuels. These critics dispute many of the above claims and also stress that, at present, microalgae production consumes more energy than it uses. As has been noted, this does not matter much for small amounts of specialty chemicals, but must be changed if microalgae are to be used to supply major amounts of energy to the economy.

It is assumed here that very large-scale production will occur, as would be necessary if microalgae were to provide the bulk of the world's transportation fuels or a large share of the present global primary energy demand of over 500 exajoules (EJ = 10^{18} J) (BP 2011, International Energy Agency (IEA) 2011), a figure which could double by 2050 (Moriarty and Honnery 2009). Indeed, Chisti (2008) argues that microalgae are the only biomass feedstock with the potential to replace petroleum-based fuels completely for transportation uses. Globally, transportation fuels from petroleum have an energy content of approximately 90 EJ (Reijnders 2009). The rest of this section first describes the various methods and stages of microalgae production and harvesting and then examines the future prospects of algal production by looking at likely monetary and energy costs.

26.2.1 Description of Microalgae Production and Harvesting

Compared with the production of biofuels (and food or fiber) from agricultural systems, annual microalgae production is tiny. Most of the 5000-tonne output comes from shallow (10–30 cm depth) open ponds in a circular or "raceway" configuration, usually constructed in concrete and lined with plastic. Mechanical circulation by paddle wheels is necessary for efficient production. To suppress competing strains of algae and grazers (zooplankton) only algae species that can tolerate extreme conditions (e.g., very high salinity and pH) are used (Rosch et al. 2009). Open ponds are also a relatively cheap means of producing algal biomass.

Because of the limited range of algal species that can be grown in open ponds, there has been rising interest in closed systems or "photobioreactors" (Darzins et al. 2010). Basic designs for these use flat plates, tubular reactors, or bags made from transparent plastics or glass. One problem with polyethlyene plastic is its limited durability in natural light—as little as one year (Wijffels and Barbosa 2010). Again, it is necessary to circulate the algal cultures using either pumps or air-lift systems. Light is necessary for photosynthesis of algae, but if the intensity is too high, photoinhibition or overheating of algae can occur. With closed reactors, it is much easier to control light intensity, temperature, and other variables and to suppress contamination, but the sophisticated systems needed greatly increase monetary and energy costs.

All algae production methods need inputs of light, water, carbon dioxide, and nutrients, especially nitrogen and phosphorus, along with other trace elements. Shallow open ponds in high-insolation semiarid areas will have very high evaporation rates—as much as one cm per day—so water availability could be a problem. The possibility of using brackish water (or even salt water in coastal regions) could ameliorate this problem. In closed reactors, water loss is not an issue.

Production of each tonne of dry algal biomass requires 1.8 tonnes of CO_2 (Wijffels and Barbosa 2010). Several researchers (e.g., Chisti 2008, Rosch et al. 2009, Stephenson et al. 2010) have therefore proposed the use of flue gas from fossil fuel plants (with CO_2 concentrations several hundred times ambient levels) as a source of CO_2. However, flue gas CO_2 would also be targeted by carbon capture and storage schemes as a means of climate-change mitigation. Furthermore, if the intention is both to mitigate climate change and to avoid fossil fuel depletion, a reduction in fossil fuel power generation will be needed, and with it availability of this source of CO_2 (Moriarty and Honnery 2010a). Scott et al. (2010) have pointed to a potential problem: it may be necessary to pretreat the CO_2 flue gas to remove algal growth inhibitors or toxic chemicals. On the other hand, if the nonoil residue is anaerobically digested to produce biogas, with the biogas used to generate electricity to provide energy for production and harvesting, the resulting flue gas CO_2 could be recycled (Rosch et al. 2009).

Costs and energy inputs can also be reduced by the use of nutrient-rich wastewaters, especially those that are in need of remediation. Algal biomass has a far higher need for mineral nutrients than land-based plants; approximately 7% by weight of algae is nitrogen and 1% phosphorus (Wijffels and Barbosa 2010). There is also scope for internal recycling of nutrients.

The microalgae must be harvested before extraction of oil or conversion into various products. The energy costs of harvesting are high because of the low concentration of microalgae (<1.0 g/L in ponds and 3–15.0 g/L in photobioreactors) (Rosch et al. 2009, Mata et al. 2010). Harvesting can be done by flocculation, centrifugation, sedimentation, and filtration, including ultrafiltration. Flocculation is usually the first step in harvesting and is done to aggregate algae cells together to increase effective particle size. Flocculation is then followed by one of the other methods, often centrifugation, despite its high energy costs (Pittman et al. 2011). However, some cultures that show promise for high oil yields, such as *Nannochloropsis*, have small spherical cells with thick cell walls, making extraction of the oil difficult (Wijffels and Barbosa 2010). Finally, the extracted oil has to be processed to produce biodiesel.

26.2.2 Present and Future Costs

Because transportation fuels sell at a much higher price per MJ than power plant fuels, they represent the best fuel market for algal biomass. Chisti (2008) has examined the economics of producing transportation fuels from algae. He calculated that for crude oil at US$100 per barrel and microalgae with an oil content of 55% of dry biomass, the algal biomass would need to be produced at <US$340 per tonne. This figure assumed that conversion of algae biomass to transportation fuels will cost about the same as converting crude oil to transportation fuels. The analysis also made no allowance for by-product sales.

As discussed above, dry algal biomass presently costs as much as US$3000/tonne to produce. Costs will therefore need to be reduced by nearly an order of magnitude; alternatively, crude oil prices will need to rise greatly. The problem for algal fuels is that land-based biofuels for transportation use are already close to competitive with petroleum-based fuels. On the other hand, as algae supporters stress (Chisti 2008), land-based biofuels cannot be produced in sufficient quantities to replace more than a small fraction of petroleum-based transport fuels. However, if algal (or even land-based biomass) transportation fuels were to become a substantial replacement for petroleum, the cost of conventional oil, at least, would fall, because in some Middle-East oil fields, it can be produced at present for as little as US$10 per barrel (Moriarty and Honnery 2010a). Lower production rates would also greatly extend the life of oil reserves. If algal fuels, whether for transport or other uses, are to have any future, their production costs must fall dramatically. Finally, Mata et al. (2010) reported that algal production costs in photobioreactors may be an order of magnitude higher than for open ponds because their higher productivity cannot compensate for their much higher capital and operating costs.

26.2.3 Energy Analysis and Greenhouse-Gas Emissions

As might be expected, energy production costs and monetary production costs are closely related. At present, energy input and carbon costs in kg CO_2-equivalent ($CO_{2\text{-e}}$) for microalgal production and harvesting are very high. The energy/carbon footprint for microalgae production can be greatly reduced if the algae are grown in open ponds rather than in bioreactors because of the far higher frictional energy costs of pumping the algal culture around perspex tubes and the lower construction energy costs (Rosch et al. 2009). Even so, Lardon et al. (2009) examined various options for open-pond production from *Chorella vulgaris*, a freshwater algal species, and found that little or no net energy would be produced. They also found that more conventional bio-oil production methods (see Section 26.3) would be superior from a net energy viewpoint.

However, growing algae in open ponds has its own drawbacks. It is both land-intensive and water-intensive; because sunlight cannot penetrate more than a few centimeters into the algae mass, it is necessary to have very shallow ponds of large areal extent. For example, a large hypersaline open-pond plant in coastal South Australia uses an area of 520 ha to produce (by bioenergy standards) very small amounts of β-carotene from *Dunaliella Salina* (Darzins et al. 2010). Provision of freshwater can have very high energy costs, which are likely to rise as the need for tapping deeper groundwater or for interbasin water transfers increases (Moriarty and Honnery 2007).

Stephenson et al. (2010) also performed a comprehensive LCA on production of oil from *Chorella vulgaris*. Cultivation both in raceway ponds and in air-lift tubular bioreactors was investigated, as well as different biomass processing methods. Flue gas from a gas-fired power plant was assumed as the CO_2 source. Results of their model showed that if annual algal yields were 40 tonne/ha of oil, raceway ponds would have a kg $CO_{2\text{-e}}$/tonne of biodiesel and a fossil fuel energy requirement both roughly 80% less than that for petroleum-based diesel fuel. For bioreactor production, however, the fossil fuel energy inputs and kg $CO_{2\text{-e}}$/tonne were respectively 362% and 273% greater than those for conventional diesel.

These researchers found that both the $CO_{2\text{-e}}$ emissions and the fossil fuel energy inputs needed per unit of output were very sensitive to the following four factors:

- Oil yield (tonne/ha/year)
- Velocity of algal circulation
- Recyclability of the culture media
- Flue gas CO_2 concentration

Clearly, if the yield is (say) halved, the energy ratio (defined as the ratio of energy output to input) will also be nearly halved. If circulation velocity is increased, friction losses will rise nonlinearly. If the culture media cannot be recycled, or if it needs pretreatment, energy inputs will likewise rise. The problem of future provision of flue CO_2 to algal cultures has already been discussed. If, instead, elevated CO_2 concentrations were supplied by air capture, energy costs would be very high (Moriarty and Honnery 2010b). Similarly, if open freshwater ponds were located in semiarid areas, energy input costs for water replacement could even match the gross energy output (Moriarty and Honnery 2007). In conclusion, the input energy costs, and with them the $CO_{2\text{-e}}$ emissions per unit of output (and the unit monetary costs), could be far higher than the values derived in the studies just reviewed.

26.2.4 Discussion

Extravagant claims have been made for microalgal production, with estimates of potential oil production yields as high as 900,000 liters/ha/year, which is, according to one commentator, "thermodynamically impossible" (Mascarelli 2009). Similarly, it has been argued (Chisti 2007) that microalgal oil yield would be over 200 times greater than that of the best oil-yielding land-based plants. With claims like these, it is clear why some think that land-based production of biomass is an outdated technology. However, these comparisons are not useful because global land-based production of biofuels (biodiesel, ethanol) in 2009 totaled over 85 billion liters (Renewable Fuels Association (RFA) 2010), whereas biodiesel production from microalgae was some six orders of magnitude lower, with no commercial-scale plants.

Compared with microalgae, land-based plants may appear to be an inefficient method of producing biomass. True, land plants divert much of their metabolic energy into maintaining a structure, producing a rooting system for plant stability, achieving uptake of water and nutrients from the soil, and developing toxins to fight off various pathogens. Plant structure also enables all leaves to have access to insolation. Oxygen poisoning is not a problem, and higher levels of CO_2 than ambient, while useful, are not needed. In contrast, microalgae need to have many of the functions performed "naturally" by more advanced plants performed by artificial, mechanical means. Lighting, mixing of nutrients, temperature control, removal of toxic levels of oxygen, and provision of adequate levels of CO_2 must all be carried out for the microalgae by energy-intensive means. It is easy to be seduced by the high outputs obtained in laboratory test setups, but these high yields come at a heavy energy and economic cost.

26.3 Pyrolysis Biorefinery for Fuels and Chemicals

Pyrolysis is the process of heating in the absence of oxygen. For biomass, the purpose is to break up and depolymerize its three major components: cellulose, hemicellulose, and lignin. Although noncondensable gases and solids (char) are produced, liquid in the form of a complex mixture of polar and nonpolar oxygenated compounds is the major product. Typical mass conversion rates from feedstock to liquid (termed bio-oil) are approximately 60–75%. Detail on pyrolysis systems, bio-oil composition, and thermochemical properties can be found in Bridgwater et al. (1999). A typical system is shown in Figure 26.1.

Almost all forms of biomass can be converted through pyrolysis to produce bio-oil. The properties and composition of bio-oil do, however, strongly depend on biomass feedstock

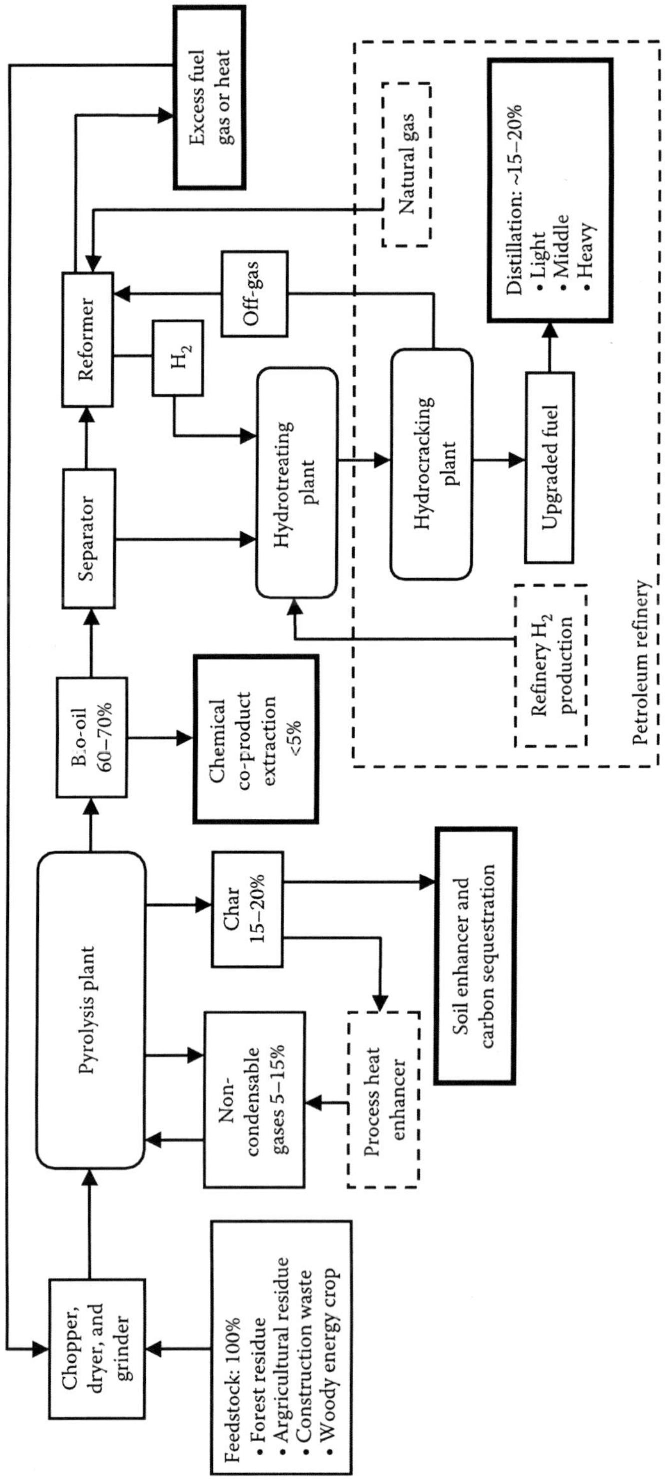

FIGURE 26.1
Pyrolysis BRF with upgraded fuel and chemical coproduct extraction. Percentages shown are typical feedstock mass conversion rates. Dashed boxes represent possible alternative processes.

composition and moisture content. They also depend on pyrolysis reactor type and on process conditions such as residence time, temperature, and heating rate (Bridgwater et al. 1999). Higher reactor temperature favors the formation of gases, lower temperature (as in traditional charcoal production) favors solids. Although the composition varies, among the many chemicals identified in bio-oil are phenol, furfural, guaiacol, and levoglucosan (Elliot 2004). Compared to the major constituent, water, these chemicals are generally found in low volumes, and once extracted, represent a possible high-value coproduct in addition to the high-volume, but low-value use of bio-oil as a fuel. In addition to liquid fuels and chemicals, other possibilities include production of electric power from any excess noncondensable gases and the possible use of char as a bio-oil fuel enhancer or as an agricultural soil supplement.

The additional steps necessary to upgrade bio-oil into a more readily usable fuel are illustrated in Figure 26.1. Following pyrolysis, the bio-oil undergoes hydrotreating and hydrocracking to produce a fuel consisting mostly of naphthenes, aromatics, and alkanes (Elliot et al. 2009). The proportions of these groups within the upgraded fuel can vary with feedstock. Nevertheless, distillation provides the possibility of producing gasoline, diesel, and jet-fuel-like grades, although the diesel-fuel equivalent has been found to have a cetane number considerably lower than most automotive-grade diesel fuels, while the gasoline component has octane numbers slightly lower than most modern low-octane-number fuels (Elliot et al. 2009). Despite these differences, these fuels can be used directly in unmodified engines, or as blends with existing fuels.

Consistent with the definition of a BRF, it has been suggested that the hydrogen needed for the upgrading processes can be supplied by reforming a portion of the bio-oil (Anex et al. 2010, Wright et al. 2010). Although this process will reduce the overall quantity of upgraded liquid fuel produced, it does result in a greater production of gases, which, although low in heating value, could be used to produce power or to supply process heat.

There are at present no commercial-scale plants to upgrade fuel produced by pyrolysis. Relatively large-scale pyrolysis plants capable of processing up to 200 tonnes of biomass per day exist, but production is limited to electricity, flavorings, and bio-oil, either directly or enhanced through blending with char. Upgrading is currently limited to laboratory-scale experimentation, although there is considerable experience in the petroleum industry with processes similar to those proposed to upgrade bio-oil. Performance and cost assessments of these systems to date are therefore theoretical and, as is to be expected, strongly dependent on the assumptions made.

26.3.1 Technoeconomic Assessments and Environmental Flows

There have been ongoing attempts to assess the product value (PV) of upgraded fuels from pyrolysis plants since at least the early 1990s (e.g., Cottam and Bridgwater 1994). Early attempts focused on bio-oil production itself rather than upgraded fuels. As Wright et al. (2010) noted in their own study, previous estimates for bio-oil PV have ranged from US$0.11 to US$0.65 per liter, reflecting mostly the large variability in estimating the capital cost of the equipment needed, rather than the feedstock itself. Recent studies have seen this variability narrow as experience has grown with pyrolysis and now also with upgrading, albeit at the laboratory scale. Additionally, the strong focus in the United States on extracting as much benefit from corn production as possible has resulted in a number of studies which examine the use of corn stover as a potential feedstock, thus enabling comparison of the methodologies for similar inputs and process assumptions. Some of the findings of these studies are listed in Table 26.1.

TABLE 26.1

Summary of Recent Technoeconomic and LCA Studies of Pyrolysis Biorefineries Producing Upgraded Fuels (Ethanol Also Shown)

Study	Feedstock	Feedstock $US/Tonne	PV $US/Liter	PV Ethanol Equivalent $US/Liter	Liquid Fuel Conversion $MJ/MJ_{biomass}$	CO_{2e} kg/MJ (% Petroleum Saving)	Process
Kalnes et al. (2009)	Logging residues	—	—	—	—	29.0 (−68.3)	Full fuel-cycle examination with H_2 production from biomass. Biomass transport and production costs included
	Poplar	—	—	—	—	22.8 (−75.0)	
	Willow	—	—	—	—	21.2 (−76.8)	
Holmgren et al. (2008)	Wood	35	0.53	0.34	0.65	—	H_2 gas supply not stated
	Corn stover	35	0.48	0.30	0.61	—	H_2 gas supply not stated
Jones et al. (2009)	Poplar	56	0.54	0.35	0.50	—	*N*th plant design costs analysis. H_2 gas production from biomass
	Poplar	56	0.46	0.30	0.76	—	*N*th plant design costs analysis. Reduced costs because of petroleum colocation. Conversion efficiency increased by natural gas purchase to produce H_2
Wright et al. (2010)	Corn stover	83	0.82	0.52	0.38	—	*N*th plant design costs analysis. H_2 gas production from biomass
	Corn stover	83	0.56	0.36	0.62	—	*N*th plant design costs analysis. H_2 gas purchase
Anex et al. (2010)	Corn stover	83	0.80	0.50	0.41	—	*N*th plant design costs analysis. H_2 gas production from biomass
	Corn stover	83	0.53	0.33	0.68		*N*th plant design costs analysis. H_2 gas purchase
Searchinger et al. (2008)	Cellulose to ethanol	—	—	—	—	28 (−70)	Without land use changes included
		—	—	—	—	138 (+50)	With land use changes included
Holmgren et al. (2008)	Corn stover to ethanol	35	—	0.57	0.32	—	Biochemical conversion from stover to ethanol
	Wood to ethanol	35	—	0.43	0.29	—	Thermal conversion from wood to ethanol
Jones and Zhu (2009)	Corn stover to ethanol	46	—	0.27	~0.45	—	Cellulose and hemicellulose fermented to ethanol, lignin used to produce heat and steam

Apart from the differing biomass feedstock types listed in Table 26.1, two major process variations are apparent: (1) hydrogen production by steam-reforming a portion of the bio-oil and (2) hydrogen supply from a refinery (or from natural gas). Product value and feedstock energy-conversion efficiency run contrary to each other for these two cases. Production of hydrogen within the BRF reduces the ultimate amount of upgraded fuel produced, typically by 35–40 %, compared with the case of hydrogen supply, but upgraded fuel PV would be reduced by 15% to as much as 50%, depending on the feedstock. An important PV marker for the upgraded fuel is the United States. DOE target price of US$0.28 per liter for cellulose-to-ethanol production by 2012 (Holmgren et al. 2008). This price is yet to be reached for the fuels assessed in what are essentially theoretical studies of pyrolysis BRFs. However, the prices determined appear to be competitive to those quoted for ethanol production from the same feedstocks, with the hydrogen supply case PVs being nearest to those of ethanol.

The gap between current PV estimates and the DOE ethanol target price needs to be considered further in view of input and process cost sensitivities. For the case of hydrogen production, Wright et al. (2010) in their detailed analysis list in order of PV sensitivity: bio-oil fuel yield, feedstock price, capital cost, gas credit values, catalyst costs, and char credit value. Their assessment suggests that there is greater negative sensitivity than positive, particularly to fuel yields and feedstock costs. A drop in fuel yield from 25% to 20% combined with a feedstock price rise from US$83 to US$100 per tonne would see the upgraded fuel PV rise from US$0.82 to approximately US$1.1 per liter, or approximately US$0.74 per liter on an ethanol-fuel energy basis. While feedstock prices depend on the market, fuel yield factors relate to the performance of the fuel conversion technology. In what they term a pessimistic view of system costs and performance, Anex et al. (2010), arrived at a cost of US$ 2.16 per liter on an ethanol fuel energy basis. Achieving the US$ 0.28/liter DOE ethanol price in the Wright et al. (2010) model would require fuel yield to rise to 30% and feedstock price to fall to less than approximately US$65 per tonne. Both the Wright et al. (2010) and Anex et al. (2010) studies show lower PV sensitivity to input variables for the hydrogen supply case. In all cases and all studies, sensitivity to the cost of the catalysts used in the upgrading processes is low.

Another important factor is the size of the plant. Most studies have examined plants which are able to process 2000 dry tonnes of feedstock per day. These plants could be expected to yield approximately 100–200 million liters/year of upgraded fuel, depending on how hydrogen is produced or supplied. Anex et al. (2010) have estimated that 34 plants of this size would be required to achieve the United States RFS2 cellulosic biofuel goal of 3 billion gallons/year (11.4 billion liters (GL)/year) by 2015 (EPA 2009), provided that the hydrogen was available for purchase. This number could rise by 60% if the plants produced their own hydrogen. These feedstock rates represent a 10-fold increase in plant size relative to the largest pyrolysis plants currently in operation, but based on output, such a plant would be less than 4% the size of a typical 100,000 barrel/day petroleum refinery. Size is, however, important in determining the PV of the upgraded fuel. Jones et al. (2009) have estimated that reducing plant size to 500 dry tonnes/day could raise PV by as much as 35%, while increasing plant size to 5000 dry tonnes/day could reduce PV by approximately 12%. Plant size will, however, ultimately depend on feedstock availability and cost, with these factors likely to become more significant if the BRF were to be located at or near a petroleum refinery, because these are generally located near a port or oil source rather than near land-based biomass sources.

As previously observed, the energy efficiency of these plants, as measured by the conversion of dry biomass energy into fuel energy, depends most significantly on whether hydrogen is produced or supplied and on feedstock type. Roughly 40–50% of the biomass input energy is expected to be captured by the upgraded fuel for the hydrogen production case. For corn stover, this compares favorably with ethanol production from both biochemical and thermochemical conversion pathways. For wood, pyrolysis systems appear to provide a significant advantage over conversion to ethanol.

As indicated in Figure 26.1, there are many internal energy flows within BRFs, most of which are driven by heat supplied by combustion of the char and noncondensable gases produced from the pyrolysis reaction. Importantly, this energy is used to provide heat to the feedstock dyer and to drive the pyrolysis reactor. An important variable in this process is feedstock moisture content. Cottam and Bridgwater (1994) have shown that decreasing char formation with increasing reactor temperature resulted in reductions in system self-sufficiency with increasing feedstock moisture content. At reactor temperatures above 600°C, there is insufficient energy in the char and noncondensable gases to provide heat for the feedstock dryer and pyrolysis reactor if the moisture content of the feedstock is above 5%. For typical reactor temperatures of 450–500°C, although excess energy drops with increasing moisture, heat is in excess of that required for the 5–25% moisture content range examined, leading to process self-sufficiency.

Electricity is another important energy input. In the study by Wright et al. (2010), electrical energy input is estimated to be 324 MWh/day (1.2 GJ) for their 2000 dry tonne/day plant. Based on the typical U.S. electricity generation efficiency, this energy could be supplied by cofiring in a power station an amount equivalent to approximately 10% of the feedstock mass.

Along with the relatively small energy inputs, the claimed process heat self-sufficiency means that production of upgraded fuels from these BRFs should result in net reductions in greenhouse-gas emissions through fossil fuel substitution. Full life cycle emission studies such as that by Kalnes et al. (2009) suggested reductions of approximately 75% relative to use of petroleum. This study, which included feedstock transport, cultivation, harvesting, and processing, found that most emissions emanated from fuel production and were independent of feedstock type. Feedstock-related emissions were found to be highest for logging residues because of the high collection-energy inputs resulting from the relatively low per-hectare yields.

26.3.2 Discussion

Although pyrolysis is already well known within the research community, wider recognition that biomass pyrolysis can form the basis of a high-quality fuel production plant is relatively new. A consequence of this is that, other than detailed technical investigations of specific processes, there are few studies that examine the broader aspects of these systems, and fewer still which provide a critique. An alternative interpretation of this last point is that, unlike other cellulose-to-liquid-fuel systems, there is little to critique. Many of the studies critical of cellulose to ethanol processes are, however, relevant to pyrolysis systems, particularly as they relate to land use, feedstocks, overall energy efficiency, and greenhouse-gas emissions. Additional points are consideration of the benefits or otherwise of coproduct production, petroleum colocation, and hydrogen supply.

26.3.2.1 Colocation

Most of the analyses of pyrolysis BRFs have been done based on the operation of one biorefinery plant imposed on rather than integrated into an existing or future economic or ecological system. The data produced from the accompanying analyses are then extended across various systems by extrapolation typically based on feedstock availability or on the number of BRFs needed to achieve a particular target. For example, as noted above, consideration has been given to the number of pyrolysis plants needed to achieve the United States RFS2 cellulosic biofuels target of 3 billion gallons/year (11.4 GL/year) by 2015. If this was to be achieved by colocating pyrolysis BRFs at petroleum refineries to take advantage of lower BRF operating costs, this would mean locating a 2000-tonne/day BRF at approximately one-quarter of all U.S. petroleum refineries. This fraction would likely rise to more than one-third if those refineries not in or near agricultural regions were excluded. Moreover, extending the RFS2 target to 16 billion gallons (60.6 GL) by 2022 (EPA 2009) would require colocation at almost all these refineries, assuming ethanol production made up approximately half this target.

Production of fuels at this level could be expected to have significant effects on petroleum-refinery operations. For example, the difficulty of feedstock storage and handling operations alone could mean that few existing petroleum refineries would be able to integrate BRFs of the proposed size successfully. In addition, the increased demand for hydrogen could itself push up the price of production, thus reducing the benefit gained by colocation. It is likely, therefore, that most BRFs would have to be located away from refineries. The economics of hydrogen supply would then have to be weighed against the additional benefits of improved greenhouse-gas emission outcomes, which along with energy security issues, drive the need for BRFs.

26.3.2.2 Land Use and Feedstocks

Location near feedstock sources raises the issue of whether it is better to transport feedstock, bio-oil, or upgraded fuel. A number of studies have attempted to develop models to explore the various options available. For example, Holmgren et al. (2008) have developed what they call a "conceptual PyOil-based refinery" which involves radial connection of pyrolysis plants to a central refinery hub. Depending on the degree of processing done at the local pyrolysis plant, the central refinery could be bio- or petroleum-based. Apart from inclusion of transport distances (e.g., Kalnes et al. 2009) and examination of the effect of moisture content on transportation of either wood chips or bio-oil (e.g., Faix et al. 2009), little effort has been made to date to analyze the energy and greenhouse-gas emission profiles of large-scale integrated systems.

Land use and feedstock are related issues unless feedstocks are based on waste from construction and manufacturing, but these sources could only ever be expected to provide a very low level of production, even at a global scale. Large-scale operation will require feedstocks based on specific energy crops or residues from farming, agriculture, or forestry. For these feedstocks, land management practices can play a significant role in determining overall greenhouse-gas emissions and energy efficiency. An example of this is given in Table 26.1, which shows the effect of inclusion of land use change on greenhouse-gas emissions for conversion of cellulose to ethanol. In the case of cellulose supply from a dedicated woody energy crop, Searchinger et al. (2008) have shown that the emission factors typical of those generated by the GREET model (Wang 2005) can shift from providing a net reduction to a net increase relative to use of fossil fuels, depending on the assumptions made about crop displacement and the related energy and material inputs. Life cycle studies of pyrolysis plants have yet to include these effects because they consider only factors related to cultivation and harvesting.

In the case of waste feedstocks, the considerations become more complex. Removal of logging residues from forests and agricultural waste from croplands could be expected to alter moisture and nutrient transfer between the soil and atmosphere, resulting in increased use of nitrogen-based fertilizers (Crutzen et al. 2008, Cherubini et al. 2009). Moreover, similarly to construction waste which no longer goes to landfill, any disturbance or alteration to existing decomposition processes could be expected to reduce methane emission profiles because these residues would no longer undergo anaerobic-like decomposition. In the case of construction wastes, this could provide an important additional benefit to overall greenhouse-gas accounting, but as already noted, the availability of this feedstock is limited.

26.3.2.3 Coproducts

Successful production of coproducts will depend on bio-oil composition, the availability of extraction technology, and the size of the coproduct market. Assessing the sensitivity of the BRF to economic, energy, and greenhouse-gas balances is therefore difficult, and this is reflected in the absence of any study of this type. The complexity of the task can be illustrated by considering the extraction of a ligno-polyphenolic coproduct. Extraction of this group could enable the replacement of phenols currently produced from crude oil via the Cumene process. In addition, use of the BRF coproduct could result in a reduction in the use of benzene, propylene, acetone, and other similar solvents. A comparative analysis of the two alternative production pathways could then be undertaken to assess the benefits, provided that the extraction technology had been developed to a level of maturity to enable a fair comparison, that the size of the market was known, and that the ability of the coproduct to act as a substitute was understood.

26.3.2.4 Process Sensitivities

Despite apparent reliance on existing or similar systems, the need to undertake additional research into the processes which underpin the claims being made for pyrolysis BRF performance is reinforced by the observations made by Jones et al. (2009) on research sensitivities. They argue that although none of the individual processes that make up the BRF is uneconomic (or technically challenging), there are few data on catalyst costs, life, and composition, component space velocity, and the degree to which the entire molecular weight spectrum of bio-oil requires a two-stage hydrocracking process. Furthermore, the effect of scale on upgrade-system performance will require greater consideration before the benefits claimed from larger plant sizes can be obtained.

26.4 Pulp and Paper

The pulp and paper industry is arguably the oldest and most common type of BRF. Paper has been produced by a continuous process for over two centuries, and mills are found in most countries. Operating as a Kraft mill, production of pulp and paper can be considered as a BRF as it combines the production of materials (pulp, papers), chemicals (turpentine, resin, and fatty acids), and energy (heat and steam from black liquor and electricity and steam from cogeneration using bark).

At present, paper is used mostly as a communication medium (fine paper, newsprint, etc.), packaging material, personal care substrate (bath, facial and towel tissues), and in specialty

TABLE 26.2

Comparison of 2005 Global and Australian Production of Paper by Grade

Product	World (%)	Australia (%)
Printing/writing	34	17
Corrugated material	34	48
Carton board	12	6
Tissue	8	7
Newsprint	12	22
Total (Mt/year)	340	3.8

Source: Adapted from Appita 2005. *Appita's Guide to the Australian and New Zealand Pulp and Paper Industry.* 70p.

grades (plasterboard liner, tea bags, filtration media, security, etc.). Of the 340 megatonnes (Mt) of paper produced worldwide in 2003, 46% was used for communication; this compares with 3.8 Mt and 39% for Australia, as shown in Table 26.2 (Appita 2005). However, with the electronic media revolution, the equivalence between paper and communication has increasingly been challenged, resulting in a constant decrease in demand and price for newsprint, fine paper, and envelopes. These factors have combined with increasing energy and water costs and rising public concern about logging practices and mill effluent management practices to bring about a crisis in the industry. As a result, pulp and paper mill closures have outnumbered new mill openings in the last decade. In an effort to ensure survival and improve profitability, the search for what has been termed a second-generation pulp and paper biorefinery has become the focus of many parts of the industry. Before the potential for this to occur can be examined, process requirements of pulp and paper production must first be addressed. Following this, a roadmap for the second generation can be proposed.

26.4.1 First-Generation Biorefinery

There are six requirements for a pulp and paper mill:

1. Fiber for paper
2. Process water
3. Process energy
4. A qualified workforce
5. A paper market
6. Capital

A pulp and paper mill consumes large amounts of lignocellulosic fibers, typically between 500 and 3000 tonnes per day (dry basis). These fibers are either virgin or recycled. Virgin fibers are generated mostly from wood (hardwoods such as eucalyptus, maple, or aspen; softwoods such as pine or spruce) or herbaceous plants (bagasse, flax, hemp, bamboo). The moisture content of these feedstocks is typically approximately 50%. Ideally, the fiber supply is located within 200 km of the pulp mill. Large amounts of water are required to manufacture paper. To insure good paper fiber and basis weight distribution (formation), pulp fibers are diluted to between 0.1% wt/vol (tissue) and 1% (packaging papers) for dispersion on the paper machine's moving wire. Water is also needed for pulp cleaning and cooling. Although water is recycled within the mill, 1.0–40 m^3 of water per tonne of paper is typically required; fiber bleaching typically consumes half the water required for pulping (Gullichsen and Fogelholm 1999a).

Papermaking can either generate or consume energy, depending upon the pulping process selected. There are two main pulping-process categories: chemical and mechanical. In the chemical process, cellulosic pulp fibers are generated by dissolving the lignin and hemicellulose fractions of the lumen lamella separating the fibers, using either an alkaline solvent (as in Kraft pulping) or an acidic system (the sulfite process). The spent pulping liquor is then regenerated by combustion into fresh pulping liquor. A tonne of dry wood yields approximately 450 kg of chemical pulp fibers, while the remaining 300 kg of lignin and 250 kg of hemicelluloses are dissolved, then gasified, to produce most of the energy supplying the process (e.g., steam and electricity). The enthalpy of the black liquor is approximately 15 MJ/kg of solids (Gullichsen and Fogelholm 1999b). The heating value of hemicelluloses and lignin are approximately 13.6 and 27 MJ/kg, respectively (van Heiningen 2007).

In mechanical pulping, pulp fibers are generated by mechanical shear in a refiner; steam is often introduced to decrease the glass transition temperature of lignin and provide longer fibers. Chemical additives (e.g., SO_2) are also sometimes used to enhance this process. Although the pulp yield is very high (96% for refiner mechanical pulp, 90%+ for thermomechanical pulp), mechanical pulping requires a large amount of energy (Sundholm 1999). The energy requirement for pulping for newsprint production ranges from 2000 to 3500 kWh/tonne. Papermaking consumes 2000–3000 kWh/tonne, most of which (75%) is for drying (Sundholm 1999). Table 26.3 presents the energy requirements in steam and electricity for the different types of paper.

The complexity and scale of the pulp and paper process and the control and environmental challenges which must be met require the presence of a highly qualified workforce. Paper mills must also have easy access to large markets and a good transportation infrastructure. With a relatively low PV of US$500–US$1000 per tonne, efficient and low-cost transportation is required for economic viability. Large amounts of initial capital are required to construct a typical mill. For example, a new 2000 tonne per day bleached Kraft pulp mill costs more than US$2 billion; assuming that the pulp sells at US$750 per tonne and the mill runs at 98% efficiency, the annual value of the product is US$536 million, or one-quarter of the capital investment. Pulping is accordingly a very capital-intensive industry with a long payback time compared with other manufacturing processes. Component replacement is also highly expensive, with a paper machine costing upward

TABLE 26.3

Papermaking Energy Requirements

Paper Grade	Steam (kWh/ Tonne)	Electricity (kWh/ Tonne)	Total (kWh/ Tonne)	Steam % of Total
Newsprint	1500	630	2130	70
Wood-free, uncoated	2000	670	2670	75
Wood-free, coated	2139	900	3039	70
Tissue	1056	1050	2106	50

Source: Adapted from Sundholm, J. (ed.) 1999. *Mechanical Pulping*. Book 5, Finnish Paper Engineers Association and TAPPI, 427p; Karlsson, M.J. (ed.) 2000. *Papermaking Part 2: Drying*. Book 9, Finnish Paper Engineers Association and TAPPI, 496p.

of US$500 million. Such levels of capital are justifiable only over long investment time frames supported by a stable and sustainable return on investment.

26.4.2 Second-Generation Biorefinery

Despite the problems facing the global pulp and paper industry, opportunities exist through the creation of new markets and new applications for existing products. Bioactive papers for health and environmental applications, novel cellulosic packaging, and nanocellulose crystals are examples of innovation. As well as these new products, there are also new directions that the industry can take through the development of what are called by the sector *second-generation biorefineries* (Fernando et al. 2006, Kamm and Kamm 2007, Huang et al. 2008). However, few studies have critically examined whether these new BRFs can achieve the outcomes claimed.

Pulp and paper BRF economics require a balance between the PV and input costs such as fibers, chemicals, energy, water, effluent treatment, and the workforce. Moreover, although it is true that the environmental impact of any new plant must always be lower than that of any existing plant, existing plants are required to reduce their impact through a progressive tightening of the regulations governing effluent handling and control. There is therefore a strong link between environmental impact and plant economics.

In the case of newsprint, there is a single product, with both demand and price decreasing, in a context of increasing input costs. The balance can be significantly perturbed should the needed electricity originate from, for example, a coal power plant which requires additional payments to account for CO_2 emissions through an emissions trading scheme, or if regulations accounting for the cost of the removal of inorganics such as sodium and calcium from water effluents become more stringent. Energy-intensive industries producing large amounts of low-value products are particularly vulnerable. Although major management and engineering efforts have driven down newsprint production cost, little success has been achieved in increasing PV. This is the objective of the proposed second-generation biorefinery.

Because of the link between cost and environmental impact, a major component of the second-generation BRF design philosophy is cost minimization. This in turn requires minimization of unwanted impacts and the integration of existing operations with new ones, using the feedstock as the integrating medium (van Heiningen 2007, Figure 26.2). Examples of how this can be done are

1. Conversion of unused reactants into valuable products. Extraction and valorization of the hemicelluloses and wood extractives (pitch) before or during mechanical or chemical pulping is one example. Hemicelluloses can provide either polysaccharides for water-soluble polymers or a source of pentose and hexose sugars after complete hydrolysis (enzymatic, acid) (Hansen and Plackett 2008). Burning wood bark or dewatered sludge in a cogenerator for steam or electricity generation is another example. This also has the benefit of CO_2 reduction.
2. Replacement of expensive reactants or a low-value product by a more favorable alternative. One example is to precipitate the lignin from the black liquor obtained from the Kraft process. This lignin, which is a network of propylphenols, can be depolymerized and used as a renewable reactant for phenol or adhesive production. This would require replacement of the energy currently generated by black liquor through an alternative GHG-neutral source such as a combustible fuel (forest or agriculture residues from the fiber supplier) or by renewable energy. A second

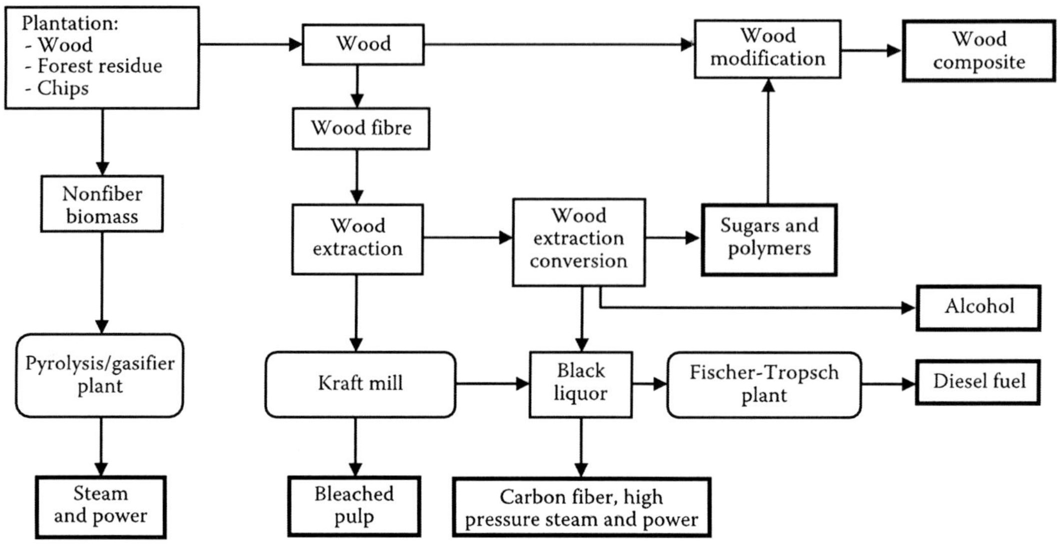

FIGURE 26.2
Simplified integrated second-generation pulp and paper BRF. (Adapted from Van Heiningen, A. 2007. *TAPPSA Technical Articles*, http://www.tappsa.co.za/archive3/Journal_papers/On_the_management_of_odour/on_the_management_of_odour.html. Accessed September 2, 2010.)

example is to modify newsprint into biodegradable mulch covers for horticultural production of high-value agricultural products such as tomatoes and strawberries. A third example is production of ethanol from the extracted hemicelluloses.

3. Production of novel products from the same reactants. This could be achieved by integrating new processes into an existing site. Wood chips could be converted into biofuels by fast pyrolysis/hydrocracking distillation processes or by gasification /liquefaction using the Fischer–Tropsch or the dimethyl ether (DME) process. Hydrolyzing wood into lignin, hexoses (C6), and pentoses (C5) for the fermentation of monomers such as succinic acid, 1–3-dipropanol, or lactic acid and their polymerization into biodegradable thermoplastics to replace those produced from fossil sources is another possibility.

26.4.3 Discussion

Given the diversity of outputs, an important aspect of this type of operation is the size of individual product markets. As an example, global paper production is relatively small compared to global transportation fuel use (<6–7% on an energy-equivalent basis) and electricity generation (Melendez 2008). Therefore, outputs such as power and fuels will have little influence on energy markets provided that they are cost-competitive. As with the coproducts from pyrolysis BRFs, markets for specialty chemicals are much smaller, so large-scale production could be seen to be self-defeating, particularly if the economic viability of the BRF is sensitive to the value of these products.

Because Kraft mills are generally net producers of energy, the net energy output from these systems depends on the balances obtained from production of the energy-related coproducts, particularly diesel fuels and ethanol. In a detailed study of these systems undertaken by van Heiningen (2007), conversion rates for hemicellulose to ethanol and black liquor to diesel were 45% and 40%, respectively, with wood yields of 10% and 43%. Product values were estimated to be US$0.33–US$0.31 per liter on an ethanol fuel equivalent basis, thus making them competitive with the PVs listed in Table 26.1. This is to be expected because the conversion processes are very similar, even though the feedstocks are different. Similar comments can be made about the environmental impact of these plants, although differences do occur because of the need to handle and control (among other things) inorganic effluent from the Kraft mill. Furthermore, the number of integrated BRFs is likely to be much lower than the number of pyrolysis BRFs, primarily because of the size of their major markets; therefore, environmental impacts will occur at a local rather than a global scale. Pulp and paper BRFs are also more likely to be more widely distributed at the global scale. However, comments previously made regarding the impact of biomass production and harvesting in relation to land use practices are also relevant here. Although researchers such as van Heiningen (2007) have undertaken detailed economic assessments of performance, their studies have yet to include these wider environmental impacts.

26.5 Conclusions

The biorefinery concept has much in common with other ideas being advanced to meet the twin challenges of global resource depletion and environmental degradation; these include green design, design for the environment, green chemistry, and industrial ecology. BRFs are of course concerned only with utilization of various forms of biomass, in contrast to the other, more general approaches. Advantages claimed for BRFs over more conventional uses of biomass include greater utilization of the entire biomass feedstock, greater product income, and less waste generated. Wider advantages claimed include greater national energy security because of replacement of imported fossil fuels, lower greenhouse-gas emissions, economic and social benefits to primary producers, and even the creation of a new industry.

Three possible BRF types have been examined in this chapter: production of fuels and chemicals from microalgae; pyrolysis of woody lignocellulose; and the pulp and paper industry. The first, though initially proposed half a century ago, is furthest from realization. The second is perhaps on the brink of large-scale production, while the third is an established technology.

The scale of global operation is crucial in evaluating both the economic viability and environmental impact of a given BRF type. Those based on pyrolysis of woody lignocellulose or microalgae production could well have as their main product liquid fuels for transportation. However, given that the global market for liquid transportation fuels is today already more than 90 EJ and growing, any attempt to meet most of this expanding market through BRFs would mean that by-product markets (e.g., for specialty chemicals) would soon be saturated, lowering their price and thus threatening the viability of the BRF.

Similarly, the scale of inputs to BRFs must be considered. Large-scale pyrolysis plants would need more than locally available biomass wastes from construction, agriculture, or forestry. Dedicated plantations would be needed to supply most of the inputs, and their energy needs and GHG releases should be evaluated from an ESS perspective, because such large-scale plantations can have significant effects on soil carbon and, because of fertilizer inputs, on N_2O releases from the soil. Through competition for land and water, they would also have downstream effects on agriculture. Similarly, for microalgae production in bioreactors, availability of flue gas CO_2 could be a problem at very high output rates. Underlying this is the need to have consistent and comprehensive data and process descriptions so that this type of analysis can be undertaken. Although progress is being made in this regard, there is still much to do.

The main biorefinery product will likely be liquid transportation fuels, given the very large global demand and the high prices they command. Although this chapter has stressed the challenges facing BRFs, it must be emphasized that they have one key advantage over petroleum-based fuels. Progressive depletion of conventional oil reserves will necessitate a shift to unconventional reserves such as oil sands and shale oil. Production costs and releases of GHGs per unit of these transportation fuels delivered are far higher than for conventional oil, so that the monetary and environmental costs of conventional transportation fuels will steadily increase. It is against these rising costs that BRFs must be evaluated.

References

Anex, R., Aden, A., Kazi, F., Fortman, J. et al. 2010. Techno-economic comparison of biomass-to-transport fuels via pyrolysis, gasification, and biochemical pathways. *Fuel* 89: S29-S35.

Appita 2005. *Appita's Guide to the Australian and New Zealand Pulp and Paper Industry*. 70p. http://www.appita.com.au/what-we-do/extensions/appita-guide

BP 2011. *BP Statistical Review of World Energy 2011*. www.bp.com/statisticalreview

Bridgwater, A., Meier, D., Radlein, D. 1999. An overview of fast pyrolysis of biomass. *Organic Geochemistry* 30: 479–493.

Cherubini, F., Bird, N., Cowie, A., Jungmeier, G., Schlamadinger, B., Woess-Gallasch, S. 2009. Energy- and greenhouse gas-based LCA of biofuel and bioenergy systems: Key issues, ranges and recommendation. *Resources, Conservation, and Recycling* 53: 434–447.

Chisti, Y. 2007. Biodiesel from microalgae. *Biotechnology Advances* 25: 294–306.

Chisti, Y. 2008. Biodiesel from microalgae beats bioethanol. *Trends in Biotechnology* 26(3): 126–131.

Clarens, A.F., Resurrection, E.P., White, M.A., Colosi, L.M. 2010. Environmental life cycle comparison of algae to other bioenergy feedstocks. *Environmental Science and Technology* 44: 1813–1819.

Cottam, M., Bridgwater, A. 1994. Techno-economic modelling of biomass flash pyrolysis and upgrading systems. *Biomass and Bioenergy* 7: 267–273.

Crutzen, P.J., Mosier, A.R., Smith, K.A., Winiwarter, W. 2008. N_2O release from agro-biofuel production negates global warming reduction by replacing fossil fuels. *Atmospheric Chemistry and Physics* 8: 389–395.

Darzins, A., Pienkos, P., Edye, L. 2010. *Current Status and Potential for Algal Biofuels Production: A Report to the IEA Bioenergy Task Force*. IEA Report T39-T2 6.

Demirbas, A. 2010. Use of algae as biofuel sources. *Energy Conversion and Management* 51: 2738–2749.

Elliot, D. 2004. Chemicals from Biomass. *Encyclopaedia of Energy* 1: 163–174.

Elliot, D., Hart, T., Neuenschwander, G., Rotness, L., Zacher, A. 2009. Catalytic hydroprocessing of biomass fast pyrolysis bio-oil to produce hydrocarbon products. *Environmental Progress and Sustainable Energy (2009)* 28: 441–449.

Faix, A., Schweinle, J., Scholl, S., Becker, G., Meier, D. 2009. Life-cycle assessment of the biomass-to-oil (BTO) process with combined heat and power (CHP) generation. *International Conference on Thermochemical Biomass Conversion Science*, September, Chicago IL.

Fernando, S., Adhikari, S., Chandrapal, C., Murali, N. 2006. Biorefineries: Current status, challenges, and future direction. *Energy and Fuels* 20: 1727–1737.

Gullichsen, J. and Fogelholm, C. (eds.) 1999a. *Chemical Pulping–A*, book 6A. Finnish Paper Engineers Association and TAPPI, 497p.

Gullichsen, J. and Fogelholm C. (eds.) 1999b. *Chemical Pulping–A*, book 6B. Finnish Paper Engineers Association and TAPPI, 693p.

Hansen, N.M.L. and Plackett, D. 2008. Sustainable films and coatings from hemicelluloses: A review. *Biomacromolecules* 9(6): 1493–1505.

Holmgren, J., Marinangeli, R., Nair, P., Elliot, D., Bain, R. 2008. Converting pyrolysis oils to renewable transport fuels: Processing challenges and opportunities. *National Petrochemical and Refiners Association Annual Meeting*, San Diego CA: pp. 9–11.

Huang, H.-J., Ramaswamy, S., Tschirner, U.W., Ramarao, B.V. 2008. A review of separation technologies in current and future biorefineries. *Separation and Purification Technology* 62: 1–21.

International Energy Agency (IEA) 2011. *Key World Energy Statistics 2011*. IEA/OECD, Paris.

Jones, B., Zhu, Y. 2009. *Preliminary Economics for the Production of Pyrolysis Oil from Lignin in a Cellulosic Ethanol Biorefinery*. PNNL-18401, Richland WA: Pacific Northwest National Laboratory (PNNL).

Jones, S., Valkenburg, C., Walton, C., Elliot, D., Holladay, J.E., Stevens, D.J., Kinchin, C., Czernik, S. 2009. *Production of Gasoline and Diesel from Biomass via Fast Pyrolysis, Hydrotreating, and Hydrocracking: A Design Case*. PNNL-18284, Rev. 1, Richland WA: Pacific Northwest National Laboratory (PNNL).

Kalnes, T., Hopkins, G., Reno, M., Shonnard, D. 2009. Emergent biofuel technology and life cycle assessment. *International Conference on Thermochemical Biomass Conversion Science*, September, Chicago, IL.

Kamm, B. and Kamm, M. 2007. Biorefineries—Multi-product processes. *Advances in Biochemical Engineering/Biotechnology* 105: 175–204.

Karlsson, M.J. (ed.) 2000. *Papermaking Part 2: Drying*. Book 9, Finnish Paper Engineers Association and TAPPI, 496p.

Lardon, L., Helias, A., Sialve, B., Steyer, J.-P., Bernard, O. 2009. Life-cycle assessment of biodiesel production from microalgae. *Environmental Science & Technology* 43(17): 6475–6481.

Marsh, G. 2009. Small wonders: Biomass from algae. *Renewable Energy Focus* (Jan/Feb): 74–78.

Mascarelli, A.L. 2009. Gold rush for algae. *Nature* 461: 460–461.

Mata, T.M., Martins, A.A., Caetano, N.S. 2010. Microalgae for biodiesel production and other applications: A review. *Renewable and Sustainable Energy Reviews* 14: 217–232.

Melendez, J. 2008. *Thermochemical Production of Biofuels from Australian Forestry Resources: Literature Review and Theoretical Case Study*. M.Eng. Thesis, Monash University, Clayton, Australia 2008.

Moriarty, P. and Honnery, D. 2007. Global bioenergy: Problems and prospects. *International Journal of Global Energy Issues* 27(2): 231–249.

Moriarty, P. and Honnery, D. 2009.What energy levels can the Earth sustain? *Energy Policy* 37: 2469–2474.

Moriarty, P. and Honnery, D. 2010a. *Rise and Fall of the Carbon Civilisation*. London: Springer.

Moriarty, P. and Honnery, D. 2010b. A human needs approach to reducing atmospheric carbon. *Energy Policy* 38: 695–700.

Pittman, J.K., Dean, A.P., Osundeko, O. 2011. The potential of sustainable algal biofuel production using wastewater resources. *Bioresource Technology* 102: 17–25.

Ragauskas, A.J., Williams, C.K., Davison, B.H., Britovsek, G., Cairney, J., Eckert, C.A., Frederick, W.J. Jr. 2006. The path forward for biofuels and biomaterials. *Science* 311: 484–489.

Reijnders, L. 2009. Microalgal and terrestrial transport biofuels to displace fossil fuels. *Energies* 2: 48–56; doi:10.3390/en20100048.

Renewable Fuels Association (RFA) 2010. *Ethanol Industry Outlook 2010*. http://www.ethanolrfa.org/page/-/objects/pdf/outlook/RFAoutlook2010_fin.pdf?nocdn=1. Accessed September 1, 2010.

Rosch, C., Skarka, J., Patyk, A. 2009. Microalgae—opportunities and challenges of an innovative energy source. *Proceedings, 17th European Biomass Conference and Exhibition*, Hamburg, Germany.

Scott, S.A., Davey, M.P., Dennis, J.S. 2010. Biodiesel from algae: Challenges and prospects. *Current Opinion in Biotechnology* 21: 277–286.

Searchinger, T., Heimlich, R., Houghton, R., Dong, F., Elobeid, A., Fabiosa, J., Tokgoz, S., Hayes, D., Yu, T.-H. 2008. Use of U.S. croplands for biofuels increases greenhouse gases through emissions from land-use change. *Science* 319(58–67): 1238–1240.

Stephenson, A.L., Kazamia, E., Dennis, J.S., Howe, C.J., Scott, S.A., Smith, A.G. 2010. Life-cycle assessment of potential algal biodiesel production in the United Kingdom: A comparison of raceways and air-lift tubular bioreactors. *Energy Fuels* 24: 4062–4077; DOI:10.1021/ef1003123.

Sundholm, J. (ed.) 1999. *Mechanical Pulping*. Book 5, Finnish Paper Engineers' Association and TAPPI, 427p.

U.S. Environmental Protection Agency (EPA) 2009. *EPA Proposes New Regulations for the National Renewable Fuel Standard Program for 2010 and Beyond*. Published by U.S. Office of Transportation and Air Quality, EPA-420-F-09-023, May.

Van Heiningen, A. 2007. Converting a kraft pulp mill into an integrated forest products refinery. *TAPPSA Technical Articles*, http://www.tappsa.co.za/archive3/Journal_papers/On_the_management_of_odour/on_the_management_of_odour.html. Accessed September 2, 2010.

Walker, D.A. 2009. Biofuels: Facts, fantasy, and feasibility. *Journal of Applied Phycology* 21: 509–517.

Wang, M. 2005. Updated energy and greenhouse gas emission results of fuel ethanol. *15th Symposium on Alcohol Fuels*, San Diego CA, September.

Wijffels, R.H., Barbosa, M.J. 2010. An outlook on microalgal biofuels. *Science* 329: 796–799.

Wright, M., Daugaard, D., Satrio, J., Brown, R. 2010. Techno-economic analysis of biomass fast pyrolysis to transportation fuels. *Fuel* 89: S2–S10.

27

LCA-Based Environmental Evaluation of Biorefinery Projects

Shabnam Sanaei, Matty Janssen, and Paul R. Stuart

CONTENTS

27.1 Introduction

Continuing fossil-based energy consumption is leading to increased depletion of fossil resources and consequently increased greenhouse-gas (GHG) emissions [1]. It has been observed that GHG emissions (carbon dioxide, methane, and nitrous oxide) are upsetting the Earth's climate due to fossil fuel combustion and land-use change as a result of human activities [2]. Using new technologies that will guarantee economic growth and environmental sustainability in the long term can overcome these issues. As a potential solution, over the past decades, attention has been paid to using renewable sources for providing energy and chemicals. Although heat and power can be produced by several renewable resource alternatives (wind, solar, hydro, biomass, etc.), biomass is the only alternative that by use of biorefinery technologies can be converted into fuels and chemicals [2]. To prove

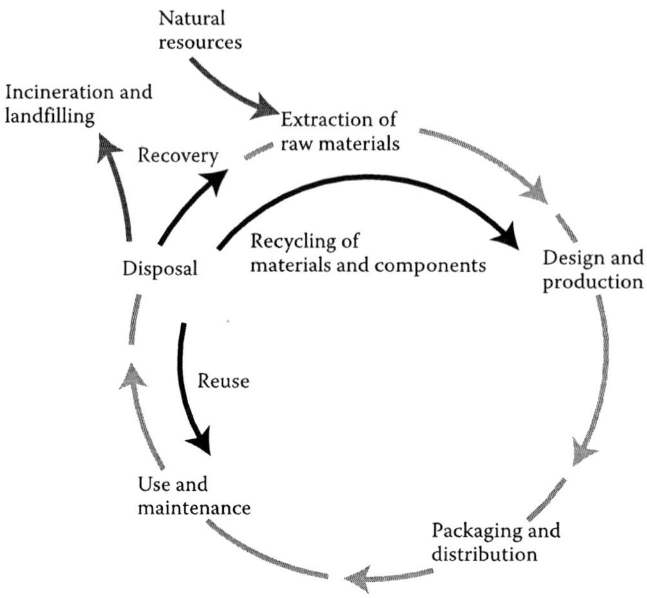

FIGURE 27.1
Product life cycle. (From Remmen, A. et al. *Life Cycle Management – A Business Guide to Sustainability*. United Nations Environment Programme, p. 1–52. With permission.)

that replacing fossil-based products by bioproducts leads to decreases in environmental impacts, several studies have performed an environmental evaluation of biorefinery systems.

Although most of these studies confirm that bioproducts have fewer environmental impacts than fossil-based fuels, a new concern is that there are several alternatives for biomass feedstock and emerging conversion technologies, from which the most promising need to be chosen for future implementation of a biorefinery strategy. Comparison of the environmental performance of possible strategies plays an important role in distinguishing them from a sustainability point of view. One of the most applicable methodologies by which the potential environmental impacts of a system can be assessed is life-cycle assessment (LCA). The main purpose of this method is to provide a quantitative assessment of the potential environmental impacts of products over their entire life cycle. A life cycle, as illustrated in Figure 27.1, consists of extracting raw materials from natural resources, production, distribution, use and eventually the recycling, reuse, recovery, or final disposal of the product. Life-cycle thinking goes beyond the traditional focus on production by including the impacts of a product, process, or service over its entire life cycle.

The following sections describe the use of LCA methodology to evaluate the environmental performance of biorefinery systems.

27.2 Overview of Life-Cycle Assessment Methodology

The LCA methodology is defined in the ISO 14040 and 14044 standards. As illustrated in Figure 27.2, the standard framework of LCA methodology has four steps: goal and scope

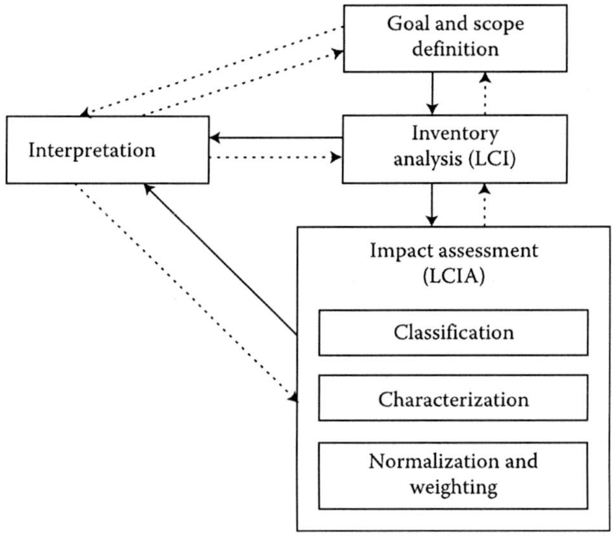

FIGURE 27.2
Main steps in the standard life-cycle assessment framework. (Adapted from Gaudreault, C., Samson, R., and Stuart, P.R. 2007. *Tappi Journal* 6, 3–9.)

definition (ISO 14040), inventory analysis (ISO 14041), impact assessment (ISO 14042), and interpretation (ISO 14043). Goal and scope definition defines the problem and system to be studied. During inventory analysis, polluting emissions, the use of renewable and nonrenewable resources, and the use of land are quantified. In the third step, these inventory data are transformed into potential impact indicators. Finally, during the interpretation step, key points are identified, and sensitivity analyses are performed to test the robustness of the results.

27.2.1 Goal and Scope Definition

Goal and scope definition consists of describing the objectives of the study, its applications, and its target audiences. Then, the function to be studied is introduced, and a representative unit for this function is selected (functional unit). The system boundary is set, and reference flows (a measure of the outputs from processes) are determined. The functional unit is the quantified performance of the product system used as a reference unit, while the reference flow is a measure of the outputs from processes in a given product system required to fulfill the functions carried out by the functional unit. The system boundary definition determines which unit processes are to be included in the study and must be consistent with the goal of the study. In general, the life-cycle stages that are included are extraction and preparation of raw materials, transportation, manufacturing, use of the product, and waste management. In the comparison of alternatives, any life-cycle stages that do not vary among the alternatives can be excluded. The LCA methodology can be applied, with proper justification, in studies that do not cover the entire life cycle of the product studied (cradle-to-gate and gate-to-gate studies). There are two approaches for determining the start and endpoint of the life cycle in LCA, which are cradle-to-gate and cradle-to-grave [5].

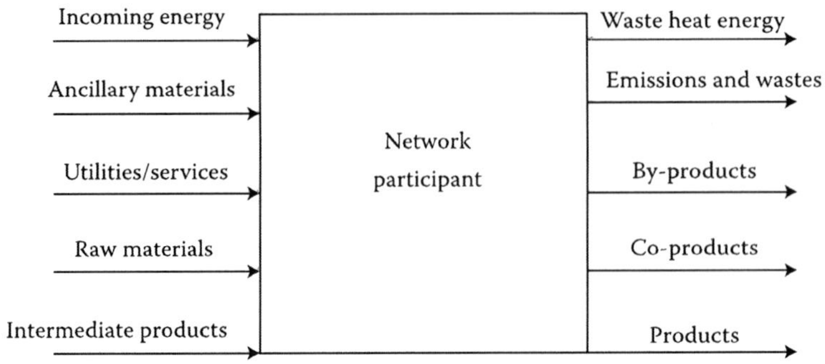

FIGURE 27.3
Mass and energy balance for LCI step. (From Halog, A. 2009. *International Journal of Global Energy Issues* 32, 83–101. With permission.)

27.2.2 Life-Cycle Inventory Analysis

Life-cycle inventory analysis (LCI) gathers information about all resource inputs and environmental outputs for each process technology. This information can be obtained by mass and energy balances. A schematic form of this step is presented in Figure 27.3.

27.2.3 Life-Cycle Impact Assessment

The aim of life-cycle impact assessment (LCIA) is to evaluate the magnitude of the potential environmental impacts of a product. It converts the results from LCI into environmental information by aggregating the LCI information into environmental indicators. When a substance is emitted into the environment, its concentration, state, or medium may change and often it is transformed into other substances. The steps that are followed by a substance before it causes an effect are called the *impact pathway* or the *cause–effect chain*. Impact characterization methods try to model the impact pathways to relate each piece of inventory data to its potential environmental impact as much as possible. Some methods, the so-called midpoint methods, stop at an intermediate level of the impact pathway, while others try to reach the endpoint and to describe environmental impacts using damage categories [5]. An example of indicators at both levels and how they relate to each other is given in Figure 27.4.

LCIA can also include optional elements such as normalization and weighting. Normalization expresses the results of LCIA relative to a reference and aims to provide a better understanding of the relative significance of the different impact category results. It also prepares the LCIA results for weighting by transforming them into dimensionless indicators. Weighting expresses the relative importance of the different impact categories by using value-based numerical factors. It enables the environmental information to be aggregated into a single indicator if desired [5].

27.2.4 Life-Cycle Interpretation

During life-cycle interpretation, the findings are evaluated to obtain conclusions and recommendations. This activity consists of two main steps: identification of significant issues and evaluation of the results. In this step, completeness, sensitivity, and consistency checks as well as uncertainty and data quality analyses can be done to ensure that the final results

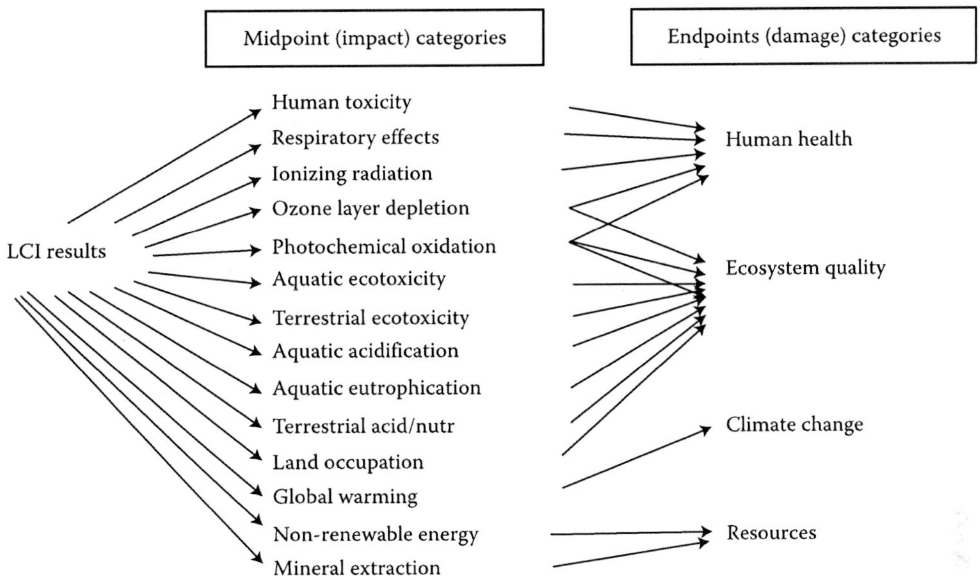

FIGURE 27.4
Example of IMPACT 2002+ indicators at midpoint and endpoint categories. (Adapted from Jolliet, O., et al. 2003. *International Journal of Life Cycle Assessment* 8, 324–330.)

are reliable. The completeness check ensures that all relevant information and data needed for the interpretation are available and complete. The sensitivity check assesses the reliability of the final results and conclusions by determining how they are affected by uncertainties in the data and the methodological choices. The consistency check determines whether the assumptions, methods, and data are consistent with the goal and scope of the analysis [5].

27.3 LCA Applications

In general, there are two main categories of LCA. The first includes all applications which aim to describe the environmental performance of systems (accounting LCA, attributional LCA), while the second includes applications which aim to describe the environmental consequences of a specific decision (change-oriented LCA, consequential LCA). On the one hand, attributional LCA is more related to the classical definition of LCA by ISO. On the other hand, in consequential LCA, any process that is affected by a given decision, whether it is in the life cycle being investigated or even in another life cycle, should be included in the analysis. This distinction is still young, and there are still debates on how and when one or the other of these approaches should be applied.

In the biorefinery context, LCA can be used to highlight the motivation for replacing fossil-based products by bioproducts, or it can be used for decision-making on possible biorefinery strategies. In both cases, the results of LCA-based environmental evaluation can be interpreted in conjunction with technical, economic, and social criteria to evaluate the sustainability of a biorefinery strategy.

27.4 LCA Challenges and Limitations

In spite of the strengths of the LCA methodology, it also has its weaknesses. The large amount of data and time needed to apply LCA can be considered as one of the main weaknesses of this methodology [8]. In addition, some of the important challenges are as follows [9]:

- Establishment of the system boundary
- Product functionality
- Choosing an allocation approach to assign environmental impacts to multiple products
- Selection of the impact assessment method
- Weighing of different impact categories against each other
- Temporal and special resolution

To increase the reliability of an LCA, the choice made in responding to each of the challenges mentioned above should be well documented [34]. In addition, the robustness and sensitivity of the results should be assessed by considering a number of possible choices.

Besides the challenges already mentioned, an unavoidable concern in LCA methodology is uncertainty. The sources of LCA uncertainty and methods to analyze them have been presented by Bjorklund [10]. Some of the sources of LCA uncertainty include data inaccuracy, data gaps, spatial and temporal variability, model uncertainty, and uncertainty of choices [11]. It is claimed that the most important source of uncertainty is site-specific information gathered in the inventory phase [11]. In general, LCA uncertainty can be categorized into three types: (1) parameter uncertainty (e.g., inventory data), (2) scenario uncertainty (e.g., choices concerning allocation), and (3) model uncertainty (e.g., lack of suitable characterization factors) [12].

Only a few studies have considered LCA uncertainty in the biorefinery context. Tan et al. [13] have used probabilistic modeling approaches for uncertainty evaluation of the life-cycle inventory of biofuels. Malca and Freire [14] have considered both parameter and scenario uncertainties for biofuel systems. Although some studies have covered one or two types of LCA uncertainty [15], there is no available LCA study in the biorefinery context that covers all three types of uncertainty simultaneously.

27.5 Environmental Concerns for Biorefinery Plants

The environmental issues which have been studied most extensively in the biorefinery context are greenhouse-gas (GHG) emissions, nonrenewable energy (NRE), and land-use change (LUC). GHG emissions may cause dangerous anthropogenic interference with the climate system and result in global warming. The most important greenhouse gases are carbon dioxide, methane, and nitrous oxide. Usually, a 500-year time horizon is chosen as a time period to represent the full impacts. With carbon dioxide as the baseline and a 500-year horizon, methane has been shown to have approximately 21 times more impact than CO_2 and nitrous oxide 310 times more [16].

Assessing GHG emissions of biorefinery systems depends strongly on the inclusion of emissions caused by land use, including direct land-use change (LUC) and indirect land-use change (ILUC) [17]. Direct land-use change (LUC) occurs when a piece of land with its previous land cover is changed to cultivated cropland to produce energy crops for biorefinery plants [18]. Fargione et al. [19] and Searchinger et al. [20] have shown the necessity of including land-use change impacts in the GHG balance for biofuels. Depending on the methodology used, the effect of land-use change on the GHG balance of a biorefinery system can be positive or negative. The conversion of forests, wetlands, and grasslands to cropland has a negative effect on GHG because of the emission of carbon from biomass and soils into the atmosphere. However, converting sparsely vegetated or disturbed lands to cropland results in a net gain in biomass production and sequestration of carbon into soil [17].

Besides the direct land-use change effects of producing bioproducts, there are also indirect effects. Because production of bioproducts competes with agricultural resources, this competition results in a price increase for agricultural products, and this price increase causes additional conversion of the world's grasslands and forests to cultivated cropland. This additional land conversion results in the release of previously sequestered carbon from grassland and forest ecosystems [21]. These emissions are an indirect result of producing bioproducts, and they should be considered in estimating the amount of GHG emissions for each biorefinery system. In addition to the agricultural-based biorefinery, indirect land-use change can also be associated with the forest-based biorefinery. For instance, using forest residues in biorefinery systems may cause feedstock shortages for energy production, which is another industrial application of forest residues. Activities to overcome this issue (increasing extraction of woody biomass to provide energy) will probably change GHG balance indirectly.

Because of the depletion of fossil resources in the world, nonrenewable energy use is considered as another important environmental concern in the biorefinery context. Among several proposed metrics for quantifying nonrenewable energy (NRE) use, the ratio of energy output of the resultant bioproduct to the fossil energy input required for its production is more representative and applicable.

The amount of water required to manufacture a bioproduct is another limiting factor for its future success. In addition, pollutants entering water sources, such as fertilizers and pesticides that are applied to the land to enhance plant growth, can affect water quality. These adverse effects can appear as eutrophication of fresh and ocean waters [22].

Among all the mentioned important environmental concerns in the biorefinery context, most available studies have focused on evaluating GHG and NRE, especially for biofuels production. However, more recent studies are also paying more attention to the impacts of direct and indirect land-use change caused by biorefinery systems. It has been confirmed that bioproducts use less NRE and release fewer GHGs during the cradle-to-gate segment of their life cycle (including feedstock preparation and manufacturing stages). However, when the product use and disposal stages are brought within the boundary of analysis (in a cradle-to-grave assessment), the outcomes become less predictable [23]. In addition, biorefinery systems sometimes have greater eutrophication and acidification impacts compared to their equivalent fossil-based pathways [23].

Because of the lack of information in the literature about possible environmental impacts of biorefinery systems (such as indirect land-use change, acidification, eutrophication, ozone depletion, and toxicity), it is difficult to conclude that bioproducts have a better environmental performance compared to their equivalent fossil-based products [23]. This difficulty underlines the importance of environmental evaluation of biorefinery systems.

27.6 LCA Studies for Biorefinery Projects

Most studies on the environmental evaluation of bioproducts have been carried out for biofuels, especially for bioethanol. Reviews of available publications during 1999–2004 and 2001–2008 which have used LCA for the environmental evaluation of biofuels production were presented by Blottnitz and Curran [24] and Ranjbar et al. [25]. In addition to biofuels, some studies have focused on other bioproduct categories. For instance, Wellisch et al. [23] reviewed 67 LCA-based environmental evaluations of several categories of bioproducts. They concluded that except for the biopolymer category, all bioproducts consume less nonrenewable energy and have lower GHG emissions compared to their fossil-based counterparts. However, they mentioned that it cannot be concluded that any particular bioproduct has the best environmental performance compared to other bioproducts.

There are several studies in the literature on LCA-based environmental evaluation of biorefinery systems. However, because they have considered various frameworks (a boundary and allocation method) and have different levels of accuracy, transparency, and consistency, it is very difficult to compare their results on a rational basis [26]. In addition, environmental evaluation of the biorefinery context is not yet mature for two main reasons: (1) not all the probable environmental issues (such as ILUC) are covered and (2) the various types of uncertainty are not considered [27]. Therefore, LCA-based environmental evaluation plays a very critical role in strategic decision-making about future implementations of biorefinery strategies.

27.7 Illustrative Example

The aim of presenting this example is to illustrate the procedure for environmental assessment of two biorefinery strategies for biofuel production using the LCA methodology and the SimaPro 7.1 software. Because of a desire for more sustainable fuel sources, recently greater attention has been paid to second-generation liquid biofuels, which can be produced through a variety of feedstocks and conversion technologies. Therefore, it is desirable to compare the environmental impacts of the various conversion processes for second-generation biofuels production. The two main types of conversion technologies by which biofuels are produced are biochemical and thermochemical technologies. The two biorefinery strategies which have been chosen for comparison using LCA are

1. Bioethanol production from a forest-residues feedstock using a thermochemical process (mixed alcohols are produced as coproducts)
2. Bioethanol production from corn-stover feedstock through a biochemical process (electricity is produced as a coproduct).

27.7.1 Goal and Scope Definition

This study intends to evaluate and compare the potential environmental impacts of two different pathways for producing bioethanol within the specified scope to determine which process has smaller potential environmental impacts. The "cradle-to-gate" impacts of the two competing processes for producing ethanol from biomass sources are assessed. This means that the potential environmental impacts are evaluated from growing the feedstock until the production stage, because the other steps are the same in both cases. Therefore, the system boundary includes the unit processes of growing, harvesting, and transporting the biomass and the production of ethanol from biomass for both processes. Incorporated into these primary unit processes are other units such as infrastructure, transportation, fuel, chemicals, and waste treatment. The primary mass balance for the processes is based on two NREL reports [28,29] which present fully developed mass balances and flowsheets for the production stage. The system boundaries as determined are shown in Figures 27.5 and 27.6. The common function of both processes is the production of ethanol, and the functional unit is defined as producing 1 kg of ethanol. However, different coproducts are produced through the two chosen pathways. ISO's recommendation is to avoid allocation by expanding the system limits to include the additional functions. However, when physical properties are not appropriate, ISO suggests using a different basis for allocation, such as the mass, energy, or economic value of the products. Although allocation is not the preferred method for handling multifunction systems, it is used in this example because of several existing multifunction issues. In the biochemical process in which electricity is produced as a coproduct, allocation is performed on an economic basis because it best represents the functionality of the process. In the

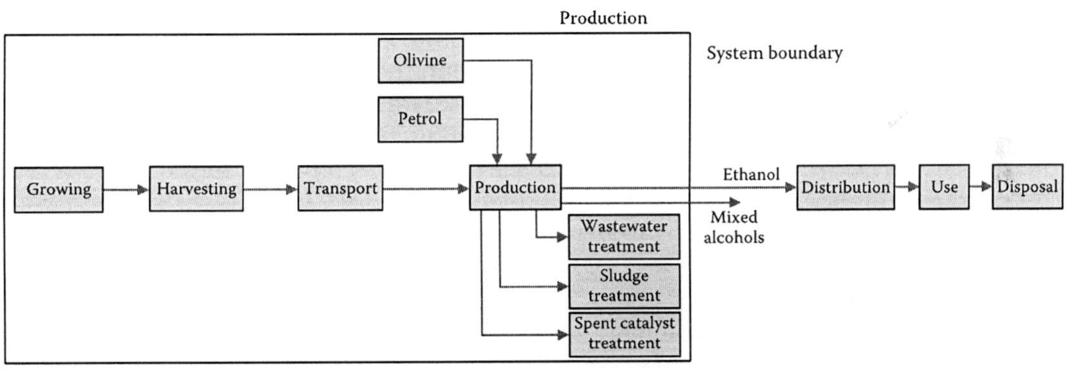

FIGURE 27.5
Thermochemical process using forest residues feedstock.

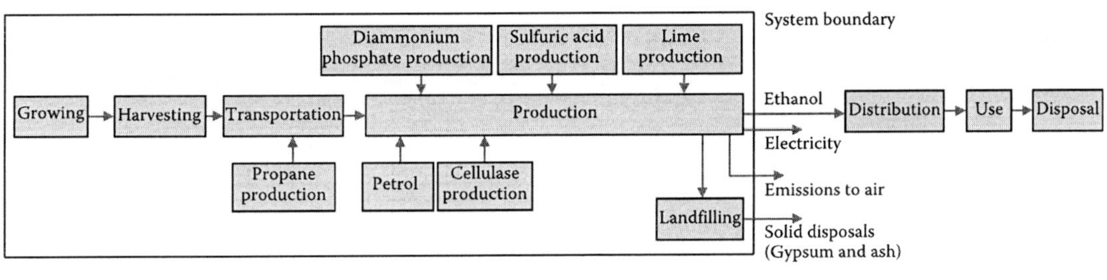

FIGURE 27.6
Biochemical process using corn stover feedstock.

TABLE 27.1
Reference Flows

Reference flows in ethanol production from forest residues through thermochemical process	**Amount**
Forest residue feedstock	9.25 kg
Ethanol to storage	1 kg
Mixed alcohols	0.181 kg
Reference flows in ethanol production from corn stover through biochemical process	**Amount**
Corn stover feedstock	3.81 kg
Ethanol to storage	1 kg
Electricity	2625.5 kJ

TABLE 27.2
Key Parameters

Parameter	Value	References
	In ethanol production from corn stover through biochemical process	
Corn/stover ratio	1	—
Harvestable stover	50%	—
Corn stover EtOH yield	98039 kg stover/25705 kg EtOH	[28]
Electricity ratio	18747kW/25705 kg/h EtOH	[28]
	In ethanol production from forest residues through thermochemical process	
Forest residue (FR) yield	367437 lb FR/39731 lb EtOH	[29]
Mixed-alcohols ratio	7204 lb mixed alcohols/39731 lb EtOH	[29]

thermochemical process in which mixed alcohols are produced as coproducts, because of the lack of reliable data about the market for mixed alcohols, allocation is performed on a mass basis. Another allocation problem exists with the feedstocks for both processes because they are both by-products of a primary function: corn stover of maize production, and forest residues of forestry. For both these cases, an economic basis is chosen because it can represent the current nature of their production. Based on the defined functional unit, the reference flows and key parameters of the systems under study are summarized in Tables 27.1 and 27.2.

27.7.2 Life-Cycle Inventory

All the assumptions made for estimating inventory data are presented in Tables 27.3 and 27.4 for the thermochemical and biochemical processes, respectively.

27.7.3 Life-Cycle Impact Assessment

The impact assessment was conducted using the Impact 2002+ method [7]. This means that the normalization data were based on European emissions. The two processes were characterized using 15 midpoint impact categories and four endpoint categories, and a comparison of the processes was undertaken. Figure 27.7 shows the comparison of midpoint

TABLE 27.3

Assumptions and their Justifications in the Thermochemical Process

Stages	Assumption	Amount	Justification
Growing	Removed 70% of fuel consumption	70%	Growing includes harvesting to side of road, so because harvesting is considered separately in this study, 70% of fuel consumption in the growing step is removed arbitrary
	Residue density	648.5 kg/m³	Same as the density of bundles
Harvesting	Tractor can carry 9998 kg (22 bundles)	9998 kg	—
	Bundle weight and volume	454 kg/0.7m³	—
Transportation	Distance from cropland to long-term storage	25 km	—
	Truck 28 t for transporting logs	—	Bundles were assumed similar enough to logs to be handled as such
Production	Infrastructure	—	A methanol plant is used as a proxy for infrastructure impacts, assuming a 20-year life and scaled to capacity
	Olivine	—	Assumed that making olivine involved the same process as sand, because it is a mineral and used as a heating medium and gasifying fluid
	Economic-based allocation	67% to EtOH	Heavy alcohol price (assumed same as butanol price) = \$5587 Ethanol price = \$11,370
	Energy-based allocation (used in sensitivity analysis)	65% to EtOH	Based on energy content: 477801 MJ to EtOH, 262,631 MJ to mixed alcohols

impacts for the processes. It can be seen that the thermochemical process (light gray bars) shows smaller potential impacts in most of the midpoint impact categories. The noncarcinogenicity and terrestrial ecotoxicity categories show net impact credits for the biochemical process, which therefore outperforms the thermochemical process in these areas. This is due to the absorption of toxic metals in the growing of corn.

Figure 27.8 shows the comparison of endpoint impacts for the two processes. As can be seen, all of the endpoint damage categories have lower scores for the thermochemical process.

The impact categories can be normalized to present the plant emissions as a fraction of the emissions in a geographical area. This is useful in understanding whether the amount of emissions from a given process is significant compared to the overall emissions of the same damage type in the area. As can be seen in Figure 27.9, only five of the impact categories present large potential impacts when normalized to European emissions. In four of the impact categories, the thermochemical process clearly has lower impacts. However, for the terrestrial ecotoxicity impact category, the biochemical process outperforms the thermochemical process because of the credits obtained from metal absorption during the growing phase.

Weighting is not necessary for this study because based on the results for the endpoint categories, it can be concluded that the thermochemical process is the environmentally more favorable process of the two examined in this study. No matter how the categories are weighted, the thermochemical process will remain the more environmentally favorable process.

TABLE 27.4

Assumptions and their Justifications in the Biochemical Process

Stages	Assumption	Amount	Justification
Growing	Mass ratio of corn to corn stover	1	—
	Harvestable stover	50%	30–50% of stover should be left on land as nutrition for the land
	Economic-based allocation	Allocating 54% of impacts to corn and 46% to corn stover	(Sensitivity analysis will be done) Corn price: 2.35 \$/bushel, corn density: 721 kg/m^3, corn stover price: 60 \$/dt
Harvesting	Two-step harvesting	—	In two-step harvesting, corn is harvested first (its environmental impacts have been included in the corn plant that has been extracted from Ecoinvent in this project), and then stovers are harvested by mowing, baling, and loading the bales
	Weight of each stover bale	700 kg	—
	Number of bales/hr for the plant with capacity of 2000 t/day feedstock	140	
Transportation	Distance from cropland to long-term storage	20 km	Plants which use agricultural feedstocks generally are located approximately 20 km from cropland
	Distance from long-term storage to short-term storage	1 km	Short-term storage should be close enough to long-term storage and also to the production process
	Truck 28 t for transporting round bales from cropland to long-term storage and also from long-term storage to short-term storage	—	Each truck can carry 17 round bales, so selecting a 28-t truck for transportation is consistent
	Number of forklifts	8	4 for unloading the trucks and 4 for putting the bales on the conveyor. Each forklift is capable of operating with a 33-lb propane tank for an 8 h shift
Production	Infrastructure	—	A methanol plant (same as in the thermochemical process (see Table 27.3))
	Chemical organics	—	Chemical organics has been chosen as a representative for enzyme in SimaPro
	Economic-based allocation	Allocation of 99% of impacts to ethanol and 1% to electricity	Economic-based allocation sensitivity analysis will be done

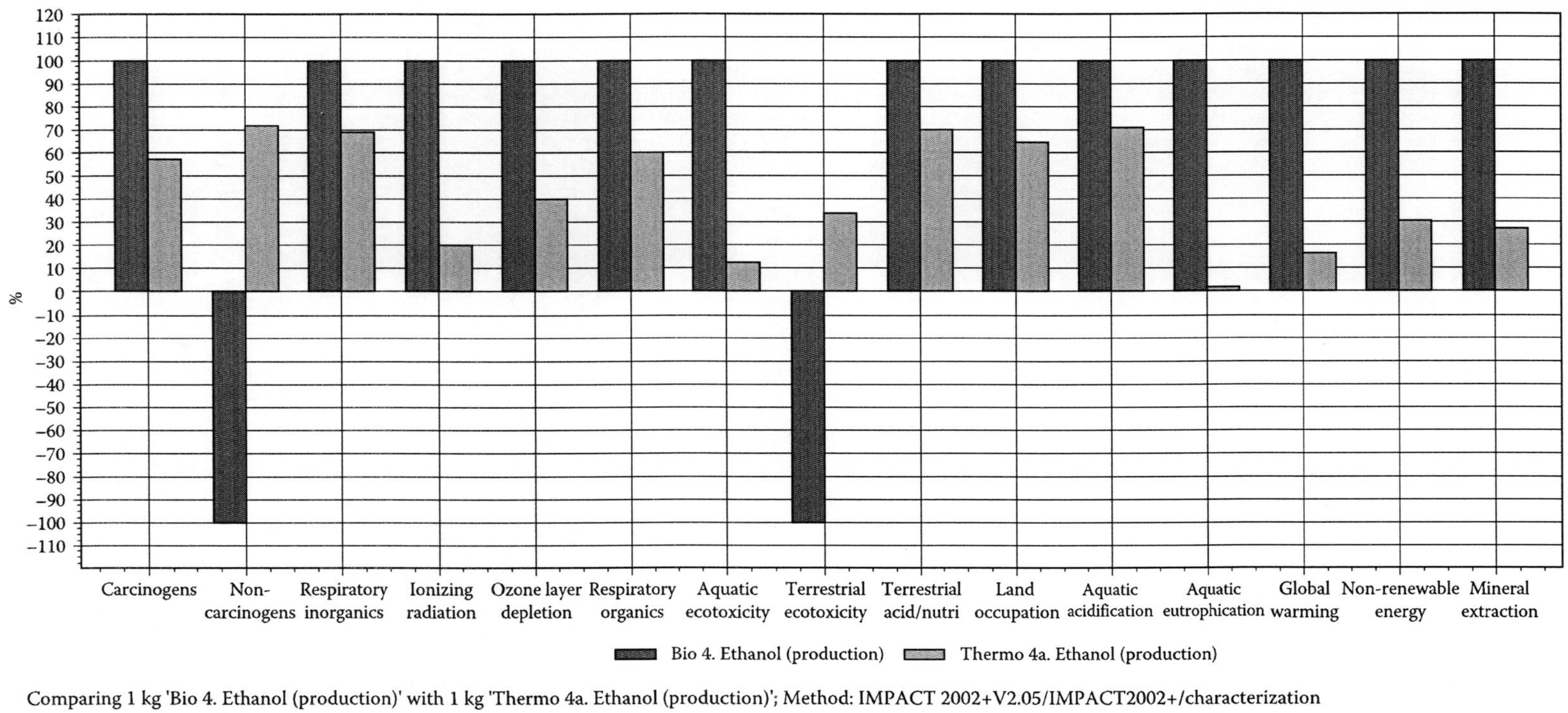

FIGURE 27.7
Midpoint characterization comparison of the thermochemical (light gray bars) and biochemical (dark gray bars) processes/1 kg ethanol.

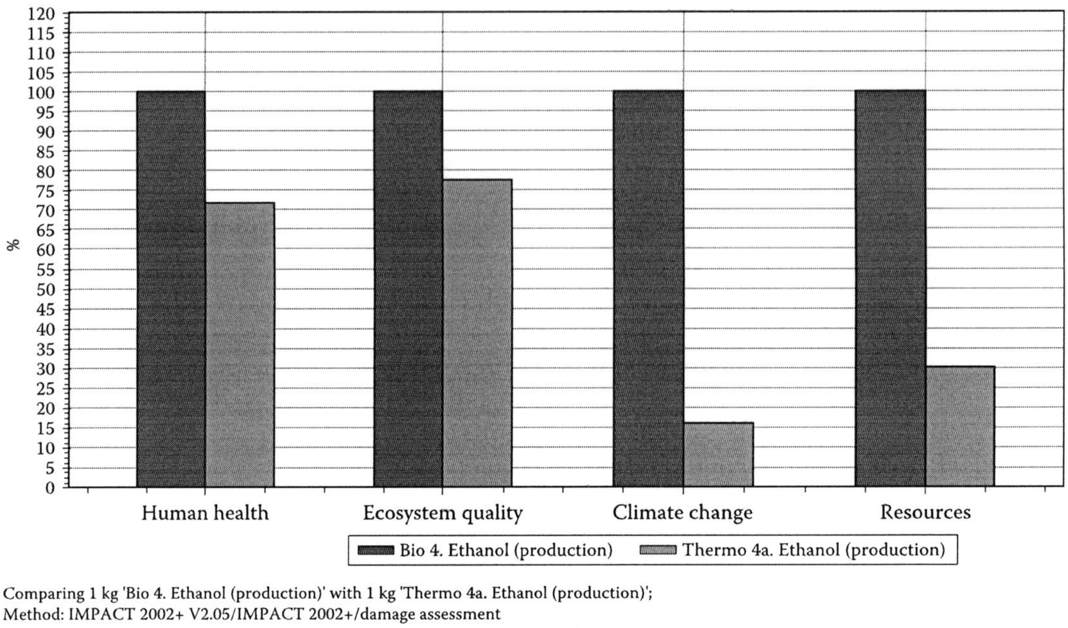

FIGURE 27.8
Endpoint characterization comparison of the thermochemical (light gray bars) and biochemical (dark gray bars) processes/1 kg ethanol.

27.7.4 Interpretation

This section aims to provide an overview of the contribution of potential damages and will enable the readers to be able to interpret the results and gain a better understanding of the cause of the impacts. Using normalization and weighting (by considering an equal weighting factor for each impact category), the contributions of each stage to the endpoint categories for both processes have been obtained and are shown in Figures 27.10 and 27.11. It is apparent that the growing and production stages are responsible for a large portion of the environmental impacts compared to harvesting and transportation. In both processes, the production stage is dominated by human health issues, which in turn is due to the midpoint impacts of respiratory inorganics and nonrenewable energy use. This is because of large emissions of NO_2 and SO_2 in the case of respiratory inorganics and the use of petrol as a denaturant in the case of nonrenewable energy. The impacts of biomass growing are dominated by ecosystem quality issues due to the occupation of land for growing trees and crops.

A substance contribution analysis was conducted by looking at the endpoints. In both cases, it was concluded that human health impacts arise mostly from nitrogen dioxide. NO_2 is generated primarily in the production process and also from the burning of fuel during the transportation stage. It can be seen that for both processes, fossil CO_2 emissions make up the largest part of the climate change category. This is due to the quantities of fossil fuels used for harvesting, transportation, and cultivation.

For each life-cycle stage, it is useful to know its contribution to the impact categories. In this section, the impact damages at both the midpoint and the endpoints are shown, and the significance of the contributions of each life-cycle stage is explored to determine the reasons for hotspots. Normalized midpoint values for both processes show that among all the midpoint impact categories, respiratory inorganics, land occupation, nonrenewable

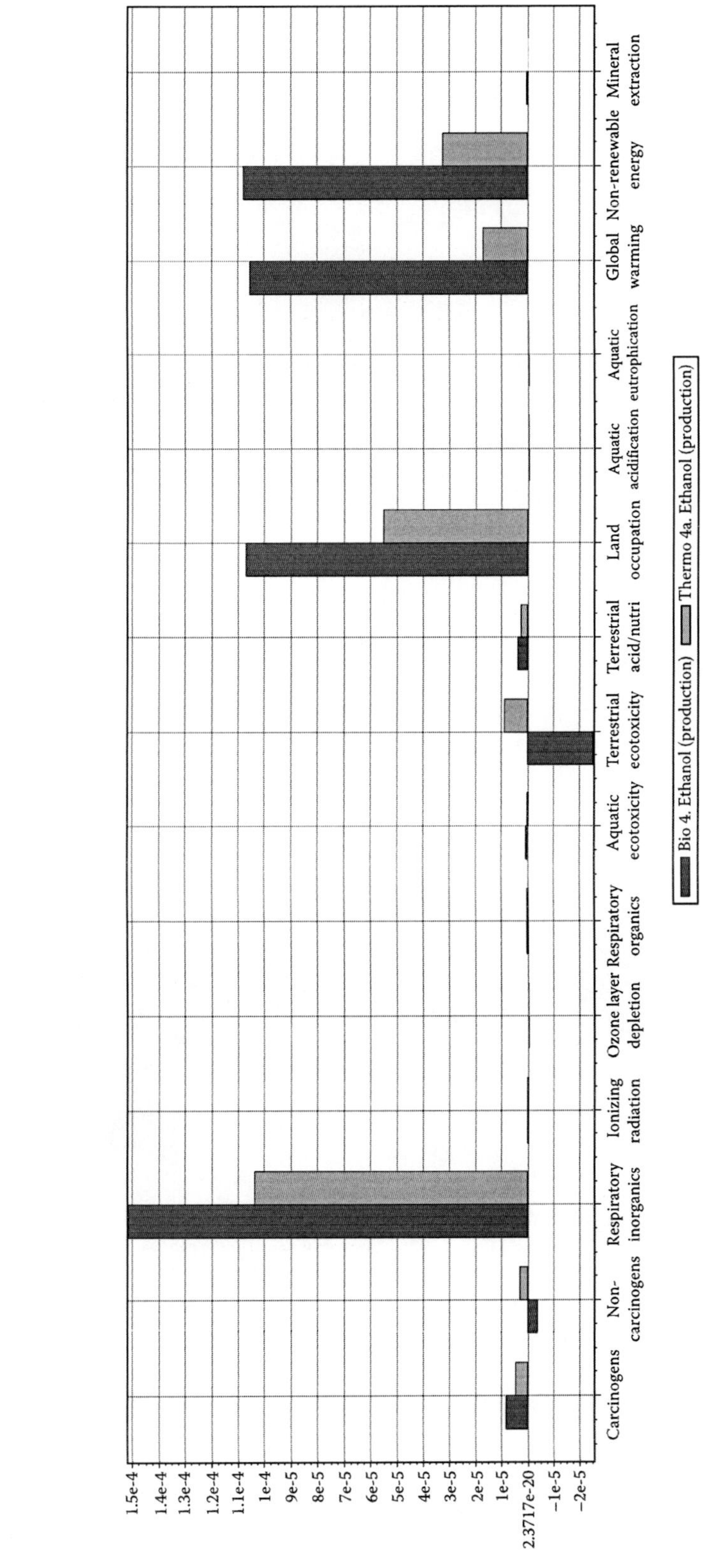

FIGURE 27.9
Midpoint normalization comparison of thermochemical (light gray bars) and biochemical (dark gray bars) processes/1 kg ethanol (Impact 2002+).

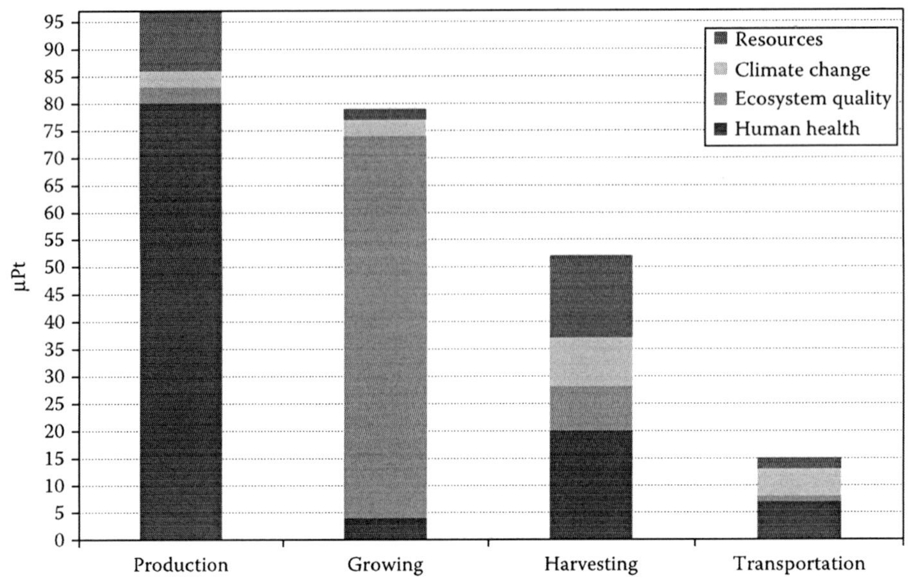

FIGURE 27.10
Thermochemical process single-score contributions/1 kg ethanol (Impact 2002+).

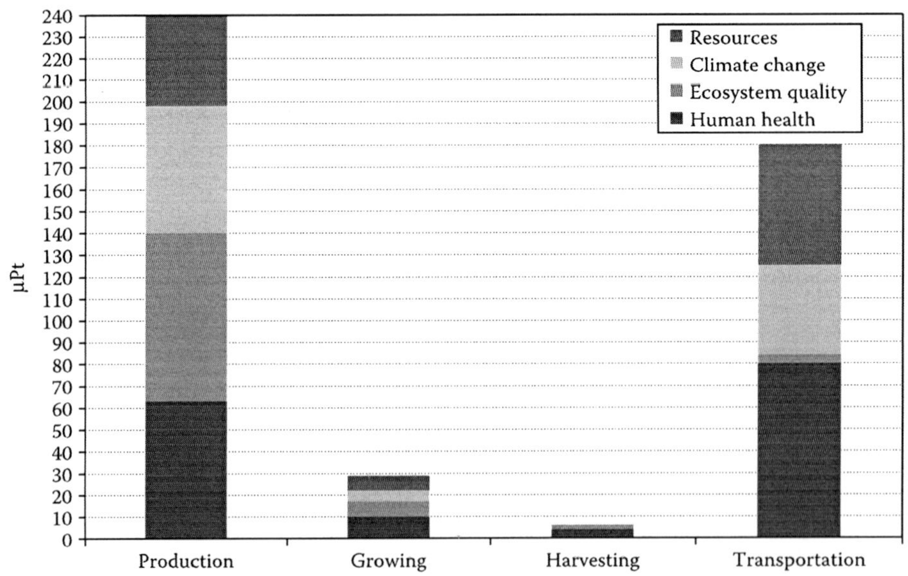

FIGURE 27.11
Biochemical process single-score contributions/1 kg ethanol (Impact 2002+) endpoint.

energy, and global warming make the largest overall contributions to potential environmental impact (Figures 27.12 and 27.13).

As can be seen in Figure 27.12 for respiratory inorganics (which cause human health impacts), this impact is dominated by the production stage, which in turn is a result of the amount of NO_2 and SO_2 emitted to the atmosphere. Land occupation (which causes

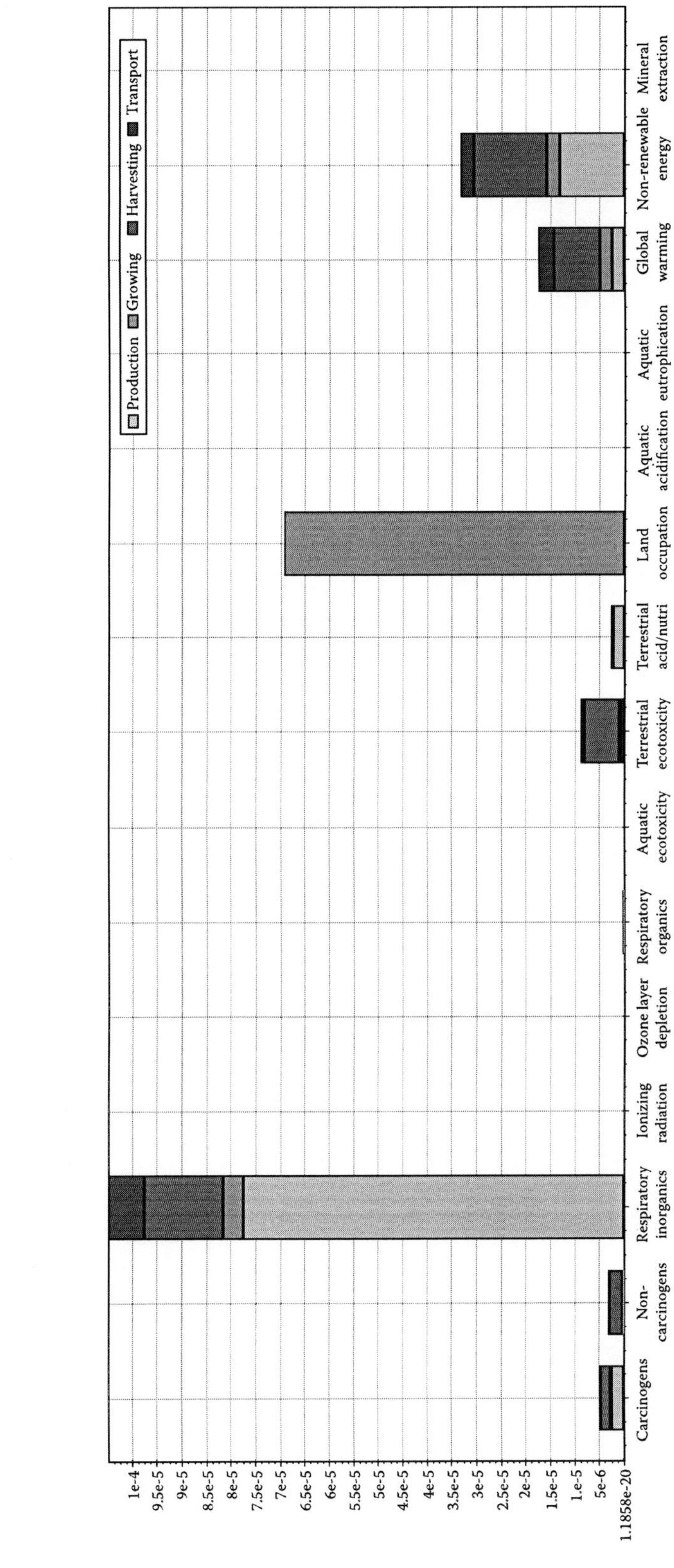

FIGURE 27.12
Thermochemical process category midpoint stage contributions (normalized)/1 kg ethanol (Impact 2002+).

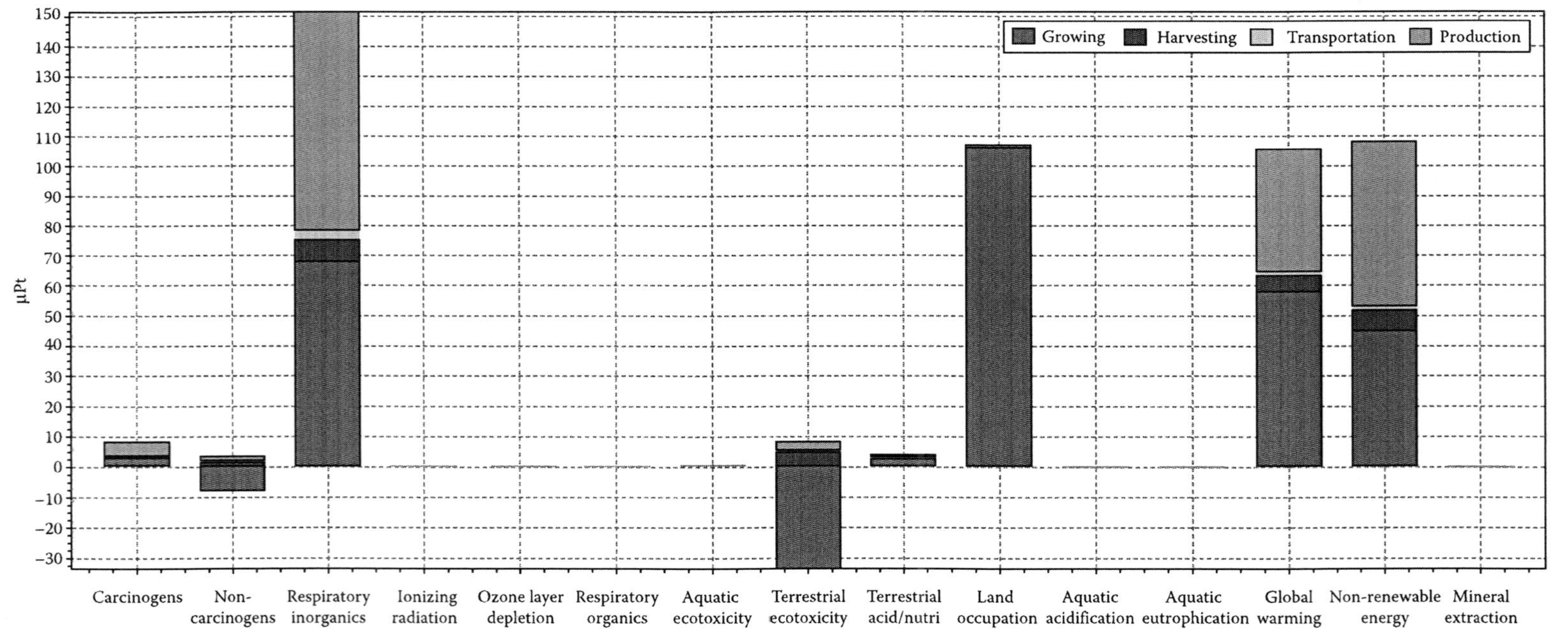

FIGURE 27.13
Biochemical process category midpoint stage contributions (normalized) / 1 kg ethanol (Impact 2002+).

ecosystem quality impacts) results almost entirely from the growing stage and is due to the amount of land required for growing biomass. For nonrenewable energy (which causes resources impacts), the impacts are caused primarily by production and harvesting and are due to fossil fuel consumption: use of gasoline as a denaturant during production and diesel consumption during harvesting. The majority of the global warming impact comes from the harvesting stage.

27.7.4.1 Sensitivity Analysis of Allocation

There are several uncertainties based on the assumptions that were made that may affect the results. To assess how much the results might possibly change based on the assumptions that were made, a sensitivity analysis was undertaken. One of the most important assumptions, to which most projects are sensitive, is related to the method of allocation. This section therefore tests the sensitivity of the final results to the allocation methods used in both feedstock and product stages. For all the parameters tested, the output was taken to be the normalized endpoint impacts, and the impacts were compared between the scenarios and also against the base case of the competing process (thermochemical or biochemical) to see how the overall results were affected.

To test the sensitivity of forest-residues allocation (vs. roundwood), the allocation was not changed directly; instead, the amount of forest-residues impact was increased and reduced to see what overall impact this would have on the comparison. Allocation had already been performed on the collected data for forest residues, and no data were available on the split used, only that it was economic. This was acceptable for the base case because it was decided that economic allocation was the most meaningful allocation method to use. To perform a sensitivity test, the impacts were therefore tested at levels of plus or minus 20% of the total impacts of the growing stage, and then the sensitivity of the overall process was measured.

Figure 27.14 shows that changing the allocation of the growing-stage products (roundwood and forest residues) would significantly affect only the ecosystem quality endpoint.

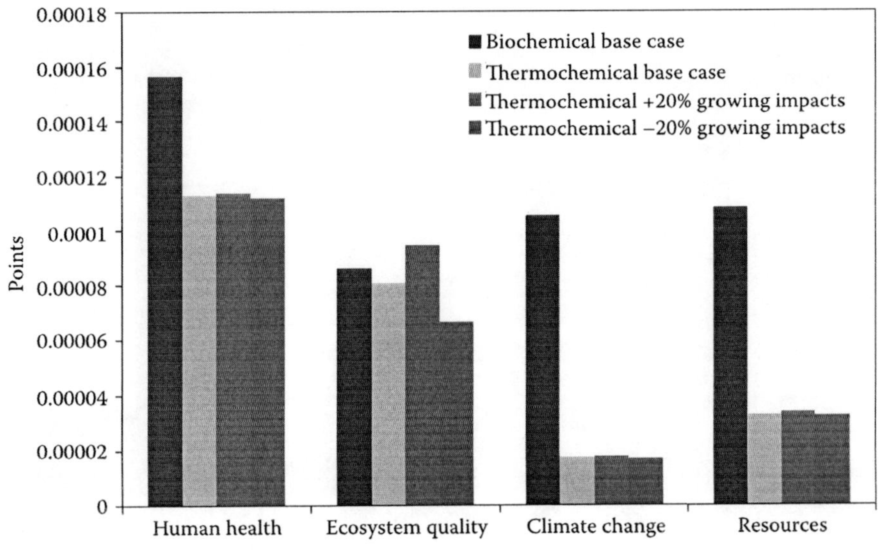

FIGURE 27.14
Thermochemical growing-stage allocation sensitivity.

This could potentially give a result where the biochemical process would be superior to the thermochemical process in this category, which means that in this case, it would be less clear which process has lower environmental impacts on ecosystem quality.

In the thermochemical process, mass allocation was selected for products in the base case. An economic allocation might have been more appropriate, but was not used in the base case due to gross uncertainties in the economic assumptions. The economic allocation was used as a comparison to reveal any differences. Furthermore, an allocation was performed using the energy content of the two products.

Figure 27.15 shows that climate change was not affected strongly by the change in allocation. Because both allocation scenarios allocate relatively few impacts to the product, as expected, these reduce the impact in the other three categories. Resource use is not significantly affected; however, human health is affected very significantly, as is ecosystem quality. Although this does not change the results, it shows that the nonselective nature of mixed-alcohol synthesis may not necessarily prove to be a significant environmental problem.

In the biochemical case, two allocations were undertaken, one in the production stage for allocating impacts to the main product and by-product, and another in the feedstock stage for allocating growth impacts to corn and corn stover. The economic allocation was used in the base case, which led to allocating 46% of impacts to corn stover and 54% to corn. To check the sensitivity of the biochemical case to the allocation strategy, two runs were done in which a mass allocation (corn:corn stover = 50:50) and an energy allocation (corn:corn stover = 46:54) were used. Because corn stover is a by-product of the corn plant, it can be assumed that all the impacts of the growing stage can be allocated to corn, which is the main product of the cropland. Therefore, one of the cases that were studied is allocating 100% of impacts to the corn produced. The results obtained for all the sensitivity analysis case studies mentioned are shown in Figure 27.16.

Because the allocation percentages in all the allocation strategies (economic, mass, and energy allocations) are similar, there is no significant difference among the results obtained. Figure 27.16 shows that the thermochemical process shows better environmental perfor-

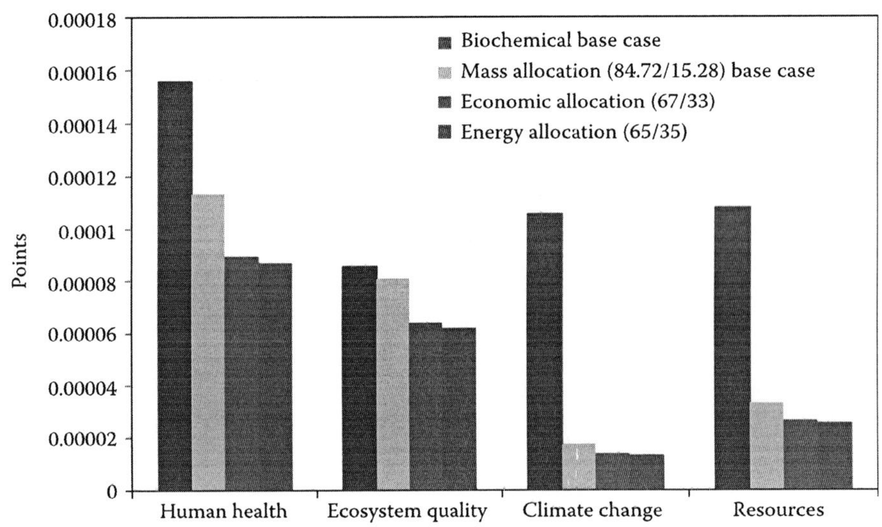

FIGURE 27.15
Thermochemical production-stage allocation sensitivity.

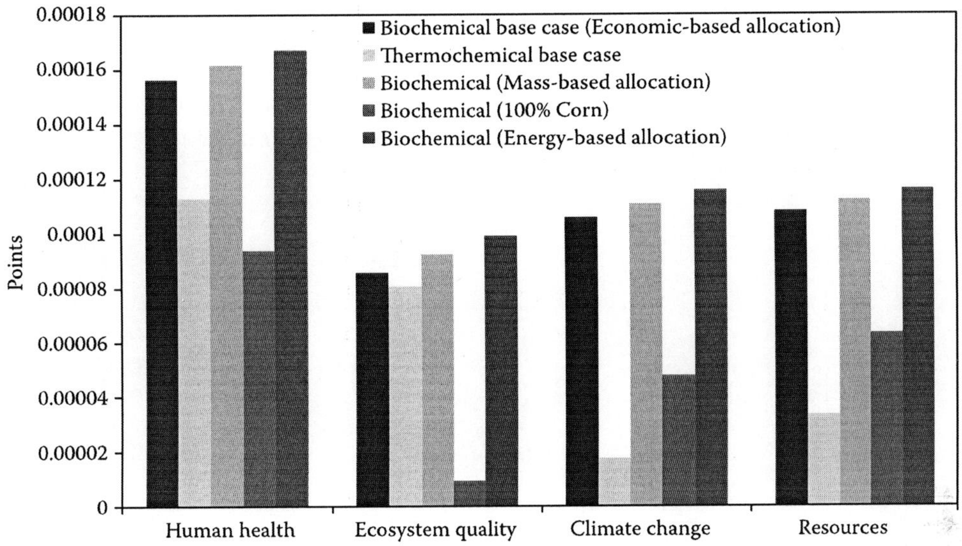

FIGURE 27.16
Sensitivity analysis of growing-stage product allocation in the biochemical project.

mance than the biochemical process when using the economic, mass, and energy allocations. However, in the case in which all the impacts were allocated to corn, a significant change was seen. In this case, for human health and ecosystem quality, the biochemical pathway will have a lower impact than the thermochemical process. Therefore, it can be concluded that the results are sensitive to allocation procedure.

Another parameter which can affect the results is the allocation of impacts to the final products. In the biochemical process, there are two products, ethanol as the main product, and electricity as the by-product. In the base case, economic allocation was used to allocate the environmental impacts to these two products. The other allocation option that can be used is energy-based allocation. In addition, because the amount of produced electricity is not significant compared to the main product, another option could be to consider electricity as an avoided burden (a method which subtracts surplus function results in a monofunctional system). The results obtained for these scenarios are presented in Figure 27.17. This figure shows that in all these cases, the thermochemical process shows better environmental performance than the biochemical process.

Based on the results presented in Figures 27.14 through 27.17, it can be concluded that both thermochemical and biochemical case studies are sensitive to allocation method in the feedstock stage; however, they are not sensitive to the chosen allocation method in the product stage.

27.7.5 Limitations of This Study

The results of this study are considered to be very preliminary and are therefore quite general and contain a significant margin of error. In general, the data quality is rather poor because the data are not based on a specific site or even an existing process. Therefore, assumptions were made and proxies used, which limit the accuracy of the study. Some of the assumptions used had the potential to affect the outcome of the study and the extent of their effects has been investigated by carrying out sensitivity analyses.

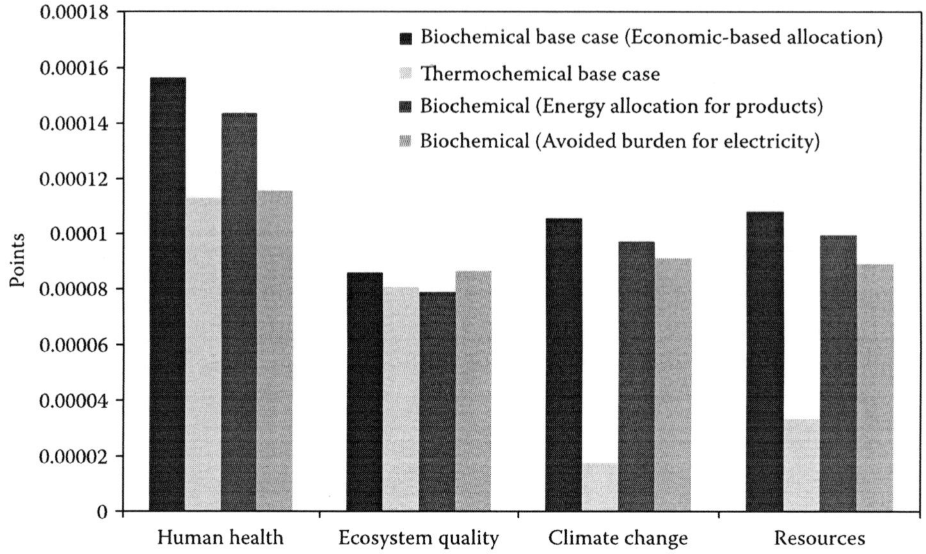

FIGURE 27.17
Sensitivity analysis of impact allocation in the production stage of the biochemical project.

In general, allocation of impacts to feedstocks was found to be the most sensitive, although not in all impact categories. Different methods of dealing with multifunctional systems should be tried in the future, including a wider variety of allocation models and expanding the system and its functions. The most widely used methodology for allocation is system expansion, after which economic allocation is the second most widely used procedure [26]. Many of the data used were based on a European database, which could limit the applicability of this work. The most highly limiting part of the methodology, however, would be the European impact assessment method, Impact 2002+, as well as the normalization step because it is based solely on European emissions.

27.8 Role of LCA in Multicriteria Decision Making

As mentioned earlier, although several possible biorefinery strategies exist, only those that at the same time can create more profit, have lower environmental impacts, and provide more social benefits in the context of a long-term vision can be considered as promising to implement in the future. This developing industry is at an early stage in which, instead of selecting the most sustainable strategies, less promising options should be screened out before any detailed engineering analysis which requires significant amounts of time and money. This goal will be achieved not only by considering environmental criteria but also by interpreting the results of LCA-based environmental evaluation in conjunction with technical, economic, and social criteria using multicriteria decision making (MCDM) methods. MCDM methods inform decision maker(s) about problem complexity and at the same time address decision uncertainty.

Cohen et al. [30] have recently defined a set of multidisciplinary criteria (including an environmental criterion) and have established an MCDM panel in which biorefinery experts were asked to elicit the importance of the defined criteria to be used to assess possible biorefinery technologies for the production of bioethanol that could be implemented at pulp and paper mills. The results of this study show that compared to the defined criteria, the environmental criterion achieved the lowest level of consensus among the panel members and finally obtained a low weight, that is, a low importance. A similar study was carried out by Quintero-Bermúdez et al. [31] with different panel members. The result for the environmental criterion was almost identical. The lower weight obtained for the environmental criterion compared to the other defined criteria can be attributed to the similarity of the environmental performance of all selected bioethanol production technologies compared to the corn-to-ethanol routes. It has been suggested by Quintero-Bermúdez et al. [31] that comparing biorefinery technologies with similar fossil-route counterparts would provide a better representation of the environmental advantages of biorefinery strategies. In addition, the low level of consensus on this criterion among panel members in both MCDM panels can be attributed to the perceived necessity of defining the environmental criterion as a "show-stopper" for screening biorefinery strategies in future studies, which would change its relative importance.

27.9 Conclusions

Biorefinery strategies are a set of potential solutions for the environmental concerns that the world faces today, including climate change and energy supply. However, among all possible biorefinery options, only those strategies that are sustainable and that can create more profit and achieve lower environmental impact and more social benefits in the long term can be considered promising for the future. LCA has been introduced as a methodology by which the environmental performance of biorefinery strategies can be assessed. To illustrate the application of the LCA methodology in the biorefinery context, an example was presented comparing two processes for bioethanol production (bioethanol production from forest residues through a thermochemical process, and bioethanol production from corn stover through a biochemical process). The results show that in both processes, four impacts are dominant: global warming, land occupation, nonrenewable energy, and respiratory inorganics. This result is consistent with available studies in the literature which claim that GHG emissions, NRE, and land use are the most important environmental concerns for biorefinery systems. In addition, the results showed that the thermochemical process using forest residues as a feedstock is more environmentally favorable than the biochemical process using corn-stover feedstock. However, it should be remembered that these results are fairly uncertain, with many errors introduced because of low data quality, regional differences in the data used and their assessment, and assumptions made because of missing data. Therefore, the results of this example can be considered suitable only for early-stage assessment or for confirming the outcomes of similar studies [32,33].

To judge the sustainability of biorefinery strategies, the results of an LCA-based environmental evaluation should be interpreted in conjunction with technical, economic, and social criteria using MCDM. The available results from MCDM panels convened to select promising biorefinery strategies show the lowest level of consensus for the environmental criterion among the panel members, which indirectly shows the necessity of defining the environmental criterion as a "show-stopper" for screening biorefinery strategies in future studies.

Acknowledgment

The authors would like to thank Matthew O'Connor for his collaboration in this study.

References

1. Singh, A., Pant, D., Korres, N.E., Nizami, A.-S., Prasad, S., and Murphy, J.D. 2010. Key issues in life cycle assessment of ethanol production from lignocellulosic biomass: Challenges and perspectives. *Bioresource Technology* 101, 5003–5012.
2. Cherubini, F. and Jungmeier, G. 2010. LCA of a biorefinery concept producing bioethanol, bioenergy, and chemicals from switchgrass. *International Journal of Life Cycle Assessment* 15, 53–66.
3. United Nations Environment Programme, Division of Technology, Industry and Economics (sustainable consumption & production branch) 2010. http://www.unep.fr/scp/lifecycle/.
4. Gaudreault, C., Samson, R., and Stuart, P.R. 2007. Life-cycle thinking in the pulp and paper industry, Part 2: LCA studies and opportunities for development. *Tappi Journal* 6, 3–9.
5. International Organization for Standardization (ISO) 14040. 2006. *Environmental Management: Life Cycle Assessment-Principles and Framework*, Switzerland.
6. Halog, A. 2009. Models for evaluating energy, environmental and sustainability performance of the biofuels value chain. *International Journal of Global Energy Issues* 32, 83–101.
7. Jolliet, O., Margni, M., Charles, R., Humber, S., Payer, J., Rebitzer, G., and Rosenbaum, R. 2003. Impact 2002+: A new life cycle impact assessment methodology. *International Journal of Life Cycle Assessment* 8, 324–330.
8. Hermann, B.G., Kroeze, C., and Jawjit, W. 2007. Assessing environmental performance by combining life cycle assessment, multi-criteria analysis, and environmental performance indicators. *Journal of Cleaner Production* 15, 1787–1796.
9. Bos, H. and Meester, K. 2008. *Sustainability evaluation of high-value-added product models.* Report 1004, Agrotechnology and Food Innovations b.v.
10. Bjorklund, A.E. 2001. Survey of approaches to improve reliability in LCA. *International Journal of Life Cycle Assessment* 7, 64–72.
11. Ross, S., Evans, D., and Webber, M. 2002. How LCA studies deal with uncertainty. *International Journal of Life Cycle Assessment* 7, 47–52.
12. Huijbregts, M.A.J. 1998. Application of uncertainty and variability in LCA. *International Journal of Life Cycle Assessment* 3, 273–280.
13. Tan, R.R., Culaba, A.B., and Purvis, M.R.I. 2002. Application of possibility theory in the life-cycle inventory assessment of biofuels. *International Journal of Energy Research* 26, 737–745.
14. Malca, J. and Freire, F. 2010. Uncertainty analysis in biofuel systems. *Journal of Industrial Ecology* 14, 322–334.
15. Huijbregts, M.A.J., Gilijamse, W., Ragas, M.J., and Reijnders, L. 2003. Evaluating uncertainty in environmental life-cycle assessment: A case study comparing two insulation options for a Dutch one-family dwelling. *Environmental Science and Technology* 37, 2600–2608.
16. Government of Canada, Statistics Canada, 2007. Catalog No.16-255-X. *Canadian Environmental Sustainability Indicators: Greenhouse Gas Emissions Indicator Data Sources and Methods*, Ottawa, Canada.
17. King, O.R.I.D. and Williams, A. 2010. *The Future of Industrial Biorefineries.* World economic forum. Geneve.
18. Rowe, R.L., Street, N.R., and Taylor, G. 2009. Identifying potential environmental impacts of large-scale deployment of dedicated bioenergy crops in the UK. *Renewable and Sustainable Energy Reviews* 13, 271–290.

19. Fargione, J.H.J., Tilman, D., Polasky, S., and Hawthorne, P. 2008. Land clearing and the biofuel carbon debt. *Science Express* 319, 1235–1238.
20. Searchinger, R.H.T., Houghton, R.A., Dong, F., Elobeid, A., Fabiosa, J., Tokgoz, S., Hayes, D., and Hg Yu, T. 2008. Use of U.S. croplands for biofuels increases greenhouse gases through emissions from land-use change. *Science Express* 319, 1238–1240.
21. Liska, A.J. and Perrin, R.K. 2009. Indirect land use emissions in the life cycle of biofuels: regulations *vs.* science. *Biofuels, Bioproducts, and Biorefineries* 3, 318–328.
22. Jacobson, M.Z. 2009. Review of solutions to global warming, air pollution, and energy security. *Energy Environmental Science* 2, 148–173.
23. Wellisch, G.J.M., Karbowski, A., Patel, M.K., and Rogulska, M. 2010. Biorefinery systems: Potential contributors to sustainable innovation. *Biofuels, Bioproducts, and Biorefining* 4, 275–285.
24. Blottnitz, H.V. and Curran, M.A. 2007. A review of assessments conducted on bio-ethanol as a transportation fuel from a net energy, greenhouse gas, and environmental life cycle perspective. *Cleaner Production* 15, 607–619.
25. Ranjbar, M., Samon, R., and Stuart, P.R. 2009. *Studying the Consequences of Different Systems Choices in LCA for Ethanol Production: An Assessment.* International Biorefinery Conference (IBC 09), New York.
26. Gnansounou, E., Dauriat, A., Villegas, J., and Panichelli, L. 2009. Life cycle assessment of biofuels: energy and greenhouse gas balances. *Bioresource Technology* 100, 4919–4930.
27. Delucchi, M.A. 2010. Impacts of biofuels on climate change, water use and land use. *Annals of New York Academy of Science* 1195, 28–45.
28. Aden, A., Ruth, M., Ibsen, K., Jechura, J., Neeves, K., Sheehan, J., Wallace, B., Montague, L., Slayton, A., and Lukas, J. 2002. *Lignocellulosic Biomass to Ethanol Process Design and Economics Utilizing Co-Current Dilute Acid Prehydrolysis and Enzymatic Hydrolysis for Corn Stover.* National Renewable Energy Laboratory (NREL), Colorado.
29. Phillips, S., Aden, A., Jechura, J., Dayton, D., and Eggeman, T. 2007. *Thermochemical ethanol via indirect gasification and mixed alcohol synthesis of lignocellulosic biomass.* NREL.
30. Cohen, J., Janssen, M., Chambost, V., and Stuart, P. 2010. Critical analysis of emerging forest biorefinery (FBR) technologies for ethanol production. *Pulp and Paper Canada* 111, 24–30.
31. Quintero-Bermúdez, M.A., Janssen, M., Cohen, J., Stuart, P. 2011. *Early Design Stage Biorefinery Process Selection.* TAPPI PEERS Conference. Portland, Oregon.
32. Mu, T.S.D., Rao, P.S., and Zhao, F. 2010. Comparative life cycle assessment of lignocellulosic ethanol production: biochemical versus thermochemical conversion. *Environmental Management* 46, 565–578.
33. Foust, A.A.T.D., Dutta, A., and Phillips, S. 2009. An economic and environmental comparison of a biochemical and a thermochemical lignocellulosic ethanol conversion process. *Cellulose* 16, 547–565.
34. Gaudreault, C. 2008. *Cadre méthodologique de gestion du cycle de vie appliqué à l'industrie papetière.* PhD Dissertation, École Polytechnique de Montréal, October 2008.

Index

E

P

T

U

V